Progress in Mathematics
Volume 244

The Unity of Mathematics

In Honor of the Ninetieth Birthday of I.M. Gelfand

Pavel Etingof
Vladimir Retakh
I.M. Singer
Editors

Birkhäuser
Boston • Basel • Berlin

Pavel Etingof
I.M. Singer
Massachusetts Institute of Technology
Department of Mathematics
Cambridge, MA 02139
U.S.A.

Vladimir Retakh
Rutgers University
Department of Mathematics
Piscataway, NJ 08854
U.S.A.

AMS Subject Classifications (2000): 14C35, 14C40, 14D20, 14D21, 14F05, 14F42, 14G22, 14J32, 14L05, 14L30, 14M17, 14N35, 14R20, 16E40, 17B67, 20E50, 20G45, 22D20, 33C35, 37K10, 43A30, 46E35, 53-02, 53D17, 53D35, 55N22, 57R17, 58-02, 58B24, 58C30, 58J42, 81-02, 81J40, 81R12, 81T13, 81T30, 81T45, 81T60, 81T70, 82B23

Library of Congress Control Number: 2005935032

ISBN-10 0-8176-4076-2 e-IBSN 0-8176-4467-9
ISBN-13 978-0-8176-4076-7

Printed on acid-free paper.

 Birkhäuser 

Printed in the United States of America. (JLS/SB)

9 8 7 6 5 4 3 2 1

www.birkhauser.com

Contents

Preface

> "Le professeur Gelfand est ainsi un pionnier qui, comme Poincaré et Hilbert, ayant défriché de nouveaux domaines, a laissé du travail à ses continuateurs pour une ou plusieurs générations."
>
> *Henri Cartan*

Israel Moiseevich Gelfand is one of the greatest mathematicians of the 20th century. His insights and ideas have helped to develop new areas in mathematics and to reshape many classical ones.

The influence of Gelfand can be found everywhere in mathematics and mathematical physics from functional analysis to geometry, algebra, and number theory. His seminar (one of the most influential in the history of mathematics) helped to create a very diverse and productive Gelfand school; indeed, many outstanding mathematicians proudly call themselves Gelfand disciples.

The width and diversity of the Gelfand school confirms one of his main ideas about the unity of the universe of mathematics, applied mathematics, and physics. The conference held in his honor reflected this unity. Talks were presented by former Gelfand students, their former students, and other outstanding mathematicians influenced by Gelfand.

The diversity of the talks and the subsequent outgrowths presented in this volume represent the diversity of Gelfand's interests. Articles by S. DeBacker and D. Kazhdan, B. Kostant and N. Wallach, G. Lusztig, and A. Vershik are devoted to various aspects of representation theory. (One cannot imagine representation theory without the fundamental works of I. M. Gelfand.) Geometry (an old love of Gelfand's) and its connections with physics are represented in the volume by the articles of M. Atiyah; D. McDuff; M. Kontsevich and Y. Soibelman; and Chien-Hao Liu, Kefeng Liu, and S.-T. Yau.

The article by A. Connes on noncommutative geometry reflects Gelfand's longtime (for more than 60 years) interests in noncommutative structures.

A majority of articles are devoted to a variety of topics in modern algebraic geometry and topology: A. Braverman, M. Finkelberg, and D. Gaitsgory; T. Coates and A. Givental; and V. Drinfeld.

Gelfand's interests in mathematical and theoretical physics are represented by papers of L. Faddeev; M. Movshev and A. Schwarz; N. Nekrasov and A. Okounkov; and A. Okounkov, N. Reshetikhin, and C. Vafa.

The article by H. Brezis connects partial differential equations and algebraic topology.

The unity of mathematics and physics cannot be separated from the life and work of Israel Gelfand. The conference and this volume are testimonies and celebrations of this unity.

P. Etingof, MIT
V. Retakh, Rutgers University
I. M. Singer, MIT

September 2005

Conference Program: An International Conference on "The Unity of Mathematics"

Sunday, August 31st

9:15am
Benedict Gross: *Opening talk* (as Dean of Harvard College)

9:30–10:30am
David Kazhdan: *Works of I. M. Gelfand on the theory of representations*

11:00am–noon
Robbert Dijkgraaf, *Random matrices, quantum geometry and integrable hierarchies*

2:00–3:00pm
Alexander Beilinson: *Around the center of a Kac–Moody algebra*

3:30–4:30pm
Vladimir Drinfeld: *Infinite-dimensional vector bundles in algebraic geometry*

4:45–5:45pm
George Lusztig: *Character sheaves and generalizations*

Monday, September 1st

9:30–10:30am
Michael Atiyah: *Some reflections on geometry and physics*

11:00am–noon
Cumrun Vafa: *Unity of topological field theories*

2:00–3:00pm
Alain Connes: *Noncommutative geometry and modular forms*

3:30–4:30pm
Albert Schwarz: *Supersymmetric gauge theories on commutative and noncommutative spaces*

4:45–5:45pm
Nathan Seiberg: *Matrix models, the Gelfand–Dikii differential polynomials, and (super) string theory*

Tuesday, September 2nd

9:30–10:30am
Shing-Tung Yau: *Mirror symmetry and localization*

11:00am–noon
Dusa McDuff: *Quantum cohomology and symplectomorphism groups*

2:00–3:00pm
Nikita Nekrasov: *Unity of instanton mathematics*

3:30–4:30pm
Ludwig Faddeev: *Algebraic lessons from quantum integrable models*

5:15–6:15pm
Israel M. Gelfand: *Mathematics as an adequate language. A few remarks*

Wednesday, September 3rd

9:30–10:30am
Alexander Givental: *Strings, loops, cobordisms and quantization*

11:00am–noon
Michael Hopkins: *Algebraic topology and modular forms*

2:00–3:00pm
Maxim Konstevich: *Integral affine structures*

3:30–4:30pm
Sergey Novikov: *Discrete complex analysis and geometry*

4:45–5:45pm
Isadore Singer: *Chiral anomalies and refined index theory*

6:30pm: Banquet at Royal East Restaurant
Israel Gelfand's talk given at dinner

Thursday, September 4th

9:30–10:30am
Peter Sarnak: *The generalized Ramanujan conjectures*

11:00am–noon
Bertram Kostant: *Macdonald's eta function formula and Peterson's Borel abelian ideal theorem*

2:00–3:00pm
Dennis Gaitsgory: *Uhlenbeck compactification of the moduli space of G-bundles on an algebraic surface*

3:30–4:30pm
Anatoly Vershik: *Gel'fand–Zetlin bases, virtual groups, harmonic analysis on infinite dimensional groups*

4:45–5:45pm
Joseph Bernstein: *Estimates of automorphic functions and representation theory*

Talk Given at the Dinner at Royal East Restaurant on September 3, 2003

Israel M. Gelfand
(Transcribed by Tatiana Alekseyevskaya)

It is a real pleasure to see all of you. I was asked many questions. I will try to answer some of them.

- The first question is, "Why at my age I can work in mathematics?"
- The second, "What must we do in mathematics?"
- And the third, "What is the future of mathematics?"

I think these questions are too specific. I will instead try to answer my own question:

- "What is mathematics?" (*Laughter.*)

Let us begin with the last question: What is mathematics?

From my point of view, mathematics is a part of our culture, like music, poetry and philosophy. I talked about this in my lecture at the conference. There, I have mentioned the closeness between the style of mathematics and the style of classical music or poetry. I was happy to find the following four common features: first, beauty; second, simplicity; third, exactness; fourth, crazy ideas. The combination of these four things: beauty, exactness, simplicity and crazy ideas is just the heart of mathematics, the heart of classical music. Classical music is not only the music of Mozart, or Bach, or Beethoven. It is also the music of Shostakovich, Schnitke, Shoenberg (the last one I understand less). All this is classical music. And I think, that all these four features are always present in it. For this reason, as I explained in my talk, it is not by chance that mathematicians like classical music. They like it because it has the same style of psychological organization.

There is also another side of the similarity between mathematics and classical music, poetry, and so on. These are languages to understand many things. For example, in my lecture I discussed a question which I will not answer now, but I have the answer: Why did great Greek philosophers study geometry? They were philosophers. They learned geometry as philosophy. Great geometers followed and follow the same tradition—to narrow the gap between vision and reasoning. For example, the works of Euclid summed up this direction in his time. But this is another topic.

An important side of mathematics is that it is an adequate language for different areas: physics, engineering, biology. Here, the most important word is adequate language. We have adequate and nonadequate languages. I can give you examples of adequate and nonadequate languages. For example, to use quantum mechanics in biology is not an adequate language, but to use mathematics in studying gene sequences is an adequate language. Mathematical language helps to organize a lot of things. But this is a serious issue, and I will not go into details.

Why this is issue important now? It is important because we have a "perestroika" in our time. We have computers which can do everything. We are not obliged to be bound by two operations—addition and multiplication. We also have a lot of other tools. I am sure that in 10 to 15 years mathematics will be absolutely different from what it was before.

The next question was: How can I work at my age? The answer is very simple. I am not a great mathematician. I speak seriously. I am just a student all my life. From the very beginning of my life I was trying to learn. And for example now, when listening to the talks and reading notes of this conference, I discover how much I still do not know and have to learn. Therefore, I am always learning. In this sense I am a student—never a "Führer."

I would like to mention my teachers. I cannot explain who all my teachers were because there were too many of them. When I was young, approximately 15–16 years old, I began studying mathematics. I did not have the formal education, I never finished any university, I "jumped" through this. At the age of 19, I became a graduate student, and I learned from my older colleagues.

At that time one of the most important teachers for me was Schnirelman, a genius mathematician, who died young. Then there were Kolmogorov, Lavrentiev, Plesner, Petrovsky, Pontriagin, Vinogradov, Lusternik. All of them were different. Some of them I liked, some of them, I understood how good they were but I did not agree with their—let us say softly—point of view. (Laughter) But they were great mathematicians. I am very grateful to all of them, and I learned a lot from them.

At the end, I want to give you an example of a short statement, not in mathematics, which combines simplicity, exactness, and other features I mentioned. This is a statement of a Nobel Prize winner, Isaac Bashevis Singer: "There will be no justice as long as man will stand with a knife or with a gun and destroy those who are weaker than he is."

Mathematics as an Adequate Language

Israel M. Gelfand

Introduction

This conference is called "The Unity of Mathematics." I would like to make a few remarks on this wonderful theme.

I do not consider myself a prophet. I am simply a student. All my life I have been learning from great mathematicians such as Euler and Gauss, from my older and younger colleagues, from my friends and collaborators, and most importantly from my students. This is my way to continue working.

Many people consider mathematics to be a boring and formal science. However, any really good work in mathematics always has in it: beauty, simplicity, exactness, and crazy ideas. This is a strange combination. I understood earlier that this combination is essential in classical music and poetry, for example. But it is also typical in mathematics. Perhaps it is not by chance that many mathematicians enjoy serious music.

This combination of beauty, simplicity, exactness, and crazy ideas is, I think, common to both mathematics and music. When we think about music, we do not divine it into specific areas as we often do in mathematics. If we ask a composer what is his profession, he will answer, "I am a composer." He is unlikely to answer, "I am a composer of quartets." Maybe this is the reason why, when I am asked what kind of mathematics I do, I just answer, "I am a mathematician."

I was lucky to meet the great Paul Dirac, with whom I spent a few days in Hungary. I learned a lot from him.

In the 1930s, a young physicist, Pauli, wrote one of the best books on quantum mechanics. In the last chapter of this book, Pauli discusses the Dirac equations. He writes that Dirac equations have weak points because they yield improbable and even crazy conclusions:

1. These equations assume that, besides an electron, there exists a positively charged particle, the positron, which no one ever observed.
2. Moreover, the electron behaves strangely upon meeting the positron. The two annihilate each other and form two photons.

And what is completely crazy:

3. Two photons can turn into an electron–positron pair.

Pauli writes that despite this, the Dirac equations are quite interesting and especially the Dirac matrices deserve attention.

I asked Dirac, "Paul, why, in spite of these comments, did you not abandon your equations and continue to pursue your results?"

"Because, they are beautiful."

Now it is time for a radical perestroika of the fundamental language of mathematics. I will talk about this later. During this time, it is especially important to remember the unity of mathematics, to remember its beauty, simplicity, exactness and crazy ideas.

It is very useful for me to remind myself than when the style of music changed in the 20th century many people said that the modern music lacked harmony, did not follow standard rules, had dissonances, and so on. However, Shoenberg, Stravinsky, Shostakovich, and Schnitke were as exact in their music as Bach, Mozart, and Beethoven.

1 Noncommutative multiplication

We may start with rethinking relations between the two simplest operations: addition and multiplication.

Traditional Arithmetic and Algebra are too restrictive. They originate from a simple counting and they describe and canonize the simplest relations between persons, groups, cells, etc. This language is sequential: to perform operations is like reading a book, and the axiomatization of this language (rings, algebras, skew-fields, categories) is too rigid. For example, a theorem by Wedderburn states that a finite-dimensional division algebra is always commutative.

1.1 Noncommutative high-school algebra

For twelve years, V. Retakh and I tried to understand associative noncommutative multiplication. This is the simplest possible operation: you operate with words in a given alphabet without any brackets and you multiply the words by concatenation. Part of these results are described in a recent survey, "Quasideterminants," by I. Gelfand, S. Gelfand, V. Retakh, and R. Wilson. I would say that noncommutative mathematics is as simple (or, even more simple) than the commutative one, but it is different. It is surprising how rich this structure is.

Take a quadratic equation

$$x^2 + px + q = 0$$

over a division algebra. Let x_1, x_2 be its left roots, i.e., $x_i^2 + px_i + q = 0, i = 1, 2$. You cannot write $-p = x_1 + x_2$, $q = x_1x_2$ as in the commutative case. To have the proper formulas, we have to give other clothes to x_1 and x_2. Namely, assume that the difference is invertible and set $x_{2,1} = (x_1 - x_2)x_1(x_1 - x_2)^{-1}$, $x_{1,2} = (x_2 - x_1)x_2(x_2 - x_1)^{-1}$. Then

$$-p = x_{1,2} + x_1 = x_{2,1} + x_2,$$
$$q = x_{1,2}x_1 = x_{2,1}x_2.$$

To generalize this theorem to polynomials of the nth degree with left roots $x_1, \dots, x_n$, we need to find "new clothes" for these roots by following the same pattern. For any subset $A \subset \{1, \dots, n\}$, $A = (i_1, \dots, i_m)$, and $i \notin A$, we introduce *pseudo-roots* $x_{A,i}$. They are given by the formula

$$x_{A,i} = v(x_{i_1}, \dots, x_{i_m}, x_i)x_i v(x_{i_1}, \dots, x_{i_m}, x_i)^{-1},$$

where $v(x_{i_1}, \dots, x_{i_m}, x_i)$ is the Vandermonde quasideterminant, $v(x_i) = 1$,

$$v(x_{i_1}, \dots, x_{i_m}, x_i) = \begin{vmatrix} x_{i_1}^m & \dots & x_{i_m}^m & \boxed{x_i^m} \\ & \dots & & \\ x_{i_1} & \dots & x_{i_m} & x_i \\ 1 & \dots & 1 & 1 \end{vmatrix}.$$

Suppose now that roots $x_1, \dots, x_n$ are multiplicity free, i.e., the differences $x_{A,i} - x_{A,j}$ are invertible for any A and $i \notin A$, $j \notin A$, $i \neq j$.

Let $x_1, \dots, x_n$ be multiplicity free roots of the equation

$$x^n + a_1x^{n-1} + \dots + a_n = 0.$$

Let $(i_1, \dots, i_n)$ be an ordering of $1, \dots, n$. Set $\tilde{x}_{i_k} = x_{\{i_1,\dots,i_{k-1}\},i_k}$, $k = 1, \dots, n$.

Theorem.

$$-a_1 = \tilde{x}_{i_n} + \dots + \tilde{x}_{i_1},$$
$$a_2 = \sum_{p>q} \tilde{x}_{i_p}\tilde{x}_{i_q},$$
$$\dots,$$
$$a_n = (-1)^n \tilde{x}_{i_n} \dots \tilde{x}_{i_1}.$$

These formulas lead to a factorization

$$P(t) = (t - \tilde{x}_{i_n})(t - \tilde{x}_{i_{n-1}}) \dots (t - \tilde{x}_{i_1}),$$

where $P(t) = t^n + a_1t^{n-1} + \dots + a_n$ and t is a central variable.

Thus, if the roots are multiplicity free, then we have $n!$ different factorizations of $P(t)$. In the commutative case we also have $n!$ factorizations of $P(t)$ but they all coincide.

The variables $x_{A,i}$ satisfy the relations

$$x_{A\cup\{i\},j} + x_{A,i} = x_{A\cup\{j\},i} + x_{A,j},$$
$$x_{A\cup\{i\},j} x_{A,i} = x_{A\cup\{j\},i} x_{A,j}$$

for $i \notin A$, $j \notin A$.

The algebra generated by these variables and these relations is called Q_n. This is a universal algebra of pseudo-roots of noncommutative polynomials. By going to quotients of this algebra, we may study special polynomials, for example, polynomials with multiple roots when $x_{A,i} = x_{A,j}$ for some i, j and A. Even to a trivial polynomial x^n there corresponds an interesting quotient algebra Q_n^0 of Q_n. For example, Q_2^0 is a nontrivial algebra with generators x_1, x_2 and relations $x_1^2 = x_2^2 = 0$.

Note that Q_n is a Koszul (i.e., "good") algebra and its dual also has an interesting structure.

1.2 Algebras with two multiplications

Sometimes a simple multiplication is a sum of two even simpler multiplications. A good example is the algebra of noncommutative symmetric functions studied by V. Retakh, R. Wilson, myself, and others. In the notation of Section 1.1, this algebra can be described as follows. Let $x_1, \dots, x_n$ be free noncommuting variables. Let $i_1, \dots, i_n$ be an ordering of $1, \dots, n$. Define elements $\tilde{x}_{i_1}, \dots, \tilde{x}_{i_n}$ as above. Let Sym be the algebra of polynomials in $\tilde{x}_{i_1}, \dots, \tilde{x}_{i_n}$ which are symmetric in $x_1, \dots, x_n$ as rational functions. The algebra Sym does not depend on an ordering of $1, \dots, n$, and we call it the algebra of noncommutative symmetric functions in variables $x_1, \dots, x_n$.

To construct a linear basis in algebra Sym, we need some notation. Let $w = a_{p_1} \dots a_{p_k}$ be a word in ordered letters $a_1 < \dots < a_n$. An integer m is called a *descent* of w if $m < k$ and $p_m > p_{m+1}$. Let $M(w)$ be the set of all descents of w.

Choose any ordering of $x_1, \dots, x_n$, say, $x_1 < x_2 < \dots < x_n$. For any set $J = (j_1, \dots, j_k)$, define

$$R_J = \sum \tilde{x}_{p_1} \dots \tilde{x}_{p_m},$$

where the sum is taken over all words $w = x_{p_1} \dots x_{p_m}$ such that $M(w) = \{j_1, j_1 + j_2, \dots, j_1 + j_2 + \dots + j_{k-1}\}$.

The polynomials R_J are called ribbon Schur functions; they are noncommutative analogues of commutative ribbon Schur functions introduced by MacMahon.

One can define two multiplications on noncommutative ribbon Schur functions. Let $I = (i_1, \dots, i_r)$, $J = (j_1, \dots, j_s)$. Set $I + J = (i_1, \dots, i_{r-1}, i_r + j_1, j_2, \dots, j_s)$, $I \cdot J = (i_1, \dots, i_{r-1}, i_r, j_1, j_2, \dots, j_s)$.

Set

$$R_I *_1 R_J = R_{I+J}, \qquad R_I *_2 R_J = R_{I \cdot J}.$$

The multiplications $*_1$ and $*_2$ are associative and their sum equals the standard multiplication in Sym. In other words,

$$R_I R_J = R_{I+J} + R_{I \cdot J}.$$

In fact, the algebra Sym is freely generated by one element $\tilde{x}_1 + \dots + \tilde{x}_n$ and two multiplications $*_1$ and $*_2$.

Two multiplications also play a fundamental role in the theory of integrable systems of Magri–Dorfman–Gelfand–Zakharevich. The theory is based on a pair of Poisson brackets such that any linear combination of them is a Poisson bracket. The Kontsevich quantization of this structure gives us a family of associative multiplications.

I think it is time to study several multiplications. It may bring a lot of new connections.

1.3 Heredity versus multiplicativity

An important problem both in pure and applied mathematics is how to deal with block-matrices. Attempts to find an adequate language for this problem go back to Frobenius and Schur. My colleagues and I think that we found an adequate language: quasideterminants. Quasideterminants do not possess the multiplicative property of determinants but unlike commutative determinants they satisfy the more important *Heredity Principle*: Let A be a square matrix over a division algebra and (A_{ij}) a block decomposition of A. Consider A_{ij}s as elements of a matrix X. Then the quasideterminant of X will be a matrix B, and (under natural assumptions) the quasideterminant of B is equal to a suitable quasideterminant of A. Maybe instead of categories, one should study structures with the Heredity Principle.

The determinants of multidimensional matrices also do not satisfy the multiplicative property. One cannot be too traditional here nor be restrained by requiring the multiplicative property of determinants. I think we have found an adequate language for dealing with multidimensional matrices. (See the book *Discriminants, Resultants and Multidimensional Determinants* by I. Gelfand, M. Kapranov, and A. Zelevinsky.) A beautiful application of this technique connecting multilinear algebra and classical number theory was given in the dissertation "Higher composition laws" by M. Bhargava. I predict that this is just a beginning.

2 Addition and multiplication

The simplicity of the relations between addition and multiplication is sometimes illusory. A free abelian group with one generator (denoted 1) and with operation of addition and a free abelian monoid with infinitely many generators and with operation of multiplication (called prime numbers) are the simplest objects one can imagine, but their "marriage" gives us the ring of integers Z.

And even Gross, Iwaniec, and Sarnak cannot answer all questions about the mysteries of the ring of integers—solving the Riemann hypothesis, for example.

The great physicist Lev Landau noticed, "I do not understand why mathematicians try to prove theorems about addition of prime numbers. Prime numbers were invented to multiply them and not to add." But for a mathematician, the nature of addition of prime numbers is a key point in understanding the relations between two operations: addition and multiplication.

Note that theories like Minkowski mixed volumes and valuations are very interesting forms of addition.

The invention of different types of canonical bases (Gelfand–Zetlin, Kazhdan–Lusztig, Lusztig, Kashiwara, Berenstein–Zelevinsky) are, in fact, attempts to relate addition and multiplication. Many good bases have a geometric nature, i.e., they are related or they should be related to triangulations of some polyhedra.

Another attempt is the invention of matroids by Whitney. Whitney tried to axiomatize a notion of linear independence for vectors. This gives interesting connections between algebra and combinatorial geometry. I will talk about this later.

Algebraic aspects of different types of matroids, including Coxeter matroids introduced by Serganova and me, are discussed in a recent book, *Coxeter Matroids*, by A. Borovik, I. Gelfand, and N. White. But this is just a beginning. In particular, we have to invent matroids in noncommutative algebra and geometry.

3 Geometry

Geometry has a different nature compared to algebra; it is based on a global perception. In geometry we operate with images like TV images. I do not understand why our students have trouble with geometry: they are watching TV all the time. We just need to think how to use it. Anyway, images play an increasingly important role in modern life, and so geometry should play a bigger role in mathematics and in education. In physics this means that we should go back to the geometrical intuition of Faraday (based on an adequate geometrical language) rather than to the calculus used by Maxwell. People were impressed by Maxwell because he used calculus, the most advanced language of his time.

Many talks in this conference (Dijkgraaf, Nekrasov, Schwarz, Seiberg, Vafa) are devoted to a search for proper geometrical language in physics. And never forget E. Cartan, and always learn from Atiyah and Singer.

3.1 Matroids and geometry

I want to mention only one part of geometry, combinatorial geometry, and give you only two examples. One is a notion of matroids. I became interested in matroids when I understood that they give an adequate language for the geometry of hypergeometric functions by S. Gelfand, M. Graev, M. Kapranov, A. Zelevinsky, and me. With R. Macpherson, I used matroids for a combinatorial description of cohomology classes of manifolds. Continuing this line, Macpherson used oriented matroids for a description of combinatorial manifolds. We should also have a similar theory based on symplectic and Lagrangian matroids.

In particular, we should have a good "matroid" description for Chern–Simon classes.

3.2 Geometry and protein design

Another example is my work with A. Kister, "Combinatorics and geometrical structures of beta-proteins." Step by step, analyzing real structures, we are trying to create an adequate language for this subject. It is a new geometry for live objects.

4 Fourier transforms and hypergeometric functions

In our search of an adequate language, we should not be afraid to challenge the classics, even such classics as Euler. Quite recently we realized that our approach to hypergeometric functions can be based on the Fourier transform of double exponents like $e^{xe^{\sqrt{-1}\omega t}}$, where x and ω are complex and t is a real number. The Fourier transforms of such functions are functionals over analytic functions. For example, let $F(x, \omega, z)$ be the Fourier transform of the double exponent $e^{xe^{\sqrt{-1}\omega t}}$. Then

$$\langle F(x, \omega, z), \phi(z) \rangle = \sum_{k=0}^{\infty} \frac{x^k}{k!} \phi(-k\omega).$$

We may define the action of $F(x, \omega, z)$ as $\phi \mapsto \sum \text{Res}[f(z)\phi(z)]$, where $f(z)$ is a meromorphic function with simple poles in $k\omega$, $k = 0, 1, 2, \dots$.

The function $f(z)$ is defined up to addition of an analytic function. As a representative of this class, we may choose the function

$$\Gamma_0(x, \omega, z) = \sum_{k=0}^{\infty} \frac{x^k}{k!} \frac{1}{z + k\omega},$$

or the function

$$(-x)^{-z/\omega} \Gamma(z/\omega).$$

We believe now that the function Γ_0 should replace the Euler function Γ in the theory of hypergeometric functions, but this work with Graev and Retakh is in progress.

5 Applied mathematics, nonlinear PDEs, and blowup

My search for an adequate language is based in part on my work in applied mathematics. Sergey Novikov called me somewhere, "an outstanding applied mathematician." I take it as a high compliment. I learned the importance of applied mathematics from Gauss. I think that the greatness of Gauss came in part because he had to deal with real-world problems like astronomy and so on and that Gauss admired computations. For example, I found recently that Gauss constructed the multiplication table for quaternions thirty years before Hamilton.

By the way, I remember my "mental conversation" with Gauss. When I discovered Fourier transforms of characters of abelian groups, I had an idea that now I can make a revolution with Gauss sums and to change number theory. I even imagined telling this to Gauss. And then I realized that Gauss probably would tell me, "You young idiot! Don't you think that I already knew it when I worked with my sums?"

5.1 PDEs and Hironaka

Working as an applied mathematician, I realized the importance of the resolution of singularities while working with nonlinear partial differential equations in the late 1950s. I understood that we have to deal with a sequence of resolutions (blowups) by changing variables and adding new ones. So, I was fully prepared to embrace the great result of Hironaka. We studied his paper for a year. Hironaka's theorem seems to have nothing to do with nonlinear PDEs. But for me it just shows the unity of mathematics.

Let me emphasize here that we still do not have a "Hironaka" theory for nonlinear PDEs.

5.2 Tricomi equation

When the books by Bourbaki started to appear in Moscow, I asked, "In which volume will a fundamental solution of the Tricomi equation be published?" Bourbaki did not publish this volume, and it is time to do it myself.

The Tricomi equation is

$$y\frac{\partial^2 u}{\partial x^2} + \frac{\partial^2 u}{\partial y^2} = f.$$

It is elliptic for $y > 0$ and hyperbolic for $y < 0$. With J. Barros-Neto, we found fundamental solutions for the Tricomi equation, continuing works by Leray, Agmon, and others.

Acknowledgments. I am grateful to Tanya Alexeevskaya and Tanya Gelfand for their help with the introduction, and to Vladimir Retakh for his help with the mathematical section.

The Unity
of
Mathematics

The Interaction between Geometry and Physics*

Michael Atiyah

School of Mathematics
University of Edinburgh
James Clerk Maxwell Building
The King's Buildings
Mayfield Road
Edinburgh EH9 3JZ
Scotland
U.K.
m.atiyah@ed.ac.uk

Dedicated to Israel Moisevich Gel'fand on his 90th birthday.

Subject Classifications: 5802, 5302, 8102

1 Introduction

The theme of this conference is "The Unity of Mathematics," embodying the attitude of Gel'fand himself as demonstrated in the wide range of his many original works. I share this outlook and am happy to describe one of the most fascinating examples, representing the unity of mathematics *and physics*. The speakers were also encouraged to look to the future and not be afraid to speculate—again, characteristics of Gel'fand. In my case, this is perhaps an unnecessary and even dangerous injunction, since my friends feel that I am already too much inclined to wild speculation, and very rash enthusiasm should be dampened down instead of being whipped up. Nevertheless, I will indulge myself and try to peer into the future, offering many hostages to fortune.

Mathematics and physics have a long and fruitful history of interaction. In fact, it is only in recent times, with the increasing tendency to specialization in knowledge, that any clear distinction was drawn between the two. Even when I was a student in Cambridge around 1950, we studied "natural philosophy," which included physics and mechanics, as part of the mathematical tripos. Going further back, it is a moot point whether mathematicians or physicists should claim that Newton was one of theirs.

* This is a slightly extended version of the lecture delivered at Harvard and takes account of some of the comments made to me afterwards.

The great theoretical breakthroughs in physics at the end of the 19th century and the beginning of the 20th century: electromagnetism, general relativity, and quantum mechanics were all highly mathematical, and it is impossible to describe modern physics in nonmathematical terms. Michael Faraday was the last great physicist who was unskilled in mathematics. So much has mathematics pervaded physics that Eugene Wigner has, in a much-quoted phrase, referred to "the unreasonable effectiveness of mathematics in physics."

The question whether this mathematical description of physics reflects "reality" or whether it is an imposition of the human mind is a perennial and fascinating problem for philosophers and scientists alike. Personally I am sure that new insights from neurophysiology, on how the human brain works, will shed much light on this age-old question and probably alter the very terms in which it is formulated.

Turning from the broad sweep of history to more contemporary events, one can, however, see some sharp oscillations in the synergy between mathematics and physics. The period (after the 1939–1945 war) of the great accelerators with their plethora of new particles, and the struggles of theorists with the infinities that plagued quantum field theories were far away from the concerns of most mathematicians. True, there were always mathematicians trying desperately to lay foundations, far behind the front line, and physicists themselves displayed great virtuosity in handling the techniques of Feynman diagrams as well as the symmetries of Lie groups that gradually brought order to the scene. But all this owed little to the broad mathematical community, unless one includes converts such as Freeman Dyson, and in turn it had little impact on mathematical research.

All this changed abruptly in the middle 1970s after the emergence of gauge theories, with their differential-geometric background, as the favored framework for the quantum field theory of elementary particles. Not only was there now a common language but it was soon discovered that some of the most delicate questions on both sides were closely related. These related the "anomalies" of quantum field theory to the index theory of elliptic differential operators.

Suddenly a new bridge was opened up or, to use a different metaphor, two groups digging tunnels from two ends suddenly joined up and found that the join fitted as beautifully as if it had been engineered by Brunel himself. This time the mathematicians were not building foundations; they were in the forefront where the action was.

I remember vividly those heady days and, in particular, a meeting I had in 1975 with the physicists at MIT, including Roman Jackiw and the young Edward Witten (who impressed me even then). I recall Jackiw asking whether this new interaction between the two sides was a short love affair or a long-term relationship!

Well, here we are, celebrating the silver wedding anniversary of this now firmly established marriage. The past 25 years has seen a really spectacular flowering, with tremendous impact both ways. The younger generation of theoretical physicists has rapidly mastered much of 20th century mathematics in the fields of algebraic geometry, differential geometry and topology. Many of them can manipulate spectral sequences with as much panache as the brightest graduate student in topology, and we mathematicians are constantly being asked the most searching and recondite questions in geometry and topology which stretch our knowledge to its limits.

On an occasion like this it seems appropriate to take a broad view and so I will try to survey rapidly the impact that the new physics has had on geometry (in the broad sense). This has usually taken the form of predictions, with great precision and detail, of some unexpected results or formulae in geometry. These predictions rarely come with any formal proof, though sometimes proofs can, with effort, be extracted from the physics. More often mathematicians are reduced to verifying these unexpected formulae by indirect and less conceptual methods.

What is surprising, beyond the wide scope of the results in question, is how successful the program has been. Despite the absence of any firm foundations, physical intuition and skillful use of techniques, has not yet led to false conclusions. I am tempted to reverse Wigner's dictum and wonder at "the unexpected effectiveness of physics in mathematics."

2 The background

It may be helpful to start by reviewing rapidly the parts of geometry and of physics which have been involved in this new interaction.

Let me begin with geometry. As indicated above the differential geometry of bundles, involving connections and curvatures, is basic. The link with physics goes back essentially to Hermann Weyl's attempt to interpret Maxwell's equations geometrically, and the later improvement by Kaluza. This was just the abelian case of $U(1)$-bundles, but the nonabelian case, involving general Lie groups G, is much more sophisticated and its full mathematical development came much later.

Another key component goes back to the pioneering work of Hodge with his theory of harmonic forms, in particular the refined theory of Kähler manifolds with application to algebraic geometry.

It was Witten who pointed out that Hodge theory should be viewed as supersymmetric quantum mechanics, thus providing an important bridge between key concepts on the two sides. Moreover, when extended to quantum field theories, it showed mathematicians that physicists were trying to make sense of Hodge theory in infinite dimensions. Their success in this venture depended on subtle ideas of physics, going back to Dirac, which mathematicians had to absorb.

In the first few decades after the 1939–1945 war, topology was taking center stage in geometry. New concepts and techniques led to a good understanding of global topological problems in differential and algebraic geometry, culminating in Hirzebruch's famous generalization of the Riemann–Roch theorem. This involved a skilled use of algebraic machinery centering around the theory of characteristic classes and the remarkable polynomials originally introduced by J. A. Todd.

All of this was motivated by internal mathematical questions derived in the main from classical algebraic geometry, now augmented by the powerful machinery of sheaf cohomology of Leray, Cartan and Serre. Any suggestion that this might have relevance to physics would have been met with disbelief.

In fact, the most remarkable fact about the new geometry–physics interface is that topology lies at its heart. In retrospect the roots of this can be seen to go back to Dirac.

His argument explaining the quantization of electric charge—where all particles have an electric charge which is an integer multiple of the charge of the electron—is essentially topological. In modern terms, he argued that a charged particle, moving in the background field of a point magnetic monopole, had a quantum-mechanical wave function which was a section of a complex line bundle (defined outside the monopole). Thus, while classical forces can be expressed purely locally by differential geometric formulae, quantum mechanics forces a global topological view and the *integer* topological invariants correspond to *quantized* charges.

The full implications of this link between quantum theory and topology only emerged when string theory appeared in physics with its Kaluza–Klein requirement for extra dimensions above the four of space–time. The geometry and topology of the extra dimensions provided a strong link with the mainstream development of contemporary geometry.

The role of Lie groups and symmetry in physics has been clear for some time and this already has geometric and topological implications, but the higher dimensions of Kaluza–Klein, as mediated by string theory, involve manifolds which are not necessarily homogeneous spaces of Lie groups. This means that algebra alone is not the answer, and that the full power of modern geometry, including Hodge theory and sheaf cohomology, is required.

As mentioned briefly in Section 1, a key connection between the geometry and the physics came from "anomalies" and their relation to index problems. These were a natural extension of Hirzebruch's work on the Riemann–Roch theorem and the famous Todd polynomials, and variants of them now appeared as having important physical significance. In fact, Hirzebruch's work had been generalized by Grothendieck with his introduction of K-theory and, in its topological version, this turned out to be a very refined tool for investigating anomalies in physics. Some of these are purely global, having no local integral formulation, and K-theory detects such torsion invariants. These, and other clues, indicate that K-theory plays a fundamental role in quantum physics but the deeper meaning of this remains obscure.

Finally, I should say a word about the role of spinors. Ever since they arose in physics with the work of Dirac they have played a fundamental part, providing the fermions of the theory. In mathematics spinors are well understood algebraically (going back to Hamilton and Clifford) and their role in the representation theory of the orthogonal group provides the link with physics. However, in global geometry, spinors are much less understood. The Dirac operator can be defined on spinor fields and its square is similar to the Hodge–Laplace operator. Its index is given by the topological formula referred to above in connection with anomalies. However, while the geometric significance of differential forms (as integrands) is clear, the geometric meaning of spinor fields is still mysterious. The only case where they can be interpreted geometrically is for complex Kähler manifolds where holomorphic function theory essentially extracts the "square root of the geometry" that is needed. Gauss is reputed to have said that the true metaphysics of $\sqrt{-1}$ is not simple. The same could be said for spinors, which are also a mysterious kind of square root. Perhaps this remains the deepest mystery on the geometry–physics frontier.

3 Dimensional hierarchy

Although string theory may require higher dimensions at a fundamental level, at normal energy scales we operate in a space–time of four dimensions, and the extra Kaluza–Klein dimensions merely determine the kinds of fields and particles that we have to deal with.

Since four-dimensional theories present many serious problems it is useful to study simpler "toy models" in low dimensions. We can then think of a dimensional hierarchy where the theory gets more complicated as we increase the dimensions. In general, we write $D = d + 1$ for the space–time dimension, d being the space dimension.

For $d = 0$, we just have quantum mechanics and the associated mathematics of (finite-dimensional) manifolds, Lie groups, etc. For $d = 1$, we get the first level of quantum field theory, which involves things like loop spaces and loop groups. Much of this can now be treated by mathematically rigorous methods, but it is still a large and sophisticated area. For $d = 2$, the quantum field theory becomes more serious and rigour, for the most part, has to be left behind. This is even more so with the case $d = 3$ of the real world.

The increasing complexity as D increases is reflected by (and perhaps due to) the increasing complexity of the Riemannian curvature. Thus for $D = 1$ there is no curvature, for $D = 2$ we have only the scalar curvature, and for $D = 3$ we have the Ricci curvature, while only for $D = 4$ do we have the full Riemann curvature tensor. At the level of the Einstein equations, for classical relativity, this is related to the increasing difficulty of geometric structures in dimensions $D \leq 4$. For $D = 2$, we have the classical theory of Riemann surfaces (or surfaces of constant curvature). For $D = 3$ the theory of 3-manifolds is already much deeper as is made clear by the work of Thurston (and more recently Perelman). For $D = 4$ the situation is vastly different, as has been shown by Donaldson, using ideas coming from physics as we shall discuss later.

In the subsequent sections, we shall review the ways in which physics has impacted on mathematics, organizing it according to this dimensional hierarchy. However, before proceeding, we should make one general remark which applies throughout. Typically, in the applications, there are formulae which depend on some integer parameter such as a degree. From the physics point of view, what naturally emerges is something like a generating function involving a sum over all values of the parameter. Traditionally this is not the way geometers would have looked for the answer, and one of the remarkable insights arising from the physics is that the generating functions are very natural objects, sometimes being solutions of differential equations.

4 Space–time dimension 2

4.1 Rigidity theorems

The space V of solutions of an elliptic differential operator on a compact manifold M is finite dimensional. If a compact group G acts on the manifold, preserving the operator, then V becomes a representation of G. In fact, all representations of compact Lie groups arise in this way. In very special circumstances we may have *rigidity*, meaning that the representation on V is trivial. For example, if V is the space of harmonic forms, of degree p, and if G is connected, then by Hodge theory the action is trivial. A different example arises if we take the Dirac operator D and spin manifold M (of dimension $4k$), then D and its adjoint D^* have solution spaces V^+, V^- and the index of D is defined as

$$\text{index } D = \dim V^+ - \dim V^-.$$

If G acts on M preserving the metric, then it commutes with D and so V^+, V^- become representations of G and the index becomes a virtual representation or character. The rigidity theorem (of Atiyah and Hirzebruch) says that this character is trivial, i.e., a constant. (In fact, more is true—it is zero for a nontrivial action.)

Arguments from quantum field theory (for maps of space–time into M) led to the discovery of a whole sequence of rigidity theorems for the Dirac operator coupled to certain bundles. Moreover, the generating function turns out to be a modular form, something predicted by the relativistic invariance of the quantum field theory.

This discovery stimulated a whole new branch of topology, called "elliptic cohomology" with fascinating connections to number theory as explained by Michael Hopkins [6].

This subject is an application of physics to differential topology, but the remaining subjects of this section will be concerned with algebraic geometry.

4.2 Moduli spaces of bundles

The Jacobian of an algebraic curve classifies all holomorphic line bundles over it with degree zero. It can also be described as the moduli space of flat $U(1)$-bundles. Its study was a major feature of 19th century mathematics in the context of theta functions. It has a natural generalization to vector bundles of higher rank, the study of which emerged in the middle of the 20th century and is much more involved. In particular, not much was even known about the topology of these moduli spaces.

Again quantum field theory in two dimensions has led to beautiful formulae relating to the cohomology of these moduli spaces. Rigorous mathematical proofs of these results are now available, inspired by the physics.

4.3 Moduli spaces of curves

Somewhat analogous, but deeper than the moduli spaces of Section 4.2, is the moduli space of curves of genus g. The classical theory of the elliptic modular function deals with the case $g = 1$, but for higher genus the moduli space remained rather unknown.

As in Section 4.1, physics has again produced remarkable formulae for the cohomology of these moduli spaces. This time the physics is related to gravity rather than gauge theory and is important for string theory.

4.4 Quantum cohomology

Classical geometry led to many enumerative problems, the simplest of which was to count how many points were common to a number of subvarieties of a given algebraic variety. This led to *intersection theory*, which, in the hands of Lefschetz, was developed as an aspect of homology theory. Subsequently, this was viewed as the ring structure of cohomology theory.

A deeper class of enumerative problems arises when we want to count not *points* but *curves*. How many curves of given type (degree, genus, singularity structure) lie on a given algebraic variety. This was a difficult unsolved problem even for curves in the plane.

In quantum field theory, holomorphic curves appear as "instantons" of a two-dimensional field theory, measuring important nonperturbative features of the theory.

Taking into account instantons of genus zero leads in particular to a ring associated with the target manifold which depends on a parameter t and in the classical limit $t \to 0$ reduces to the cohomology ring. This new ring is called the quantum cohomology ring and it encodes information about numbers of rational curves.

The quantum cohomology rings of various varieties have been calculated, thus leading to explicit enumerative formulae. An important point to mention is that the quantum cohomology ring only has a grading into odd and even parts, not an integer grading like the classical cohomology.

4.5 Mirror symmetry

This subject, which has now grown into a large industry, is related to Section 4.4 and is one of the ways in which the enumerative problems have been solved.

It was discovered by physicists that certain algebraic varieties come in pairs M and M^*, called mirror pairs. The most interesting case is when M and M^* are three-dimensional complex algebraic varieties with vanishing first Chern class (Calabi–Yau manifolds). The remarkable thing about mirror symmetry is that M and M^* have quite different topologies. In fact, the ranks of the odd and even Betti numbers switch, so that

$$\chi(M^*) = -\chi(M),$$

where χ is the Euler number.

For physicists, M and M^* give rise to the same two-dimensional quantum field theory, but quantum invariants involving instanton calculations on M can be calculated by classical invariants involving periods of integrals on M^*. This is what leads to effectively computable formulae and gives the theory its power.

The mathematical study of Mirror Symmetry has now progressed quite far involving symplectic as well as complex geometry. Recent work is formulated in the language of derived categories and surprisingly such extremely abstract mathematical techniques appear in return to be relevant to the physics of string theory.

5 Three-dimensional space–time

The most striking application of quantum field theory in three dimensions was undoubtedly Witten's interpretation of the polynomial knot invariants discovered by Vaughan Jones. It was already clear, from the work of Jones, that his invariants were essentially new and very powerful. Old conjectures were quickly disposed of. What Witten did was to show how the Jones invariants could be easily understood (and generalized) in terms of the quantum field theory defined by the *Chern–Simons Lagrangian*. One immediate benefit of this was that it worked for any oriented 3-manifold, not just S^3. In particular, taking the empty knot one obtained numerical invariants for compact 3-manifolds.

These developments have stimulated a great deal of work by geometers. In particular, there are combinatorial treatments which are fully rigorous and mimic much of the physics.

In three dimensions, we are in the odd situation of having two completely different theories. One the one hand there are the quantum invariants just discussed, while on the other hand there is the deep work of Thurston on geometric structures, including the important special case of hyperbolic 3-manifolds. It has been a long-standing and embarrassing situation that there was little or no connection between these two theories. For example, given an explicit compact hyperbolic 3-manifold, how does one compute its quantum invariants? Some answers were available for the simpler structures (positive curvature or fibrations) but not for the hyperbolic case.

Recently conjectures have been made proposing a general link, with hyperbolic volumes appearing as limits of Jones invariants. In particular Gukov [5] has attempted to establish a basis for this link using Chern–Simons theory for the noncompact groups $SL(2, \mathbb{C})$ which ties in to three-dimensional gravity. This follows earlier work of Witten and others. It looks very promising and one might hope to connect it ultimately to the recent work of Perelman [9].

Perhaps, looking at current research and peering into the future, I can make a few further comments.

In the first place, while quantum field theory gives (at least heuristically) a very satisfying explanation of the Jones theory and most of its properties, it fails in one important respect. It does not explain why the coefficients of the Jones polynomials are *integers*. In Witten's description, the values of the Jones polynomials at certain roots of unity are expectation values and the physics gives no indication of their arithmetic nature.

A really fundamental treatment should provide such an explanation, while preserving the elegance of the quantum field theory approach.

After my lecture, I was reminded that recent work of Khovanov [7] does give a direct explanation for the integer coefficients in the Jones polynomial. Khovanov constructs, from a knot, certain homology groups as invariants and the Jones coefficients appear as Euler characteristics. While this explains their integrality it does not explain what relations these Khovanov homology groups have to the physics.

There is a somewhat parallel situation with respect to the Casson invariant of a homology 3-space. On the one hand Witten has shown that it is given by a variant of Chern–Simons theory. On the other hand it can also be interpreted as the Euler characteristic of the Floer cohomology groups, which are the Hilbert spaces of the Donaldson quantum field theory in four dimensions (as we shall discuss in the next section). This might suggest that the Khovanov homology groups should simply be interpreted as the Hilbert spaces of some four-dimensional quantum field theory. No such theory appears at present to be known.

Speculating in another direction, I note that the Jones polynomial is naturally a character of the circle, the integers being the multiplicities of the irreducible representations. One may ask where the circle comes from. Now the knots studied by Jones are traditional ones in $\mathbb{R}^3$ and we have an S^2 at ∞, on which $SO(3)$ acts. Moreover, the equivariant K-theory of S^2 is given by the character ring of the circle

$$K_{SO(3)}(S^2) \cong R(S^1).$$

Here S^1 appears as the isotropy group of the action (and is unique up to conjugation). Since K-theory appears to play a special role in quantum field theory it is tempting to interpret the Jones polynomial as an element of the equivariant K-theory of S^2, where we think of S^2 as any large 2-space enclosing the knot. This idea receives some encouragement from the fact that the Jones polynomial for links (generalizing knots) involves integer series in $t^{\frac{1}{2}}$, which corresponds to characters of the double cover of our original S^1. But this is natural if we replace $SO(3)$ by Spin(3), as a physicist would do.

Such equivariant K-groups have appeared in connection not with knots in $\mathbb{R}^3$ but in connection with finite configurations of distinct points in $\mathbb{R}^3$ [2] and this might provide some link. This idea is reinforced by the further speculation made in [3] relating to Hecke algebras, which provide the original Jones approach to knot invariants.

As explained in [2] the 2-sphere involved there is naturally the complex 2-sphere, which occurs as the base of the light-cone in Minkowski space. The connection with quantum theory that is postulated is in the spirit of the ideas of Roger Penrose as mentioned later in Section 9.

6 Four-dimensional space–time

I have already alluded several times to Donaldson theory, on which I shall now elaborate.

For any compact oriented 4-manifold X, any compact Lie group G and any positive integer k, Donaldson studies the moduli space M of k-instantons. These are anti-self-dual connections for the G-bundle (with topology fixed by k). For this he has first to choose a Riemannian metric (or rather a conformal structure), but he then computes some intersection numbers on M and shows these are independent of the metric. In this way, Donaldson defines invariants of X, which are just polynomials on the second homology of X.

As is now well known, these Donaldson invariants proved spectacularly successful in distinguishing between 4-manifolds and they opened up the whole subject of smooth 4-manifolds just as Freedman had closed the subject of topological 4-manifolds.

While the idea of using instantons came from physics, Donaldson was just using the classical equations of Yang–Mills theory. But Witten subsequently explained that Donaldson's theory could be interpreted as a suitable quantum field theory in four dimensions. Moreover, this was just a slight variant on a standard theory known as $N = 2$ supersymmetric Yang–Mills.

This physical interpretation of the Donaldson theory was interesting for physicists but it was not clear what the mathematical benefit was. However, a few years later, the benefit became abundantly clear. As part of some very general ideas of duality in quantum fields theories, Seiberg and Witten produced a quite different theory which was expected to be equivalent to Donaldson theory. This has now been essentially confirmed by mathematicians, though a rigorous proof of the equivalence is not yet complete. Moreover, the Seiberg–Witten equations are technically easier to handle and so they have proved more powerful in many cases. In particular, they have led to a proof of the old conjecture of René Thom about the genus of surfaces embedded in $\mathbb{CP}_2$.

I should emphasize that the equivalence between the Donaldson and Seiberg–Witten theories is one between generating functions. Each theory has its instantons, but there is no simple relation between instantons of separate degrees, only between the total sums over all degrees. This should be compared with the classical Poisson summation formula which expresses a sum over one lattice in terms of the sum of Fourier transforms over the dual lattice. Thus these dualities of quantum field theories should be viewed as some kind of nonlinear analogues of the Fourier transfom. I shall return to this theme at the end of my lecture.

One surprising feature of the Seiberg–Witten theory, and the classical equations they lead to, is that they deal with a $U(1)$ theory coupled nonlinearly to spinors. Thus spinors appear explicitly here, while they do not appear in the $SU(2)$ Donaldson theory. This only increases the mystery of spinors and emphasizes my earlier remarks about our lack of any deep understanding of them.

At present it is not clear whether Donaldson theory, with various refinements, will explain all geometric phenomena in four dimensions. It may do so, but it is also possible that it may take another 100 years to fully understand the geometry of four dimensions, just as it has taken a century to move from Riemann surfaces to an equivalent understanding of three dimensions. If so, this may accompany a similar period for a proper understanding of the physics of space–time, a topic to which I will return in the last section.

7 Topological quantum theories

As I have quickly outlined, there are a large number of important areas where quantum theories yield topological results. All of these are, in fact, topological field theories. They are especially simple theories in which the only output is topological. The Hamiltonian of such a theory is zero, so there is no continuous dynamics. However, the theory has nontrivial content related to topological phenomena. This makes this area much simpler and hence more tractable mathematically. In [1] I gave an axiomatic description of a topological quantum field theory, analogous to the classical axiomatization of homology by Eilenberg and Steenrod. The key part of such a theory is its construction by some explicit method which could in principle (as with homology) be either combinatorial or analytic.

It might appear that, for a real physicist, such purely topological theories could have no serious interest. But this is wrong for two reasons. In the first place the complexity of a really physical quantum field theory can rise from one of two sources. First there is the analytical study of small fluctuations, but this is to a great extent based on standard examples and perturbation theory. Then there are nonperturbative phenomena and these are illustrated very well by purely topological theories.

But not only can topological theories play the role of toy models to study nonperturbative effects, they can also arise from a physical theory in some limiting regime. As a simple illustration consider Hodge theory, or supersymmetric quantum mechanics. The full theory requires us to know all the eigenvalues of the Hodge Laplacian. But under rescaling we can consider the limit when all eigenvalues get very large, so that only the zero eigenvalues survive. This recovers the homology (as the harmonic forms).

Although I have spoken only about quantum field theories, the connection between geometry and physics also extends to string theories. In particular, Witten has shown that Chern–Simons theory for $U(N)$, in its perturbative form, is a topological string theory for open strings on T^*S^3, the cotangent bundle of S^3, with the 0-section as a brane of multiplicity N where the string must end.

More recently Vafa and others have argued that this theory is dual to the (topological) theory of closed strings on a rank 2 vector bundle over $\mathbb{CP}_1$. In this duality, one switches from a perturbative expansion valid for *large level*, of Chern–Simons $U(N)$ gauge theory, to a perturbative expansion of the closed string theory for *large* N. The geometry behind this duality is best understood in M-theory terms involving a suitable 7-manifold of G_2-holonomy [4].

This duality of Vafa leads to explicit formulae for every genus of the world-sheet and these have now been verified by mathematical computations using fixed-point methods on moduli spaces [8]. Interestingly, in the end, everything boils down to purely combinatorial formulae. On the one hand string theory arises from the Feynman diagrams of perturbation theory, while on the other hand we have combinatorial data associated with degenerate algebraic curves. Riemann surfaces and analysis interpolate between these two different combinatorial schemes, but with sufficient skill a direct computation can be made. However, this is not very enlightening.

In his lecture at this conference, Vafa [10] discusses many aspects of topological theories in much greater detail. I refer to his text for more information.

8 The significance for mathematics

It should be clear from my rapid survey that quantum theory, in its modern form, has had profound consequences for mathematics and in particular for geometry. But it is hard to grasp the real significance of all this and to predict what its future will be.

While physics can inject new ideas and techniques into mathematics, it cannot in the end provide a foundation for it. Even if, one day, we can develop a completely rigorous quantum field theory or string theory, it would be bizarre if this had to be the pillar on which mathematics, or large parts of it, rested.

A historical perspective may help us get a glimpse of the future in this respect. Fourier analysis emerged, in the 18th century, from physics, specifically the study of heat conduction. But in due course it was absorbed into a purely mathematical theory, and was fundamental in the subsequent development of Linear Analysis. Later, in the 20th century, this theory was generalized to the noncommutative situation centering around group representation theory. In fact, one could say that noncommutative Fourier analysis has been one of the central theories of 20th century mathematics.

As I have mentioned earlier, the dualities of quantum field theory and string theory which lie behind some of the most striking applications of physics to geometry can be viewed as some kind of nonlinear Fourier Transform. In special finite-dimensional cases these are now understood mathematically, and are related to classical ideas of integral geometry. These include the Penrose Transform, the Mukai Transform, the Nahm Transform, and the inverse scattering transform in soliton theory. In fact, solitons are a prominent part of all these dualities. However, the full dualities of string theory (or QFT) are infinite dimensional and nonlinear.

All this suggests that a prominent theme of 21st century mathematics might be the development of a fully-fledged nonlinear Fourier Transform theory for function spaces. Of course, there are too many kinds of nonlinearity to be encompassed in any nontrivial way by a single theory. Clearly, physics appears to be singling out a type of nonlinearity for which a deep but tractable duality will hold. The key feature of this nonlinearity appears to be supersymmetry, which in some way extends the symmetry arising in group theory. In geometric terms this means that we deal not just with homogeneous spaces of Lie groups but also with Riemannian manifolds having special holonomy, such as Kähler manifolds, Calabi–Yau manifolds or G_2-manifolds. The Lie groups are still there, but only at the (integrable) infinitesimal level. These ideas are, in fact, not far removed from the original ideas of Lie, who moved on from finite-dimensional Lie groups to the infinite-dimensional structures occurring for example in complex manifolds. Lie himself was disappointed that his fundamental ideas did not appear to be given their due credit. The 20th century certainly rectified this omission, but perhaps the 21st century will take it even further.

9 The significance for physics

In the previous section, I tried to peer into the future to see what kind of mathematics might emerge from the current geometry–physics interface. Trying to forecast the physics is even harder, and I am less qualified, but perhaps an outsider can offer a different perspective.

As we know, the holy grail in current fundamental physics is how to combine Einstein's Theory of General Relativity with Quantum Theory. These two theories operate very effectively but at quite different scales, GR at cosmic distances and QM at subatomic scales.

The difficulty in combining the two theories is both conceptual and technical. As is well known, Einstein dreamed of a unified geometric theory, extending GR, and he never accepted the philosophical foundations of QM, with its uncertainty principle. In the long debate on this controversy between Einstein and Bohr the general verdict of the physics community was that Einstein lost and that his idea of a unified field theory was a hopeless pipe dream.

With the remarkable success of the standard model of elementary particles, incorporating geometrically the electromagnetic, the weak force and the strong force, Einstein's ideas were given new life. But the framework remained that of QM, and GR remained strictly outside the scope of the unification. Now, with string theory offering the hope of the ultimate unification it might appear that the old controversy between Einstein and Bohr has been resolved, with the honours more equally split. Unification is perhaps being achieved, but QM has persisted.

This is the orthodox view of string theorists and they have impressive evidence in their favor. The only fly in the ointment is that no one yet has any real idea of what their ultimate M-theory is. Perhaps in the coming years this will be clarified and we will learn to live with the mysterious world of 11 dimensions and its hidden supersymmetries. Perhaps only a few technical obstacles remain to complete the structure.

But it is at least worth exploring alternative scenarios. There are in particular two attractive ideas that have their devotees. The first (in historical precedence) is Roger Penrose's twistor theory. On the one hand this has, as a technical mathematical tool, proved its worth in a number of problems. It is also related to supersymmetry and duality. Links with string theory are being explored. But beyond these mathematical technicalities there lies a deeper philosophical idea. Penrose is an Einsteinian who believes that in the hoped-for marriage between GR and QM it is the latter that must give the most, adapting itself to the beauty of GR. Twistors are thought of as a first step to achieving this goal. Moreover, Penrose speculates that the mysterious role of complex numbers in QM should ultimately have a geometric origin in the natural complex structure of the base of the light-cone in Minkowski space. So far, it has to be conceded that the weight of evidence is not in Penrose's favor, but that does not mean that he may not ultimately be vindicated.

A completely different scenario is offered by Alain Connes' noncommutative geometry, a theory with a rich mathematical background and a promising future. Links with physics exist and new ones are being discovered. In a sense Connes takes off from the Heisenberg commutation relations, in a definitely non-Einsteinian direction. However, he tries to keep the geometric spirit by using the same concepts and terminology. It is certainly possible that the final version of M-theory may use Connes' framework for its formulation.

Perhaps I can end by indulging in some wild speculation of my own, not I hope totally unrelated to the other ideas above.

I start, further back, by asking some philosophical or metaphysical questions. If we end up with a coherent and consistent unified theory of the universe, involving extremely complicated mathematics, do we believe that this represents "reality"? Do we believe that the laws of nature are laid down using the elaborate algebraic machinery that is now emerging in string theory? Or is it possible that nature's laws are much deeper, simple yet subtle, and that the mathematical description we use is simply the best we can do with the tools we have? In other words, perhaps we have not yet found the right language or framework to see the ultimate simplicity of nature.

To get a better idea of what I am trying to say, let us consider GR as a description of gravity. To a mathematician this theory is beautifully simple but yet subtle. Moreover, it is highly nonlinear so that it is extremely complicated in its detailed implications. This is no doubt why it appeals to both Einstein and Penrose as a model theory. Is it not possible that something having the same inherent simplicity (and nonlinearity) can explain all of nature?

While everyone might agree that this would be an ideal philosophical ambition, there appears to be the insuperable obstacle presented by QM. To get round this will require some conceptual leap, and such leaps have in the past only come when one is prepared to sacrifice some accepted dogma, such as Einstein did with the separation of space and time.

Let me, in such a speculative mood, raise one possibility of a dogma to be sacrificed. Ever since Newton, it has been a cardinal principle of physical sciences that we can predict the future from the present (given complete knowledge). This even holds in QM, where the state at time zero evolves by a Hamiltonian flow to give the state at future times. This assumption, which may have seemed rash to some, has abundantly proved its worth. But is it really true? Perhaps all we can say is that a knowledge of present *and past* enables us to predict the future? After all, this, in a sense, is true in the biological world where our DNA represents our past.

Of course, to explain the remarkable success of the standard dogma, the effect of the past would have to be minute and only noticeable at very short time-scales. But this is precisely where QM comes into play. So perhaps the uncertainty in QM is really a reflection of the fact that we (the observers) do not know our past. Perhaps the Hilbert Space state at the present time is determined by our past.

This metaphysical idea would have to be embodied in precise mathematical form consistent with GR. In particular, the fundamental equations would not be differential equations but integrodifferential equations, involving integration over the past. The nonlinearity of GR, together with the effect of past history, would be difficult to solve mathematically. But very good approximations might be obtained by using high precision mathematical tools of the kind appearing in string theory. The various dualities might appear from alternative ways of making the necessary approximations.

A theory on these lines would have satisfied Einstein and it seems at least worth exploring. The dream of all mathematical physicists is to find ultimate explanations which are inherently simple in mathematical form and yet can explain the fascinating diversity of nature. We should not settle for less.

References

[1] M. F. Atiyah, *Topological Quantum Field Theories*, Publications Mathématiques 68, Institut des Hautes Études Scientifiques, Bures-sur-Yvette, France, 1989, 175–186.

[2] M. F. Atiyah, Configurations of points, *Philos. Trans. Roy. Soc. London Ser.* A, **359** (2001), 1375–1387.

[3] M. F. Atiyah and R. Bielawski, Nahm's equations, configuration spaces and flag manifolds, *Bull. Brazil Math. Soc. N. S.*, **33** (2002), 157–176.

[4] M. F. Atiyah, J. Maldacena, and C. Vafa, An M-theory flop as a large $\mathbb{N}$ duality, *J. Math. Phys.*, **6** (2002), 3209–3220.

[5] S. Gukov, Three-dimensional quantum gravity Chern-Simons theory and the A-polynomial, 2003, arXiv:hep-th/0306165.

[6] M. J. Hopkins, Algebraic topology and modular forms, in *Proceedings of the ICM, Beijing* 2002, Vol. 1, Higher Education Press, Beijing, 2002, 283–309; also in *Comm. Math. Phys.*, **255**-3 (2005), 577–627.

[7] M. Khovanov. A categorification of the Jones polynomial, *Duke Math. J.*, **101**-3 (2000), 359–426.

[8] K. Liu, Mathematical results inspired by physics, in L. I. Tatsien, ed., *Proceedings of the International Congress of Mathematicians, Beijing* 2002, Vol. 3, Higher Education Press, Beijing, 2003, 457–466.

[9] G. Perelman, The entropy formula for the Ricci flow and its geometric applications, 2002, arXiv:math.DG/0211159.

[10] C. Vafa, Unity of topological field theories, lecture given at *An International Conference on "The Unity of Mathematics,"* Harvard University, Cambridge, MA, 2003.

Uhlenbeck Spaces via Affine Lie Algebras

Alexander Braverman[1], Michael Finkelberg[2], and Dennis Gaitsgory[3]

[1] Department of Mathematics
Brown University
Providence, RI 02912
USA
braval@math.brown.edu
[2] Independent Moscow University
11 Bolshoj Vlasjevskij per.
Moscow 119002
Russia
fnklberg@mccme.ru
[3] Department of Mathematics
The University of Chicago
Chicago, IL 60637
USA
gaitsgde@math.uchicago.edu

Abstract. Let G be an almost simple simply connected group over $\mathbb{C}$, and let $\mathrm{Bun}_G^a(\mathbb{P}^2, \mathbb{P}^1)$ be the moduli scheme of principal G-bundles on the projective plane $\mathbb{P}^2$, of second Chern class a, trivialized along a line $\mathbb{P}^1 \subset \mathbb{P}^2$.

We define the Uhlenbeck compactification $\mathfrak{U}_G^a$ of $\mathrm{Bun}_G^a(\mathbb{P}^2, \mathbb{P}^1)$, which classifies, roughly, pairs $(\mathcal{F}_G, D)$, where D is a 0-cycle on $\mathbb{A}^2 = \mathbb{P}^2 - \mathbb{P}^1$ of degree b, and $\mathcal{F}_G$ is a point of $\mathrm{Bun}_G^{a-b}(\mathbb{P}^2, \mathbb{P}^1)$, for varying b.

In addition, we calculate the stalks of the Intersection Cohomology sheaf of $\mathfrak{U}_G^a$. To do that we give a geometric realization of Kashiwara's crystals for affine Kac–Moody algebras.

Subject Classifications: 14D20, 14D21, 17B67

Introduction

0.1

Let G be an almost simple simply connected group over $\mathbb{C}$, with Lie algebra $\mathfrak{g}$, and let $\mathbf{S}$ be a smooth projective surface.

Let us denote by $\mathrm{Bun}_G^a(\mathbf{S})$ the moduli space (stack) of principal G-bundles on $\mathbf{S}$ of second Chern class a. It is easy to see that $\mathrm{Bun}_G^a(\mathbf{S})$ cannot be compact, and the source of the noncompactness can be explained as follows:

By checking the valuative criterion of properness, we arrive at the following situation: we are given a G-bundle $\mathcal{F}_G$ on a three-dimensional variety $\mathcal{X}$ defined away from a point, and we would like to extend it to the entire $\mathcal{X}$. However, such an extension does not always exist, and the obstruction is given by a positive integer, which one can think of as the second Chern class of the restriction of $\mathcal{F}_G$ to a suitable 4-sphere corresponding to the point x.

However, this immediately suggests what a compactification of $\mathrm{Bun}_G^a(\mathbf{S})$ could look like: it should be a union

$$\bigcup_{b\in\mathbb{N}} \mathrm{Bun}_G^{a-b}(\mathbf{S}) \times \mathrm{Sym}^b(\mathbf{S}). \tag{1}$$

In the differential-geometric framework of moduli spaces of K-instantons on Riemannian 4-manifolds (where K is the maximal compact subgroup of G) such a compactification was introduced in the pioneering work [U]. Therefore, we shall call its algebro-geometric version the Uhlenbeck space, and denote it by $\mathfrak{U}_G^a(\mathbf{S})$.

Unfortunately, one still does not know how to construct the spaces $\mathfrak{U}_G^a(\mathbf{S})$ for a general group G and an arbitrary surface $\mathbf{S}$. More precisely, one would like to formulate a moduli problem, to which $\mathfrak{U}_G^a(\mathbf{S})$ would be the answer, and so far this is not known. In this formulation the question of constructing the Uhlenbeck spaces has been posed (to the best of our knowledge) by V. Ginzburg. He and V. Baranovsky (cf. [BaGi]) have made the first attempts to solve it, as well as indicated the approach adopted in this paper.

A significant simplification occurs for $G = SL_n$. Let us note that when $G = SL_n$, there exists another natural compactification of the stack $\mathrm{Bun}_n^a(\mathbf{S}) := \mathrm{Bun}_{SL_n}^a(\mathbf{S})$, by torsion-free sheaves of generic rank n and of second Chern class a, called the Gieseker compactification, which in this paper we will denote by $\widetilde{\mathfrak{N}}_n^a(\mathbf{S})$. One expects that there exists a proper map $\mathsf{f} : \widetilde{\mathfrak{N}}_n^a(\mathbf{S}) \to \mathfrak{U}_{SL_n}^a(\mathbf{S})$, described as follows:

A torsion-free sheaf $\mathcal{M}$ embeds into a short exact sequence

$$0 \to \mathcal{M} \to \mathcal{M}' \to \mathcal{M}_0 \to 0,$$

where $\mathcal{M}'$ is a vector bundle (called the saturation of $\mathcal{M}$), and $\mathcal{M}_0$ is a finite-length sheaf. The map should send a point of $\widetilde{\mathfrak{N}}_n^a(\mathbf{S})$ corresponding to $\mathcal{M}$ to the pair $(\mathcal{M}', \mathrm{cycle}(\mathcal{M}_0)) \in \mathrm{Bun}_n^{a-b}(\mathbf{S}) \times \mathrm{Sym}^b(\mathbf{S})$, where b is the length of $\mathcal{M}_0$, and $\mathrm{cycle}(\mathcal{M}_0)$ is the cycle of $\mathcal{M}_0$. In other words, the map must "collapse" the information of the quotient $\mathcal{M}' \twoheadrightarrow \mathcal{M}_0$ to just the information of the length of $\mathcal{M}_0$ at various points of $\mathbf{S}$.

Since the spaces $\widetilde{\mathfrak{N}}_n^a(\mathbf{S})$, being a solution of a moduli problem, are easy to construct, one may attempt to construct the Uhlenbeck spaces $\mathfrak{U}_{SL_n}^a(\mathbf{S})$ by constructing an explicit blowdown of the Gieseker spaces $\widetilde{\mathfrak{N}}_n^a(\mathbf{S})$. This has indeed been performed in the works of J. Li (cf. [Li]) and J. W. Morgan (cf. [Mo]).

The problem simplifies even further, when we put $\mathbf{S} = \mathbb{P}^2$, the projective plane, and consider bundles trivialized along a fixed line $\mathbb{P}^1 \subset \mathbb{P}^2$. In this case, the sought-for space $\mathfrak{U}_n^a(\mathbf{S})$ has been constructed by S. Donaldson (cf. [DK, Chapter 3]) and thoroughly studied by H. Nakajima (cf., e.g., [Na]) in his works on quiver varieties.

In the present paper, we will consider the case of an arbitrary group G, but the surface equal to $\mathbb{P}^2$ (and we will be interested in bundles trivialized along $\mathbb{P}^1 \subset \mathbb{P}^2$, i.e., we will work in the Donaldson–Nakajima setup.)

We will be able to construct the Uhlenbeck spaces $\mathfrak{U}^a_G$, but only up to nilpotents. In other words, we will have several definitions, two of which admit modular descriptions, and which produce the same answer on the level of reduced schemes. We do not know whether the resulting schemes actually coincide when we take the nilpotents into account. And neither do we know whether the resulting reduced scheme is normal.

We should say that the problem of constructing the Uhlenbeck spaces can be posed over a base field of any characteristic. However, the proof of one of the main results of this paper, Theorem 4.8, which ensures that our spaces $\mathfrak{U}^a_G$ are invariantly defined, uses the char. $= 0$ assumption. It is quite possible that in order to treat the char. $= p$ case, one needs a finer analysis.

0.2

The construction of $\mathfrak{U}^a_G$ used in this paper is a simplification of a suggestion of Drinfeld's (the latter potentially works for an arbitrary surface **S**).

We are trying to express points of $\mathfrak{U}^a_G$ (one may call them quasi-bundles) by replacing the original problem for the surface $\mathbb{P}^2$, or rather for a rationally equivalent surface $\mathbb{P}^1 \times \mathbb{P}^1$, by another problem for the curve $\mathbb{P}^1$.

As a motivation, let us consider the following simpler situation. Let $\mathcal{M}^0$ be the trivial rank-2 bundle on a curve $\mathbf{C}$. A flag in $\mathcal{M}^0$ is by definition a line subbundle $\mathcal{L} \subset \mathcal{M}^0$, or equivalently a map from $\mathbf{C}$ to the flag variety of GL_2, i.e., $\mathbb{P}^1$.

However, there is a natural generalization of a notion of a flag, also suggested by Drinfeld: instead of line *subbundles*, we may consider all pairs $(\mathcal{L}, \kappa : \mathcal{L} \to \mathcal{M}^0)$, where $\mathcal{L}$ is still a line bundle, but κ need not be a bundle map, just an embedding of coherent sheaves. We define a *quasi-map* $\mathbf{C} \to \mathbb{P}^1$ (generalizing the notion of a *map*) to be such a pair $(\mathcal{L}, \kappa)$.

In fact, one can introduce the notion of a quasi-map from a curve (or any projective variety) $\mathbf{C}$ to another projective variety $\mathcal{T}$. When $\mathcal{T}$ is the flag variety of a semisimple group G, the corresponding quasi-map spaces have been studied in [FFKM, BG1, BFGM].

Our construction of the Uhlenbeck space is based on considering quasi-maps from $\mathbb{P}^1$ (thought of as a "horizontal" component of $\mathbb{P}^1 \times \mathbb{P}^1$) to various flag varieties associated to the loop group of G; among them the most important are Kashiwara's thick Grassmannian and the Beilinson–Drinfeld Grassmannian Gr^{BD}_G.

The spaces of maps and quasi-maps from a projective curve $\mathbf{C}$ to Kashiwara's flag schemes are of independent interest and have been another major source of motivation for us.

In [FFKM, FKMM] it was shown that if one considers the space of (based) maps, of multidegree μ, from $\mathbf{C}$ to the flag variety of a finite-dimensional group G, one obtains an affine scheme, which we denote by $\mathrm{Maps}^\mu(\mathbf{C}, \mathcal{B}_\mathfrak{g})$, endowed with a symplectic structure, and which admits a Lagrangian projection to the space of colored divisors on $\mathbf{C}$, denoted $\mathbf{C}^\mu$.

Moreover, the irreducible components of the *central fiber* $\mathfrak{F}^\mu_{\mathfrak{g}}$ of this projection (i.e., the fiber over $\mu \cdot \mathbf{c} \in \mathbf{C}^\mu$, for some point $\mathbf{c} \in \mathbf{C}$) form in a natural way a basis for the μ-weight piece of $U(\check{\mathfrak{n}})$, where $\check{\mathfrak{g}} \supset \check{\mathfrak{n}}$ are the Langlands dual Lie algebra and its maximal nilpotent subalgebra, respectively.

One may wonder if this picture can be generalized for an arbitrary Kac–Moody $\mathfrak{g}'$, instead of the finite-dimensional algebra $\mathfrak{g}$. We discuss such a generalization in Parts I and IV of this paper. We formulate Conjecture 2.27, which is subsequently proven for $\mathfrak{g}'$ affine (and, of course, finite), which allows one to define on the set of irreducible components of the central fibers $\underset{\mu}{\cup}\mathfrak{F}^\mu_{\mathfrak{g}'}$ a structure of Kashiwara's crystal, and thereby link it to the combinatorics of the Langlands dual Lie algebra $\check{\mathfrak{g}}'$.

In particular, when $\mathfrak{g}'$ is affine, the space $\mathrm{Maps}^\mu(\mathbf{C}, \mathcal{B}_{\mathfrak{g}'})$ turns out to be closely related to the space of bundles on $\mathbb{P}^2$, and the space of quasi-maps $\mathrm{QMaps}^\mu(\mathbf{C}, \mathcal{B}_{\mathfrak{g}'})$ to the corresponding Uhlenbeck space. The relation between the irreducible components of $\mathfrak{F}^\mu$ and the Lie algebra $\check{\mathfrak{g}}'$ mentioned above allows us to explicitly compute the Intersection Cohomology sheaf on $\mathfrak{U}^a_G$, and express it in terms of $\check{\mathfrak{g}}'$ (in this case $\mathfrak{g}' = \mathfrak{g}_{\mathrm{aff}}$, the affinization of $\mathfrak{g}$.)

0.3

The two main results of this paper are construction of the scheme $\mathfrak{U}^a_G$, so that it has the stratification as in (1) (Theorem 7.2), and the explicit description of the Intersection Cohomology sheaf of $\mathfrak{U}^a_G$ (Theorem 7.10). Let us now explain the logical structure of the paper and the main points of each of the parts.

Part I is mostly devoted to the preliminaries. In Section 1 we introduce the notion of a quasi-map (in rather general circumstances) and prove some of its basic properties. The reader familiar with any of the works [FFKM], [BG1], or [BFGM] may skip Section 1 and return for proofs of statements referred to in the subsequent sections.

In Section 2 we collect some facts about Kashiwara's flag schemes $\mathcal{G}_{\mathfrak{g}',\mathfrak{p}}$ for a general Kac–Moody Lie algebra $\mathfrak{g}'$, and study the quasi-map spaces from a curve $\mathbf{C}$ to $\mathcal{G}_{\mathfrak{g}',\mathfrak{p}}$. In the main body of the paper, we will only use the cases when $\mathfrak{g}'$ is the affine algebra $\mathfrak{g}_{\mathrm{aff}}$, or the initial finite-dimensional Lie algebra $\mathfrak{g}$.

In Section 3 we collect some basic facts about G-bundles on a surface $\mathbb{P}^1 \times \mathbb{P}^1$ trivialized along the divisor at infinity.

In Part II we introduce the Uhlenbeck space $\mathfrak{U}^a_G$ and study its properties. In Section 4 we give three definitions of $\mathfrak{U}^a_G$, of which two are almost immediately equivalent; the equivalence with the (most invariant) third one is established later.

Section 5 contains a proof of Theorem 5.12, (conjectured in [FGK]) about the isomorphism of our definition of the Uhlenbeck space $\mathfrak{U}^a_{SL_n}$ with Donaldson's definition. The proof follows the ideas indicated by Drinfeld.

In Section 6 we prove some additional functoriality properties of $\mathfrak{U}^a_G$, which, combined with Theorem 5.12 of the previous section, yields Theorem 4.8 about the equivalence of all three definitions from Section 4. In addition, in Section 6 we establish the factorization property of $\mathfrak{U}_G^a$ with respect to the projection on the symmetric power $\mathbb{A}^{(a)}$ of the "horizontal" line, which will be one of the principal technical tools in the study of the geometry of $\mathfrak{U}_G^a$, and in particular, for the computation of the IC sheaf.

In Section 7 we prove Theorem 7.2 saying that $\mathfrak{U}_G^a$ indeed has a stratification as in (1). In addition, we formulate Theorem 7.10 describing the stalks of the IC sheaf on the various strata.

In Section 8 we present two moduli problems, whose solutions provide two more variants of the definition of $\mathfrak{U}_G^a$, and which coincide with the original one on the level of reduced schemes.

In Part III we define the "parabolic" version of Uhlenbeck spaces, $\mathfrak{U}_{G,P}^\theta$ and $\widetilde{\mathfrak{U}}_{G,P}^\theta$, which are two different compactifications of the space $\mathrm{Bun}_{G;P}(\mathbf{S}, \mathbf{D}_\infty; \mathbf{D}_0)$, classifying G-bundles on $\mathbb{P}^2$ with a trivialization along a divisor $\mathbb{P}^1 \simeq \mathbf{D}_\infty \subset \mathbb{P}^2$, and a reduction to a parabolic P along another divisor $\mathbb{P}^1 \simeq \mathbf{D}_0 \subset \mathbb{P}^2$. Introducing these more general spaces is necessary for our calculation of stalks of the IC sheaf.

Thus in Section 9 we give the definition of $\mathfrak{U}_{G,P}^\theta$ and $\widetilde{\mathfrak{U}}_{G,P}^\theta$, and establish the corresponding factorization properties.

In Section 10, we prove Theorem 10.2, which describes the stratifications of $\mathfrak{U}_{G,P}^\theta$ and $\widetilde{\mathfrak{U}}_{G,P}^\theta$ parallel to those of $\mathfrak{U}_G^a$.

In Section 11, we prove an important geometric property of $\widetilde{\mathfrak{U}}_{G,P}^\theta$ when $P = B$ saying that its *boundary* (in a natural sense) is a Cartier divisor.

In Part IV we make a digression and discuss a construction of crystals (in the sense of Kashiwara), using the quasi-map spaces $\mathrm{QMaps}(\mathbf{C}, \mathcal{G}_{\mathfrak{g}',\mathfrak{p}})$ introduced in Section 2, for an arbitrary Kac–Moody algebra $\mathfrak{g}'$. Unfortunately, to make this construction work one has to assume a certain geometric property of the $\mathrm{QMaps}(\mathbf{C}, \mathcal{G}_{\mathfrak{g}',\mathfrak{p}})$ spaces, Conjecture 2.27, which we verify in Section 15.6 for $\mathfrak{g}'$ of affine type, using some geometric properties of the parabolic Uhlenbeck spaces. As was mentioned above, it is via Kashiwara's crystals—more precisely, using Theorem 12.8—that we relate the IC stalks on $\mathfrak{U}_G^a$ and the Lie algebra $\check{\mathfrak{g}}_{\mathrm{aff}}$.

In Section 12 we recollect some general facts about Kashiwara's crystals. In particular, we review what properties are necessary to prove that a given crystal $\mathsf{B}_{\mathfrak{g}'}$ is isomorphic to the standard crystal $\mathsf{B}_{\mathfrak{g}'}^\infty$ of [Ka5].

In Section 13 we take our Lie algebra $\mathfrak{g}'$ to be finite dimensional and spell out our "new" construction of crystals using the affine Grassmannian of the corresponding group G.

In Section 14 we consider the case of a general Kac–Moody algebra, and essentially repeat the construction of the previous section using the scheme $\mathrm{QMaps}(\mathbf{C}, \mathcal{G}_{\mathfrak{g}',\mathfrak{p}})$ instead of the affine Grassmannian.

In Section 15 we verify Conjecture 2.27 and Conjecture 15.3 for finite-dimensional and affine Lie algebras, which ensures that in these cases the crystal of the previous section is well defined and can be identified with the standard crystal $\mathsf{B}_{\mathfrak{g}}^\infty$.

In Part V we perform the calculation of the IC sheaf on the schemes $\mathfrak{U}^a_G$, $\widetilde{\mathfrak{U}}^\theta_{G,P}$ and $\mathfrak{U}^\theta_{G,P}$.

In Section 16 we formulate four theorems, which describe the behavior of the IC sheaf, and in Section 17 we prove all four statements by an inductive argument borrowed from [BFGM].

Finally, in the appendix we reproduce a theorem of A. Joseph, formulated in Part I, Section 2, which says that the space of based maps $\mathbf{C} \to \mathcal{G}_{\mathfrak{g}',\mathfrak{p}}$ of given degree is a scheme of finite type for any Kac–Moody Lie algebra $\mathfrak{g}'$.

Part I: Preliminaries on Quasi-Maps

1 Maps and quasi-maps

1.1

The simplest framework in which one defines the notion of quasi-map is the following:

Let $\mathcal{Y}$ be a projective scheme, $\mathcal{E}$ a vector space, and $\mathcal{T} \subset \mathbb{P}(\mathcal{E})$ a closed subscheme.

There exists a scheme, which we will denote $\mathrm{Maps}(\mathcal{Y}, \mathcal{T})$ that represents the functor which assigns to a test scheme S the set of maps $\mathcal{Y} \times S \to \mathcal{T}$, which commute with the natural projection of both sides to $\mathcal{Y}$.

To show the representability, it is enough to assume that $\mathcal{T}$ is the entire $\mathbb{P}(\mathcal{E})$ (since in general $\mathrm{Maps}(\mathcal{Y}, \mathcal{T})$ is evidently a closed subfunctor in $\mathrm{Maps}(\mathcal{Y}, \mathbb{P}(\mathcal{E}))$), and in the latter case our functor can be rewritten as pairs $(\mathcal{L}, \kappa)$, where $\mathcal{L}$ is a line bundle on $\mathcal{Y} \times S$, and κ is an injective bundle map

$$\kappa : \mathcal{L} \hookrightarrow \mathcal{O}_{\mathcal{Y} \times S} \otimes \mathcal{E}.$$

Therefore, we are dealing with an open subset of a suitable Hilbert scheme. The scheme $\mathrm{Maps}(\mathcal{Y}, \mathcal{T})$ splits as a disjoint union of subschemes, denoted $\mathrm{Maps}^a(\mathcal{Y}, \mathcal{T})$, and indexed by the set of connected components of the Picard stack $\mathrm{Pic}(\mathcal{Y})$ of $\mathcal{Y}$ (our normalization is such that $\sigma = (\mathcal{L}, \kappa) \in \mathrm{Maps}^a(\mathcal{Y}, \mathcal{T})$ if $\mathcal{L}^{-1} \in \mathrm{Pic}^a(\mathcal{Y})$), and each $\mathrm{Maps}^a(\mathcal{Y}, \mathcal{T})$ is a quasi-projective scheme.

We will introduce a bigger scheme, denoted $\mathrm{QMaps}^a(\mathcal{Y}, \mathcal{T}; \mathcal{E})$, which contains $\mathrm{Maps}^a(\mathcal{Y}, \mathcal{T})$ as an open subscheme. First, we will consider the case of $\mathcal{T} = \mathbb{P}(\mathcal{E})$.

By definition, $\mathrm{QMaps}^a(\mathcal{Y}, \mathbb{P}(\mathcal{E}); \mathcal{E})$ represents the functor that assigns to a scheme S the set of pairs $(\mathcal{L}, \kappa)$, where $\mathcal{L}$ is a line bundle on the product $\mathcal{Y} \times S$ belonging to the connected component $\mathrm{Pic}^{-a}(\mathcal{Y})$ of the Picard stack, and κ is an *injective map of coherent sheaves*

$$\kappa : \mathcal{L} \hookrightarrow \mathcal{O}_{\mathcal{Y} \times S} \otimes \mathcal{E},$$

such that the quotient is S-flat. (The latter condition is equivalent to the fact that for every geometric point $s \in S$, the restriction of κ to $\mathcal{Y} \times s$ is injective.) This functor is also representable by a quasi-projective scheme, for the same reason.

Now let $\mathcal{T}$ be arbitrary, and let $\mathcal{I}_{\mathcal{T}} = \underset{n \geq 0}{\oplus} \mathcal{I}^n_{\mathcal{T}} \subset \mathrm{Sym}(\mathcal{E}^*)$ be the corresponding graded ideal. We define the closed subscheme $\mathrm{QMaps}^a(\mathcal{Y}, \mathcal{T}; \mathcal{E}) \subset \mathrm{QMaps}^a(\mathcal{Y}, \mathbb{P}(\mathcal{E}); \mathcal{E})$ by the condition that for every n the composition

$$\mathcal{O}_{\mathcal{Y}\times S} \otimes \mathcal{I}^n_{\mathcal{T}} \hookrightarrow \mathcal{O}_{\mathcal{Y}\times S} \otimes \mathrm{Sym}^n(\mathcal{E}^*) \to (\mathcal{L}^{-1})^{\otimes n}$$

vanishes.

We will denote by $\mathrm{QMaps}(\mathcal{Y}, \mathcal{T}; \mathcal{E})$ the union of $\mathrm{QMaps}^a(\mathcal{Y}, \mathcal{T}; \mathcal{E})$ over all connected components of $\mathrm{Pic}(\mathcal{Y})$.

In the main body of the present paper the scheme $\mathcal{Y}$ will be a smooth algebraic curve, but for completeness in this section we will consider the general case. Note that in the case of curves, the parameter a amounts to an integer, normalized to be the negative of $\deg(\mathcal{L})$.

In the rest of this section we will study various properties and generalizations of the notion of a quasi-map introduced above.

1.2 Variant

Assume for a moment that $\mathcal{Y}$ is integral. Observe that if $\sigma = (\mathcal{L}, \kappa)$ is an S-point of $\mathrm{QMaps}^a(\mathcal{Y}, \mathbb{P}(\mathcal{E}); \mathcal{E})$, there exists an open dense subset $U \subset \mathcal{Y} \times S$ over which κ is a bundle map.

Consider the (automatically closed) subfunctor of $\mathrm{QMaps}^a(\mathcal{Y}, \mathbb{P}(\mathcal{E}); \mathcal{E})$, corresponding to the condition that the resulting map $U \to \mathbb{P}(\mathcal{E})$ factors through $\mathcal{T} \subset \mathbb{P}(\mathcal{E})$. It is easy to see that this subfunctor coincides with $\mathrm{QMaps}^a(\mathcal{Y}, \mathcal{T}; \mathcal{E})$.

The above definition can be also spelled out as follows. Let $C(\mathcal{T}; \mathcal{E})$ be the affine cone over $\mathcal{T}$, i.e., the closure in $\mathcal{E}$ of the preimage of $\mathcal{T}$ under the natural map $(\mathcal{E} - 0) \to \mathbb{P}(\mathcal{E})$. The multiplicative group $\mathbb{G}_m$ acts naturally on $C(\mathcal{T}; \mathcal{E})$ and we can form the stack $C(\mathcal{T}; \mathcal{E})/\mathbb{G}_m$, which contains $\mathcal{T}$ as an open substack.

It is easy to see that a map $S \to \mathrm{QMaps}(\mathcal{Y}, \mathcal{T}; \mathcal{E})$ is the same as a map $\sigma : \mathcal{Y} \times S \to C(\mathcal{T}; \mathcal{E})/\mathbb{G}_m$ such that for every geometric $s \in S$, the map $\mathcal{Y} \simeq \mathcal{Y} \times s \to C(\mathcal{T}; \mathcal{E})/\mathbb{G}_m$ sends the generic point of $\mathcal{Y}$ into $\mathcal{T}$.

We will now introduce a still bigger scheme, $\mathrm{QQMaps}^p(\mathcal{Y}, \mathcal{T}; \mathcal{E})$. First, $\mathrm{QQMaps}^p(\mathcal{Y}, \mathbb{P}(\mathcal{E}); \mathcal{E})$ is the scheme classifying pairs $(\mathcal{L}, \kappa)$, where $\mathcal{L}$ is an S-flat coherent sheaf on $\mathcal{Y} \times S$ whose restriction to every geometric fiber $\mathcal{Y} \times s$ is of generic rank 1, and κ is a map of coherent sheaves, injective at each $Y \times s$ (as above $s \in S$ denotes a geometric point). The superscript p signifies that the Hilbert polynomial p of $\mathcal{L}$ (with respect to some ample line bundle on $\mathcal{Y}$) is fixed. Omitting the superscript means that we are taking the union over all possible Hilbert polynomials.

As in the case of $\mathrm{QMaps}^p(\mathcal{Y}, \mathcal{T}; \mathcal{E})$, given an S-point $(\mathcal{L}, \kappa)$ of $\mathrm{QQMaps}^p(\mathcal{Y}, \mathbb{P}(\mathcal{E}); \mathcal{E})$, there exists an open dense subset $U \subset \mathcal{Y} \times S$, over which $\mathcal{L}$ is a line bundle, and κ is a bundle map.

For a closed subscheme $\mathcal{T} \subset \mathbb{P}(\mathcal{E})$, we defined the closed subfunctor $\mathrm{QQMaps}^p(\mathcal{Y}, \mathcal{T}; \mathcal{E})$ of $\mathrm{QQMaps}^p(\mathcal{Y}, \mathbb{P}(\mathcal{E}); \mathcal{E})$ by the condition that the resulting map $U \to \mathbb{P}(\mathcal{E})$ factors through $\mathcal{T}$.

Since $\mathrm{QQMaps}^p(\mathcal{Y}, \mathbb{P}(\mathcal{E}); \mathcal{E})$ is actually the entire Hilbert scheme, we obtain the following.

Lemma 1.3. *The scheme* $\mathrm{QQMaps}^p(\mathcal{Y}, \mathbb{P}(\mathcal{E}); \mathcal{E})$ *is proper.*

Note that by assumption, for every geometric point $s \in S$, the restriction $\mathcal{L}_s := \mathcal{L}|_{\mathcal{Y}\times s}$ embeds into $\mathcal{E} \otimes \mathcal{O}_{\mathcal{Y}}$; therefore, $\mathcal{L}$ is torsion-free. In particular, when $\mathcal{Y}$ is a smooth curve, $\mathrm{QMaps}(\mathcal{Y}, \mathcal{T}; \mathcal{E})$ coincides with $\mathrm{QQMaps}(\mathcal{Y}, \mathcal{T}; \mathcal{E})$. But in general we have an open embedding $\mathrm{QMaps}(\mathcal{Y}, \mathcal{T}; \mathcal{E}) \hookrightarrow \mathrm{QQMaps}(\mathcal{Y}, \mathcal{T}; \mathcal{E})$.

Proposition 1.4. *Suppose $\mathcal{Y}$ is smooth. Then the scheme $\mathrm{QMaps}^a(\mathcal{Y}, \mathcal{T}; \mathcal{E})$ is proper as well.*

Proof. Let us check the valuative criterion of properness. Using Lemma 1.3 we can assume that we have the following setup:

Let $\mathbf{X}$ be a curve with a marked point $0_{\mathbf{X}} \in \mathbf{X}$, and let $\mathcal{L}$ be a torsion-free coherent sheaf on $\mathcal{Y} \times \mathbf{X}$, embedded into $\mathcal{E} \otimes \mathcal{O}_{\mathcal{Y}\times\mathbf{X}}$, and such that $\mathcal{L}|_{\mathcal{Y}\times(\mathbf{X}-0_{\mathbf{X}})}$ is a line bundle.

We claim that $\mathcal{L}$ itself is a line bundle. Indeed, $\mathcal{L}$ is locally free outside of codimension 2, and since $\mathcal{Y} \times \mathbf{X}$ is smooth, it admits a saturation, i.e., there exists a unique line bundle $\mathcal{L}^0$ containing $\mathcal{L}$, such that $\mathcal{L}^0/\mathcal{L}$ is supported in codimension 2. Moreover, it is easy to see that the map $\mathcal{L} \to \mathcal{E} \otimes \mathcal{O}_{\mathcal{Y}\times\mathbf{X}}$ extends to a map $\mathcal{L}^0 \to \mathcal{E} \otimes \mathcal{O}_{\mathcal{Y}\times\mathbf{X}}$. But since the cokernel of κ was $\mathbf{X}$-flat, we obtain that $\mathcal{L} \to \mathcal{L}^0$ must be an isomorphism. □

In particular, we see that when $\mathcal{Y}$ is smooth, $\mathrm{QMaps}(\mathcal{Y}, \mathcal{T}; \mathcal{E})$ is the union of certain of the connected components of $\mathrm{QQMaps}(\mathcal{Y}, \mathcal{T}; \mathcal{E})$.

1.5 Based quasi-maps

Let $\mathcal{Y}' \subset \mathcal{Y}$ be a closed subscheme, and $\sigma' : \mathcal{Y}' \to \mathcal{T}$ a fixed map. We introduce the scheme $\mathrm{QMaps}^a(\mathcal{Y}, \mathcal{T}; \mathcal{E})_{\mathcal{Y}',\sigma'}$ as a (locally closed) subfunctor of $\mathrm{QMaps}^a(\mathcal{Y}, \mathcal{T}; \mathcal{E})$ defined by the following two conditions:

- The restriction of the map κ to $\mathcal{Y}' \times S$ is a *bundle map*. (Equivalently, the quotient $\mathcal{O}_{\mathcal{Y}\times S} \otimes \mathcal{E}/\mathcal{L}$ is locally free in a neighborhood of $\mathcal{Y}' \times S$.)
- The resulting map $\mathcal{Y}' \times S \to \mathcal{T}$ equals $\mathcal{Y}' \times S \to \mathcal{Y}' \overset{\sigma'}{\to} \mathcal{T}$.

When $\mathcal{Y}$ is integral, one defines in a similar way the subscheme

$$\mathrm{QQMaps}^a(\mathcal{Y}, \mathcal{T}; \mathcal{E})_{\mathcal{Y}',\sigma'} \subset \mathrm{QQMaps}^a(\mathcal{Y}, \mathcal{T}; \mathcal{E}).$$

The following assertion will be needed in the main body of the paper: Assume that $\mathcal{Y}$ is a smooth curve $\mathbf{C}$ and $\mathcal{Y}'$ is a point $\mathbf{c}$, so that $\sigma' : pt \to \mathcal{T}$ corresponds to some point of $\mathcal{T}$.

Lemma 1.6. *There exists an affine map $\mathrm{QMaps}^a(\mathbf{C}, \mathcal{T}; \mathcal{E})_{\mathbf{c},\sigma'} \to (\mathbf{C} - \mathbf{c})^{(a)}$. In particular, the scheme $\mathrm{QMaps}^a(\mathbf{C}, \mathcal{T}; \mathcal{E})_{\mathbf{c},\sigma'}$ is affine.*

Proof. We can assume that $\mathcal{T} = \mathbb{P}(\mathcal{E})$, so that σ' corresponds to a line $\ell \subset \mathcal{E}$. Let us choose a splitting $\mathcal{E} \simeq \mathcal{E}' \oplus \ell$.

Then we have a natural map $\mathrm{QMaps}^a(\mathbf{C}, \mathcal{E})_{\mathbf{c},\sigma'} \to (\mathbf{C} - \mathbf{c})^{(a)}$, that assigns to a pair $(\mathcal{L}, \kappa : \mathcal{L} \to \mathcal{O} \otimes \mathcal{E})$ the divisor of zeroes of the composition $\mathcal{L} \to \mathcal{O} \otimes \mathcal{E} \to \mathcal{O} \otimes \ell \simeq \mathcal{O}$.

Since $\mathbf{C} - \mathbf{c}$ is affine, the symmetric power $(\mathbf{C} - \mathbf{c})^{(a)}$ is affine as well. We claim that the above morphism $\mathrm{QMaps}^a(\mathbf{C}, \mathcal{E})_{\mathbf{c},\sigma'} \to (\mathbf{C} - \mathbf{c})^{(a)}$ is also affine.

Indeed, given a divisor $D \in (\mathbf{C} - \mathbf{c})^{(a)}$, the fiber of $\mathrm{QMaps}^a(\mathbf{C}, \mathcal{E})_{\mathbf{c},\sigma'}$ over it is the vector space $\mathrm{Hom}(\mathcal{O}(-D), \mathcal{E}')$. □

1.7 The relative version

Suppose now that $\mathcal{Y}$ itself is a flat family of projective schemes over some base $\mathcal{X}$, and $\mathcal{E}$ is a vector bundle on $\mathcal{Y}$, with $\mathcal{T} \subset \mathbb{P}(\mathcal{E})$ a closed subscheme.

The scheme of maps $\mathrm{Maps}(\mathcal{Y}, \mathcal{T})$ assigns to a test scheme S over $\mathcal{X}$ the set of maps $\sigma : \mathcal{Y} \underset{\mathcal{X}}{\times} S \to \mathcal{T}$, such that the composition $\mathcal{Y} \underset{\mathcal{X}}{\times} S \to \mathcal{T} \to \mathcal{Y}$ is the projection on the first factor. By definition, $\mathrm{Maps}(\mathcal{Y}, \mathcal{T})$ is also a scheme over $\mathcal{X}$.

By essentially repeating the construction of Section 1.1, we obtain a scheme $\mathrm{QMaps}(\mathcal{Y}, \mathcal{T}; \mathcal{E})$, which is quasi-projective over $\mathcal{X}$.

As before, if $\mathcal{Y}' \subset \mathcal{Y}$ is a closed subscheme, flat over $\mathcal{X}$, and $\sigma' : \mathcal{Y}' \to \mathcal{T}$ is a $\mathcal{X}$-map, we can define a locally closed subscheme $\mathrm{QMaps}^a(\mathcal{Y}, \mathcal{T}; \mathcal{E})_{\mathcal{Y}',\sigma'}$ of based maps. When $\mathcal{Y} \to \mathcal{X}$ has integral fibers, one can define the relative versions of $\mathrm{QQMaps}(\mathcal{Y}, \mathcal{T}; \mathcal{E})$ and $\mathrm{QQMaps}(\mathcal{Y}, \mathcal{T}; \mathcal{E})_{\mathcal{Y}',\sigma'}$ in a similar fashion.

In what follows, we will mostly discuss the "absolute" case, leaving the (straightforward) modifications required in the relative and based cases to the reader.

1.8

Next we will study some functorial properties of the quasi-map spaces. From now on we will assume that $\mathcal{Y}$ is integral.

For $\mathcal{T} \subset \mathbb{P}(\mathcal{E})$ as above, let $\mathcal{P}$ be the very ample line bundle $\mathcal{O}(1)|_{\mathcal{T}}$. Note that we have a map $H^0(\mathcal{T}, \mathcal{P})^* \to \mathcal{E}$.

Now let $\mathcal{T}_1 \subset \mathbb{P}(\mathcal{E}_1)$ and $\mathcal{T}_2 \subset \mathbb{P}(\mathcal{E}_2)$ be two projective schemes, and $\phi : \mathcal{T}_1 \to \mathcal{T}_2$ be a map, such that $\phi^*(\mathcal{P}_2) \simeq \mathcal{P}_1$. Assume, moreover, that the corresponding map $H^0(\mathcal{T}_2, \mathcal{P}_2) \to H^0(\mathcal{T}_1, \mathcal{P}_1)^*$ extends to a commutative diagram

$$\begin{array}{ccc} H^0(\mathcal{T}_1, \mathcal{P}_1)^* & \longrightarrow & H^0(\mathcal{T}_2, \mathcal{P}_2)^* \\ \downarrow & & \downarrow \\ \mathcal{E}_1 & \longrightarrow & \mathcal{E}_2. \end{array}$$

Lemma 1.9. *Under the above circumstances the morphism* $\mathrm{Maps}(\mathcal{Y}, \mathcal{T}_1) \to \mathrm{Maps}(\mathcal{Y}, \mathcal{T}_2)$ *extends to a Cartesian diagram:*

$$\begin{array}{ccccc} \mathrm{Maps}(\mathcal{Y}, \mathcal{T}_1) & \longrightarrow & \mathrm{QMaps}(\mathcal{Y}, \mathcal{T}_1, \mathcal{E}_1) & \longrightarrow & \mathrm{QQMaps}(\mathcal{Y}, \mathcal{T}_1, \mathcal{E}_1) \\ \downarrow & & \downarrow & & \downarrow \\ \mathrm{Maps}(\mathcal{Y}, \mathcal{T}_2) & \longrightarrow & \mathrm{QMaps}(\mathcal{Y}, \mathcal{T}_2, \mathcal{E}_2) & \longrightarrow & \mathrm{QQMaps}(\mathcal{Y}, \mathcal{T}_2, \mathcal{E}_2). \end{array}$$

Proof. The existence of the map $\mathrm{QMaps}(\mathcal{Y}, \mathcal{T}_1, \mathcal{E}_1) \to \mathrm{QMaps}(\mathcal{Y}, \mathcal{T}_2, \mathcal{E}_2)$ is nearly obvious from the interpretation of quasi-maps as maps to the affine cone.

Note that the condition of the lemma is equivalent to the fact that the map $\mathcal{T}_1 \to \mathcal{T}_2$ extends to a commutative diagram of the corresponding affine cones:

$$\begin{array}{ccccccc} \mathcal{T}_1 & \longleftarrow & C(\mathcal{T}_1;\mathcal{E}_1) - 0 & \longrightarrow & C(\mathcal{T}_1;\mathcal{E}_1) & \longrightarrow & \mathcal{E}_1 \\ \downarrow & & \downarrow & & \downarrow & & \downarrow \\ \mathcal{T}_2 & \longleftarrow & C(\mathcal{T}_2;\mathcal{E}_2) - 0 & \longrightarrow & C(\mathcal{T}_2;\mathcal{E}_2) & \longrightarrow & \mathcal{E}_2. \end{array}$$

Therefore, we can just compose a map $\mathcal{Y} \times S \to C(\mathcal{T}_1;\mathcal{E}_1)/\mathbb{G}_m$ with the map $C(\mathcal{T}_1;\mathcal{E}_1)/\mathbb{G}_m \to C(\mathcal{T}_2;\mathcal{E}_2)/\mathbb{G}_m$.

Let us construct the map $\mathrm{QQMaps}(\mathcal{Y}, \mathcal{T}_1, \mathcal{E}_1) \to \mathrm{QQMaps}(\mathcal{Y}, \mathcal{T}_2, \mathcal{E}_2)$. Given a scheme S, and a coherent sheaf of generic rank 1 $\mathcal{L}$ on $\mathcal{Y} \times S$ with a map $\kappa_1 : \mathcal{L} \to \mathcal{O}_{\mathcal{Y}\times S} \otimes \mathcal{E}_1$, let us consider the composition

$$\kappa_2 : \mathcal{L} \to \mathcal{O}_{\mathcal{Y}\times S} \otimes \mathcal{E}_1 \to \mathcal{O}_{\mathcal{Y}\times S} \otimes \mathcal{E}_2.$$

We claim that the restriction of κ_2 to every fiber $\mathcal{Y} \times s$ is still injective. Indeed, we are dealing with a map $\mathcal{L}_s \to \mathcal{O}_{\mathcal{Y}} \otimes \mathcal{E}_2$ which is injective over an open subset of $\mathcal{Y}$ (because on some open subset U, κ_2 corresponds to a *map* $U \to \mathcal{T}_2$), and we know that $\mathcal{L}_s$ is torsion-free.

The square

$$\begin{array}{ccc} \mathrm{QMaps}(\mathcal{Y}, \mathcal{T}_1, \mathcal{E}_1) & \longrightarrow & \mathrm{QQMaps}(\mathcal{Y}, \mathcal{T}_1, \mathcal{E}_1) \\ \downarrow & & \downarrow \\ \mathrm{QMaps}(\mathcal{Y}, \mathcal{T}_2, \mathcal{E}_2) & \longrightarrow & \mathrm{QQMaps}(\mathcal{Y}, \mathcal{T}_2, \mathcal{E}_2). \end{array}$$

is Cartesian, because being in QMaps is just the condition that $\mathcal{L}$ is locally free.

To prove that the square

$$\begin{array}{ccc} \mathrm{Maps}(\mathcal{Y}, \mathcal{T}_1) & \longrightarrow & \mathrm{QMaps}(\mathcal{Y}, \mathcal{T}_1, \mathcal{E}_1) \\ \downarrow & & \downarrow \\ \mathrm{Maps}(\mathcal{Y}, \mathcal{T}_2) & \longrightarrow & \mathrm{QMaps}(\mathcal{Y}, \mathcal{T}_2, \mathcal{E}_2). \end{array}$$

is Cartesian, we have to show that if $\mathrm{coker}(\kappa_2)$ is locally free, then so is $\mathrm{coker}(\kappa_1)$. But this follows from the four-term exact sequence

$$\begin{aligned} 0 \to \mathcal{O}_{\mathcal{Y}\times S} \otimes \ker(\mathcal{E}_1 \to \mathcal{E}_2) \to \mathrm{coker}(\kappa_1) \\ \to \mathrm{coker}(\kappa_2) \to \mathcal{O}_{\mathcal{Y}\times S} \otimes \mathrm{coker}(\mathcal{E}_1 \to \mathcal{E}_2) \to 0. \end{aligned}$$ □

This lemma has a straightforward generalization to the based case:

If we fix $\mathcal{Y}'$ and $\sigma_1' : \mathcal{Y}' \to \mathcal{T}_1$ with $\sigma_2' := \phi \circ \sigma_1'$, we have a Cartesian square

$$\begin{array}{ccc} \mathrm{QMaps}(\mathcal{Y}, \mathcal{T}_1, \mathcal{E}_1)_{\mathcal{Y}',\sigma'_1} & \longrightarrow & \mathrm{QMaps}(\mathcal{Y}, \mathcal{T}_1, \mathcal{E}_1) \\ \downarrow & & \downarrow \\ \mathrm{QMaps}(\mathcal{Y}, \mathcal{T}_2, \mathcal{E}_2)_{\mathcal{Y}',\sigma'_2} & \longrightarrow & \mathrm{QMaps}(\mathcal{Y}, \mathcal{T}_2, \mathcal{E}_2), \end{array}$$

and similarly for QQMaps.

Assume now that we are in the situation of Lemma 1.9, and that the map $\mathcal{T}_1 \to \mathcal{T}_2$ is a closed embedding.

Proposition 1.10. *Under the above circumstances, the resulting morphisms*

$$\mathrm{QMaps}(\mathcal{Y}, \mathcal{T}_1, \mathcal{E}_1) \to \mathrm{QMaps}(\mathcal{Y}, \mathcal{T}_2, \mathcal{E}_2)$$

and

$$\mathrm{QQMaps}(\mathcal{Y}, \mathcal{T}_1, \mathcal{E}_1) \to \mathrm{QQMaps}(\mathcal{Y}, \mathcal{T}_2, \mathcal{E}_2)$$

are closed embeddings.

If $\mathcal{T}_1 \to \mathcal{T}_2$ is an isomorphism and $\mathcal{Y}$ is normal, then the morphism $\mathrm{QMaps}(\mathcal{Y}, \mathcal{T}_1, \mathcal{E}_1) \to \mathrm{QMaps}(\mathcal{Y}, \mathcal{T}_2, \mathcal{E}_2)$ *induces an isomorphism on the level of reduced schemes.*

(Due to the Cartesian diagram above, the same assertions will hold in the based situation.)

Proof. If $\mathcal{E}_1 \to \mathcal{E}_2$ is an isomorphism, then the assertion of the lemma is obvious, as both schemes are closed subschemes in $\mathrm{QMaps}(\mathcal{Y}, \mathbb{P}(\mathcal{E}_2), \mathcal{E}_2)$ (respectively, $\mathrm{QQMaps}(\mathcal{Y}, \mathbb{P}(\mathcal{E}_2), \mathcal{E}_2)$). Therefore, we can assume that $\mathcal{T}_1 \simeq \mathcal{T}_2 =: \mathcal{T}$.

First, we claim that the map

$$\mathrm{Hom}(S, \mathrm{QQMaps}(\mathcal{Y}, \mathcal{T}, \mathcal{E}_1)) \to \mathrm{Hom}(S, \mathrm{QQMaps}(\mathcal{Y}, \mathcal{T}, \mathcal{E}_2))$$

is an injection for any scheme S.

This follows from the fact that if $\mathcal{L}$ is a torsion-free coherent sheaf on $\mathcal{Y} \times S$, then any two maps $\mathcal{L} \to \mathcal{O}_{\mathcal{Y} \times S} \otimes \mathcal{E}_1$ that coincide over a dense subset $U \subset \mathcal{Y} \times S$ must coincide globally.

Since the scheme $\mathrm{QQMaps}(\mathcal{Y}, \mathcal{T}, \mathcal{E}_1)$ is proper, the map

$$\mathrm{QQMaps}(\mathcal{Y}, \mathcal{T}, \mathcal{E}_1) \to \mathrm{QQMaps}(\mathcal{Y}, \mathcal{T}, \mathcal{E}_2)$$

is proper. Combined with the previous statement, we obtain that it is a closed embedding. Using the Cartesian diagrams of Lemma 1.9, we obtain that $\mathrm{QMaps}(\mathcal{Y}, \mathcal{T}, \mathcal{E}_1) \to \mathrm{QMaps}(\mathcal{Y}, \mathcal{T}, \mathcal{E}_2)$ is a closed embedding as well.

To prove the last assertion, we claim that for any normal scheme S, the map

$$\mathrm{Hom}(S, \mathrm{QMaps}(\mathcal{Y}, \mathcal{T}, \mathcal{E}_1)) \to \mathrm{Hom}(S, \mathrm{QMaps}(\mathcal{Y}, \mathcal{T}, \mathcal{E}_2))$$

is in fact a bijection. Indeed, let us consider the map between the cones $C(\mathcal{T}; \mathcal{E}_1) \to C(\mathcal{T}; \mathcal{E}_2)$. This map is finite, and is an isomorphism away from $0 \in C(\mathcal{T}; \mathcal{E}_2)$.

For a given element in $\mathrm{Hom}(S, \mathrm{QMaps}(\mathcal{Y}, \mathcal{T}, \mathcal{E}_2))$, consider the Cartesian product

$$(\mathcal{Y} \times S) \underset{C(\mathcal{T};\mathcal{E}_2)/\mathbb{G}_m}{\times} C(\mathcal{T}; \mathcal{E}_1)/\mathbb{G}_m.$$

This is a scheme, finite and generically isomorphic to $\mathcal{Y} \times S$. Since $\mathcal{Y} \times S$ is normal, we obtain that it admits a section, i.e., we have a map $\mathcal{Y} \times S \to C(\mathcal{T}, \mathcal{E}_1)/\mathbb{G}_m$. □

1.11

The definition of quasi-maps given above depends on the projective embedding of $\mathcal{T}$. In fact, one can produce another scheme, which depends only on the corresponding very ample line bundle $\mathcal{P} = \mathcal{O}(1)|_{\mathcal{T}}$. Note that for the discussion below it is crucial that we work with QMaps and not QQMaps, because we will be taking tensor products of the corresponding line bundles.

For a test scheme S, we let $\mathrm{Hom}(S, \mathrm{QMaps}(\mathcal{Y}, \mathcal{T}; \mathcal{P}))$ be the set consisting of pairs $(\mathcal{L}, \kappa)$, where $\mathcal{L}$ is, as before, a line bundle on $\mathcal{Y} \times S$, and κ is a map $\mathcal{L} \to \mathcal{O}_{\mathcal{Y}\times S} \otimes \Gamma(\mathcal{T}, \mathcal{P})^*$ which extends to a map of algebras

$$\bigoplus_n H^0(\mathcal{T}, \mathcal{P}^{\otimes n}) \to \bigoplus_n (\mathcal{L}^*)^{\otimes n},$$

and such that κ is injective over every geometric point $s \in S$. (This definition makes sense for an arbitrary, i.e., not necessarily integral scheme $\mathcal{Y}$.)

Adopting again the assumption that $\mathcal{Y}$ is integral, we can spell out the above definition as follows. Let $C(\mathcal{T}; \mathcal{P})$ be the affine closure of the total space of $(\mathcal{P}^{-1} - 0)$, i.e., $C(\mathcal{T}; \mathcal{P})$ is the spectrum of the algebra $\bigoplus_n H^0(\mathcal{T}, \mathcal{P}^{\otimes n})$. We can form the stack $C(\mathcal{T}; \mathcal{P})/\mathbb{G}_m$, which contains $\mathcal{T}$ as an open subset.

As before, we can consider the functor $\mathrm{QMaps}(\mathcal{Y}, \mathcal{T}; \mathcal{P})$ that assigns to a scheme S the set of maps $\mathcal{Y} \times S \to C(\mathcal{T}; \mathcal{P})/\mathbb{G}_m$, such that for every geometric point $s \in S$, the map $\mathcal{Y} \simeq \mathcal{Y} \times s \to C(\mathcal{T}; \mathcal{P})/\mathbb{G}_m$ sends the generic point of $\mathcal{Y}$ into $\mathcal{T}$.

Proposition 1.12. *For $\mathcal{T} \subset \mathbb{P}(\mathcal{E})$ and $\mathcal{P}$ as above we have the following:*

(a) *There is a canonical morphism* $\mathrm{QMaps}(\mathcal{Y}, \mathcal{T}; \mathcal{P}) \to \mathrm{QMaps}(\mathcal{Y}, \mathcal{T}; \mathcal{E})$. *When $\mathcal{Y}$ is smooth, this map is a closed embedding, which induces an isomorphism on the level of reduced schemes.*
(b) *For $n \in \mathbb{N}$, we have a morphism* $\mathrm{QMaps}(\mathcal{Y}, \mathcal{T}; \mathcal{P}) \to \mathrm{QMaps}(\mathcal{Y}, \mathcal{T}; \mathcal{P}^{\otimes n})$. *When $\mathcal{Y}$ is smooth, this map is also a closed embedding, which induces an isomorphism on the level of reduced schemes.*
(c) *If $\phi : \mathcal{T}_1 \to \mathcal{T}_2$ is a map of projective schemes, such that $\phi^*(\mathcal{P}_2) \simeq \mathcal{P}_1$, we have a morphism* $\mathrm{QMaps}(\mathcal{Y}, \mathcal{T}; \mathcal{P}_1) \to \mathrm{QMaps}(\mathcal{Y}, \mathcal{T}; \mathcal{P}_2)$, *which is a closed embedding, whenever ϕ is.*

Proof. Observe that we have a finite map $C(\mathcal{T};\mathcal{P}) \to C(\mathcal{T};H^0(\mathcal{T},\mathcal{P})^*)$, and moreover, for any positive integer n, a finite map $C(\mathcal{T};\mathcal{P}) \to C(\mathcal{T};\mathcal{P}^{\otimes n})$. This makes the existence of the maps $\mathrm{QMaps}(\mathcal{Y},\mathcal{T};\mathcal{P}) \to \mathrm{QMaps}(\mathcal{Y},\mathcal{T};\mathcal{E})$ and $\mathrm{QMaps}(\mathcal{Y},\mathcal{T};\mathcal{P}) \to \mathrm{QMaps}(\mathcal{Y},\mathcal{T};\mathcal{P}^{\otimes n})$ obvious.

For k large enough, the map of $C(\mathcal{T};\mathcal{P})$ into the product $\prod_{i=1,\dots,k} C(\mathcal{T};H^0(\mathcal{T},\mathcal{P}^{\otimes i})^*)$ is a closed embedding. Hence $\mathrm{QMaps}(\mathcal{Y},\mathcal{T};\mathcal{P})$ is representable by a closed subscheme in the product $\prod_{i=1,\dots,k} \mathrm{QMaps}(\mathcal{Y},\mathcal{T};H^0(\mathcal{T},\mathcal{P}^{\otimes i})^*)$.

In particular, when $\mathcal{Y}$ is smooth, we obtain that $\mathrm{QMaps}(\mathcal{Y},\mathcal{T};\mathcal{P})$ is proper.

To finish the proof of the proposition, it would suffice to show that for a test scheme S, the maps $\mathrm{Hom}(S,\mathrm{QMaps}(\mathcal{Y},\mathcal{T};\mathcal{P})) \to \mathrm{Hom}(S,\mathrm{QMaps}(\mathcal{Y},\mathcal{T};\mathcal{E}))$ and $\mathrm{Hom}(S,\mathrm{QMaps}(\mathcal{Y},\mathcal{T};\mathcal{P})) \to \mathrm{Hom}(S,\mathrm{QMaps}(\mathcal{Y},\mathcal{T};\mathcal{P}^{\otimes n}))$ are injective and are, moreover, bijective if S is normal. This is done exactly as in the proof of Lemma 1.10.

Assertion (c) of the Proposition follows from Lemma 1.9 and Lemma 1.10. □

Point (c) of the above proposition allows to define $\mathrm{QMaps}(\mathcal{Y},\mathcal{T};\mathcal{P})$ as a strict ind-scheme, when $\mathcal{T}$ is a strict ind-projective ind-scheme endowed with a very ample line bundle.

Indeed, if $\mathcal{T} = \varinjlim \mathcal{T}_i$, where $\mathcal{T}_i$ are projective schemes, and $\mathcal{T}_i \to \mathcal{T}_j$ are closed embeddings, and $\mathcal{P}_i$ is a compatible system of line bundles on $\mathcal{T}_i$, we have a system of closed embeddings

$$\mathrm{QMaps}(\mathcal{Y},\mathcal{T}_i;\mathcal{P}_i) \to \mathrm{QMaps}(\mathcal{Y},\mathcal{T}_j;\mathcal{P}_j).$$

In this section we will discuss the quasi-map spaces in their $\mathrm{Hom}(S,\mathrm{QMaps}(\mathcal{Y},\mathcal{T};\mathcal{E}))$ incarnation. The $\mathrm{Hom}(S,\mathrm{QMaps}(\mathcal{Y},\mathcal{T};\mathcal{P}))$-versions of the corresponding results are obtained similarly.

1.13

Suppose now that we have an embedding $\mathcal{T} \subset \mathbb{P}(\mathcal{E}_1) \times \cdots \times \mathbb{P}(\mathcal{E}_k)$. In this situation, for a k-tuple $\overline{a}$ of parameters a_i, we can introduce a scheme of quasi-maps:

We first define $\mathrm{QMaps}^{\overline{a}}(\mathcal{Y},\mathbb{P}(\mathcal{E}_1)\times\cdots\times\mathbb{P}(\mathcal{E}_k);\mathcal{E}_1,\dots,\mathcal{E}_k)$ simply as the product

$$\prod_i \mathrm{QMaps}^{a_i}(\mathcal{Y},\mathbb{P}(\mathcal{E}_i)).$$

For an arbitrary $\mathcal{T}$, which corresponds to a multigraded ideal

$$\bigoplus_{n_1,\dots,n_k} \mathcal{I}_{\mathcal{T}}^{n_1,\dots,n_k} \subset \mathrm{Sym}(\mathcal{E}_1^*) \otimes \cdots \otimes \mathrm{Sym}(\mathcal{E}_n^*),$$

we set $\mathrm{QMaps}^{\overline{a}}(\mathcal{Y},\mathcal{T};\mathcal{E}_1,\dots,\mathcal{E}_k)$ to be the closed subscheme in $\prod_i \mathrm{QMaps}^{a_i}(\mathcal{Y},\mathbb{P}(\mathcal{E}_i))$, defined by the condition that for each n-tuple $n_1,\dots,n_k$ of nonnegative integers, the composition

$$\mathcal{O}_{\mathcal{Y}\times S} \otimes (\mathcal{I}_{\mathcal{T}})^{n_1,\dots,n_k} \hookrightarrow \mathcal{O}_{\mathcal{Y}\times S} \otimes \mathrm{Sym}^{n_1}(\mathcal{E}_1^*) \otimes \cdots \otimes \mathrm{Sym}^{n_k}(\mathcal{E}_k^*)$$

$$\to (\mathcal{L}_1^{-1})^{\otimes n_1} \otimes \cdots \otimes (\mathcal{L}^{-1})_k^{\otimes n_k}$$

vanishes.

For $\mathcal{Y}$ integral, this condition is equivalent to demanding that the generic point of $\mathcal{Y}$ gets mapped to $\mathcal{T}$, just as in the definition of $\mathrm{QMaps}^a(\mathcal{Y}, \mathcal{T}; \mathcal{E})$.

Alternatively, let $C(\mathcal{T}; \mathcal{E}_1, \ldots, \mathcal{E}_k) \subset \mathcal{E}_1 \times \cdots \times \mathcal{E}_k$ be the affine cone over $\mathcal{T}$. We have an action of $\mathbb{G}_m^k$ on $C(\mathcal{T}; \mathcal{E}_1, \ldots, \mathcal{E}_k)$, and the stack-quotient $C(\mathcal{T}; \mathcal{E}_1, \ldots, \mathcal{E}_k)/\mathbb{G}_m^k$ again contains $\mathcal{T}$ as an open subset. By definition, $\mathrm{Hom}(S, \mathrm{QMaps}^{\overline{a}}(\mathcal{Y}, \mathcal{T}; \mathcal{E}_1, \ldots, \mathcal{E}_k))$ is the set of maps from $\mathcal{Y} \times S$ to the stack $C(\mathcal{T}, \mathcal{E}_1, \ldots, \mathcal{E}_k)/\mathbb{G}_m^k$, such that for every geometric $s \in S$, the map $\mathcal{Y} \times s \to C(\mathcal{T}; \mathcal{E}_1, \ldots, \mathcal{E}_k)/\mathbb{G}_m^k$ sends the generic point of $\mathcal{Y}$ to $\mathcal{T}$.

The above definition has an obvious QQMaps version, and Lemma 1.9, Proposition 1.10 with its based analogues, generalize in a straightforward way. Note, however, that for the next proposition it is essential that we work with QMaps, and not with QQMaps, as we will be taking tensor products of $\mathcal{L}_i$s.

Observe that $\mathcal{T}$ can be naturally embedded into $\mathbb{P}(\mathcal{E}_1 \otimes \cdots \otimes \mathcal{E}_k)$ via the Segre embedding $\mathbb{P}(\mathcal{E}_1) \times \cdots \times \mathbb{P}(\mathcal{E}_k) \hookrightarrow \mathbb{P}(\mathcal{E}_1 \otimes \cdots \otimes \mathcal{E}_k)$.

Lemma 1.14. *There exists a map*

$$\mathrm{QMaps}^{\overline{a}}(\mathcal{Y}, \mathcal{T}; \mathcal{E}_1, \ldots, \mathcal{E}_k)) \to \mathrm{QMaps}^{a_1+\cdots+a_k}(\mathcal{Y}, \mathcal{T}; \mathcal{E}_1 \otimes \cdots \otimes \mathcal{E}_k),$$

extending the identity map on $\mathrm{Maps}(\mathcal{Y}, S)$. *Moreover, when $\mathcal{Y}$ is smooth, this map is finite.*

Proof. We have a map of stacks

$$C(\mathcal{T}; \mathcal{E}_1 \otimes \cdots \otimes \mathcal{E}_k)/\mathbb{G}_m \to C(\mathcal{T}; \mathcal{E}_1, \ldots, \mathcal{E}_k)/\mathbb{G}_m^k,$$

which makes the existence of the map $\mathrm{QMaps}(\mathcal{Y}, \mathcal{T}; \mathcal{E}_1, \ldots, \mathcal{E}_k)) \to \mathrm{QMaps}(\mathcal{Y}, \mathcal{T}; \mathcal{E}_1 \otimes \cdots \otimes \mathcal{E}_k)$ obvious.

We know that when $\mathcal{Y}$ is smooth, $\mathrm{QMaps}(\mathcal{Y}, \mathcal{T}; \mathcal{E}_1, \ldots, \mathcal{E}_k)$ is proper; therefore, to check that our map is finite, it is enough to show that the fiber over every geometric point of the scheme $\mathrm{QMaps}(\mathcal{Y}, \mathcal{T}; \mathcal{E}_1 \otimes \cdots \otimes \mathcal{E}_k)$ is finite.

Suppose that we have a line bundle $\mathcal{L}$ on $\mathcal{Y}$ with an injective map $\kappa : \mathcal{L} \to \mathcal{O}_\mathcal{Y} \otimes (\mathcal{E}_1 \otimes \cdots \otimes \mathcal{E}_k)$, such that there exists an open subset $U \subset \mathcal{Y}$ such that $\kappa|_U$ is a bundle map corresponding to a map $U \to \mathcal{T}$. In particular, we have the line bundles $\mathcal{L}_1^U, \ldots, \mathcal{L}_k^U$, defined on U, such that $\mathcal{L}_1^U \otimes \cdots \otimes \mathcal{L}_k^U \simeq \mathcal{L}|_U$.

Let $\mathcal{L}'_i$ be the (unique) extension of $\mathcal{L}_i^U$ such that the map $\mathcal{L}'_i \to \mathcal{O}_\mathcal{Y} \otimes \mathcal{E}_i$ is regular and is a bundle map away from codimension 2. Clearly, $\mathcal{L}$ is a subsheaf in $\mathcal{L}'_1 \otimes \cdots \otimes \mathcal{L}'_k$.

Then the fiber of $\mathrm{QMaps}(\mathcal{Y}, \mathcal{T}; \mathcal{E}_1, \ldots, \mathcal{E}_k)) \to \mathrm{QMaps}(\mathcal{Y}, \mathcal{T}; \mathcal{E}_1 \otimes \cdots \otimes \mathcal{E}_k)$ is the scheme classifying subsheaves $\mathcal{L}_i \subset \mathcal{L}'_i$ such that $\mathcal{L}_1 \otimes \cdots \otimes \mathcal{L}_k \simeq \mathcal{L}$ and the maps $\mathcal{L}_i \to \mathcal{O}_\mathcal{Y} \otimes \mathcal{E}_i$ continue to be regular. This scheme is clearly finite. □

An example of the situation of Section 1.13 is this: Suppose that $\mathcal{T} = \mathcal{T}_1 \times \cdots \times \mathcal{T}_k$, and each $\mathcal{T}_i$ embeds into the corresponding $\mathcal{E}_i$. Then from Lemma 1.14 we obtain that there exists a finite morphism

$$\prod_i \mathrm{QMaps}^{a_i}(\mathcal{Y}, \mathcal{T}_i; \mathcal{E}_i) \simeq \mathrm{QMaps}^{\overline{a}}(\mathcal{Y}, \mathcal{T}; \mathcal{E}_1, \ldots, \mathcal{E}_k)$$
$$\to \mathrm{QMaps}^{a_1+\cdots+a_k}(\mathcal{Y}, \mathcal{T}; \mathcal{E}_1 \otimes \cdots \otimes \mathcal{E}_k).$$

In the main body of the paper, we will need the following technical statement. Let $\mathcal{T}_0 \subset \mathcal{E}_0$ and $\mathcal{T}'_0 \subset \mathcal{E}'_0$ be two projective varieties.

Let $\mathcal{T}$ be another projective variety, embedded into $\mathbb{P}(\mathcal{E}_0) \times \mathbb{P}(\mathcal{E}_1) \times \cdots \times \mathbb{P}(\mathcal{E}_n)$, and endowed with a map $\mathcal{T} \to \mathcal{T}_0$, such that the two maps from $\mathcal{T}$ to $\mathbb{P}(\mathcal{E}_0)$ coincide.

Lemma 1.15. *The (a priori finite) map*

$$\mathrm{QMaps}(\mathcal{Y}, \mathcal{T}; \mathcal{E}_0, \mathcal{E}_1, \ldots, \mathcal{E}_k) \times \mathrm{QMaps}(\mathcal{Y}, \mathcal{T}'_0; \mathcal{E}'_0)$$
$$\to \mathrm{QMaps}(\mathcal{Y}, \mathcal{T} \times \mathcal{T}'_0; \mathcal{E}_0 \otimes \mathcal{E}'_0, \mathcal{E}_1, \ldots, \mathcal{E}_k) \underset{\mathrm{QMaps}(\mathcal{Y}, \mathcal{T}_0 \times \mathcal{T}'_0; \mathcal{E}_0 \otimes \mathcal{E}'_0)}{\times} \mathrm{QMaps}(\mathcal{Y}, \mathcal{T}_0 \times \mathcal{T}'_0; \mathcal{E}_0, \mathcal{E}'_0).$$

is an isomorphism.

The proof is straightforward.

1.16

Let $\overset{\circ}{\mathrm{QQMaps}}(\mathcal{Y}, \mathcal{T}; \mathcal{E})$ denote the open subset of $\mathrm{QQMaps}(\mathcal{Y}, \mathcal{T}; \mathcal{E})$ corresponding to the condition that $\mathrm{coker}(\kappa)$ is a torsion-free coherent sheaf on $\mathcal{Y}$ (over every geometric point s in the corresponding test scheme S). Let us denote by $\partial(\mathrm{QQMaps}(\mathcal{Y}, \mathcal{T}; \mathcal{E}))$ the complement to $\overset{\circ}{\mathrm{QQMaps}}(\mathcal{Y}, \mathcal{T}; \mathcal{E})$ in $\mathrm{QQMaps}(\mathcal{Y}, \mathcal{T}; \mathcal{E})$.

For a Hilbert polynomial p, let $\mathrm{TFree}_1^p(\mathcal{Y})$ denote the corresponding connected component of the stack of torsion-free coherent sheaves of generic rank 1 on $\mathcal{Y}$. For two parameters p, p', let us denote by $\mathcal{Y}^{Q;(p,p')}$ the stack classifying triples $(\mathcal{L}, \mathcal{L}', \beta : \mathcal{L} \hookrightarrow \mathcal{L}')$, where $\mathcal{L}, \mathcal{L}'$ are points of $\mathrm{TFree}_1^p(\mathcal{Y})$ and $\mathrm{TFree}_1^{p'}(\mathcal{Y})$, respectively, and β is a nonzero map, which is automatically an embedding. Note that for $p = p'$, $\mathcal{Y}^{Q;(p,p')}$ projects isomorphically onto $\mathrm{TFree}_1^p(\mathcal{Y})$.

Proposition 1.17. *There is a natural proper morphism*

$$\bar{\iota}_{p',p} : \mathrm{QQMaps}^{p'}(\mathcal{Y}, \mathcal{T}; \mathcal{E}) \underset{\mathrm{TFree}_1^{p'}(\mathcal{Y})}{\times} \mathcal{Y}^{Q;(p,p')} \to \mathrm{QQMaps}^p(\mathcal{Y}, \mathcal{T}; \mathcal{E}).$$

The composition

$$\iota_{p',p} : \overset{\circ}{\mathrm{QQMaps}}{}^{p'}(\mathcal{Y}, \mathcal{T}; \mathcal{E}) \underset{\mathrm{TFree}_1^{p'}(\mathcal{Y})}{\times} \mathcal{Y}^{Q;(p,p')}$$
$$\hookrightarrow \mathrm{QQMaps}^{p'}(\mathcal{Y}, \mathcal{T}; \mathcal{E}) \underset{\mathrm{TFree}_1^{p'}(\mathcal{Y})}{\times} \mathcal{Y}^{Q;(p,p')} \to \mathrm{QQMaps}^p(\mathcal{Y}, \mathcal{T}; \mathcal{E})$$

is a locally closed embedding. Moreover, every geometric point of $\mathrm{QQMaps}^p(\mathcal{Y}, \mathcal{T}; \mathcal{E})$ *belongs to the image of* $\iota_{p',p}$ *for exactly one* p'.

Proof. The map $\bar{\iota}_{p',p}$ is constructed in a most straightforward way: given an S-point of $\mathrm{QQMaps}^{p'}(\mathcal{Y}, \mathcal{T}, \mathcal{E})$, i.e., a sheaf $\mathcal{L}'$ on $\mathcal{Y} \times S$, with an embedding $\kappa' : \mathcal{L}' \to \mathcal{O}_{\mathcal{Y}\times S} \otimes \mathcal{E}$, and an S-point of $\mathcal{Y}^{Q,(p,p')}$, i.e., another sheaf $\mathcal{L}$ on $\mathcal{Y} \times S$ with an embedding $\beta : \mathcal{L} \to \mathcal{L}'$, we define a new point of $\mathrm{QMaps}(\mathcal{Y}, \mathcal{T}, \mathcal{E})$, by setting κ to be the composition $\kappa' \circ \beta$.

This map is proper because the scheme $\mathrm{QQMaps}^{p'}(\mathcal{Y}, \mathcal{T}, \mathcal{E})$ is proper, and so is the projection $\mathcal{Y}^{Q;(p,p')} \to \mathrm{TFree}_1^p(\mathcal{Y})$.

Observe now that the map

$$\bigcup_{p'} \overset{\circ}{\mathrm{QQMaps}}{}^{p'}(\mathcal{Y}, \mathcal{T}; \mathcal{E}) \underset{\mathrm{TFree}_1^{p'}(\mathcal{Y})}{\times} \mathcal{Y}^{Q;(p,p')} \to \mathrm{QQMaps}^p(\mathcal{Y}, \mathcal{T}; \mathcal{E})$$

is injective on the level of S-points for any S.

Indeed, for $(\mathcal{L}, \mathcal{L}', \kappa', \beta)$ as above with $(\mathcal{L}', \kappa') \in \overset{\circ}{\mathrm{QQMaps}}{}^{p'}(\mathcal{Y}, \mathcal{T}; \mathcal{E})$, the sheaf $\mathcal{L}'$ is reconstructed as the unique subsheaf in $\mathcal{O}_{\mathcal{Y}\times S} \otimes \mathcal{E}$ containing $\mathcal{L}$, such that the quotient $\mathcal{O}_{\mathcal{Y}\times S} \otimes \mathcal{E}/\mathcal{L}'$ is torsion-free over every geometric point of S. (Moreover, for every geometric point $(\mathcal{L}, \kappa)$ of $\mathrm{QQMaps}^p(\mathcal{Y}, \mathcal{T}; \mathcal{E})$, such $\mathcal{L}'$ exists and equals the preimage in $\mathcal{O}_{\mathcal{Y}} \otimes \mathcal{E}$ of the maximal torsion subsheaf in $\mathrm{coker}(\kappa)$.)

Thus we obtain that the open subset in $\mathrm{QQMaps}^{p'}(\mathcal{Y}, \mathcal{T}; \mathcal{E}) \underset{\mathrm{TFree}_1^{p'}(\mathcal{Y})}{\times} \mathcal{Y}^{Q;(p,p')}$ equal to

$$(\bar{\iota}_{p',p})^{-1}(\mathrm{QQMaps}^p(\mathcal{Y}, \mathcal{T}; \mathcal{E}) - \bar{\iota}_{p',p}(\partial(\mathrm{QQMaps}^{p'}(\mathcal{Y}, \mathcal{T}; \mathcal{E}))))$$

coincides with $\overset{\circ}{\mathrm{QQMaps}}{}^{p'}(\mathcal{Y}, \mathcal{T}; \mathcal{E}) \underset{\mathrm{TFree}_1^{p'}(\mathcal{Y})}{\times} \mathcal{Y}^{Q;(p,p')}$. Hence the map

$$\bar{\iota}_{p',p} : \overset{\circ}{\mathrm{QQMaps}}{}^{p'}(\mathcal{Y}, \mathcal{T}; \mathcal{E}) \underset{\mathrm{TFree}_1^{p'}(\mathcal{Y})}{\times} \mathcal{Y}^{Q;(p,p')} \to (\mathrm{QQMaps}^p(\mathcal{Y}, \mathcal{T}; \mathcal{E}) - \bar{\iota}_{p',p}(\partial(\mathrm{QQMaps}^{p'}(\mathcal{Y}, \mathcal{T}; \mathcal{E}))))$$

is also proper, and being an injection on the level of S-points, it is a closed embedding. □

Now set $\overset{\circ}{\mathrm{QMaps}}(\mathcal{Y}, \mathcal{T}; \mathcal{E}) := \overset{\circ}{\mathrm{QQMaps}}(\mathcal{Y}, \mathcal{T}; \mathcal{E}) \cap \mathrm{QMaps}(\mathcal{Y}, \mathcal{T}; \mathcal{E})$, and $\partial(\mathrm{QMaps}(\mathcal{Y}, \mathcal{T}; \mathcal{E})) := \partial(\mathrm{QQMaps}(\mathcal{Y}, \mathcal{T}; \mathcal{E})) \cap \mathrm{QMaps}(\mathcal{Y}, \mathcal{T}; \mathcal{E})$.

For example, when $\mathcal{Y}$ is a smooth curve (in which case there is no difference between QMaps and QQMaps), the locus $\overset{\circ}{\mathrm{QMaps}}(\mathcal{Y}, \mathcal{T}; \mathcal{E})$ coincides with $\mathrm{Maps}(\mathcal{Y}, \mathcal{T}; \mathcal{E})$. Indeed, the condition that the quotient is torsion-free is equivalent in this case to its being a vector bundle.

Recall that by $\mathrm{Pic}^a(\mathcal{Y})$ we denote connected components of the Picard stack of $\mathcal{Y}$. Observe that if the parameters a, a' are such that $\mathrm{Pic}^{-a}(\mathcal{Y}) \times \mathrm{Pic}^{-a'}(\mathcal{Y}) \subset \mathrm{TFree}_1^p(\mathcal{Y}) \times \mathrm{TFree}_1^{p'}(\mathcal{Y})$, then

$$\mathcal{Y}^{Q;(p,p')} \underset{\mathrm{TFree}_1^p(\mathcal{Y})\times \mathrm{TFree}_1^{p'}(\mathcal{Y})}{\times} (\mathrm{Pic}^{-a}(\mathcal{Y}) \times \mathrm{Pic}^{-a'}(\mathcal{Y})) \simeq \mathrm{Pic}^{-a'}(\mathcal{Y}) \times \mathcal{Y}^{(a-a')},$$

where by $\mathcal{Y}^{(b)}$ we denote the scheme that classifies pairs $(\mathcal{L}, s)$, where $\mathcal{L} \in \mathrm{Pic}^b(\mathcal{Y})$ and $s : \mathcal{O}_{\mathcal{Y}} \to \mathcal{L}$ is a nonzero section. Note that when $\mathcal{Y}$ is a smooth curve, $\mathcal{Y}^{(b)}$ is just the corresponding symmetric power.

Thus from the previous proposition, we have a map

$$\bar{\iota}_{a',a} : \mathrm{QMaps}^{a'}(\mathcal{Y}, \mathcal{T}; \mathcal{E}) \times \mathcal{Y}^{(a-a')} \to \mathrm{QMaps}^a(\mathcal{Y}, \mathcal{T}; \mathcal{E}), \tag{2}$$

and a locally closed embedding $\iota_{a',a} : \overset{\circ}{\mathrm{QMaps}}{}^{a'}(\mathcal{Y}, \mathcal{T}; \mathcal{E}) \times \mathcal{Y}^{(a-a')} \to \mathrm{QMaps}^a(\mathcal{Y}, \mathcal{T}; \mathcal{E})$.

Assume now that $\mathcal{Y}$ is smooth. We have the following.

Lemma 1.18. *The map* $\bar{\iota}_{a',a} : \mathrm{QMaps}^{a'}(\mathcal{Y}, \mathcal{T}; \mathcal{E}) \times \mathcal{Y}^{(a-a')} \to \mathrm{QMaps}^a(\mathcal{Y}, \mathcal{T}; \mathcal{E})$ *above is proper and finite. Every geometric point of* $\mathrm{QMaps}^a(\mathcal{Y}, \mathcal{T}; \mathcal{E})$ *belongs to the image of exactly one* $\iota_{a',a}$.

Proof. The properness assertion follows from the fact that the schemes $\mathrm{QMaps}^{a'}(\mathcal{Y}, \mathcal{T}; \mathcal{E})$ and $\mathcal{Y}^{(a-a')}$ are proper when $\mathcal{Y}$ is smooth. The finiteness follows from the fact that given two line bundles $\mathcal{L} \subset \mathcal{L}'$ on $\mathcal{Y}$, the scheme classifying line bundles $\mathcal{L}''$, squeezed between the two, is finite.

To prove the last assertion, it suffices to observe that if $\mathcal{L}$ is a torsion-free sheaf of generic rank 1 on a smooth variety, embedded into $\mathcal{O}_{\mathcal{Y}} \otimes \mathcal{E}$, such that the quotient is torsion-free, then $\mathcal{L}$ is a line bundle. □

The above assertions have an obvious based analogue. The only required modification is that instead of $\mathcal{Y}^{Q;(p,p')}$ (respectively, $\mathcal{Y}^{(a-a')}$) we need to consider its open substack corresponding to the condition that $\mathcal{L} \to \mathcal{L}'$ is an isomorphism in a neighborhood of our subscheme $\mathcal{Y}'$ (respectively, the section s has no zeroes on $\mathcal{Y}'$).

1.19 Meromorphic quasi-maps

In what follows, we will have to consider the following generalization of the notion of quasi-map. We will formulate it in the relative situation.

Let $\mathcal{Y}$ be a flat projective scheme over a base $\mathcal{X}$, and let $\mathcal{Y}_1 \subset \mathcal{Y}$ be a relative Cartier divisor. Let $\mathcal{E}$ be a vector bundle defined over $\mathcal{Y} - \mathcal{Y}_1$, and $\mathcal{T} \subset \mathbb{P}(\mathcal{E})$ a closed subscheme.

We define the functor ${}_{\infty\cdot\mathcal{Y}_1}\mathrm{QMaps}(\mathcal{Y}, \mathcal{T}; \mathcal{E})$ on the category of schemes over $\mathcal{X}$, to assign to a test scheme S the set of pairs $(\mathcal{L}, \kappa)$, where $\mathcal{L}$ is a line bundle over $\mathcal{Y} \underset{\mathcal{X}}{\times} S$, and κ is a map of coherent sheaves on $(\mathcal{Y} - \mathcal{Y}_1) \underset{\mathcal{X}}{\times} S$

$$\kappa : \mathcal{L}|_{(\mathcal{Y}-\mathcal{Y}_1)\underset{\mathcal{X}}{\times} S} \to \mathcal{E} \underset{\mathcal{O}_{\mathcal{X}}}{\otimes} \mathcal{O}_S,$$

such that for an open dense subset $U \subset (\mathcal{Y} - \mathcal{Y}_1) \underset{\mathcal{X}}{\times} S$, over which κ is a bundle map, the resulting map $U \to \mathbb{P}(\mathcal{E})$ factors through $\mathcal{T}$.

Proposition 1.20. *The functor* ${}_{\infty\cdot\mathcal{Y}_1}\mathrm{QMaps}(\mathcal{Y},\mathcal{T};\mathcal{E})$ *is representable by a strict ind-scheme of ind-finite type over* $\mathcal{X}$.

Proof. Suppose that $\mathcal{Y}_2$ is another relative Cartier divisor containing $\mathcal{Y}_1$. We have a closed embedding of functors ${}_{\infty\cdot\mathcal{Y}_1}\mathrm{QMaps}(\mathcal{Y},\mathcal{T};\mathcal{E}) \hookrightarrow {}_{\infty\cdot\mathcal{Y}_2}\mathrm{QMaps}(\mathcal{Y},\mathcal{T};\mathcal{E})$.

Therefore, the question of ind-representability being local on $\mathcal{X}$, by enlarging $\mathcal{Y}_1$ we can assume that $\mathcal{E}$ is actually trivial; we will denote by the same symbol its extension to the entire $\mathcal{Y}$.

Let $\mathcal{L}_1$ be a line bundle on $\mathcal{Y}$ and $s:\mathcal{O}\to\mathcal{Y}$ be a section such that $\mathcal{Y}_1$ is its set of zeroes. If $\mathcal{L}_1\in\mathrm{Pic}^{a_1}_{\mathcal{X}}(\mathcal{Y})$, we have a natural closed embedding

$$\mathrm{QMaps}^a(\mathcal{Y},\mathcal{T};\mathcal{E})\to\mathrm{QMaps}^{a+a_1}(\mathcal{Y},\mathcal{T};\mathcal{E}),$$

as in (2), which sends a point $(\mathcal{L},\kappa)\in\mathrm{QMaps}^a(\mathcal{Y},\mathcal{T};\mathcal{E})$ to $(\mathcal{L}\otimes\mathcal{L}_1^{-1},\mathcal{L}\otimes\mathcal{L}_1^{-1}\xrightarrow{s^{-1}}\mathcal{L}\xrightarrow{\kappa}\mathcal{E})$.

It is now easy to see that ${}_{\infty\cdot\mathcal{Y}_1}\mathrm{QMaps}(\mathcal{Y},\mathcal{T};\mathcal{E})$ is representable by the inductive limit $\varinjlim\mathrm{QMaps}^{a+n\cdot a_1}(\mathcal{Y},\mathcal{T};\mathcal{E})$. □

If $\mathcal{Y}'\subset\mathcal{Y}$ is another closed subscheme, disjoint from $\mathcal{Y}_1$, endowed with a section $\sigma':\mathcal{Y}'\to\mathcal{T}$, in the same manner we define the corresponding based version ${}_{\infty\cdot\mathcal{Y}_1}\mathrm{QMaps}(\mathcal{Y},\mathcal{T};\mathcal{E})_{\mathcal{Y}',\sigma'}$, which is also a strict ind-scheme of ind-finite type.

1.21

Now let $\mathcal{E}$ be a pro-finite-dimensional vector space, i.e., $\mathcal{E}=\varprojlim\mathcal{E}_i$. (In the relative situation, we will assume that $\mathcal{E}$ is an inverse limit of vector bundles on $\mathcal{Y}$.)

Let us recall the definition of the scheme $\mathbb{P}(\mathcal{E})$ in this case. By definition, for a test scheme S, a map $S\to\mathbb{P}(\mathcal{E})$ is a line bundle $\mathcal{L}$ on S together with a map $S\to\mathcal{E}$ (which by definition means a compatible system of maps $S\to\mathcal{E}_i$), such that locally on S, starting from some index i, the map $S\to\mathcal{E}_i$ is an injective bundle map.

By construction, $\mathbb{P}(\mathcal{E})$ is a union of open subschemes "$\mathbb{P}(\mathcal{E})-\mathbb{P}(\ker(\mathcal{E}\to\mathcal{E}_i))$," where each such open is the inverse limit of finite-dimensional schemes $\mathbb{P}(\mathcal{E}_j)-\mathbb{P}(\ker(\mathcal{E}_j\to\mathcal{E}_i))$, $j\geq i$.

When $\mathcal{Y}$ is a projective scheme, the scheme $\mathrm{Maps}(\mathcal{Y},\mathbb{P}(\mathcal{E}))$ makes sense, and it is also a union of schemes that are projective limits of schemes of finite type.

The definition of $\mathrm{QQMaps}(\mathcal{Y},\mathbb{P}(\mathcal{E}))$ proceeds in exactly the same way as when $\mathcal{E}$ is finite dimensional. Namely, for a test scheme $\mathrm{Hom}(S,\mathrm{QQMaps}(\mathcal{Y},\mathbb{P}(\mathcal{E}))$ is the set of pairs $\sigma=(\mathcal{L},\kappa)$, where $\mathcal{L}$ a coherent sheaf of generic rank 1 on $\mathcal{Y}\times S$, and κ is a compatible system of maps $\mathcal{L}\to\mathcal{O}_{\mathcal{Y}\times S}\otimes\mathcal{E}_i$, such that for every geometric point $s\in S$ starting from some index i the map $\mathcal{L}_s\to\mathcal{O}_{\mathcal{Y}}\otimes\mathcal{E}_i$ is injective.

It is easy to show that $\mathrm{QQMaps}(\mathcal{Y},\mathbb{P}(\mathcal{E}))$ is a union of open subschemes, each of which is a projective limit of schemes of finite type. However, $\mathrm{QQMaps}(\mathcal{Y},\mathbb{P}(\mathcal{E}))$ itself is generally not of finite type.

If now $\mathcal{T}$ is a closed subscheme on $\mathbb{P}(\mathcal{E})$, we define $\mathrm{QQMaps}(\mathcal{Y},\mathcal{T};\mathcal{E})$ as the corresponding closed subscheme of $\mathrm{QQMaps}(\mathcal{Y},\mathbb{P}(\mathcal{E}))$. As before, the open subscheme

$\mathrm{QMaps}(\mathcal{Y}, \mathcal{T}; \mathcal{E}) \subset \mathrm{QQMaps}(\mathcal{Y}, \mathcal{T}; \mathcal{E})$ is defined by the condition that $\mathcal{L}$ is a line bundle.

The following easily follows from Lemma 1.9 and Section 1.11. Let $\mathcal{T}$ be a projective scheme of finite type embedded into $\mathbb{P}(\mathcal{E})$, where $\mathcal{E}$ is a pro-finite-dimensional vector space. Set $\mathcal{P} := \mathcal{O}(1)_{\mathcal{T}}$.

Lemma 1.22. *We have a natural closed embedding* $\mathrm{QMaps}(\mathcal{Y}, \mathcal{T}; \mathcal{P}) \to \mathrm{QMaps}(\mathcal{Y}, \mathbb{P}(\mathcal{E}))$.

2 Quasi-maps into flag schemes and Zastava spaces

In this section $\mathfrak{g}$ will be an arbitrary Kac–Moody algebra. Below we list some facts related to the flag scheme (and certain partial flag schemes) attached to $\mathfrak{g}$. These facts are established in [Ka] when $\mathfrak{g}$ is symmetrizable, but according to [Ka1], they hold in general. In the main body of the paper we will only need the cases when $\mathfrak{g}$ is either finite dimensional or nontwisted affine.

2.1

We will work in the following setup:

Let $A = (A_{ij})_{i,j \in I}$ be a finite Cartan matrix. We fix a root datum, that is, two finitely generated free abelian groups $\Lambda, \check{\Lambda}$ with a perfect bilinear pairing $\langle , \rangle : \Lambda \times \check{\Lambda} \to \mathbb{Z}$, and embeddings $I \subset \check{\Lambda}, i \mapsto \check{\alpha}_i$ (simple roots), $I \subset \Lambda, i \mapsto \alpha_i$ (simple coroots) such that $\langle \alpha_i, \check{\alpha}_j \rangle = A_{ij}$.

We also assume that the subsets $I \subset \Lambda$, $I \subset \check{\Lambda}$ are both linearly independent, and that the subgroup of Λ generated by α_i is saturated (i.e., the quotient is torsion-free). In the finite type case, the latter condition is equivalent to the fact that our root datum is simply connected.

We denote by $\mathfrak{g}(A) = \mathfrak{g}$ (for short) the completed Kac–Moody Lie algebra associated to the above datum (see [Ka, 3.2]). In particular, the Cartan subalgebra $\mathfrak{h}$ identifies with $\Lambda \underset{\mathbb{Z}}{\otimes} \mathbb{C}$. Sometimes we will denote the corresponding root datum by $\Lambda_{\mathfrak{g}}, \check{\Lambda}_{\mathfrak{g}}$ to stress the relation with $\mathfrak{g}$.

The semigroup of dominant coweights (respectively, weights) will be denoted by $\Lambda^+_{\mathfrak{g}} \subset \Lambda_{\mathfrak{g}}$ (respectively, $\check{\Lambda}^+_{\mathfrak{g}} \subset \check{\Lambda}_{\mathfrak{g}}$). We say that $\lambda \leq \mu$ if $\mu - \lambda$ is an integral nonnegative linear combination of simple coroots α_i, $i \in I$. The semigroup of coweights λ, which are ≥ 0 in this sense will be denoted $\Lambda^{\mathrm{pos}}_{\mathfrak{g}}$.

To a dominant weight $\check{\lambda} \in \check{\Lambda}^+_{\mathfrak{g}}$, one attaches the integrable $\mathfrak{g}$-module, denoted $\mathcal{V}_{\check{\lambda}}$, with a fixed highest-weight vector $v_{\check{\lambda}} \in \mathcal{V}_{\check{\lambda}}$. For a pair of weights $\check{\lambda}_1, \check{\lambda}_2 \in \check{\Lambda}^+_{\mathfrak{g}}$, there is a canonical map $\mathcal{V}_{\check{\lambda}_1+\check{\lambda}_2} \to \mathcal{V}_{\check{\lambda}_1} \otimes \mathcal{V}_{\check{\lambda}_2}$ that sends $v_{\check{\lambda}_1+\check{\lambda}_2}$ to $v_{\check{\lambda}_1} \otimes v_{\check{\lambda}_2}$.

The Lie algebra $\mathfrak{g}$ has a triangular decomposition $\mathfrak{g} = \mathfrak{n} \oplus \mathfrak{h} \oplus \mathfrak{n}^-$ (here $\mathfrak{n}$ is a pro-finite-dimensional vector space), and we have standard Borel subalgebras $\mathfrak{b} = \mathfrak{n} \oplus \mathfrak{h}$, $\mathfrak{b}^- = \mathfrak{h} \oplus \mathfrak{n}^-$.

If $\mathfrak{p} \subset \mathfrak{g}$ is a standard parabolic, we will denote by $\mathfrak{n}(\mathfrak{p}) \subset \mathfrak{n}$ its unipotent radical, by $\mathfrak{p}^- \subset \mathfrak{g}$ (respectively, $\mathfrak{n}(\mathfrak{p}^-) \subset \mathfrak{n}^-$) the corresponding opposite parabolic (respectively, its unipotent radical), and by $\mathfrak{m}(\mathfrak{p}) := \mathfrak{p} \cap \mathfrak{p}^-$ (or just $\mathfrak{m}$) the Levi factor. We will write $\mathfrak{n}(\mathfrak{m})$ (respectively, $\mathfrak{n}^-(\mathfrak{m})$) for the intersections $\mathfrak{m} \cap \mathfrak{n}$ and $\mathfrak{m} \cap \mathfrak{n}^-$, respectively.

If i is an element of I, we will denote the corresponding subminimal parabolic by $\mathfrak{p}_i$, and by $\mathfrak{p}_i^-$, $\mathfrak{n}_i := \mathfrak{n}(\mathfrak{p}_i)$, $\mathfrak{n}_i^- := \mathfrak{n}(\mathfrak{p}_i^-)$, $\mathfrak{m}_i$, respectively, the corresponding associated subalgebras. For a standard Levi subalgebra $\mathfrak{m}$, we will write that $i \in \mathfrak{m}$ if the corresponding simple root $\check{\alpha}_i$ belongs to $\mathfrak{n}(\mathfrak{m})$.

From now on we will only work with parabolics corresponding to subdiagrams of I of *finite type*. In particular, the Levi subalgebra $\mathfrak{m}$ is finite dimensional, and there exists a canonically defined reductive group M, such that $\mathfrak{m}$ is its Lie algebra. (When $\mathfrak{p} = \mathfrak{b}$, the corresponding Levi subalgebra is the Cartan torus T, whose set of cocharacters is our lattice $\Lambda_{\mathfrak{g}}$.) By the assumption on the root datum, the derived group of M is simply connected.

We will denote by $\Lambda_{\mathfrak{g},\mathfrak{p}}$ the quotient of $\Lambda = \Lambda_{\mathfrak{g}}$ by the subgroup generated by $\mathrm{Span}(\alpha_i, i \in \mathfrak{m})$. In other words, $\Lambda_{\mathfrak{g},\mathfrak{p}}$ is the group of cocharacters of $M/[M,M]$. By $\Lambda^{\mathrm{pos}}_{\mathfrak{g},\mathfrak{p}}$ we will denote the subsemigroup of $\Lambda_{\mathfrak{g},\mathfrak{p}}$ equal to the positive span of the images of $\alpha_i \in \Lambda_{\mathfrak{g}} \twoheadrightarrow \Lambda_{\mathfrak{g},\mathfrak{p}}$ for $i \notin \mathfrak{m}$.

In addition, there exists a pro-algebraic group P, with Lie algebra $\mathfrak{p}$, which projects onto M, and the kernel $N(P)$ is pro-unipotent.

The entire group G associated to the Lie algebra $\mathfrak{g}$, along with its subgroup P^-, exists as a group ind-scheme. Of course, if $\mathfrak{g}$ is finite dimensional, G is the corresponding reductive group. In the untwisted affine case, i.e., for $\mathfrak{g}_{\mathrm{aff}} = \mathfrak{g}((x)) \oplus K \cdot \mathbb{C} \oplus d \cdot \mathbb{C}$ being the affinization of a finite-dimensional simple $\underline{\mathfrak{g}}$ (cf. Section 3.7), the corresponding group ind-scheme is $G_{\mathrm{aff}} := \widehat{G} \times \mathbb{G}_m$, where $\widehat{G}$ is the canonical central extension $1 \to \mathbb{G}_m \to \widehat{G} \to G((x)) \to 1$ of the loop group ind-scheme $G((x))$, corresponding to the minimal $ad_{\mathfrak{g}}$-invariant scalar product on $\mathfrak{g}$, such that the induced bilinear form on $\Lambda_{\mathfrak{g}}$ is integral-valued.

2.2 Kashiwara's flag schemes

Let $\mathfrak{p} \subset \mathfrak{g}$ be a standard parabolic, corresponding to a subdiagram of I of finite type. Following Kashiwara (cf. [Ka]), one defines the (partial) flag schemes $\mathcal{G}_{\mathfrak{g},\mathfrak{p}}$, equipped with the action of G. Each $\mathcal{G}_{\mathfrak{g},\mathfrak{p}}$ has a unit point $1_{\mathcal{G}_{\mathfrak{g},\mathfrak{p}}}$, and its stabilizer in G equals P^-, so that $\mathcal{G}_{\mathfrak{g},\mathfrak{p}}$ should be thought of as the quotient "G/P^-."

The scheme $\mathcal{G}_{\mathfrak{g},\mathfrak{p}}$ comes equipped with a closed embedding (called the *Plücker embedding*)

$$\mathcal{G}_{\mathfrak{g},\mathfrak{p}} \hookrightarrow \prod \mathbb{P}(\mathcal{V}^*_{\check{\lambda}}),$$

where the product is being taken over $\check{\Lambda}^+_{\mathfrak{g},\mathfrak{p}} := \check{\Lambda}^+_{\mathfrak{g}} \cap \check{\Lambda}_{\mathfrak{g},\mathfrak{p}} \subset \check{\Lambda}_{\mathfrak{g}}$, and where each $\mathcal{V}^*_{\check{\lambda}}$ is the pro-finite-dimensional vector space dual to the integrable module $\mathcal{V}_{\check{\lambda}}$. The subscheme $\mathcal{G}_{\mathfrak{g},\mathfrak{p}} \hookrightarrow \prod \mathbb{P}(\mathcal{V}^*_{\check{\lambda}})$ is cut out by the so-called *Plücker equations*.

A collection of lines $(\ell_{\check{\lambda}} \subset \mathcal{V}^*_{\check{\lambda}})_{\check{\lambda} \in \check{\Lambda}^+_{\mathfrak{g},\mathfrak{p}}}$ satisfies the Plücker equations if

(a) for the canonical morphism $\mathcal{V}_{\check\lambda_1+\check\lambda_2} \to \mathcal{V}_{\check\lambda_1} \otimes \mathcal{V}_{\check\lambda_2}$, the map $\mathcal{V}^*_{\check\lambda_1} \hat\otimes \mathcal{V}^*_{\check\lambda_2} \to \mathcal{V}^*_{\check\lambda_1+\check\lambda_2}$ sends $\ell_{\check\lambda_1} \otimes \ell_{\check\lambda_2}$ to $\ell_{\check\lambda_1+\check\lambda_2}$;
(b) for any $\mathfrak{g}$-morphism $\mathcal{V}_{\check\mu} \to \mathcal{V}_{\check\lambda_1} \otimes \mathcal{V}_{\check\lambda_2}$ with $\check\mu \neq \check\lambda_1+\check\lambda_2$, the map $\mathcal{V}^*_{\check\lambda_1} \hat\otimes \mathcal{V}^*_{\check\lambda_2} \to \mathcal{V}^*_{\check\mu}$ sends $\ell_{\check\lambda_1} \otimes \ell_{\check\lambda_2}$ to 0.

The point $1_{\mathcal{G}_{\mathfrak{g},\mathfrak{p}}}$ corresponds to the system of lines $(\ell^0_{\check\lambda} \subset \mathcal{V}^*_{\check\lambda})$, where $\ell^0_{\check\lambda} = (\mathcal{V}^*_{\check\lambda})^{\mathfrak{n}_-}$. The inverse image of the line bundle $\mathcal{O}(1)$ on $\mathbb{P}(\mathcal{V}^*_{\check\lambda})$ is the line bundle on $\mathcal{G}_{\mathfrak{g},\mathfrak{p}}$ denoted by $\mathcal{P}^{\check\lambda}_{\mathfrak{g},\mathfrak{p}}$. We have $\Gamma(\mathcal{G}, \mathcal{P}^{\check\lambda}_{\mathfrak{g},\mathfrak{p}}) = \mathcal{V}_\lambda$.

The orbits of the action of the pro-algebraic group $N(P)$ on $\mathcal{G}_{\mathfrak{g},\mathfrak{p}}$ are parametrized by the double-quotient $\mathcal{W}_{\mathfrak{m}}\backslash\mathcal{W}_{\mathfrak{g}}/\mathcal{W}_{\mathfrak{m}}$, where $\mathcal{W}_{\mathfrak{g}}$ and $\mathcal{W}_{\mathfrak{m}}$ are the Weyl groups of $\mathfrak{g}$ and $\mathfrak{m}$, respectively. For $w \in \mathcal{W}_{\mathfrak{m}}\backslash\mathcal{W}_{\mathfrak{g}}/\mathcal{W}_{\mathfrak{m}}$, we will denote by $\mathcal{G}_{\mathfrak{g},\mathfrak{p},w}$ the corresponding orbit (in our normalization, the unit point $1_{\mathcal{G}_{\mathfrak{g},\mathfrak{p}}}$ belongs to $\mathcal{G}_{\mathfrak{g},\mathfrak{p},e}$, where $e \in \mathcal{W}_{\mathfrak{g}}$ is the unit element). By $\overline{\mathcal{G}}_{\mathfrak{g},\mathfrak{p},w}$ we will denote the closure of $\mathcal{G}_{\mathfrak{g},\mathfrak{p},w}$ (e.g., $\overline{\mathcal{G}}_{\mathfrak{g},\mathfrak{p},e}$ equals the entire $\mathcal{G}_{\mathfrak{g},\mathfrak{p}}$), and by $\mathcal{G}^w_{\mathfrak{g},\mathfrak{p}}$ the open subscheme equal to union of orbits with parameters $w' \leq w$ in the sense of the usual Bruhat order. For each $w \in \mathcal{W}_{\mathfrak{m}}\backslash\mathcal{W}_{\mathfrak{g}}/\mathcal{W}_{\mathfrak{m}}$ there exists a canonical subgroup $N(P)_w \subset N(P)$, of finite codimension, such that $N(P)_w$ acts freely on $\mathcal{G}^w_{\mathfrak{g},\mathfrak{p}}$ with a finite-dimensional quotient.

For every simple reflection $s_i \in \mathcal{W}_{\mathfrak{g}}$ with $\alpha_i \notin \mathfrak{m}$, we have a codimension-1 subscheme $\mathcal{G}_{\mathfrak{g},\mathfrak{p},s_i}$, and its closure $\overline{\mathcal{G}}_{\mathfrak{g},\mathfrak{p},s_i}$ is an effective Cartier divisor. The union $\bigcup_i \overline{\mathcal{G}}_{\mathfrak{g},\mathfrak{p},s_i}$ is called *the Schubert divisor* and it equals the complement $\mathcal{G}_{\mathfrak{g},\mathfrak{p}} - \mathcal{G}^e_{\mathfrak{g},\mathfrak{p}}$.

Consider the affine cone $C(\mathcal{G}_{\mathfrak{g},\mathfrak{p}})$ over $\mathcal{G}_{\mathfrak{g},\mathfrak{p}}$ corresponding to the Plücker embedding. This is a closed subscheme of $\prod_{\check\lambda \in \check\Lambda^+_{\mathfrak{g},\mathfrak{p}}} \mathcal{V}^*_{\check\lambda}$ consisting of collections of vectors $(u_{\check\lambda} \subset \mathcal{V}^*_{\check\lambda})_{\check\lambda \in \check\Lambda^+_{\mathfrak{g},\mathfrak{p}}}$ satisfying the Plücker equations:

(a) For the canonical morphism $\mathcal{V}_{\check\lambda_1+\check\lambda_2} \to \mathcal{V}_{\check\lambda_1} \otimes \mathcal{V}_{\check\lambda_2}$, the map $\mathcal{V}^*_{\check\lambda_1} \hat\otimes \mathcal{V}^*_{\check\lambda_2} \to \mathcal{V}^*_{\check\lambda_1+\check\lambda_2}$ sends $u_{\check\lambda_1} \otimes u_{\check\lambda_2}$ to $u_{\check\lambda_1+\check\lambda_2}$.
(b) For any $\mathfrak{g}$-morphism $\mathcal{V}_{\check\mu} \to \mathcal{V}_{\check\lambda_1} \otimes \mathcal{V}_{\check\lambda_2}$ with $\check\mu \neq \check\lambda_1+\check\lambda_2$, the map $\mathcal{V}^*_{\check\lambda_1} \hat\otimes \mathcal{V}^*_{\check\lambda_2} \to \mathcal{V}^*_{\check\mu}$ sends $u_{\check\lambda_1} \otimes u_{\check\lambda_2}$ to 0.

We have a natural action of the torus $M/[M, M]$ on $C(\mathcal{G}_{\mathfrak{g},\mathfrak{p}})$. Let $\overset{\circ}{C}(\mathcal{G}_{\mathfrak{g},\mathfrak{p}})$ be the open subset corresponding to the condition that all $u_{\check\lambda} \neq 0$. Then $\overset{\circ}{C}(\mathcal{G}_{\mathfrak{g},\mathfrak{p}})$ is a principal $M/[M, M]$-bundle over $\mathcal{G}_{\mathfrak{g},\mathfrak{p}}$.

2.3 Quasi-maps into flag schemes

Let $\mathbf{C}$ be a smooth projective curve, with a marked point $\infty_{\mathbf{C}} \in \mathbf{C}$, called "infinity." We will denote by $\overset{\circ}{\mathbf{C}}$ the complement $\mathbf{C} - \infty_{\mathbf{C}}$.

We will use a shorthand notation $\text{QMaps}(\mathbf{C}, \mathcal{G}_{\mathfrak{g},\mathfrak{p}})$ for the scheme of *based quasi-maps* $\text{QMaps}(\mathbf{C}, \mathcal{G}_{\mathfrak{g},\mathfrak{p}}; \mathcal{V}_{\check\lambda}, \check\lambda \in \check\Lambda^+_{\mathfrak{g},\mathfrak{p}})_{\infty_{\mathbf{C}},\sigma_{const}}$, where $\sigma_{const} : \infty_{\mathbf{C}} \to \mathcal{G}_{\mathfrak{g},\mathfrak{p}}$ is the constant map corresponding to $1_{\mathcal{G}_{\mathfrak{g},\mathfrak{p}}} \in \mathcal{G}_{\mathfrak{g},\mathfrak{p}}$. In other words, $\mathrm{QMaps}(\mathbf{C}, \mathcal{G}_{\mathfrak{g},\mathfrak{p}})$ classifies based maps from $\mathbf{C}$ to the stack $C(\mathcal{G}_{\mathfrak{g},\mathfrak{p}})/(M/[M, M])$. By $\mathrm{Maps}(\mathbf{C}, \mathcal{G}_{\mathfrak{g},\mathfrak{p}})$ we will denote the open subscheme of maps.

The scheme $\mathrm{QMaps}(\mathbf{C}, \mathcal{G}_{\mathfrak{g},\mathfrak{p}})$ splits as a disjoint union according the degree, which in our case is given by elements $\theta \in \Lambda^{\mathrm{pos}}_{\mathfrak{g},\mathfrak{p}}$.

For a fixed parameter θ as above, let $\overset{\circ}{\mathbf{C}}{}^{\theta}$ be the corresponding partially symmetrized power of $\overset{\circ}{\mathbf{C}}$. In other words, a point of $\overset{\circ}{\mathbf{C}}{}^{\theta}$ assigns to every $\check{\lambda} \in \check{\Lambda}^{+}_{\mathfrak{g},\mathfrak{p}}$ an element of the usual symmetric power $\overset{\circ}{\mathbf{C}}{}^{(n)}$, where $n = \langle \theta, \check{\lambda} \rangle$, and this assignment must be linear in $\check{\lambda}$ in a natural sense. Explicitly, if $\theta = \underset{i \in I}{\Sigma} n_i \cdot \alpha_i$, for $i \notin \mathfrak{m}$, then $\overset{\circ}{\mathbf{C}}{}^{\theta} \simeq \prod\limits_{i \in I} \overset{\circ}{\mathbf{C}}{}^{(n_i)}$.

From Section 1.16, we obtain that for each $\theta', \theta - \theta' \in \Lambda^{\mathrm{pos}}_{\mathfrak{g},\mathfrak{p}}$, we have a finite map

$$\bar{\iota}_{\theta'} : \mathrm{QMaps}^{\theta-\theta'}(\mathbf{C}, \mathcal{G}_{\mathfrak{g},\mathfrak{p}}) \times \overset{\circ}{\mathbf{C}}{}^{\theta'} \to \mathrm{QMaps}^{\theta}(\mathbf{C}, \mathcal{G}_{\mathfrak{g},\mathfrak{p}}), \tag{3}$$

such that the corresponding map

$$\iota_{\theta'} : \mathrm{Maps}^{\theta-\theta'}(\mathbf{C}, \mathcal{G}_{\mathfrak{g},\mathfrak{p}}) \times \overset{\circ}{\mathbf{C}}{}^{\theta'} \to \mathrm{QMaps}^{\theta}(\mathbf{C}, \mathcal{G}_{\mathfrak{g},\mathfrak{p}})$$

is a locally closed embedding. Moreover, the images of $\iota_{\theta'}$ define a decomposition of $\mathrm{QMaps}^{\theta}(\mathbf{C}, \mathcal{G}_{\mathfrak{g},\mathfrak{p}})$ into locally closed pieces. We will denote the embedding of the deepest stratum $\overset{\circ}{\mathbf{C}}{}^{\theta} \to \mathrm{QMaps}^{\theta}(\mathbf{C}, \mathcal{G}_{\mathfrak{g},\mathfrak{p}})$ by $\mathfrak{s}^{\theta}_{\mathfrak{p}}$.

Given a quasi-map $\sigma \in \mathrm{QMaps}^{\theta}(\mathbf{C}, \mathcal{G}_{\mathfrak{g},\mathfrak{p}})$ which is the image of

$$(\sigma' \in \mathrm{QMaps}^{\theta-\theta'}(\mathbf{C}, \mathcal{G}_{\mathfrak{g},\mathfrak{p}}), \theta_k \cdot \mathbf{c}_k \in \overset{\circ}{\mathbf{C}}{}^{\theta'}),$$

we will say that (a) the defect of σ is concentrated in $\underset{k}{\cup} \mathbf{c}_k$, (b) the defect of σ at $\mathbf{c}_k$ equals θ_k, (c) the total defect equals θ', and (d) σ' is the saturation of σ. Sometimes, we will assemble statements (a)–(c) into one by saying that the defect of σ is the colored divisor $\underset{k}{\Sigma} \theta_k \cdot \mathbf{c}_k \in \overset{\circ}{\mathbf{C}}{}^{\theta'}$.

2.4 The finite-dimensional case

Assume for a moment that $\mathfrak{g}$ is finite dimensional. Let $\mathrm{Bun}_G(\mathbf{C})$ (respectively, $\mathrm{Bun}_G(\mathbf{C}, \infty_{\mathbf{C}})$) denote the stack of G-bundles on $\mathbf{C}$ (respectively, G-bundles trivialized at $\infty_{\mathbf{C}}$).

We can consider a relative situation over $\mathrm{Bun}_G(\mathbf{C})$ (respectively, over $\mathrm{Bun}_G(\mathbf{C}, \infty_{\mathbf{C}})$), when one takes maps or quasi-maps (respectively, based maps or quasi-maps) from $\mathbf{C}$ to $\mathcal{F}_G \overset{G}{\times} \mathcal{G}_{\mathfrak{g},\mathfrak{p}}$, where $\mathcal{F}_G$ is a point of $\mathrm{Bun}_G(\mathbf{C})$ (respectively, $\mathrm{Bun}_G(\mathbf{C}, \infty_{\mathbf{C}})$).

The corresponding stacks of maps identify, respectively, with the stack $\mathrm{Bun}_{P^-}(\mathbf{C})$ of P^--bundles on $\mathbf{C}$, and $\mathrm{Bun}_{P^-}(\mathbf{C}, \infty_{\mathbf{C}})$–the stack P^--bundles on $\mathbf{C}$ trivialized at $\infty_{\mathbf{C}}$.

In the nonbased case, the stack of relative quasi-maps is denoted by $\overline{\mathrm{Bun}}_{P^-}(\mathbf{C})$, and it was studied in [BG1] and [BFGM]. The corresponding stack of relative based quasi-maps will be denoted by $\overline{\mathrm{Bun}}_{P^-}(\mathbf{C}, \infty_{\mathbf{C}})$.

2.5 Some properties of quasi-map spaces

Let $\nu \in \Lambda_{\mathfrak{g}}$ be a coweight space of quasi-maps such that $\langle \nu, \check{\alpha}_i \rangle = 0$ for $i \in \mathfrak{m}$, and $\langle \nu, \check{\alpha}_i \rangle > 0$ if $i \notin \mathfrak{m}$. Consider the corresponding 1-parameter subgroup $\mathbb{G}_m \to T$. Since the point $1_{\mathcal{G}_{\mathfrak{g},\mathfrak{p}}}$ is T-stable, we obtain a $\mathbb{G}_m$-action on the scheme $\mathrm{QMaps}^\theta(\mathbf{C}, \mathcal{G}_{\mathfrak{g},\mathfrak{p}})$.

In what follows, if a group $\mathbb{G}_m$ acts on a scheme $\mathcal{Y}$, we will say that the action contracts $\mathcal{Y}$ to a subscheme $\mathcal{Y}' \subset \mathcal{Y}$, if: (a) the action on $\mathcal{Y}'$ is trivial, and (b) the action map extends to a morphism $\mathbb{A}^1 \times \mathcal{Y} \to \mathcal{Y}$, such that $0 \times \mathcal{Y}$ is mapped to $\mathcal{Y}'$.

Proposition 2.6. *The above* $\mathbb{G}_m$*-action contracts* $\mathrm{QMaps}^\theta(\mathbf{C}, \mathcal{G}_{\mathfrak{g},\mathfrak{p}})$ *to* $\overset{\circ}{\mathbf{C}}{}^\theta \overset{\mathfrak{s}_{\mathfrak{p}}^\theta}{\subset} \mathrm{QMaps}^\theta(\mathbf{C}, \mathcal{G}_{\mathfrak{g},\mathfrak{p}})$.

Proof. The above T-action on $\mathrm{QMaps}^\theta(\mathbf{C}, \mathcal{G}_{\mathfrak{g},\mathfrak{p}})$ corresponds to an action of T on the cone $C(\mathcal{G}_{\mathfrak{p},\mathfrak{g}})$ that takes a collection of vectors $(u_{\check{\lambda}} \in \mathcal{V}^*_{\check{\lambda}})_{\check{\lambda} \in \check{\Lambda}^+_{\mathfrak{g},\mathfrak{p}}}$ to $\check{\lambda}(t) \cdot t \cdot u_{\check{\lambda}}$, where $t \cdot u$ denotes the T-action on the representation $\mathcal{V}^*_{\check{\lambda}}$.

It is clear now that a 1-parameter subgroup corresponding to ν as in the proposition contracts each $\mathcal{V}^*_{\check{\lambda}}$ to the line $\ell^0_{\check{\lambda}}$.

Moreover, the subscheme of quasi-maps which map $\mathbf{C}$ to $(\ell^0_{\check{\lambda}})_{\check{\lambda} \in \check{\Lambda}^+_{\mathfrak{g},\mathfrak{p}}} \subset C(\mathcal{G}_{\mathfrak{p},\mathfrak{g}})$ coincides with $\mathfrak{s}_{\mathfrak{p}}(\overset{\circ}{\mathbf{C}}{}^\theta)$. □

Proposition 2.7. *The scheme* $\mathrm{Maps}^\theta(\mathbf{C}, \mathcal{G}_{\mathfrak{g},\mathfrak{p}})$ *is a union of open subschemes of finite type.*

Remark. One can prove that the scheme $\mathrm{Maps}^\theta(\mathbf{C}, \mathcal{G}_{\mathfrak{g},\mathfrak{b}})$ is in fact globally of finite type, at least when $\mathfrak{g}$ is symmetrizable (cf. the appendix), but the proof is more involved. Of course, when $\mathfrak{g}$ is finite dimensional, this is obvious. When $\mathfrak{g}$ is (untwisted) affine, another proof will be given in the next section, using a modular interpretation of $\mathcal{G}_{\mathfrak{g},\mathfrak{b}}$ via bundles on the projective line.

Proof. Let us first remark that if $\mathcal{Y}$ and $\mathcal{T}$ are projective schemes, and $\mathcal{T}^0 \subset \mathcal{T}$ is an open subscheme, the subfunctor of $\mathrm{Maps}(\mathcal{Y}, \mathcal{T})$ consisting of maps landing in $\mathcal{T}^0$ is in fact an open subscheme.

For an element $w \in \mathcal{W}_{\mathfrak{m}} \backslash \mathcal{W} / \mathcal{W}_{\mathfrak{m}}$, let $\mathrm{Maps}^\theta(\mathbf{C}, \mathcal{G}^w_{\mathfrak{g},\mathfrak{p}})$ be the open subset in $\mathrm{Maps}^\theta(\mathbf{C}, \mathcal{G}_{\mathfrak{g},\mathfrak{p}})$ of maps that land in the open subset $\mathcal{G}^w_{\mathfrak{g},\mathfrak{p}} \subset \mathcal{G}_{\mathfrak{g},\mathfrak{p}}$. Evidently, $\mathrm{Maps}^\theta(\mathbf{C}, \mathcal{G}_{\mathfrak{g},\mathfrak{p}}) \simeq \bigcup_w \mathrm{Maps}^\theta(\mathbf{C}, \mathcal{G}^w_{\mathfrak{g},\mathfrak{p}})$. We claim that each $\mathrm{Maps}^\theta(\mathbf{C}, \mathcal{G}^w_{\mathfrak{g},\mathfrak{p}})$ is of finite type.

Recall that the subgroup $N(P)_w \subset N(P)$ acts freely on $\mathcal{G}^w_{\mathfrak{g},\mathfrak{p}}$ and that the quotient $\mathcal{G}^w_{\mathfrak{g},\mathfrak{p}}/N(P)_w$ is a quasi-projective scheme of finite type. Hence the scheme (of based maps) $\mathrm{Maps}(\mathbf{C}, \mathcal{G}^w_{\mathfrak{g},\mathfrak{p}}/N(P)_w)$ is of finite type.

We now claim that the natural projection map

$$\mathrm{Maps}^{\theta}(\mathbf{C}, \mathcal{G}^w_{\mathfrak{g},\mathfrak{p}}) \to \mathrm{Maps}(\mathbf{C}, \mathcal{G}^w_{\mathfrak{g},\mathfrak{p}})/N(P)_w$$

is a closed embedding. (We will show, moreover, that this map is actually an isomorphism if $\mathbf{C}$ is of genus 0.)

First, let us show that this map is injective on the level of S-points for any S. Let σ_1, σ_2 be two based maps $\mathbf{C} \times S \to \mathcal{G}^w_{\mathfrak{g},\mathfrak{p}}$ which project to the same map to $\mathcal{G}^w_{\mathfrak{g},\mathfrak{p}}/N(P)_w$. Then by definition we obtain a map $\widetilde{\sigma} : \mathbf{C} \times S \to N(P)_w$ with $\widetilde{\sigma}|_{\infty_{\mathbf{C}} \times S} \equiv 1$. Since $\mathbf{C}$ is projective and $N(P)_w$ is an inverse limit of groups which are extensions of $\mathbb{G}_a$, we obtain that $\widetilde{\sigma} \equiv 1$.

Now let us show that the map $\mathrm{Maps}^{\theta}(\mathbf{C}, \mathcal{G}^w_{\mathfrak{g},\mathfrak{p}}) \to \mathrm{Maps}(\mathbf{C}, \mathcal{G}^w_{\mathfrak{g},\mathfrak{p}})/N(P)_w$ is proper by checking the valuative criterion.

Let σ' be a based map $\mathbf{C} \times \mathbf{X} \to \mathcal{G}^w_{\mathfrak{g},\mathfrak{p}}/N(P)_w$, where $\mathbf{X}$ is an affine curve such that the restriction $\sigma'|_{\mathbf{C}\times(\mathbf{X}-0_{\mathbf{X}})}$ lifts to a map to $\mathcal{G}^w_{\mathfrak{g},\mathfrak{p}}$. Let us filter $N(P)_w$ by normal subgroups $N(P)_w = N_0 \supset N_1 \supset \dots$ with associated graded quotients isomorphic to the additive group $\mathbb{G}_a$. Set $\sigma'_0 = \sigma'$, and assume that we have found a lifting of σ' to a based map $\sigma'_i : \mathbf{C} \times \mathbf{X} \to \mathcal{G}^w_{\mathfrak{g},\mathfrak{p}}/N_i$. Then the obstruction to lifting it to the next level lies in $H^1(\mathbf{C}\times\mathbf{X}, \mathcal{O}(-\infty_{\mathbf{C}}) \boxtimes \mathcal{O}_{\mathbf{X}})$. Since $\mathbf{X}$ is affine, the latter group is isomorphic to the space of $H^1(\mathbf{C}, \mathcal{O}(-\infty_{\mathbf{C}}))$-valued functions on $\mathbf{X}$. By assumption, the function corresponding to the obstruction class vanishes on the open subset $\mathbf{X} - 0_{\mathbf{X}}$; therefore, it vanishes.

Note that when $\mathbf{C}$ is of genus 0, and for an arbitrary affine base S, the same argument shows that any based map $\sigma' : \mathbf{C} \times S \to \mathcal{G}^w_{\mathfrak{g},\mathfrak{p}}/N(P)_w$ lifts to a map $\mathbf{C} \times \mathbf{X} \to \mathcal{G}^w_{\mathfrak{g},\mathfrak{p}}$, because in this case $H^1(\mathbf{C}, \mathcal{O}(-\infty_{\mathbf{C}})) = 0$. □

Let $\mathfrak{p} \subset \mathfrak{p}'$ be a pair of standard parabolics, and set $\mathfrak{p}(\mathfrak{m}') = \mathfrak{m}' \cap \mathfrak{p}$ to be the corresponding parabolic in $\mathfrak{m}'$. Note that we have an exact sequence

$$0 \to \Lambda_{\mathfrak{m}',\mathfrak{p}(\mathfrak{m}')} \to \Lambda_{\mathfrak{g},\mathfrak{p}} \to \Lambda_{\mathfrak{g},\mathfrak{p}'} \to 0.$$

Lemma 2.8. *We have a natural map* $\mathrm{QMaps}^{\theta}(\mathbf{C}, \mathcal{G}_{\mathfrak{g},\mathfrak{p}}) \to \mathrm{QMaps}^{\theta'}(\mathbf{C}, \mathcal{G}_{\mathfrak{g},\mathfrak{p}'})$*, where* θ' *is the image of* θ *under the above map of lattices. If* $\theta' = 0$*, there is an isomorphism* $\mathrm{QMaps}^{\theta}(\mathbf{C}, \mathcal{G}_{\mathfrak{g},\mathfrak{p}}) \simeq \mathrm{QMaps}^{\theta}(\mathbf{C}, \mathcal{G}_{\mathfrak{m}',\mathfrak{p}(\mathfrak{m}')})$.

Proof. The existence of the map $\mathrm{QMaps}^{\theta}(\mathbf{C}, \mathcal{G}_{\mathfrak{g},\mathfrak{p}}) \to \mathrm{QMaps}^{\theta'}(\mathbf{C}, \mathcal{G}_{\mathfrak{g},\mathfrak{p}'})$ is immediate from the definitions.

Note that if $\theta' = 0$, $\mathrm{QMaps}^{\theta'}(\mathbf{C}, \mathcal{G}_{\mathfrak{g},\mathfrak{p}'})$ is a point-scheme that corresponds to the constant map $\mathbf{C} \to 1_{\mathcal{G}_{\mathfrak{g},\mathfrak{p}'}} \in \mathcal{G}_{\mathfrak{g},\mathfrak{p}'}$. Note also that the preimage of $1_{\mathcal{G}_{\mathfrak{g},\mathfrak{p}'}}$ in $\mathcal{G}_{\mathfrak{g},\mathfrak{p}}$ identifies naturally with $\mathcal{G}_{\mathfrak{m}',\mathfrak{m}'(\mathfrak{p})}$. In particular, for θ as above, we have a closed embedding $\mathrm{QMaps}^{\theta}(\mathbf{C}, \mathcal{G}_{\mathfrak{m}',\mathfrak{p}(\mathfrak{m}')}) \to \mathrm{QMaps}^{\theta}(\mathbf{C}, \mathcal{G}_{\mathfrak{g},\mathfrak{p}})$.

To see that it is an isomorphism, note that for any $\sigma \in \mathrm{QMaps}^{\theta}(\mathbf{C}, \mathcal{G}_{\mathfrak{g},\mathfrak{p}})$ there exists an open dense subset $U \subset \mathbf{C} \times S$ such that the map $\sigma|_U$ projects to the constant map $U \to \mathcal{G}_{\mathfrak{g},\mathfrak{p}'}$, and hence has its image in $\mathcal{G}_{\mathfrak{m}',\mathfrak{p}(\mathfrak{m}')}$. But this implies that σ itself is a quasi-map into $\mathcal{G}_{\mathfrak{m}',\mathfrak{p}(\mathfrak{m}')}$. □

Assume for example that $\mathfrak{p} = \mathfrak{b}$ and that $\mu \in \Lambda_{\mathfrak{g}}^{\text{pos}} = \Lambda_{\mathfrak{g},\mathfrak{b}}^{\text{pos}}$ equals α_i for some $i \notin \mathfrak{m}$. By putting $\mathfrak{p}' = \mathfrak{p}_i$, from the previous lemma we obtain that the space $\text{QMaps}^\theta(\mathbf{C}, \mathcal{G}_{\mathfrak{g},\mathfrak{b}})$ in this case identifies with the space of based quasi-maps $\mathbf{C} \to \mathbb{P}^1$ of degree 1. In particular, it is empty unless $\mathbf{C}$ is of genus 0, and for $\mathbf{C} \simeq \mathbb{P}^1$, it is isomorphic to $\mathbb{A}^2$.

Note that we have a natural map $\text{Maps}(\mathbf{C}, \mathcal{G}_{\mathfrak{g},\mathfrak{p}}) \to \text{Bun}_M(\mathbf{C}, \infty_{\mathbf{C}})$. Indeed, for a based map $\mathbf{C} \times S \to \mathcal{G}_{\mathfrak{g},\mathfrak{p}}$ we define an M-bundle on $\mathbf{C} \times S$ by taking the Cartesian product

$$(\mathbf{C} \times S) \underset{\mathcal{G}_{\mathfrak{g},\mathfrak{p}}}{\times} \overset{\circ}{\widetilde{\mathcal{G}}}_{\mathfrak{g},\mathfrak{p}},$$

where $\overset{\circ}{\widetilde{\mathcal{G}}}_{\mathfrak{g},\mathfrak{p}}$ is as in Section 2.29.

The corresponding schemes $\text{Maps}(\mathbf{C}, \mathcal{G}_{\mathfrak{g},\mathfrak{p}})$ and $\text{Maps}(\mathbf{C}, \mathcal{G}_{\mathfrak{g},\mathfrak{p}'})$ are related in the following explicit way:

Let $P^-(M')$ be the parabolic in M' corresponding to $\mathfrak{p}^-(\mathfrak{m}')$. Let $\text{Bun}_{M'}(\mathbf{C}, \infty_{\mathbf{C}})$ be the stack as in Section 2.4, and let $\text{Bun}_{P^-(M')}(\mathbf{C}, \infty_{\mathbf{C}}) \subset \overline{\text{Bun}}_{P^-(M')}(\mathbf{C}, \infty_{\mathbf{C}})$ be the corresponding stacks of $P^-(M')$-bundles and generalized $P^-(M')$-bundles, respectively.

The following is straightforward from the definitions.

Lemma 2.9. *There exists an open embedding*

$$\text{Maps}(\mathbf{C}, \mathcal{G}_{\mathfrak{g},\mathfrak{p}'}) \underset{\text{Bun}_{M'}(\mathbf{C},\infty_{\mathbf{C}})}{\times} \overline{\text{Bun}}_{P^-(M')}(\mathbf{C}, \infty_{\mathbf{C}}) \hookrightarrow \text{QMaps}(\mathbf{C}, \mathcal{G}_{\mathfrak{g},\mathfrak{p}}),$$

whose image is the union of $\text{Im}(\iota_{\theta'})$ *over* $\theta' \in \Lambda^{\text{pos}}_{\mathfrak{m}',\mathfrak{p}(\mathfrak{m}')}$. *Moreover, the map*

$$\text{Maps}(\mathbf{C}, \mathcal{G}_{\mathfrak{g},\mathfrak{p}'}) \underset{\text{Bun}_{M'}(\mathbf{C},\infty_{\mathbf{C}})}{\times} \text{Bun}_{P^-(M')}(\mathbf{C}, \infty_{\mathbf{C}}) \hookrightarrow \text{Maps}(\mathbf{C}, \mathcal{G}_{\mathfrak{g},\mathfrak{p}})$$

is an isomorphism.

For $\mathfrak{p} = \mathfrak{b}$ and a vertex $i \in I$, let us denote by $\partial(\text{QMaps}(\mathbf{C}, \mathcal{G}_{\mathfrak{g},\mathfrak{b}}))_i$ the locally closed subset equal to $\text{Im}(\iota_{\alpha_i})$.

The above lemma combined with Proposition 2.7 yields the following.

Corollary 2.10. *The open subscheme* $\text{Maps}(\mathbf{C}, \mathcal{G}_{\mathfrak{g},\mathfrak{b}}) \bigcup (\underset{i \in I}{\cup} \partial(\text{QMaps}(\mathbf{C}, \mathcal{G}_{\mathfrak{g},\mathfrak{b}}))_i)$ *is (locally) of finite type.*

2.11 Projection to the configuration space

We claim that there exists a canonical map $\varrho^\theta_{\mathfrak{p}} : \text{QMaps}^\theta(\mathbf{C}, \mathcal{G}_{\mathfrak{g},\mathfrak{p}}) \to \overset{\circ}{\mathbf{C}}{}^\theta$. Indeed, let σ be an S-point of $\text{QMaps}^\theta(\mathbf{C}, \mathcal{G}_{\mathfrak{g},\mathfrak{p}})$, i.e., for every $\check{\lambda} \in \check{\Lambda}^+_{\mathfrak{g},\mathfrak{p}}$ we have a line bundle $\mathcal{L}^{\check{\lambda}}$ on $\mathbf{C} \times S$ and a map $\kappa^{\check{\lambda}} : \mathcal{L}^{\check{\lambda}} \to \mathcal{O}_{\mathbf{C}\times S} \otimes \mathcal{V}^*_{\check{\lambda}}$. Recall that the choice of the standard parabolic $\mathfrak{p}$ defines a highest-weight vector in $\mathcal{V}_{\check{\lambda}}$; hence by composing $\kappa^{\check{\lambda}}$ with the corresponding map $\mathcal{O}_{\mathbf{C}\times S} \otimes \mathcal{V}^*_{\check\lambda} \to \mathcal{O}_{\mathbf{C}\times S}$, we obtain a map $\mathcal{L}^{\check\lambda} \to \mathcal{O}$, i.e., a point of $\mathbf{C}^{(\langle\theta,\check\lambda\rangle)}$. The obtained divisor avoids $\infty_{\mathbf{C}}$ because the composition $\ell^0_{\check\lambda} \to \mathcal{V}^*_{\check\lambda} \to \mathbb{C}$ is nonzero.

The above construction of $\varrho^\theta_{\mathfrak{p}}$ can be alternatively viewed as follows. Recall the cone $C(\mathcal{G}_{\mathfrak{g},\mathfrak{p}})$, and observe that for each vertex $i \in I$ with $i \notin \mathfrak{m}$, we have a function on $C(\mathcal{G}_{\mathfrak{g},\mathfrak{p}})$ corresponding to the map $\mathcal{V}_{\check\omega_i} \to \mathbb{C}$, where $\check\omega_i$ is the fundamental weight. Let us denote by $C(\overline{\mathcal{G}}_{\mathfrak{g},\mathfrak{p},s_i})$ the corresponding Cartier divisor. Note that the intersection $\overset{\circ}{C}(\mathcal{G}_{\mathfrak{g},\mathfrak{p}}) \cap C(\overline{\mathcal{G}}_{\mathfrak{g},\mathfrak{p},s_i})$ is the preimage of the Cartier divisor $\overline{\mathcal{G}}_{\mathfrak{g},\mathfrak{p},s_i} \subset \mathcal{G}_{\mathfrak{g},\mathfrak{p}}$.

An S-point of $\mathrm{QMaps}^\theta(\mathbf{C}, \mathcal{G}_{\mathfrak{g},\mathfrak{p}})$ is a map $\sigma : \mathbf{C} \times S \to C(\mathcal{G}_{\mathfrak{g},\mathfrak{p}})/(M/[M,M])$, and by pulling back $C(\overline{\mathcal{G}}_{\mathfrak{g},\mathfrak{p},s_i})$, we obtain a Cartier divisor on $\mathbf{C} \times S$, which by the conditions on σ is in fact a relative Cartier divisor over S, i.e., it gives rise to a map from S to the suitable symmetric power of $\mathbf{C}$.

Note that the composition of the map $\mathfrak{s}^\theta_{\mathfrak{p}}$ and $\varrho^\theta_{\mathfrak{p}}$ is the identity map on $\overset{\circ}{\mathbf{C}}{}^\theta$. More generally, for $\theta' \in \Lambda^{\mathrm{pos}}_{\mathfrak{g},\mathfrak{p}}$, the composition

$$\varrho^\theta_{\mathfrak{p}} \circ \iota_{\theta'} : \mathrm{QMaps}^{\theta-\theta'}(\mathbf{C}, \mathcal{G}_{\mathfrak{g},\mathfrak{p}}) \times \overset{\circ}{\mathbf{C}}{}^{\theta'} \to \overset{\circ}{\mathbf{C}}{}^\theta \tag{4}$$

covers the addition map $\overset{\circ}{\mathbf{C}}{}^{\theta-\theta'} \times \overset{\circ}{\mathbf{C}}{}^{\theta'} \to \overset{\circ}{\mathbf{C}}{}^\theta$.

2.12 Zastava spaces

We will introduce twisted versions of the schemes $\mathrm{QMaps}^\theta(\mathbf{C}, \mathcal{G}_{\mathfrak{g},\mathfrak{p}})$, called *Zastava spaces* $\mathcal{Z}^\theta_{\mathfrak{g},\mathfrak{p}}(\mathbf{C})$; cf. [BFGM]. We will first define $\mathcal{Z}^\theta_{\mathfrak{g},\mathfrak{p}}(\mathbf{C})$ as a functor, and later show that it is representable by a scheme.

Let $\mathbf{C}$ be a smooth curve (not necessarily complete). An S-point of $\mathcal{Z}^\theta_{\mathfrak{g},\mathfrak{p}}(\mathbf{C})$ is given by data $(D^\theta, \mathcal{F}_{N(P)}, \kappa)$, where we have the following:

- D^θ is an S-point of $\mathbf{C}^\theta$. In particular, we obtain a principal $M/[M,M]$-bundle $\mathcal{F}_{M/[M,M]}$ on $\mathbf{C} \times S$, such that for every $\check\lambda \in \check\Lambda_{\mathfrak{g},\mathfrak{p}}$ the associated line bundle denoted $\mathcal{L}^{\check\lambda}_{\mathcal{F}_{M/[M,M]}}$ equals $\mathcal{O}_{\mathbf{C}\times S}(-\check\lambda(D^\theta))$.
- $\mathcal{F}_{N(P)}$ is a principal $N(P)$-bundle on $\mathbf{C} \times S$. Note that this makes sense, since $N(P)$ is a pro-algebraic group. In particular, for every $N(P)$-integrable $\mathfrak{g}$-module $\mathcal{V}$, we can form a pro-vector bundle $(\mathcal{V}^*)_{\mathcal{F}_{N(P)}}$ on $\mathbf{C} \times S$. If $v \in \mathcal{V}$ is an $N(P)$-invariant vector, it gives rise to a map $(\mathcal{V}^*)_{\mathcal{F}_{N(P)}} \to \mathcal{O}_{\mathbf{C}\times S}$.
- κ is a system of maps $\kappa^{\check\lambda} : \mathcal{L}^{\check\lambda}_{\mathcal{F}_{M/[M,M]}} \to (\mathcal{V}^*_{\check\lambda})_{\mathcal{F}_{N(P)}}$ which satisfy the Plücker relations, and such that the composition of $\kappa^{\check\lambda}$ and the projection $(\mathcal{V}^*_{\check\lambda})_{\mathcal{F}_{N(P)}} \to \mathcal{O}_{\mathbf{C}\times S}$ corresponding to the highest-weight vector $v_{\check\lambda} \in \mathcal{V}_{\check\lambda}$ is the tautological map $\mathcal{O}_{\mathbf{C}\times S}(-\check\lambda(D^\theta)) \to \mathcal{O}_{\mathbf{C}\times S}$.

We will denote by $\overset{\circ}{\mathcal{Z}}{}^\theta_{\mathfrak{g},\mathfrak{p}}(\mathbf{C})$ the open subfunctor in $\mathcal{Z}^\theta_{\mathfrak{g},\mathfrak{p}}(\mathbf{C})$ corresponding to the condition that the $\kappa^{\check\lambda}$s are injective bundle maps.

Another way to spell out the definition of $\mathcal{Z}^{\theta}_{\mathfrak{g},\mathfrak{p}}(\mathbf{C})$ is as follows: Let us consider the quotient $N(P)\backslash C(\mathcal{G}_{\mathfrak{g},\mathfrak{p}})/(M/[M,M])$. This is a nonalgebraic stack (i.e., all the axioms, except for the one about covering by a scheme, hold). It contains as an open substack $N(P)\backslash\mathcal{G}^{e}_{\mathfrak{g},\mathfrak{p}} \simeq pt$, and the Cartier divisors $N(P)\backslash C(\overline{\mathcal{G}}_{\mathfrak{g},\mathfrak{p},s_i})/(M/[M,M])$ for $i \in I$.

It is easy to see that an S-point of $\mathcal{Z}^{\theta}_{\mathfrak{g},\mathfrak{p}}(\mathbf{C})$ is the same as a map

$$\mathbf{C} \times S \to N(P)\backslash C(\mathcal{G}_{\mathfrak{g},\mathfrak{p}})/(M/[M,M]),$$

such that for every geometric point $s \in S$, the map

$$\mathbf{C} \simeq \mathbf{C} \times s \to N(P)\backslash C(\mathcal{G}_{\mathfrak{g},\mathfrak{p}})/(M/[M,M])$$

sends the generic point of $\mathbf{C}$ to $N(P)\backslash\mathcal{G}^{e}_{\mathfrak{g},\mathfrak{p}} \simeq pt$, and the intersection

$$\mathbf{C} \cap (N(P)\backslash C(\overline{\mathcal{G}}_{\mathfrak{g},\mathfrak{p},s_i})/M/[M,M])$$

is a divisor of degree $\langle \theta, \check{\omega}_i \rangle$ on $\mathbf{C}$.

From this description of $\mathcal{Z}^{\theta}_{\mathfrak{g},\mathfrak{p}}(\mathbf{C})$, we obtain the following.

Lemma 2.13. *For an S-point $(D^{\theta}, \mathcal{F}_{N(P)}, \kappa)$ of $\mathcal{Z}^{\theta}_{\mathfrak{g},\mathfrak{p}}(\mathbf{C})$, the $N(P)$-bundle $\mathcal{F}_{N(P)}$ admits a canonical trivialization away from the support of D^{θ}. Moreover, in terms of this trivialization, over this open subset the maps $\kappa^{\check{\lambda}}$ are constant maps corresponding to $v_{\check{\lambda}} \in \ell^{0}_{\check{\lambda}} \subset \mathcal{V}^{*}_{\check{\lambda}}$.*

For the proof it suffices to observe that for a map $\mathbf{C} \times S \to N(P)\backslash C(\mathcal{G}_{\mathfrak{g},\mathfrak{p}})/(M/[M,M])$, the open subset $\mathbf{C} \times S - \mathrm{supp}(D^{\theta})$ is exactly the preimage of $pt = N(P)\backslash\mathcal{G}^{e}_{\mathfrak{g},\mathfrak{p}}$.

Let us now consider the following setup, suggested in this generality by Drinfeld. Let $\mathcal{T}$ be a (not necessarily algebraic) stack, which contains an open substack $\mathcal{T}^0$ isomorphic to pt. Let S be a scheme, embedded as an open subscheme into a scheme S_1, and let $\sigma : S \to \mathcal{T}$ be a map such that $\sigma^{-1}(\mathcal{T} - \mathcal{T}^0)$ is closed in S_1.

Lemma 2.14. *There is a canonical bijection between the set of maps σ as above and the set of maps $\sigma_1 : S_1 \to \mathcal{T}$ such that $\sigma_1^{-1}(\mathcal{T} - \mathcal{T}^0)$ is contained in S.*

Proof. Of course, starting from σ_1, we get the corresponding σ by restriction to S.

Conversely, for $\sigma : \mathcal{Y} \times S \to \mathcal{T}$ as above, consider the two open subsets S and $S_1 - (\sigma^{-1}(\mathcal{T} - \mathcal{T}^0))$, which cover S_1.

By setting $\sigma_1|_S = \sigma$, $\sigma_1|_{S_1 - (\sigma^{-1}(\mathcal{T} - \mathcal{T}^0))}$ to be the constant map to $pt = \mathcal{T}^0$, we have gluing data for σ_1. □

We apply this lemma in the following situation.

Corollary 2.15. *Let $\mathbf{C} \to \mathbf{C}_1$ be an open embedding of curves. We have an isomorphism*

$$\mathcal{Z}^{\theta}_{\mathfrak{g},\mathfrak{p}}(\mathbf{C}) \simeq \mathcal{Z}^{\theta}_{\mathfrak{g},\mathfrak{p}}(\mathbf{C}_1) \underset{\mathbf{C}_1^{\theta}}{\times} \mathbf{C}^{\theta}.$$

The proof follows from the fact that for an S-point of $\mathcal{Z}^{\theta}_{\mathfrak{g},\mathfrak{p}}(\mathbf{C})$, the support of the colored divisor D^{θ}, which is the same as the preimage of $N(P)\backslash(C(\mathcal{G}_{\mathfrak{g},\mathfrak{p}}))/(M/[M,M]) - pt$, is finite over S.

2.16 Factorization principle

Let $pt = \mathcal{T}^0 \subset \mathcal{T}$ be an embedding of stacks as before, and suppose now that the complement $\mathcal{T} - \mathcal{T}^0$ is a union of Cartier divisors $\mathcal{T}_i$, for i belonging to some set of indices I.

For a set of nonnegative integers $\overline{n} = n_i, i \in I$, consider the functor $\mathrm{Maps}^{\overline{n}}(\mathbf{C}, \mathcal{T})$ that assigns to a scheme S the set of maps

$$\sigma : \mathbf{C} \times S \to \mathcal{T},$$

such that each $\sigma^{-1}(\mathcal{T}_i) \subset \mathbf{C} \times S$ is a relative (over S) Cartier divisor of degree n_i. It is easy to see that $\mathrm{Maps}^{\overline{n}}(\mathbf{C}, \mathcal{T})$ is a sheaf in the faithfully flat topology on the category of schemes.[1]

By construction, we have a map (of functors) $\mathrm{Maps}^{\overline{n}}(\mathbf{C}, \mathcal{T}) \to \prod_{i \in I} \mathbf{C}^{(n_i)}$. The following factorization principle is due to Drinfeld.

Proposition 2.17. *Let $n_i = n_i' + n_i''$ be a decomposition, and let*

$$\left(\prod_{i \in I} \mathbf{C}^{(n_i')} \times \prod_{i \in I} \mathbf{C}^{(n_i'')} \right)_{\mathrm{disj}}$$

be the open subset corresponding to the condition that the divisors $D_i' \in \mathbf{C}^{(n_i')}$ and $D_i'' \in \mathbf{C}^{(n_i'')}$ have disjoint supports. We have a canonical isomorphism

$$\mathrm{Maps}^{\overline{n}}(\mathbf{C}, \mathcal{T}) \underset{\prod_{i \in I} \mathbf{C}^{(n_i)}}{\times} \left(\prod_{i \in I} \mathbf{C}^{(n_i')} \times \prod_{i \in I} \mathbf{C}^{(n_i'')} \right)_{\mathrm{disj}}$$

$$\simeq (\mathrm{Maps}^{\overline{n}'}(\mathbf{C}, \mathcal{T}) \times \mathrm{Maps}^{\overline{n}''}(\mathbf{C}, \mathcal{T})) \underset{\prod_{i \in I} \mathbf{C}^{(n_i')} \times \prod_{i \in I} \mathbf{C}^{(n_i'')}}{\times} \left(\prod_{i \in I} \mathbf{C}^{(n_i')} \times \prod_{i \in I} \mathbf{C}^{(n_i'')} \right)_{\mathrm{disj}}.$$

Proof. Given an S-point (σ, D', D'') of

$$\mathrm{Maps}^{\overline{n}}(\mathbf{C}, \mathcal{T}) \underset{\prod_{i \in I} \mathbf{C}^{(n_i)}}{\times} \left(\prod_{i \in I} \mathbf{C}^{(n_i')} \times \prod_{i \in I} \mathbf{C}^{(n_i'')} \right)_{\mathrm{disj}}$$

we produce the S-points σ' and σ'' of $\mathrm{Maps}^{\overline{n}'}(\mathbf{C}, \mathcal{T})$ and $\mathrm{Maps}^{\overline{n}''}(\mathbf{C}, \mathcal{T})$ as follows:

As a map $\mathbf{C} \times S \to \mathcal{T}$, σ' is set to be equal to σ on the open subset $\mathbf{C} \times S - \mathrm{supp}(D'')$, and to the constant map $\mathbf{C} \times S \to \mathcal{T}^0 = pt$ on the open subset $\mathbf{C} \times S - \mathrm{supp}(D')$. By assumption, this gives well-defined gluing data for σ'. The map σ'' is defined by interchanging the roles of $'$ and $''$.

[1] According to Drinfeld, one can formulate a general hypothesis on $\mathcal{T}$, under which the functor $\mathrm{Maps}^{\overline{n}}(\mathbf{C}, \mathcal{T})$ is representable by a scheme.

Conversely, given σ' and σ'', with the corresponding colored divisors D' and D'', respectively, we define σ as follows. On the open subset $\mathbf{C} \times S - \mathrm{supp}(D'')$, σ is set to be equal to σ', and on the open subset $\mathbf{C} \times S - \mathrm{supp}(D')$, σ is set to be equal to σ''. Since σ' and σ'' agree on $\mathbf{C} \times S - \mathrm{supp}(D' + D'')$, this gives well-defined gluing data for σ. □

The above proposition admits the following generalization. Let $p : \widetilde{\mathbf{C}} \to \mathbf{C}$ be an étale cover. Let $\widetilde{\mathbf{C}}^{(n)}_{\mathrm{disj}} \subset \widetilde{\mathbf{C}}^{(n)}$ be the open subset that consists of divisors $\widetilde{D}$ on $\widetilde{\mathbf{C}}$ such that the divisor $p^*(p_*(\widetilde{D})) - \widetilde{D}$ is disjoint from $\widetilde{D}$.

Essentially the same proof gives the following.

Proposition 2.18. *There is a canonical isomorphism*

$$\mathrm{Maps}^{\overline{n}}(\mathbf{C}, \mathcal{T}) \underset{\prod_{i\in I} \mathbf{C}^{(n_i)}}{\times} \prod_{i\in I} \widetilde{\mathbf{C}}^{(n_i)}_{\mathrm{disj}} \simeq \mathrm{Maps}^{\overline{n}}(\widetilde{\mathbf{C}}, \mathcal{T}) \underset{\prod_{i\in I} \widetilde{\mathbf{C}}^{(n_i)}}{\times} \prod_{i\in I} \widetilde{\mathbf{C}}^{(n_i)}_{\mathrm{disj}}.$$

As an application, we take $\mathcal{T} = N(P)\backslash(C(\mathcal{G}_{\mathfrak{g},\mathfrak{p}}))/(M/[M, M])$, and we obtain the following factorization property of the Zastava spaces.

Proposition 2.19. *For an étale map $p : \widetilde{\mathbf{C}} \to \mathbf{C}$, we have a canonical isomorphism*

$$\mathcal{Z}^\theta_{\mathfrak{g},\mathfrak{p}}(\mathbf{C}) \underset{\mathbf{C}^\theta}{\times} \widetilde{\mathbf{C}}^\theta_{\mathrm{disj}} \simeq \mathcal{Z}^\theta_{\mathfrak{g},\mathfrak{p}}(\widetilde{\mathbf{C}}) \underset{\widetilde{\mathbf{C}}^\theta}{\times} \widetilde{\mathbf{C}}^\theta_{\mathrm{disj}}.$$

In particular, for $\theta = \theta_1 + \theta_2$, we have a canonical isomorphism

$$\mathcal{Z}^\theta_{\mathfrak{g},\mathfrak{p}}(\mathbf{C}) \underset{\mathbf{C}^\theta}{\times} (\mathbf{C}^{\theta_1} \times \mathbf{C}^{\theta_2})_{\mathrm{disj}} \simeq (\mathcal{Z}^{\theta_1}_{\mathfrak{g},\mathfrak{p}}(\mathbf{C}) \times \mathcal{Z}^{\theta_2}_{\mathfrak{g},\mathfrak{p}}(\mathbf{C})) \underset{\mathbf{C}^{\theta_1} \times \mathbf{C}^{\theta_2}}{\times} (\mathbf{C}^{\theta_1} \times \mathbf{C}^{\theta_2})_{\mathrm{disj}}.$$

Moreover, the above isomorphisms preserve the loci $\overset{\circ}{\mathcal{Z}}{}^\theta_{\mathfrak{g},\mathfrak{p}}(\mathbf{C}) \subset \mathcal{Z}^\theta_{\mathfrak{g},\mathfrak{p}}(\mathbf{C})$, $\overset{\circ}{\mathcal{Z}}{}^\theta_{\mathfrak{g},\mathfrak{p}}(\widetilde{\mathbf{C}}) \subset \mathcal{Z}^\theta_{\mathfrak{g},\mathfrak{p}}(\widetilde{\mathbf{C}})$.

This proposition together with Lemma 2.14 expresses the locality property of the Zastava spaces $\mathcal{Z}^\theta_{\mathfrak{g},\mathfrak{p}}(\mathbf{C})$ with respect to $\mathbf{C}$.

Another application of Proposition 2.17 is Proposition 6.6, where $\mathcal{T}$ is taken to be the stack of coherent sheaves of generic rank n on $\mathbb{P}^1$ with a trivialization at ∞.

2.20 The case of genus 0

Now let $\mathbf{C}$ be a projective line, and $\overset{\circ}{\mathbf{C}} = \mathbf{C} - \infty_{\mathbf{C}}$ the corresponding affine line.

The following proposition will play a key role in this paper.

Proposition 2.21. *The functor represented by* $\mathrm{QMaps}^\theta(\mathbf{C}, \mathcal{G}_{\mathfrak{g},\mathfrak{p}})$ *is naturally isomorphic to $\mathcal{Z}^\theta_{\mathfrak{g},\mathfrak{p}}(\overset{\circ}{\mathbf{C}})$.*

Proof. By the very definition of $\mathcal{Z}^\theta_{\mathfrak{g},\mathfrak{p}}(\overset{\circ}{\mathbf{C}})$, the assertion of the proposition amounts to the following: For an S-point $(D^\theta, \mathcal{F}_{N(P)}, \kappa)$ of $\mathcal{Z}^\theta_{\mathfrak{g},\mathfrak{p}}(\overset{\circ}{\mathbf{C}})$, the $N(P)$-bundle $\mathcal{F}_{N(P)}$ on $\overset{\circ}{\mathbf{C}} \times S$ can be canonically trivialized.

Note that according to Corollary 2.15, $\mathcal{F}_{N(P)}$ extends to a principal $N(P)$-bundle on the entire $\mathbf{C} \times S$, with a trivialization on $\infty_\mathbf{C} \times S$. But since $N(P)$ is pro-unipotent, and $\mathbf{C}$ is of genus 0, such a trivialization extends uniquely to $\mathbf{C} \times S$.

(Note, however, that this trivialization and the one coming from Lemma 2.13 *do not* agree, but they do coincide on $\infty_\mathbf{C} \times S$.) □

Remark. For a curve of arbitrary genus, it is easy to see that the natural map $\mathrm{QMaps}^\theta(\mathbf{C}, \mathcal{G}_{\mathfrak{g},\mathfrak{p}}) \to \mathcal{Z}^\theta_{\mathfrak{g},\mathfrak{p}}(\overset{\circ}{\mathbf{C}})$ is a closed embedding.

Indeed, if the $N(P)$-bundle $\mathcal{F}_{N(P)}$ trivialized at $\infty_\mathbf{C}$ can be trivialized globally, this can be done in a unique fashion (since $H^0(\mathbf{C}, \mathcal{O}_\mathbf{C}(-\infty_\mathbf{C})) = 0$), and the property that it admits such a trivialization is a closed condition.

As a corollary of Proposition 2.21, we obtain the following statement.

Corollary 2.22. *The functor $\mathcal{Z}^\theta_{\mathfrak{g},\mathfrak{p}}(\mathbf{C})$ is representable by a scheme for any curve $\mathbf{C}$. The corresponding open subscheme $\overset{\circ}{\mathcal{Z}}{}^\theta_{\mathfrak{g},\mathfrak{p}}(\mathbf{C})$ is locally of finite type.*

Proof. Using Corollary 2.15 and Proposition 2.19, we reduce the assertions of the proposition to the case when the curve in question is the projective line. The representability now follows from Proposition 2.21. The fact that $\overset{\circ}{\mathcal{Z}}{}^\theta_{\mathfrak{g},\mathfrak{p}}(\mathbf{C})$ is of finite type follows from Proposition 2.7. □

2.23 Further properties of spaces of maps

In this subsection (until Section 2.29) we will assume that $\mathbf{C}$ is of genus 0, and establish certain properties of the scheme $\mathrm{Maps}^\theta(\mathbf{C}, \mathcal{G}_{\mathfrak{g},\mathfrak{p}})$. Using Proposition 2.21, the same assertions will hold for the space $\overset{\circ}{\mathcal{Z}}{}^\theta_{\mathfrak{g},\mathfrak{p}}(\mathbf{C})$ on any curve $\mathbf{C}$.

First, note that Proposition 2.21 and Proposition 2.19 yield the following factorization property of $\mathrm{QMaps}^\theta_{\mathfrak{g},\mathfrak{p}}(\mathbf{C})$ with respect to the projection $\varrho^\theta_\mathfrak{p}$:

$$\begin{aligned} &\mathrm{QMaps}^\theta_{\mathfrak{g},\mathfrak{p}}(\mathbf{C}) \underset{\mathbf{C}^\theta}{\times} (\mathbf{C}^{\theta_1} \times \mathbf{C}^{\theta_2})_{\mathrm{disj}} \\ &\quad \simeq (\mathrm{QMaps}^{\theta_1}_{\mathfrak{g},\mathfrak{p}}(\mathbf{C}) \times \mathrm{QMaps}^{\theta_2}_{\mathfrak{g},\mathfrak{p}}(\mathbf{C})) \underset{\mathbf{C}^{\theta_1} \times \mathbf{C}^{\theta_2}}{\times} (\mathbf{C}^{\theta_1} \times \mathbf{C}^{\theta_2})_{\mathrm{disj}}. \end{aligned} \tag{5}$$

Proposition 2.24. *The scheme $\mathrm{Maps}^\theta(\mathbf{C}, \mathcal{G}_{\mathfrak{g},\mathfrak{p}})$ is smooth.*

Proof. Let $\sigma : \mathbf{C} \times S \to \mathcal{G}_{\mathfrak{g},\mathfrak{p}}$ be a based map, where S is an Artinian scheme, and let $S' \supset S$ be a bigger Artinian scheme. We must show that σ extends to a based map $\sigma' : \mathbf{C} \times S' \to \mathcal{G}_{\mathfrak{g},\mathfrak{p}}$.

Since $\mathcal{G}_{\mathfrak{g},\mathfrak{p}}$ is a union of open subschemes, each of which is a projective limit of smooth schemes under smooth maps, locally on $\mathbf{C}$ there is no obstruction to extending σ. Therefore, by induction on the length of S', we obtain that the obstruction to the existence of σ' lies in $H^1(\mathbf{C}, \sigma^*(T\mathcal{G}_{\mathfrak{g},\mathfrak{p}})(-\infty_{\mathbf{C}}))$, where $T\mathcal{G}_{\mathfrak{g},\mathfrak{p}}$ is the tangent sheaf of $\mathcal{G}_{\mathfrak{g},\mathfrak{p}}$.

However, the Lie algebra $\mathfrak{g}$ surjects onto the tangent space to $\mathcal{G}_{\mathfrak{g},\mathfrak{p}}$ at every point. Therefore, we have a surjection $\mathcal{O}_{\mathbf{C}} \otimes \mathfrak{g} \twoheadrightarrow \sigma^*(T\mathcal{G}_{\mathfrak{g},\mathfrak{p}})$. Hence we have a surjection on the level of H^1, but $H^1(\mathbf{C}, \mathcal{O}_{\mathbf{C}} \otimes \mathfrak{g}(-\infty_{\mathbf{C}})) \simeq H^1(\mathbf{C}, \mathcal{O}_{\mathbf{C}}(-\infty_{\mathbf{C}})) \otimes \mathfrak{g} = 0$. □

Now let $\mathfrak{p} = \mathfrak{b}$, and for an element $\mu \in \Lambda_{\mathfrak{g}}^{\mathrm{pos}}$ equal to $\mu = \sum_i n_i \cdot \alpha_i$ let us define the length of μ as $|\mu| = \sum n_i$.

Proposition 2.25. *The scheme* $\mathrm{Maps}^{\mu}(\mathbf{C}, \mathcal{G}_{\mathfrak{g},\mathfrak{b}})$ *is connected.*

Proof. The proof proceeds by induction on the length of μ. If $|\mu| = 1$, or more generally, if μ is the image of $n \cdot \alpha_i$ for some $i \in I$, the scheme $\mathrm{Maps}^{\mu}(\mathbf{C}, \mathcal{G}_{\mathfrak{g},\mathfrak{b}})$ is isomorphic to the scheme of based maps $\mathbf{C} \to \mathbb{P}^1$ of degree n, and hence is connected.

Let μ be an element of minimal length for which $\mathrm{Maps}^{\mu}(\mathbf{C}, \mathcal{G}_{\mathfrak{g},\mathfrak{b}})$ is disconnected. By what we said above, we can assume that μ is not a multiple of one simple coroot.

By the factorization property, equation (5), and the minimality assumption, $\mathrm{Maps}^{\mu}(\mathbf{C}, \mathcal{G}_{\mathfrak{g},\mathfrak{b}})$ contains a connected component $\mathbf{K}$, which projects under $\varrho_{\mathfrak{b}}^{\mu}$ to the main diagonal $\Delta(\overset{\circ}{\mathbf{C}}) \subset \overset{\circ}{\mathbf{C}}{}^{\mu}$. We claim that this leads to a contradiction.

Indeed, by the definition of $\varrho_{\mathfrak{b}}^{\mu}$, for any map $\sigma \in \mathbf{K}$ there exists a unique point $\mathbf{c} \in \mathbf{C}$ with $\sigma(\mathbf{c}) \in \mathcal{G}_{\mathfrak{g},\mathfrak{b}} - \mathcal{G}^{e}_{\mathfrak{g},\mathfrak{b}}$. Let $w \in W$ be minimal with the property that there exists $\sigma \in \mathbf{K}$ as above such that $\sigma(\mathbf{c}) \in \mathcal{G}_{\mathfrak{g},\mathfrak{b},w}$.

If w is a simple reflection s_i, then $\sigma(\mathbf{C})$ intersects only $\mathcal{G}_{\mathfrak{g},\mathfrak{b},s_i}$, i.e., by the definition of $\varrho_{\mathfrak{b}}^{\mu}$, θ is a multiple of α_i, which is impossible.

Hence we can assume that $\ell(w) > 1$, and let w' be such that $w = w' \cdot s_i$, $\ell(w') = \ell(w) - 1$. We claim that we will be able to find $\sigma' \in \mathbf{K}$ such that $\sigma'(\mathbf{C}) \cap \mathcal{G}_{\mathfrak{g},\mathfrak{b},w'} \neq \emptyset$.

Indeed, the group SL_2 corresponding to $i \in I$ acts on $\mathcal{G}_{\mathfrak{g},\mathfrak{b}}$, and the corresponding $N_i^- \subset SL_2$ preserves the point $1_{\mathcal{G}_{\mathfrak{g},\mathfrak{b}}}$. Hence N_i^- acts on $\mathrm{Maps}^{\theta}(\mathbf{C}, \mathcal{G}_{\mathfrak{g},\mathfrak{b}})$, and, being connected, it preserves the connected component $\mathbf{K}$. However, for any nontrivial element $u \in N_i^-$, $u(\mathcal{G}_{\mathfrak{g},\mathfrak{b},w}) \subset \mathcal{G}_{\mathfrak{g},\mathfrak{b},w'}$. Hence if we define σ' as $u(\sigma)$, for the same point $\mathbf{c} \in \mathbf{C}$, $\sigma'(\mathbf{c}) \in \mathcal{G}_{\mathfrak{g},\mathfrak{b},w'}$. □

Remark. For finite-dimensional $\mathfrak{g}$ the above proposition is not new; see [Th, FFKM, KiP, Pe]. However, all the proofs avoiding factorization use Kleiman's theorem on generic transversality, unavailable in the infinite-dimensional setting.

Corollary 2.26. *The dimension of* $\mathrm{Maps}^{\mu}(\mathbf{C}, \mathcal{G}_{\mathfrak{g},\mathfrak{b}})$ *equals* $2|\mu|$.

Proof. Using Proposition 2.25 and Proposition 2.24, we know that $\mathrm{Maps}^{\mu}(\mathbf{C}, \mathcal{G}_{\mathfrak{g},\mathfrak{b}})$ is irreducible. Recall the map $\varrho_{\mathfrak{b}}^{\mu} : \mathrm{Maps}^{\mu}(\mathbf{C}, \mathcal{G}_{\mathfrak{g},\mathfrak{b}}) \to \overset{\circ}{\mathbf{C}}{}^{\mu}$, and consider the open subset in $\mathrm{Maps}^{\mu}(\mathbf{C}, \mathcal{G}_{\mathfrak{g},\mathfrak{b}})$ equal to the preimage of the locus of $\overset{\circ}{\mathbf{C}}{}^{\mu}$ corresponding to multiplicity-free divisors.

Using (5), this reduces us to the case when μ is a simple coroot α_i, $i \in I$. But as we have seen before, the scheme $\mathrm{Maps}^\mu(\mathbf{C}, \mathcal{G}_{\mathfrak{g},\mathfrak{b}})$ in this case is isomorphic to $\mathbb{A}^2$. □

Note that since the base $\overset{\circ}{\mathbf{C}}{}^\mu$ is smooth, from the above corollary it follows that for any point $\mathbf{c} \in \overset{\circ}{\mathbf{C}}$ every irreducible component of the preimage of the corresponding point $\mu \cdot \mathbf{c} \in \overset{\circ}{\mathbf{C}}{}^\mu$ in $\mathrm{Maps}^\mu(\mathbf{C}, \mathcal{G}_{\mathfrak{g},\mathfrak{b}})$ is of dimension $\geq |\mu|$.

The following conjecture will be established in the case when $\mathfrak{g}$ is of finite and affine type.

Conjecture 2.27. The projection $\varrho^\mu_{\mathfrak{b}} : \mathrm{Maps}^\mu(\mathbf{C}, \mathcal{G}_{\mathfrak{g},\mathfrak{b}}) \to \overset{\circ}{\mathbf{C}}{}^\mu$ is flat. Equivalently, the preimage of the point $\mu \cdot \mathbf{c} \in \overset{\circ}{\mathbf{C}}{}^\mu$ (for any point $\mathbf{c} \in \overset{\circ}{\mathbf{C}}$) is equidimensional of dimension $|\mu|$.

For a parabolic $\mathfrak{p}$ and $\theta \in \Lambda^{\mathrm{pos}}_{\mathfrak{g},\mathfrak{p}}$ equal to the projection of $\widetilde{\theta} = \sum n_i \cdot \alpha_i \in \Lambda^{\mathrm{pos}}_{\mathfrak{g}}$ with $i \notin \mathfrak{m}$, define $|\theta|$ as $\sum n_i$, and $|\theta|' = |\theta| - \langle \widetilde{\theta}, \check{\rho}_M \rangle$, where $\check{\rho}_M$ is half the sum of the positive roots of $\mathfrak{m}$.

Corollary 2.28. *The dimension of* $\mathrm{Maps}^\theta(\mathbf{C}, \mathcal{G}_{\mathfrak{g},\mathfrak{p}})$ *equals* $2|\theta|'$.

Proof. Pick an element $\check{\mu} \in \Lambda^{\mathrm{pos}}_{\mathfrak{g}}$ which projects to θ under $\Lambda_{\mathfrak{g}} \to \Lambda_{\mathfrak{g},\mathfrak{p}}$, and which is sufficiently dominant with respect to $\mathfrak{m}$, so that the map $\mathrm{Bun}^\mu_{B^-(M)}(\mathbf{C}, \infty_{\mathbf{C}}) \to \mathrm{Bun}_M(\mathbf{C}, \mathbf{c})$ is smooth.

It is sufficient to show that for any such μ,

$$\dim((\mathrm{Maps}^\theta(\mathbf{C}, \mathcal{G}_{\mathfrak{g},\mathfrak{p}}) \underset{\mathrm{Bun}_M(\mathbf{C},\mathbf{c})}{\times} \mathrm{Bun}^\mu_{B^-(M)}(\mathbf{C}, \infty_{\mathbf{C}}))$$
$$= 2|\theta|' + \text{rel. dim.}(\mathrm{Bun}^\mu_{B^-(M)}(\mathbf{C}, \infty_{\mathbf{C}}), \mathrm{Bun}_M(\mathbf{C}, \mathbf{c})).$$

However, according to Lemma 2.9 and Corollary 2.26, the left-hand side of the above expression equals $2|\mu|$, and rel. dim.$(\mathrm{Bun}^\mu_{B^-(M)}(\mathbf{C}, \infty_{\mathbf{C}}), \mathrm{Bun}_M(\mathbf{C}, \mathbf{c}))$ is readily seen to equal $\langle \mu, 2\check{\rho}_M \rangle$. Together, this yields the desired result. □

2.29 Enhanced quasi-maps

For an element $\check{\lambda} \in \check{\Lambda}^+_{\mathfrak{g}}$, let $\mathcal{U}_{\check{\lambda}}$ denote the corresponding integrable module over the Levi subalgebra M. Note that each such $\mathcal{U}_{\check{\lambda}}$ can be realized as $(\mathcal{V}_{\check{\lambda}})_{\mathfrak{p}^-}$. Therefore, every map of $\mathfrak{g}$-modules $\mathcal{V}_{\check{\nu}} \to \mathcal{V}_{\check{\mu}} \otimes \mathcal{V}_{\check{\lambda}}$ gives rise to a map $\mathcal{U}_{\check{\nu}} \to \mathcal{U}_{\check{\mu}} \otimes \mathcal{U}_{\check{\lambda}}$.

Consider the subscheme

$$\widetilde{C}(\mathcal{G}_{\mathfrak{g},\mathfrak{p}}) \subset \prod_{\check{\Lambda}^+_{\mathfrak{g}}} \mathrm{Hom}(\mathcal{V}_{\check{\lambda}}, \mathcal{U}_{\check{\lambda}})$$

given by the following equations:

A system of maps $\varphi_{\check{\lambda}} \in \operatorname{Hom}(\mathcal{V}_{\check{\lambda}}, \mathcal{U}_{\check{\lambda}})$ belongs to $\widetilde{C}(\mathcal{G}_{\mathfrak{g},\mathfrak{p}})$ if, for every $\check{\nu}, \check{\lambda}, \check{\mu} \in \check{\Lambda}^+_{\mathfrak{g}}$ and a map $\mathcal{V}_{\check{\nu}} \to \mathcal{V}_{\check{\mu}} \otimes \mathcal{V}_{\check{\lambda}}$, the diagram

$$\begin{array}{ccc} \mathcal{V}_{\check{\nu}} & \longrightarrow & \mathcal{V}_{\check{\mu}} \otimes \mathcal{V}_{\check{\lambda}} \\ {\scriptstyle \varphi_{\check{\nu}}}\big\downarrow & & {\scriptstyle \varphi_{\check{\mu}} \otimes \varphi_{\check{\lambda}}}\big\downarrow \\ \mathcal{U}_{\check{\nu}} & \longrightarrow & \mathcal{U}_{\check{\mu}} \otimes \mathcal{U}_{\check{\lambda}} \end{array} \tag{6}$$

is commutative.

There is a natural map from $\widetilde{C}(\mathcal{G}_{\mathfrak{g},\mathfrak{p}})$ to $C(\mathcal{G}_{\mathfrak{g},\mathfrak{p}})$ which remembers the data of $\varphi_{\check{\lambda}}$ for $\check{\lambda} \in \Lambda^+_{\mathfrak{g},\mathfrak{p}}$. Let $\overset{\circ}{\widetilde{C}}(\mathcal{G}_{\mathfrak{g},\mathfrak{p}})$ be the open subscheme of $\widetilde{C}(\mathcal{G}_{\mathfrak{g},\mathfrak{p}})$ corresponding to the condition that all the $\varphi_{\check{\lambda}}$ are surjections. We have a natural map

$$\overset{\circ}{\widetilde{C}}(\mathcal{G}_{\mathfrak{g},\mathfrak{p}}) \to \mathcal{G}_{\mathfrak{g},\mathfrak{p}},$$

and the former is a principal M-bundle over the latter.

For a (not necessarily complete) curve $\mathbf{C}$, we shall now define a certain scheme $\widetilde{\mathcal{Z}}^\theta_{\mathfrak{g},\mathfrak{p}}(\mathbf{C})$, called the enhanced version of the Zastava space.

By definition, $\widetilde{\mathcal{Z}}^\theta_{\mathfrak{g},\mathfrak{p}}(\mathbf{C})$ classifies quadruples $(D^\theta, \mathcal{F}_{N(P)}, \mathcal{F}_M, \kappa)$, where

- $(D^\theta, \mathcal{F}_{N(P)})$ are as in the definition of $\mathcal{Z}^\theta_{\mathfrak{g},\mathfrak{p}}(\mathbf{C})$,
- $\mathcal{F}_M$ is a principal M-bundle on $\mathbf{C}$, such that the induced $M/[M,M]$-bundle $\mathcal{F}_{M/[M,M]}$ is identified with the one coming from D^θ,
- κ is a system of generically surjective maps

$$\kappa^{\check{\lambda}} : (\mathcal{V}_{\check{\lambda}})_{\mathcal{F}_{N(P)}} \to (\mathcal{U}_{\check{\lambda}})_{\mathcal{F}_M}, \check{\lambda} \in \check{\Lambda}^+_{\mathfrak{g}},$$

such that

- the $\kappa^{\check{\lambda}}$s satisfy the Plücker relations (cf. equation (6)) and
- for $\check{\lambda} \in \check{\Lambda}^+_{\mathfrak{g},\mathfrak{p}}$, the composition

$$\mathcal{O}_{\mathbf{C}} \to (\mathcal{V}_{\check{\lambda}})_{\mathcal{F}_{N(P)}} \xrightarrow{\kappa^{\check{\lambda}}} \mathcal{L}^{\check{\lambda}}_{\mathcal{F}_{M/[M,M]}}$$

  equals the tautological embedding $\mathcal{O}_{\mathbf{C}} \to \mathcal{O}(\check{\lambda}(D^\theta))$.

The proof that $\widetilde{\mathcal{Z}}^\theta_{\mathfrak{g},\mathfrak{p}}(\mathbf{C})$ is indeed representable by a scheme is given below.

One can reformulate the definition of $\widetilde{\mathcal{Z}}^\theta_{\mathfrak{g},\mathfrak{p}}(\mathbf{C})$ as follows: it classifies maps from $\mathbf{C}$ to the stack $N(P)\backslash\widetilde{C}(\mathcal{G}_{\mathfrak{g},\mathfrak{p}})/M$ which send the generic point of $\mathbf{C}$ to $N(P)\backslash\mathcal{G}^e_{\mathfrak{g},\mathfrak{p}} \simeq pt$.

From this it is easy to see that the analogues of Lemma 2.13, Corollary 2.14, and Proposition 2.19 hold. In particular, for an S-point of $\widetilde{\mathcal{Z}}^\theta_{\mathfrak{g},\mathfrak{p}}(\mathbf{C})$, on the open subset $\mathbf{C} \times S - D^\theta$ the bundles $\mathcal{F}_{N(P)}$ and $\mathcal{F}_M$ admit canonical trivializations, such that the maps $\kappa^{\check{\lambda}}$ become the projections $\mathcal{O}_{\mathbf{C}} \otimes \mathcal{V}_{\check{\lambda}} \to \mathcal{O}_{\mathbf{C}} \otimes \mathcal{U}_{\check{\lambda}}$.

If $\mathbf{C}$ is a complete curve with a marked point $\infty_{\mathbf{C}}$, we define the scheme of based enhanced quasi-maps $\widetilde{\mathrm{QMaps}}^\theta(\mathbf{C}, \mathcal{G}_{\mathfrak{g},\mathfrak{p}})$ as

$$\widetilde{\mathcal{Z}}^\theta_{\mathfrak{g},\mathfrak{p}}(\mathbf{C}) \underset{\mathcal{Z}^\theta_{\mathfrak{g},\mathfrak{p}}(\mathbf{C})}{\times} \mathrm{QMaps}^\theta(\mathbf{C}, \mathcal{G}_{\mathfrak{g},\mathfrak{p}}).$$

In other words, $\widetilde{\mathrm{QMaps}}^\theta(\mathbf{C}, \mathcal{G}_{\mathfrak{g},\mathfrak{p}})$ classifies maps from $\mathbf{C}$ to the stack $\widetilde{C}(\mathcal{G}_{\mathfrak{g},\mathfrak{p}})/M$ which send a neighborhood of $\infty_\mathbf{C}$ to $\overset{\circ}{\widetilde{C}}(\mathcal{G}_{\mathfrak{g},\mathfrak{p}})/M \simeq \mathcal{G}_{\mathfrak{g},\mathfrak{p}}$, and such that $\infty_\mathbf{C}$ gets sent to $1_{\mathcal{G}_{\mathfrak{g},\mathfrak{p}}} \in \mathcal{G}_{\mathfrak{g},\mathfrak{p}}$.

Just as in Proposition 2.21, when $\mathbf{C}$ is of genus 0 we have an isomorphism between $\widetilde{\mathcal{Z}}^\theta_{\mathfrak{g},\mathfrak{p}}(\overset{\circ}{\mathbf{C}})$ and $\widetilde{\mathrm{QMaps}}^\theta(\mathbf{C}, \mathcal{G}_{\mathfrak{g},\mathfrak{p}})$, and in general the latter is a closed subscheme of the former.

When $\mathfrak{g}$ is finite dimensional, we introduce the corresponding (relative over $\mathrm{Bun}_G(\mathbf{C}, \infty_\mathbf{C})$) version of $\widetilde{\mathrm{QMaps}}^\theta(\mathbf{C}, \mathcal{G}_{\mathfrak{g},\mathfrak{p}})$, denoted $\widetilde{\mathrm{Bun}}_{P^-}(\mathbf{C}, \infty_\mathbf{C})$.

2.30

To formulate the next assertion, we need to recall some notation related to affine Grassmannians.

If M is a reductive group and $\mathbf{C}$ a curve, we will denote by $\mathrm{Gr}_{M,\mathbf{C}}$ the corresponding affine Grassmannian. By definition, this is an ind-scheme classifying triples $(\mathbf{c}, \mathcal{F}_M, \beta)$, where $\mathbf{c}$ is a point of $\mathbf{C}$, $\mathcal{F}_M$ is an M-bundle, and β is a trivialization of $\mathcal{F}_M$ on $\mathbf{C} - \mathbf{c}$. When the point $\mathbf{c}$ is fixed, we will denote the corresponding subscheme of $\mathrm{Gr}_{M,\mathbf{C}}$ by $\mathrm{Gr}_{M,\mathbf{c}}$, and sometimes simply by Gr_M.

More generally, for $a \in \mathbb{N}$, we have the Beilinson–Drinfeld version of the affine Grassmannian, denoted $\mathrm{Gr}^{BD,a}_{M,\mathbf{C}}$, which is now an ind-scheme over $\mathbf{C}^{(a)}$. For a fixed divisor $D \in \mathbf{C}^{(a)}$, we will denote by $\mathrm{Gr}^{BD,a}_{M,\mathbf{C},D}$ the fiber of $\mathrm{Gr}^{BD,a}_{M,\mathbf{C}}$ over D. By definition, the latter scheme classifies pairs $(\mathcal{F}_M, \beta)$, where $\mathcal{F}_M$ is as before an M-bundle on $\mathbf{C}$, and β is its trivialization off the support of D.

Assume now that M is realized as the reductive group corresponding to a Levi subalgebra $\mathfrak{m} \subset \mathfrak{g}$, for a Kac–Moody algebra $\mathfrak{g}$, and let θ be an element of $\Lambda^+_{\mathfrak{g},\mathfrak{p}}$.

We define the (finite-dimensional) scheme $\mathrm{Mod}^{\theta,+}_{M,\mathbf{C}}$ to classify triples $(D^\theta, \mathcal{F}_M, \beta)$, where D^θ is a point of $\mathbf{C}^\theta$, $\mathcal{F}_M$ is a principal M-bundle on $\mathbf{C}$, and β is a trivialization of $\mathcal{F}_M$ off the support of D^θ, such that the following conditions hold:

(1) The trivialization given by β of the induced $M/[M, M]$-bundle $\mathcal{F}_{M/[M,M]}$ is such that $\mathcal{L}^{\check{\lambda}}_{\mathcal{F}_{M/[M,M]}} \simeq \mathcal{O}(\check{\lambda}(D^\theta))$, $\check{\lambda} \in \Lambda^+_{\mathfrak{g},\mathfrak{p}}$, where $\mathcal{L}^{\check{\lambda}}_{\mathcal{F}_{M/[M,M]}}$ is the line bundle associated with $\mathcal{F}_{M/[M,M]}$ and the character $\check{\lambda} : M/[M, M] \to \mathbb{G}_m$.
(2) For an integrable $\mathfrak{g}$-module $\mathcal{V}$, and the corresponding $\mathfrak{m}$-module $\mathcal{U} := \mathcal{V}^{n(\mathfrak{p})}$, the (a priori meromorphic) map $\mathcal{O}_\mathbf{C} \otimes \mathcal{U} \to \mathcal{U}_{\mathcal{F}_M}$ induced by β is regular.

Since M admits a faithful representation of the form $\mathcal{V}^{n(\mathfrak{p})}$, where $\mathcal{V}$ is an integrable $\mathfrak{g}$-module, we obtain that $\mathrm{Mod}^{\theta,+}_{M,\mathbf{C}}$ is indeed finite dimensional.

Let us describe more explicitly the fibers of $\mathrm{Mod}^{\theta,+}_{M,\mathbf{C}}$ over $\mathbf{C}^\theta$. For simplicity, let us take a point in $\mathbf{C}^\theta$ equal to $\theta \cdot \mathbf{c}$, where $\mathbf{c}$ is some point of $\mathbf{C}$. Recall that the entire affine Grassmannian $\mathrm{Gr}_{M,\mathbf{c}}$ is the union of Schubert cells, denoted $\mathrm{Gr}^\mu_{M,\mathbf{c}}$, where μ runs over the set of dominant coweights of M. It is easy to see that the fiber of $\mathrm{Mod}^{\theta,+}_{M,\mathbf{C}}$ over $\theta\cdot\mathbf{c}$, being a closed subscheme in $\mathrm{Gr}_{M,\mathbf{c}}$, contains (equivalently, intersects) $\mathrm{Gr}^{\mu}_{M,\mathbf{c}}$ if and only if the followinng two conditions hold:

(1) The projection of μ under $\Lambda_{\mathfrak{m}}=\Lambda_{\mathfrak{g}}\to\Lambda_{\mathfrak{g},\mathfrak{p}}$ equals θ.
(2) $w_0^M(\mu)\in\Lambda^{\mathrm{pos}}_{\mathfrak{g}}$, where w_0^M is the longest element of the Weyl group of M.

In particular, for a given θ, there are only finitely many such μ, a fact that follows alternatively from the above finite-dimensionality statement.

Note that there is a canonical map $\mathfrak{r}_{\mathfrak{p}}:\widetilde{\mathcal{Z}}^{\theta}_{\mathfrak{g},\mathfrak{p}}(\mathbf{C})\to\mathcal{Z}^{\theta}_{\mathfrak{g},\mathfrak{p}}(\mathbf{C})$ which remembers the data of $\mathcal{F}_{M/[M,M]}$, and $\kappa^{\check{\lambda}}$ for $\check{\lambda}\in\check{\Lambda}^{+}_{\mathfrak{g},\mathfrak{p}}$. It is easy to see that $\mathfrak{r}_{\mathfrak{p}}$ is an isomorphism over the open subset $\overset{\circ}{\mathcal{Z}}{}^{\theta}_{\mathfrak{g},\mathfrak{p}}(\mathbf{C})$.

Lemma 2.31. *We have a closed embedding (of functors)* $\widetilde{\mathcal{Z}}^{\theta}_{\mathfrak{g},\mathfrak{p}}(\mathbf{C})\to\mathcal{Z}^{\theta}_{\mathfrak{g},\mathfrak{p}}(\mathbf{C})\underset{\mathbf{C}^{\theta}}{\times}\mathrm{Mod}^{\theta,+}_{M,\mathbf{C}}$.

This lemma implies, in particular, that both $\widetilde{\mathcal{Z}}^{\theta}_{\mathfrak{g},\mathfrak{p}}(\mathbf{C})$ and $\widetilde{\mathrm{QMaps}}^{\theta}(\mathbf{C},\mathcal{G}_{\mathfrak{g},\mathfrak{p}})$ are representable, being closed subfunctors of representable functors.

Proof. For an S-point of $\widetilde{\mathcal{Z}}^{\theta}_{\mathfrak{g},\mathfrak{p}}(\mathbf{C})$, we already know that the corresponding M-bundle admits a trivialization on $\mathbf{C}\times S-D^{\theta}$.

Moreover, if $\mathcal{U}^{\check{\lambda}}$ is an $\mathfrak{m}$-module with $\check{\lambda}\in\check{\Lambda}^{+}_{\mathfrak{g}}$, the corresponding map $\beta^{\check{\lambda}}:\mathcal{O}_{\mathbf{C}}\otimes\mathcal{U}_{\check{\lambda}}\to(\mathcal{U}_{\check{\lambda}})_{\mathcal{F}_M}$ equals the composition

$$\mathcal{O}_{\mathbf{C}}\otimes\mathcal{U}_{\check{\lambda}}\to\mathcal{O}_{\mathbf{C}}\otimes\mathcal{V}_{\check{\lambda}}\to(\mathcal{U}_{\check{\lambda}})_{\mathcal{F}_M},\tag{7}$$

where the first arrow comes from the embedding $\mathcal{U}_{\check{\lambda}}\simeq(\mathcal{V}_{\check{\lambda}})^{\mathfrak{n}(\mathfrak{p})}\hookrightarrow\mathcal{V}_{\check{\lambda}}$.

This proves that $\mathcal{F}_M$ with its trivialization indeed defines a point of $\mathrm{Mod}^{\theta,+}_{M,\mathbf{C}}$.

Conversely, given a point of $\mathcal{Z}^{\theta}_{\mathfrak{g},\mathfrak{p}}(\mathbf{C})$, and an M-bundle $\mathcal{F}_M$ trivialized on $\mathbf{C}\times S-D^{\theta}$, from Lemma 2.13 we obtain that there is a meromorphic map $\kappa^{\check{\lambda}}:\mathcal{O}_{\mathbf{C}}\otimes\mathcal{V}_{\check{\lambda}}\to(\mathcal{U}_{\check{\lambda}})_{\mathcal{F}_M}$. Our data defines a point of $\widetilde{\mathcal{Z}}^{\theta}_{\mathfrak{g},\mathfrak{p}}(\mathbf{C})$ if and only if $\kappa^{\check{\lambda}}$ is regular, which is a closed condition. □

We will denote by ϱ^{θ}_{M} the projection $\widetilde{\mathcal{Z}}^{\theta}_{\mathfrak{g},\mathfrak{p}}(\mathbf{C})\to\mathrm{Mod}^{\theta,+}_{M,\mathbf{C}}$.

2.32

Let us once again assume that $\mathbf{C}$ is complete. We will introduce yet two more versions of $\mathcal{Z}^{\theta}_{\mathfrak{g},\mathfrak{p}}(\mathbf{C})$ (denoted $\mathbf{Z}^{\theta}_{\mathfrak{g},\mathfrak{p}}(\mathbf{C})$, and $\widetilde{\mathbf{Z}}^{\theta}_{\mathfrak{g},\mathfrak{p}}(\mathbf{C})$, respectively, and called the twisted Zastava spaces), which will be fibered over the stack $\mathrm{Bun}_M(\mathbf{C})$ classifying M-bundles on $\mathbf{C}$.

By definition, $\widetilde{\mathbf{Z}}^{\theta}_{\mathfrak{g},\mathfrak{p}}(\mathbf{C})$ classifies the data of $(D^{\theta},\mathcal{F}_P,\mathcal{F}_M,\kappa)$, where $(D^{\theta},\mathcal{F}_M)$ are as before, $\mathcal{F}_P$ is a principal P-bundle on $\mathbf{C}$, and κ is a collection of maps

$$\kappa^{\check{\lambda}} : (\mathcal{V}_{\check{\lambda}})_{\mathcal{F}_P} \to (\mathcal{U}_{\check{\lambda}})_{\mathcal{F}_M}, \check{\lambda} \in \check{\Lambda}^+_{\mathfrak{g}},$$

which are generically surjective and satisfy the Plücker relations, and for $\check{\lambda} \in \check{\Lambda}^+_{\mathfrak{g},\mathfrak{p}}$ the compositions $\mathcal{L}^{\check{\lambda}}_{\mathcal{F}_P} \to (\mathcal{V}_{\check{\lambda}})_{\mathcal{F}_P} \to \mathcal{L}^{\check{\lambda}}_{\mathcal{F}_M}$ induce isomorphisms $\mathcal{L}^{\check{\lambda}}_{\mathcal{F}_M} \simeq \mathcal{L}^{\check{\lambda}}_{\mathcal{F}_P}(\check{\lambda}(D^\theta))$.

The stack $\mathbf{Z}^\theta_{\mathfrak{g},\mathfrak{p}}(\mathbf{C})$ classifies triples $(D^\theta, \mathcal{F}_P, \kappa)$, where $(D^\theta, \mathcal{F}_P)$ are as above, and κ are collections of maps $\kappa^{\check{\lambda}} : (\mathcal{V}_{\check{\lambda}})_{N(P)} \to \mathcal{O}(\check{\lambda}(D^\theta))$, defined for $\check{\lambda} \in \check{\Lambda}^+_{\mathfrak{g},\mathfrak{p}}$. The forgetful map $\widetilde{\mathbf{Z}}^\theta_{\mathfrak{g},\mathfrak{p}}(\mathbf{C}) \to \mathbf{Z}^\theta_{\mathfrak{g},\mathfrak{p}}(\mathbf{C})$ will be denoted by the same symbol $\mathfrak{r}_\mathfrak{p}$.

We have an open substack $\overset{\circ}{\mathbf{Z}}{}^\theta_{\mathfrak{g},\mathfrak{p}}(\mathbf{C}) \subset \mathbf{Z}^\theta_{\mathfrak{g},\mathfrak{p}}(\mathbf{C})$ that corresponds to the condition that the maps $\kappa^{\check{\lambda}}$ are surjective. Over it, $\mathfrak{r}_\mathfrak{p}$ is an isomorphism.

We have a projection, which we will call $\mathfrak{q}_\mathfrak{p}$, from both $\mathbf{Z}^\theta_{\mathfrak{g},\mathfrak{p}}(\mathbf{C})$ and $\widetilde{\mathbf{Z}}^\theta_{\mathfrak{g},\mathfrak{p}}(\mathbf{C})$ to the stack $\mathrm{Bun}_M(\mathbf{C})$ that "remembers" the data of the M-bundle $\mathcal{F}'_M := N(P)\backslash\mathcal{F}_P$. In the case of $\widetilde{\mathbf{Z}}^\theta_{\mathfrak{g},\mathfrak{p}}(\mathbf{C})$ we also have the other projection to $\mathrm{Bun}_M(\mathbf{C})$, denoted $\mathfrak{q}_{\mathfrak{p}^-}$, that "remembers" the data of $\mathcal{F}_M$. Of course, the preimage of the trivial bundle $\mathcal{F}^0_M \in \mathrm{Bun}_M(\mathbf{C})$ under $\mathfrak{q}_\mathfrak{p}$ identifies with the schemes $\mathcal{Z}^\theta_{\mathfrak{g},\mathfrak{p}}(\mathbf{C})$ and $\widetilde{\mathcal{Z}}^\theta_{\mathfrak{g},\mathfrak{p}}(\mathbf{C})$, respectively.

It is easy to see that Lemma 2.13, Corollary 2.14, and Proposition 2.19 generalize to the context of twisted Zastava spaces. In particular, we have the following assertion.

Lemma 2.33. *Let $\phi_1, \phi_2 : S \to \mathrm{Bun}_M(\mathbf{C})$ be two arrows such that the corresponding M-bundles on $\mathbf{C} \times S$ are identified over an open subset $U \subset \mathbf{C} \times S$. In addition let D^θ be the graph of a map $S \to \mathbf{C}^\theta$, such that $D^\theta \subset U$. Then the two Cartesian products $S \underset{\mathrm{Bun}_M(\mathbf{C})\times\mathbf{C}^\theta}{\times} \widetilde{\mathbf{Z}}^\theta_{\mathfrak{g},\mathfrak{p}}(\mathbf{C})$, taken with respect to either ϕ_1 or ϕ_2 are naturally isomorphic, and a similar assertion holds for $\mathbf{Z}^\theta_{\mathfrak{g},\mathfrak{p}}(\mathbf{C})$.*

This lemma implies that $\widetilde{\mathbf{Z}}^\theta_{\mathfrak{g},\mathfrak{p}}(\mathbf{C})$ and $\mathbf{Z}^\theta_{\mathfrak{g},\mathfrak{p}}(\mathbf{C})$ are algebraic stacks, such that the morphism $\mathfrak{q}_\mathfrak{p}$ is representable.

As in the case of $\mathcal{Z}^\theta_{\mathfrak{g},\mathfrak{p}}(\mathbf{C})$, we have a stratification of $\mathbf{Z}^\theta_{\mathfrak{g},\mathfrak{p}}(\mathbf{C})$ by locally closed substacks of the form $\overset{\circ}{\mathbf{Z}}{}^{\theta-\theta'}_{\mathfrak{g},\mathfrak{p}}(\mathbf{C}) \times \mathbf{C}^{\theta'}$ for $\theta', \theta-\theta' \in \Lambda^{\mathrm{pos}}_{\mathfrak{g},\mathfrak{p}}$. To describe the preimages of these strata in $\widetilde{\mathbf{Z}}^\theta_{\mathfrak{g},\mathfrak{p}}(\mathbf{C})$ we need to introduce some notation:

Let $\mathcal{H}_{M,\mathbf{C}}$ be the Hecke stack, i.e., a relative over $\mathrm{Bun}_M(\mathbf{C})$ version of $\mathrm{Gr}_{M,\mathbf{C}}$, which classifies quadruples $(\mathbf{c}, \mathcal{F}_M, \mathcal{F}'_M, \beta)$, where $\mathbf{c}$ is a point of $\mathbf{C}$, $\mathcal{F}_M, \mathcal{F}'_M$ are principal M-bundles on $\mathbf{C}$, and β is an identification $\mathcal{F}_M \simeq \mathcal{F}'_M|_{\mathbf{C}-\mathbf{c}}$. Let $\mathcal{H}^{BD,a}_{M,\mathbf{C}}$, $\mathcal{H}^{+,\theta}_{M,\mathbf{C}}$ be the corresponding relative versions of $\mathrm{Gr}^{BD,a}_{M,\mathbf{C}}$ and $\mathrm{Mod}^{+,\theta}_{M,\mathbf{C}}$, respectively. We will denote by $\overleftarrow{h}, \overrightarrow{h}$ the two projections from any of the stacks $\mathcal{H}_{M,\mathbf{C}}, \mathcal{H}^{BD,a}_{M,\mathbf{C}}$ or $\mathrm{Mod}^{+,\theta}_{M,\theta}$ to $\mathrm{Bun}_M(\mathbf{C})$ that "remember" the data of $\mathcal{F}_M$ and $\mathcal{F}'_M$, respectively.

Note that we have a canonical (convolution) map $\mathcal{H}_{M,\mathbf{C}} \underset{\mathrm{Bun}_M(\mathbf{C})\times\mathbf{C}}{\times} \mathcal{H}_{M,\mathbf{C}} \to \mathcal{H}_{M,\mathbf{C}}$, which sends $(\mathbf{c}, \mathcal{F}_M, \mathcal{F}'_M, \beta) \times (\mathbf{c}, \mathcal{F}'_M, \mathcal{F}''_M, \beta')$ to $(\mathbf{c}, \mathcal{F}_M, \mathcal{F}''_M, \beta'')$, where $\beta'' = \beta' \circ \beta$. Similarly, we have a map $\mathcal{H}^{+,\theta}_{M,\mathbf{C}} \underset{\mathrm{Bun}_M(\mathbf{C})}{\times} \mathcal{H}^{+,\theta'}_{M,\mathbf{C}} \to \mathcal{H}^{+,\theta+\theta'}_{M,\mathbf{C}}$ that covers the addition map $\mathbf{C}^\theta \times \mathbf{C}^{\theta'} \to \mathbf{C}^{\theta+\theta'}$.

As in the case of $\mathcal{Z}^\theta_{\mathfrak{g},\mathfrak{p}}(\mathbf{C})$, we have a map $\varrho^\theta_M : \widetilde{\mathsf{Z}}^\theta_{\mathfrak{g},\mathfrak{p}}(\mathbf{C}) \to \mathcal{H}^{+,\theta}_{M,\mathbf{C}}$, such that the map

$$\widetilde{\mathsf{Z}}^\theta_{\mathfrak{g},\mathfrak{p}}(\mathbf{C}) \hookrightarrow \mathsf{Z}^\theta_{\mathfrak{g},\mathfrak{p}} \underset{\mathrm{Bun}_M(\mathbf{C})}{\times} \mathcal{H}^{+,\theta}_{M,\mathbf{C}}$$

is a closed embedding, and such that the maps $\mathfrak{q}_{\mathfrak{p}}$, $\mathfrak{q}_{\mathfrak{p}^-}$ are the compositions of ϱ^θ_M with $\overleftarrow{h}$ and $\overrightarrow{h}$, respectively.

Proposition 2.34. *The preimage of a stratum* $\overset{\circ}{\mathsf{Z}}{}^{\theta-\theta'}_{\mathfrak{g},\mathfrak{p}}(\mathbf{C}) \times \mathbf{C}^{\theta'}$ *in* $\widetilde{\mathsf{Z}}^\theta_{\mathfrak{g},\mathfrak{p}}(\mathbf{C})$ *is isomorphic to* $\overset{\circ}{\mathsf{Z}}{}^{\theta-\theta'}_{\mathfrak{g},\mathfrak{p}}(\mathbf{C}) \underset{\mathrm{Bun}_M(\mathbf{C})}{\times} \mathcal{H}^{+,\theta'}_{M,\mathbf{C}}$, *where the projections from* $\overset{\circ}{\mathsf{Z}}{}^{\theta-\theta'}_{\mathfrak{g},\mathfrak{p}}(\mathbf{C})$ *and* $\mathcal{H}^{+,\theta'}_{M,\mathbf{C}}$ *to* $\mathrm{Bun}_M(\mathbf{C})$ *are* $\mathfrak{q}_{\mathfrak{p}^-}$ *and* $\overleftarrow{h}$, *respectively.*

By fixing the P-bundle $\mathcal{F}_P$ (respectively, the M-bundle $\widetilde{N(P)}\backslash\mathcal{F}_P$) to be trivial, we obtain the description of the corresponding strata in $\mathrm{QMaps}^\theta(\mathbf{C}, \mathcal{G}_{\mathfrak{g},\mathfrak{p}})$ and $\widetilde{\mathcal{Z}}^\theta_{\mathfrak{g},\mathfrak{p}}(\mathbf{C})$, respectively.

Proof. Given a point of $\mathsf{Z}^\theta_{\mathfrak{g},\mathfrak{p}}(\mathbf{C})$ factoring through $\overset{\circ}{\mathsf{Z}}{}^{\theta-\theta'}_{\mathfrak{g},\mathfrak{p}}(\mathbf{C}) \times \mathbf{C}^{\theta'}$, we have a triple $(\mathcal{F}_P, \mathcal{F}^1_M, \kappa)$:

$$(\mathcal{V}_{\check{\lambda}})_{\mathcal{F}_P} \xrightarrow{\kappa^\theta} (\mathcal{U}_{\check{\lambda}})_{\mathcal{F}^1_M},$$

where the maps κ^θ are surjections.

Then it is clear that the scheme of possible $\mathcal{F}_M$s with $\beta : \mathcal{F}_M \simeq \mathcal{F}^1_M|_{\mathbf{C}-D^{\theta'}}$, such that the (a priori meromorphic) maps

$$(\mathcal{V}_{\check{\lambda}})_{\mathcal{F}_P} \xrightarrow{\kappa^\theta} (\mathcal{U}_{\check{\lambda}})_{\mathcal{F}_M}$$

continue to be regular, identifies with the fiber of $\mathcal{H}^{+,\theta'}_{M,\mathbf{C}}$ over $\mathcal{F}^1{}_M \in \mathrm{Bun}_M(\mathbf{C})$. □

We will denote by $\widetilde{\mathfrak{s}}^\theta_{\mathfrak{p}}$ the embedding of the last stratum $\mathcal{H}^{\theta,+}_{M,\mathbf{C}} \to \widetilde{\mathsf{Z}}^\theta_{\mathfrak{g},\mathfrak{p}}(\mathbf{C})$. Note that in terms of the above proposition, the composition

$$\overset{\circ}{\mathsf{Z}}{}^{\theta-\theta'}_{\mathfrak{g},\mathfrak{p}}(\mathbf{C}) \underset{\mathrm{Bun}_M(\mathbf{C})}{\times} \mathcal{H}^{+,\theta'}_{M,\mathbf{C}} \to \widetilde{\mathsf{Z}}^\theta_{\mathfrak{g},\mathfrak{p}}(\mathbf{C}) \xrightarrow{\varrho^\theta_M} \mathcal{H}^{+,\theta}_{M,\mathbf{C}}$$

equals

$$\overset{\circ}{\mathsf{Z}}{}^{\theta-\theta'}_{\mathfrak{g},\mathfrak{p}}(\mathbf{C}) \underset{\mathrm{Bun}_M(\mathbf{C})}{\times} \mathcal{H}^{+,\theta'}_{M,\mathbf{C}} \xrightarrow{\varrho^{\theta-\theta'}_M \times \mathrm{id}} \mathcal{H}^{+,\theta-\theta'}_{M,\mathbf{C}} \underset{\mathrm{Bun}_M(\mathbf{C})}{\times} \mathcal{H}^{+,\theta'}_{M,\mathbf{C}} \to \mathcal{H}^{+,\theta}_{M,\mathbf{C}}.$$

2.35 The meromorphic case

Let $\mathfrak{X}$ be a scheme, and $D_{\mathfrak{X}} \subset \mathbf{C} \times \mathfrak{X}$ a relative Cartier divisor disjoint from $\infty_{\mathbf{C}} \times \mathfrak{X}$. We will now develop a notion of a meromorphic quasi-map parallel to Section 1.19. These will appear in the following two contexts.

First, let $\mathfrak{g}$ be finite dimensional, and let $\mathcal{F}_G$ be a principal G-bundle defined on $\mathbf{C} \times \mathfrak{X} - D_{\mathfrak{X}}$, trivialized along $\infty_{\mathbf{C}} \times \mathfrak{X}$. Let ${}_{\infty \cdot D_{\mathfrak{X}}}\mathrm{QMaps}(\mathbf{C}, \mathcal{F}_G \overset{G}{\times} \mathcal{G}_{\mathfrak{g},\mathfrak{p}})$ be the functor on the category of schemes over $\mathfrak{X}$ that assigns to a test scheme S the data of $(\mathcal{F}_{M/[M,M]}, \kappa^{\check{\lambda}}, \check{\lambda} \in \Lambda^+_{\mathfrak{g},\mathfrak{p}})$, where $\mathcal{F}_{M/[M,M]}$ is a principal $M/[M, M]$-bundle defined on $\mathbf{C} \times S$, and the $\kappa^{\check{\lambda}}$s are maps

$$\mathcal{L}^{\check{\lambda}}_{\mathcal{F}_{M/[M,M]}}|_{(\mathbf{C}\times\mathfrak{X}-D_{\mathfrak{X}})\underset{\mathfrak{X}}{\times} S} \to (\mathcal{V}^*_{\check{\lambda}})_{\mathcal{F}_G}, \tag{8}$$

satisfying the Plücker equations, with a prescribed value at $\infty_{\mathbf{C}} \times S$ corresponding to the trivialization of $\mathcal{F}_G$. The functor ${}_{\infty \cdot D_{\mathfrak{X}}}\mathrm{QMaps}^{\theta}(\mathbf{C}, \mathcal{G}_{\mathfrak{g},\mathfrak{p}})$ is representable by a strict ind-scheme of ind-finite type, by Proposition 1.20.

Similarly, we define the ind-scheme ${}_{\infty \cdot D_{\mathfrak{X}}}\widetilde{\mathrm{QMaps}}(\mathbf{C}, \mathcal{F}_G \overset{G}{\times} \mathcal{G}_{\mathfrak{g},\mathfrak{p}})$, where instead of an $M/[M, M]$-bundle $\mathcal{F}_{M/[M,M]}$, we have an M-bundle $\mathcal{F}_M$, and the maps $\kappa^{\check{\lambda}}$: $(\mathcal{V}_{\check{\lambda}})_{\mathcal{F}_G} \to (\mathcal{U}_{\check{\lambda}})_{\mathcal{F}_M}$ are defined for all $\check{\lambda} \in \Lambda^+_{\mathfrak{g}}$ and satisfy the Plücker relations in the sense of (6).

Now let $\mathfrak{g}$ be arbitrary. We define ${}_{\infty \cdot D_{\mathfrak{X}}}\mathrm{QMaps}(\mathbf{C}, \mathcal{G}_{\mathfrak{g},\mathfrak{p}})$ to be the ind-scheme which is the union over $\nu \in \Lambda^{\mathrm{pos}}_{\mathfrak{g},\mathfrak{p}}$ of schemes classifying the data of $(\mathcal{F}_{M/[M,M]}, \kappa^{\check{\lambda}}, \check{\lambda} \in \Lambda^+_{\mathfrak{g},\mathfrak{p}})$, where $\mathcal{F}_{M/[M,M]}$ is a principal $M/[M, M]$-bundle on $\mathbf{C} \times S$, and the $\kappa^{\check{\lambda}}$s are maps

$$\mathcal{L}^{\check{\lambda}}_{\mathcal{F}_{M/[M,M]}}(-\langle \nu, \check{\lambda}\rangle \cdot D_{\mathfrak{X}}|_{\mathbf{C}\times S}) \to \mathcal{V}^*_{\check{\lambda}},$$

satisfying the Plücker relations, with the prescribed value at $\infty_{\mathbf{C}} \times S$. The ind-scheme ${}_{\infty \cdot D_{\mathfrak{X}}}\widetilde{\mathrm{QMaps}}(\mathbf{C}, \mathcal{G}_{\mathfrak{g},\mathfrak{p}})$ is defined similarly, but where the union goes over $\nu \in \Lambda^{\mathrm{pos}}_{\mathfrak{g}}$.

3 Bundles on $\mathbb{P}^1 \times \mathbb{P}^1$

In this section $\mathbf{C}$ will be a projective curve of genus 0 with a marked infinity $\infty_{\mathbf{C}}$, and $\infty_{\mathbf{X}} \in \mathbf{X}$ will be another such curve. We will be interested in the surface $\mathbf{S}' := \mathbf{C} \times \mathbf{X}$, and we will call $\mathbf{D}'_{\infty} := \mathbf{C} \times \infty_{\mathbf{X}} \cup \infty_{\mathbf{C}} \times \mathbf{X} \subset \mathbf{S}'$ the divisor at infinity.

Throughout this section $\mathfrak{g}$ will be a finite-dimensional simple Lie algebra, and G the corresponding simply connected group.

3.1

Consider the stack $\mathrm{Bun}_G(\mathbf{S}', \mathbf{D}'_{\infty})$ that classifies G-bundles on $\mathbf{S}'$ with a trivialization on $\mathbf{D}'_{\infty}$. We will see shortly that $\mathrm{Bun}_G(\mathbf{S}', \mathbf{D}'_{\infty})$ is in fact a scheme. We have

$$\mathrm{Bun}_G(\mathbf{S}', \mathbf{D}'_\infty) = \underset{a \in \mathbb{N}}{\cup} \mathrm{Bun}_G^a(\mathbf{S}', \mathbf{D}'_\infty),$$

where $\mathrm{Bun}_G^a(\mathbf{S}', \mathbf{D}'_\infty)$ corresponds to G-bundles with second Chern class equal to a. (It is easy to see that $\mathrm{Bun}_G(\mathbf{S}', \mathbf{D}'_\infty)$ contains no points with negative Chern class.)

Recall now that $\mathrm{Bun}_G(\mathbf{X}, \infty_\mathbf{X})$ denotes the stack classifying G-bundles on $\mathbf{X}$ with a trivialization at $\infty_\mathbf{X}$. The stack $\mathrm{Bun}_G(\mathbf{X}, \infty_\mathbf{X})$ contains an open subset corresponding to the *trivial bundle*, which is isomorphic to pt. Since $\mathrm{Bun}_G(\mathbf{X}, \infty_\mathbf{X})$ is smooth, the complement $\mathrm{Bun}_G(\mathbf{X}, \infty_\mathbf{X}) - pt$, being of codimension 1, is a Cartier divisor. That is, we are in the situation of Section 2.16. We will denote the corresponding line bundle on $\mathrm{Bun}_G(\mathbf{X}, \infty_\mathbf{X})$ by $\mathcal{P}_{\mathrm{Bun}_G(\mathbf{X}, \infty_\mathbf{X})}$.

We will use the shorthand notation of $\mathrm{Maps}(\mathbf{C}, \mathrm{Bun}_G(\mathbf{X}, \infty_\mathbf{X}))$ for the functor of based maps from $\mathbf{C}$ to $\mathrm{Bun}_G(\mathbf{X}, \infty_\mathbf{X})$ that send $\infty_\mathbf{C}$ to $pt \subset \mathrm{Bun}_G(\mathbf{X}, \infty_\mathbf{X})$. By definition,

$$\mathrm{Maps}(\mathbf{C}, \mathrm{Bun}_G(\mathbf{X}, \infty_\mathbf{X})) = \underset{a \in \mathbb{N}}{\cup} \mathrm{Maps}^a(\mathbf{C}, \mathrm{Bun}_G(\mathbf{X}, \infty_\mathbf{X})),$$

where each $\mathrm{Maps}^a(\mathbf{C}, \mathrm{Bun}_G(\mathbf{X}, \infty_\mathbf{X})$ corresponds to maps σ such that $\sigma^*(\mathcal{P}_{\mathrm{Bun}_G(\mathbf{X}, \infty_\mathbf{X})})$ is of degree a. Since $\mathcal{P}_{\mathrm{Bun}_G(\mathbf{X}, \infty_\mathbf{X})}$ comes equipped with a section, we obtain a map $\varpi_h^a : \mathrm{Maps}^a(\mathbf{C}, \mathrm{Bun}_G(\mathbf{X}, \infty_\mathbf{X})) \to \overset{\circ}{\mathbf{C}}{}^{(a)}$.

We have an obvious isomorphism of functors:

$$\mathrm{Bun}_G(\mathbf{S}', \mathbf{D}'_\infty) \simeq \mathrm{Maps}(\mathbf{C}, \mathrm{Bun}_G(\mathbf{X}, \infty_\mathbf{X})). \tag{9}$$

The following assertion is left to the reader.

Lemma 3.2. *Under the above isomorphism,* $\mathrm{Bun}_G^a(\mathbf{S}', \mathbf{D}'_\infty)$ *maps to* $\mathrm{Maps}^a(\mathbf{C}, \mathrm{Bun}_G(\mathbf{X}, \infty_\mathbf{X}))$.

In particular, we obtain a map $\varpi_h^a : \mathrm{Bun}_G^a(\mathbf{S}', \mathbf{D}'_\infty) \to \overset{\circ}{\mathbf{C}}{}^{(a)}$. By interchanging the roles of $\mathbf{X}$ and $\mathbf{C}$, we obtain a map $\varpi_v^a : \mathrm{Bun}_G^a(\mathbf{S}', \mathbf{D}'_\infty) \to \overset{\circ}{\mathbf{X}}{}^{(a)}$.

3.3

Let $\mathcal{G}_{G,\mathbf{X}}$ denote the "thick" Grassmannian corresponding to G and the curve $\mathbf{X}$. In other words, $\mathcal{G}_{G,\mathbf{X}}$ is the scheme classifying pairs, $(\mathcal{F}_G, \beta)$, where $\mathcal{F}_G$ is a G-bundle on $\mathbf{X}$, and β is a trivialization of $\mathcal{F}_G$ over the formal neighborhood of $\mathcal{D}_{\infty_\mathbf{X}}$ of $\infty_\mathbf{X}$.

As we shall recall later, $\mathcal{G}_{G,\mathbf{X}}$ is one of the partial flag schemes associated to the affine Kac–Moody algebra corresponding to $\mathfrak{g}$. The unit point $1_{\mathcal{G}_{G,\mathbf{X}}}$ is the pair $(\mathcal{F}_G^0, \beta_{taut})$, where $\mathcal{F}_G^0$ is the trivial bundle and β_{taut} is its tautological trivialization. In particular, the scheme of based maps $\mathrm{Maps}^a(\mathbf{C}, \mathcal{G}_{G,\mathbf{X}})$ makes sense. (And we know from Proposition 2.7 that it is locally of finite type and smooth.)

Moreover, $\mathcal{G}_{G,\mathbf{X}}$ carries a very ample line bundle that will be denoted by $\mathcal{P}_{\mathcal{G}_{G,\mathbf{X}}}$, which is in fact the pullback of the line bundle $\mathcal{P}_{\mathrm{Bun}_G(\mathbf{X}, \infty_\mathbf{X})}$ under the natural projection $\mathcal{P}_{\mathcal{G}_{G,\mathbf{X}}} \to \mathrm{Bun}_G(\mathbf{X}, \infty_\mathbf{X})$. This enables us to define the scheme of based quasi-maps $\mathrm{QMaps}^a(\mathbf{C}, \mathcal{G}_{G,\mathbf{X}})$, studied in the previous section.

Recall also that over the symmetric power $\overset{\circ}{\mathbf{X}}{}^{(a)}$ we have the ind-scheme $\mathrm{Gr}^{BD,a}_{G,\mathbf{X}}$, classifying pairs $(D, \mathcal{F}_G, \beta)$, where $D \in \overset{\circ}{\mathbf{X}}{}^{(a)}$, $\mathcal{F}_G$ is a G-bundle on $\mathbf{X}$, and β is its trivialization off the support of D. We have a section $\overset{\circ}{\mathbf{X}}{}^{(a)} \to \mathrm{Gr}^{BD,a}_{G,\mathbf{X}}$, which also corresponds to the trivial bundle with a tautological trivialization. We will denote by $\mathrm{Maps}(\mathbf{C}, \mathrm{Gr}^{BD,a}_{G,\mathbf{X}})$ the (ind-)scheme of based relative maps $\mathbf{C} \to \mathrm{Gr}^{BD,a}_{G,\mathbf{X}}$. By definition, this is an ind-scheme over $\overset{\circ}{\mathbf{X}}{}^{(a)}$.

There exists a natural map $\mathrm{Gr}^{BD,a}_{G,\mathbf{X}} \to \mathcal{G}_{G,\mathbf{X}} \times \overset{\circ}{\mathbf{X}}{}^{(a)}$, which corresponds to the restriction of the trivialization on $\mathbf{X} - D$ to $\mathcal{D}_{\infty_\mathbf{X}}$. It is easy to see that this map is in fact a closed embedding. Moreover, the restriction $\mathcal{P}_{\mathrm{Gr}^{BD,a}_{G,\mathbf{X}}}$ of the line bundle $\mathcal{P}_{\mathcal{G}_{G,\mathbf{X}}}$ to $\mathrm{Gr}^{BD,a}_{G,\mathbf{X}}$ is relatively very ample. Therefore, we can introduce the ind-scheme of based quasi-maps $\mathrm{QMaps}(\mathbf{C}, \mathrm{Gr}^{BD,a}_{G,\mathbf{X}})$.

Proposition 3.4. *The natural morphisms*

$$\begin{aligned}\mathrm{Maps}^a(\mathbf{C}, \mathrm{Gr}^{BD,a}_{G,\mathbf{X}}) &\to \mathrm{Maps}^a(\mathbf{C}, \mathcal{G}_{G,\mathbf{X}})\\ &\to \mathrm{Maps}^a(\mathbf{C}, \mathrm{Bun}_G(\mathbf{X}, \infty_\mathbf{X})) \simeq \mathrm{Bun}^a_G(\mathbf{S}', \mathbf{D}'_\infty)\end{aligned}$$

are all isomorphisms. The resulting map $\mathrm{Bun}^a_G(\mathbf{S}', \mathbf{D}'_\infty) \to \overset{\circ}{\mathbf{X}}{}^{(a)}$ *coincides with the map* ϖ^a_v.

Proof. Let us show first that any map $S \to \mathrm{Bun}^a_G(\mathbf{S}', \mathbf{D}'_\infty)$ (for any test scheme S) lifts to a map $S \to \mathrm{Maps}(\mathbf{C}, \mathrm{Gr}^{BD,a}_{G,\mathbf{X}})$. Indeed, given a G-bundle $\mathcal{F}_G$ on $\mathbf{S}' \times S$ and using the map ϖ^a_v we obtain a divisor $D^v \subset \overset{\circ}{\mathbf{X}} \times S$, such that $\mathcal{F}_G$ is trivialized on $\mathbf{C} \times (\mathbf{X} \times S - D^v)$. But this by definition means that we are dealing with a based map $\mathbf{C} \times S \to \mathrm{Gr}^{BD,a}_{G,\mathbf{X}}$, which covers the map $S \to \overset{\circ}{\mathbf{X}}{}^{(a)}$ corresponding to D^v.

To prove the proposition it remains to show that if $\mathcal{F}_G$ is a G-bundle in $S \times \mathbf{S}'$ equipped with two trivializations on $\mathbf{C} \times \mathcal{D}_{\infty_\mathbf{X}} \times S$ which agree on $\infty_\mathbf{C} \times \mathcal{D}_{\infty_\mathbf{X}} \times S$, then these two trivializations coincide.

Indeed, the difference between the trivializations is a map from $\mathbf{C} \times S$ to the group of automorphisms of the trivial G-bundle on $\mathcal{D}_{\infty_\mathbf{X}}$. And since $\mathbf{C}$ is complete and the group in question is pro-affine, any such map is constant along the $\mathbf{C}$ factor. □

3.5

Using Proposition 2.24 we obtain that the scheme $\mathrm{Bun}^a_G(\mathbf{S}', \mathbf{D}'_\infty)$ is smooth. Moreover, we claim that $\dim(\mathrm{Bun}^a_G(\mathbf{S}', \mathbf{D}'_\infty)) = 2 \cdot \check{h} \cdot a$, where $\check{h}$ is the dual Coxeter number. One way to see this is via Corollary 2.28, and another way is as follows:

The tangent space to $\mathcal{F}_G \subset \mathrm{Bun}^a_G(\mathbf{S}', \mathbf{D}'_\infty)$ at a point corresponding to a G-bundle $\mathcal{F}_G$ equals $H^1(\mathbf{S}', \mathfrak{g}_{\mathcal{F}_G}(-\mathbf{D}'))$. Note that the vector bundle corresponding $\mathfrak{g}_{\mathcal{F}_G}$ has zero first Chern class, and the second Chern class equals $2 \cdot \check{h} \cdot a$.

Since points of $\mathrm{Bun}^a_G(\mathbf{S}', \mathbf{D}'_\infty)$ have no automorphisms, we obtain that $H^0(\mathbf{S}', \mathfrak{g}_{\mathcal{F}_G}(-\mathbf{D}')) = 0$, and by Serre duality we obtain that $H^2(\mathbf{S}', \mathfrak{g}_{\mathcal{F}_G}(-\mathbf{D}')) \simeq H^0(\mathbf{S}', \mathfrak{g}^*_{\mathcal{F}_G}(-\mathbf{D}'))^* = 0$. Hence the dimension of $H^1(\mathbf{S}', \mathfrak{g}_{\mathcal{F}_G}(-\mathbf{D}'))$ can be calculated by the Riemann–Roch formula, which yields $2\cdot\check{h}\cdot a$. Note that this calculation in fact re-proves that $\mathrm{Bun}^a_G(\mathbf{S}', \mathbf{D}'_\infty)$ is smooth, since we have shown that all the tangent spaces have the same dimension.

Let $0_\mathbf{X} \subset \mathbf{X}$ be another point, and let us denote by $\mathbf{D}'_0 \subset \mathbf{S}'$ the divisor $\mathbf{C} \times 0_\mathbf{X}$. By taking restriction of G-bundles we obtain a map from $\mathrm{Bun}^a_G(\mathbf{S}', \mathbf{D}'_\infty)$ to $\mathrm{Bun}_G(\mathbf{C}, \infty_\mathbf{C})$.

Lemma 3.6. *The above map* $\mathrm{Bun}^a_G(\mathbf{S}', \mathbf{D}'_\infty) \to \mathrm{Bun}_G(\mathbf{C}, \infty_\mathbf{C})$ *is smooth.*

Proof. Since both $\mathrm{Bun}^a_G(\mathbf{S}', \mathbf{D}'_\infty)$ and $\mathrm{Bun}_G(\mathbf{C}, \infty_\mathbf{C})$ are smooth, it suffices to check the surjectivity of the corresponding map on the level of tangent spaces:

$$H^1(\mathbf{S}', \mathfrak{g}_{\mathcal{F}_G}(-\mathbf{D}')) \to H^1(\mathbf{C}, \mathfrak{g}_{\mathcal{F}_G}|_{\mathbf{D}'_0}(-\infty_\mathbf{C})).$$

By the long exact sequence, the cokernel is given by $H^2(\mathbf{S}', \mathfrak{g}_{\mathcal{F}_G}(-\mathbf{D}' - \mathbf{D}'_0))$, and we claim that this cohomology group vanishes. Indeed, by Serre duality, it suffices to show that $H^0(\mathbf{S}', \mathfrak{g}_{\mathcal{F}_G}(-\infty_\mathbf{C} \times \mathbf{X})) = 0$, and we have an exact sequence

$$\begin{aligned} 0 &\to H^0(\mathbf{S}', \mathfrak{g}_{\mathcal{F}_G}(-\mathbf{D}')) \to H^0(\mathbf{S}', \mathfrak{g}_{\mathcal{F}_G}(-\infty_\mathbf{C} \times \mathbf{X})) \\ &\to H^0(\mathbf{C}, \mathfrak{g}_{\mathcal{F}_G}|_{\mathbf{C}\times\infty_\mathbf{X}}(-\infty_\mathbf{C}))\cdots. \end{aligned}$$

We know already that the first term vanishes (since $\mathrm{Bun}^a_G(\mathbf{S}', \mathbf{D}'_\infty)$ is a scheme) and the last term vanishes too (since $\mathfrak{g}_{\mathcal{F}_G}|_{\mathbf{C}\times\infty_\mathbf{X}}$ is trivial). □

As another corollary of Proposition 3.4 we obtain the following factorization property of $\mathrm{Bun}^a_G(\mathbf{S}', \mathbf{D}'_\infty)$ with respect to the projection ϖ^a_h (and by symmetry, with respect to ϖ^a_v):

$$\begin{aligned} &\mathrm{Bun}^a_G(\mathbf{S}', \mathbf{D}'_\infty) \underset{\overset{\circ}{\mathbf{C}}{}^{(a)}}{\times} (\overset{\circ}{\mathbf{C}}{}^{(a_1)} \times \overset{\circ}{\mathbf{C}}{}^{(a_2)})_{\mathrm{disj}} \\ &\quad\simeq (\mathrm{Bun}^{a_1}_G(\mathbf{S}', \mathbf{D}'_\infty) \times \mathrm{Bun}^{a_2}_G(\mathbf{S}', \mathbf{D}'_\infty)) \underset{\overset{\circ}{\mathbf{C}}{}^{(a_1)}\times\overset{\circ}{\mathbf{C}}{}^{(a_2)}}{\times} (\overset{\circ}{\mathbf{C}}{}^{(a_1)} \times \overset{\circ}{\mathbf{C}}{}^{(a_2)})_{\mathrm{disj}}. \end{aligned} \tag{10}$$

3.7 Relation with affine Lie algebras

Let $0_\mathbf{X} \in \mathbf{X}$ be another chosen point, and let x be a coordinate on $\mathbf{X}$ with $x(0_\mathbf{X}) = \infty$, $x(\infty_\mathbf{X}) = 0$. Let $\mathfrak{g}_{\mathrm{aff}} \simeq \mathfrak{g}((x)) \oplus K\cdot\mathbb{C} \oplus d\cdot\mathbb{C}$ be the corresponding untwisted affine Kac–Moody algebra, such that the element d acts via the derivation $x\cdot\partial_x$. We will denote by $\widehat{\mathfrak{g}}$ the derived algebra of $\mathfrak{g}_{\mathrm{aff}}$, i.e., $\mathfrak{g}((x)) \oplus K\cdot\mathbb{C}$.

The lattice $\Lambda_{\mathfrak{g}_{\mathrm{aff}}}$ is by definition the direct sum $\Lambda_\mathfrak{g} \oplus \delta\cdot\mathbb{Z} \oplus \mathbb{Z}$, and we will denote by $\widehat{\Lambda}_\mathfrak{g}$ the direct sum of the first two factors, i.e., the cocharacter lattice of the corresponding derived group. (The other factor can be largely ignored, since, for example, the semigroup $\Lambda^{pos}_{\mathfrak{g}_{aff}}$ is contained in $\widehat{\Lambda}_{\mathfrak{g}}$, and we will sometimes write $\widehat{\Lambda}^{pos}_{\mathfrak{g}}$ instead of $\Lambda^{pos}_{\mathfrak{g}_{aff}}$.) We will write an element $\mu \in \widehat{\Lambda}_{\mathfrak{g}}$ as $(\overline{\mu}, a)$ for $\overline{\mu} \in \Lambda_{\mathfrak{g}}, a \in \mathbb{Z}$. By definition, the element $\delta \in \widehat{\Lambda}_{\mathfrak{g}}$ equals $(0, 1)$.

Let $\overline{\alpha}_0$ denote the positive coroot of $\mathfrak{g}$ dual to the long dominant root. The simple affine coroot α_0 equals $(-\overline{\alpha}_0, 1)$. Note that $(\overline{\mu}, a) \in \widehat{\Lambda}^{pos}_{\mathfrak{g}}$ if and only if $a \geq 0$ and $\overline{\mu} + a \cdot \overline{\alpha}_0 \in \Lambda^{pos}_{\mathfrak{g}}$.

We will denote by $\mathfrak{g}^+_{aff}$ (respectively, $\mathfrak{g}^-_{aff}$) the subalgebra $\mathfrak{g}[[x]] \oplus K \cdot \mathbb{C} \oplus d \cdot \mathbb{C}$ (respectively, $\mathfrak{g}[x^{-1}] \oplus K \cdot \mathbb{C} \oplus d \cdot \mathbb{C}$). These algebras are the maximal parabolic and its opposite corresponding to $I \subset I_{aff}$. The corresponding partial flag scheme $\mathcal{G}_{\mathfrak{g}_{aff}, \mathfrak{g}^+_{aff}}$ identifies with $\mathcal{G}_{G,\mathbf{X}}$. Note that the latter does not depend on the choice of the point $0_{\mathbf{X}} \in \mathbf{X}$.

If $P \subset G$ is a parabolic and P^- is the corresponding opposite parabolic, let us denote by $\mathcal{G}_{G,P,\mathbf{X}}$ the scheme classifying triples $(\mathcal{F}_G, \beta, \gamma)$, where $(\mathcal{F}_G, \beta)$ are as in the definition of $\mathcal{G}_{G,\mathbf{X}}$, and γ is the data of a reduction to P of the fiber of $\mathcal{F}_G$ at $0_{\mathbf{X}}$.

We will denote by $\mathfrak{p}^+_{aff}$ the subalgebra of $\mathfrak{g}^+_{aff}$ consisting of elements whose value modulo x belongs to $\mathfrak{p}$. Similarly, we will denote by $\mathfrak{p}^-_{aff}$ the subalgebra of $\mathfrak{g}^-_{aff}$ consisting of elements whose value modulo x^{-1} belongs to $\mathfrak{p}^-$. The corresponding Lie algebras $\mathfrak{p}^+_{aff}, \mathfrak{p}^-_{aff}$ are a parabolic and its opposite in $\mathfrak{g}_{aff}$, and $\mathcal{G}_{\mathfrak{g}_{aff}, \mathfrak{p}^+_{aff}} \simeq \mathcal{G}_{G,P,\mathbf{X}}$. The Levi subgroup corresponding to $\mathfrak{p}^+_{aff}$ is $M \times \mathbb{G}_m \times \mathbb{G}_m$; we will denote by M_{aff} the group $M \times \mathbb{G}_m$ corresponding to the first $\mathbb{G}_m$-factor.

The lattice $\Lambda_{\mathfrak{g}_{aff}, \mathfrak{p}^+_{aff}}$ is the direct sum $\Lambda_{\mathfrak{g},\mathfrak{p}} \oplus \delta \cdot \mathbb{Z} \oplus \mathbb{Z}$, and we will denote by $\widehat{\Lambda}_{\mathfrak{g},\mathfrak{p}} \subset \Lambda_{\mathfrak{g}_{aff}, \mathfrak{p}_{aff}}$ the direct sum of the first two factors. For $\theta \in \widehat{\Lambda}^{pos}_{\mathfrak{g},\mathfrak{p}} := \Lambda^{pos}_{\mathfrak{g}_{aff}, \mathfrak{p}^+_{aff}}$, we have the corresponding scheme $\mathrm{Mod}^{\theta,+}_{M_{aff},\mathbf{C}}$.

Thus for a projective curve $\mathbf{C}$, we can consider the schemes of based maps $\mathrm{Maps}^\theta(\mathbf{C}, \mathcal{G}_{G,P,\mathbf{X}})$ for $\theta \in \widehat{\Lambda}^{pos}_{\mathfrak{g},\mathfrak{p}}$. For $\theta = (\overline{\theta}, a)$, consider the stack $\mathrm{Bun}^\theta_{G;P}(\mathbf{S}', \mathbf{D}'_\infty; \mathbf{D}'_0)$ that classifies the data of a G-bundle $\mathcal{F}_G \in \mathrm{Bun}^a_G(\mathbf{S}', \mathbf{D}'_\infty)$, and equipped with a reduction to P on $\mathbf{D}'_0$ of weight $\overline{\theta}$, compatible with the above trivialization on $\mathbf{D}'_\infty \cap \mathbf{D}'_0 = \infty_{\mathbf{C}} \times \infty_{\mathbf{X}}$.

From Proposition 3.4 we obtain an isomorphism

$$\mathrm{Bun}^\theta_{G;P}(\mathbf{S}', \mathbf{D}'_\infty; \mathbf{D}'_0) \simeq \mathrm{Maps}^\theta(\mathbf{C}, \mathcal{G}_{G,P,\mathbf{X}}). \tag{11}$$

We can also consider the scheme of based quasi-maps $\mathrm{QMaps}^\theta(\mathbf{C}, \mathcal{G}_{G,P,\mathbf{X}})$. We will denote by $\varrho^\theta_{\mathfrak{p}^+_{aff}}$ the map $\mathrm{QMaps}^\theta(\mathbf{C}, \mathcal{G}_{G,P,\mathbf{X}}) \to \overset{\circ}{\mathbf{C}}{}^\theta$.

In addition, we have the scheme of enhanced based quasi-maps $\widetilde{\mathrm{QMaps}}^\theta(\mathbf{C}, \mathcal{G}_{G,P,\mathbf{X}})$. We have a projection denoted $\mathfrak{r}_{\mathfrak{p}^+_{aff}} : \widetilde{\mathrm{QMaps}}^\theta(\mathbf{C}, \mathcal{G}_{G,P,\mathbf{X}}) \to \mathrm{QMaps}^\theta(\mathbf{C}, \mathcal{G}_{G,P,\mathbf{X}})$, and a map $\varrho^\theta_{M_{aff}} : \widetilde{\mathrm{QMaps}}^\theta(\mathbf{C}, \mathcal{G}_{G,P,\mathbf{X}}) \to \mathrm{Mod}^{\theta,+}_{M_{aff},\mathbf{X}}$, such that

$$\widetilde{\mathrm{QMaps}}^\theta(\mathbf{C}, \mathcal{G}_{G,P,\mathbf{X}}) \to \mathrm{QMaps}^\theta(\mathbf{C}, \mathcal{G}_{G,P,\mathbf{X}}) \underset{\overset{\circ}{\mathbf{C}}{}^\theta}{\times} \mathrm{Mod}^{\theta,+}_{M_{aff},\mathbf{C}}$$

is a closed embedding.

Note that the map $\mathfrak{r}_{\mathfrak{p}^+_{\text{aff}}}$ is not an isomorphism even when $P = G$. But it is, of course, an isomorphism when $\mathfrak{p} = \mathfrak{b}$.

3.8

Finally, let us show that the scheme $\text{Maps}^a(\mathbf{C}, \mathcal{G}_{G,\mathbf{X}}) = \text{Bun}^a_G(\mathbf{S}', \mathbf{D}'_\infty)$ is globally of finite type. Using equation (11), this would imply that the schemes $\text{Maps}^\theta(\mathbf{C}, \mathcal{G}_{\mathfrak{g}_{\text{aff}}, \mathfrak{p}^+_{\text{aff}}})$ are also of finite type.

Using Proposition 3.4, it suffices to show that any map $\sigma : \mathbf{C} \to \text{Gr}^{BD,a}_{G,\mathbf{X}}$ of degree a has its image in a fixed finite-dimensional subscheme of $\text{Gr}^{BD,a}_{G,\mathbf{X}}$. To simplify the notation, we will fix a divisor $D \in \overset{\circ}{\mathbf{X}}{}^{(a)}$ and consider based maps $\mathbf{C} \to \text{Gr}^{BD,a}_{G,\mathbf{X},D}$.

Write $D = \sum_k n_k \cdot \mathbf{x}_k$ with $\mathbf{x}_k$ pairwise distinct. Then $\text{Gr}^{BD,a}_{G,\mathbf{X},D}$ is the product of the affine Grassmannians $\prod_k \text{Gr}_{G,\mathbf{x}_k}$, and consider the subscheme $\prod_k \overline{\text{Gr}}^{n_k \cdot \overline{\alpha}_0}_{G,\mathbf{x}_k} \subset \text{Gr}^{BD,a}_{G,\mathbf{X},D}$, where for a dominant coweight $\overline{\lambda}$ of G, $\overline{\text{Gr}}^{\overline{\lambda}}_{G,\mathbf{x}}$ denotes the corresponding finite-dimensional subscheme of $\text{Gr}_{G,\mathbf{x}}$.

Lemma 3.9. *For $D \in \overset{\circ}{\mathbf{X}}{}^{(a)}$ as above, any based map $\sigma : \mathbf{C} \to \text{Gr}^{BD,a}_{G,\mathbf{X},D}$ of degree a has its image in the subscheme $\prod_k \overline{\text{Gr}}^{n_k \cdot \overline{\alpha}_0}_{G,\mathbf{x}_k}$.*

Of course, an analogous statement holds globally, i.e., when D moves along $\overset{\circ}{\mathbf{X}}{}^{(a)}$.

Proof. For a fixed point $\mathcal{F}_G \in \text{Bun}^a_G(\mathbf{S}', \mathbf{D}'_\infty)$, let $\sigma_\mathbf{C}$ be the corresponding based map $\mathbf{C} \to \text{Gr}^{BD,a}_{G,\mathbf{X}}$, and let $\sigma_\mathbf{X}$ be the corresponding based map $\mathbf{X} \to \mathcal{G}_{G,\mathbf{C}}$.

Let us fix a point $\mathbf{c} \in \mathbf{C}$, which we may as well call $0_\mathbf{C}$. We must show that the value of $\sigma_\mathbf{C}$ at $0_\mathbf{C}$ belongs to $\prod_k \overline{\text{Gr}}^{n_k \cdot \overline{\alpha}_0}_{G,\mathbf{x}_k} \subset \text{Gr}^{BD,a}_{G,\mathbf{X},D}$.

From Section 3.7, we have a map $\text{Maps}^a(\mathbf{X}, \mathcal{G}_{G,\mathbf{C}}) \to \text{Mod}^{a,+}_{G_{\text{aff}},\mathbf{X}}$ covering the map $\varpi^a_v : \text{Maps}^a(\mathbf{X}, \mathcal{G}_{G,\mathbf{C}}) \to \overset{\circ}{\mathbf{X}}{}^{(a)}$.

Now, it is easy to see that for our map $\sigma_\mathbf{X}$, $\varpi^a_v(\sigma_\mathbf{X}) = D$ and the fiber of $\text{Mod}^{a,+}_{G_{\text{aff}},\mathbf{X}}$ at D is a closed subscheme of $\prod_k \overline{\text{Gr}}^{n_k \cdot \overline{\alpha}_0}_{G,\mathbf{x}_k}$ (cf. [BFGM, Proposition 1.7]); moreover, this embedding induces an isomorphism on the level of reduced schemes. The resulting point of $\prod_k \overline{\text{Gr}}^{n_k \cdot \overline{\alpha}_0}_{G,\mathbf{x}_k}$ equals the value of $\sigma_\mathbf{C}$ at $0_\mathbf{C}$, which is what we had to show. □

Part II: Uhlenbeck Spaces

Throughout Part II, G will be a simple simply connected group and $\mathfrak{g}$ its Lie algebra. When $G = SL_n$, the subscript "SL_n" will often be replaced by just "n."

4 Definition of Uhlenbeck spaces

4.1 Rational surfaces

As was explained in the introduction, Uhlenbeck spaces $\mathfrak{U}^a_G$ are attached to the surface $\mathbf{S} \simeq \mathbb{P}^2$ with a distinguished "infinity" line $\mathbf{D}_\infty \simeq \mathbb{P}^1 \subset \mathbb{P}^2$. However, in order to define $\mathfrak{U}^a_G$, we will need to replace $\mathbf{S}$ by all possible rationally equivalent surfaces isomorphic to $\mathbb{P}^1 \times \mathbb{P}^1$.

Let $\overset{\circ}{\mathbf{S}} \subset \mathbf{S}$ denote the affine plane $\mathbf{S} - \mathbf{D}_\infty$. For two distinct points $\mathbf{d}_v, \mathbf{d}_h \in \mathbf{D}_\infty$ we obtain a decomposition of $\overset{\circ}{\mathbf{S}}$ as a product of two affine lines (horizontal and vertical):

$$\overset{\circ}{\mathbf{S}} \simeq \overset{\circ}{\mathbf{C}} \times \overset{\circ}{\mathbf{X}},$$

where $\mathbf{d}_v$ corresponds to the class of parallel lines in $\overset{\circ}{\mathbf{S}}$ that project to a single point in $\overset{\circ}{\mathbf{C}}$, and similarly for $\mathbf{d}_h$. Let $\mathbf{C} := \overset{\circ}{\mathbf{C}} \cup \infty_{\mathbf{C}}$, $\mathbf{X} := \overset{\circ}{\mathbf{X}} \cup \infty_{\mathbf{X}}$ be the corresponding projective lines.

Let us denote by $\mathbf{S}'$ the surface $\mathbf{C} \times \mathbf{X}$, and by π_v, π_h the projections from $\mathbf{S}'$ to $\mathbf{C}$ and $\mathbf{X}$, respectively. Let $\mathbf{D}'_\infty := \infty_{\mathbf{C}} \times \mathbf{X} \cup \mathbf{C} \times \infty_{\mathbf{X}}$ be the corresponding divisor "at infinity" in $\mathbf{S}'$. The surfaces $\mathbf{S}$ and $\mathbf{S}'$ are connected by a flip-flop. Namely, let $\mathbf{S}''$ be the blow-up of $\mathbf{S}$ at the two points $\mathbf{d}_v, \mathbf{d}_h$. Then $\mathbf{S}'$ is obtained from $\mathbf{S}''$ by blowing down the proper transform of $\mathbf{D}_\infty$.

In particular, it is easy to see that (a family of) G-bundles on $\mathbf{S}$ trivialized along $\mathbf{D}_\infty$ is the same as (a family of) G-bundles on $\mathbf{S}'$ trivialized along $\mathbf{D}'_\infty$.

4.2

Let O denote the variety $\mathbf{D}_\infty \times \mathbf{D}_\infty - \Delta(\mathbf{D}_\infty)$, i.e., the variety classifying pairs of distinct points $(\mathbf{d}_v, \mathbf{d}_h) \in \mathbf{D}_\infty$, and let us consider the "relative over O" versions of the varieties discussed above.

In particular, we have the relative affine (respectively, projective) lines $\overset{\circ}{\mathbf{C}}_{\mathsf{O}}, \overset{\circ}{\mathbf{X}}_{\mathsf{O}}$ (respectively, $\mathbf{C}_{\mathsf{O}}, \mathbf{X}_{\mathsf{O}}$), and the relative surface $\mathbf{S}'_{\mathsf{O}}$. Let $\pi_{v,\mathsf{O}}$ (respectively, $\pi_{h,\mathsf{O}}$) denote the projection $\mathbf{S}'_{\mathsf{O}} \to \mathbf{C}_{\mathsf{O}}$ (respectively, $\mathbf{S}'_{\mathsf{O}} \to \mathbf{X}_{\mathsf{O}}$.) We will denote by $\overset{\circ}{\mathbf{C}}{}^{(a)}_{\mathsf{O}}$, $\overset{\circ}{\mathbf{X}}{}^{(a)}_{\mathsf{O}}$ the corresponding fibrations into symmetric powers.

Let us recall the following general construction. Suppose that $\mathcal{Y}_1$ is a scheme (of finite type), and $\mathcal{Y}_2 \to \mathcal{Y}_1$ is an affine morphism (also of finite type). Then the functor on the category of schemes that sends a test scheme S to the set of sections (i.e., $\mathcal{Y}_1$-maps) $S \times \mathcal{Y}_1 \to \mathcal{Y}_2$ is representable by an affine ind-scheme of ind-finite type, which we will denote by $\mathrm{Sect}(\mathcal{Y}_1, \mathcal{Y}_2)$.

(To show the representability, it is enough to assume that $\mathcal{Y}_2$ is the total space of a vector bundle $\mathcal{E}$, in which case $\mathrm{Sect}(\mathcal{Y}_1, \mathcal{Y}_2)$ is representable by the vector space $\Gamma(\mathcal{Y}_1, \mathcal{E})$.)

For example, by applying this construction to $\mathcal{Y}_1 = \mathsf{O}$, we obtain the ind-schemes $\mathrm{Sect}(\mathsf{O}, \overset{\circ}{\mathbf{C}}{}^{(a)}_{\mathsf{O}})$, $\mathrm{Sect}(\mathsf{O}, \overset{\circ}{\mathbf{X}}{}^{(a)}_{\mathsf{O}})$, $\mathrm{Sect}(\mathsf{O}, \overset{\circ}{\mathbf{C}}{}^{(a)}_{\mathsf{O}} \underset{\mathsf{O}}{\times} \overset{\circ}{\mathbf{X}}{}^{(a)}_{\mathsf{O}})$.

4.3

We are now ready to give the first definition of the Uhlenbeck space $\mathfrak{U}_G^a$.

For a fixed pair of directions $(\mathbf{d}_v, \mathbf{d}_h) \in \mathbf{D}_\infty$, i.e., a point of O, and a divisor $D_v \in \overset{\circ}{\mathbf{X}}{}^{(a)}$, consider the scheme of based quasi-maps $\mathrm{QMaps}^a(\mathbf{C}, \mathrm{Gr}_{G,\mathbf{X},D_v}^{BD,a})$; cf. Section 1.11. This ind-scheme is ind-affine, of ind-finite type; cf. Lemma 1.6.

By making $D_v \in \overset{\circ}{\mathbf{X}}{}^{(a)}$ a parameter, we obtain an ind-affine ind-scheme $\mathrm{QMaps}^a(\mathbf{C}, \mathrm{Gr}_{G,\mathbf{X}}^{BD,a})$ fibered over $\overset{\circ}{\mathbf{X}}{}^{(a)}$. Finally, by letting $(\mathbf{d}_v, \mathbf{d}_h) \in \mathsf{O}$ move, we obtain ind-affine fibrations

$$\mathrm{QMaps}^a(\mathbf{C}, \mathrm{Gr}_{G,\mathbf{X}}^{BD,a})_\mathsf{O} \to \overset{\circ}{\mathbf{X}}{}_\mathsf{O}^{(a)} \to \mathsf{O}.$$

Thus we can consider the ind-scheme $\mathrm{Sect}(\mathsf{O}, \mathrm{QMaps}^a(\mathbf{C}, \mathrm{Gr}_{G,\mathbf{X}}^{BD,a})_\mathsf{O})$. By construction, we have a natural map

$$\varpi_{v,\mathsf{O}}^a : \mathrm{Sect}(\mathsf{O}, \mathrm{QMaps}^a(\mathbf{C}, \mathrm{Gr}_{G,\mathbf{X}}^{BD,a})_\mathsf{O}) \to \mathrm{Sect}(\mathsf{O}, \overset{\circ}{\mathbf{X}}{}_\mathsf{O}^{(a)}).$$

When a pair of directions $(\mathbf{d}_v, \mathbf{d}_h)$ is fixed, by further evaluation we obtain the map $\varpi_v^a : \mathrm{Sect}(\mathsf{O}, \mathrm{QMaps}^a(\mathbf{C}, \mathrm{Gr}_{G,\mathbf{X}}^{BD,a})_\mathsf{O}) \to \overset{\circ}{\mathbf{X}}{}^{(a)}$.

Now, we claim that we have a natural map

$$\mathrm{Bun}_G^a(\mathbf{S}, \mathbf{D}_\infty) \to \mathrm{Sect}(\mathsf{O}, \mathrm{QMaps}^a(\mathbf{C}, \mathrm{Gr}_{G,\mathbf{X}}^{BD,a})_\mathsf{O}).$$

Indeed, constructing such a map amounts to giving a map $\mathrm{Bun}_G^a(\mathbf{S}, \mathbf{D}_\infty) \simeq \mathrm{Bun}_G(\mathbf{S}', \mathbf{D}'_\infty) \to \mathrm{QMaps}^a(\mathbf{C}, \mathrm{Gr}_{G,\mathbf{X}}^{BD,a})$ for every pair of directions $(\mathbf{d}_v, \mathbf{d}_h) \in \mathsf{O}$, but this has been done in the previous section, Proposition 3.4.

Since $\mathrm{Bun}_G^a(\mathbf{S}, \mathbf{D}_\infty)$ is a scheme, its image in $\mathrm{Sect}(\mathsf{O}, \mathrm{QMaps}^a(\mathbf{C}, \mathrm{Gr}_{G,\mathbf{X}}^{BD,a})_\mathsf{O})$ is contained in a closed subscheme of finite type. In fact, it is contained in the subscheme described in Section 3.8.

Definition 4.4. *We define* $\mathfrak{U}_G^a$ *to be the closure of the image of* $\mathrm{Bun}_G^a(\mathbf{S}, \mathbf{D}_\infty)$ *in the ind-scheme* $\mathrm{Sect}(\mathsf{O}, \mathrm{QMaps}^a(\mathbf{C}, \mathrm{Gr}_{G,\mathbf{X}}^{BD,a})_\mathsf{O})$.

By construction, $\mathfrak{U}_G^a$ is an affine scheme of finite type, which functorially depends on the pair $(\mathbf{S}, \mathbf{D}_\infty)$.

Lemma 4.5. *The map* $\mathrm{Bun}_G^a(\mathbf{S}, \mathbf{D}_\infty) \to \mathfrak{U}_G^a$ *is an open embedding.*

Proof. Let us fix a pair of directions $(\mathbf{d}_v, \mathbf{d}_h)$ and consider the corresponding evaluation map $\mathrm{Sect}(\mathsf{O}, \mathrm{QMaps}^a(\mathbf{C}, \mathrm{Gr}_{G,\mathbf{X}}^{BD,a})_\mathsf{O}) \to \mathrm{QMaps}^a(\mathbf{C}, \mathrm{Gr}_{G,\mathbf{X}}^{BD,a})$. We have a composition

$$\mathrm{Bun}_G^a(\mathbf{S}, \mathbf{D}_\infty) \to \mathfrak{U}_G^a \underset{\mathrm{QMaps}^a(\mathbf{C}, \mathrm{Gr}_{G,\mathbf{X}}^{BD,a})}{\times} \mathrm{Maps}^a(\mathbf{C}, \mathrm{Gr}_{G,\mathbf{X}}^{BD,a})$$
$$\to \mathrm{Maps}^a(\mathbf{C}, \mathrm{Gr}_{G,\mathbf{X}}^{BD,a}) \simeq \mathrm{Bun}_G^a(\mathbf{S}, \mathbf{D}_\infty).$$

Since $\mathrm{Bun}^a_G(\mathbf{S}, \mathbf{D}_\infty)$ is dense in $\mathfrak{U}^a_G$, we obtain that all arrows in the above formula are isomorphisms. The assertion of the lemma follows since

$$\mathfrak{U}^a_G \underset{\mathrm{QMaps}^a(\mathbf{C}, \mathrm{Gr}^{BD,a}_{G,\mathbf{X}})}{\times} \mathrm{Maps}^a(\mathbf{C}, \mathrm{Gr}^{BD,a}_{G,\mathbf{X}})$$

is clearly open in $\mathfrak{U}^a_G$. □

4.6

We will now give two more definitions of the space $\mathfrak{U}^a_G$, which, on the one hand, are more economical, but on the other hand possess less symmetry. Of course, later we will establish the equivalence of all the definitions.

Let us fix a point $(\mathbf{d}_v, \mathbf{d}_h) \in \mathbf{D}_\infty$ and consider the space of quasi-maps $\mathrm{QMaps}^a(\mathbf{C}, \mathcal{G}_{G,\mathbf{X}})$. As we have seen in the previous section, this is an affine scheme of infinite type.

We have a natural map

$$\mathrm{Bun}^a_G(\mathbf{S}, \mathbf{D}_\infty) \to \mathrm{QMaps}^a(\mathbf{C}, \mathcal{G}_{G,\mathbf{X}}) \times \mathrm{Sect}(\mathcal{O}, \overset{\circ}{\mathbf{X}}{}^{(a)}_{\mathcal{O}}),$$

constructed as in the previous section.

We set $'\mathfrak{U}^a_G$ to be the closure of the image of $\mathrm{Bun}^a_G(\mathbf{S}, \mathbf{D}_\infty)$ in the above product. From this definition, it is not immediately clear that $'\mathfrak{U}^a_G$ is of finite type.

Again, for a fixed pair of directions $(\mathbf{d}_v, \mathbf{d}_h) \in \mathbf{D}_\infty$, consider the fibration $\mathrm{Gr}^{BD,a}_{G,\mathbf{X}} \to \overset{\circ}{\mathbf{X}}{}^{(a)}$, and consider the corresponding fibration of quasi-map spaces $\mathrm{QMaps}^a(\mathbf{C}, \mathrm{Gr}^{BD,a}_{G,\mathbf{X}}) \to \overset{\circ}{\mathbf{X}}{}^{(a)}$. This is an affine ind-scheme of ind-finite type.

We have a natural map

$$\mathrm{Bun}^a_G(\mathbf{S}, \mathbf{D}_\infty) \to \mathrm{QMaps}^a(\mathbf{C}, \mathrm{Gr}^{BD,a}_{G,\mathbf{X}}) \underset{\overset{\circ}{\mathbf{X}}{}^{(a)}}{\times} \mathrm{Sect}(\mathcal{O}, \overset{\circ}{\mathbf{X}}{}^{(a)}_{\mathcal{O}}),$$

where the projection $\mathrm{Sect}(\mathcal{O}, \overset{\circ}{\mathbf{X}}{}^{(a)}_{\mathcal{O}}) \to \overset{\circ}{\mathbf{X}}{}^{(a)}$ corresponds to the evaluation at our fixed point $(\mathbf{d}_v, \mathbf{d}_h)$.

We set $''\mathfrak{U}^a_G$ to be the closure of its image. By construction, $''\mathfrak{U}^a_G$ is an affine scheme of finite type.

Proposition 4.7. *There is a canonical isomorphism* $'\mathfrak{U}^a_G \simeq {}''\mathfrak{U}^a_G$.

Proof. Recall that we have a natural map $\mathrm{Gr}^{BD,a}_{G,\mathbf{X}} \to \mathcal{G}_{G,\mathbf{X}}$, such that the canonical line bundle on $\mathrm{Gr}^{BD,a}_{G,\mathbf{X}}$ is the restriction of that on $\mathcal{G}_{G,\mathbf{X}}$; moreover, the map $\mathrm{Gr}^{BD,a}_{G,\mathbf{X}} \to \mathcal{G}_{G,\mathbf{X}} \times \overset{\circ}{\mathbf{X}}{}^{(a)}$ is a closed embedding.

Therefore, using Lemma 1.22, we obtain a closed embedding

$$\mathrm{QMaps}^a(\mathbf{C}, \mathrm{Gr}^{BD,a}_{G,\mathbf{X}}) \to \mathrm{QMaps}^a(\mathbf{C}, \mathcal{G}_{G,\mathbf{X}}) \times \overset{\circ}{\mathbf{X}}{}^{(a)}.$$

This defines a map

$$\mathrm{QMaps}^a(\mathbf{C}, \mathrm{Gr}_{G,\mathbf{X}}^{BD,a}) \underset{\overset{\circ}{\mathbf{X}}{}^{(a)}}{\times} \mathrm{Sect}(\mathsf{O}, \overset{\circ}{\mathbf{X}}{}^{(a)}_{\mathsf{O}}) \to \mathrm{QMaps}^a(\mathbf{C}, \mathcal{G}_{G,\mathbf{X}}) \times \mathrm{Sect}(\mathsf{O}, \overset{\circ}{\mathbf{X}}{}^{(a)}_{\mathsf{O}}),$$

which is also a closed embedding. This proves the proposition. □

We also have a natural map $\mathfrak{U}^a_G \to {}''\mathfrak{U}^a_G$, which corresponds to the evaluation map $\mathrm{Sect}(\mathsf{O}, \mathrm{QMaps}^a(\mathbf{C}, \mathrm{Gr}_{G,\mathbf{X}}^{BD,a})_{\mathsf{O}}) \to \mathrm{QMaps}^a(\mathbf{C}, \mathrm{Gr}_{G,\mathbf{X}}^{BD,a})$.

Theorem 4.8. *The map $\mathfrak{U}^a_G \to {}''\mathfrak{U}^a_G$ is an isomorphism.*

This theorem will be proved in the next section for $G = SL_n$, and in Section 6.3 for G general.

4.9

We conclude this section with the following observation:

Let $(\mathbf{d}_v, \mathbf{d}_h)$ be a fixed pair of directions. Using Section 2.11, we obtain a map

$$\mathrm{QMaps}^a(\mathbf{C}, \mathcal{G}_{G,\mathbf{X}}) \to \overset{\circ}{\mathbf{C}}{}^{(a)}.$$

By composing it with $\mathrm{QMaps}^a(\mathbf{C}, \mathrm{Gr}_{G,\mathbf{X}}^{BD,a}) \to \mathrm{QMaps}^a(\mathbf{C}, \mathcal{G}_{G,\mathbf{X}})$, we also obtain the map $\varpi^a_h : \mathrm{QMaps}^a(\mathbf{C}, \mathrm{Gr}_{G,\mathbf{X}}^{BD,a}) \to \overset{\circ}{\mathbf{C}}{}^{(a)}$. By making $(\mathbf{d}_v, \mathbf{d}_h)$ vary along O, we have, therefore, a morphism:

$$\varpi^a_{h,\mathsf{O}} : \mathrm{Sect}(\mathsf{O}, \mathrm{QMaps}^a(\mathbf{C}, \mathrm{Gr}_{G,\mathbf{X}}^{BD,a})_{\mathsf{O}}) \to \mathrm{Sect}(\mathsf{O}, \overset{\circ}{\mathbf{C}}{}^{(a)}_{\mathsf{O}}).$$

Now note that the space O carries a natural involution, which interchanges the roles of $\mathbf{d}_v$ and $\mathbf{d}_h$. In particular, we have a map $\tau : \mathrm{Sect}(\mathsf{O}, \overset{\circ}{\mathbf{X}}{}^{(a)}_{\mathsf{O}}) \to \mathrm{Sect}(\mathsf{O}, \overset{\circ}{\mathbf{C}}{}^{(a)}_{\mathsf{O}})$.

We will denote by $\mathrm{Sect}(\mathsf{O}, \mathrm{QMaps}^a(\mathbf{C}, \mathrm{Gr}_{G,\mathbf{X}}^{BD,a})_{\mathsf{O}})^{\tau}$ the equalizer of the two maps

$$\tau \circ \varpi_{v,\mathsf{O}} \quad \text{and} \quad \varpi_{h,\mathsf{O}} : \mathrm{Sect}(\mathsf{O}, \mathrm{QMaps}^a(\mathbf{C}, \mathrm{Gr}_{G,\mathbf{X}}^{BD,a})_{\mathsf{O}}) \to \mathrm{Sect}(\mathsf{O}, \overset{\circ}{\mathbf{C}}{}^{(a)}_{\mathsf{O}}). \quad (12)$$

We have the following.

Lemma 4.10. $\mathfrak{U}^a_G \subset \mathrm{Sect}(\mathsf{O}, \mathrm{QMaps}^a(\mathbf{C}, \mathrm{Gr}_{G,\mathbf{X}}^{BD,a})_{\mathsf{O}})^{\tau}$.

The proof follows from the fact that the maps $\tau \circ \varpi_{v,\mathsf{O}}$ and $\varpi_{h,\mathsf{O}}$ coincide on $\mathrm{Bun}^a_G(\mathbf{S}, \mathbf{D}_\infty)$; cf. Proposition 3.4.

5 Comparison of the definitions: The case of SL_n

5.1

In order to prove Theorem 4.8 for SL_n, we will recall yet one more definition of Uhlenbeck spaces, essentially due to S. Donaldson. To simplify the notation, we will write $\mathfrak{U}_n^a$ instead of $\mathfrak{U}_{SL_n}^a$.

In the case $G = SL_n$, the moduli space $\mathrm{Bun}_n(\mathbf{S}, \mathbf{D}_\infty)$ admits the following linear algebraic ADHM description going back to Barth (see a modern exposition in [Na]). We consider vector spaces $V = \mathbb{C}^a$, $W = \mathbb{C}^n$, and consider the affine space

$$\mathrm{End}(V) \oplus \mathrm{End}(V) \oplus \mathrm{Hom}(W, V) \oplus \mathrm{Hom}(V, W),$$

a typical element of which will be denoted $(B_1, B_2, \imath, \jmath)$. We define a subscheme $M_n^a \subset \mathrm{End}(V) \oplus \mathrm{End}(V) \oplus \mathrm{Hom}(W, V) \oplus \mathrm{Hom}(V, W)$ by the equation $[B_1, B_2] + \imath\jmath = 0$.

A quadruple $(B_1, B_2, \imath, \jmath) \in M_n^a$ is called *stable* if V has no proper subspace containing the image of $\imath$ and invariant with respect to B_1, B_2. A quadruple $(B_1, B_2, \imath, \jmath) \in M_n^a$ is called *costable* if $\mathrm{Ker}(\jmath)$ contains no nonzero subspace invariant with respect to B_1, B_2. We have the open subscheme ${}^sM_n^a \subset M_n^a$ (respectively, ${}^cM_n^a \subset M_n^a$) formed by stable (respectively, costable) quadruples. Their intersection ${}^sM_n^a \cap {}^cM^c$ is denoted by ${}^{sc}M_n^a$.

According to [Na, 2.1], the natural action of $\mathrm{GL}(V)$ on ${}^sM_n^a$ (respectively, ${}^cM_n^a$) is free, and the GIT quotient $\widetilde{\mathfrak{N}}_n^a := {}^sM_n^a/\mathrm{GL}(V)$ is canonically isomorphic to the fine moduli space of torsion-free sheaves of rank n and second Chern class a on $\mathbf{S}$ equipped with a trivialization on $\mathbf{D}_\infty$. The open subset ${}^{sc}M_n^a/\mathrm{GL}(V) \subset {}^sM_n^a/\mathrm{GL}(V)$ corresponds, under this identification, to the locus of vector bundles $\mathrm{Bun}_n^a(\mathbf{S}, \mathbf{D}_\infty) \subset \widetilde{\mathfrak{N}}_n^a$.

Finally, consider the categorical quotient $\mathfrak{N}_n^a := M_n^a//\mathrm{GL}(V)$. The natural projective morphism $\widetilde{\mathfrak{N}}_n^a \to \mathfrak{N}_n^a$ is the affinization of $\widetilde{\mathfrak{N}}_n^a$, i.e., $\mathfrak{N}_n^a$ is the spectrum of the algebra of regular functions on $\widetilde{\mathfrak{N}}_n^a$. Moreover, $\mathfrak{N}_n^a$ is reduced and irreducible and the natural map $\mathrm{Bun}_n(\mathbf{S}, \mathbf{D}_\infty) \to \mathfrak{N}_n^a$ is an open embedding.

Our present goal is to construct a map $\mathfrak{N}_n^a \to \mathfrak{U}_n^a$ and show that the composition $\mathfrak{N}_n^a \to \mathfrak{U}_n^a \to {}'\mathfrak{U}_n^a$ is an isomorphism. The proof given below was indicated by Drinfeld, and the main step is to interpret $\mathfrak{N}_n^a$ as a coarse moduli space of *coherent perverse sheaves* (cf. [B]) on $\mathbf{S}$.

5.2

Let $\mathbf{S}$ be a smooth surface.

Definition 5.3. *We will call a complex* $\mathcal{M} \in \mathrm{DCoh}^{\geq 0, \leq 1}(\mathbf{S})$ *a coherent perverse sheaf if*

- $h^1(\mathcal{M})$ *is a finite-length sheaf,*
- $h^0(\mathcal{M})$ *is a torsion-free coherent sheaf.*

Coherent perverse sheaves obviously form an additive subcategory of $\mathrm{DCoh}(X)$, denoted $\mathrm{PCoh}(X)$. It is easy to see that Serre–Grothendieck duality maps $\mathrm{PCoh}(X)$ to itself.

Lemma 5.4. *For* $\mathcal{M}_1, \mathcal{M}_2 \in \mathrm{PCoh}(X)$*, the inner Hom satisfies* $\underline{\mathrm{RHom}}^i(\mathcal{M}_1, \mathcal{M}_2) = 0$ *for* $i < 0$.

If S is a scheme, an S-family of coherent perverse sheaves on $\mathbf{S}$ is an object $\mathcal{M}$ of $\mathrm{DCoh}(\mathbf{S} \times S)$ such that for every geometric point $s \in S$ the (derived) restriction $\mathcal{M}_s \in \mathrm{DCoh}(\mathbf{S})$ belongs to $\mathrm{PCoh}(\mathbf{S})$.

Lemma 5.5. *The functor Schemes* $\to$ *Groupoids, that assigns to a test scheme* S *the full subgroupoid of* $\mathrm{DCoh}(\mathbf{S} \times S)$ *consisting of* S*-families of coherent perverse sheaves on* $\mathbf{S}$*, is a sheaf of categories in the faithfully flat topology.*

The lemma is proved in exactly the same manner as the usual faithfully flat descent theorem for sheaves, using the fact that for two S-families of coherent perverse sheaves $\mathcal{M}_1$ and $\mathcal{M}_2$, the cone of any arrow $\mathcal{M}_1 \to \mathcal{M}_2$ is a *canonically* defined object of $\mathrm{DCoh}(\mathbf{S} \times S)$, which follows from Lemma 5.4.

5.6

For $\mathbf{S} = \mathbb{P}^2$, consider the functor Schemes $\to$ Groupoids, denoted $\mathrm{Perv}^a_n(\mathbf{S}, \mathbf{D}_\infty)$, that assigns to a test scheme S the groupoid whose objects are S-families of coherent perverse sheaves $\mathcal{M}$ on $\mathbf{S}$, such that

- $\mathcal{M}$ is of generic rank n, $ch_2(\mathcal{M}) = -a$;
- in a neighborhood of $\mathbf{D}_\infty \times S \subset \mathbf{S} \times S$, $\mathcal{M}$ is a vector bundle, and its restriction to the divisor $\mathbf{D}_\infty \times S$ is trivialized.

Morphisms in this category are isomorphisms between coherent perverse sheaves (as objects in $\mathrm{DCoh}(\mathbf{S} \times S)$), which respect the trivialization at $\mathbf{D}_\infty$.

The following theorem, due to Drinfeld, is a generalization of Donaldson–Nakajima theory.

Theorem 5.7. *The functor* $\mathrm{Perv}^a_n(\mathbf{S}, \mathbf{D}_\infty)$ *is representable by the stack* $M^a_n/\mathrm{GL}(V)$.

Proof. The proof is a modification of Nakajima's argument in [Na, Chapter 2]. Let us choose homogeneous coordinates z_0, z_1, z_2 on $\mathbb{P}^2$, so that the line $\mathbf{D}_\infty \subset \mathbf{S}$ is given by equation $z_0 = 0$. In particular, this defines a pair of directions $(\mathbf{d}_v, \mathbf{d}_h) \in \mathsf{O}$.

Lemma 5.8. *For a coherent perverse sheaf* $\mathcal{M}$ *on* $\mathbf{S}$ *trivialized at* $\mathbf{D}_\infty$*, we have*

$$H^{\pm 1}(\mathbf{S}, \mathcal{M}(-1)[1]) = H^{\pm 1}(\mathbf{S}, \mathcal{M}(-2)[1]) = H^{\pm 1}(\mathbf{S}, \mathcal{M} \otimes \Omega^1[1]) = 0.$$

The proof for a torsion-free sheaf (in cohomological degree 0) is given in [Na, Chapter 2]. The statement of the lemma obviously also holds for a finite-length sheaf in cohomological degree 1, and an arbitrary $\mathcal{M}$ is an extension of two such perverse sheaves. □

Let p_1, p_2 be the two projections $\mathbf{S} \times \mathbf{S} \to \mathbf{S}$, and let $C^\bullet$ be the Koszul complex on $\mathbf{S} \times \mathbf{S}$, i.e., the complex

$$0 \to \mathcal{O}(-2) \boxtimes \Omega^2(2) \to \mathcal{O}(-1) \boxtimes \Omega^1(1) \to \mathcal{O} \boxtimes \mathcal{O} \to 0,$$

which is known to be quasi-isomorphic to $\mathcal{O}_{\Delta(\mathbf{S})}$. Then the complex of perverse coherent sheaves $p_2^*(\mathcal{M}(-1)) \otimes C^\bullet$ looks like

$$0 \to \mathcal{O}(-2) \boxtimes (\mathcal{M}(-1) \otimes \Omega^2(2)) \to \mathcal{O}(-1) \boxtimes (\mathcal{M}(-1) \otimes \Omega^1(1)) \to \mathcal{O} \boxtimes \mathcal{M}(-1) \to 0$$

The first term of the Beilinson spectral sequence for $Rp_{1*}(p_2^*(\mathcal{M}(-1)) \otimes C^\bullet) \simeq \mathcal{M}(-1)$ reduces to

$$\mathcal{O}(-2) \otimes H^1(\mathbf{S}, \mathcal{M}(-2)) \to \mathcal{O}(-1) \otimes H^1(\mathbf{S}, \mathcal{M} \otimes \Omega^1) \to \mathcal{O} \otimes H^1(\mathbf{S}, \mathcal{M}(-1))$$

(in degrees $-1, 0, 1$). Hence $\mathcal{M}$ is canonically quasi-isomorphic to the complex (monad)

$$\mathcal{O}(-1) \otimes H^1(\mathbf{S}, \mathcal{M}(-2)) \stackrel{d}{\to} \mathcal{O} \otimes H^1(\mathbf{S}, \mathcal{M} \otimes \Omega^1) \stackrel{b}{\to} \mathcal{O}(1) \otimes H^1(\mathbf{S}, \mathcal{M}(-1))$$

and d is injective.

Now we are able to go from perverse sheaves to the linear algebraic data and back. We set $V = H^1(\mathbf{S}, \mathcal{M}(-2))$, $W' = H^1(\mathbf{S}, \mathcal{M} \otimes \Omega^1)$, $V' = H^1(\mathbf{S}, \mathcal{M}(-1))$. We have $\dim V = \dim V' = a$ and $\dim W' = 2a + n$. Since $H^0(\mathbf{S}, \mathcal{O}(1))$ has a basis $\{z_0, z_1, z_2\}$, we may write in the above monad $d = z_0 d_0 + z_1 d_1 + z_2 d_2$, $b = z_0 b_0 + z_1 b_1 + z_2 b_2$, where $d_i \in \mathrm{Hom}(V, W')$, $b_i \in \mathrm{Hom}(W', V')$. Nakajima checks in [Na] that $b_1 d_2 = -b_2 d_1$ is an isomorphism from V to V', and identifies V' with V via this isomorphism. Nakajima defines $W \subset W'$ as $W := \mathrm{Ker}(b_1) \cap \mathrm{Ker}(b_2)$, and identifies W' with $V \oplus V \oplus W$ via $(d_1, d_2) : V \oplus V \leftrightarrows W' : (-b_2, b_1)$. Note that $\dim W = n$. Under these identifications, we write $V \stackrel{d_0}{\to} V \oplus V \oplus W \stackrel{b_0}{\to} V$ as $d_0 = (B_1, B_2, \jmath)$, $b_0 = (-B_2, B_1, \imath)$, where $B_1, B_2 \in \mathrm{End}(V)$, $\imath \in \mathrm{Hom}(W, V)$, $\jmath \in \mathrm{Hom}(V, W)$ satisfy the relation $[B_1, B_2] + \imath\jmath = 0$.

Conversely, given $V, W, B_1, B_2, \imath, \jmath$ as above, we define $\mathcal{M}$ as a monad

$$V \otimes \mathcal{O}(-1) \stackrel{d}{\to} (V \oplus V \oplus W) \otimes \mathcal{O} \stackrel{b}{\to} V \otimes \mathcal{O}(1)$$

(in cohomological degrees $-1, 0, 1$), where $d = (z_0 B_1 - z_1, z_0 B_2 - z_2, z_0 \jmath)$, $b = (-z_0 B_2 + z_2, z_0 B_1 - z_1, z_0 \imath)$. Evidently, $\mathcal{M}|_{\mathbf{D}_\infty} = W \otimes \mathcal{O}_{\mathbf{D}_\infty}$. □

If $\mathcal{M}$ is a coherent perverse sheaf on $\mathbf{S} = \mathbb{P}^2$ trivialized along $\mathbf{D}_\infty \simeq \mathbb{P}^1$, we will denote by the same symbol $\mathcal{M}$ the corresponding coherent perverse sheaf on $\mathbf{S}' = \mathbf{C} \times \mathbf{X}$, trivialized along the divisor $\mathbf{D}'_\infty$.

The following assertion can be deduced from the above proof of Theorem 5.7

Lemma 5.9. *The maps ϖ_h^a, ϖ_v^a from* $\mathrm{Bun}_n^a(\mathbf{S}, \mathbf{D}_\infty)$ *to* $(\mathbb{A}^1)^{(a)} \simeq \overset{\circ}{\mathbf{C}}{}^{(a)}$ *and* $(\mathbb{A}^1)^{(a)} \simeq \overset{\circ}{\mathbf{X}}{}^{(a)}$, *respectively* (*cf. Proposition* 3.4), *are given by the characteristic polynomials of B_1 and B_2.*

For the proof one has to observe that if $\mathcal{M}$ is a point of $\mathrm{Bun}_n^a(\mathbf{S}, \mathbf{D}_\infty)$, corresponding to a quadruple $(B_1, B_2, \imath, \jmath)$, and D_h (respectively, D_v) is the divisor on $\mathbb{A}^1$ given by the characteristic polynomial of B_1 (respectively, B_2), then as a bundle on $\mathbf{S}'$, $\mathcal{M}$ will be trivialized on $(\mathbf{C} - D_h) \times \mathbf{X}$ (respectively, $\mathbf{C} \times (\mathbf{X} - D_v)$).

5.10

We will now construct a map $\mathrm{Perv}_n^a(\mathbf{S}, \mathbf{D}_\infty) \to \mathfrak{U}_n^a$. First, for a fixed pair of directions $(\mathbf{d}_v, \mathbf{d}_h)$, we will construct a map

$$\mathrm{Perv}_n^a(\mathbf{S}, \mathbf{D}_\infty) \to \mathrm{QMaps}^a(\mathbf{C}, \mathcal{G}_{SL_n,\mathbf{X}}). \tag{13}$$

Thus let $\mathcal{M}$ be an S-point of $\mathrm{Perv}_n^a(\mathbf{S}, \mathbf{D}_\infty)$ or $\mathrm{Perv}_n^a(\mathbf{S}', \mathbf{D}'_\infty)$. Over the open subscheme $(\mathbf{C} \times S)_0$ of $\mathbf{C} \times S$, over which $\mathcal{M}$ is a vector bundle, we obtain a genuine map $(\mathbf{C} \times S)_0 \to \mathcal{G}_{SL_n,\mathbf{X}}$, and we have to show that this map extends as a quasi-map to the entire $\mathbf{C} \times S$.

Let us recall the description of the fundamental representation of $\widehat{\mathfrak{sl}}_n$ as a Clifford module Cliff_n; cf. [FGK, Section 2.2]. Thus we have to attach to $\mathcal{M}$ a line bundle $\mathcal{L}_\mathcal{M}$ on $\mathbf{C} \times S$, and a map

$$\mathcal{L}_\mathcal{M} \to \mathcal{O}_{\mathbf{C}\times S} \otimes \Lambda^\bullet(x^{-d_1}\mathbb{C}^n[[x]]/x^{d_2}\mathbb{C}^n[[x]]) \otimes (\det(x^{-d_1}\mathbb{C}^n[[x]]/\mathbb{C}^n[[x]]))^{-1}$$

for each pair of positive integers d_1 and d_2.

Consider the (derived) direct image

$$\mathcal{N}_d := (\pi_v \times \mathrm{id})_*(\mathcal{M}(d \cdot (\mathbf{C} \times \infty_\mathbf{X} \times S))).$$

This is a complex on $\mathbf{C} \times S$, whose fiber at every geometric point of $\mathbf{C} \times S$ lies in cohomological degrees 0 and 1. Therefore, locally on $\mathbf{C} \times S$, $\mathcal{N}$ can be represented by a length-2 complex of vector bundles $\mathcal{F}_d^0 \to \mathcal{F}_d^1$. We set $\mathcal{L}_\mathcal{M} := \det(\mathcal{F}_d^0) \otimes \det(\mathcal{F}_d^1)^{-1} \otimes \det(x^{-d}\mathbb{C}^n[[x]]/\mathbb{C}^n[[x]])$.

We have also a canonical map on $\mathbf{S}' \times S$:

$$\mathcal{M} \to \mathcal{O}^{\oplus n}/\mathcal{O}^{\oplus n}(-d_2 \cdot (\mathbf{C} \times \infty_\mathbf{X} \times S)),$$

which comes from a trivialization of $\mathcal{M}$ around the divisor $\mathbf{D}'_h \times S$. Therefore, we obtain a map in the derived category $\mathcal{N}_{d_1} \to \mathcal{O}_{\mathbf{C}\times S} \otimes (x^{-d_1}\mathbb{C}^n[[x]]/x^{d_2}\mathbb{C}^n[[x]])$.

Moreover, by replacing $\mathcal{F}_{d_1}^0 \to \mathcal{F}_{d_1}^1$ by a quasi-isomorphic complex of vector bundles, we can assume that the above map $\mathcal{N}_{d_1} \to \mathcal{O}_{\mathbf{C}\times S} \otimes (x^{-d_1}\mathbb{C}^n[[x]]/x^{d_2}\mathbb{C}^n[[x]])$ comes from a map $\mathcal{F}_{d_1}^0 \to \mathcal{O}_{\mathbf{C}\times S} \otimes (x^{-d_1}\mathbb{C}^n[[x]]/x^{d_2}\mathbb{C}^n[[x]])$.

Thus we have a map $\mathcal{F}_{d_1}^0 \to \mathcal{F}_{d_1}^1 \oplus (\mathcal{O}_{\mathbf{C}\times S} \otimes (x^{-d_1}\mathbb{C}^n[[x]]/x^{d_2}\mathbb{C}^n[[x]]))$ and, hence, a map

$$\begin{aligned}\Lambda^{\mathrm{rk}(\mathcal{F}_{d_1}^0)}(\mathcal{F}_{d_1}^0) &\to \Lambda^{\mathrm{rk}(\mathcal{F}_{d_1}^0)}(\mathcal{F}_{d_1}^1 \oplus \mathcal{O}_{\mathbf{C}\times S} \otimes (x^{-d_1}\mathbb{C}^n[[x]]/x^{d_2}\mathbb{C}^n[[x]]))\\ &\twoheadrightarrow \Lambda^{\mathrm{rk}(\mathcal{F}_{d_1}^1)}(\mathcal{F}_{d_1}^1) \otimes \Lambda^{\mathrm{rk}(\mathcal{F}_{d_1}^0)-\mathrm{rk}(\mathcal{F}_{d_1}^1)}(\mathcal{O}_{\mathbf{C}\times S} \otimes (x^{-d_1}\mathbb{C}^n[[x]]/x^{d_2}\mathbb{C}^n[[x]])).\end{aligned}$$

By tensoring both sides by $(\det(x^{-d_1}\mathbb{C}^n[[x]]/\mathbb{C}^n[[x]]))^{-1}$, we obtain a map $\mathcal{L}_{\mathcal{M}} \to \mathcal{O}_{\mathbf{C}\times S} \otimes \Lambda^\bullet(x^{-d_1}\mathbb{C}^n[[x]]/x^{d_2}\mathbb{C}^n[[x]]) \otimes (\det(x^{-d_1}\mathbb{C}^n[[x]]/\mathbb{C}^n[[x]]))^{-1}$, as required.

It is easy to check that the definition of this morphism does not depend on a particular choice of representing complex, and over $(\mathbf{C} \times S)_0$, this is the same map as the one defining the map $(\mathbf{C} \times S)_0 \to \mathcal{G}_{SL_n,\mathbf{X}}$.

5.11

Thus we have a map $\mathrm{Perv}_n^a(\mathbf{S}, \mathbf{D}_\infty) \to \mathrm{QMaps}^a(\mathbf{C}, \mathcal{G}_{SL_n,\mathbf{X}})$ for every fixed pair of directions $(\mathbf{d}_v, \mathbf{d}_h)$. In particular, we have a map $\mathrm{Perv}_n^a(\mathbf{S}, \mathbf{D}_\infty) \to \overset{\circ}{\mathbf{C}}{}^{(a)}$, and by letting $(\mathbf{d}_v, \mathbf{d}_h)$ vary, we obtain a map $\mathrm{Perv}_n^a(\mathbf{S}, \mathbf{D}_\infty) \to \mathrm{Sect}(\mathsf{O}, \overset{\circ}{\mathbf{C}}{}^{(a)}_{\mathsf{O}})$. By interchanging the roles of $\mathbf{d}_h$ and $\mathbf{d}_v$, we also obtain a map $\mathrm{Perv}_n^a(\mathbf{S}, \mathbf{D}_\infty) \to \mathrm{Sect}(\mathsf{O}, \overset{\circ}{\mathbf{X}}{}^{(a)}_{\mathsf{O}})$.

We now claim that for any fixed pair of directions $(\mathbf{d}_v, \mathbf{d}_h)$, the map

$$\mathrm{Perv}_n^a(\mathbf{S}, \mathbf{D}_\infty) \to \overset{\circ}{\mathbf{X}}{}^{(a)} \times \mathrm{QMaps}^a(\mathbf{C}, \mathcal{G}_{SL_n,\mathbf{X}})$$

factors through the closed subscheme $\mathrm{QMaps}^a(\mathbf{C}, \mathrm{Gr}^{BD,a}_{SL_n,\mathbf{X}}) \subset \overset{\circ}{\mathbf{X}}{}^{(a)} \times \mathrm{QMaps}^a(\mathbf{C}, \mathcal{G}_{SL_n,\mathbf{X}})$. Indeed, this is so because the corresponding fact is true over the open dense substack $\mathrm{Bun}_n^a(\mathbf{S}, \mathbf{D}_\infty) \subset \mathrm{Perv}_n^a(\mathbf{S}, \mathbf{D}_\infty)$. (The density assertion is a corollary of Theorem 5.7).

Therefore, we obtain a map

$$\mathrm{Perv}_n^a(\mathbf{S}, \mathbf{D}_\infty) \to \mathrm{Sect}(\mathsf{O}, \mathrm{QMaps}^a(\mathbf{C}, \mathrm{Gr}^{BD,a}_{SL_n,\mathbf{X}})_{\mathsf{O}}). \tag{14}$$

The image of this map lies in $\mathfrak{U}_n^a$ because the open dense substack $\mathrm{Bun}_n^a(\mathbf{S}, \mathbf{D}_\infty)$ does map there. In other words, we obtain a map $\mathrm{Perv}_n^a(\mathbf{S}, \mathbf{D}_\infty) \to \mathfrak{U}_n^a$.

In particular, from Theorem 5.7, we have a $\mathrm{GL}(V)$-invariant map $M_n^a \to \mathfrak{U}_n^a$, and since $\mathfrak{U}_n^a$ is affine, we thus have a map $\mathfrak{N}_n^a \to \mathfrak{U}_n^a$.

Moreover, from Lemma 5.9 we obtain that for a fixed pair of directions $(\mathbf{d}_v, \mathbf{d}_h)$, the composition $M_n^a \to \mathfrak{U}_n^a \overset{\varpi_h^a}{\to} \overset{\circ}{\mathbf{C}}{}^{(a)} \simeq (\mathbb{A}^1)^{(a)}$ is given by the map sending B_1 to its characteristic polynomial, and similarly for ϖ_v^a and B_2. Indeed, this is so because the corresponding fact holds for the open part ${}^{sc}M_n^a$.

Remark. Above we have shown that the map $\mathrm{Bun}_n^a(\mathbf{S}, \mathbf{D}_\infty) \to \mathfrak{U}_n^a$ extends to a $\mathrm{GL}(V)$-invariant map $M_n^a \to \mathfrak{U}_n^a$. However, if we used the results of [FGK], the existence of the latter map could be proved differently:

In [FGK] we constructed a map $\widetilde{\mathfrak{N}}_n^a \to \mathfrak{U}_n^a$, using the interpretation of $\widetilde{\mathfrak{N}}_n^a$ as the moduli space of torsion-free coherent sheaves on $\mathbf{S}$. Since $\mathfrak{N}_n^a$ is the affinization of $\widetilde{\mathfrak{N}}_n^a$, and $\mathfrak{U}_n^a$ is affine, we do obtain a map $\mathfrak{N}_n^a \to \mathfrak{U}_n^a$ and hence a map $M_n^a/\mathrm{GL}(V) \to \mathfrak{N}_n^a$.

Theorem 5.12. *The maps $\mathfrak{N}_n^a \to \mathfrak{U}_n^a \to {}'\mathfrak{U}_n^a$ are isomorphisms.*

The rest of this section is devoted to the proof of this theorem. Since all three varieties that appear in Theorem 5.12 have a common dense open piece, namely, $\mathrm{Bun}_n^a(\mathbf{S}, \mathbf{D}_\infty)$, it is sufficient to prove that the map $\mathsf{f} : \mathfrak{N}_n^a \to {}'\mathfrak{U}_n^a$ is an isomorphism.

5.13

Consider the map $\widetilde{\mathfrak{N}}^a_n \to {}'\mathfrak{U}^a_n$. It was shown in [FGK] that this map is proper. Since $\mathfrak{N}^a_n$ is an affinization of $\widetilde{\mathfrak{N}}^a_n$, we obtain that $\mathsf{f} : \mathfrak{N}^a_n \to {}'\mathfrak{U}^a_n$ is finite.

Choose a point $0_{\mathbf{S}} \in \overset{\circ}{\mathbf{S}}$, and let $0_{\mathbf{C}}$, $0_{\mathbf{X}}$ be the corresponding points on $\mathbf{C}$ and $\mathbf{X}$, respectively. Consider the corresponding action of $\mathbb{G}_m$ by dilations. By transport of structure, we obtain a $\mathbb{G}_m$-action on the scheme $\mathrm{QMaps}(\mathbf{C}, \mathrm{Gr}^{BD,a}_{\mathbf{X},SL_n})$ and on $\mathrm{Sect}(\mathsf{O}, \overset{\circ}{\mathbf{X}}{}^{(a)}_{\mathsf{O}})$.

It is easy to see that the $\mathbb{G}_m$-action contracts both these varieties to a single point. Namely, $\mathrm{Sect}(\mathsf{O}, \overset{\circ}{\mathbf{X}}{}^{(a)}_{\mathsf{O}})$ is contracted to the section that assigns to every $(\mathbf{d}^1_v, \mathbf{d}^1_h) \in \mathbf{D}_\infty$ the point $a \cdot 0_{\mathbf{X}}$ in the symmetric power of $\overset{\circ}{\mathbf{X}}$. Using Proposition 2.6, we obtain that $\mathrm{QMaps}(\mathbf{C}, \mathrm{Gr}^{BD,a}_{SL_n,\mathbf{X}})$ is contracted to the quasi-map whose saturation is the constant map $\mathbf{C} \to \mathrm{Gr}^{BD,a}_{SL_n,\mathbf{X}}$, corresponding to the trivial bundle and tautological trivialization, with defect of order a at $0_{\mathbf{C}}$. Let us denote by $\sigma^{\mathfrak{U}}$ the attracting point of ${}'\mathfrak{U}^a_n$ described above.

The above action of $\mathbb{G}_m$ on ${}'\mathfrak{U}^a_n$ is covered by a natural $\mathbb{G}_m$-action on the stack $\mathrm{Perv}^a_n(\mathbf{S}, \mathbf{D}_\infty)$. Moreover, if we identify $\overset{\circ}{\mathbf{S}}$ with $\mathbb{A}^2$, such that $0_{\mathbf{S}}$ corresponds to the origin, the above action on $\mathrm{Perv}^a_n(\mathbf{S}, \mathbf{D}_\infty)$ corresponds to the canonical $\mathbb{G}_m$-action on the variety $\mathbb{A}^2$ by homotheties.

The induced $\mathbb{G}_m$-action on $\mathfrak{N}^a_n \simeq M^a_n//\mathrm{GL}(V)$ contracts this variety to a single point, which we will denote by $\sigma^{\mathfrak{N}}$.

Thus it would be sufficient to show that the scheme-theoretic preimage $\mathsf{f}^{-1}(\sigma^{\mathfrak{U}})$ is in fact a point-scheme corresponding to $\sigma^{\mathfrak{N}}$.

5.14

Let $\mathsf{g}^{\mathfrak{N}}$ (respectively, $\mathsf{g}^{\mathfrak{U}}$) denote the canonical map from $\mathrm{Perv}^a_n(\mathbf{S}, \mathbf{D}_\infty)$ to $\mathfrak{N}^a_n$ (respectively, $\mathfrak{U}^a_n$). We claim that it is sufficient to show that the inclusion $(\mathsf{g}^{\mathfrak{N}})^{-1}(z^{\mathfrak{N}}_0) \hookrightarrow (\mathsf{g}^{\mathfrak{U}})^{-1}(z^{\mathfrak{U}}_0)$ is an equality. This follows from the next general observation.

Lemma 5.15. *Let $\mathcal{Y}_1$ be an affine algebraic variety (in char 0) with an action of a reductive group G. Let $\mathcal{Y}_2$ be another affine variety and $\mathsf{g} : \mathcal{Y}_1 \to \mathcal{Y}_2$ be a G-invariant map, and let us denote by $\mathsf{f} : \mathcal{Y}_1//G \to \mathcal{Y}_2$, $\mathsf{g}' : \mathcal{Y}_1 \to \mathcal{Y}_1//G$ the corresponding maps. Then if for some $z' \in \mathcal{Y}_1//G$, $z \in \mathcal{Y}_2$ with $\mathsf{f}(z') = z$ the inclusion $(\mathsf{g}')^{-1}(z) \subset (\mathsf{g})^{-1}(z')$ is an isomorphism, then $(\mathsf{f})^{-1}(z)$ is a point-scheme.*

Consider now the following ind-stack $\mathrm{Perv}^a_n(\mathbf{S}, \mathbf{S} - 0_{\mathbf{S}})$: For a scheme S, its S-points are S-families of coherent perverse sheaves on $\mathbf{S}$ (with $c_1 = 0$, $ch_2 = -a$), equipped with a trivialization on $\mathbf{S} - 0_{\mathbf{S}}$ and such that for every pair of directions $(\mathbf{d}_v, \mathbf{d}_h) \in \mathsf{O}$, the composition

$$S \to \mathrm{Perv}^a_n(\mathbf{S}) \to \mathrm{QMaps}^a(\mathbf{C}, \mathcal{G}_{SL_n,\mathbf{X}}) \to \overset{\circ}{\mathbf{C}}{}^{(a)}$$

maps to the point $a \cdot 0_{\mathbf{C}} \in \overset{\circ}{\mathbf{C}}{}^{(a)}$.

Proposition 5.16. *The composition* $\mathrm{Perv}^a_n(\mathbf{S}, \mathbf{S} - 0_{\mathbf{S}}) \to \mathrm{Perv}^a_n(\mathbf{S}, \mathbf{D}_\infty) \to \mathfrak{N}^a_n$ *is the constant map to the point* $\sigma^{\mathfrak{N}}$.

Proof. For a triple $(B_1, B_2, \imath, \jmath)$, representing a point of $\mathrm{Perv}^a_n(\mathbf{S}, \mathbf{D}_\infty)$, let us denote by T_W any endomorphism of the vector space W obtained by composing the maps $B_1, B_2, \imath, \jmath$, and by T_V any similarly obtained endomorphism of V.

It is easy to see that the space of regular functions on $\mathfrak{N}^a_n$ is obtained by taking matrix coefficients of all possible T_Ws and traces of all possible T_Vs.

Let $\mathcal{M}$ be an S-point of $\mathrm{Perv}^a_n(\mathbf{S}, \mathbf{S} - 0_{\mathbf{S}})$. For an integer m, let $\mathcal{M}'$ be the (constant) S-family of coherent perverse sheaves on $\mathbf{S}$, corresponding to the torsion-free sheaf

$$\ker(\mathcal{O}^{\oplus n} \to (\mathcal{O}/\mathfrak{m}^m_{0_{\mathbf{S}}})^{\oplus n}),$$

where $\mathfrak{m}_{0_{\mathbf{S}}}$ is the maximal ideal of the point $0_{\mathbf{S}}$. Then, when m is large enough, we can find a map $\mathcal{M}' \to \mathcal{M}$, which respects the trivializations of both sheaves on $\mathbf{S} - 0_{\mathbf{S}}$. The cone of this map is set-theoretically supported at $0_{\mathbf{S}}$ and has cohomologies in degrees 0 and 1.

Let $(V, W, B_1, B_2, \imath, \jmath)$ and $(V', W', B'_1, B'_2, \imath', \jmath')$ denote the linear algebra data corresponding to $\mathcal{M}$ and $\mathcal{M}'$, respectively. By unraveling the proof of Theorem 5.7, we obtain that there are maps $V' \to V$ and $W' \simeq W$ which commute with all the endomorphisms.

By the definition of $\mathcal{M}'$, $\jmath' = 0$. From this we obtain that all the matrices T_W vanish, and the only nonzero T_V-matrices are of the form $B_1^{k_1} \circ B_2^{k_2} \circ \cdots \circ B_1^{k_{m-1}}$. It remains to show that any such matrix is traceless.

Note that for any matrices T^1_V and T^2_V as above, the trace of $T^1_V \circ i \circ j \circ T^2_V \in \mathrm{End}(V)$ equals the trace of the corresponding endomorphism of V', and hence vanishes. The relation $[B_1, B_2] + \imath \circ \jmath = 0$ implies that the trace of a matrix $B_1^{k_1} \circ B_2^{k_2} \circ \cdots \circ B_1^{k_{m-1}}$ does not depend on the order of the factors. Therefore, it is sufficient to show that the characteristic polynomial of a matrix $B_1 + c \cdot B_2$ vanishes for all $c \in \mathbb{C}$.

However, using [FGK, Lemma 3.5], for any such c, we can find a pair of directions $(\mathbf{d}_v, \mathbf{d}_h) \in \mathsf{O}$, such that this characteristic polynomial equals the value of $M^a_n \to \mathfrak{U}^a_n \overset{\varpi^a_h}{\to} \overset{\circ}{\mathbf{C}}{}^{(a)} \simeq (\mathbb{A}^1)^{(a)}$ at our point of M^a_n. □

Using this proposition, we obtain that in order to prove Theorem 5.12 it would be sufficient to show that any S-point of the stack $(\mathsf{g}^{\mathfrak{U}})^{-1}(\sigma^{\mathfrak{U}})$ factors through an S-point of $\mathrm{Perv}^a_n(\mathbf{S}, \mathbf{S} - 0_{\mathbf{S}})$.

5.17

Let $\mathcal{M}$ be an arbitrary S-family of coherent perverse sheaves on $\mathbf{S}'$ corresponding to an S-point of $\mathrm{Perv}^a_n(\mathbf{S}, \mathbf{D}_\infty)$. Let $(\mathbf{d}_v, \mathbf{d}_h)$ be a fixed configuration, and let us denote by $D_h \subset \overset{\circ}{\mathbf{C}} \times S$, $D_v \subset \overset{\circ}{\mathbf{X}} \times S$ the corresponding divisors.

First, by unraveling the definition of the map

$$\mathrm{Perv}^a_n(\mathbf{S}, \mathbf{D}_\infty) \to \mathrm{QMaps}(\mathbf{C}, \mathcal{G}_{SL_n,\mathbf{X}}) \to \overset{\circ}{\mathbf{C}}{}^{(a)},$$

we obtain that $\mathcal{M}$ is a *vector bundle* away from the divisors $D_h \times \mathbf{X}$ and $\mathbf{C} \times D_v$. (In other words, a quasi-map necessarily acquires a defect at some point of $\mathbf{C}$ if $\mathcal{M}$ has a singularity on the vertical divisor over this point.)

Moreover, when we view $\mathcal{M}|_{\mathbf{S}' \times S - \mathbf{C} \times D_v}$ as an $(\mathbf{X} \times S - D_v)$-family of bundles on $\mathbf{C}$, this family is canonically trivialized.

Consider now the restriction of $\mathcal{M}$ to the complement of $D_h \times \mathbf{X} \cup \mathbf{C} \times D_v \subset \mathbf{S} \times S$. This is a vector bundle equipped with two trivializations. One comes from the fact that we are dealing with a *map* $\mathbf{C} \times S - D_h \to \mathrm{Gr}^{BD,a}_{SL_n,\mathbf{X},D_v}$, and the other one comes from the trivialization of $\mathcal{M}|_{\mathbf{S}' \times S - \mathbf{C} \times D_v}$ mentioned above.

We claim that these two trivializations actually coincide. This is so because the corresponding fact is true over the dense substack $\mathrm{Bun}^a_n(\mathbf{S}, \mathbf{D}_\infty) \subset \mathrm{Perv}^a_n(\mathbf{S}, \mathbf{D}_\infty)$.

Going back to the proof of the theorem, assume that $\mathcal{M}$ corresponds to an S-point of $(\mathsf{g}^{\mathfrak{U}})^{-1}(\sigma^{\mathfrak{U}})$. Then, first of all, the divisors D_h, D_v are $(a \cdot 0_{\mathbf{C}}) \times S$ and $(a \cdot 0_{\mathbf{X}}) \times S$, respectively.

Moreover, $\mathcal{M}$ is a vector bundle away from $0_{\mathbf{S}} \times S$, and it is trivialized away from $\mathbf{C} \times 0_{\mathbf{X}} \times S \subset \mathbf{S}' \times S$. Therefore, we only have to show that this trivialization extends across the divisor $\mathbf{C} \times 0_{\mathbf{X}} \times S$ over $(\mathbf{C} - 0_{\mathbf{C}}) \times S$.

However, by the definition of $\sigma^{\mathfrak{U}}$, the map $(\mathbf{C} - 0_{\mathbf{C}}) \times S \to \mathrm{Gr}_{SL_n,\mathbf{X}} \to \mathcal{G}_{SL_n}$ is the constant map corresponding to the trivial bundle. Therefore, the trivialization does extend.

6 Properties of Uhlenbeck spaces

6.1

Let $\phi : G_1 \to G_2$ be a homomorphism of simple simply connected groups, and let $\phi_{\mathbb{Z}}$ denote the corresponding homomorphism $\phi_{\mathbb{Z}} : \mathbb{Z} \simeq H_3(G_1, \mathbb{Z}) \to H_3(G_2, \mathbb{Z}) \simeq \mathbb{Z}$. Observe that for a curve $\mathbf{X}$, the pullback of the canonical line bundle $\mathcal{P}_{\mathrm{Bun}_{G_2}(\mathbf{X})}$ under the induced map $\mathrm{Bun}_{G_1}(\mathbf{X}) \to \mathrm{Bun}_{G_2}(\mathbf{X})$ is $\mathcal{P}^{\otimes \phi_{\mathbb{Z}}(1)}_{\mathrm{Bun}_{G_1}(\mathbf{X})}$. In particular, the map $\mathrm{Bun}_{G_1}(\mathbf{S}, \mathbf{D}_\infty) \to \mathrm{Bun}_{G_2}(\mathbf{S}, \mathbf{D}_\infty)$ sends $\mathrm{Bun}^a_{G_1}(\mathbf{S}, \mathbf{D}_\infty)$ to $\mathrm{Bun}^{\phi_{\mathbb{Z}}(a)}_{G_2}(\mathbf{S}, \mathbf{D}_\infty)$.

Lemma 6.2. *The map* $\mathrm{Bun}^a_{G_1}(\mathbf{S}, \mathbf{D}_\infty) \to \mathrm{Bun}^{\phi_{\mathbb{Z}}(a)}_{G_2}(\mathbf{S}, \mathbf{D}_\infty)$ *extends to morphisms* $\mathfrak{U}^a_{G_1} \to \mathfrak{U}^{\phi_{\mathbb{Z}}(a)}_{G_2}$ *and* $''\mathfrak{U}^a_{G_1} \to {}''\mathfrak{U}^{\phi_{\mathbb{Z}}(a)}_{G_2}$. *When* ϕ *is injective, both these maps are closed embeddings.*

Proof. First, for a curve $\mathbf{X}$, from $\phi_{\mathbb{Z}}(a)$ we obtain a map $\mathbf{X}^{(a)} \to \mathbf{X}^{(\phi_{\mathbb{Z}}(a))}$, and the corresponding map $\mathrm{Gr}^{BD,a}_{G_1,\mathbf{X}} \to \mathrm{Gr}^{BD,\phi_{\mathbb{Z}}(a)}_{G_2,\mathbf{X}}$. The latter morphism is a closed embedding if ϕ is.

Hence according to Proposition 1.12(c), for a curve $\mathbf{C}$ with a marked point $\mathbf{c} \in \mathbf{C}$ we obtain a morphism of the corresponding based quasi-map spaces:

$$\phi_{\mathrm{QMaps}} : \mathrm{QMaps}^a(\mathbf{C}, \mathrm{Gr}^{BD,a}_{G_1,\mathbf{X}}) \to \mathrm{QMaps}^{\phi_{\mathbb{Z}}(a)}(\mathbf{C}, \mathrm{Gr}^{BD,\phi_{\mathbb{Z}}(a)}_{G_2,\mathbf{X}}).$$

The existence of the maps $\mathfrak{U}^a_{G_1} \to \mathfrak{U}^{\phi_{\mathbb{Z}}(a)}_{G_2}$ and $''\mathfrak{U}^a_{G_1} \to {}''\mathfrak{U}^{\phi_{\mathbb{Z}}(a)}_{G_2}$ now follows from the definition of $\mathfrak{U}^a_G$ and $''\mathfrak{U}^a_G$. □

6.3 Comparison of the definitions: The general case

The above lemma allows us to establish the equivalence of the two definitions of $\mathfrak{U}^a_G$.

Proof of Theorem 4.8. Let us choose a faithful representation $\phi : G \to SL_n$. By the previous lemma, $\mathfrak{U}^a_G$ and $''\mathfrak{U}^a_G$ are isomorphic to the closures of $\mathrm{Bun}^a_G(\mathbf{S}, \mathbf{D}_\infty)$ in $\mathfrak{U}^{\phi_{\mathbb{Z}}(a)}_n(\mathbf{S}, \mathbf{D}_\infty)$ and $''\mathfrak{U}^{\phi_{\mathbb{Z}}(a)}_n(\mathbf{S}, \mathbf{D}_\infty)$, respectively.

However, by Theorem 5.12, the map $\mathfrak{U}^{\phi_{\mathbb{Z}}(a)}_n(\mathbf{S}, \mathbf{D}_\infty) \to {}''\mathfrak{U}^{\phi_{\mathbb{Z}}(a)}_n(\mathbf{S}, \mathbf{D}_\infty)$ is an isomorphism. Hence $\mathfrak{U}^a_G \to {}''\mathfrak{U}^a_G$ is an isomorphism as well. □

6.4 Factorization property

Next, we will establish the factorization property of Uhlenbeck compactifications. Let us fix a pair of nonparallel lines $(\mathbf{d}_v, \mathbf{d}_h)$, and consider the corresponding projection $\varpi^a_h : \mathfrak{U}^a_G \to \overset{\circ}{\mathbf{C}}{}^{(a)}$.

Proposition 6.5. *For $a = a_1 + a_2$, there is a natural isomorphism*

$$(\overset{\circ}{\mathbf{C}}{}^{(a_1)} \times \overset{\circ}{\mathbf{C}}{}^{(a_2)})_{\mathrm{disj}} \underset{\overset{\circ}{\mathbf{C}}{}^{(a)}}{\times} \mathfrak{U}^a_G \simeq (\overset{\circ}{\mathbf{C}}{}^{(a_1)} \times \overset{\circ}{\mathbf{C}}{}^{(a_2)})_{\mathrm{disj}} \underset{\overset{\circ}{\mathbf{C}}{}^{(a_1)} \times \overset{\circ}{\mathbf{C}}{}^{(a_2)}}{\times} (\mathfrak{U}^{a_1}_G \times \mathfrak{U}^{a_2}_G).$$

To prove this proposition, we will first consider the case of SL_n.

Proposition 6.6. *For $a = a_1 + a_2$, there are natural isomorphisms*

$$(\overset{\circ}{\mathbf{C}}{}^{(a_1} \times \overset{\circ}{\mathbf{C}}{}^{(a_2)})_{\mathrm{disj}} \underset{\overset{\circ}{\mathbf{C}}{}^{(a)}}{\times} \widetilde{\mathfrak{N}}^a_n \simeq (\overset{\circ}{\mathbf{C}}{}^{(a_1)} \times \overset{\circ}{\mathbf{C}}{}^{(a_2)})_{\mathrm{disj}} \underset{\overset{\circ}{\mathbf{C}}{}^{(a_1)} \times \overset{\circ}{\mathbf{C}}{}^{(a_2)}}{\times} (\widetilde{\mathfrak{N}}^{a_1}_n \times \widetilde{\mathfrak{N}}^{a_2}_n),$$

$$(\overset{\circ}{\mathbf{C}}{}^{(a_1} \times \overset{\circ}{\mathbf{C}}{}^{(a_2)})_{\mathrm{disj}} \underset{\overset{\circ}{\mathbf{C}}{}^{(a)}}{\times} \mathfrak{N}^a_n \simeq (\overset{\circ}{\mathbf{C}}{}^{(a_1)} \times \overset{\circ}{\mathbf{C}}{}^{(a_2)})_{\mathrm{disj}} \underset{\overset{\circ}{\mathbf{C}}{}^{(a_1)} \times \overset{\circ}{\mathbf{C}}{}^{(a_2)}}{\times} (\mathfrak{N}^{a_1}_n \times \mathfrak{N}^{a_2}_n).$$

Proof. The second factorization isomorphism follows from the first one since $\mathfrak{N}^a_n$ is the affinization of $\widetilde{\mathfrak{N}}^a_n$.

Let $\mathsf{Coh}_n(\mathbf{X}, \infty_{\mathbf{X}})$ be the stack of coherent sheaves on $\mathbf{X}$ with a trivialization at $\infty_{\mathbf{X}}$. This is a smooth stack which contains an open subset isomorphic to the point scheme, that corresponds to the trivial rank-n vector bundle. The complement of this open subset is of codimension 1 and hence is a Cartier divisor.

Observe now that the scheme $\widetilde{\mathfrak{N}}^a_n$ represents the functor of maps, $\mathrm{Maps}^a(\overset{\circ}{\mathbf{C}}, \mathsf{Coh}_n(\mathbf{X}, \infty_{\mathbf{X}}))$, where the latter is as in Section 2.16.

Indeed, an S-point of $\mathrm{Maps}^a(\overset{\circ}{\mathbf{C}}, \mathsf{Coh}_n(\mathbf{X}, \infty_{\mathbf{X}}))$, is according to Lemma 2.14 the same as coherent sheaf $\mathcal{M}$ on $\mathbf{X} \times \mathbf{C} \times S$, trivialized over $\infty_{\mathbf{X}} \times \mathbf{C} \times S$ and $\mathbf{X} \times \infty_{\mathbf{C}} \times S$, and which is $\mathbf{C} \times S$-flat. But this implies that for every geometric point $s \in S$, the restriction $\mathcal{M}|_{\mathbf{S} \times s}$ is torsion-free.

Hence the assertion about the factorization of $\widetilde{\mathfrak{N}}^a_n$ follows from Proposition 2.17. □

Now we can prove Proposition 6.5.

Proof. Let us choose a faithful representation $\phi : G \to SL_n$ and consider the corresponding closed embedding

$$\mathfrak{U}_G^a \to \mathrm{QMaps}^a(\mathbf{C}, \mathcal{G}_{G,\mathbf{X}}) \times \mathfrak{U}_n^{\phi_{\mathbb{Z}}(a)}.$$

The image of this map lies in the closed subscheme

$$\mathrm{QMaps}^a(\mathbf{C}, \mathcal{G}_{G,\mathbf{X}}) \underset{\overset{\circ}{\mathbf{C}}^{(\phi_{\mathbb{Z}}(a))}}{\times} \mathfrak{U}_n^{\phi_{\mathbb{Z}}(a)} \simeq \mathrm{QMaps}^a(\mathbf{C}, \mathcal{G}_{G,\mathbf{X}}) \underset{\overset{\circ}{\mathbf{C}}^{(a)}}{\times} (\overset{\circ}{\mathbf{C}}^{(a)} \underset{\overset{\circ}{\mathbf{C}}^{(\phi_{\mathbb{Z}}(a))}}{\times} \mathfrak{U}_n^{\phi_{\mathbb{Z}}(a)}).$$

Now, $\mathrm{QMaps}^a(\mathbf{C}, \mathcal{G}_{G,\mathbf{X}})$ factorizes over $\overset{\circ}{\mathbf{C}}^{(a)}$ according to Proposition 2.19, and the fiber product $\overset{\circ}{\mathbf{C}}^{(a)} \underset{\overset{\circ}{\mathbf{C}}^{(\phi_{\mathbb{Z}}(a))}}{\times} \mathfrak{U}_n^{\phi_{\mathbb{Z}}(a)}$ factorizes over $\overset{\circ}{\mathbf{C}}^{(a)}$ because $\mathfrak{U}_n^{\phi_{\mathbb{Z}}(a)}$ does so over $\overset{\circ}{\mathbf{C}}^{(\phi_{\mathbb{Z}}(a))}$. This implies the proposition in view of the isomorphism (10). □

6.7

Let us fix a point $\mathbf{d}_v \in \mathbf{D}_\infty$ and observe that the curve $\mathbf{C}$ is well defined without the additional choice of $\mathbf{d}_h \in \mathbf{D}_\infty - \mathbf{d}_v$. Indeed, $\mathbf{C}$ is canonically identified with the projectivization of the tangent space $T\mathbf{S}_{\mathbf{d}_v}$. Thus the map $\varpi^a_{h,\mathrm{O}}$ gives rise to a map

$$\varpi^a_{h,\mathbf{d}_v} : (\mathbf{D}_\infty - \mathbf{d}_v) \times \mathfrak{U}_G^a \to \overset{\circ}{\mathbf{C}}^{(a)}.$$

Proposition 6.8. *The above map $\varpi^a_{h,\mathbf{d}_v}$ is independent of the variable $\mathbf{D}_\infty - \mathbf{d}_v$. Moreover, the corresponding factorization isomorphisms of Proposition* 6.5 *are also independent of the choice of $\mathbf{d}_h$.*

Proof. Since $\mathrm{Bun}_G^a(\mathbf{S}, \mathbf{D}_\infty)$ is dense in $\mathfrak{U}_G^a$, it is enough to show that the corresponding map $(\mathbf{D}_\infty - \mathbf{d}_v) \times \mathrm{Bun}_G^a(\mathbf{S}, \mathbf{D}_\infty) \to \overset{\circ}{\mathbf{C}}^{(a)}$ is independent of the first $\mathbf{d}_h$, and similarly for the factorization isomorphisms.

Consider the surface $\mathbf{S}^{\mathbf{d}_v}$ obtained by blowing up $\mathbf{S}$ at $\mathbf{d}_v$. Then the exceptional divisor identifies canonically with $\mathbf{C}$, and we have a projection $\pi_v^{\mathbf{d}_v} : \mathbf{S}^{\mathbf{d}_v} \to \mathbf{C}$. We can consider the corresponding stack $\mathrm{Bun}_G^a(\mathbf{S}_{\mathbf{d}_v}, \mathbf{D}_\infty^{\mathbf{d}_v})$, which classifies G-bundles with a trivialization along $\mathbf{D}_\infty^{\mathbf{d}_v} := (\mathbf{C} \cup (\pi_v^{\mathbf{d}_v})^{-1}(\infty_{\mathbf{C}})) \subset \mathbf{S}^{\mathbf{d}_v}$.

In addition, $\mathbf{S}^{\mathbf{d}_v}$ can be regarded as a relative curve $\mathbf{X}_\mathbf{C}$ with a marked infinity over $\mathbf{C}$, and we can consider the relative stack $\mathrm{Bun}_G(\mathbf{X}_\mathbf{C}, \infty_{\mathbf{X}_\mathbf{C}})$. This stack contains an open substack (corresponding to the trivial bundle) and its complement is a Cartier divisor.

The corresponding space of sections $\mathbf{C} \to \mathrm{Bun}_G(\mathbf{X}_\mathbf{C}, \infty_{\mathbf{X}_\mathbf{C}})$ identifies with $\mathrm{Bun}_G^a(\mathbf{S}_{\mathbf{d}_v}, \mathbf{D}_\infty^{\mathbf{d}_v})$, and thus gives rise to a map

$$\mathrm{Bun}_G^a(\mathbf{S}_{\mathbf{d}_v}, \mathbf{D}_\infty^{\mathbf{d}_v}) \to \overset{\circ}{\mathbf{C}}^{(a)},$$

and the corresponding factorization isomorphisms; cf. Proposition 2.17. This makes the assertion of the proposition manifest. □

6.9

Let us now fix a point $\mathbf{d}_h \in \mathbf{D}_\infty$, and consider the family of curves $\mathbf{C}_{\mathbf{D}_\infty - \mathbf{d}_h}$ corresponding to the moving point $\mathbf{d}_v \in \mathbf{D}_\infty - \mathbf{d}_h$.

We have a natural forgetful map

$$\mathrm{Sect}(\mathsf{O}, \overset{\circ}{\mathbf{C}}{}^{(a)}_{\mathsf{O}}) \to \mathrm{Sect}(\mathbf{D}_\infty - \mathbf{d}_h, \overset{\circ}{\mathbf{C}}{}^{(a)}_{\mathbf{D}_\infty - \mathbf{d}_h}),$$

and by composing with $\varpi^a_{h,\mathsf{O}}$ we obtain a map

$$\mathfrak{U}^a_G \to \mathrm{Sect}(\mathbf{D}_\infty - \mathbf{d}_h, \overset{\circ}{\mathbf{C}}{}^{(a)}_{\mathbf{D}_\infty - \mathbf{d}_h}).$$

From Proposition 6.8, we obtain the following corollary.

Corollary 6.10. *The map*

$$\mathfrak{U}^a_G \to \mathrm{QMaps}^a(\mathbf{C}, \mathcal{G}_{G,\mathbf{X}}) \times \mathrm{Sect}(\mathbf{D}_\infty - \mathbf{d}_h, \overset{\circ}{\mathbf{C}}{}^{(a)}_{\mathbf{D}_\infty - \mathbf{d}_h})$$

is a closed embedding, where $\mathbf{C}$ *corresponds to some fixed point of* $\mathbf{D}_\infty - \mathbf{d}_h$.

Proof. We know that $\mathfrak{U}^a_G \to \mathrm{Maps}^a(\mathbf{C}, \mathcal{G}_{G,\mathbf{X}}) \times \mathrm{Sect}(\mathsf{O}, \overset{\circ}{\mathbf{C}}{}^{(a)}_{\mathsf{O}})$ is a closed embedding, and if we consider the open subset $\overset{\circ}{\mathsf{O}} := \mathsf{O} - \mathbf{d}_h \times (\mathbf{D}_\infty - \mathbf{d}_h)$, the map

$$\mathfrak{U}^a_G \to \mathrm{Maps}^a(\mathbf{C}, \mathcal{G}_{G,\mathbf{X}}) \times \mathrm{Sect}(\overset{\circ}{\mathsf{O}}, \overset{\circ}{\mathbf{C}}{}^{(a)}_{\overset{\circ}{\mathsf{O}}})$$

would be a closed embedding as well. However, from Proposition 6.8, we obtain that the map $\mathfrak{U}^a_G \to \mathrm{Sect}(\overset{\circ}{\mathsf{O}}, \overset{\circ}{\mathbf{C}}{}^{(a)}_{\overset{\circ}{\mathsf{O}}})$ factors as

$$\mathfrak{U}^a_G \to \mathrm{Sect}(\mathbf{D}_\infty - \mathbf{d}_h, \overset{\circ}{\mathbf{C}}{}^{(a)}_{\mathbf{D}_\infty - \mathbf{d}_h}) \to \mathrm{Sect}(\overset{\circ}{\mathsf{O}}, \overset{\circ}{\mathbf{C}}{}^{(a)}_{\overset{\circ}{\mathsf{O}}}),$$

where the last arrow comes from the projection to the first factor $\overset{\circ}{\mathsf{O}} \to (\mathbf{D}_\infty - \mathbf{d}_h)$.

This establishes the proposition. □

7 Stratifications and IC stalks

7.1

In this section we will introduce a stratification of $\mathfrak{U}^a_G$ and formulate a theorem describing the intersection cohomology sheaf $\mathrm{IC}_{\mathfrak{U}^a_G}$.

Let b be an integer $0 \le b \le a$, and set by definition $\mathfrak{U}^{a;b}_G := \mathrm{Bun}^{a-b}_G(\mathbf{S}, \mathbf{D}_\infty) \times \mathrm{Sym}^b(\overset{\circ}{\mathbf{S}})$.

Theorem 7.2. *There exists a canonical locally closed embedding* $\iota_b : \mathfrak{U}^{a;b}_G \to \mathfrak{U}^a_G$. *Moreover,* $\mathfrak{U}^a_G = \bigcup_b \mathfrak{U}^{a;b}_G$.

7.3 Proof of Theorem 7.2

Let $(\mathcal{F}'_G, D_{\mathbf{S}})$ be a point of $\mathrm{Bun}_G^{a-b}(\mathbf{S}, \mathbf{D}_\infty) \times \mathrm{Sym}^b(\overset{\circ}{\mathbf{S}})$. For every pair of directions $(\mathbf{d}_v, \mathbf{d}_h)$, the data of $\mathcal{F}'_G$ defines a based *map* $\sigma' : \mathbf{C} \to \mathrm{Gr}_{G,\mathbf{X}}^{BD,a-b}$ of degree $a - b$.

Consider the embedding $\mathrm{Gr}_{G,\mathbf{X}}^{BD,a-b} \hookrightarrow \mathrm{Gr}_{G,\mathbf{X}}^{BD,a}$ obtained by adding to the divisor $D_v \in \overset{\circ}{\mathbf{X}}{}^{(a-b)}$ the divisor $\pi_h(D_{\mathbf{S}})$. We define the sought-for quasi-map $\sigma : \mathbf{C} \to \mathrm{Gr}_{G,\mathbf{X}}^{BD,a}$ by adding to the map

$$\mathbf{C} \xrightarrow{\sigma'} \mathrm{Gr}_{G,\mathbf{X}}^{BD,a-b} \to \mathrm{Gr}_{G,\mathbf{X}}^{BD,a} \tag{15}$$

the defect equal to $\pi_v(D_{\mathbf{S}})$; cf. Sections 1.16 and 2.3, (3). Since this can be done for any pair of directions $(\mathbf{d}_v, \mathbf{d}_h)$, we obtain a point of $\mathrm{Sect}(\mathsf{O}, \mathrm{QMaps}(\mathbf{C}, \mathrm{Gr}_{G,\mathbf{X}}^{BD,a})_{\mathsf{O}})$. This morphism is a locally closed embedding due to the corresponding property of the quasi-map spaces.

Note that the above construction defines in fact a map $\bar{\iota}_b : \mathfrak{U}_G^{a-b} \times \mathrm{Sym}^b(\overset{\circ}{\mathbf{S}}) \to \mathrm{Sect}(\mathsf{O}, \mathrm{QMaps}(\mathbf{C}, \mathrm{Gr}_{G,\mathbf{X}}^{BD,a})_{\mathsf{O}})$.

Now our goal is to prove that the image of ι_b belongs to $\mathfrak{U}_G^a$, and that every geometric point of $\mathfrak{U}_G^a$ belongs to (exactly) one of the subschemes $\mathfrak{U}_G^{a;b}$. First, we shall do this for $G = SL_n$.

Let $\widetilde{\mathfrak{N}}_n^{a;b}$ be the scheme classifying pairs $\mathcal{M} \subset \mathcal{M}'$, where $\mathcal{M} \in \mathfrak{N}_n^a$, and $\mathcal{M}'$ is a vector bundle on $\mathbf{S}$, such that $\mathcal{M}'/\mathcal{M}$ is a torsion sheaf of length b. We have a natural proper and surjective map

$$\widetilde{\mathfrak{N}}_n^{a;b} \to \mathrm{Bun}_n^{a-b}(\mathbf{S}, \mathbf{D}_\infty) \times \mathrm{Sym}^b(\overset{\circ}{\mathbf{S}}),$$

and a locally closed embedding $\widetilde{\mathfrak{N}}_n^{a;b} \hookrightarrow \widetilde{\mathfrak{N}}_n^a$. Moreover, it is easy to see that the square

$$\begin{array}{ccc} \widetilde{\mathfrak{N}}_n^{a;b} & \longrightarrow & \widetilde{\mathfrak{N}}_n^a \\ \downarrow & & \downarrow \\ \mathrm{Bun}_n^{a-b}(\mathbf{S}, \mathbf{D}_\infty) \times \mathrm{Sym}^b(\overset{\circ}{\mathbf{S}}) & \xrightarrow{\iota_b} & \mathrm{Sect}(\mathsf{O}, \mathrm{QMaps}(\mathbf{C}, \mathrm{Gr}_{SL_n,\mathbf{X}}^{BD,a})_{\mathsf{O}}) \end{array} \tag{16}$$

is commutative, where in the above diagram the right vertical arrow is the composition of $\widetilde{\mathfrak{N}}_n^a \to \mathfrak{N}_n^a \simeq \mathfrak{U}_n^a$ and $\mathfrak{U}_n^a \hookrightarrow \mathrm{Sect}(\mathsf{O}, \mathrm{QMaps}(\mathbf{C}, \mathrm{Gr}_{SL_n,\mathbf{X}}^{BD,a})_{\mathsf{O}})$.

This readily shows that the image of ι_b belongs to $\mathfrak{U}_G^a$ for $G = SL_n$. Moreover, since $\widetilde{\mathfrak{N}}_n^a \simeq \bigcup_b \widetilde{\mathfrak{N}}_n^{a;b}$, we obtain that $\mathfrak{U}_n^a \simeq \bigcup_b \mathfrak{U}_n^{a;b}$.

Now let us treat the case of an arbitrary G. To show that the image of ι_b belongs to $\mathfrak{U}_G^a$ consider the open subset in the product $\mathrm{Bun}_G^{a-b}(\mathbf{S}, \mathbf{D}_\infty) \times \mathrm{Sym}^b(\overset{\circ}{\mathbf{S}})$ corresponding to pairs $(\mathcal{F}_G, D_{\mathbf{S}})$ such that $\pi_v(D_{\mathbf{S}})$ is a multiplicity-free divisor, disjoint from $\varpi_h^{a-b}(\mathcal{F}_G)$. It would be enough to show that the image of this open subset in $\mathrm{Sect}(\mathsf{O}, \mathrm{QMaps}(\mathbf{C}, \mathrm{Gr}_{G,\mathbf{X}}^{BD,a})_{\mathsf{O}})$ is contained in $\mathfrak{U}_G^a$.

Note that from equation (16) we obtain the following compatibility relation of the map ι_b with the factorization isomorphisms of Proposition 6.5 for SL_n. For $b = b_1 + b_2$ the map

$$(\mathrm{Bun}_n^{a-b}(\mathbf{S}, \mathbf{D}_\infty) \times \mathrm{Sym}^b(\overset{\circ}{\mathbf{S}})) \underset{\overset{\circ}{\mathbf{C}}{}^{(a)}}{\times} (\overset{\circ}{\mathbf{C}}{}^{(a-b)} \times \overset{\circ}{\mathbf{C}}{}^{(b_1)} \times \overset{\circ}{\mathbf{C}}{}^{(b_2)})_{\mathrm{disj}}$$

$$\Big\downarrow \iota_b$$

$$\mathfrak{U}_n^a \underset{\overset{\circ}{\mathbf{C}}{}^{(a)}}{\times} (\overset{\circ}{\mathbf{C}}{}^{(a-b)} \times \overset{\circ}{\mathbf{C}}{}^{(b_1)} \times \overset{\circ}{\mathbf{C}}{}^{(b_2)})_{\mathrm{disj}}$$

coincides with the composition

$$(\mathrm{Bun}_n^{a-b}(\mathbf{S}, \mathbf{D}_\infty) \times \mathrm{Sym}^b(\overset{\circ}{\mathbf{S}})) \underset{\overset{\circ}{\mathbf{C}}{}^{(a)}}{\times} (\overset{\circ}{\mathbf{C}}{}^{(a-b)} \times \overset{\circ}{\mathbf{C}}{}^{(b_1)} \times \overset{\circ}{\mathbf{C}}{}^{(b_2)})_{\mathrm{disj}}$$

$$\Big\downarrow \sim$$

$$(\mathrm{Bun}_n^{a-b}(\mathbf{S}, \mathbf{D}_\infty) \times \mathrm{Sym}^{b_1}(\overset{\circ}{\mathbf{S}}) \times \mathrm{Sym}^{b_2}(\overset{\circ}{\mathbf{S}})) \underset{\overset{\circ}{\mathbf{C}}{}^{(a-b)} \times \overset{\circ}{\mathbf{C}}{}^{(b_1)} \times \overset{\circ}{\mathbf{C}}{}^{(b_2)}}{\times} (\overset{\circ}{\mathbf{C}}{}^{(a-b)} \times \overset{\circ}{\mathbf{C}}{}^{(b_1)} \times \overset{\circ}{\mathbf{C}}{}^{(b_2)})_{\mathrm{disj}}$$

$$\Big\downarrow \mathrm{id} \times \iota_{b_1} \times \iota_{b_2}$$

$$(\mathfrak{U}_n^{a-b} \times \mathfrak{U}_n^{b_1} \times \mathfrak{U}_n^{b_2}) \underset{\overset{\circ}{\mathbf{C}}{}^{(a-b)} \times \overset{\circ}{\mathbf{C}}{}^{(b_1)} \times \overset{\circ}{\mathbf{C}}{}^{(b_2)}}{\times} (\overset{\circ}{\mathbf{C}}{}^{(a-b)} \times \overset{\circ}{\mathbf{C}}{}^{(b_1)} \times \overset{\circ}{\mathbf{C}}{}^{(b_2)})_{\mathrm{disj}}$$

$$\Big\downarrow \sim$$

$$\mathfrak{U}_n^a \underset{\overset{\circ}{\mathbf{C}}{}^{(a)}}{\times} (\overset{\circ}{\mathbf{C}}{}^{(a-b)} \times \overset{\circ}{\mathbf{C}}{}^{(b_1)} \times \overset{\circ}{\mathbf{C}}{}^{(b_2)})_{\mathrm{disj}}.$$

Hence by embedding G into SL_n and using Lemma 6.2, we reduce our assertion to the case when $b = a = 1$. Using the action of the group of affine-linear transformations, we reduce the assertion further to the fact that the image of the point-scheme, thought of as $\mathrm{Bun}_G^0(\mathbf{S}, \mathbf{D}_\infty) \times 0_{\mathbf{S}}$, belongs to $\mathfrak{U}_G^1$.

However, as in the proof of Theorem 5.12, the above point-scheme is the only attractor of the $\mathbb{G}_m$-action on $\mathrm{Sect}(\mathsf{O}, \mathrm{QMaps}(\mathbf{C}, \mathcal{G}_{G,\mathbf{X}})_{\mathsf{O}})$. In particular, it is contained in the closure of $\mathrm{Bun}_G^1(\mathbf{S}, \mathbf{D}_\infty)$, which is $\mathfrak{U}_G^1$.

To finish the proof of the theorem, we have to show that every geometric point of $\mathfrak{U}_G^a$ belongs to (exactly) one of the subschemes $\mathfrak{U}_G^{a;b}$. That these subschemes are mutually disjoint is clear from the fact that b can be recovered as a total defect of a quasi-map.

Let σ_{O} be a point of $\mathfrak{U}_G^a$, thought of as a collection of quasi-maps $\sigma : \mathbf{C} \to \mathrm{Gr}_{G,\mathbf{X}}^{BD,a}$ for every pair of directions $(\mathbf{d}_v, \mathbf{d}_h)$. We have to show that there exists a point $\sigma'_{\mathsf{O}} \in \mathrm{Bun}_G^{a-b}(\mathbf{S}, \mathbf{D}_\infty)$, and a 0-cycle $D_{\mathbf{S}} = \Sigma b_i \cdot \mathbf{s}_i$ on $\mathbf{S}$, such that for every pair of directions $(\mathbf{d}_v, \mathbf{d}_h)$, the corresponding quasi-map $\sigma : \mathbf{C} \to \mathrm{Gr}_{G,\mathbf{X}}^{BD,a}$ has as its saturation the composition

$$\mathbf{C} \xrightarrow{\sigma'} \mathrm{Gr}_{G,\mathbf{X}}^{BD,a-b} \hookrightarrow \mathrm{Gr}_{G,\mathbf{X}}^{BD,a},$$

(the last arrow is as in (15)), with defect $\pi_v(D_{\mathbf{S}})$. (In what follows we will say that the point $\sigma_{\mathbf{O}} \in \mathfrak{U}^a_G$ has saturation equal to the G-bundle $\mathcal{F}'_G$ corresponding to $\sigma'_{\mathbf{O}}$, and defect given by the 0-cycle $D_{\mathbf{S}}$.)

For a faithful representation $\phi : G \to SL_n$, let $\sigma_{\mathbf{O},n}$ be the corresponding point of $\mathfrak{U}_n^{\phi_{\mathbb{Z}}(a)}$. We know that $\sigma_{\mathbf{O},n}$ can be described as a pair $(\sigma'_{\mathbf{O},n}, D_{\mathbf{S},n})$, where $\sigma'_{\mathbf{O},n}$ is the saturation of $\sigma_{\mathbf{O},n}$, i.e., a point of $\mathrm{Bun}_n^{\phi_{\mathbb{Z}}(a-b)}(\mathbf{S}, \mathbf{D}_\infty)$, and $D_{\mathbf{S},n} = \Sigma b_{i,n} \cdot \mathbf{s}_i$ is a 0-cycle on $\mathbf{S}$.

For a fixed pair of directions $(\mathbf{d}_v, \mathbf{d}_h)$, let σ'' be the saturation of σ. In particular, σ'' defines a G-bundle on $\mathbf{S}$, with a trivialization along $\mathbf{D}_\infty$, such that the induced SL_n-bundle is isomorphic to $\sigma'_{\mathbf{O},n}$. Since the morphism $\mathrm{Bun}_G(\mathbf{S}, \mathbf{D}_\infty) \to \mathrm{Bun}_n(\mathbf{S}, \mathbf{D}_\infty)$ is an embedding, we obtain that this G-bundle on $\mathbf{S}$ is well defined, i.e., is independent of $(\mathbf{d}_v, \mathbf{d}_h)$. We set $\sigma'_{\mathbf{O}}$ to be the corresponding point of $\mathrm{Bun}_G(\mathbf{S}, \mathbf{D}_\infty)$.

It remains to show that the integers $b_{i,n}$ are divisible by $\phi_{\mathbb{Z}}(1)$. For that we choose a pair of directions $(\mathbf{d}_v, \mathbf{d}_h)$ so that the points $\pi_v(\mathbf{s}_i)$ are pairwise distinct. Then the corresponding b_is are reconstructed as the defect of the quasi-map σ.

This completes the proof of the theorem.

7.4

Note that in the course of the proof of Theorem 7.2 we have established that the morphisms $\bar{\iota} : \mathfrak{U}^{a-b}_G \times \mathrm{Sym}(\overset{\circ}{\mathbf{S}}) \to \mathfrak{U}^a_G$ are compatible in a natural sense with the factorization isomorphisms of Proposition 6.5. This is so because the corresponding property is true for SL_n.

As a corollary of Theorem 7.2, we obtain, in particular, the following.

Corollary 7.5. *The variety* $\mathrm{Bun}^a_G(\mathbf{S}, \mathbf{D}_\infty)$ *is quasi-affine, and its affine closure is isomorphic to the normalization of* $\mathfrak{U}^a_G$.

Proof. Indeed, we know that $\dim \mathrm{Bun}^a_G(\mathbf{S}, \mathbf{D}_\infty) = 2\check{h}a$ and from Theorem 7.2, we obtain that the complement to $\mathrm{Bun}^a_G(\mathbf{S}, \mathbf{D}_\infty)$ inside the normalization of $\mathfrak{U}^a_G$ is of dimension $2\check{h}(a-1)+2$. In other words, it is of codimension $2(\check{h}-1) \geq 2$. □

We do not know whether $\mathfrak{U}^a_G$ is in general a normal variety, but this is true for $G = SL_n$ since $\mathfrak{U}^a_n \simeq \mathfrak{M}^a_n$; cf. [CB].

7.6

For a fixed pair of directions $(\mathbf{d}_v, \mathbf{d}_h)$ and a point $0_{\mathbf{X}} \in \mathbf{X}$, consider the $\mathbb{G}_m$-action on $\mathbf{S}$ which acts as the identity on $\mathbf{C}$ and by dilations on $\mathbf{X}$, fixing $0_{\mathbf{X}}$. By transport of structure, we obtain a $\mathbb{G}_m$-action on $\mathfrak{U}^a_G$.

Proposition 7.7. *The above* $\mathbb{G}_m$*-action on* $\mathfrak{U}^a_G$ *contracts this space to*

$$(\overset{\circ}{\mathbf{C}} \times 0_{\mathbf{X}})^{(a)} \subset \overset{\circ}{\mathbf{S}}{}^{(a)} \simeq \mathfrak{U}^{a;a}_G \hookrightarrow \mathfrak{U}^a_G.$$

The contraction map $\mathfrak{U}^a_G \to \overset{\circ}{\mathbf{C}}{}^{(a)}$ *coincides with* ϖ^a_h.

Proof. We will use the description of $\mathfrak{U}_G^a$ via Corollary 6.10, i.e., we realize it as a closed subset in

$$\mathrm{QMaps}(\mathbf{C}, \mathcal{G}_{G,\mathbf{X}}) \times \mathrm{Sect}(\mathbf{D}_\infty - \mathbf{d}_h, \overset{\circ}{\mathbf{C}}{}^{(a)}_{\mathbf{D}_\infty - \mathbf{d}_h}).$$

Note that the family of curves $\mathbf{C}_{\mathbf{D}_\infty - \mathbf{d}_h}$ can be naturally identified with the constant family $(\mathbf{D}_\infty - \mathbf{d}_h) \times \mathbf{C}$ via the projection $\mathbf{C} \times 0_{\mathbf{X}} \overset{\pi_v}{\to} \mathbf{C}_{\mathbf{d}_v}$ for each $\mathbf{d}_v \in \mathbf{D}_\infty - \mathbf{d}_h$. In terms of this trivialization, the above $\mathbb{G}_m$-action contracts the space $\mathrm{Sect}(\mathbf{D}_\infty - \mathbf{d}_h, \overset{\circ}{\mathbf{C}}{}^{(a)}_{\mathbf{D}_\infty - \mathbf{d}_h})$ to the space of constant sections, which can be identified with $\overset{\circ}{\mathbf{C}}{}^{(a)}$.

In addition, from Proposition 2.6 we obtain that the above $\mathbb{G}_m$-action contracts the space $\mathrm{QMaps}^a(\mathbf{C}, \mathcal{G}_{G,\mathbf{X}})$ to $\overset{\circ}{\mathbf{C}}{}^{(a)} \subset \mathrm{QMaps}^a(\mathbf{C}, \mathcal{G}_{G,\mathbf{X}})$.

By comparing it with our description of the map $i_b : \mathfrak{U}_G^{a;b} \to \mathfrak{U}_G^a$ for $b = a$ we obtain the assertion of the proposition. □

7.8 An example

Let us describe explicitly the Uhlenbeck space $\mathfrak{U}_G^a$ for $a = 1$. (In particular, using Proposition 6.5, this will imply the description of the singularities of $\mathfrak{U}_G^a$ for general a along the strata $\mathfrak{U}_G^{a;b}$ for $b = 1$.)

Choose a pair of directions $(\mathbf{d}_v, \mathbf{d}_h)$, and consider the corresponding surface $\mathbf{S}' = \mathbf{C} \times \mathbf{X}$. Let us choose points $0_{\mathbf{C}} \in \mathbf{C} - \infty_{\mathbf{C}}$, $0_{\mathbf{X}} \in \mathbf{X} - \infty_{\mathbf{X}}$, and consider the affine Grassmannian $\mathrm{Gr}_G := \mathrm{Gr}_{G,\mathbf{X},0_{\mathbf{X}}}$. Let $\overline{\mathrm{Gr}}_G^{\bar{\alpha}_0} \subset \mathrm{Gr}_G$ be the corresponding closed subscheme. It contains the unit point $\mathbf{1}_{\mathrm{Gr}} \in \mathrm{Gr}_G$, which is its only singularity.

By definition, we have a map $\overline{\mathrm{Gr}}_G^{\bar{\alpha}_0} \to \mathrm{Bun}_G(\mathbf{X})$, and, as we know, the stack $\mathrm{Bun}_G(\mathbf{X})$ carries a line bundle $\mathcal{P}_{\mathrm{Bun}_G(\mathbf{X})}$ with a section whose set of zeroes is the locus of nontrivial bundles. Let us denote by $D_{\overline{\mathrm{Gr}}_G^{\bar{\alpha}_0}}$ the corresponding Cartier divisor in $\overline{\mathrm{Gr}}_G^{\bar{\alpha}_0}$. Note that the point $\mathbf{1}_{\mathrm{Gr}}$ belongs to $\overline{\mathrm{Gr}}_G^{\bar{\alpha}_0} - D_{\overline{\mathrm{Gr}}_G^{\bar{\alpha}_0}}$.

We claim that there is a natural isomorphism

$$\mathfrak{U}_G^1 \simeq \overset{\circ}{\mathbf{C}} \times \overset{\circ}{\mathbf{X}} \times (\overline{\mathrm{Gr}}_G^{\bar{\alpha}_0} - D_{\overline{\mathrm{Gr}}_G^{\bar{\alpha}_0}}).$$

We have the projection

$$\varpi_h^1 \times \varpi_v^1 : \mathfrak{U}_G^1 \to \overset{\circ}{\mathbf{C}} \times \overset{\circ}{\mathbf{X}},$$

and the group $\mathbb{A}^2$ acts on $\mathfrak{U}_G^1$ via its action on $\overset{\circ}{\mathbf{S}}$ by shifts. Therefore, it is enough to show that

$$(\varpi_h^1)^{-1}(0_{\mathbf{C}}) \cap (\varpi_v^1)^{-1}(0_{\mathbf{X}}) \simeq \overline{\mathrm{Gr}}_G^{\bar{\alpha}_0} - D_{\overline{\mathrm{Gr}}_G^{\bar{\alpha}_0}}. \tag{17}$$

Note that the intersection $(\varpi_h^1)^{-1}(0_{\mathbf{C}}) \cap (\varpi_v^1)^{-1}(0_{\mathbf{X}})$ contains only one geometric point, which is not in $\mathrm{Bun}_G(\mathbf{S}', \mathbf{D}'_\infty)$, namely, the point $0_{\mathbf{C}} \times 0_{\mathbf{X}}$ from the stratum $\mathfrak{U}_G^{1;1}$.

Let $\mathbf{1}_\mathbf{C} \in \mathbf{C}$ be a point different from $0_\mathbf{C}$ and $\infty_\mathbf{C}$. Given a point in $(\varpi_h^1)^{-1}(0_\mathbf{C}) \cap (\varpi_v^1)^{-1}(0_\mathbf{X})$, we have a quasi-map $\sigma : \mathbf{C} \to \overline{\mathrm{Gr}}_G^{\overline{\alpha}_0}$ (cf. Section 3.8), such that σ, restricted to $\mathbf{C} - 0_\mathbf{C}$, is a map whose image belongs to $\overline{\mathrm{Gr}}_G^{\overline{\alpha}_0} - D_{\mathrm{Gr}_G^{\overline{\alpha}_0}}$. Hence $\sigma \mapsto \sigma(\mathbf{1}_\mathbf{C})$ defines a map in one direction $(\varpi_h^1)^{-1}(0_\mathbf{C}) \cap (\varpi_v^1)^{-1}(0_\mathbf{X}) \to (\overline{\mathrm{Gr}}_G^{\overline{\alpha}_0} - D_{\mathrm{Gr}_G^{\overline{\alpha}_0}})$.

Let us first show that this map is one-to-one. From the proof of Theorem 7.2, it is clear that it sends the unique point in $(\varpi_h^1)^{-1}(0_\mathbf{C}) \cap (\varpi_v^1)^{-1}(0_\mathbf{X}) \cap \mathfrak{U}_G^{1;1}$ to the point $\mathbf{1}_{\mathrm{Gr}} \in \overline{\mathrm{Gr}}_G^{\overline{\alpha}_0} - D_{\mathrm{Gr}_G^{\overline{\alpha}_0}}$.

Now, let σ_1, σ_2 be two elements of $(\varpi_h^1)^{-1}(0_\mathbf{C}) \cap (\varpi_v^1)^{-1}(0_\mathbf{X})$. Since both σ_1 and σ_2 are of degree 1, if we had $\mathbf{1}_{\mathrm{Gr}} = \sigma_1(\infty_\mathbf{C}) = \sigma_2(\infty_\mathbf{C})$, and $\sigma_1(\mathbf{1}_\mathbf{C}) = \sigma_2(\mathbf{1}_\mathbf{C})$, this would imply $\sigma_1 = \sigma_2$. For the same reason, no *map* σ can send $\mathbf{1}_\mathbf{C}$ to $\mathbf{1}_{\mathrm{Gr}}$.

Hence the map $(\varpi_h^1)^{-1}(0_\mathbf{C}) \cap (\varpi_v^1)^{-1}(0_\mathbf{X}) \to (\overline{\mathrm{Gr}}_G^{\overline{\alpha}_0} - D_{\mathrm{Gr}_G^{\overline{\alpha}_0}})$ is one-to-one on the level of geometric points. Since the left-hand side is reduced, and the right-hand side is normal, we obtain that it is an open embedding. Recall now the action of $\mathbb{G}_m$ by dilations along the $\mathbf{X}$-factor. We claim that it contracts $(\overline{\mathrm{Gr}}_G^{\overline{\alpha}_0} - D_{\mathrm{Gr}_G^{\overline{\alpha}_0}})$ to the point $\mathbf{1}_{\mathrm{Gr}}$. Indeed, we have a closed embedding $\mathrm{Gr}_G \hookrightarrow \mathcal{G}_{G,\mathbf{X}}$, such that $(\overline{\mathrm{Gr}}_G^{\overline{\alpha}_0} - D_{\mathrm{Gr}_G^{\overline{\alpha}_0}})$ gets mapped to the corresponding open cell $\mathcal{G}_{\mathfrak{g}_{\mathrm{aff}}, \mathfrak{g}_{\mathrm{aff}}^+, e}$ (cf. Section 2.2), and for Schubert cells the contraction assertion is evident.

Using Proposition 7.7, this implies that $(\varpi_h^1)^{-1}(0_\mathbf{C}) \cap (\varpi_v^1)^{-1}(0_\mathbf{X}) \to (\overline{\mathrm{Gr}}_G^{\overline{\alpha}_0} - D_{\mathrm{Gr}_G^{\overline{\alpha}_0}})$ is in fact an isomorphism.

Remark. It is well known that $(\overline{\mathrm{Gr}}_G^{\overline{\alpha}_0} - D_{\mathrm{Gr}_G^{\overline{\alpha}_0}})$ is isomorphic to the minimal nilpotent orbit closure $\overline{\mathbb{O}}_{\min}$ in $\mathfrak{g}$. It was noticed by V. Drinfeld in 1998 that the transversal slice to the singularity of $\mathfrak{U}_G^1$ should look like $\overline{\mathbb{O}}_{\min}$.

7.9

For a fixed b, $0 \leq b \leq a$, let $\mathfrak{P}(b)$ be a partition, i.e., $b = \sum_k n_k \cdot d_k$, with $d_k > 0$ and pairwise distinct. Let us denote by $\mathrm{Sym}^{\mathfrak{P}(b)}(\overset{\circ}{\mathbf{S}})$ the variety $\prod_k \mathrm{Sym}^{n_k}(\overset{\circ}{\mathbf{S}})$, and by $\overset{\circ}{\mathrm{Sym}}{}^{\mathfrak{P}(b)}(\overset{\circ}{\mathbf{S}})$ the open subset of $\mathrm{Sym}^{\mathfrak{P}(b)}(\overset{\circ}{\mathbf{S}})$ obtained by removing all the diagonals.

The symmetric power $\mathrm{Sym}^b(\overset{\circ}{\mathbf{S}})$ is the union $\bigcup_{\mathfrak{P}(b)} \overset{\circ}{\mathrm{Sym}}{}^{\mathfrak{P}(b)}(\overset{\circ}{\mathbf{S}})$ over all the partitions of b. Let $\mathfrak{U}_G^{a;\mathfrak{P}(b)} = \mathrm{Bun}_G^{a-b}(\mathbf{S}, \mathbf{D}_\infty) \times \overset{\circ}{\mathrm{Sym}}{}^{\mathfrak{P}(b)}(\overset{\circ}{\mathbf{S}})$ be the corresponding subscheme in $\mathfrak{U}_G^{a;\mathfrak{P}(b)}$.

The next theorem, which can be regarded as the main result of this paper, describes the restriction of the intersection cohomology sheaf of $\mathfrak{U}_G^a$ to the strata $\mathfrak{U}_G^{a;\mathfrak{P}}$.

Let $\check{\mathfrak{g}}_{\mathrm{aff}}$ denote the Langlands dual Lie algebra to $\mathfrak{g}_{\mathrm{aff}}$. Consider the (maximal) parabolic $\mathfrak{g}[[t]] \oplus K \cdot \mathbb{C} \oplus d \cdot \mathbb{C} \subset \mathfrak{g}_{\mathrm{aff}}$; then we obtain the corresponding parabolic inside $\check{\mathfrak{g}}_{\text{aff}}$, whose unipotent radical $\mathfrak{V}$ is an (integrable) module over the corresponding dual Levi subalgebra, i.e., $\mathfrak{V}$ is a representation of the group $\check{G}_{\text{aff}} = \check{G} \times \mathbb{G}_m$. In particular, we can consider the space $\mathfrak{V}^f$, where $f \in \check{\mathfrak{g}}$ is a principal nilpotent element. We regard $\mathfrak{V}^f$ as a bi-graded vector space, $\mathfrak{V}^f = \underset{m,l}{\oplus} (\mathfrak{V}^f)_l^m$: where the "first" grading, corresponding to the upper index m, is the principal grading coming from the Jacobson–Morozov triple containing f, and the "second" grading, corresponding to the lower index l, comes from the action of $\mathbb{G}_m \subset G_{\text{aff}}$.

For a partition $\mathfrak{P}(b)$ as above, we will denote by $\text{Sym}^{\mathfrak{P}(b)}(\mathfrak{V}^f)$ the graded vector space

$$\bigotimes_k \big(\underset{i \geq 0}{\oplus} (\text{Sym}^i(\mathfrak{V}^f))_{d_k}[2i] \big)^{\otimes n_k},$$

where the subscript d_k means that we are taking the graded subspace of the corresponding index with respect to the "second" grading, and the notation $[j]$ means the shift with respect to the "first" grading. By declaring the "first" grading to be cohomological we obtain a semisimple complex of vector spaces.

Theorem 7.10. *The restriction* $\text{IC}_{\mathfrak{U}_G^a}|_{\mathfrak{U}_G^{a;\mathfrak{P}(b)}}$ *is locally constant and is isomorphic to the IC sheaf on this scheme tensored by the (constant) complex* $\text{Sym}^{\mathfrak{P}(b)}(\mathfrak{V}^f)$.

8 Functorial definitions

8.1

An obvious disadvantage of our definition of $\mathfrak{U}_G^a$ is that we were not able to say what functor it represents. In this section, we will give two more variants of the definition of the Uhlenbeck space, we will call them ${}^{\text{Tann}}\mathfrak{U}_G^a$ and ${}^{\text{Drinf}}\mathfrak{U}_G^a$, respectively, and which will be defined as functors. Moreover, ${}^{\text{Tann}}\mathfrak{U}_G^a$ will be a scheme, whereas ${}^{\text{Drinf}}\mathfrak{U}_G^a$ is only an ind-scheme.

We will have a sequence of closed embeddings

$$\mathfrak{U}_G^a \to {}^{\text{Tann}}\mathfrak{U}_G^a \to {}^{\text{Drinf}}\mathfrak{U}_G^a,$$

which induce isomorphisms on the level of the corresponding reduced schemes. (In other words, up to nilpotents, one can give a functorial definition of the Uhlenbeck space.)

8.2 A Tannakian definition

In this subsection, we will express the Uhlenbeck space for an arbitrary G in terms of that of SL_n.

For two integers n_1 and n_2, consider the homomorphisms

$$SL_{n_1} \times SL_{n_2} \to SL_{n_1+n_2} \quad \text{and} \quad SL_{n_1} \times SL_{n_2} \to SL_{n_1 \cdot n_2},$$

and the corresponding morphisms

$$\mathrm{Bun}_{n_1}(\mathbf{X}) \times \mathrm{Bun}_{n_2}(\mathbf{X}) \to \mathrm{Bun}_{n_1+n_2}(\mathbf{X})$$

and

$$\mathrm{Bun}_{n_1}(\mathbf{X}) \times \mathrm{Bun}_{n_2}(\mathbf{X}) \to \mathrm{Bun}_{n_1 \cdot n_2}(\mathbf{X}).$$

Note that the pullbacks of the line bundles $\mathcal{P}_{\mathrm{Bun}_{n_1+n_2}(\mathbf{X})}$ and $\mathcal{P}_{\mathrm{Bun}_{n_1 \cdot n_2}(\mathbf{X})}$ are isomorphic to $\mathcal{P}_{\mathrm{Bun}_{n_1}(\mathbf{X})} \boxtimes \mathcal{P}_{\mathrm{Bun}_{n_2}(\mathbf{X})}$ and $\mathcal{P}^{\otimes n_2}_{\mathrm{Bun}_{n_1}(\mathbf{X})} \boxtimes \mathcal{P}^{\otimes n_1}_{\mathrm{Bun}_{n_2}(\mathbf{X})}$, respectively.

As in Lemma 6.2, we obtain the closed embeddings $\mathfrak{U}^{a_1}_{n_1} \times \mathfrak{U}^{a_2}_{n_2} \to \mathfrak{U}^{a_1+a_2}_{n_1+n_2}$ and $\mathfrak{U}^{a_1}_{n_1} \times \mathfrak{U}^{a_2}_{n_2} \to \mathfrak{U}^{a_1 \cdot n_2 + a_2 \cdot n_1}_{n_1 \cdot n_2}$, which extend the natural morphisms between the corresponding moduli spaces $\mathrm{Bun}^a_n(\mathbf{S}, \mathbf{D}_\infty)$.

Moreover, it is easy to deduce from the proof of Theorem 7.2 that if $\sigma_{\mathcal{O},1}$ (respectively, $\sigma_{\mathcal{O},2}$) is a collection of based quasi-maps $\mathbf{C} \to \mathrm{Gr}^{BD,a}_{SL_{n_1},\mathbf{X}}$ describing a point of $\mathfrak{U}^{a_1}_{n_1}$, whose saturation is $\sigma'_{\mathcal{O},1}$ and defect $D_{1,\mathbf{S}}$ (respectively, $\sigma'_{\mathcal{O},2}$, $D_{2,\mathbf{S}}$), then the corresponding point of $\mathfrak{U}^{a_1+a_2}_{n_1+n_2}$ (respectively, $\mathfrak{U}^{a_1 \cdot n_2 + a_2 \cdot n_1}_{n_1 \cdot n_2}$) will have saturation $\sigma'_{\mathcal{O},1} \oplus \sigma'_{\mathcal{O},2}$ (respectively, $\sigma'_{\mathcal{O},1} \otimes \sigma'_{\mathcal{O},2}$) and defect $D_{1,\mathbf{S}} + D_{2,\mathbf{S}}$ (respectively, $n_2 \cdot D_{1,\mathbf{S}} + n_1 \cdot D_{2,\mathbf{S}}$).

8.3

We set ${}^{\mathrm{Tann}}\mathfrak{U}^a_G$ to represent the functor that assigns to a test scheme S the following data:

- an S-point of $\mathrm{Sect}(\mathcal{O}, \mathrm{QMaps}^a(\mathbf{C}, \mathrm{Gr}^{BD,a}_{G,\mathbf{X}})_{\mathcal{O}})$;
- for every representation $\phi : G \to SL_n$, a point of $\mathfrak{U}^{\phi_{\mathbb{Z}}(a)}_n$,

such that we have the following:

- The corresponding S-points of $\mathrm{Sect}(\mathcal{O}, \mathrm{QMaps}^{\phi_{\mathbb{Z}}(a)}(\mathbf{C}, \mathrm{Gr}^{BD,\phi_{\mathbb{Z}}(a)}_{SL_n,\mathbf{X}})_{\mathcal{O}})$, coming from

  $$\mathrm{Sect}(\mathcal{O}, \mathrm{QMaps}^a(\mathbf{C}, \mathrm{Gr}^{BD,a}_{G,\mathbf{X}})_{\mathcal{O}}) \to \mathrm{Sect}(\mathcal{O}, \mathrm{QMaps}^{\phi_{\mathbb{Z}}(a)}(\mathbf{C}, \mathrm{Gr}^{BD,\phi_{\mathbb{Z}}(a)}_{SL_n,\mathbf{X}})_{\mathcal{O}}) \leftarrow \mathfrak{U}^{\phi_{\mathbb{Z}}(a)}_n,$$

  coincide.
- For $\phi = \phi_1 \otimes \phi_2$, $n = n_1 \cdot n_2$ (respectively, $\phi = \phi_1 \oplus \phi_2$, $n = n_1 + n_2$), the corresponding S-point of $\mathfrak{U}^{\phi_{\mathbb{Z}}(a)}_n$ equals the image of the corresponding point of $\mathfrak{U}^{\phi_{1\mathbb{Z}}(a)}_{n_1} \times \mathfrak{U}^{\phi_{2\mathbb{Z}}(a)}_{n_2}$ under

  $$\mathfrak{U}^{\phi_{1\mathbb{Z}}(a)}_{n_1} \times \mathfrak{U}^{\phi_{2\mathbb{Z}}(a)}_{n_2} \to \mathfrak{U}^{\phi_{\mathbb{Z}}(a)}_n.$$

Since for a faithful representation $\phi : G \to SL_n$, the map $\mathrm{Sect}(\mathsf{O}, \mathrm{QMaps}^a(\mathbf{C}, \mathrm{Gr}^{BD,a}_{G,\mathbf{X}})_{\mathsf{O}}) \to \mathrm{Sect}(\mathsf{O}, \mathrm{QMaps}^{\phi_{\mathbb{Z}}(a)}(\mathbf{C}, \mathrm{Gr}^{BD,\phi_{\mathbb{Z}}(a)}_{SL_n,\mathbf{X}})_{\mathsf{O}})$ is a closed embedding, it is easy to see that ${}^{\mathrm{Tann}}\mathfrak{U}^a_G$ is in fact a closed subfunctor in $\mathfrak{U}^{\phi_{\mathbb{Z}}(a)}_n$; in particular, ${}^{\mathrm{Tann}}\mathfrak{U}^a_G$ is indeed representable by an affine scheme of finite type. Moreover, by construction, we have a closed embedding $\mathfrak{U}^a_G \to {}^{\mathrm{Tann}}\mathfrak{U}^a_G$.

Proposition 8.4. *The above map* $\mathfrak{U}^a_G \to {}^{\mathrm{Tann}}\mathfrak{U}^a_G$ *induces an isomorphism on the level of reduced schemes.*

We do not know whether it is in general true that ${}^{\mathrm{Tann}}\mathfrak{U}^a_G$ is actually isomorphic to $\mathfrak{U}^a_G$.

Proof. Since the map in question is a closed embedding, we only have to check that it defines a surjection on the level of geometric points.

Thus let $\sigma_{\mathsf{O}} : \mathbf{C} \to \mathrm{Gr}^{BD,a}_{G,\mathbf{X}}$ be the collection of quasi-maps representing a point of $\mathrm{Sect}(\mathsf{O}, \mathrm{QMaps}^a(\mathbf{C}, \mathrm{Gr}^{BD,a}_{G,\mathbf{X}})_{\mathsf{O}})$, and $\sigma_{\mathsf{O},\phi}$ be a compatible system of points of $\mathfrak{U}^{\phi_{\mathbb{Z}}(a)}_n$ for every representation $\phi : G \to SL_n$.

Using Theorem 7.2, we have to show that there exists a point $\mathcal{F}'_G \in \mathrm{Bun}_G(\mathbf{S}, \mathbf{D}_\infty)$ (corresponding to a system of maps σ'_{O}) and a 0-cycle $D_{\mathbf{S}}$, such that for every $(\mathbf{d}_v, \mathbf{d}_h) \in \mathsf{O}$, the corresponding quasi-map σ has as its saturation the map of (15) and $\pi_v(D_{\mathbf{S}})$ as its defect.

Let $\mathcal{F}'_\phi$ be the SL_n bundle corresponding to the saturation of $\sigma_{\mathsf{O},\phi}$. From our description of the maps $\mathfrak{U}^{a_1}_{n_1} \times \mathfrak{U}^{a_2}_{n_2} \to \mathfrak{U}^{a_1+a_2}_{n_1+n_2}$ and $\mathfrak{U}^{a_1}_{n_1} \times \mathfrak{U}^{a_2}_{n_2} \to \mathfrak{U}^{a_1\cdot n_2+a_2\cdot n_1}_{n_1\cdot n_2}$, it follows that $\mathcal{F}'_{\phi_1\oplus\phi_2} \simeq \mathcal{F}'_{\phi_1} \oplus \mathcal{F}'_{\phi_2}$ and $\mathcal{F}'_{\phi_1\otimes\phi_2} \simeq \mathcal{F}'_{\phi_1} \otimes \mathcal{F}'_{\phi_2}$.

Therefore, the collection $\{\mathcal{F}'_\phi\}$ defines a G-bundle on $\mathbf{S}$, trivialized along $\mathbf{D}_\infty$, which we set to be our $\mathcal{F}'_G$. The 0-cycle $D_{\mathbf{S}}$ is reconstructed using just one faithful representation of G, as in the proof of Theorem 7.2. □

8.5 Drinfeld's approach

Consider the functor {Schemes of finite type} $\to$ {Groupoids}; we will call it ${}^{\mathrm{Drinf}}\mathfrak{U}^a_G$ since the definition below follows a suggestion of Drinfeld. The category corresponding to a test scheme S has as objects the data of the following:

- A principal G-bundle $\mathcal{F}_G$ defined on an open subscheme of $U \subset \mathbf{S} \times S$, such that $\mathbf{S} \times S - U$ is finite over S, and $U \supset \mathbf{D}_\infty \times S$, and $\mathcal{F}_G$ is equipped with a trivialization along $\mathbf{D}_\infty \times S$. (We will denote by U_{O} the corresponding open subset in $\mathbf{S}'_{\mathsf{O}} \times S$, and by $\mathcal{F}_{G,\mathsf{O}}$ the corresponding G-bundle over it.)
- A map $S \to \mathrm{Sect}(\mathsf{O}, \mathrm{QMaps}^a(\mathbf{C}, \mathrm{Gr}^{BD,a}_{G,\mathbf{X}})_{\mathsf{O}})^\tau$. (This expression makes use of the superscript τ, which was introduced in Lemma 4.10.) We will denote by σ_{O} the corresponding relative quasi-map $\mathbf{C}_{\mathsf{O}} \times S \to \mathrm{Gr}^{BD,a}_{G,\mathbf{X}_{\mathsf{O}}}$. Let $V_{\mathsf{O}} \subset \mathbf{C}_{\mathsf{O}} \times S$ denote the open subset over which σ_{O} is defined as a map, and $D_{v,\mathsf{O}} \subset \mathbf{X}_{\mathsf{O}} \times S$ the corresponding relative divisor. Let $U^{\sigma_{\mathsf{O}}} \subset \mathbf{S}'_{\mathsf{O}} \times S$ be the open subset $V_{\mathsf{O}} \underset{\mathsf{O}}{\times} \mathbf{X}_{\mathsf{O}} \cup$

$\mathbf{C}_{\mathsf{O}} \underset{\mathsf{O}}{\times} (\mathbf{X}_{\mathsf{O}} - D_{v,\mathsf{O}})$, and $\mathcal{F}_G^{\sigma_{\mathsf{O}}}$ be the corresponding principal G-bundle defined on $U^{\sigma_{\mathsf{O}}}$.

- An isomorphism α between $\mathcal{F}_{G,\mathsf{O}}|_{U^{\sigma_{\mathsf{O}}} \cap U_{\mathsf{O}}}$ and $\mathcal{F}_G^{\sigma_{\mathsf{O}}}|_{U^{\sigma_{\mathsf{O}}} \cap U_{\mathsf{O}}}$, respecting the trivializations along $\mathbf{D}'_{\infty,\mathsf{O}} \times S$.

Morphisms between an object $(\mathcal{F}_G, U, \sigma_{\mathsf{O}}, \alpha)$ and another object $(\mathcal{F}_G^1, U^1, \sigma_{\mathsf{O}}^1, \alpha^1)$ are G-bundle isomorphisms $\mathcal{F}_G \simeq \mathcal{F}_G^1|_{U_{\mathsf{O}}^1 \cap U_{\mathsf{O}}}$, which commute with other pieces of data (in particular, the data of σ_{O} must be the same). In particular, we see that objects in our category have no nontrivial automorphisms, i.e., ${}^{\mathrm{Drinf}}\mathfrak{U}_G^a$ is a functor {Schemes of finite type} $\to$ {Sets}.

Lemma 8.6. *The functor* ${}^{\mathrm{Drinf}}\mathfrak{U}_G^a$ *is representable by an affine ind-scheme of ind-finite type.*

(Later we will show that ${}^{\mathrm{Drinf}}\mathfrak{U}_G^a \to \mathrm{Sect}(\mathsf{O}, \mathrm{QMaps}^a(\mathbf{C}, \mathrm{Gr}_{G,\mathbf{X}}^{BD,a})_{\mathsf{O}})$ is, in fact, a closed embedding.)

Proof. It is enough to show that the forgetful map ${}^{\mathrm{Drinf}}\mathfrak{U}_G^a \to \mathrm{Sect}(\mathsf{O}, \mathrm{QMaps}^a(\mathbf{C}, \mathrm{Gr}_{G,\mathbf{X}}^{BD,a})_{\mathsf{O}})$ is ind-representable (and ind-affine). Thus for a test-scheme S let σ_{O} be a relative quasi-map $\mathbf{C}_{\mathsf{O}} \times S \to \mathrm{Gr}_{G,\mathbf{X}_{\mathsf{O}}}^{BD,a}$, and let $U^{\sigma_{\mathsf{O}}}$, $\mathcal{F}_G^{\sigma_{\mathsf{O}}}$ be as above.

Let $(\mathbf{d}_v, \mathbf{d}_h) \in \mathsf{O}$ be some fixed pair of directions, and set $U \subset \mathbf{S}' \times S$ (respectively, $\mathcal{F}_G$) to be the fiber of $U^{\sigma_{\mathsf{O}}}$ (respectively, $\mathcal{F}_G^{\sigma_{\mathsf{O}}}$) over the corresponding point of O. We will denote by the same character U (respectively, $\mathcal{F}_G$) the corresponding open subset in $\mathbf{S} \times S$ (respectively, the G-bundle over it).

Consider the group ind-scheme over S, call it $\mathrm{Aut}_U(\mathcal{F}_G)$, of automorphisms of $\mathcal{F}_G$, respecting the trivialization at the divisor of infinity. Consider now the group ind-scheme of S-maps $S \times \mathsf{O} \to \mathrm{Aut}_U(\mathcal{F}_G)$. Then it is easy to see that the fiber product

$$S \underset{\mathrm{Sect}(\mathsf{O}, \mathrm{QMaps}^a(\mathbf{C}, \mathrm{Gr}_G^a)_{\mathsf{O}})}{\times} {}^{\mathrm{Drinf}}\mathfrak{U}_G^a$$

is (ind-)represented by the above ind-scheme $\mathrm{Maps}_S(S \times \mathsf{O}, \mathrm{Aut}_U(\mathcal{F}_G))$. □

The rest of this section will be devoted to the proof of the following result.

Theorem 8.7. *We have a closed embedding* ${}^{\mathrm{Tann}}\mathfrak{U}_G^a \to {}^{\mathrm{Drinf}}\mathfrak{U}_G^a$, *which induces an isomorphism on the level of reduced schemes.*

8.8

Our first task is to construct the map ${}^{\mathrm{Tann}}\mathfrak{U}_G^a \to {}^{\mathrm{Drinf}}\mathfrak{U}_G^a$. The first case to consider is $G = SL_n$. In this case, by the definition of ${}^{\mathrm{Tann}}\mathfrak{U}_G^a$, we have $\mathfrak{U}_n^a \simeq {}^{\mathrm{Tann}}\mathfrak{U}_n^a$.

Since ${}^{\mathrm{Drinf}}\mathfrak{U}_n^a$ is ind-affine, and $\mathfrak{U}_n^a \simeq \mathfrak{N}_n^a$ is the affinization of the scheme $\widetilde{\mathfrak{N}}_n^a$, to construct a map $\mathfrak{U}_n^a \to {}^{\mathrm{Drinf}}\mathfrak{U}_n^a$, it suffices to construct a map $\widetilde{\mathfrak{N}}_n^a \to {}^{\mathrm{Drinf}}\mathfrak{U}_G^a$.

Given an S-point $\mathcal{M}$ of $\widetilde{\mathfrak{N}}_n^a$, i.e., an S-family of torsion-free sheaves on $\mathbf{S}$, there exists an open subset $U \subset \mathbf{S} \times S$ whose complement is finite over S, such that $\mathcal{M}$

is a vector bundle when restricted to U, i.e., we obtain a principal SL_n-bundle $\mathcal{F}_{SL_n}$ on U.

The data of a relative quasi-map σ_{O} are obtained from the composition

$$\widetilde{\mathfrak{N}}_n^a \to \mathfrak{U}_n^a \to \mathrm{Sect}(\mathsf{O}, \mathrm{QMaps}^a(\mathbf{C}, \mathrm{Gr}_G^a)_{\mathsf{O}}).$$

The isomorphisms $\alpha : \mathcal{F}_{SL_n,\mathsf{O}}|_{U^{\sigma_{\mathsf{O}}} \cap U_{\mathsf{O}}} \simeq \mathcal{F}_{SL_n}^{\sigma_{\mathsf{O}}}|_{U^{\sigma_{\mathsf{O}}} \cap U_{\mathsf{O}}}$ follow from the construction of the map (13).

Thus we have a morphism $\mathfrak{U}_n^a \to {}^{\mathrm{Drinf}}\mathfrak{U}_n^a$, and moreover, for $n = n_1 + n_2$ (respectively, $n = n_1 \cdot n_2$) and $a = a_1 + a_2$ (respectively, $a = n_1 \cdot a_2 + n_2 \cdot a_1$), the natural morphism

$${}^{\mathrm{Drinf}}\mathfrak{U}_{n_1}^{a_1} \times {}^{\mathrm{Drinf}}\mathfrak{U}_{n_2}^{a_2} \to {}^{\mathrm{Drinf}}\mathfrak{U}_n^a$$

is compatible with the corresponding morphism $\mathfrak{U}_{n_1}^{a_1} \times \mathfrak{U}_{n_2}^{a_2} \to \mathfrak{U}_n^a$. Hence we have the morphism ${}^{\mathrm{Tann}}\mathfrak{U}_G^a \to {}^{\mathrm{Drinf}}\mathfrak{U}_G^a$ for an arbitrary G.

Next, we will show that the composition $\mathfrak{U}_G^a \to {}^{\mathrm{Tann}}\mathfrak{U}_G^a \to {}^{\mathrm{Drinf}}\mathfrak{U}_G^a$ is a closed embedding. Let $\overline{\mathfrak{U}}_G^a$ be the closure of $\mathrm{Bun}_G(\mathbf{S}, \mathbf{D}_\infty)$ in ${}^{\mathrm{Drinf}}\mathfrak{U}_G^a$.

Then under the projection ${}^{\mathrm{Drinf}}\mathfrak{U}_G^a \to \mathrm{Sect}(\mathsf{O}, \mathrm{QMaps}^a(\mathbf{C}, \mathrm{Gr}_{G,\mathbf{X}}^{BD,a})_{\mathsf{O}})$, $\overline{\mathfrak{U}}_G^a$ gets mapped to $\mathfrak{U}_G^a$, and we obtain a pair of arrows

$$\mathfrak{U}_G^a \leftrightarrows \overline{\mathfrak{U}}_G^a,$$

such that both compositions induce the identity map on $\mathrm{Bun}_G(\mathbf{S}, \mathbf{D}_\infty)$. Since the latter is dense, we obtain that the map $\mathfrak{U}_G^a \to \overline{\mathfrak{U}}_G^a$ is in fact an isomorphism.

Thus $\mathfrak{U}_G^a \to {}^{\mathrm{Drinf}}\mathfrak{U}_G^a$ is a closed embedding, and let us show now that it induces an isomorphism from $\mathfrak{U}_G^a$ to the reduced (ind-)scheme underlying ${}^{\mathrm{Drinf}}\mathfrak{U}_G^a$. For that we have to check that this map is surjective on the level of geometric points.

Let $(\mathcal{F}_G, \sigma_{\mathsf{O}}, \alpha)$ be a geometric point of ${}^{\mathrm{Drinf}}\mathfrak{U}_G^a$, and let $D_{v,\mathsf{O}}$ (respectively, $D_{h,\mathsf{O}}$) be the relative Cartier divisors in $\mathbf{X}_{\mathsf{O}}$ (respectively, $\mathbf{C}_{\mathsf{O}}$) obtained from σ_{O} via $\varpi_{v,\mathsf{O}}^a$ (respectively, $\varpi_{h,\mathsf{O}}^a$). Let us denote by $\mathcal{F}'_G$ the (unique) extension of $\mathcal{F}_G$ to the entire $\mathbf{S}$. Let σ'_{O} be the map $\mathbf{C}_{\mathsf{O}} \times S \to \mathrm{Gr}_{G,\mathbf{X}_{\mathsf{O}}}^{BD,a-b}$ corresponding to $\mathcal{F}'_G$. Let $D'_{v,\mathsf{O}}$ (respectively, $D'_{h,\mathsf{O}}$) be the relative divisors in $\mathbf{X}_{\mathsf{O}}$ (respectively, $\mathbf{C}_{\mathsf{O}}$) corresponding to σ'_{O}.

Set $\mathbf{S} - U = \underset{i}{\cup}\mathbf{s}_i$. Our task is to show that there exist integers b_i such that for every pair of directions $(\mathbf{d}_v, \mathbf{d}_h) \in \mathsf{O}$, we have the following:

(1) $D_v = D'_v + b_i \cdot \pi_h(\mathbf{s}_i)$, $D_h = D'_h + b_i \cdot \pi_v(\mathbf{s}_i)$.

(2) The quasi-map σ has the saturation equal to $\mathbf{C} \xrightarrow{\sigma'} \mathrm{Gr}_{G,\mathbf{X},D'_v}^{BD,a-b} \to \mathrm{Gr}_{G,\mathbf{X},D_v}^{BD,a}$.

(3) The defect of σ equals $\underset{i}{\Sigma} b_i \cdot \pi_v(\mathbf{s}_i)$.

Let $\sigma'' : \mathbf{C} \to \mathrm{Gr}_{G,\mathbf{X},D_v}^{BD,a}$ be a based map of degree $a - b$ which is the saturation of σ. We know that σ is obtained from σ'' by adding to it a divisor of degree b supported on $\underset{i}{\cup}\pi_v(\mathbf{s}_i)$. Hence by letting $(\mathbf{d}_v, \mathbf{d}_h)$ move along O, we obtain a section $\mathsf{O} \to \overset{\circ}{\mathbf{X}}{}_{\mathsf{O}}^{(b)}$

with the above support property. However, it is easy to see that any such section is of the form

$$(\mathbf{d}_v, \mathbf{d}_h) \mapsto \sum_i b_i \cdot \pi_v(\mathbf{s}_i)$$

for some integers b_i.

Now, since the maps σ' and σ'' give rise to the same bundle on $\mathbf{S}'$, the two compositions

$$\mathbf{C} \xrightarrow{\sigma'} \mathrm{Gr}^{BD,a-b}_{G,\mathbf{X},D'_v} \to \mathcal{G}_{G,\mathbf{X}} \quad \text{and} \quad \mathbf{C} \xrightarrow{\sigma''} \mathrm{Gr}^{BD,a}_{G,\mathbf{X},D_v} \to \mathcal{G}_{G,\mathbf{X}}$$

coincide. Hence σ' and σ'' have the same degree $a-b$ and $\varpi_h^{a-b}(\sigma') = \varpi_h^{a-b}(\sigma'')$. Since by the property of saturations (cf. (4)) we have $\varpi_h^a(\sigma) = \varpi_v^{a-b}(\sigma'') + \sum_i b_i \cdot \pi_v(\mathbf{s}_i)$, we obtain $D_h = D'_h + b_i \cdot \pi_v(\mathbf{s}_i)$. By combining this with the condition that $\sigma_{\mathcal{O}} \in \mathrm{Sect}(\mathcal{O}, \mathrm{QMaps}^a(\mathbf{C}, \mathrm{Gr}^{BD,a}_{G,\mathbf{X}})_{\mathcal{O}})^\tau$ (cf. Lemma 4.10), we also obtain that $D'_v = D_v + \sum_i b_i \cdot \pi_h(\mathbf{s}_i)$. Moreover, we obtain that σ'' is the composition of σ' and the embedding $\mathrm{Gr}^{BD,a-b}_{G,\mathbf{X},D'_v} \to \mathrm{Gr}^{BD,a}_{G,\mathbf{X},D_v}$, which is what we had to show.

To complete the proof of the theorem, it would be enough to show that the projection ${}^{\mathrm{Drinf}}\mathfrak{U}^a_G \to \mathrm{Sect}(\mathcal{O}, \mathrm{QMaps}^a(\mathbf{C}, \mathrm{Gr}^{BD,a}_{G,\mathbf{X}})_{\mathcal{O}})$ is a closed embedding. We know this for the closed subscheme $\mathfrak{U}^a_G$, and hence, we obtain a priori that the map in question is ind-finite. Therefore, it would suffice to show that any tangent vector to ${}^{\mathrm{Drinf}}\mathfrak{U}^a_G$, which is vertical with respect to the map ${}^{\mathrm{Drinf}}\mathfrak{U}^a_G \to \mathrm{Sect}(\mathcal{O}, \mathrm{QMaps}^a(\mathbf{C}, \mathrm{Gr}^{BD,a}_{G,\mathbf{X}})_{\mathcal{O}})$, is in fact zero.

Let $\mathrm{Spec}(\mathbb{C}[\epsilon]/\epsilon^2) \to {}^{\mathrm{Drinf}}\mathfrak{U}^a_G$ be such a tangent vector. Then the relative quasi-map $\sigma_{\mathcal{O}} : \mathbf{C}_{\mathcal{O}} \times \mathrm{Spec}(\mathbb{C}[\epsilon]/\epsilon^2) \to \mathrm{Gr}^{BD,a}_{G,\mathbf{X}_{\mathcal{O}}}$ is actually independent of ϵ, and as we have seen in the proof of Lemma 8.6, we can assume that the generically defined G-bundle $\mathcal{F}_G$ is also independent of ϵ. Thus we are dealing with an infinitesimal automorphism of $\mathcal{F}_G$, and we must show that it is zero.

However, $\mathcal{F}_G$ extends canonically to a G-bundle on the entire $\mathbf{S}$, and so does our infinitesimal automorphism. Since this automorphism preserves the trivialization along $\mathbf{D}_\infty$, our assertion reduces to the fact that points of $\mathrm{Bun}_G(\mathbf{S}, \mathbf{D}_\infty)$ have no automorphisms.

Part III: Parabolic Versions of Uhlenbeck Spaces

Throughout this part, $\mathfrak{g}$ will be a simple finite-dimensional Lie algebra with the corresponding simply connected group denoted G. By $\mathfrak{g}_{\mathrm{aff}}$ we will denote the corresponding untwisted affine Kac–Moody algebra, i.e., $\mathfrak{g}_{\mathrm{aff}} = \mathfrak{g}((x)) \oplus K \cdot \mathbb{C} \oplus d \cdot \mathbb{C}$, as in Section 3.7.

9 Parabolic Uhlenbeck spaces

9.1

Let $\mathbb{P}^1 \simeq \mathbf{D}_0 \subset \mathbf{S} \simeq \mathbb{P}^2$ be a divisor different from $\mathbf{D}_\infty$; we will denote by $\overset{\circ}{\mathbf{D}}_0$ the intersection $\mathbf{D}_0 \cap \overset{\circ}{\mathbf{S}}$.

Let us denote by $\mathrm{Bun}_{G;P}(\mathbf{S}, \mathbf{D}_\infty; \mathbf{D}_0)$ the scheme classifying the data consisting of a G-bundle $\mathcal{F}_G$ on $\mathbf{S}$, a trivialization of $\mathcal{F}_G|_{\mathbf{D}_\infty}$, and a reduction to P of $\mathcal{F}_G|_{\mathbf{D}_0}$, compatible with the above trivialization at the point $\mathbf{D}_\infty \cap \mathbf{D}_0$. In other words, if we denote by $\mathrm{Bun}_P(\mathbf{D}_0; \mathbf{D}_\infty \cap \mathbf{D}_0)$ the moduli stack classifying P-bundles on the curve $\mathbf{D}_0$ with a trivialization at $\mathbf{D}_\infty \cap \mathbf{D}_0$, we have an identification

$$\mathrm{Bun}_{G;P}(\mathbf{S}, \mathbf{D}_\infty; \mathbf{D}_0) \simeq \mathrm{Bun}_G(\mathbf{S}, \mathbf{D}_\infty) \underset{\mathrm{Bun}_G(\mathbf{D}_0;\mathbf{D}_\infty\cap\mathbf{D}_0)}{\times} \mathrm{Bun}_P(\mathbf{D}_0; \mathbf{D}_\infty \cap \mathbf{D}_0). \quad (18)$$

From this description, we see that $\mathrm{Bun}_{G;P}(\mathbf{S}, \mathbf{D}_\infty; \mathbf{D}_0)$ splits into connected components indexed by $\theta \in \widehat{\Lambda}_{\mathfrak{g},\mathfrak{p}} \simeq \Lambda_{\mathfrak{g},\mathfrak{p}} \oplus \delta \cdot \mathbb{Z}$, where for $\theta = (\overline{\theta}, a)$, the index a indexes the component of $\mathrm{Bun}_G(\mathbf{S}, \mathbf{D}_\infty)$, and $\overline{\theta}$ that of $\mathrm{Bun}_P(\mathbf{D}_0; \mathbf{D}_\infty \cap \mathbf{D}_0)$.

Recall the scheme $\mathcal{G}_{G,P,\mathbf{X}}$ (cf. Section 3.7), and observe that from Proposition 3.4 it follows that the scheme $\mathrm{Bun}^\theta_{G;P}(\mathbf{S}, \mathbf{D}_\infty; \mathbf{D}_0) \simeq \mathrm{Bun}^\theta_{G;P}(\mathbf{S}', \mathbf{D}'_\infty; \mathbf{D}'_0)$ is isomorphic to the scheme of based maps, $\mathrm{Maps}^\theta(\mathbf{C}, \mathcal{G}_{G,P,\mathbf{X}})$. This description implies that $\mathrm{Bun}^\theta_{G;P}(\mathbf{S}, \mathbf{D}_\infty; \mathbf{D}_0)$ is empty unless $\theta \in \widehat{\Lambda}^{\mathrm{pos}}_{\mathfrak{g},\mathfrak{p}}$.

Our present goal is to introduce parabolic versions of the Uhlenbeck spaces, $\mathfrak{U}^\theta_{G;P}$ and $\widetilde{\mathfrak{U}}^\theta_{G;P}$, both of which contain $\mathrm{Bun}_{G;P}(\mathbf{S}, \mathbf{D}_\infty; \mathbf{D}_0)$ as an open subset.

Let us first choose a pair of directions $(\mathbf{d}_v, \mathbf{d}_h) \in \mathbf{D}_\infty$, but with the condition that $\mathbf{d}_h = \mathbf{D}_\infty \cap \mathbf{D}_0$. In other words, we have a rational surface $\mathbf{S}' \simeq \mathbf{C} \times \mathbf{X}$, and in addition to $\infty_\mathbf{C} \in \mathbf{C}$ and $\infty_\mathbf{X} \in \mathbf{X}$, we have a distinguished point $0_\mathbf{X} \in \mathbf{X}$, such that $\mathbf{D}'_0 = \mathbf{C} \times 0_\mathbf{X}$ is the proper transform of $\mathbf{D}_0$. Henceforth, we will identify $\mathbf{D}_0 \simeq \mathbf{D}'_0 \simeq \mathbf{C}$.

We will first introduce $\mathfrak{U}^\theta_{G;P}$ and $\widetilde{\mathfrak{U}}^\theta_{G;P}$ using a choice of $\mathbf{d}_v$, and then show that the definition is in fact independent of that choice.

Definition 9.2. *For $\theta = (\overline{\theta}, a)$, the parabolic Uhlenbeck space $\mathfrak{U}^\theta_{G;P}$ is defined as the closure of* $\mathrm{Bun}^\theta_{G;P}(\mathbf{S}', \mathbf{D}'_\infty; \mathbf{D}'_0)$ *in the product* $\mathrm{QMaps}^\theta(\mathbf{C}, \mathcal{G}_{G,P,\mathbf{X}}) \underset{\mathrm{QMaps}^a(\mathbf{C},\mathcal{G}_{G,\mathbf{X}})}{\times} \mathfrak{U}^a_G$.

Note that for $P = G$, the scheme $\mathfrak{U}^\theta_{G;P}$ is nothing but $\mathfrak{U}^a_G$.

Definition 9.3. *The enhanced parabolic Uhlenbeck space $\widetilde{\mathfrak{U}}^\theta_{G;P}$ is defined as the closure of* $\mathrm{Bun}^\theta_{G;P}(\mathbf{S}', \mathbf{D}'_\infty; \mathbf{D}'_0)$ *in the product* $\widetilde{\mathrm{QMaps}}^\theta(\mathbf{C}, \mathcal{G}_{G,P,\mathbf{X}}) \underset{\mathrm{QMaps}^a(\mathbf{C},\mathcal{G}_{G,\mathbf{X}})}{\times} \mathfrak{U}^a_G$.

As we shall see later, both $\mathfrak{U}^\theta_{G;P}$ and $\widetilde{\mathfrak{U}}^\theta_{G;P}$ are schemes of finite type. Note also that from the definition of $\mathfrak{U}^a_G$ and Theorem 4.8, it follows that $\mathfrak{U}^\theta_{G;P}$ (respectively, $\widetilde{\mathfrak{U}}^\theta_{G;P}$) could be equivalently defined as the closure of $\mathrm{Bun}^\theta_{G;P}(\mathbf{S}', \mathbf{D}'_\infty; \mathbf{D}'_0)$ inside the product $\mathrm{QMaps}^\theta(\mathbf{C}, \mathcal{G}_{G,P,\mathbf{X}}) \times \mathrm{Sect}(\mathsf{O}, \overset{\circ}{\mathbf{C}}{}^{(a)}_{\mathsf{O}})$ (respectively, $\widetilde{\mathrm{QMaps}}^\theta(\mathbf{C}, \mathcal{G}_{G,P,\mathbf{X}}) \times \mathrm{Sect}(\mathsf{O}, \overset{\circ}{\mathbf{C}}{}^{(a)}_{\mathsf{O}})$). Note also that $\widetilde{\mathfrak{U}}^\theta_{G;P}$ is a closed subscheme in the product $\mathfrak{U}^\theta_{G;P} \underset{\overset{\circ}{\mathbf{C}}{}^\theta}{\times} \mathrm{Mod}^{\theta,+}_{M_{\mathrm{aff}},\mathbf{C}}$ (this is so because the corresponding property holds on the level of quasi-map spaces). We will denote the natural projection $\widetilde{\mathfrak{U}}^\theta_{G;P} \to \mathfrak{U}^\theta_{G;P}$ by $\mathfrak{r}_{\mathfrak{p}^+_{\mathrm{aff}}}$. Obviously, for $P = B$, the map $\mathfrak{r}_{\mathfrak{p}^+_{\mathrm{aff}}}$ is an isomorphism, but it is NOT an isomorphism for $P = G$.

We will denote by $\varrho^\theta_{\mathfrak{p}^+_{\mathrm{aff}}}$ and $\varrho^\theta_{M_{\mathrm{aff}}}$ the natural maps $\mathfrak{U}^\theta_{G;P} \to \overset{\circ}{\mathbf{C}}{}^\theta$ and $\widetilde{\mathfrak{U}}^\theta_{G;P} \to \mathrm{Mod}^{\theta,+}_{M_{\mathrm{aff}},\mathbf{C}}$.

The following *horizontal factorization property* of $\mathfrak{U}^\theta_{G;P}$ and $\widetilde{\mathfrak{U}}^\theta_{G;P}$ follows from the corresponding factorization properties of quasi-map spaces (Proposition 2.19) and of the Uhlenbeck space (Proposition 6.5).

Corollary 9.4. *For $\theta = \theta_1 + \theta_2$, we have canonical isomorphisms*

$$\mathfrak{U}^\theta_{G;P} \underset{\overset{\circ}{\mathbf{C}}{}^\theta}{\times} (\overset{\circ}{\mathbf{C}}{}^{\theta_1} \times \overset{\circ}{\mathbf{C}}{}^{\theta_2})_{\mathrm{disj}} \simeq (\mathfrak{U}^{\theta_1}_{G;P} \times \mathfrak{U}^{\theta_2}_{G;P}) \underset{\overset{\circ}{\mathbf{C}}{}^{\theta_1} \times \overset{\circ}{\mathbf{C}}{}^{\theta_1}}{\times} (\overset{\circ}{\mathbf{C}}{}^{\theta_1} \times \overset{\circ}{\mathbf{C}}{}^{\theta_2})_{\mathrm{disj}}$$

and

$$\widetilde{\mathfrak{U}}^\theta_{G;P} \underset{\overset{\circ}{\mathbf{C}}{}^\theta}{\times} (\overset{\circ}{\mathbf{C}}{}^{\theta_1} \times \overset{\circ}{\mathbf{C}}{}^{\theta_2})_{\mathrm{disj}} \simeq (\widetilde{\mathfrak{U}}^{\theta_1}_{G;P} \times \widetilde{\mathfrak{U}}^{\theta_2}_{G;P}) \underset{\overset{\circ}{\mathbf{C}}{}^{\theta_1} \times \overset{\circ}{\mathbf{C}}{}^{\theta_2}}{\times} (\overset{\circ}{\mathbf{C}}{}^{\theta_1} \times \overset{\circ}{\mathbf{C}}{}^{\theta_2})_{\mathrm{disj}}.$$

9.5

Now our goal is to prove that the schemes $\mathfrak{U}^\theta_{G;P}$ and $\widetilde{\mathfrak{U}}^\theta_{G;P}$ are in fact canonically attached to the surface $\mathbf{S}$ with the divisor $\mathbf{D}_0$, i.e., that they do not depend on the choice of the direction $\mathbf{d}_v$. We will consider the case of $\widetilde{\mathfrak{U}}^\theta_{G;P}$, since that of $\mathfrak{U}^\theta_{G;P}$ is analogous (and simpler).

By Theorem 8.7, for $S = \mathfrak{U}^a_G$, the product $\mathbf{S} \times S$ contains an open subset U, such that $\mathbf{S} \times S - U$ is finite over S, and over which we have a well-defined G-bundle $\mathcal{F}_G$. Let D be a divisor on $\mathbf{C} \times S$, such that $\mathbf{C} \times S - D \subset U$. We set ${}^{\mathrm{Huge}}\widetilde{\mathfrak{U}}^\theta_{G;P}$ to be the ind-scheme of meromorphic enhanced quasi-maps

$${}_{D\cdot\infty}\widetilde{\mathrm{QMaps}}(\mathbf{C}, \mathcal{F}_G|_{\mathbf{C}\times S-D} \overset{G}{\times} \mathcal{G}_{\mathfrak{g},\mathfrak{p}});$$

cf. Section 2.35 (recall that $\mathcal{G}_{\mathfrak{g},\mathfrak{p}}$ is the finite-dimensional flag variety G/P).

Clearly, ${}^{\mathrm{Huge}}\widetilde{\mathfrak{U}}^\theta_{G;P}$, is defined in a way independent of the choice of $\mathbf{d}_v$. We now claim that for any $\mathbf{d}_v$ we have a closed embedding $\widetilde{\mathfrak{U}}^\theta_{G;P} \hookrightarrow {}^{\mathrm{Huge}}\widetilde{\mathfrak{U}}^\theta_{G;P}$.

Indeed, the scheme $\widetilde{\mathrm{QMaps}}(\mathbf{C}, \mathcal{G}_{G,P,\mathbf{X}})$ embeds as a closed subscheme into the corresponding ind-scheme of meromorphic enhanced quasi-maps ${}_{D\cdot\infty}\widetilde{\mathrm{QMaps}}(\mathbf{C}, \mathcal{G}_{G,P,\mathbf{X}})$. Now

$$ {}_{D\cdot\infty}\widetilde{\mathrm{QMaps}}(\mathbf{C}, \mathcal{G}_{G,P,\mathbf{X}}) \underset{{}_{D\cdot\infty}\mathrm{QMaps}(\mathbf{C},\mathcal{G}_{G,\mathbf{X}})}{\times} S \simeq {}^{\mathrm{Huge}}\widetilde{\mathfrak{U}}^\theta_{G;P}. $$

This is because the data of an enhanced quasi-map $(\mathbf{C} \times S - D) \to \mathcal{G}_{G,P,\mathbf{X}}$ that projects to a given map $(\mathbf{C} \times S - D) \to \mathcal{G}_{G,\mathbf{X}}$ is equivalent to a data of an enhanced quasi-map

$$ (\mathbf{C} \times S - D) \to \mathcal{F}_G|_{\mathbf{C}\times S-D} \overset{G}{\times} \mathcal{G}_{\mathfrak{g},\mathfrak{p}}. $$

Note that the composition

$$ \mathrm{Bun}^\theta_{G;P}(\mathbf{S}', \mathbf{D}'_\infty; \mathbf{D}'_0) \to \widetilde{\mathfrak{U}}^\theta_{G;P} \to {}^{\mathrm{Huge}}\widetilde{\mathfrak{U}}^\theta_{G;P} $$

is also independent of $\mathbf{d}_\upsilon$. This shows that we can identify $\widetilde{\mathfrak{U}}^\theta_{G;P}$ with the closure of $\mathrm{Bun}^\theta_{G;P}(\mathbf{S}', \mathbf{D}'_\infty; \mathbf{D}'_0)$ inside ${}^{\mathrm{Huge}}\widetilde{\mathfrak{U}}^\theta_{G;P}$, which makes it manifestly independent of $\mathbf{d}_\upsilon$. Note also that in the course of the proof, we also established the following.

Corollary 9.6. *The map* $\widetilde{\mathfrak{U}}^\theta_{G;P} \to \mathrm{Bun}_M(\mathbf{C}, \infty_{\mathbf{C}})$*, which is the composition of* $\varrho^\theta_{M_{\mathrm{aff}}} : \widetilde{\mathfrak{U}}^\theta_{G;P} \to \mathrm{Mod}^{\theta,+}_{M_{\mathrm{aff}},\mathbf{C}}$ *and the natural projection* $\mathrm{Mod}^{\theta,+}_{M_{\mathrm{aff}},\mathbf{C}} \to \mathrm{Bun}_M(\mathbf{C}, \infty_{\mathbf{C}})$*, is independent of the choice of* $\mathbf{d}_\upsilon$.

(Of course, the maps $\varrho^\theta_{M_{\mathrm{aff}}}$ and $\varrho^\theta_{\mathfrak{p}^+_{\mathrm{aff}}} : \mathfrak{U}^\theta_{G;P} \to \overset{\circ}{\mathbf{C}}{}^\theta$ do depend on the choice of $\mathbf{d}_\upsilon$.)

Corollary 9.7. *The schemes* $\mathfrak{U}^\theta_{G;P}$ *and* $\widetilde{\mathfrak{U}}^\theta_{G;P}$ *are of finite type.*

(Indeed, the ind-scheme ${}^{\mathrm{Huge}}\mathfrak{U}^\theta_{G;P}$ is of ind-finite type, and $\widetilde{\mathfrak{U}}^\theta_{G;P}$ is the closure of $\mathrm{Bun}^a_{G;P}(\mathbf{S}, \mathbf{D}_\infty, \mathbf{D}_0)$ inside ${}^{\mathrm{Huge}}\mathfrak{U}^\theta_{G;P}$.)

9.8 Beilinson–Drinfeld–Kottwitz flag space

Consider the ind-scheme $\mathrm{Gr}^{BD,a}_{G,P,\mathbf{X},0_{\mathbf{X}}}$ fibered over over $\overset{\circ}{\mathbf{X}}{}^{(a)}$, which classifies the data of

- a divisor $D \in \overset{\circ}{\mathbf{X}}{}^{(a)}$,
- a principal G-bundle $\mathcal{F}_G$ on $\mathbf{X}$,
- a trivialization $\mathcal{F}_G \simeq \mathcal{F}^0_G|_{\mathbf{X}-D}$, and
- a reduction to P of the G-torsor $\mathcal{F}_G|_{0_{\mathbf{X}}}$.

We have an evident map $\mathrm{Gr}^{BD,a}_{G,P,\mathbf{X},0_{\mathbf{X}}} \to \mathcal{G}_{G,P,\mathbf{X}}$, which remembers of the trivialization of $\mathcal{F}_G$ on $\mathbf{X} - D$ its restriction to the formal neighborhood of $\infty_{\mathbf{X}}$. Moreover, it is easy to see that the map $\mathrm{Gr}^{BD,a}_{G,P,\mathbf{X},0_{\mathbf{X}}} \to \mathcal{G}_{G,P,\mathbf{X}} \times \overset{\circ}{\mathbf{X}}{}^{(a)}$ is, in fact, a closed embedding.

Thus we can consider the relative based quasi-map space $\mathrm{QMaps}(\mathbf{C}, \mathrm{Gr}^{BD,a}_{G,P,\mathbf{X},0_{\mathbf{X}}})$, fibered over $\overset{\circ}{\mathbf{X}}{}^{(a)}$, which is a closed subscheme inside $\mathrm{QMaps}(\mathbf{C}, \mathcal{G}_{G,P,\mathbf{X}}) \times \overset{\circ}{\mathbf{X}}{}^{(a)}$.

As in Proposition 4.7, we obtain that the parabolic Uhlenbeck space $\mathfrak{U}^\theta_{G;P}$ can be alternatively defined as the closure of $\mathrm{Bun}^\theta_{G;P}(\mathbf{S}', \mathbf{D}'_\infty; \mathbf{D}'_0)$ in the product

$$\mathrm{QMaps}^\theta(\mathbf{C}, \mathrm{Gr}^{BD,a}_{G,P,\mathbf{X},0_\mathbf{X}}) \underset{\mathrm{QMaps}^a(\mathbf{C},\mathrm{Gr}^{BD,a}_{G,\mathbf{X}})}{\times} \mathfrak{U}^a_G,$$

where a is the projection of θ under $\widehat{\Lambda}_{\mathfrak{g},\mathfrak{p}} \to \mathbb{Z}$.

The above realization of $\mathfrak{U}^\theta_{G;P}$ gives an alternative proof that $\mathfrak{U}^\theta_{G;P}$ is a scheme of finite type, since $\mathrm{Gr}^{BD,a}_{G,P,\mathbf{X},0_\mathbf{X}}$ is of ind-finite type.

We will establish another important property of parabolic Uhlenbeck spaces (going back to the work [Va] of G. Valli), called the *vertical factorization property*.

Proposition 9.9. *Let $\theta = (\overline{\theta}, a)$ be an element of $\widehat{\Lambda}^{\mathrm{pos}}_{\mathfrak{g},\mathfrak{p}}$. If $a = a_1 + a_2$ is such that $\theta_1 = \theta - a_2 \cdot \delta$ also lies in $\widehat{\Lambda}^{\mathrm{pos}}_{\mathfrak{g},\mathfrak{p}}$, then there are canonical isomorphisms*

$$\begin{aligned}&\mathfrak{U}^\theta_{G;P} \underset{\overset{\circ}{\mathbf{X}}{}^{(a)}}{\times} (\overset{\circ}{\mathbf{X}}{}^{(a_1)} \times (\overset{\circ}{\mathbf{X}} - 0_\mathbf{X})^{(a_2)})_{\mathrm{disj}} \\ &\quad\simeq (\mathfrak{U}^{\theta_1}_{G;P} \times \mathfrak{U}^{a_2}_G) \underset{\overset{\circ}{\mathbf{X}}{}^{(a_1)} \times (\overset{\circ}{\mathbf{X}} - 0_\mathbf{X})^{(a_2)}}{\times} (\overset{\circ}{\mathbf{X}}{}^{(a_1)} \times (\overset{\circ}{\mathbf{X}} - 0_\mathbf{X})^{(a_2)})_{\mathrm{disj}},\end{aligned}$$

and

$$\begin{aligned}&\widetilde{\mathfrak{U}}^\theta_{G;P} \underset{\overset{\circ}{\mathbf{X}}{}^{(a)}}{\times} (\overset{\circ}{\mathbf{X}}{}^{(a_1)} \times (\overset{\circ}{\mathbf{X}} - 0_\mathbf{X})^{(a_2)})_{\mathrm{disj}} \\ &\quad\simeq (\widetilde{\mathfrak{U}}^{\theta_1}_{G;P} \times \mathfrak{U}^{a_2}_G) \underset{\overset{\circ}{\mathbf{X}}{}^{(a_1)} \times (\overset{\circ}{\mathbf{X}} - 0_\mathbf{X})^{(a_2)}}{\times} (\overset{\circ}{\mathbf{X}}{}^{(a_1)} \times (\overset{\circ}{\mathbf{X}} - 0_\mathbf{X})^{(a_2)})_{\mathrm{disj}}.\end{aligned}$$

If $\theta - \delta \notin \Lambda^{\mathrm{pos}}_{aff,\mathfrak{g},\mathfrak{p}}$, then the composition $\mathfrak{U}^\theta_{G;P} \to \mathfrak{U}^a_G \xrightarrow{\varpi^a_v} \overset{\circ}{\mathbf{X}}{}^{(a)}$ maps to the point $a \cdot 0_\mathbf{X} \in \overset{\circ}{\mathbf{X}}{}^{(a)}$.

Proof. First, the usual Uhlenbeck space $\mathfrak{U}^a_G$ has the vertical factorization property

$$\mathfrak{U}^a_G \underset{\overset{\circ}{\mathbf{X}}{}^{(a)}}{\times} (\overset{\circ}{\mathbf{X}}{}^{(a_1)} \times \overset{\circ}{\mathbf{X}}{}^{(a_2)})_{\mathrm{disj}} \simeq (\mathfrak{U}^{a_1}_G \times \mathfrak{U}^{a_2}_G) \underset{\overset{\circ}{\mathbf{X}}{}^{(a_1)} \times \overset{\circ}{\mathbf{X}}{}^{(a_2)}}{\times} (\overset{\circ}{\mathbf{X}}{}^{(a_1)} \times \overset{\circ}{\mathbf{X}}{}^{(a_2)})_{\mathrm{disj}}, \quad (19)$$

because $\mathbf{C}$ and $\mathbf{X}$ play symmetric roles in the definition of $\mathfrak{U}^a_G$.

Using (18) we obtain that the corresponding factorization property for the open subscheme $\mathrm{Bun}^\theta_{G;P}(\mathbf{S}, \mathbf{D}_\infty; \mathbf{D}_0)$ follows from that of $\mathrm{Bun}^a_G(\mathbf{S}, \mathbf{D}_\infty)$.

To establish the factorization property of $\mathfrak{U}^\theta_{G;P}$ we will use the fact (cf. [Ga]) that

$$\mathrm{Gr}^{BD,a}_{G,P,\mathbf{X},0_\mathbf{X}} \underset{\overset{\circ}{\mathbf{X}}{}^{(a)}}{\times} (\overset{\circ}{\mathbf{X}}{}^{(a_1)} \times (\overset{\circ}{\mathbf{X}} - 0_\mathbf{X})^{(a_2)})_{\mathrm{disj}}$$

$$\simeq (\mathrm{Gr}^{BD,a_1}_{G,P,\mathbf{X},0_{\mathbf{X}}} \times \mathrm{Gr}^{BD,a_2}_{G,\mathbf{X}}) \underset{\overset{\circ}{\mathbf{X}}{}^{(a_1)} \times \overset{\circ}{\mathbf{X}}{}^{(a_2)}}{\times} (\mathbf{X}^{(a_1)} \times (\overset{\circ}{\mathbf{X}} - 0_{\mathbf{X}})^{(a_2)})_{\mathrm{disj}}. \tag{20}$$

Moreover, this decomposition is compatible with the corresponding line bundles. Therefore, (20) combined with (19) and Lemma 1.15 yield an isomorphism:

$$(\mathrm{QMaps}^{\theta}(\mathbf{C}, \mathrm{Gr}^{BD,a}_{G,P,\mathbf{X},0_{\mathbf{X}}}) \underset{\mathrm{QMaps}^a(\mathbf{C},\mathrm{Gr}^{BD,a}_{G,\mathbf{X}})}{\times} \mathfrak{U}^a_G) \underset{\overset{\circ}{\mathbf{X}}{}^{(a)}}{\times} (\overset{\circ}{\mathbf{X}}{}^{(a_1)} \times (\overset{\circ}{\mathbf{X}} - 0_{\mathbf{X}})^{(a_2)})_{\mathrm{disj}}$$

$$\simeq (\mathrm{QMaps}^{\theta_1}(\mathbf{C}, \mathrm{Gr}^{BD,a_1}_{G,P,\mathbf{X},0_{\mathbf{X}}}) \times \mathrm{QMaps}^{a_2}(\mathbf{C}, \mathrm{Gr}^{BD,a_2}_{G,\mathbf{X}}))$$

$$\underset{\mathrm{QMaps}^{a_1}(\mathbf{C},\mathrm{Gr}^{BD,a_1}_{G,\mathbf{X}}) \times \mathrm{QMaps}^{a_2}(\mathbf{C},\mathrm{Gr}^{BD,a_2}_{G,\mathbf{X}})}{\times} (\mathfrak{U}^{a_1}_G \times \mathfrak{U}^{a_2}_G)$$

$$\underset{\overset{\circ}{\mathbf{X}}{}^{(a_1)} \times (\overset{\circ}{\mathbf{X}} - 0_{\mathbf{X}})^{(a_2)}}{\times} (\overset{\circ}{\mathbf{X}}{}^{(a_1)} \times (\overset{\circ}{\mathbf{X}} - 0_{\mathbf{X}})^{(a_2)})_{\mathrm{disj}}.$$

Since $\mathfrak{U}^{\theta}_{G;P}$ is the closure of $\mathrm{Bun}^{\theta}_{G;P}(\mathbf{S}', \mathbf{D}'_{\infty}; \mathbf{D}'_0)$ in the product

$$\mathrm{QMaps}^{\theta}(\mathbf{C}, \mathrm{Gr}^{BD,a}_{G,P,\mathbf{X},0_{\mathbf{X}}}) \underset{\mathrm{QMaps}^a(\mathbf{C},\mathrm{Gr}^{BD,a}_{G,\mathbf{X}})}{\times} \mathfrak{U}^a_G,$$

we conclude that it factorizes in the required fashion.

To prove the assertion for $\widetilde{\mathfrak{U}}^{\theta}_{G;P}$, recall that it is the closure of $\mathrm{Bun}^{\theta}_{G;P}(\mathbf{S}', \mathbf{D}'_{\infty}; \mathbf{D}'_0)$ in the product $\mathfrak{U}^{\theta}_{G;P} \times \mathrm{Mod}^{\theta,+}_{M_{\mathrm{aff}}}$.

Note also that we have a natural closed embedding $\mathrm{Mod}^{\theta_1,+}_{M_{\mathrm{aff}}} \to \mathrm{Mod}^{\theta,+}_{M_{\mathrm{aff}}}$, such that we have a commutative square

$$\begin{array}{ccc}
(\overset{\circ}{\mathbf{X}}{}^{(a_1)} \times \overset{\circ}{\mathbf{X}}{}^{(a_2)})_{\mathrm{disj}} \underset{\overset{\circ}{\mathbf{X}}{}^{(a_1)} \times \overset{\circ}{\mathbf{X}}{}^{(a_2)}}{\times} (\mathrm{Bun}^{\theta_1}_{G;P}(\mathbf{S}', \mathbf{D}'_{\infty}; \mathbf{D}'_0) \times \mathrm{Bun}^{a_2}_G(\mathbf{S}', \mathbf{D}'_{\infty})) & \longrightarrow & \mathrm{Mod}^{\theta_1,+}_{M_{\mathrm{aff}}} \\
\sim\downarrow & & \downarrow \\
(\overset{\circ}{\mathbf{X}}{}^{(a_1)} \times \overset{\circ}{\mathbf{X}}{}^{(a_2)})_{\mathrm{disj}} \underset{\overset{\circ}{\mathbf{X}}{}^{(a)}}{\times} \mathrm{Bun}^{\theta}_{G;P}(\mathbf{S}', \mathbf{D}'_{\infty}; \mathbf{D}'_0) & \longrightarrow & \mathrm{Mod}^{\theta,+}_{M_{\mathrm{aff}}}.
\end{array}$$

Therefore, the factorization property for $\widetilde{\mathfrak{U}}^{\theta}_{G;P}$ follows from that for $\mathfrak{U}^{\theta}_{G;P}$.

Suppose now that $\theta - \delta \notin \widehat{\Lambda}^{\mathrm{pos}}_{\mathfrak{g},\mathfrak{p}}$. Since $\mathrm{Bun}^{\theta}_{G;P}(\mathbf{S}, \mathbf{D}_{\infty}; \mathbf{D}_0)$ is dense in $\mathfrak{U}^{\theta}_{G;P}$, it is enough to prove that

$$\mathrm{Bun}^{\theta}_{G;P}(\mathbf{S}, \mathbf{D}_{\infty}; \mathbf{D}_0) \to \mathrm{Bun}^a_G(\mathbf{S}, \mathbf{D}_{\infty}) \overset{\varpi^a_v}{\longrightarrow} \overset{\circ}{\mathbf{X}}{}^{(a)}$$

is the constant map to $a \cdot 0_{\mathbf{X}} \in \overset{\circ}{\mathbf{X}}{}^{(a)}$.

If this were not the case, there would exist $a_1, a_2 \neq 0$ with $a_1 + a_2 = a$ and a based map

$$\mathbf{C} \to \mathrm{Gr}^{BD,a_1}_{G,P,\mathbf{X},0_{\mathbf{X}}} \times \mathrm{Gr}^{BD,a_2}_{G,\mathbf{X}}$$

of total degree θ, such that its projection onto the second factor, i.e., $\mathbf{C} \to \mathrm{Gr}^{BD,a_2}_{G,\mathbf{X}}$ was not the constant map. However, the space of such maps is the union over $b \in \mathbb{Z}$ of

$$\mathrm{Maps}^{\theta-b\cdot\delta}(\mathbf{C}, \mathrm{Gr}^{BD,a_1}_{G,P,\mathbf{X},0_\mathbf{X}}) \times \mathrm{Maps}^{b}(\mathbf{C}, \mathrm{Gr}^{BD,a_2}_{G,\mathbf{X}}),$$

which is nonempty only when $b \geq 0$ and $\theta - b \cdot \delta \in \Lambda^{\mathrm{pos}}_{\mathfrak{g},\mathfrak{p}}$. But this forces $b = 0$, i.e., our map $\mathbf{C} \to \mathrm{Gr}^{BD,a_2}_{G,\mathbf{X}}$ must actually be the constant map. $\square$

10 Stratification of parabolic Uhlenbeck spaces

10.1

For an element $\theta \in \widehat{\Lambda}^{\mathrm{pos}}_{\mathfrak{g},\mathfrak{p}}$ consider a decomposition $\theta = \theta_1 + \theta_2 + b \cdot \delta$ with θ_1, $\theta_2 \in \widehat{\Lambda}^{\mathrm{pos}}_{\mathfrak{g},\mathfrak{p}} - 0$, $b \in \mathbb{N}$. Let us denote by $\overset{\circ\circ}{\mathbf{S}}$ the complement to $\mathbf{D}_0$ in $\overset{\circ}{\mathbf{S}}$, and let us denote by $\mathfrak{U}^{\theta;\theta_2,b}_{G;P}$ the following scheme:

$$\mathrm{Bun}^{\theta_1}_{G;P}(\mathbf{S}, \mathbf{D}_\infty; \mathbf{D}_0) \times \overset{\circ}{\mathbf{C}}{}^{\theta_2} \times \mathrm{Sym}^b(\overset{\circ\circ}{\mathbf{S}}).$$

The following is a generalization of Theorem 7.2.

Theorem 10.2. *There exists a finite map $\bar{\iota}_{\theta_2,b} : \mathfrak{U}^{\theta_1}_{G;P} \times \overset{\circ}{\mathbf{C}}{}^{\theta_2} \times \mathrm{Sym}^b(\overset{\circ}{\mathbf{S}}) \to \mathfrak{U}^{\theta}_{G;P}$, compatible with the factorization isomorphisms of Corollary 9.4 and Proposition 9.9. Moreover, the composition*

$$\iota_{\theta_2,b} : \mathfrak{U}^{\theta;\theta_2,b}_{G;P} \hookrightarrow \mathfrak{U}^{\theta_1}_{G;P} \times \overset{\circ}{\mathbf{C}}{}^{\theta_2} \times \mathrm{Sym}^b(\overset{\circ}{\mathbf{S}}) \to \mathfrak{U}^{\theta}_{G;P}$$

is a locally closed embedding, and $\mathfrak{U}^{\theta}_{G;P} \simeq \underset{\theta_2,b}{\cup} \mathfrak{U}^{\theta;\theta_2,b}_{G;P}$.

Proof. We construct a map

$$\mathfrak{U}^{\theta_1}_{G;P} \times \overset{\circ}{\mathbf{C}}{}^{\theta_2} \times \mathrm{Sym}^b(\overset{\circ}{\mathbf{S}}) \to \mathrm{QMaps}^{\theta}(\mathbf{C}, \mathcal{G}_{G,P,\mathbf{X}}) \underset{\mathrm{QMaps}^a(\mathbf{C},\mathcal{G}_{G,\mathbf{X}})}{\times} \mathfrak{U}^a_G$$

as follows:

The projection $\mathfrak{U}^{\theta_1}_{G;P} \times \overset{\circ}{\mathbf{C}}{}^{\theta_2} \times \mathrm{Sym}^b(\overset{\circ}{\mathbf{S}}) \to \mathfrak{U}^a_G$ is the map

$$\mathfrak{U}^{\theta_1}_{G;P} \times \overset{\circ}{\mathbf{C}}{}^{\theta_2} \times \mathrm{Sym}^d(\overset{\circ}{\mathbf{S}}) \to \mathfrak{U}^{a_1}_G \times \overset{\circ}{\mathbf{C}}{}^{(a_2)} \times \mathrm{Sym}^b(\overset{\circ}{\mathbf{S}}) \to \mathfrak{U}^{a_1}_G \times \mathrm{Sym}^{a_2+b}(\overset{\circ}{\mathbf{S}}) \xrightarrow{\bar{\iota}_{a_2+b}} \mathfrak{U}^a_G,$$

where $\bar{\iota}_{a_2+b}$ is as in Theorem 7.2.

The projection $\mathfrak{U}^{\theta_1}_{G;P} \times \overset{\circ}{\mathbf{C}}{}^{\theta_2} \times \mathrm{Sym}^b(\overset{\circ}{\mathbf{S}}) \to \mathrm{QMaps}^{\theta}(\mathbf{C}, \mathcal{G}_{G,P,\mathbf{X}})$ is

$$\begin{aligned}\mathfrak{U}^{\theta_1}_{G;P} \times \overset{\circ}{\mathbf{C}}{}^{\theta_2} \times \operatorname{Sym}^b(\overset{\circ}{\mathbf{S}}) &\to \operatorname{QMaps}^{\theta_1}(\mathbf{C}, \mathcal{G}_{G,P,\mathbf{X}}) \times \overset{\circ}{\mathbf{C}}{}^{\theta_2} \times \overset{\circ}{\mathbf{C}}{}^{(b)} \\ &\to \operatorname{QMaps}^{\theta_1}(\mathbf{C}, \mathcal{G}_{G,P,\mathbf{X}}) \times \overset{\circ}{\mathbf{C}}{}^{\theta_2+b\cdot\delta} \\ &\to \operatorname{QMaps}^{\theta}(\mathbf{C}, \mathcal{G}_{G,P,\mathbf{X}}),\end{aligned}$$

where the first arrow in the above composition uses the map $\pi_v : \operatorname{Sym}^b(\overset{\circ}{\mathbf{S}}) \to \overset{\circ}{\mathbf{C}}{}^{(b)}$, the second arrow corresponds to the addition of divisors, and the third arrow comes from (3). It is easy to see that the resulting map $\mathfrak{U}^{\theta_1}_{G;P} \times \overset{\circ}{\mathbf{C}}{}^{\theta_2} \times \operatorname{Sym}^b(\overset{\circ}{\mathbf{S}}) \to \operatorname{QMaps}^{\theta}(\mathbf{C}, \mathcal{G}_{G,P,\mathbf{X}}) \times \mathfrak{U}^a_G$ indeed maps to the fiber product $\operatorname{QMaps}^{\theta}(\mathbf{C}, \mathcal{G}_{G,P,\mathbf{X}}) \underset{\operatorname{QMaps}^a(\mathbf{C},\mathcal{G}_{G,\mathbf{X}})}{\times} \mathfrak{U}^a_G$ and that the map $\iota_{\theta_2,b}$ is a locally closed embedding.

Moreover, it is easy to see that the image of $\bar{\iota}_{\theta_2,b}$ belongs in fact to the subscheme

$$\operatorname{QMaps}^{\theta}(\mathbf{C}, \operatorname{Gr}^{BD,a}_{G,P,\mathbf{X},0_{\mathbf{X}}}) \underset{\operatorname{QMaps}^a(\mathbf{C},\operatorname{Gr}^{BD,a}_{G,\mathbf{X}})}{\times} \mathfrak{U}^a_G,$$

and that this map is compatible with the isomorphisms of Proposition 9.9.

The map $\bar{\iota}_{\theta_2,b}$ is also compatible with the factorization isomorphisms (9.4) (due to the corresponding property of the map $\bar{\iota}_b$ for the Uhlenbeck space $\mathfrak{U}^a_G$). Therefore, to prove that the image of $\bar{\iota}_{\theta_2,b}$ belongs to $\mathfrak{U}^{\theta}_{G;P}$ (i.e., to the closure of $\operatorname{Bun}^{\theta}_{G;P}(\mathbf{S}', \mathbf{D}'_\infty; \mathbf{D}'_0)$ in the product $\operatorname{QMaps}^{\theta}(\mathbf{C}, \mathcal{G}_{G,P,\mathbf{X}}) \underset{\operatorname{QMaps}^a(\mathbf{C},\mathcal{G}_{G,\mathbf{X}})}{\times} \mathfrak{U}^a_G$), it suffices to analyze separately the following cases: (a) $\theta_2 = 0, b = 0$; (b) $\theta_1 = 0, \theta_2 = 0$; (c) $\theta_1 = 0, b = 0$. Of course, case a) is trivial, since it corresponds to the open stratum, i.e., $\operatorname{Bun}^{\theta}_{G;P}(\mathbf{S}', \mathbf{D}'_\infty; \mathbf{D}'_0)$ itself.

The second case, when $\theta_1 = 0, \theta_2 = 0$, follows immediately from Proposition 9.9. Indeed, the preimage of $(\overset{\circ}{\mathbf{X}} - 0_{\mathbf{X}})^{(b)}$ under $\mathfrak{U}^{b\cdot\delta}_{G;P} \to \mathfrak{U}^b_G \xrightarrow{\varpi^b_h} \overset{\circ}{\mathbf{X}}{}^{(b)}$ in this case is isomorphic to

$$\mathfrak{U}^b_G \underset{\overset{\circ}{\mathbf{X}}{}^{(b)}}{\times} (\overset{\circ}{\mathbf{X}} - 0_{\mathbf{X}})^{(b)},$$

which contains $\operatorname{Sym}^b(\overset{\circ\circ}{\mathbf{S}})$ as a closed subscheme, according to Theorem 7.2.

Thus it remains to treat the case $\theta_1 = 0, b = 0$. We will use again the horizontal factorization property, i.e., Corollary 9.4, which allows us to assume that θ_2 is the projection of a simple coroot of $\mathfrak{g}_{\mathrm{aff}}$.

To show that the image of $\overset{\circ}{\mathbf{C}} \simeq \operatorname{Bun}^0_{G;P}(\mathbf{S}, \mathbf{D}_\infty; \mathbf{D}_0) \times \overset{\circ}{\mathbf{C}} \times \operatorname{Sym}^0(\overset{\circ\circ}{\mathbf{S}})$ belongs to $\mathfrak{U}^{\theta}_{G;P}$, we will consider a certain $\mathbb{G}_m$-action on $\operatorname{QMaps}^{\theta}(\mathbf{C}, \mathcal{G}_{G,P,\mathbf{X}}) \underset{\operatorname{QMaps}^a(\mathbf{C},\mathcal{G}_{G,\mathbf{X}})}{\times} \mathfrak{U}^a_G$.

We choose a direction $\mathbf{d}_v \in \mathbf{D}_\infty - \mathbf{d}_h$ and consider the $\mathbb{G}_m$-action on $\mathbf{S}$ as in Proposition 7.7. This action induces a $\mathbb{G}_m$-action on $\operatorname{QMaps}^{\theta}(\mathbf{C}, \mathcal{G}_{G,P,\mathbf{X}}) \underset{\operatorname{QMaps}^a(\mathbf{C},\mathcal{G}_{G,\mathbf{X}})}{\times} \mathfrak{U}^a_G$, which we shall call "the action of the first kind." Note that the corresponding action on the first factor, i.e., $\operatorname{QMaps}^{\theta}(\mathbf{C}, \mathcal{G}_{G,P,\mathbf{X}})$, corresponds to the canonical homomorphism $\mathbb{G}_m \to M_{\mathrm{aff}} \times \mathbb{G}_m$, where the latter is thought of as the Levi subgroup corresponding to the parabolic $\mathfrak{p}^+_{\mathrm{aff}} \subset \mathfrak{g}_{\mathrm{aff}}$.

Another $\mathbb{G}_m$-action, which we shall call "the action of the second kind," corresponds to a dominant central cocharacter $\overline{\nu} : \mathbb{G}_m \to Z(M)$, as in Proposition 2.6.

Note that the actions of the first and the second kind commute with one another, and the compound $\mathbb{G}_m$-action, which we shall call "new," corresponds to a cocharacter $\mathbb{G}_m \to M_{\mathrm{aff}} \times \mathbb{G}_m$, which satisfies the dominance condition of Proposition 2.6. Using this proposition, combined with Proposition 7.7, we obtain that the "new" action contracts the space $\mathrm{QMaps}^\theta(\mathbf{C}, \mathcal{G}_{G,P,\mathbf{X}}) \underset{\mathrm{QMaps}^a(\mathbf{C},\mathcal{G}_{G,\mathbf{X}})}{\times} \mathfrak{U}^a_G$ to $\iota_{\theta,0}(\overset{\circ}{\mathbf{C}})$. This contraction map is dominant, due to its equivariance with respect to the action of $\mathbb{A}^1 \simeq \overset{\circ}{\mathbf{C}}$ by horizontal shifts.

In particular, we obtain that $\iota_{\theta,0}(\overset{\circ}{\mathbf{C}})$ lies in the closure of $\mathrm{Bun}^\theta_{G;P}(\mathbf{S}, \mathbf{D}_\infty; \mathbf{D}_0)$, which is what we had to show.

Now let us show that every geometric point of $\mathfrak{U}^\theta_{G;P}$ belongs to one of the locally closed subschemes $\mathfrak{U}^{\theta;\theta_2,b}_{G;P}$. We will argue by induction on the length of θ.

Consider the map $\mathfrak{U}^\theta_{G;P} \to \mathfrak{U}^a_G$, and let $z \in \mathfrak{U}^a_G$ be the image of our geometric point. If z has a singularity at $\mathbf{s} \in \overset{\circ\circ}{\mathbf{S}}$, then $\varpi^a_\upsilon(z) \neq a \cdot 0_{\mathbf{X}}$, and by Proposition 9.9, $\theta - \delta \in \widehat{\Lambda}^{\mathrm{pos}}_{\mathfrak{g},\mathfrak{p}}$. Hence our geometric point is contained in the image of $\bar{\imath}_{0,1}$, and our assertion follows from the induction hypothesis.

If z has a singularity at $\mathbf{s} \in \overset{\circ}{\mathbf{C}} \times 0_{\mathbf{X}}$, then our geometric point is contained in the image of $\bar{\imath}_{\alpha_0,0}$, and again the assertion follows by the induction hypothesis.

Hence it remains to analyze the case when z belongs to $\mathrm{Bun}^a_G(\mathbf{S}, \mathbf{D}_\infty)$, i.e., we are dealing with the locus of $\mathfrak{U}^\theta_{G;P}$ isomorphic to

$$\mathrm{Bun}_G(\mathbf{S}, \mathbf{D}_\infty) \underset{\mathrm{Bun}_G(\mathbf{C},\infty_{\mathbf{C}})}{\times} \overline{\mathrm{Bun}}^\theta_P(\mathbf{C}, \infty_{\mathbf{C}}),$$

where $\overline{\mathrm{Bun}}^\theta_P(\mathbf{C}, \infty_{\mathbf{C}})$ is the corresponding version of Drinfeld's stack of generalized P-bundles on $\mathbf{C}$ with a trivialization at $\infty_{\mathbf{C}} \in \mathbf{C}$.

The assertion in this case follows from the corresponding property of Drinfeld's compactifications; cf. [BG1]. □

10.3

In the course of the proof of Theorem 10.2, we have established the following.

Corollary 10.4. *The "new" $\mathbb{G}_m$-action on* $\mathrm{QMaps}^\theta(\mathbf{C}, \mathcal{G}_{G,P,\mathbf{X}}) \underset{\mathrm{QMaps}^a(\mathbf{C},\mathcal{G}_{G,\mathbf{X}})}{\times} \mathfrak{U}^a_G$ *preserves* $\mathfrak{U}^\theta_{G;P}$ *and contracts it to the subscheme* $\overset{\circ}{\mathbf{C}}{}^\theta \overset{\iota_{\theta,0}}{\hookrightarrow} \mathfrak{U}^\theta_{G;P}$. *The contraction map* $\mathfrak{U}^\theta_{G;P} \to \overset{\circ}{\mathbf{C}}{}^\theta$ *coincides with the map* $\varrho^\theta_{\mathfrak{p}^+_{\mathrm{aff}}}$.

In what follows, we will sometimes use the notation $\mathfrak{s}_{\mathfrak{p}^+_{\mathrm{aff}}}$ for the map $\overset{\circ}{\mathbf{C}}{}^\theta \to \mathfrak{U}^\theta_{G;P}$.

10.5

Now we turn to the space $\widetilde{\mathfrak{U}}^{\theta}_{G;P}$. For $\theta = \theta_1 + \theta_2 + b \cdot \delta$, let us denote by $\widetilde{\mathfrak{U}}^{\theta;\theta_2,b}_{G;P}$ the preimage in $\widetilde{\mathfrak{U}}^{\theta}_{G;P}$ of the locally closed subset $\mathfrak{U}^{\theta;\theta_2,b}_{G;P} \subset \mathfrak{U}^{\theta}_{G;P}$.

Proposition 10.6. *We have an isomorphism on the level of reduced schemes:*

$$\widetilde{\mathfrak{U}}^{\theta;\theta_2,b}_{G;P} \simeq (\mathrm{Bun}^{\theta_1}_{G;P}(\mathbf{S}, \mathbf{D}_\infty; \mathbf{D}_0) \underset{\mathrm{Bun}_M(\mathbf{C},\infty_\mathbf{C})}{\times} \mathcal{H}^{\theta_2,+}_{M_{\mathrm{aff}},\mathbf{C}}) \times \mathrm{Sym}^b(\overset{\circ\circ}{\mathbf{S}}).$$

Proof. Recall the map $\mathfrak{r}_{\mathfrak{p}^+_{\mathrm{aff}}} : \widetilde{\mathrm{QMaps}}^\theta(\mathbf{C}, \mathcal{G}_{G,P,\mathbf{X}}) \to \mathrm{QMaps}^\theta(\mathbf{C}, \mathcal{G}_{G,P,\mathbf{X}})$, and note that we have a natural closed embedding

$$\widetilde{\mathfrak{U}}^{\theta}_{G;P} \to \mathfrak{U}^{\theta}_{G;P} \underset{\mathrm{QMaps}^\theta(\mathbf{C},\mathcal{G}_{G,P,\mathbf{X}})}{\times} \widetilde{\mathrm{QMaps}}^\theta(\mathbf{C}, \mathcal{G}_{G,P,\mathbf{X}}).$$

Hence we a priori have a closed embedding

$$\widetilde{\mathfrak{U}}^{\theta;\theta_2,b}_{G;P} \to (\mathrm{Bun}^{\theta_1}_{G;P}(\mathbf{S}, \mathbf{D}_\infty; \mathbf{D}_0) \underset{\mathrm{Bun}_M(\mathbf{C},\infty_\mathbf{C})}{\times} \mathcal{H}^{\theta_2+b\cdot\delta,+}_{M_{\mathrm{aff}},\mathbf{C}}) \times \mathrm{Sym}^b(\overset{\circ\circ}{\mathbf{S}}),$$

by Proposition 2.34. However, using the vertical factorization property, Proposition 9.9, we obtain that its image is in fact contained in $(\mathrm{Bun}^{\theta_1}_{G;P}(\mathbf{S}, \mathbf{D}_\infty; \mathbf{D}_0) \underset{\mathrm{Bun}_M(\mathbf{C},\infty_\mathbf{C})}{\times} \mathcal{H}^{\theta_2,+}_{M_{\mathrm{aff}},\mathbf{C}}) \times \mathrm{Sym}^b(\overset{\circ\circ}{\mathbf{S}})$. Thus we have to show that the above inclusion is in fact an equality on the level of reduced schemes.

Using again the vertical factorization property, we can assume that $b = 0$. We know (cf. Lemma 3.6) that the map $\mathrm{Bun}_G(\mathbf{S}, \mathbf{D}_\infty) \simeq \mathrm{Bun}_G(\mathbf{S}', \mathbf{D}'_\infty) \to \mathrm{Bun}_G(\mathbf{C}, \infty_\mathbf{C})$, given by restriction of G-bundles under $\mathbf{C} \simeq \mathbf{D}_0 \to \mathbf{S}$, is smooth. Hence the map

$$\begin{aligned}\mathrm{Bun}_{G;P}(\mathbf{S}, \mathbf{D}_\infty; \mathbf{D}_0) \simeq \mathrm{Bun}_G(\mathbf{S}, \mathbf{D}_\infty) &\underset{\mathrm{Bun}_G(\mathbf{C},\infty_\mathbf{C})}{\times} \mathrm{Bun}_P(\mathbf{C}, \infty_\mathbf{C}) \\ &\to \mathrm{Bun}_P(\mathbf{C}, \infty_\mathbf{C}) \to \mathrm{Bun}_M(\mathbf{C}, \infty_\mathbf{C})\end{aligned}$$

is smooth as well. Therefore, the scheme $\mathrm{Bun}^{\theta_1}_G(\mathbf{S}, \mathbf{D}_\infty; \mathbf{D}_0) \underset{\mathrm{Bun}_M(\mathbf{C},\infty_\mathbf{C})}{\times} \mathcal{H}^{\theta_2,+}_{M_{\mathrm{aff}},\mathbf{C}}$ is irreducible, and it is enough to check our surjectivity assertion at the generic point.

Consider the map

$$\mathrm{Bun}^{\theta_1}_{G;P}(\mathbf{S}, \mathbf{D}_\infty; \mathbf{D}_0) \underset{\mathrm{Bun}_M(\mathbf{C},\infty_\mathbf{C})}{\times} \mathcal{H}^{\theta_2,+}_{M_{\mathrm{aff}},\mathbf{C}} \to (\overset{\circ}{\mathbf{C}}{}^{\theta_1} \times \overset{\circ}{\mathbf{C}}{}^{\theta_2}).$$

It is sufficient to analyze the locus which projects to pairs of simple and mutually disjoint divisors in $\overset{\circ}{\mathbf{C}}{}^{\theta_1} \times \overset{\circ}{\mathbf{C}}{}^{\theta_2}$. Moreover, using the horizontal factorization property, we reduce the assertion to the case when $\theta_1 = 0$, and $\theta_2 = \theta$ is the projection of a simple coroot in $\mathfrak{g}_{\mathrm{aff}}$.

For θ of the form $(\overline{\theta}, 0)$ we have an isomorphism $\widetilde{\mathfrak{U}}^{\theta}_{G;P} \simeq \widetilde{\mathrm{Bun}}^{\overline{\theta}}_P(\mathbf{C}, \infty_{\mathbf{C}})$. Therefore, if θ is the projection of a coroot belonging to $\mathfrak{g}$, our assertion follows from the corresponding fact for $\widetilde{\mathrm{Bun}}^{\overline{\theta}}_P(\mathbf{C}, \infty_{\mathbf{C}})$; cf. [BFGM]. Hence it remains to consider the case when θ is the projection of the simple affine coroot α_0.

Consider the "new" $\mathbb{G}_m$-action on $\widetilde{\mathfrak{U}}^{\theta}_{G;P}$ as in Corollary 10.4. In the same way, we obtain a $\mathbb{G}_m$-action on $\widetilde{\mathfrak{U}}^{\theta}_{G;P}$, which contracts this space to

$$\mathrm{Mod}^{\theta,+}_{M_{\mathrm{aff}},\mathbf{C}} \simeq \overset{\circ}{\mathbf{C}}{}^{\theta} \underset{\overset{\circ}{\mathbf{C}}{}^{\theta}}{\times} \mathrm{Mod}^{\theta,+}_{M_{\mathrm{aff}},\mathbf{C}} \subset \mathfrak{U}^{\theta}_{G;P} \underset{\overset{\circ}{\mathbf{C}}{}^{\theta}}{\times} \mathrm{Mod}^{\theta,+}_{M_{\mathrm{aff}},\mathbf{C}}.$$

Thus it is enough to show that the map $\varrho^{\theta}_{M_{\mathrm{aff}}} : \widetilde{\mathfrak{U}}^{\theta}_{G;P} \to \mathrm{Mod}^{\theta,+}_{M_{\mathrm{aff}},\mathbf{C}}$ is dominant. Of course, it is sufficient to prove that the restriction of $\varrho^{\theta}_{M_{\mathrm{aff}}}$ to $\mathrm{Bun}^{\theta}_{G;P}(\mathbf{S}, \mathbf{D}_\infty; \mathbf{D}_0)$ is dominant.

Let us fix a point $\mathbf{c} \in \overset{\circ}{\mathbf{C}}$ and consider the preimage of $a \cdot \mathbf{c} \in \mathbf{C}$ under

$$\mathrm{Bun}^{\theta}_{G;P}(\mathbf{S}, \mathbf{D}_\infty; \mathbf{D}_0) \to \mathrm{Bun}^{a}_{G}(\mathbf{S}, \mathbf{D}_\infty) \xrightarrow{\varpi^a_h} \overset{\circ}{\mathbf{C}}{}^{(a)},$$

which we will denote by $\overset{\circ}{\mathfrak{F}}{}^{\theta}_{\mathfrak{g}_{\mathrm{aff}},\mathfrak{p}^+_{\mathrm{aff}}}$; cf. Section 14.2. The map $\varrho^{\theta}_{M_{\mathrm{aff}}}$ gives rise to a map $\overset{\circ}{\mathfrak{F}}{}^{\theta}_{\mathfrak{g}_{\mathrm{aff}},\mathfrak{p}^+_{\mathrm{aff}}} \to \mathrm{Gr}^{\theta,+}_{M_{\mathrm{aff}}} := \mathrm{Gr}_{M_{\mathrm{aff}}} \cap \mathrm{Mod}^{\theta,+}_{M_{\mathrm{aff}},\mathbf{C}}$.

Recall (cf. Section 13.7) that the affine Grassmannian Gr_M is a union of locally closed subsets Gr^{ν}_M, $\nu \in \Lambda^+_{\mathfrak{m}}$, and $\overline{\mathrm{Gr}}^{\nu}_M = \underset{\nu' \leq \nu}{\cup} \mathrm{Gr}^{\nu'}_M$. Note that on the level of reduced schemes, $(\mathrm{Gr}^{\theta,+}_{M_{\mathrm{aff}}})_{\mathrm{red}}$ is isomorphic to $\overline{\mathrm{Gr}}^{-w^M_0(\overline{\theta})}_M$, where w^M_0 is the longest element in the Weyl group of M.

Recall now that we have fixed θ to be α_0. Therefore, the complement to $\mathrm{Gr}^{-w^M_0(\overline{\alpha}_0)}_M$ in $\overline{\mathrm{Gr}}^{-w^M_0(\overline{\alpha}_0)}_M$ is a point-orbit corresponding to $\mathrm{Gr}^{0,+}_{M_{\mathrm{aff}}} \xrightarrow{\delta} \mathrm{Gr}^{\alpha_0,+}_{M_{\mathrm{aff}}}$ (unless $M = T$). Therefore, using Corollary 14.4, we conclude that if the map $\overset{\circ}{\mathfrak{F}}{}^{\alpha_0}_{\mathfrak{g}_{\mathrm{aff}},\mathfrak{p}^+_{\mathrm{aff}}} \to \mathrm{Gr}^{\alpha_0,+}_{M_{\mathrm{aff}}}$ were not dominant for some (and hence, every) point $\mathbf{c} \in \mathbf{C}$, we would obtain that for every geometric point of $\mathrm{Bun}^{\theta}_{G;P}(\mathbf{S}, \mathbf{D}_\infty; \mathbf{D}_0)$, the corresponding P-bundle on $\mathbf{C}$ is such that the induced M-bundle is trivial. However, since $\mathbf{C}$ is of genus 0, this would mean the G-bundle on $\mathbf{C}$ is trivial. But this is a contradiction: we know (cf. Proposition 3.4) that there must exist a horizontal line $\mathbf{C} \times \mathbf{x}$ such that the restriction of the G-bundle to it is nontrivial, but according to the second part of Proposition 9.9, this point $\mathbf{x}$ must equal $0_{\mathbf{X}}$. □

In what follows, we will denote by $\widetilde{\mathfrak{s}}_{\mathfrak{p}^+_{\mathrm{aff}}}$ the map $\mathrm{Mod}^{\theta,+}_{M_{\mathrm{aff}},\mathbf{C}} \to \widetilde{\mathfrak{U}}^{\theta}_{G;P}$.

10.7 Functorial definitions in the parabolic case

Let us briefly discuss alternative definitions of $\mathfrak{U}^{\theta}_{G;P}$ $\widetilde{\mathfrak{U}}^{\theta}_{G;P}$, which will be functorial in the spirit of ${}^{\mathrm{Drinf}}\mathfrak{U}^{a}_{G}$. We will consider $\widetilde{\mathfrak{U}}^{\theta}_{G;P}$, since the case of $\mathfrak{U}^{\theta}_{G;P}$ is analogous (and simpler).

Consider the functor ${}^{\mathrm{Funct}}\widetilde{\mathfrak{U}}^{\theta}_{G;P}$: {Schemes} $\to$ {Sets} that assigns to a scheme S the data of

- an S-point of $\mathfrak{U}^a_G$ (in particular, we obtain a G-bundle $\mathcal{F}_G$ defined on an open subset $U \subset \mathbf{S} \times S$ such that $\mathbf{S} \times S - U$ is finite over S; let $D \subset \mathbf{C} \times S$ be a relative Cartier divisor such that $\mathbf{C} \times S - D \subset U \cap \mathbf{C} \times S$);
- a meromorphic enhanced quasi-map $\sigma \in {}_{\infty \cdot D_h}\widetilde{\mathrm{QMaps}}^{\theta}(\mathbf{C}, \mathcal{F}_G \overset{G}{\times} \mathcal{G}_{\mathfrak{g},\mathfrak{p}})$ (in particular, we have an M-bundle $\mathcal{F}_M$ defined on $\mathbf{C} \times S$),

such that the following condition is satisfied:

For any direction $\mathbf{d}_v \in \mathbf{D}_\infty - \mathbf{d}_h$, *and the corresponding decomposition* $\mathbf{S}' \simeq \mathbf{C} \times \mathbf{X}$, *the meromorphic enhanced quasi-map*

$$\widehat{\sigma} : \mathbf{C} \to \mathcal{G}_{G,P,\mathbf{X}},$$

obtained from $\mathcal{F}_G$ *and* σ, *is, in fact, regular.*

As before one shows that the functor ${}^{\mathrm{Funct}}\widetilde{\mathfrak{U}}^{\theta}_{G;P}$ is representable by an ind-scheme of ind-finite type.

Theorem 10.8. *There exists a canonical closed embedding* $\widetilde{\mathfrak{U}}^{\theta}_{G;P} \to {}^{\mathrm{Funct}}\widetilde{\mathfrak{U}}^{\theta}_{G;P}$, *which induces an isomorphism on the level of geometric points.*

Proof. The closed embedding $\widetilde{\mathfrak{U}}^{\theta}_{G;P} \to {}^{\mathrm{Funct}}\widetilde{\mathfrak{U}}^{\theta}_{G;P}$ has been constructed in Section 9.5.

Consider a geometric point of ${}^{\mathrm{Funct}}\widetilde{\mathfrak{U}}^{\theta}_{G;P}$, and let $\mathbf{s}_1, \ldots, \mathbf{s}_k$ be the points of $\overset{\circ\circ}{\mathbf{S}}$, where the corresponding point $z \in \mathfrak{U}^a_G$ has a singularity, and let $b_1, \ldots, b_k$ be the corresponding multiplicities. Let also $\mathbf{c}_1, \ldots, \mathbf{c}_n$ be the points on $\mathbf{D}_0 \simeq \mathbf{C}$ where the enhanced meromorphic quasi-map σ (as in the definition of ${}^{\mathrm{Funct}}\widetilde{\mathfrak{U}}^{\theta}_{G;P}$) is either nondefined or has a singularity.

Let us choose a direction $\mathbf{d}_v$ so that the points $\mathbf{c}_i$ are disjoint from $\pi_v(\mathbf{s}_j)$. Then the enhanced quasi-map $\widehat{\sigma}$ (as in the definition of ${}^{\mathrm{Funct}}\widetilde{\mathfrak{U}}^{\theta}_{G;P}$) has singularities only at the above points $\mathbf{c}_i$ and $\pi_v(\mathbf{s}_j)$. Let $\theta_i \in \widehat{\Lambda}^{\mathrm{pos}}_{\mathfrak{g},\mathfrak{p}}$ denote the defect of $\widehat{\sigma}$ at $\mathbf{c}_i$. The defect of $\widehat{\sigma}$ at $\pi_v(\mathbf{s}_j)$ is automatically equal to $b_j \cdot \delta$.

Set $\theta' = \theta - \underset{i}{\Sigma}\theta_i - \underset{j}{\Sigma} b_j \cdot \delta$. According to Proposition 2.34, the quasi-map $\widehat{\sigma}$ can be described as its saturation $\widehat{\sigma}' \in \mathrm{Bun}^{\theta'}_{G;P}(\mathbf{S}, \mathbf{D}_\infty; \mathbf{D}_0)$ and the corresponding positive modification, i.e., a point in

$$pt \underset{\mathrm{Bun}_{M_{\mathrm{aff}}}(\mathbf{C},\infty_{\mathbf{C}})}{\times} (\prod_j \mathcal{H}^{b_j \cdot \delta,+}_{M_{\mathrm{aff}},\mathbf{C},\pi_v(\mathbf{s}_j)} \times \prod_i \mathcal{H}^{\theta_i,+}_{M_{\mathrm{aff}},\mathbf{C},\mathbf{c}_i}),$$

where $pt \in \mathrm{Bun}_{M_{\mathrm{aff}}}$ corresponds to the pair $\mathcal{F}_M$ (cf. the definition of ${}^{\mathrm{Funct}}\widetilde{\mathfrak{U}}^{\theta}_{G;P}$) and the line bundle corresponding to the divisor $\varpi^a_h(z)$ (for $z \in \mathfrak{U}^a_G$ as above).

The theorem now follows from the description of points of $\widetilde{\mathfrak{U}}^{\theta}_G$ given by Proposition 10.6. □

11 The boundary

11.1

Let $\mathcal{Y}$ be a scheme of finite type whose underlying reduced subscheme $\mathcal{Y}_{red}$ is irreducible; let $\mathcal{Y}_n \to \mathcal{Y}_{red}$ be its normalization. Let $\mathcal{Y}' \subset \mathcal{Y}$ be a closed subset, and let $\mathcal{Y}'_n$ be the preimage of $\mathcal{Y}'$ in $\mathcal{Y}_n$.

Definition 11.2. *$\mathcal{Y}'$ will be called a quasi-effective Cartier divisor if there exists a line bundle $\mathcal{P}$ on $\mathcal{Y}$, and a meromorphic (i.e., defined over a dense open subset) section $s : \mathcal{O}_{\mathcal{Y}} \to \mathcal{P}$, such that*

- *the section s is an isomorphism on $\mathcal{Y} - \mathcal{Y}'$;*
- *the section $s_n : \mathcal{O}_{\mathcal{Y}_n} \to \mathcal{P} \underset{\mathcal{O}_{\mathcal{Y}}}{\otimes} \mathcal{O}_{\mathcal{Y}_n}$ is regular;*
- *the set of zeroes of s_n coincides with $\mathcal{Y}'_n$.*

The notion of quasi-effective Cartier divisor is useful because of the following lemma.

Lemma 11.3. *Let $\mathcal{Y}' \subset \mathcal{Y}$ be a quasi-effective Cartier divisor, and let $Z \subset \mathcal{Y}'$ be a nonempty irreducible subvariety, not contained in $\mathcal{Y}'$. Then* $\dim(Z \cap \mathcal{Y}') = \dim(Z) - 1$.

Proof. Let Z_n be the preimage of Z in $\mathcal{Y}_n$. It is sufficient to show that $\dim(Z_n \cap \mathcal{Y}'_n) = \dim(Z_n) - 1$; but this is true because $\mathcal{Y}'_n$ is a Cartier divisor in $\mathcal{Y}_n$. □

Here is a simple criterion for how to show that a subset $\mathcal{Y}'$ is a quasi-effective Cartier divisor. Suppose that we are given a line bundle $\mathcal{P}$ on $\mathcal{Y}$ with a trivialization $s_n : \mathcal{O} \to \mathcal{P}$ over $\mathcal{Y} - \mathcal{Y}'$. Suppose also that $\mathcal{Y}'$ contains a dense subset of the form $\underset{i}{\cup} \mathcal{Y}'_i$, where $\mathcal{Y}'_i \subset \mathcal{Y}$ are subvarieties of codimension 1, such that s has a zero at the generic point of each $\mathcal{Y}'_i$.

Lemma 11.4. *Under the above circumstances, $\mathcal{Y}'$ is a quasi-effective Cartier divisor.*

11.5 The finite-dimensional case

Recall the space $\overline{\mathrm{Bun}}_B(\mathbf{C})$ (cf. Section 2.4) defined for any projective curve $\mathbf{C}$. By definition, the boundary of $\overline{\mathrm{Bun}}_B(\mathbf{C})$ is the closed subset $\partial(\overline{\mathrm{Bun}}_B(\mathbf{C})) := \overline{\mathrm{Bun}}_B(\mathbf{C}) - \mathrm{Bun}_B(\mathbf{C})$.

Theorem 11.6. *The boundary $\partial(\overline{\mathrm{Bun}}_B(\mathbf{C}))$ is a quasi-effective Cartier divisor in $\overline{\mathrm{Bun}}_B(\mathbf{C})$.*

The idea of the proof given below is borrowed from Faltings' approach to the construction of the determinant line bundle on the moduli space of bundles on a curve; cf. [Fa].

Recall that the moduli space $\mathrm{Bun}_G(\mathbf{C})$ is equipped with the line bundle $\mathcal{P}_{\mathrm{Bun}_G(\mathbf{C})}$. Its $2\check{h}$th tensor power can be identified with the determinant line bundle $\mathcal{P}_{\mathrm{Bun}_G(\mathbf{C}),\det}$, whose fiber at $\mathcal{F}_G \in \mathrm{Bun}_G(\mathbf{C})$ is $\det(R\Gamma(\mathbf{C}, \mathfrak{g}_{\mathcal{F}_G}))$.

Let us consider the line bundle $\mathcal{P}_{\mathrm{Bun}_T(\mathbf{C})}$ on the stack $\mathrm{Bun}_T(\mathbf{C})$ whose fiber at $\mathcal{F}_T \in \mathrm{Bun}_T(\mathbf{C})$ is

$$\bigotimes_{\check{\alpha}} \det(R\Gamma(\mathbf{C}, \mathcal{L}^{\check{\alpha}}_{\mathcal{F}_T})),$$

where $\check{\alpha}$ runs over the set of all roots of $\mathfrak{g}$.

We have the canonical projections $p_G : \overline{\mathrm{Bun}}_B(\mathbf{C}) \to \mathrm{Bun}_G(\mathbf{C})$ and $p_T : \overline{\mathrm{Bun}}_B(\mathbf{C}) \to \mathrm{Bun}_T(\mathbf{C})$, and we define the line bundle $\mathcal{P}_{\overline{\mathrm{Bun}}_B(\mathbf{C})}$ on $\overline{\mathrm{Bun}}_B(\mathbf{C})$ as $p_G^*((\mathcal{P}_{\mathrm{Bun}_G(\mathbf{C}),\det})^{-1}) \otimes p_T^*(\mathcal{P}_{\mathrm{Bun}_T(\mathbf{C})})$.

Observe that over the open part $\mathrm{Bun}_B(\mathbf{C}) \subset \overline{\mathrm{Bun}}_B(\mathbf{C})$ we have an (almost canonical) trivialization of $\mathcal{P}_{\overline{\mathrm{Bun}}_B(\mathbf{C})}$. Indeed, for a point $\mathcal{F}_B \in \mathrm{Bun}_B(\mathbf{C})$, with the induced G- and T- bundle being $\mathcal{F}_G$ and $\mathcal{F}_T$, respectively, the vector bundle $\mathfrak{g}_{\mathcal{F}_G}$ has a filtration with the successive quotients being $\mathcal{L}^{\check{\alpha}}_{\mathcal{F}_T}$, and the trivial bundle $\mathfrak{h} \otimes \mathcal{O}$. Hence

$$\det(R\Gamma(\mathbf{C}, \mathfrak{g}_{\mathcal{F}_G})) \simeq \bigotimes_{\check{\alpha}} \det(R\Gamma(\mathbf{C}, \mathcal{L}^{\check{\alpha}}_{\mathcal{F}_T})) \otimes \det(R\Gamma(\mathbf{C}, \mathcal{O}))^{\dim(T)}.$$

In order to be able to apply Lemma 11.4, we need to analyze the behavior of the section $s : \mathcal{O} \to \mathcal{P}_{\overline{\mathrm{Bun}}_B(\mathbf{C})}$ just constructed at the generic point at each irreducible component of $\partial(\overline{\mathrm{Bun}}_B(\mathbf{C}))$.

It is well known (cf. [BG1]) that $\partial(\overline{\mathrm{Bun}}_B(\mathbf{C}))$ contains as a dense subset the following codimension-1 locus: it is the union of connected components indexed by the vertices of the Dynkin diagram of $\mathfrak{g}$; each such component $\partial(\overline{\mathrm{Bun}}_B(\mathbf{C}))_i$ is isomorphic to $\mathbf{C} \times \mathrm{Bun}_B(\mathbf{C})$, and we have an identification

$$\mathrm{Bun}_B(\mathbf{C}) \cup \partial(\overline{\mathrm{Bun}}_B(\mathbf{C}))_i \simeq \mathrm{Bun}_{P_i}(\mathbf{C}) \underset{\mathrm{Bun}_{M_i}(\mathbf{C})}{\times} \overline{\mathrm{Bun}}_{B_i}(\mathbf{C}),$$

where P_i (respectively, M_i) is the corresponding subminimal parabolic (respectively, its Levi quotient), and $\overline{\mathrm{Bun}}_{B_i}(\mathbf{C})$ is the corresponding space for the group M_i; cf. Lemma 2.9 and Corollary 2.10.

This reduces our problem to the following calculation. Let G be a reductive group of semisimple rank 1 and the derived group isomorphic to SL_2, and let $\mathcal{V}$ be an irreducible G-module of dimension $n+1$. Let $\mathcal{P}_{\mathrm{Bun}_G(\mathbf{C}),\mathcal{V}}$ be the line bundle on $\mathrm{Bun}_G(\mathbf{C})$ given by

$$\mathcal{F}_G \mapsto \det(R\Gamma(\mathbf{C}, \mathcal{V}_{\mathcal{F}_G})),$$

and $\mathcal{P}_{\mathrm{Bun}_T(\mathbf{C}),\mathcal{V}}$ be the line bundle on $\mathrm{Bun}_T(\mathbf{C})$ given by

$$\mathcal{F}_T \mapsto \bigotimes_{\check{\mu}} \det(R\Gamma(\mathbf{C}, \mathcal{L}^{\check{\mu}}_{\mathcal{F}_T})),$$

where $\check{\mu}$ runs over the set of weights of $\mathcal{V}$. Let $\mathcal{P}_{\overline{\mathrm{Bun}}_B(\mathbf{C}),\mathcal{V}}$ be the corresponding line bundle on $\overline{\mathrm{Bun}}_B(\mathbf{C})$, and s its trivialization over $\mathrm{Bun}_B(\mathbf{C})$.

Lemma 11.7. *The section s has a zero of order $\frac{n(n+1)(n+2)}{6}$ along the boundary $\partial(\overline{\mathrm{Bun}}_B(\mathbf{C}))$.*

Proof. Recall that for groups of semisimple rank 1, the stack $\overline{\mathrm{Bun}}_B(\mathbf{C})$ is smooth, and the variety $\partial(\overline{\mathrm{Bun}}_B(\mathbf{C}))$ is irreducible.

First, it is easy to see that our initial group can be replaced by GL_2, such that $\mathcal{V}$ is the nth symmetric power of the standard representation. We will construct a map of the projective line $\mathbb{P}^1$ into $\overline{\mathrm{Bun}}_B(\mathbf{C})$, such that $\mathbb{A}^1 \subset \mathbb{P}^1$ maps to the open part $\mathrm{Bun}_B(\mathbf{C})$ and the intersection $\mathbb{P}^1 \cap \partial(\overline{\mathrm{Bun}}_B(\mathbf{C}))$ is transversal, and such that the pullback of $\mathcal{P}_{\overline{\mathrm{Bun}}_B(\mathbf{C}),\mathcal{V}}$ to $\mathbb{P}^1$ has degree $\frac{n(n+1)(n+2)}{6}$.

Fix a point $\mathbf{c} \in \mathbf{C}$ and a point $\mathcal{F}_B \in \mathrm{Bun}_B(\mathbf{C})$, i.e., a vector bundle $\mathcal{M}$ on $\mathbf{C}$ and a short exact sequence

$$0 \to \mathcal{L}_1 \to \mathcal{M} \to \mathcal{L}_2 \to 0.$$

Consider the projective line of all elementary lower modifications of $\mathcal{M}$ at $\mathbf{c}$. This $\mathbb{P}^1$ maps to $\overline{\mathrm{Bun}}_B(\mathbf{C})$ by setting the new T-bundle to be $(\mathcal{L}_1(-\mathbf{c}), \mathcal{L}_2)$. The required properties of the embedding $\mathbb{P}^1 \to \overline{\mathrm{Bun}}_B(\mathbf{C})$ are easy to verify. By construction, the composition $\mathbb{P}^1 \to \overline{\mathrm{Bun}}_B(\mathbf{C}) \overset{p_T}{\to} \mathrm{Bun}_T(\mathbf{C})$ is the constant map; therefore, it suffices to calculate the degree of the pullback of $\mathcal{P}_{\mathrm{Bun}_G(\mathbf{C}),\mathcal{V}}$ under $\mathbb{P}^1 \to \overline{\mathrm{Bun}}_B(\mathbf{C}) \overset{p_G}{\to} \mathrm{Bun}_G(\mathbf{C})$.

But for a lower modification $\mathcal{M}' \subset \mathcal{M}$, the quotient $\mathrm{Sym}^n(\mathcal{M})/\mathrm{Sym}^n(\mathcal{M}')$ has a canonical n-step filtration with successive quotients isomorphic to

$$(\mathcal{M}/\mathcal{M}')^{\otimes j} \otimes \mathrm{Sym}^{n-j}(\mathcal{M}/\mathcal{M}(-x)).$$

Therefore, on the level of determinant lines, $\det(R\Gamma(\mathbf{C}, \mathrm{Sym}^n(\mathcal{M}')))$ is isomorphic to a fixed one-dimensional vector space times $(\mathcal{M}/\mathcal{M}')^{\otimes \frac{n(n+1)(n+2)}{6}}$. □

This implies the assertion of the theorem.

11.8

We will now formulate and prove an analogue of Theorem 11.6 for Uhlenbeck spaces on $\mathbb{P}^2$.

For $\mu \in \widehat{\Lambda}_{\mathfrak{g}}^{\mathrm{pos}} = \widehat{\Lambda}_{\mathfrak{g},\mathfrak{b}}^{\mathrm{pos}}$, consider the open subset $\overset{\circ}{\mathfrak{U}}{}^{\mu}_{G;B}$ in $\mathfrak{U}^{\mu}_{G;B}$ equal to the union of the strata $\mathfrak{U}^{\mu;0,b}_{G;B}$. In other words, $\overset{\circ}{\mathfrak{U}}{}^{\mu}_{G;B}$ consists of points which have no singularities along $\mathbf{D}_0$. (This means that the induced point of $\mathfrak{U}^a_G$ gives rise to a bundle $\mathcal{F}_G$ defined in a neighborhood of $\mathbf{D}_0$, and the corresponding quasi-map $\mathbf{C} \to \mathcal{F}_G|_{\mathbf{C}} \overset{G}{\times} \mathcal{G}_{\mathfrak{g},\mathfrak{b}}$ is in fact a map.) Of course, $\overset{\circ}{\mathfrak{U}}{}^{\mu}_{G;B}$ contains $\mathrm{Bun}^{\mu}_{G;B}(\mathbf{S}, \mathbf{D}_\infty; \mathbf{D}_0)$ as an open subset, and from Theorem 10.2 we obtain that the complement is of codimension at least 2.

We define the boundary $\partial(\mathfrak{U}^{\mu}_{G;B}) \subset \mathfrak{U}^{\mu}_{G;B}$ as the complement $\mathfrak{U}^{\mu}_{G;B} - \overset{\circ}{\mathfrak{U}}{}^{\mu}_{G;B}$. From Theorem 10.2, we obtain that $\partial(\mathfrak{U}^{\mu}_{G;B})$ contains a dense subset of codimension 1 equal to the disjoint union over $i \in I_{\mathrm{aff}}$ of

$$\partial(\mathfrak{U}^{\mu}_{G;B})_i := \mathfrak{U}^{\mu;\alpha_i,0}_{G;B} \simeq \mathrm{Bun}^{\mu-\alpha_i}_{G;B}(\mathbf{S}, \mathbf{D}_\infty; \mathbf{D}_0) \times \overset{\circ}{\mathbf{C}}.$$

Remark. Note also that $\mathfrak{U}^{\mu}_{G;B}$ is regular at the generic point of each $\partial(\mathfrak{U}^{\mu}_{G;B})_i$ (in particular, $\mathfrak{U}^{\mu}_{G;B}$ is regular in codimension 1.) Indeed, from Theorem 10.2, it is easy to deduce that the open subset

$$\mathrm{Bun}^{\mu}_{G;B}(\mathbf{S}, \mathbf{D}_\infty; \mathbf{D}_0) \bigcup (\underset{i \in I_{\mathrm{aff}}}{\cup} \mathfrak{U}^{\mu;\alpha_i,0}_{G;B})$$

projects one-to-one on the open subset in $\mathrm{QMaps}^{\mu}(\mathbf{C}, \mathcal{G}_{G,P,\mathbf{X}})$ equal to

$$\mathrm{Maps}(\mathbf{C}, \mathcal{G}_{\mathfrak{g}_{\mathrm{aff}}, \mathfrak{b}^+_{\mathrm{aff}}}) \bigcup (\underset{i \in I_{\mathrm{aff}}}{\cup} \partial(\mathrm{QMaps}(\mathbf{C}, \mathcal{G}_{G,P,\mathbf{X}}))_i),$$

and the latter is smooth, by Lemma 2.9.

Theorem 11.9. *The boundary $\partial(\mathfrak{U}^{\mu}_{G;B})$ is a quasi-effective Cartier divisor in $\mathfrak{U}^{\mu}_{G;B}$.*

The rest of this section is devoted to the proof of this theorem.

11.10

Let us choose a pair of directions $(\mathbf{d}_v, \mathbf{d}_h)$, and consider the corresponding decomposition $\mathbf{S}' \simeq \mathbf{C} \times \mathbf{X}$. From the map $\varpi^a_v : \mathfrak{U}^a_G \to \overset{\circ}{\mathbf{X}}{}^{(a)}$, we obtain a relative divisor on $\mathfrak{U}^a_G \times \mathbf{X}$ and the associated line bundle.

We define the line bundle $\mathcal{P}_{\mathfrak{U},G}$ on $\mathfrak{U}^a_G$ as the restriction of the above line bundle to $\mathfrak{U}^a_G \times 0_{\mathbf{X}} \subset \mathfrak{U}^a_G \times \mathbf{X}$.

Let $\overset{\circ}{\mathfrak{U}}{}^a_G$ be the open subset in $\mathfrak{U}^a_G$ corresponding to points with no singularity along $\mathbf{D}_0$. We have a canonical projection $\overset{\circ}{\mathfrak{U}}{}^a_G \to \mathrm{Bun}_G(\mathbf{C}, \infty_{\mathbf{C}})$.

The next assertion follows from the construction of the map ϖ^a_v.

Lemma 11.11. *The restriction of $\mathcal{P}_{\mathfrak{U},G}$ to $\overset{\circ}{\mathfrak{U}}{}^a_G$ can be canonically identified with the pullback of $\mathcal{P}_{\mathrm{Bun}_G(\mathbf{C})}$ under $\overset{\circ}{\mathfrak{U}}{}^a_G \to \mathrm{Bun}_G(\mathbf{C})$.*

We have a canonical projection $p_{T,aff}$ from $\mathfrak{U}^{\mu}_{G;B}$ to $\mathrm{Bun}_{T_{\mathrm{aff}}}(\mathbf{C})$. We define the line bundle $\mathcal{P}_{\mathfrak{U},B}$ on $\mathfrak{U}^{\mu}_{G;B}$ as the tensor product of the pullback of $\mathcal{P}_{\mathrm{Bun}_T(\mathbf{C})}$ under

$$\mathfrak{U}^{\mu}_{G;B} \overset{p_{T,aff}}{\longrightarrow} \mathrm{Bun}_{T_{\mathrm{aff}}}(\mathbf{C}) \to \mathrm{Bun}_T(\mathbf{C})$$

and the inverse of the pullback of $\mathcal{P}^{\otimes 2 \cdot h^\vee}_{\mathfrak{U},G}$ under $\mathfrak{U}^{\mu}_{G;B} \to \mathfrak{U}^a_G$.

We claim that over $\overset{\circ}{\mathfrak{U}}{}^{\mu}_{G;B}$ there is a canonical trivialization of $s : \mathcal{O} \to \mathcal{P}_{\mathfrak{U},B}$. Indeed, we have a projection $\overset{\circ}{\mathfrak{U}}{}^{\mu}_{G;B} \to \mathrm{Bun}_B(\mathbf{C})$, and the restriction of $\mathcal{P}_{\mathfrak{U},B}$ is the pullback of $\mathcal{P}_{\overline{\mathrm{Bun}}_B(\mathbf{C})}$ under this map. Therefore, the existence of the trivialization follows from the corresponding fact for $\mathrm{Bun}_B(\mathbf{C})$, using the fact that $\mathcal{P}_{\mathrm{Bun}_G(\mathbf{C}),\det} \simeq \mathcal{P}^{\otimes 2 \cdot h^\vee}_{\mathrm{Bun}_G(\mathbf{C})}$.

Thus to prove Theorem 11.9, it remains to show that the above meromorphic section s has a zero at the generic point of each $\partial(\mathfrak{U}^{\mu}_{G;B})_i$, $i \in I_{\mathrm{aff}}$.

When $i \in I \subset I_{\mathrm{aff}}$, this readily follows from Theorem 11.6, since in this case $\partial(\mathfrak{U}^{\mu}_{G;B})_i$ is contained in the preimage of $\overset{\circ}{\mathfrak{U}}{}^{a}_{G}$ in $\mathfrak{U}^{\mu}_{G;B}$.

Therefore, it remains to analyze the behavior of s on the open subvariety

$$\mathfrak{U}^{\mu}_{G;B} - \overline{\underset{i\in I}{\cup}\partial(\mathfrak{U}^{\mu}_{G;B})_i}$$

at the generic point of $\partial(\mathfrak{U}^{\mu}_{G;B})_{i_0}$, where i_0 corresponds to the affine root.

Lemma 11.12. *Let $\mathcal{Y}$ be a variety (regular in codimension 1), acted on by the group $\mathbb{G}_m$; let $\mathcal{P}$ be an equivariant line bundle, and $s : \mathcal{O} \to \mathcal{P}$ be an equivariant nowhere vanishing section defined outside an irreducible subvariety $\mathcal{Y}'$ of codimension 1.*

Suppose that there exists a point $y \in \mathcal{Y} - \mathcal{Y}'$ such that the action map $\mathbb{G}_m \times y \to \mathcal{Y}$ extends to a map $\mathbb{A}^1 \to \mathcal{Y}$, such that the image of 0, call it y', belongs to $\mathcal{Y}'$. Assume, moreover, that the $\mathbb{G}_m$-action on the fiber of $\mathcal{P}$ at y' (note that y' is automatically $\mathbb{G}_m$-stable) is given by a positive power of the standard character. Then s vanishes at the generic point of $\mathcal{Y}'$.

Proof. If we lift y to a geometric point in the normalization of $\mathcal{Y}$, the same conditions will hold; therefore, we can assume that $\mathcal{Y}$ is normal.

To prove the lemma we have to exclude the possibility that s has a pole of order ≥ 0 on $\mathcal{Y}'$. If it did, the same would be true for the pullback of the pair $(\mathcal{P}, s)$ to $\mathbb{A}^1$, endowed with the standard $\mathbb{G}_m$-action. In the latter case, to have a pole of nonnegative order means that $\mathcal{P} \simeq \mathcal{O}_{\mathbb{A}^1}(-n \cdot 0)$ with $n \geq 0$, but the action of $\mathbb{G}_m$ on the fiber of $\mathcal{P} \simeq \mathcal{O}_{\mathbb{A}^1}(-n \cdot 0)$ at 0 is given by the character $-n$, which is a contradiction. □

Thus to prove the theorem, we need to construct a $\mathbb{G}_m$-action as in the lemma and check its properties. Consider the $\mathbb{G}_m$-action "of the first kind" on $\mathfrak{U}^{\mu}_{G;B}$, as in the proof of Theorem 10.2. The line bundle $\mathcal{P}_{\mathfrak{U},B}$ and the section s are $\mathbb{G}_m$-equivariant, by construction.

Let us now construct a point $y \in \mathfrak{U}^{\mu}_{G;B}$ as in the lemma. Let us write $\mu = (\overline{\mu}, a)$, and recall the projection

$$\varrho^{\mu}_{\mathfrak{b}^+_{\mathrm{aff}}} : \mathfrak{U}^{\mu}_{G;B} \to \overset{\circ}{\mathbf{C}}{}^{\mu} \simeq \overset{\circ}{\mathbf{C}}{}^{\overline{\mu - a\cdot\delta}} \times \overset{\circ}{\mathbf{C}}{}^{(a)}.$$

Let y be any point of $\overset{\circ}{\mathfrak{U}}{}^{\mu}_{G;B}$ which projects to a multiplicity-free point in $\overset{\circ}{\mathbf{C}}{}^{\mu}$. We claim that the $\mathbb{G}_m$ action will contract this point to a point y' on the subscheme

$$\mathfrak{U}^{\mu;a\cdot\alpha_0,0}_{G;B} \subset \overline{\partial(\mathfrak{U}^{\mu}_{G;B})_{i_0}} \cap (\mathfrak{U}^{\mu}_{G;B} - \overline{\underset{i\in I}{\cup}\partial(\mathfrak{U}^{\mu}_{G;B})_i}).$$

Indeed, using the horizontal factorization property, Proposition 9.4, it suffices to analyze separately the cases when $a = 0$, and when and $\mu = \alpha_0$. In the former case, the assertion is trivial, since in this case $\mathfrak{U}^{\mu}_{G;B} \simeq \overline{\mathrm{Bun}}^{\overline{\mu}}_B(\mathbf{C}, \mathbf{c})$, and the $\mathbb{G}_m$-action is trivial.

In the latter case, from Theorem 10.2, we obtain that the projection

$$\mathfrak{U}^{\alpha_0}_{G;B} \to \mathrm{QMaps}^{\alpha_0}(\mathbf{C}, \mathcal{G}_{G,B,\mathbf{X}})$$

is one-to-one. From the remark following Lemma 2.8, we obtain that $\mathrm{QMaps}^{\alpha_0}(\mathbf{C}, \mathcal{B}_{\mathfrak{g}_{\mathrm{aff}}}) \simeq \overset{\circ}{\mathbf{C}} \times \mathbb{A}^1$, so that the $\mathbb{G}_m$ action comes from the standard $\mathbb{G}_m$-action on $\mathbb{A}^1$. Hence $\mathfrak{U}^{\alpha_0}_{G;B} \to \mathrm{QMaps}^{\alpha_0}(\mathbf{C}, \mathcal{B}_{\mathfrak{g}_{\mathrm{aff}}})$ is an isomorphism, and the contraction statement in this case is manifest as well.

Thus it remains to check that for any point $y' \in \mathfrak{U}^{\mu;a\cdot\alpha_0,0}_{G;B}$, the $\mathbb{G}_m$-action on the fiber of $\mathcal{P}_{\mathfrak{U},B}$ at y' is given by a positive character. However, by construction this action is the same as the $\mathbb{G}_m$-action on the fiber of $\mathcal{O}(a \cdot 0_{\mathbf{X}})^{\otimes 2\cdot h^\vee}$ at $0_{\mathbf{X}} \in \mathbf{X}$, i.e., corresponds to the $2 \cdot a \cdot h^\vee$th power of the standard character.

Thus the theorem is proved.

11.13 Remark

Let us give another interpretation of the line bundle $\mathcal{P}_{\mathfrak{U},B}$ for $G = SL_n$ in terms of the scheme $\widetilde{\mathfrak{N}}^a_n$ classifying torsion-free sheaves on $\mathbf{S}$.

For $\mu = (\overline{\mu}, a)$, and $\overline{\mu} = (\mu_1, \ldots, \mu_n)$, consider the scheme $\widetilde{\mathfrak{N}}^\mu_{n,flag}$ that classifies the data of

- a torsion-free sheaf of generic rank n, denoted $\mathcal{M}$, on $\mathbf{S}$, which is locally free near $\mathbf{D}_\infty$, and with $ch_2(\mathcal{M}) = -a$;
- a trivialization $\mathcal{M}|_{\mathbf{D}_\infty} \simeq \mathcal{O}^{\oplus n}$;
- a flag of locally free subsheaves

$$\mathcal{M} = \mathcal{M}_0 \subset \mathcal{M}_1 \subset \cdots \subset \mathcal{M}_{n-1} \subset \mathcal{M}_n = \mathcal{M}(\mathbf{D}_0),$$

such that each $\mathcal{M}_i/\mathcal{M}_{i-1}$ is a coherent sheaf on $\mathbf{D}_0$ of generic rank 1 and of degree μ_i.

On $\widetilde{\mathfrak{N}}^\mu_{n,flag}$ there is a natural line bundle $\mathcal{P}_{\widetilde{\mathfrak{N}},B}$, whose fiber at a point $\mathcal{M} \subset \mathcal{M}_1 \subset \cdots \subset \mathcal{M}_{n-1} \subset \mathcal{M}(\mathbf{D}_0)$ is the line

$$\bigotimes_{i=1,\ldots,n} (\det(R\Gamma(\det(\mathcal{M}_i/\mathcal{M}_{i-1}))) \otimes \det(R\Gamma(\mathcal{M}_i/\mathcal{M}_{i-1}))^{-1}).$$

In [FGK] it was shown that there exists a natural proper map $\widetilde{\mathfrak{N}}^\mu_{n,flag} \to \mathfrak{U}^\mu_{SL_n,B}$, which is, in fact, a semismall resolution of singularities. One can show that the pullback of $\mathcal{P}_{\mathfrak{U},B}$ to $\widetilde{\mathfrak{N}}^\mu_{n,flag}$ is isomorphic to $\mathcal{P}^{\otimes 2n}_{\widetilde{\mathfrak{N}},B}$.

Part IV: Crystals

In this part, $\mathfrak{g}$ will be an arbitrary Kac–Moody Lie algebra, with the exception of Sections 13, 15.5, and 15.6.

12 Crystals in general

12.1

We refer to [KaS, Section 3.1] for the definition of crystals associated to a Kac–Moody algebra. By a slight abuse of language, by crystal over $\mathfrak{g}$ we will actually mean a crystal over the Langlands dual algebra of $\mathfrak{g}$, in particular, the weight function will take values in the lattice Λ.

Crystals form a category, where the morphisms from B_1 to B_2 are the maps of sets $\mathsf{B}_1 \to \mathsf{B}_2$ which preserve the e_i, f_i operations and the ϵ_i, ϕ_i–functions for $i \in I$.

If $\mathfrak{m}$ is a Levi subalgebra in $\mathfrak{g}$, it corresponds to a subdiagram of the Dynkin graph; therefore, it makes sense to talk about $\mathfrak{m}$-crystals. If B is a $\mathfrak{g}$-crystal, we will denote by the same symbol the underlying $\mathfrak{m}$-crystal. Similarly, for every $\mathfrak{m}$-crystal B we define a $\mathfrak{g}$-crystal structure on the same underlying set, by leaving $e_i, f_i, \epsilon_i, \phi_i$ for $i \in \mathfrak{m}$ unchanged, and declaring that for $j \notin \mathfrak{m}$, e_j, f_j send everything to 0, and $\epsilon_j = \phi_j = -\infty$.

12.2

Let $\mathsf{B}_\mathfrak{g}$ be a crystal. Our goal in this section is to review some extra structures and properties of $\mathsf{B}_\mathfrak{g}$ which allow one to identify it with the standard crystal $\mathsf{B}_\mathfrak{g}^\infty$ that parametrizes the canonical basis in $U(\check{\mathfrak{n}})$; cf. [KaS, Section 3.2].

For $i \in I$, let $\mathfrak{m}_i$ be the corresponding subminimal Levi subalgebra. Let B_i be the $\mathfrak{g}$-crystal obtained in the above way from the "standard" $\mathfrak{sl}_2$-crystal. In other words, as a set it is $\mathbb{Z}^{\geq 0}$ and consists of elements that we will denote $\mathsf{b}_i(n), n \geq 0$ with $wt_i(\mathsf{b}_i(n)) = n \cdot \check{\alpha}_i$, $f_i(\mathsf{b}_i(n)) = \mathsf{b}_i(n-1)$, $\phi_i(\mathsf{b}_i(n)) = n$, and all the operations for $j \neq i$ are trivial.

According to [KaS, Proposition 3.2.3], we need to understand what kind of structure on $\mathsf{B}_\mathfrak{g}$ allows one to construct maps $\Psi_i : \mathsf{B}_\mathfrak{g} \to \mathsf{B}_\mathfrak{g} \otimes \mathsf{B}_i$, satisfying conditions (1)–(7) of loc. cit.

Our crystal $\mathsf{B}_\mathfrak{g}$ will automatically satisfy conditions (1)–(4), and it will be ϕ-normal, i.e.,

$$\text{for } \mathsf{b} \in \mathsf{B}_\mathfrak{g}, \quad \phi_i(\mathsf{b}) = \max\{n | f_i^n(\mathsf{b}) \neq 0\}.$$

Then the ϵ-functions are defined by the rule

$$\phi_i(\mathsf{b}) = \epsilon_i(\mathsf{b}) + \langle wt(\mathsf{b}), \check{\alpha}_i \rangle.$$

12.3

We suppose that the set $\mathsf{B}_\mathfrak{g}$ carries *another* crystal structure (also assumed ϕ-normal), given by operations e_i^* and f_i^*, $i \in I$ with the same weight function wt. Also assume the following:

(a) The operations e_i, f_i commute with e_j^* and f_j^* whenever $i \neq j$.

(b) For every vertex $i \in I$, we have a set decomposition

$$\mathsf{B}_{\mathfrak{g}} = \underset{n \in \mathbb{N}}{\cup} \mathsf{C}^n_{\mathfrak{m}_i} \times \mathsf{B}^n_{\mathfrak{m}_i} \times \mathsf{B}_i, \tag{21}$$

where we have the following:

- $\mathsf{C}^n_{\mathfrak{m}_i}$ is an abstract set, $\mathsf{B}^n_{\mathfrak{m}_i}$ is the set underlying the finite $\mathfrak{sl}_2$-crystal of highest weight n, and B_i is as above.
- This decomposition respects both the e_i, f_i and the e_i^*, f_i^*-operations.
- The e_i, f_i–action on each $\mathsf{C}^n_{\mathfrak{m}_i} \times \mathsf{B}^n_{\mathfrak{m}_i} \times \mathsf{B}_i$ is the trivial action on $\mathsf{C}^n_{\mathfrak{m}_i}$, times the action on $\mathsf{B}^n_{\mathfrak{m}_i} \times \mathsf{B}_i$ as on the tensor product of crystals.
- The e_i^*, f_i^*–action on each $\mathsf{C}^n_{\mathfrak{m}_i} \times \mathsf{B}^n_{\mathfrak{m}_i} \times \mathsf{B}_i$ is trivial on the first two factors times the standard action on B_i.

12.4

We claim that under the above circumstances, we do have canonical maps $\Psi_i : \mathsf{B}_{\mathfrak{g}} \to \mathsf{B}_{\mathfrak{g}} \times \mathsf{B}_i$, satisfying [KaS, condition (6) of Proposition 3.2.3].

Indeed, such maps obviously exist for $\mathfrak{g} = \mathfrak{sl}_2$, i.e., for each i we have a map of $\mathfrak{m}_i$-crystals

$$\Psi_i^{can} : \mathsf{B}_i \to \mathsf{B}_i \otimes \mathsf{B}_i.$$

We define the map Ψ_i on $\mathsf{B}_{\mathfrak{g}}$ in terms of the decomposition (21). Namely, on each $\mathsf{C}^n_{\mathfrak{m}_i} \times \mathsf{B}^n_{\mathfrak{m}_i} \times \mathsf{B}_i \subset \mathsf{B}_{\mathfrak{g}}$, Ψ_i is the identity map on the first two factors times Ψ_i^{can} on the third one.

In other words, each $\mathsf{b} \in \mathsf{B}_{\mathfrak{g}}$ can be uniquely written in the form $(e_i^*)^n(\mathsf{b}')$, where $f_i^*(\mathsf{b}') = 0$, and

$$\Psi_i(\mathsf{b}) = \mathsf{b}' \otimes \mathsf{b}_i(n). \tag{22}$$

We claim that Ψ_i indeed respects the operations e_j and f_j. First, when $j \neq i$, the assertion follows immediately from (22), since f_i^*, e_i^* commute with e_j and f_j.

For $j = i$ the assertion is also clear, since we are dealing with the map of $\mathfrak{m}_i$-crystals

$$\mathsf{B}^n_{\mathfrak{m}_i} \otimes \mathsf{B}_i \overset{\mathrm{id} \otimes \Psi_i^{can}}{\longrightarrow} \mathsf{B}^n_{\mathfrak{m}_i} \otimes \mathsf{B}_i \otimes \mathsf{B}_i.$$

12.5

Finally, in order to apply the uniqueness theorem of [KaS], the crystal $\mathsf{B}_{\mathfrak{g}}$ must satisfy the "highest-weight property" (i.e., [KaS, condition (7) of Proposition 3.2.3]). It can be stated in either of the two ways:

For $\mathsf{b} \in \mathsf{B}_{\mathfrak{g}}$ different from the canonical highest-weight vector $\mathsf{b}(0)$,

- there exists $i \in I$ such that $f_i(\mathsf{b}) \neq 0$;
- there exists $i \in I$ such that $f_i^*(\mathsf{b}) \neq 0$.

Let us denote by $\mathsf{B}_{\mathfrak{g}}^{\mathfrak{m}_i}$ the subset of $\mathsf{B}_{\mathfrak{g}}$ consisting of elements annihilated by f_i^*. We claim that it also has a natural crystal structure. Indeed, the operations e_j and f_j obtained by restriction from $\mathsf{B}_{\mathfrak{g}}$ preserve $\mathsf{B}_{\mathfrak{g}}^{\mathfrak{m}_i}$, since they commute with e_i^*.

The operations e_i and f_i are defined in terms of the decomposition (21). Namely,

$$\mathsf{B}_{\mathfrak{g}}^{\mathfrak{m}_i} \cap (\mathsf{C}_{\mathfrak{m}_i}^n \times \mathsf{B}_{\mathfrak{m}_i}^n \times \mathsf{B}_i) = (\mathsf{C}_{\mathfrak{m}_i}^n \times \mathsf{B}_{\mathfrak{m}_i}^n) \times \mathsf{b}_i(0),$$

and we set e_i, f_i to act along the $\mathsf{B}_{\mathfrak{m}_i}^n$–factor.

Moreover, note that the map Ψ_i constructed above maps $\mathsf{B}_{\mathfrak{g}}$ to $\mathsf{B}_{\mathfrak{g}}^{\mathfrak{m}_i} \otimes \mathsf{B}_i$, and we claim that this is a map of crystals with respect to the crystal structure on $\mathsf{B}_{\mathfrak{g}}^{\mathfrak{m}_i}$ introduced above. Moreover, one easily checks that this map is an isomorphism.

12.6

Thus if $\mathsf{B}_{\mathfrak{g}}$ satisfies the highest-weight property, and admits decompositions as in (21) for every i, it can be identified with the standard crystal $\mathsf{B}_{\mathfrak{g}}^{\infty}$ of [KaS]. In what follows, we will need several additional properties of $\mathsf{B}_{\mathfrak{g}}^{\infty}$.

Let $\mathfrak{m}$ be an arbitrary Levi subalgebra of $\mathfrak{g}$. (In our applications we will only consider Levi subalgebras that correspond to subdiagrams of finite type.) The following result is a generalization of what we have constructed for $\mathfrak{m} = \mathfrak{m}_i$; cf. [Ka4, Ka5].

Theorem 12.7. *For every Levi subalgebra $\mathfrak{m}$ there exists an isomorphism of $\mathfrak{g}$-crystals (with respect to the $e_i, f_i, \epsilon_i, \phi_i, i \in I$)*

$$\Psi_{\mathfrak{m}} : \mathsf{B}_{\mathfrak{g}}^{\infty} \simeq \mathsf{B}_{\mathfrak{g}}^{\mathfrak{m}} \otimes \mathsf{B}_{\mathfrak{m}}^{\infty},$$

(here $\mathsf{B}_{\mathfrak{m}}^{\infty}$ is viewed as a $\mathfrak{g}$-crystal), such that

(a) *the operations e_i^*, f_i^* for $i \in \mathfrak{m}$ on $\mathsf{B}_{\mathfrak{g}}^{\infty}$ go over to the e_i^*, f_i^* operations along the second factor;*
(b) *the $\mathfrak{m}$-crystal structure on $\mathsf{B}_{\mathfrak{g}}^{\mathfrak{m}}$ is normal.*

In particular, from this theorem we obtain that, as a set, $\mathsf{B}_{\mathfrak{g}}^{\mathfrak{m}}$ can be realized as a subset of $\mathsf{B}_{\mathfrak{g}}^{\infty}$ consisting of elements annihilated by the f_i^*, $i \in \mathfrak{m}$. In terms of this set-theoretic embedding $\mathsf{B}_{\mathfrak{g}}^{\mathfrak{m}} \to \mathsf{B}_{\mathfrak{g}}^{\infty}$, the operations $e_j, f_j, \epsilon_j, \phi_j$ for $j \notin \mathfrak{m}$ are induced from those on $\mathsf{B}_{\mathfrak{g}}^{\infty}$. For $i \in \mathfrak{m}$, the e_i and f_i operators on an element $\mathsf{b} \in \mathsf{B}_{\mathfrak{g}}^{\mathfrak{m}}$ can be explicitly described as follows:

The element $f_i(\mathsf{b})$ (for the action that comes from $\mathsf{B}_{\mathfrak{g}}^{\mathfrak{m}}$-crystal structure) equals $f_i(\mathsf{b})$ (for the action that is induced by the $\mathsf{B}_{\mathfrak{g}}^{\infty}$-crystal structure). The element $e_i(\mathsf{b})$ (for the action coming from the $\mathsf{B}_{\mathfrak{g}}^{\mathfrak{m}}$-crystal structure) equals $e_i(\mathsf{b})$ (for the action induced by the $\mathsf{B}_{\mathfrak{g}}^{\infty}$-crystal structure) if the latter belongs to $\mathsf{B}_{\mathfrak{g}}^{\mathfrak{m}}$, and 0 otherwise.

In addition, we have the following assertion; cf. [Ka3, Theorem 3].

Theorem 12.8. *The set $\mathsf{B}_{\mathfrak{g}}^{\mathfrak{m}}$ parametrizes the canonical basis for $U(\mathfrak{n}(\check{\mathfrak{p}}))$, where $\check{\mathfrak{p}}$ is the corresponding parabolic subalgebra, and $\mathfrak{n}(\check{\mathfrak{p}})$ is its unipotent radical. Moreover, as an $\mathfrak{m}$-crystal, $\mathsf{B}_{\mathfrak{g}}^{\mathfrak{m}}$ splits as a disjoint union*

$$\mathsf{B}_{\mathfrak{g}}^{\mathfrak{m}} = \bigcup_{\nu} \mathsf{C}_{\mathfrak{m}}^{\nu} \otimes \mathsf{B}_{\mathfrak{m}}^{\nu},$$

where ν *runs over the set of dominant integral weights of* $\mathfrak{m}$, $\mathsf{B}_{\mathfrak{m}}^{\nu}$ *is the crystal associated to the integrable* $\check{\mathfrak{m}}$*-module* V_M^{ν}, *and* $\mathsf{C}_{\mathfrak{m}}^{\nu}$ *parametrizes a basis in* $\mathrm{Hom}_{\check{\mathfrak{m}}}(V_M^{\nu}, U(\mathfrak{n}(\check{\mathfrak{p}})))$.

Note that the set $\mathsf{C}_{\mathfrak{m}}^{\nu}$ of the above theorem embeds into $\mathsf{B}_{\mathfrak{g}}^{\infty}$ as the set of those $\mathsf{b} \in \mathsf{B}_{\mathfrak{g}}^{\infty}(\nu)$ which are annihilated by f_i and f_i^* for $i \in \mathfrak{m}$.

13 Crystals via the affine Grassmannian

13.1

In this section we will take $\mathfrak{g}$ to be finite dimensional, and give a construction of a crystal $\mathsf{B}_{\mathfrak{g}}$ using the affine Grassmannian Gr_G of G, in the spirit of [BG]. We will assume that the reader is familiar with the notation of [BG].

In the next section we will generalize this construction for an arbitrary Kac–Moody algebra $\mathfrak{g}$, using the space $\mathrm{QMaps}(\mathbf{C}, \mathcal{G}_{\mathfrak{g},\mathfrak{b}})$ instead of the (nonexistent in the general case) affine Grassmannian.

Recall that for a standard parabolic P, we denote by $N(P)$ (respectively, M, P^-, $N(P^-)$) its unipotent radical (respectively, the Levi subgroup, the corresponding opposite parabolic, etc.) By $B(M)$ (respectively, $N(M)$, $B^-(M)$, $N^-(M)$) we will denote the standard Borel subgroup in M (respectively, unipotent radical of $B(M)$, etc.) Finally, recall that T denotes the Cartan subgroup of G.

Let $\mathbf{C}$ be a (smooth, but not necessarily complete) curve, and $\mathbf{c} \in \mathbf{C}$ a point. We will choose a local parameter on $\mathbf{C}$ at $\mathbf{c}$ and call it t. Recall that if H is an algebraic group, we can consider the group-scheme $H[[t]]$, the group ind-scheme $H((t))$ and the ind-scheme of ind-finite type $\mathrm{Gr}_H := H((t))/H[[t]]$, called the affine Grassmannian. In what follows we will consider the affine Grassmannians corresponding to the groups G, M, P, $N(P)$, etc.

For a coweight λ, we have a canonical point in $T((t))$, denoted t^{λ}. We will denote by the same symbol its image in Gr_T and Gr_G via the embedding of $\mathrm{Gr}_T \to \mathrm{Gr}_G$.

We have the maps $i_P : \mathrm{Gr}_P \to \mathrm{Gr}_G$, $i_{P^-} : \mathrm{Gr}_{P^-} \to \mathrm{Gr}_G$, which are locally closed embeddings on every connected component, and projections $\mathfrak{q}_P : \mathrm{Gr}_P \to \mathrm{Gr}_M$, $\mathfrak{q}_{P^-} : \mathrm{Gr}_{P^-} \to \mathrm{Gr}_M$. The projections $\mathfrak{q}_P$, $\mathfrak{q}_{P^-}$ induce a bijection on the set of connected components. For a coweight $\lambda \in \Lambda$, we will denote by Gr_T^{λ} (respectively, Gr_B^{λ}, $\mathrm{Gr}_{B^-}^{\lambda}$) the corresponding connected component of Gr_T (respectively, Gr_B, Gr_{B^-}).

13.2

It is known that the intersection $\mathrm{Gr}_B^{\lambda_1} \cap \mathrm{Gr}_{B^-}^{\lambda_2}$ is of pure dimension $\langle \lambda_1 - \lambda_2, \check{\rho} \rangle$. Let us denote by $\mathsf{B}_{\mathfrak{g}}(\lambda)$ the set of irreducible components of the above intersection with parameters $\lambda_1 - \lambda_2 = \lambda$. (It is easy to see that the action of $t^{\lambda'} \in T((t))$ identifies $\mathrm{Gr}_B^{\lambda_1} \cap \mathrm{Gr}_{B^-}^{\lambda_2}$ with $\mathrm{Gr}_B^{\lambda_1+\lambda'} \cap \mathrm{Gr}_{B^-}^{\lambda_2+\lambda'}$.)

Our goal now is to show that the set $\mathsf{B}_\mathfrak{g} := \underset{\lambda}{\cup} \mathsf{B}_\mathfrak{g}(\lambda)$ has a natural structure of $\mathfrak{g}$-crystal. We need to define the weight function $wt : \mathsf{B}_\mathfrak{g} \to \Lambda$, the functions $\epsilon_i, \phi_i : \mathsf{B}_\mathfrak{g} \to \mathbb{Z}$ for $i \in I$, and the operations $e_i, f_i : \mathsf{B}_\mathfrak{g} \to \mathsf{B}_\mathfrak{g} \cup 0$.

Of course, the function wt is defined so that $wt(\mathsf{B}_\mathfrak{g}(\lambda)) = \lambda$. The function ϕ_i is defined in terms of f_i by the normality condition

$$\text{for } \mathsf{b} \in \mathsf{B}_\mathfrak{g}, \quad \phi_i(\mathsf{b}) = max\{n | f_i^n(\mathsf{b}) \neq 0\}.$$

The function ϵ_i will be defined by the rule

$$\epsilon_i(\mathsf{b}) = \phi_i(\mathsf{b}) - \langle \check{\alpha}_i, wt(\mathsf{b}) \rangle.$$

Thus we have to define the operations e_i and f_i.

13.3

First, let us assume that G is of semisimple rank 1. Then we can identify the coroot lattice with $\mathbb{Z}$, and it is easy to see that $\mathrm{Gr}_B^\lambda \cap \mathrm{Gr}_{B^-}^0$ is nonempty if and only if $\lambda = n \cdot \alpha$ (where α is the unique positive coroot), with $n \geq 0$. In the latter case, this intersection is an irreducible variety, isomorphic to $\mathbb{A}^n - \mathbb{A}^{n-1}$. Thus $\mathsf{B}_\mathfrak{g}$ in this case is naturally $\mathbb{Z}^{\geq 0}$, and the definition of e_i and f_i is evident: they are the raising and the lowering operators, respectively.

In the case of a general G, the operations will be defined by reduction to the rank-1 case.

For a parabolic P and $\lambda, \mu \in \Lambda$, consider the intersection

$$\mathfrak{q}_P^{-1}(\mathrm{Gr}_{B^-(M)}^\mu) \cap \mathrm{Gr}_{B^-}^\lambda \subset \mathrm{Gr}_G,$$

projecting by means of $\mathfrak{q}_P$ on $\mathrm{Gr}_{B^-(M)}^\mu$.

Lemma 13.4. *The above intersection splits as a direct product*

$$\mathrm{Gr}_{B^-(M)}^\mu \times ((\mathfrak{q}_P)^{-1}(g) \cap \mathrm{Gr}_{B^-}^\lambda)$$

for any $g \in \mathrm{Gr}_{B^-(M)}^\mu$.

Proof. The group $N^-(M)((t))$ acts transitively on $\mathrm{Gr}_{B^-(M)}^\mu$, and it also acts on Gr_G preserving the intersection $(\mathfrak{q}_P)^{-1}(\mathrm{Gr}_{B^-(M)}^\mu) \cap \mathrm{Gr}_{B^-}^\lambda$.

Moreover, for $g' \in \mathrm{Gr}_P \subset \mathrm{Gr}_G$ and $g = \mathfrak{q}_P(g')$, the inclusion

$$\mathrm{Stab}_{N^-(M)((t))}(g', \mathrm{Gr}_G) \subset \mathrm{Stab}_{N^-(M)((t))}(g, \mathrm{Gr}_M)$$

is an equality, since $N^-(M) \cap N(P) = 1$.

Therefore, for any g as above, the action of $N^-(M)((t)) / \mathrm{Stab}_{N^-(M)((t))}(g, \mathrm{Gr}_M)$ on the fiber $(\mathfrak{q}_P)^{-1}(g) \cap \mathrm{Gr}_{B^-}^\lambda$ defines an isomorphism

$$(\mathfrak{q}_P)^{-1}(\mathrm{Gr}_{B^-(M)}^\mu) \cap \mathrm{Gr}_{B^-}^\lambda \simeq \mathrm{Gr}_{B^-(M)}^\mu \times ((\mathfrak{q}_P)^{-1}(g) \cap \mathrm{Gr}_{B^-}^\lambda). \qquad \square$$

We will use this lemma as follows. For $\lambda_1, \lambda_2, \mu \in \Lambda$, let us note that the intersection

$$\mathrm{Gr}_B^{\lambda_1} \cap (\mathfrak{q}_P)^{-1}(\mathrm{Gr}_{B^-(M)}^{\mu}) \cap \mathrm{Gr}_{B^-}^{\lambda_2}$$

is the same as

$$(\mathfrak{q}_P)^{-1}(\mathrm{Gr}_{B(M)}^{\lambda_1} \cap \mathrm{Gr}_{B^-(M)}^{\mu}) \cap \mathrm{Gr}_{B^-}^{\lambda_2}.$$

In particular, by putting $\lambda_1 = \mu$ we obtain that the variety $(\mathfrak{q}_P)^{-1}(g) \cap \mathrm{Gr}_{B^-}^{\lambda}$ of Lemma 13.4 is finite dimensional of dim $\leq \langle \mu - \lambda, \check{\rho} \rangle$.

Of course, the variety $(\mathfrak{q}_P)^{-1}(\mathrm{Gr}_{B^-(M)}^{\mu}) \cap \mathrm{Gr}_{B^-}^{\lambda}$ depends up to isomorphism only on the difference $\mu' := \mu - \lambda$. Let us denote by $\mathsf{B}_{\mathfrak{g}}^{\mathfrak{m},*}(\mu')$ the set of irreducible components of $(\mathfrak{q}_P)^{-1}(g) \cap \mathrm{Gr}_{B^-}^{\lambda}$ of (the maximal possible) dimension $\langle \mu', \check{\rho} \rangle$.

Since the group $N^-(B)((t))$ is ind-pro-unipotent, we obtain that the stabilizer

$$\mathrm{Stab}_{N^-(M)((t))}(g, \mathrm{Gr}_M)$$

appearing in the proof of Lemma 13.4 is connected. Therefore, the set $\mathsf{B}_{\mathfrak{g}}^{\mathfrak{m},*}(\mu')$ is well defined, i.e., is independent of the choice of the point g.

Clearly, for every irreducible component $\mathbf{K}$ of $\mathrm{Gr}_B^{\lambda_1} \cap \mathrm{Gr}_{B^-}^{\lambda_2}$ there exists a unique $\mu \in \Lambda$ such that the intersection $\mathbf{K} \cap (\mathfrak{q}_P)^{-1}(\mathrm{Gr}_{B^-(M)}^{\mu})$ is dense in $\mathbf{K}$.

Using Lemma 13.4, we obtain that the set $\mathsf{B}_{\mathfrak{g}}(\lambda)$ can be canonically decomposed as a union

$$\mathsf{B}_{\mathfrak{g}}(\lambda) = \underset{\mu}{\cup} \mathsf{B}_{\mathfrak{g}}^{\mathfrak{m},*}(\mu) \times \mathsf{B}_{\mathfrak{m}}(\lambda - \mu). \tag{23}$$

Finally, we are able to define the operations e_i and f_i. For each $i \in I$ take P to be the corresponding subminimal parabolic P_i and consider the decomposition

$$\mathsf{B}_{\mathfrak{g}}(\lambda) = \underset{\mu}{\cup} \mathsf{B}_{\mathfrak{g}}^{\mathfrak{m}_i,*}(\mu) \times \mathsf{B}_{\mathfrak{m}_i}(\lambda - \mu).$$

For an element $\mathsf{b}_{\mathfrak{g}} \in \mathsf{B}_{\mathfrak{g}}(\lambda)$ of the form

$$\mathsf{b}' \times \mathsf{b}_{\mathfrak{m}_i}, \mathsf{b}' \in \mathsf{B}_{\mathfrak{g}}^{\mathfrak{m}_i,*}(\mu), \mathsf{b}_{\mathfrak{m}_i} \in \mathsf{B}_{\mathfrak{m}_i}(\lambda - \mu),$$

we set $e_i(\mathsf{b}_G)$ to be

$$\mathsf{b}' \times e_i(\mathsf{b}_{\mathfrak{m}_i}) \in \mathsf{B}_{\mathfrak{g}}^{\mathfrak{m}_i,*}(\mu) \times \mathsf{B}_{\mathfrak{m}_i}(\lambda + \alpha_i - \mu) \cup 0 \subset \mathsf{B}_{\mathfrak{g}}(\lambda + \alpha_i) \cup 0,$$

and similarly for f_i.

This is well defined, because M_i is of semisimple rank 1, and we know how the operators e_i and f_i act on $\mathsf{B}_{\mathfrak{m}_i}$.

13.5

Let us now define the operations e_i^* and f_i^*. First, when G is of semisimple rank 1, we set $e_i^* = e_i$ and $f_i^* = f_i$. To treat the general case, we simply interchange the roles of the projections $\mathfrak{q}_P$ and $\mathfrak{q}_{P^-}$.

In more detail, for any λ and μ as before, let us consider the intersection

$$\mathrm{Gr}^{\lambda}_{B} \cap (\mathfrak{q}_{P^-})^{-1}(\mathrm{Gr}^{\mu}_{B(M)}).$$

As in Lemma 13.4, this intersection can be represented as a product $\mathrm{Gr}^{\mu}_{B(M)} \times ((\mathfrak{q}_{P^-})^{-1}(g) \cap \mathrm{Gr}^{\lambda}_{B})$ for any $g \in \mathrm{Gr}^{\mu}_{B(M)}$. For $\mu' = \lambda - \mu$, let us denote by $\mathsf{B}^{\mathfrak{m}}_{\mathfrak{g}}(\mu')$ the set of irreducible components of top dimension of $(\mathfrak{q}_{P^-})^{-1}(g) \cap \mathrm{Gr}^{\lambda}_{B}$.

By looking at the intersections of irreducible components of $\mathrm{Gr}^{\lambda_1}_{B} \cap \mathrm{Gr}^{\lambda_2}_{B^-}$ with the various $(\mathfrak{q}_{P^-})^{-1}(\mathrm{Gr}^{\mu}_{B(M)})$, we obtain a decomposition

$$\mathsf{B}_{\mathfrak{g}}(\lambda) = \underset{\mu}{\cup} \mathsf{B}^{\mathfrak{m}}_{\mathfrak{g}}(\mu) \times \mathsf{B}_{\mathfrak{m}}(\lambda - \mu). \tag{24}$$

By taking $M = M_i$, for $\mathsf{b}_{\mathfrak{g}} \in \mathsf{B}_{\mathfrak{g}}(\lambda)$ of the form

$$\mathsf{b}' \times \mathsf{b}_{\mathfrak{m}_i}, \mathsf{b}' \in \mathsf{B}^{\mathfrak{m}_i}_{\mathfrak{g}}(\mu), \mathsf{b}_{\mathfrak{m}_i} \in \mathsf{B}_{\mathfrak{m}_i}(\lambda - \mu),$$

we set $e_i^*(\mathsf{b}_{\mathfrak{g}})$ to be

$$\mathsf{b}' \times e_i^*(\mathsf{b}_{\mathfrak{m}_i}) \in \mathsf{B}^{\mathfrak{m}_i}_{\mathfrak{g}}(\mu) \times \mathsf{B}_{\mathfrak{m}_i}(\lambda + \alpha_i - \mu) \cup 0 \subset \mathsf{B}_{\mathfrak{g}}(\lambda + \alpha_i) \cup 0,$$

and similarly for f_i^*.

Remark. Let us explain the consistency of the notation $\mathsf{B}^{\mathfrak{m}}_{\mathfrak{g}}(\mu)$. As we shall see later, the crystals $\mathsf{B}_{\mathfrak{g}}$ (and in particular $\mathsf{B}_{\mathfrak{m}}$) have the highest-weight property. If we assume this, we will obtain that the set $\mathsf{B}^{\mathfrak{m}}_{\mathfrak{g}}(\mu)$ is precisely the subset of $\mathsf{B}_{\mathfrak{g}}(\mu)$ consisting of elements annihilated by f_i^*, $i \in \mathfrak{m}$. Indeed, the latter elements correspond to the irreducible components of

$$\mathrm{Gr}^{0}_{B} \cap (\mathfrak{q}_{P^-})^{-1}(\mathrm{Gr}^{\mu}_{B(M)} \cap \mathrm{Gr}^{\mu}_{B^-(M)})$$

of dimension $\langle \mu, \check{\rho} \rangle$, which is the same as the set $\mathsf{B}^{\mathfrak{m}}_{\mathfrak{g}}(\mu)$.

Proposition 13.6. *For $i \neq j$, the operations e_i, f_i commute with the operations e_j^*, f_j^*.*

Proof. Consider the intersection $(\mathfrak{q}_{P_i})^{-1}(\mathrm{Gr}^{\mu_1}_{B^-(M_i)}) \cap (\mathfrak{q}_{P_j^-})^{-1}(\mathrm{Gr}^{\mu_2}_{B(M_j)}) \subset \mathrm{Gr}_G$. We claim that it decomposes as

$$(\mathrm{Gr}^{\mu_1}_{B^-(M_i)} \times \mathrm{Gr}^{\mu_2}_{B(M_j)}) \times ((\mathfrak{q}_{P_i})^{-1}(g_1) \cap (\mathfrak{q}_{P_j^-})^{-1}(g_2))$$

for any $g_1 \times g_2 \in \mathrm{Gr}^{\mu_1}_{B^-(M_i)} \times \mathrm{Gr}^{\mu_2}_{B(M_j)}$.

Indeed, as in the proof of Lemma 13.4, the group $N^-(M_i)((t)) \times N(M_j)((t))$ acts transitively on the base $\mathrm{Gr}^{\mu_1}_{B^-(M_i)} \times \mathrm{Gr}^{\mu_2}_{B(M_j)}$, and since the subgroups $N^-(M_i)$, $N(M_j) \subset G$ commute with one another, we obtain the desired decomposition as in Lemma 13.4.

For $\mu = \mu_1 - \mu_2$, let us denote by $\mathsf{C}_{i,j}(\mu)$ the set of irreducible components of top $(= \langle \mu, \check{\rho} \rangle)$ dimension of $(q_{P_i})^{-1}(g_1) \cap (q_{P_j^-})^{-1}(g_2)$.

By intersecting $\mathrm{Gr}_B^{\lambda_1} \cap \mathrm{Gr}_{B^-}^{\lambda_2}$ with $(\mathfrak{q}_{P_i})^{-1}(\mathrm{Gr}^{\mu_1}_{B^-(M_i)}) \cap (\mathfrak{q}_{P_j^-})^{-1}(\mathrm{Gr}^{\mu_2}_{B(M_j)})$ for the various μ_1 and μ_2, we obtain a decomposition

$$\mathsf{B}_{\mathfrak{g}}(\lambda) \simeq \bigcup_{\mu_1, \mu_2} \mathsf{C}_{i,j}(\mu_1 - \mu_2) \times \mathsf{B}_{\mathfrak{m}_i}(\lambda_1 - \mu_1) \times \mathsf{B}_{\mathfrak{m}_j}(\mu_2 - \lambda_2).$$

This decomposition is a refinement of (23) and (24). Therefore, for an element $\mathsf{b}_G \in \mathsf{B}_{\mathfrak{g}}(\lambda)$ of the form $\mathbf{c} \times \mathsf{b}_i \times \mathsf{b}_j$ with

$$\mathsf{c} \in \mathsf{C}_{i,j}(\mu_1 - \mu_2), \mathsf{b}_i \in \mathsf{B}_{\mathfrak{m}_i}(\lambda - \mu_1), \mathsf{b}_j \in \mathsf{B}_{\mathfrak{m}_j}(\lambda - \mu_2),$$

the e_i and f_i operations act via their action on b_i, and the e_j^* and f_j^* act via b_j, respectively. This makes the required commutativity property manifest. □

13.7

Finally, we are going to define the decompositions of $\mathsf{B}_{\mathfrak{g}}$ as in (21). In fact, we will define the decompositions as in Theorem 12.7 for any Levi subalgebra $\mathfrak{m}$. First, we need to discuss certain $\mathfrak{m}$-crystals associated with the convolution diagram of M.

Let us recall some notation related to the affine Grassmannian Gr_M. The group-scheme $M[[t]]$ (being a subgroup of the group ind-scheme $M((t)))$ acts naturally on Gr_M, and its orbits are parametrized by the set $\Lambda^+_{\mathfrak{m}}$ of dominant coweights of M; for $\nu \in \Lambda^+_{\mathfrak{m}}$ we will denote the corresponding orbit by Gr^ν_M.

Let Conv_M denote the convolution diagram for M. It is by definition of the ind-scheme parametrizing the data of $(\mathcal{F}_M, \mathcal{F}'_M, \beta, \widetilde{\beta}')$, where $\mathcal{F}_M, \mathcal{F}'_M$ are principal M-bundles on $\mathbf{C}$, β is a trivialization of $\mathcal{F}_M$ on $\mathbf{C} - \mathbf{c}$, and $\widetilde{\beta}^1$) is an isomorphism $\mathcal{F}_M|_{\mathbf{C}-\mathbf{c}} \to \mathcal{F}'_M|_{\mathbf{C}-\mathbf{c}}$. We will think of Conv_M as a fibration over Gr_M with typical fiber Gr_M; more precisely,

$$\mathrm{Conv}_M \simeq M((t)) \underset{M[[t]]}{\times} \mathrm{Gr}_M .$$

Sometimes, we will write $\mathrm{Conv}_M \simeq \mathrm{Gr}_M \star \mathrm{Gr}_M$, so that the first factor is perceived as the base, and the second one as the fiber. In particular, if $\mathcal{Y} \subset \mathrm{Gr}_M$ is a subscheme, and $\mathcal{Y}' \subset \mathrm{Gr}_M$ is an $M[[t]]$-invariant subscheme, it makes sense to consider the subscheme $\mathcal{Y} \star \mathcal{Y}' \subset \mathrm{Gr}_M \star \mathrm{Gr}_M \simeq \mathrm{Conv}_M$. For $\nu \in \Lambda^+_{\mathfrak{m}}$, let $\mathrm{Conv}^\nu_M \subset \mathrm{Conv}_M$ be equal to $\mathrm{Gr}_M \star \mathrm{Gr}^\nu_M$.

Note that for a point $(\mathcal{F}_M, \mathcal{F}'_M, \beta, \widetilde{\beta}') \in \mathrm{Conv}_M$, by taking the composition $\beta^1 := \widetilde{\beta}^1 \circ \beta$ we obtain a trivialization of $\mathcal{F}'_M$ on $\mathbf{C} - \mathbf{c}$, and thus a map $\mathrm{Conv}_M \to \mathrm{Gr}_M \times \mathrm{Gr}_M$, which is easily seen to be an isomorphism. We will denote by p and p' the two projections $\mathrm{Conv}_M \to \mathrm{Gr}_M$, which remember the data of $(\mathcal{F}_M, \beta)$ and $(\mathcal{F}'_M, \beta')$, respectively. Note that the locally closed subsets Conv^ν_M introduced above are exactly the orbits of the diagonal $M((t))$ action on $\mathrm{Conv}_M \simeq \mathrm{Gr}_M \times \mathrm{Gr}_M$.

13.8

For another pair of coweights λ_1 and λ_2, consider the intersection

$$(\mathrm{Gr}^{\lambda_1}_{B(M)} \times \mathrm{Gr}^{\lambda_2}_{B^-(M)})^\nu := (\mathrm{Gr}^{\lambda_1}_{B(M)} \times \mathrm{Gr}^{\lambda_2}_{B^-(M)}) \cap \mathrm{Conv}^\nu_M .$$

Again, up to isomorphism, the above variety depends only on the difference $\lambda = \lambda_1 - \lambda_2$. Let us denote by $\mathsf{D}^\nu_{\mathfrak{m}}(\lambda)$ the set of irreducible components of the above intersection of the (maximal possible) dimension $\langle \nu + \lambda, \check{\rho} \rangle$.

We will define on the set $\mathsf{D}^\nu_{\mathfrak{m}} := \underset{\lambda}{\cup} \mathsf{D}^\nu_{\mathfrak{m}}(\lambda)$ two structures of a (ϕ-normal) $\mathfrak{m}$-crystal.

First, let us define the operations e_i, f_i. For this let us fix $\lambda_1, \lambda_2 \in \Lambda$ and consider the ind-scheme

$$(\mathrm{Gr}^{\lambda_1}_{B^-(M)} \times \mathrm{Gr}^{\lambda_2}_{B^-(M)})^\nu := (\mathrm{Gr}^{\lambda_1}_{B^-(M)} \times \mathrm{Gr}^{\lambda_2}_{B^-(M)}) \cap \mathrm{Conv}^\nu_M$$

projecting by means of p onto $\mathrm{Gr}^{\lambda_1}_{B^-(M)}$. As in Lemma 13.4, we obtain that there is an isomorphism

$$(\mathrm{Gr}^{\lambda_1}_{B^-(M)} \times \mathrm{Gr}^{\lambda_2}_{B^-(M)})^\nu \simeq \mathrm{Gr}^{\lambda_1}_{B^-(M)} \times ((g \times \mathrm{Gr}^{\lambda_2}_{B^-(M)}) \cap \mathrm{Conv}^\nu_M)$$

for any $g \in \mathrm{Gr}^{\lambda_1}_{B^-(M)}$.

For $\lambda = \lambda_1 - \lambda_2$ and $g \in \mathrm{Gr}^{\lambda_1}_{B^-(M)}$, let $\mathsf{B}^{\nu,*}_{\mathfrak{m}}(\lambda)$ denote the set of irreducible components of the intersection $(g \times \mathrm{Gr}^{\lambda_2}_{B^-(M)}) \cap \mathrm{Conv}^\nu_M$ of (the maximal possible) dimension $\langle \nu + \lambda, \check{\rho} \rangle$. (As in Lemma 13.4, this set does not depend on the choice of $g \in \mathrm{Gr}_M$).

Thus we have a set-theoretic decomposition

$$\mathsf{D}^\nu_{\mathfrak{m}}(\lambda) \simeq \underset{\lambda'}{\cup} \mathsf{B}^{\nu,*}_{\mathfrak{m}}(\lambda') \times \mathsf{B}_{\mathfrak{m}}(\lambda - \lambda').$$

By taking $M = M_i$, we define the operations e_i and f_i to act along the second multiple of this decomposition. It is easy to see that this definition agrees with the one in Section 13.3.

The operations e^*_i, f^*_i are defined in a similar fashion. We fix λ_1 and λ_2 and consider the ind-scheme

$$(\mathrm{Gr}^{\lambda_1}_{B(M)} \times \mathrm{Gr}^{\lambda_2}_{B(M)})^\nu := (\mathrm{Gr}^{\lambda_1}_{B(M)} \times \mathrm{Gr}^{\lambda_2}_{B(M)}) \cap \mathrm{Conv}^\nu_M$$

projecting by means of p' onto $\mathrm{Gr}^{\lambda_2}_{B(M)}$. It also splits as a product

$$(\mathrm{Gr}^{\lambda_1}_{B(M)} \times \mathrm{Gr}^{\lambda_2}_{B(M)})^\nu \simeq \mathrm{Gr}^{\lambda_2}_{B(M)} \times ((\mathrm{Gr}^{\lambda_1}_{B(M)} \times g) \cap \mathrm{Conv}^\nu_M)$$

for any $g \in \mathrm{Gr}^{\lambda_2}_{B(M)}$.

Let $\mathsf{B}^\nu_{\mathfrak{m}}(\lambda)$ denote the set of irreducible components of (the maximal possible) dimension $\langle \nu + \lambda, \check{\rho} \rangle$ of $(\mathrm{Gr}^{\lambda_1}_{B(M)} \times g) \cap (\mathrm{Gr}_M \times \mathrm{Gr}_M)^\nu$, and as before, we obtain a set-theoretic decomposition

$$\mathsf{D}^{\nu}_{\mathfrak{m}}(\lambda) \simeq \underset{\lambda'}{\cup} \mathsf{B}^{\nu}_{\mathfrak{m}}(\lambda') \times \mathsf{B}_{\mathfrak{m}}(\lambda - \lambda'). \tag{25}$$

We define the operations e_i^*, f_i^* to act along the second multiple of this decomposition for $M = M_i$, and again this definition is easily seen to coincide with the one of Section 13.5.

Note that by taking g to be the unit element in Gr_M, we obtain that $\mathsf{B}^{\nu}_{\mathfrak{m}}(\lambda)$ are exactly the elements of the crystal associated to the integrable $\check{M}$-module with highest weight ν of weight λ. Set $\mathsf{B}^{\nu}_{\mathfrak{m}} := \underset{\lambda}{\cup} \mathsf{B}^{\nu}_{\mathfrak{m}}(\lambda)$, and recall from [BG] that this set has a canonical structure of a (normal) $\mathfrak{m}$-crystal.

The following is a generalization of [BG, Theorem 3.2] when one of the finite crystals is replaced by $\mathsf{B}_{\mathfrak{m}}$.

Proposition 13.9. *In terms of the above set-theoretic decomposition $\mathsf{D}^{\nu}_{\mathfrak{m}} \simeq \mathsf{B}^{\nu}_{\mathfrak{m}} \times \mathsf{B}_{\mathfrak{m}}$, the crystal structure on $\mathsf{D}^{\nu}_{\mathfrak{m}}$, given by e_i and f_i, corresponds to the tensor product crystal structure on the right-hand side.*

We omit the proof, since the argument is completely analogous to the corresponding proof in loc. cit.

13.10

Finally, we are ready to define the decompositions of (21).

Consider the projection

$$(\mathfrak{q}_P \times \mathfrak{q}_{P^-}) : \mathrm{Gr}_P \underset{\mathrm{Gr}_G}{\times} \mathrm{Gr}_{P^-} \to \mathrm{Gr}_M \times \mathrm{Gr}_M \simeq \mathrm{Conv}_M .$$

This projection is $M((t))$-equivariant. For a fixed M-dominant coweight ν let us denote by $(\mathrm{Gr}_P \underset{\mathrm{Gr}_G}{\times} \mathrm{Gr}_{P^-})^{\nu}$ the preimage of Conv^{ν}_M under this map. For a point $g \in \mathrm{Conv}^{\nu}_M$ consider the scheme $(\mathfrak{q}_P \times \mathfrak{q}_{P^-})^{-1}(g) \subset (\mathrm{Gr}_P \underset{\mathrm{Gr}_G}{\times} \mathrm{Gr}_{P^-})^{\nu}$. The set of its irreducible components of (the maximal possible) dimension $\langle \nu, \check{\rho} \rangle$ will be denoted by $\mathsf{C}^{\nu}_{\mathfrak{m}}$.

Thus we obtain a decomposition

$$\mathsf{B}_{\mathfrak{g}} = \underset{\nu}{\cup} \mathsf{C}^{\nu}_{\mathfrak{m}} \times \mathsf{D}^{\nu}_{\mathfrak{m}}.$$

Moreover, by unraveling the construction of the operations, we obtain that the above decomposition respects the action of both e_i, f_i, and e_i^*, f_i^*, such that all of the operations act via the second multiple (i.e., $\mathsf{D}^{\nu}_{\mathfrak{m}}$) in the way specified above.

This defines the required decomposition in view of Proposition 13.9.

Finally, by combining what we said above with Theorem 12.8 (and assuming the highest-weight property of the crystal $\mathsf{B}_{\mathfrak{g}}$), we obtain that the set $\mathsf{C}^{\nu}_{\mathfrak{m}}$, which enumerates irreducible components of $(\mathfrak{q}_P \times \mathfrak{q}_{P^-})^{-1}(g)$ for $g \in \mathrm{Conv}^{\nu}_M$, parametrizes a basis of $\mathrm{Hom}_M(V^{\nu}_M, U(\mathfrak{n}(\check{\mathfrak{p}})))$.

14 Crystals via quasi-map spaces

14.1

In this section $\mathfrak{g}$ will again be an arbitrary Kac–Moody algebra. Recall that $\mathcal{G}_{\mathfrak{g},\mathfrak{b}}$ denotes Kashiwara's flag scheme. We will fix $\mathbf{C}$ to be an arbitrary smooth (but not necessarily complete) curve, with a distinguished point $\mathbf{c} \in \mathbf{C}$. As in the previous section, we will denote by t a local coordinate on $\mathbf{C}$ defined near $\mathbf{c}$.

Recall that for $\lambda \in \Lambda_{\mathfrak{g}}^{\text{pos}}$ in Section 2.12 we introduced the scheme $\mathcal{Z}^{\lambda}_{\mathfrak{g},\mathfrak{b}}(\mathbf{C})$ and its open subset $\overset{\circ}{\mathcal{Z}}{}^{\lambda}_{\mathfrak{g},\mathfrak{b}}(\mathbf{C})$. We showed in Proposition 2.21 that when $\mathbf{C} = \mathbb{A}^1$, then $\mathcal{Z}^{\lambda}_{\mathfrak{g},\mathfrak{b}}(\mathbf{C})$ can be identified with the scheme of based quasi-maps $\mathrm{QMaps}^{\lambda}(\mathbb{P}^1, \mathcal{G}_{\mathfrak{g},\mathfrak{b}})$, such that $\overset{\circ}{\mathcal{Z}}{}^{\lambda}_{\mathfrak{g},\mathfrak{b}}(\mathbf{C})$ corresponds to the locus of *maps* inside quasi-maps.

We have a natural projection $\varrho^{\lambda}_{\mathfrak{b}} : \mathcal{Z}^{\lambda}_{\mathfrak{g},\mathfrak{b}}(\mathbf{C}) \to \mathbf{C}^{\lambda}$, and we set $\mathfrak{F}^{\lambda}_{\mathfrak{g},\mathfrak{b}} := (\varrho^{\lambda}_{\mathfrak{b}})^{-1}(\lambda \cdot 0_{\mathbf{C}})$ (respectively, $\overset{\circ}{\mathfrak{F}}{}^{\lambda}_{\mathfrak{g},\mathfrak{b}} := \mathfrak{F}^{\lambda}_{\mathfrak{g},\mathfrak{b}} \cap \overset{\circ}{\mathcal{Z}}{}^{\lambda}_{\mathfrak{g}}(\mathbf{C})$).

It was shown in [FFKM] (cf. also [BFGM]) that when $\mathfrak{g}$ is finite dimensional, the scheme $\overset{\circ}{\mathfrak{F}}{}^{\lambda}_{\mathfrak{g},\mathfrak{b}}$ can be identified with the intersection $\mathrm{Gr}^{0}_{B} \cap \mathrm{Gr}^{-\lambda}_{B^-}$ considered in the previous section. Thus the contents of the previous section amount to defining a crystal structure on the union of the sets of irreducible components of the union of $\overset{\circ}{\mathfrak{F}}{}^{\lambda}_{\mathfrak{g},\mathfrak{b}}$ (over all $\lambda \in \Lambda_{\mathfrak{g}}^{\text{pos}}$).

For an arbitrary Kac–Moody algebra $\mathfrak{g}$, set $\mathsf{B}_{\mathfrak{g}}(\lambda)$ to be the set of irreducible components of $\overset{\circ}{\mathfrak{F}}{}^{\lambda}_{\mathfrak{g},\mathfrak{b}}$, and set $\mathsf{B}_{\mathfrak{g}} = \underset{\lambda}{\cup}\mathsf{B}_{\mathfrak{g}}(\lambda)$. In this section we will generalize the construction of the previous section to define on $\mathsf{B}_{\mathfrak{g}}$ a structure of a crystal. However, since in the general case we do not have the affine Grassmannian picture, and we will have to spell out the definitions using quasi-map spaces.

To define the crystal structure on $\mathsf{B}_{\mathfrak{g}}$ we will need to assume Conjecture 2.27. To identify it with the standard crystal of [KaS], we will need to assume one more conjecture, Conjecture 15.3 (cf. Section 15). Both these conjectures are verified when $\mathfrak{g}$ is of affine (and, of course, finite) type (cf. Sections 15.5 and 15.6).

14.2

For a standard parabolic $\mathfrak{p} \subset \mathfrak{g}$, and an element $\theta \in \Lambda^{\text{pos}}_{\mathfrak{g},\mathfrak{p}}$, let $\mathfrak{F}^{\theta}_{\mathfrak{g},\mathfrak{p}}$ (respectively, $\widetilde{\mathfrak{F}}^{\theta}_{\mathfrak{g},\mathfrak{p}}$, $\overset{\circ}{\mathfrak{F}}{}^{\theta}_{\mathfrak{g},\mathfrak{p}}$) denote the preimage of $\theta \cdot \mathbf{c}$ under $\varrho^{\theta}_{\mathfrak{b}} : \mathcal{Z}^{\theta}_{\mathfrak{g},\mathfrak{p}}(\mathbf{C}) \to \mathbf{C}^{\theta}$ (respectively, in $\widetilde{\mathcal{Z}}^{\theta}_{\mathfrak{g},\mathfrak{p}}(\mathbf{C})$, $\overset{\circ}{\mathcal{Z}}{}^{\theta}_{\mathfrak{g},\mathfrak{p}}(\mathbf{C})$).

Assuming for a moment that $\mathbf{C}$ is complete (cf. the proof of Proposition 14.3, where we get rid of this assumption), recall the stacks $\mathsf{Z}^{\theta}_{\mathfrak{g},\mathfrak{p}}(\mathbf{C})$ (respectively, $\widetilde{\mathsf{Z}}^{\theta}_{\mathfrak{g},\mathfrak{p}}(\mathbf{C})$, $\overset{\circ}{\mathsf{Z}}{}^{\theta}_{\mathfrak{g},\mathfrak{p}}(\mathbf{C})$), introduced in Section 2.32. Recall also the forgetful map $\mathrm{Gr}_M \to \mathrm{Bun}_M(\mathbf{C})$, where Gr_M is $\mathrm{Gr}_{M,\mathbf{c}}$. We now define three ind-schemes

$$\mathbb{S}^{\theta}_{\mathfrak{g},\mathfrak{p}} := \mathsf{Z}^{\theta}_{\mathfrak{g},\mathfrak{p}}(\mathbf{C}) \underset{\mathrm{Bun}_M(\mathbf{C}) \times \mathbf{C}^{\theta}}{\times} (\mathrm{Gr}_M \times pt),$$

$$\widetilde{\mathbb{S}}^\theta_{\mathfrak{g},\mathfrak{p}} := \widetilde{\mathbf{Z}}^\theta_{\mathfrak{g},\mathfrak{p}}(\mathbf{C}) \underset{\mathrm{Bun}_M(\mathbf{C})\times\mathbf{C}^\theta}{\times} (\mathrm{Gr}_M \times pt)$$

and

$$\overset{\circ}{\mathbb{S}}{}^\theta_{\mathfrak{g},\mathfrak{p}} := \overset{\circ}{\mathbf{Z}}{}^\theta_{\mathfrak{g},\mathfrak{p}}(\mathbf{C}) \underset{\mathrm{Bun}_M(\mathbf{C})\times\mathbf{C}^\theta}{\times} (\mathrm{Gr}_M \times pt),$$

where we are using the projection $\mathfrak{q}_\mathfrak{p} : \mathbf{Z}^\theta_{\mathfrak{g},\mathfrak{p}}(\mathbf{C}) \to \mathrm{Bun}_M(\mathbf{C})$ (and similarly for $\widetilde{\mathbf{Z}}^\theta_{\mathfrak{g},\mathfrak{p}}(\mathbf{C})$, $\overset{\circ}{\mathbf{Z}}{}^\theta_{\mathfrak{g},\mathfrak{p}}(\mathbf{C})$) to define the fiber product, and $pt \to \mathbf{C}^\theta$ corresponds to the point $\theta \cdot \mathbf{c}$. By definition, $\mathfrak{F}^\theta_{\mathfrak{g},\mathfrak{p}} \simeq \mathbb{S}^\theta_{\mathfrak{g},\mathfrak{p}} \underset{\mathrm{Gr}_M}{\times} pt$, where $pt \hookrightarrow \mathrm{Gr}_M$ is the unit point (and similarly for $\widetilde{\mathfrak{F}}^\theta_{\mathfrak{g},\mathfrak{p}}$, $\overset{\circ}{\mathfrak{F}}{}^\theta_{\mathfrak{g},\mathfrak{p}}$).

Using the map $\varrho^\theta_M : \widetilde{\mathbf{Z}}^\theta_{\mathfrak{g},\mathfrak{p}}(\mathbf{C}) \to \mathcal{H}_{M,\mathbf{C}}$, we can rewrite

$$\widetilde{\mathbb{S}}^\theta_{\mathfrak{g},\mathfrak{p}} \simeq \widetilde{\mathbf{Z}}^\theta_{\mathfrak{g},\mathfrak{p}}(\mathbf{C}) \underset{\mathcal{H}_{M,\mathbf{C}}}{\times} \mathrm{Conv}_M \quad \text{and} \quad \overset{\circ}{\mathbb{S}}{}^\theta_{\mathfrak{g},\mathfrak{p}}(\mathbf{C}) \simeq \overset{\circ}{\mathbf{Z}}{}^\theta_{\mathfrak{g},\mathfrak{p}} \underset{\mathcal{H}_{M,\mathbf{C}}}{\times} \mathrm{Conv}_M .$$

Proposition 14.3. *There exists a canonical action of $M((t))$ on $\widetilde{\mathbb{S}}^\theta_{\mathfrak{g},\mathfrak{p}}$ (respectively, $\mathbb{S}^\theta_{\mathfrak{g},\mathfrak{p}}$, $\overset{\circ}{\mathbb{S}}{}^\theta_{\mathfrak{g},\mathfrak{p}}$) compatible with its action on* Conv_M *(respectively,* Gr_M, Conv_M*).*

Proof. We will give a proof for $\widetilde{\mathbb{S}}^\theta_{\mathfrak{g},\mathfrak{p}}$, since the corresponding facts for $\mathbb{S}^\theta_{\mathfrak{g},\mathfrak{p}}$ and $\overset{\circ}{\mathbb{S}}{}^\theta_{\mathfrak{g},\mathfrak{p}}$ are analogous (and simpler).

Let $\mathcal{D}_\mathbf{c}$ (respectively, $\mathcal{D}^*_\mathbf{c}$) be the formal (respectively, formal punctured) disc in $\mathbf{C}$ around $\mathbf{c}$. Using Lemma 2.13, the data defining a point of $\widetilde{\mathbb{S}}^\theta_{\mathfrak{g},\mathfrak{p}}$ can be rewritten using the formal disc $\mathcal{D}_\mathbf{c}$ instead of the curve $\mathbf{C}$ as follows:

It consists of a principal P-bundle $\mathcal{F}_P$ on $\mathcal{D}_\mathbf{c}$, a principal M-bundle $\mathcal{F}_M$ on $\mathcal{D}_\mathbf{c}$ with a trivialization $\beta : \mathcal{F}_M|_{\mathcal{D}^*_\mathbf{c}} \to \mathcal{F}^0_M|_{\mathcal{D}^*_\mathbf{c}}$ on the formal punctured disc $\mathcal{D}^*_\mathbf{c}$, and a collection of maps

$$\kappa^{\check{\lambda}} : (\mathcal{V}_{\check{\lambda}})_{\mathcal{F}_P} \to (\mathcal{U}_{\check{\lambda}})_{\mathcal{F}_M},$$

satisfying the Plücker equations, such that for $\check{\lambda} \in \check{\Lambda}^+_{\mathfrak{g},\mathfrak{p}}$ the composed map of line bundles

$$\mathcal{L}^{\check{\lambda}}_{\mathcal{F}_P} \to (\mathcal{V}_{\check{\lambda}})_{\mathcal{F}_P} \to \mathcal{L}^{\check{\lambda}}_{\mathcal{F}_M}$$

has a zero of order $\langle\theta, \check{\lambda}\rangle$ at $\mathbf{c}$.

(In particular, this description makes it clear that the schemes $\widetilde{\mathbb{S}}^\theta_{\mathfrak{g},\mathfrak{p}}$ etc., do not depend on the global curve $\mathbf{C}$, but rather on the formal neighborhood $\mathcal{D}_\mathbf{c}$ of $\mathbf{c}$.)

In these terms, the action of $M((t))$ leaves the data of $(\mathcal{F}_P, \mathcal{F}_M, \kappa^{\check{\lambda}})$ intact, and only acts on the data of the trivialization β. □

For $\nu \in \Lambda^+_\mathfrak{m}$, recall the subscheme $\mathrm{Conv}^\nu_M \subset \mathrm{Conv}_M$. Let us denote by $\mathbb{S}^\nu_{\mathfrak{g},\mathfrak{p}} \subset \overset{\circ}{\mathbb{S}}{}^\theta_{\mathfrak{g},\mathfrak{p}}$ the preimage of Conv^ν_M under the map ϱ^θ_M. The resulting map $\mathbb{S}^\nu_{\mathfrak{g},\mathfrak{p}} \to \mathrm{Conv}^\nu_M$ will be denoted by ϱ^ν_M. From Proposition 14.3, we obtain the following.

Corollary 14.4. *The map* $\varrho^\nu_M : \mathbb{S}^\nu_{\mathfrak{g},\mathfrak{p}} \to \mathrm{Conv}^\nu_M$ *is a fibration, locally trivial in the smooth topology.*

The proof is immediate from the fact that the $M((t))$-action on Conv^ν_M is transitive.

14.5

For $\lambda_1, \lambda_2 \in \Lambda_\mathfrak{g}$, $\lambda = \lambda_1 - \lambda_2$, and θ being the projection of λ under $\Lambda_\mathfrak{g} \to \Lambda_{\mathfrak{g},\mathfrak{p}}$, recall the subscheme $(\mathrm{Gr}^{\lambda_1}_{B(M)} \times \mathrm{Gr}^{\lambda_2}_{B^-(M)})^\nu \subset \mathrm{Conv}_M$.

By unraveling the definitions of $\overset{\circ}{\mathbb{S}}{}^\theta_{\mathfrak{g},\mathfrak{p}}$ and $\overset{\circ}{\mathbb{S}}{}^\lambda_{\mathfrak{g},\mathfrak{p}}$, we obtain an isomorphism:

$$\overset{\circ}{\mathbb{S}}{}^\theta_{\mathfrak{g},\mathfrak{p}} \underset{\mathrm{Conv}_M}{\times} (\mathrm{Gr}^{\lambda_1}_{B(M)} \times \mathrm{Gr}^{\lambda_2}_{B^-(M)}) \simeq \overset{\circ}{\mathbb{S}}{}^\lambda_{\mathfrak{g},\mathfrak{b}} \underset{\mathrm{Conv}_T}{\times} (t^{\lambda_1} \times t^{\lambda_2}). \tag{26}$$

(Here we denote by t^λ the point-scheme, identified with the reduced subscheme of the corresponding connected component of Gr_T.)

The isomorphism (26), combined with Conjecture 2.27 which we assume, implies the following dimension estimate.

Corollary 14.6. *The fibers of the projection* $\varrho^\nu_M : \mathbb{S}^\nu_{\mathfrak{g},\mathfrak{p}} \to \mathrm{Conv}^\nu_M$ *are of dimension at most* $\langle \nu, \check{\rho} - 2\check{\rho}_M \rangle$.

Proof. Pick $\lambda^1 = \lambda = \nu$ and $\lambda^2 = 0$ so that $(\mathrm{Gr}^{\lambda_1}_{B(M)} \times \mathrm{Gr}^{\lambda_2}_{B^-(M)}) \cap \mathrm{Conv}^\nu_M$ is nonempty, and is in fact of pure dimension $\langle \nu, 2\check{\rho}_M \rangle$. Therefore, it would suffice to show that $\overset{\circ}{\mathbb{S}}{}^\theta_{\mathfrak{g},\mathfrak{p}} \underset{\mathrm{Conv}_M}{\times} (\mathrm{Gr}^{\lambda_1}_{B(M)} \times \mathrm{Gr}^{\lambda_2}_{B^-(M)})$ is of dimension $\leq \langle \nu, \check{\rho} \rangle = \langle \lambda_1 - \lambda_2, \check{\rho} \rangle$.

However, from Conjecture 2.27 (which we assume to hold for our $\mathfrak{g}$), we obtain that

$$\begin{aligned} \dim(\overset{\circ}{\mathbb{S}}{}^\theta_{\mathfrak{g},\mathfrak{p}} \underset{\mathrm{Conv}_M}{\times} (\mathrm{Gr}^{\lambda_1}_{B(M)} \times \mathrm{Gr}^{\lambda_2}_{B^-(M)})) &= \dim(\overset{\circ}{\mathbb{S}}{}^\lambda_{\mathfrak{g},\mathfrak{b}} \underset{\mathrm{Conv}_T}{\times} (t^{\lambda_1} \times t^{\lambda_2})) \\ &= \dim(\overset{\circ}{\mathfrak{F}}{}^\lambda_{\mathfrak{g},\mathfrak{b}}) = \langle \lambda_1 - \lambda_2, \check{\rho} \rangle \end{aligned}$$

which implies the desired dimension estimate. □

Since the stabilizers of the $M((t))$-action on Conv^ν_M are connected, we obtain that irreducible components of $\overset{\circ}{\mathbb{S}}{}^\nu_{\mathfrak{g},\mathfrak{p}}$ are in one-to-one correspondence with irreducible components of any fiber of $\varrho^\nu_M : \mathbb{S}^\nu_{\mathfrak{g},\mathfrak{p}} \to \mathrm{Conv}^\nu_M$. Let us denote by $\mathsf{C}^\nu_\mathfrak{m}$ the set of irreducible components of any such fiber of (the maximal possible) dimension $\langle \nu, \check{\rho} - 2\check{\rho}_M \rangle$.

Thus our set $\mathsf{B}_\mathfrak{g}(\lambda)$ can be identified with the set of irreducible components of the fibers of

$$\overset{\circ}{\mathbb{S}}{}^\lambda_{\mathfrak{g},\mathfrak{b}} \to \mathrm{Conv}^\lambda_T = \bigcup_{\lambda_1 - \lambda_2 = \lambda} t^{\lambda_1} \times t^{\lambda_2}.$$

Using the isomorphism (26), we obtain a decomposition

$$\mathsf{B}_{\mathfrak{g}} \simeq \underset{\nu}{\cup} \mathsf{C}^{\nu}_{\mathfrak{m}} \times \mathsf{D}^{\nu}_{\mathfrak{m}}, \tag{27}$$

where $\mathsf{D}^{\nu}_{\mathfrak{m}}$ is as in the previous section.

This already allows one to introduce the operations e_i, f_i, e_i^*, f_i^*. Indeed, by choosing $\mathfrak{m}$ to be the subminimal Levi subalgebra $\mathfrak{m}_i$, we let our operations act along the second factor in the decomposition

$$\mathsf{B}_{\mathfrak{g}} \simeq \underset{n}{\cup} \mathsf{C}^{n}_{\mathfrak{m}_i} \times \mathsf{D}^{n}_{\mathfrak{m}_i},$$

as in Section 13.7. Moreover, the properties of (21) hold due to Proposition 13.9.

To be able to apply the uniqueness theorem of [KaS], it remains to do two things: to check that the operations e_i, f_i commute with e_j^*, f_j^* whenever $i \neq j$, and to establish the highest-weight property of $\mathsf{B}_{\mathfrak{g}}$. Then the uniqueness theorem of [KaS] would guarantee the isomorphism $\mathsf{B}_{\mathfrak{g}} \simeq \mathsf{B}^{\infty}_{\mathfrak{g}}$.

14.7

The above definition of the operations e_i, etc. mimics the definition of Section 13.7. We will now give another (of course, equivalent) construction in the spirit of Section 13.3, which would enable us to prove the commutation relation.

Again, for a standard parabolic $\mathfrak{p}$, let us consider the ind-scheme $\mathbb{S}^{\mu}_{\mathfrak{g},\mathfrak{p},\mathfrak{b}}$ equal to

$$\overset{\circ}{\mathbb{S}}{}^{\theta}_{\mathfrak{g},\mathfrak{p}} \underset{\mathrm{Conv}_M}{\times} (\mathrm{Gr}^{\mu}_{B^-(M)} \times \mathrm{Gr}^{0}_{B^-(M)}),$$

where θ is the image of μ under $\Lambda_{\mathfrak{g}} \to \Lambda_{\mathfrak{g},\mathfrak{p}}$.

In other words, $\mathbb{S}^{\mu}_{\mathfrak{g},\mathfrak{p},\mathfrak{b}}$ classifies the data of

- a principal P-bundle $\mathcal{F}_P$ (as usual, the induced bundle will be denoted by $\mathcal{F}'_M$);
- regular bundle maps $(\mathcal{V}_{\check{\lambda}})_{\mathcal{F}_P} \to \mathcal{L}^{\check{\lambda}}_{\mathcal{F}^0_T}$; and
- meromorphic maps $\mathcal{L}^{\check{\nu}}_{\mathcal{F}^0_T} \to (\mathcal{U}_{\check{\nu}})_{\mathcal{F}'_M}$,

such that

- $\mathcal{L}^{\check{\lambda}}_{\mathcal{F}^0_T} \to (\mathcal{U}_{\check{\lambda}})_{\mathcal{F}'_M} \to (\mathcal{V}_{\check{\lambda}})_{\mathcal{F}_P} \to \mathcal{L}^{\check{\lambda}}_{\mathcal{F}^0_T}$ are the identity maps and
- the (a priori meromorphic) compositions $(\mathcal{U}_{\check{\lambda}})_{\mathcal{F}'_M} \to \mathcal{L}^{\check{\lambda}}_{\mathcal{F}^0_T}(\langle -\mu, \check{\lambda}\rangle \cdot \mathbf{c})$ are regular and surjective.

Using Proposition 14.3, we obtain that the natural projection

$$\mathbb{S}^{\mu}_{\mathfrak{g},\mathfrak{p},\mathfrak{b}} \to (\mathrm{Gr}^{\mu}_{B^-(M)} \times \mathrm{Gr}^{0}_{B^-(M)}) \to \mathrm{Gr}^{\mu}_{B^-(M)}$$

splits as a direct product, as in Lemma 13.4.

For another element $\lambda \in \Lambda_{\mathfrak{g}}$, consider the preimage (call it $\mathbb{S}^{\lambda,\mu}_{\mathfrak{g},\mathfrak{p},\mathfrak{b}}$) of $\mathrm{Gr}^{\lambda}_{B(M)} \cap \mathrm{Gr}^{\mu}_{B^-(M)} \subset \mathrm{Gr}^{\mu}_{B^-(M)}$ in $\mathbb{S}^{\mu}_{\mathfrak{g},\mathfrak{p},\mathfrak{b}}$. Using (26), we conclude that $\mathbb{S}^{\lambda,\mu}_{\mathfrak{g},\mathfrak{p},\mathfrak{b}}$ is isomorphic to a locally closed subset inside $\overset{\circ}{\mathbb{S}}{}^{\lambda}_{\mathfrak{g},\mathfrak{b}} \underset{\mathrm{Conv}_T}{\times} (t^{\lambda} \times t^{0})$.

Therefore, as in Section 13.3, we obtain that the fibers of $\mathbb{S}^{\mu}_{\mathfrak{g},\mathfrak{p},\mathfrak{b}} \to \mathrm{Gr}^{\mu}_{B^-(M)}$ are of dimensions $\leq \langle \mu, \check{\rho} \rangle$, and if we denote by $\mathsf{B}^{\mathfrak{m},*}_{\mathfrak{g}}(\mu)$ the set of irreducible components of any such fiber, we obtain a decomposition

$$\mathsf{B}_{\mathfrak{g}}(\lambda) = \underset{\mu}{\cup} \mathsf{B}^{\mathfrak{m},*}_{\mathfrak{g}}(\mu) \times \mathsf{B}_{\mathfrak{m}}(\lambda - \mu). \tag{28}$$

Moreover, by unraveling our definition of the e_is and f_is we obtain that they act along the second multiple of the above decomposition when we choose $\mathfrak{p} = \mathfrak{p}_i$.

14.8

We will now perform a similar procedure for the e_i^* and f_i^*. Let $\mathbb{S}^{\mu}_{\mathfrak{g},\mathfrak{b},\mathfrak{p}}$ be the ind-scheme equal to

$$\overset{\circ}{\mathbb{S}}{}^{\theta}_{\mathfrak{g},\mathfrak{p}} \underset{\mathrm{Conv}_M}{\times} (\mathrm{Gr}^{0}_{B(M)} \times \mathrm{Gr}^{\mu}_{B(M)}).$$

In other words, $\mathbb{S}^{\mu}_{\mathfrak{g},\mathfrak{b},\mathfrak{p}}$ classifies the data of

- a B-bundle $\mathcal{F}_B$, such that the induced T-bundle $\mathcal{F}_T$ is trivial;
- an M-bundle $\mathcal{F}_M$;
- regular bundle maps $(\mathcal{V}_{\check{\lambda}})_{\mathcal{F}_B} \to (\mathcal{U}_{\check{\lambda}})_{\mathcal{F}_M}$; and
- meromorphic maps $(\mathcal{U}_{\check{\nu}})_{\mathcal{F}_M} \to \mathcal{L}^{\check{\nu}}_{\mathcal{F}^0_T}$,

such that

- the compositions $\mathcal{L}^{\check{\lambda}}_{\mathcal{F}^0_T} \to (\mathcal{V}_{\check{\lambda}})_{\mathcal{F}_B} \to (\mathcal{U}_{\check{\lambda}})_{\mathcal{F}_M} \to \mathcal{L}^{\check{\lambda}}_{\mathcal{F}^0_T}$ are the identity maps and
- the $\kappa^{\check{\lambda},-}_{\mathfrak{p}}$s induce regular bundle maps $\mathcal{L}^{\check{\lambda}}_{\mathcal{F}^0_T} \to (\mathcal{U}_{\check{\lambda}})_{\mathcal{F}_M}(\langle \mu, \check{\lambda} \rangle \cdot \mathbf{c})$.

As above, we have a projection $\mathbb{S}^{\mu}_{\mathfrak{g},\mathfrak{b},\mathfrak{p}} \to \mathrm{Gr}^{\mu}_{B(M)}$, which splits as a direct product and defines a decomposition

$$\mathsf{B}_{\mathfrak{g}}(\lambda) = \underset{\mu}{\cup} \mathsf{B}^{\mathfrak{m}}_{\mathfrak{g}}(\mu) \times \mathsf{B}_{\mathfrak{m}}(\lambda - \mu), \tag{29}$$

where $\mathsf{B}^{\mathfrak{m}}_{\mathfrak{g}}(\mu)$ is the set of irreducible components of dimension $\langle \mu, \check{\rho} \rangle$ of any fiber of $\varrho^{\mu}_{\mathfrak{b},M}$.

And as before, the operations e_i^*, f_i^* introduced earlier coincide with those defined in terms of the above decomposition for $M = M_i$.

14.9

At last, we are ready to check the required commutation property of e_i and f_i with e_j^* and f_j^* for $i \neq j \in I$.

Consider the scheme $\mathbb{S}^{\mu_1,\mu_2}_{\mathfrak{g},i,j}$ classifying the data of

- a P_i-bundle $\mathcal{F}_{P_i}$ (with the induced M_i-bundle $\mathcal{F}'_{M_i}$);
- an M_j-bundle $\mathcal{F}_{M_j}$;
- regular maps $(\mathcal{V}_{\check{\nu}})_{\mathcal{F}_{P_i}} \to (\mathcal{U}_{\check{\nu}})_{\mathcal{F}_{M_j}}$;
- meromorphic maps : $\mathcal{L}^{\check{\nu}}_{\mathcal{F}^0_T} \to (\mathcal{U}_{\check{\nu}})_{\mathcal{F}'_{M_i}}$; and
- meromorphic maps $(\mathcal{U}_{\check{\nu}})_{\mathcal{F}_{M_j}} \to \mathcal{L}^{\check{\nu}}_{\mathcal{F}^0_T}$,

such that

- the compositions $\mathcal{L}^{\check{\lambda}}_{\mathcal{F}^0_T} \to (\mathcal{U}_{\check{\lambda}})_{\mathcal{F}'_{M_i}} \to (\mathcal{V}_{\check{\lambda}})_{\mathcal{F}_{P_i}} \to (\mathcal{U}^{\check{\lambda}})_{\mathcal{F}_{M_j}} \to \mathcal{L}^{\check{\lambda}}_{\mathcal{F}^0_T}$ are the identity maps;
- the (a priori meromorphic) compositions

$$(\mathcal{U}_{\check{\lambda}})_{\mathcal{F}'_{M_i}} \to (\mathcal{V}_{\check{\lambda}})_{\mathcal{F}_{P_i}} \to (\mathcal{U}_{\check{\lambda}})_{\mathcal{F}_{M_j}} \to \mathcal{L}^{\check{\lambda}}_{\mathcal{F}^0_T}(-\langle \mu_1, \check{\lambda}\rangle \cdot \mathbf{c})$$

  are regular bundle maps; and
- the (a priori meromorphic) compositions

$$\mathcal{L}^{\check{\lambda}}_{\mathcal{F}^0_T}(-\langle \mu_2, \check{\lambda}\rangle \cdot \mathbf{c}) \to (\mathcal{U}_{\check{\lambda}})_{\mathcal{F}'_{M_i}} \to (\mathcal{V}_{\check{\lambda}})_{\mathcal{F}_{P_i}} \to (\mathcal{U}_{\check{\lambda}})_{\mathcal{F}_{M_j}}$$

  are regular bundle maps as well.

As before, we have a map

$$\mathbb{S}^{\mu_1,\mu_2}_{\mathfrak{g},i,j} \to \mathrm{Gr}^{\mu_1}_{B^-(M_i)} \times \mathrm{Gr}^{\mu_2}_{B(M_j)}.$$

We claim that $\mathbb{S}^{\mu_1,\mu_2}_{\mathfrak{g},i,j}$ over $\mathrm{Gr}^{\mu_1}_{B^-(M_i)} \times \mathrm{Gr}^{\mu_2}_{B(M_j)}$ also splits as a direct product. This follows from the fact that we can make the group $(N^-(M_i) \times N(M_j))((t))$ act on $\mathbb{S}^{\mu_1,\mu_2}_{\mathfrak{g},i,j}$ lifting its action on $\mathrm{Gr}^{\mu_1}_{B^-(M_i)} \times \mathrm{Gr}^{\mu_2}_{B(M_j)}$.

The isomorphism class of $\mathbb{S}^{\mathfrak{g},\mu_1,\mu_2}_{i,j}$ also depends only on the difference $\mu = \mu_1 - \mu_2$, and let us denote by $\mathsf{C}_{i,j}(\mu)$ the set of irreducible components of the top dimension of any fiber of $\mathbb{S}^{\mathfrak{g},\mu_1,\mu_2}_{i,j}$ over $\mathrm{Gr}^{\mu_1}_{B^-(M_i)} \times \mathrm{Gr}^{\mu_2}_{B(M_j)}$.

As before, the preimage of $(\mathrm{Gr}^{\lambda_1}_{B(M_i)} \cap \mathrm{Gr}^{\mu_1}_{B^-(M_i)}) \times (\mathrm{Gr}^{\mu_2}_{B(M_j)} \times \mathrm{Gr}^{\lambda_2}_{B^-(M_j)})$ in $\mathbb{S}^{\mathfrak{g},\mu_1,\mu_2}_{i,j}$ is naturally a locally closed subset inside $\overset{\circ}{\mathbb{S}}{}^{\lambda_1-\lambda_2}_{\mathfrak{g},\mathfrak{b}}$ and we obtain a decomposition

$$\mathsf{B}_{\mathfrak{g}}(\lambda) \simeq \bigcup_{\mu_1,\mu_2} \mathsf{B}_{\mathfrak{m}_i}(\lambda_1 - \mu_1) \times \mathsf{B}_{\mathfrak{m}_j}(\mu_2 - \lambda_2) \times \mathsf{C}_{i,j}(\mu_1 - \mu_2). \tag{30}$$

Moreover, this decomposition refines those of (28) and (29). Hence the e_i, f_i, e_j^*, f_j^* operations preserve the decomposition of (30), and e_i, f_i act along the first factor, whereas e_j^*, f_j^* act only along the second factor, and hence they commute.

15 The highest-weight property

In this section (with the exception of Sections 15.5 and 15.6), $\mathfrak{g}$ will still be a general Kac–Moody algebra. We will reduce Conjecture 2.27 to Conjecture 15.3 and verify the latter in the finite and affine cases.

15.1

Observe that for any $i \in I$, using (3), by adding a defect at $\mathbf{c} \in \mathbf{C}$, we can realize $\mathcal{Z}^{\lambda-\alpha_i}_{\mathfrak{g},\mathfrak{b}}(\mathbf{C})$ as a closed subscheme of $\mathcal{Z}^{\lambda}_{\mathfrak{g},\mathfrak{b}}(\mathbf{C})$ of codimension 2. Similarly, $\mathfrak{F}^{\lambda-\alpha_i}_{\mathfrak{g},\mathfrak{b}}$ is a closed subscheme of $\mathfrak{F}^{\lambda}_{\mathfrak{g},\mathfrak{b}}$.

Let $\mathbf{K}$ be an irreducible component of $\overset{\circ}{\mathfrak{F}}{}^{\lambda}_{\mathfrak{g},\mathfrak{b}}$, and let $\overline{\mathbf{K}}$ be its closure in $\mathfrak{F}^{\lambda}_{\mathfrak{g},\mathfrak{b}}$. Consider the intersection

$$\overline{\mathbf{K}} \cap \overset{\circ}{\mathfrak{F}}{}^{\lambda-\alpha_i}_{\mathfrak{g},\mathfrak{b}} \subset \mathfrak{F}^{\lambda}_{\mathfrak{g},\mathfrak{b}}.$$

Proposition 15.2. *If the above intersection is nonempty, it consists of one irreducible component, whose dimension is* $\dim(\mathbf{K}) - 1$.

Proof of Proposition 15.2. Recall the stack $\mathbb{S}^{\mu}_{\mathfrak{g},\mathfrak{b},\mathfrak{p}_i}$ introduced in the previous section. Let $\mathbb{S}^{\mu,\leq\lambda}_{\mathfrak{g},\mathfrak{b},\mathfrak{p}_i}$ and $\mathbb{S}^{\mu,\lambda}_{\mathfrak{g},\mathfrak{b},\mathfrak{p}_i}$ be its closed (respectively, locally closed) subschemes equal to

$$\overset{\circ}{\mathbb{S}}{}^{\theta}_{\mathfrak{g},\mathfrak{p}_i} \underset{\mathrm{Conv}_{M_i}}{\times} (\mathrm{Gr}^0_{B(M_i)} \times (\mathrm{Gr}^{\mu}_{B(M_i)} \cap \overline{\mathrm{Gr}}{}^{\lambda}_{B^-(M_i)})),\ \overset{\circ}{\mathbb{S}}{}^{\theta}_{\mathfrak{g},\mathfrak{p}_i} \underset{\mathrm{Conv}_{M_i}}{\times} (\mathrm{Gr}^0_{B(M_i)} \times (\mathrm{Gr}^{\mu}_{B(M_i)} \cap \mathrm{Gr}^{\lambda}_{B^-(M_i)})),$$

respectively, where $\overline{\mathrm{Gr}}{}^{\lambda}_{B^-(M_i)}$ is the closure of $\mathrm{Gr}^{\lambda}_{B^-(M_i)}$ in Gr_{M_i}. (In other words, we impose the condition that the maps $(\mathcal{U}^{\nu})_{\mathcal{F}_{M_i}} \to \mathcal{L}_{\mathcal{F}^0_T}(\langle\lambda,\nu\rangle\cdot\mathbf{c})$ are regular (respectively, regular bundle maps).)

We have $\mathbb{S}^{\mu,\leq\lambda}_{\mathfrak{g},\mathfrak{b},\mathfrak{p}_i} = \bigcup\limits_{0\leq\lambda'\leq\lambda} \mathbb{S}^{\mu,\lambda'}_{\mathfrak{g},\mathfrak{b},\mathfrak{p}_i}$, and $\mathbb{S}^{\mu,\leq\lambda}_{\mathfrak{g},\mathfrak{b},\mathfrak{p}_i}$ identifies naturally with a locally closed subset of $\mathfrak{F}^{\lambda}_{\mathfrak{g},\mathfrak{b}}$ (of the same dimension), and

$$\mathfrak{F}^{\lambda}_{\mathfrak{g},\mathfrak{b}} \simeq \bigcup_{\mu} \mathbb{S}^{\mu,\leq\lambda}_{\mathfrak{g},\mathfrak{b},\mathfrak{p}_i},\ \overset{\circ}{\mathfrak{F}}{}^{\lambda}_{\mathfrak{g},\mathfrak{b}} \simeq \bigcup_{\mu} \mathbb{S}^{\mu,\lambda}_{\mathfrak{g},\mathfrak{b},\mathfrak{p}_i}.$$

We have a pair of Cartesian squares:

$$\begin{array}{ccccc} \mathbb{S}^{\mu,\lambda}_{\mathfrak{g},\mathfrak{b},\mathfrak{p}_i} & \longrightarrow & \mathbb{S}^{\mu,\leq\lambda}_{\mathfrak{g},\mathfrak{b},\mathfrak{p}_i} & \longleftarrow & \mathbb{S}^{\mu,\lambda-\alpha_i}_{\mathfrak{g},\mathfrak{b},\mathfrak{p}_i} \\ \downarrow & & \downarrow & & \downarrow \\ \overset{\circ}{\mathfrak{F}}{}^{\lambda}_{\mathfrak{g},\mathfrak{b}} & \longrightarrow & \mathfrak{F}^{\lambda}_{\mathfrak{g},\mathfrak{b}} & \overset{\bar{\iota}_{\alpha_i}}{\longleftarrow} & \overset{\circ}{\mathfrak{F}}{}^{\lambda-\alpha_i}_{\mathfrak{g},\mathfrak{b}}. \end{array}$$

Therefore, it is enough to show that for any irreducible component $\mathbf{K}$ of $\mathbb{S}^{\mu,\lambda}_{\mathfrak{g},\mathfrak{b},\mathfrak{p}_i}$, the intersection

$$\overline{\mathbf{K}} \cap \mathbb{S}^{\mu,\lambda-\alpha_i}_{\mathfrak{g},\mathfrak{b},\mathfrak{p}_i}$$

consists of one irreducible component, whose dimension is $\dim(\mathbf{K}) - 1$.

However, since $\mathbb{S}^{\mu}_{\mathfrak{g},\mathfrak{b},\mathfrak{p}_i} \to \mathrm{Gr}^{\mu}_{B(M_i)}$ splits as a product, our assertion follows from the fact that $\mathrm{Gr}^{\mu}_{B(M_i)} \cap \mathrm{Gr}^{\lambda}_{B^-(M_i)}$ and $\mathrm{Gr}^{\mu}_{B(M_i)} \cap \mathrm{Gr}^{\lambda-\alpha_i}_{B^-(M_i)}$ are irreducible varieties, since M_i is a reductive group of rank 1. □

In order to be able to apply the uniqueness theorem of [KaS], we need the following conjecture to be satisfied for our $\mathfrak{g}$.

Conjecture 15.3. If $\lambda \neq 0$, for every irreducible component $\mathbf{K}$ of $\overset{\circ}{\mathfrak{F}}{}^{\lambda}_{\mathfrak{g},\mathfrak{b}}$, the intersection $\overline{\mathbf{K}} \cap \overset{\circ}{\mathfrak{F}}{}^{\lambda-\alpha_i}_{\mathfrak{g},\mathfrak{b}} \subset \mathfrak{F}^{\lambda}_{\mathfrak{g},\mathfrak{b}}$ is nonempty for at least one $i \in I$.

Note that Conjecture 15.3 implies Conjecture 2.27.

Proof. We will argue by induction on the length of λ. The dimension estimate is obvious for $\lambda = 0$, and let us suppose that it is verified for all $\lambda' < \lambda$.

Let $\mathbf{K}$ be an irreducible component of $\overset{\circ}{\mathfrak{F}}{}^{\lambda}_{\mathfrak{g},\mathfrak{b}}$, and let $i \in I$ be such that $\mathbf{K}' := \overline{\mathbf{K}} \cap \overset{\circ}{\mathfrak{F}}{}^{\lambda-\alpha_i}_{\mathfrak{g},\mathfrak{b}}$ is nonempty.

We know that $\dim(\mathbf{K}') \leq \langle \lambda - \alpha_i, \check{\rho} \rangle$. Hence by Proposition 15.2, $\dim(\mathbf{K}) \leq \langle \lambda, \check{\rho} \rangle$, which is what we had to prove. □

Note also that if we assume Conjecture 15.3, and thus obtain a well-defined crystal structure on $\mathsf{B}_{\mathfrak{g}}$, the operation

$$\mathbf{K} \mapsto \overline{\mathbf{K}} \cap \overset{\circ}{\mathfrak{F}}{}^{\lambda-\alpha_i}_{\mathfrak{g},\mathfrak{b}},$$

as in Proposition 15.2, viewed as a map $\mathsf{B}_{\mathfrak{g}}(\lambda) \to \mathsf{B}_{\mathfrak{g}}(\lambda - \alpha_i) \cup 0$, equals f_i^*.

Hence if Conjecture 15.3 is verified, we obtain, according to Section 12.5, that our crystal $\mathsf{B}_{\mathfrak{g}}$ can be identified with Kashiwara's crystal $\mathsf{B}^{\infty}_{\mathfrak{g}}$.

In particular, we obtain the following corollary.

Corollary 15.4. *If Conjecture* 15.3 *is verified, then for a parabolic subalgebra* $\mathfrak{p} \subset \mathfrak{g}$ *such that its Levi subalgebra is of finite type, the dimension of the fibers of the map* $\mathbb{S}^{\nu}_{\mathfrak{g},\mathfrak{p}} \to \mathrm{Conv}^{\nu}_{M}$ *is* $\leq \langle \nu, \check{\rho} - 2\check{\rho}_M \rangle$, *and the set of irreducible components of dimension* $\langle \nu, \check{\rho} - 2\check{\rho}_M \rangle$ *of (any) such fiber parametrizes a basis of the space* $\mathrm{Hom}_{\mathfrak{m}}(V^{\mu}_{M}, U(n(\check{\mathfrak{p}})))$.

15.5

Although Conjecture 15.3 is well known for $\mathfrak{g}$ of finite type, we will give here yet another proof, using Theorem 11.6. Then we will modify this argument to prove Conjecture 15.3 in the affine case.

We will argue by contradiction. Let λ be a minimal element in $\Lambda_{\mathfrak{g}}^{pos}$, for which there exists a component $\mathbf{K}$ such that all $\overline{\mathbf{K}} \cap \overset{\circ}{\mathfrak{F}}{}^{\lambda-\alpha_i}_{\mathfrak{g},\mathfrak{b}}$ are empty. As in the proof of Conjecture 2.27 above, we obtain in particular that for all $\lambda' < \lambda$, the dimension estimate $\dim(\mathfrak{F}^{\lambda'}_{\mathfrak{g},\mathfrak{b}}) = |\lambda'|$ holds.

Let $\mathbf{C}$ be a genus 0 curve. According to [BFGM], we have a smooth map $\mathfrak{Z}^{\lambda}_{\mathfrak{g},\mathfrak{b}}(\mathbf{C}) \to \overline{\mathrm{Bun}}_B(\mathbf{C})$, and let $\partial(\mathfrak{Z}^{\lambda}_{\mathfrak{g},\mathfrak{b}}(\mathbf{C}))$ be the preimage of $\partial(\overline{\mathrm{Bun}}_B(\mathbf{C}))$ in $\mathfrak{Z}^{\lambda}_{\mathfrak{g},\mathfrak{b}}(\mathbf{C})$. According to Theorem 11.6, $\partial(\mathfrak{Z}^{\lambda}_{\mathfrak{g},\mathfrak{b}}(\mathbf{C}))$ is a quasi-effective Cartier divisor in $\mathfrak{Z}^{\lambda}_{\mathfrak{g},\mathfrak{b}}(\mathbf{C})$. Let us consider the intersection $\overline{\mathbf{K}} \cap \partial(\mathfrak{Z}^{\lambda}_{\mathfrak{g},\mathfrak{b}}(\mathbf{C}))$. This intersection is nonempty, since it contains the point-scheme, corresponding to $\mathfrak{Z}^{0}_{\mathfrak{g},\mathfrak{b}}(\mathbf{C}) \simeq \mathfrak{F}^{0}_{\mathfrak{g},\mathfrak{b}} \subset \mathfrak{Z}^{\lambda}_{\mathfrak{g},\mathfrak{b}}(\mathbf{C})$.

On the one hand, from Lemma 11.3, we obtain that $\dim(\overline{\mathbf{K}} \cap \partial(\mathfrak{Z}^{\lambda}_{\mathfrak{g}})) \geq \dim(\mathbf{K}) - 1 \geq |\lambda| - 1$. On the other hand, this intersection is contained in

$$\bigcap_{\lambda'<\lambda} \overline{\mathbf{K}} \cap \mathfrak{F}^{\lambda'}_{\mathfrak{g},\mathfrak{b}}.$$

But, according to our assumption, all the λ' appearing in the above expression have the property that $|\lambda'| \leq |\lambda| - 2$, and hence $\dim(\overline{\mathbf{K}} \cap \partial(\mathfrak{Z}^{\lambda}_{\mathfrak{g},\mathfrak{b}}(\mathbf{C})))$ is of dimension $\leq |\lambda| - 2$, which is a contradiction.

15.6

In this subsection we will prove Conjecture 15.3 for affine Lie algebras. We will change our notation, and from now on $\mathfrak{g}$ will denote a finite-dimensional semisimple Lie algebra, and $\mathfrak{g}_{\mathrm{aff}}$ will be the corresponding affine Kac–Moody algebra.

As above, we will argue by contradiction. Let μ be a minimal element of $\widehat{\Lambda}_{\mathfrak{g}}^{pos}$ for which there exists a component $\mathbf{K}$ violating Conjecture 15.3. In particular, we can assume that the dimension estimate $\dim(\overset{\circ}{\mathfrak{F}}{}^{\lambda'}_{\mathfrak{g}_{\mathrm{aff}},\mathfrak{b}^+_{\mathrm{aff}}}) = |\lambda'|$ holds for all $\lambda' < \lambda$.

Let us consider the corresponding Uhlenbeck space $\mathfrak{U}^{\lambda}_{G;B}$. We can view our component $\mathbf{K}$ as a subset in $\mathrm{Bun}^{\lambda}_{G;B}(\mathbf{S}, \mathbf{D}_\infty; \mathbf{D}_0)$; let $\overline{\overline{\mathbf{K}}}$ be the closure of $\mathbf{K}$ inside $\mathfrak{U}^{\lambda}_{G;B}$.

Consider the intersection $\overline{\overline{\mathbf{K}}} \cap \partial(\mathfrak{U}^{\lambda}_{G;B})$. As before, it is nonempty, since it contains the point $a \cdot \mathbf{c} \in \mathfrak{U}^{\mu;\mu,0}_{G,B} \subset \mathfrak{U}^{\lambda}_{G;B}$, and hence is of dimension at least $|\lambda| - 1$. On the other hand we will show that if $\mathbf{K}$ violates Conjecture 15.3, this intersection will be of dimension $\leq |\lambda| - 2$.

Indeed, consider the fiber of $\varrho^{\lambda}_{\mathfrak{b}^+_{\mathrm{aff}}} : \mathfrak{U}^{\lambda}_{G;B} \to \overset{\circ}{\mathbf{C}}{}^{\lambda}$ over $\lambda \cdot \mathbf{c} \in \overset{\circ}{\mathbf{C}}{}^{\lambda}$. According to Theorem 10.2, it is the union over decompositions $\lambda = \lambda_1 + \lambda_2 + b \cdot \delta$ of varieties

$$\overset{\circ}{\mathfrak{F}}{}^{\lambda_1}_{\mathfrak{g}_{\mathrm{aff}},\mathfrak{b}^+_{\mathrm{aff}}} \times \mathrm{Sym}^b(\overset{\circ}{\mathbf{X}} - 0_{\mathbf{X}}),$$

and $\partial(\overset{\circ}{\mathfrak{U}}{}^{\lambda}_{G;B})$ corresponds to the locus where $\lambda_2 \neq 0$. Since $|\delta|$ is the dual Coxeter number, and is > 1, the intersection

$$\overline{\overline{\mathbf{K}}} \cap (\overset{\circ}{\mathfrak{F}}{}^{\lambda_1}_{\mathfrak{g}_{\text{aff}}, \mathfrak{b}^+_{\text{aff}}} \times \text{Sym}^b(\overset{\circ}{\mathbf{X}} - 0_{\mathbf{X}})) \tag{31}$$

is of dimension $\leq |\lambda| - 2$ unless $b = 0$ and $|\lambda_1| = |\lambda| - 1$.

However, under the projection $\mathfrak{U}^{\mu}_{G;B} \to \mathfrak{Z}^{\mu}_{\mathfrak{g}_{\text{aff}}, \mathfrak{b}^+_{\text{aff}}}(\mathbf{C})$, the intersection (31) gets mapped to $\overline{\mathbf{K}} \cap \overset{\circ}{\mathfrak{F}}{}^{\lambda_1}_{\mathfrak{g}_{\text{aff}}, \mathfrak{b}^+_{\text{aff}}}$, and by assumption the latter is empty for $\lambda_1 = \lambda - \alpha_i$ for all $i \in I_{\text{aff}}$.

This completes the proof of Conjecture 15.3 for untwisted affine Lie algebras. The proof for a twisted affine Lie algebra follows by realizing it (as well as all the related moduli spaces) as the fixed-point set of a finite order automorphism of the corresponding untwisted affine Lie algebra. We will not use the result for the twisted affine Lie algebras, and we leave the details to the interested reader.

Part V: Computation of the IC Sheaf

16 IC stalks: Statements

16.1

The purpose of this section is to formulate some statements about the behavior of the IC-sheaves on the parabolic Uhlenbeck spaces $\widetilde{\mathfrak{U}}^{\theta}_{G,P}$ and $\mathfrak{U}^{\theta}_{G,P}$. Since the material here is largely parallel to that of [BFGM], we will assume that the reader is familiar with the notation of [BFGM].

Recall that we have a natural map

$$\mathfrak{r}_{\mathfrak{p}^+_{\text{aff}}} : \widetilde{\mathfrak{U}}^{\theta}_{G;P} \to \mathfrak{U}^{\theta}_{G;P}$$

and the decompositions into locally closed subschemes

$$\mathfrak{U}^{\theta}_{G;P} = \bigcup_{\theta_2, b} \mathfrak{U}^{\theta;\theta_2,b}_{G;P} \quad \text{and} \quad \widetilde{\mathfrak{U}}^{\theta}_{G;P} = \bigcup_{\theta_2, b} \widetilde{\mathfrak{U}}^{\theta;\theta_2,b}_{G;P},$$

where $\widetilde{\mathfrak{U}}^{\theta;\theta_2,b}_{G;P} = \mathfrak{r}^{-1}_{\mathfrak{p}^+_{\text{aff}}}(\mathfrak{U}^{\theta;\theta_2,b}_{G;P})$. Recall also that we have the natural sections

$$\mathfrak{s}_{\mathfrak{p}^+_{\text{aff}}} : \overset{\circ}{\mathbf{C}}{}^{\theta} \to \mathfrak{U}^{\theta}_{G;P} \quad \text{and} \quad \widetilde{\mathfrak{s}}_{\mathfrak{p}^+_{\text{aff}}} : \text{Mod}^{\theta,+}_{M_{\text{aff}}} \to \widetilde{\mathfrak{U}}^{\theta}_{G;P}.$$

16.2

For a standard parabolic $\mathfrak{p} \subset \mathfrak{g}$, consider the corresponding parabolic $\mathfrak{p}^+_{\text{aff}} \subset \mathfrak{g}_{\text{aff}}$, and let $\mathfrak{V}_{\mathfrak{p}}$ be the unipotent radical of the parabolic attached to it in the Langlands dual Lie algebra $\check{\mathfrak{g}}_{\text{aff}}$. (Recall that by $\mathfrak{V}$ we denote $\mathfrak{V}_{\mathfrak{p}}$ for $\mathfrak{p} = \mathfrak{g}$.)

The corresponding Levi subgroup contains the subgroup $\check{M}_{\text{aff}} \simeq \check{M} \times \mathbb{G}_m$ (we ignore the other $\mathbb{G}_m$-factor because it is central), and consider $\mathfrak{V}_{\mathfrak{p}}$ as a $\check{M}_{\text{aff}}$-module.

Note that the lattice $\widehat{\Lambda}_{\mathfrak{g},\mathfrak{p}}$ is naturally the lattice of central characters of $\check{M}_{\text{aff}}$, and $\widehat{\Lambda}^{\text{pos}}_{\mathfrak{g},\mathfrak{p}} \subset \widehat{\Lambda}_{\mathfrak{g},\mathfrak{p}}$ corresponds to those that appear in $\text{Sym}(\mathfrak{V}_{\mathfrak{p}})$.

Let us denote by $\text{Loc}^{M_{\text{aff}}}$ the canonical equivalence of categories between $\text{Rep}(\check{M}_{\text{aff}})$ and the category of spherical perverse sheaves on the affine Grassmannian of M_{aff}. More generally, for a scheme (or stack S) mapping to $\text{Bun}_{M_{\text{aff}}}$, we will denote by $\text{Loc}^{M_{\text{aff}}}_{S,\mathbf{C}}$ the corresponding functor from $\text{Rep}(\check{M}_{\text{aff}})$ to the category of perverse sheaves on $S \underset{\text{Bun}_{M_{\text{aff}}}}{\times} \mathcal{H}_{M_{\text{aff}},\mathbf{C}}$; cf. [BFGM, Theorem 1.12].

For an element $\theta \in \widehat{\Lambda}^{\text{pos}}_{\mathfrak{g},\mathfrak{p}}$, we will denote by $\mathfrak{P}(\theta)$ elements of the set of partitions of θ as a sum $\theta = \sum_k n_k \cdot \theta_k$, $\theta_k \in \widehat{\Lambda}^{\text{pos}}_{\mathfrak{g},\mathfrak{p}}$; let $|\mathfrak{P}(\theta)|$ be the sum $\sum_k n_k$. For a partition $\mathfrak{P}(\theta)$, let $\text{Sym}^{\mathfrak{P}(\theta)}(\overset{\circ}{\mathbf{C}})$ denote the corresponding partially symmetrized power of $\overset{\circ}{\mathbf{C}}$, i.e.,

$$\text{Sym}^{\mathfrak{P}(\theta)}(\overset{\circ}{\mathbf{C}}) := \prod_k \overset{\circ}{\mathbf{C}}{}^{\theta_k}.$$

Let $\mathcal{H}^{\mathfrak{P}(\theta),+}_{M_{\text{aff}},\mathbf{C}}$ be the appropriate version of the Hecke stack over $\text{Sym}^{\mathfrak{P}(\theta)}(\overset{\circ}{\mathbf{C}})$.

Recall (cf. [BFGM, Section 4.1]) that we have a natural finite map

$$\widetilde{\text{Norm}}_{\mathfrak{P}(\theta)} : \mathcal{H}^{\mathfrak{P}(\theta),+}_{M_{\text{aff}},\mathbf{C}} \to \mathcal{H}^{\theta,+}_{M_{\text{aff}},\mathbf{C}},$$

covering the map $\text{Norm}_{\mathfrak{P}(\theta)} : \mathbf{C}^{\mathfrak{P}(\theta)} \to \mathbf{C}^{\theta}$.

As in [BFGM, Theorem 1.12], we have the functor $\text{Loc}^{M_{\text{aff}},\mathfrak{P}(\theta)}_{S,\mathbf{C}}$ from $\text{Rep}(\check{M}_{\text{aff}})$ to the category of perverse sheaves on $S \underset{\text{Bun}_{M_{\text{aff}}}}{\times} \mathcal{H}^{\mathfrak{P}(\theta),+}_{M_{\text{aff}},\mathbf{C}}$.

Theorem 16.3. *The object* $(\mathfrak{s}_{\mathfrak{p}^+_{\text{aff}}})^!(\text{IC}_{\widetilde{\mathfrak{U}}^\theta_{G;P}})$ *is isomorphic to the direct sum over all* $\mathfrak{P}(\theta)$ *of*

$$(\widetilde{\text{Norm}}_{\mathfrak{P}(\theta)})_*(\text{Loc}^{M_{\text{aff}},\mathfrak{P}(\theta)}_{pt,\mathbf{C}}(\mathfrak{V}_{\mathfrak{p}}))[-|\mathfrak{P}(\theta)|],$$

where $pt \to \text{Bun}_M(\mathbf{C})$ *is the map corresponding to the trivial bundle.*

16.4

Next we will state the theorem that describes the stalks of $\text{IC}_{\widetilde{\mathfrak{U}}^\theta_{G;P}}$.

Let $\theta \in \widehat{\Lambda}^{\text{pos}}_{\mathfrak{g},\mathfrak{p}}$ be decomposed as $\theta = \theta_1 + \theta_2 + b\cdot\delta$, and recall that the corresponding stratum $\widetilde{\mathfrak{U}}^{\theta;\theta_2,b}_{G;P}$ is isomorphic to

$$(\text{Bun}^{\theta_1}_{G;P}(\mathbf{S}', \mathbf{D}'_\infty; \mathbf{D}'_0) \underset{\text{Bun}_M(\mathbf{C},\infty_{\mathbf{C}})}{\times} \mathcal{H}^{\theta_2,+}_{M_{\text{aff}},\mathbf{C}}) \times \text{Sym}^b(\overset{\circ\circ}{\mathbf{S}}).$$

Let us fix partitions $\mathfrak{P}(b)$ and $\mathfrak{P}(\theta_2)$ of b and θ_2, respectively. Let $\text{Sym}^{\mathfrak{P}(\theta_2)}(\overset{\circ}{\mathbf{C}}) \subset \text{Sym}^{\mathfrak{P}(\theta_2)}(\overset{\circ}{\mathbf{C}})$ and $\text{Sym}^{\mathfrak{P}(b)}(\overset{\circ}{\mathbf{S}}) \subset \text{Sym}^{\mathfrak{P}(b)}(\overset{\circ}{\mathbf{S}})$ be the open subsets obtained by removing all the diagonals. Let us denote by $\widetilde{\mathfrak{U}}^{\theta;\mathfrak{P}(\theta_2),\mathfrak{P}(b)}_{G;P}$ the corresponding locally closed subvariety in $\widetilde{\mathfrak{U}}^{\theta;\theta_2,b}_{G;P}$. Note that it is isomorphic to

$$(\mathrm{Bun}^{\theta_1}_{G;P}(\mathbf{S}', \mathbf{D}'_\infty; \mathbf{D}'_0) \underset{\mathrm{Bun}_M(\mathbf{C},\infty_{\mathbf{C}})}{\times} (\mathcal{H}^{\mathfrak{P}(\theta_2),+}_{M_{\mathrm{aff}},\mathbf{C}} \underset{\mathrm{Sym}^{\mathfrak{P}(\theta_2)}(\mathbf{C})}{\times} \overset{\circ}{\mathrm{Sym}}{}^{\mathfrak{P}(\theta_2)}(\overset{\circ}{\mathbf{C}})) \times \overset{\circ}{\mathrm{Sym}}{}^{\mathfrak{P}(b)}(\overset{\circ\circ}{\mathbf{S}}).$$

Theorem 16.5. *The *-restriction of* $\mathrm{IC}_{\widetilde{\mathfrak{U}}^\theta_{G;P}}$ *to the stratum* $\widetilde{\mathfrak{U}}^{\theta;\mathfrak{P}(\theta_2),\mathfrak{P}(b)}_{G;P}$ *is isomorphic to*

$$\mathrm{Loc}^{M_{\mathrm{aff}},\mathfrak{P}(\theta_2)}_{\mathrm{Bun}^{\theta_1}_{G;P}(\mathbf{S},\mathbf{D}_\infty;\mathbf{D}_0),\mathbf{C}}(\bigoplus_{i\geq 0} \mathrm{Sym}^i(\mathfrak{V}_{\mathfrak{p}})[2i]) \boxtimes (\mathrm{Sym}^{\mathfrak{P}(b)}(\mathfrak{V}^f) \otimes \mathrm{IC}_{\overset{\circ}{\mathrm{Sym}}{}^{\mathfrak{P}(b)}(\overset{\circ\circ}{\mathbf{S}})}).$$

16.6

We will now formulate theorems parallel to Theorems 16.3 and 16.5 for the schemes $\mathfrak{U}^\theta_{G;P}$.

Let V be a representation of the group $\check{M}_{\mathrm{aff}}$, and let $\mathfrak{P}(\theta) : \theta = \sum_k n_k \cdot \theta_k$ be a partition. We will denote by $\overline{\mathrm{Loc}}^{\mathfrak{P}(\theta)}(V)$ the semisimple complex of sheaves on $\mathrm{Sym}^{\mathfrak{P}(\theta)}(\mathbf{C})$ which is the IC-sheaf tensored by the complex of vector spaces equal to

$$\bigotimes_k (V^f_{\theta_k})^{\otimes n_k},$$

where V^f denotes the kernel of the principal nilpotent element $f \in \check{\mathfrak{m}}$ acting on V^f endowed with a principal grading, which we declare cohomological.

Note that the same procedure makes sense when V is actually a semisimple complex of $\check{M}_{\mathrm{aff}}$-representations.

Theorem 16.7. *The object* $(\mathfrak{s}_{\mathfrak{p}^+_{\mathrm{aff}}})^*(\mathrm{IC}_{\mathfrak{U}^\theta_{G;P}})$ *is isomorphic to the direct sum over all* $\mathfrak{P}(\theta)$ *of*

$$(\mathrm{Norm}_{\mathfrak{P}(\theta)})_*(\overline{\mathrm{Loc}}^{\mathfrak{P}(\theta)}(\mathfrak{V}_{\mathfrak{p}}))[|\mathfrak{P}(\theta)|].$$

And finally, we have the following.

Theorem 16.8. *The *-restriction of* $\mathrm{IC}_{\mathfrak{U}^\theta_{G;P}}$ *to the stratum*

$$\mathfrak{U}^{\theta;\mathfrak{P}(\theta_2),\mathfrak{P}(b)}_{G;P} \simeq \mathrm{Bun}^{\theta_1}_{G;P}(\mathbf{S}, \mathbf{D}_\infty; \mathbf{D}_0) \times \overset{\circ}{\mathrm{Sym}}{}^{\mathfrak{P}(\theta_2)}(\overset{\circ}{\mathbf{C}}) \times \overset{\circ}{\mathrm{Sym}}{}^{\mathfrak{P}(b)}(\overset{\circ\circ}{\mathbf{S}})$$

is isomorphic to

$$\mathrm{IC}_{\mathrm{Bun}_{G;P}(\mathbf{S},\mathbf{D}_\infty;\mathbf{D}_0)} \boxtimes \overline{\mathrm{Loc}}^{\mathfrak{P}(\theta_2)}(\bigoplus_{i\geq 0} \mathrm{Sym}^i(\mathfrak{V}_{\mathfrak{p}})[2i]) \boxtimes (\mathrm{Sym}^{\mathfrak{P}(b)}(\mathfrak{V}^f) \otimes \mathrm{IC}_{\overset{\circ}{\mathrm{Sym}}{}^{\mathfrak{P}(b)}(\overset{\circ\circ}{\mathbf{S}})}).$$

Corollary 16.9. *Theorem* 7.10 *holds.*

To prove Corollary 16.9 it is enough to note that it is a particular case of Theorem 16.8 when $P = G$.

17 IC stalks: Proofs

17.1 Logic of the proof

To prove all five theorems—i.e., Theorems 16.3, 16.5, 16.7, 16.8, and 7.10—we will proceed by induction on the length $|\theta|$ of the parameter θ.

Thus let θ be $(\overline{\theta}, a)$, and we assume that the assertions of Theorems 16.3, 16.5, 16.7, and 16.8 hold for all $\theta' < \theta$ and that the assertion of Theorem 7.10 holds for all $a' < a$.

As we shall see, the induction hypothesis will give a description of the restrictions of $\mathrm{IC}_{\mathfrak{U}^\theta_{G;P}}$ and (respectively, $\mathrm{IC}_{\widetilde{\mathfrak{U}}^\theta_{G;P}}$) to almost all the strata $\mathrm{IC}_{\mathfrak{U}^{\theta;\mathfrak{P}(\theta_2),\mathfrak{P}(b)}_{G;P}}$ (respectively, $\mathrm{IC}_{\widetilde{\mathfrak{U}}^{\theta;\mathfrak{P}(\theta_2),\mathfrak{P}(b)}_{G;P}}$).

This will allow us to perform the induction step and prove Theorem 16.3 for the parameter equal to θ. Then we will deduce Theorems 16.5, 16.7, and 16.8 from Theorem 16.3. Finally, as was noted before, Theorem 7.10 is a particular case of Theorem 16.8.

17.2

Consider a decomposition $\theta = \theta_1 + \theta_2 + b \cdot \delta$, and let $\widetilde{\mathfrak{U}}^{\theta;\mathfrak{P}(\theta_2),\mathfrak{P}(b)}_{G;P}$ be the corresponding stratum. We claim that if

$$\Big((\theta_2 < \theta) \bigwedge (b < a)\Big) \bigvee (|\mathfrak{P}(\theta_2)| > 1) \bigvee (|\mathfrak{P}(b)| > 1)$$

then the restriction of $\mathrm{IC}_{\widetilde{\mathfrak{U}}^\theta_{G;P}}$ to this stratum is known and given by the expression of Theorem 16.5, by the induction hypothesis.

We will first establish the required isomorphism locally in the étale topology on $\widetilde{\mathfrak{U}}^{\theta;\mathfrak{P}(\theta_2),\mathfrak{P}(b)}_{G;P}$ and then argue that it holds globally. An absolutely similar argument establishes the description of the restriction of $\mathrm{IC}_{\mathfrak{U}^\theta_{G;P}}$ to the corresponding stratum in $\mathfrak{U}^\theta_{G;P}$ under the same assumption on $\mathfrak{P}(\theta_2)$, $\mathfrak{P}(b)$.

We claim that every geometric point belonging to the stratum $\widetilde{\mathfrak{U}}^{\theta;\mathfrak{P}(\theta_2),\mathfrak{P}(b)}_{G;P}$ has an étale neighborhood, which is smoothly equivalent to an étale neighborhood of a point in $\widetilde{\mathfrak{U}}^{\theta';\theta_2,b}_{G;P}$, where $\theta' < \theta$.

Let us first consider the case when $\theta_1 \neq 0$. Recall the map $\varrho^\theta_{\mathfrak{p}^+_{\mathrm{aff}}} : \mathfrak{U}^\theta_{G;P} \to \overset{\circ}{\mathbf{C}}{}^\theta$ and the map $\mathfrak{r}_{\mathfrak{p}^+_{\mathrm{aff}}} : \widetilde{\mathfrak{U}}^\theta_{G;P} \to \mathfrak{U}^\theta_{G;P}$. Using Proposition 10.6, let us write a point σ of $\widetilde{\mathfrak{U}}^{\theta;\theta_2,b}_{G;P}$ as a triple

$$(\sigma', D^{\theta_2}, (\mathcal{F}_{M_{\mathrm{aff}}}, \beta), (\Sigma b_i \cdot \mathbf{s}_i)),$$

where we have the following:

- σ' is a based map of degree $\mathbf{C} \to \mathrm{Bun}_{G;P}(\mathbf{X}, \infty_{\mathbf{X}}; 0_{\mathbf{X}})$. Let us denote by $\mathcal{F}'_G$ the corresponding principal G-bundle on $\mathbf{S}'$, endowed with a trivialization along $\mathbf{D}'_\infty$ and a reduction to P along $\mathbf{D}'_0$; let $\mathcal{F}'_M$ denote the corresponding M-bundle on $\mathbf{D}'_0 \simeq \mathbf{C}$.

- D^{θ_2} is an element of $\overset{\circ}{\mathbf{C}}{}^{\theta_2}$.
- $(\mathcal{F}'_M, D^{\theta_2}, \mathcal{F}_{M_{\mathrm{aff}}}, \beta)$ is a point of $\mathcal{H}^{\theta_2,+}_{M_{\mathrm{aff}},\mathbf{C}}$.
- $\Sigma b_i \cdot \mathbf{s}_i$ is a 0-cycle of degree b on $\overset{\circ\circ}{\mathbf{S}}$.

Assume for a moment that the support of the divisor $\varrho^{\theta}_{\mathfrak{p}^+_{\mathrm{aff}}}(\sigma') \in \overset{\circ}{\mathbf{C}}{}^{\theta}$ is disjoint from both $\pi_v(\Sigma b_i \cdot \mathbf{s}_i)$ and D^{θ_2}. Then, according to the factorization property Proposition 9.4, our point of $\widetilde{\mathfrak{U}}^{\theta;\theta_2,b}_{G;P}$ is smoothly equivalent to a point in the product $\widetilde{\mathfrak{U}}^{\theta-\theta_1;\theta_2,b}_{G;P} \times \widetilde{\mathfrak{U}}^{\theta_1;0,0}_{G;P}$, and our assertion about smooth equivalence follows since $\widetilde{\mathfrak{U}}^{\theta_1;0,0}_{G;P} \simeq \mathrm{Bun}^{\theta_1}_{G;P}(\mathbf{S};, \mathbf{D}'_\infty; \mathbf{D}'_0)$ and the latter is smooth.

It remains to analyze the case when the support of $\varrho^{\theta}_{\mathfrak{p}^+_{\mathrm{aff}}}(\sigma')$ does intersect the supports of D^{θ_2} and $\pi_v(\Sigma b_i \cdot \mathbf{s}_i)$. By applying again Proposition 9.4, we can assume that there exists a point $\mathbf{c} \in \overset{\circ}{\mathbf{C}}$, such that $D^{\theta_2} = \theta_2 \cdot \mathbf{c}$ and $\pi_v(\Sigma b_i \cdot \mathbf{s}_i) = b \cdot \mathbf{c}$.

Consider the fiber product

$$(\widetilde{\mathfrak{U}}^{\theta}_{G;P} \times \mathrm{Bun}^{a_1}_G(\mathbf{S}', \mathbf{D}'_\infty)) \underset{\mathbf{X}^{(a)} \times \mathbf{X}^{(a_1)}}{\times} (\mathbf{X}^{(a)} \times (\overset{\circ}{\mathbf{X}} - 0_{\mathbf{X}})^{(a_1)})_{\mathrm{disj}},$$

which, according to Proposition 9.9, maps smoothly to $\widetilde{\mathfrak{U}}^{\theta+a_1\cdot\delta}_{G;P}$.

We can view this morphism as a convolution of a point $\sigma \in \widetilde{\mathfrak{U}}^{\theta}_{G;P}$ with a point $\sigma_1 \in \mathrm{Bun}^{a_1}_G(\mathbf{S}', \mathbf{D}'_\infty)$ such that $\varpi_v(\sigma)$ and $\varpi_v(\sigma_1)$ have disjoint supports. We will denote the resulting point of $\widetilde{\mathfrak{U}}^{\theta+a_1\cdot\delta}_{G;P}$ by $\sigma \circ \sigma_1$.

It is easy to see that if $\sigma \in \widetilde{\mathfrak{U}}^{\theta;\theta_2,b}_{G;P}$, then $\sigma \circ \sigma_1 \in \widetilde{\mathfrak{U}}^{\theta+a_1\cdot\delta;\theta_2,b}_{G;P}$; moreover, $\sigma \circ \sigma_1$ corresponds to the triple

$$(\sigma' \circ \sigma_1, D^{\theta_2}, (\mathcal{F}_{M_{\mathrm{aff}}}, \beta), (\Sigma b_i \cdot \mathbf{s}_i)),$$

where all but the first piece of data remain unchanged.

Therefore, it is sufficient to show that there exists an integer a_1 large enough and $\sigma_1 \in \mathrm{Bun}^{a_1}_G(\mathbf{S}', \mathbf{D}'_\infty)$, such that the support of $\varrho^{\theta_1+a_1\cdot\delta}_{\mathfrak{p}^+_{\mathrm{aff}}}(\sigma \circ \sigma_1)$ is disjoint from $\mathbf{c}$. The latter means that the G-bundle obtained from $\sigma' \circ \sigma_1$ on the line $\mathbf{c} \times \mathbf{X}$ should be trivial and its reduction to P at $\mathbf{c} \times 0_{\mathbf{X}}$ in the generic position with respect to the trivialization at $\mathbf{c} \times \infty_{\mathbf{X}}$.

We can view σ_1 as a pair consisting of a divisor $D \in (\overset{\circ}{\mathbf{X}} - 0_{\mathbf{X}})^{(a_1)}$ and a based map $\mathbf{C} \to \mathrm{Gr}^{BD,a_1}_{G,\mathbf{X},D}$. Then the restriction of $\sigma \circ \sigma_1$ to $\mathbf{c} \times \mathbf{X}$ is obtained from the restriction of σ to $\mathbf{c} \times \mathbf{X}$ by a Hecke transformation corresponding to the value of the above map $\mathbf{C} \to \mathrm{Gr}^{BD,a_1}_{G,\mathbf{X},D}$ at $\mathbf{c}$.

Now, it is clear that by choosing a_1 to be large enough we can always find a based map $\mathbf{C} \to \mathrm{Gr}^{BD,a_1}_{G,\mathbf{X},D}$ which will bring $\sigma'|_{\mathbf{c}\times\mathbf{X}}$ to the desired generic position.

17.3

Now, let us assume that $\theta_1 = 0$, but $(0 < \theta_2 < \theta) \bigwedge (0 < b)$. Write θ_2 as $(\overline{\theta}_2, b_2)$ and let us consider separately the cases $b_2 \neq 0$ and $b_2 = 0$.

If $b_2 \neq 0$, the assertion about the smooth equivalence follows from Proposition 9.9.

If $b_2 = 0$, the composition $\widetilde{\mathfrak{U}}^{\theta;\theta_2,b}_{G;P} \hookrightarrow \widetilde{\mathfrak{U}}^{\theta}_{G;P} \to \mathfrak{U}^a_G \stackrel{\varpi^a_h}{\to} \mathbf{X}^{(a)}$ maps to $(\overset{\circ}{\mathbf{X}} - 0_{\mathbf{X}})^{(a)}$, and the assertion once again follows from Proposition 9.9.

Thus it remains to analyze the cases when either $\theta_2 = \theta$, but $|\mathfrak{P}(\theta_2)| > 1$ or $b \cdot \delta = \theta$, but $|\mathfrak{P}(b)| > 1$.

In the former case the assertion about the smooth equivalence follows immediately from Proposition 9.4. In the latter case, given a point

$$(\Sigma b_i \cdot \mathbf{s}_i) \in \mathrm{Sym}^b(\overset{\circ\circ}{\mathbf{S}}) \simeq \widetilde{\mathfrak{U}}^{b\cdot\delta;0,b}_{G;P},$$

which belongs to the stratum $\overset{\circ}{\mathrm{Sym}}{}^{\mathfrak{P}(b)}(\overset{\circ\circ}{\mathbf{S}})$ with $|\mathfrak{P}(b)| > 1$, its projection with respect to at least one of the projections ϖ^b_h or ϖ^b_v will be a divisor supported in more than one point. In other words, we deduce the smooth equivalence assertion from either Proposition 9.9 or Proposition 9.4.

To summarize, we obtain that there are only two types of strata not covered by the induction hypothesis. One is when $\theta_2 = \theta$, and the corresponding stratum is isomorphic to $\overset{\circ}{\mathbf{C}}$, which is contained in the image of $\mathfrak{s}_{\mathfrak{p}^+_{\mathrm{aff}}}$. The restriction is locally constant, because of the equivariance with respect to the group $\mathbb{A}_1$ action on the pair $(\mathbf{S}', \mathbf{D}'_0)$ by "horizontal" shifts (i.e., along the $\mathbf{C}$-factor).

The other type of stratum occurs only when $\theta = a \cdot \delta$; the stratum itself is isomorphic to $\overset{\circ\circ}{\mathbf{S}}$, and it is contained in an open subset of $\widetilde{\mathfrak{U}}^{\theta}_{G;P}$ that projects isomorphically onto an open subset of $\mathfrak{U}^a_G$, via Proposition 9.9.

Note that the restriction of $\mathrm{IC}_{\widetilde{\mathfrak{U}}^{\theta}_{G;P}}$ to this last stratum is automatically locally constant. Indeed, it is enough to show that $\mathrm{IC}_{\mathfrak{U}^a_G}|_{\overset{\circ}{\mathbf{S}}}$ is locally constant, but this follows from the fact that $\mathrm{IC}_{\mathfrak{U}^a_G}$ is equivariant with respect to the group of affine-linear transformations acting on $\mathbf{S}$.

Thus we have established that the restriction of $\mathrm{IC}_{\widetilde{\mathfrak{U}}^{\theta}_{G;P}}$ to $\widetilde{\mathfrak{U}}^{\theta;\mathfrak{P}(\theta_2),\mathfrak{P}(b)}_{G;P}$ locally has the required form.

Let us now show that the isomorphism in fact holds globally. First, we claim that the complex $\mathrm{IC}_{\widetilde{\mathfrak{U}}^{\theta}_{G;P}}|_{\widetilde{\mathfrak{U}}^{\theta;\mathfrak{P}(\theta_2),\mathfrak{P}(b)}_{G;P}}$ is semisimple. This follows from the fact that by retracing our calculation in the category of mixed Hodge modules, we obtain that $\mathrm{IC}_{\widetilde{\mathfrak{U}}^{\theta}_{G;P}}|_{\widetilde{\mathfrak{U}}^{\theta;\mathfrak{P}(\theta_2),\mathfrak{P}(b)}_{G;P}}$ is pure. Hence the semisimplicity assertion follows from the decomposition theorem.

Therefore, it remains to see that the cohomologically shifted perverse sheaves that appear as direct summands in Theorem 16.5 have no monodromy. This is shown in the same way as the corresponding assertion in [BFGM, Section 5.11].

17.4 The induction step

We will now perform the induction step and prove Theorem 16.3 for the value of the parameter equal to θ. The proof essentially mimics the argument of [BFGM, Section 5].

Recall the map $\varrho^\theta_{M_{\mathrm{aff}}} : \widetilde{\mathfrak{U}}^\theta_{G;P} \to \mathrm{Mod}^{\theta,+}_{M_{\mathrm{aff}}}$, and its section $\widetilde{\mathfrak{s}}_{\mathfrak{p}^+_{\mathrm{aff}}} : \mathrm{Mod}^{\theta,+}_{M_{\mathrm{aff}}} \to \widetilde{\mathfrak{U}}^\theta_{G;P}$. Recall also (cf. Corollary 10.4 and Proposition 10.6) that we have the "new" $\mathbb{G}_m$-action on $\widetilde{\mathfrak{U}}^\theta_{G;P}$ that contracts it onto the image of $\widetilde{\mathfrak{s}}_{\mathfrak{p}^+_{\mathrm{aff}}}$.

Under these circumstances, we have teh following, as in [BFGM, Lemma 5.2 and Proposition 5.3].

Lemma 17.5. *There is a canonical isomorphism*

$$(\widetilde{\mathfrak{s}}_{\mathfrak{p}^+_{\mathrm{aff}}})^!(\mathrm{IC}_{\widetilde{\mathfrak{U}}^\theta_{G;P}}) \simeq (\varrho^\theta_{M_{\mathrm{aff}}})_!(\mathrm{IC}_{\widetilde{\mathfrak{U}}^\theta_{G;P}}),$$

and $(\widetilde{\mathfrak{s}}_{\mathfrak{p}^+_{\mathrm{aff}}})^!(\mathrm{IC}_{\widetilde{\mathfrak{U}}^\theta_{G;P}})$ *is a semisimple complex.*

By the induction hypothesis and the factorization property, we may assume that the restriction of $(\widetilde{\mathfrak{s}}_{\mathfrak{p}^+_{\mathrm{aff}}})^!(\mathrm{IC}_{\widetilde{\mathfrak{U}}^\theta_{G;P}})$ to the open subset in $\mathrm{Mod}^{\theta,+}_{M_{\mathrm{aff}}}$ equal to the preimage of the complement of the main diagonal $\overset{\circ}{\mathbf{C}}{}^\theta - \Delta(\overset{\circ}{\mathbf{C}})$ has the desired form. In other words,

$$(\widetilde{\mathfrak{s}}_{\mathfrak{p}^+_{\mathrm{aff}}})^!(\mathrm{IC}_{\widetilde{\mathfrak{U}}^\theta_{G;P}}) \simeq \bigoplus_{\mathfrak{P}(\theta),|\mathfrak{P}(\theta)|>1} \widetilde{(\mathrm{Norm}_{\mathfrak{P}(\theta)})_*}(\mathrm{Loc}^{M_{\mathrm{aff}},\mathfrak{P}(\theta)}_{pt,\mathbf{C}}(\mathfrak{V}_{\mathfrak{p}}))[-|\mathfrak{P}(\theta)|] \bigoplus \mathcal{K}^\theta,$$

where $\mathcal{K}^\theta$ is a semisimple complex on $\mathrm{Gr}^{\theta,+}_{M_{\mathrm{aff}},\mathbf{C}} \subset \mathrm{Mod}^{\theta,+}_{M_{\mathrm{aff}}}$, and our goal is to show that

$$\mathcal{K}^\theta \simeq \mathrm{Loc}^{M_{\mathrm{aff}}}_{pt,\mathbf{C}}((\mathfrak{V}_{\mathfrak{p}})_\theta)[-1], \tag{32}$$

where the subscript θ means that we are taking the direct summand of $\mathfrak{V}_{\mathfrak{p}}$ corresponding to the central character θ.

We pick a point in $\mathbf{C}$ that we will call $\mathbf{c}$, and consider the fiber of $\widetilde{\mathfrak{U}}^\theta_{G;P}$ over it, which we denote by ${}^{\mathfrak{U}}\mathfrak{F}^\theta_{G;P}$. By definition, we have a projection

$$\varrho^\theta_{M_{\mathrm{aff}}} : {}^{\mathfrak{U}}\mathfrak{F}^\theta_{G;P} \to \mathrm{Gr}^{\theta,+}_{M_{\mathrm{aff}}}.$$

As in [BFGM, Section 5.12], it is sufficient to prove that $(\varrho^\theta_{M_{\mathrm{aff}}})_!(\mathrm{IC}_{\widetilde{\mathfrak{U}}^\theta_{G;P}}|_{{}^{\mathfrak{U}}\mathfrak{F}^\theta_{G;P}})$ is concentrated in the (perverse) cohomological degrees ≤ 0 and that its 0th cohomology is isomorphic to

$$\mathrm{Loc}^{M_{\mathrm{aff}}}((U(\mathfrak{V}_{\mathfrak{p}}))_\theta).$$

17.6

We calculate $(\varrho^\theta_{M_{\text{aff}}})_!(\mathrm{IC}_{\widetilde{\mathfrak{U}}^\theta_{G;P}}|_{{}^{\mathfrak{U}}\mathfrak{F}^\theta_{G;P}})$ by intersecting ${}^{\mathfrak{U}}\mathfrak{F}^\theta_{G;P}$ with the various strata $\widetilde{\mathfrak{U}}^{\theta;\mathfrak{P}(\theta_2),\mathfrak{P}(b)}_{G;P}$ of $\widetilde{\mathfrak{U}}^\theta_{G;P}$.

Case 1: *The open stratum.* Set $\overset{\circ}{\mathfrak{F}}{}^\theta_{G;P} = {}^{\mathfrak{U}}\mathfrak{F}^\theta_{G;P} \cap \mathrm{Bun}^\theta_{G;P}(\mathbf{S}, \mathbf{D}_0; \mathbf{D}_\infty)$. (Of course, $\overset{\circ}{\mathfrak{F}}{}^\theta_{G;P} \simeq \overset{\circ}{\mathfrak{F}}{}^\theta_{\mathfrak{g}_{\text{aff}},\mathfrak{p}^+_{\text{aff}}}$, in the terminology of Part IV.) The projection of the latter scheme onto $\mathrm{Gr}^{\theta,+}_{M_{\text{aff}}}$ was studied in Section 15.

For an M_{aff}-dominant coweight ν, let $\overset{\circ}{\mathfrak{F}}{}^\nu_{G;P}$ denote the preimage in $\overset{\circ}{\mathfrak{F}}{}^\theta_{G;P}$ of $\mathrm{Gr}^\nu_{M_{\text{aff}}}$, and we know from Corollary 15.4 that $\overset{\circ}{\mathfrak{F}}{}^\nu_{G;P}$ is of dimension $\leq \langle \nu, \check{\rho}_{\text{aff}} \rangle$. Hence

$$(\varrho^\theta_{M_{\text{aff}}})_!(\mathbb{C}_{\overset{\circ}{\mathfrak{F}}{}^\nu_{G;P}})$$

lies in the cohomological degrees $\leq \langle \theta, 2(\check{\rho}_{\text{aff}} - \check{\rho}_{M_{\text{aff}}}) \rangle$. Moreover, by Corollary 15.4 its top (i.e., $\langle \theta, 2(\check{\rho}_{\text{aff}} - \check{\rho}_{M_{\text{aff}}}) \rangle$) perverse cohomology is isomorphic to

$$\mathrm{IC}_{\mathrm{Gr}^\nu_{M_{\text{aff}}}} \otimes \mathrm{Hom}_{\check{M}_{\text{aff}}}(V^\nu_{M_{\text{aff}}}, U(\mathfrak{n}(\check{\mathfrak{p}})).$$

Since $\mathrm{Bun}^\theta_{G;P}(\mathbf{S}', \mathbf{D}'_\infty; \mathbf{D}'_0)$ is smooth of dimension $\langle \theta, 2(\check{\rho}_{\text{aff}} - \check{\rho}_{M_{\text{aff}}}) \rangle$, we obtain that $(\varrho^\theta_{M_{\text{aff}}})_!(\mathrm{IC}_{\widetilde{\mathfrak{U}}^\theta_{G;P}}|_{\overset{\circ}{\mathfrak{F}}{}^\theta_{G;P}})$ indeed lies in the cohomological degrees ≤ 0, and its 0th perverse cohomology is $\mathrm{Loc}(U(\mathfrak{n}(\check{\mathfrak{p}}))_\theta)$.

Next, we will show that all other strata do not contribute to the cohomological degrees ≥ 0.

Case 2: *The intermediate strata, when* $\theta_2 \neq \theta$, $b \cdot \delta \neq \theta$. Note that if the intersection of $\widetilde{\mathfrak{U}}^{\theta;\mathfrak{P}(\theta_2),\mathfrak{P}(b)}_{G;P}$ with ${}^{\mathfrak{U}}\mathfrak{F}^\theta_{G;P}$ is nonempty, then necessarily $|\mathfrak{P}(\theta_2)| = 1$. In this case, we will denote this intersection by $\mathfrak{F}^{\theta;\theta_2,\mathfrak{P}(b)}_{G;P}$. As a scheme it is isomorphic to

$$(\overset{\circ}{\mathfrak{F}}{}^{\theta_1}_{G;P} \underset{\mathrm{Bun}_M(\mathbf{C})}{\times} \mathcal{H}^{\theta_2,+}_{M_{\text{aff}}}) \times \mathrm{Sym}^{\mathfrak{P}(b)}(\overset{\circ}{\mathbf{X}} - 0_{\mathbf{X}}).$$

Note also that

$$\overset{\circ}{\mathfrak{F}}{}^{\theta_1}_{G;P} \underset{\mathrm{Bun}_M}{\times} \mathcal{H}^{\theta_2,+}_{M_{\text{aff}}} \simeq \overset{\circ}{\mathfrak{F}}{}^{\theta_1}_{G;P} \underset{\mathrm{Gr}^{\theta_1,+}_{M_{\text{aff}}}}{\times} (\mathrm{Gr}^{\theta_1,+}_{M_{\text{aff}}} \star \mathrm{Gr}^{\theta_2,+}_{M_{\text{aff}}}),$$

where $\mathrm{Gr}^{\theta_1,+}_{M_{\text{aff}}} \star \mathrm{Gr}^{\theta_2,+}_{M_{\text{aff}}} \subset \mathrm{Conv}_{M_{\text{aff}}}$ is the corresponding subscheme in the convolution diagram; cf. Section 13.7. We shall view $\mathfrak{F}^{\theta;\theta_2,\mathfrak{P}(b)}_{G;P}$ as a fibration over $\overset{\circ}{\mathfrak{F}}{}^{\theta_1}_{G;P} \times \mathrm{Sym}^{\mathfrak{P}(b)}(\overset{\circ}{\mathbf{X}} - 0_{\mathbf{X}})$ with the typical fiber $\mathrm{Gr}^{\theta_2,+}_{M_{\text{aff}}}$.

In terms of the above identifications, the projection $\varrho^\theta_{M_{\text{aff}}} : {}^{\mathfrak{U}}\mathfrak{F}^\theta_{G;P} \to \mathrm{Gr}^{\theta,+}_M$ is equal to the composition

$$(\overset{\circ}{\mathfrak{F}}{}^{\theta_1}_{G;P} \underset{\mathrm{Gr}^{\theta_1,+}_{M_{\mathrm{aff}}}}{\times} (\mathrm{Gr}^{\theta_1,+}_{M_{\mathrm{aff}}} \star \mathrm{Gr}^{\theta_2,+}_{M_{\mathrm{aff}}})) \times \overset{\circ}{\mathrm{Sym}}{}^{\mathfrak{P}(b)}(\overset{\circ}{\mathbf{X}} - 0_{\mathbf{X}})$$
$$\to \mathrm{Gr}^{\theta_1,+}_{M_{\mathrm{aff}}} \star \mathrm{Gr}^{\theta_2,+}_{M_{\mathrm{aff}}} \to \mathrm{Gr}^{\theta_1+\theta_2,+}_{M_{\mathrm{aff}}} \to \mathrm{Gr}^{\theta,+}_{M_{\mathrm{aff}}},$$

where the latter map is induced by the central cocharacter $b \cdot \delta$.

Now, using the assumption that $\theta_2 \neq \theta$ and $b \cdot \delta \neq \theta$, we can use Section 17.2 to write down the restriction of $\mathrm{IC}_{\widetilde{\mathfrak{U}}^{\theta}_{G;P}}$ to $\mathfrak{F}^{\theta;\theta_2,\mathfrak{P}(b)}_{G;P}$.

We obtain that it is equal to the external product of the complex

$$(\mathbb{C}_{\overset{\circ}{\mathfrak{F}}{}^{\theta_1}_{G;P}}[\langle \theta_1, 2(\check{\rho}_{\mathrm{aff}} - \check{\rho}_{M_{\mathrm{aff}}})\rangle]) \boxtimes (\mathrm{Sym}^{\mathfrak{P}(b)}(\mathfrak{V}^f)[2|\mathfrak{P}(b)|] \otimes \underline{\mathbb{C}}_{\overset{\circ}{\mathrm{Sym}}{}^{\mathfrak{P}(b)}(\overset{\circ}{\mathbf{X}}-0_{\mathbf{X}})})$$

along the base $\overset{\circ}{\mathfrak{F}}{}^{\theta_1}_{G;P} \times \overset{\circ}{\mathrm{Sym}}{}^{\mathfrak{P}(b)}(\overset{\circ}{\mathbf{X}} - 0_{\mathbf{X}})$, and the perverse complex

$$\mathrm{Loc}^{M_{\mathrm{aff}}}(\underset{i \geq 0}{\oplus} (\mathrm{Sym}^i(\mathfrak{V}_{\mathfrak{p}}))_{\theta_2}[2i])$$

along the fiber $\mathrm{Gr}^{\theta_2,+}_{M_{\mathrm{aff}}}$.

By Corollary 15.4, we obtain that the !-direct image of the restriction $\mathrm{IC}_{\widetilde{\mathfrak{U}}^{\theta}_{G;P}}|_{\mathfrak{F}^{\theta;\theta_2,\mathfrak{P}(b)}_{G;P}}$ onto $\mathrm{Gr}^{\theta_1,+}_{M_{\mathrm{aff}}} \star \mathrm{Gr}^{\theta_2,+}_{M_{\mathrm{aff}}}$ is a complex of sheaves lying in strictly negative cohomological degrees with spherical perverse cohomology. By the exactness of convolution, its further direct image onto $\mathrm{Gr}^{\theta,+}_{M_{\mathrm{aff}}}$ also lies in strictly negative cohomological degrees, which is what we had to show.

Case 3: *The strata with* $\theta = b \cdot \delta$. The intersection $\widetilde{\mathfrak{U}}^{b\cdot\delta;0,b}_{G;P} \cap {}^{\mathfrak{U}}\mathfrak{F}^{\theta}_{G;P}$ is isomorphic to $(\overset{\circ}{\mathbf{X}} - 0_{\mathbf{X}})^{(b)}$, which we further subdivide according to partitions $\mathfrak{P}(b)$ of b. The map $\widetilde{\mathfrak{U}}^{b\cdot\delta;0,b}_{G;P} \cap {}^{\mathfrak{U}}\mathfrak{F}^{\theta}_{G;P} \to \mathrm{Gr}^{\theta,+}_{M_{\mathrm{aff}}}$ is the composition of the projection $\widetilde{\mathfrak{U}}^{b\cdot\delta;0,b}_{G;P} \cap {}^{\mathfrak{U}}\mathfrak{F}^{\theta}_{G;P} \to pt$ and the embedding $pt \to \mathrm{Gr}^{\theta,+}_{M_{\mathrm{aff}}}$ corresponding to the central cocharacter $b \cdot \delta$.

We will consider separately two cases: (a) when $|\mathfrak{P}(b)| > 1$ and (b) when $|\mathfrak{P}(b)| = 1$.

In case (a), the restriction of $\mathrm{IC}_{\widetilde{\mathfrak{U}}^{\theta}_{G;P}}$ to the corresponding stratum $\overset{\circ}{\mathrm{Sym}}{}^{\mathfrak{P}(b)}(\overset{\circ\circ}{\mathbf{S}})$ is known, according to Section 17.3. In particular, we know that this complex lives in the perverse cohomological degrees $\leq -|\mathfrak{P}(b)|$. Hence when we further restrict it to $\overset{\circ}{\mathrm{Sym}}{}^{\mathfrak{P}(b)}(\overset{\circ}{\mathbf{X}} - 0_{\mathbf{X}}) \subset \overset{\circ}{\mathrm{Sym}}{}^{\mathfrak{P}(b)}(\overset{\circ\circ}{\mathbf{S}})$, it lives in the perverse cohomological degrees $\leq -2|\mathfrak{P}(b)| < -1 - |\mathfrak{P}(b)|$. Therefore, when we take its direct image along $\overset{\circ}{\mathrm{Sym}}{}^{\mathfrak{P}(b)}(\overset{\circ}{\mathbf{X}} - 0_{\mathbf{X}})$ we obtain a complex in the cohomological degrees < 0.

In case (b), by the definition of intersection cohomology, the restriction of $\mathrm{IC}_{\widetilde{\mathfrak{U}}^{\theta}_{G;P}}$ to $\overset{\circ\circ}{\mathbf{S}}$ lives in strictly negative cohomological degrees. According to Section 17.3, this restriction is locally constant; therefore, its further restriction to $\mathbf{X} - 0_{\mathbf{X}} \subset \overset{\circ\circ}{\mathbf{S}}$ lives in the cohomological degrees < -1. Hence its cohomology along $\mathbf{X} - 0_{\mathbf{X}}$ lives in the cohomological degrees < 0.

Case 4: *The stratum* $\theta_2 = \theta$. According to Proposition 10.2, the intersection $\widetilde{\mathfrak{U}}^{\theta;\theta,0}_{G;P} \cap {}^{\mathfrak{U}}\mathfrak{F}^{\theta}_{G;P}$ is in fact isomorphic to $\mathrm{Gr}^{\theta,+}_{M_{\mathrm{aff}}}$, such that the map $\varrho^{\theta}_{M_{\mathrm{aff}}}$ is the identity map.

Hence the required assertion follows from the fact that the restriction of $\mathrm{IC}_{\widetilde{\mathfrak{U}}^{\theta}_{G;P}}$ to $\mathrm{Gr}^{\theta,+}_{M_{\mathrm{aff}}}$ lives in the negative cohomological degrees, by the definition of the IC sheaf.

17.7

Thus the induction step has been performed and we have established Theorem 16.3 for the parameter equal to θ. Now Theorem 16.7 is deduced from Theorem 16.3 is the same way as [BFGM, Theorem 7.2] is deduced from [BFGM, Theorem 4.5].

Let us prove now Theorem 16.8. As was explained in Section 17.2, it remains to identify the restriction of $\mathrm{IC}_{\mathfrak{U}^{\theta}_{G;P}}$ to strata of two types.

Type 1. $\mathfrak{U}^{\theta;\mathfrak{P}(\theta),0}_{G;P}$, where $\mathfrak{P}(\theta)$ is a 1-element partition.

Note that $\mathfrak{U}^{\theta;\mathfrak{P}(\theta),0}_{G;P}$ is a closed subset of $\mathfrak{U}^{\theta;\theta,0}_{G;P}$, and the latter is exactly the image of the map $\mathfrak{s}_{\mathfrak{p}^+_{\mathrm{aff}}}$. Therefore, we have to calculate the $*$-restriction of

$$\bigoplus_{\mathfrak{P}(\theta)} (\mathrm{Norm}_{\mathfrak{P}(\theta)})_*(\overline{\mathrm{Loc}}^{\mathfrak{P}(\theta)}(\mathfrak{V}_{\mathfrak{p}}))[|\mathfrak{P}(\theta)|]$$

to $\mathrm{Gr}^{\theta,+}_{M_{\mathrm{aff}},\mathbf{C}} \subset \mathrm{Mod}^{\theta,+}_{M_{\mathrm{aff}},\mathbf{C}}$, and as in the proof of [BFGM, Theorem 1.12], we obtain the required answer for Theorem 16.8.

Type 2. We can assume that $P = G$ and we are dealing with the stratum $\overset{\circ}{\mathbf{S}} \subset \mathrm{Sym}^a(\overset{\circ}{\mathbf{S}}) \subset \mathfrak{U}^a_G$.

By Theorem 16.7 we know the restriction of $\mathrm{IC}_{\mathfrak{U}^a_G}$ to $\overset{\circ}{\mathbf{C}} \subset \overset{\circ}{\mathbf{S}}$, and we obtain

$$\bigoplus_i (\mathrm{Sym}^i(\mathfrak{V}^f))_a[2i] \otimes \underline{\mathbb{C}}_{\overset{\circ}{\mathbf{C}}}[2].$$

In particular, by passing to the category of mixed Hodge modules and retracing the proof of Theorem 16.7, we obtain that $\mathrm{IC}_{\mathfrak{U}^a_G}|_{\overset{\circ}{\mathbf{C}}}$ is pure.

However, since $\mathrm{IC}_{\mathfrak{U}^a_G}|_{\overset{\circ}{\mathbf{S}}}$ is equivariant with respect to the group of affine-linear transformations of $\overset{\circ}{\mathbf{S}}$, we obtain that $\mathrm{IC}_{\mathfrak{U}^a_G}|_{\overset{\circ}{\mathbf{S}}}$ is also pure, and hence semisimple. Therefore, it is isomorphic to $\bigoplus_i (\mathrm{Sym}^i(\mathfrak{V}^f))_a[2i] \otimes \underline{\mathbb{C}}_{\overset{\circ}{\mathbf{S}}}[2]$, as required.

Finally, let us prove Theorem 16.5. In this case there are also two types of strata, not covered by the induction hypothesis: $\widetilde{\mathfrak{U}}^{\theta;\mathfrak{P}(\theta),0}_{G;P}$ and $\widetilde{\mathfrak{U}}^{a\cdot\delta;0,a}_{G;P}$.

The assertion for the stratum of the first type follows from Theorem 16.3 just as the corresponding assertion for $\mathfrak{U}^{\theta}_{G,P}$ follows from Theorem 16.7. The assertion for the stratum of the second type follows from the corresponding fact for $\mathfrak{U}^a_G$, which has been established above.

18 Appendix

We recall the setup of Proposition 2.7. Assume that $\mathfrak{g}$ is symmetrizable. Then a stronger statement holds.

Theorem 18.1. *The scheme* $\mathrm{Maps}^\theta(\mathbf{C}, \mathcal{G}_{\mathfrak{g},\mathfrak{p}})$ *is of finite type.*

We reproduce here the proof due to V. Drinfeld and A. Joseph. Recall that according to Proposition 2.7, $\mathrm{Maps}^\theta(\mathbf{C}, \mathcal{G}_{\mathfrak{g},\mathfrak{p}})$ is a union of finite type open subschemes $\mathrm{Maps}^\theta(\mathbf{C}, \mathcal{G}^w_{\mathfrak{g},\mathfrak{p}})$, $w \in \mathcal{W}$. We have to check that the above union is actually finite, i.e., for w big enough we have $\mathrm{Maps}^\theta(\mathbf{C}, \mathcal{G}_{\mathfrak{g},\mathfrak{p}}) = \mathrm{Maps}^\theta(\mathbf{C}, \mathcal{G}^w_{\mathfrak{g},\mathfrak{p}})$. In other words, we have to find $w \in \mathcal{W}$ such that any map $\sigma \in \mathrm{Maps}^\theta(\mathbf{C}, \mathcal{G}_{\mathfrak{g},\mathfrak{p}})$ lands in the open subscheme $\mathcal{G}^w_{\mathfrak{g},\mathfrak{p}} \subset \mathcal{G}_{\mathfrak{g},\mathfrak{p}}$. We will give a proof for $\mathfrak{p} = \mathfrak{b}$, and the general case follows immediately.

More precisely, for $\mu = \sum\limits_{i\in I} n_i \cdot \alpha_i \in \Lambda^{\mathrm{pos}}_{\mathfrak{g}}$ we set $|\mu| := \sum\limits_{i\in I} n_i$. We will prove that the image of $\sigma(\mathbf{C})$ in $\mathcal{G}_{\mathfrak{g},\mathfrak{b}}$ never intersects the Schubert subvariety $\overline{\mathcal{G}}_{\mathfrak{g},\mathfrak{b},w}$ for $\ell(w) > 2|\mu|$ (the reduced length).

To this end note that $|\mu|$ is the intersection multiplicity of the curve $\sigma(\mathbf{C})$ and the Schubert divisor $\bigcup\limits_{i\in I} \overline{\mathcal{G}}_{\mathfrak{g},\mathfrak{b},s_i}$ in $\mathcal{G}_{\mathfrak{g},\mathfrak{b}}$. If $\sigma(\mathbf{C})$ passes through a point of $\overline{\mathcal{G}}_{\mathfrak{g},\mathfrak{b},w}$, this intersection multiplicity cannot be smaller than the multiplicity of the Schubert divisor at this point. Thus it suffices to prove that the multiplicity of the Schubert divisor $\bigcup\limits_{i\in I} \overline{\mathcal{G}}_{\mathfrak{g},\mathfrak{b},s_i}$ at a point of the Schubert variety $\overline{\mathcal{G}}_{\mathfrak{g},\mathfrak{b},w}$ is at least $\frac{1}{2}\ell(w)$.

We choose $\check{\rho} \in \mathfrak{h}^*$ such that $\langle \alpha_i, \check{\rho}\rangle = 1 \ \forall\, i \in I$. Let $\mathcal{V}_{\check{\rho}}$ denote the simple $\mathfrak{g}$-module with highest weight $\check{\rho}$. For each $w \in \mathcal{W}$, let $v_{w\check{\rho}}$ denote an extremal vector in $\mathcal{V}_{\check{\rho}}$ of weight $w\check{\rho}$; let $v^*_{w\check{\rho}}$ be an extremal vector in $\mathcal{V}^*_{\check{\rho}}$ of weight $-w\check{\rho}$.

Recall that under the projective embedding $\mathcal{G}_{\mathfrak{g},\mathfrak{b}} \hookrightarrow \mathbb{P}(\mathcal{V}^*_{\check{\rho}})$ the Schubert divisor $\bigcup_{i\in I} \overline{\mathcal{G}}_{\mathfrak{g},\mathfrak{b},s_i}$ is cut out by the equation $v_{\check{\rho}}$. There is a transversal slice to $\overline{\mathcal{G}}_{\mathfrak{g},\mathfrak{b},w}$ through the T-fixed point w isomorphic to $w(\mathfrak{n}) \cap \mathfrak{n}^-$ (and the isomorphism is given by the action of the corresponding nilpotent Lie group), and in the coordinates $f \in w(\mathfrak{n}) \cap \mathfrak{n}^-$ the above equation reads

$$\langle v_{\check{\rho}}, \exp(f) v^*_{w\check{\rho}}\rangle = 0,$$

where exp is the isomorphism between the nilpotent Lie algebra $w(\mathfrak{n}) \cap \mathfrak{n}^-$ and the corresponding Lie group.

Hence the multiplicity m_w of the Schubert divisor at the point $w \in \overline{\mathcal{G}}_{\mathfrak{g},\mathfrak{b},w}$ is the maximal integer m such that $\langle f_1 \dots f_m v_{\check{\rho}}, v^*_{w\check{\rho}}\rangle = 0$ for any $f_1, \dots, f_m \in w(\mathfrak{n}) \cap \mathfrak{n}^-$. Let F^n denote the canonical filtration on the universal enveloping algebra $U(\mathfrak{n}^-)$. Now the desired estimate $m_w \geq \frac{1}{2}\ell(w)$ is a consequence of the following lemma belonging to A. Joseph. We are grateful to him for the permission to reproduce it here.

Lemma 18.2. *Suppose* $v_{w\check{\rho}} \in F^m(U(\mathfrak{n}^-))v_{\check{\rho}}$. *Then* $2m \geq \ell(w)$.

Proof. The proof uses the Chevalley–Kostant construction of $\mathcal{V}_{\check{\rho}}$ as extended in [J] to the affine case. The extension generalizes with no significant change for $\mathfrak{g}$ as above. The extension from the semisimple case (of Chevalley–Kostant) needs a little care as infinite sums are involved. A slightly more streamlined analysis is given in [GrJ]. Some details are given below.

Recall that $\mathfrak{g}$ admits a nondegenerate symmetric invariant bilinear form $(\, , \,)$. From this one may construct the Clifford algebra $C(\mathfrak{g})$ of $\mathfrak{g}$ defined as a quotient of the tensor algebra $T(\mathfrak{g})$ of $\mathfrak{g}$ by the ideal generated by the elements $x \otimes y - y \otimes x - 2(x, y)$ $\forall\, x, y \in \mathfrak{g}$.

One defines a Lie algebra homomorphism φ of $\mathfrak{g}$ to a subspace of certain infinite sums of quadratic elements of $C(\mathfrak{g})$. (See [J, 4.11] or [GrJ, 4.7]). In this one checks that only finitely many commutators contribute to a given term in $[\varphi(x), \varphi(y)]$, which therefore makes sense and by the construction equals $\varphi[x, y]$ (cf. [GrJ, 4.7]).

It turns out that we need to reorder the expression for $\varphi(x)$ so that only finitely many of the negative root vectors lie to the right. This process is well defined for root vectors x of weight $\pm\check{\alpha}_i$, $i \in I$. This is extended to $\mathfrak{g}$ via commutation and the Jacobi identities (see [J, 4.11]). Notably,

$$\varphi(h) = \check{\rho}(h) - \frac{1}{2} \sum_{\check{\alpha} \in \Delta^+} \sum_{r} \check{\alpha}(h) e^r_{-\check{\alpha}} e^r_{\check{\alpha}} \quad \forall\, h \in \mathfrak{h}, \tag{33}$$

where Δ^+ is the set of positive roots, $e^r_{\check{\alpha}}$ is a basis for $\mathfrak{g}_{\check{\alpha}}$ and $e^r_{-\check{\alpha}}$ a dual basis for $\mathfrak{g}_{-\check{\alpha}}$.

Through the diamond lemma one shows that $C(\mathfrak{g})$ admits a triangular decomposition, that is to say there is a vector space isomorphism

$$C(\mathfrak{n}^-) \otimes C(\mathfrak{h}) \otimes C(\mathfrak{n}) \xrightarrow{\sim} C(\mathfrak{g})$$

given by multiplication. (The left-hand side is defined by restricting the bilinear form: in particular $C(\mathfrak{n}^-) = \Lambda(\mathfrak{n}^-)$).

View $\mathfrak{n}$ as a subspace of $C(\mathfrak{g})$ and let $I(\mathfrak{g})$ denote the left ideal it generates. Through multiplication by $\varphi(x)$, $x \in \mathfrak{g}$, and the above ordering, it follows that $C(\mathfrak{g})/I(\mathfrak{g})$ becomes a $\mathfrak{g}$-module. By [J, 4.12] it is a direct sum of $\dim C(\mathfrak{h})$ copies of $\mathcal{V}_{\check{\rho}}$. In particular if $v_{\check{\rho}}$ is the image of the identity of $C(\mathfrak{g})$ in $C(\mathfrak{g})/I(\mathfrak{g})$, then $U(\varphi(\mathfrak{g}))v_{\check{\rho}}$ is isomorphic to $\mathcal{V}_{\check{\rho}}$. (However, it does not lie in $C(\mathfrak{n}^-) \pmod{I(\mathfrak{g})}$, only in $C(\mathfrak{n}^-) \otimes C(\mathfrak{h}) \pmod{I(\mathfrak{g})}$). (Note by (33) above, $v_{\check{\rho}}$ has weight $\check{\rho}$.)

Given $w \in \mathcal{W}$, set $S(w^{-1}) = \{\check{\alpha} \in \Delta^+ \text{ such that } w^{-1}\check{\alpha} \in -\Delta^+\}$. Since $\check{\beta} \in S(w^{-1})$ is a real root there is a unique up to scalars root vector $e_{-\check{\beta}} \in \mathfrak{n}^-$ of weight $-\check{\beta}$. Since $C(\mathfrak{g})/I(\mathfrak{g}) \approx \Lambda(\mathfrak{n}^-) \otimes C(\mathfrak{h})$ as an $\mathfrak{h}$-module it follows that the subspace of $C(\mathfrak{g})/I(\mathfrak{g})$ of weight $w\check{\rho}$ is just

$$\left(\prod_{\check{\beta} \in S(w^{-1})} e_{-\check{\beta}} \right) C(\mathfrak{h}).$$

With respect to the canonical filtration F^n of $C(\mathfrak{g})$, such an element has degree between $\ell(w)$ and $\ell(w) + \dim \mathfrak{h}$. Yet $\varphi(\mathfrak{g}) \subset F^2C(\mathfrak{g})$ and $v_{\check{\rho}} \in F^0C(\mathfrak{g})$. The assertion of the lemma follows. □

This completes the proof of Theorem 18.1.

Acknowledgments. It is clear from the above discussion that the present paper owes its existence to V. Drinfeld's generous explanations. We have also benefited strongly from discussions with V. Baranovsky and V. Ginzburg. In fact, the idea that the Uhlenbeck space must be related to the space of quasi-maps into the affine Grassmannian was formulated by V. Ginzburg back in 1997.

In the course of our study of Uhlenbeck spaces, M. F. has enjoyed the hospitality and support of the Hebrew University of Jerusalem, the University of Chicago, and the University of Massachusetts at Amherst. His research was conducted for the Clay Mathematical Institute, and partially supported by CRDF Award RM1-2545-MO-03.

D.G. is a prize fellow of the Clay Mathematics Institute. He also wishes to thank the Hebrew University of Jerusalem where the major part of this paper was written.

We would also like to thank the referee for helpful remarks.

References

[B] R. Bezrukavnikov, *Perverse Coherent Sheaves (after Deligne)*, preprint, 2000; math.AG/0005152.

[BBD] A. Beilinson, J. Bernstein, and P. Deligne, Faisceaux pervers, in *Analyse et topologie sur les espaces singulières* I (*Luminy*, 1981), Astérisque 100, Société Mathématique de France, Paris, 1982, 5–171.

[BaGi] V. Baranovsky and V. Ginzburg, *Algebraic Construction of the Uhlenbeck Moduli Space*, manuscript, 1998.

[BFGM] A. Braverman, M. Finkelberg, D. Gaitsgory, and I. Mirković, Intersection Cohomology of Drinfeld's compactifications, *Selecta Math.* (*N.S.*), **8** (2002), 381–418.

[BG] A. Braverman and D. Gaitsgory, Crystals via the affine Grassmannian, *Duke Math. J.*, **107** (2001), 561–575.

[BG1] A. Braverman and D. Gaitsgory, Geometric Eisenstein series, *Invent. Math.*, **150** (2002), 287–384.

[CB] W. Crawley-Boevey, Normality of Marsden-Weinstein reductions for representations of quivers, *Math. Ann.*, **325**-1 (2003), 55–79.

[Do] S. K. Donaldson, Connections, cohomology and the intersection forms of four-manifolds, *J. Differential Geom.*, **24** (1986), 275–341.

[DK] S. K. Donaldson and P. Kronheimer, *The Geometry of Four-Manifolds*, Oxford Mathematical Monographs, Oxford University Press, London, 1990.

[Fa] G. Faltings, Algebraic loop groups and moduli spaces of bundles, *J. European Math. Soc.*, **5** (2003), 41–68.

[FFKM] B. Feigin, M. Finkelberg, A. Kuznetsov, and I. Mirković, Semi-infinite flags, *AMS Transl.*, **194** (1999), 81–148.

[FGK] M. Finkelberg, D. Gaitsgory, and A. Kuznetsov, Uhlenbeck spaces for $\mathbb{A}^2$ and affine Lie algebra $\widehat{\mathfrak{sl}}_n$, *Publ. RIMS Kyoto Univ.*, **39**-4 (2003), 721–766.

[FKMM] M. Finkelberg, A. Kuznetsov, N. Markarian, and I. Mirković, A note on a symplectic structure on the space of G-monopoles, *Comm. Math. Phys.*, **201** (1999), 411–421.

[Ga] D. Gaitsgory, Construction of central elements in the affine Hecke algebra via nearby cycles, *Invent. Math.*, **144** (2001), 253–280.

[Gi] D. Gieseker, On the moduli of vector bundles on an algebraic surface, *Ann. Math.* (2), **106** (1977), 45–60.

[GrJ] J. Greenstein and A. Joseph, A Chevalley-Kostant presentation of basic modules for $\widehat{sl(2)}$ and the associated affine KPRV determinants at $q = 1$, *Bull. Sci. Math.*, **125** (2001), 85–108.

[J] A. Joseph, On an affine quantum KPRV determinant at $q = 1$, *Bull. Sci. Math.*, **125** (2001), 23–48.

[Ka] M. Kashiwara, The flag manifold of Kac-Moody Lie algebra, in *Algebraic Analysis, Geometry, and Number Theory: Proceedings of the JAMI Inaugural Conference*, Johns Hopkins University Press, Baltimore, 1989, 161–190.

[Ka1] M. Kashiwara, private communication.

[Ka2] M. Kashiwara, Kazhdan-Lusztig conjecture for a symmetrizable Kac-Moody Lie algebra, *Progr. Math.*, **87** (1990), 407–433.

[Ka3] M. Kashiwara, Crystallizing the q-analogue of universal enveloping algebras, *Duke Math. J.* **63** (1991), 465–516.

[Ka4] M. Kashiwara, Crystal base and Littelmann's refined Demazure character formula, *Duke Math. J.*, **71** (1993), 839–858.

[Ka5] M. Kashiwara, On crystal bases, *CMS Conf. Proc.*, **16** (1995), 155–197.

[KaS] M. Kashiwara and Y. Saito, Geometric construction of crystal bases, *Duke Math. J.*, **89** (1997), 9–36.

[KiP] B. Kim and R. Pandharipande, *The Connectedness of the Moduli Space of Maps to Homogeneous Spaces*, preprint, 2000; math.AG/0003168.

[Li] J. Li, Algebraic geometric interpretation of Donaldson's polynomial invariants, *J. Differential Geom.*, **37** (1993), 417–466.

[Mo] J. W. Morgan, Comparison of the Donaldson polynomial invariants with their algebro-geometric analogues, *Topology*, **32** (1993), 449–488.

[Na] H. Nakajima, *Lectures on Hilbert Schemes of Points on Surfaces*, AMS University Lecture Series 18, American Mathematical Society, Providence, 1999.

[Pe] N. Perrin, Courbes rationelles sur les variétés homogènes, *Ann. Inst. Fourier*, **52** (2002), 105–132.

[Th] J. F. Thomsen, Irreducibility of $\overline{M}_{0,n}(G/P, \beta)$, *Internat. J. Math.*, **9** (1998), 367–376.

[U] K. K. Uhlenbeck, Connections with L^p bounds on curvature, *Comm. Math. Phys.*, **83** (1982), 31–42.

[Va] G. Valli, Interpolation theory, loop groups and instantons, *J. Reine Angew. Math.*, **446** (1994), 137–163.

New Questions Related to the Topological Degree

Haïm Brezis[1–3]

[1] Laboratoire J.-L. Lions
Université Pierre et Marie Curie
B.C. 187
4 Place Jussieu
75252 Paris Cedex 05 France
brezis@ccr.jussieu.fr
[2] Institut Universitaire de France
[3] Department of Mathematics
Rutgers University
Hill Center, Busch Campus
110 Frelinghuysen Road
Piscataway, NJ 08854
USA
brezis@math.rutgers.edu

To I. M. Gelfand with admiration.

Subject Classifications: 58C30, 46E35

1 Topological degree and VMO

Degree theory for continuous maps has a long history and has been extensively studied, both from the point of view of analysis and topology. If $f \in C^0(S^n, S^n)$, $\deg f$ is a well-defined element of $\mathbb{Z}$, which is stable under continuous deformation. Starting in the early 1980s, the need to define a degree for some classes of discontinuous maps emerged from the study of some nonlinear PDEs (related to problems in liquid crystals and superconductors). These examples involved Sobolev maps in the limiting case of the Sobolev embedding; see Sections 2 and 3 below. (Topological questions for Sobolev maps strictly below the limiting exponent have been investigated in [15] and [14].) In these cases, the Sobolev embedding asserts only that such maps belong to the space VMO (see below) and *need not be continuous*.

In connection with degree for $H^{1/2}(S^1, S^1)$, L. Boutet de Monvel and O. Gabber suggested a concept of degree for maps in VMO(S^1, S^1) (see [2] and Section 3 below). In our joint work with L. Nirenberg [16], we followed up on their suggestion and established on firm grounds a degree theory for maps in VMO(S^n, S^n). Here is a brief summary of our contribution.

First, recall the definition of BMO (bounded mean oscillation), a concept originally introduced by F. John and L. Nirenberg in 1961. Let Ω be a smooth bounded open domain in $\mathbb{R}^n$, or a smooth, compact, n-dimensional Riemannian manifold (with or without boundary). An integrable function $f : \Omega \to \mathbb{R}$ belongs to BMO if

$$|f|_{\mathrm{BMO}} = \operatorname*{Sup}_{B\subset\Omega} ⨍_B ⨍_B |f(x) - f(y)| dx dy < \infty,$$

where the Sup is taken over all (geodesic) balls in Ω. It is easy to see that an equivalent seminorm is given by

$$\operatorname*{Sup}_{B\subset\Omega} ⨍_B \left| f(x) - ⨍_B f(y) dy \right| dx.$$

A very important subspace of BMO, introduced by L. Sarason, consists of VMO (vanishing mean oscillation) functions in the sense that

$$\lim_{|B|\to 0} ⨍_B ⨍_B |f(x) - f(y)| dx dy = 0.$$

It is easy to see that

$$\mathrm{VMO}(\Omega, \mathbb{R}) = \overline{C^0(\overline{\Omega}, \mathbb{R})}^{\mathrm{BMO}}.$$

The space VMO is equipped with the BMO seminorm $|f|_{\mathrm{BMO}}$. Clearly, $L^\infty \subset \mathrm{BMO}$. It is well known that BMO is strictly bigger than L^∞ (a standard example is $f(x) = |\log|x||$); however, as a consequence of the classical John–Nirenberg inequality,

$$\mathrm{BMO} \subset \bigcap_{p<\infty} L^p.$$

Thus BMO is "squeezed" between L^∞ and $\cap_{p<\infty} L^p$ and for many purposes serves as an interesting "substitute" for L^∞.

Concerning VMO, it is easy to see that $L^\infty \not\subset \mathrm{VMO}$, but, of course, $C^0 \subset \mathrm{VMO}$. A useful example showing that the inclusion is strict is the function

$$f(x) = |\log|x||^\alpha,$$

which belongs to VMO for every $\alpha < 1$. In some sense, VMO serves as a "substitute" for C^0. The Sobolev space $W^{1,n}$ provides an important class of VMO functions. Recall that for every $1 \le p < \infty$,

$$W^{1,p}(\Omega, \mathbb{R}) = \{f \in L^p(\Omega); \nabla f \in L^p(\Omega)\}.$$

Poincaré's inequality asserts that

$$\int_B \left| f - ⨍_B f \right| \le C|B|^{1/n} \int_B |\nabla f|,$$

from which we deduce, using Hölder, that

$$\fint_B \left| f - \fint_B f \right| \leq C \left[\int_B |\nabla f|^n \right]^{1/n}$$

and thus $W^{1,n} \subset \text{VMO}$.

Similarly, the fractional Sobolev space $W^{s,p}(\Omega)$ is contained in VMO for all $0 < s < 1$ and all $1 < p < \infty$ with $sp = n$ (the limiting case of the Sobolev embedding). Indeed, in the Gagliardo characterization, we have

$$W^{s,p}(\Omega) = \{f \in L^p(\Omega); \int_\Omega \int_\Omega \frac{|f(x) - f(y)|^p}{|x - y|^{n+sp}} dxdy < \infty\}. \tag{1.1}$$

Clearly,

$$\begin{aligned}\int_B \int_B |f(x) - f(y)| dxdy &= \int_B \int_B \frac{|f(x) - f(y)|}{|x - y|^{(n/p)+s}} |x - y|^{(n/p)+s} dxdy \\ &\leq C|B|^{(1/p)+(s/n)} \int_B \int_B \frac{|f(x) - f(y)|}{|x - y|^{(n/p)+s}} dxdy.\end{aligned}$$

Using Hölder, we deduce that

$$\begin{aligned}&\int_B \int_B |f(x) - f(y)| dxdy \\ &\quad \leq C|B|^{(1/p)+(s/n)+2-(2/p)} \left[\int_B \int_B \frac{|f(x) - f(y)|^p}{|x - y|^{n+sp}} dxdy \right]^{1/p},\end{aligned}$$

and thus when $sp = n$,

$$\fint_B \fint_B |f(x) - f(y)| dxdy \leq C \left[\int_B \int_B \frac{|f(x) - f(y)|^p}{|x - y|^{n+sp}} dxdy \right]^{1/p},$$

which implies that $W^{s,p} \subset \text{VMO}$.

One of the basic results in [16] is the following.

Theorem 1 (H. Brezis and L. Nirenberg [16]). *Every map* $f \in \text{VMO}(S^n, S^n)$ *has a well-defined degree. Moreover,*

(a) *this degree coincides with the standard degree when* f *is continuous;*
(b) *the map* $f \mapsto \deg f$ *is continuous on* $\text{VMO}(S^n, S^n)$ *under BMO-convergence.*

It is quite easy to define the VMO-degree. For any given measurable map $f : S^n \to S^n$ and $0 < \varepsilon < 1$, set

$$\bar{f}_\varepsilon(x) = \fint_{B_\varepsilon(x)} f(y) dy.$$

Next, we present an elementary lemma that is extremely useful.

Lemma 1. *If* $f \in \text{VMO}(S^n, S^n)$, *then*

$$|\bar{f}_\varepsilon(x)| \to 1 \quad \text{as } \varepsilon \to 0, \quad \text{uniformly in } x \in S^n.$$

Proof. Set

$$\rho_\varepsilon(x) = \fint_{B_\varepsilon(x)} \fint_{B_\varepsilon(x)} |f(y) - f(z)| dy dz,$$

so that $\rho_\varepsilon(x) \to 0$ as $\varepsilon \to 0$, uniformly in $x \in S^n$, since $f \in \text{VMO}$. Then observe that

$$1 - \rho_\varepsilon(x) \leq |\bar{f}_\varepsilon(x)| \leq 1. \qquad \square$$

If $f \in \text{VMO}(S^n, S^n)$, we may now set

$$f_\varepsilon(x) = \frac{\bar{f}_\varepsilon(x)}{|\bar{f}_\varepsilon(x)|}, \quad x \in S^n, \quad 0 < \varepsilon < \varepsilon_0(f).$$

Using ε as a homotopy parameter, we see that deg f_ε is well defined and *independent of* ε for $\varepsilon > 0$ sufficiently small. This integer is, by definition, VMO-deg f. The proof of Theorem 1(a) is straightforward. For the proof of (b), we refer to [16].

The space $\text{VMO}(S^n, S^n)$ is larger than $C^0(S^n, S^n)$. However, its structure, from the point of view of connected (or, equivalently, path-connected) components, is similar to $C^0(S^n, S^n)$. More precisely, there is a VMO version of the celebrated Hopf result.

Theorem 2. *The homotopy classes (i.e., the path-connected components) of* $\text{VMO}(S^n, S^n)$ *are characterized by their VMO-degree.*

Remark 1. By contrast, it is *not* possible to define a degree for maps in $L^\infty(S^n, S^n)$. In fact, the space $L^\infty(S^n, S^n)$ is path-connected (see [16, Section I.5]).

2 Degree for $H^1(S^2, S^2)$ and beyond

In my earlier paper with J. M. Coron [12] (see also [9, 10]), we were led to a concept of degree for maps in $H^1(S^2, S^2)$. Our original motivation came from solving a nonlinear elliptic system, proposed in [17], which amounts to finding critical points of the Dirichlet integral

$$E(u) = \int_\Omega |\nabla u|^2$$

subject to the constraint

$$u \in H^1_\varphi(\Omega, S^2) = \{u \in H^1(\Omega; S^2); u = \varphi \text{ on } \partial\Omega\},$$

where Ω denotes the unit disc in $\mathbb{R}^2$ and $\varphi : \partial\Omega \to S^2$ is given (smooth). In the process of finding critical points, it is natural to investigate the connected components of $H^1_\varphi(\Omega, S^2)$, a question which is closely related to the study of the components of $H^1(S^2, S^2)$. The way we defined a degree for $H^1(S^2, S^2)$ was with the help of an *integral formula*. Recall that if $f \in C^1(S^n, S^n)$, Kronecker's formula asserts that

$$\deg f = \int_{S^n} \det(\nabla f), \tag{2.1}$$

where $\det(\nabla f)$ denotes the $n \times n$ Jacobian determinant of f. When $n = 2$, the right-hand side of (2.1) still makes sense when f is not C^1, but merely in $H^1(S^2, S^2)$ because $\det(\nabla f) \in L^1$. We were able to prove (via a density argument) that the RHS in (2.1) belongs to $\mathbb{Z}$ and we took it as a definition of the H^1-degree of f. Similarly, one may use (2.1) to define a degree for every map $f \in W^{1,n}$. In view of the discussion in Section 1, we know that $W^{1,n} \subset \text{VMO}$ and thus any $f \in W^{1,n}(S^n, S^n)$ admits a VMO-degree in the sense of Section 1. Fortunately, the two definitions coincide. In fact, we have the following.

Lemma 2. *For every $f \in W^{1,n}(S^n, S^n)$,*

$$W^{1,n}\text{-deg } f = \text{VMO-deg } f.$$

Moreover, the components of $W^{1,n}(S^n, S^n)$ are characterized by their degree.

Using this concept of degree, we managed to prove in [12] that if φ is not a constant, then E achieves its minimum on two distinct components of $H^1_\varphi(\Omega, S^2)$. A very interesting question remains open.

Open Problem 1. *Does E admit a critical point in each component of $H^1_\varphi(\Omega, S^2)$ when φ is not a constant?*

Even the special case

$$\varphi(x, y) = (Rx, Ry, \sqrt{1 - R^2}), \quad 0 < R < 1, \quad x^2 + y^2 = 1,$$

is open.

It is also interesting to study the homotopy structure of $W^{1,p}(S^n, S^n)$ for values of $p \neq n$. This was done in my joint paper with Y. Li [14].

Theorem 3. *When $p > n$, the standard (C^0) degree of maps in $W^{1,p}$ is well defined and the components of $W^{1,p}$ are characterized by their degree. When $1 \leq p < n$, $W^{1,p}$ is path-connected.*

Following the earlier paper [15], we started to investigate with Y. Li [14] the homotopy structure of $W^{1,p}(M, N)$ when M and N are general Riemannian manifolds (M possibly with boundary, while $\partial N = \emptyset$). When $p \geq \dim M$, the homotopy structure of $W^{1,p}(M, N)$ is identical to that of $C^0(M, N)$. When $\dim M > 1$ and $1 \leq p < 2$, we proved in [14] that $W^{1,p}(M, N)$ is *always* path-connected. When p decreases from $\dim M$ to 2, the set $W^{1,p}(M, N)$ becomes larger and larger while various surprising phenomena may occur:

(a) Some homotopy classes persist below the Sobolev threshold $p = \dim M$, where maps need not belong to VMO.

(b) As p decreases, the set $W^{1,p}(M, N)$ increases, and in this process some of the homotopy classes "coalesce" as p crosses distinguished *integer* values—and usually there is a cascade of such levels where the homotopy structure undergoes "dramatic" jumps.

(c) As p decreases, new homotopy classes may "suddenly" appear at some (integral) levels; every map in these new classes must have "robust" singularities: they cannot be erased via homotopy.

We refer the interested reader to [14] and to the subsequent remarkable paper by F. B. Hang and F. H. Lin [18].

3 Degree for $H^{1/2}(S^1, S^1)$. Can one hear the degree of continuous maps?

Another important example that motivated my work with L. Nirenberg [16] was the concept of degree for maps in $H^{1/2}(S^1, S^1)$ due to L. Boutet de Monvel and O. Gabber (presented in [2, appendix]). The motivation in [2] came from a Ginzburg–Landau model arising in superconductivity. This $H^{1/2}$-degree also plays an important role in our study of the Ginzburg–Landau vortices with F. Bethuel and F. Hélein (see [1]). For example, it is at the heart of the proof of the following.

Lemma 3. *Let Ω be the unit disc in $\mathbb{R}^2$ and let φ be a smooth map from $\partial\Omega = S^1$ into S^1. Then*

$$[H^1_\varphi(\Omega, S^1) \neq \emptyset] \Leftrightarrow [\deg \varphi = 0].$$

The way Boutet de Monvel and Gabber originally defined a degree for $H^{1/2}(S^1, S^1)$ went as follows. First, observe that if $f \in C^1(S^1, \mathbb{C} \setminus \{0\})$, then the Cauchy formula asserts that

$$\deg f = \frac{1}{2i\pi} \int_{S^1} \frac{\dot{f}}{f}. \tag{3.1}$$

In particular, if $f \in C^1(S^1, S^1)$ we may write (3.1) as

$$\deg f = \frac{1}{2i\pi} \int_{S^1} \bar{f}\dot{f} = \frac{1}{2\pi} \int_{S^1} \det(f, \dot{f}) \tag{3.2}$$

(which is the simplest form of Kronecker's formula (2.1)). Then Boutet de Monvel and Gabber observed that the right-hand side of (3.2) still makes when f is not C^1, but merely in $H^{1/2}$. To do so, they interpret the RHS in (3.2) as a scalar product in the duality $H^{1/2} - H^{-1/2}(\bar{f} \in H^{1/2}, \dot{f} \in H^{-1/2})$. Using a density argument, they prove that the RHS in (3.2) belongs to $\mathbb{Z}$ and they take it as definition for the $H^{1/2}$-degree of f. On the other hand, recall (see Section 1) that $H^{1/2}(S^1) \subset \mathrm{VMO}(S^1)$. Therefore, any $f \in H^{1/2}(S^1, S^1)$ admits a VMO-degree in the sense of Section 1, and, in fact, we have the following.

Lemma 4. *For every* $f \in H^{1/2}(S^1, S^1)$,

$$H^{1/2}\text{-deg } f = \text{VMO-deg } f.$$

Lemmas 2 and 4 show the unifying character of the VMO-degree, putting various concepts of degree (for continuous maps, for $W^{1,n}(S^n, S^n)$ maps, for $H^{1/2}(S^1, S^1)$ maps, etc.) under a common roof.

In 1996, I. M. Gelfand invited me to present at his seminar the VMO-degree theory we had just developed with Louis Nirenberg. He asked me to elaborate on the special case of the $H^{1/2}(S^1, S^1)$-degree. I wrote down Gagliardo's characterization of $H^{1/2}$ which, in this special case, takes the form

$$H^{1/2}(S^1) = \left\{ f \in L^2(S^1);\ \int_{S^1}\int_{S^1} \frac{|f(x) - f(y)|^2}{|x - y|^2} dx dy < \infty \right\}.$$

Since I. M. Gelfand was not fully satisfied with Gagliardo's formulation, I also wrote down the characterization of $H^{1/2}$ in terms of the Fourier coefficients (a_n) of f:

$$H^{1/2}(S^1) = \left\{ f \in L^2(S^1);\ \sum_{n=-\infty}^{+\infty} |n||a_n|^2 < \infty \right\}$$

(see also Lemma 5 below). At that point, I. M. Gelfand asked whether there is a connection between the degree and the Fourier coefficients. At first, I was surprised by his question, but I realized shortly afterwards that if one inserts the Fourier expansion

$$f(\theta) = \sum_{n=-\infty}^{+\infty} a_n e^{in\theta}$$

into (3.2), one finds

$$\deg f = \sum_{n=-\infty}^{+\infty} n|a_n|^2. \tag{3.3}$$

Formula (3.3) is easily justified when $f \in C^1(S^1, S^1)$. The density of $C^1(S^1, S^1)$ into $H^{1/2}(S^1, S^1)$ and the stability of degree under VMO-convergence (and thus under $H^{1/2}$-convergence) yield the following.

Theorem 4. *For every* $f \in H^{1/2}(S^1, S^1)$,

$$\text{VMO-deg } f = \sum_{n=-\infty}^{+\infty} n|a_n|^2. \tag{3.4}$$

Formula (3.4) raises some intriguing questions. First, however, we present a consequence of Theorem 4.

Corollary 1. *Let (a_n) be a sequence of complex numbers satisfying*

$$\sum_{n=-\infty}^{+\infty} |n||a_n|^2 < \infty, \tag{3.5}$$

$$\sum_{n=-\infty}^{+\infty} |a_n|^2 = 1, \tag{3.6}$$

and

$$\sum_{n=-\infty}^{+\infty} a_n \bar{a}_{n+k} = 0 \quad \forall k \neq 0. \tag{3.7}$$

Then

$$\sum_{n=-\infty}^{+\infty} n|a_n|^2 \in \mathbb{Z}. \tag{3.8}$$

Proof. Set

$$f(\theta) = \sum_{n=-\infty}^{+\infty} a_n e^{in\theta},$$

so that $f \in H^{1/2}(S^1, \mathbb{C})$. Moreover, we have

$$\int_{S^1} (|f(\theta)|^2 - 1)e^{ik\theta} d\theta = 0 \quad \forall k. \tag{3.9}$$

Indeed, for $k = 0$, (3.9) follows from (3.6), and for $k \neq 0$, (3.9) follows from (3.7). Thus we obtain

$$|f(\theta)| = 1 \quad \text{a.e.} \tag{3.10}$$

Applying Theorem 4, we find (3.8). □

Pedagogical Question. Is there an elementary proof of Corollary 1 that does not rely on Theorem 4?

Suppose now $f \in C^0(S^1, S^1)$ and $f \notin H^{1/2}$. Then the series

$$\sum_{n=-\infty}^{+\infty} |n||a_n|^2$$

is divergent. The LHS in (3.4) is well defined, but the RHS is not. It is natural to ask whether $\deg f$ may still be computed as a "principal value" of the series $\sum_{n=-\infty}^{+\infty} n|a_n|^2$ (which is not absolutely convergent). In [11] we raised the question of whether standard summation processes can be used to compute the degree of a general $f \in C^0(S^1, S^1)$. Let, for example,

$$\sigma_N = \sum_{n=-N}^{+N} n|a_n|^2$$

or

$$P_r = \sum_{n=-\infty}^{+\infty} n|a_n|^2 r^{|n|}, \quad 0 < r < 1.$$

Is it true that, for any $f \in C^0(S^1, S^1)$,

$$\deg f = \lim_{N\to+\infty} \sigma_N \quad \text{or} \quad \deg f = \lim_{r\downarrow 1} P_r?$$

J. Korevaar [20] has shown that the answer is negative. He has constructed interesting examples of maps $f \in C^0(S^1, S^1)$ of degree 0 such that σ_N (respectively, P_r) need not have a limit as $N \to \infty$ (respectively, $r \to 1$) or may converge to any given real number $\lambda \neq 0$, including $\pm\infty$. In view of this fact, we now propose a more "modest" question: Do the absolute values of the Fourier coefficients determine the degree? More precisely, we have the following.

Open Problem 2 (Can one hear the degree of continuous maps?). *Let $f, g \in C^0(S^1, S^1)$ and let (a_n), (b_n) denote the Fourier coefficients of f and g, respectively. Assume that*

$$|a_n| = |b_n| \quad \forall n \in \mathbb{Z}. \tag{3.11}$$

Can one conclude that

$$\deg f = \deg g?$$

Answer the same question if one assumes only that $f, g \in \mathrm{VMO}(S^1, S^1)$.

Of course, the answer to Open Problem 2 is positive if, in addition, $f, g \in H^{1/2}(S^1, S^1)$. This is a consequence of Theorem 4. The answer is still positive in a class of functions strictly larger than $H^{1/2}$. The proof is based on the following.

Theorem 5. *For every $f \in W^{1/3,3}(S^1, S^1)$, we have*

$$\text{VMO-deg}\, f = \lim_{\varepsilon\downarrow 0} \frac{1}{\varepsilon^2} \sum_{\substack{n\in\mathbb{Z}\\ n\neq 0}} |a_n|^2 \frac{\sin^2 n\varepsilon}{n}. \tag{3.12}$$

Corollary 2. *Assume that $f, g \in W^{1/3,3}(S^1, S^1)$ satisfy* (3.11). *Then*

$$\text{VMO-deg}\, f = \text{VMO-deg}\, g.$$

Corollary 3 (J. P. Kahane [19]). *Assume that $f, g \in C^{0,\alpha}(S^1, S^1)$, with $\alpha > 1/3$, satisfy* (3.11). *Then*

$$\deg f = \deg g.$$

Note that $C^{0,\alpha} \subset W^{1/3,3}$ $\forall \alpha > 1/3$. (This is an obvious consequence of Gagliardo's characterization (1.1)). Thus Corollary 2 implies Corollary 3. Our proof of Theorem 5 is a straightforward adaptation of the ingenious argument of J. P. Kahane [19] for $C^{0,\alpha}$, $\alpha > 1/3$.

Remark 2. The conclusion of Theorem 5 holds if $f \in W^{1/p,p}(S^1, S^1)$ with $1 < p \leq 3$ (since $W^{1/p,p} \cap L^\infty \subset W^{1/3,3}$ $\forall p \leq 3$). (Note that when $1 < p \leq 2$ the conclusion of Theorem 5 is an immediate consequence of Theorem 4 since $\sum |n||a_n|^2 < \infty$. However, in the range $2 < p \leq 3$, the conclusion is far from obvious since the series $\sum |n||a_n|^2$ may be divergent.) It is interesting to point out that formula (3.12) *fails* if one assumes only $f \in W^{1/p,p}(S^1, S^1)$ with $p > 3$. In fact, J. P. Kahane [19] has constructed an example of a function $f \in C^{0,1/3}(S^1, S^1)$ such that $\deg f = 0$ while

$$\lim_{\varepsilon \downarrow 0} \frac{1}{\varepsilon^2} \sum_{\substack{n \in \mathbb{Z} \\ n \neq 0}} |a_n|^2 \frac{\sin^2 n\varepsilon}{n} = \lambda,$$

where λ could be any real number $\lambda \neq 0$. The heart of the matter is the existence of a 2π-periodic function $\varphi \in C^{0,1/3}(\mathbb{R}, \mathbb{R})$ such that

$$\int_0^{2\pi} (\varphi(\theta + h) - \varphi(\theta))^3 d\theta = \sin h \quad \forall h.$$

This still leaves open the question whether Corollary 2 holds when $W^{1/3,3}$ is replaced by $W^{1/p,p}$, $p > 3$.

Taking $p \to 1$ in Remark 2 suggests that Theorem 5 holds for $f \in W^{1,1}$. This is indeed true, and there is even a stronger statement.

Theorem 6. *For every* $f \in C^0(S^1, S^1) \cap \mathrm{BV}(S^1, S^1)$, *we have*

$$\deg f = \lim_{\varepsilon \downarrow 0} \frac{1}{\varepsilon} \sum_{n=-\infty}^{+\infty} |a_n|^2 \sin n\varepsilon.$$

Consequently, we also have the following.

Corollary 4. *Assume* $f, g \in C^0(S^1, S^1) \cap \mathrm{BV}(S^1, S^1)$ *satisfy* (3.11). *Then*

$$\deg f = \deg g.$$

Remark 3. It was already observed by J. Korevaar in [20] that for every $f \in C^0 \cap \mathrm{BV}$, one has

$$\deg f = \lim_{N \to \infty} \sum_{n=-N}^{+N} n|a_n|^2,$$

which also implies Corollary 4.

Proof of Theorem 5. We follow the argument of J. P. Kahane [19], except that we work in the fractional Sobolev space $W^{1/3,3}$ instead of the smaller Hölder space $C^{0,\alpha}$, $\alpha > 1/3$. Set

$$d = \text{VMO-deg } f.$$

By [16, Theorem 3 (and Remark 10)], we may write

$$f(\theta) = e^{i(\varphi(\theta)+d\theta)}$$

for some $\varphi \in \text{VMO}(S^1, \mathbb{R})$. Applying [3, Theorem 1] and the uniqueness of the lifting in VMO, we know that $\varphi \in W^{1/3,3}$.

Write

$$\int_0^{2\pi} f(\theta+h)\bar{f}(\theta)d\theta = 2\pi \sum_{n=-\infty}^{+\infty} |a_n|^2 e^{inh} = \int_0^{2\pi} e^{idh} e^{i(\varphi(\theta+h)-\varphi(\theta))} d\theta, \tag{3.13}$$

$$e^{idh} = 1 + idh + O(|h|^2), \tag{3.14}$$

and

$$\begin{aligned} e^{i(\varphi(\theta+h)-\varphi(\theta))} &= 1 + i(\varphi(\theta+h)-\varphi(\theta)) - \frac{1}{2}(\varphi(\theta+h)-\varphi(\theta))^2 \\ &\quad + O(|\varphi(\theta+h)-\varphi(\theta)|^3). \end{aligned} \tag{3.15}$$

Thus

$$\begin{aligned} \text{Im}[e^{idh}e^{i(\varphi(\theta+h)-\varphi(\theta))}] &= \text{Im}[(1+idh)e^{i(\varphi(\theta+h)-\varphi(\theta))}] + O(|h|^2) \\ &= (\varphi(\theta+h)-\varphi(\theta)) + dh + O|h|^2) \\ &\quad + O(|h||\varphi(\theta+h)-\varphi(\theta)|^2) \\ &\quad + O(|\varphi(\theta+h)-\varphi(\theta)|^3). \end{aligned} \tag{3.16}$$

Integrating (3.16) with respect to θ yields

$$\left| \sum_{n=-\infty}^{+\infty} |a_n|^2 \sin nh - dh \right| \le C|h|^2 + C\int_0^{2\pi} |\varphi(\theta+h)-\varphi(\theta)|^3 d\theta. \tag{3.17}$$

Next, integrating (3.17) with respect to h on $(0, 2\varepsilon)$ gives

$$\begin{aligned} &\left| \sum_{\substack{n\in\mathbb{Z} \\ n\neq 0}} |a_n|^2 \left(\frac{1-\cos 2n\varepsilon}{n}\right) - 2d\varepsilon^2 \right| \\ &\qquad \le C\varepsilon^3 + C\int_0^{2\varepsilon} dh \int_0^{2\pi} |\varphi(\theta+h)-\varphi(\theta)|^3 d\theta \end{aligned}$$

and therefore

$$\begin{aligned}&\left|\frac{1}{\varepsilon^2}\sum_{\substack{n\in\mathbb{Z}\\ n\neq 0}}|a_n|^2\frac{\sin^2 n\varepsilon}{n}-d\right|\\ &\quad\le C\varepsilon+\frac{C}{\varepsilon^2}\int_0^{2\varepsilon}\int_0^{2\pi}|\varphi(\theta+h)-\varphi(\theta)|^3 dhd\theta\\ &\quad\le C\varepsilon+C\int_0^{2\varepsilon}\int_0^{2\pi}\frac{|\varphi(\theta+h)-\varphi(\theta)|^3}{|h|^2}dhd\theta,\end{aligned}\tag{3.18}$$

which implies (3.12) since $\varphi \in W^{1/3,3}$. □

Proof of Theorem 6. Since $f \in C^0 \cap \mathrm{BV}$, the corresponding φ satisfies $\varphi \in C^0 \cap \mathrm{BV}$. We return to (3.17) with $h = \varepsilon$,

$$\left|\frac{1}{\varepsilon}\sum_{h=-\infty}^{+\infty}|a_n|^2\sin n\varepsilon-d\right|\le C\varepsilon+\frac{C}{\varepsilon}\int_0^{2\pi}|\varphi(\theta+\varepsilon)-\varphi(\theta)|^3 d\theta. \tag{3.19}$$

Next, we have

$$\int_0^{2\pi}|\varphi(\theta+\varepsilon)-\varphi(\theta)|d\theta\le\varepsilon\|\varphi\|_{\mathrm{BV}}. \tag{3.20}$$

Inserting (3.2) into (3.19) gives

$$\left|\frac{1}{\varepsilon}\sum_{n=-\infty}^{+\infty}|a_n|^2\sin n\varepsilon-d\right|\le C\varepsilon+C\,\mathop{\mathrm{Sup}}_{\theta}\|\varphi(\theta+\varepsilon)-\varphi(\theta)\|^2_{L^\infty}, \tag{3.21}$$

and the conclusion follows since $\varphi \in C^0$. □

Remark 4. It has been pointed out to me by J. P. Kahane that a slightly stronger conclusion holds in Theorem 5.

Theorem 5′. *For every $f \in W^{1/3,3}(S^1, S^1)$, we have*

$$\text{VMO-deg}\, f=\lim_{\varepsilon\downarrow 0}\frac{1}{\varepsilon}\sum_{n=-\infty}^{+\infty}|a_n|^2\sin n\varepsilon. \tag{3.22}$$

Proof. Returning to (3.17), it suffices to verify that

$$\lim_{h\downarrow 0}\frac{1}{h}\int_0^{2\pi}|\varphi(\theta+h)-\varphi(\theta)|^3 d\theta=0. \tag{3.23}$$

Set

$$I(t)=\int_0^{2\pi}|\varphi(\theta+t)-\varphi(\theta)|^3 d\theta$$

so that

$$I^{1/3}(t_1 + t_2) \leq I^{1/3}(t_1) + I^{1/3}(t_2).$$

Thus

$$I(t_1 + t_2) \leq 4(I(t_1) + I(t_2)).$$

Consequently,

$$I(h) \leq \frac{8}{h} \int_{h/2}^{h} (I(s) + I(h-s))ds = \frac{8}{h} \int_0^h I(s)ds \leq 8h \int_0^h \frac{I(s)}{s^2} ds.$$

Since $\varphi \in W^{1/3,3}$, we know that

$$\int_0^{2\pi} \frac{I(s)}{s^2} ds < \infty$$

and (3.23) follows.

4 New estimates for the degree

Going back to (3.3), we see that for every $f \in C^1(S^1, S^1)$,

$$|\deg f| \leq \sum |n||a_n|^2. \tag{4.1}$$

Combining (4.1) with Gagliardo's characterization (1.1) of $H^{1/2}$, we find that

$$|\deg f| \leq C \int_{S^1} \int_{S^1} \frac{|f(x) - f(y)|^2}{|x-y|^2} dx dy. \tag{4.2}$$

In fact, the sharp estimate

$$|\deg f| \leq \frac{1}{4\pi^2} \int_{S^1} \int_{S^1} \frac{|f(x) - f(y)|^2}{|x-y|^2} dx dy \tag{4.3}$$

is an immediate consequence of (4.1) and the following.

Lemma 5. *For every $f \in H^{1/2}$, one has*

$$\int_{S^1} \int_{S^1} \frac{|f(x) - f(x)|^2}{|x-y|^2} dx dy = 4\pi^2 \sum_{n=-\infty}^{+\infty} |n||a_n|^2. \tag{4.4}$$

Proof. Write

$$\int_{S^1} \int_{S^1} \frac{|f(x) - f(y)|^2}{|x-y|^2} dx dy = \int_0^{2\pi} \int_0^{2\pi} \frac{|\sum a_n e^{in\theta} - \sum a_n e^{in\psi}|^2}{|e^{i\theta} - e^{i\psi}|^2} d\theta d\psi$$

$$= \int_0^{2\pi} \frac{d\gamma}{|e^{i\gamma} - 1|^2} \int_0^{2\pi} \left| \sum a_n (1 - e^{in\gamma}) e^{in\theta} \right|^2 d\theta$$
$$= 2\pi \sum |a_n|^2 \int_0^{2\pi} \frac{|e^{in\gamma} - 1|^2}{|e^{i\gamma} - 1|^2} d\gamma.$$

However, for $|n| \geq 1$,

$$\frac{|e^{in\gamma} - 1|^2}{|e^{i\gamma} - 1|^2} = (e^{i(n-1)\gamma} + \cdots + 1)(e^{-i(n-1)\gamma} + \cdots + 1),$$

and thus

$$\int_0^{2\pi} \frac{|e^{in\gamma} - 1|^2}{|e^{i\gamma} - 1|^2} d\gamma = 2\pi |n|.$$

Inserting this into the previous equality yields (4.4). □

Remark 5. Inequality (4.3) can be viewed as an estimate for the "least amount of $H^{1/2}$-energy" necessary to produce a map $f : S^1 \to S^1$ with assigned degree. More precisely, we have

$$\operatorname*{Inf}_{\substack{f:S^1 \to S^1 \\ \deg f = n}} \int_{S^1} \int_{S^1} \frac{|f(x) - f(y)|^2}{|x - y|^2} dx\, dy = 4\pi^2 |n|, \tag{4.5}$$

and the Inf in (4.5) is achieved when $f(\theta) = e^{in\theta}$. The existence of a minimizer for similar problems where the standard $H^{1/2}$ norm is replaced by equivalent norms (e.g., the trace of an H^1 norm on the disc with variable coefficients) is a very delicate question because of "lack of compactness"; we refer to [21].

Remark 6. Estimate (4.2) serves as a building block in the study of the least $H^{1/2}$-energy of maps $u : S^2 \to S^1$ with prescribed singularities. Such a question has been investigated in [5]. More precisely, recall that

$$\|u\|^2_{H^{1/2}(S^2)} = \int_{S^2} \int_{S^2} \frac{|u(x) - u(y)|^2}{|x - y|^3} dx dy.$$

Given points $\Sigma = \{p_1, p_2, \ldots, p_k\} \cup \{n_1, n_2, \ldots, n_k\}$, consider the class of maps

$$A = \{u \in C^1(S^2 \setminus \Sigma, S^1); \deg(u, p_i) = +1 \text{ and } \deg(u, n_i) = -1 \ \forall i\}.$$

Theorem 7 (Bourgain–Brezis–Mironescu [5]). *There exist absolute constants C_1, $C_2 > 0$ such that*

$$C_1 L(\Sigma) \leq \operatorname*{Inf}_{u \in A} \|u\|^2_{H^{1/2}(S^2)} \leq C_2 L(\Sigma), \tag{4.6}$$

where $L(\Sigma)$ is the length of a minimal connection connecting the points (p_i) to the points (n_i).

Theorem 7 is the $H^{1/2}$-version of an earlier result [13] concerning H^1 maps from S^3 into S^2 with singularities that had been motivated by questions arising in liquid crystals with point defects, while the analysis in [5] has its source in the Ginzburg–Landau model for superconductors. It is the LHS inequality in (4.6), which is related to (4.2). The RHS inequality in (4.6) comes from a "brute force" construction called the "dipole construction."

Remark 7. An immediate consequence of (4.3) is the estimate

$$|\deg f| \leq \frac{1}{2\pi^2}\int_{S^1}\int_{S^1}\frac{|f(x)-f(y)|^p}{|x-y|^2} \quad \forall f \in C^1(S^1,S^1), \quad \forall p \in (1,2). \quad (4.7)$$

Estimate (4.7) deteriorates as $p \downarrow 1$ since the RHS in (4.7) tends to $+\infty$ unless f is constant (see [4]). It would be desirable to improve the constant $(1/2\pi^2)$ and establish that

$$|\deg f| \leq C_p \int_{S^1}\int_{S^1}\frac{|f(x)-f(y)|^p}{|x-y|^2}dxdy \quad \forall f \in C^1(S^1,S^1), \quad \forall p \in (1,2). \quad (4.8)$$

with a constant $C_p \sim (p-1)$ as $p \downarrow 1$. In the limit as $p \downarrow 1$, one should be able to recover (in the spirit of [4]) the obvious inequality

$$|\deg f| \leq \frac{1}{2\pi}\int |\dot{f}|. \quad (4.9)$$

Inequality (4.8) is also valid for $p > 2$, but it cannot be deduced from (4.3) and its proof requires much work.

Theorem 8 (Bourgain–Brezis–Mironescu [6]). *For every $p > 1$, there is a constant C_p such that for any (smooth) $f : S^1 \to S^1$,*

$$|\deg f| \leq C_p \int_{S^1}\int_{S^1}\frac{|f(x)-f(y)|^p}{|x-y|^2} = C_p\|f\|^p_{W^{1/p,p}}. \quad (4.10)$$

There is an estimate stronger than (4.10).

Theorem 9 (Bourgain–Brezis–Mironescu [7]). *For any $\delta > 0$ sufficiently small, there is a constant C_δ such that, $\forall f \in C^0(S^1,S^1)$,*

$$|\deg f| \leq C_\delta \int_{S^1}\int_{S^1}\limits_{[|f(x)-f(y)|>\delta]}\frac{1}{|x-y|^2}dxdy. \quad (4.11)$$

Remark 8. In Bourgain–Brezis–Nguyen [8], it was proved that (4.11) holds for any $\delta < 2^{1/2}$. This was later improved by H.-M. Nguyen [22], who established the bound

$$|\deg f| \leq C \int_{S^1}\int_{S^1}\limits_{[|f(x)-f(y)|>3^{1/2}]}\frac{1}{|x-y|^2}dxdy; \quad (4.12)$$

Nguyen [22] has also constructed examples showing that (4.11) fails for any $\delta > 3^{1/2}$.

Open Problem 3. *What is the behavior of the best constant C_δ in (4.11) as $\delta \downarrow 0$? Is there a more precise estimate of the form*

$$|\deg f| \leq C\delta \int_{S^1}\int_{S^1}\limits_{[|f(x)-f(y)|>\delta]} \frac{1}{|x-y|^2} dxdy \tag{4.13}$$

with C independent of δ, for all $\delta < 3^{1/2}$?

In the spirit of [4], one might then be able to recover (4.9) as $\delta \to 0$.

Higher-dimensional analogues

Theorem 9 can be extended to higher dimensions.

Theorem 10 (Bourgain–Brezis–Mironescu [8]). *Let $n \geq 1$. For any $\delta \in (0, 2^{1/2})$, there is a constant C_δ such that $\forall f \in C^0(S^n, S^n)$,*

$$|\deg f| \leq C \int_{S^n}\int_{S^n}\limits_{[|f(x)-f(y)|>\delta]} \frac{1}{|x-y|^{2n}} dxdy. \tag{4.14}$$

A more refined version of Theorem 10 was obtained by H.-M. Nguyen [22]. He proved that (4.14) holds for any $\delta < [2 + 2/(n+1)]^{1/2}$ and that this range of δs is optimal for all dimensions n. From Theorem 10 we may, of course, recover the earlier estimate of Bourgain–Brezis–Mironescu [6]: $\forall n \geq 1$, $\forall p > n$, $\forall f \in W^{n/p,p}(S^n, S^n)$,

$$|\deg f| \leq C(p,n) \int_{S^n}\int_{S^n} \frac{|f(x)-f(y)|^p}{|x-y|^{2n}} dxdy = C(p,n)\|f\|^p_{W^{n/p,p}}. \tag{4.15?}$$

In a different direction, it might be interesting to estimate other topological invariants in terms of fractional Sobolev norms. One of the simplest examples could be the following.

Open Problem 4. *Does one have*

$$|\,\text{Hopf-degree } f| \leq C_p \int_{S^3}\int_{S^3} \frac{|f(x)-f(y)|^p}{|x-y|^6} \quad \forall p > 3, \quad \forall f \in C^1(S^3, S^2)?$$

Acknowledgments. I am very grateful to J. P. Kahane for enlightening discussions which have led me to Theorem 5 (and Corollary 2). Special thanks to H. Furstenberg and P. Mironescu for useful conversations. This work was partially supported by an EC grant through the RTN program "Front-Singularities," HPRN-CT-2002-00274.

References

[1] F. Bethuel, H. Brezis, and F. Hélein, *Ginzburg–Landau Vortices*, Birkhäuser Boston, Cambridge, MA, 1994.

[2] A. Boutet de Monvel-Berthier, V. Georgescu, and R. Purice, A boundary value problem related to the Ginzburg-Landau model, *Comm. Math. Phys*, **141** (1991), 1–23.

[3] J. Bourgain, H. Brezis, and P. Mironescu, Lifting in Sobolev spaces, *J. Anal. Math.*, **80** (2000), 37–86.

[4] J. Bourgain, H. Brezis, and P. Mironescu, Another look at Sobolev spaces, in J. L. Menaldi, E. Rofman, and A. Sulem, eds., *Optimal Control and Partial Differential Equations: A Volume in Honour of A. Bensoussan's 60th Birthday*, IOS Press, Amsterdam, 2001, 439–455.

[5] J. Bourgain, H. Brezis, and P. Mironescu, $H^{1/2}$ maps into the circle: Minimal connections, lifting, and the Ginzburg–Landau equation, *Publ. Math. IHES*, **99** (2004), 1–115.

[6] J. Bourgain, H. Brezis, and P. Mironescu, Lifting, degree and distributional Jacobian revisited, *Comm. Pure Appl. Math.*, **58**-4 (2005), 529–551.

[7] J. Bourgain, H. Brezis, and P. Mironescu, Complements to the paper "Lifting, degree and distributional Jacobian revisited," available online from http://www.ann.jussieu.fr/publications.

[8] J. Bourgain, H. Brezis, and H.-M. Nguyen, A new estimate for the topological degree, *C. R. Acad. Sci. Paris*, **340** (2005), 787–791.

[9] H. Brezis, Large harmonic maps in two dimensions, in A. Marino, L. Modica, S. Spagnolo, and M. Degiovanni, eds., *Nonlinear Variational Problems* (*Isola d'Elba*, 1983), Research Notes in Mathematics 127, Pitman, Boston, 1985, 33–46.

[10] H. Brezis, Metastable harmonic maps, in S. Antman, J. Ericksen, D. Kinderlehrer, and I. Müller, eds., *Metastability and Incompletely Posed Problems*, IMA Volumes in Mathematics and Its Applications 3, Springer-Verlag, Berlin, New York, Heidelberg, 1987, 35–42.

[11] H. Brezis, Degree theory: Old and new, in M. Matzeu and A. Vignoli, eds., *Topological Nonlinear Analysis* II: *Degree, Singularity and Variations*, Birkhäuser Boston, Cambridge, MA, 1997, 87–108.

[12] H. Brezis and J. M. Coron, Large solutions for harmonic maps in two dimensions, *Comm. Math. Phys.*, **92** (1983), 203–215.

[13] H. Brezis, J. M. Coron, and E. Lieb, Harmonic maps with defects, *Comm. Math. Phys.*, **107** (1986), 649–705.

[14] H. Brezis and Y. Li, Topology and Sobolev spaces, *J. Functional Anal.*, **183** (2001), 321–369.

[15] H. Brezis, Y. Li, P. Mironescu, and L. Nirenberg, Degree and Sobolev spaces, *Topological Methods Nonlinear Anal.*, **13** (1999), 181–190.

[16] H. Brezis and L. Nirenberg, Degree theory and BMO, Part I: Compact manifolds without boundaries, *Selecta Math.*, **1** (1995), 197–263.

[17] M. Giaquinta and S. Hildebrandt, A priori estimates for harmonic mappings, *J. Reine Angew. Math.*, **336** (1982), 124–164.

[18] F.-B. Hang and F.-H. Lin, Topology of Sobolev mappings II, *Acta Math.*, **191** (2003), 55–107.

[19] J. P. Kahane, Sur l'équation fonctionnelle $\int_{\mathbb{T}}(\psi(t+s)-\psi(s))^3 ds = \sin t$, *C. R. Acad. Sci. Paris*, to appear.

[20] J. Korevaar, On a question of Brezis and Nirenberg concerning the degree of circle maps, *Selecta Math.*, **5** (1999), 107–122.

[21] P. Mironescu and A. Pisante, A variational problem with lack of compactness for $H^{1/2}(S^1, S^1)$ maps of prescribed degree, *J. Functional Anal.*, **217** (2004), 249–279.

[22] H.-M. Nguyen, Optimal constant in a new estimate for the degree, to appear.

Quantum Cobordisms and Formal Group Laws

Tom Coates[1] and Alexander Givental[2]

[1] Department of Mathematics
Harvard University
Cambridge, MA 02138
USA
tomc@math.harvard.edu

[2] Department of Mathematics
University of California at Berkeley
Berkeley, CA 94720
USA
givental@math.berkeley.edu

To I. M. Gelfand, who teaches us the unity of mathematics.

Subject Classifications: 14D20, 55N22, 14C40, 14L05

The present paper is closely based on the lecture given by the second author at *The Unity of Mathematics* symposium and is based on our joint work in progress on Gromov–Witten invariants with values in complex cobordisms. We will mostly consider here only the simplest example, elucidating one of the key aspects of the theory. We refer the reader to [9] for a more comprehensive survey of the subject and to [5] for all further details. Consider $\overline{\mathcal{M}}_{0,n}$, $n \geq 3$, the Deligne–Mumford compactification of the moduli space of configurations of n distinct ordered points on the Riemann sphere $\mathbf{C}P^1$. Obviously, $\overline{\mathcal{M}}_{0,3} = \text{pt}$, $\overline{\mathcal{M}}_{0,4} = \mathbf{C}P^1$, while $\overline{\mathcal{M}}_{0,5}$ is known to be isomorphic to $\mathbf{C}P^2$ blown up at four points. In general, $\overline{\mathcal{M}}_{0,n}$ is a compact complex manifold of dimension $n-3$, and it makes sense to ask what is the complex cobordism class of this manifold. The Thom complex cobordism ring, after tensoring with $\mathbf{Q}$, is known to be isomorphic to $U^* = \mathbf{Q}[\mathbf{C}P^1, \mathbf{C}P^2, \ldots]$, the polynomial algebra with generators $\mathbf{C}P^k$ of degree $-2k$. Thus our question is to express $\overline{\mathcal{M}}_{0,n}$, modulo the relation of complex cobordism, as a polynomial in complex projective spaces.

This problem can be generalized in the following three directions.

First, one can develop intersection theory for complex cobordism classes from the complex cobordism ring $U^*(\overline{\mathcal{M}}_{0,n})$. Such intersection numbers take values in the coefficient algebra $U^* = U^*(\text{pt})$ of complex cobordism theory.

Second, one can consider the Deligne–Mumford moduli spaces $\overline{\mathcal{M}}_{g,n}$ of stable n-pointed genus-g complex curves. They are known to be compact complex *orbifolds*, and for an orbifold, one can mimic (as explained below) cobordism-valued intersection theory using cohomological intersection theory over $\mathbf{Q}$ against a certain characteristic class of the tangent orbibundle.

Third, one can introduce [12, 3] more general moduli spaces $\overline{\mathcal{M}}_{g,n}(X,d)$ of degree-d *stable maps* from n-pointed genus-g complex curves to a compact Kähler (or almost-Kähler) target manifold X. One defines Gromov–Witten invariants of X using *virtual intersection theory* in these spaces (see [2, 8, 13, 15, 16]). Furthermore, using *virtual tangent bundles* of the moduli spaces of stable maps and their characteristic classes, one can extend Gromov–Witten invariants to take values in the cobordism ring U^*.

The *Quantum Hirzebruch–Riemann–Roch Theorem* (see [5]) expresses cobordism-valued Gromov–Witten invariants of X in terms of cohomological ones. Cobordism-valued intersection theory in Deligne–Mumford spaces is included as the special case $X = \mathrm{pt}$. In these notes, we will mostly be concerned with this special case, and with curves of genus zero, i.e., with cobordism-valued intersection theory in the manifolds $\overline{\mathcal{M}}_{0,n}$. The cobordism classes of $\overline{\mathcal{M}}_{0,n}$ that we seek are then interpreted as the self-intersections of the fundamental classes.

1 Cohomological intersection theory on $\overline{\mathcal{M}}_{0,n}$

Let L_i, $i = 1, \dots, n$, denote the line bundle over $\overline{\mathcal{M}}_{0,n}$ formed by the cotangent lines to the complex curves at the marked point with the index i. More precisely, consider the *forgetful map* $\mathrm{ft}_{n+1} : \overline{\mathcal{M}}_{0,n+1} \to \overline{\mathcal{M}}_{0,n}$ defined by forgetting the last marked point. Let a point $p \in \overline{\mathcal{M}}_{0,n}$ be represented by a stable genus-zero complex curve Σ equipped with the marked points $\sigma_1, \dots, \sigma_n$. Then the fiber $\mathrm{ft}_{n+1}^{-1}(p)$ can be canonically identified with Σ. In particular, the map ft_{n+1} has n canonical sections σ_i defined by the marked points, and the diagram formed by the forgetful map and the sections can be considered as the universal family of stable n-pointed curves of genus zero. The L_is are defined as the conormal bundles to the sections $\sigma_i : \overline{\mathcal{M}}_{0,n} \to \overline{\mathcal{M}}_{0,n+1}$ and are often called *universal cotangent lines* at the marked points.

Put $\psi_i = c_1(L_i)$ and define the *correlator* $\langle \psi_1^{k_1}, \dots, \psi_n^{k_n} \rangle_{0,n}$ to be the intersection index $\int_{\overline{\mathcal{M}}_{0,n}} \psi_1^{k_1} \dots \psi_n^{k_n}$. These correlators are not too hard to compute (see [17, 11]). Moreover, it turns out that intersection theory in the spaces $\overline{\mathcal{M}}_{0,n}$ is governed by the "universal monotone function." Namely, one introduces the *genus-0 potential*

$$\mathcal{F}_0(t_0, t_1, t_2, \dots) := \sum_{n \geq 3} \sum_{k_1, \dots, k_n \geq 0} \langle \psi_1^{k_1}, \dots, \psi_n^{k_n} \rangle_{0,n} \frac{t_{k_1} \dots t_{k_n}}{n!}.$$

It is a formal function of $t_0, t_1, t_2, \dots$ whose Taylor coefficients are the correlators. Then

$$\mathcal{F}_0 = \text{critical value of } \left\{ \frac{1}{2} \int_0^\tau Q^2(x)dx \right\}, \tag{1}$$

where

$$Q(x) := -x + t_0 + t_1 x + \cdots + t_n \frac{x^n}{n!} + \cdots.$$

More precisely, the "monotone function" of the variable τ depending on the parameters $t_0, t_1, \ldots$ has the critical point $\tau = \tau(t_0, t_1, \ldots)$. It can be computed as a formal function of the t_is from the relation $Q(\tau) = 0$ which has the form of the "universal fixed-point equation":

$$\tau = t_0 + t_1\tau + \cdots + t_n\frac{\tau^n}{n!} + \cdots .$$

Then termwise integration in (1) yields the critical value as a formal function of the t_is which is claimed to coincide with $\mathcal{F}_0$.

The formula (1) describing $\mathcal{F}_0$ can be easily derived by the method of characteristics applied to the following PDE (called the *string equation*):

$$\partial_0\mathcal{F}_0 - t_1\partial_{t_0}\mathcal{F}_0 - t_2\partial_{t_1}\mathcal{F}_0 - \cdots = t_0^2/2.$$

The initial condition $\mathcal{F}_0 = 0$ at $t_0 = 0$ holds for dimensional reasons: $\dim_{\mathbf{R}} \overline{\mathcal{M}}_{0,n} < 2n = \deg(\psi_1 \ldots \psi_n)$. The string equation itself expresses the well-known fact that for $i = 1, \ldots, n$, the class ψ_i on $\overline{\mathcal{M}}_{0,n+1}$ does not coincide with the pull-back $\mathrm{ft}_{n+1}^*(\psi_i)$ of its counterpart from $\overline{\mathcal{M}}_{0,n}$ but differs from it by the class of the divisor $\sigma_i(\overline{\mathcal{M}}_{0,n})$.

The family of "monotone functions" in (1) can alternatively be viewed as a family of quadratic forms in Q depending on one parameter τ. This leads to the following description of $\mathcal{F}_0$ in terms of linear symplectic geometry.

Let $\mathcal{H}$ denote the space of Laurent series $\mathbf{Q}((z^{-1}))$ in one indeterminate z^{-1}. Given two such Laurent series $f, g \in \mathcal{H}$, we put

$$\Omega(f, g) := \frac{1}{2\pi i}\oint f(-z)g(z)dz = -\Omega(g, f).$$

This pairing is a symplectic form on $\mathcal{H}$, and

$$f = \cdots - p_2z^{-3} + p_1z^{-2} - p_0z^{-1} + q_0z^0 + q_1z^2 + q_2z^2 + \cdots$$

is a Darboux coordinate system.

The subspaces $\mathcal{H}_+ := \mathbf{Q}[z]$ and $\mathcal{H}_- := z^{-1}\mathbf{Q}[[z^{-1}]]$ form a Lagrangian polarization of $(\mathcal{H}, \Omega)$ and identify the symplectic space with $T^*\mathcal{H}_+$. Next, we consider $\mathcal{F}_0$ as a formal function on $\mathcal{H}_+$ near the shifted origin $-z$ by putting

$$q_0 + q_1z + \cdots + q_nz^n + \cdots = -z + t_0 + t_1z + \cdots + t_nz^n + \cdots . \tag{2}$$

This convention—called the *dilaton shift*—makes many ingredients of the theory homogeneous, as is illustrated by the following examples: the vector field on the LHS of the string equation is $-\sum q_{k+1}\partial_{q_k}$; $Q(x)$ in (1) becomes $\sum q_kx^k/k!$; $\mathcal{F}_0$ becomes homogeneous of degree 2, i.e., $\sum q_k\partial_{q_k}\mathcal{F}_0 = 2\mathcal{F}_0$. Furthermore, we define a (formal germ of a) Lagrangian submanifold $\mathcal{L} \subset \mathcal{H}$ as the graph of the differential of $\mathcal{F}_0$:

$$\mathcal{L} = \{(\mathbf{p}, \mathbf{q}) \mid \mathbf{p} = d_{z+\mathbf{q}}\mathcal{F}_0\} \subset \mathcal{H} \cong T^*\mathcal{H}_+.$$

Then $\mathcal{L} \subset \mathcal{H}$ is the Lagrangian cone

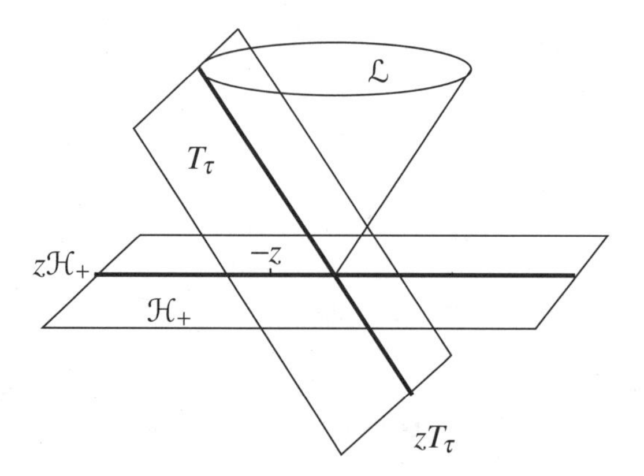

Fig. 1. The Lagrangian cone $\mathcal{L}$.

$$\mathcal{L} = \cup_\tau e^{\tau/z} z\mathcal{H}_+ = \{ze^{\tau/z}\mathbf{q}(z) \mid \mathbf{q} \in \mathbf{Q}[z], \tau \in \mathbf{Q}\}.$$

We make a few remarks about this formula. The tangent spaces to the cone $\mathcal{L}$ form a one-parameter family $T_\tau = e^{\tau/z}\mathcal{H}_+$ of semi-infinite Lagrangian subspaces. These are graphs of the differentials of the quadratic forms from (1). The subspaces T_τ are invariant under multiplication by z and form a *variation of semi-infinite Hodge structures* in the sense of S. Barannikov [1]. The graph $\mathcal{L}$ of $d\mathcal{F}_0$ is therefore the envelope to such a variation. Moreover, *the tangent spaces T_τ of $\mathcal{L}$ are tangent to $\mathcal{L}$ exactly along zT_τ*. All these facts, properly generalized, remain true in genus-zero Gromov–Witten theory with a nontrivial target space X (see [9]).

2 Complex cobordism theory

Complex cobordism is an extraordinary cohomology theory $U^*(\cdot)$ defined in terms of homotopy classes Π of maps to the spectrum $MU(k)$ of the Thom spaces of universal $U_{k/2}$-bundles:

$$U^n(B) = \lim_{k\to\infty} \Pi(\Sigma^k B, MU(n+k)).$$

The dual homology theory, called *bordism*, can be described geometrically as

$$U_n(B) := \{\text{maps } Z^n \to B\} \,/\, \text{bordism},$$

where Z^n is a compact stably almost complex manifold of real dimension n, and the bordism manifold should carry a stably almost complex structure compatible in the obvious sense with that of the boundary. See Figure 2.

When B itself is a compact stably almost complex manifold of real dimension m, the celebrated Pontryagin–Thom construction identifies $U^n(B)$ with $U_{m-n}(B)$ and plays the role of the Poincaré isomorphism.

One can define characteristic classes of complex vector bundles which take values in complex cobordism. The splitting principle in the theory of vector bundles identifies

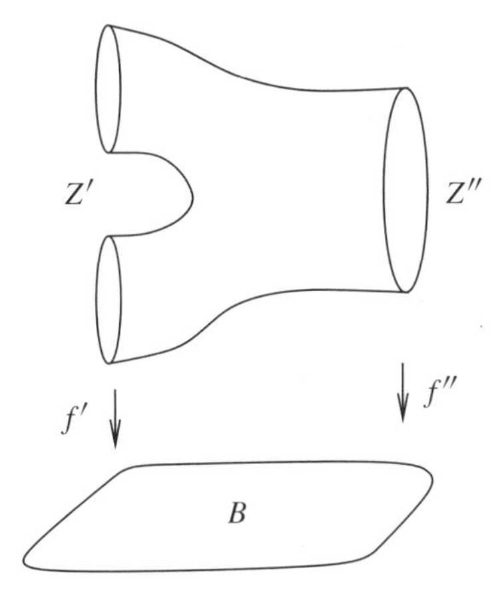

Fig. 2. A bordism between (Z', f') and (Z'', f'').

$U^*(BU_n)$ with the symmetric part of $U^*((BU_1)^n)$. Thus to define cobordism-valued Chern classes, it suffices to describe the first Chern class $u \in U^2(\mathbf{C}P^\infty)$ of the universal complex line bundle. By definition, the first Chern class of $\mathcal{O}(1)$ over $\mathbf{C}P^N$ is Poincaré-dual to the embedding $\mathbf{C}P^{N-1} \to \mathbf{C}P^N$ of a hyperplane section.

The operation of tensor product of line bundles defines a formal commutative group law on the line with coordinate u. Namely, the tensor product of the line bundles with first Chern classes v and w has the first Chern class $u = F(v, w) = v + w + \cdots$. The group properties follow from associativity of tensor product and invertibility of line bundles.

Henceforth we use the notation $U^*(\cdot)$ for the cobordism theory *tensored with* $\mathbf{Q}$.

Much as in complex K-theory, there is the *Chern–Dold character* which provides natural multiplicative isomorphisms

$$\mathrm{Ch} : U^*(B) \to H^*(B, U^*).$$

Here $U^* = U^*(pt)$ is the coefficient ring of the theory and is isomorphic to the polynomial algebra on the generators of degrees $-2k$ Poincaré-dual to the bordism classes $[\mathbf{C}P^k]$. We refer to [4] for the construction. The Chern–Dold character applies, in particular, to the universal cobordism-valued first Chern class of complex line bundles:

$$\mathrm{Ch}(u) = u(z) = z + a_1 z^2 + a_2 z^3 + \cdots, \tag{3}$$

where z is the cohomological first Chern of the universal line bundle $\mathcal{O}(1)$ over $\mathbf{C}P^\infty$, and $\{a_k\}$ is another set of generators in U^*. The series $u(z)$ can be interpreted as an isomorphism between the formal group corresponding to complex cobordism and the additive group $(x, y) \mapsto x + y$:

$$F(v, w) = u(z(v) + z(w)),$$

where $z(\cdot)$ is the series inverse to $u(z)$. This is known as the *logarithm* of the formal group law and takes the form

$$z = u + [\mathbf{C}P^1]\frac{u^2}{2} + [\mathbf{C}P^2]\frac{u^3}{3} + [\mathbf{C}P^3]\frac{u^4}{4} + \cdots. \tag{4}$$

Much as in K-theory, one can compute push-forwards in complex cobordism in terms of cohomology theory. In particular, for a map $\pi : B \to \mathrm{pt}$ from a stably almost complex manifold B to a point, we have the *Hirzebruch–Riemann–Roch formula*

$$\pi_*^U(c) = \int_B \mathrm{Ch}(c)\,\mathrm{Td}(T_B) \in U^* \quad \forall c \in U^*(B), \tag{5}$$

where $\mathrm{Td}(T_B)$ is the *Todd genus* of the tangent bundle. By definition, the push-forward π_*^U of the cobordism class c represented by the Poincaré-dual bordism class represented by $Z \to B$ is the class of the manifold Z in U^*. In cobordism theory, the Todd genus is the *universal* cohomology-valued stable multiplicative characteristic class of complex vector bundles:

$$\mathrm{Td}(\cdot) = \exp \sum_{m=1}^{\infty} s_m \,\mathrm{ch}_m(\cdot),$$

where $\mathbf{s} = (s_1, s_2, s_3, \dots)$ is yet another set of generators of the coefficient algebra U^*. To find out which one, use the fact that on the universal line bundle $\mathrm{Td}(\mathcal{O}(1)) = \frac{z}{u(z)}$ and so

$$\exp \sum_{m=1}^{\infty} s_m \frac{z^m}{m!} = \sum_{k \geq 0} [\mathbf{C}P^k] \frac{u^k(z)}{k+1}.$$

3 Cobordism-valued intersection theory of $\overline{\mathcal{M}}_{0,n}$

Let $\Psi_i \in U^*(\overline{\mathcal{M}}_{0,n})$, $i = 1, \dots, n$, denote the cobordism-valued first Chern classes of the universal cotangent lines L_i over $\overline{\mathcal{M}}_{0,n}$. We introduce the correlators

$$\langle \Psi_1^{k_1}, \dots, \Psi_n^{k_n} \rangle_{0,n}^U := \pi_*^U(\Psi_1^{k_1} \dots \Psi_n^{k_n})$$

and the genus-0 potential

$$\mathcal{F}_0^U(t_0, t_1, \dots) := \sum_{n \geq 3} \sum_{k_1, \dots, k_n \geq 0} \langle \Psi_1^{k_1}, \dots, \Psi_n^{k_n} \rangle_{0,n} \frac{t_{k_1} \dots t_{k_n}}{n!}$$

which take values in the coefficient ring U^* of complex cobordism theory. In particular, the generating function $F_0^U := \sum_{n \geq 3} [\overline{\mathcal{M}}_{0,n}] t_0^n / n!$ for the bordism classes of $\overline{\mathcal{M}}_{0,n}$ is obtained from $\mathcal{F}_0^U$ by putting $t_1 = t_2 = \cdots = 0$. Our present goal is to compute $\mathcal{F}_0^U$.

According to the Hirzebruch–Riemann–Roch formula (5), the computation can be reduced to cohomological intersection theory in $\overline{\mathcal{M}}_{0,n}$ between ψ-classes and characteristic classes of the tangent bundle. In the more general context of (higher genus)

Gromov–Witten theory, one can consider intersection numbers involving characteristic classes of the virtual tangent bundles of moduli spaces of stable maps as a natural generalization of "usual" Gromov–Witten invariants. It turns out that it is possible to express these generalized Gromov–Witten invariants in terms of the usual ones, but the explicit formulas seem unmanageable *unless* one interprets the generalized invariants as the RHS of the Hirzebruch–Riemann–Roch formula in complex cobordism theory. This interpretation of the (cohomological) intersection theory problem in terms of cobordism theory dictates a change of the symplectic formalism described in Section 1, and this change alone miraculously provides a radical simplification of the otherwise unmanageable formulas. In this section, we describe how this happens in the example of the spaces $\overline{\mathcal{M}}_{0,n}$.

Let U denote the formal series completion of the algebra U^* in the topology defined by the grading. Introduce the symplectic space[1] $(\mathcal{U}, \Omega^U)$ defined over U. Let $\mathcal{U}$ denote the space of Laurent series $\sum_{k\in\mathbf{Z}} f_k u^k$ with coefficients $f_k \in U$ which are possibly infinite in both directions but satisfy the condition $\lim_{k\to+\infty} f_k = 0$ in the topology of U. We will call such series *convergent* and write $\mathcal{U} = U\{\{u^{-1}\}\}$. For $f, g \in \mathcal{U}$, we define $\Omega^U(f, g) \in U$ by

$$\Omega^U(f, g) := \frac{1}{2\pi i} \sum_{n=0}^{\infty} [\mathbf{C}P^n] \oint f(u^*) g(u) u^n du,$$

where u^* is the inverse to u in the formal group law described in Section 2. Using the formal group isomorphism $u = u(z)$ we find $u^*(z) = u(-z(u))$. The formula (4) for the logarithm $z = z(u)$ shows that our integration measure $\sum [\mathbf{C}P^n] u^n du$ coincides with dz. Thus the following *quantum Chern character* qCh is a symplectic isomorphism, $\mathrm{qCh}^* \Omega = \Omega^U$:

$$\mathrm{qCh} : \mathcal{U} \mapsto \mathcal{H} \hat{\otimes} U, \qquad \sum_{k\in\mathbf{Z}} f_k u^k \mapsto \sum_{k\in\mathbf{Z}} f_k u^k(z). \tag{6}$$

The "hat" over the tensor product sign indicates the necessary completion by convergent Laurent series.

The next step is to define a Lagrangian submanifold $\mathcal{L}^U \subset \mathcal{U}$ as the graph of the differential of the generating function $\mathcal{F}_0^U$. This requires a Lagrangian polarization $\mathcal{U} = \mathcal{U}_+ \oplus \mathcal{U}_-$. We define $\mathcal{U}_+ = U\{u\}$ to be the space of convergent power series. However, there is no reason for the opposite subspace $u^{-1}U\{\{u^{-1}\}\}$ to be Lagrangian relative to Ω^U. We need a more conceptual construction of the polarization. This is provided by the following residue formula:

$$\frac{1}{2\pi i} \oint \frac{dz}{u(z-x)u(-z-y)} = \begin{cases} +\frac{1}{u(-x-y)} & \text{if } |x| < |z| < |y|, \\ -\frac{1}{u(-x-y)} & \text{if } |y| < |z| < |x|, \\ 0 & \text{otherwise.} \end{cases}$$

The integrand here has first-order poles at $z = x$ and $z = -y$, and the integral depends on the property of the contour to enclose both, neither, or one of them.

[1] We continue to call it "space" although it is a U-module.

One can pick a topological basis $\{u_k(z) \mid k = 0, 1, 2, \dots\}$ in the U-module $U\{u\} =_{\mathrm{qCh}} U\{z\}$ (e.g., $1, u(z), u(z)^2, \dots$ or $1, z, z^2, \dots$) and expand $u(x+y)$ in the region $|x| < |y|$ as

$$\frac{1}{u(-x-y)} = \sum_{k=0}^{\infty} u_k(x) v_k(y), \tag{7}$$

where the v_k are convergent Laurent series in y^{-1} (or, equivalently, in u^{-1}). The residue formulas then show that

$$\sum_{l,m \geq 0} \Omega^U(v_l, u_m) u_l(x) v_m(y) = \sum_{k \geq 0} u_k(x) v_k(y),$$
$$\sum_{l,m \geq 0} \Omega^U(u_l, u_m) v_l(x) v_m(y) = 0,$$
$$\sum_{l,m \geq 0} \Omega^U(v_l, v_m) u_l(x) u_m(y) = 0,$$

or, in other words, $\{v_0, u_0; v_1, u_1; \dots\}$ is a Darboux basis in $(\mathcal{U}, \Omega^U)$. For example, in cohomology theory (which we recover by setting $a_1 = a_2 = \dots = 0$), the Darboux basis $\{-z^{-1}, 1; z^{-2}, z; -z^{-3}, z^2; \dots\}$ is obtained from the expansion

$$\frac{1}{-x-y} = \sum_{k \geq 0} x^k (-y)^{-1-k}.$$

We define $\mathcal{U}_-$ to be the Lagrangian subspace spanned by $\{v_k\}$:

$$\mathcal{U}_- := \left\{ \sum_{k \geq 0} f_{-k} v_k, \ f_{-k} \in U \right\}.$$

A different choice $u'_k = \sum c_{kl} u_l$ of basis in $\mathcal{U}_+$ yields

$$\sum_l u_l(x) v_l(y) = \sum_k u'_k(x) v'_k(y) = \sum_{kl} u_l(x) c_{kl} v'_k(y),$$

so that $v_l = \sum_l c_{kl} v'_k$. Thus the subspace $\mathcal{U}_-$ spanned by v'_k remains the same.

As before, the Lagrangian polarization $\mathcal{U} = \mathcal{U}_+ \oplus \mathcal{U}_-$ identifies $(\mathcal{U}, \Omega^U)$ with $T^*\mathcal{U}_+$. We define a (formal germ of a) Lagrangian section

$$\mathcal{L}^U := \{(\mathbf{p}, \mathbf{q}) \in T^*\mathcal{U}_+ \mid \mathbf{p} = d_{-u^* + \mathbf{q}} \mathcal{F}^U\} \subset \mathcal{U}.$$

Note that the *dilaton shift* is defined by the formula

$$q_0 + q_1 u + q_2 u^2 + \dots = u^*(u) + t_0 + t_1 u + t_2 u^2 + \dots \tag{8}$$

involving the inversion $u^*(u)$ in the formal group. This reduces to the dilaton shift $-z$ in the cohomological theory when we set $a_1 = a_2 = \dots = 0$.

Our goal is to express the image $\mathrm{qCh}(\mathcal{L}^U)$ in terms of the Lagrangian cone

$$\mathcal{L} = \cup_{\tau \in U} z e^{\tau/z} (\mathcal{H}_+ \hat{\otimes} U) \subset \mathcal{H} \hat{\otimes} U$$

defined by cohomological intersection theory in $\overline{\mathcal{M}}_{0,n}$. The answer is given by the following theorem (see [5]).

Theorem. $\mathrm{qCh}(\mathcal{L}^U) = \mathcal{L}$.

Corollary. *$\mathcal{L}^U$ is a Lagrangian cone with the property that its tangent spaces T are tangent to $\mathcal{L}^U$ exactly along uT.*

4 Extracting intersection indices

In this section, we unpack the information hidden in the abstract formulation of the Theorem to compute the fundamental classes $[\overline{\mathcal{M}}_{0,n}]$ in U^*. We succeed for small values of n, but things soon become messy. A lesson to learn is that there is no conceptual advantage in doing this, and that the Theorem as stated provides a better way of representing the answers.

The Theorem together with our conventions (7), (8) on the polarization and dilaton shift encode cobordism-valued intersection indices in $\overline{\mathcal{M}}_{0,n}$. We may view $\mathrm{qCh}(\mathcal{U}_-)$ as a family of Lagrangian subspaces depending on the parameters $(a_1, a_2, \dots)$. The dilaton shift $u(-z)$ can be interpreted in a similar parametric sense. First, the value of $\mathcal{F}_0^U$ considered as a function on the conical graph $\mathcal{L}^U$ of $d\mathcal{F}_0^U$ is equal at a point $f = \sum (p_k v_k + q_k u_k)$ to the value of the quadratic form

$$\frac{1}{2} \sum_{k \geq 0} p_k q_k = \frac{1}{2} \sum_{k \geq 0} q_k \Omega^U (f, u_k).$$

The projection $\sum q_k u_k$ of f to $\mathcal{U}_+$ along $\mathcal{U}_-$ is

$$\sum_{k \geq 0} q_k u_k(x) = \sum_{k \geq 0} u_k(x) \Omega^U \left(\sum v_k, f \right) = \frac{1}{2\pi i} \oint_{|x|<|z|} \frac{f(z)dz}{u(z-x)}.$$

Combining, we find that the value of $\mathcal{F}_0^U$ at the point $f = -z e^{\tau/z} \mathbf{q}(z) \in \mathcal{L}$ is given by the double residue

$$\frac{1}{8\pi^2} \oint \oint_{|x|<|z|} \frac{dz dx}{u(z+x)} xz \mathbf{q}(x) \mathbf{q}(z) e^{[\tau(\frac{1}{x}+\frac{1}{z})]}.$$

The integral vanishes at $\tau = 0$. Differentiating in τ brings down the factor $(z+x)/zx$. We expand $(z+x)/u(z+x) = 1 + \sum_{k>0} b_k (z+x)^k$, where $(b_1, b_2, \dots)$ is one more set of generators of U^*—"the Bernoulli polynomials." Each factor $(z+x)$ can be replaced by zx and differentiation in τ. Using this we express the double integrals via the product of single integrals. After some elementary calculations, we obtain the result in the form

$$-\frac{1}{2}\int_0^\tau [Q^{(-1)}(y)]^2 dy - \frac{1}{2}\sum_{k>0} b_k \frac{d^{k-1}}{d\tau^{k-1}}[Q^{(-1-k)}(\tau)]^2, \tag{9}$$

$$\text{where } Q^{(-l)}(\tau) = \left(\frac{d}{d\tau}\right)^{-l} Q(\tau) := \sum_{m\geq 0} q_m \frac{\tau^{m+l}}{(m+l)!}.$$

Thus (9) gives the value of the potential $\mathcal{F}_0^U$ at the point $(t_0, t_1, \dots)$, which is computed as

$$t_0 + t_1 x + t_2 x^2 + \cdots = -u(-x) - \frac{1}{2\pi i}\oint_{|x|<|z|} \frac{z e^{\tau/z}\mathbf{q}(z)dz}{u(z-x)}.$$

To evaluate the latter expression, we write $1/(z-x) = \sum_{r>0} x^{r-1}/z^r$, expand $(z-x)/u(z-x)$ as before and use the binomial formula for $(z-x)^r$. The resulting expression is

$$-u(-x) - \sum_{k\geq 0} x^k Q^{(k-1)}(\tau) - \sum_{k\geq 0}(-x)^k \sum_{l\geq 0} Q^{(-2-l)}(\tau)\binom{k+l}{k} b_{k+l+1}.$$

It leads to the following sequence of equations (we put $a_{-1} = 0, a_0 = b_0 = 1$):

$$t_k + (-1)^k a_{k-1} + \sum_{l\geq 0} b_l Q^{(k-1-l)}(\tau)\frac{(1-l)(2-l)\dots(k-l)}{k!} = 0. \tag{10}$$

These equations are homogeneous relative to the grading

$$\deg \tau = 1, \qquad \deg t_k = 1-k, \qquad \deg a_k = \deg b_k = \deg q_k = -k$$

and need to be solved for $q_0, q_1, \dots \in U^*$ and $\tau \in U^*[[t_0, t_1, \dots]]$.

With the aim of computing $F_0^U(t) := \mathcal{F}_0^U|_{t_0=t, t_1=t_2=\cdots=0}$ modulo (t^7), we work modulo elements in U^* of degree < -3 and set up the equations (10) with $k = 0, 1, 2, 3, 4$:

$$\begin{aligned}
t + q_0\tau + q_1\tau^2/2 &+ q_2\tau^3/6 + q_3\tau^4/24 \\
+ b_1(q_0\tau^2/2 &+ q_1\tau^3/6 + q_2\tau^4/24) \\
&+ b_2(q_0\tau^3/6 + q_1\tau^4/24) \\
&+ b_3 q_0\tau^4/24 = 0, \\
q_0 - 1 + q_1\tau &+ q_2\tau^2/2 + q_3\tau^3/6 \\
&- b_2(q_0\tau^2/2 + q_1\tau^3/6) \\
&- 2b_3 q_0\tau^3/6 = 0, \\
a_1 + q_1 &+ q_2\tau + q_3\tau^2/2 \\
&+ b_3 q_0\tau^2/2 = 0 \\
&- a_2 + q_2 + q_3\tau = 0 \\
&a_3 + q_3 = 0.
\end{aligned}$$

We solve this system consecutively modulo elements in U^* of degree $n < 0, -1, -2, -3$. The relation $\sum b_k x^k = (\sum a_k x^k)^{-1}$ implies

$$b_1 = -a_1, \qquad b_2 = a_1^2 - a_2, \qquad b_3 = 2a_1a_2 - a_3 - a_1^3.$$

Taking this into account, we find (using MAPLE)

$$\begin{array}{llll} \tau = -t & & +(a_1^2/3 - a_2/2)t^3 & +(7a_3 - 11a_1a_2 + 5a_1^3)t^4/12 \\ q_0 = 1 & -a_1t & +a_1^2t^2/2 & +(4a_3 - 11a_1a_2 + 5a_1^3)t^3/6 \\ q_1 = & -a_1 & +a_2t & +(-a_1a_2 + a_1^3/2)t^2 \\ q_2 = & & a_2 & -a_3t \\ q_3 = & & & -a_3. \end{array}$$

From (9), we find $F_0^U(t)$ modulo (t^7):

$$\begin{aligned} F_0^U =_{\mathrm{mod}(t^7)} & -\frac{q_0^2\tau^3}{6} - \frac{q_0q_1\tau^4}{8} - \frac{q_1^2\tau^5}{40} - \frac{q_0q_2\tau^5}{30} - \frac{q_1q_2\tau^6}{72} \\ & - \frac{q_0q_3\tau^6}{144} - \frac{b_1}{2}\left(\frac{q_0\tau^2}{2} + \frac{q_1\tau^3}{6} + \frac{q_2\tau^4}{24}\right)^2 \\ & - b_2\left(\frac{q_0\tau^2}{2} + \frac{q_1\tau^3}{6}\right)\left(\frac{q_0\tau^3}{6} + \frac{q_1\tau^4}{24}\right) - \frac{7b_3q_0^2\tau^6}{144}. \end{aligned}$$

Substituting the previous formulas into this mess, we compute

$$F_0^U = \frac{t^3}{6} - a_1\frac{t^4}{12} + (9a_2 - 2a_1^2)\frac{t^5}{120} + (10a_1^3 - 10a_1a_2 - 34a_3)\frac{t^6}{720} + O(t^7).$$

Expressing $u(x) = x + a_1x^2 + a_2x^3 + a_3x^4 + \cdots$ as the inverse function to $x(u) = u + p_1u^2/2 + p_2u^3/3 + p_3u^4/4 + \cdots$, we find

$$a_1 = -p_1/2, \qquad a_2 = p_1^2/2 - p_2/3, \qquad a_3 = -5p_1^3/8 + 5p_1p_2/6 - p_3/4$$

and finally arrive at

$$F_0^U = \frac{t^3}{6} + p_1\frac{t^4}{24} + (4p_1^2 - 3p_2)\frac{t^5}{120} + \left(\frac{45}{2}p_1^3 - 30p_1p_2 + \frac{17}{2}p_3\right)\frac{t^6}{720} + O(t^7).$$

The coefficients in this series mean that $[\overline{\mathcal{M}}_{0,3}] = [\mathrm{pt}]$, $[\overline{\mathcal{M}}_{0,4}] = [\mathbf{C}P^1]$, that each blowup of a complex surface M ($= \mathbf{C}P^2$ in our case) adds $[\mathbf{C}P^1 \times \mathbf{C}P^1] - [\mathbf{C}P^2]$ to its cobordism class,[2] and that

$$[\overline{\mathcal{M}}_{0,6}] = \frac{45}{2}[\mathbf{C}P^1 \times \mathbf{C}P^1 \times \mathbf{C}P^1] - 30[\mathbf{C}P^1 \times \mathbf{C}P^2] + \frac{17}{2}[\mathbf{C}P^3].$$

The coefficient sum $(45/2 - 30 + 17/2)$ yields the arithmetical genus of $[\overline{\mathcal{M}}_{0,6}]$ (which is equal to 1 for all $\overline{\mathcal{M}}_{0,n}$ since they are rational manifolds).

[2] This is not hard to verify by studying the behavior under blowups of the Chern characteristic numbers $\int_M c_1(T_M)^2$ and $\int_M c_2(T_M)$, or simply using the fact that $\mathbf{C}P^1 \times \mathbf{C}P^1$ is obtained from $\mathbf{C}P^2$ by two blowups and one blowdown.

5 Quantum Hirzebruch–Riemann–Roch theorem

Assuming that the reader is familiar with generalities of Gromov–Witten theory, we briefly explain below how the Theorem of Section 3 generalizes to higher genera and arbitrary target spaces.

Let X be a compact almost Kähler manifold. Gromov–Witten invariants of X with values in complex cobordism are defined by

$$\langle \phi_1 \Psi_1^{k_1}, \dots, \phi_n \Psi_n^{k_n} \rangle_{g,n}^{X,d} := \int_{[\overline{\mathcal{M}}_{g,n}(X,d)]} \mathrm{Td}(T_{g,n}^{X,d}) \prod_{i=1}^{n} \mathrm{ev}_i^*(\mathrm{Ch}(\phi_i)) u(\psi_i)^{k_i}. \quad (11)$$

Here $[\overline{\mathcal{M}}_{g,n}(X,d)]$ is the virtual fundamental class of the moduli space of degree-d stable pseudoholomorphic maps to X from genus-g curves with n marked points, $\phi_i \in U^*(X)$ are cobordism classes on X, $\mathrm{ev}_i : \overline{\mathcal{M}}_{g,n}(X,d) \to X$ are evaluation maps at the marked points, ψ_i are (cohomological) first Chern classes of the universal cotangent line bundles over $\overline{\mathcal{M}}_{g,n}(X,d)$, and $T_{g,n}^{X,d}$ is the *virtual tangent bundle* of $\overline{\mathcal{M}}_{g,n}(X,d)$.

The *genus-g* potential of X is the generating function

$$\mathcal{F}_{g,X}^U := \sum_{n,d} \frac{Q^d}{n!} \sum_{k_1,\dots,k_n \geq 0} \sum_{\alpha_1,\dots,\alpha_n} \langle \phi_{\alpha_1} \Psi_1^{k_1}, \dots, \phi_{\alpha_n} \Psi_n^{k_n} \rangle_{g,n}^{X,d} t_{k_1}^{\alpha_1} \dots t_{k_n}^{\alpha_n},$$

where $\{\phi_\alpha\}$ is a basis of the free $U^*(\mathrm{pt})$-module $U^*(X)$, Q^d is the monomial in the Novikov ring representing the degree $d \in H_2(X)$, and $\{t_k^\alpha, k = 0, 1, 2, \dots\}$ are formal variables. The *total potential* of X is the expression

$$\mathcal{D}_X^U := \exp\left(\sum_{g=0}^{\infty} \hbar^{g-1} \mathcal{F}_{g,X}^U \right). \quad (12)$$

The exponent is a formal function (with values in a coefficient ring that includes $U^*(\mathrm{pt})$, the Novikov ring and Laurent series in $\hbar$) on the space of polynomials $\mathbf{t} := t_0 + t_1 \Psi + t_2 \Psi^2 + \cdots$ with vector coefficients $t_k = \sum_\alpha t_k^\alpha \phi_\alpha$.

We prefer to consider $\mathcal{D}_X^U$ as a family of formal expressions depending on the sequence of parameters $\mathbf{s} = (s_1, s_2, \dots)$—generators of $U^*(\mathrm{pt})$ featuring in the definition of the Todd genus $\mathrm{Td}(\cdot) = \exp \sum s_k \, \mathrm{ch}_k(\cdot)$. The specialization $\mathbf{s} = 0$ yields the total potential $\mathcal{D}_X$ of cohomological Gromov–Witten theory on X. Our goal is to express $\mathcal{D}_X^U$ in terms of $\mathcal{D}_X$.

To formulate our answer, we interpret $\mathcal{D}_X^U$ as an *asymptotic element* of some Fock space, the quantization of an appropriate symplectic space defined as follows. Let U now denote the superspace $U^*(X)$ tensored with $\mathbf{Q}$ and with the Novikov ring and completed, as before, in the formal series topology of the coefficient ring $U^*(\mathrm{pt})$. Let $(a, b)^U$ denote the cobordism-valued Poincaré pairing

$$(a, b)^U := \pi_*^U(ab) \in U,$$

where $\pi : X \to \mathrm{pt}$. We put

$$\mathcal{U} := U\{\{u^{-1}\}\}, \qquad \Omega^U(f, g) = \frac{1}{2\pi i} \sum_{n \geq 0} [\mathbf{C}P^n] \oint (f(u^*), f(u))^U u^n du$$

and define a Lagrangian polarization of $(\mathcal{U}, \Omega^U)$ using (7):

$$\mathcal{U}_+ = U\{u\}, \qquad \mathcal{U}_- := \left\{ \sum_{k \geq 0} p_k v_k \mid p_k \in U \right\}.$$

Using the dilaton shift $\mathbf{q}(u) = 1u^* + \mathbf{t}(u)$ (where 1 is the unit element of U), we identify $\mathcal{D}_X$ with an *asymptotic function* of $\mathbf{q} \in \mathcal{U}_+$. By definition, this means that the exponent in $\mathcal{D}_X$ is a formal function of $\mathbf{q}$ near the shifted origin. The Heisenberg Lie algebra of $(\mathcal{U}, \Omega^U)$ acts on functions on $\mathcal{U}$ invariant under translations by $\mathcal{U}_-$, and this action extends to the (nonlinear) space of asymptotic functions.

In the specialization $\mathbf{s} = 0$, the above structure degenerates into its cohomological counterpart introduced in [10]: $(\mathcal{H}, \Omega) = T^*\mathcal{H}_+$ where $\mathcal{H} := H((z^{-1}))$ consists of Laurent series in z^{-1} with vector coefficients in the cohomology superspace $H = H^*(X, \mathbf{Q})$ tensored with the Novikov ring, $\mathcal{H}_+ = H[z]$, $\mathcal{H}_- = z^{-1}H[[z^{-1}]]$, $\Omega(f, g) = (2\pi i)^{-1} \oint (f(-z), g(z)) dz$, and $(a, b) := \int_X ab$ is the cohomological Poincaré pairing. The total potential $\mathcal{D}_X$ is identified via the dilaton shift $\mathbf{q}(z) = \mathbf{t}(z) - z$ with an asymptotic element of the corresponding Fock space.

The quantum Chern–Dold character of Section 3 is generalized in the following way:

$$\mathrm{qCh}\left(\sum_{k \in z} f_k u^k \right) := \sqrt{\mathrm{Td}(T_X)} \sum_{k \in \mathbf{Z}} \mathrm{Ch}(f_k) u^k(z).$$

The factor $\sqrt{\mathrm{Td}(T_X)}$ is needed to match the cobordism-valued Poincaré pairing $\pi^U_*(ab) = \int_X \mathrm{Td}(T_X)\,\mathrm{Ch}(a)\,\mathrm{Ch}(b)$ with the cohomological one. The quantum Chern–Dold character provides a symplectic isomorphism of $(\mathcal{U}, \Omega^U)$ with $(\mathcal{H}\hat{\otimes} U, \Omega)$. This isomorphism identifies the Heisenberg Lie algebras and thus—due to the Stone–von Neumann theorem—gives a projective identification of the corresponding Fock spaces. Understanding the quantum Chern–Dold character in this sense, we obtain a family of "quantum states"

$$\langle \mathcal{D}^{\mathbf{s}}_X \rangle := \mathrm{qCh} \langle \mathcal{D}^U_X \rangle.$$

These are one-dimensional spaces depending formally on the parameter $\mathbf{s} = (s_1, s_2, \ldots)$ and spanned by asymptotic elements of the Fock space associated with $(\mathcal{H}, \Omega)$. We have $\langle \mathcal{D}^{\mathbf{s}}_X \rangle|_{\mathbf{s}=\mathbf{0}} = \langle \mathcal{D}_X \rangle$.

Introduce the virtual bundle $E = T_X \ominus \mathbf{C}$ over X, and identify $z \in H^*(\mathbf{C}P^\infty)$ with the equivariant first Chern class of the trivial line bundle L over X equipped with the standard fiberwise S^1-action.

Theorem (see [5]). *Let $\hat{\Delta}$ be the quantization of the linear symplectic transformation Δ given by multiplication (in the algebra $\mathcal{H}$) by the asymptotic expansion of the infinite product*

$$\sqrt{\mathrm{Td}(E)}\prod_{m=1}^{\infty}\mathrm{Td}(E\otimes L^{-m}).$$

Then

$$\langle\mathcal{D}_X^{\mathbf{s}}\rangle=\hat{\Delta}(\mathbf{s})\langle\mathcal{D}_X\rangle.$$

We need to add to this formulation the following comments. The asymptotic expansion in question is obtained using the famous formula relating integration with its finite-difference version via the Bernoulli numbers:

$$\frac{s(x)}{2}+\sum_{m=1}^{\infty}s(x-mz)=\frac{1+e^{-zd/dx}}{1-e^{-zd/dx}}\frac{s(x)}{2}$$

$$\sim\sum_{m=0}^{\infty}\frac{B_{2m}}{(2m)!}z^{2m-1}\frac{d^{2m-1}s(x)}{dx^{2m-1}}.$$

Taking $s(x)=\sum_{k\geq 0}s_k x^k/k!$ and letting x run over the Chern roots of E, we find that

$$\ln\Delta=\sum_{m=0}^{\infty}\sum_{l=0}^{D}s_{2m-1+l}\frac{B_{2m}}{(2m)!}\,\mathrm{ch}_l(E)z^{2m-1},$$

where $D=\dim_{\mathbb{C}}(X)$. The operators A on $\mathcal{H}$ defined as multiplication by $\mathrm{ch}_l(E)z^{2m-1}$ are infinitesimal symplectic transformations—they are antisymmetric with respect to Ω—and so define quadratic Hamiltonians $\Omega(Af,f)/2$ on $\mathcal{H}$. We use the quantization rule of quadratic Hamiltonians written in a Darboux coordinate system $\{p_\alpha,q_\alpha\}$:

$$(q_\alpha q_\beta)\hat{}:=\frac{q_\alpha q_\beta}{\hbar},\qquad (q_\alpha p_\beta)\hat{}:=q_\alpha\frac{\partial}{\partial q_\beta},\qquad (p_\alpha p_\beta)\hat{}:=\hbar\frac{\partial^2}{\partial p_\alpha\partial p_\beta}.$$

The rule defines the quantization $\hat{A}$ and the action on the quantum state $\langle\mathcal{D}_X\rangle$ of $\hat{\Delta}:=\exp(\ln\Delta)\hat{}$.

A more precise version of this Quantum–Hirzebruch–Riemann–Roch theorem also provides the proportionality coefficient between $\hat{\Delta}\mathcal{D}_X$ and $\mathcal{D}_X^{\mathbf{s}}$. Namely, let us write $\Delta=\Delta_1\Delta_2$, where $\ln\Delta_2$ consists of the z^{-1}-terms in (12) and $\ln\Delta_1$ contains the rest. Let us agree that $\hat{\Delta}\mathcal{D}_X$ means $\hat{\Delta}_1\hat{\Delta}_2\mathcal{D}_X$ (this is important since the two operators commute only up to scalar factor). Then (see [5])

$$\mathcal{D}_X^{\mathbf{s}}=\left(\mathrm{sdet}\sqrt{\mathrm{Td}(E)}\right)^{1/24}e^{\frac{1}{24}\sum_{l>0}s_{l-1}\int_X\mathrm{ch}_l(E)c_{D-1}(T_X)}\hat{\Delta}(\mathbf{s})\mathcal{D}_X,\tag{13}$$

where c_{D-1} is the $(D-1)$st Chern class, and sdet stands for the Berezinian.

Taking the quasi-classical limit $\hbar\to 0$, we obtain the genus-zero version of the theorem. As in Section 3, the graph of the differential $d\mathcal{F}_{0,X}^U$ of the cobordism-valued genus-zero potential defines a (formal germ of a) Lagrangian submanifold $\mathcal{L}_X^U$ of $\mathcal{U}$.

Corollary. *The image* $\mathrm{qCh}(\mathcal{L}_X^U)$ *of the Lagrangian submanifold* $\mathcal{L}_X^U \subset \mathcal{U}$ *is obtained from the Lagrangian cone* $\mathcal{L}_X \subset \mathcal{H}$ *representing* $d\mathcal{F}_X^0$ *by the family of linear symplectic transformations* $\Delta(\mathbf{s})$*:*

$$\mathrm{qCh}(\mathcal{L}_X^U) = \Delta \mathcal{L}_X.$$

In particular, $\mathcal{L}_X^U$ *is a Lagrangian cone with the property that its tangent spaces* T *are tangent to* $\mathcal{L}_X^U$ *exactly along* uT*.*

When $X = \mathrm{pt}$, we have $\ln \Delta = -\sum_{m>0} s_{2m-1} B_{2m} z^{2m-1}/(2m)!$. In the case $g = 0$, since $\mathcal{L}_{\mathrm{pt}}$ is invariant under multiplication by z, we see that $\mathrm{qCh}(\mathcal{L}_{\mathrm{pt}}^U) = \mathcal{L}_{\mathrm{pt}}$. This is our Theorem from Section 3.

6 Outline of the proof of the QHRR theorem

The proof proceeds by showing that the derivatives in s_k, $k = 1, 2, \ldots$, of the LHS and the RHS of (13) are equal. Differentiating the LHS in s_k brings down a factor of $\mathrm{ch}_k(T_{g,n}^{X,d})$ inside all the correlators (11). Since the RHS contains the generating function $\mathcal{D}_X$ for intersection numbers involving the classes ψ_i, our main problem now is to express the Chern character $\mathrm{ch}(T_{g,n}^{X,d})$ in terms of ψ-classes.

One can view the moduli space $\overline{\mathcal{M}}_{g,n}(X, d)$ as fibered over the moduli stack $\overline{\mathfrak{M}}_{g,n}$ of marked nodal curves $(\Sigma; \sigma_1, \ldots, \sigma_n)$. Thus the virtual tangent bundle $T_{g,n}^{X,d}$ falls into three parts:

$$T_{g,n}^{X,d} = T' + T'' + T''',$$

where T' is the virtual tangent bundle to the fibers, T'' is the (pullback of the) tangent sheaf to the moduli stack $\overline{\mathfrak{M}}_{g,n}$ *logarithmic* with respect to the divisor of nodal curves, and T''' is a sheaf supported on the divisor. The subbundle T'—which is the *index bundle* of the Cauchy–Riemann operator describing infinitesimal variations of pseudoholomorphic maps to X from a fixed complex curve Σ with a fixed configuration of marked points—can alternatively be described in terms of the *twisting bundles* considered in [6]. Namely, the diagram formed by the forgetful map $\mathrm{ft}_{n+1} : \overline{\mathcal{M}}_{g,n+1}(X, d) \to \overline{\mathcal{M}}_{g,n}(X, d)$ and the evaluation map $\mathrm{ev}_{n+1} : \overline{\mathcal{M}}_{g,n+1}(X, d) \to X$ can be considered as the universal family of genus-g, n-pointed stable maps to X of degree d. Let E be a complex bundle (or virtual bundle) over X. The K-theoretic pull-back/push-forward $E_{g,n}^{X,d} := (\mathrm{ft}_{n+1})_* \mathrm{ev}_{n+1}^*(E)$ is an element, called a *twisting bundle*, of the Grothendieck group of orbibundles $K^*(\overline{\mathcal{M}}_{g,n}(X, d))$. The virtual bundle T' coincides with the twisting bundle $(TX)_{g,n}^{X,d}$.

Intersection numbers in $\overline{\mathcal{M}}_{g,n}(X, d)$ against characteristic classes of twisting bundles $E_{g,n}^{X,d}$ are called Gromov–Witten invariants of X *twisted by* E. The "Quantum Riemann–Roch" theorem of [6] expresses such twisted Gromov–Witten invariants in terms of untwisted ones. The key to this is an application of the Grothendieck–Riemann–Roch theorem to the universal family $\mathrm{ft}_{n+1} : \overline{\mathcal{M}}_{g,n+1}(X, d) \to \overline{\mathcal{M}}_{g,n}(X, d)$, analogous to Mumford's famous computation [14] of the Hodge classes in $\overline{\mathcal{M}}_{g,0}$ (and to its generalization to $\overline{\mathcal{M}}_{g,n}(X,d)$ by Faber and Pandharipande [7]). Applying the same idea here allows us to express the classes $\mathrm{ch}_k(T')$ in terms of ψ-classes on the universal family $\overline{\mathcal{M}}_{g,n+1}(X,d)$.

Next, fibers of the logarithmic tangent bundle T'' can be viewed as dual to spaces of quadratic differentials on Σ twisted appropriately at the marked points and nodes. More precisely, $T'' = -(\mathrm{ft}_{n+1})_*(L_{n+1}^{-1})$, where L_{n+1} is the universal cotangent line at the $(n+1)$st marked point and $(\mathrm{ft}_{n+1})_*$ is the K-theoretic push-forward. Thus by applying the Grothendieck–Riemann–Roch formula again, we can express $\mathrm{ch}_k(T'')$ in terms of ψ-classes on the universal family $\overline{\mathcal{M}}_{g,n+1}(X,d)$.

Finally, the sheaf T''' can be expressed as the K-theoretic push-forward $(\mathrm{ft}_{n+1})_*\mathcal{O}_Z$ of the structure sheaf of the locus $Z \subset \overline{\mathcal{M}}_{g,n+1}(X,d)$ of nodes of the curves Σ. This has (virtual) complex codimension 2 in the universal family $\overline{\mathcal{M}}_{g,n+1}(X,d)$. It is parametrized by certain pairs of stable maps with genera g_1 and g_2 where $g_1+g_2=g$, each of which carry an extra marked point, and by stable maps of genus $g-1$ which carry two extra marked points; the extra marked points are glued to form the node. Intersection numbers involving the classes $\mathrm{ch}_k(T''')$ can therefore be expressed in terms of intersections against ψ-classes in (products of) "simpler" moduli spaces of stable maps.

Together, the previous three paragraphs give recursive formulas which reduce intersection numbers involving the classes $\mathrm{ch}_k(T_{g,n}^{X,d})$ to those involving only ψ-classes. Processing the s_k-derivative of the LHS of (13) in this way one finds, after some 20 pages of miraculous cancellations and coincidences, that it is equal to the s_k-derivative of the RHS. We do not have any conceptual explanation for these cancellations, which often look quite surprising. For example, it turns out to be vital that the orbifold Euler characteristic $\chi(\overline{\mathcal{M}}_{1,1})$ is equal to $5/12$ (or, equivalently, that $c_1(T_{1,1}^{\mathrm{pt},0}) = 10\psi_1$). Were this not the case, a delicate cancellation involving the cocycle coming from the projective representation of the Heisenberg Lie algebra would not have occurred, and the multiplicativity of (13) with respect to the group $\mathrm{Td}(\cdot) = \exp\sum s_k\,\mathrm{ch}_k(\cdot)$ of characteristic classes would have been destroyed.

The three summands T', T'', and T''' play differing roles in the ultimate formula, as we now explain. Comparing the QHRR theorem with the Quantum Riemann–Roch theorem from [6], one sees that the potentials $\mathcal{D}_X^{\mathbf{s}}$ coincide with the total potentials of X for cohomological Gromov–Witten theory twisted by the characteristic class $\mathrm{Td}(\cdot) = \exp\sum s_k\,\mathrm{ch}_k(\cdot)$ and the bundle $E = TX - \mathbf{C}$. But the total potential $\mathcal{D}_X^U$ for cobordism-valued Gromov–Witten theory of X differs from $\mathcal{D}_X^{\mathbf{s}}$ precisely because of the additional $\mathbf{s}$-dependence coming from the quantum Chern–Dold character, i.e., through the $\mathbf{s}$-dependence of the dilaton shift $u(-z)$ and of the polarization $\mathcal{H}_+ \oplus \mathrm{qCh}(\mathcal{U}_-)$. These effects are compensated for by $(\mathrm{ft}_{n+1})_*(\mathbf{C} - L_{n+1}^{-1})$ and $(\mathrm{ft}_{n+1})_*\mathcal{O}_Z$ respectively. Thus, roughly speaking, the quantized symplectic transformation $\hat{\Delta}$ accounts for variations T' of maps $\Sigma \to X$, while the symplectic formalism of Section 3 based on formal groups accounts for variations $T'' + T'''$ of complex structures on Σ. This suggests that there is an intrinsic relationship between formal group laws and the moduli space of Riemann surfaces. However, the precise nature of this relationship remains mysterious.

Acknowledgments. We would like to congratulate Israel Moiseevich Gelfand on the occasion of his 90th birthday and to thank the organizers P. Etingof, V. Retakh, and I. Singer for the opportunity to take part in the celebration.

This material is based upon work supported by National Science Foundation grant DMS-0306316.

References

[1] S. Barannikov, Quantum periods I: Semi-infinite variations of Hodge structures, *Internat. Math. Res. Notices*, **23** (2001), 1243–1264.

[2] K. Behrend and B. Fantechi, The intrinsic normal cone, *Invent. Math.*, **128** (1997), 45–88.

[3] K. Behrend and Yu. Manin, Stacks of stable maps and Gromov-Witten invariants, *Duke Math. J.*, **85** (1996), 1–60.

[4] V. M. Bukhshtaber. The Chern-Dold character in cobordisms, *Mat. Sb. (N.S.)*, **83** (1970), 575–595.

[5] T. Coates. *Riemann–Roch Theorems in Gromov–Witten Theory*, Ph.D. thesis, University of California at Berkeley, Berkeley, CA, 2003; availble online from http://abel.math.harvard.edu/~tomc/.

[6] T. Coates and A. Givental, *Quantum Riemann–Roch, Lefschetz and Serre*, 2001; arXiv: math.AG/0110142.

[7] C. Faber and R. Pandharipande, Hodge integrals and Gromov–Witten theory, *Invent. Math.*, **139** (2000), 173–199.

[8] K. Fukaya and K. Ono, Arnold conjecture and Gromov-Witten invariants, *Topology*, **38**-5 (1999), 933–1048.

[9] A. Givental, Symplectic geometry of Frobenius structures, in C. Hertling and M. Marcolli, eds., *Frobenius Manifolds: Quantum Cohomology and Singularities*, Aspects of Mathematics E 36,Vieweg, Braunschweig, Germany, 2004, 91–112.

[10] A. Givental, Gromov–Witten invariants and quantization of quadratic hamiltonians, *Moscow Math. J.*, **1**-4 (2001), 551–568.

[11] M. Kontsevich, Intersection theory on the moduli space of curves and the matrix Airy function, *Comm. Math. Phys.*, **147** (1992), 1–23.

[12] M. Kontsevich, Enumeration of rational curves via toric actions, in R, Dijkgraaf, C. Faber, and G. van der Geer, eds., *The Moduli Space of Curves*, Progress in Mathematics 129, Birkhäuser Boston, Cambridge, MA, 1995, 335–368.

[13] J. Li and G. Tian, Virtual moduli cycles and Gromov-Witten invariants of algebraic varieties, *J. Amer. Math. Soc.*, **11**-1(1998), 119–174.

[14] D. Mumford, Towards enumerative geometry on the moduli space of curves, in M. Artin and J. Tate, eds., *Arithmetics and Geometry*, Vol. 2, Birkhäuser Boston, Cambridge, MA, 1983, 271–328.

[15] Y. Ruan, Virtual neighborhoods of pseudo-holomorphic curves, *Turkish J. Math.*, **23**-1 (1999) (Proceedings of the 6th Gökova Geometry-Topology Conference), 161–231.

[16] B. Siebert, Algebraic and symplectic Gromov–Witten invariants coincide, *Ann. Inst. Fourier (Grenoble)*, **49**-6 (1999), 1743–1795.

[17] E. Witten, Two-dimensional gravity and intersection theory on moduli space, *Surveys Differential Geom.*, **1** (1991), 243–310.

On the Foundations of Noncommutative Geometry

A. Connes[1–3]

[1] Collège de France
3, rue d'Ulm
75231 Paris cedex 05
France
[2] I.H.E.S.
Le Bois-Marie
35, route de Chartres
91440 Bures-sur-Yvette
France
[3] Department of Mathematics
Vanderbilt University
2201 West End Avenue
Nashville, TN 37235
USA
alain@connes.org

Subject Classifications: 58B24, 58J42

Israel Gelfand is one of the handful of mathematicians who really shaped the mathematics of the twentieth century. Even among them he stands out by the fecundity of the concepts he created and the astonishing number of new fields he originated.

One characteristic feature of his mathematics is that, while working at a high level of conceptual breadth, it never loses contact with concrete computations and applications, including those to theoretical physics, a subject in which his influence is hard to match.

Although mathematicians before Gelfand had studied normed rings, it was he who created the tools that got the theory really started. In his thesis he brought to light the fundamental concept of maximal ideal and proved that the quotient of a commutative Banach algebra by a maximal ideal is always the field $\mathbb{C}$ of complex numbers. This easily implied, for instance, Wiener's well-known result that the inverse of a function with no zeros and absolutely convergent Fourier expansion also has absolutely convergent Fourier expansion. The fundamental result in the commutative case characterized the rings of continuous functions on a (locally) compact space in a purely algebraic manner. Dropping the commutativity assumption led Gelfand and Naimark to the theory of C^*-algebras, again proving the fundamental result that any such ring can be realized as an involutive norm closed subalgebra of the algebra of operators in Hilbert space. The key step, known as the "Gelfand–Naimark–Segal" construction, plays a basic role in quantum field theory, and was used early on by Gelfand and Raikov to show that any locally compact group admits enough irreducible Hilbert space representations. These and many others of Gelfand's results were so influential that it is hard for us to imagine mathematics without them.

They played a decisive role in the foundations of Noncommutative Geometry, a subject to which I have devoted most of my mathematical work. I refer to the survey [35] for a thorough presentation of the subject and will describe here, after a brief introduction, a few of the open frontiers and problems which are actively being explored at this point.

I The framework of noncommutative geometry

As long as we consider geometry as intimately related to our model of space-time, Einstein's general relativity clearly vindicated the ideas of Gauss and Riemann, allowing for variable curvature, and formulating the intrinsic geometry of a curved space independently of its embedding in Euclidean space. The two key notions are those of manifold of arbitrary dimension, whose points are locally labeled by finitely many real numbers x^μ, and that of the line element, i.e., the infinitesimal unit of length, which, when transported, allows one to measure distances. The infinitesimal calculus encodes the geometry by the formula for the line element ds in local terms

$$ds^2 = g_{\mu\nu} dx^\mu dx^\nu,$$

and allows one to generalize most of the concepts which were present either in Euclidean or non-Euclidean geometry, while considerably enhancing the number of available interesting examples.

Riemann was sufficiently cautious in his lecture on the foundation of geometry to question the validity of his hypotheses in the infinitely small. He explicitly proposed to "gradually modify the foundations under the compulsion of facts which cannot be explained by it" in case physics would find new unexplained phenomena in the exploration of smaller scales.

The origin of noncommutative geometry can be traced back to the discovery of such unexplained phenomena in the phase space of the microscopic mechanical system describing an atom. This system manifests itself through its interaction with radiation and the basic laws of spectroscopy, as found in particular by Ritz and Rydberg, are in contradiction with the "manifold" picture of the phase space.

The very bare fact, which came directly from experimental findings in spectroscopy and was unveiled by Heisenberg (and then understood at a more mathematical level by Born, Jordan, Dirac and the physicists of the late 1920s), is the following. Whereas when you are dealing with a manifold you can parameterize (locally) its points x by real numbers $x_1, x_2, \ldots$, which specify completely the situation of the system, when you turn to the phase space of a microscopic mechanical system, even of the simplest kind, the coordinates, namely, the real numbers $x_1, x_2, \ldots$ that you would like to use to parameterize points, actually do not commute.

This means that the classical geometrical framework is too narrow to describe in a faithful manner many physical spaces of great interest. In noncommutative geometry one replaces the usual notion of manifold formed of points labeled by coordinates with spaces of a more general nature, as we shall see shortly. Usual geometry is just a particular case of this new theory, in the same way as Euclidean and non-Euclidean geometry are particular cases of Riemannian geometry. Many of the familiar geometrical concepts do survive in the new theory, but they also carry a new unexpected meaning.

Before describing the novel notion of space, it is worthwhile to explain in simple terms how noncommutative geometry modifies the measurement of distances. Such a simple description is possible because the evolution between the Riemannian way of measuring distances and the new (noncommutative) way exactly parallels the improvement of the standard of length[1] in the metric system. The original definition of the meter at the end of the 18th century was based on a small portion (one forty millionth part) of the size of the largest available macroscopic object (here the earth's circumference). Moreover, this "unit of length" became concretely represented in 1799 as "mètre des archives" by a platinum bar localized near Paris. The international prototype was a more stable copy of the "mètre des archives" which served to define the meter. The most drastic change in the definition of the meter occurred in 1960 when it was redefined as a multiple of the wavelength of a certain orange spectral line in the light emitted by isotope 86 of krypton. This definition was then replaced in 1983 by the current definition which, using the speed of light as a conversion factor, is expressed in terms of inverse frequencies rather than wavelength, and is based on a hyperfine transition in the cesium atom. The advantages of the new standard are obvious. No comparison to a localized "mètre des archives" is necessary, the uncertainties are estimated as 10^{-15} and for most applications a commercial cesium beam is sufficiently accurate. Also we could (if any communication were possible) communicate our choice of unit of length to aliens, and uniformize length units in the galaxy without having to send out material copies of the "mètre des archives"!

As we shall see below, the concept of "metric" in noncommutative geometry is precisely based on such a spectral data.

Let us now come to "spaces." What the discovery of Heisenberg showed is that the familiar duality of algebraic geometry between a space and its algebra of coordinates (i.e., the algebra of functions on that space) is too restrictive to model the phase space of microscopic physical systems. The basic idea then is to extend this duality, so that the algebra of coordinates on a space is no longer required to be commutative. Gelfand's work on C^*-algebras provides the right framework to define noncommutative topological spaces. They are given by their algebra of continuous functions which can be an arbitrary, not necessarily commutative, C^*-algebra.

It turns out that there is a wealth of examples of spaces, which have obvious geometric meaning but which are best described by a noncommutative algebra of coordinates. The first examples came, as we saw above, from phase space in quantum mechanics but there are many others, such as

[1] Or, equivalently, of time using the speed of light as a conversion factor.

- space of leaves of foliations,
- space of irreducible representations of discrete groups,
- space of Penrose tilings of the plane,
- Brillouin zone in the quantum Hall effect,
- phase space in quantum mechanics,
- space time,
- space of $\mathbb{Q}$-lattices in $\mathbb{R}^n$.

This last class of examples [44, 45] appears to be of great relevance in number theory and will be discussed at the end of this short survey. The space of $\mathbb{Q}$-lattices [45] is a natural geometric space, with an action of the scaling group providing a spectral interpretation of the zeros of the L-functions of number theory and an interpretation of the Riemann explicit formulas as a trace formula [31]. Another rich class of examples arises from deformation theory, such as deformation of Poisson manifolds, quantum groups and their homogeneous spaces. Moduli spaces also generate very interesting new examples as in [32, 72], as well as the fiber at ∞ in arithmetic geometry [46].

Thus there is no shortage of examples of noncommutative spaces that beg our understanding but which are very difficult to comprehend. Among them the noncommutative tori were fully analyzed at a very early stage of the theory in 1980 [15] and a beginner might be tempted to be happy with the understanding of such simple examples ignoring the wild diversity of the general landscape. The common feature of many of these spaces is that when one tries to analyze them from the usual set theoretic point of view, the usual tools break down for the following simple reason. Even though as a set they have the cardinality of the continuum, it is impossible to distinguish their points by a finite (or countable) set of explicit functions. In other words, any explicit countable family of *invariants* fails to separate points.

Here is the general principle that allows one to nevertheless encode them by a function algebra, which will no longer be commutative. The above spaces are obtained as quotients from a larger classical space Y gifted with an equivalence relation $\mathcal{R}$. The usual algebra of functions associated to the quotient is

$$\mathcal{A} = \{f \mid f(a) = f(b) \;\; \forall\,(a,b) \in \mathcal{R}\}. \tag{1}$$

This algebra is, by construction, a *subalgebra* of the original function algebra on Y and remains commutative. There is, however, a much better way to encode in an algebraic manner the above quotient operation. It consists, instead of taking the subalgebra given by (1), of adjoining to the algebra of functions the identification of a with b, whenever $(a, b) \in \mathcal{R}$. The algebra obtained in this way,

$$\mathcal{B} = \{f = [f_{ab}],\; (a,b) \in \mathcal{R}\}, \tag{2}$$

is the convolution algebra of the groupoid associated to $\mathcal{R}$ and is, of course, no longer commutative in general, nor Morita equivalent to a commutative algebra. By encoding the *dynamics* underlying the identification of points by the relation $\mathcal{R}$, it bypasses the problem created by the lack of constructible *static* invariants labeling points in the quotient.

The first operation (1) is of a cohomological flavor, while the second (2) always gives a satisfactory answer, which keeps a close contact with the quotient space. One then recovers the "naive" function spaces generated by the first operation (1) from the cyclic cohomology of the noncommutative algebra obtained from the second operation (2).

The second vital ingredient of the theory is the extension of geometric ideas to the noncommutative framework. It may seem at first sight that it is a simple matter to rewrite algebraically the usual geometric concepts but, in fact, the extension of geometric thinking imposed by passing to noncommutative spaces forces one to rethink most of our familiar notions. The most interesting part comes from totally unexpected new features, such as the canonical dynamics of noncommutative measure spaces, which have no counterpart in the classical geometric setup.

As we shall see below, far reaching extensions of classical concepts have been obtained, with variable degrees of perfection, for measure theory, topology, differential geometry, and Riemannian geometry:

- metric geometry,
- differential geometry,
- topology,
- measure theory.

II Measure theory

One compelling reason to start working in noncommutative geometry is that, even at the very coarse level of measure theory, the general noncommutative case is becoming highly nontrivial. When one looks at an ordinary space and measures theory, one uses the Lebesgue theory, which is a beautiful theory, but all spaces are the same. There is nothing really happening as far as classification is concerned. This is not at all the case in noncommutative measure theory. What happens there is very surprising. It is an absolutely fascinating fact that, when one takes a noncommutative algebra M from the measure theory point of view, such an algebra evolves with time!

More precisely, it admits a god-given time evolution, given by a canonical group homomorphism [10, 11]

$$\delta : \mathbb{R} \to \mathrm{Out}(M) = \mathrm{Aut}(M)/\,\mathrm{Int}(M) \tag{1}$$

from the additive group $\mathbb{R}$ to the center of the group of automorphism classes of M modulo inner automorphisms.

This homomorphism is provided by the uniqueness of the, a priori state dependent, modular automorphism group of a state. Together with the earlier work of Powers, Araki–Woods, and Krieger, it was the beginning of a long story that eventually led to the complete classification [11, 86, 87, 68, 29, 12, 13, 19, 57] of approximately finite-dimensional factors (also called hyperfinite).

They are classified by their module,

$$\mathrm{Mod}(M) \subset_{\sim} \mathbb{R}_+^*, \tag{2}$$

which is a virtual closed subgroup of $\mathbb{R}_+^*$ in the sense of G. Mackey, i.e., an ergodic action of $\mathbb{R}_+^*$, called the flow of weights [29]. This invariant was first defined and used in my thesis [11] to show in particular the existence of hyperfinite factors which are not isomorphic to Araki–Woods factors.

There is a striking analogy, which I described in [27], between the above classification and the Brauer theory of central simple algebras. It has taken new important steps recently, since noncommutative manifolds [40, 41] give examples of construction of the hyperfinite II_1 factor as the crossed product of the field K_q of elliptic functions by a subgroup of its Galois group, in perfect analogy with the Brauer theory.

Thus we see that noncommutative measure theory is already highly nontrivial; hence we have many reasons to believe that if one goes further in the natural hierarchy of features of a space, one will discover really interesting new phenomena.

III Topology

The development of the topological ideas was prompted by the work of Israel Gelfand, whose C^*-algebras give the required framework for noncommutative topology. The two main driving forces in the development of noncommutative topology were the Novikov conjecture on homotopy invariance of higher signatures of ordinary manifolds as well as the Atiyah–Singer index theorem. It has led, through the work of Atiyah, Singer, Brown, Douglas, Fillmore, Miščenko, and Kasparov [1, 83, 5, 75, 62], to the realization that not only the Atiyah–Hirzebruch K-theory but more importantly the dual K-homology admits Hilbert space techniques and functional analysis as their natural framework. The cycles in the K-homology group $K_*(X)$ of a compact space X are indeed given by Fredholm representations of the C^*-algebra A of continuous functions on X. The central tool is the Kasparov bivariant K-theory. A basic example of C^*-algebra, to which the theory applies, is the group ring of a discrete group, and restricting oneself to commutative algebras is an obviously undesirable assumption.

For a C^*-algebra A, let $K_0(A)$, $K_1(A)$ be its K-theory groups. Thus $K_0(A)$ is the algebraic K_0 theory of the ring A and $K_1(A)$ is the algebraic K_0 theory of the ring $A \otimes C_0(\mathbb{R}) = C_0(\mathbb{R}, A)$. If $A \to B$ is a morphism of C^*-algebras, then there are induced homomorphisms of abelian groups $K_i(A) \to K_i(B)$. Bott periodicity provides a six term K-theory exact sequence for each exact sequence $0 \to J \to A \to B \to 0$ of C^*-algebras, and excision shows that the K-groups involved in the exact sequence only depend on the respective C^*-algebras.

Discrete groups, Lie groups, group actions, and foliations give rise, through their convolution algebra, to a canonical C^*-algebra, hence to K-theory groups. The analytical meaning of these K-theory groups is clear as a receptacle for indices of elliptic operators. However, these groups are difficult to compute. For instance, in the case of semisimple Lie groups, the free abelian group with one generator for each irreducible discrete series representation is contained in $K_0(C_r^*(G))$, where $C_r^*(G)$ is the reduced C^*-algebra of G. Thus an explicit determination of the K-theory in this case in particular involves an enumeration of the discrete series.

We introduced with P. Baum [3] a geometrically defined K-theory, which specializes to discrete groups, Lie groups, group actions, and foliations. Its main features are its computability and the simplicity of its definition. In essence, it is the group of topological data labeling the symbols of elliptic operators. It does not involve the difficult quotient spaces, but replaces them (up to homotopy) by the familiar homotopy quotient (replacing free actions by proper actions).

In the case of semisimple Lie groups, it elucidates the role of the homogeneous space G/K (K the maximal compact subgroup of G) in the Atiyah–Schmid geometric construction of the discrete series [2]. Using elliptic operators, we constructed a natural map μ from our geometrically defined K-theory groups to the above analytic (i.e., C^*-algebra) K-theory groups.

Much progress has been made in the past years to determine the range of validity of the isomorphism between the geometrically defined K-theory groups and the above analytic (i.e., C^*-algebra) K-theory groups. We refer to the three Bourbaki seminars [84] for an update on this topic and for a precise account of the various contributions. Among the most important contributions are those of Kasparov and Higson, who showed that the conjectured isomorphism holds for amenable groups. It also holds for real semisimple Lie groups, thanks in particular to the work of A. Wassermann. Moreover, the recent work of V. Lafforgue [69] crossed the barrier of property T, showing that it holds for cocompact subgroups of rank-one Lie groups and also of $\mathrm{SL}(3,\mathbb{R})$ or of p-adic Lie groups. He also gave the first general conceptual proof of the isomorphism for real or p-adic semisimple Lie groups. The proof of the isomorphism, for all connected locally compact groups, based on Lafforgue's work, has been obtained by J. Chabert, S. Echterhoff, and R. Nest [7]. The proof by G. Yu of the analogue (due to J. Roe) of the conjecture in the context of coarse geometry for metric spaces that are uniformly embeddable in Hilbert space and the work of G. Skandalis, J. L. Tu, J. Roe, and N. Higson on the groupoid case has very striking consequences, such as the injectivity of the map μ for exact $C_r^*(\Gamma)$, due to Kaminker, Guentner and Ozawa.

Finally, the independent results of Lafforgue and Mineyev–Yu [74] show that the conjecture holds for arbitrary hyperbolic groups (most of which have property T), and P. Julg was even able to prove the conjecture with coefficients for rank-one groups. This was the strongest existing positive result until its extension to arbitrary hyperbolic groups which has recently been achieved by Vincent Lafforgue. On the negative side, recent progress due to Gromov, Higson, Lafforgue, and Skandalis gives counterexamples to the general conjecture for locally compact groupoids, for the simple reason that the functor $G \to K_0(C_r^*(G))$ is not half-exact, unlike the functor given by the geometric group. This makes the general problem of computing $K(C_r^*(G))$ really interesting. It shows that besides determining the large class of locally compact groups, for which the original conjecture is valid, one should understand how to take homological algebra into account to deal with the correct general formulation.

It also raises many integrality questions in cyclic cohomology of both discrete groups and foliations since a number of natural cyclic cocycles take integral values on the range of the map μ from the geometric group to the analytic group [20].

In summary, the above gives in any of the listed examples a natural construction, based on index problems, of K-theory classes in the relevant algebra, and tools to decide if this construction exhausts all the K-theory. It also provides a classifying space, which gives a rough approximation "up to homotopy" of the singular quotient encoded by the noncommutative geometric description.

IV Differential geometry

The development of differential geometric ideas, including de Rham homology, connections and curvature of vector bundles, etc. took place during the 1980s thanks to cyclic cohomology, which I introduced in 1981, including the spectral sequence relating it to Hochschild cohomology [16]. This led Loday and Quillen to their interpretation of cyclic homology in terms of the homology of the Lie algebra of matrices, which was also obtained independently by Tsygan in [88]. My papers appeared in preprint form in 1982 [17] and were quoted by Loday and Quillen [71] (see also [18] and [6]).

The first role of cyclic cohomology was to obtain index formulas computing an index, by the pairing of a K-theory class with a cyclic cocycle. (See [15] for a typical example with a cyclic 2-cocycle on the algebra $C^\infty(\mathbb{T}^2_\theta)$ of smooth functions on the noncommutative 2-torus.) This pairing is a simple extension of the Chern–Weil theory of characteristic classes, using the following dictionary to relate geometrical notions to their algebraic counterpart in such a way that the latter is meaningful in the general noncommutative situation.

Space	Algebra
Vector bundle	Finite projective module
Differential form	(Class of) Hochschild cycle
de Rham current	(Class of) Hochschild cocycle
de Rham homology	Cyclic cohomology
Chern–Weil theory	Pairing $\langle K(\mathcal{A}), HC(\mathcal{A})\rangle$

The pairing $\langle K(\mathcal{A}), HC(\mathcal{A})\rangle$ has a very concrete form and we urge the reader to prove the following simple lemma to get the general flavor of these computations of differential geometric nature.

Lemma 1. *Let $\mathcal{A}$ be an algebra and φ a trilinear form on $\mathcal{A}$ such that*

- $\varphi(a_0, a_1, a_2) = \varphi(a_1, a_2, a_0)\ \forall\, a_j \in \mathcal{A}$;
- $\varphi(a_0a_1, a_2, a_3) - \varphi(a_0, a_1a_2, a_3) + \varphi(a_0, a_1, a_2a_3) - \varphi(a_2a_0, a_1, a_2) = 0$ $\forall\, a_j \in \mathcal{A}$.

Then the scalar $\varphi_n(E, E, E)^2$ *is invariant under homotopy for projectors (idempotents)* $E \in M_n(\mathcal{A})$.

In the example of the noncommutative torus, the cyclic 2-cocycle representing the fundamental class [15] gives an integrality theorem, which J. Bellissard showed to be the integrality of the Hall conductivity in the quantum Hall effect, when applied to a specific spectral projection of the Hamiltonian (see [24] for an account of the work of J. Bellissard).

Basically, by extending the Chern–Weil characteristic classes to the general framework, the theory allows for many concrete computations of a differential geometric nature on noncommutative spaces. Indeed, the purely K-theoretic description of the Atiyah–Singer index formula would be of little practical use if it were not supplemented by the explicit local formulas in terms of characteristic classes and of the Chern character. This is achieved in the general noncommutative framework by the "local index formula," which will be described below in more detail.

Cyclic cohomology was used at a very early stage [20] to obtain index theorems, whose implications could be formulated independently of the whole framework of noncommutative geometry. A typical example is the following strengthening of a well-known result of A. Lichnerowicz [70].

Theorem 2 ([20]). *Let* M *be a compact oriented manifold and assume that the* $\hat{A}$*-genus* $\hat{A}(M)$ *is nonzero (since* M *is not assumed to be a Spin manifold* $\hat{A}(M)$ *need not be an integer). Let then* F *be an integrable Spin subbundle of* TM*. There exists no metric on* F *for which the scalar curvature (of the leaves) is strictly positive* $(\geq \varepsilon > 0)$ *on* M.

The proof is based on the construction of cyclic cohomology classes (on the algebra of the foliation) associated to Gelfand–Fuchs cohomology. The main difficulty is in extending the cocycles to a subalgebra stable under the holomorphic functional calculus.

The reason for working with cyclic cocycles rather than with the (obviously dual) cyclic homology can be understood easily in the above example. Cyclic cocycles are functionals of a differential geometric nature and as such are, of course, not everywhere defined on the algebra of all continuous functions. There is, however, in general a strong compatibility between differentiability and continuity, which reflects itself in the *closability* of the densely defined operators of differential geometry. It is precisely this *closability* that is exploited in [20] to construct the *smooth* domain of the cocycles. In that way cyclic cocycles are closely related to unbounded operators in Hilbert space, each defining its own independent *smooth* domain.

The theory also showed, early on, the depth of the relation between the above classification of factors and the geometry of foliations. In a remarkable series of papers (see [60] for references), J. Heitsch and S. Hurder have analyzed the interplay between the vanishing of the Godbillon–Vey invariant of a compact foliated manifold (V, F) and the type of the von Neumann algebra of the foliation. Their work

[2] Note that φ has been uniquely extended to $M_n(\mathcal{A})$ using the trace on $M_n(\mathbb{C})$, i.e., $\varphi_n = \varphi \otimes \text{Trace}$.

culminates in the following beautiful result of S. Hurder [60]. If the von Neumann algebra is *semifinite*, then the Godbillon–Vey invariant *vanishes*. We have shown, in fact, that cyclic cohomology yields a stronger result, proving that if $\mathrm{GV} \neq 0$, then the central decomposition of M necessarily contains factors M, whose virtual modular spectrum is of finite covolume in $\mathbb{R}_+^*$.

Theorem 3 ([20]). *Let (V, F) be an oriented, transversally oriented, compact, foliated manifold* (codim $F = 1$). *Let M be the associated von Neumann algebra, and* $\mathrm{Mod}(M)$ *its flow of weights. Then, if the Godbillon–Vey class of (V, F) is different from* 0, *there exists an invariant probability measure for the flow* $\mathrm{Mod}(M)$.

One actually constructs an invariant measure for the flow $\mathrm{Mod}(M)$, exploiting the following remarkable property of the natural cyclic 1-cocycle τ on the algebra $\mathcal{A}$ of the transverse 1-jet bundle for the foliation. When viewed as a linear map δ from $\mathcal{A}$ to its dual, δ is an unbounded derivation, which is *closable*, and whose domain extends to the center Z of the von-Neumann algebra generated by $\mathcal{A}$. Moreover, δ vanishes on this center, whose elements $h \in Z$ can then be used to obtain new cyclic cocycles τ_h on $\mathcal{A}$. The pairing

$$L(h) = \langle \tau_h, \mu(x) \rangle$$

with the K-theory classes $\mu(x)$ obtained from the assembly map μ, which we had constructed with P. Baum (cf. the topology section), then gives a measure on Z, whose invariance under the flow of weights follows from the discreteness of the K-group. To show that it is nonzero, one uses an index formula that evaluates the cyclic cocycles, associated as above to the Gelfand–Fuchs classes, on the range of the assembly map μ.

Cyclic cohomology led H. Moscovici and myself to the first proof of the Novikov conjecture for hyperbolic groups.

Theorem 4 ([36]). *Any hyperbolic discrete group satisfies the Novikov conjecture.*

The proof is based on the higher index theorem for discrete groups and the analysis[3] of a natural dense subalgebra stable under holomorphic functional calculus in the reduced group C^*-algebra. Cyclic cohomology has had many other applications, which I will not describe here. The very important general result of excision was obtained in the work of Cuntz and Quillen [48, 49].

In summary, the basic notions of differential geometry extend to the noncommutative framework and, starting from the noncommutative algebra $\mathcal{A}$ of "coordinates," the first task is to compute both its Hochschild and cyclic cohomologies in order to get the relevant tools before proceeding further to its "geometric" structure.

One should keep in mind the new subtleties that arise from noncommutativity. For instance, unlike the de Rham cohomology, its noncommutative replacement, which is cyclic cohomology, is not graded but filtered. Also it inherits from the Chern character map a natural integral lattice. These two features play a basic role in the description of

[3] Due to Haagerup and Jolissaint.

the natural moduli space (or, more precisely, its covering Teichmüller space, together with a natural action of SL(2, $\mathbb{Z}$) on this space) for the noncommutative tori $\mathbb{T}^2_\theta$. The discussion parallels the description of the moduli space of elliptic curves, but involves the even cohomology instead of the odd cohomology [32].

V Quantized calculus

The infinitesimal calculus is built on the tension expressed in the basic formula

$$\int_a^b df = f(b) - f(a)$$

between the integral and the infinitesimal variation df. One comes to terms with this tension by developing the Lebesgue integral and the notion of differential form. At the intuitive level, the naive picture of the "infinitesimal variation" df as the increment of f for very nearby values of the variable is good enough for most purposes, so that there is no need in trying to create a theory of infinitesimals.

The scenario is different in noncommutative geometry, where quantum mechanics provides a natural stage for the calculus [24, 35]. It is, of course, a bit hard to pass from the classical stage, where one just deals with functions, to the new one, in which operators in a Hilbert space $\mathcal{H}$ play the central role, but one basic input of quantum mechanics is precisely that the intuitive notion of a real variable quantity should be modeled as a self-adjoint operator in $\mathcal{H}$. One gains a lot in doing so. The set of values of the variable is the spectrum of the operator, and the number of times a value is reached is the spectral multiplicity. Continuous variables (operators with continuous spectrum) coexist happily with discrete variables precisely because of noncommutativity of operators. Furthermore, we now have a perfect home for infinitesimals, namely, for variables that are smaller than ϵ for any ϵ, without being zero. Of course, requiring that the operator norm is smaller than ϵ for any ϵ is too strong, but one can be more subtle and ask that for any positive ϵ, one can condition the operator by a finite number of linear conditions, so that its norm becomes less than ϵ. This is a well-known characterization of compact operators in Hilbert space, and they are the obvious candidates for infinitesimals. The basic rules of infinitesimals are easy to check, for instance, the sum of two compact operators is compact, the product compact times bounded is compact and they form a two-sided ideal $\mathcal{K}$ in the algebra of bounded operators in $\mathcal{H}$.

The size of the infinitesimal $\epsilon \in \mathcal{K}$ is governed by the rate of decay of the decreasing sequence of its characteristic values $\mu_n = \mu_n(\epsilon)$ as $n \to \infty$. (By definition, $\mu_n(\epsilon)$ is the nth eigenvalue of the absolute value $|\epsilon| = \sqrt{\epsilon^*\epsilon}$.) In particular, for all real positive α, the following condition defines infinitesimals of order α:

$$\mu_n(\epsilon) = O(n^{-\alpha}) \quad \text{when } n \to \infty. \tag{1}$$

Infinitesimals of order α also form a two-sided ideal and, moreover,

$$\epsilon_j \text{ of order } \alpha_j \Rightarrow \epsilon_1\epsilon_2 \text{ of order } \alpha_1 + \alpha_2. \tag{2}$$

The other key ingredient in the new calculus is the integral

$$⨍.$$

It has the usual properties of additivity and positivity of the ordinary integral, but it allows one to recover the power of the usual infinitesimal calculus, by automatically neglecting the ideal of infinitesimals of order > 1

$$⨍ \epsilon = 0 \quad \forall \epsilon \ \mu_n(\epsilon) = o(n^{-1}). \tag{3}$$

By filtering out these operators, one passes from the original stage of the quantized calculus described above to a *classical* stage where, as we shall see later, the notion of locality finds its correct place.

Using (3), one recovers the above mentioned tension of the ordinary differential calculus, which allows one to neglect infinitesimals of higher order (such as $(df)^2$) in an integral expression.

We refer to [24] for the construction of the integral in the required generality, obtained by the analysis, mainly due to Dixmier [51], of the logarithmic divergence of the ordinary trace for an infinitesimal of order 1.

The first interesting concrete example is provided by pseudodifferential operators k on a differentiable manifold M. When k is of order 1 in the above sense, it is measurable and $⨍ k$ is the noncommutative residue of k [89]. It has a local expression in terms of the distribution kernel $k(x, y)$, $x, y \in M$. For k of order 1 in the above sense, the kernel $k(x, y)$ diverges logarithmically near the diagonal,

$$k(x, y) = -a(x) \log |x - y| + O(1) \quad (\text{for } y \to x), \tag{4}$$

where $a(x)$ is a 1-density independent of the choice of Riemannian distance $|x - y|$. Then one has (up to normalization),

$$⨍ k = \int_M a(x). \tag{5}$$

The right-hand side of this formula makes sense for all pseudodifferential operators (cf. [89]), since one can easily see that the kernel of such an operator is asymptotically of the form

$$k(x, y) = \sum a_n(x, x - y) - a(x) \log |x - y| + O(1), \tag{6}$$

where $a_n(x, \xi)$ is homogeneous of degree $-n$ in ξ, and the 1-density $a(x)$ is defined intrinsically, since the logarithm does not mix with rational terms under a change of local coordinates.

What is quite remarkable is that this allows one to extend the domain of the integral $⨍$ to infinitesimals that are of order < 1, hence to obtain a computable answer to questions that would be meaningless in the ordinary calculus—the prototype being "what is the area of a four manifold?"—which we shall discuss below. The same principle of extension of $⨍$ to infinitesimals of order < 1 turns out to work in much greater generality. It works, for instance, for hypoelliptic operators and more generally for spectral triples, whose dimension spectrum is simple, as we shall see below.

VI Metric geometry and spectral action

With the above calculus as a tool, we now have a home for infinitesimals and can come back to the two basic notions introduced by Riemann in the classical framework, those of *manifold* and of *line element* [82]. We have shown that both of these notions adapt remarkably well to the noncommutative framework and lead to the notion of spectral triple, on which noncommutative geometry is based (cf. [35] for an overall presentation and [26, 25, 55] for the more technical aspects). This definition is entirely spectral: the elements of the algebra are operators, the points, if they exist, come from the joint spectrum of operators, and the line element is an operator. In a *spectral triple*

$$(\mathcal{A}, \mathcal{H}, D), \tag{1}$$

the algebra $\mathcal{A}$ of coordinates is concretely represented on the Hilbert space $\mathcal{H}$ and the operator D is an unbounded self-adjoint operator, which is the inverse of the line element,

$$ds = 1/D. \tag{2}$$

The basic properties of such spectral triples are easy to formulate and do not make any reference to the commutativity of the algebra $\mathcal{A}$. They are

$$[D, a] \text{ is bounded} \quad \text{for any } a \in \mathcal{A}, \tag{3}$$

$$D = D^* \quad \text{and} \quad (D + \lambda)^{-1} \text{ is a compact operator} \quad \forall \lambda \notin \mathbb{C}. \tag{4}$$

There is a simple formula for the distance in the general noncommutative case. It measures the distance between states,[4]

$$d(\varphi, \psi) = \text{Sup}\{|\varphi(a) - \psi(a)|;\ a \in \mathcal{A},\ \|[D, a]\| \leq 1\}. \tag{5}$$

The significance of D is twofold. On the one hand it defines the metric by the above equation, on the other hand its homotopy class represents the K-homology fundamental class of the space under consideration. In the classical geometric case, both the fundamental cycle in K-homology and the metric are encoded in the *spectral triple* $(\mathcal{A}, \mathcal{H}, D)$, where $\mathcal{A}$ is the algebra of functions acting in the Hilbert space $\mathcal{H}$ of spinors, while D is the Dirac operator. In some sense this encoding of Riemannian geometry takes a square root of the usual ansatz giving ds^2 as $g_{\mu\nu}dx^\mu dx^\nu$, the point

[4] Recall that a state is a normalized positive linear form on $\mathcal{A}$.

being that the Spin structure allows for the extraction of the square root of ds^2. (As is well known, Dirac found the corresponding operator as a differential square root of a Laplacian.)

The first thing one checks is that in the classical Riemannian case the geodesic distance $d(x, y)$ between two points is reobtained by

$$d(x, y) = \mathrm{Sup}\{|f(x) - f(y)|;\ f \in \mathcal{A},\ \|[D, f]\| \leq 1\}, \tag{6}$$

with $D = ds^{-1}$ as above, and where $\mathcal{A}$ is the algebra of smooth functions. Note that ds has the dimension of a length L, D has dimension L^{-1}, and the above expression for $d(x, y)$ also has the dimension of a length. It is also important to notice that we do not have to give the algebra of smooth functions. Indeed, imagine we are just given the von Neumann algebra $\mathcal{A}''$ in $\mathcal{H}$, weak closure of $\mathcal{A}$. How do we recover the subalgebra $C^\infty(M)$ of smooth functions? This is hopeless without using D, since the pair $(\mathcal{A}'', \mathcal{H})$ contains no more information than the multiplicity of this representation of a Lebesgue measure space. (Recall that they are all isomorphic in the measure category.) Using D, however, the answer is as follows. We shall say that an operator T in $\mathcal{H}$ is *smooth* iff the following map is smooth:

$$t \to F_t(T) = e^{it|D|}Te^{-it|D|} \in C^\infty(\mathbb{R}, \mathcal{L}(\mathcal{H})). \tag{7}$$

We let

$$OP^0 = \{T \in \mathcal{L}(\mathcal{H});\ T \text{ is smooth}\}. \tag{8}$$

It is then an exercise to show, in the Riemannian context, that

$$C^\infty(M) = OP^0 \cap L^\infty(M),$$

where $L^\infty(M) = \mathcal{A}''$ is the von Neumann algebra weak closure in $\mathcal{H}$.

In the general context, the flow (7) plays the role of the geodesic flow, assuming the following regularity hypothesis on $(\mathcal{A}, \mathcal{H}, D)$:

$$a \text{ and } [D, a] \in \cap \operatorname{Dom} \delta^k \quad \forall a \in \mathcal{A}, \tag{9}$$

where δ is the derivation $\delta(T) = [|D|, T]$, for any operator T. This derivation is the generator of the geodesic flow.

The usual notion of *dimension* of a space is replaced by the *dimension spectrum*, which is the subset Σ of $\{z \in \mathbb{C}, \mathrm{Re}(z) \geq 0\}$ of singularities of the analytic functions

$$\zeta_b(z) = \mathrm{Trace}(b|D|^{-z}), \quad \mathrm{Re}\, z > p, \quad b \in \mathcal{B}, \tag{10}$$

where we let $\mathcal{B}$ denote the algebra generated by $\delta^k(a)$ and $\delta^k([D, a])$, for $a \in \mathcal{A}$. The dimension spectrum Σ is, of course, bounded above by the crude dimension provided by the growth of eigenvalues of D, or equivalently by the order of the infinitesimal ds. In essence, the dimension spectrum is the set of complex numbers where the space under consideration becomes visible from the classical standpoint of the integral $⨍$.

The dimension spectrum of an ordinary manifold M is the set $\{0, 1, \ldots, n\}$, $n = \dim M$; it is simple. Multiplicities appear for singular manifolds. Cantor sets provide examples of complex points $z \notin \mathbb{R}$ in the dimension spectrum.

Going back to the usual Riemannian case, one checks that one recovers the volume form of the Riemannian metric by the equality (valid up to a normalization constant [24])

$$⨍ f |ds|^n = \int_{M_n} f \sqrt{g} d^n x, \tag{11}$$

but the first interesting point is that besides this coherence with the usual computations, there are new simple questions we can ask now, such as "what is the two-dimensional measure of a four-manifold?" or, in other words, "what is its area?" Thus one should compute

$$⨍ ds^2. \tag{12}$$

From invariant theory, this should be proportional to the Hilbert–Einstein action. The direct computation has been done in [63], the result being

$$⨍ ds^2 = \frac{-1}{24\pi^2} \int_{M_4} r \sqrt{g} d^4 x, \tag{13}$$

where, as above, $dv = \sqrt{g} d^4 x$ is the volume form, $ds = D^{-1}$ the length element, i.e., the inverse of the Dirac operator, and r is the scalar curvature.

A spectral triple is, in effect, a fairly minimal set of data allowing one to start doing quantum field theory. First, the inverse

$$ds = D^{-1} \tag{14}$$

plays the role of the propagator for Euclidean fermions and allows one to start writing the contributions of Feynman graphs whose internal lines are fermionic. The gauge bosons then appear as derived objects through the simple issue of Morita equivalence. Indeed, to define the analogue of the operator D for the algebra of endomorphisms of a finite projective module over $\mathcal{A}$

$$\mathcal{B} = \mathrm{End}_{\mathcal{A}}(\mathcal{E}), \tag{15}$$

where $\mathcal{E}$ is a finite, projective, Hermitian right $\mathcal{A}$–module, requires the choice of a *Hermitian connection* on $\mathcal{E}$. Such a connection ∇ is a linear map $\nabla : \mathcal{E} \to \mathcal{E} \otimes_{\mathcal{A}} \Omega^1_D$, satisfying the rules [24]

$$\nabla(\xi a) = (\nabla \xi) a + \xi \otimes da \qquad \forall \xi \in \mathcal{E}, \quad a \in \mathcal{A}, \tag{16}$$

$$(\xi, \nabla \eta) - (\nabla \xi, \eta) = d(\xi, \eta) \qquad \forall \xi, \eta \in \mathcal{E}, \tag{17}$$

where $da = [D, a]$ and where $\Omega^1_D \subset \mathcal{L}(\mathcal{H})$ is the $\mathcal{A}$–bimodule of operators of the form

$$A = \Sigma a_i [D, b_i], \quad a_i, b_i \in \mathcal{A}. \tag{18}$$

Any algebra $\mathcal{A}$ is Morita equivalent to itself (with $\mathcal{E} = \mathcal{A}$), and when one applies the construction above in this context, one gets the inner deformations of the spectral geometry. (We ignore the real structure and refer to [26] for the full story.) These replace the operator D by

$$D \to D + A \tag{19}$$

where $A = A^*$ is an arbitrary self-adjoint operator of the form (18), where we disregard the real structure for simplicity. Analyzing the divergences of the simplest diagrams with fermionic internal lines, as proposed early on in [23], provides perfect candidates for the counterterms, and hence the bosonic self-interactions. Such terms are readily expressible as residues or Dixmier traces and are gauge invariant by construction. The basic results are the following:

- In the above general context of NCG and in dimension 4, the obtained counterterms are a sum of a Chern–Simons action associated to a cyclic 3-cocycle on the algebra $\mathcal{A}$ and a Yang–Mills action expressed from a Dixmier trace, along the lines of [23] and [24]. The main additional hypothesis is the vanishing of the "tadpole," which expresses that one expands around an extremum.
- In the above generality exactly the same terms appear in the spectral action $\langle N(\Lambda)\rangle$ as the terms independent of the cutoff parameter Λ.

The spectral action was defined in [9] and computed there for the natural spectral triple describing the standard model. We refer to [64] for the detailed calculation. The overall idea of the approach is to use the above more flexible geometric framework to model the geometry of space-time, starting from the observed Lagrangian of gravity coupled with matter. The usual paradigm guesses space-time from the Maxwell part of the Lagrangian concludes that it is Minkowski space and then adds more and more particles to account for new terms in the Lagrangian. We start instead from the full Lagrangian and derive the geometry of space-time directly from this *empirical* data. The only rule is that we want a theory that is pure gravity, with the action functional given by the spectral action $\langle N(\Lambda)\rangle$ explained below, with an added fermionic term

$$S = \langle N(\Lambda)\rangle + \langle \psi, D\psi\rangle.$$

Note that D here stands for the Dirac operator with all its decorations, such as the inner fluctuations A explained above. Thus D stands for $D + A$, but this decomposition is an artifact of the standard distinction between gravity and matter, which is irrelevant in our framework.[5] The gauge bosons appear as the *inner* part of the metric, in the same way as the invariance group, which is the noncommutative geometry analogue of the group of diffeomorphisms, contains inner automorphisms as a normal subgroup (corresponding to the internal symmetries in physics).

The phenomenological Lagrangian of physics is the Einstein Lagrangian plus the minimally coupled standard model Lagrangian. The fermionic part of this action is used to determine a spectral triple $(\mathcal{A}, \mathcal{H}, D)$, where the algebra $\mathcal{A}$ determines the effective space-time from the internal symmetries and yields an answer which differs from the usual space-time (coming from QED). The Hilbert space $\mathcal{H}$ encodes not only the ordinary spinors (coming from QED) but all quarks and leptons, and the operator D encodes not only the ordinary Dirac operator but also the Yukawa coupling matrix.

[5] The spectral action is clearly superior to the Dixmier trace (residue) version of the Yang–Mills action in that it does not use this artificial splitting.

Then one recovers the bosonic part as follows. The Hilbert–Einstein action functional for the Riemannian metric, the Yang–Mills action for the vector potentials, the self-interaction, and the minimal coupling for the Higgs fields all appear with the correct signs in the asymptotic expansion for large Λ of the number $N(\Lambda)$ of eigenvalues of D that are $\leq \Lambda$ (cf. [9]),

$$N(\Lambda) = \#\text{ eigenvalues of } D \text{ in } [-\Lambda, \Lambda]. \tag{20}$$

This step function $N(\Lambda)$ is the superposition of two terms,

$$N(\Lambda) = \langle N(\Lambda) \rangle + N_{\text{osc}}(\Lambda).$$

The oscillatory part $N_{\text{osc}}(\Lambda)$ is the same as for a random matrix, governed by the statistic dictated by the symmetries of the system, and does not concern us here. The average part $\langle N(\Lambda) \rangle$ is computed by a semiclassical approximation from local expressions involving the familiar heat equation expansion and delivers the correct terms. Other nonzero terms in the asymptotic expansion are cosmological, Weyl gravity and topological terms. In general, the average part $\langle N(\Lambda) \rangle$ is given as a sum of residues. Assuming that the dimension spectrum Σ is simple, it is given by

$$\langle N(\Lambda) \rangle := \sum_{k>0} \frac{\Lambda^k}{k} \operatorname{Res}_{s=k} \zeta_D(s) + \zeta_D(0), \tag{21}$$

where the sum is over $k \in \Sigma$ and

$$\zeta_D(s) = \operatorname{Trace}(|D|^{-s}).$$

For instance, for Spec $D \subset \mathbb{Z}$ and $P(n)$ the total multiplicity of $\{\pm n\}$, for a polynomial P, one has

$$\langle N(\Lambda) \rangle = \int_0^\Lambda P(u)du + \text{cst},$$

which smoothly interpolates through the irregular step function

$$N(\Lambda) = \sum_0^\Lambda P(n).$$

As explained above, the Yang–Mills action appears in general as a part of the spectral action, but can also be defined directly using the calculus. This analogue of the Yang–Mills action functional and the classification of Yang–Mills connections on noncommutative tori were developed in [30], with the primary goal of finding a "manifold shadow" for these noncommutative spaces. These moduli spaces turned out indeed to fit this purpose perfectly, allowing us, for instance, to find the usual Riemannian space of gauge equivalence classes of Yang–Mills connections as an invariant of the noncommutative metric. We refer to [24] for the construction of the metrics on noncommutative tori from the conceptual point of view and to [25] for the verification that all natural axioms of noncommutative geometry are fulfilled in that case.

Gauge theory on noncommutative tori was shown to be relevant in string theory compactifications in [32]. Indeed, both the noncommutative tori and the components ∇_j of the Yang–Mills connections occur naturally in the classification of the BPS states in M-theory [32]. In the matrix formulation of M-theory, the basic equations to obtain periodicity of two of the basic coordinates X_i turn out to be the following:

$$U_i X_j U_i^{-1} = X_j + a\delta_i^j, \quad i = 1, 2, \tag{22}$$

where the U_i are unitary gauge transformations. The multiplicative commutator

$$U_1 U_2 U_1^{-1} U_2^{-1}$$

is then central and, in the irreducible case, its scalar value $\lambda = \exp 2\pi i\theta$ brings in the algebra of coordinates on the noncommutative torus. The X_j are then the components of the Yang–Mills connections. It is quite remarkable that the same picture emerged from the other information one has about M-theory, concerning its relation with 11-dimensional supergravity, and that string theory dualities could be interpreted using Morita equivalence. The latter [79] relates the values of θ on an orbit of $\mathrm{SL}(2, \mathbb{Z})$, and this type of relation, which is obvious from the foliation point of view [15], would be invisible in a purely deformation theoretic perturbative expansion like the one given by the Moyal product. The gauge theories on noncommutative 4-space were used very successfully in [76] to give a conceptual meaning to the compactifications of moduli spaces of instantons on $\mathbb{R}^4$ in terms of instantons on noncommutative $\mathbb{R}^4$. The corresponding spectral triple has been shown to fit in the general framework of NCG in [52] and the spectral action has been computed in [53]. These constructions apply to flat spaces, but were greatly generalized to isospectral deformations of Riemannian geometries of rank > 1 in [39, 40]. We shall not review here the renormalization of QFT on noncommutative spaces, but we simply refer to a recent remarkable positive result by H. Grosse and R. Wulkenhaar [56].

In summary, we now have at our disposal an operator theoretic analogue of the "calculus" of infinitesimals and a general framework of geometry.

In general, given a noncommutative space with coordinate algebra $\mathcal{A}$, the determination of corresponding geometries $(\mathcal{A}, \mathcal{H}, D)$ is obtained in two independent steps:

1. The first step consists of presenting the algebraic relations between coordinates $x \in \mathcal{A}$ and the inverse line element D, the simplest instance being the equation

   $$U^{-1}[D, U] = 1$$

   fixing the geometry of the one-dimensional circle. We refer to [39, 40] for noncommutative versions of this equation in dimension 3 and 4. Basically, this fixes the volume form v as a Hochschild cocycle and then allows for arbitrary metrics with v as volume form as solutions.
2. Once the algebraic relations between coordinates and the inverse line element have been determined, the second step is to find irreducible representations of

these relations in Hilbert space. Different metrics will correspond to the various inequivalent irreducible representations of the pair $(\mathcal{A}, D)$, fulfilling the prescribed relations. One guiding principle is that the homotopy class of such a representation should yield a nontrivial K-homology class on $\mathcal{A}$, playing the role of the "fundamental class." The corresponding index problem yields the analogue of the Pontrjagin classes and of "curvature," as discussed below. In order to compare noncommutative metrics, it is very natural to use spectral invariants, such as the spectral action mentioned above.

In many ways, the above two steps parallel the description of particles as irreducible representations of the Poincaré group. We thus view a given geometry as an irreducible representation of the algebraic relations between the coordinates and the line element, while the choice of such representations breaks the natural invariance group of the theory. The simplest instance of this view of geometry as a symmetry breaking phenomenon is what happens in the Higgs sector of the standard model.

VII Metric geometry, the local index formula

In the spectral noncommutative framework, the next appearance of the notion of curvature (besides the above spectral one) comes from the local computation of the analogue of Pontrjagin classes, i.e., of the components of the cyclic cocycle that is the Chern character of the K-homology class of D, which make sense in general. This result allows us, using the infinitesimal calculus, to go from local to global in the general framework of spectral triples $(\mathcal{A}, \mathcal{H}, D)$.

The Fredholm index of the operator D determines (we only look at the odd case for simplicity but there are similar formulas in the even case) an additive map $K_1(\mathcal{A}) \xrightarrow{\varphi} \mathbb{Z}$, given by the equality

$$\varphi([u]) = \text{Index}(PuP), \quad u \in GL_1(\mathcal{A}), \tag{1}$$

where P is the projector $P = \frac{1+F}{2}$, $F = \text{Sign}(D)$.

It is an easy fact that this map is computed by the pairing of $K_1(\mathcal{A})$ with the cyclic cocycle

$$\tau(a^0, \ldots, a^n) = \text{Trace}(a^0[F, a^1] \ldots [F, a^n]) \quad \forall a^j \in \mathcal{A}, \tag{2}$$

where $F = \text{Sign}\, D$, and we assume that the dimension p of our space is finite, which means that $(D + i)^{-1}$ is of order $1/p$, also $n \geq p$ is an odd integer. There are similar formulas involving the grading γ in the even case, and it is quite satisfactory [21, 61] that both cyclic cohomology and the Chern character formula adapt to the infinite-dimensional case, in which the only hypothesis is that $\exp(-D^2)$ is a trace class operator.

The cocycle τ is, however, nonlocal in general, because the formula (2) involves the ordinary trace instead of the local trace $⨍$ and it is crucial to obtain a local form of the above cocycle.

This problem is solved by the "local index formula" [37], under the regularity hypothesis (9) on $(\mathcal{A}, \mathcal{H}, D)$.

We assume that the dimension spectrum Σ is discrete and simple and refer to [37] for the case of a spectrum with multiplicities. Let $(\mathcal{A}, \mathcal{H}, D)$ be a regular spectral triple with simple dimension spectrum; the local index theorem is the following [37].

Theorem 5.

- *The equality*

$$\int\!\!\!\!\!\!- P = \mathrm{Res}_{z=0}\, \mathrm{Trace}(P|D|^{-z})$$

defines a trace on the algebra generated by $\mathcal{A}$, $[D, \mathcal{A}]$, and $|D|^z$, where $z \in \mathbb{C}$.
- *There is only a finite number of nonzero terms in the following formula defining the odd components $(\varphi_n)_{n=1,3,\ldots}$ of a cocycle in the bicomplex (b, B) of $\mathcal{A}$,*

$$\varphi_n(a^0, \ldots, a^n) = \sum_k c_{n,k} \int\!\!\!\!\!\!- a^0[D, a^1]^{(k_1)} \ldots [D, a^n]^{(k_n)}|D|^{-n-2|k|} \quad \forall a^j \in \mathcal{A}$$

where the following notations are used: $T^{(k)} = \nabla^k(T)$ and $\nabla(T) = D^2T - TD^2$, k is a multi-index, $|k| = k_1 + \cdots + k_n$,

$$c_{n,k} = (-1)^{|k|}\sqrt{2i}\,(k_1! \ldots k_n!)^{-1}((k_1+1) \ldots (k_1+k_2+\cdots+k_n+n))^{-1}\Gamma\left(|k| + \frac{n}{2}\right).$$

- *The pairing of the cyclic cohomology class $(\varphi_n) \in HC^*(\mathcal{A})$ with $K_1(\mathcal{A})$ gives the Fredholm index of D with coefficients in $K_1(\mathcal{A})$.*

We refer to [59] for a user friendly account of the proof. The first test of this general local index formula was the computation of the local Pontrjagin classes in the case of foliations. Their transverse geometry is, as explained in [37], encoded by a spectral triple. The answer for the general case [38] was obtained thanks to a Hopf algebra $\mathcal{H}(n)$ only depending on the codimension n of the foliation. It allowed us to organize the computation and to encode algebraically the noncommutative curvature. It also dictated the correct generalization of cyclic cohomology for Hopf algebras [38, 42]. This extension of cyclic cohomology has been pursued with great success recently [65, 66, 58] and we shall use it later. In the above context of foliations, the index computation transits through the cyclic cohomology of the Hopf algebra $\mathcal{H}(n)$ and we showed that it coincides with the Gelfand–Fuchs cohomology [54].

Another case of great interest came recently from quantum groups, where the local index formula works fine and yields quite remarkable formulas involving a sequence of rational approximations to the logarithmic derivative of the Dedekind eta function, even in the simplest case of $SU_q(2)$ (cf. [8, 43]). What is very striking in this example is that it displays the meaning of *locality* in the noncommutative framework. This notion requires no definition in the usual topological framework but would appear far more elusive in the noncommutative case without such concrete examples. What it means is that one works at ∞ in momentum space, but with very precise rules that allow one to strip all formulas from irrelevant details that won't have any effect in the computation of residues. We urge the reader to look at the concrete computations of [43] to really appreciate this point.

After the breakthrough of [50] showing that, contrary to a well-established negative belief, one could get nontrivial spectral geometries associated to quantum homogeneous spaces, it was shown in [77] how the above local index formula should be adapted to deal with the situation when the *principal symbol* of D^2 is no longer scalar because of a q-twist. Finally, recent work in the general case of quantum flag manifolds opens the way to a large class of examples, on which the above machinery should be tested and improved [67].

An open question of great relevance in the general framework of the analysis of spectral triples is to associate to any spectral triple $(\mathcal{A}, \mathcal{H}, D)$ the coarse geometry of its "momentum space" P. This space should be an ordinary metric space, with growth exactly governed by the spectrum of $|D|$ (which would give the set of distances to the origin). When viewed in the sense of the coarse geometry of John Roe, the space P should give an accurate description of the "infinitesimal" structure of the noncommutative space given by the spectral triple $(\mathcal{A}, \mathcal{H}, D)$. In particular, the classification by M. Gromov of discrete groups with polynomial growth should be extended to show that the "local structure" of noncommutative finite-dimensional manifolds is essentially of nilpotent nature. (The map from discrete groups G to spectral triples is the action of the group ring in $l^2(\Gamma)$. With $|D|$ being the multiplication by the word metric, it only takes care of the absolute value of $|D|$.)

VIII Renormalization, residues, and locality

At about the same time as the Hopf algebra $\mathcal{H}(n)$, another Hopf algebra was independently discovered by Dirk Kreimer, as the organizing concept in the computations of renormalization in quantum field theory.

His Hopf algebra is commutative as an algebra, and we showed in [33] that it is the dual Hopf algebra of the enveloping algebra of a Lie algebra $\underline{G}$, whose basis is labeled by the one-particle irreducible Feynman graphs. The Lie bracket of two such graphs is computed from insertions of one graph in the other and vice versa. The corresponding Lie group G is the group of characters of $\mathcal{H}$.

We showed that the group G is a semidirect product of an easily understood abelian group by a highly nontrivial group closely tied up with groups of diffeomorphisms.

Our joint work shows that the essence of the concrete computations performed by physicists using the renormalization technique is conceptually understood as a special case of a general principle of multiplicative extraction of finite values coming from the Birkhoff decomposition in the Riemann–Hilbert problem. The Birkhoff decomposition is the factorization

$$\gamma(z) = \gamma_-(z)^{-1}\gamma_+(z), \quad z \in C, \tag{1}$$

where we let $C \subset P_1(\mathbb{C})$ be a smooth simple curve, C_- the component of the complement of C containing $\infty \notin C$ and C_+ the other component. Both γ and $\gamma_\pm$ are loops with values in G,

$$\gamma(z) \in G \quad \forall z \in \mathbb{C}$$

and $\gamma_\pm$ are boundary values of holomorphic maps (still denoted by the same symbol),

$$\gamma_\pm : C_\pm \to G. \tag{2}$$

The normalization condition $\gamma_-(\infty) = 1$ ensures that if it exists, the decomposition (2) is unique (under suitable regularity conditions).

When G is a simply connected nilpotent complex Lie group, the existence (and uniqueness) of the Birkhoff decomposition (2) is valid for any γ. When the loop $\gamma : C \to G$ extends to a holomorphic loop: $C_+ \to G$, the Birkhoff decomposition is given by $\gamma_+ = \gamma$, $\gamma_- = 1$. In general, for $z \in C_+$, the evaluation

$$\gamma \to \gamma_+(z) \in G \tag{3}$$

is a natural principle to extract a finite value from the singular expression $\gamma(z)$. This extraction of finite values coincides with the removal of the pole part when G is the additive group $\mathbb{C}$ of complex numbers and the loop γ is meromorphic inside C_+ with z as its only singularity. It is convenient, in fact, to use the decomposition relative to an infinitesimal circle C_+ around z.

The main result of our joint work [34] is that the renormalized theory is just the evaluation at $z = D$ of the holomorphic part γ_+ of the Birkhoff decomposition of the loop γ with values in G provided by the dimensional regularization.

In fact, the relation that we uncovered in [34] between the Hopf algebra of Feynman graphs and the Hopf algebra of coordinates on the group of formal diffeomorphisms of the dimensionless coupling constants of the theory allows us to prove the following result, which for simplicity deals with the case of a single dimensionless coupling constant.

Theorem 6 ([34]). *Let the unrenormalized effective coupling constant* $g_{\text{eff}}(\varepsilon)$ *be viewed as a formal power series in* g *and let* $g_{\text{eff}}(\varepsilon) = g_{\text{eff}_+}(\varepsilon)(g_{\text{eff}_-}(\varepsilon))^{-1}$ *be its (opposite) Birkhoff decomposition in the group of formal diffeomorphisms. Then the loop* $g_{\text{eff}_-}(\varepsilon)$ *is the bare coupling constant and* $g_{\text{eff}_+}(0)$ *is the renormalized effective coupling.*

This allows us, using the relation between the Birkhoff decomposition and the classification of holomorphic bundles, to encode geometrically the operation of renormalization. It also signals a very clear analogy between the renormalization group as an "ambiguity" group of physical theories and the missing Galois theory at Archimedean places alluded to above. We refer to [28] for more information.

Note also that the residue, which is the cornerstone of our integral calculus (Section V), plays a key role in the Birkhoff decomposition. Indeed, we showed in [34] that the negative part in the Birkhoff decomposition (the part coming from divergences and giving the counterterms) is entirely determined by its residue (the term in $1/\epsilon$) in the dimensional regularization, a strong form of the 't Hooft relations in QFT. This then gives much weight to the idea that *local functionals* are best expressed in the general context of NCG as noncommutative integrals, i.e., residues.

With this conceptual understanding of renormalization at hand, the next obvious question is to match it with the above framework of NCG and apply it to the spectral action. One very nice feature of the framework of NCG is that it allows for the *dressing* of the geometry. Indeed, in quantum field theory the Dirac propagator undergoes a whole series of quantum corrections that provide a formal power series in powers of $\hbar$. In our framework, these corrections mean that the whole geometry is affected by the quantum field theory. This directly fine tunes its fundamental ingredient, which is the "line element" $ds = \times\!\!-\!\!\!-\!\!\times$. Of course, it is necessary, in order to really formulate things coherently, to pass to the second quantized Hilbert space, rather than staying in the one particle Hilbert space. This remains to be done in the general framework of NCG, but one can already appreciate a direct benefit of passing to the second quantized Hilbert space. Indeed, if we concentrate on space (versus space-time), its "line element" $ds = \times\!\!-\!\!\!-\!\!\times$ becomes positive, as the inverse of the Dirac Hamiltonian, thanks to the "Dirac sea" construction, which makes a preferred choice of the spin representation of the infinite-dimensional Clifford algebra. What remains at the second quantized level of the cohomological significance of ds (as a generator of Poincaré duality in K-homology) should be captured by a "regulator" pairing with algebraic K-theory along the lines of [22].

IX Modular forms and the space of $\mathbb{Q}$-lattices

We shall end this short presentation of the subject with examples of noncommutative spaces, which appeared in our joint work with H. Moscovici [44] and M. Marcolli [45] and have obvious relevance in number theory. Modular forms already appeared in noncommutative geometry in the classification of noncommutative three-spheres [40, 41], where hard computations with the noncommutative analogue of the Jacobian, involving the ninth power of the Dedekind eta function, were necessary in order to analyze the relation between such spheres and noncommutative nilmanifolds.

The coexistence of two a priori unrelated structures on modular forms, namely, the algebra structure given by the pointwise product on the one hand and the action of the Hecke operators on the other, led us in [44] to associate to any congruence subgroup Γ of $\mathrm{SL}(2, \mathbb{Z})$ a crossed product algebra $\mathcal{A}(\Gamma)$, the *modular Hecke algebra* of level Γ, which is a direct extension of both the ring of classical Hecke operators and of the algebra $\mathcal{M}(\Gamma)$ of Γ-modular forms. With $\mathcal{M}$ denoting the algebra of modular forms of arbitrary level, the elements of $\mathcal{A}(\Gamma)$ are maps with finite support

$$F : \Gamma\backslash \mathrm{GL}_2^+(\mathbb{Q}) \to \mathcal{M}, \qquad \alpha \mapsto F_\alpha \in \mathcal{M},$$

satisfying the covariance condition

$$F_{\alpha\gamma} = F_\alpha|\gamma \quad \forall \alpha \in \mathrm{GL}_2^+(\mathbb{Q}), \quad \gamma \in \Gamma \tag{1}$$

and their product is given by convolution

$$(F^1 * F^2)_\alpha := \sum_{\beta\in\Gamma\backslash \mathrm{GL}_2^+(\mathbb{Q})} F^1_\beta \cdot F^2_{\alpha\beta^{-1}}|\beta.$$

In the simplest case $\Gamma(1) = \mathrm{SL}(2,\mathbb{Z})$, the elements of $\mathcal{A}(\Gamma(1))$ are encoded by a finite number of modular forms $f_N \in \mathcal{M}(\Gamma_0(N))$ of arbitrary high level and the product operation is nontrivial.

Our starting point is the basic observation that the Hopf algebra $\mathcal{H}_1 = \mathcal{H}(1)$ of transverse geometry in codimension 1 mentioned above admits a natural action on the modular Hecke algebras $\mathcal{A}(\Gamma)$. As an algebra, $\mathcal{H}_1$ coincides with the universal enveloping algebra of the Lie algebra with basis $\{X, Y, \delta_n; n \geq 1\}$ and brackets

$$[Y, X] = X, \quad [Y, \delta_n] = n\delta_n, \quad [X, \delta_n] = \delta_{n+1}, \quad [\delta_k, \delta_\ell] = 0, \quad n, k, \ell \geq 1, \tag{2}$$

while the coproduct, which endows it with the Hopf algebra structure, is determined by the identities

$$\Delta Y = Y \otimes 1 + 1 \otimes Y, \qquad \Delta\delta_1 = \delta_1 \otimes 1 + 1 \otimes \delta_1,$$
$$\Delta X = X \otimes 1 + 1 \otimes X + \delta_1 \otimes Y,$$

together with the property that $\Delta : \mathcal{H}_1 \to \mathcal{H}_1 \otimes \mathcal{H}_1$ is an algebra homomorphism. The action of X on $\mathcal{A}(\Gamma)$ is given by a classical operator going back to Ramanujan, which corrects the usual differentiation by the logarithmic derivative of the Dedekind eta function $\eta(z)$. The action of Y is given by the standard grading by the weight (the Euler operator) on modular forms. Finally, δ_1 and its higher "derivatives" δ_n act by generalized cocycles on $\mathrm{GL}_2^+(\mathbb{Q})$ with values in modular forms.

One can then analyze this action of $\mathcal{H}_1$ through its cyclic cohomology. The latter admits three basic generators, with one of them cobounding in the periodized theory. We describe them and their meaning for foliations.

- *Godbillon–Vey cocycle*. Its class is represented by δ_1 which is a primitive element of $\mathcal{H}_1$, i.e., fulfills
$$\Delta\delta_1 = \delta_1 \otimes 1 + 1 \otimes \delta_1.$$
It is cyclic, so that $B(\delta_1) = 0$, where B is the boundary operator in cyclic cohomology, whose definition involves the antipode, the product and the coproduct. One shows that the class of δ_1,
$$[\delta_1] \in HC^1_{\mathrm{Hopf}}(\mathcal{H}_1),$$
is the generator of $PHC^{\mathrm{odd}}_{\mathrm{Hopf}}(\mathcal{H}_1)$ and corresponds to the Godbillon–Vey class in the isomorphism with Gelfand–Fuchs cohomology. It is this class which is responsible for the type III property of codimension one foliations explained above.
- *Schwarzian derivative*. The element $\delta_2' := \delta_2 - \frac{1}{2}\delta_1^2 \in \mathcal{H}_1$ is a Hopf cyclic cocycle, whose action in the foliation case corresponds to the multiplication by the Schwarzian derivative of the holonomy and whose class,
$$[\delta_2'] \in HC^1_{\mathrm{Hopf}}(\mathcal{H}_1),$$
is equal to $B(c)$, where c is the Hochschild 2-cocycle
$$c := \delta_1 \otimes X + \frac{1}{2}\delta_1^2 \otimes Y.$$

- *Fundamental class*. The generator of the even group $PHC^{\text{even}}_{\text{Hopf}}(\mathcal{H}_1)$ is the class of the cyclic 2-cocycle

$$F := X \otimes Y - Y \otimes X - \delta_1 Y \otimes Y,$$

which for foliations represents the transverse fundamental class.

We showed in [44] that each of the above cocycles admits a beautiful interpretation in its action on the modular Hecke algebras. First, the action of δ'_1, coupled with modular symbols, yields a rational representative for the Euler class of $\mathrm{GL}_2^+(\mathbb{Q})$. Next the action of δ'_2 is an inner derivation implemented by the modular form ω_4,

$$\omega_4 = -\frac{E_4}{72}, \quad E_4(q) := 1 + 240 \sum_1^\infty n^3 \frac{q^n}{1-q^n}.$$

Moreover, there is no way to perturb the action of the Hopf algebra $\mathcal{H}_1$ on the modular Hecke algebras so that δ'_2 vanishes, and the obstruction exactly agrees with the obstruction found by D. Zagier in his work on Rankin–Cohen algebras [90].

Next, the action of the fundamental class gives a Hochschild 2-cocycle which is the natural extension to modular Hecke algebras of the first Rankin–Cohen bracket. This led us to the following general result, which provides the correct notion of one-dimensional projective structure for noncommutative spaces and extends the Rankin–Cohen deformation in that generality. Let the Hopf algebra $\mathcal{H}_1$ act on an algebra $\mathcal{A}$, in such a way that the derivation δ'_2 is inner, implemented by an element $\Omega \in \mathcal{A}$,

$$\delta'_2(a) = \Omega a - a\Omega \quad \forall a \in \mathcal{A} \tag{3}$$

with

$$\delta_k(\Omega) = 0 \quad \forall k \in \mathbb{N}. \tag{4}$$

Such an action of $\mathcal{H}_1$ on an algebra $\mathcal{A}$ will be said to define a *projective structure* on $\mathcal{A}$, and the element $\Omega \in \mathcal{A}$ implementing the inner derivation δ'_2 will be called its *quadratic differential*.

Our main result in [44] is the construction of a sequence of brackets RC_*, which, applied to any algebra $\mathcal{A}$ endowed with a projective structure, yields a family of formal associative deformations of $\mathcal{A}$. The formulas for RC_n are completely explicit, but they rapidly become quite involved, as witnessed by the complexity of the following formula for RC_3, which gives the *bidifferential operator* expression for the third bracket. (The expression of RC_4 is much longer, it would occupy several pages.)

$$\begin{aligned} RC_3 = {} & -2X \otimes X^2 - 2X \otimes X^2.Y + 2X \otimes \alpha[\Omega].Y + 2X \otimes \alpha[\Omega].Y^2 + 2X^2 \otimes X \\ & + 6X^2 \otimes X.Y \\ & + 4X^2 \otimes X.Y^2 - \frac{2X^3 \otimes Y}{3} - 2X^3 \otimes Y^2 - \frac{4X^3 \otimes Y^3}{3} + \frac{2Y \otimes X^3}{3} \\ & - \frac{2}{3} Y \otimes \alpha[\Omega].X \end{aligned}$$

$$
\begin{aligned}
&-\frac{2}{3}Y\otimes\alpha[X[\Omega]].Y-2Y\otimes\alpha[\Omega].X.Y+2Y^2\otimes X^3-2Y^2\otimes\alpha[\Omega].X\\
&\quad-2Y^2\otimes\alpha[X[\Omega]].Y\\
&-6Y^2\otimes\alpha[\Omega].X.Y+\frac{4Y^3\otimes X^3}{3}-\frac{4}{3}Y^3\otimes\alpha[\Omega].X-\frac{4}{3}Y^3\otimes\alpha[X[\Omega]].Y\\
&\quad-4Y^3\otimes\alpha[\Omega].X.Y\\
&-6X.Y\otimes X^2-6X.Y\otimes X^2.Y+6X.Y\otimes\alpha[\Omega].Y+6X.Y\otimes\alpha[\Omega].Y^2\\
&\quad-4X.Y^2\otimes X^2\\
&-4X.Y^2\otimes X^2.Y+4X.Y^2\otimes\alpha[\Omega].Y+4X.Y^2\otimes\alpha[\Omega].Y^2+2X^2.Y\otimes X\\
&\quad+6X^2.Y\otimes X.Y\\
&+4X^2.Y\otimes X.Y^2-2\delta_1.X\otimes X-6\delta_1.X\otimes X.Y-4\delta_1.X\otimes X.Y^2\\
&\quad+2\delta_1.X^2\otimes Y+6\delta_1.X^2\otimes Y^2\\
&+4\delta_1.X^2\otimes Y^3+2\delta_1.Y\otimes X^2+2\delta_1.Y\otimes X^2.Y-2\delta_1.Y\otimes\alpha[\Omega].Y\\
&\quad-2\delta_1.Y\otimes\alpha[\Omega].Y^2\\
&+6\delta_1.Y^2\otimes X^2+6\delta_1.Y^2\otimes X^2.Y-6\delta_1.Y^2\otimes\alpha[\Omega].Y-6\delta_1.Y^2\otimes\alpha[\Omega].Y^2\\
&\quad+4\delta_1.Y^3\otimes X^2\\
&+4\delta_1.Y^3\otimes X^2.Y-4\delta_1.Y^3\otimes\alpha[\Omega].Y-4\delta_1.Y^3\otimes\alpha[\Omega].Y^2-\delta_1^2.X\otimes Y\\
&\quad-3\delta_1^2.X\otimes Y^2\\
&-2\delta_1^2.X\otimes Y^3+\delta_1^2.Y\otimes X+3\delta_1^2.Y\otimes X.Y+2\delta_1^2.Y\otimes X.Y^2+3\delta_1^2.Y^2\otimes X\\
&\quad+9\delta_1^2.Y^2\otimes X.Y\\
&+6\delta_1^2.Y^2\otimes X.Y^2+2\delta_1^2.Y^3\otimes X+6\delta_1^2.Y^3\otimes X.Y+4\delta_1^2.Y^3\otimes X.Y^2\\
&\quad+\frac{1}{3}\delta_1^3.Y\otimes Y+\delta_1^3.Y\otimes Y^2\\
&+\frac{2}{3}\delta_1^3.Y\otimes Y^3+\delta_1^3.Y^2\otimes Y+3\delta_1^3.Y^2\otimes Y^2+2\delta_1^3.Y^2\otimes Y^3+\frac{2}{3}\delta_1^3.Y^3\otimes Y\\
&\quad+2\delta_1^3.Y^3\otimes Y^2\\
&+\frac{4}{3}\delta_1^3.Y^3\otimes Y^3+\frac{2}{3}\alpha[\Omega].X\otimes Y+2\alpha[\Omega].X\otimes Y^2+\frac{4}{3}\alpha[\Omega].X\otimes Y^3\\
&\quad-2\alpha[\Omega].Y\otimes X\\
&-6\alpha[\Omega].Y\otimes X.Y-4\alpha[\Omega].Y\otimes X.Y^2-2\alpha[\Omega].Y^2\otimes X\\
&\quad-6\alpha[\Omega].Y^2\otimes X.Y-4\alpha[\Omega].Y^2\otimes X.Y^2\\
&+\frac{2}{3}\alpha[X[\Omega]].Y\otimes Y+2\alpha[X[\Omega]].Y\otimes Y^2+\frac{4}{3}\alpha[X[\Omega]].Y\otimes Y^3\\
&\quad-6\delta_1.X.Y\otimes X-18\delta_1.X.Y\otimes X.Y\\
&-12\delta_1.X.Y\otimes X.Y^2-4\delta_1.X.Y^2\otimes X-12\delta_1.X.Y^2\otimes X.Y
\end{aligned}
$$

$$
\begin{aligned}
&\quad - 8\delta_1.X.Y^2 \otimes X.Y^2 + 2\delta_1.X^2.Y \otimes Y \\
&+ 6\delta_1.X^2.Y \otimes Y^2 + 4\delta_1.X^2.Y \otimes Y^3 - 3\delta_1^2.X.Y \otimes Y - 9\delta_1^2.X.Y \otimes Y^2 \\
&\quad - 6\delta_1^2.X.Y \otimes Y^3 \\
&- 2\delta_1^2.X.Y^2 \otimes Y - 6\delta_1^2.X.Y^2 \otimes Y^2 - 4\delta_1^2.X.Y^2 \otimes Y^3 + 2\alpha[\Omega].X.Y \otimes Y \\
&\quad + 6\alpha[\Omega].X.Y \otimes Y^2 \\
&+ 4\alpha[\Omega].X.Y \otimes Y^3 - 2\alpha[\Omega].\delta_1.Y \otimes Y - 6\alpha[\Omega].\delta_1.Y \otimes Y^2 \\
&\quad - 4\alpha[\Omega].\delta_1.Y \otimes Y^3 - 2\alpha[\Omega].\delta_1.Y^2 \otimes Y \\
&- 6\alpha[\Omega].\delta_1.Y^2 \otimes Y^2 - 4\alpha[\Omega].\delta_1.Y^2 \otimes Y^3.
\end{aligned}
$$

The modular Hecke algebras turn out to be intimately related to the analysis of a very natural noncommutative space, which arose in a completely different context [45], having to do with the interplay between number theory and phase transitions with spontaneous symmetry breaking in quantum statistical mechanics, as initiated in [4]. The search for a two-dimensional analogue of the statistical system of [4] was obtained in [45], by first reinterpreting the latter from the geometry of the space of $\mathbb{Q}$-lattices in dimension one and then passing to two dimensions.

An n-dimensional $\mathbb{Q}$-lattice consists of an ordinary lattice Λ in $\mathbb{R}^n$ and a homomorphism

$$\phi : \mathbb{Q}^n/\mathbb{Z}^n \to \mathbb{Q}\Lambda/\Lambda.$$

Two such $\mathbb{Q}$-lattices are called *commensurable* if and only if the corresponding lattices are commensurable and the maps agree modulo the sum of the lattices.

The space $\mathcal{L}_n$ of commensurability classes of $\mathbb{Q}$-lattices in $\mathbb{R}^n$ turns out to be a very complicated noncommutative space, which appears to be of great number theoretical significance because of its relation to both the Riemann zeta function (for $n = 1$) and modular forms (for $n = 2$). In physics language, what emerges is that the zeros of zeta appear as an absorption spectrum in the L^2 space of the space of commensurability classes of one-dimensional $\mathbb{Q}$-lattices as in [31]. The noncommutative geometry description of the space of one-dimensional $\mathbb{Q}$-lattices modulo scaling recovers the quantum statistical mechanical system of [4], which exhibits a phase transition with spontaneous symmetry breaking. In a similar manner, we showed in [45] that the space of $\mathbb{Q}$-lattices in $\mathbb{C}$ modulo scaling generates a very interesting quantum statistical mechanical system whose ground states are parameterized by

$$\mathrm{GL}_2(\mathbb{Q})\backslash \mathrm{GL}_2(\mathbb{A})/\mathbb{C}^*,$$

while the natural symmetry group of the system is the quotient

$$S = \mathbb{Q}^*\backslash \mathrm{GL}_2(\mathbb{A}_f).$$

The values of a ground state φ on the natural *rational* subalgebra $\mathcal{A}_{\mathbb{Q}}$ of rational observables generates, in the generic case, a specialization $F_\varphi \subset \mathbb{C}$ of the modular field F. The state φ then intertwines the symmetry group S of the system with the Galois group of the modular field and there exists an isomorphism θ of S with $\mathrm{Gal}(F_\varphi/\mathbb{Q})$, such that

$$\alpha \circ \varphi = \varphi \circ \theta^{-1}(\alpha) \quad \forall \alpha \in \mathrm{Gal}(F_\varphi/\mathbb{Q}).$$

In general, while the zeros of zeta and L-functions appear at the critical temperature, the analysis of the low temperature equilibrium states concentrates on the Langlands space

$$\mathrm{GL}_n(\mathbb{Q})\backslash \mathrm{GL}_n(\mathbb{A}).$$

The subalgebra $\mathcal{A}_{\mathbb{Q}}$ of rational observables turns out to be intimately related to the modular Hecke algebras, which leads us to suspect that many of the results of [44] will survive in the context of [45], and will therefore be relevant in the analysis of the higher-dimensional analogue of the trace formula of [31].

References

[1] M. F. Atiyah, Global theory of elliptic operators, in *Proceedings of the International Conference on Functional Analysis and Related Topics* (*Tokyo*, 1969), University of Tokyo Press, Tokyo, 1970, 21–30.

[2] M. F. Atiyah and W. Schmid, A geometric construction of the discrete series for semisimple Lie groups, *Invent. Math.*, **42** (1977), 1–62.

[3] P. Baum and A. Connes, Geometric K-theory for Lie groups and foliations. *Enseign. Math.*, **46** (2000), 1–35.

[4] J.-B. Bost and A. Connes, Hecke Algebras, type III factors and phase transitions with spontaneous symmetry breaking in number theory, *Selecta Math. N.S.*, **1**-3 (1995), 411–457.

[5] L. G. Brown, R. G. Douglas, and P. A. Fillmore, Extensions of C^*-algebras and K-homology, *Ann. Math.* 2, **105** (1977), 265–324.

[6] P. Cartier, Homologie cyclique: Rapport sur les travaux récents de Connes, Karoubi, Loday, Quillen, in *Seminaire Bourbaki*, Vol. 1983/84, Astérisque 121–122, Société Mathématique de France, Paris, 1985, exposé 621, 123–146.

[7] J. Chabert, S. Echterhoff, and R. Nest, The Connes-Kasparov conjecture for almost connected groups and for linear p-adic groups, *Publ. Math. IHES*, **97** (2003), 239–278.

[8] P. S. Chakraborty and A. Pal, Equivariant spectral triple on the quantum $SU(2)$-group, *K-Theory*, **28**-2 (2003), 107–126.

[9] A. Chamseddine and A. Connes, Universal formulas for noncommutative geometry actions, *Phys. Rev. Lett.*, **77**-24 (1996), 4868–4871.

[10] A. Connes, Groupe modulaire d'une algèbre de von Neumann, *C. R. Acad. Sci. Paris Sér.* A–B, **274** (1972), 1923–1926.

[11] A. Connes, Une classification des facteurs de type III, *Ann. Sci. Ecole Norm. Sup.*, **6**-4 (1973), 133–252.

[12] A. Connes, Classification of injective factors, *Ann. Math.*, **104**-2 (1976), 73–115.

[13] A. Connes, Outer conjugacy classes of automorphisms of factors, *Ann. Sci. Ecole Norm. Sup.*, **8**-4 (1975), 383–419.

[14] A. Connes, The von Neumann algebra of a foliation, in *Mathematical Problems in Theoretical Physics* (*Proceedings of the International Conference, University of Rome, Rome*, 1977), Lecture Notes in Physics 80, Springer-Verlag, Berlin-New York, 1978, 145–151.

[15] A. Connes, C^* algèbres et géométrie différentielle, *C. R. Acad. Sci. Paris Sér.* A–B, **290**-13 (1980), A599–A604.
[16] A. Connes, Spectral sequence and homology of currents for operator algebras, *Math. Forschungsinst. Oberwolfach Tagungsbericht*, **41** (1981); *Funktionalanal. C^*-Algebren*, **27**-9 (1981), 3–10.
[17] A. Connes, Noncommutative differential geometry, *Inst. Hautes Etudes Sci. Publ. Math.*, **62** (1985), 257–360.
[18] A. Connes, Cohomologie cyclique et foncteur Ext^n, *C. R. Acad. Sci. Paris Ser.* I *Math.*, **296** (1983), 953–958.
[19] A. Connes, Factors of type III_1, property L'_λ and closure of inner automorphisms, *J. Operator Theory*, **14** (1985), 189–211.
[20] A. Connes, Cyclic cohomology and the transverse fundamental class of a foliation, in *Geometric Methods in Operator Algebras* (*Kyoto*, 1983), Pitman Research Notes in Mathematics 123, Longman, Harlow, UK, 1986, 52–144.
[21] A. Connes, Entire cyclic cohomology of Banach algebras and characters of θ-summable Fredholm modules, *K-Theory*, **1** (1988), 519–548.
[22] A. Connes and M. Karoubi, Caractère multiplicatif d'un module de Fredholm, *K-Theory*, **2**-3 (1988), 431–463.
[23] A. Connes, Essay on physics and noncommutative geometry, in *The Interface of Mathematics and Particle Physics* (*Oxford*, 1988), Instiitute of Mathematics and Its Applications Conference Series (New Series) 24, Oxford University Press, New York, 1990, 9–48.
[24] A. Connes, *Noncommutative Geometry*, Academic Press, New York, 1994.
[25] A. Connes, Gravity coupled with matter and foundations of noncommutative geometry, *Comm. Math. Phys.*, **182** (1996), 155–176.
[26] A. Connes, Noncommutative geometry and reality, *J. Math. Physics*, **36**-11 (1995), 6194–6231.
[27] A. Connes, Noncommutative geometry and the Riemann zeta function, in V. Arnold, M. Atiyah, P. Lax, and B. Mazur, eds., *Mathematics: Frontiers and Perspectives* 2000, American Mathematical Society, Providence, 2000, 35–55.
[28] A. Connes, Symétries Galoisiennes et renormalisation, in B. Duplantier and V. Rivasseau, eds., *Poincaré Seminar* 2002: *Vacuum Energy, Renormalization*, Progress in Mathematical Physics 30, Birkhäuser Boston, Cambridge, MA, 2003.
[29] A. Connes and M. Takesaki, The flow of weights on factors of type III, *Tohoku Math. J.*, **29** (1977), 473–575.
[30] A. Connes and M. Rieffel, Yang–Mills for noncommutative two tori, in P. E. T. Jorgensen and P. S. Muhly, eds., *Operator Algebras and Mathematical Physics: Proceedings*, Contemporary Mathematics 62, American Mathematical Society, Providence, 1987, 237–266.
[31] A. Connes, Trace formula in noncommutative geometry and the zeros of the Riemann zeta function, *Selecta Math.* (*N.S.*), **5** (1999), 29–106.
[32] A. Connes, M. Douglas, and A. Schwarz, Noncommutative geometry and matrix theory: Compactification on tori, *J. High Energy Phys.*, **2** (1998), 003.
[33] A. Connes and D. Kreimer, Hopf algebras, renormalization and noncommutative geometry, *Comm. Math. Phys.*, **199** (1998), 203–242.
[34] A. Connes and D. Kreimer, Renormalization in quantum field theory and the Riemann-Hilbert problem I: The Hopf algebra structure of graphs and the main theorem, II: The β function, diffeomorphisms and the renormalization group, *Comm. Math. Phys.*, **210** (2000), 249–273, **216** (2001), 215–241.
[35] A. Connes, A short survey of noncommutative geometry, *J. Math. Phys.*, **41** (2000), 3832–3866.

[36] A. Connes and H. Moscovici, Cyclic cohomology, the Novikov conjecture and hyperbolic groups, *Topology*, **29** (1990), 345–388.

[37] A. Connes and H. Moscovici, The local index formula in noncommutative geometry, *Geom. Functional Anal.*, **5** (1995), 174–243.

[38] A. Connes and H. Moscovici, Hopf algebras, cyclic cohomology and the transverse index theorem, *Comm. Math. Phys.*, **198** (1998), 199–246.

[39] A. Connes and G. Landi, Noncommutative manifolds, the instanton algebra and isospectral deformations, *Comm. Math. Phys.*, **221** (2001), 141–159.

[40] A. Connes and M. Dubois-Violette, Noncommutative finite-dimensional manifolds I: Spherical manifolds and related examples, preprint, 2001; math.QA/0107070.

[41] A. Connes and M. Dubois-Violette, Moduli space and structure of noncommutative 3-spheres, *Lett. Math. Phys.*, **66** (2003), 91–121.

[42] A. Connes and H. Moscovici, Cyclic cohomology and Hopf algebra symmetry, *Lett. Math. Phys.*, **52**-1 (2000), 1–28.

[43] A. Connes, Cyclic cohomology, quantum group symmetries and the local index formula for $SU_q(2)$, *J. Inst. Math. Jussieu*, **3**-1 (2004), 17–68.

[44] A. Connes and H. Moscovici, Modular Hecke algebras and their Hopf symmetry, *Moscow Math. J.*, **4**-1 (2003), 67–109.

A. Connes and H. Moscovici, Rankin-Cohen brackets and the Hopf algebra of transverse geometry, *Moscow Math. J.*, **4**-1 (2003), 111–139.

[45] A. Connes and M. Marcolli, *From physics to number theory via noncommutative geometry, Part* I: *Quantum statistical mechanics of* $\mathbb{Q}$*-lattices*, preprint, 2004; math.NT/0404128.

[46] C. Consani and M. Marcolli, Non-commutative geometry, dynamics, and infinity-adic Arakelov geometry, *Selecta Math.* (*N.S.*), **10**-2 (2004), 167–251.

[47] J. Cuntz and D. Quillen, On excision in periodic cyclic cohomology I, II, *C. R. Acad. Sci. Paris Ser.* I *Math.*, **317** (1993), 917–922, **318** (1994), 11–12.

[48] J. Cuntz and D. Quillen, Cyclic homology and nonsingularity, *J. Amer. Math. Soc.*, **8** (1995), 373–442.

[49] J. Cuntz and D. Quillen, Operators on noncommutative differential forms and cyclic homology, in S.-T. Yau, ed., *Geometry, Topology, and Physics for Raoul Bott*, Conference Proceedings and Lecture Notes in Geometry and Topology, International Press, Boston, 1995, 77–111.

[50] L. Dabrowski and A. Sitarz, Dirac operator on the standard Podles quantum sphere, in W. Pusz and P. M. Hajac, eds., *Noncommutative Geometry and Quantum Groups*, Banach Center Publications 61, Polish Academy of Sciences, Warsaw, 2003, 49–58.

[51] J. Dixmier, Existence de traces non normales, *C. R. Acad. Sci. Paris Ser.* A–B, **262** (1966), A1107–A1108.

[52] V. Gayral, J. M. Gracia-Bondía, B. Iochum, T. Schucker, and J. C. Varilly, Moyal planes are spectral triples, *Comm. Math. Phys.*, **246**-3 (2004), 569–623.

[53] V. Gayral and B. Iochum, The spectral action for Moyal planes, *J. Math. Phys.*, **46** (2005), 043503.

[54] I. M. Gelfand and D. B. Fuchs, Cohomology of the Lie algebra of formal vector fields, *Izv. Akad. Nauk SSSR*, **34** (1970), 322–337.

[55] J. M. Gracia-Bondia, J. C. Varilly, and H. Figueroa, *Elements of Noncommutative Geometry*, Birkhäuser Boston, Cambridge, MA, 2000.

[56] H. Grosse and R. Wulkenhaar, Renormalisation of ϕ^4-theory on noncommutative $\mathbb{R}^4$ in the matrix base, *Comm. Math. Phys.*, **256** (2005), 305–374.

[57] U. Haagerup, Connes' bicentralizer problem and uniqueness of the injective factor of type III_1, *Acta Math.*, **158** (1987), 95–148.

[58] P. M. Hajac, M. Khalkhali, B. Rangipour, and Y. Sommerhaeuser, Hopf-cyclic homology and cohomology with coefficients, *C. R. Acad. Sci. Paris Ser.* I *Math.*, **338** (2004), 667–672.

[59] N. Higson, The local index formula in noncommutative geometry, in *Lectures Given at the School and Conference on Algebraic K-Theory and Its Applications*, Trieste, Italy, 2002; available online from http://www.math.psu.edu/higson/ResearchPapers.html.

[60] S. Hurder, *Secondary Classes and the von Neumann Algebra of a Foliation*, preprint, Mathematical Sciences Research Institute, Berkeley, CA, 1983.

[61] A. Jaffe, A. Lesniewski, and K. Osterwalder, Quantum K-theory I: The Chern character, *Comm. Math. Phys.*, **118** (1988), 1–14.

[62] G. G. Kasparov, The operator K-functor and extensions of C^* algebras, *Izv. Akad. Nauk SSSR Ser. Mat.*, **44** (1980), 571–636; *Math. USSR Izv.*, **16** (1981), 513–572.

[63] D. Kastler, The Dirac operator and gravitation, *Comm. Math. Phys.*, **166** (1995), 633–643.

[64] D. Kastler, Noncommutative geometry and fundamental physical interactions: The lagrangian level, *J. Math. Phys.*, **41** (2000), 3867–3891.

[65] M. Khalkhali and B. Rangipour, A note on cyclic duality and Hopf algebras, *Comm. Algebra*, **33**-3 (2005), 763–773.

[66] M. Khalkhali and B. Rangipour, Cyclic cohomology of Hopf algebras and Hopf algebroids, preprint, 2003; math.KT/0303069.

[67] U. Kraehmer, Dirac operators on quantum flag manifolds, *Lett. Math. Phys.*, **67**-1 (2004), 49–59.

[68] W. Krieger, On ergodic flows and the isomorphism of factors, *Math. Ann.*, **223** (1976), 19–70.

[69] V. Lafforgue, K-théorie bivariante pour les algèbres de Banach et conjecture de Baum-Connes, *Invent. Math.*, **149** (2002), 1–97.

[70] A. Lichnerowicz, Spineurs harmoniques, *C. R. Acad. Sci. Ser.* A–B, **257** (1963), 7–9.

[71] J. L. Loday and D. Quillen, Homologie cyclique et homologie de l'algèbre de Lie des matrices, *C. R. Acad. Sci. Paris Ser.* A–B, **296** (1983), 295–297.

[72] Yu. I. Manin and M. Marcolli, Continued fractions, modular symbols, and noncommutative geometry, *Selecta Math.* (*N.S.*), **8**-3 (2002), 475–521.

[73] J. Milnor and D. Stasheff, *Characteristic Classes*, Annals of Mathematics Studies 76, Princeton University Press, Princeton, NJ, 1974.

[74] I. Mineyev and G. Yu, The Baum-Connes conjecture for hyperbolic groups, *Invent. Math.*, **149** (2002), 97–123.

[75] A. S. Miščenko, C^* algebras and K theory, in *Algebraic Topology: Aarhus* (1978), Lecture Notes in Mathematics 763, Springer-Verlag, Berlin, New York, Heidelberg, 1979, 262–274.

[76] N. Nekrasov and A. Schwarz, Instantons in noncommutative $\mathbb{R}^4$ and (2, 0) superconformal six dimensional theory, *Comm. Math. Phys.*, **198** (1998), 689–703.

[77] S. Neshveyev and L. Tuset, A local index formula for the quantum sphere, *Comm. Math. Phys.*, **254**-2 (2005), 323–341.

[78] M. Pimsner and D. Voiculescu, Exact sequences for K groups and Ext group of certain crossed product C^*-algebras, *J. Operator Theory*, **4** (1980), 93–118.

[79] M. A. Rieffel, Morita equivalence for C^*-algebras and W^*-algebras, *J. Pure Appl. Algebra*, **5** (1974), 51–96.

[80] M. A. Rieffel, C^*-algebras associated with irrational rotations, *Pacific J. Math.*, **93** (1981), 415–429.

[81] M. A. Rieffel, The cancellation theorem for projective modules over irrational rotation C^*-algebras, *Proc. London Math. Soc.*, **47** (1983), 285–302.

[82] B. Riemann, *Gesammelte mathematische Werke und wissenschaftlicher Nachlass*, Dover, New York, 1953.

[83] I. M. Singer, Future extensions of index theory and elliptic operators, *Ann. Math. Stud.*, **70** (1971), 171–185.

[84] G. Skandalis, Approche de la conjecture de Novikov par la cohomologie cyclique, in *Seminaire Bourbaki*, Vol. 1990/91, Astérisque 201–203, Société Mathématique de France, Paris, 1992, exposé 739, 299–316.

P. Julg, Travaux de N. Higson et G. Kasparov sur la conjecture de Baum-Connes, in *Seminaire Bourbaki*, Vol. 1997/98, Astérisque 252, Société Mathématique de France, Paris, 1998, exposé 841, 151–183.

G. Skandalis, Progres recents sur la conjecture de Baum-Connes, contribution de Vincent Lafforgue, in *Seminaire Bourbaki*, Vol. 1999/2000, Astérisque 276, Société Mathématique de France, Paris, 2002, exposé 869.

[85] D. Sullivan, *Geometric Periodicity and the Invariants of Manifolds*, Lecture Notes in Mathematics 197, Springer-Verlag, Berlin, New York, Heidelberg, 1971.

[86] M. Takesaki, *Tomita's Theory of Modular Hilbert Algebras and Its Applications*, Lecture Notes in Mathematics 128, Springer-Verlag, Berlin, New York, Heidelberg, 1970.

[87] M. Takesaki, Duality for crossed products and the structure of von Neumann algebras of type III, *Acta Math.*, **131** (1973), 249–310.

[88] B. L. Tsygan, Homology of matrix Lie algebras over rings and the Hochschild homology, *Uspekhi Math. Nauk.*, **38** (1983), 217–218.

[89] M. Wodzicki, *Noncommutative Residue, Part* I: *Fundamentals K-Theory, Arithmetic and Geometry*, Lecture Notes in Mathematics 1289, Springer-Verlag, Berlin, 1987, 320–399.

[90] D. Zagier, Modular forms and differential operators, *Proc. Indian Acad. Sci. Math. Sci.*, **104**-1 (1994) (K. G. Ramanathan memorial issue), 57–75.

Stable Distributions Supported on the Nilpotent Cone for the Group G_2

Stephen DeBacker[1] and David Kazhdan[2]

[1] Department of Mathematics
University of Michigan
2074 East Hall
530 Church Street
Ann Arbor, MI 48109-1043
USA
smdbackr@umich.edu

[2] Institute of Mathematics
Hebrew University
Givat-Ram
Jerusalem, 91904
Israel
kazhdan@math.huji.ac.il

Abstract. Assuming that p is sufficiently large, we describe the stable distributions supported on the set of nilpotent elements for p-adic G_2.

Key words: stability, harmonic analysis, reductive p-adic group

Subject Classifications: Primary 20E50

1 Introduction

We describe the stable distributions supported on the nilpotent cone for p-adic G_2.

1.1 The problem

Traditionally, the study of harmonic analysis on $\mathfrak{g}$, the Lie algebra of a reductive p-adic group G, was concerned with understanding invariant distributions (for example, orbital integrals and their Fourier transforms) for a single group. Many of the modern problems in harmonic analysis (for example, the fundamental lemma) are concerned with establishing identities between distributions for two different groups. For these problems, it is often more natural to consider stable distributions.

Unfortunately, our understanding of stable distributions is quite limited. In the early 1970s it was realized that a better understanding of invariant distributions could be gained by studying the distributions supported on the nilpotent cone. It is natural to expect that a similar approach will be useful in the study of stable distributions. However, this will require a description of the stable distributions supported on the nilpotent cone. Thanks to Waldspurger [28], for unramified classical groups we have such a description. In this paper we provide a description for p-adic G_2.

1.1.1 Stable distributions

To motivate the definition of stability, we begin by recalling a result of Harish-Chandra. Let k denote a characteristic zero nonarchimedean local field and suppose that $G = \mathrm{G}_2(k)$.

A *distribution* is an element of $C_c^\infty(\mathfrak{g})^*$, the linear dual of the space of locally constant, compactly supported functions on $\mathfrak{g}$. A distribution T is said to be *invariant* provided that $T(f^g) = T(f)$ for all $f \in C_c^\infty(\mathfrak{g})$ and all $g \in G$. Here $f^g(X) = f(\mathrm{Ad}(g)X)$ for $X \in \mathfrak{g}$ and $g \in G$.

Suppose $\mathcal{D}^{\mathrm{ann}}$ is the subspace of $C_c^\infty(\mathfrak{g})$ consisting of functions for which all regular semisimple orbital integrals are equal to zero (see Section 2.7). In [15], Harish-Chandra showed that a distribution $T \in C_c^\infty(\mathfrak{g})^*$ is invariant if and only if $\mathrm{res}_{\mathcal{D}^{\mathrm{ann}}} T = 0$. (Here $\mathrm{res}_{\mathcal{D}^{\mathrm{ann}}} T$ denotes the restriction of T to the subspace $\mathcal{D}^{\mathrm{ann}}$.) In other words, regular semisimple orbital integrals are dense in the space of invariant distributions.

We now define $J^{\mathrm{st}}(\mathfrak{g})$, the space of stable distributions on $\mathfrak{g}$, in a similar way. Suppose $X \in \mathfrak{g}$ is regular semisimple. There is a finite set $\{X_1, X_2, \ldots, X_n\}$ of regular semisimple elements of $\mathfrak{g}$ so that ${}^{\mathrm{G}_2(K)}X \cap \mathfrak{g}$ can be written as a disjoint union

$$ {}^{\mathrm{G}_2(K)}X \cap \mathfrak{g} = {}^G X_1 \sqcup {}^G X_2 \sqcup \cdots \sqcup {}^G X_n. $$

(Here K is a fixed maximal unramified extension of k and ${}^{\mathrm{G}_2(K)}X$ denotes the $\mathrm{G}_2(K)$-orbit of X in the Lie algebra of $\mathrm{G}_2(K)$.) After normalizing measures (see Section 2.7), we set

$$ S\mu_X = \sum_{\ell=1}^{n} \mu_{X_\ell}, $$

where μ_{X_ℓ} is the orbital integral associated to X_ℓ. We call $S\mu_X$ a *stable orbital integral*. The role of $\mathcal{D}^{\mathrm{ann}}$ is now played by $\mathcal{D}^{\mathrm{stann}}$, the space of functions that vanish on every stable orbital integral. We define

$$ J^{\mathrm{st}}(\mathfrak{g}) := \{T \in C_c^\infty(\mathfrak{g})^* \mid \mathrm{res}_{\mathcal{D}^{\mathrm{stann}}} T = 0\}. $$

Note that each stable distribution is an invariant distribution.

1.1.2 The main result

Let $J(\mathcal{N})$ denote the subspace of $C_c^\infty(\mathfrak{g})^*$ consisting of the invariant distributions supported on $\mathcal{N}$, the set of nilpotent elements in $\mathfrak{g}$. From Harish-Chandra [15], we know that the set of nilpotent orbital integrals form a basis for $J(\mathcal{N})$.

In Theorem 7.0.1 we describe a basis for $J^{\mathrm{st}}(\mathcal{N}) := J(\mathcal{N}) \cap J^{\mathrm{st}}(\mathfrak{g})$, the set of stable distributions supported on $\mathcal{N}$. More precisely, we show that a distribution belongs to $J^{\mathrm{st}}(\mathcal{N})$ if and only if it is a linear combination of special nilpotent orbital integrals. (We call a nilpotent $\mathrm{G}_2(k)$-orbit in $\mathfrak{g}$ *special* provided that it is either trivial or k-distinguished (see Section 5.2.1).)

1.2 A guide to this paper

One goal of this paper is to describe some of the "machinery" occurring in [28] in a uniform way via Bruhat–Tits theory; to this end, Sections 2 and 3 introduce the notations and normalizations necessary to carry this out. In particular, we discuss how to attach various data to a facet in the Bruhat–Tits building. For example, to a facet F we can attach lattices $\mathfrak{g}_F^+ \subset \mathfrak{g}_F$ of $\mathfrak{g}$ so that the quotient $\mathsf{L}_F(\mathfrak{f}) := \mathfrak{g}_F/\mathfrak{g}_F^+$ is isomorphic to the Lie algebra of a finite group of Lie type.

In Section 4 we examine the interaction between invariant distributions and functions of depth zero. A function of depth zero is an element of $C_c^\infty(\mathfrak{g})$ that can be obtained (after extending by zero) by inflating a function on $\mathsf{L}_F(\mathfrak{f})$, for some facet F, to a function on $\mathfrak{g}_F$. In particular, we find a basis for $J(\mathcal{N})^*$ consisting of functions of depth zero associated to generalized Green functions. That we can find such a basis is a byproduct of the proof of a homogeneity result of Waldspurger [28] and DeBacker [11].

This homogeneity result lies at the core of our considerations. For any alcove C in the Bruhat–Tits building of G, we have a lattice $\mathfrak{g}_C$ in $\mathfrak{g}$. We set

$$\mathcal{D}_0 := \sum_C C_c(\mathfrak{g}/\mathfrak{g}_C),$$

where the sum is over the set of alcoves in the building. The subspace $\mathcal{D}_0$ of $C_c^\infty(\mathfrak{g})$ may be though of as a kind of Lie algebra version of the Iwahori–Hecke algebra. We define

$$\mathfrak{g}_0 := \bigcup_F \mathfrak{g}_F,$$

where the indexing set is the set of facets in the building. The set $\mathfrak{g}_0$ is a (very large) subset of $\mathfrak{g}$ which is closed, open, invariant, and contains $\mathcal{N}$. The homogeneity result states that

$$\operatorname{res}_{\mathcal{D}_0} J(\mathfrak{g}_0) = \operatorname{res}_{\mathcal{D}_0} J(\mathcal{N}), \tag{$*$}$$

where $J(\mathfrak{g}_0)$ denotes the set of invariant distributions supported on $\mathfrak{g}_0$.

We use equation $(*)$ in the following way. Suppose $T \in J(\mathfrak{g}_0)$ and D is the unique distribution in $J(\mathcal{N})$ for which

$$\operatorname{res}_{\mathcal{D}_0} T = \operatorname{res}_{\mathcal{D}_0} D.$$

Waldspurger [28] has shown that if T is stable, then D is stable. Thus we have two problems: (1) Find a basis for $\operatorname{res}_{\mathcal{D}_0} J(\mathfrak{g}_0) \cap \operatorname{res}_{\mathcal{D}_0} J^{\mathrm{st}}(\mathfrak{g})$, and (2) use this basis to find a basis for $J^{\mathrm{st}}(\mathcal{N})$. The second problem is addressed in Section 7.

In Section 5 we produce a basis for $\operatorname{res}_{\mathcal{D}_0} J(\mathfrak{g}_0)$. This basis is dual to the basis for $\operatorname{res}_{\mathcal{D}_0} J(\mathfrak{g}_0)^*$ coming from the depth zero functions attached to generalized Green functions. Thanks to Kazhdan [19], to a generalized Green function which comes from a Deligne–Lusztig generalized character, we can associate an "orbital integral" in $\mathsf{L}_F(\mathfrak{f})$. The corresponding distribution in $J(\mathfrak{g}_0)$ should be thought of as a lift of this orbital integral. For G_2, the remaining generalized Green functions are all associated to cuspidal local systems. The idea is to associate to a cuspidal local system an invariant distribution which is supported on the set of regular semisimple elements. This association is related (as Waldspurger has noticed) to the Kazhdan–Lusztig map [20].

In Section 6 we obtain a new basis for $\operatorname{res}_{\mathcal{D}_0} J(\mathfrak{g}_0)$ by taking combinations of the basis elements produced in Section 5. This new basis has the property that each element is either stable or a combination of "κ-orbital integrals". We conclude this section by repeating an argument of Waldspurger to show that the stable elements of this new basis are a basis for $\operatorname{res}_{\mathcal{D}_0} J(\mathfrak{g}_0) \cap \operatorname{res}_{\mathcal{D}_0} J^{\mathrm{st}}(\mathfrak{g})$.

2 Notation and normalizations

2.1 Basic notation

Let k be a characteristic zero nonarchimedean local field. Let ν be a valuation on k so that $\nu(k^\times) = \mathbb{Z}$. We denote the (finite) residue field of k by $\mathfrak{f}$. We denote by R the ring of integers in k, by $\wp$ its prime ideal, and by ϖ a fixed uniformizer. (So $\mathfrak{f} = R/\wp$ and $\wp = \varpi R$.) We let Λ denote an additive character of k which is nontrivial on R and trivial on $\wp$. We shall call the corresponding additive character on $\mathfrak{f}$ by Λ as well.

Let K be a fixed maximal unramified extension of k. We denote by R_K the ring of integers of K. Let $\mathfrak{F}$ denote the residue field of K. Note that $\mathfrak{F}$ is an algebraic closure of $\mathfrak{f}$.

Let $\Gamma = \operatorname{Gal}(K/k)$. Let σ denote a topological generator for Γ.

Let $\mathbf{G}$ be the connected reductive algebraic k-split group with root system Φ of type G_2. Fix a maximal k-split torus $\mathbf{T}$ in $\mathbf{G}$. We fix a basis Δ for the root system and let β denote the long root and α the short root in this basis (see Figure 1). We denote by $\mathbf{g}$ the Lie algebra of $\mathbf{G}$, and, with respect to $\mathbf{T}$, we fix a Chevalley basis

$$\{H_\delta, X_\phi \mid \delta \in \Delta \text{ and } \phi \in \Phi\}$$

for $\mathbf{G}$. We let $G = \mathbf{G}(k)$, the group of k-rational points of $\mathbf{G}$, and, similarly, we set $T = \mathbf{T}(k)$ and $\mathfrak{g} = \mathbf{g}(k)$.

2.2 Buildings and associated notation

Let $\mathcal{B}(G)$ denote the Bruhat–Tits building of G. We identify $\mathcal{B}(G)$ with the Γ-fixed points of $\mathcal{B}(\mathbf{G}, K)$, the Bruhat–Tits building of $\mathbf{G}(K)$.

For every maximal k-split torus $\mathbf{S}$ in $\mathbf{G}$, we can associate an apartment $\mathcal{A}(\mathbf{S}, k) = \mathcal{A}(\mathbf{S}(k))$ in $\mathcal{B}(G)$. Each apartment of $\mathcal{B}(G)$ can be carried to $\mathcal{A}(T)$ via the action of

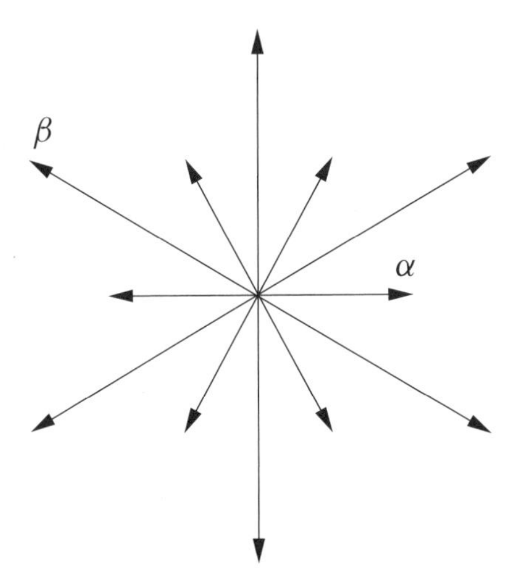

Fig. 1. The root system G_2.

G on $\mathcal{B}(G)$. Since we have fixed a Chevalley basis, we may identify $\mathcal{A}(T)$ with the vector space $\mathbf{X}_*(\mathbf{T}) \otimes \mathbb{R}$. Via the natural pairing between co-roots and roots, $\mathcal{A}(T)$ also carries a natural simplicial decomposition: Each element of

$$\Psi := \{\phi + n \mid \phi \in \Phi \text{ and } n \in \mathbb{Z}\},$$

the set of affine roots with respect to $\mathbf{T}$, $\mathbf{G}$, and ν, can be thought of as a function on $\mathbf{X}_*(\mathbf{T}) \otimes \mathbb{R}$ by setting

$$(\phi + n)(r \cdot \lambda) := r\langle \phi, \lambda \rangle + n$$

for $\lambda \in \mathbf{X}_*(\mathbf{T})$ and $r \in \mathbb{R}$. For each $\psi \in \Psi$ we let H_ψ denote the corresponding hyperplane, and, in the natural way, these hyperplanes gives us a simplicial decomposition of $\mathcal{A}(T)$. In our case, this simplicial decomposition may be thought of as a tiling of the plane by $(\frac{\pi}{6}, \frac{\pi}{3}, \frac{\pi}{2})$-right triangles. The normalizer in G of T acts transitively on the set of maximal simplices, called alcoves, in $\mathcal{A}(T)$. For this paper, we will focus on that alcove in $\mathcal{A}(T)$ bounded by the hyperplanes $H_{\alpha+0}$, $H_{\beta+0}$, and $H_{-(2\beta+3\alpha)+1}$ (see Figure 2). If $\psi = \phi + n$ is an affine root, then $\dot{\psi} := \phi$ is called the *gradient* of ψ.

Suppose $x \in \mathcal{B}(\mathbf{G}, K)$. We denote the parahoric subgroup of $\mathbf{G}(K)$ corresponding to x by $\mathbf{G}(K)_x$. Since $\mathbf{G}$ is simply connected, $\mathbf{G}(K)_x$ is the fixator in $\mathbf{G}(K)$ of x. We denote the pro-unipotent radical of $\mathbf{G}(K)_x$ by $\mathbf{G}(K)_x^+$. The subgroups $\mathbf{G}(K)_x$ and $\mathbf{G}(K)_x^+$ depend only on the facet of $\mathcal{B}(\mathbf{G}, K)$ to which x belongs. If F is a facet in $\mathcal{B}(\mathbf{G}, K)$ and $x \in F$, then we define $\mathbf{G}(K)_F := \mathbf{G}(K)_x$ and $\mathbf{G}(K)_F^+ := \mathbf{G}(K)_x^+$.

Example 2.2.1. In Figure 3 we have placed a label on each of the facets occurring in the closure of our preferred chamber. (Recall that $\tilde{\mathrm{A}}_1$ signifies a subdiagram consisting of a short root.) These labels will be used extensively later in the paper. It is a bit notationally messy to explicitly describe the groups $\mathbf{G}(K)_F$ and $\mathbf{G}(K)_F^+$, but here is

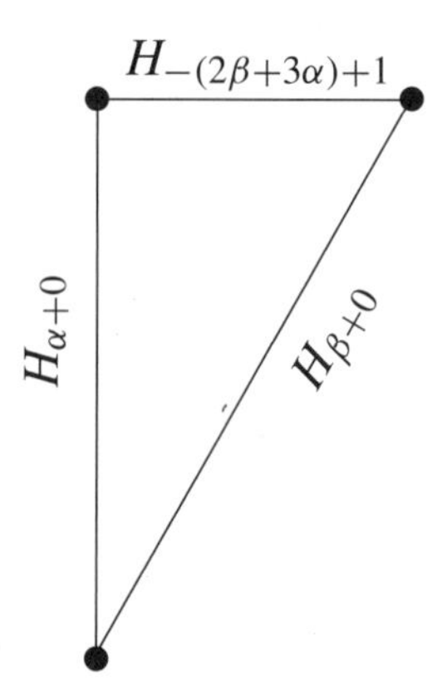

Fig. 2. Our fixed alcove in $\mathcal{B}(G)$.

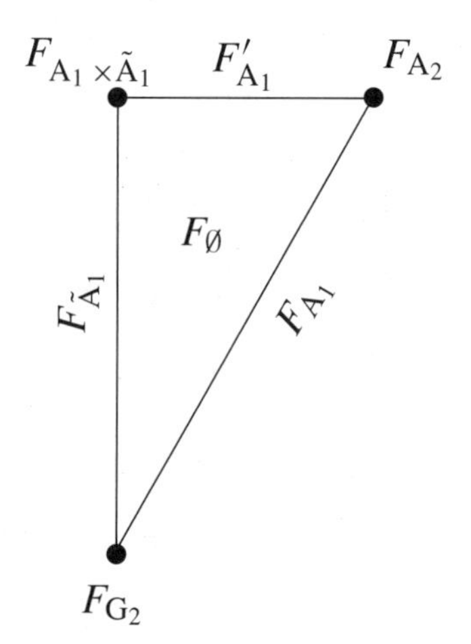

Fig. 3. A labeling of the facets.

how to construct them: For each root $\phi \in \Phi$, we let U_ϕ denote the corresponding root group in $\mathbf{G}(K)$. Our choice of a Chevalley basis determines an isomorphism from U_ϕ to K. Thus U_ϕ carries a natural filtration indexed by the integers and hence by the affine roots of the form $\phi + n$ with $n \in \mathbb{Z}$. We set $U_{\phi+0} := U_\phi \cap \mathbf{G}(R_K)$ and define $U_{\phi+1}$ to be the first filtration subgroup of $U_{\phi+0}$. For every $n \in \mathbb{Z}$ this uniquely determines a subgroup $U_{\phi+n}$ of U_ϕ. Thus for each $\psi \in \Psi$, we have a subgroup U_ψ of $\mathbf{G}(K)$. For a facet F in $\mathcal{A}(\mathbf{T}, K)$ the parahoric $\mathbf{G}(K)_F$ is the group generated by $\mathbf{T}(R_K)$ and the groups U_ψ such that $\psi \in \Psi$ and ψ is nonnegative on F. The prounipotent radical $\mathbf{G}(K)_F^+$ is the subgroup of $\mathbf{G}(K)_F$ generated by the first filtration subgroup of $\mathbf{T}(R_K)$ and the groups U_ψ such that $\psi \in \Psi$ is positive on F. Note that $\mathbf{G}(K)_{F_{\mathrm{G}_2}} = \mathrm{G}_2(R_K)$.

For a facet F in $\mathcal{B}(\mathbf{G}, K)$, the quotient $\mathbf{G}(K)_F/\mathbf{G}(K)_F^+$ is the group of $\mathfrak{F}$-rational points of a connected, reductive $\mathfrak{F}$-group G_F.

Example 2.2.2. In Figure 4 each facet in our preferred chamber has been labeled with the corresponding connected reductive $\mathfrak{f}$-group. Note that every vertex occurring in the closure of $F_\emptyset$ corresponds to a node in the extended Dynkin diagram of type G_2 (see Figure 5). So, for example, F_{A_2} corresponds to the node labeled $\alpha + 0$ (by deleting this node we obtain a subdiagram of type A_2).

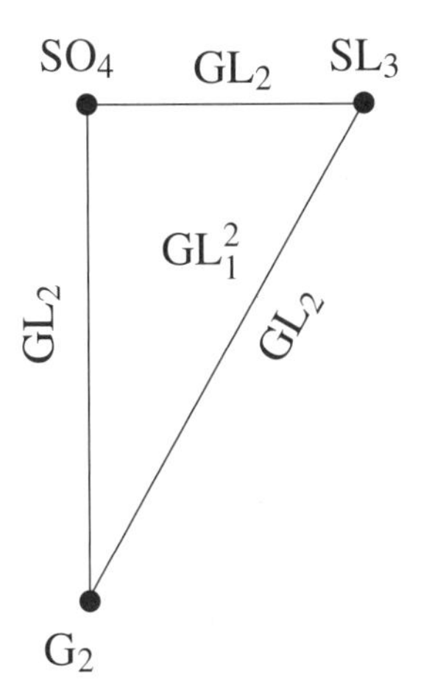

Fig. 4. The groups G_F.

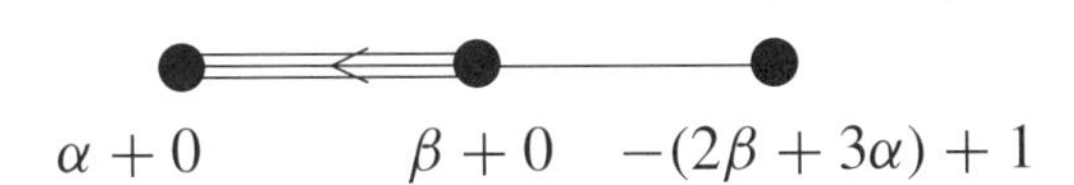

Fig. 5. The extended Dynkin diagram of type G_2.

Similarly, we define (see [2, Section 2.2]) lattices $\mathfrak{g}(K)_x$, $\mathfrak{g}(K)_x^+$, $\mathfrak{g}(K)_F$, and $\mathfrak{g}(K)_F^+$ in $\mathfrak{g}(K)$. If we denote the Lie algebra of G_F by L_F, then $\mathsf{L}_F(\mathfrak{F}) \cong \mathfrak{g}(K)_F/\mathfrak{g}(K)_F^+$. If $\psi \in \Psi$ and $x \in \mathcal{A}(\mathbf{T}, K)$ such that $\psi(x) = 0$, then we define $\mathfrak{g}(K)_\psi := \mathfrak{g}_{\dot{\psi}} \cap \mathfrak{g}(K)_x$ and $\mathfrak{g}(K)_\psi^+ := \mathfrak{g}_{\dot{\psi}} \cap \mathfrak{g}(K)_x^+$. This definition is independent of the choice of x.

Suppose now that x is Γ-fixed (that is, $x \in \mathcal{B}(G)$). In this case the parahoric subgroup of G attached to x is $G_x := \mathbf{G}(K)_x^\Gamma$, and the prounipotent radical of G_x is $G_x^+ := (\mathbf{G}(K)_x^+)^\Gamma$. As above, G_x and G_x^+ depend only on the facet of $\mathcal{B}(G)$ to which x belongs. If F is a facet in $\mathcal{B}(G)$ and $x \in F$, then we define the parahoric subgroup $G_F := G_x$, the associated pro-unipotent radical $G_F^+ := G_x^+$, and the associated connected reductive $\mathfrak{f}$-group $\mathsf{G}_F = \mathsf{G}_x$. For a facet F in $\mathcal{B}(G)$ we have $\mathsf{G}_F(\mathfrak{f}) = G_F/G_F^+$. Similarly, by taking Γ-fixed points, we can define lattices $\mathfrak{g}_x$, $\mathfrak{g}_x^+$, $\mathfrak{g}_F$, and $\mathfrak{g}_F^+$ in $\mathfrak{g}$, and we have $\mathsf{L}_F(\mathfrak{f}) \cong \mathfrak{g}_F/\mathfrak{g}_F^+$.

2.3 Levis and facets

If M is the group of k-rational points of a Levi k-subgroup of a parabolic k-subgroup, then we let (M) denote the conjugacy class of M. The set of such conjugacy classes is partially ordered: $(L) \leq (M)$ if and only if there exist $L_1 \in (L)$ and $M_1 \in (M)$ such that $L_1 \leq M_1$. We write $(L) < (M)$ provided that $(L) \leq (M)$ and $(L) \neq (M)$.

To each facet F in $\mathcal{B}(G)$ we can associate a conjugacy class (M_F). Without loss of generality, we suppose F is in $\mathcal{A}(T)$. Let M_F be the subgroup of G generated by T and the root groups U_ϕ, where ϕ runs over the set of roots for which the affine roots $\phi + n \in \Psi$ are constant on F. The group M_F is the group of k-rational points of a Levi k-subgroup of a parabolic k-subgroup.

Example 2.3.1. If F is a vertex, then (M_F) is $\{G\}$. If F is an alcove, then $(M_F) = (T)$. We have that $(M_{F_{\tilde{A}_1}})$ and $(M_{F_{A_1}})$ correspond to the distinct conjugacy classes of $\mathrm{GL}_2(k)$-Levi subgroups of G.

2.4 Compact and topologically unipotent elements

We define

$$G_0 := \bigcup_{x \in \mathcal{B}(G)} G_x$$

and

$$G_{0^+} := \bigcup_{x \in \mathcal{B}(G)} G_x^+.$$

The set G_0 is often referred to as the set of compact elements in G, and the set G_{0^+} is called the set of topologically unipotent elements in G. Since $\mathbf{G}$ is simply connected, the set of compact elements is the union of all compact subgroups of G. The set of topologically unipotent elements has the property that if $g \in G_{0^+}$, then

$$\lim_{n \to \infty} g^{p^n} = 1,$$

where p denotes the characteristic of $\mathfrak{f}$.

One can also describe the set of topologically unipotent elements as follows. If $g \in G$, then $g \in G_{0^+}$ provided that there exists a facet F in $\mathcal{B}(G)$ such that $g \in G_F$ and the image of g in $\mathsf{G}_F(\mathfrak{f})$ is unipotent.

We define, in a completely analogous manner, the set $\mathfrak{g}_0$ of compact elements in $\mathfrak{g}$ and the set $\mathfrak{g}_{0^+}$ of topologically nilpotent elements in $\mathfrak{g}$.

2.5 An assumption on the characteristic of $\mathfrak{f}$

We will require that p, the characteristic of $\mathfrak{f}$, be sufficiently large. Taking $p > 16$ is sufficient, but probably not necessary. We briefly state the facts we use which require some restrictions on p.

We require that every k-torus in $\mathbf{G}$ splits over a tamely ramified extension of k.

We let B denote the Killing form on $\mathfrak{g}$—that is, $B(X, Y) = \mathrm{tr}(\mathrm{ad}(X) \cdot \mathrm{ad}(Y))$. It is a G-invariant, nondegenerate, symmetric bilinear form on $\mathfrak{g}$. We assume that for each x in $\mathcal{B}(G)$, the form B induces a $\mathsf{G}_x(\mathfrak{f})$-invariant, nondegenerate, symmetric bilinear form on $\mathsf{L}_x(\mathfrak{f})$.

For every facet F in $\mathcal{B}(G)$ we assume that there is a $\mathsf{G}_F(\mathfrak{F})$-equivariant bijection from $\mathcal{N}_F$, the set of nilpotent elements in $\mathsf{L}_F(\mathfrak{F})$, to $\mathcal{U}_F$, the set of unipotent elements in $\mathsf{G}_F(\mathfrak{F})$.

We assume that Theorem 4.4.1 is valid (see [12, Section 4.2]).

We assume that the Lie algebra of a torus over the finite field contains regular semisimple elements. An element of a Lie algebra is called regular semisimple provided that its centralizer is a torus.

If $e \in \mathcal{N}_F(\mathfrak{f})$, then we assume that we can complete e to an $\mathfrak{sl}_2(\mathfrak{F})$-triple (f, h, e) with f and h both Γ-fixed. Moreover, we assume that there exists an $\mathfrak{sl}_2(K)$-triple (Y, H, X) lifting (f, h, e) with Y, H, and X each Γ-fixed.

2.6 The Fourier transform and attendant normalizations

Suppose V is a finite-dimensional k-vector space, B' is a symmetric, nondegenerate bilinear two-form on V, and dv is a Haar measure on V. Let $C_c^\infty(V)$ denote the space of compactly supported locally constant functions on V. We define the Fourier transform, denoted $\hat{f}$ or $\mathcal{F}(f)$, of $f \in C_c^\infty(V)$ by

$$\hat{f}(v') = \int_V \Lambda(B'(v', v)) \cdot f(v)\, dv$$

for $v' \in V$. Unless explicitly stated to the contrary, we shall always assume that dv is normalized so that $\hat{\hat{f}}(v') = f(-v')$ for $f \in C_c^\infty(V)$ and $v' \in V$. In particular, when our vector space is $\mathfrak{g}$, we shall take B' to be the Killing form and, therefore, uniquely pin down a Haar measure.

Suppose F is a facet in $\mathcal{B}(G)$. For a complex-valued function f on $\mathsf{L}_F(\mathfrak{f})$, we define its Fourier transform by

$$\hat{f}(\bar{Z}) = |\mathfrak{f}|^{-\dim(\mathsf{L}_F)/2} \sum_{\bar{X} \in \mathsf{L}_F(\mathfrak{f})} \Lambda(B(X, Z)) \cdot f(\bar{X})$$

for $Z \in \mathfrak{g}_F$ with image $\bar{Z} \in \mathsf{L}_F(\mathfrak{f})$. (Recall that we are identifying $\mathsf{L}_F(\mathfrak{f})$ with $\mathfrak{g}_F/\mathfrak{g}_F^+$.)

Definition 2.6.1. *If F is a facet in $\mathcal{B}(G)$ and f is a function on $\mathsf{L}_F(\mathfrak{f})$, then we let f_F denote the natural inflation of f to a function in $C(\mathfrak{g}_F/\mathfrak{g}_F^+) \subset C_c^\infty(\mathfrak{g})$.*

Suppose dX is the Haar measure on $\mathfrak{g}$ and F is a facet in $\mathcal{B}(G)$. We have

$$\mathrm{meas}_{dX}(\mathfrak{g}_F^+) = |\mathsf{L}_F(\mathfrak{f})|^{-1/2} = |\mathfrak{f}|^{-\dim(\mathsf{L}_F)/2}.$$

We also have

$$(\hat{f})_F(Z) = \widehat{f_F}(Z)$$

for all $Z \in \mathfrak{g}$ and for each complex-valued function f on $\mathsf{L}_F(\mathfrak{f})$; for this reason we will not distinguish between $(\hat{f})_F$ and $\widehat{f_F}$.

We will always assume that the Haar measure dg on G is chosen so that

$$\mathrm{meas}_{dg}(G_F^+) = |\mathsf{L}_F(\mathfrak{f})|^{-1/2}.$$

This normalization is independent of F.

2.7 Semisimple orbital integrals

Suppose $Y \in \mathfrak{g}$ is semisimple. Let $\mathcal{O}_Y$ denote the G-orbit of Y in $\mathfrak{g}$; this orbit carries an invariant measure, unique up to scaling, which we denote by $d\mu_Y$. Since $\mathcal{O}_Y$ is closed, if $f \in C_c^\infty(\mathfrak{g})$, then the restriction of f to $\mathcal{O}_Y$ belongs to $C_c^\infty(\mathcal{O}_Y)$. Consequently, we may define an invariant distribution, μ_Y, on $\mathfrak{g}$ by

$$\mu_Y(f) := \int_{\mathcal{O}_Y} f(Z)\, d\mu_Y(Z)$$

for $f \in C_c^\infty(\mathfrak{g})$.

For our calculations, it is important to specify how the measure $d\mu_Y$ is normalized. The centralizer in G of Y, denoted $C_G(Y)$, is the group of k-rational points of a connected reductive k-group; in Section 2.6 we specified a choice of a Haar measure for such a group (all the material there applies to the group of k-rational points of any reductive k-group). We identify the G-orbit of Y with $G/C_G(Y)$ and define μ_Y by taking the quotient measure.

Suppose that $X, Y \in \mathfrak{g}$ are regular semisimple and stably-conjugate; that is, there is a $g \in \mathbf{G}(K)$ such that $\mathrm{Ad}(g)X = {}^gX = Y$. In this case, we have ${}^g(C_G(X)) = C_G(Y)$. Consequently, $\mathrm{Ad}(g)_*(\mu_X) = \mu_Y$.

3 Induction, restriction, and cuspidality

3.1 Induction, restriction, and the pairing over $\mathfrak{f}$

We begin by considering any connected reductive $\mathfrak{f}$-group G. We let L_G denote the Lie algebra of G.

We let $C(\mathsf{L}_\mathsf{G}(\mathfrak{f}))$ denote the space of complex-valued functions on $\mathsf{L}_\mathsf{G}(\mathfrak{f})$. For functions $h, h' \in C(\mathsf{L}_\mathsf{G}(\mathfrak{f}))$ we define

$$(h, h')_{\mathsf{L}_\mathsf{G}} = \frac{1}{|\mathsf{G}(\mathfrak{f})|} \sum_{\bar{X} \in \mathsf{L}_\mathsf{G}(\mathfrak{f})} \overline{h}(\bar{X}) \cdot h'(\bar{X}).$$

Here $\overline{h}(\bar{X})$ denotes the complex conjugate of $h(\bar{X})$.

Suppose P is a parabolic $\mathfrak{f}$-subgroup of G with unipotent radical U and a Levi $\mathfrak{f}$-subgroup M so that P has a Levi decomposition $\mathsf{P} = \mathsf{MU}$. If $f \in C(\mathsf{L_M}(\mathfrak{f}))$, we define $\mathrm{Ind}_{\mathsf{P}}^{\mathsf{G}} f \in C((\mathsf{L_G})(\mathfrak{f}))$ by

$$\mathrm{Ind}_{\mathsf{P}}^{\mathsf{G}} f(\bar{X}) := \frac{1}{|\mathsf{P}(\mathfrak{f})|} \sum_{(\bar{x},\bar{Z},\bar{Y})} f(\bar{Y}) \cdot [\bar{X}]({}^{\bar{x}}(\bar{Y} + \bar{Z}))$$

for $\bar{X} \in \mathsf{L_G}(\mathfrak{f})$. Here $[\bar{X}]$ denotes the characteristic function associated to the set $\{\bar{X}\}$, and the sum is over triples

$$(\bar{x}, \bar{Z}, \bar{Y}) \in \mathsf{G}(\mathfrak{f}) \times \mathsf{L_U}(\mathfrak{f}) \times \mathsf{L_M}(\mathfrak{f}).$$

In the opposite direction, if $h \in C(\mathsf{L_G}(\mathfrak{f}))$, then we define $\mathrm{r}_{\mathsf{P}}^{\mathsf{G}}\, h \in C(\mathsf{L_M}(\mathfrak{f}))$ by

$$(\mathrm{r}_{\mathsf{P}}^{\mathsf{G}}\, h)(\bar{Y}) := \frac{1}{|\mathsf{G}(\mathfrak{f})| \cdot |\mathsf{U}(\mathfrak{f})|} \sum_{(\bar{x},\bar{Z})} h({}^{\bar{x}}(\bar{Y} + \bar{Z}))$$

for $\bar{Y} \in \mathsf{L_M}(\mathfrak{f})$. Here the sum is over pairs

$$(\bar{x}, \bar{Z}) \in \mathsf{G}(\mathfrak{f}) \times \mathsf{L_U}(\mathfrak{f}).$$

For $f \in C(\mathsf{L_M}(\mathfrak{f}))$ and $h \in C(\mathsf{L_G}(\mathfrak{f}))$ we have a version of Frobenius Reciprocity:

$$(\mathrm{r}_{\mathsf{P}}^{\mathsf{G}}\, h, f)_{\mathsf{L_M}} = (h, \mathrm{Ind}_{\mathsf{P}}^{\mathsf{G}} f)_{\mathsf{L_G}}.$$

Finally, we call a function $h \in C(\mathsf{L_G}(\mathfrak{f}))$ *cuspidal* provided that $\mathrm{r}_{\mathsf{P}}^{\mathsf{G}}\, h = 0$ for each proper parabolic $\mathfrak{f}$-subgroup $\mathsf{P} = \mathsf{MU}$ of G.

3.2 Extension to our situation

Suppose that H is a facet in $\mathcal{B}(G)$. For functions $h, h' \in C(\mathsf{L}_H(\mathfrak{f}))$, we define

$$(h, h')_H := (h, h')_{\mathsf{L}_H}.$$

Suppose F is another facet in $\mathcal{B}(G)$ for which $H \subset \bar{F}$; that is, H is contained in the closure of F. We have

$$G_H^+ \subset G_F^+ \subset G_F \subset G_H$$

with G_F/G_H^+ the group of $\mathfrak{f}$-rational points of a parabolic $\mathfrak{f}$-subgroup P of G_H. Moreover, if U denotes the unipotent radical of P, then the group of $\mathfrak{f}$-rational points of U is G_F^+/G_H^+. Finally, the Levi $\mathfrak{f}$-subgroup factor of P is isomorphic to G_F. Similarly, we have

$$\mathfrak{g}_H^+ \subset \mathfrak{g}_F^+ \subset \mathfrak{g}_F \subset \mathfrak{g}_H$$

with $\mathfrak{g}_F/\mathfrak{g}_H^+$ the vector space of $\mathfrak{f}$-rational points of the parabolic subalgebra $\mathsf{L_P}$ of L_H, and so on.

Example 3.2.1. For example, in the notation introduced in Figure 3, $G_{F_\emptyset}/G_{F_{\mathrm{G}_2}}$ is a Borel subgroup of $\mathrm{G}_2(\mathfrak{f}) = G_{F_{\mathrm{G}_2}}/G^+_{F_{\mathrm{G}_2}}$. The reader is encouraged to spend some time working out the details.

For $f \in C(\mathsf{L}_F(\mathfrak{f}))$, we define

$$\mathrm{Ind}_F^H f := \mathrm{Ind}_{\mathsf{P}}^{\mathsf{G}_H} f \in C(\mathsf{L}_H(\mathfrak{f})),$$

and for $h \in C(\mathsf{L}_H(\mathfrak{f}))$ we define $\mathrm{r}_F^H h \in C(\mathsf{L}_F(\mathfrak{f}))$ by

$$\mathrm{r}_F^H h := \mathrm{r}_{\mathsf{P}}^{\mathsf{G}_H} h.$$

Of course, for $f \in C(\mathsf{L}_F(\mathfrak{f}))$ and $h \in C(\mathsf{L}_H(\mathfrak{f}))$, we have

$$(\mathrm{r}_F^H h, f)_F = (h, \mathrm{Ind}_F^H f)_H,$$

and $h \in C(\mathsf{L}_H(\mathfrak{f}))$ is cuspidal if and only if $\mathrm{r}_F^H h = 0$ whenever $F \neq H$ is a facet in $\mathcal{B}(G)$ which contains H in its closure.

3.3 Cuspidality over $\mathfrak{f}$ and k

Suppose P is the group of k-rational points of a parabolic k-subgroup $\mathbf{P}$ of $\mathbf{G}$. Suppose M is the group of k-rational points of a Levi k-subgroup of $\mathbf{P}$ and N is the group of k-rational points of the unipotent radical of $\mathbf{P}$. Let $\mathfrak{p} = \mathfrak{m} + \mathfrak{n}$ denote the associated Lie algebras in $\mathfrak{g}$. For $f \in C_c^\infty(\mathfrak{g})$, we define

$$f_P(Y) := \int_K dk \int_{\mathfrak{n}} f({}^k(Y+Z))\, dZ,$$

where dZ is a Haar measure on $\mathfrak{n}$ and dk is the normalized Haar measure on the compact open subgroup $K = G_{F_{\mathrm{G}_2}}$.

Lemma 3.3.1. *We use the notation introduced above and suppose H is a facet in $\mathcal{B}(G)$ and $h \in C(\mathsf{L}_H(\mathfrak{f}))$ is cuspidal. If $(M) < (M_H)$, then $(h_H)_P = 0$.*

Proof. Fix $k \in G_{F_{\mathrm{G}_2}}$. It will be enough to show that

$$0 = \int_{\mathfrak{n}} (h_H)({}^{k^{-1}}(Y+Z))\, dZ$$

for $Y \in \mathfrak{m}$. From [2, Corollary 2.4.3], there is an $n \in N$ such that nkH is a facet in $\mathcal{B}(M)$. Since

$${}^{n^{-1}}Y = Y \text{ modulo } \mathfrak{n}$$

and dZ is a Haar measure, we have

$$\int_{\mathfrak{n}} (h_H)({}^{k^{-1}}(Y+Z))\, dZ = \int_{\mathfrak{n}} (({}^{nk}h)_{nkH})(Y+Z)\, dZ.$$

Here ${}^{nk}h \in C(L_{nkH}(\mathfrak{f}))$ is defined by

$$ {}^{nk}h(\bar{X}) = h(\overline{{}^{k^{-1}n^{-1}}X}), $$

where $X \in \mathfrak{g}_{nkH}$ is any lift of $\bar{X}$. If $(Y + \mathfrak{n}) \cap \mathfrak{g}_{nkH} = \emptyset$, then the integral

$$ \int_{\mathfrak{n}} (({}^{nk}h)_{nkH})(Y + Z)\, dZ $$

is zero. Suppose now that $(Y + \mathfrak{n}) \cap \mathfrak{g}_{nkH} \neq \emptyset$. Since $nkH \subset \mathcal{B}(M)$, we have

$$ \mathfrak{g}_{nkH} \cap (\mathfrak{m} + \mathfrak{n}) = (\mathfrak{g}_{nkH} \cap \mathfrak{m}) + (\mathfrak{g}_{nkH} \cap \mathfrak{n}). $$

Thus, since $(Y + \mathfrak{n}) \cap \mathfrak{g}_{nkH} \neq \emptyset$, we conclude that $Y \in \mathfrak{g}_{nkH} \cap \mathfrak{m} \subset \mathfrak{g}_{nkH}$. Since $(M) < (M_H)$, the image of $N \cap G_{nkH}$ in $\mathsf{G}_{nkH}(\mathfrak{f})$ is the group of $\mathfrak{f}$-rational points of the unipotent radical of a proper parabolic subgroup of G_{nkH}. Thus, since ${}^{nk}h$ is a cuspidal function in $\mathsf{L}_{nkH}(\mathfrak{f})$, the integral is zero. □

4 Some comments on invariant distributions on the Lie algebra

Recall that a distribution on $\mathfrak{g}$ is an element of the linear dual of $C_c^\infty(\mathfrak{g})$. We let $J(\mathfrak{g})$ denote the space of $\mathrm{Ad}(G)$-invariant distributions on $\mathfrak{g}$.

4.1 An equivalence relation on functions of depth zero

We begin by introducing the indexing set I^f.

Definition 4.1.1.

$$ I^f := \{(F, f)\colon F \textit{ is a facet in } \mathcal{B}(G) \textit{ and } f \in C(\mathsf{L}_F(\mathfrak{f}))\}. $$

The set I^f carries a natural equivalence relation. However, before introducing this equivalence relation, we must recall some notation and facts. If $\mathcal{A}$ is an apartment in $\mathcal{B}(G)$ and F is a facet in $\mathcal{A}$, then we denote by $A(\mathcal{A}, F)$ the smallest affine subspace of $\mathcal{A}$ that contains F. If F and F' are two facets in $\mathcal{A}$, then the condition

$$ \emptyset \neq A(\mathcal{A}, gF) = A(\mathcal{A}, F') $$

implies that the natural maps

$$ \mathbf{g}(K)_{gF} \cap \mathbf{g}(K)_{F'} \to \mathsf{L}_{gF}(\mathfrak{F}) \quad \text{and} \quad \mathbf{g}(K)_{gF} \cap \mathbf{g}(K)_{F'} \to \mathsf{L}_{F'}(\mathfrak{F}) $$

are surjective, Γ-equivariant maps with kernel equal to

$$ \mathbf{g}(K)^+_{gF} \cap \mathbf{g}(K)^+_{F'}. $$

In this way, we get a natural $\mathfrak{f}$-isomorphism between L_{gF} and $\mathsf{L}_{F'}$, which we write as $\mathsf{L}_{gF} \overset{i}{=} \mathsf{L}_{F'}$. Whenever we want to identify objects (see Definition 4.1.2 and Remarks 4.1.3 and 4.1.4) via this isomorphism, we will use the "$\overset{i}{=}$" notation.

Definition 4.1.2. *For (F, f) and (F', f') in I^f we write $(F, f) \sim (F', f')$ if and only if there exist a $g \in G$ and an apartment $\mathcal{A}$ in $\mathcal{B}(G)$ such that*

1. $\emptyset \neq A(\mathcal{A}, gF) = A(\mathcal{A}, F')$ *and*
2. ${}^g f \stackrel{i}{=} f'$ *in* $C(\mathsf{L}_{gF}(\mathfrak{f})) \stackrel{i}{=} C(\mathsf{L}_{F'}(\mathfrak{f}))$.

Here ${}^g f$ is the function defined by ${}^g f(\bar{X}) := f_F({}^{g^{-1}}X)$, where $X \in \mathfrak{g}_{gF}$ is any lift of $\bar{X}$.

Remark 4.1.3. Instead of considering pairs (F, f) with f a function on $C(\mathsf{L}_F(\mathfrak{f}))$, we could (and will) consider an equivalence relation on pairs of the form (F, e), where e is an element of $\mathsf{L}_F(\mathfrak{f})$ and the second condition in Definition 4.1.2 is replaced by

$${}^g e \stackrel{i}{=} e' \quad \text{in } \mathsf{L}_{gF}(\mathfrak{f}) \stackrel{i}{=} \mathsf{L}_{F'}(\mathfrak{f}).$$

Remark 4.1.4. There is a group analogue of the above with G_F replacing L_F, etc. We shall use it in the following context: Instead of considering pairs (F, f) as above, we consider pairs (F, S) with S a maximal $\mathfrak{f}$-minisotropic torus in G_F. A torus is $\mathfrak{f}$-*minisotropic* in G_F provided that its maximal $\mathfrak{f}$-split subtorus lies in the center of G_F. We then say that (F, S) is equivalent to (F', S') provided that the first condition of Definition 4.1.2 is true and the second is replaced by

$${}^g\mathsf{S} \stackrel{i}{=} \mathsf{S}' \quad \text{in } \mathsf{G}_{gF} \stackrel{i}{=} \mathsf{G}_{F'}.$$

Example 4.1.5. Since our preferred alcove is a fundamental domain for the action of G on $\mathcal{B}(G)$, in order to determine representatives for the above equivalence relation, it is enough to look at our alcove. The set of pairs $(F, f) \in I^f$, where F runs over the nonprimed facets in Figure 3, form a set of representatives for $I^f/\sim$. (The two facets F_{A_1} and F'_{A_1} correspond to two "$\mathrm{GL}_2(\mathfrak{f})$-Levi subgroups" of $\mathrm{SL}_3(\mathfrak{f})$, all such Levi subgroups are $\mathrm{SL}_3(\mathfrak{f})$-conjugate.)

4.2 Functions of depth zero and invariant distributions

Lemma 4.2.1. *Let H and F be facets in $\mathcal{B}(G)$. Suppose $H \subset \bar{F}$. Let P denote the parabolic $\mathfrak{f}$-subgroup of G_H corresponding to F. If $f \in C(\mathsf{L}_F(\mathfrak{f}))$ and $D \in J(\mathfrak{g})$, then*

$$D(f_F) = \frac{|\mathsf{P}(\mathfrak{f})|}{|\mathsf{G}_H(\mathfrak{f})|} \cdot D((\mathrm{Ind}_F^H f)_H).$$

Proof. Suppose $f \in C(\mathsf{L}_F(\mathfrak{f}))$. Let U denote the unipotent radical of P. Since

$$\mathfrak{g}_H^+ \subset \mathfrak{g}_F^+ \subset \mathfrak{g}_F \subset \mathfrak{g}_H$$

we have $f_F \in C(\mathfrak{g}_H/\mathfrak{g}_H^+)$. Choose $h \in C(\mathsf{L}_H(\mathfrak{f}))$ such that $f_F = h_H$.

Since $D \in J(\mathfrak{g})$, there exists a $\mathsf{G}_H(\mathfrak{f})$-invariant $d \in C(\mathsf{L}_H(\mathfrak{f}))$ such that

$$D(h'_H) = (d, h')_H$$

for all $h' \in C(\mathsf{L}_H(\mathfrak{f}))$. We have

$$\begin{aligned}
D((\operatorname{Ind}_F^H f)_H) &= (d, \operatorname{Ind}_F^H f)_H = (\mathrm{r}_F^H d, f)_F \\
&= \frac{1}{|\mathsf{G}_F(\mathfrak{f})|} \sum_{\bar{Y} \in \mathsf{L}_F(\mathfrak{f})} \overline{\mathrm{r}_F^H(d)(\bar{Y})} \cdot f(\bar{Y}) \\
&= \frac{1}{|\mathsf{G}_H(\mathfrak{f})| \cdot |\mathsf{G}_F(\mathfrak{f})| \cdot |\mathsf{U}(\mathfrak{f})|} \sum_{\bar{Y} \in \mathfrak{g}_F/\mathfrak{g}_F^+} f_F(Y) \sum_{\substack{\bar{x} \in G_H/G_H^+ \\ \bar{Z} \in \mathfrak{g}_F^+/\mathfrak{g}_H^+}} \overline{d}_H({}^x(Y+Z)) \\
&= \frac{1}{|\mathsf{P}(\mathfrak{f})|} \sum_{\bar{Y} \in \mathfrak{g}_F/\mathfrak{g}_H^+} \overline{d}_H(Y) \cdot f_F(Y) \\
&= \frac{|\mathsf{G}_H(\mathfrak{f})|}{|\mathsf{P}(\mathfrak{f})|} \cdot \frac{1}{|\mathsf{G}_H(\mathfrak{f})|} \sum_{\bar{Y} \in \mathfrak{g}_H/\mathfrak{g}_H^+} h_H(Y) \cdot \overline{d}_H(Y) \\
&= \frac{|\mathsf{G}_H(\mathfrak{f})|}{|\mathsf{P}(\mathfrak{f})|} \cdot (d, h)_H = \frac{|\mathsf{G}_H(\mathfrak{f})|}{|\mathsf{P}(\mathfrak{f})|} \cdot D(h_H) \\
&= \frac{|\mathsf{G}_H(\mathfrak{f})|}{|\mathsf{P}(\mathfrak{f})|} \cdot D(f_F).
\end{aligned}$$

□

Corollary 4.2.2. *Suppose* $D \in J(\mathfrak{g})$. *If* $(F_i, f_i) \in I^f$ *and* $\operatorname{supp}((f_i)_{F_i}) \subset \mathfrak{g}_{0^+}$; *then*

$$(F_1, f_1) \sim (F_2, f_2) \Rightarrow D({f_1}_{F_1}) = D({f_2}_{F_2}).$$

Proof. The condition $\operatorname{supp}((f_i)_{F_i}) \subset \mathfrak{g}_{0^+}$ implies that the support of f_i is contained in $\mathcal{N}_{F_i}(\mathfrak{f})$.

Since D is G-invariant, without loss of generality, we may assume there is an apartment $\mathcal{A}$ such that

$$A(\mathcal{A}, F_1) = A(\mathcal{A}, F_2) \neq \emptyset$$

and

$$f_1 \stackrel{i}{=} f_2 \quad \text{in } C(\mathsf{L}_{F_1}(\mathfrak{f})) \stackrel{i}{=} C(\mathsf{L}_{F_2}(\mathfrak{f})).$$

Moreover, we may assume that f_i is a $\mathsf{G}_{F_i}(\mathfrak{f})$-invariant function.

There is a sequence of pairs $(F(i), f(i))$ with $1 \le i \le m$ so that

1. each $F(i)$ is maximal in $A(\mathcal{A}, F_1)$ for each i,
2. $f(j) \stackrel{i}{=} f(\ell)$ in $C(\mathsf{L}_{F(j)}(\mathfrak{f})) \stackrel{i}{=} C(\mathsf{L}_{F(\ell)}(\mathfrak{f}))$ for $1 \le j, \ell \le m$,
3. $(F(1), f(1)) = (F_1, f_1)$ and $(F(m), f(m)) = (F_2, f_2)$,
4. $\bar{F}(i) \cap \bar{F}(i+1) \neq \emptyset$ for $1 \le i < m$.

Thus, without loss of generality, we assume there is a facet H in $\mathcal{A}$ such that $H \subset \bar{F}_1 \cap \bar{F}_2$. From Lemma 4.2.1, we have

$$D({f_i}_{F_i}) = \frac{|G_{F_i}/G_H^+|}{|\mathsf{G}_H(\mathfrak{f})|} \cdot D(\operatorname{Ind}_{F_i}^H (f_i)_H).$$

Since the unipotent radicals of any two parabolics sharing a common Levi factor have the same dimension, we have that $|G_{F_1}/G_H^+| = |G_{F_2}/G_H^+|$. Therefore, we need only check that

$$\mathrm{Ind}_{F_1}^H f_1 = \mathrm{Ind}_{F_2}^H f_2$$

in $C(\mathsf{L}_H(\mathfrak{f}))$. Note that both functions are supported on $\mathcal{N}_H(\mathfrak{f})$ and both are induced from a nilpotently supported function on the Levi subalgebra. After identifying $\mathcal{N}_H(\mathfrak{f})$ with $\mathcal{U}_H(\mathfrak{f})$, the statement we want follows from the fact that for finite groups of Lie type, parabolic induction is independent of the parabolic containing the Levi (see, for example, [17]). □

4.3 Some indexing sets

We define I^G to be the subset of I^f consisting of pairs $(F, \mathcal{G})$, where $\mathcal{G}$ is a generalized Green function in $C(\mathsf{L}_F(\mathfrak{f}))$. We shall describe these functions in Section 5. Note that by extending by zero, we are thinking of the generalized Green function as a function on all of $\mathsf{L}_F(\mathfrak{f})$, not just the set of nilpotent elements in $\mathsf{L}_F(\mathfrak{f})$. We let I^c denote the subset of I^G consisting of pairs $(F, \mathcal{G})$, where $\mathcal{G}$ is a cuspidal generalized Green function on $\mathsf{L}_F(\mathfrak{f})$; that is, $\mathcal{G}$ is a generalized Green function on $\mathsf{L}_F(\mathfrak{f})$ which is cuspidal in the sense that $\mathrm{r}_H^F\,\mathcal{G} = 0$ for each facet H with $F \subsetneq \bar{H}$.

Finally, we define I^n to be the set of pairs (F, e), where F is a facet in $\mathcal{B}(G)$ and $e \in \mathcal{N}_F(\mathfrak{f})$. To each such pair, we associate the function $[(F, e)]$, the characteristic function of the coset e.

Example 4.3.1. In Figure 6, we tally the number of cuspidal generalized Green functions supported on $\mathsf{L}_F(\mathfrak{f})$; we will give a more precise description in Section 5. Note that the cuspidal generalized Green functions associated to F_{A_1} and F'_{A_1} are equivalent.

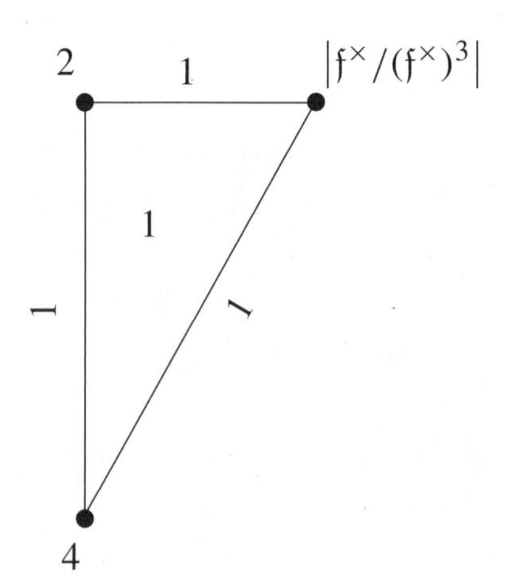

Fig. 6. A tally of the cuspidal generalized Green functions.

4.4 A homogeneity result

We will need a very precise version of Howe's finiteness conjecture for the Lie algebra [16]. Such results were first proved by Waldspurger for unramified classical groups [29]. Waldspurger's work has been generalized in [11].

We let $J(\mathfrak{g}_0)$ denote the subspace of $J(\mathfrak{g})$ consisting of distributions having support in $\mathfrak{g}_0$. For example, if $Y \in \mathfrak{g}_0$, then the associated orbital integral, μ_Y, is an element of $J(\mathfrak{g}_0)$.

We let $\mathcal{D}_0$ denote the (invariant version of the) Lie algebra analogue of the Iwahori–Hecke algebra. More precisely, we have

$$\mathcal{D}_0 = \sum_{C'} C_c(\mathfrak{g}/\mathfrak{g}_{C'}),$$

where the sum runs over the set of alcoves in $\mathcal{B}(G)$. Finally, we define

$$\mathcal{D}_0^0 := \sum_{F \subset \bar{F}_\emptyset} C(\mathfrak{g}_F/\mathfrak{g}_{F_\emptyset}),$$

where $F_\emptyset$ is our fixed alcove.

Theorem 4.4.1 (Waldspurger, DeBacker). *We have*

1.
$$\operatorname{res}_{\mathcal{D}_0} J(\mathfrak{g}_0) = \operatorname{res}_{\mathcal{D}_0} J(\mathcal{N}).$$

2. *Suppose $D \in J(\mathfrak{g}_0)$. We have*
$$\operatorname{res}_{\mathcal{D}_0} D = 0 \quad \textit{if and only if } \operatorname{res}_{\mathcal{D}_0^0} D = 0.$$

Note that the second statement in the theorem says that a dual basis for $\operatorname{res}_{\mathcal{D}_0} J(\mathfrak{g}_0)$ may be found in $\mathcal{D}_0^0$. The following corollary is a generalization of a result of Waldspurger [28, Corollaire III.10(i)].

Corollary 4.4.2. *Suppose $D \in J(\mathfrak{g}_0)$. We have*

$$\operatorname{res}_{\mathcal{D}_0} D = 0 \quad \textit{if and only if } D(\hat{\mathcal{G}}_F) = 0 \quad \textit{for all } (F, \mathcal{G}) \in I^c/\sim.$$

Remark 4.4.3. Thanks to Corollary 4.2.2, the right-hand side makes sense.

Proof. "$\Rightarrow$" If $(F, \mathcal{G}) \in I^c$, then $\mathcal{G}_F \in C(\mathfrak{g}_F/\mathfrak{g}_F^+) \cap C_c^\infty(\mathfrak{g}_{0+})$. Therefore, from [2, Lemma 4.2.3] we have $\hat{\mathcal{G}}_F \in C(\mathfrak{g}_F/\mathfrak{g}_F^+) \cap \mathcal{D}_0$. Thus for all $(F, \mathcal{G}) \in I^c$ we have $\hat{\mathcal{G}}_F \in \mathcal{D}_0$ and so $D(\hat{\mathcal{G}}_F) = 0$.

"$\Leftarrow$" Recall that for $f \in C_c^\infty(\mathfrak{g})$ we have $\hat{D}(f) := D(\hat{f})$. It therefore follows, from Lemma 4.2.1, Corollary 4.2.2, and the fact that generalized Green functions behave well with respect to parabolic induction,[1] that we may assume that $\hat{D}(\mathcal{G}_F) = 0$ for all $(F, \mathcal{G}) \in I^G$. Since D is G-invariant and since the characteristic function of

[1] Sometimes called Harish-Chandra induction.

each nilpotent orbit in $\mathsf{L}_F(f)$ can be written as a combination of generalized Green functions, we conclude that $\hat{D}([(F,e)]) = 0$ for all $(F,e) \in I^n$.

Fix an alcove C' in $\mathcal{B}(G)$. Since for all facets $F \subset \bar{C}'$, we have that $\mathfrak{g}^+_{C'}/\mathfrak{g}^+_F$ is the nilradical of the Borel subgroup $G_{C'}/G^+_F$ in $\mathsf{G}_F(\mathfrak{f}) \cong G_F/G^+_F$, it follows from the above paragraph that for all facets $F \subset \bar{C}'$, we have

$$\operatorname{res}_{C(\mathfrak{g}^+_{C'}/\mathfrak{g}^+_F)} \hat{D} = 0.$$

Consequently, since the Fourier transform maps $C(\mathfrak{g}^+_{C'}/\mathfrak{g}^+_F)$ bijectively to $C(\mathfrak{g}_F/\mathfrak{g}_{C'})$, for all facets $F \subset \bar{C}'$, we have

$$\operatorname{res}_{C(\mathfrak{g}_F/\mathfrak{g}_{C'})} D = 0.$$

But this means that $\operatorname{res}_{\mathcal{D}^0_0} D = 0$. The result now follows from Theorem 4.4.1. □

5 A basis for $\operatorname{res}_{\mathcal{D}_0} J(\mathfrak{g}_0)$

From Corollary 4.4.2 we have a particularly nice spanning set for $(\operatorname{res}_{\mathcal{D}_0} J(\mathfrak{g}_0))^*$; namely, the set of functions

$$\{\hat{\mathcal{G}}_F\},$$

where $(F,\mathcal{G}) \in I^c$ runs over a set of representatives for $I^c/\sim$. (In fact, this set is a basis; see below.) The goal of this section is to produce a dual basis in $\operatorname{res}_{\mathcal{D}_0} J(\mathfrak{g}_0)$ with good properties. In particular, for every pair $(F,\mathcal{G}) \in I^c$ we will construct an invariant distribution which is supported on the set of regular semisimple elements. The restriction to $\mathcal{D}_0$ of these distributions will constitute our dual basis.

Since $\mathbf{G}$ is of type G_2, for a facet F in $\mathcal{B}(G)$ the cuspidal generalized Green functions on $\mathsf{L}_F(\mathfrak{f})$ come in two flavors: they are either toric Green functions, i.e., the restriction to the unipotent set (identified with the nilpotent set) of Deligne–Lusztig generalized characters, or they are attached to cuspidal local systems (these occur only if F is a vertex). In general, the situation is more complicated. Note that, in all cases, cuspidal generalized Green functions are real-valued functions.

5.1 Distributions associated to the toric Green functions

Fix a facet F in $\mathcal{B}(G)$ and let S denote a $\mathfrak{f}$-minisotropic torus in G_F (From, for example, [10, Section 3.3] the set of conjugacy classes of $\mathfrak{f}$-minisotropic tori in G_F corresponds to the set of conjugacy classes in the (absolute) Weyl group W_F of G_F which do not intersect a proper parabolic subgroup of W_F.) Let Q^F_{S} denote the associated toric Green function (see, for example, [10, Section 7.6]).

Example 5.1.1. In Figure 7 we enumerate the cuspidal toric Green functions by listing (in the notation of Carter—see Table 1) the elliptic conjugacy classes in W_F. (In Table 1, the elements w_α and w_β are the simple reflections corresponding to our simple roots α and β.)

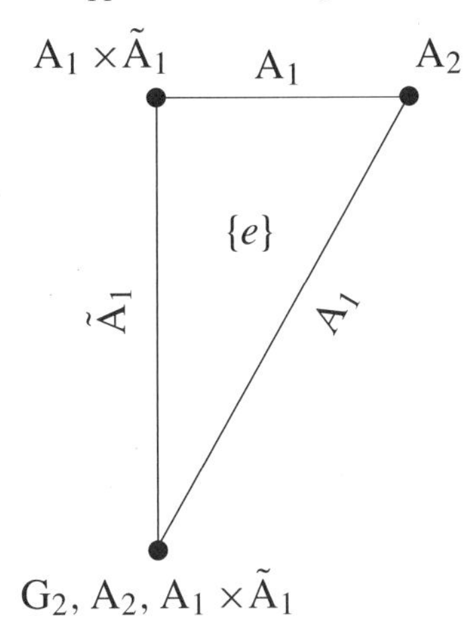

Fig. 7. A list of the cuspidal toric Green functions.

Table 1. A tabulation of data for the Weyl group of G_2.

Weyl group element	conjugacy class	image of α	image of β
1	$\{e\}$	α	β
w_α	$\tilde{A}_1$	$-\alpha$	$\beta+3\alpha$
w_β	A_1	$\beta+\alpha$	$-\beta$
$w_\alpha w_\beta$	G_2	$\beta+2\alpha$	$-\beta-3\alpha$
$w_\beta w_\alpha$	G_2	$-\beta-\alpha$	$2\beta+3\alpha$
$w_\beta w_\alpha w_\beta$	$\tilde{A}_1$	$\beta+2\alpha$	$-2\beta-3\alpha$
$w_\alpha w_\beta w_\alpha$	A_1	$-\beta-2\alpha$	$2\beta+3\alpha$
$w_\alpha w_\beta w_\alpha w_\beta$	A_2	$\beta+\alpha$	$-2\beta-3\alpha$
$w_\beta w_\alpha w_\beta w_\alpha$	A_2	$-\beta-2\alpha$	$\beta+3\alpha$
$w_\alpha w_\beta w_\alpha w_\beta w_\alpha$	$\tilde{A}_1$	$-\beta-\alpha$	β
$w_\beta w_\alpha w_\beta w_\alpha w_\beta$	A_1	α	$-\beta-3\alpha$
$w_\alpha w_\beta w_\alpha w_\beta w_\alpha w_\beta$	$A_1 \times \tilde{A}_1$	$-\alpha$	$-\beta$

From [19], we have

$$\left(\mathcal{F}\left(\sum_{g\in\mathsf{G}_F(\mathfrak{f})}[\bar{X}_{\mathsf{S}}]^g\right)\right)(\bar{X}) = \frac{(-1)^{\mathrm{rk}(\mathsf{S})}\cdot|\mathsf{S}(\mathfrak{f})|}{|\mathsf{L}_{\mathsf{S}}(\mathfrak{f})|^{1/2}}\cdot Q^F_{\mathsf{S}}(\bar{X}) \tag{1}$$

for $\bar{X}\in\mathcal{N}_F(\mathfrak{f})$. Here $\bar{X}_{\mathsf{S}}$ is any element of $\mathsf{L}_F(\mathfrak{f})$ whose centralizer in G_F is S and $\mathrm{rk}(\mathsf{S})$ denotes the dimension of the maximal $\mathfrak{f}$-split torus in S. One should think of the left-hand side as the function (up to a constant) that represents the Fourier transform of the (finite) orbital integral corresponding to $\bar{X}_{\mathsf{S}}$. We also note that for all $\bar{X}\in\mathcal{N}_F(\mathfrak{f})$, we have

$$Q^F_{\mathsf{S}}(\bar{X}) = (-1)^{\mathrm{rk}(\mathsf{S})}\cdot|\mathsf{L}_F(\mathfrak{f})|^{1/2}\cdot\hat{Q}^F_{\mathsf{S}}(\bar{X}).$$

(From [21], the two sides are proportional. The constant of proportionality may be computed explicitly by evaluating both sides at zero.)

Finally, we let $\mathbf{S}$ denote a lift of (F, S) to a maximal torus of $\mathbf{G}$ (that is, $\mathbf{S}$ has the property that $F \subset \mathcal{A}(\mathbf{S}, K)$ and the image of $\mathbf{S}(K) \cap \mathbf{G}(K)_F$ in $\mathsf{G}_F(\mathfrak{F})$ is $\mathsf{S}(\mathfrak{F})$—see [13]). From [13], the torus $\mathbf{S}$ is unique up to conjugation by G_F^+. We let S denote the group of k-rational points of $\mathbf{S}$.

Lemma 5.1.2. *Fix $(F, Q_{\mathsf{S}}^F) \in I^c$. Suppose $X_{\mathsf{S}} \in \mathfrak{g}_F$ is such that the centralizer of its image in $\mathsf{L}_F(\mathfrak{f})$ is S. If $(F', \mathcal{G}) \in I^c$, then we have*

$$\mu_{X_{\mathsf{S}}}(\hat{\mathcal{G}}_{F'}) = \begin{cases} \dfrac{(-1)^{\mathrm{rk}(\mathsf{S})} \cdot |\mathsf{G}_F(\mathfrak{f})| \cdot |N_G(S)/S|}{|\mathsf{L}_F(\mathfrak{f})|^{1/2} \cdot |\mathsf{S}(\mathfrak{f})|} & \text{if } (F', \mathcal{G}) \sim (F, Q_{\mathsf{S}}^F), \\ 0 & \text{otherwise.} \end{cases}$$

Proof. Since the centralizer of X_{S} in $\mathbf{G}$ is a torus lifting (F, S), we may assume that this torus is $\mathbf{S}$. Working through the definitions, we have

$$\mu_{X_{\mathsf{S}}}(\hat{\mathcal{G}}_{F'}) = \int_{G/S} \hat{\mathcal{G}}_{F'}({}^g X_{\mathsf{S}})\, dg^*,$$

where dg^* is the quotient measure $\frac{dg}{ds}$. Fix a $g \in G$. Since $\hat{\mathcal{G}}_{F'} \in C(\mathfrak{g}_{F'}/\mathfrak{g}_{F'}^+)$, we have that $\hat{\mathcal{G}}_{F'}({}^g X_{\mathsf{S}}) \neq 0$ implies that ${}^g X_{\mathsf{S}} \in \mathfrak{g}_{F'}$. This, in turn, implies that $X_{\mathsf{S}} \in \mathfrak{g}_{g^{-1}F'}$, which, from, for example, a slight modification of [12, Section 4.4], implies that $g^{-1}F' \subset \mathcal{B}(S) := \mathcal{A}(\mathbf{S}, K)^\Gamma \hookrightarrow \mathcal{B}(\mathbf{G}, K)^\Gamma = \mathcal{B}(G)$.

Let $\mathcal{F}$ denote the set of G-facets in the intersection of $\mathcal{B}(S)$ with the G-orbit of F'. From [13] we can find an apartment $\mathcal{A}$ in $\mathcal{B}(G)$ such that $\mathcal{B}(S) \subset \mathcal{A}$ and, in fact, $\mathcal{B}(S) = A(\mathcal{A}, F)$; thus without loss of generality, we suppose F is in $\mathcal{A}(T)$. Since $S = C_G(X_{\mathsf{S}})$ and $H \subset \mathcal{B}(S)$, for each facet H in $\mathcal{F}$ the centralizer in G_H of the image of X_{S} in $\mathsf{L}_H(\mathfrak{f}) = \mathfrak{g}_H/\mathfrak{g}_H^+$ is naturally isomorphic to S; we denote the corresponding toric Green function by Q_{S}^H. We remark that $Q_{\mathsf{S}}^H = \mathrm{Ind}_F^H Q_{\mathsf{S}}^F$.

We let $\mathcal{F}^{\mathrm{rep}}$ denote a set of representatives for $\mathcal{F}$ modulo the action of S. For each $H \in \mathcal{F}^{\mathrm{rep}}$, we fix $g_H \in G$ for which $g_H^{-1}F' = H$. Since ${}^{g_H}G_H = G_{F'}$ and $\hat{\mathcal{G}}_{F'}$ is $G_{F'}$-invariant, we have

$$\begin{aligned} \mu_{X_{\mathsf{S}}}(\hat{\mathcal{G}}_{F'}) &= \sum_{H \in \mathcal{F}^{\mathrm{rep}}} \int_{G_H S/S} \hat{\mathcal{G}}_{F'}({}^{g_H g} X_{\mathsf{S}})\, dg^* \\ &= \sum_{H \in \mathcal{F}^{\mathrm{rep}}} \mathrm{meas}_{dg^*}(G_H S/S) \cdot \hat{\mathcal{G}}_{F'}({}^{g_H} X_{\mathsf{S}}). \end{aligned} \tag{2}$$

Fix $H \in \mathcal{F}^{\mathrm{rep}}$. We are interested in the term

$$\hat{\mathcal{G}}_{F'}({}^{g_H} X_{\mathsf{S}})$$

occurring in the sum above. For notational convenience we set $\mathcal{G}_H := {}^{g_H^{-1}}\mathcal{G}_{F'}$. We first observe that

$$\hat{\mathcal{G}}_H(X_{\mathsf{S}}) = \int_{\mathfrak{g}_H} \Lambda(B(X_{\mathsf{S}}, Y)) \cdot \mathcal{G}_H(Y)\, dY. \tag{3}$$

On the other hand, we have

$$\begin{aligned}(Q_{\mathsf{S}}^F, \mathrm{r}_F^H({}^{g_H^{-1}}\mathcal{G}))_{\mathsf{L}_F} &= (Q_{\mathsf{S}}^H, {}^{g_H^{-1}}\mathcal{G})_{\mathsf{L}_H} \\ &= \frac{1}{|\mathsf{G}_H(\mathfrak{f})|} \cdot \sum_{\bar{X} \in \mathsf{L}_H(\mathfrak{f})} Q_{\mathsf{S}}^H(\bar{X}) \cdot \mathcal{G}_H(X).\end{aligned}$$

Now, since $\mathcal{G}$ is supported on $\mathcal{N}_{F'}(\mathfrak{f})$, we can apply equation (1) to arrive at

$$\begin{aligned}&\frac{(-1)^{\mathrm{rk}(\mathsf{S})} \cdot |\mathsf{S}(\mathfrak{f})|}{|\mathsf{L}_{\mathsf{S}}(\mathfrak{f})|^{1/2}} \cdot (Q_{\mathsf{S}}^F, \mathrm{r}_F^H({}^{g_H^{-1}}\mathcal{G}))_{\mathsf{L}_F} \\ &\quad = \frac{1}{|\mathsf{G}_H(\mathfrak{f})|} \cdot \sum_{\bar{X} \in \mathsf{L}_H(\mathfrak{f})} \left(\mathcal{F}\left(\sum_{g \in \mathsf{G}_H(\mathfrak{f})} [\bar{X}_{\mathsf{S}}]^g\right)\right)(\bar{X}) \cdot \mathcal{G}_H(X).\end{aligned}$$

Expanding the right-hand side yields

$$\frac{1}{|\mathsf{G}_H(\mathfrak{f})| \cdot |\mathsf{L}_H(\mathfrak{f})|^{1/2}} \cdot \sum_{\bar{X} \in \mathsf{L}_H(\mathfrak{f})} \sum_{\bar{Y} \in \mathsf{L}_H(\mathfrak{f})} \sum_{g \in \mathsf{G}_H(\mathfrak{f})} [\bar{X}_{\mathsf{S}}]({}^g\bar{Y}) \cdot \Lambda(B(X, Y)) \cdot \mathcal{G}_H(X).$$

By moving the sum over $\mathsf{G}_H(\mathfrak{f})$ in front of the other two sums and making the changes of variables $(X \mapsto {}^{g^{-1}}X)$ and $(Y \mapsto {}^{g^{-1}}Y)$, we can take advantage of the invariance properties of B and $\mathcal{G}_H$ to arrive at

$$\begin{aligned}&\frac{(-1)^{\mathrm{rk}(\mathsf{S})} \cdot |\mathsf{S}(\mathfrak{f})|}{|\mathsf{L}_{\mathsf{S}}(\mathfrak{f})|^{1/2}} \cdot (Q_{\mathsf{S}}^F, \mathrm{r}_F^H({}^{g_H^{-1}}\mathcal{G})_{\mathsf{L}_F} \\ &\quad = \mathrm{meas}_{dX}(\mathfrak{g}_H^+) \cdot \sum_{\bar{X} \in \mathsf{L}_H(\mathfrak{f})} \Lambda(B(X, X_{\mathsf{S}})) \cdot \mathcal{G}_H(X).\end{aligned} \tag{4}$$

Combining equations (2), (3), and (4), we arrive at

$$\mu_{X_{\mathsf{S}}}(\hat{\mathcal{G}}_{F'}) = (-1)^{\mathrm{rk}(\mathsf{S})} \cdot \sum_{H \in \mathcal{F}^{\mathrm{rep}}} \frac{\mathrm{meas}_{dg^*}(G_H S/S) \cdot |\mathsf{S}(\mathfrak{f})|}{|\mathsf{L}_{\mathsf{S}}(\mathfrak{f})|^{1/2}} \cdot (Q_{\mathsf{S}}^F, \mathrm{r}_F^H({}^{g_H^{-1}}\mathcal{G}))_{\mathsf{L}_F}.$$

Since for the pair $(F', \mathcal{G})$ the generalized Green function $\mathcal{G}$ was assumed to be cuspidal, in order for the sum above to be nonzero, it must be the case that there is an apartment $\mathcal{A}$ and a $g \in G$ such that

$$A(\mathcal{A}, F) = A(\mathcal{A}, gF').$$

We therefore assume that this is true. Consequently, since generalized Green functions are orthogonal, in order for the sum above to be nonzero we must have $(F, Q_{\mathsf{S}}^F) \sim (F', \mathcal{G})$.

To complete the proof, we consider the situation when $(F, Q^F_{\mathsf{S}}) \sim (F', \mathcal{G})$. Without loss of generality, from Corollary 4.2.2 we may assume that $F' = F$ and $\mathcal{G} = Q^F_{\mathsf{S}}$. The orthogonality relations for toric Green functions [10, Proposition 7.6.2] give us

$$\mu_{X_{\mathsf{S}}}(\hat{\mathcal{G}}_{F'}) = \sum_{H \in \mathcal{F}^{\text{rep}}} \frac{(-1)^{\text{rk}(\mathsf{S})} \cdot \text{meas}_{dg^*}(G_F S/S) \cdot \left|N_{\mathsf{G}_F}(\mathsf{S})(\mathfrak{f})/\mathsf{S}(\mathfrak{f})\right|}{|\mathsf{L}_{\mathsf{S}}(\mathfrak{f})|^{1/2}}.$$

We first note that

$$\begin{aligned} \text{meas}_{dg^*}(G_F S/S) &= \frac{\text{meas}_{dg}(G_F)}{\text{meas}_{ds}(S \cap G_F)} \\ &= \frac{|\mathsf{G}_F(\mathfrak{f})| \cdot |\mathsf{L}_{\mathsf{S}}(\mathfrak{f})|^{1/2}}{|\mathsf{L}_F(\mathfrak{f})|^{1/2} \cdot |\mathsf{S}(\mathfrak{f})|}. \end{aligned} \tag{5}$$

So we have

$$\mu_{X_{\mathsf{S}}}(\hat{\mathcal{G}}_{F'}) = \frac{(-1)^{\text{rk}(\mathsf{S})} \cdot |\mathcal{F}^{\text{rep}}| \cdot |\mathsf{G}_F(\mathfrak{f})| \cdot \left|N_{\mathsf{G}_F}(\mathsf{S})(\mathfrak{f})/\mathsf{S}(\mathfrak{f})\right|}{|\mathsf{L}_F(\mathfrak{f})|^{1/2} \cdot |\mathsf{S}(\mathfrak{f})|}.$$

To complete the proof, we now show that

$$\left|\mathcal{F}^{\text{rep}}\right| = \frac{|N_G(S)/S|}{\left|N_{\mathsf{G}_F}(\mathsf{S})(\mathfrak{f})/\mathsf{S}(\mathfrak{f})\right|}.$$

Suppose $H \in \mathcal{F}^{\text{rep}}$ and $g \in G$ such that $gF = H \subset \mathcal{B}(S)$. Let ${}^g\mathsf{S}$ denote the $\mathfrak{f}$-torus in G_H whose group of $\mathfrak{F}$-rational points agrees with the image of ${}^g\mathbf{S}(K) \cap \mathbf{G}(K)_H$ in G_H. Via the identification $\mathsf{G}_H \stackrel{i}{=} \mathsf{G}_F$ we see that there is a $k \in G_H \cap G_F$ such that ${}^{kg}\mathsf{S} \stackrel{i}{=} \mathsf{S}$ in $\mathsf{G}_H \stackrel{i}{=} \mathsf{G}_F$. Since $\mathbf{S}$ is a lift of ${}^{kg}\mathsf{S}$, from [13] there is an element $k' \in G_H^+$ such that ${}^{k'kg}\mathbf{S} = \mathbf{S}$. Thus every element of $\mathcal{F}^{\text{rep}}$ uniquely determines, up to right multiplication by $N_{G_F}(S)/S$, an element of $N_G(S)/S$. The desired equality follows. □

Definition 5.1.3. *Suppose $(F, Q^F_{\mathsf{S}}) \in I^c$. Choose $X_{\mathsf{S}} \in \mathfrak{g}_F$ such that the image of X_{S} in $\mathsf{L}_F(\mathfrak{f})$ has centralizer S in G_F. We define*

$$D_{(F, Q^F_{\mathsf{S}})} := \mu_{X_{\mathsf{S}}}.$$

Remark 5.1.4. As a distribution on $\mathfrak{g}$, the distribution $D_{(F,Q^F_{\mathsf{S}})}$ is not independent of our choice of X_{S}. However, thanks to Lemma 5.1.2 and Corollary 4.4.2, we have that $\text{res}_{\mathcal{D}_0} D_{(F,Q^F_{\mathsf{S}})}$, the restriction of $D_{(F,Q^F_{\mathsf{S}})}$ to $\mathcal{D}_0$, is independent of the choice of X_{S}.

5.2 Distributions associated to cuspidal local systems

There are four (or two, depending on whether or not $\mathfrak{f}^\times$ has cubic roots of unity) classes in $I^c/\!\sim$ which are not covered by the material in Section 5.1. Each of these pairs is of the form $(F, \mathcal{G})$, where F is a vertex in $\mathcal{B}(G)$, and $\mathcal{G}$ is a cuspidal generalized Green function associated to a cuspidal local system. The goal of this section is to associate to each such pair a particularly good element of $\text{res}_{\mathcal{D}_0} J(\mathfrak{g}_0)$. As Waldspurger has noticed, this association is related to the Kazhdan–Lusztig [20] map from nilpotent orbits in Lie $\mathbf{G}(\mathbb{C})$ to conjugacy classes of maximal tori in $\mathbf{G}(\mathbb{C}((t)))$.

5.2.1 Distinguished elements

Suppose $\mathfrak{k}$ is a field and $\mathfrak{h}$ is the Lie algebra of H, the group of $\mathfrak{k}$-rational points of a reductive $\mathfrak{k}$-group. A nilpotent H-orbit in $\mathfrak{h}$ is said to be $\mathfrak{k}$-*distinguished* provided that it does not intersect a Levi $\mathfrak{k}$-subalgebra of a proper parabolic $\mathfrak{k}$-subgroup. An element of a $\mathfrak{k}$-distinguished orbit is said to be $\mathfrak{k}$-distinguished.

5.2.2 Cuspidal local systems

Suppose F is a vertex. A cuspidal local system for $\mathsf{L}_F(\mathfrak{f})$ is specified by a $\mathfrak{F}$-distinguished nilpotent orbit $\bar{\mathcal{O}}$ in $\mathsf{L}_F(\mathfrak{F})$ and a "cuspidal" character χ of an irreducible representation of the (twisted by σ) component group associated to the orbit $\bar{\mathcal{O}}$. Choose a Γ-fixed element $e \in \bar{\mathcal{O}}$. From Lang–Steinberg [10, Section 1.17], for every c in the component group $C(F, e) := (\mathsf{G}_F)_e/(\mathsf{G}_F)_e^\circ$ there is a $g_c \in \mathsf{G}_F(\mathfrak{F})$ such that the image of $\sigma(g_c)^{-1}g_c$ in $C(F, e)$ is c. We have

$$\bar{\mathcal{O}}^\Gamma = \coprod_{c \in C(F,e)} {}^{\mathsf{G}_F(\mathfrak{f})g_c}e.$$

The generalized Green function associated to this cuspidal local system is

$$\mathcal{G} := \sum_{c \in C(F,e)/\sim} \chi(c) \cdot [{}^{\mathsf{G}_F(\mathfrak{f})g_c}e],$$

where $[{}^{\mathsf{G}_F(\mathfrak{f})g_c}e]$ denotes the characteristic function of the orbit ${}^{\mathsf{G}_F(\mathfrak{f})g_c}e$ and the equivalence relation is σ-conjugacy. We now describe the $(|\mu_3(\mathfrak{f})| + 1)$ elements of $I^c/\sim$ which arise from cuspidal local systems.

Independent of the status of the cubic roots of unity, we shall always have the classes in $I^c/\sim$ represented by

$$(F_{\mathrm{G}_2}, \mathcal{G}_{\mathrm{sgn}}) \quad \text{and} \quad (F_{\mathrm{A}_1 \times \tilde{\mathrm{A}}_1}, \mathcal{G}_{\mathrm{sgn}}).$$

For the facet F_{G_2}, the generalized Green function $\mathcal{G}_{\mathrm{sgn}}$ comes from the cuspidal local system supported on the $\mathfrak{F}$-distinguished nonregular orbit in $\mathsf{L}_{F_{\mathrm{G}_2}}(\mathfrak{F})$; we take the sign character, sgn, on the associated component group, which is S_3. For the facet $F_{\mathrm{A}_1 \times \tilde{\mathrm{A}}_1}$ the generalized Green function $\mathcal{G}_{\mathrm{sgn}}$ arises from the cuspidal local system supported on the regular orbit in $\mathsf{L}_{F_{\mathrm{A}_1 \times \tilde{\mathrm{A}}_1}}(\mathfrak{F})$; we consider the sign character, sgn, on $\mathbb{Z}/2\mathbb{Z}$, the associated component group. If $\mathfrak{f}^\times \neq (\mathfrak{f}^\times)^3$, then we have the additional classes in $I^c/\sim$ represented by

$$(F_{\mathrm{A}_2}, \mathcal{G}_{\chi'}) \quad \text{and} \quad (F_{\mathrm{A}_2}, \mathcal{G}_{\chi''}),$$

where, in each case, the generalized Green function is coming from the cuspidal local system supported on the regular nilpotent orbit in $\mathsf{L}_{F_{\mathrm{A}_2}}(\mathfrak{F})$. The characters χ' and χ'' are the two nontrivial characters on the associated component group, $\mathbb{Z}/3\mathbb{Z}$. We summarize this notation in Figure 8.

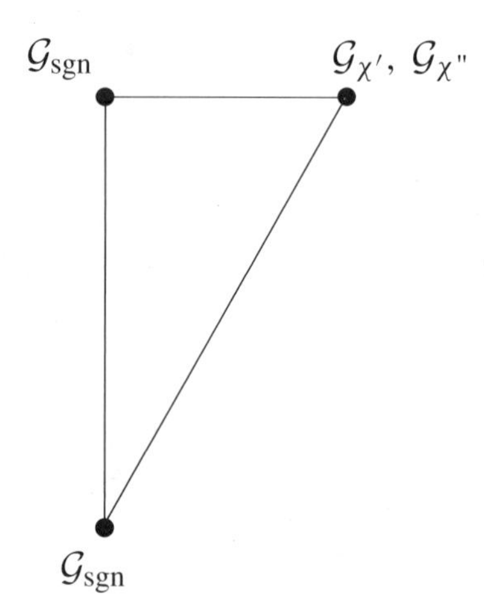

Fig. 8. Cuspidal generalized Green functions associated to cuspidal local systems.

In order to continue to use our preferred chamber $F_\emptyset$ as a way to visualize information, we must be somewhat careful about how e gets chosen in each case. Since we will also eventually need to choose a lift X of e, we also do this now. In Table 2 we list our choices. The superscript 1 on e_1^1 and X_1^1 denotes the identity element of $\mu_3(\mathfrak{f})$.

Table 2. Choices for e and X for cuspidal local systems.

$(F, \mathcal{G})$	X	e
$(F_{\mathrm{G}_2}, \mathcal{G}_{\mathrm{sgn}})$	$X_1 := X_\beta - X_{\beta+3\alpha} \in \mathfrak{g}_{F_{\mathrm{G}_2}}$	e_1, the image of X_1 in $\mathsf{L}_{F_{\mathrm{G}_2}}$
$(F_{\mathrm{A}_1 \times \tilde{\mathrm{A}}_1}, \mathcal{G}_{\mathrm{sgn}})$	$X_1^+ := X_\alpha + \varpi X_{-(2\beta+3\alpha)} \in \mathfrak{g}_{F_{\mathrm{A}_1 \times \tilde{\mathrm{A}}_1}}$	e_1^+, the image of X_1^+ in $\mathsf{L}_{F_{\mathrm{A}_1 \times \tilde{\mathrm{A}}_1}}$
$(F_{\mathrm{A}_2}, \mathcal{G}_{\chi'})$	$X_1^1 := X_\beta + \varpi X_{-(2\beta+3\alpha)} \in \mathfrak{g}_{F_{\mathrm{A}_2}}$	e_1^1, the image of X_1^1 in $\mathsf{L}_{F_{\mathrm{A}_2}}$
$(F_{\mathrm{A}_2}, \mathcal{G}_{\chi''})$	X_1^1	e_1^1

5.2.3 Component groups over $\mathfrak{F}$ and K

Before we can continue, we need a better understanding of the connection between the component groups associated to distinguished nilpotent orbits over $\mathfrak{F}$ and K.

Suppose F is a vertex and $\bar{\mathcal{O}}$ is a $\mathfrak{F}$-distinguished orbit in $\mathsf{L}_F(\mathfrak{F})$. Choose a Γ-fixed $e \in \bar{\mathcal{O}}$.

Recall that $\mathbf{T}$ is our fixed maximal k-split torus in $\mathbf{G}$, and our alcove $F_\emptyset$ belongs to $\mathcal{A}(T)$. Let T be the maximal $\mathfrak{f}$-torus in G_F corresponding to $\mathbf{T}$.

We complete e to an $\mathfrak{sl}_2(\mathfrak{F})$-triple (f, h, e) with f and h both Γ-fixed. We suppose that the associated one-parameter subgroup $\bar{\mu}$ belongs to $\mathbf{X}_*^{\mathfrak{f}}(\mathsf{T})$ and that there exists an $\mathfrak{sl}_2(K)$-triple (Y, H, X) lifting (f, h, e) with Y, H, and X each Γ-fixed and such that the lift $\mu \in \mathbf{X}_*^k(\mathbf{T})$ of $\bar{\mu}$ is the associated one-parameter subgroup.

Let $\mathsf{L}_F(i)$ denote the i-eigenspace for the action of the one parameter subgroup $\bar{\mu}$ and set

$$\mathsf{L}_F(\geq j) = \bigoplus_{i \geq j} \mathsf{L}_F(i).$$

Let P denote the parabolic $\mathfrak{f}$-subgroup with Lie algebra $\mathsf{L}_F(\geq 0)$ and let M denote the $\mathfrak{f}$-Levi subgroup of P with Lie algebra $\mathsf{L}_F(0)$. The unipotent radical of P will be denoted N. (The parabolic P is the distinguished parabolic associated to e.) Let $(\mathsf{G}_F)_e$ (respectively, M_e, N_e) denote the centralizer of e in G_F (respectively, M, N). We have [10, Section 5.7]

$$(\mathsf{G}_F)_e = \mathsf{M}_e\mathsf{N}_e$$

and, since e is $\mathfrak{F}$-distinguished,

$$\mathsf{M}_e \cong (\mathsf{G}_F)_e/(\mathsf{G}_F)_e^\circ,$$

where $(\mathsf{G}_F)_e^\circ$ denotes the connected component of $(\mathsf{G}_F)_e$. (Note that $\mathsf{M}_e = C_{\mathsf{G}_F}(e) \cap C_{\mathsf{G}_F}(f) \cap C_{\mathsf{G}_F}(h)$.) From Lang–Steinberg, for all $g \in \mathsf{M}_e(\mathfrak{F})$, there exists $m \in \mathsf{M}(\mathfrak{F})$ such that $\sigma(m)^{-1}m = g$.

Similarly, from μ we construct a k-parabolic subgroup $\mathbf{P}$ of $\mathbf{G}$, a k-Levi $\mathbf{M}$ of $\mathbf{P}$, and the unipotent radical $\mathbf{N}$ of $\mathbf{P}$ such that $\mathbf{P}$ has a Levi decomposition $\mathbf{P} = \mathbf{MN}$. (The k-parabolic $\mathbf{P}$ is the distinguished parabolic associated to X.) Let $\mathbf{G}_X$ (respectively, $\mathbf{M}_X$, $\mathbf{N}_X$) denote the centralizer of X in $\mathbf{G}$ (respectively, $\mathbf{M}$, $\mathbf{N}$). As before, we have

$$\mathbf{G}_X = \mathbf{M}_X\mathbf{N}_X.$$

It follows that $\mathbf{G}_X(K) = \mathbf{M}_X(K)\mathbf{N}_X(K)$.

Example 5.2.1. We describe some of these objects for the e we chose in Table 2 of Section 5.2.2. We let $\check{\alpha}$ and $\check{\beta}$ denote the coroots of α and β, respectively. For the vertex F_{G_2} we have $\mu = 2\check{\alpha} + 4\check{\beta}$ and P is the image of the parahoric $\mathbf{G}(K)_{F_{\tilde{A}_1}}$ in $\mathrm{G}_2(\mathfrak{F}) = \mathsf{G}_{F_{\mathrm{G}_2}}(\mathfrak{F})$. The Levi subgroup M is isomorphic to $\mathrm{GL}_2 = \mathsf{G}_{F_{\tilde{A}_1}}$ and M_{e_1} is S_3 embedded in GL_2 as the group generated by

$$\check{\alpha}(\mu_3) \quad \text{and} \quad \bar{n}_\alpha.$$

Here $\bar{n}_\alpha$ is the image of an involution $n_\alpha \in N_G(T)$ which represents the Weyl group element corresponding to α. We note that $\mathbf{M}(K)$ acts on the root space $\mathfrak{g}(K)_{-(2\beta+3\alpha)}$ by $\det^{-1}$.

For the vertex $F_{\mathrm{A}_1 \times \tilde{\mathrm{A}}_1}$ we have $\mu = -2\check{\beta}$ and P is the Borel subgroup realized as the image of the Iwahori subgroup $\mathbf{G}(K)_{F_\emptyset}$ in $\mathrm{SO}_4(\mathfrak{F}) = \mathsf{G}_{F_{\mathrm{A}_1 \times \tilde{\mathrm{A}}_1}}(\mathfrak{F})$. The Levi subgroup M is isomorphic to $(\mathrm{GL}_1)^2 = \mathsf{G}_{F_{\tilde{A}_1}}$, and $\mathsf{M}_{e_1^+}$ is μ_2 embedded in $(\mathrm{GL}_1)^2$ as $\check{\alpha}(\mu_2)$.

Finally, for the vertex F_{A_2} we have $\mu = -2(\check{\alpha} + \check{\beta})$, and P is the image of the Iwahori subgroup $\mathbf{G}(K)_{F_\emptyset}$ in $\mathrm{SL}_3(\mathfrak{F}) = \mathsf{G}_{F_{\mathrm{A}_2}}(\mathfrak{F})$. The Levi subgroup M is isomorphic to $(\mathrm{GL}_1)^2 = \mathsf{G}_{F_{\tilde{A}_1}}$ and $\mathsf{M}_{e_1^1}$ is μ_3 embedded in $(\mathrm{GL}_1)^2$ as $\check{\alpha}(\mu_3)$.

Now, since F is a vertex, X is K-distinguished in $\mathfrak{g}(K)$; moreover, F is the unique facet in $\mathcal{B}(G)$ for which X, Y, and H belong to $\mathfrak{g}(K)_F$ (see, for example, [3, Corollary 4.4]). Since X is K-distinguished, $\mathbf{M}_X(K)$ is finite. Since

$$\mathbf{M}_X = C_{\mathbf{G}}(Y) \cap C_{\mathbf{G}}(H) \cap C_{\mathbf{G}}(X),$$

and F is the unique facet in $\mathcal{B}(G)$ for which X, H, and Y belong to $\mathfrak{g}(K)_F$, it follows that

$$\mathbf{M}_X(K) \subset \mathbf{G}(K)_F.$$

Indeed, if $m \in \mathbf{M}_X(K)$, then X, H, and Y belong to $\mathfrak{g}(K)_{mF}$. But F is the unique facet for which X, H, and Y belong to $\mathfrak{g}(K)_F$. Hence $mF = F$.

Lemma 5.2.2. *The natural surjective map*

$$\mathbf{G}(K)_F \to \mathbf{G}(K)_F/\mathbf{G}(K)_F^+ \cong \mathsf{G}_F(\mathfrak{F})$$

induces an isomorphism $\mathbf{M}_X(K) \cong \mathsf{M}_e(\mathfrak{F})$.

Proof. We first show that the map is surjective. Suppose $\bar{m} \in \mathsf{M}_e(\mathfrak{F})$. Let $p \in \mathbf{G}(K)_F$ be any lift of $\bar{m}$. From [12, Corollary 5.2.3], since

$${}^pX \in (X + \mathfrak{g}(K)_F^+) \cap {}^{\mathbf{G}(K)}X,$$

there exists $k \in \mathbf{G}(K)_F^+$ such that ${}^{kp}X = X$. So without loss of generality, $p \in \mathbf{G}_X(K) \cap \mathbf{G}(K)_F$. Since $\mathbf{G}_X(K) \subset \mathbf{M}_X(K)\mathbf{N}_X(K)$ and $\mathbf{P}(K) \cap \mathbf{G}(K)_F = (\mathbf{M}(K) \cap \mathbf{G}(K)_F) \cdot (\mathbf{N}(K) \cap \mathbf{G}(K)_F)$ (with uniqueness of decompostion), we can write $p = mn$ with

$$m \in \mathbf{M}_X(K) \cap \mathbf{G}(K)_F$$

and

$$n \in \mathbf{N}_X(K) \cap \mathbf{G}(K)_F.$$

Since $\mathsf{M}(\mathfrak{F}) = \mathbf{M}(K)_F/\mathbf{M}(K)_F^+$, we conclude that $m \in \mathbf{M}_X(K) \cap \mathbf{G}(K)_F$ and $n \in \mathbf{N}_X(K) \cap \mathbf{G}(K)_F^+$. Thus the image of m in $\mathsf{M}(\mathfrak{F})$ is $\bar{m}$.

We now show that the map is injective. Suppose $m \in \mathbf{M}_X(K) \leq \mathbf{G}(K)_F$ such that the image of m in $\mathsf{M}_e(\mathfrak{F})$ is trivial. We have $m \in \mathbf{G}(K)_F^+$ and m is of finite order. Therefore, m is trivial. □

Corollary 5.2.3. *If* $g \in \mathbf{M}_X(K)$, *then there exists* $m \in \mathbf{M}(K)_F$ *such that* $\sigma(m)^{-1}m = g$.

Proof. Let $\bar{g}$ denote the image of g in M_e. From Lang–Steinberg, we can choose $\bar{m} \in \mathsf{M}(\mathfrak{F})$ such that $\sigma(\bar{m})^{-1}\bar{m} = \bar{g}$. We have that ${}^{\bar{m}}e$ is an element of $\mathsf{L}_F(\mathfrak{f})$. By completing ${}^{\bar{m}}e$ to an $\mathfrak{sl}_2(\mathfrak{f})$-triple, we see from [12, Lemma 5.3.3] that we may choose a Γ-fixed lift X' of ${}^{\bar{m}}e$ so that the nilpotent orbit ${}^{\mathbf{G}(K)}X'$ is the unique nilpotent orbit of minimal dimension intersecting ${}^{\bar{m}}e$ nontrivially. Let $m' \in \mathbf{M}(K)_F$ be any lift of $\bar{m}$. It follows from [12, Corollary 5.2.3], that there exists a $j \in \mathbf{G}(K)_F^+$ such that ${}^{jm'}X = X'$. By construction, we have $\sigma(jm')^{-1}jm' \in \mathbf{G}_X(K) = \mathbf{M}_X(K)\mathbf{N}_X(K)$, and the image of $\sigma(jm')^{-1}jm'$ in $\mathsf{G}_F(\mathfrak{F})$ is $\bar{g} \in \mathsf{M}_e$. Thanks to the Iwahori decomposition, we can write $j = \bar{n}''n''m''$ with $m'' \in \mathbf{G}(K)_F^+ \cap \mathbf{M}(K)$, $n'' \in \mathbf{G}(K)_F^+ \cap \mathbf{N}(K)$, and $\bar{n}'' \in \bar{\mathbf{N}}(K) \cap \mathbf{G}(K)_F^+$, where $\bar{\mathbf{N}}$ denotes the unipotent radical of the parabolic subgroup opposite $\mathbf{P} = \mathbf{MN}$. We have

$$\sigma(jm')^{-1}jm' = \sigma(m')^{-1}\sigma(m'')^{-1}\sigma(n'')^{-1}\sigma(\bar{n}'')^{-1}\bar{n}''n''m''m' \in \mathbf{G}_X(K) \leq \mathbf{P}(K).$$

Thus $\sigma(\bar{n}'') = \bar{n}''$. Consequently, $\bar{n}'' \in \bar{\mathbf{N}}(k) \cap \mathbf{G}(K)_F^+$. Without loss of generality, we may replace X' with ${}^{\bar{n}''}X'$. We now have $j = n''m''$. Let $m = m''m'$. Then ${}^{n''m}X = X'$. Since

$$\sigma(m)^{-1}\sigma(n'')^{-1}n''m = [\sigma(m)^{-1}m][{}^{m^{-1}}(\sigma(n'')n'')] \in \mathbf{M}_X(K)\mathbf{N}_X(K),$$

we have $\sigma(m)^{-1}m \in \mathbf{M}_X(K)$, and the image of $\sigma(m)^{-1}m$ in $\mathsf{G}_F(\mathfrak{F})$ is $\bar{g}$. The corollary now follows from Lemma 5.2.2. □

5.2.4 Identifying regular semisimple elements

We now associate to each of the $(|\mu_3(\mathfrak{f})| + 1)$ pairs $(F, \mathcal{G})$ a subset of the regular semisimple elements in $\mathfrak{g}$. More specifically, recall that we have associated to $(F, \mathcal{G})$ a nilpotent element e. For each $m \in \mathsf{M}_e$ we will choose a subset of the regular semisimple elements in $\mathfrak{g}$. Essentially, these are certain elements of a torus associated to e by the Kazhdan–Lusztig map. Although there are explicit lists [25] describing this map, because we need to associate to the pair (m, e) a set of topologically nilpotent regular semisimple elements of $\mathfrak{g}$ and not just $\mathbf{g}(K)$, we must do this part of the proof "by hand." To this end, we offer a *caveat*: although our approach seems very general, it only works for G_2.

Fix one of the pairs $(F, \mathcal{G}) \in I^c$ with $\mathcal{G}$ corresponding to a cuspidal local system. Recall that in Table 2 and Example 5.2.1 we have associated to the pair $(F, \mathcal{G})$ a nilpotent $e \in \mathsf{L}_F(\mathfrak{f})$, a nilpotent $X \in \mathfrak{g}_F$, subgroups M_e and $\mathbf{M}_X(K)$, etc.

Recall (see Example 2.2.2) that each vertex occurring in the closure of $F_\emptyset$ corresponds to a node in the extended Dynkin diagram of type G_2. Let ψ_F denote the affine root labeling the node corresponding to F.

We let H denote the fixator in $\mathbf{M}_X(K)$ of $\mathbf{g}(K)_{\dot{\psi}_F}$; that is

$$H = \{m \in \mathbf{M}_X(K) \mid {}^mZ = Z \text{ for all } Z \in \mathbf{g}(K)_{\dot{\psi}_F}\}.$$

The group H is finite, and its centralizer, $\mathbf{G}(X)$, is a connected reductive subgroup of $\mathbf{G}$ which contains $\mathbf{T}$. The centralizer (in $\mathbf{G}$) of each element of the set

$$B(F, \mathcal{G}, X) := X + \mathbf{g}(K)_{\psi_F} \setminus \mathbf{g}(K)^+_{\psi_F} \subset \mathrm{Lie}(\mathbf{G}(X))(K)$$

is a K-elliptic torus in $\mathbf{G}(X)$ which splits over a totally ramified extension of degree h. Here h denotes the Coxeter[2] number of $\mathbf{G}(X)$. In Table 3 we describe all these

[2] The symbol h is also used for the semisimple element of an $\mathfrak{sl}_2$-triple over $\mathfrak{f}$. However, this should not cause any confusion.

objects for our four specific pairs; in the column labeled "class" we identify the Weyl group conjugacy class corresponding to the K-elliptic torus (see [25, Table I]). In fact, if $\mathbf{S}$ is such a torus, then our assumptions on p imply that that the building of $\mathbf{S}(K)$ embeds in the building of $\mathbf{G}(X)(K)$ which, in turn, embeds into that of $\mathbf{G}(K)$ as the point

$$y = F + \frac{\hat{\mu}}{2h}.$$

More specifically, the building of a torus corresponding to the Weyl conjugacy class A_2 occurs in the interior of the facet $F_{\tilde{A}_1}$, and the building of a torus corresponding to the Weyl conjugacy class G_2 occurs in the interior of the facet $F_{\emptyset}$. In all cases, the quotient $\mathbf{G}(K)_y/\mathbf{G}(K)_y^+$ is isomorphic to $\mathsf{M}(\mathfrak{F})$.

Table 3. Some data associated to cuspidal local systems.

$(F, \mathcal{G})$	H	$\mathbf{G}(X)$	h	class
$(F_{\mathrm{G}_2}, \mathcal{G}_{\mathrm{sgn}})$	$\check{\alpha}(\mu_3)$	SL_3	3	A_2
$(F_{\mathrm{A}_1 \times \tilde{\mathrm{A}}_1}, \mathcal{G}_{\mathrm{sgn}})$	1	G_2	6	G_2
$(F_{\mathrm{A}_2}, \mathcal{G}_{\chi'})$	1	G_2	6	G_2
$(F_{\mathrm{A}_2}, \mathcal{G}_{\chi''})$	1	G_2	6	G_2

5.2.5 Two distributions associated to $(F, \mathcal{G})$

Let $(F, \mathcal{G})$ be one of our $(|\mu_3(\mathfrak{f})| + 1)$ pairs. We will associate two distributions, denoted $T_{(F,\mathcal{G})}$ and $D_{(F,\mathcal{G})}$, to this pair. While the first distribution is most naturally associated to the pair $(F, \mathcal{G})$, it is the second which will play a fundamental role in the remainder of the paper. In Section 5.2.8 we show that the restrictions to $\mathcal{D}_0$ of these two distributions agree (up to an explicit constant).

We begin by defining $D_{(F,\mathcal{G})}$. Let e and X be the nilpotent elements associated to $(F, \mathcal{G})$ and let χ denote the associated character of M_e. For $m \in \mathsf{M}_e$, we fix $g_m \in \mathbf{M}_X(K) \cap \mathbf{G}(K)_F$ such that $\sigma(g_m)^{-1} g_m = m$ and set

$$A(F, \mathcal{G}, X, g_m) = \{{}^{g_m}Y \mid Y \in B(F, \mathcal{G}, X) \text{ and } {}^{m}Y = \sigma(Y)\}.$$

In all cases, $A(F, \mathcal{G}, X, g_m)$ is a subset of $\mathfrak{g}$ which is naturally topologically isomorphic to $R^\times$ and consists entirely of regular semisimple elements for which the centralizer is a K-elliptic k-torus; that is, an elliptic maximal k-torus which splits over a totally ramified extension of k. Moreover, since $\mathbf{g}_{\psi_F}(K)$ is $\mathbf{M}(K)$-invariant, we have

$$A(F, \mathcal{G}, X, g_m)_{\psi_F} = \mathfrak{g}_{\psi_F} \smallsetminus \mathfrak{g}_{\psi_F}^+,$$

where $A(F, \mathcal{G}, X, g_m)_{\psi_F}$ denotes the image of $A(F, \mathcal{G}, X, g_m)$ under the projection map $\mathfrak{g} \to \mathfrak{g}_{\psi_F}$.

We let dY denote the measure on $A(F, \mathcal{G}, X, g_m)$ which arises from the normalized Haar measure on $R^{\times}$. We shall study the distribution

$$D_{(F,\mathcal{G})} := \sum_{\mathsf{M}_e/\sim} \frac{\chi(m) \cdot |S_m/S_m^{\circ}|}{|C_{\mathsf{G}_F(\mathfrak{f})}({}^{g_m}e)|} \int_{A(F,\mathcal{G},X,g_m)} \mu_Y \, dY.$$

Here S_m is the centralizer in G of any element of $A(F, \mathcal{G}, X, g_m)$ and S_m° is the parahoric subgroup of S_m.

We now take up the definition of the distribution $T_{(F,\mathcal{G})}$.

Lemma 5.2.4. *Suppose $f \in C_c^{\infty}(\mathfrak{g})$. The function from G to $\mathbb{C}$ defined by*

$$g \mapsto \int_{\mathfrak{g}} f({}^g Z) \cdot \mathcal{G}_F(Z) \, dZ$$

is locally constant and compactly supported.

Proof. This is a standard result which dates back to Harish-Chandra's notes [14]. We sketch the main idea: Fix $f \in C_c^{\infty}(\mathfrak{g})$. Choose $n \in \mathbb{Z}_{>0}$ so that f is locally constant with respect to $\mathfrak{g}_{F,n} := \varpi^n \mathfrak{g}_F$. For g outside of a compact subset of G we have that the image of ${}^g\mathfrak{g}_{F,n} \cap \mathfrak{g}_F$ in $\mathsf{L}_F(\mathfrak{f})$ is the nilradical of a proper parabolic subgroup of $\mathsf{G}_F(\mathfrak{f})$. Since $\mathcal{G}$ is a cuspidal function, the result follows. □

Thanks to Lemma 5.2.4, it makes sense to define the invariant distribution $T_{(F,\mathcal{G})}$ by

$$T_{(F,\mathcal{G})}(f) := \int_G \int_{\mathfrak{g}} f({}^g Z) \cdot \mathcal{G}_F(Z) \, dZ \, dg$$

for $f \in C_c^{\infty}(\mathfrak{g})$. (Here dg (respectively, dZ) is the Haar measure on G (respectively, $\mathfrak{g}$).)

5.2.6 An orthogonality result for $T_{(F,\mathcal{G})}$

We continue to use the notation introduced in the previous section.

Lemma 5.2.5. *If $(F', \mathcal{G}') \in I^c$, then we have*

$$T_{(F,\bar{\mathcal{G}})}(\hat{\mathcal{G}}'_{F'}) = \begin{cases} \dfrac{|\mathsf{G}_F(\mathfrak{f})|^2}{|\mathsf{L}_F(\mathfrak{f})|} (\hat{\mathcal{G}}, \mathcal{G})_{\mathsf{L}_F} & \text{if } (F', \mathcal{G}') \sim (F, \mathcal{G}), \\ 0 & \text{otherwise}. \end{cases}$$

Proof. We fix $g \in G$ and consider the integral

$$\int_{\mathfrak{g}} \hat{\mathcal{G}}'_{F'}({}^g Z) \cdot \bar{\mathcal{G}}_F(Z) \, dZ.$$

If ${}^g Z \in \mathfrak{g}_{F'}$, then, since the support of $\mathcal{G}_F(Z)$ lies in $\mathfrak{g}_{0^+}$, we have that the image of ${}^g Z$ in $\mathsf{L}_{F'}(\mathfrak{f})$ is nilpotent. Consequently, since, on the set of nilpotent elements, a generalized Green function agrees with its Fourier transform up to a constant, we may replace $\hat{\mathcal{G}}'_{F'}$ in the above integral with $\mathrm{const}(\mathcal{G}', F') \cdot \mathcal{G}'_{F'}$. Thus we are interested in the integral

$$\mathrm{const}(\mathcal{G}', F') \cdot \int_{\mathfrak{g}} \mathcal{G}'_{F'}({}^{g}Z) \cdot \bar{\mathcal{G}}_F(Z)\, dZ.$$

We first show that this integral is zero if $g^{-1}F' \neq F$. Recall that F is a vertex. If $g^{-1}F' \neq F$, then the above integral is, up to a constant, equal to the integral

$$\int_{\mathfrak{g}} \mathcal{G}'_{F'}({}^{g}Z) \int_{\mathfrak{g}_F \cap \mathfrak{g}^{+}_{g^{-1}F'}} \bar{\mathcal{G}}_F(Z + Z')\, dZ'\, dZ.$$

However, since $\mathcal{G}$ is a cuspidal function on $\mathsf{L}_F(\mathfrak{f})$ and the image of $\mathfrak{g}_F \cap \mathfrak{g}^{+}_{g^{-1}F'}$ in $\mathsf{L}_F(\mathfrak{f})$ is the nilradical of a proper parabolic subgroup of G_F, we conclude that the above double integral is zero.

We now suppose that ${}^{g^{-1}}F' = F$. If $\mathcal{G} \neq {}^{g^{-1}}\mathcal{G}'$, then the integral is zero since nonequivalent generalized Green functions are orthogonal.

We have shown that $T_{(F,\mathcal{G})}(\hat{\mathcal{G}}'_{F'})$ is zero unless $(F, \mathcal{G}) \sim (F', \mathcal{G}')$. We now take up the case when $(F, \mathcal{G}) \sim (F', \mathcal{G}')$. Without loss of generality, we may assume that $(F, \mathcal{G}) = (F', \mathcal{G}')$. We have

$$T_{(F,\mathcal{G})}(\hat{\mathcal{G}}_F) = \mathrm{const}(\mathcal{G}, F) \cdot \int_{G_F} \int_{\mathfrak{g}_F} \mathcal{G}_F({}^{g}Z) \cdot \bar{\mathcal{G}}_F(Z)\, dZ\, dg.$$

Using the invariance of $\mathcal{G}$, we obtain the desired formula. □

5.2.7 Some auxiliary functions and vanishing results

Unfortunately, we have not been able to produce a notationally simple proof of the relationship between $T_{(F,\mathcal{G})}$ and $D_{(F,\mathcal{G})}$; consequently, the reader is encouraged to skip the proofs in this subsection.

Let $\mathfrak{g}_F(i)$ denote the intersection of $\mathfrak{g}_F$ with the i-eigenspace in $\mathfrak{g}$ for μ; similar notation applies to L_F, etc. For G_2, we have $\mathfrak{g}_F(2i+1) = 0$ for all i; that is, all nilpotent orbits are even. We will not take advantage of this fact.

We denote by $\mathfrak{g}_{y,-(1/h)}$ (using the notation of Moy and Prasad [24]) the unique lattice in $\mathfrak{g}$ for which

$$\mathfrak{g}_y \subset \mathfrak{g}_{y,-(1/h)} \subset \mathfrak{g}_F + \mathfrak{g}_{\psi_F}$$

and

$$\mathfrak{g}_{y,-(1/h)}/\mathfrak{g}_y \cong \mathsf{L}_F(-2) \oplus \mathfrak{g}_{-\psi_F}/\mathfrak{g}^{+}_{-\psi_F}.$$

Thus

$$\mathsf{L}_F(-2) \cong \mathfrak{g}_{y,-(1/h)}(-2)/\mathfrak{g}_y(-2), \tag{6}$$

while

$$\mathfrak{g}_{-\psi_F} \not\subset \mathfrak{g}_F \quad \text{and} \quad \mathfrak{g}^{+}_{-\psi_F} \subset \mathfrak{g}_y \subset \mathfrak{g}_F. \tag{7}$$

Definition 5.2.6. *Let dj denote the Haar measure on G. For $Z \in \mathfrak{g}_0$, we set*

$$I(Z) := \int_{G_y} \sum_{\bar{m} \in \mathsf{M}_e/\sim} \frac{\chi(m)}{\left|C_{\mathsf{G}_F(\mathfrak{f})}({}^{g_m}e)\right|} \int_{A(F,\mathcal{G},X,g_m)} \Lambda(B(Z, {}^{j}Y))\, dY\, dj.$$

Since $G_F^+ \leq G_y$, we have $I(Z) = I({}^{\ell}Z)$ for all $\ell \in G_F^+$. Consequently, it makes sense to define the following.

Definition 5.2.7.

$$J(Z) := \sum_{\bar{i} \in \mathsf{G}_F(\mathfrak{f})} I({}^{i}Z).$$

We now prove a vanishing result for the function $Z \mapsto I(Z)$. As usual for such results, the statement is simple yet the proof is technically demanding.

Lemma 5.2.8. *If $Z \in \mathfrak{g}_0 \setminus \mathfrak{g}_F$, then $I(Z) = 0$.*

Proof. Fix $Z \in \mathfrak{g}_0 \setminus \mathfrak{g}_F$.

First, suppose that $Z \notin \mathfrak{g}_{y,-(1/h)}$. For all $m \in \mathbf{M}_X(K)$ every element of $A(F, \mathcal{G}, X, g_m)$ is "good" of depth $1/h$ in the sense of [4]. Thus for all $Y \in A(F, \mathcal{G}, X, g_m)$,

$$\int_{G_y} \Lambda(B(Z, {}^{j}Y))\, dj = 0$$

from [4, Lemma 6.3.3]. Consequently, when $Z \notin \mathfrak{g}_{y,-(1/h)}$, we have $I(Z) = 0$.

Now suppose $Z \in \mathfrak{g}_{y,-(1/h)} \setminus \mathfrak{g}_F$. Since $Z \in \mathfrak{g}_{y,-(1/h)}$, we can write $I(Z)$ as

$$\frac{\mathrm{meas}_{dj}(G_y^+)}{|\mathfrak{f}^\times|} \cdot \sum_{\bar{\ell} \in \mathsf{M}(\mathfrak{f})} \sum_{\bar{m} \in \mathbf{M}_X(K)/\sim} \frac{\chi(m)}{\left|C_{\mathsf{G}_F(\mathfrak{f})}({}^{g_m}e)\right|} \cdot \Lambda(B(Z, {}^{\ell g_m}X)) \cdot \sum_{\bar{W}} \Lambda(B(Z, {}^{\ell}W)),$$

where the sum is over $\bar{W}$ in

$$(\mathfrak{g}_{\psi_F} \setminus \mathfrak{g}_{\psi_F}^+)/\mathfrak{g}_{\psi_F}^+.$$

Note that the last sum is independent of $\bar{\ell} \in \mathsf{M}(\mathfrak{f})$, so it will be enough to show that

$$\sum_{\bar{\ell} \in \mathsf{M}(\mathfrak{f})} \sum_{\bar{m} \in \mathbf{M}_X(K)/\sim} \frac{\chi(m)}{\left|C_{\mathsf{G}_F(\mathfrak{f})}({}^{g_m}e)\right|} \cdot \Lambda(B(Z, {}^{\ell g_m}X)) \tag{8}$$

is zero.

Since $Z \notin \mathfrak{g}_F$, from equation (7) we have that the element Z is congruent to $Z_{-\dot{\psi}_F}$ modulo $\mathfrak{g}_F$. Here

$$Z_{-\dot{\psi}_F} \in \mathfrak{g}_{-\psi_F} \setminus \mathfrak{g}_{-\psi_F}^+ \subset \mathfrak{g}_{y,-(1/h)} \setminus \mathfrak{g}_y \tag{9}$$

denotes the image of Z under the projection $\mathfrak{g} \to \mathfrak{g}_{-\dot{\psi}_F}$.

Since ${}^{\ell g_m}X \in \mathfrak{g}_F(2)$, for purposes of evaluating equation (8) we may restrict our attention to $Z_{-2} \in \mathfrak{g}_F$, the image of Z under the projection $\mathfrak{g} \to \mathfrak{g}_{-2}$. Since the Fourier transform of a function on $\mathsf{L}_F(2)(\mathfrak{f})$ is a function on $\mathsf{L}_F(-2)(\mathfrak{f})$, in order to evaluate equation (8), we need to compute the Fourier transform of

$$\sum_{\bar{m} \in \mathsf{M}_e/\sim} \frac{\chi(m) \cdot \left|C_{\mathsf{M}(\mathfrak{f})}({}^{g_m}e)\right|}{\left|C_{\mathsf{G}_F(\mathfrak{f})}({}^{g_m}e)\right|} \cdot [{}^{\mathsf{M}(\mathfrak{f}) g_m}e].$$

From [20, Proposition 10.6] and [21], the Fourier transform of the above function is supported on the $\mathfrak{f}$-rational points of the unique Zariski dense M-orbit in $\mathsf{L}_F(-2)$. Thus it will be enough to show that the image of Z_{-2} in $\mathsf{L}_F(-2)$ cannot lie in this orbit.

Suppose the image of Z_{-2} lies in this orbit. We would then have that $Z_{-2} + Z_{-\dot{\psi}_F}$ is not nilpotent—in fact, it is good in the sense of [4] and its centralizer is a maximal k-torus which splits over a totally ramified extension. However, since $Z_{-\dot{\psi}_F} \notin \mathfrak{g}_y$ (see equation (9)) and $Z_{-2} \notin \mathfrak{g}_y$ (see equation (6)), we conclude that the coset $Z + \mathfrak{g}_y = Z_{-2} + Z_{-\dot{\psi}_F} + \mathfrak{g}_y$ is nondegenerate, that is, it contains no nilpotent elements. However, $Z \in \mathfrak{g}_0$ and $\mathfrak{g}_0 \subset \mathcal{N} + \mathfrak{g}_y$ from [2, Corollary 3.3.2]. □

With the above vanishing result, we can prove the following.

Lemma 5.2.9. *For $Z \in \mathfrak{g}_0$, we have*

$$J(Z) = \frac{\operatorname{meas}(G_y)}{\operatorname{meas}(G_F^+)} \cdot \hat{\mathcal{G}}_F(Z).$$

Proof. From Lemma 5.2.8 we may assume that $Z \in \mathfrak{g}_F$. We then have

$$\begin{aligned} J(Z) &= \operatorname{meas}(G_y) \cdot \sum_{\bar{i} \in G_F/G_F^+} \sum_{\bar{m} \in \mathsf{M}_e/\sim} \frac{\chi(m)}{\left|C_{\mathsf{G}_F(\mathfrak{f})}({}^{g_m}e)\right|} \int_{A(F,\mathcal{G},X,g_m)} \Lambda(B(Z, {}^iY))\, dY \\ &= \operatorname{meas}(G_y) \cdot \sum_{\bar{m} \in \mathsf{M}_e/\sim} \chi(m) \sum_{\bar{W} \in \mathsf{L}_F(\mathfrak{f})} \Lambda(B(Z, W)) \cdot [{}^{\mathsf{G}_F(\mathfrak{f}) g_m}e](\bar{W}) \\ &\qquad \text{(we now switch the order of summation to arrive at)} \\ &= \frac{\operatorname{meas}(G_y)}{\operatorname{meas}(G_F^+)} \cdot \hat{\mathcal{G}}_F(Z). \end{aligned}$$

□

5.2.8 The relation between $D_{(F,\mathcal{G})}$ and $T_{(F,\mathcal{G})}$

Lemma 5.2.10. *If $(F', \mathcal{G}') \in I^c$, then we have*

$$D_{(F,\bar{\mathcal{G}})}(\hat{\mathcal{G}}'_{F'}) = \begin{cases} \dfrac{|\mathsf{G}_F(\mathfrak{f})|}{|\mathsf{L}_F(\mathfrak{f})|^{1/2}} \cdot (\hat{\mathcal{G}}, \mathcal{G})_{\mathsf{L}_F} & \text{if } (F', \mathcal{G}') \sim (F, \mathcal{G}), \\ 0 & \text{otherwise.} \end{cases}$$

Proof. From Lemma 5.2.5 it will be enough to show that for $Z \in \mathfrak{g}_0$, we have

$$\hat{D}_{(F,\mathcal{G})}(Z) = \frac{|\mathsf{L}_F(\mathfrak{f})|^{1/2}}{|\mathsf{G}_F(\mathfrak{f})|} \cdot \hat{T}_{(F,\mathcal{G})}(Z).$$

Since the Fourier transform bijectively maps $C_c^\infty(\mathfrak{g}_0)$ to

$$\mathcal{D}_{0+} := \sum_{x \in \mathcal{B}(G)} C_c^\infty(\mathfrak{g}/\mathfrak{g}_x^+),$$

it will be enough to show

$$D_{(F,\mathcal{G})}(f) = \frac{|\mathsf{L}_F(\mathfrak{f})|^{1/2}}{|\mathsf{G}_F(\mathfrak{f})|} \cdot T_{(F,\mathcal{G})}(f)$$

for all $f \in \mathcal{D}_{0+}$. Fix $f \in \mathcal{D}_{0+}$. We have

$$D_{(F,\mathcal{G})}(f) = \sum_{\bar{m} \in \mathsf{M}_e/\sim} \frac{\chi(m) \cdot |S_m/S_m^\circ|}{|C_{\mathsf{G}_F(\mathfrak{f})}({}^{g_m}e)|} \int_{A(F,\mathcal{G},X,g_m)} \mu_Y(f)\, dY$$

(since the volume of $C_G(Y)$ is $|S_m/S_m^\circ|$, this becomes)

$$= \sum_{\bar{m} \in \mathsf{M}_e/\sim} \frac{\chi(m)}{|C_{\mathsf{G}_F(\mathfrak{f})}({}^{g_m}e)|} \int_{A(F,\mathcal{G},X,g_m)} \int_G f({}^jY)\, dj\, dY$$

$$= \sum_{\bar{m} \in \mathsf{M}_e/\sim} \frac{\chi(m)}{|C_{\mathsf{G}_F(\mathfrak{f})}({}^{g_m}e)|} \int_{A(F,\mathcal{G},X,g_m)} \sum_{\bar{j} \in G/G_F} \int_{G_F} f({}^{ji}Y)\, di\, dY$$

(Here di is the restriction of the Haar measure dj to G_F.)

$$= \frac{\operatorname{meas}(G_F^+)}{\operatorname{meas}(G_y)} \cdot \sum_{\bar{j} \in G/G_F} \int_{\mathfrak{g}} \hat{f}(-{}^jZ) \cdot J(Z)\, dZ$$

(since $\hat{f}$ is supported in $\mathfrak{g}_0$, from Lemma 5.2.9 we derive)

$$= \sum_{\bar{j} \in G/G_F} \int_{\mathfrak{g}} f({}^jZ) \cdot \mathcal{G}_F(Z)\, dZ$$

$$= (\operatorname{meas}(G_F))^{-1} \cdot \int_G \int_{\mathfrak{g}} f({}^jZ) \cdot \mathcal{G}_F(Z)\, dZ\, dj$$

$$= \frac{|\mathsf{L}_F(\mathfrak{f})|^{1/2}}{|\mathsf{G}_F(\mathfrak{f})|} \cdot T_{(F,\mathcal{G})}(f).$$

□

Remark 5.2.11. For a pair $(F, \mathcal{G}) \in I^c$ with $\mathcal{G}$ coming from a cuspidal local system, we have shown that

$$\operatorname{res}_{\mathcal{D}_{0+}} T_{(F,\mathcal{G})} = \frac{|\mathsf{G}_F(\mathfrak{f})|}{|\mathsf{L}_F(\mathfrak{f})|^{1/2}} \cdot \operatorname{res}_{\mathcal{D}_{0+}} D_{(F,\mathcal{G})}.$$

Table 4. A listing of distributions.

distribution name	description
$D^{\mathrm{st}}_{G_2}$	$D_{(F_{G_2}, \mathcal{Q}_{\mathsf{S}_{G_2}}^{F_{G_2}})}$
$D^{\mathrm{st}}_{A_2}$	$\left(D_{(F_{G_2}, \mathcal{Q}_{\mathsf{S}_{A_2}}^{F_{G_2}})} + 2 \cdot D_{(F_{A_2}, \mathcal{Q}_{\mathsf{S}_{A_2}}^{F_{A_2}})}\right)$
$D^{\mathrm{unst}}_{A_2}$	$\left(D_{(F_{G_2}, \mathcal{Q}_{\mathsf{S}_{A_2}}^{F_{G_2}})} - D_{(F_{A_2}, \mathcal{Q}_{\mathsf{S}_{A_2}}^{F_{A_2}})}\right)$
$D^{\mathrm{st}}_{A_1 \times \tilde{A}_1}$	$\left(D_{(F_{G_2}, \mathcal{Q}_{\mathsf{S}_{A_1 \times \tilde{A}_1}}^{F_{G_2}})} + 3 \cdot D_{(F_{A_1 \times \tilde{A}_1}, \mathcal{Q}_{\mathsf{S}_{A_1 \times \tilde{A}_1}}^{F_{A_1 \times \tilde{A}_1}})}\right)$
$D^{\mathrm{unst}}_{A_1 \times \tilde{A}_1}$	$(D_{(F_{G_2}, \mathcal{Q}_{\mathsf{S}_{A_1 \times \tilde{A}_1}}^{F_{G_2}})} - D_{(F_{A_1 \times \tilde{A}_1}, \mathcal{Q}_{\mathsf{S}_{A_1 \times \tilde{A}_1}}^{F_{A_1 \times \tilde{A}_1}})})$
$D^{\mathrm{st}}_{\tilde{A}_1}$	$D_{(F_{\tilde{A}_1}, \mathcal{Q}_{\mathsf{S}_{\tilde{A}_1}}^{F_{\tilde{A}_1}})}$
$D^{\mathrm{st}}_{A_1}$	$D_{(F_{A_1}, \mathcal{Q}_{\mathsf{S}_{A_1}}^{F_{A_1}})}$
$D^{\mathrm{st}}_{\{e\}}$	$D_{(F_{\{e\}}, \mathcal{Q}_{\mathsf{S}_{\{e\}}}^{F_{\{e\}}})}$
$D^{\mathrm{st}}_{(F_{G_2}, \mathcal{G}_{\mathrm{sgn}})}$	$D_{(F_{G_2}, \mathcal{G}_{\mathrm{sgn}})}$
$D^{\mathrm{st}}_{(F_{A_1 \times \tilde{A}_1}, \mathcal{G}_{\mathrm{sgn}})}$	$D_{(F_{A_1 \times \tilde{A}_1}, \mathcal{G}_{\mathrm{sgn}})}$
$D^{\mathrm{st}}_{(F_{A_2}, \mathcal{G}_{\chi'})}$	$D_{(F_{A_2}, \mathcal{G}_{\chi'})}$
$D^{\mathrm{st}}_{(F_{A_2}, \mathcal{G}_{\chi''})}$	$D_{(F_{A_2}, \mathcal{G}_{\chi''})}$

6 A basis for $\mathrm{res}_{\mathcal{D}_0} J^{\mathrm{st}}(\mathfrak{g}) \cap \mathrm{res}_{\mathcal{D}_0} J(\mathfrak{g}_0)$

In this section, we produce a basis for $\mathrm{res}_{\mathcal{D}_0} J^{\mathrm{st}}(\mathfrak{g}) \cap \mathrm{res}_{\mathcal{D}_0} J(\mathfrak{g}_0)$.

6.1 A new basis for $\mathrm{res}_{\mathcal{D}_0} J(\mathfrak{g}_0)$

Using the notation of Figures 3, 7, and 8 we produce a new basis for $\mathrm{res}_{\mathcal{D}_0} J(\mathfrak{g}_0)$ which is more amenable to our purposes. If $\mathfrak{f}^\times \neq (\mathfrak{f}^\times)^3$, then we let $\mathfrak{B}$ denote the twelve distributions listed in Table 4. If $\mathfrak{f}^\times = (\mathfrak{f}^\times)^3$, then we let $\mathfrak{B}$ denote the first ten distributions listed in Table 4. We let $\mathfrak{B}^{\mathrm{st}}$ denote the subset of $\mathfrak{B}$ consisting of those elements of $\mathfrak{B}$ whose label has a superscript st. Similarly, we let $\mathfrak{B}^{\text{unst}}$ denote the two element subset of $\mathfrak{B}$ consisting of those elements of $\mathfrak{B}$ whose label has a superscript unst.

It follows from Lemma 5.2.10 and Corollary 4.4.2 that the set

$$\{\operatorname{res}_{\mathcal{D}_0} D \mid D \in \mathfrak{B}\}$$

is a basis for $\operatorname{res}_{\mathcal{D}_0} J(\mathfrak{g}_0)$. Moreover, for each $D \in \mathfrak{B}$, there exists a unique combination, which we shall call $\mathcal{G}_D$, of the functions $\mathcal{G}_F$ (with $(F, \mathcal{G}) \in I^c$ our chosen representatives for the equivalence classes in I^c) for which

$$D(\hat{\mathcal{G}}_{D'}) = \begin{cases} 1 & \text{if } D = D', \\ 0 & \text{if } D \neq D' \end{cases}$$

for all $D, D' \in \mathfrak{B}$.

Recall that from Remark 5.1.4 the distributions associated to toric Green functions are not independent of the choice of X_{S} (as distributions on $\mathfrak{g}$). In Sections 6.2 and 6.3 we show that for a suitable choice of the X_{S}, the distributions in $\mathfrak{B}^{\text{st}}$ are stable and the distributions in $\mathfrak{B}^{\text{unst}}$ are the images under endoscopic induction of elliptic (but not G_2) endoscopic groups in the sense of Waldspurger [27]. In Section 6.4, we present a result of Waldspurger which shows that the elements of the set

$$\{\operatorname{res}_{\mathcal{D}_0} D \mid D \in \mathfrak{B}^{\text{st}}\}$$

form a basis for $\operatorname{res}_{\mathcal{D}_0} J(\mathfrak{g}_0) \cap \operatorname{res}_{\mathcal{D}_0} J^{\text{st}}(\mathfrak{g})$.

6.2 The distributions associated to unramified tori

Let C denote a conjugacy class in the Weyl group of G_2. We now turn our attention to the distributions

$$D_C^*$$

of Table 4.

Suppose $(F, Q_C^F) \in I^c$. Let S_C be a torus in G_F corresponding to C. Let $\mathbf{S}$ be a a maximal K-split k-torus in G which lifts the pair (F, S_C). Choose a regular semisimple element $X_{\mathsf{S}_C} \in \operatorname{Lie}(\mathbf{S})(k) \subset \mathfrak{g}_F$ for which the centralizer in G_F of the image of X_{S_C} in $\mathsf{L}_F(\mathfrak{f})$ is S_C. Since G_2 is simply connected, the short exact sequence

$$1 \to \mathbf{S}(K) \to \mathbf{G}(K) \to {}^{\mathbf{G}(K)}X_{\mathsf{S}_C} \to 1$$

yields the exact sequence (of pointed sets)

$$1 \to \mathbf{S}(k) \to \mathbf{G}(k) \to ({}^{\mathbf{G}(K)}X_{\mathsf{S}_C})^{\operatorname{Gal}(K/k)} \to \mathrm{H}^1(\operatorname{Gal}(K/k), \mathbf{S}(K)) \to 1.$$

Thus the number of rational conjugacy classes in

$${}^{\mathbf{G}(K)}X_{\mathsf{S}_C} \cap \mathfrak{g}$$

Table 5. A tabulation of $\mathrm{tor}[\mathbf{X}_*(\mathbf{T})/(1-w)\mathbf{X}_*(\mathbf{T})]$.

Class of w	$\mathrm{tor}[\mathbf{X}_*(\mathbf{T})/(1-w)\mathbf{X}_*(\mathbf{T})]$
$\{e\}$	trivial group
A_1	trivial group
$\tilde{A}_1$	trivial group
A_2	$\mathbb{Z}/3\mathbb{Z}$
$A_1 \times \tilde{A}_1$	$\mathbb{Z}/2\mathbb{Z} \times \mathbb{Z}/2\mathbb{Z}$
G_2	trivial group

is in bijective correspondence with the elements of the group $\mathrm{H}^1(\mathrm{Gal}(K/k), \mathbf{S}(K))$. Thanks to Tate–Nakayama duality, we have $\mathrm{H}^1(\mathrm{Gal}(K/k), \mathbf{S}(K))$ is isomorphic to $\mathrm{tor}[\mathbf{X}_*(\mathbf{T})/(1-w)\mathbf{X}_*(\mathbf{T})]$, the group of torsion points of $\mathbf{X}_*(\mathbf{T})/(1-w)\mathbf{X}_*(\mathbf{T})$. (Here w is any element of C.) The groups $\mathrm{tor}[\mathbf{X}_*(\mathbf{T})/(1-w)\mathbf{X}_*(\mathbf{T})]$ are listed in Table 5.

For each character κ of $\mathrm{tor}[\mathbf{X}_*(\mathbf{T})/(1-w)\mathbf{X}_*(\mathbf{T})]$, we have a distribution

$$T_C(\kappa) := \sum_{\rho} \kappa(\rho) \cdot \mu_{X^{\rho}_{\mathsf{S}_C}},$$

where the sum if over ρ in $\mathrm{tor}[\mathbf{X}_*(\mathbf{T})/(1-w)\mathbf{X}_*(\mathbf{T})]$ and $X^{\rho}_{\mathsf{S}_C}$ belongs to the G-conjugacy class in $({}^{\mathbf{G}(K)}X_{\mathsf{S}_C} \cap \mathfrak{g})$ indexed by ρ. Note that $T_C(1)$ is stable.

Suppose first that $C \notin \{A_2, A_1 \times \tilde{A}_1\}$. From Lemma 5.1.2 and Corollary 4.4.2 the restrictions to $\mathcal{D}_0$ of the distributions D^{st}_C are independent of the choice of X_{S_C}. Hence we may and do assume that

$$D^{\mathrm{st}}_C = T_C(1).$$

That is, $D^{\mathrm{st}}_C \in J^{\mathrm{st}}(\mathfrak{g})$.

Now suppose that $C \in \{A_2, \mathrm{A}_1 \times \tilde{\mathrm{A}}_1\}$. We analyze these cases in two steps: first, we examine the G-conjugacy classes of maximal K-split k-tori corresponding to C. We then examine how the set $({}^{\mathbf{G}(K)}X_{\mathsf{S}_C} \cap \mathfrak{g})$ interacts with these conjugacy classes of tori.

According to [13], the G-conjugacy classes of maximal K-split k-tori are parameterized by $I^t/\sim$. Here I^t is the set of pairs (F, S), where F is a facet in $\mathcal{B}(G)$ and S is a maximal $\mathfrak{f}$-minisotropic torus in G_F and $\sim$ is the equivalence relation introduced in Remark 4.1.4. Moreover, a maximal K-split k-torus lifting (F_1, S_1) is $\mathrm{G}(K)$-conjugate to a maximal K-split k-torus lifting (F_2, S_2) if and only if the Weyl group conjugacy classes corresponding to S_1 and S_2 are the same. Thus the distribution D^*_C is associated to a single "$\mathbf{G}(K)$-conjugacy class" of maximal K-split k-tori. If $C = \mathrm{A}_2$, then the two corresponding G-conjugacy classes of maximal K-split k-tori correspond to the pairs $(F_{\mathrm{G}_2}, \mathsf{S}_{\mathrm{A}_2})$ and $(F_{\mathrm{A}_2}, \mathsf{S}_{\mathrm{A}_2})$. On the other hand, if

$C = \mathrm{A}_1 \times \tilde{A}_1$, then the two corresponding G-conjugacy classes of maximal K-split k-tori correspond to the pairs $(F_{\mathrm{G}_2}, \mathsf{S}_{\mathrm{A}_1 \times \tilde{\mathrm{A}}_1})$ and $(F_{\mathrm{A}_1 \times \tilde{\mathrm{A}}_1}, \mathsf{S}_{\mathrm{A}_1 \times \tilde{\mathrm{A}}_1})$.

The rational classes in ${}^{\mathbf{G}(K)}X$ that intersect $\mathrm{Lie}(\mathbf{S})(k)$ are parameterized by the quotient

$$N(F, \mathsf{S}_C) := [(N_{\mathbf{G}(K)}(\mathbf{S}(K)))/(\mathbf{S}(K))]^{\Gamma}/[N_G(S)/S].$$

In Table 6 we describe the cardinality of these quotients for the classes of interest.

Table 6. The quotient $N(F, \mathsf{S})$.

Class of w	vertex	$\lvert N(F, \mathsf{S})\rvert$
A_2	F_{G_2}	1
A_2	F_{A_2}	2
$A_1 \times \tilde{A}_1$	F_{G_2}	1
$A_1 \times \tilde{A}_1$	$F_{A_1 \times \tilde{A}_1}$	3

Combining the previous two paragraphs with Lemma 5.1.2 and Corollary 4.4.2 , we have that, up to scaling,

$$\begin{aligned}
\mathrm{res}_{\mathcal{D}_0} T_{\mathrm{A}_1 \times \tilde{\mathrm{A}}_1}(1) &= \mathrm{res}_{\mathcal{D}_0} D^{\mathrm{st}}_{A_1 \times \tilde{A}_1}, \\
\mathrm{res}_{\mathcal{D}_0} T_{\mathrm{A}_2}(1) &= \mathrm{res}_{\mathcal{D}_0} D^{\mathrm{st}}_{A_2}, \\
\mathrm{res}_{\mathcal{D}_0} T_{\mathrm{A}_1 \times \tilde{\mathrm{A}}_1}(\kappa) &= \mathrm{res}_{\mathcal{D}_0} D^{\mathrm{unst}}_{A_1 \times \tilde{A}_1},
\end{aligned}$$

and

$$\mathrm{res}_{\mathcal{D}_0} T_{\mathrm{A}_2}(\kappa) = \mathrm{res}_{\mathcal{D}_0} D^{\mathrm{unst}}_{A_2},$$

where κ is any nontrivial character of $\mathrm{tor}[\mathbf{X}_*(\mathbf{T})/(1-w)\mathbf{X}_*(\mathbf{T})]$ with $w \in C$. Thus we may assume that $D^{\mathrm{st}}_C = T_C(1) \in J^{\mathrm{st}}(\mathfrak{g})$ and $D^{\mathrm{unst}}_C = T_C(\kappa)$ with κ nontrivial.

6.3 The distributions associated to cuspidal local systems

The remaining distributions are all associated to tori which split over totally ramified extensions.

Fix a maximal k-torus $\mathbf{S}$ which splits over a totally ramified extension. Let E be the extension of K over which $\mathbf{S}$ splits. Since we are assuming that every maximal k-torus splits over a tame extension, $\mathrm{Gal}(E/K)$ is cyclic. Hence, if $\mathbf{S} = {}^{g}\mathbf{T}$ with $g \in \mathbf{G}(E)$, then $\tau(g)^{-1}g \in N_{\mathbf{G}(E)}(\mathbf{T}(E))$, where τ generates $\mathrm{Gal}(E/K)$. Let C denote the conjugacy class in W corresponding to $\tau(g)^{-1}g$. Let w denote the image of $\tau(g)^{-1}g$ in W.

Fix a regular semisimple element $X \in \mathfrak{g}$ which lies in the the Lie algebra of $\mathbf{S}(k)$. As above, the set of rational conjugacy classes in

$$^{\mathbf{G}(K)}X \cap \mathfrak{g}$$

is controlled by the group $\mathrm{H}^1(\mathrm{Gal}(K/k), \mathbf{S}(K))$. Thanks to a result of Bruhat and Tits [9], we have that $\mathrm{H}^1(\mathrm{Gal}(K/k), \mathbf{S}(K))$ is isomorphic to $(\mathbf{S}(K)/\mathbf{S}(K)_0)/(1-\sigma)(\mathbf{S}(K)/\mathbf{S}(K)_0)$, where $\mathbf{S}(K)_0$ denotes the parahoric subgroup of $\mathbf{S}(K)$. Since $\mathrm{H}^1(\mathrm{Gal}(E/K), \mathbf{S}(E)_{0+})$ is trivial, $\mathbf{S}(K)_0 = \mathbf{S}(K)_{0+}$, and, from [4], $\mathbf{S}(K)_{0+} = \mathbf{S}(E)_{0+}^{\mathrm{Gal}(E/K)}$, we have

$$\mathbf{S}(K)/\mathbf{S}(K)_0 \cong (\mathbf{S}(E)/\mathbf{S}(E)_{0+})^{\mathrm{Gal}(E/K)}.$$

Since $\mathbf{S}$ is K-elliptic, we have

$$(\mathbf{S}(E)/\mathbf{S}(E)_{0+})^{\mathrm{Gal}(E/K)} = (\mathbf{S}(E)_0/\mathbf{S}(E)_{0+})^{\mathrm{Gal}(E/K)}$$

which is isomorphic to $(\mathbf{T}(E)_0/\mathbf{T}(E)_{0+})^{w\circ\tau}$, the subgroup of $\mathbf{T}(E)_0/\mathbf{T}(E)_{0+}$ fixed by $w \circ \tau$. Since τ acts trivially on the quotient $\mathbf{T}(E)_0/\mathbf{T}(E)_{0+}$, we need only compute the group of w-fixed points in $\mathbf{T}(E)_0/\mathbf{T}(E)_{0+}$. For $w \in \mathrm{G}_2$, this group is trivial. It follows that the distributions $D_{(F_{\mathrm{A}_1 \times \tilde{\mathrm{A}}_1}, \mathcal{G}_{\mathrm{sgn}})}$, $D_{(F_{\mathrm{A}_2}, \mathcal{G}_{\chi'})}$, and $D_{(F_{\mathrm{A}_2}, \mathcal{G}_{\chi''})}$ are stable.

We are left to consider why the distribution $D_{(F_{\mathrm{G}_2}, \mathcal{G}_{\mathrm{sgn}})}$ is stable. The torus $\mathbf{S}$ associated to this distribution corresponds to the Weyl group conjugacy class A_2; a computation shows that the group $\mathbf{S}(K)/\mathbf{S}(K)_0$ is isomorphic to $\mathbb{Z}/3\mathbb{Z}$. We have

$$D_{(F_{\mathrm{G}_2}, \mathcal{G}_{\mathrm{sgn}})} := \sum_{\mathsf{M}_e/\sim} \frac{\mathrm{sgn}(m) \cdot |S_m/S_m^\circ|}{|C_{\mathsf{G}_F(\mathfrak{f})}(^{g_m}e)|} \int_{A(F,\mathcal{G},X,g_m)} \mu_Y \, dY$$

and $\mathsf{M}_e = S_3 = \langle \check{\alpha}(\mu_3), \bar{n}_\alpha \rangle$. We consider two cases.

6.3.1 Case I: The cubic roots of unity belong to $\mathfrak{f}^\times$

In this case, σ acts trivially on M_e. If the conjugacy class of m is represented by an element of $\check{\alpha}(\mu_3)$, then, by construction, the centralizer $\mathbf{S}_m$ of an element of $A(F_{\mathrm{G}_2}, \mathcal{G}_{\mathrm{sgn}}, X_1, g_m)$ belongs to a copy of SL_3 in G_2 which is defined over k. In this case S_m/S_m° has cardinality three and $\mathrm{H}^1(\mathrm{Gal}(K/k), \mathbf{S}_m(K)) = \mathbb{Z}/3\mathbb{Z}$. On the other hand, if m belongs to the conjugacy class containing $\bar{n}_\alpha$, then the centralizer $\mathbf{S}_m$ of an element of $A(F_{\mathrm{G}_2}, \mathcal{G}_{\mathrm{sgn}}, X_1, g_m)$ belongs to a copy of SU_3 in G_2 which is defined over k. In this case, S_m/S_m° is trivial and $\mathrm{H}^1(\mathrm{Gal}(K/k), \mathbf{S}_m(K))$ is trivial.

We have that $D_{(F_{\mathrm{G}_2}, \mathcal{G}_{\mathrm{sgn}})}$ is equal to

$$\frac{3}{6} \int_{A(F_{\mathrm{G}_2}, \mathcal{G}_{\mathrm{sgn}}, X_1, g_{\check{\alpha}(1)})} \mu_Y \, dY + \frac{3}{3} \int_{A(F_{\mathrm{G}_2}, \mathcal{G}_{\mathrm{sgn}}, X_1, g_{\check{\alpha}(\xi)})} \mu_Y \, dY + \frac{-1}{2} \int_{A(F_{\mathrm{G}_2}, \mathcal{G}_{\mathrm{sgn}}, X_1, g_{\bar{n}_\alpha})} \mu_Y \, dY,$$

where ξ is a nontrivial element of μ_3. As noted above, $\mathrm{H}^1(\mathrm{Gal}(K/k), \mathbf{S}_{\bar{n}_\alpha}(K))$ is trivial, so the last term in the sum is stable. The first two terms combine to give a stable distribution. More precisely, if

$$Y = X_\beta - X_{\beta+3\alpha} + \gamma\varpi X_{-2\beta-3\alpha} \in A(F_{\mathrm{G}_2}, \mathcal{G}_{\mathrm{sgn}}, X_1, g_e)$$

($\gamma \in R^\times$), then Y, ${}^{g_{\check{\alpha}(\xi)}}Y$, and ${}^{g_{\check{\alpha}(\xi^2)}}Y$ are representatives for the G-conjugacy classes in $({}^{\mathbf{G}(K)}Y \cap \mathfrak{g})$. We have $\check{\alpha}(\xi^2) = {}^{n_\alpha}(\check{\alpha}(\xi))$ and $g_{\check{\alpha}(\xi^2)} = {}^{n_\alpha}g_{\check{\alpha}(\xi)}$. Since $\check{\alpha}(\xi)$ acts trivially on $\mathfrak{g}_{\dot{\psi}_F}$, it follows that

$${}^{g_{\check{\alpha}(\xi)}}Y \in A(F_{\mathrm{G}_2}, \mathcal{G}_{\mathrm{sgn}}, X_1, g_{\check{\alpha}(\xi)})$$

and

$$\begin{aligned} {}^{g_{\check{\alpha}(\xi^2)}}Y &= {}^{n_\alpha g_{\check{\alpha}(\xi)} n_\alpha^{-1}}(X_\beta - X_{\beta+3\alpha} + \gamma\varpi X_{-2\beta-3\alpha}) \\ &= {}^{n_\alpha g_{\check{\alpha}(\xi)}}(X_\beta - X_{\beta+3\alpha} - \gamma\varpi X_{-2\beta-3\alpha}) \\ &\in {}^{n_\alpha}A(F_{\mathrm{G}_2}, \mathcal{G}_{\mathrm{sgn}}, X_1, g_{\check{\alpha}(\xi)}) = A(F_{\mathrm{G}_2}, \mathcal{G}_{\mathrm{sgn}}, X_1, g_{\check{\alpha}(\xi)}). \end{aligned}$$

Thus, for each $Y \in A(F_{\mathrm{G}_2}, \mathcal{G}_{\mathrm{sgn}}, X_1, g_{\check{\alpha}(1)})$, representatives for both of the remaining rational conjugacy classes in $({}^{\mathbf{G}(K)}Y \cap \mathfrak{g})$ occur in $A(F_{\mathrm{G}_2}, \mathcal{G}_{\mathrm{sgn}}, X_1, g_{\check{\alpha}(\xi)})$.

6.3.2 Case II: The cubic roots of unity do not belong to $\mathfrak{f}^\times$

In this case, since σ acts nontrivially on M_e, we consider σ-conjugacy classes in M_e. If m represents the σ-conjugacy class $\check{\alpha}(\mu_3)$, then, by construction, the centralizer $\mathbf{S}_m$ of an element of $A(F_{\mathrm{G}_2}, \mathcal{G}_{\mathrm{sgn}}, X_1, g_m)$ belongs to a copy of SL_3 in G_2 which is defined over k. In this case S_m/S_m° has cardinality one and $\mathrm{H}^1(\mathrm{Gal}(K/k), \mathbf{S}_m(K))$ is trivial. On the other hand, if m belongs to either the σ-conjugacy class

$$\{\bar{n}_\alpha\}$$

or the σ-conjugacy class

$$\{\bar{n}_\alpha\check{\alpha}(t) \mid t \in \mu_3 \setminus \{1\}\},$$

then the centralizer $\mathbf{S}_m$ of an element of $A(F_{\mathrm{G}_2}, \mathcal{G}_{\mathrm{sgn}}, X_1, g_m)$ belongs to a copy of SU_3 in G_2 which is defined over k. In this case, S_m/S_m° has three elements and $\mathrm{H}^1(\mathrm{Gal}(K/k), \mathbf{S}_m(K)) = \mathbb{Z}/3\mathbb{Z}$.

We have $D_{(F_{\mathrm{G}_2}, \mathcal{G}_{\mathrm{sgn}})}$ is equal to

$$\frac{-3}{6}\int_{A(F_{\mathrm{G}_2}, \mathcal{G}_{\mathrm{sgn}}, X_1, g_{\bar{n}_\alpha})} \mu_Y\, dY + \frac{-3}{3}\int_{A(F_{\mathrm{G}_2}, \mathcal{G}_{\mathrm{sgn}}, X_1, g_{\bar{n}_\alpha\check{\alpha}(\xi)})} \mu_Y\, dY + \frac{1}{2}\int_{A(F_{\mathrm{G}_2}, \mathcal{G}_{\mathrm{sgn}}, X_1, g_{\check{\alpha}(1)})} \mu_Y\, dY,$$

where ξ is a nontrivial element of μ_3. Since $\mathrm{H}^1(\mathrm{Gal}(K/k), \mathbf{S}_{\check{\alpha}(1)}(K))$ is trivial, the last term in the sum is stable. The first two terms combine to give a stable distribution. More precisely, if

$$Y = (X_\beta - X_{\beta+3\alpha} + \gamma\varpi X_{-2\beta-3\alpha}) \in B(F_{\mathrm{G}_2}, \mathcal{G}_{\mathrm{sgn}}, X_1)$$

($\gamma \in R_K^\times$ and $\sigma(\gamma) = -\gamma$), then ${}^{g_{\bar{n}_\alpha}}Y$, ${}^{g_{\bar{n}_\alpha\check{\alpha}(\xi)}}Y$, and ${}^{g_{\bar{n}_\alpha\check{\alpha}(\xi^2)}}Y$ represent the three G-conjugacy classes in $({}^{\mathbf{G}(K)}Y \cap \mathfrak{g})$. We have $n_\alpha\check{\alpha}(\xi^2) = {}^{n_\alpha}(n_\alpha\check{\alpha}(\xi))$ and $g_{\bar{n}_\alpha\check{\alpha}(\xi^2)} = {}^{n_\alpha}g_{\bar{n}_\alpha\check{\alpha}(\xi)}$. Since $\check{\alpha}(\mu_3)$ acts trivially on $\mathbf{g}(K)_{\dot{\psi}_F}$, we have that

$$\begin{aligned}
{}^{g_{\bar{n}_\alpha}}Y &\in A(F_{\mathrm{G}_2}, \mathcal{G}_{\mathrm{sgn}}, X_1, g_{\bar{n}_\alpha}),\\
{}^{g_{\bar{n}_\alpha\check{\alpha}(\xi)}}Y &\in A(F_{\mathrm{G}_2}, \mathcal{G}_{\mathrm{sgn}}, X_1, g_{\bar{n}_\alpha\check{\alpha}(\xi)}),
\end{aligned}$$

and

$$\begin{aligned}
{}^{g_{\bar{n}_\alpha\check{\alpha}(\xi^2)}}Y &= {}^{n_\alpha g_{\bar{n}_\alpha\check{\alpha}(\xi)}}(X_\beta - X_{\beta+3\alpha} - \delta\varpi X_{-2\beta-3\alpha})\\
&\in {}^{n_\alpha}A(F_{\mathrm{G}_2}, \mathcal{G}_{\mathrm{sgn}}, X_1, g_{\bar{n}_\alpha\check{\alpha}(\xi)}) = A(F_{\mathrm{G}_2}, \mathcal{G}_{\mathrm{sgn}}, X_1, g_{\bar{n}_\alpha\check{\alpha}(\xi)}).
\end{aligned}$$

Thus, for each $Z \in A(F_{\mathrm{G}_2}, \mathcal{G}_{\mathrm{sgn}}, X_1, g_{\bar{n}_\alpha})$, representatives for both of the remaining rational conjugacy classes in $({}^{\mathbf{G}(K)}Z \cap \mathfrak{g})$ occur in $A(F_{\mathrm{G}_2}, \mathcal{G}_{\mathrm{sgn}}, X_1, g_{\bar{n}_\alpha\check{\alpha}(\xi)})$.

6.4 A result of Waldspurger

The proof of the following lemma is a straightforward adaptation of a result of Waldspurger [28, Théorème IV.13].

Lemma 6.4.1. *The elements of the set*

$$\{\mathrm{res}_{\mathcal{D}_0} D \mid D \in \mathfrak{B}^{\mathrm{st}}\}$$

form a basis for $\mathrm{res}_{\mathcal{D}_0} J(\mathfrak{g}_0) \cap \mathrm{res}_{\mathcal{D}_0} J^{\mathrm{st}}(\mathfrak{g})$.

Proof (*Waldspurger*). Suppose $T \in J^{\mathrm{st}}(\mathfrak{g})$ such that $\mathrm{res}_{\mathcal{D}_0} T$ is an element of $\mathrm{res}_{\mathcal{D}_0} J(\mathfrak{g}_0)$. It is enough to show that if

$$\mathrm{res}_{\mathcal{D}_0} T$$

is in the span of the elements of the set

$$\{\mathrm{res}_{\mathcal{D}_0} D \mid D \in \mathfrak{B}^{\mathrm{unst}}\},$$

then $\mathrm{res}_{\mathcal{D}_0} T = 0$.

Since $\mathrm{res}_{\mathcal{D}_0} T \in \mathrm{res}_{\mathcal{D}_0} J(\mathfrak{g}_0)$, we have that the Fourier transform $\hat{T}$ of T is represented by a locally integrable function on $C_c^\infty(\mathfrak{g}_{0+})$. We shall denote by $\hat{T}$ the extension (by zero) of this function to $\mathfrak{g}$.

Suppose

$$\mathrm{res}_{\mathcal{D}_0} T = \sum_{D \in \mathfrak{B}^{\mathrm{unst}}} c(D) \cdot \mathrm{res}_{\mathcal{D}_0} D$$

with $c(D) \in \mathbb{C}$ and at least one of the $c(D)$ nonzero. Since for each $D \in \mathfrak{B}$ we have $\hat{\mathcal{G}}_D \in \mathcal{D}_0$ and $\mathcal{G}_D \in C_c^\infty(\mathfrak{g}_{0+})$, we conclude that

$$\begin{aligned} c(D) &= T(\hat{\mathcal{G}}_D) = \hat{T}(\mathcal{G}_D) \\ &= \int_{\mathfrak{g}} \hat{T}(X) \cdot \mathcal{G}_D(X)\, dX \\ &= \sum c_{\mathfrak{h}} \cdot \int_{\mathfrak{h}} |\eta(H)| \cdot \hat{T}(H) \cdot \mu_H(\mathcal{G}_D)\, dH. \end{aligned}$$

The last displayed line is Weyl's integration formula; the sum is over conjugacy classes of Cartan subalgebras of $\mathfrak{g}$, the $c_{\mathfrak{h}}$ are positive constants, and dH is a Haar measure on $\mathfrak{h}$. Thus there is a regular semisimple H in $\mathfrak{g}_{0^+}$ for which $\mu_H(\mathcal{G}_D) \neq 0$. Fix such an H.

An inspection of the elements of $\mathfrak{B}^{\text{unst}}$ reveals that $(\mathcal{G}_D)_P = 0$ for all proper parabolic subgroups P of G. Hence, from Lemma 3.3.1 and [15, Lemma 1.5] we conclude that H is elliptic. Consequently, there is an elliptic regular semisimple H in $\mathfrak{g}_{0^+}$ for which

$$0 \neq \hat{T}(H) = \sum_{D \in \mathfrak{B}^{\text{unst}}} c(D) \cdot \hat{D}(H). \tag{10}$$

However, from [27, I, Proposition] the function $\hat{T}$ is stable—that is, it is constant on the rational points in the orbit

$$^{\mathbf{G}(\bar{k})}H.$$

On the other hand, from [27, I, Proposition A], the right-hand side of equation (10) cannot be stable unless it is zero, a contradiction. □

7 Stable distributions supported on the nilpotent cone

In this section, we explicitly describe the stable distributions supported on the nilpotent cone. Thanks to Harish-Chandra, we know that the set of nilpotent orbital integrals is a basis for the set of invariant distributions supported on the nilpotent cone. We therefore need to describe a basis for $J^{\text{st}}(\mathcal{N})$, the set of stable distributions supported on $\mathcal{N}$, in terms of nilpotent orbital integrals. For $\mathrm{G}_2(k)$, we call a nilpotent orbit *special* provided that it is either k-distinguished or trivial. We shall prove the following.

Theorem 7.0.1. *Suppose $\mathcal{O}$ is a nilpotent orbit. The orbital integral $\mu_{\mathcal{O}}$ is stable if and only if $\mathcal{O}$ is special. Moreover, the set*

$$\{\mu_{\mathcal{O}} \mid \mathcal{O} \text{ is special}\}$$

is a basis for $J^{\text{st}}(\mathcal{N})$.

Remark 7.0.2. We shall assume that the measure on a nilpotent orbit is normalized as in [28].

7.1 A parameterization of nilpotent orbits

We let I^d denote the subset of I^n consisting of pairs (F, e), where F is a facet in $\mathcal{B}(G)$ and e is a $\mathfrak{f}$-distinguished nilpotent element of $\mathsf{L}(\mathfrak{f})$; that is, e is nilpotent and does not belong to the Levi subalgebra of a proper parabolic $\mathfrak{f}$-subgroup. The set I^d carries the equivalence relation discussed in Remark 4.1.3 of Section 4.1.

Thanks to [6], if $(F, e) \in I^d$, then there is a unique nilpotent orbit $\mathcal{O}(F, e)$ of minimal dimension which intersects the coset e nontrivially. The map $(F, e) \mapsto \mathcal{O}(F, e)$ induces a bijective correspondence between $I^d/\sim$ and $\mathcal{O}(0)$, the set of nilpotent orbits in $\mathfrak{g}$ [12]. Thus we can use the set I^d to keep track of the nilpotent orbits. To do this, we introduce some additional notation. For each facet F, the number of pairs $(F, e) \in I^d$ listed below agrees (as it should) with the numbers given in Figure 6. Some of this notation has been defined previously.

Let $e_0 \in \mathsf{L}_{F_{G_2}}(\mathfrak{f})$ be an element of the regular nilpotent $\mathsf{G}_{F_{G_2}}(\mathfrak{f})$-orbit. We denote by $\mathcal{O}_0$ the regular G-orbit $\mathcal{O}(F_{G_2}, e_0)$.

Let $e_1, e_1', e_1'' \in \mathsf{L}_{F_{G_2}}(\mathfrak{f})$ be elements of the the subregular nilpotent $\mathsf{G}_{F_{G_2}}(\mathfrak{f})$-orbits with $|\mathsf{M}_{e_1}(\mathfrak{f})| = 6$, $|\mathsf{M}_{e_1'}(\mathfrak{f})| = 3$, and $|\mathsf{M}_{e_1''}(\mathfrak{f})| = 2$. We let $\mathcal{O}_1$, $\mathcal{O}_1'$ and $\mathcal{O}_1''$ denote the nilpotent orbits $\mathcal{O}(F_{G_2}, e_1)$, $\mathcal{O}(F_{G_2}, e_1')$, and $\mathcal{O}(F_{G_2}, e_1'')$, respectively.

Let $e_1^{\pm} \in \mathsf{L}_{F_{A_1 \times \tilde{A}_1}}(\mathfrak{f})$ be elements of the the regular nilpotent $\mathsf{G}_{F_{A_1 \times \tilde{A}_1}}(\mathfrak{f})$-orbits; we assume e_1^+ and e_1^- lie in distinct $\mathsf{G}_{F_{A_1 \times \tilde{A}_1}}(\mathfrak{f})$-orbits. We let $\mathcal{O}_1^{\pm}$ denote the nilpotent orbit $\mathcal{O}(F_{A_1 \times \tilde{A}_1}, e_1^{\pm})$.

Let $e_1^{\delta} \in \mathsf{L}_{F_{A_2}}(\mathfrak{f})$, with $\delta \in \mu_3(\mathfrak{f})$, be representatives for the regular nilpotent $\mathsf{G}_{F_{A_2}}(\mathfrak{f})$-orbits. We assume that e_1^{δ} and $e_1^{\delta'}$ lie in the same $\mathsf{G}_{F_{A_2}}(\mathfrak{f})$-orbit if and only if $\delta = \delta'$. We let $\mathcal{O}_1^{\delta}$ denote the nilpotent orbit $\mathcal{O}(F_{A_2}, e_1^{\delta})$.

We have now labeled each of the $6 + |\mathfrak{f}^{\times}/(\mathfrak{f}^{\times})^3|$ k-distinguished G-orbits in $\mathfrak{g}$. They are all special. There are three more nilpotent orbits in $\mathfrak{g}$, these are parameterized by the pairs (F_{A_1}, e_2), $(F_{\tilde{A}_1}, e_3)$, and $(F_{\emptyset}, e_4)$. Here e_2 (respectively, e_3) is a regular nilpotent element in $\mathsf{L}_{F_{A_1}}(\mathfrak{f})$ (respectively, $\mathsf{L}_{F_{\tilde{A}_1}}(\mathfrak{f})$). Finally, e_4 is the zero element in $\mathsf{L}_{F_{\emptyset}}(\mathfrak{f})$, and $\mathcal{O}(F_{\emptyset}, e_4) = \{0\}$.

7.2 Generalized Gelfand–Graev characters

Suppose $(F, e) \in I^n$. Let $(f, h, e) \in \mathsf{L}_F(\mathfrak{f})$ denote an $\mathfrak{sl}_2(\mathfrak{f})$-completion of e. Recall that

$$\mathsf{L}_F(\leq 1) = \sum_{j \leq 1} \mathsf{L}_F(j),$$

where $\mathsf{L}_F(j)$ is the j-eigenspace for the action of h on $\mathsf{L}_F(\mathfrak{f})$.

We let $\Gamma_{(F,e)}$ denote the generalized Gelfand–Graev character for (f, h, e); it is defined by

$$\Gamma_{(F,e)}(\bar{Z}) = \frac{|\mathsf{L}_F(-1)|^{1/2}}{|\mathsf{L}_F(\leq -1)|} \sum_{\bar{g} \in \mathsf{G}_F(\mathfrak{f}); {}^{\bar{g}}\bar{Z} \in \mathsf{L}_F(\leq -2)} \Lambda(B(X, {}^{g}Z)),$$

where X is any lift of e and $\bar{Z} \in \mathsf{L}_F(\mathfrak{f})$.

Let $h(F,e) \in C(\mathfrak{g}_F/\mathfrak{g}_F^+)$ denote the characteristic function of the lift of the subset

$$e + \mathsf{L}_F(\leq 1)$$

of $\mathsf{L}_F(\mathfrak{f})$. Note that $h(F,e) \in \mathcal{D}_0$. From, for example, the proof of [23, Lemma 2.2] we have that the map $(\bar{g}, \bar{X}) \mapsto {}^{\bar{g}}\bar{X}$ from $\mathsf{G}_F(\leq -1) \times (e + C_{\mathsf{L}_F(\mathfrak{f})}(f))$ to $e + \mathsf{L}_F(\leq 1)$ is bijective. Here $\mathsf{G}_F(\leq -1)$ is the unipotent radical of the parabolic subgroup in $\mathsf{G}_F(\mathfrak{f})$ with Lie algebra $\mathsf{L}_F(\leq 0)$.

The functions $\Gamma_{(F,e)}$ and $h(F,e)$ are related by the fact (see [23, Section 2] and [28, p. 283, (2)]) that

$$\sum_{\bar{g} \in \mathsf{G}_F(\mathfrak{f})} \hat{h}(F,e)({}^{g}\bar{Z}) = |\mathsf{L}_F(1)| \cdot |\mathsf{L}_F(\mathfrak{f})|^{1/2} \cdot \Gamma_{(F,e)}(\bar{Z}) \tag{11}$$

for all $\bar{Z} \in \mathsf{L}_F(\mathfrak{f})$.

7.3 The distributions of interest evaluated at the functions $h(F,e)$

We now evaluate various distributions at the functions $h(F,e)$. We perform these calculations in increasing order of difficulty (from the authors' perspective).

7.3.1 Nilpotent orbital integrals

For nilpotent orbital integrals, the result we need is a straightforward generalization of a result of Waldspurger [28, Section IX.4].

Lemma 7.3.1. *Suppose $(F,e), (F',e') \in I^n$. We have*

$$\mu_{\mathcal{O}(F,e)}(h(F',e'))$$

is zero unless the (p-adic) closure of $\mathcal{O}(F',e')$ is contained in the (p-adic) closure of $\mathcal{O}(F,e)$. Moreover,

$$\mu_{\mathcal{O}(F,e)}(h(F,e)) = |\mathsf{L}_F(1)|^{1/2}.$$

7.3.2 Distributions not associated to toric Green functions

For a pair $(F,\mathcal{G}) \in I^c$, where $\mathcal{G}$ is not a toric Green function, the result we need is not difficult to obtain.

Lemma 7.3.2. *Fix $(F,\mathcal{G}) \in I^c$. Suppose that $\mathcal{G}$ is not a toric Green function. If x is a vertex in $\mathcal{B}(G)$ and $(x,e) \in I^d$, then $D^{\mathrm{st}}_{(F,\mathcal{G})}(h(x,e))$ is zero unless F lies in the G-orbit of x. In this case, we may assume $x = F$. We then have*

$$D^{\mathrm{st}}_{(x,\mathcal{G})}(h(x,e)) = |\mathsf{L}_x(0)|^{-1/2} \cdot \mathcal{G}(e).$$

Proof. In all cases, F is a vertex. We have

$$
\begin{aligned}
&D^{\mathrm{st}}_{(F,\mathcal{G})}(h(x,e)) \\
&= \frac{|\mathsf{L}_F(\mathfrak{f})|^{1/2}}{|\mathsf{G}_F(\mathfrak{f})|} \cdot \int_G \int_{\mathfrak{g}} h(x,e)({}^g Z) \cdot \mathcal{G}_F(Z)\, dZ\, dg \\
&= \frac{|\mathsf{L}_F(\mathfrak{f})|^{1/2}}{|\mathsf{G}_F(\mathfrak{f})|} \cdot \int_G \int_{\mathfrak{g}_F} h(x,e)({}^g Z) \cdot \mathcal{G}_F(Z)\, dZ\, dg \\
&= \frac{|\mathsf{L}_F(\mathfrak{f})|^{1/2}}{|\mathsf{G}_F(\mathfrak{f})|} \cdot \int_G \sum_{\bar{Z} \in \mathfrak{g}_F/(\mathfrak{g}_F \cap \mathfrak{g}^+_{g^{-1}x})} h(x,e)({}^g Z) \int_{\mathfrak{g}_F \cap \mathfrak{g}^+_{g^{-1}x}} \mathcal{G}_F(Z+Z')\, dZ'\, dg.
\end{aligned}
$$

If $g^{-1}x \neq F$, then the image of $\mathfrak{g}_F \cap \mathfrak{g}^+_{g^{-1}x}$ in $\mathsf{L}_F(\mathfrak{f})$ is the nilradical of a proper parabolic subgroup of $\mathsf{G}_F(\mathfrak{f})$. Hence, as $\mathcal{G}$ is a cuspidal function, the inner integral is zero. Thus if x and F do not lie in the same G-orbit of vertices, then $D^{\mathrm{st}}_{(F,\mathcal{G})}(h(x,e)) = 0$. On the other hand, if they do lie in the same orbit, then we may assume $F = x$. We then have

$$
\begin{aligned}
D^{\mathrm{st}}_{(x,\mathcal{G})}(h(x,e)) &= \frac{|\mathsf{L}_x(\mathfrak{f})|^{1/2}}{|\mathsf{G}_x(\mathfrak{f})|} \cdot \int_{G_x} \int_{\mathfrak{g}_x} h(x,e)({}^g Z) \cdot \mathcal{G}_x(Z)\, dZ\, dg \\
&= \frac{|\mathsf{L}_x(\mathfrak{f})|^{1/2}}{|\mathsf{G}_x(\mathfrak{f})|} \cdot \mathrm{meas}_{dg}(G_x) \cdot \mathrm{meas}_{dZ}(\mathfrak{g}^+_x) \cdot |\mathsf{L}_x(\leq -1)| \cdot \mathcal{G}(e) \\
&= |\mathsf{L}_x(0)|^{-1/2} \cdot \mathcal{G}(e).
\end{aligned}
$$

□

7.3.3 Distributions associated to toric Green functions

Finally, we consider the distributions associated to toric Green functions. These calculations are rather lengthy.

Lemma 7.3.3. *Fix $(F, \mathcal{Q}^F_{\mathsf{S}}) \in I^c$. Suppose $X_{\mathsf{S}} \in \mathfrak{g}_F$ such that the image of X_{S} in $\mathsf{L}_F(\mathfrak{f})$ has centralizer S in G_F. Let S be a lift of (F, S). If x is a vertex in $\mathcal{B}(G)$ and $(x,e) \in I^d$, then*

$$\mu_{X_{\mathsf{S}}}(h(x,e)) = 0$$

unless there exists a $g \in G$ for which $gx \in \mathcal{B}(S)$; in this case, we may assume that $x \in \mathcal{B}(S)$ and we have

$$\mu_{X_{\mathsf{S}}}(h(x,e)) = \frac{(-1)^{\mathrm{rk}(\mathsf{S})} \cdot |\mathsf{L}_x(1)|^{1/2} \cdot |N_G(S)/S|}{\left|N_{\mathsf{G}_x}(\mathsf{S})(\mathfrak{f})/\mathsf{S}(\mathfrak{f})\right|} \cdot (\mathrm{Ind}^x_F\, \mathcal{Q}^F_{\mathsf{S}}, \Gamma_{(x,e)})_{\mathsf{L}_x}.$$

Proof. Since the centralizer of X_{S} in G is a torus lifting (F, S), we may assume that this torus is S. We have

$$\mu_{X_{\mathsf{S}}}(h(x,e)) = \int_{G/S} h(x,e)({}^g X_{\mathsf{S}})\, dg^* = \int_{G/S} h_x({}^g X_{\mathsf{S}})\, dg^*,$$

where dg^* is the quotient measure $\frac{dg}{ds}$ and

$$h_x(Y) = \frac{1}{|\mathsf{G}_x(\mathfrak{f})|} \cdot \sum_{\bar{g} \in \mathsf{G}_x(\mathfrak{f})} h(x,e)({}^{g}Y)$$

for $Y \in \mathfrak{g}$.

Fix a $g \in G$. Since $h(x,e) \in C(\mathfrak{g}_x/\mathfrak{g}_x^+)$, we have that $h(x,e)({}^{g}X_{\mathsf{S}}) \neq 0$ implies that ${}^{g}X_{\mathsf{S}} \in \mathfrak{g}_x$. This, in turn, implies that $X_{\mathsf{S}} \in \mathfrak{g}_{g^{-1}x}$, which, from, for example, a slight modification of [12, Section 4.4], implies that $g^{-1}x \in \mathcal{B}(S) := \mathcal{A}(\mathbf{S}, K)^{\Gamma} \hookrightarrow \mathcal{B}(\mathbf{G}, K)^{\Gamma} = \mathcal{B}(G)$.

Let $\mathcal{F}$ denote the set of vertices in the intersection of $\mathcal{B}(S)$ with the G-orbit of x. Note that for each vertex y in $\mathcal{F}$ the centralizer in G_y of the image of X_{S} in $\mathsf{L}_y(\mathfrak{f}) = \mathfrak{g}_y/\mathfrak{g}_y^+$ is naturally isomorphic to S; we denote the corresponding toric Green function by Q_{S}^{y}. We remark that $Q_{\mathsf{S}}^{y} = \mathrm{Ind}_F^y\, Q_{\mathsf{S}}^{F}$.

We let $\mathcal{F}^{\mathrm{rep}}$ denote a set of representatives for $\mathcal{F}$ modulo the action of S. Without loss of generality, $x \in \mathcal{F}$. For each $y \in \mathcal{F}^{\mathrm{rep}}$, we fix $g_y \in G$, for which $g_y^{-1}x = y$. Since ${}^{g_y}G_y = G_x$ and h_x is G_x-invariant, we have

$$\begin{aligned}
\mu_{X_{\mathsf{S}}}(h_x) &= \sum_{y \in \mathcal{F}^{\mathrm{rep}}} \int_{G_y S/S} h_x({}^{g_y g}X_{\mathsf{S}})\, dg^* \\
&= \sum_{y \in \mathcal{F}^{\mathrm{rep}}} \mathrm{meas}_{dg^*}(G_y S/S) \cdot h_x({}^{g_y}X_{\mathsf{S}}) \qquad (12)\\
&\quad \text{(from equation (5))} \\
&= \frac{|\mathsf{G}_x(\mathfrak{f})| \cdot |\mathsf{L}_{\mathsf{S}}(\mathfrak{f})|^{1/2}}{|\mathsf{L}_x(\mathfrak{f})|^{1/2} \cdot |\mathsf{S}(\mathfrak{f})|} \cdot \sum_{y \in \mathcal{F}^{\mathrm{rep}}} h_x({}^{g_y}X_{\mathsf{S}}).
\end{aligned}$$

Fix $y \in \mathcal{F}^{\mathrm{rep}}$. We are interested in the term

$$h_x({}^{g_y}X_{\mathsf{S}})$$

occurring in the sum above. For notational convenience we set $h_y := h_x^{g_y}$ and let h denote the corresponding element of $C(\mathsf{L}_y(\mathfrak{f}))$. We first observe that

$$h_y(X_{\mathsf{S}}) = \int_{\mathfrak{g}_y} \Lambda(B(X_{\mathsf{S}}, Y)) \cdot \hat{h}_y(-Y)\, dY. \qquad (13)$$

On the other hand, we have

$$\begin{aligned}
(Q_{\mathsf{S}}^{F}, \mathrm{r}_F^y \hat{h})_{\mathsf{L}_F} &= (Q_{\mathsf{S}}^{y}, \hat{h})_{\mathsf{L}_y} \\
&= \frac{1}{|\mathsf{G}_y(\mathfrak{f})|} \cdot \sum_{\bar{X} \in \mathsf{L}_y(\mathfrak{f})} Q_{\mathsf{S}}^{y}(\bar{X}) \cdot \hat{h}_y(X).
\end{aligned}$$

Now, since $\hat{h}$ is supported on the nilpotent cone, we can apply equation (1) to arrive at

$$\begin{aligned}\frac{(-1)^{\mathrm{rk}(\mathsf{S})}\cdot|\mathsf{S}(\mathfrak{f})|}{|\mathsf{L}_{\mathsf{S}}(\mathfrak{f})|^{1/2}}&\cdot(Q^F_{\mathsf{S}},\mathrm{r}^y_F\hat{h})_{\mathsf{L}_F}\\ &=\frac{1}{\left|\mathsf{G}_y(\mathfrak{f})\right|}\cdot\sum_{\bar{X}\in\mathsf{L}_y(\mathfrak{f})}\left(\mathcal{F}\left(\sum_{g\in\mathsf{G}_y(\mathfrak{f})}[\bar{X}_{\mathsf{S}}]^g\right)\right)(\bar{X})\cdot\hat{h}_y(X).\end{aligned}$$

Expanding the right-hand side yields

$$\frac{1}{\left|\mathsf{G}_y(\mathfrak{f})\right|\cdot\left|\mathsf{L}_y(\mathfrak{f})\right|^{1/2}}\cdot\sum_{\bar{X}\in\mathsf{L}_y(\mathfrak{f})}\sum_{\bar{Y}\in\mathsf{L}_y(\mathfrak{f})}\sum_{g\in\mathsf{G}_y(\mathfrak{f})}[\bar{X}_{\mathsf{S}}]({}^g\bar{Y})\cdot\Lambda(B(X,Y))\cdot\hat{h}_y(X).$$

By moving the sum over $\mathsf{G}_y(\mathfrak{f})$ in front of the other two sums and making the changes of variables $(X\mapsto {}^{g^{-1}}X)$ and $(Y\mapsto {}^{g^{-1}}Y)$, we can take advantage of the invariance properties of B and h_y to arrive at

$$\begin{aligned}\frac{(-1)^{\mathrm{rk}(\mathsf{S})}\cdot|\mathsf{S}(\mathfrak{f})|}{|\mathsf{L}_{\mathsf{S}}(\mathfrak{f})|^{1/2}}&\cdot(Q^F_{\mathsf{S}},\mathrm{r}^y_F\hat{h})_{\mathsf{L}_F}\\ &=\mathrm{meas}_{dX}(\mathfrak{g}_y^+)\cdot\sum_{\bar{X}\in\mathsf{L}_y(\mathfrak{f})}\Lambda(B(X,X_{\mathsf{S}}))\cdot\hat{h}_y(X)\\ &\qquad\text{(since the left-hand side is independent of the choice of } X_{\mathsf{S}})\\ &=\mathrm{meas}_{dX}(\mathfrak{g}_y^+)\cdot\sum_{\bar{X}\in\mathsf{L}_y(\mathfrak{f})}\Lambda(B(X,X_{\mathsf{S}}))\cdot\hat{h}_y(-X).\end{aligned}\tag{14}$$

Combining equations (12), (13), and (14), we arrive at

$$\begin{aligned}\mu_{X_{\mathsf{S}}}(h_x)&=(-1)^{\mathrm{rk}(\mathsf{S})}\cdot\sum_{y\in\mathcal{F}^{\mathrm{rep}}}\frac{|\mathsf{G}_x(\mathfrak{f})|}{|\mathsf{L}_x(\mathfrak{f})|^{1/2}}\cdot(Q^F_{\mathsf{S}},\mathrm{r}^y_F\hat{h})_{\mathsf{L}_F}\\ &=(-1)^{\mathrm{rk}(\mathsf{S})}\cdot\frac{|\mathcal{F}^{\mathrm{rep}}|\cdot|\mathsf{G}_x(\mathfrak{f})|}{|\mathsf{L}_x(\mathfrak{f})|^{1/2}}\cdot(Q^x_{\mathsf{S}},\hat{h})_{\mathsf{L}_x}\\ &\qquad\text{(from equation (11))}\\ &=(-1)^{\mathrm{rk}(\mathsf{S})}\cdot\left|\mathcal{F}^{\mathrm{rep}}\right|\cdot|\mathsf{L}_x(1)|^{1/2}\cdot(\mathrm{Ind}^x_F\,Q^F_{\mathsf{S}},\Gamma_{(x,e)})_{\mathsf{L}_x}.\end{aligned}$$

To complete the proof, we now show that

$$\left|\mathcal{F}^{\mathrm{rep}}\right|=\frac{|N_G(S)/S|}{\left|N_{\mathsf{G}_x}(\mathsf{S})(\mathfrak{f})/\mathsf{S}(\mathfrak{f})\right|}.$$

From [13, Lemma 2.2.1], we can find a maximal k-split torus $\mathbf{T}'$ of $\mathbf{G}$ such that $\mathcal{B}(S)\subset\mathcal{A}(\mathbf{T}',k)$ and, in fact, $\mathcal{B}(S)=A(\mathcal{A}(\mathbf{T}',k),F)$ (see Section 4.1 for notation). Suppose $y\in\mathcal{F}^{\mathrm{rep}}$ and $g\in G$ such that $gx=y\in\mathcal{B}(S)$. Let ${}^g\mathsf{S}$ denote the $\mathfrak{f}$-torus in G_y whose group of $\mathfrak{F}$-rational points agrees with the image of ${}^g\mathbf{S}(K)\cap\mathbf{G}(K)_y$ in

G_y. Via the natural identification of G_F in G_y we see that there is a $k \in G_y$ such that ${}^{kg}\mathsf{S} \stackrel{i}{=} \mathsf{S}$ in G_F in G_y. Since S is a lift of ${}^{kg}\mathsf{S}$, from [13, Lemma 2.2.2] there is an element $k' \in G_y^+$ such that ${}^{k'kg}S = S$. Thus every element of $\mathcal{F}^{\mathrm{rep}}$ uniquely determines, up to right multiplication by $N_{G_x}(S)/S$, an element of $N_G(S)/S$, and *vice-versa*. The desired equality follows. □

Suppose x is a vertex in $\mathcal{B}(G)$. For each pair $(x, e) \in I^d$ and each $(F, \mathcal{Q}_{\mathsf{S}}^F) \in I^c$ we set $X_{(F,\mathsf{S})}^{(x,e)}$ equal to zero if the G-orbit of x does not intersect the building $\mathcal{B}(S)$, where S is any lift of (F, S). Otherwise, we assume x lies in $\mathcal{B}(S)$ and set $X_{(F,\mathsf{S})}^{(x,e)}$ equal to the Green polynomial associated to $(\mathrm{Ind}_F^x\, \mathcal{Q}_{\mathsf{S}}^F)(e)$ evaluated at q^{-1}. These polynomials can be explicitly described: If e is regular in $\mathsf{L}_x(\mathfrak{f})$ and the G-orbit of x intersects $\mathcal{B}(S)$, then $X_{(F,\mathsf{S})}^{(x,e)} = 1$. Otherwise, either $X_{(F,\mathsf{S})}^{(x,e)} = 0$ or e is subregular in $\mathsf{L}_{F_{G_2}}(\mathfrak{f})$ and we have (see [26, Section 7])

$$X_{(F,\mathsf{S})}^{(F_{G_2},e_1)} = 1 + q^{-1}(\chi(w) + 2\tau(w)),$$
$$X_{(F,\mathsf{S})}^{(F_{G_2},e_1')} = 1 + q^{-1}(\chi(w) - \tau(w)),$$

and

$$X_{(F,\mathsf{S})}^{(F_{G_2},e_1'')} = 1 + q^{-1}(\chi(w)),$$

where χ and τ are characters of W (see Table 7) and $w \in W$ is a representative of the conjugacy class associated to S.

Table 7. Characters for the Weyl group of G_2.

	1	ε	τ	$\varepsilon\tau$	χ	$\chi\tau$
$\{e\}$	1	1	1	1	2	2
$\tilde{A}_1$	1	−1	1	−1	0	0
A_1	1	−1	−1	1	0	0
G_2	1	1	−1	−1	1	−1
A_2	1	1	1	1	−1	−1
$A_1 \times \tilde{A}_1$	1	1	−1	−1	−2	2

Corollary 7.3.4. *Fix $(F, \mathcal{Q}_{\mathsf{S}}^F) \in I^c$. Suppose $X_{\mathsf{S}} \in \mathfrak{g}_F$ such that the image of X_{S} in $\mathsf{L}_F(\mathfrak{f})$ has centralizer S in G_F. Let S be a lift of (F, S). If x is a vertex in $\mathcal{B}(G)$ and $(x, e) \in I^d$; then*

$$\mu_{X_{\mathsf{S}}}(h(x, e)) = 0$$

unless there exists a $g \in G$ *for which* $gx \in \mathcal{B}(S)$*; in this case, we may assume that* $x \in \mathcal{B}(S)$ *and we have*

$$D_{(F,\mathcal{Q}^F_{\mathsf{S}})}(h(x,e)) = \frac{|\mathsf{L}_x(0)|^{1/2} \cdot |\mathsf{L}_x(1)| \cdot |N_G(S)/S| \cdot X^{(x,e)}_{(F,\mathsf{S})}}{|W_x| \cdot q},$$

where W_x *denotes the absolute Weyl group of* G_x *and the constants* $X^{(x,e)}_{(F,\mathsf{S})}$ *were described above.*

Proof. From Lemma 7.3.3 we have

$$\begin{aligned}
\mu_{X_{\mathsf{S}}}(h(x,e)) &= \frac{(-1)^{\mathrm{rk}(\mathsf{S})} \cdot |\mathsf{L}_x(1)|^{1/2} \cdot |N_G(S)/S|}{\left|N_{\mathsf{G}_x}(\mathsf{S})(\mathfrak{f})/\mathsf{S}(\mathfrak{f})\right|} \cdot (\mathrm{Ind}^x_F \, \mathcal{Q}^F_{\mathsf{S}}, \Gamma_{(x,e)})_{\mathsf{L}_x} \\
&\quad \text{(from [18, 2.3.2])} \\
&= \frac{|\mathsf{L}_x(1)| \cdot |\mathsf{L}_x(0)|^{1/2} \cdot |N_G(S)/S| \cdot |\mathsf{S}(\mathfrak{f})| \cdot X^{(x,e)}_{(F,\mathsf{S})}}{\left|N_{\mathsf{G}_x}(\mathsf{S})(\mathfrak{f})/\mathsf{S}(\mathfrak{f})\right| \cdot |W_x| \cdot q} \\
&\quad \cdot (\mathrm{Ind}^x_F \, \mathcal{Q}^F_{\mathsf{S}}, \mathrm{Ind}^x_F \, \mathcal{Q}^F_{\mathsf{S}})_{\mathsf{L}_x} \\
&\quad \text{(from [10, Proposition 7.6.2])} \\
&= \frac{|\mathsf{L}_x(0)|^{1/2} \cdot |\mathsf{L}_x(1)| \cdot |N_G(S)/S| \cdot X^{(x,e)}_{(F,\mathsf{S})}}{|W_x| \cdot q}.
\end{aligned}$$

□

Recall that if C is a conjugacy class in W, then, after scaling, we may assume

$$D^{\mathrm{st}}_C = \sum_{(F,\mathcal{Q}^F_{\mathsf{S}}) \in I^c/\sim} \frac{D_{(F,\mathcal{Q}^F_{\mathsf{S}})}}{|N_G(S)/S|},$$

where the sum is over the set of equivalence classes of pairs $(F, \mathcal{Q}^F_{\mathsf{S}})$ for which S corresponds to C. For notational ease, if $w \in C$, then $D^{\mathrm{st}}_w := D^{\mathrm{st}}_C$. For each character κ of W we define

$$D^{\mathrm{st}}_\kappa = \sum_{w \in W} \kappa(w) \cdot D^{\mathrm{st}}_w.$$

The distribution D^{st}_κ is stable. Thanks to Corollary 7.3.4, if x is vertex and $(x, e) \in I^d$, then we can compute $D^{\mathrm{st}}_\kappa(h(x,e))$. Indeed, we have

$$D^{\mathrm{st}}_\kappa(h(x,e)) = \frac{|\mathsf{L}_x(0)|^{1/2} \cdot |\mathsf{L}_x(1)| \cdot Y_\kappa(x,e)}{q \cdot |W_x|}, \tag{15}$$

where the value of $Y_\kappa(x,e)$ is given in Table 8.

7.4 A basis for $J^{\mathrm{st}}(\mathcal{N})$

We now prove Theorem 7.0.1.

Table 8. The values of $Y_\kappa(x, e)$.

	(F_{G_2}, e_0)	(F_{G_2}, e_1)	(F_{G_2}, e_1')	(F_{G_2}, e_1'')	$(F_{A_1\times\tilde{A}_1}, e_1^{\pm})$	(F_{A_2}, e_1^{δ})
1	12	12	12	12	8	6
ε	0	0	0	0	-4	0
τ	0	$24q^{-1}$	$-12q^{-1}$	0	0	0
$\varepsilon\tau$	0	0	0	0	0	6
χ	0	$12q^{-1}$	$12q^{-1}$	$12q^{-1}$	0	0
$\chi\tau$	0	0	0	0	4	0

A proof of Theorem 7.0.1. If $T \in J(\mathfrak{g}_0)$, then from Theorem 4.4.1, the restriction of T to $\mathcal{D}_0$ has a *local expansion*. That is, there exist constants $c_{\mathcal{O}}(T)$, indexed by $\mathcal{O} \in \mathcal{O}(0)$, so that for all $f \in \mathcal{D}_0$, we have

$$T(f) = \sum_{\mathcal{O}\in\mathcal{O}(0)} c_{\mathcal{O}}(T) \cdot \mu_{\mathcal{O}}(f).$$

Thanks to Waldspurger [28, Lemme IV.15], we know that if T is stable, then $\sum c_{\mathcal{O}}(T) \cdot \mu_{\mathcal{O}}$ is also stable. Moreover, for each i, the distribution

$$\sum_{\dim(\mathcal{O})=i} c_{\mathcal{O}}(T) \cdot \mu_{\mathcal{O}}$$

is stable.

Since $\mu_{\{0\}}$ is stable, from Lemma 6.4.1 we need to produce $6 + |\mathfrak{f}^\times/(\mathfrak{f}^\times)^3|$ additional combinations of nilpotent orbital integrals which are stable. That each of the remaining combinations can be taken to be of the form $\mu_{\mathcal{O}}$ with $\mathcal{O}$ k-distinguished follows directly from Lemma 7.3.1, Lemma 7.3.2, and equation (15). We now explain how this happens.

We first examine the regular nilpotent orbital integral $\mu_{\mathcal{O}_0}$. From Theorem 4.4.1, the restriction of D_1^{st} to $\mathcal{D}_0$ has a local expansion:

$$D_1^{\mathrm{st}}(f) = \sum c_{\mathcal{O}}(D_1^{\mathrm{st}}) \cdot \mu_{\mathcal{O}}(f) \tag{16}$$

for all $f \in \mathcal{D}_0$. By evaluating both sides of equation (16) at $h(F_{G_2}, e_0)$, we conclude from Lemma 7.3.1, equation (15), and Table 8 that $c_{\mathcal{O}_0}(D_1^{\mathrm{st}}) \neq 0$. Hence, thanks to [28, Lemme IV.15], the distribution

$$\mu_{\mathcal{O}_0}$$

is stable.

We next look at the three k-distinguished nilpotent orbital integrals $\mu_{\mathcal{O}_1}$, $\mu_{\mathcal{O}_1'}$, and $\mu_{\mathcal{O}_1''}$. The restrictions to $\mathcal{D}_0$ of the stable distributions D_τ^{st}, D_χ^{st}, and $D_{F_{G_2}}^{\mathrm{st}} := D_{(F_{G_2},\mathcal{G}_{\mathrm{sgn}})}^{\mathrm{st}}$ have local expansions:

$$D^{\rm st}_\tau(f)=\sum c_{\mathcal{O}}(D^{\rm st}_\tau)\cdot\mu_{\mathcal{O}}(f),\qquad D^{\rm st}_\chi(f)=\sum c_{\mathcal{O}}(D^{\rm st}_\chi)\cdot\mu_{\mathcal{O}}(f),$$

and

$$D^{\rm st}_{F_{G_2}}(f)=\sum c_{\mathcal{O}}(D^{\rm st}_{F_{G_2}})\cdot\mu_{\mathcal{O}}(f)$$

for all $f\in\mathcal{D}_0$. From equation (15) and Table 8, we conclude that

$$D^{\rm st}_\chi(h(F_{G_2},e_0))=D^{\rm st}_\chi(h(F_{\mathrm{A}_1\times\tilde{\mathrm{A}}_1},e_1^{\pm}))=D^{\rm st}_\chi(h(F_{\mathrm{A}_2},e_1^{\delta}))=0 \tag{17}$$

and

$$D^{\rm st}_\tau(h(F_{G_2},e_0))=D^{\rm st}_\tau(h(F_{\mathrm{A}_1\times\tilde{\mathrm{A}}_1},e_1^{\pm}))=D^{\rm st}_\tau(h(F_{\mathrm{A}_2},e_1^{\delta}))=0. \tag{18}$$

Similarly, from Lemma 7.3.2 we have

$$D^{\rm st}_{F_{G_2}}(h(F_{G_2},e_0))=D^{\rm st}_{F_{G_2}}(h(F_{\mathrm{A}_1\times\tilde{\mathrm{A}}_1},e_1^{\pm}))=D^{\rm st}_{F_{G_2}}(h(F_{\mathrm{A}_2},e_1^{\delta}))=0. \tag{19}$$

Therefore, we conclude from Lemma 7.3.1 that $c_{\mathcal{O}}(D^{\rm st}_\tau)=c_{\mathcal{O}}(D^{\rm st}_\chi)=c_{\mathcal{O}}(D^{\rm st}_{F_{G_2}})=0$ for all k-distinguished nilpotent orbits $\mathcal{O}$ other than $\mathcal{O}_1$, $\mathcal{O}_1'$, and $\mathcal{O}_1''$.

By evaluating the distribution $D^{\rm st}_\chi$ at the functions $h(F_{G_2},e_1)$, $h(F_{G_2},e_1')$, and $h(F_{G_2},e_1'')$ we conclude from equation (17), Lemma 7.3.1, and equation (15) that

$$c_{\mathcal{O}_1}(D^{\rm st}_\chi)=c_{\mathcal{O}_1'}(D^{\rm st}_\chi)=c_{\mathcal{O}_1''}(D^{\rm st}_\chi)\neq 0.$$

Hence, thanks to [28, Lemme IV.15], the distribution

$$\mu_{\mathcal{O}_1}+\mu_{\mathcal{O}_1'}+\mu_{\mathcal{O}_1''} \tag{20}$$

is stable.

By evaluating the distribution $D^{\rm st}_\tau$ at the functions $h(F_{G_2},e_1)$, $h(F_{G_2},e_1')$, and $h(F_{G_2},e_1'')$ we conclude from Lemma 7.3.1, equation (15), and equation (18) that

$$c_{\mathcal{O}_1}(D^{\rm st}_\chi)=-2c_{\mathcal{O}_1'}(D^{\rm st}_\chi)\neq 0\quad\text{and}\quad c_{\mathcal{O}_1''}(D^{\rm st}_\chi)=0.$$

Hence, thanks to [28, Lemme IV.15], the distribution

$$2\mu_{\mathcal{O}_1}-\mu_{\mathcal{O}_1'} \tag{21}$$

is stable.

By evaluating the distribution $D^{\rm st}_{F_{G_2}}$ at the functions $h(F_{G_2},e_1)$, $h(F_{G_2},e_1')$, and $h(F_{G_2},e_1'')$ we conclude from Lemma 7.3.1, Lemma 7.3.2, and equation (19) that

$$c_{\mathcal{O}_1}(D^{\rm st}_{F_{G_2}})=c_{\mathcal{O}_1'}(D^{\rm st}_{F_{G_2}})=-c_{\mathcal{O}_1''}(D^{\rm st}_{F_{G_2}})\neq 0.$$

Hence, thanks to [28, Lemme IV.15], the distribution

$$\mu_{\mathcal{O}_1} + \mu_{\mathcal{O}_1'} - \mu_{\mathcal{O}_1''} \tag{22}$$

is stable.

Consequently, from equations (20), (21), and (22), we conclude that the distributions $\mu_{\mathcal{O}_1}$, $\mu_{\mathcal{O}_1'}$, and $\mu_{\mathcal{O}_1''}$ are stable.

We now examine the two k-distinguished nilpotent orbital integrals $\mu_{\mathcal{O}_1^+}$ and $\mu_{\mathcal{O}_1^-}$. The restrictions to $\mathcal{D}_0$ of the stable distributions $D^{\mathrm{st}}_{\chi\tau}$ and $D^{\mathrm{st}}_{F_{\mathrm{A}_1\times\tilde{\mathrm{A}}_1}} := D^{\mathrm{st}}_{(F_{\mathrm{A}_1\times\tilde{\mathrm{A}}_1}, \mathcal{G}_{\mathrm{sgn}})}$ have local expansions:

$$D^{\mathrm{st}}_{\chi\tau}(f) = \sum c_{\mathcal{O}}(D^{\mathrm{st}}_{\chi\tau}) \cdot \mu_{\mathcal{O}}(f)$$

and

$$D^{\mathrm{st}}_{F_{\mathrm{A}_1\times\tilde{\mathrm{A}}_1}}(f) = \sum c_{\mathcal{O}}(D^{\mathrm{st}}_{F_{\mathrm{A}_1\times\tilde{\mathrm{A}}_1}}) \cdot \mu_{\mathcal{O}}(f)$$

for all $f \in \mathcal{D}_0$. From equation (15) and Table 8, we have

$$\begin{aligned} D^{\mathrm{st}}_{\chi\tau}(h(F_{G_2}, e_0)) &= D^{\mathrm{st}}_{\chi\tau}(h(F_{G_2}, e_1)) = D^{\mathrm{st}}_{\chi\tau}(h(F_{G_2}, e_1')) \\ &= D^{\mathrm{st}}_{\chi\tau}(h(F_{G_2}, e_1'')) = D^{\mathrm{st}}_{\chi\tau}(h(F_{A_2}, e_1^{\delta})) = 0. \end{aligned} \tag{23}$$

Similarly, from Lemma 7.3.2,

$$\begin{aligned} D^{\mathrm{st}}_{F_{\mathrm{A}_1\times\tilde{\mathrm{A}}_1}}(h(F_{G_2}, e_0)) &= D^{\mathrm{st}}_{F_{\mathrm{A}_1\times\tilde{\mathrm{A}}_1}}(h(F_{G_2}, e_1)) = D^{\mathrm{st}}_{F_{\mathrm{A}_1\times\tilde{\mathrm{A}}_1}}(h(F_{G_2}, e_1')) \\ &= D^{\mathrm{st}}_{F_{\mathrm{A}_1\times\tilde{\mathrm{A}}_1}}(h(F_{G_2}, e_1'')) = D^{\mathrm{st}}_{F_{\mathrm{A}_1\times\tilde{\mathrm{A}}_1}}(h(F_{A_2}, e_1^{\delta})) = 0. \end{aligned} \tag{24}$$

Therefore, we conclude from Lemma 7.3.1 that $c_{\mathcal{O}}(D^{\mathrm{st}}_{\chi\tau}) = c_{\mathcal{O}}(D^{\mathrm{st}}_{F_{\mathrm{A}_1\times\tilde{\mathrm{A}}_1}}) = 0$ for all k-distinguished nilpotent orbits $\mathcal{O}$ other than $\mathcal{O}_1^+$ and $\mathcal{O}_1^-$.

By evaluating the distribution $D^{\mathrm{st}}_{\chi\tau}$ at the functions $h(F_{\mathrm{A}_1\times\tilde{\mathrm{A}}_1}, e_1^{\pm})$, we conclude from equation (23), Lemma 7.3.1, and equation (15) that

$$c_{\mathcal{O}_1^+}(D^{\mathrm{st}}_{\chi\tau}) = c_{\mathcal{O}_1^-}(D^{\mathrm{st}}_{\chi\tau}) \neq 0.$$

Hence, thanks to [28, Lemme IV.15], the distribution

$$\mu_{\mathcal{O}_1^+} + \mu_{\mathcal{O}_1^-} \tag{25}$$

is stable.

By evaluating the distribution $D^{\mathrm{st}}_{F_{\mathrm{A}_1\times\tilde{\mathrm{A}}_1}}$ at the functions $h(F_{\mathrm{A}_1\times\tilde{\mathrm{A}}_1}, e_1^{\pm})$, we conclude from Lemma 7.3.1, equation (15), and equation (24) that

$$c_{\mathcal{O}_1^+}(D^{\mathrm{st}}_{\chi\tau}) = -c_{\mathcal{O}_1^-}(D^{\mathrm{st}}_{\chi\tau}) \neq 0.$$

thanks to [28, Lemme IV.15], the distribution

$$\mu_{\mathcal{O}_1^+} - \mu_{\mathcal{O}_1^-} \tag{26}$$

is stable.

From equations (25) and (26), we conclude that the distributions $\mu_{\mathcal{O}_1^+}$ and $\mu_{\mathcal{O}_1^-}$ are stable.

Finally, we examine the k-distinguished nilpotent orbital integrals $\mu_{\mathcal{O}_1^\delta}$ for $\delta \in \mu_3(\mathfrak{f})$. The restriction to $\mathcal{D}_0$ of the stable distribution $D^{\mathrm{st}}_{\varepsilon\tau}$ has local expansion

$$D^{\mathrm{st}}_{\varepsilon\tau}(f) = \sum c_{\mathcal{O}}(D^{\mathrm{st}}_{\varepsilon\tau}) \cdot \mu_{\mathcal{O}}(f)$$

for $f \in \mathcal{D}_0$. If $|\mu_3(\mathfrak{f})| \neq 1$, then the restrictions to $\mathcal{D}_0$ of the stable distributions $D^{\mathrm{st}}_{\chi'} := D^{\mathrm{st}}_{(F_{A_2},\mathcal{G}_{\chi'})}$ and $D^{\mathrm{st}}_{\chi''} := D^{\mathrm{st}}_{(F_{A_2},\mathcal{G}_{\chi''})}$ have local expansions

$$D^{\mathrm{st}}_{\chi'}(f) = \sum c_{\mathcal{O}}(D^{\mathrm{st}}_{\chi'}) \cdot \mu_{\mathcal{O}}(f) \quad \text{and} \quad D^{\mathrm{st}}_{\chi''}(f) = \sum c_{\mathcal{O}}(D^{\mathrm{st}}_{\chi''}) \cdot \mu_{\mathcal{O}}(f)$$

for all $f \in \mathcal{D}_0$. From equation (15) and Table 8, we have

$$\begin{aligned} D^{\mathrm{st}}_{\varepsilon\tau}(h(F_{G_2}, e_0)) = D^{\mathrm{st}}_{\varepsilon\tau}(h(F_{G_2}, e_1)) = D^{\mathrm{st}}_{\varepsilon\tau}(h(F_{G_2}, e_1')) = D^{\mathrm{st}}_{\varepsilon\tau}(h(F_{G_2}, e_1'')) \\ = D^{\mathrm{st}}_{\varepsilon\tau}(h(F_{\mathrm{A}_1 \times \tilde{\mathrm{A}}_1}, e_1^{\pm})) = 0. \end{aligned} \tag{27}$$

When $|\mu_3(\mathfrak{f})| \neq 1$, similar statements can be made for $D^{\mathrm{st}}_{\chi'}$ and $D^{\mathrm{st}}_{\chi''}$ (using Lemma 7.3.2). Therefore, we conclude from Lemma 7.3.1 that $c_{\mathcal{O}}(D^{\mathrm{st}}_{\varepsilon\chi}) = 0$ for all k-distinguished nilpotent orbits $\mathcal{O}$ other than $\mathcal{O}_1^\delta$. Similarly, if $|\mu_3(\mathfrak{f})| \neq 1$, then $c_{\mathcal{O}}(D^{\mathrm{st}}_{\chi'}) = c_{\mathcal{O}}(D^{\mathrm{st}}_{\chi''}) = 0$ for all k-distinguished nilpotent orbits $\mathcal{O}$ other than $\mathcal{O}_1^\delta$.

By evaluating the distribution $D^{\mathrm{st}}_{\varepsilon\tau}$ at the function(s) $h(F_{A_2}, e_1^\delta)$ we conclude from equation (27), Lemma 7.3.1, equation (15), and [28, Lemme IV.15] that the distribution

$$\sum_{\delta \in \mu_3(\mathfrak{f})} \mu_{\mathcal{O}_1^\delta}$$

is stable. If $|\mu_3(\mathfrak{f})| \neq 1$, then by evaluating the distributions $D^{\mathrm{st}}_{\chi'}$, and $D^{\mathrm{st}}_{\chi''}$ at the functions $h(F_{G_2}, e_1^\delta)$, we conclude from Lemma 7.3.1, Lemma 7.3.2, the $D^{\mathrm{st}}_{F_{G_2}}$-analogue of equation (27), and [28, Lemme IV.15] that the distributions

$$\sum_{\delta \in \mu_3(\mathfrak{f})} \chi'(\delta) \cdot \mu_{\mathcal{O}_1^\delta}$$

and

$$\sum_{\delta \in \mu_3(\mathfrak{f})} \chi''(\delta) \cdot \mu_{\mathcal{O}_1^\delta}$$

are stable. Consequently, independent of the order of $|\mu_3(\mathfrak{f})|$, each element of the set

$$\{\mu_{\mathcal{O}_1^\delta} \mid \delta \in \mu_3(\mathfrak{f})\}$$

is a stable distribution. □

List of symbols

Symbol	Description	Page
$(\,,\,)_{\mathsf{L}_{\mathsf{G}}}$	An invariant pairing for functions on a finite reductive Lie algebra.	214
$\stackrel{i}{=}$		217
$A(\mathcal{A}, F)$	The smallest affine subspace of $\mathcal{A}$ containing F.	217
$\check{\alpha}$	The coroot of α.	229
$A(F, \mathcal{G}, X, g_m)$	$\{{}^{g_m}Y \mid Y \in B(F, \mathcal{G}, X) \text{ and } {}^{m}Y = \sigma(Y)\}$.	232
A_1	A conjugacy class in the Weyl group.	222
$\tilde{\mathrm{A}}_1$	A conjugacy class in the Weyl group.	222
A_2	A conjugacy class in the Weyl group.	222
$\mathrm{A}_1 \times \tilde{\mathrm{A}}_1$	A conjugacy class in the Weyl group.	222
B	The Killing form on $\mathfrak{g}$.	213
$\mathcal{B}(G)$	The Bruhat–Tits building of G.	208
$\mathfrak{B}$	The $(9 + \lvert\mu_3(\mathfrak{f})\rvert)$ distributions listed in Table 4.	238
$\mathfrak{B}^{\mathrm{st}}$	Those elements of $\mathfrak{B}$ whose label has a superscript st.	238
$\mathfrak{B}^{\mathrm{unst}}$	Those elements of $\mathfrak{B}$ whose label has a superscript unst.	239
$B(F, \mathcal{G}, X)$	$X + \mathbf{g}(K)_{\psi_F} \setminus \mathbf{g}(K)^{+}_{\psi_F}$.	231
$C_c^{\infty}(V)$	The space of complex valued compactly supported locally constant functions on V.	213
cuspidal		215
Δ	A basis for Φ.	208
$D_{(F,\mathcal{G})}$	$\sum_{\mathsf{M}_e/\sim} \frac{\chi(m)\cdot\lvert S_m/S_m^{\circ}\rvert}{\lvert C_{\mathsf{G}_F(\mathfrak{f})}({}^{g_m}e)\rvert} \int_{A(F,\mathcal{G},X,g_m)} \mu_Y \, dY$.	233
$D_{(F,\mathcal{Q}^F_{\mathsf{S}})}$	$\mu_{X_{\mathsf{S}}}$, the distribution associated to a toric Green function.	226
D_C^{st}	An element of $\mathfrak{B}$ associated to the Weyl conjugacy class C.	238
D_C^{unst}	An element of $\mathfrak{B}$ associated to a Weyl conjugacy class C.	238
$D^{\mathrm{st}}_{(F,\mathcal{G})}$	An element of $\mathfrak{B}$ associated to a Green function arising from a cuspidal local system.	238
$\mathcal{D}_0$	A subspace of $C_c^{\infty}(\mathfrak{g})$.	221
$\mathcal{D}_0^0$	A subspace of $\mathcal{D}_0$.	221
$\mathcal{D}_{0^+}$	A subspace of $C_c^{\infty}(\mathfrak{g})$.	237
distinguished		227
ε	A character of the Weyl group.	251
e_0	A regular nilpotent element of $\mathsf{L}_{F_{G_2}}(\mathfrak{f})$.	246
e_1	A subregular nilpotent element of $\mathsf{L}_{F_{G_2}}(\mathfrak{f})$.	228
e_1'	A subregular nilpotent element of $\mathsf{L}_{F_{G_2}}(\mathfrak{f})$.	246
e_1''	A subregular nilpotent element of $\mathsf{L}_{F_{G_2}}(\mathfrak{f})$.	246
e_1^{+}	A regular nilpotent element of $\mathsf{L}_{F_{\mathrm{A}_1 \times \tilde{\mathrm{A}}_1}}(\mathfrak{f})$.	228
$e_1^{\pm}$	A subregular nilpotent element of $\mathsf{L}_{F_{\mathrm{A}_1 \times \tilde{\mathrm{A}}_1}}(\mathfrak{f})$.	246
e_1^{1}	A regular nilpotent element of $\mathsf{L}_{F_{\mathrm{A}_2}}(\mathfrak{f})$.	228
e_1^{δ}	A regular nilpotent element of $\mathsf{L}_{F_{\mathrm{A}_2}}(\mathfrak{f})$.	246
$\mathfrak{f}$	The residue field of k.	208
$\mathfrak{F}$	The residue field of K.	208

$F_\emptyset$	A fixed alcove in $\mathcal{A}(T)$.	210
F_{G_2}	The hyperspecial vertex in the closure of our preferred alcove $F_\emptyset$.	210
$F_{\mathrm{A}_1\times\tilde{\mathrm{A}}_1}$	A vertex in the closure of $F_\emptyset$.	210
F_{A_2}	A vertex in the closure of $F_\emptyset$.	210
F_{A_1}	A facet in the closure of $F_\emptyset$.	210
$F_{\tilde{\mathrm{A}}_1}$	A facet in the closure of $F_\emptyset$.	210
$\bar{F}$	The closure of the facet F.	215
f_P	The constant term of f along P.	216
$\hat{f}$	The Fourier transform of f.	213
$\mathcal{F}(f)$	The Fourier transform of f.	213
$[(F,e)]$	The characteristic function of the coset e.	220
G_2	A conjugacy class in the Weyl group.	222
$\mathcal{G}_{\mathrm{sgn}}$	A generalized Green function arising from a cuspidal local system.	227
$\mathcal{G}_{\chi'}$	A generalized Green function on $\mathsf{L}_{\mathrm{A}_2}(\mathfrak{f})$ arising from a cuspidal local system.	227
$\mathcal{G}_{\chi''}$	A generalized Green function on $\mathsf{L}_{\mathrm{A}_2}(\mathfrak{f})$ arising from a cuspidal local system.	227
$\mathbf{G}$	The connected reductive algebraic k-split group of type G_2.	208
G	The group of k-rational points of $\mathbf{G}$.	208
G_0	The set of compact elements in G.	212
G_{0+}	The set of topologically unipotent elements in G.	212
$\mathbf{G}(K)_x$	The parahoric subgroup of $\mathbf{G}(K)$ corresponding to x.	209
$\mathbf{G}(K)_x^+$	The pro-unipotent radical of $\mathbf{G}(K)_x$.	209
G_x	The parahoric subgroup attached to x.	211
G_x^+	The prounipotent radical of the parahoric G_x.	211
G_F	The connected reductive F group attached to the facet F.	210
$\mathbf{g}$	The Lie algebra of $\mathbf{G}$.	208
$\mathfrak{g}$	The vector space of k-rational points of $\mathbf{g}$.	208
$\mathfrak{g}_0$	The set of compact elements in $\mathfrak{g}$.	212
$\mathfrak{g}_{0+}$	The set of topologically nilpotent elements in $\mathfrak{g}$.	212
$\mathbf{g}(K)_x$	A lattice in $\mathbf{g}(K)$ associated to x.	211
$\mathbf{g}(K)_x^+$	A sublattice of $\mathbf{g}(K)_x$.	211
$\mathfrak{g}_x$	A lattice in $\mathfrak{g}$ associated to x.	211
$\mathfrak{g}_x^+$	A sublattice of $\mathfrak{g}_x$.	211
$\mathfrak{g}_F(i)$	The intersection of $\mathfrak{g}_F$ with the i-eigenspace in $\mathfrak{g}$ for μ.	234
Γ	The Galois group $\mathrm{Gal}(K/k)$.	208
$\Gamma_{(F,e)}$	The generalized Gelfand–Graev character for (f,h,e).	246
h	The Coxeter number.	231
H_δ	A Chevalley basis element.	208
I^c	A subset of I^G.	220
I^d	A subset of I^n.	246
I^f	The set of pairs (F,f), where F is a facet and $f \in C(\mathsf{L}_F(\mathfrak{f}))$.	217

I^G	The set of pairs $(F, \mathcal{G})$ with $\mathcal{G} \in C(\mathsf{L}_F(\mathfrak{f}))$ a generalized Green function.	220
I^n	The set of pairs (F, e), where F is a facet and $e \in \mathsf{L}_F(\mathfrak{f})$ is nilpotent.	220
Ind	The induction map.	215
invariant		206
$J(\mathfrak{g})$	The space of invariant distributions on $\mathfrak{g}$.	217
$J(\mathfrak{g}_0)$	The subspace of $J(\mathfrak{g})$ consisting of distributions supported on $\mathfrak{g}_0$.	221
$J(\mathcal{N})$	The span of the nilpotent orbital integrals.	206
$J^{\mathrm{st}}(\mathcal{N})$	The set of stable distributions supported on $\mathcal{N}$.	207
$J^{\mathrm{st}}(\mathfrak{g})$	The space of stable distributions on $\mathfrak{g}$.	206
k	A characteristic zero nonarchimedean local field.	208
K	A fixed maximal unramified extension of k.	208
L_F	The Lie algebra of G_F.	211
L_G	The Lie algebra of G.	214
Λ	An additive character of k that descends to an additive character of $\mathfrak{f}$.	208
lifts		224
local expansion		253
(M_F)	The conjugacy class of Levi subgroups of G corresponding to F.	212
minisotropic		218
$\mu_\mathcal{O}$	The orbital integral attached to the orbit $\mathcal{O}$.	245
μ	The one-parameter subgroup associated to an $\mathfrak{sl}_2(k)$-triple.	228
μ_Y	The orbital integral associated to the semisimple element Y.	214
$\bar{\mu}$	The one-parameter subgroup associated to an $\mathfrak{sl}_2(\mathfrak{f})$-triple.	228
$\mathcal{N}$	The set of nilpotent elements in $\mathfrak{g}$.	206
$\mathcal{N}_F$	The set of nilpotent elements in $\mathsf{L}_F(\mathfrak{F})$.	213
ν	A nontrivial discrete valuation.	208
n_α	An involution in $N_G(T)$ corresponding to the root α.	229
$\bar{n}_\alpha$	The image in the Weyl group of the involution $n_\alpha \in N_G(T)$.	229
$\mathcal{O}(F, e)$	The nilpotent orbit attached to $(F, e) \in I^d$.	246
$\mathcal{O}_0$	$\mathcal{O}(F_{\mathrm{G}_2}, e_0)$.	246
$\mathcal{O}_1$	$\mathcal{O}(F_{\mathrm{G}_2}, e_1)$.	246
$\mathcal{O}_1'$	$\mathcal{O}(F_{\mathrm{G}_2}, e_1')$.	246
$\mathcal{O}_1''$	$\mathcal{O}(F_{\mathrm{G}_2}, e_1'')$.	246
$\mathcal{O}_1^\pm$	$\mathcal{O}(F_{\mathrm{A}_1 \times \tilde{\mathrm{A}}_1}, e_1^\pm)$.	246
$\mathcal{O}_1^\delta$	$\mathcal{O}(F_{\mathrm{A}_2}, e_1^\delta)$.	246
$\wp$	The prime ideal of R.	208
Φ	The root system of type G_2.	208
Ψ	The set of affine roots with respect to $\mathbf{T}$, $\mathbf{G}$, and ν.	209
$\dot{\psi}$	The gradient of the affine root ψ.	209
Q_S^F	A toric Green function.	222
r	The restriction map.	215
R	The ring of integers of k.	208

R_K	The ring of integers of K.	208
res	Restriction.	221
regular semisimple		213
σ	A fixed topological generator for Γ.	208
special orbit		245
stable		206
$\mathbf{T}$	A fixed maximal k-split torus in $\mathbf{G}$.	208
T	The group of k-rational points of $\mathbf{T}$.	208
$T_C(\kappa)$	A combination of toric Green functions.	240
τ	A character of the Weyl group.	251
tor	Torsion.	240
$T_{(F,\mathcal{G})}$	$\int_G \int_{\mathfrak{g}} f({}^g Z) \cdot \mathcal{G}_F(Z)\, dZ\, dg$.	233
U_ϕ	The root group in $\mathbf{G}(K)$ corresponding to the root ϕ.	210
U_ψ	The subgroup of $U_{\dot{\psi}}$ corresponding to the affine root ψ.	210
$\mathcal{U}_F$	The set of unipotent elements in $\mathsf{G}_F(\mathfrak{F})$.	213
ϖ	A uniformizer for k.	208
W_x	The absolute Weyl group of G_x.	252
X_ϕ	A Chevalley basis element.	208
$X_{(F,\mathsf{S})}^{(x,e)}$	Values related to Green polynomials.	251
X_{S}	An element of $\mathfrak{g}_F$ such that its image in $\mathsf{L}_F(\mathfrak{f})$ has centralizer S in G_F.	226
X_1	A subregular nilpotent element of $\mathfrak{g}$.	228
X_1^+	A subregular nilpotent element of $\mathfrak{g}$.	228
X_1^1	A subregular nilpotent element of $\mathfrak{g}$.	228
χ	A character of the Weyl group.	251
ξ	A nontrivial element of μ_3.	243
$Y_\kappa(x,e)$		252

Acknowledgments. The first author was supported by National Science Foundation grant 0345121. The second author was supported by the ISF. Part of this work was performed while the first author was at the Department of Mathematics of Harvard University.

References

[1] J. D. Adler, Refined anisotropic K-types and supercuspidal representations, *Pacific J. Math.*, **185** (1998), 1–32.

[2] J. Adler and S. DeBacker, Some applications of Bruhat-Tits theory to harmonic analysis on the Lie algebra of a reductive p-adic group, *Michigan Math. J.*, **50**-2 (2002), 263–286.

[3] J. Adler and S. DeBacker, A generalization of a result of Kazhdan and Lusztig, *Proc. Amer. Math. Soc.*, **132**-6 (2004), 1861–1868 (electronic).

[4] J. Adler and S. DeBacker, Murnaghan-Kirillov theory for supercuspidal representations of tame general linear groups, *J. Reine Angew. Math.*, **575** (2004), 1–35.

[5] J. Adler and A. Roche, An intertwining result for p-adic groups, *Canad. J. Math.*, **52**-3 (2000), 449–467.

[6] D. Barbasch and A. Moy, Local character expansions, *Ann. Sci. École Norm. Sup.* (4), **30**-5 (1997), 553–567.
[7] A. Borel, *Linear Algebraic Groups*, 2nd enlarged ed., Graduate Texts in Mathematics 126, Springer-Verlag, New York, 1991.
[8] F. Bruhat and J. Tits, Groupes réductifs sur un corps local II: Schémas en groupes. Existence d'une donnée radicielle valuée, *Inst. Hautes Études Sci. Publ. Math.*, **60** (1984), 197–376.
[9] F. Bruhat and J. Tits, Groupes algébriques sur un corps local, Chapitre III: Compléments et applications la cohomologie galoisienne, *J. Fac. Sci. Univ. Tokyo Sect.* IA *Math.*, **34**-3 (1987), 671–698.
[10] R. Carter, *Finite Groups of Lie Type: Conjugacy Classes and Complex Characters*, reprint of the 1985 original, Wiley Classics Library, Wiley, Chichester, UK, 1993.
[11] S. DeBacker, Homogeneity results for invariant distributions of a reductive p-adic group, *Ann. Sci. École Norm. Sup.* (4), **35**-3 (2002), 391–422.
[12] S. DeBacker, Parametrizing nilpotent orbits via Bruhat-Tits theory, *Ann. Math.* (2), **156**-1 (2002), 295–332.
[13] S. DeBacker, Parametrizing conjugacy classes of maximal unramified tori, *Michigan Math. J*, to appear.
[14] Harish-Chandra, *Harmonic Analysis on Reductive p-Adic Groups*, notes by G. van Dijk, Lecture Notes in Mathematics 162, Springer-Verlag, Berlin, New York, 1970.
[15] Harish-Chandra, *Admissible Invariant Distributions on Reductive p-Adic Groups*, preface and notes by S. DeBacker and P. J. Sally, Jr., University Lecture Series 16, American Mathematical Society, Providence, RI, 1999.
[16] R. Howe, Two conjectures about reductive p-adic groups, in C. C. Moore, ed., *Harmonic Analysis on Homogeneous Spaces*, Proceedings of Symposia in Pure Mathematics 26, American Mathematical Society, Providence, RI, 1973, 377–380.
[17] R. Howlett and G. Lehrer, On Harish-Chandra induction and restriction for modules of Levi subgroups, *J. Algebra*, **165**-1 (1994), 172–183.
[18] N. Kawanaka, Shintani lifting and Gelfand-Graev representations, in P. Fong, ed., *The Arcata Conference on Representations of Finite Groups*, Proceedings of Symposia in Pure Mathematics 47, Part 1, American Mathematical Society, Providence, RI, 1987, 147–163.
[19] D. Kazhdan, Proof of Springer's hypothesis, *Israel J. Math.*, **28** (1977), 272–286.
[20] D. Kazhdan and G. Lusztig, Fixed point varieties on affine flag manifolds, *Israel J. Math.*, **62**-2 (1988), 129–168.
[21] G. Lusztig, Fourier transforms on a semisimple Lie algebra over F_q, in W. H. Hesselink, W. L. J. Van Der Kallen, A. M. Cohen, T. A. Springer, J. R. Strooker, and A. M. Cohen, eds., *Algebraic Groups: Utrecht* 1986, Lecture Notes in Mathematics 1271, Springer-Verlag, Berlin, 1987, 177–188.
[22] G. Lusztig, Green functions and character sheaves, *Ann. Math.*, **131** (1990), 355–408.
[23] G. Lusztig, A unipotent support for irreducible representations, *Adv. Math.*, **94**-2 (1992), 139–179.
[24] A. Moy and G. Prasad, Unrefined minimal K-types for p-adic groups, *Invent. Math.*, **116** (1994), 393–408.
[25] N. Spaltenstein, On the Kazhdan-Lusztig map for exceptional Lie algebras, *Adv. Math.*, **83**-1 (1990), 48–74.
[26] T. Springer, Trigonometric sums, Green functions of finite groups and representations of Weyl groups, *Invent. Math.*, **36** (1976), 173–207.
[27] J.-L. Waldspurger, Transformation de Fourier et endoscopie, *J. Lie Theory*, **10**-1 (2000), 195–206.

[28] J.-L. Waldspurger, *Intégrales orbitales nilpotentes et endoscopie pour les groupes classiques non ramifiés*, Astérisque 269, Société Mathématique de France, Paris, 2001.
[29] J.-L. Waldspurger, Quelques résultats de finitude conçernant les distributions invariantes sur les algèbres de Lie p-adiques, preprint, 1993.

Infinite-Dimensional Vector Bundles in Algebraic Geometry

An Introduction

Vladimir Drinfeld

Department of Mathematics
University of Chicago
5734 University Avenue
Chicago, IL 60637
USA
drinfeld@math.uchicago.edu

To Izrail Moiseevich Gelfand with deepest gratitude and admiration.

> "According to the Lamb conjecture, the key to the future development of Quantum Field Theory is probably buried in some forgotten paper published in the 30's. Attempts to follow up this conjecture, however, will probably be unsuccessful because of the Peierls-Jensen paradox; namely, that even if one finds the right paper, the point will probably be missed until it is found independently and accidentally by experiment."
>
> "The Future of Field Theory," by Pure Imaginary Observer [PI]

Subject Classifications: 14F05, 14F42, 14C35

1 Introduction

1.1 Subject of the article

The goal of this work is to show that there is a reasonable algebro-geometric notion of vector bundle with infinite-dimensional locally linearly compact fibers and that these objects appear "in nature." Our approach is based on some results and ideas discovered in algebra during the period 1958–1972 by H. Bass, L. Gruson, I. Kaplansky, M. Karoubi, and M. Raynaud.

This article contains definitions and formulations of the main theorems, but practically no proofs. A detailed exposition will appear in [Dr].

1.2 Conventions

We use the words *S-family of vector spaces* as shorthand for "vector bundle on a scheme S" and *Tate space* as shorthand for "locally linearly compact vector space."

1.3 Overview of the results and structure of the article

1.3.1 General theory

In Section 2 we recall the Raynaud–Gruson theorem on the local nature of projectivity, which shows that there is a good notion of family of discrete infinite-dimensional vector spaces.

In Section 3 we introduce the notion of a *Tate module* over an arbitrary ring R and show that if R is commutative one thus gets a reasonable notion of S-family of Tate spaces, $S = \operatorname{Spec} R$. One has to take into account that K_0 of the additive category of Tate R-modules may be nontrivial. In fact, it equals $K_{-1}(R)$. We show that $K_{-1}(R) = 0$ if R is Henselian. We give a proof of this fact because it explains the fundamental role of the Nisnevich topology in this work. We discuss the notions of dimension torsor and determinant gerbe of a family of Tate spaces.

At least technically, the theory of Tate R-modules is based on the notion of *almost projective* module, which is introduced in Section 4. Roughly speaking, a module is almost projective if it is projective up to finitely generated modules. Unlike Tate modules, almost projective modules are discrete. Any Tate module can be represented as the projective limit of a filtering projective system of almost projective modules with surjective transition maps.

Section 5 is devoted to the canonical central extension of the automorphism groups of almost projective and Tate R-modules. In Section 5.5 we discuss an interesting (although slightly vague) picture, which I learned from A. Beilinson.

1.3.2 Application to the space of formal loops

In Section 6 we define a class of *Tate-smooth* ind-schemes (morally, these are smooth infinite-dimensional algebraic manifolds modeled on Tate spaces). According to Theorem 6.3, the ind-scheme of formal loops of a smooth affine manifold Y over the local field $k((t))$ is Tate-smooth over k. This is one of our main results. In Section 6.10 we use it to define a "refined" version of the motivic integral of a differential form on Y with no zeros over the intersection of Y with a polydisk. Unlike the usual motivic integral, the "refined" one is an object of a triangulated category rather than an element of a group.

1.3.3 Application to vector bundles on a manifold with punctures

In Section 7 we first show that almost projective and Tate modules appear naturally in the study of the cohomology of a family of finite-dimensional vector bundles on a punctured smooth manifold. Then we briefly explain how the canonical central extension that comes from this cohomology allows one (in the case of $GL(n)$-bundles) to interpret the "Uhlenbeck compactification" constructed in [FGK, BFG] as the fine moduli space of a certain type of generalized vector bundles on $\mathbb{P}^2$ (we call them *gundles*). In fact, the application to the "Uhlenbeck compactification" was one of the main motivations of this work.

2 Families of discrete infinite-dimensional vector spaces (after Kaplansky, Raynaud, and Gruson)

Is there a reasonable notion of not necessarily finite-dimensional vector bundle on a scheme? We know due to Serre [S] that a finite-dimensional vector bundle on an affine scheme Spec R is the same as a finitely generated projective R-module. So it is natural to give the following definition.

Definition. A vector bundle on a scheme X is a quasicoherent sheaf of $\mathcal{O}_X$-modules $\mathcal{F}$ such that for every open affine subset Spec $R \subset X$ the R-module $H^0(\operatorname{Spec} R, \mathcal{F})$ is projective.

Key Question. is this a local notion? More precisely, the question is as follows. Let Spec $R = \bigcup_i U_i$, $U_i =$ Spec R_i. Let M be a (not necessarily finitely generated) R-module such that $M \otimes_R R_i$ is projective for all i. Does it follow that M is projective?

The question is difficult: the arguments used in the case that M is finitely generated fail for modules of infinite type. Nevertheless, Grothendieck [Gr2, Remark 9.5.8] conjectured that the answer is positive. This was proved by Raynaud and Gruson [RG] (in Chapter 1 for countably generated modules and in Chapter 2 for arbitrary ones). Moreover, they proved the following theorem, which says that projectivity is a local property for the fpqc topology (not only for Zariski).

Theorem 2.1. *Let M be a module over a commutative ring R and R' be a flat commutative R-algebra such that the morphism* Spec $R' \to$ Spec R *is surjective. If $R' \otimes_R M$ is projective, then M is.*

In fact, they derived it as an easy corollary of the following remarkable and nontrivial theorem due to Kaplansky [Ka] and Raynaud–Gruson [RG], which explains what projectivity really is. Theorem 2.1 follows from the fact that for commutative rings properties (a)–(c) below are local.

Theorem 2.2. *Let R be a (not necessarily commutative) ring. An R-module M is projective if and only if the following properties hold:*

(a) *M is flat;*
(b) *M is a direct sum of countably generated modules;*
(c) *M is a Mittag–Leffler module.*

The fact that a projective module can be represented as a direct sum of countably generated ones was proved by Kaplansky [Ka].

The remaining part of Theorem 2.2 is due to Raynaud and Gruson [RG]. The key notion of Mittag–Leffler module was introduced in [RG]. Here I prefer only to explain what a flat Mittag–Leffler module is. By the Govorov–Lazard lemma [Gov, Laz], a flat R-module M can be represented as the inductive limit of a directed family of finitely generated projective modules P_i. According to [RG], in this situation M is Mittag–Leffler if and only if the projective system formed by the dual (right) R-modules $P_i^* := \operatorname{Hom}_R(P_i, R)$ satisfies the Mittag–Leffler condition: for every i there exists $j \geq i$ such that $\operatorname{Im}(P_j^* \to P_i^*) = \operatorname{Im}(P_k^* \to P_i^*)$ for all $k \geq j$.

Remarks.

(i) One gets a slightly different definition of not necessarily finite-dimensional vector bundle on a scheme if one replaces projectivity by the property of being a flat Mittag–Leffler module. The product of infinitely many copies of $\mathbb{Z}$ is an example of a flat Mittag–Leffler $\mathbb{Z}$-module which is not projective (it is due to Baer; see [Ka2, pp. 48 and 82]). Unlike projectivity, the property of M being a flat Mittag–Leffler module is a first-order property (in the sense of mathematical logic) of $R^{(\mathbb{N})} \otimes_R M$ viewed as a module over $\operatorname{End}_R R^{(\mathbb{N})}$ (here $R^{(\mathbb{N})}$ is the right R-module freely generated by $\mathbb{N}$). Let me also mention that one does not need AC (the axiom of choice) to prove that a vector space over a field is a flat Mittag–Leffler module, but in set theory without AC one cannot prove that $\mathbb{R}$ is a direct summand of a free $\mathbb{Q}$-module[1] (one cannot even prove the existence of a $\mathbb{Q}$-linear embedding of $\mathbb{R}$ into a free $\mathbb{Q}$-module F, for given such an embedding and using a $\mathbb{Q}$-linear retraction $F \to \mathbb{Q}$ one would get a splitting $s : \mathbb{R}/\mathbb{Q} \to \mathbb{R}$ of the exact sequence $0 \to \mathbb{Q} \to \mathbb{R} \to \mathbb{R}/\mathbb{Q} \to 0$ and therefore a nonmeasurable subset $s(\mathbb{R}/\mathbb{Q}) \subset \mathbb{R}$, but it is known [So] that the existence of such a subset cannot be proved in set theory without AC).

(ii) Instead of property (c) from Theorem 2.2 the authors of [RG] used a slightly different one, which is harder to formulate. Probably their property has some technical advantages.

(iii) Here are some more comments regarding the work [RG]. First, there is no evidence that the authors of [RG] knew that Theorem 2.1 had been conjectured by Grothendieck. Second, their notion of Mittag–Leffler module and their results on infinitely generated projective modules were probably largely forgotten (even though they deserve to be mentioned in algebra textbooks). Probably they were "lost" among many other powerful and important results of [RG] (mostly in the spirit of EGA IV).

3 Families of Tate vector spaces and the K_{-1}-functor

3.1 A class of topological vector spaces

We consider topological vector spaces over a discrete field k.

[1] Without AC, it is not true that any free module F is projective, i.e., every epimorphism $M \to F$ has a section. So without AC, projectivity is not equivalent to being a direct summand of a free module.

Definition. A topological vector space is *linearly compact* if it is the topological dual of a discrete vector space.

Example. $k[[t]] \simeq k \times k \times \cdots = (k \oplus k \oplus \dots)^*$.

A topological vector space V is linearly compact if and only if it has the following three properties:

1. V is complete and Hausdorff,
2. V has a base of neighborhoods of 0 consisting of vector subspaces,
3. each open subspace of V has finite codimension.

Definition. A *Tate space* is a topological vector space isomorphic to $P \oplus Q^*$, where P and Q are discrete.

A topological vector space T is a Tate space if and only if it has an open linearly compact subspace.

Example. $k((t))$ equipped with its usual topology (the subspaces $t^n k[[t]]$ form a base of neighborhoods of 0). This is a Tate space because it is a direct sum of the linearly compact space $k[[t]]$ and the discrete space $t^{-1}k[t^{-1}]$, or because $k[[t]] \subset k((t))$ is an open linearly compact subspace.

Tate spaces play an important role in the algebraic geometry of curves (e.g., the ring of adèles corresponding to an algebraic curve is a Tate space) and also in the theory of ∞-dimensional Lie algebras and Conformal Field Theory. In fact, they were introduced by Lefschetz [L, pp. 78–79], under the name of locally linearly compact spaces. The name "Tate space" was introduced by Beilinson because these spaces are implicit in Tate's remarkable work [T]. In fact, the approach to residues on curves developed in [T] can be most naturally interpreted in terms of the canonical central extension of the endomorphism algebra of a Tate space, which is also implicit in [T].

3.2 What is a family of Tate spaces?

Probably this question has not been considered. We suggest the following answer. In the category of topological modules over a (not necessarily commutative) ring R we define a full subcategory of *Tate R-modules*. If R is commutative, then we suggest considering Tate R-modules as "families of Tate spaces." This viewpoint is justified by Theorems 3.3 and 3.4 below.

3.2.1 Definitions

An *elementary Tate R-module* is a topological R-module isomorphic to $P \oplus Q^*$, where P, Q are discrete projective R-modules (P is a left module, Q is a right one). A *Tate R-module* is a direct summand of an elementary Tate R-module. A Tate R-module M is *quasi-elementary* if $M \oplus R^n$ is elementary for some $n \in \mathbb{N}$.

By definition, a morphism of Tate modules is a continuous homomorphism. The following lemma is very easy.

Lemma 3.1. *Let P, Q be as in the definition of Tate R-module. Then every morphism $Q^* \to P$ has finitely generated image.* □

3.2.2 Examples

1. $R((t))^n$ is an elementary Tate R-module.
2. A finitely generated projective $R((t))$-module M has a unique structure of topological $R((t))$-module such that every $R((t))$-linear morphism $M \to R((t))$ is continuous. This topology is called the *standard topology* of M. Clearly, M equipped with its standard topology is a Tate R-module. In general, it is not quasi-elementary. For example, let k be a field, $R := \{f \in k[x] | f(0) = f(1)\}$ and

$$M := \{u = u(x,t) \in k[x]((t)) \mid u(1,t) = tu(0,t)\}. \tag{3.1}$$

Then M is a finitely generated projective $R((t))$-module which is not quasi-elementary as a Tate R-module (see Section 3.5.3).

Remark. The precise relation between finitely generated projective $R((t))$-modules and Tate R-modules is explained in Theorem 3.10 below.

3.2.3 Lattices and bounded submodules

A submodule L of a topological R-module M is said to be a *lattice* if it is open and L/U is finitely generated for every open submodule $U \subset L$. A subset of a Tate R-module M is *bounded* if it is contained in some lattice. A lattice L in a Tate module M is *coprojective* if M/L is projective.

Remarks.

(i) One can show that a lattice L in a Tate module M is coprojective if and only if M/L is flat.
(ii) In every Tate R-module lattices exist and, moreover, form a base of neighborhoods of 0. On the other hand, a Tate R-module M has a coprojective lattice if and only if M is elementary.

Theorem 3.2. *A Tate R-module M has the following properties:*

(a) *M is complete and Hausdorff;*
(b) *lattices in M form a base of neighborhoods of* 0;
(c) *the functor that associates to a discrete R-module N the group* $\mathrm{Hom}(M, N)$ *of continuous homomorphisms $M \to N$ is exact.*

If a topological R-module M has a countable base of neighborhoods of 0 *and satisfies* (a)–(c), *then it is a Tate R-module.*

Only the last statement of the theorem is nontrivial. The countability assumption is essential in its proof.

Remark. If a topological R-module M satisfies (b), then (c) is equivalent to the following property: for every lattice L there is a lattice $L' \subset L$ such that the morphism $M/L' \to M/L$ admits a factorization $M/L' \to P \to M/L$ for some projective module P.

3.2.4 Duality

The *dual* of a Tate R-module M is defined to be the right R-module M^* of continuous homomorphisms $M \to R$ equipped with the topology whose base is formed by orthogonal complements of open bounded submodules $L \subset M$. Then M^* is again a Tate module, and $M^{**} = M$ (it suffices to check this for elementary Tate modules).

3.2.5 Tate modules as families of Tate spaces

Theorem 3.3. *The notion of Tate module over a commutative ring R is local for the flat topology, i.e., for every faithfully flat commutative R-algebra R' the category of Tate R-modules is canonically equivalent to that of Tate R'-modules equipped with a descent datum.*

The proof is based on the Raynaud–Gruson technique.

Theorem 3.4. *Let R be a commutative ring. Then every Tate R-module M is Nisnevich-locally elementary; in other words, there exists a Nisnevich covering* $\operatorname{Spec} R' \to \operatorname{Spec} R$ *such that $R'\hat{\otimes}_R M$ has a coprojective lattice L'. Moreover, for every lattice $L \subset M$ one can choose R' and L' so that $L' \supset R'\hat{\otimes}_R L$.*

The proof is not hard. A closely related statement (Theorem 3.7) will be proved in Section 3.4.

Let me give the definition of Nisnevich covering. A morphism $\pi : X \to \operatorname{Spec} R$ is said to be a *Nisnevich covering* if it is étale and there exist closed subschemes $\operatorname{Spec} R = F_0 \supset F_1 \supset \cdots \supset F_n = \emptyset$ such that each F_i is defined by finitely many equations and π admits a section over $F_{i-1} \setminus F_i$, $i = 1, \ldots, n$. A morphism $\pi : X \to Y$ is a Nisnevich covering if for every open affine $U \subset Y$ the morphism $\pi^{-1}(U) \to U$ is a Nisnevich covering. (If Y is locally Noetherian, then an étale morphism $X \to Y$ is a Nisnevich covering if and only if it admits a section over each point of Y; this is the usual definition.) The Nisnevich topology is weaker than étale but stronger than Zariski. The following table may be helpful:

Topology	Stalks of $\mathcal{O}_X$, $X = \operatorname{Spec} R$
Zariski	Localizations of R
Nisnevich	Henselizations of R
Ètale	Strict Henselizations of R

3.2.6 Remarks on Theorem 3.4

(i) In Theorem 3.4, *one cannot replace "Nisnevich" by "Zariski."* For example, we will see in Section 3.5.3 that the Tate module (3.1) is not Zariski-locally elementary.

(ii) It is easy to show that every quasi-elementary Tate module over a commutative ring R is Zariski-locally elementary.

3.3 Tate R-modules and $K_{-1}(R)$

How does one see that a Tate R-module is not quasi-elementary? We will assign to each Tate R-module M a class $[M] \in K_{-1}(R)$ such that $[M] = 0$ if and only if M is quasi-elementary. It is easy to define $[M]$ if one uses the following definition of $K_{-1}(R)$.

3.3.1 K_{-1} via Calkin category

First, introduce the following category $\mathcal{C}^{\text{all}} = \mathcal{C}_R^{\text{all}}$: its objects are all R-modules and the group $\operatorname{Hom}_{\mathcal{C}^{\text{all}}}(M, M')$ of $\mathcal{C}^{\text{all}}$-morphisms $M \to M'$ is defined by

$$\operatorname{Hom}_{\mathcal{C}^{\text{all}}}(M, M') := \operatorname{Hom}(M, M')/\operatorname{Hom}_f(M, M'),$$

where $\operatorname{Hom}_f(M, M')$ is the group of R-linear maps $A : M \to M'$ whose image is contained in a finitely generated submodule of M'. Let $\mathcal{C} \subset \mathcal{C}^{\text{all}}$ be the full subcategory whose objects are *projective* modules. The idempotent completion[2] of $\mathcal{C}$ (also known as the Karoubi envelope of $\mathcal{C}$) will be denoted by $\mathcal{C}^{\text{Kar}}$ or $\mathcal{C}_R^{\text{Kar}}$ and will be called the *Calkin category* of R. Let $\mathcal{C}_{\aleph_0} \subset \mathcal{C}$ be the full subcategory of countably generated projective R-modules and $\mathcal{C}_{\aleph_0}^{\text{Kar}}$ its idempotent completion.

Proposition 3.5. *Every object of $\mathcal{C}^{\text{Kar}}$ is stably equivalent[3] to an object of $\mathcal{C}_{\aleph_0}^{\text{Kar}}$. Two objects of $\mathcal{C}_{\aleph_0}^{\text{Kar}}$ are stably equivalent in $\mathcal{C}^{\text{Kar}}$ if and only if they are stably equivalent in $\mathcal{C}_{\aleph_0}^{\text{Kar}}$.*

As a corollary, we see that $K_0(\mathcal{C}^{\text{Kar}})$ is well defined[4] (even though $\mathcal{C}^{\text{Kar}}$ is not equivalent to a small category), and the morphism $K_0(\mathcal{C}_{\aleph_0}^{\text{Kar}}) \to K_0(\mathcal{C}^{\text{Kar}})$ is an isomorphism. Now define $K_{-1}(R)$ by

$$K_{-1}(R) := K_0(\mathcal{C}_R^{\text{Kar}}). \tag{3.2}$$

Remarks.

(i) The above definition of K_{-1} is slightly nonstandard but equivalent to the standard ones.

---

[2] The idempotent completion of a category $\mathcal{B}$ is the category $\mathcal{B}^{\text{Kar}}$ in which an object is a pair $(B, p : B \to B)$ with $B \in \mathcal{B}$ and $p^2 = p$, and a morphism $(B_1, p_1) \to (B_2, p_2)$ is a $\mathcal{B}$-morphism $\varphi : B_1 \to B_2$ such that $p_2\varphi p_1 = \varphi$. This construction was explained by P. Freyd in [Fr, Chapter 2, Exercise B2] a few years before Karoubi.

[3] Objects X, Y of an additive category $\mathcal{A}$ are said to be *stably equivalent* if $X \oplus Z \simeq Y \oplus Z$ for some $Z \in \mathcal{A}$.

[4] K_0 of an additive category $\mathcal{A}$ is defined by the usual universal property. It may exist even if $\mathcal{A}$ is not equivalent to a small category, e.g., K_0 of the category of *all* vector spaces equals 0.

(ii) Define the *algebraic Calkin ring* by

$$\mathrm{Calk}(R) := \mathrm{End}_{\mathcal{C}} R^{(\mathbb{N})} := \mathrm{End}\, R^{(\mathbb{N})} / \mathrm{End}_f R^{(\mathbb{N})}, \quad R^{(\mathbb{N})} := R \oplus R \oplus \cdots .$$

(Calk(R) is an algebraic version of the analysts' Calkin algebra, which is defined to be the quotient of the ring of continuous endomorphisms of a Banach space by the ideal of compact operators). If $P \in \mathcal{C}_{\aleph_0}^{\mathrm{Kar}}$, then $\mathrm{Hom}_{\mathcal{C}^{\mathrm{Kar}}}(P, R^{(\mathbb{N})})$ is a finitely generated projective module over Calk(R). Thus one gets an antiequivalence between $\mathcal{C}_{\aleph_0}^{\mathrm{Kar}}$ and the category of finitely generated projective Calk(R)-modules, which induces an isomorphism

$$K_{-1}(R) \xrightarrow{\sim} K_0(\mathrm{Calk}(R)).$$

3.3.2 The class of a Tate R-module

Let $\mathcal{T}_R$ denote the additive category of Tate R-modules. We will define a functor

$$\Phi : \mathcal{T}_R \to \mathcal{C}_R^{\mathrm{Kar}}. \tag{3.3}$$

Let $E_R \subset \mathcal{T}_R$ be the full subcategory of elementary Tate modules. One gets a functor $\Psi : E_R \to \mathcal{C}_R$ by setting $\Psi(P \oplus Q^*) := P$ (here P, Q are discrete projective modules) and defining $\Psi(f) \in \mathrm{Hom}_{\mathcal{C}}(P, P_1)$, $f : P \oplus Q^* \to P_1 \oplus Q_1^*$, to be the image of the composition $P \hookrightarrow P \oplus Q^* \xrightarrow{f} P_1 \oplus Q_1^* \twoheadrightarrow P_1$ in $\mathrm{Hom}_{\mathcal{C}}(P, P_1)$. (The equality $\Psi(f'f) = \Psi(f')\Psi(f)$ follows from Lemma 3.1.) The functor (3.3) is defined to be the extension of $\Psi : E_R \to \mathcal{C}_R \subset \mathcal{C}_R^{\mathrm{Kar}}$ to $\mathcal{T}_R = E_R^{\mathrm{Kar}}$.

Now define the class $[M]$ of a Tate R-module M by $[M] := [\Phi(M)] \in K_0(\mathcal{C}_R^{\mathrm{Kar}}) = K_{-1}(R)$.

3.3.3 K_0 of the category of Tate R-modules

Theorem 3.6.

(i) *A Tate R-module has zero class in $K_{-1}(R)$ if and only if it is quasi-elementary.*
(ii) *$K_0(\mathcal{T}_R)$ is well defined (even though $\mathcal{T}_R$ is not equivalent to a small category).*
(iii) *The morphism $K_0(\mathcal{T}_R) \to K_0(\mathcal{C}_R^{\mathrm{Kar}}) = K_{-1}(R)$ induced by* (3.3) *is an isomorphism.*
(iv) *Every element of $K_0(\mathcal{T}_R) = K_{-1}(R)$ can be represented as the class of $R((t)) \otimes_{R[t,t^{-1}]} P$ for some finitely generated projective $R[t,t^{-1}]$-module P.*

Remark. The only nontrivial point of the proof is the surjectivity of the composition

$$K_0(R[t,t^{-1}]) \to K_0(R((t))) \to K_0(\mathcal{T}_R) \to K_{-1}(R), \tag{3.4}$$

which is used in the proof of (iii) and (iv) (in fact, to prove (iii) it suffices to use Theorem 4.1(a) below). The surjectivity of (3.4) is a standard fact[5] from K-theory. It is proved by noticing that there is a canonical section $K_{-1}(R) \to K_0(R[t,t^{-1}])$, namely, multiplication by the canonical element of $K_1(\mathbb{Z}[t,t^{-1}])$.

[5] The surjectivity of (3.4) is a tautology if one uses the definition of K_{-1} given by H. Bass [Ba]. But it is a theorem if one defines K_{-1} by (3.2).

3.4 Nisnevich-local vanishing of K_{-1}

Theorem 3.4 is closely related[6] to the following theorem, which I was unable to find in the literature.

Theorem 3.7. *Let R be a commutative ring. Then every element of $K_{-1}(R)$ vanishes Nisnevich-locally.*

Remarks.

(i) According to [We2, Example 8.5] (which goes back to L. Reid's work [Re]), it is *not true* that every element of $K_i(R)$, $i < -1$, vanishes Nisnevich-locally.
(ii) It is known that K_{-1} commutes with filtering inductive limits. So Theorem 3.7 is equivalent to vanishing of $K_{-1}(R)$ for commutative Henselian rings R. I prefer the above formulation of the theorem because commutation of K_{-1} with filtering inductive limits is not immediate if one defines K_{-1} by (3.2), i.e., via the Calkin category.

In the proof of Theorem 3.7 given below we use the definition of K_{-1} from Section 3.3.1, but it is also easy to prove the theorem using the definition of K_{-1} given by H. Bass [Ba].

Proof. It suffices to show that if P is an R-module,[7] $F \subset P$ is a finitely generated submodule, and $\pi \in \operatorname{End} P$ is such that $\operatorname{Im}(\pi^2 - \pi) \subset F$ then after Nisnevich localization there exists $\widetilde{\pi} \in \operatorname{End} P$ such that $\widetilde{\pi}^2 = \widetilde{\pi}$ and $\operatorname{Im}(\widetilde{\pi} - \pi) \subset F$.

The idea is to look at the spectrum of π. There exists a monic $f \in R[\lambda]$ such that $f(\pi^2 - \pi)$ annihilates F. Then $f(\pi^2 - \pi)(\pi^2 - \pi) = 0$. Put $g(\lambda) := (\lambda^2 - \lambda)f(\lambda^2 - \lambda)$; then there is a unique morphism $R[\lambda]/(g) \to \operatorname{End} P$ such that $\lambda \mapsto \pi$. Put $S := \operatorname{Spec} R[\lambda]/(g) \subset \operatorname{Spec} R \times \mathbb{A}^1$; then $S \supset \underline{0} \cup \underline{1}$, where $\underline{0} = \operatorname{Spec} R \times \{0\}$ and $\underline{1} = \operatorname{Spec} R \times \{1\}$.

Suppose we have a decomposition

$$S = S_0 \amalg S_1, \quad S_i \text{ open}, \quad S_0 \supset \underline{0}, \quad S_1 \supset \underline{1}. \tag{3.5}$$

Then we can define $e \in R[\lambda]/(g) = H^0(S, \mathcal{O}_S)$ by $e|_{S_0} = 0$, $e|_{S_1} = 1$ and define $\widetilde{\pi}$ to be the image of e in $\operatorname{End} P$.

Claim: A decomposition (3.5) *exists Nisnevich-locally on* $\operatorname{Spec} R$. Indeed, according to the table at the end of Section 3.2.5, it suffices to show that this decomposition exists if R is Henselian. Let $\bar{g} \in (R/m)[\lambda]$ be the reduction of g modulo the maximal ideal $m \subset R$. To get (3.5) it suffices to choose a factorization $\bar{g} = \bar{g}_0\bar{g}_1$ such that $\bar{g}_0, \bar{g}_1$ are coprime, $\bar{g}_0(0) = 0$, $\bar{g}_1(1) = 0$ and then lift it to a factorization $g = g_0 g_1$. □

[6] More precisely: Theorem 3.7 follows from Theorems 3.4 and 3.6(iii); Theorem 3.4 follows from Theorems 3.7 and 4.2(iii).

[7] We need only the case that P is projective, but projectivity is not used in what follows.

3.5 The dimension torsor

Let R be commutative. Then it follows from [We2, Theorem 8.5] that there is a canonical epimorphism $K_{-1}(R) \to H^1_{\text{ét}}(\operatorname{Spec} R, \mathbb{Z})$, so a Tate R-module M should define $\alpha_M \in H^1_{\text{ét}}(\operatorname{Spec} R, \mathbb{Z})$. We will define α_M explicitly as a class of a certain $\mathbb{Z}$-torsor Dim_M on $\operatorname{Spec} R$ canonically associated to M. Dim_M is called "the torsor of dimension theories" or "dimension torsor."

3.5.1 The case that R is a field

If M is a Tate vector space over a field R the notion of dimension torsor is well known.[8] Notice that if $L \subset M$ is open and linearly compact, then usually $\dim L = \infty$ and $\dim(M/L) = \infty$. But for any open linearly compact $L, L' \subset M$ one has the *relative dimension* $d_L^{L'} := \dim(L'/L' \cap L) - \dim(L/L' \cap L) \in \mathbb{Z}$.

Definition. A *dimension theory* on a Tate vector space M is a function

$$d : \{\text{open linearly compact subspaces } L \subset M\} \to \mathbb{Z}$$

such that $d(L') - d(L) = d_L^{L'}$.

A dimension theory exists and is unique up to adding $n \in \mathbb{Z}$. So dimension theories on a Tate space form a $\mathbb{Z}$-torsor. This is Dim_M.

Example. Let T be a $\mathbb{Z}$-torsor, let $R^{(T)}$ be the vector space over a field R freely generated by T. Then $\mathbb{Z}$ acts on $R^{(T)}$, so $R^{(T)}$ becomes a $R[z, z^{-1}]$-module (multiplication by z coincides with the action of $1 \in \mathbb{Z}$). Put $M := R((z)) \otimes_{R[z,z^{-1}]} R^{(T)}$. Then one has a canonical isomorphism

$$\operatorname{Dim}_M \xrightarrow{\sim} T : \tag{3.6}$$

to $t \in T$ one associates the dimension theory d_t such that $d_t(L_t) = 0$, where $L_t \subset M$ is the $R[[z]]$-subspace generated by t.

3.5.2 The general case

If M is a Tate module and $L \subset L' \subset M$ are coprojective lattices then L'/L is a finitely generated projective R-module, so if R is commutative then $d_L^{L'} := \operatorname{rank}(L'/L) \in H^0(\operatorname{Spec} R, \mathbb{Z})$ is well defined.

Definition. Let M be a Tate module over a commutative ring R. A *dimension theory* on M is a rule that associates to each R-algebra R' and each coprojective lattice $L \subset R' \hat{\otimes}_R M$ a locally constant function $d_L : \operatorname{Spec} R' \to \mathbb{Z}$ in a way compatible with base change and so that $d_{L_2} - d_{L_1} = \operatorname{rank}(L_2/L_1)$ for any pair of coprojective lattices $L_1 \subset L_2 \subset R' \hat{\otimes}_R M$. Here $R' \hat{\otimes}_R M$ denotes the completed tensor product.

[8] I copied the definition below from [Ka3], but the notion goes back at least to the physical concept of "Dirac sea," which many years later became the "infinite wedge construction" in the representation theory of infinite-dimensional Lie algebras.

Theorem 3.4 implies that if the functions d_L with the above properties are defined for all étale R-algebras, then there exists a unique way to extend the definition to all R-algebras. It also shows that dimension theories form a $\mathbb{Z}$-torsor for the Nisnevich topology.[9] It is called the *dimension torsor* and denoted by Dim_M.

One has a canonical isomorphism

$$\mathrm{Dim}_{M_1 \oplus M_2} \xrightarrow{\sim} \mathrm{Dim}_{M_1} + \mathrm{Dim}_{M_2} . \tag{3.7}$$

So one gets a morphism $K_0(\mathcal{T}_R) = K_{-1}(R) \to H^1_{\text{ét}}(\operatorname{Spec} R, \mathbb{Z})$. It is surjective. Indeed, let T be a $\mathbb{Z}$-torsor on $S := \operatorname{Spec} R$. Then the free $\mathcal{O}_S$-module $\mathcal{O}_S^{(T)}$ generated by the sheaf of sets T is equipped with an action of $\mathbb{Z}$, so it is a module over $\mathcal{O}_S[z, z^{-1}]$ (multiplication by z coincides with the action of $1 \in \mathbb{Z}$). This module is locally free of rank one, so its global sections form a projective $R[z, z^{-1}]$-module $R^{(T)}$ of rank 1. Therefore, $R((z)) \otimes_{R[z,z^{-1}]} R^{(T)}$ is a Tate R-module. Its dimension torsor is canonically isomorphic to T (cf. (3.6)).

3.5.3 Example

Let M be the Tate module (3.1) over $R := \{f \in k[x] | f(0) = f(1)\}$. Then the $\mathbb{Z}$-torsor Dim_M is nontrivial (its pullback to $S := \operatorname{Spec}(R \otimes_k \bar{k})$ corresponds to the universal covering of S). So the class of M in $K_0(\mathcal{T}_R) = K_{-1}(R)$ is nontrivial and therefore M is not quasi-elementary. Moreover, it does not become quasi-elementary after Zariski localization.

3.5.4 The kernel of the morphism $K_{-1}(R) \to H^1_{\text{ét}}(\operatorname{Spec} R, \mathbb{Z})$ may be nonzero

Moreover, this can happen even if R is local. Examples can be found in [We3]. More precisely, [We3, Section 6] contains examples of algebras R over a field k such that $H^1_{\text{ét}}(\operatorname{Spec} R, \mathbb{Z}) = 0$ but $K_{-1}(R) \neq 0$. In each of these examples $\operatorname{Spec} R$ is a normal surface with one singular point x. Let R_x denote the local ring of x. According to [We1], the map $K_{-1}(R) \to K_{-1}(R_x)$ is an isomorphism, so $K_{-1}(R_x) \neq 0$.

3.6 The determinant gerbe

Given a Tate space M over a field, Kapranov [Ka3] defines its *groupoid of determinant theories*. The definition is based on the notion of relative determinant of two lattices in a Tate space and goes back to J.-L. Brylinski [Br] (and further back to the Japanese school and [ACK]). If M is a Tate module over a commutative ring R, then rephrasing the definition from [Ka3] in the obvious way one gets a sheaf of groupoids on the Nisnevich topology of $S := \operatorname{Spec} R$ (details will be explained in Section 5). This sheaf of groupoids is, in fact, an $\mathcal{O}_S^\times$-gerbe. We call it *the determinant gerbe of M*. Associating the class of this gerbe to a Tate R-module M one gets a morphism

[9] In fact, the categories of $\mathbb{Z}$-torsors for the Nisnevich, étale, fppf, and fpqc topologies are equivalent.

$$K_0(\mathcal{T}_R) = K_{-1}(R) \to H^2_{\mathrm{Nis}}(S, \mathcal{O}_S^\times). \tag{3.8}$$

Probably this is well known to K-theorists. One can get the restriction of (3.8) to $\mathrm{Ker}(K_{-1}(R) \to H^1_{\text{ét}}(\mathrm{Spec}\, R, \mathbb{Z}))$ (and possibly the morphism (3.8) itself) from the Brown–Gersten–Thomason spectral sequence [TT, Section 10.8]. More details on the determinant gerbes will be given in Section 5.

3.7 Co-Sato Grassmannian

Let M be a Tate module over a commutative ring R. The *co-Sato Grassmannian* of M is the following functor Gras_M from the category of commutative R-algebras R' to that of sets: $\mathrm{Gras}_M(R')$ is the set of coprojective lattices in $R'\widehat{\otimes}_R M$. Given lattices $L \subset M$ and $\tilde{L} \subset M^*$ let $\mathrm{Gras}_M^{L,\tilde{L}}(R') \subset \mathrm{Gras}_M(R')$ be the set of coprojective lattices in $R'\widehat{\otimes}_R M$ containing $R'\widehat{\otimes}_R L$ and orthogonal to $R'\widehat{\otimes}_R \tilde{L}$. The functor Gras_M is the inductive limit of the subfunctors $\mathrm{Gras}_M^{L,\tilde{L}}$, and these subfunctors form a filtering family. Theorem 3.4 easily implies the following proposition.

Proposition 3.8.

(i) $\mathrm{Gras}_M^{L,\tilde{L}}$ *is an algebraic space proper and of finite presentation over* $\mathrm{Spec}\, R$. *Locally for the Nisnevich topology of* $\mathrm{Spec}\, R$ *it is a projective scheme over* $\mathrm{Spec}\, R$.
(ii) Gras_M *is an ind-algebraic space ind-proper over* $\mathrm{Spec}\, R$.

Remarks.

(a) A standard argument based on the Plücker embedding (see Section 5.4.3) shows that if the determinant gerbe of M is trivial then $\mathrm{Gras}_M^{L,\tilde{L}}$ is projective over $\mathrm{Spec}\, R$ and Gras_M is an ind-projective ind-scheme.
(b) Using Proposition 3.8 it is easy to prove ind-representability and ind-properness of *the $\mathcal{F}$-twisted affine Grassmannian $\mathcal{GR}_{\mathcal{F}}$* of a reductive group scheme G over R. Here $\mathcal{F}$ is a G-torsor on $\mathrm{Spec}\, R((t))$ and $\mathcal{GR}_{\mathcal{F}}$ is the functor that sends a commutative R-algebra R' to the set of extensions of $\mathcal{F} \otimes_{R((z))} R'((z))$ to a G-torsor over $\mathrm{Spec}\, R'[[z]]$ (up to isomorphisms whose restriction to $\mathcal{F} \otimes_{R((z))} R'((z))$ equals the identity).

3.8 Finitely generated projective $R((t))$-modules from the Tate viewpoint

Theorem 3.10 below says that a finitely generated projective $R((t))$-module is the same as a Tate R-module equipped with a topologically nilpotent automorphism. An endomorphism (in particular, an automorphism) of a Tate R-module M is said to be *topologically nilpotent* if it satisfies the equivalent conditions of the next lemma.

Lemma 3.9. *Let M be a Tate R-module, $T \in \mathrm{End}\, M$. Then the following conditions are equivalent:*

(i) $T^n \to 0$ *for* $n \to 0$ *(which means that for every pair of lattices $L, L' \subset M$ there exists N such that $T^n L' \subset L$ for all $n > N$).*

(ii) *There exists a (unique) structure of topological* $R[[t]]$*-module on* M *such that* T *acts as multiplication by* t.

If M is a finitely generated projective $R((t))$-module equipped with its standard topology, then multiplication by t is a topologically nilpotent automorphism of M. The next theorem says that the converse statement is also true.

Theorem 3.10. *Let* M *be a Tate* R*-module and* $T : M \to M$ *be a topologically nilpotent automorphism. Equip* M *with the topological* $R((t))$*-module structure such that* $tm = T(m)$ *for* $m \in M$. *Then* M *is a finitely generated projective* $R((t))$*-module, and the topology on* M *is the standard one.*

Theorem 3.11. *Let* R *be commutative. Then the notion of finitely generated projective* $R((t))$*-module is local for the fpqc topology of* Spec R. *More precisely, let* R' *be a faithfully flat commutative* R*-algebra,* $R'' := R' \otimes_R R'$*, and let* $f, g : R'((t)) \to R''((t))$ *be defined by* $f(a) := 1 \otimes a$, $g(a) := a \otimes 1$*; then the category of finitely generated projective* $R((t))$*-modules is canonically equivalent to that of finitely generated projective* $R'((t))$*-modules* M' *equipped with an isomorphism* $R''((t)) \otimes_f M' \xrightarrow{\sim} R''((t)) \otimes_g M'$ *satisfying the usual cocycle condition.*

This is an immediate corollary of Theorems 3.3 and 3.10.

Remark. If R is of finite type over a field k and the morphism Spec $R' \to$ Spec R is a Zariski covering, then Theorem 3.11 is well known from the theory of nonarchimedian analytic spaces [BGR, Be], which is applicable because $R((t))$ is an affinoid $k((t))$-algebra in the sense of Section 6.5.

3.9 The dimension torsor of a projective $R((t))$-module

Let R be a commutative ring. Let M be a finitely generated projective $R((t))$-module equipped with an isomorphism $\varphi : \det M \xrightarrow{\sim} R((t))$. If R is a field, then M has an $R[[t]]$-stable lattice; moreover, there is a lattice $L \subset M$ such that

$$R[[t]]L \subset L, \qquad \varphi(\det L) = R[[t]]. \tag{3.9}$$

So it is easy to see that if R is a field, then there is a unique dimension theory d_φ on M such that $d_\varphi(L) = 0$ for all lattices $L \subset M$ satisfying (3.9). Therefore, if R is any commutative ring, then the $\mathbb{Z}$-torsor Dim_M is trivialized over each point of Spec R.

Proposition 3.12. *These trivializations come from a (unique) trivialization* d_φ *of the* $\mathbb{Z}$*-torsor* Dim_M.

By Proposition 3.12 the morphism $K_0(R((t))) \to H^1_{\text{ét}}(\mathrm{Spec}\, R, \mathbb{Z})$ that sends the class of a projective $R((t))$-module M to the class of Dim_M annihilates the kernel of the epimorphism $\det : K_0(R((t))) \twoheadrightarrow \mathrm{Pic}\, R((t))$, so we get a morphism

$$f : \mathrm{Pic}\, R((t)) \to H^1_{\text{ét}}(\mathrm{Spec}\, R, \mathbb{Z}) \tag{3.10}$$

such that the diagram

$$\begin{array}{ccc} K_0(R((t))) & \xrightarrow{\det} & \text{Pic } R((t)) \\ \downarrow & \swarrow & \\ H^1_{\text{ét}}(\text{Spec } R, \mathbb{Z}) & & \end{array} \tag{3.11}$$

commutes. The composition

$$g : \text{Pic } R[t, t^{-1}]) \to \text{Pic } R((t)) \xrightarrow{f} H^1_{\text{ét}}(\text{Spec } R, \mathbb{Z}) \tag{3.12}$$

was studied in [We2].

Remarks.

(i) As explained in [We2], the kernels of (3.10) and (3.12) may be nontrivial (even if R is Henselian). *Example*: If k is a field and R is either $k[x^2, x^3] \subset k[x]$ or the Henselization of $k[x^2, x^3]$ at the singular point of its spectrum, then Ker $f \simeq k((t))/k[[t]]$, Ker $g \simeq k[t, t^{-1}]/k[[t]]$ (e.g., to show that Ker $f \simeq k((t))/k[[t]]$ for $R = k[x^2, x^3]$ notice that a line bundle on Spec $R((t))$ is the same as a triple consisting of a line bundle on Spec $k[x]((t))$, a line bundle on Spec $k((t))$ and an isomorphism between their pullbacks to Spec $k[x]((t))/(x^2)$). It is also explained in [We2] that g has a splitting (and therefore f has). Indeed, Pic $R[t, t^{-1}] = H^1_{\text{ét}}(\text{Spec } R, C)$, where C is the derived direct image of the étale sheaf of invertible functions on Spec $R[t, t^{-1}]$, and the morphism $\mathbb{Z} \to C$ defined by $n \mapsto t^n$ gives a splitting.

(ii) The interested reader can easily lift the diagram (3.11) of abelian groups to a commutative diagram of appropriate Picard groupoids (in the sense of [Del, Section 1.4]).

4 Almost projective and 2-almost projective modules

4.1 Main definitions and results

Recall that every Tate R-module has a lattice but not necessarily a coprojective one. If M is a Tate R-module and $L \subset M$ is a lattice (respectively, a bounded open submodule) then M/L is 2-almost projective (respectively, almost projective) in the sense of the following definitions.

Definitions. An *elementary almost projective R-module* is a module isomorphic to a direct sum of a projective R-module and a finitely generated one. An *almost projective R-module* is a direct summand of an elementary almost projective module. An almost projective R-module M is *quasi-elementary* if $M \oplus R^n$ is elementary for some $n \in \mathbb{N}$.

Definition. An R-module M is 2-almost projective if it can be represented as a direct summand of $P \oplus F$ with P a projective R-module and F an R-module of finite presentation.

In fact, there is a reasonable notion of n-almost projectivity for any positive n; see Remark (3) at the end of this subsection.

Remark. It is easy to show that an almost projective module M is quasi-elementary if and only if it can be represented as P/N with P projective and $N \subset P$ a submodule of a finitely generated submodule of P. It is also easy to show that for P and N as above P/N is 2-almost projective if and only if N is finitely generated.

Theorem 4.1.

(a) *Every almost projective R-module M_0 can be represented as M/L with M being a Tate R-module and $L \subset M$ a bounded open submodule.*
(b) *If M_0 is 2-almost projective, then in such a representation L is a lattice.*

Theorem 4.2.

(i) *The notion of almost projective module over a commutative ring R is local for the flat topology, i.e., for every faithfully flat commutative R-algebra R' almost projectivity of an R-module M is equivalent to almost projectivity of the R'-module $R' \otimes_R M$. The same is true for 2-almost projectivity.*
(ii) *For every almost projective module M over a commutative ring R there exists a Nisnevich covering* Spec $R' \to$ Spec R *such that $R' \otimes_R M$ is elementary.*
(iii) *For every quasi-elementary almost projective module M over a commutative ring R there exists a Zariski covering* $\operatorname{Spec} R = \bigcup_i \operatorname{Spec} R_{f_i}$ *such that $R_{f_i} \otimes_R M$ is elementary for all i.*

The proof of (i) is based on the Raynaud–Gruson technique. The proofs of (ii) and (iii) are much easier. In particular, (iii) easily follows from Kaplansky's Theorem [Ka], which says that a projective module over a local field is free (even if it is not finitely generated!).

Remarks.

(1) In statement (ii) of the theorem *one cannot replace "Nisnevich" by "Zariski."* For example, the quotient of the Tate R-module (3.1) by any open bounded submodule is an almost projective module which is not Zariski-locally elementary (because the Tate module (3.1) is not; see Section 3.5.3).

(2) My impression is that statement (ii) is more important than (i) even though it is much easier to prove. Statement (i) gives you peace of mind (without it one would have two candidates for the notion of almost projectivity), but in the examples of almost projective modules that I know one can prove almost projectivity directly rather than showing that the property holds locally. The roles of Theorems 3.3 and 3.4 in the theory of Tate R-modules are similar.

(3) Although we do not need it in the rest of this work, let us define the notion of n-almost projectivity for any $n \in \mathbb{N}$: an R-module M is *n-almost projective* if in the derived category of R-modules M can be represented as a direct summand of $P \oplus F^{\cdot}$ with P being a projective R-module and $F^{\cdot}$ being a complex of projective

R-modules such that $F^i = 0$ for $i > 0$ and F^i is finitely generated for $i > -n$.[10] One can show that for $n = 1, 2$ this is equivalent to the above definitions of almost projectivity and 2-almost projectivity and that if $n > 2$ then an R-module M is n-almost projective if and only if it is 2-almost projective and for some (or for any) epimorphism $f : P \twoheadrightarrow M$ with P projective $\operatorname{Ker} f$ is $(n-1)$-almost projective. One can also show that a module M over a commutative ring is n-almost projective if and only if it can be Nisnevich-locally represented as a direct sum of a projective module and a module M' having a resolution $P_{n-1} \to P_{n-2} \to \dots P_0 \to M' \to 0$ by finitely generated projective modules.

4.2 Class of an almost projective module in K_{-1}

In Section 3.3.1 we defined the category $\mathcal{C}^{\mathrm{all}}$ and its full subcategory $\mathcal{C}$ formed by projective modules. Let $\mathcal{C}^{\mathrm{ap}} \subset \mathcal{C}^{\mathrm{all}}$ denote the full subcategory of almost projective modules. By definition, an almost projective module M is a direct summand of $F \oplus P$ with F finitely generated and P projective, so M viewed as an object of $\mathcal{C}^{\mathrm{ap}}$ becomes a direct summand of $P \in \mathcal{C}$. So we get a fully faithful functor $\Phi : \mathcal{C}^{\mathrm{ap}} \to \mathcal{C}^{\mathrm{Kar}}$ (in fact, it is not hard to prove that Φ is an equivalence). To an almost projective R-module M one associates an element $[M] \in K_{-1}(R) := K_0(\mathcal{C}^{\mathrm{Kar}})$, namely, $[M]$ is the class of $\Phi(M) \in \mathcal{C}^{\mathrm{Kar}}$.

Let T be a Tate R-module and $L \subset T$ an open bounded submodule (so T/L is almost projective). Then $[T/L] = [T]$.

4.3 The dimension torsor of an almost projective module

To an almost projective module one associates its *dimension torsor*. The definition is given below. It is parallel to the definition of the dimension torsor of a Tate R-module, but there is one new feature: the dimension torsor of an almost projective module is equipped with a *canonical upper semicontinuous section*.

A submodule L of an almost projective R-module M is said to be a *lattice* if it is finitely generated. In this case M/L is also almost projective. A lattice $L \subset M$ is said to be *coprojective* if M/L is projective.[11] One shows that in this case M/L is projective and L has finite presentation, so coprojective lattices exist if and only if M is elementary.

Now let R be commutative. We define a *dimension theory* (respectively, *upper semicontinuous dimension theory*) on an almost projective R-module M to be a rule that associates to each R-algebra R' and each coprojective lattice $L \subset R' \otimes_R M$ a locally constant (respectively, an upper semicontinuous) function $d_L : \operatorname{Spec} R' \to \mathbb{Z}$ in a way compatible with base change and such that $d_{L_2} - d_{L_1} = \operatorname{rank}(L_2/L_1)$ for

[10] One can show that this definition is equivalent to the following one: an R-module M is *n-almost projective* if a projective resolution of M viewed as a complex in the Calkin category $\mathcal{C}_R^{\mathrm{Kar}}$ from Section 3.3.1 is homotopy equivalent to a direct sum of an object of $\mathcal{C}_R^{\mathrm{Kar}}$ and a complex $C^{\cdot}$ in $\mathcal{C}_R^{\mathrm{Kar}}$ such that $C^i \neq 0$ only for $i \leq -n$.

[11] One can show that if M is 2-almost projective this is equivalent to M/L being flat.

any pair of coprojective lattices $L_1 \subset L_2 \subset R' \otimes_R M$. The notion of dimension theory (or upper semicontinuous dimension theory) does not change if one considers only étale R-algebras instead of arbitrary ones. Dimension theories on an almost projective R-module M form a $\mathbb{Z}$-torsor for the Nisnevich topology of Spec R, which is denoted by Dim_M. One defines the *canonical upper semicontinuous dimension theory* d^{can} on M by $d_L^{\mathrm{can}}(x) := \dim_{K_x}(K_x \otimes_{R'} L)$, where R' is an R-algebra, $L \subset R' \otimes_R M$ is a coprojective lattice, $x \in \mathrm{Spec}\, R'$, and K_x is the residue field of x. An upper semicontinuous dimension theory on M is the same as an *upper semicontinuous section* of Dim_M, by which we mean a $\mathbb{Z}$-antiequivariant morphism from the $\mathbb{Z}$-torsor Dim_M to the sheaf of upper semicontinuous $\mathbb{Z}$-valued functions on Spec R. Clearly, d^{can} is a true (i.e., locally constant) section of Dim_M if and only if the quotient of M modulo the nilradical $I \subset R$ is projective over R/I. In this case d^{can} defines a trivialization of Dim_M.

If N is a Tate R-module and $L \subset N$ is an open bounded submodule then the dimension torsor of the almost projective module N/L canonically identifies with that of N.

5 Finer points: Determinants and the canonical central extension

Section 5.6 (in which we discuss the canonical central extension of the automorphism group of an almost projective module) is the only part of this section used in the rest of the article, namely, in Section 7. Therefore, some readers (especially those interested primarily in spaces of formal loops and refined motivic integration) may prefer to skip this section. But it contains an interesting (though slightly vague) picture, which I learned from A. Beilinson (see Section 5.5).

In Sections 5.1–5.4 we follow [BBE, Section 2]. In particular, we combine the dimension torsor and the determinant gerbe into a single object, which is a Torsor over a certain Picard groupoid (these notions are defined below). The reason why it is convenient and maybe necessary to do this is explained in Section 5.3. Our terminology is slightly different from that of [BBE], and our determinant Torsor is inverse to that of [BBE].

5.1 Terminology

According to [Del, Section 1.4], a *Picard groupoid* is a symmetric monoidal category $\mathcal{A}$ such that all the morphisms of $\mathcal{A}$ are invertible and the semigroup of isomorphism classes of the objects of $\mathcal{A}$ is a group. A Picard groupoid is said to be *strictly commutative* if for every $a \in \mathrm{Ob}\,\mathcal{A}$ the commutativity isomorphism $a \otimes a \xrightarrow{\sim} a \otimes a$ equals id_a. As explained in [Del, Section 1.4], there is also a notion of *sheaf of Picard groupoids* (champ de catégories de Picard) on a site.

We will work with the following simple examples.

Examples. For a commutative ring R we have the Picard groupoid $\mathcal{P}ic_R$ of invertible R-modules and the Picard groupoid $\mathcal{P}ic_R^{\mathbb{Z}}$ of $\mathbb{Z}$-graded invertible R-modules (the latter is not strictly commutative because we use the "super" commutativity constraint $a \otimes b \mapsto (-1)^{p(a)p(b)} b \otimes a$). For a scheme S denote by $\mathcal{P}ic_S^{\mathbb{Z}}$ (respectively, $\mathcal{P}ic_S$) the sheaf of Picard groupoids on the Nisnevich site of S formed by $\mathbb{Z}$-graded invertible $\mathcal{O}_S$-modules (respectively, plain invertible $\mathcal{O}_S$-modules, also known as $\mathcal{O}_S^\times$-torsors).

We need more terminology. An *action* of a monoidal category $\mathcal{A}$ on a category $\mathcal{C}$ is a monoidal functor from $\mathcal{A}$ to the monoidal category $\operatorname{Funct}(\mathcal{C}, \mathcal{C})$ of functors $\mathcal{C} \to \mathcal{C}$. Suppose $\mathcal{A}$ acts on $\mathcal{C}$ and $\mathcal{C}'$, i.e., one has monoidal functors $\Phi : \mathcal{A} \to \operatorname{Funct}(\mathcal{C}, \mathcal{C})$ and $\Phi' : \mathcal{A} \to \operatorname{Funct}(\mathcal{C}', \mathcal{C}')$. Then an *$\mathcal{A}$-functor* $\mathcal{C} \to \mathcal{C}'$ is a functor $F : \mathcal{C} \to \mathcal{C}'$ equipped with isomorphisms $F\Phi(a) \xrightarrow{\sim} \Phi'(a)F$ satisfying the natural compatibility condition (the two ways of constructing an isomorphism

$$F\Phi(a_1 \otimes a_2) \xrightarrow{\sim} \Phi'(a_1 \otimes a_2)F$$

must give the same result). An *$\mathcal{A}$-equivalence* $\mathcal{C} \to \mathcal{C}'$ is an $\mathcal{A}$-functor $\mathcal{C} \to \mathcal{C}'$ which is an equivalence.

There is also an obvious notion of action of a sheaf of monoidal categories $\mathcal{A}$ on a sheaf of categories $\mathcal{C}$, and given an action of $\mathcal{A}$ on $\mathcal{C}$ and $\mathcal{C}'$ there is an obvious notion of $\mathcal{A}$-functor $\mathcal{C} \to \mathcal{C}'$ and $\mathcal{A}$-equivalence $\mathcal{C} \to \mathcal{C}'$.

Definition. Let $\mathcal{A}$ be a sheaf of Picard groupoids on a site. A sheaf of categories $\mathcal{C}$ equipped with an action of $\mathcal{A}$ is an *$\mathcal{A}$-Torsor* if it is locally $\mathcal{A}$-equivalent to $\mathcal{A}$.

Remark. The notion of Torsor makes sense even if $\mathcal{A}$ is nonsymmetric. But $\mathcal{A}$ has to be symmetric if we want to have a notion of product of $\mathcal{A}$-Torsors.

5.2 The determinant Torsor

Let R be a commutative ring, $S := \operatorname{Spec} R$. Slightly modifying the construction of [Ka3], we will associate a Torsor over $\mathcal{P}ic_S^{\mathbb{Z}}$ to an almost projective R-module M. Recall that a coprojective lattice $L \subset M$ is a finitely generated submodule such that M/L is projective. The set of coprojective lattices $L \subset M$ will be denoted by $G(M)$. In general, $G(M)$ may be empty, and it is not clear if every $L_1, L_2 \in G(M)$ are contained in some $L \in G(M)$. But it follows from Theorem 4.2(ii) that these properties hold after Nisnevich localization (to show that every $L_1, L_2 \in G(M)$ are Nisnevich-locally contained in some coprojective lattice apply statement (ii) or (iii) of Theorem 4.2 to $M/(L_1 + L_2)$). In other words, for every $x \in \operatorname{Spec} R$ the inductive limit of $G(R' \otimes_R M)$ over the filtering category of all étale R-algebras R' equipped with an R-morphism $x \to \operatorname{Spec} R'$ is a nonempty directed set.

For each pair $L_1 \subset L_2$ in $G(M)$ one has the invertible R-module $\det(L_2/L_1)$. It is equipped with a $\mathbb{Z}$-grading (the determinant of an n-dimensional vector space has grading n).

Definition. A *determinant theory* on M (respectively, a *weak determinant theory* on M) is a rule Δ which associates to each R-algebra R' and each $L \in G(R' \otimes_R M)$ an invertible graded R'-module $\Delta(L)$ (respectively, an invertible R'-module $\Delta(L)$) to each pair $L_1 \subset L_2$ in $G(R' \otimes_R M)$ an isomorphism

$$\Delta_{L_1L_2} : \Delta(L_1) \otimes \det(L_2/L_1) \xrightarrow{\sim} \Delta(L_2), \tag{5.1}$$

and to each morphism $f : R' \to R''$ of R-algebras a collection of base change morphisms $\Delta_f = \Delta_{f,L'} : \Delta(L') \to \Delta(R''L')$, $L' \in G(R' \otimes_R M)$. These data should satisfy the following conditions:

(i) every $\Delta_{f,L'}$ induces an isomorphism $R'' \otimes_{R'} \Delta(L') \xrightarrow{\sim} \Delta(R''L')$;
(ii) $\Delta_{f_2 f_1} = \Delta_{f_2}\Delta_{f_1}$;
(iii) the isomorphisms (5.1) commute with base change;
(iv) for any triple $L_1 \subset L_2 \subset L_3$ in $G(R' \otimes_R M)$, the obvious diagram

$$\begin{array}{ccc} \Delta(L_1) \otimes \det(L_2/L_1) \otimes \det(L_3/L_3) & \xrightarrow{\sim} & \Delta(L_1) \otimes \det(L_3/L_1) \\ \downarrow & & \downarrow \\ \Delta(L_2) \otimes \det(L_3/L_2) & \xrightarrow{\sim} & \Delta(L_3) \end{array}$$

commutes.

Remark. It follows from Theorem 4.2(ii) that the notion of (weak) determinant theory does not change if one considers only étale R-algebras instead of arbitrary ones.

The groupoid of all determinant theories on M is equipped with an obvious action of the Picard groupoid $\mathcal{P}ic_R^{\mathbb{Z}}$ of invertible $\mathbb{Z}$-graded R-modules: $P \in \mathcal{P}ic_R^{\mathbb{Z}}$ sends Δ to $P\Delta$, where $(P\Delta)(L) := P \otimes_R \Delta(L)$.

Determinant theories on $R' \otimes_R M$ for all étale R algebras R' form a sheaf of groupoids Det_M on the Nisnevich site of $S := \mathrm{Spec}\, R$, which is equipped with an action of the sheaf of Picard groupoids $\mathcal{P}ic_S^{\mathbb{Z}}$. It follows from Theorem 4.2(ii) that Det_M is a Torsor over $\mathcal{P}ic_S^{\mathbb{Z}}$. We call it the *determinant Torsor* of M.

If M is a Tate module (rather than an almost projective one), then the above definition of determinant theory and determinant Torsor still applies (of course, in this case the words "coprojective lattice" should be understood in the sense of Section 3.2 and $\otimes$ should be replaced by $\widehat{\otimes}$). If M is an almost projective or Tate module and $L \subset M$ is a lattice, then M/L is almost projective and $\mathrm{Det}_{M/L}$ canonically identifies with Det_M.

Remark. Consider the category whose set of objects is $\mathbb{Z}$ and whose only morphisms are the identities. We will denote it simply by $\mathbb{Z}$. Addition of integers defines a functor $\mathbb{Z} \times \mathbb{Z} \to \mathbb{Z}$, so $\mathbb{Z}$ becomes a Picard groupoid. We have a canonical Picard functor from $\mathcal{P}ic_S^{\mathbb{Z}}$ to the constant sheaf $\mathbb{Z}$ of Picard groupoids: an invertible $\mathcal{O}_S$-module placed in degree n goes to n. The $\mathbb{Z}$-torsor corresponding to the $\mathcal{P}ic_S^{\mathbb{Z}}$-Torsor Det_M is the dimension torsor Dim_M from Sections 3.5 and 4.3.

5.3 On the notion of determinant gerbe

We also have the forgetful functor from the category of $\mathbb{Z}$-graded invertible R-modules to that of plain invertible R-modules and the corresponding functor $F : \mathcal{P}ic_S^{\mathbb{Z}} \to \mathcal{P}ic_S$. Notice that F is a monoidal functor, but *not* a Picard functor. Applying

F to the $\mathcal{P}ic_S^{\mathbb{Z}}$-Torsor Det_M one gets a $\mathcal{P}ic_S$-Torsor, which is the same as an $\mathcal{O}_S^{\times}$-gerbe.[12] This is the *determinant gerbe* considered by Kapranov [Ka3] and mentioned in Section 3.6. Its sections are weak determinant theories. As F does not commute with the commutativity constraint, there is no canonical equivalence between the $\mathcal{O}_S^{\times}$-gerbe corresponding to a direct sum of almost projective modules M_i, $i \in I$, Card $I < \infty$, and the product of the $\mathcal{O}_S^{\times}$-gerbes corresponding to $M_i, i \in I$ (but there is an equivalence which depends on the choice of an ordering of I). This is the source of the numerous signs in [ACK] and the reason why we prefer to consider Torsors over $\mathcal{P}ic_S^{\mathbb{Z}}$ rather than pairs consisting of an $\mathcal{O}_S^{\times}$-gerbe and a $\mathbb{Z}$-torsor (as Kapranov does in [Ka3]).

5.4 Fermion modules, determinant theories, and co-Sato Grassmannian

We follow [BBE, Sections 2.14–2.15] (in particular, see [BBE, Remark (iii) at the end of Section 2.15]).

5.4.1 Fermion modules and weak determinant theories

Fix a Tate R-module M. Let $\mathrm{Cl}(M \oplus M^*)$ denote the Clifford algebra of $M \oplus M^*$. Define a *Clifford module* to be a module V over $\mathrm{Cl}(M \oplus M^*)$ such that for any $v \in V$ the set $\{a \in M \oplus M^* | av = 0\}$ is open in $M \oplus M^*$. A Clifford module V is said to be a *fermion module*[13] if V is fiberwise irreducible and projective over R.

If V is a fermion module and $L \subset M$ is a coprojective lattice, let $\Delta_V(L)$ denote the annihilator of $L \oplus L^{\perp} \subset M \oplus M^*$ in V. As explained in [BBE], $\Delta_V(L)$ is a *line* in V (i.e., a direct summand of V which is an invertible R-module) and Δ_V is a weak determinant theory: if $L_1 \subset L_2 \subset M$ are coprojective lattices, then the isomorphism (5.1) comes from the composition $\bigwedge^r L_2 \to \bigwedge M \to \mathrm{Cl}(M \oplus M^*)$, where r is the rank of L_2/L_1 and $\bigwedge M$ is the exterior algebra of M. Thus one gets a functor $V \mapsto \Delta_V$ from the groupoid of fermion modules to that of weak determinant theories. As explained in [BBE], it is an equivalence: to construct the inverse functor $\Delta \mapsto V_\Delta$ one first constructs V_Δ Nisnevich-locally, then glues the results of the local constructions, and finally uses Theorem 2.1 to prove that V_Δ is a projective R-module.

The equivalences $V \mapsto \Delta_V$ and $\Delta \mapsto V_\Delta$ are compatible with the actions of the groupoid Pic_R of invertible R-modules.

5.4.2 Graded fermion modules and determinant theories

As explained in [BBE], the fermion module V_Δ corresponding to a weak determinant theory Δ is equipped with a T-grading, where T is the dimension torsor of M. Given a

[12] This follows from the definitions, but also from the Grothendieck-Deligne dictionary mentioned in Section 5.5 (the complex of sheaves of abelian groups corresponding via this dictionary to the sheaf of Picard categories $\mathcal{P}ic_S$ is $\mathcal{O}_S^{\times}[1]$, i.e., $\mathcal{O}_S^{\times}$ placed in degree -1).

[13] Motivation of the name: if M is a discrete projective R-module then fermion modules have the form $(\bigwedge M) \otimes_R \mathcal{L}$ with $\mathcal{L}$ being an invertible R-module. Here $\bigwedge M$ is the exterior algebra of M.

determinant theory on M rather than a weak determinant theory one gets a $\mathbb{Z}$-grading on the fermion module compatible with the $\mathbb{Z}$-grading of $\mathrm{Cl}(M \oplus M^*)$ for which M has degree 1 and M^* has degree -1. Thus Det_M identifies with the groupoid of $\mathbb{Z}$-graded fermion modules.

5.4.3 The Plücker embedding of the co-Sato Grassmannian

The co-Sato Grassmannian Gras_M of a Tate R-module M was defined in Section 3.7. Now suppose that the determinant gerbe of M is trivial and fix a weak determinant theory Δ on M. Then we get a line bundle $\mathcal{A}_\Delta$ on Gras_M whose fiber over a coprojective lattice L equals $\Delta(L)$.

On the other hand, we have the fermion module $V = V_\Delta$ such that Δ=Δ_V (see Section 5.4.1). Assigning to a coprojective lattice L the line $\Delta_V(L)$ one gets a morphism $i : \mathrm{Gras}_M \to \mathbb{P}$, where $\mathbb{P}$ is the ind-scheme of lines in V. As explained by Plücker, i is a closed embedding.

Clearly, $\mathcal{A}_\Delta = i^*\mathcal{O}(-1)$.

5.5 A somewhat vague picture

5.5.1 The picture I learned from Beilinson

Let S be a spectrum in the sense of algebraic topology. We put $\pi^i(S) := \pi_{-i}(S)$ and define $\tau^{\leq k} S$ to be the spectrum equipped with a morphism $\tau^{\leq k} S \to S$ such that $\pi^i(\tau^{\leq k} S) = 0$ for $i > k$ and the morphism $\pi^i(\tau^{\leq k} S) \to \pi^i(S)$ is an isomorphism for $i \leq k$. There is a notion of torsor over a spectrum S, which depends only on $\tau^{\leq 1} S$. Namely, an S-torsor is a point of the infinite loop space L corresponding to $(\tau^{\leq 1} S)[1]$ (or equivalently, a morphism from the spherical spectrum to $S[1]$). A homotopy equivalence between torsors is a path connecting the corresponding points of L, so equivalence classes are parametrized by $\pi^1(S) := \pi_{-1}(S)$.

Beilinson's first remark: an object of the Calkin category $\mathcal{C}_R^{\mathrm{Kar}}$ (see Section 3.3.1) defines a point of the infinite loop space corresponding to the K-theory spectrum $K(\mathcal{C}_R^{\mathrm{Kar}})$, and as $K(\mathcal{C}_R^{\mathrm{Kar}}) = K(R))[1]$ it defines a $K(R)$-torsor. In particular, an almost projective R-module M defines a $K(R)$-torsor, whose class in $\pi_{-1}(K(R)) = K_{-1}(R)$ is the class $[M]$ considered in Section 3.3. If R is commutative then by Thomason's localization theorem [TT, Section 10.8], $K(R) = R\Gamma(S, \mathcal{K})$, where $\mathcal{K}$ is the sheaf of K-theories of $\mathcal{O}_S$ (this is a sheaf of spectra on the Nisnevich site of S). So the notion of $K(R)$-torsor should[14] coincide with that of $\mathcal{K}$-torsor. Both of them should coincide with that of $\tau^{\leq 1}\mathcal{K}$-torsor. By Theorem 3.7, $\mathcal{K}^1 := \mathcal{K}_{-1} = 0$, so $\tau^{\leq 1}\mathcal{K} = \tau^{\leq 0}\mathcal{K}$ and therefore we get a morphism $\tau^{\leq 1}\mathcal{K} = \tau^{\leq 0}\mathcal{K} \to \mathcal{K}_{[0,1]} := \mathcal{K}^{[-1,0]} := \tau^{\geq -1}\tau^{\leq 0}\mathcal{K}$. So to an almost projective R-module M there should correspond a $\mathcal{K}_{[0,1]}$-torsor Δ_M. According to Beilinson, $\mathcal{K}_{[0,1]}$ *and* Δ_M *should*

[14] Here and in what follows, I use the word "should" to indicate the parts of the picture that I do not quite understand (probably due to the fact that I have not learned the theory of sheaves of spectra).

identify with $\mathcal{P}ic_S^{\mathbb{Z}}$ *and the Torsor* Det_M *from Section* 5.2 *via the following dictionary*, which goes back to A. Grothendieck and was used in [Del, Sections 1.4–1.5] and in [Del87, Section 4].

5.5.2 Grothendieck's dictionary

According to it, *a Picard groupoid is essentially the same as a spectrum* X *with* $\pi_i(X) = 0$ *for* $i \neq 0, 1$. More precisely, the following two constructions become essentially inverse to each other if the first one is applied only to infinite loop spaces X with $\pi_i(X) = 0$ for $i > 1$:

(i) To an infinite loop space X one associates its fundamental groupoid $\Pi(X)$ viewed as a Picard groupoid.[15]
(ii) To a Picard groupoid one associates its classifying space viewed as an infinite loop space.[16]

For strictly commutative Picard groupoids there is a similar dictionary and, moreover, a precise reference, namely, [Del, Corollary 1.4.17]. The statement from [Del] is formulated in a more general context of sheaves. It says that *a sheaf of strictly commutative Picard groupoids is essentially the same as a complex of sheaves of abelian groups with cohomology concentrated in degrees* 0 *and* -1.

Hopefully, there is also a sheafified version of the dictionary in the nonstrictly commutative case. It should say that *a sheaf* $\mathcal{A}$ *of Picard groupoids is essentially the same as a sheaf of spectra* $\mathcal{S}$ *whose sheaves of homotopy groups* π_i *vanish for* $i \neq 0, 1$ and that *the notion of* $\mathcal{A}$*-Torsor from Section* 5.1 *is equivalent to that of* $\mathcal{S}$*-torsor*.

5.5.3 Problem: Make the above somewhat vague picture precise

The notion of determinant Torsor is very useful, and its rigorous interpretation in the standard homotopy-theoretic language of algebraic K-theory would be helpful.

5.6 The central extension for almost projective modules

Let M be an almost projective module over a commutative ring R and $\widetilde{M}$ be the corresponding quasicoherent sheaf on the Nisnevich topology of $S := \operatorname{Spec} R$. Then the sheaf $\mathcal{A}ut\, M := \mathcal{A}ut\, \widetilde{M}$ has a canonical central extension

$$0 \to \mathcal{O}_S^\times \to \widehat{\mathcal{A}ut}\, M \to \mathcal{A}ut\, M \to 0. \tag{5.2}$$

[15] If $X = \Omega Y$ the group structure on $\pi_0(X) = \pi_1(Y)$ lifts to a monoidal category structure on $\Pi(X)$. If $X = \Omega^2 Z$ the proof of the commutativity of $\pi_0(X) = \pi_2(Z)$ "lifts" to a braiding on $\Pi(X)$, the "square of the braiding" map $t : \pi_0(X) \times \pi_0(X) \to \pi_1(X)$ equals the Whitehead product $\pi_2(Z) \times \pi_2(Z) \to \pi_3(Z)$, and therefore t vanishes if Z is a loop space.

[16] The classifying space $B\mathcal{A}$ of any symmetric monoidal category $\mathcal{A}$ is a Γ-space (see [Seg]) or if you prefer, an E_∞ space (see [M]). So if every object of $\mathcal{A}$ is invertible (i.e., if $\pi_0(B\mathcal{A})$ is a group) then $B\mathcal{A}$ is an infinite loop space.

Its definition is similar to that of the Tate central extension of the automorphism group of a Tate vector space, also known as "Japanese" extension (see [Br, Ka3, PS, BBE]). Namely, if M has a determinant theory Δ, then $\mathcal{A}ut\, M$ is the sheaf of automorphism of (M, Δ). This sheaf does not depend (up to canonical isomorphism) on the choice of Δ. This allows one to define $\widetilde{\mathcal{A}ut}\, M$ even if Δ exists only locally.

Now suppose that a group R-scheme G acts on M (i.e., one has a compatible collection of morphisms $G(R') \to \mathrm{Aut}(R' \otimes_R M)$ for all R-algebras R'). Then (5.2) induces a canonical central extension of group schemes

$$0 \to \mathbb{G}_m \to \widehat{G} \to G \to 0. \tag{5.3}$$

(One first defines $\widehat{G}$ as a functor $\{R\text{-algebras}\} \to \{\text{groups}\}$ and then notices that $\widehat{G}$ is representable because it is a $\mathbb{G}_m$-torsor over G.) If G is abelian we get the commutator map $G \times G \to \mathbb{G}_m$. If M is projective (in particular, if k is a field) the extension (5.3) canonically splits because in this case there is a canonical determinant theory on M defined by $\Delta(L) = \det L$. The following example shows that in general the extension (5.3) can be nontrivial.

Example (*A. Beilinson*). Let k be a field, $R = k[\varepsilon]/(\varepsilon^2)$. Fix $g \in k((t))$; then $L_g := R[[t]] + \varepsilon g R[[t]]$ is a lattice in the Tate R-module $R((t))$. Put $M := R((t))/L_g$. Let G_1 denote the multiplicative group of $R[[t]]$ viewed as a group scheme over R. On M we have the natural action of G_1 and also the action of $\mathbb{G}_a$ such that $c \in \mathbb{G}_a$ acts as multiplication by $1 + c\varepsilon g$. So $G := G_1 \times \mathbb{G}_a$ acts on M. The theory of the Tate extension (see, e.g., [BBE, Section 3]) tells us that in the corresponding central extension (5.3) the commutator of $c \in \mathbb{G}_a$ and $u \in R[[t]]^\times$ equals $c\varepsilon \cdot \mathrm{res}(u \cdot dg)$. So the extension (5.3) is not commutative if $dg \neq 0$.

Remarks.

(i) To define the central extension (5.3), it suffices to have an action of G on M as an object of the Calkin category $\mathcal{C}^{\mathrm{Kar}}$ defined in Section 3.3.1 (see [BBE, Section 2] for more details). Of course, in this setting the extension (5.3) may be nontrivial even if R is a field.

(i′) One can define the extension (5.3) if G is any group-valued functor on the category of R-algebras (e.g, a group ind-scheme).

(ii) As explained in [BBE], the canonical central extension of the automorphism group of a Tate vector space should rather be considered as a "superextension" (this is necessary to formulate the compatibility between the extensions corresponding to the Tate spaces T_1, T_2 and $T_1 \oplus T_2$). The same is true for the canonical central extension of the automorphism group of an object of the Calkin category $\mathcal{C}^{\mathrm{Kar}}$. But in the case of an almost projective module M "super" is unnecessary because any automorphism of M has degree 0, i.e., preserves the dimension torsor (this follows from the existence of the canonical upper semicontinuous dimension theory on M; see Section 4.3).

6 Applications to spaces of formal loops. "Refined" motivic integration

In this section all rings and algebras are assumed to be commutative. We fix a ring k. Starting from Section 6.4 we suppose that k is a field.

6.1 A class of schemes

We will use the following notation for affine spaces:

$$\mathbb{A}^I := \operatorname{Spec} k[x_i]_{i \in I}, \qquad \mathbb{A}^\infty := \mathbb{A}^{\mathbb{N}}.$$

We say that a k-scheme is *nice* if it is isomorphic to $X \times \mathbb{A}^I$, where X is of finite presentation over k (the set I may be infinite). An affine scheme is nice if and only if it can be defined by finitely many equations in a (not necessarily finite-dimensional) affine space.

Definition. A k-scheme X is *locally nice* (respectively, *Zariski-locally nice, étale-locally nice*) if it becomes nice after Nisnevich localization (respectively, Zariski or étale localization). X is *differentially nice* if for every open affine $\operatorname{Spec} R \subset X$ the R-module $\Omega_R^1 := \Omega_{R/k}^1$ is 2-almost projective.

By Theorem 4.2(i) étale-local niceness implies differential niceness. I do not know if étale-local niceness implies local niceness. Local niceness does not imply Zariski-local niceness (see Section 6.2 below).

For a differentially nice k-scheme X one defines the *dimension torsor* Dim_X: if X is an affine scheme $\operatorname{Spec} R$, then Dim_X is the dimension torsor of the almost projective R-module Ω_R^1, and for a general X one defines Dim_X by gluing together the torsors Dim_U for all open affine $U \subset X$. If $X' \subset X$ is a closed subscheme defined by finitely many equations and X is differentially nice, then X' is; in this situation $\operatorname{Dim}_{X'}$ canonically identifies with the restriction of Dim_X to X'.

In the next subsection we will see that the dimension torsor of a locally nice k-scheme may be nontrivial, and on the other hand, there exists a locally nice k-scheme with trivial dimension torsor which is not Zariski-locally nice.

6.2 Examples

(i) Define $i : \mathbb{A}^\infty \to \mathbb{A}^\infty$ by $i(x_1, x_2, \dots) := (0, x_1, x_2, \dots)$. Take $\mathbb{A}^1 \times \mathbb{A}^\infty$ and then glue $(0, x) \in \mathbb{A}^1 \times \mathbb{A}^\infty$ with $(1, i(x)) \in \mathbb{A}^1 \times \mathbb{A}^\infty$. Thus one gets a locally nice k-scheme X whose dimension torsor is nontrivial and even not Zariski-locally trivial.

(ii) Let M be an almost projective module over a finitely generated algebra R over a Noetherian ring k. Let X denote the spectrum of the symmetric algebra of M. Then X is locally nice. This follows from Theorem 4.2(ii) and the next theorem, which is due to H. Bass [Ba2, Corollary 4.5].

Theorem 6.1. *If R is a commutative Noetherian ring whose spectrum is connected, then every infinitely generated projective R-module is free.*

It is easy to deduce from Theorem 6.1 that X is Zariski-locally nice if and only if the class of M in $K_{-1}(R)$ vanishes locally for the Zariski topology (to prove the "only if" statement consider the restriction of Ω^1_X to the zero section $\operatorname{Spec} R \hookrightarrow X$). If R and M are as in (3.1), then we get the above Example (i).

(iii) There exists a locally nice scheme X over a field k which is not Zariski-locally nice but has trivial dimension torsor.[17] According to (ii), to get such an example it suffices to find a finitely generated k-algebra R and an almost projective R-module M such that $H^1_{\text{ét}}(\operatorname{Spec} R, \mathbb{Z}) = 0$ but the class of M in $K_{-1}(R)$ is not Zariski-locally trivial. [We3, Section 6] contains examples of finitely generated normal k-algebras R with $K_{-1}(R) \neq 0$. In each of them $\operatorname{Spec} R$ has a unique singular point x, and according to [We1], the map $K_{-1}(R) \to K_{-1}(R_x)$ is an isomorphism. Now take any nonzero element of $K_{-1}(R)$ and represent it as a class of an almost projective R-module M.

6.3 Generalities on ind-schemes

The key notions introduced in this subsection are those of reasonable, T-smooth, and Tate-smooth ind-scheme (see Sections 6.3.3–6.3.7).

6.3.1 Definition of ind-scheme and formal scheme

Functors from the category of k-algebras to that of sets will be called "spaces." For example, a k-scheme can be considered as a space. *Subspace* means "subfunctor." A subspace $Y \subset X$ is said to be *closed* if for every (affine) scheme Z and every $f : Z \to X$ the subspace $Z \times_X Y \subset Z$ is a closed subscheme.

Let us agree that an *ind-scheme* is a space which can be represented as $\varinjlim X_\alpha$, where $\{X_\alpha\}$ is a directed family of quasi-compact schemes such that all the maps $i_{\alpha\beta} : X_\alpha \to X_\beta$, $\alpha \leq \beta$, are closed embeddings. (Notice that if the same space can also be represented as the inductive limit of a directed family of quasi-compact schemes X'_β then each X'_β is contained in some X_α and each X_α is contained in some X'_β.) If X can be represented as above so that the set of indices α is countable, then X is said to be an $\aleph_0$-*ind-scheme*. If P is a property of schemes stable under passage to closed subschemes then we say that X satisfies the *ind-P* property if each X_α satisfies P. For example, one has the notion of ind-affine ind-scheme and that of ind-scheme of ind-finite type.

Set $X_{\text{red}} := \varinjlim X_{\alpha\,\text{red}}$; an ind-scheme X is said to be *reduced* if $X_{\text{red}} = X$.

[17] In Section 6.13.3 we will see that one can get such X from the loop space of a smooth affine manifold.

6.3.2 $\mathcal{O}$-modules and pro-$\mathcal{O}$-modules on ind-schemes

A *pro-module* over a ring R is defined to be a pro-object[18] of the category of R-modules. We identify the category of Tate R-modules with a full subcategory of the category of pro-R-modules by associating to a Tate R-module the projective system formed by its discrete quotient modules.

An *$\mathcal{O}$-module* (respectively, a *pro-$\mathcal{O}$-module*) P on a space X is a rule that assigns to a commutative algebra A and a point $\phi \in X(A)$ an A-module (respectively, a pro-A-module) P_ϕ, and to any morphism of algebras $f : A \to B$ a B-isomorphism $f_P : B \otimes_f P_\phi \xrightarrow{\sim} P_{f\phi}$ in a way compatible with composition of fs. An $\mathcal{O}$-module on a scheme Y is the same as a quasicoherent sheaf of $\mathcal{O}_Y$-modules, and an $\mathcal{O}$-module P on an ind-scheme $X = \varinjlim X_\alpha$ is the same as a collection of $\mathcal{O}$-modules P_{X_α} on X_α together with identifications $i^*_{\alpha\beta} P_{X_\beta} = P_{X_\alpha}$ for $\alpha \leq \beta$ that satisfy the obvious transitivity property.

A pro-$\mathcal{O}$-module is said to be a *Tate sheaf* if for each ϕ as above the pro-module P_ϕ is a Tate module.

The *cotangent sheaf* Ω^1_X of an ind-scheme $X = \varinjlim X_\alpha$ is the pro-$\mathcal{O}$-module whose restriction to each X_α is defined by the projective system $i^*_{\alpha\beta}\Omega^1_{X_\beta}$, $\beta \geq \alpha$ (here $i_{\alpha\beta}$ is the embedding $X_\alpha \to X_\beta$).

6.3.3 The notion of reasonable ind-scheme

The following definitions are due to A. Beilinson. A closed quasi-compact subscheme Y of an ind-scheme X is called *reasonable* if for any closed subscheme $Z \subset X$ containing Y the ideal of Y in $\mathcal{O}_Z$ is finitely generated. Notice that reasonable subschemes of X form a directed set. An ind-scheme X is *reasonable* if X is the union of its reasonable subschemes, i.e., if it can be represented as $\varinjlim X_\alpha$, where all X_α's are reasonable.

Any scheme is a reasonable ind-scheme. A closed subspace of a reasonable ind-scheme is a reasonable ind-scheme. The product of two reasonable ind-schemes is reasonable. The completion of any ind-scheme along a reasonable closed subscheme is a reasonable ind-scheme.

6.3.4 Main example: ind-scheme of formal loops

Let Y be an affine scheme over $F := k((t))$. Define a functor $\mathcal{L}Y$ from the category of k-algebras to that of sets by $\mathcal{L}Y(R) := Y(R\hat{\otimes}F)$, $R\hat{\otimes}F := R((t))$. It is well known and easy to see that $\mathcal{L}Y$ is an ind-affine ind-subscheme. This is the *ind-scheme of formal loops* of Y. If Y is an affine scheme of *finite type* over F, then $\mathcal{L}Y$ is a reasonable $\aleph_0$-ind-scheme.

[18] A nice exposition of the theory of pro-objects and ind-objects of a category is given in [GV, Section 8]. See also [AM, Appendix].

6.3.5 Formal schemes

We define a *formal scheme* to be an ind-scheme X such that X_{red} is a scheme. An $\aleph_0$-*formal scheme* is a formal scheme which is an $\aleph_0$-ind-scheme. The *completion* of an ind-scheme Z along a closed subscheme $Y \subset Z$ is the direct limit of closed subschemes $Y' \subset Z$ such that $Y'_{\text{red}} = Y_{\text{red}}$. In the case of formal schemes we write "affine" instead of "ind-affine." A formal scheme X is affine if and only if X_{red} is affine.

Remark. As soon as you compare the above definition of formal scheme with the one from EGA I you see that they are not equivalent (even in the affine case) but the difference is not big: an $\aleph_0$-formal scheme in our sense which is reasonable in the sense of Section 6.3.3 is a formal scheme in the sense of EGA I, and on the other hand, a Noetherian formal scheme in the sense of EGA I is a formal scheme in our sense.

6.3.6 Formal smoothness

Following Grothendieck ([Gr64], [Gr]), we say that X is *formally smooth* if for every k-algebra A and every nilpotent ideal $I \subset A$ the map $X(A) \to X(A/I)$ is surjective. A morphism $X \to Y$ is said to be formally smooth if for every k-algebra k' and every morphism $\operatorname{Spec} k' \to Y$ the k'-space $X \times_Y \operatorname{Spec} k'$ is formally smooth. Clearly, formal smoothness of any ind-scheme (respectively, a reasonable ind-scheme) is equivalent to formal smoothness of its completions along all closed subschemes (respectively, all reasonable closed subschemes).

Theorem 6.2.

(i) *For reasonable formal schemes formal smoothness is an étale-local property.*
(ii) *A reasonable closed subscheme of a formally smooth ind-scheme is differentially nice.*
(iii) *If a reasonable $\aleph_0$-ind-scheme X is formally smooth then Ω^1_X is a Tate sheaf* (*the notions of cotangent sheaf of an ind-scheme, Tate sheaf, and Mittag–Leffler–Tate sheaf are defined in Section* 6.3.2).

In the case of schemes statement (i) of the theorem was proved by Grothendieck (cf. [Gr2, Remark 9.5.8]) modulo the conjecture on the local nature of projectivity (which was proved a few years later in [RG]). The proof of Theorem 6.2 in the general case is slightly more complicated but based on the same ideas.

6.3.7 T-smoothness and Tate-smoothness

We say that a reasonable ind-scheme X is *T-smooth* if

(i) every reasonable closed subscheme of X is locally nice;
(ii) X is formally smooth.

A T-smooth ind-scheme X is said to be *Tate-smooth* if its cotangent sheaf is a Tate sheaf (according to Theorem 6.2(iii), this is automatic for $\aleph_0$-ind-schemes).

Remark. In the above definitions we do not require every closed subscheme of X to be contained in a formally smooth subscheme. It is not clear if this property holds for $\mathcal{L}(SL(n))$ or for the affine Grassmannian, even though these ind-schemes are Tate-smooth. See also Remark (ii) from Section 6.4.

6.3.8 Dimension torsor

Let X be a reasonable ind-scheme such that all its reasonable closed subschemes are differentially nice (by Theorem 6.2(ii) this is true for any formally smooth reasonable ind-scheme). Then there is an obvious notion of the *dimension torsor* of X: for each reasonable closed subscheme $Y \subset X$ one has the dimension torsor Dim_Y, and if $Y' \subset Y$ are reasonable closed subschemes then $\mathrm{Dim}_{Y'}$ identifies with the restriction of Dim_Y to Y'.

6.3.9 Relation with the Kapranov–Vasserot theory

The notion of T-smooth ind-scheme is similar to the notion of "smooth locally compact ind-scheme" introduced by M. Kapranov and E. Vasserot (see [KV, Definition 4.4.4]). Neither of these classes of ind-schemes contains the other one. The theory of $\mathcal{D}$-modules on smooth locally compact ind-schemes developed in [KV] extends to the class of T-smooth ind-schemes, and the same is true for the Kapranov–Vasserot theory of de Rham complexes (which goes back to the notion of *chiral de Rham complex* from [MSV]). According to A. Beilinson (private communication), these theories, in fact, extend to the class of formally smooth reasonable ind-schemes, which contains both "smooth locally compact" ind-schemes in the sense of [KV] and T-smooth ones.

6.4 Loops of an affine manifold

From now on we assume that k is a field (I have not checked if Theorems 6.3 and 6.4 hold for any commutative ring k). So $F = k((t))$ is also a field. For any affine F-scheme Y one has the ind-scheme of formal loops $\mathcal{L}Y$ (see Section 6.3.4).

Theorem 6.3. *If an affine F-scheme Y is smooth, then $\mathcal{L}Y$ is Tate-smooth.*

Remarks.

(i) The theorem is not hard. It is only property (i) from the definition of T-smoothness (see Section 6.3.7) that requires some efforts. See Section 6.7 for more details.

(ii) If Y is a smooth affine F-scheme then by Theorem 6.3 every reasonable closed subscheme $X \subset \mathcal{L}Y$ is locally nice. But there exist Y and $X \subset \mathcal{L}Y$ as above such that X is not Zariski-locally nice. One can choose Y and X such that Dim_X is not Zariski-locally trivial. But one can also choose Y and X such that Dim_X is trivial but X is not Zariski-locally nice. See Section 6.13 for examples of these situations. According to H. Bass [Ba], K_{-1} of a regular ring is zero, so in these examples $\mathcal{L}Y$ cannot be represented (even Zariski-locally) as the union of an increasing sequence of smooth closed subschemes.

In the next subsection we formulate an analog of Theorem 6.3 for affinoid analytic spaces (this is a natural thing to do in view of Section 6.6).

6.5 Loops of an affinoid space

We will use the terminology from [BGR] (which goes back to Tate) rather than the one from [Be]. Let $F\langle z_1, \dots, z_n\rangle \subset F[[z_1, \dots z_n]]$ be the algebra of power series which converge in the polydisk $|z_i| \leq 1$. As $F = k((t))$ one has $F\langle z_1, \dots, z_n\rangle = k[z_1, \dots, z_n]((t))$. For every k-algebra R the F-algebra $R\hat{\otimes}F = R((t))$ is equipped with the norm whose unit ball is $R[[t]]$. In particular, $F\langle z_1, \dots, z_n\rangle$ is a Banach algebra. An *affinoid F-algebra* is a topological F-algebra isomorphic to a quotient of $F\langle z_1, \dots, z_n\rangle$ for some n. All morphisms between affinoid F-algebras are automatically continuous (see, e.g., [BGR, Section 6.1.3]). The category of *affinoid analytic spaces* is defined to be dual to that of affinoid F-algebras; the affinoid space corresponding to an affinoid F-algebra A will be denoted by $\mathcal{M}(A)$.

For an affinoid analytic space $Z = \mathcal{M}(A)$ and a k-algebra R denote by $\mathcal{L}Z(R)$ the set of continuous F-homomorphisms from A to the Banach F-algebra $R\hat{\otimes}F = R((t))$. It is easy to see that the functor $\mathcal{L}Z$ is a reasonable affine $\aleph_0$-formal scheme in the sense of Sections 6.3.5 and 6.3.3 (and therefore an affine formal scheme in the sense of EGA I). For example, if Z is the unit disk then $\mathcal{L}Z$ is the completion of the ind-scheme of formal Laurent series along the subscheme of formal Taylor series.

Theorem 6.4. *If an affinoid space Z is smooth then the formal scheme $\mathcal{L}Z$ is Tate-smooth. In particular, $(\mathcal{L}Z)_{\text{red}}$ is a locally nice scheme.*

6.6 Theorem 6.3 follows from Theorem 6.4

Let $Y = \operatorname{Spec} B$ be a closed subscheme of $\mathbb{A}^n = \operatorname{Spec} F[z_1, \dots, z_n]$. The ind-scheme $\mathcal{L}Y$ is the union of its closed subschemes $\mathcal{L}_N Y$ defined by $(\mathcal{L}_N Y)(R) := Y(R) \cap (t^{-N}R[[t]])^n \subset R((t))^n$ for any k-algebra R. The completion of $\mathcal{L}Y$ along $\mathcal{L}_N Y$ identifies with $\mathcal{L}Y_N$, where Y_N is the affinoid analytic space defined by

$$Y_N := \mathcal{M}(B_N), \quad B_N := B \otimes_{F[z_1,\dots,z_n]} F\langle t^N z_1, \dots, t^N z_n\rangle$$

(in other words, Y_N is the intersection of Y with the polydisk of radius r^n, $r := |t^{-1}| > 1$). Therefore, Theorem 6.3 follows from Theorem 6.4.

6.7 Sketch of the proof of Theorem 6.4

The formal smoothness of $\mathcal{L}Z$ follows immediately from the definitions. It is also easy to describe the cotangent sheaf of $\mathcal{L}Z$. Let A be the affinoid F-algebra corresponding to Z. Every finite-dimensional vector bundle E on Z defines a Tate sheaf $\mathcal{L}E$ on $\mathcal{L}Z$: if $\operatorname{Spec} R \subset \mathcal{L}Z$ is a closed affine subscheme and $f : A \to R((t))$ corresponds to the morphism $\operatorname{Spec} R \hookrightarrow \mathcal{L}Z$, then the pullback of $\mathcal{L}E$ to $\operatorname{Spec} R$ is the Tate R-module $R((t)) \otimes_A \Gamma(Z, E)$. The proof of the next lemma is straightforward.

Lemma 6.5. *The cotangent sheaf of $\mathcal{L}Z$ identifies with the Tate sheaf $\mathcal{L}\Omega^1_Z$ corresponding to the cotangent bundle Ω^1_Z of the analytic space Z.*

Corollary. *Let* Spec $R \subset \mathcal{L}Z$ *be a reasonable closed subscheme. Let M be the module of global sections of the pullback to* Spec R *of the Tate sheaf $\mathcal{L}\Omega^1_Z$. Then Ω^1_R is the quotient of the Tate R-module M by some lattice.*

It remains to show that a reasonable closed subscheme Spec $R \subset \mathcal{L}Z$ is locally nice. It easily follows from the above corollary and Theorem 3.4 that after Nisnevich localization Ω^1_R becomes a direct sum of a free module and a module of finite presentation. This is a linearized version of local niceness. To deduce local niceness from its linearized version one works with the implicit function theorem.

6.8 The renormalized dualizing complex

Fix a prime $l \neq \operatorname{char} k$. Let $D^b_c(X, \mathbb{Z}_l)$ denote the appropriately defined bounded constructible l-adic derived category on a scheme X (see [E, Ja]). For a general locally nice k-scheme X there is no natural way to define the dualizing complex $K_X \in D^b_c(X, \mathbb{Z}_l)$. Indeed, if X is the product of $\mathbb{A}^\infty$ and a k-scheme Y of finite type and if $\pi : X \to Y$ is the projection, then K_X should equal $\pi^* K_Y \otimes (\mathbb{Z}_l[2](1))^{\otimes\infty}$, which makes no sense. But suppose that the dimension $\mathbb{Z}$-torsor Dim_X is trivial and that we have chosen its trivialization η. Then one can define the *renormalized dualizing complex* $K^\eta_X \in D^b_c(X, \mathbb{Z}_l)$. The definition (which is straightforward) is given below. The reader can skip it and go directly to Section 6.9.

First assume that X is nice, i.e., there exists a morphism $\pi : X \to Y$ such that Y is a k-scheme of finite type and X is Y-isomorphic to $Y \times \mathbb{A}^I$ for some set I. Let $\mathcal{C}_X$ be the category of all such pairs (Y, π). A morphism $f : (Y, \pi) \to (Y', \pi')$ is defined to be a morphism $f : Y \to Y'$ such that $\pi' = f\pi$. Such an f is unique if it exists. The category $\mathcal{C}_X$ is equivalent to a directed set. So to define K^η_X it suffices to define a functor

$$\mathcal{C}_X \to D^b_c(X, \mathbb{Z}_l), \qquad (Y, \pi) \mapsto K^{\eta,\pi}_X \tag{6.1}$$

which sends all morphisms to isomorphisms.

If $(Y, \pi) \in \mathcal{C}_X$, then $\pi^*\Omega^1_Y \subset \Omega^1_X$ is locally of finite presentation and $\Omega^1_X/\pi^*\Omega^1_Y$ is locally free. So for every open affine $U \subset X$ one has the coprojective lattice $\Gamma(U, \pi^*\Omega^1_Y) \subset \Gamma(U, \Omega^1_X)$ and therefore a section of the torsor Dim_X over U. These sections agree with each other, so we get a global section η_π of Dim_X. Put

$$m := \eta_\pi - \eta \in H^0(X, \mathbb{Z}), \tag{6.2}$$

$$K^{\eta,\pi}_X := \pi^* K_Y \otimes (\mathbb{Z}_l[2](1))^{\otimes m}, \tag{6.3}$$

Now let $f : (Y, \pi) \to (Y', \pi')$ be a morphism. One easily shows that $f : Y \to Y'$ is smooth,[19] so one has a canonical isomorphism

[19] Choosing a section $Y \to X$ one sees that Y is Y'-isomorphic to a retract of $Y' \times \mathbb{A}^J$ for some J. So f is formally smooth and therefore smooth.

$$K_Y \xrightarrow{\sim} f^* K_{Y'} \otimes (\mathbb{Z}_l[2](1))^{\otimes d}, \tag{6.4}$$

where d is the relative dimension of Y over Y'. It is easy to see that $\pi^* d = \eta_{\pi'} - \eta_\pi$, so (6.4) induces an isomorphism $\alpha_f : K_X^{\eta,\pi} \to K_X^{\eta,\pi'}$. We define (6.1) on morphisms by $f \mapsto \alpha_f$.

So we have defined K_X^η if X is nice. The formation of K_X^η commutes with étale localization of X. It is easy to see that $\mathrm{Ext}^i(K_X^\eta, K_X^\eta) = 0$ for $i < 0$. So by [BBD, Theorem 3.2.4] there is a unique way to extend the definition of K_X^η to all étale-locally nice k-schemes X so that the formation of K_X^η still commutes with étale localization.

6.9 $R\Gamma_c$ of a locally nice scheme

Suppose we are in the situation of Section 6.8, i.e., we have a locally nice k-scheme X, a trivialization η of its dimension torsor, and a prime $l \neq \operatorname{char} k$. Assume that X is quasicompact and quasiseparated. Then we put

$$R\Gamma_c^\eta(X \otimes \bar{k}, \mathbb{Z}_l) := R\Gamma(X \otimes_k \bar{k}, K_X^\eta)^*, \tag{6.5}$$

where K_X^η is the renormalized dualizing complex defined in Section 6.8. $R\Gamma_c^\eta(X \otimes \bar{k}, \mathbb{Z}_l)$ is an object of $D_c^b(\operatorname{Spec} k, \mathbb{Z}_l)$, i.e., of the appropriately defined bounded constructible derived category of l-adic representations of $\mathrm{Gal}(k^s/k)$, where k^s is a separable closure of k.

Problems.

(1) Define an object of the triangulated category of k-motives [VSF, VMW] whose l-adic realization equals $R\Gamma_c^\eta(X \otimes \bar{k}, \mathbb{Z}_l)$ for each $l \neq \operatorname{char} k$ (Voevodsky says this can probably be done).
(2) Now suppose that the determinant gerbe of X is trivial and we have fixed its trivialization ξ. Can one canonically lift $R\Gamma_c^\eta(X \otimes \bar{k}, \mathbb{Z}_l)$ to an object of the motivic stable homotopy category depending on η and ξ? Or at least, can one canonically lift $R\Gamma_c^\eta(X \otimes \bar{k}, \mathbb{Q}_l)$ to an object of the motivic stable homotopy category tensored by $\mathbb{Q}$? (The motivic stable homotopy category, also known as $\mathbb{A}^1$ stable homotopy category, was defined in [Vo]). The reason why ξ is assumed to exist and to be fixed: if $k = \mathbb{R}$ this allows us to define $R\Gamma_c^{\eta,\xi}(X(\mathbb{R}), \mathbb{Z})$.

6.10 "Refined" motivic integration

Suppose that in the situation of Theorem 6.4 the canonical bundle $\det \Omega_Z^1$ is trivial. Choose a trivialization of $\det \Omega_Z^1$, i.e., a differential form $\omega \in H^0(Z, \det \Omega_Z^1)$ with no zeros. By Theorem 6.4, the scheme $X := (\mathcal{L}Z)_{\mathrm{red}}$ is locally nice. By Section 3.9 and the corollary of Lemma 6.5, our trivialization of $\det \Omega_Z^1$ induces a trivialization η of the dimension torsor $\operatorname{Dim} X$. We put

$$\int_Z |\omega| := R\Gamma_c^\eta(X, \mathbb{Z}_l) \in D_c^b(\operatorname{Spec} k, \mathbb{Z}_l), \tag{6.6}$$

where $R\Gamma_c^\eta(X, \mathbb{Z}_l)$ is defined by (6.5). Clearly, $\int_Z |\omega|$ does not depend on the choice of X.

6.11 Comparison with usual motivic integration

In the situation of Section 6.10 (i.e., integrating a holomorphic form with no zeros over an affinoid domain) the usual motivic integral [Lo] belongs to $M_k := M'_k[L^{-1}]$, where M'_k is the Grothendieck ring of k-varieties[20] and $L \in M'_k$ is the class of the affine line. Its definition can be reformulated as follows.

Given a connected nice k-scheme X and a trivialization η of its dimension torsor one chooses $\pi : X \to Y$ as in Section 6.8, defines $m \in H^0(X, \mathbb{Z}) = \mathbb{Z}$ by (6.2) and puts $[X]^\eta := [Y]L^m \in M_k$. If X is any quasicompact quasiseparated locally nice k-scheme, choose closed subschemes $X = F_0 \supset F_1 \supset \cdots \supset F_n = \emptyset$ so that each F_i is defined by finitely many equations and $F_i \setminus F_{i+1}$ is nice and connected; then put $[X]^\eta := \sum_i [F_i \setminus F_{i+1}]^\eta$. Finally, in the situation of Section 6.10, one puts

$$\left(\int_Z |\omega|\right)_{\text{usual}} := [X]^\eta \in M_k. \tag{6.7}$$

Clearly, (6.7) is well defined, and the images of (6.7) and (6.6) in $K_0(D_c^b(\operatorname{Spec} k, \mathbb{Z}_l))$ are equal. So (6.7) and (6.6) can be considered as different refinements of the same object of $K_0(D_c^b(\operatorname{Spec} k, \mathbb{Z}_l))$. Unless the map $M_k \to K_0(D_c^b(\operatorname{Spec} k, \mathbb{Z}_l))$ is injective (which seems unlikely), the "refined" motivic integral (6.6) cannot be considered as the refinement of the usual motivic integral (6.7). This is why I am using quotation marks.

6.12 Remark

Our definition of "refined" motivic integration works only in the case of integrating a holomorphic form with no zeros over an affinoid domain (which is probably too special for serious applications).

On the other hand, in an unpublished manuscript V. Vologodsky defined a different kind of "refined motivic integration" in the case of K3 surfaces. More precisely, let $\omega \neq 0$ be a regular differential form on a K3 surface X over $F = k((t))$, $\operatorname{char} k = 0$. Let A denote the Grothendieck ring of the category of Grothendieck motives over k, and let I_n denote the motivic integral of ω over $X \otimes_F k((t^{1/n}))$ viewed as an object of $A \otimes \mathbb{Q}$. Vologodsky defined objects M_1, M_2, M_3 of the category of Grothendieck motives so that I_n is a certain linear combination of the classes of M_1, M_2, M_3. The objects M_1, M_2, M_3 depend functorially on (X, ω). His definition of M_1, M_2, M_3 is mysterious.

6.13 Counterexamples

Here are the examples promised in Remark (ii) of Section 6.4.

[20] M'_k is generated by elements $[X]$ corresponding to isomorphism classes of k-schemes of finite type, and the defining relations are $[X] = [Y] + [X \setminus Y]$ for any k-scheme X of finite type and any closed subscheme $Y \subset X$. In particular, these relations imply that $[X]$ depends only on the reduced subscheme corresponding to X.

6.13.1 Not Zariski-locally trivial dimension torsor

Put $Y := (\mathbb{P}^1 \times \mathbb{P}^1) \setminus \Gamma_f$, where $\mathbb{P}^1$ is the projective line over $F := k((t))$ and Γ_f is the graph of a morphism $f : \mathbb{P}^1 \to \mathbb{P}^1$ of degree $n > 0$. Clearly, Y is affine, and

$$\det \Omega_Y^1 = p_1^* \mathcal{O}(-2) \otimes p_2^* \mathcal{O}(-2) = p_1^* \mathcal{O}(2n-2), \tag{6.8}$$

where $p_1, p_2 : Y \to \mathbb{P}^1$ are the projections. We claim that if $n > 1$, then the dimension torsor of $\mathcal{L}Y$ is not Zariski-locally trivial. Moreover, there exists a morphism $\phi : \operatorname{Spec} R \to \mathcal{L}Y$, $R := \{f \in k[x] | f(0) = f(1)\}$, such that $\phi^* \operatorname{Dim}_{\mathcal{L}Y}$ is not Zariski-locally trivial. One constructs ϕ as follows. Consider the $R((t))$-module M defined by (3.1). One can represent M as a direct summand of $R((t))^2$. Indeed, the $R((t))$-module

$$\{u = u(x,t) \in k[x]((t))^2 \mid u(1,t) = A(t)u(0,t)\}, \quad A(t) := \begin{pmatrix} t & 0 \\ 0 & t^{-1} \end{pmatrix}$$

is isomorphic to $R((t))^2$ because there exists $A(x,t) \in SL(2, k[x,t,t^{-1}])$ such that $A(0,t)$ is the identity matrix and $A(1,t) = A(t)$ (to find $A(x,t)$ represent $A(t)$ as a product of elementary matrices). Representing M as a direct summand of $R((t))^2$ one gets a morphism

$$g : \operatorname{Spec} R((t)) \to \mathbb{P}^1. \tag{6.9}$$

As $p_1 : Y \to \mathbb{P}^1$ is a locally trivial fibration with fiber $\mathbb{A}^1$, one can represent g as $p_1\varphi$ for some $\varphi : \operatorname{Spec} R((t)) \to Y$. Let $\phi : \operatorname{Spec} R \to \mathcal{L}Y$ be the morphism corresponding to φ. By (6.8) and the corollary of Lemma 6.5, the $\mathbb{Z}$-torsor $\phi^* \operatorname{Dim}_{\mathcal{L}Y}$ canonically identifies with the dimension torsor of $M^{\otimes(2n-2)}$. In particular, ϕ has the desired property, i.e., $\phi^* \operatorname{Dim}_{\mathcal{L}Y}$ is not Zariski-locally trivial.

The class of $\phi^* \operatorname{Dim}_{\mathcal{L}Y}$ in $H^1_{\text{ét}}(\operatorname{Spec} R, \mathbb{Z})$ is not a generator of this group (using (6.8) and the morphism (3.10) one sees that it equals $(2n-2)v$, where v is a generator). Below we construct a slightly different pair $(Y, \phi : \operatorname{Spec} R \to \mathcal{L}Y)$ such that the class of $\phi^* \operatorname{Dim}_{\mathcal{L}Y}$ in $H^1_{\text{ét}}(\operatorname{Spec} R, \mathbb{Z})$ is a generator.

6.13.2 Modification of the above example

Let Y be the space of triples (v, l, l'), where l, l' are transversal one-dimensional subspaces in F^2 and $v \in l$. Then there exists a morphism $\phi : \operatorname{Spec} R \to \mathcal{L}Y$, $R := \{f \in k[x] | f(0) = f(1)\}$, such that the class of the $\mathbb{Z}$-torsor $\phi^* \operatorname{Dim}_{\mathcal{L}Y}$ is a generator of $H^1_{\text{ét}}(\operatorname{Spec} R, \mathbb{Z})$.

More precisely, define $\pi : Y \to \mathbb{P}^1$ by $\pi(v, l, l') := l$, let $\tilde{g} : \operatorname{Spec} R((t)) \to Y$ be such that $\pi\tilde{g}$ equals (6.9), and let $\phi : \operatorname{Spec} R \to \mathcal{L}Y$ be the morphism corresponding to $\tilde{g}$. Then the class of $\phi^* \operatorname{Dim}_{\mathcal{L}Y}$ is a generator of $H^1_{\text{ét}}(\operatorname{Spec} R, \mathbb{Z})$.

6.13.3 Any "unpleasant thing" can happen

This is what the following theorem essentially says. For example, combining statement (ii) of the theorem with Weibel's examples mentioned in Section 3.5.4, one sees that for some smooth affine scheme Y over $F = k((t))$ with trivial canonical bundle there exists a reasonable closed subscheme of $\mathcal{L}Y$ which is not Zariski-locally nice (even though its dimension torsor is trivial).

Theorem 6.6. *Let R be a k-algebra and $u \in K_{-1}(R)$.*

(i) *There exists a smooth affine scheme Y over $F = k((t))$ and a morphism f : $\operatorname{Spec} R \to \mathcal{L}Y$ such that the pullback of the cotangent sheaf of $\mathcal{L}Y$ to $\operatorname{Spec} R$ has class u.*
(ii) *If the image of u in $H^1_{\text{ét}}(\operatorname{Spec} R, \mathbb{Z})$ equals 0, then one can choose Y to have trivial canonical bundle (in this case the dimension torsor of $\mathcal{L}Y$ is trivial).*

Sketch of the proof. Consider schemes Y of the following type[21]:

$$\begin{aligned} Y = Y_0 \otimes_k F, \quad Y_0 = (G \times V)/H, \quad G = \operatorname{Aut}(k^m \oplus k^n), \\ H = \operatorname{Aut} k^m \times \operatorname{Aut} k^n, \quad m, n \in \mathbb{N}, \end{aligned} \tag{6.10}$$

where G and H are viewed as algebraic groups over k and V is a suitable representation of H. To prove statement (i) of the theorem it suffices to take $V = \operatorname{Lie}[H, H] \oplus W^*$, where W is the representation of H in k^m. To prove (ii) it suffices to take $V = \operatorname{Lie}[H, H] \oplus W^* \oplus \det W$.

7 Application to finite-dimensional vector bundles on manifolds with punctures

7.1 The top cohomology

Let R be commutative,[22] $S_n := \operatorname{Spec} R[[t_1, \dots, t_n]])$, $\mathbf{0} \subset S_n$ the subset defined by the equations $t_1 = \cdots = t_n = 0$, and $S'_n := S_n \setminus \mathbf{0}$. Let Vect denote the category of vector bundles on S'_n (of finite rank). For $L \in$ Vect write $H^i(L)$ instead of $H^i(S'_n, L)$. The cohomology functors H^i : Vect $\to \{R\text{-modules}\}$ vanish for $i \geq n$ and if $n > 1$, then H^{n-1} commutes with base change $R \to \tilde{R}$, $R[[t_1, \dots, t_n]] \to \tilde{R}[[t_1, \dots, t_n]]$.

Theorem 7.1. *If $n > 1$, then for every $\mathfrak{L} \in$ Vect the R-module $H^{n-1}(S'_n, \mathfrak{L})$ is 2-almost projective.*

[21] The manifold Y from Section 6.13.2 is a particular example of (6.10), in which $m = n = 1$ and $\dim V = 1$.

[22] One can formulate and prove Theorems 7.1–7.3 without the commutativity assumption. In this case there is no S'_n, but one can define a vector bundle on S'_n to be a collection of finitely generated projective modules P_i over $\operatorname{Spec} R[[t_1, \dots, t_n]][t_i^{-1}]$ with a compatible system of isomorphisms $P_i[t_j^{-1}] \xrightarrow{\sim} P_j[t_i^{-1}]$.

7.2 Derived version

A. Beilinson explained to me that such a version should exist. Consider $D^{\text{perf}}(S'_n) = \mathrm{K}^b(\text{Vect}) := \{$homotopy category of bounded complexes in Vect$\}$. We will decompose $R\Gamma : \mathrm{K}^b(\text{Vect}) \to D(R)$ as

$$\mathrm{K}^b(\text{Vect}) \xrightarrow{(R\Gamma)_{\text{topol}}} \mathrm{K}^b(\mathcal{T}_R) \xrightarrow{\text{Forget}} D(R),$$

where $\mathcal{T}_R$ is the category of Tate R-modules. First, we have the derived functor

$$R\Gamma : \mathrm{K}^b(\text{Vect}) \to D^-(R[[t_1, \dots, t_n]]) = \mathrm{K}^-(\mathcal{P}), \tag{7.1}$$

where $\mathcal{P}$ is the category of projective $R[[t_1, \dots, t_n]]$-modules. Second, a projective module P over $R[[t_1, \dots, t_n]]$ carries a natural topology (the strongest one such that all $R[[t_1, \dots, t_n]]$-linear maps from finitely generated free $R[[t_1, \dots, t_n]]$-modules to P are continuous), so we get a functor from $\mathcal{P}$ to the additive category R-top of topological R-modules and therefore a functor

$$\mathrm{K}^-(\mathcal{P}) \to \mathrm{K}^-(R\text{-top}). \tag{7.2}$$

Theorem 7.2. *The composition of* (7.1) *and* (7.2) *belongs to the essential image of* $\mathrm{K}^b(\mathcal{T}_R)$ *in* $\mathrm{K}^-(R\text{-top})$, *so we get a triangulated functor* $(R\Gamma)_{\text{topol}} : \mathrm{K}^b(\text{Vect}) \to \mathrm{K}^b(\mathcal{T}_R)$. *If* $\mathcal{L} \in \text{Vect}$, *then* $(R\Gamma)_{\text{topol}}(\mathcal{L}) \subset \mathrm{K}^{[0,n-1]}(\mathcal{T})$.

To formulate the basic properties of $(R\Gamma)_{\text{topol}}$ we need some notation. Let $\mathcal{C}^{\text{Kar}}$ denote the Calkin category of R (see Section 3.3.1). Consider the functor $\mathrm{K}^b(\mathcal{T}_R) \to \mathrm{K}^b(\mathcal{C}^{\text{Kar}})$ induced by (3.3). The composition

$$\mathrm{K}^b(\text{Vect}) \xrightarrow{(R\Gamma)_{\text{topol}}} \mathrm{K}^b(\mathcal{T}_R) \to \mathrm{K}^b(\mathcal{C}^{\text{Kar}}) \tag{7.3}$$

will be denoted by $R\Gamma_{\text{discr}}$, because the image of a Tate R-module T in $\mathcal{C}^{\text{Kar}}$ may be viewed as the "discrete part" of T. One also has the "compact part" functor from $\mathcal{T}_R$ to the category $(\mathcal{C}^{\text{Kar}})^\circ$ dual to $\mathcal{C}^{\text{Kar}}$: this is the composition of the dualization functor $\mathcal{T}_R \to \mathcal{T}_R^\circ$ and the functor $\mathcal{T}_R^\circ \to (\mathcal{C}^{\text{Kar}})^\circ$ corresponding to (3.3). So we get $R\Gamma_{\text{comp}} : \mathrm{K}^b(\text{Vect}) \to \mathrm{K}^b((\mathcal{C}^{\text{Kar}})^\circ)$.

Theorem 7.1 is an easy consequence of statement (i) of the following theorem.

Theorem 7.3.

(i) *If* $\mathcal{L} \in \text{Vect}$, *then* $R\Gamma_{\text{discr}}(\mathcal{L})$ *is an object of* $\mathcal{C}^{\text{Kar}}$ *placed in degree* $n-1$.
(ii) *If* $\mathcal{L} \in \text{Vect}$, *then* $R\Gamma_{\text{comp}}(\mathcal{L})$ *is an object of* $(\mathcal{C}^{\text{Kar}})^\circ$ *placed in degree* 0.
(iii) *Let* ω *denote the relative (over* R*) canonical line bundle on* S'_n. *Then there is a canonical duality between* $(R\Gamma)_{\text{topol}}(\mathcal{L})$, $\mathcal{L} \in \mathrm{K}^b(\text{Vect})$, *and* $(R\Gamma)_{\text{topol}}(\mathcal{L}^* \otimes \omega[n-1])$.

7.3 The dimension torsor corresponding to a vector bundle

Let $X = \operatorname{Spec} R$ be an affine scheme of finite type over $\mathbb{C}$. Let Y denote $X(\mathbb{C})$ equipped with the usual topology. Given a vector bundle $\mathcal{L}$ on $X \times (\mathbb{A}^n \setminus \{0\})$, $n > 1$, one has the R-module $M := H^{n-1}(X \times (\mathbb{A}^n \setminus \{0\}), \mathcal{L})$, which is almost projective by Theorem 7.1. So by Section 4.3 one has the dimension torsor Dim_M (which can be viewed as a torsor on Y) and its canonical upper semicontinuous section d^{can}. Here is a geometric description of $(\mathrm{Dim}_M, d^{\mathrm{can}})$.

(i) Notice that a complex vector bundle of any rank m on the topological space $\mathbb{C}^n \setminus \{0\}$ is trivial (because $\pi_{2n-2}(GL(m, \mathbb{C})) = 0$), and the homotopy classes of its trivializations form a torsor over $\pi_{2n-1}(GL(m, \mathbb{C}))$. One has the natural morphism $\pi_{2n-1}(GL(m, \mathbb{C})) \to \pi_{2n-1}(GL(\infty, \mathbb{C})) = K_0^{\mathrm{top}}(S^{2n-2}) = \mathbb{Z}$. So $\mathcal{L}$ defines a $\mathbb{Z}$-torsor $T_{\mathcal{L}}$ on Y.

(ii) More generally, a finite complex $\mathcal{L}^{\cdot}$ of topological vector bundles on $Y \times (\mathbb{C}^n \setminus \{0\})$ defines a $\mathbb{Z}$-torsor $T_{\mathcal{L}^{\cdot}} := \sum_i (-1)^i T_{\mathcal{L}^i}$, and a homotopy equivalence $f : \mathcal{L}_1^{\cdot} \to \mathcal{L}_2^{\cdot}$ defines an isomorphism $T_{\mathcal{L}_1} \xrightarrow{\sim} T_{\mathcal{L}_2}$ (because the dimension torsor of $\mathrm{Cone}(f)$ is canonically trivialized). Of course, this isomorphism depends only on the homotopy class of f. An extension of $\mathcal{L}^{\cdot}$ to an object of the homotopy category of complexes of topological vector bundles on $Y \times \mathbb{C}^n$ defines a trivialization of $T_{\mathcal{L}^{\cdot}}$.

(iii) Let $\mathcal{L}$ be an algebraic vector bundle on $X \times (\mathbb{A}^n \setminus \{0\})$, $n > 1$. Let $j : \mathbb{A}^n \setminus \{0\} \to \mathbb{A}^n$ be the embedding. For each $x \in X$ the sheaf $j_* \mathcal{L}_x$ is coherent and has a finite locally free resolution (here $\mathcal{L}_x$ is the restriction of $\mathcal{L}$ to $\{x\} \times (\mathbb{A}^n \setminus \{0\})$). So by (ii) one gets a trivialization of $T_{\mathcal{L}_x}$ for each x, i.e., a set-theoretical section s of $T_{\mathcal{L}}$.

(iv) One can show that $(T_{\mathcal{L}}, s)$ is canonically isomorphic to $(\mathrm{Dim}_M, d^{\mathrm{can}})$ (maybe up to a sign).

7.4 Central extension

Theorem 7.4. *Let X be a smooth scheme over $S := \operatorname{Spec} R$ of pure relative dimension $n > 1$. Let $F \subset X$ be a closed subscheme which is finite over S and a locally complete intersection over S. Let $j : X \setminus F \hookrightarrow X$ denote the open embedding. Then for any vector bundle $\mathcal{L}$ on $X \setminus F$ the R-module $H^0(X, \mathrm{R}^{n-1} j_* \mathcal{L})$ is 2-almost projective.*

This easily follows from Theorems 7.1 and 4.2(i).

Now let O_F be the ring of regular functions on the formal completion of X along F. In the situation of Theorem 7.4 $H^0(X, \mathrm{R}^{n-1} j_* \mathcal{L})$ is an O_F-module, so it is equipped with an action of the group scheme $G := O_F^{\times}$. Therefore, applying (5.3), one gets a central extension

$$0 \to \mathbb{G}_m \to \widehat{G}_{\mathcal{L}} \to O_F^{\times} \to 0. \tag{7.4}$$

Remarks.

(i) Suppose that in the situation of Theorem 7.4 $\mathcal{L}$ extends to a vector bundle on X. Then the R-module $H^0(X, \mathrm{R}^{n-1} j_*\mathcal{L})$ is projective, and therefore the extension (7.4) canonically splits.
(ii) Suppose that in the situation of Theorem 7.4 $F \subset \tilde{F} \subset X$ and $\tilde{F}$ satisfies the same conditions as F. Put $\tilde{\mathcal{L}} := \mathcal{L}|_{X \setminus \tilde{F}}$. Then we have the central extension (7.4) and a similar central extension

$$0 \to \mathbb{G}_m \to \widehat{G}_{\tilde{\mathcal{L}}} \to O_{\tilde{F}}^{\times} \to 0. \tag{7.5}$$

Using the functor $(R\Gamma)_{\mathrm{topol}}$ from Section 7.2 one can construct a canonical morphism from (7.5) to (7.4) which induces the restriction map $O_{\tilde{F}}^{\times} \to O_F^{\times}$ and the identity map $\mathbb{G}_m \to \mathbb{G}_m$. If $F = \emptyset$ this amounts to Remark (i) above.

7.5 Commutativity theorem

Let $c_{\mathcal{L}} : O_F^{\times} \times O_F^{\times} \to \mathbb{G}_m$ be the commutator map of the central extension (7.4).

Theorem 7.5. *Suppose that in the situation of Theorem* 7.4 $n = 2$. *Then* $c_{\mathcal{L}} = 1$ *if and only if* $\det \mathcal{L}$ *extends to an invertible sheaf on* X.

Remarks.

(i) If an extension of $\det \mathcal{L}$ to an invertible sheaf on X exists, it equals $j_* \det \mathcal{L}$. In particular, the extension is unique.
(ii) Theorems 7.4 and 7.5 are still true for vector bundles on $(\operatorname{Spec} O_F) \setminus F$ instead of $X \setminus F$.

Question. What is the geometric meaning of $c_{\mathcal{L}}$ and the condition $c_{\mathcal{L}} = 1$ if $n > 2$?

7.6 Generalizing the notion of vector bundle on a surface

Let G be a reductive group over $\mathbb{Q}$. The moduli scheme of G-bundles on $\mathbb{P}^2_{\mathbb{Q}}$ trivialized over a fixed projective line $\mathbb{P}^1_{\mathbb{Q}} \subset \mathbb{P}^2_{\mathbb{Q}}$ has a remarkable "Uhlenbeck compactification" $\mathfrak{U}_G$ constructed in [FGK, BFG], which goes back to the physical picture of "instanton gas." It would be very important to interpret $\mathfrak{U}_G$ as a moduli scheme of some kind of geometric objects.[23] These conjectural new objects are, so to speak, "G-bundles with singularities." I suggest calling them *G-gundles*. The new word "gundle" can be considered as an abbreviation for "generalized G-bundle." On the other hand, its first three letters are also the first letters of the names of D. Gaitsgory, V. Ginzburg, K. Uhlenbeck, and H. Nakajima.[24]

---

[23] For example, such an interpretation would hopefully allow to define an analog of $\mathfrak{U}_G$ for *any* proper smooth surface.

[24] Gaitsgory is an author of [FGK, BFG], and the relation of the other three mathematicians to these articles is explained in the introductions to them.

It turns out that the central extension (7.4) allows us to give a definition of $GL(n)$-gundle on any smooth family of surfaces over any base so that $\mathfrak{U}_{GL(n)}$ identifies with the moduli scheme of $GL(n)$-gundles on $\mathbb{P}^2_{\mathbb{Q}}$ trivialized over $\mathbb{P}^1_{\mathbb{Q}}$.

Let X be a scheme smooth over S of pure relative dimension 2. The definition of $GL(n)$-gundle on X consists of several steps. I will only explain the first one and list the other steps.

7.6.1 Pre-gundles 1

Let F be as in Theorem 7.4.

Definition. A *$GL(n)$-pre-gundle on X nonsingular outside F* is a pair that consists of a rank n vector bundle $\mathcal{L}$ on $X \setminus F$ and a splitting of (7.4). The groupoid of such pairs will be denoted by $\text{Pre-gun}_F(X)$.

Remark. If (7.4) admits a splitting, then by Theorem 7.5 $\det \mathcal{L}$ extends to a line bundle on X.

7.6.2 Pre-gundles 2

If F, $\tilde{F}$ are as in Theorem 7.4 and $\tilde{F} \supset F$, then one defines a fully faithful functor $\text{Pre-gun}_F(X) \to \text{Pre-gun}_{\tilde{F}}(X)$ using Remark (ii) at the end of Section 7.4.

7.6.3 Pre-gundles 3

If X is projective over S, one defines the groupoid of pre-gundles on X to be the inductive 2-limit of $\text{Pre-gun}_F(X)$ over all closed subschemes $F \subset X$ which are finite over S and locally complete intersections over S. This groupoid is denoted by $\text{Pre-gun}(X)$, and its objects are called *$GL(n)$-pre-gundles on X*.

If X is arbitrary, one first defines $\text{Pre-gun}_F(X)$ for *any* subscheme $F \subset X$ quasi-finite over S (a standard étale or Nisnevich localization technique allows one to reduce this to the case of finite locally complete intersection considered above). Then one defines $\text{Pre-gun}(X)$ to be the inductive 2-limit of $\text{Pre-gun}_F(X)$ over *all* closed subschemes $F \subset X$ quasi-finite over S.

7.6.4 Remark

If S is the spectrum of a field, then $\text{Pre-gun}(X)$ identifies with the groupoid of pairs $(\mathcal{L}, Z)$ with $\mathcal{L}$ being a $GL(n)$-bundle on X and Z a 0-cycle on X, it being understood that an isomorphism $(\mathcal{L}_1, Z) \xrightarrow{\sim} (\mathcal{L}_2, Z)$ is the same as an isomorphism $\mathcal{L}_1 \xrightarrow{\sim} \mathcal{L}_2$ and that there are no isomorphisms $(\mathcal{L}_1, Z_1) \xrightarrow{\sim} (\mathcal{L}_2, Z_2)$ if $Z_1 \neq Z_2$.

To see this, first notice that for any finite $F \subset X$ a vector bundle on $X \setminus F$ uniquely extends to X. Second, by Remark (i) from Section 7.4, the central extension (7.4) has a canonical splitting, so *all* splittings of (7.4) are parametrized by $\text{Hom}(O_F^\times, \mathbb{G}_m)$, i.e., by the group of 0-cycles on X supported on F.

7.6.5 Pre-gundles 4

Let S again be arbitrary. Associating to an S-scheme S' the groupoid of pre-gundles on $X \times_S S'$ one gets a (nonalgebraic) S-stack Pre-gun_X. This is *the stack of* $GL(n)$-*pre-gundles on* X.

7.6.6 Gundles

One defines a closed substack $\text{Gun}_X \subset \text{Pre-gun}_X$, whose formation commutes with base change $S' \to S$. Its S-points are called $GL(n)$-*gundles on* X.

By Section 7.6.4, if $S = \operatorname{Spec} k$ with k being a field, then $GL(n)$-pre-gundles on X identify with pairs $(\mathcal{L}, Z)$ with $\mathcal{L}$ being a $GL(n)$-bundle on X and Z being a 0-cycle on X. It turns out that *such a pair* $(\mathcal{L}, Z)$ *is a* $GL(n)$-*gundle if and only if* $Z \geq 0$.

Remark. I can define the closed substack $\text{Gun}_X \subset \text{Pre-gun}_X$ using the method of [FGK, BFG], i.e., by working with various curves on X. Unfortunately, I do not know a "purely two-dimensional" way to do it.

7.6.7 Hope

If X is proper over S, then Gun_X is an algebraic stack.

7.6.8 Fact

Now let $S = \operatorname{Spec} \mathbb{Q}$ and $X = \mathbb{P}^2_{\mathbb{Q}}$. Fix a projective line $\mathbb{P}^1_{\mathbb{Q}} \subset \mathbb{P}^2_{\mathbb{Q}}$, and consider the open substack $U \subset \text{Gun}_X$ parametrizing those gundles which are nonsingular on a neighborhood of $\mathbb{P}^1$ and whose restriction to $\mathbb{P}^1$ is trivial. Then U identifies with the quotient of the "Uhlenbeck compactification" $\mathfrak{U}_{GL(n)}$ from [BFG] by the action[25] of $GL(n)$. In particular, the stack U is algebraic.

Acknowledgments. I thank A. Beilinson, D. Gaitsgory, V. Ginzburg, D. Hirschfeldt, M. Kapranov, D. Kazhdan, A. Suslin, and C. Weibel for stimulating discussions.

Speaking at the "Unity of Mathematics" conference (Harvard, 2003) was a great honor and pleasure for me, and an important stimulus to write this article. I am very grateful to the organizers of the conference.

This research was partially supported by NSF grants DMS-0100108 and DMS-0401164.

References

[ACK] E. Arbarello, C. De Concini, and V. G. Kac, Infinite wedge representation and the reciprocity law on algebraic curves, in *Theta Functions, Bowdoin* 1987, Proceedings of Symposia in Pure Mathematics 49, Part 1, American Mathematical Society, Providence, RI, 1989, 171–190.

[25] $GL(n)$ acts on $\mathfrak{U}_{GL(n)}$ by changing the trivialization over $\mathbb{P}^1$.

[AM] M. Artin and B. Mazur, *Etale Homotopy*, Lecture Notes in Mathematics 100, Springer-Verlag, Berlin, New York, Heidelberg, 1969.

[Ba] H. Bass, *Algebraic K-Theory*, Benjamin, New York, 1968.

[Ba2] H. Bass, Big projective modules are free, *Illinois J. Math.*, **7** (1963), 24–31.

[BBD] A. Beilinson, J. Bernstein, and P. Deligne, Faisceaux pervers, in *Analyse et topologie sur les espaces singuliers*, Vol. 1, Astérisque 100, Société Mathématique de France, Paris, 1982.

[BBE] A. Beilinson, S. Bloch, and H. Esnault, Epsilon-factors for Gauss-Manin determinants, *Moscow Math, J.*, **2**-3 (2002), 477–532; see also xxx.lanl.gov, e-print math.AG/0111277.

[Be] V. G. Berkovich. *Spectral Theory and Analytic Geometry over Non-Archimedean Fields*, Mathematical Surveys and Monographs 33, American Mathematical Society, Providence, RI, 1990.

[BFG] A. Braverman, M. Finkelberg, and D. Gaitsgory, Uhlenbeck spaces via affine Lie algebras, e-print AG/0301176, 2003.

[BGR] S. Bosch, U. Güntzer, and R. Remmert, *Non-Archimedean Analysis: A Systematic Approach to Rigid Analytic Geometry*, Grundlehren der Mathematischen Wissenschaften 261, Springer-Verlag, Berlin, 1984.

[Br] J.-L. Brylinski, Central extensions and reciprocity laws, *Cahiers Topol. Géom. Différentielle Catég.*, **38**-3 (1997), 193–215.

[Del] P. Deligne, La formule de dualité globale, in *Theorie des topos et cohomologie étale des schemas* (*SGA* 4), Tome 3, Lecture Notes in Mathematics 305, Springer-Verlag, Berlin, New York, Heidelberg, 1973, 481–587.

[Del87] P. Deligne, Le déterminant de la cohomologie, in K. A. Ribet, ed., *Current Trends in Arithmetical Algebraic Geometry*, Contemporary Mathematics 67, American Mathematical Society, Providence, RI, 1987, 93–177.

[Dr] V. Drinfeld, Infinite-dimensional vector bundles in algebraic geometry, to appear at xxx.lanl.gov.

[E] T. Ekedahl, On the adic formalism, in *Grothendieck Festschrift*, Vol. II, Progress in Mathematics 87, Birkhäuser Boston, Cambridge, MA, 1990, 197–218.

[FGK] M. Finkelberg, D. Gaitsgory, and A. Kuznetsov, Uhlenbeck spaces for $\mathbb{A}^2$ and affine Lie algebra $\widehat{\mathfrak{sl}}_n$, *Publ. Res. Inst. Math. Sci.*, **39**-4 (2003), 721–766.

[Fr] P. Freyd, *Abelian Categories: An Introduction to the Theory of Functors*, Harper and Row, New York, 1964.

[Gov] V. E. Govorov, On flat modules, *Siberian Math. J.*, **6** (1965), 300–304 (in Russian).

[Gr64] A. Grothendieck, *Éléments de géométrie algébrique* IV: *Étude locale des schémas et de morphismes de schémas*, première partie, Publications Mathématiques IHES 20, Institut des Hautes Études Scientifiques, Bures-sur-Yvette, France, 1964.

[Gr] A. Grothendieck, *Éléments de géométrie algébrique* IV: *Étude locale des schémas et de morphismes de schémas*, quatrième partie, Publications Mathématiques IHES 32, Institut des Hautes Études Scientifiques, Bures-sur-Yvette, France, 1967.

[Gr2] A. Grothendieck, *Catégories cofibrées additives et complexe cotangent relatif*, Lecture Notes in Mathematics 79, Springer-Verlag, Berlin, New York, Heidelberg, 1968.

[GV] A. Grothendieck and J. L. Verdier, Préfaisceaux, in *Théorie des topos et cohomologie étale des schemas* (*SGA* 4), Tome 1, Lecture Notes in Mathematics 269, Springer-Verlag, Berlin, New York, Heidelberg, 1972, 1–217.

[Ja] U. Jannsen, Continuous étale cohomology, *Math. Ann.*, **280**-2 (1988), 207–245.

[Ka] I. Kaplansky, Projective modules, *Ann. Math.*, **68** (1958), 372–377.

[Ka2] I. Kaplansky, *Infinite Abelian Groups*, University of Michigan Press, Ann Arbor, MI, 1969.

[Ka3] M. Kapranov, Semiinfinite symmetric powers, e-print math.QA/0107089, 2001.

[KV] M. Kapranov and E. Vasserot, Vertex algebras and the formal loop space, *Publ. Math. IHES*, **100** (2004), 209–269.

[Laz] D. Lazard, Autour de la platitude, *Bull. Soc. Math. France*, **97** (1969), 81–128.

[L] S. Lefschetz, *Algebraic Topology*, AMS Colloquium Publications 27, American Mathematical Society, Providence, RI, 1942.

[Lo] E. Looijenga, Motivic measures, in *Seminaire Bourbaki*, Vol. 1999/2000, Astérisque 276, Société Mathématique de France, Paris, 2002, 267–297; see also xxx.lanl.gov, e-print math.AG/0006220.

[MSV] F. Malikov, V. Schechtman, and A. Vaintrob, Chiral de Rham complex, *Comm. Math. Phys.*, **204** (1999), 439–473.

[M] J. P. May, E_∞ spaces, group completions, and permutative categories, in *New Developments in Topology*, London Mathematical Society Lecture Note Series 11, Cambridge University Press, London, 1974, 61–93.

[PI] Pure Imaginary Observer, The future of field theory, *J. Irreproducible Results*, **12**-3 (1963), 3–5 (English); in *Fiziki Prodolzhayut Shutit*, Mir, Moscow, 1968 (Russian).

[PS] A. Pressley and G. Segal, *Loop Groups*, Oxford University Press, Oxford, UK, 1986.

[Re] L. Reid, N-dimensional rings with an isolated singular point having nonzero K_{-N}, *K-Theory*, **1**-2 (1987), 197–205.

[RG] M. Raynaud and L. Gruson, Critères de platitude et de projectivité, *Invent. Math.*, **13** (1971), 1–89.

[Seg] G. Segal, Categories and cohomology theories, *Topology*, **13** (1974), 293–312.

[S] J.-P. Serre, Modules projectifs et espaces fibrés à fibre vectorielle, in *Algèbre Théorie Nombres*, Séminaire P. Dubreil, M.-L. Dubreil-Jacotin et C. Pisot 11, 1957–1958, Secrétariat Mathématique, Paris, 1958, Exposé 23.

[So] R. M. Solovay, A model of set-theory in which every set of reals is Lebesgue measurable, *Ann. Math.* (2), **92** (1970), 1–56.

[T] J. Tate, Residues of differentials on curves, *Ann. Sci. École Norm. Sup. Ser.* 4, **1** (1968), 149–159.

[TT] R. W. Thomason and T. Trobaugh, Higher algebraic K-theory of schemes and of derived categories, in *Grothendieck Festschrift*, Vol. III, Progress in Mathematics 88, Birkhäuser Boston, Cambridge, MA, 1990, 247–435.

[Vo] V. Voevodsky, $\mathbb{A}^1$-homotopy theory, *Documenta Math.*, **Extra Vol. I** (1998) (Proceedings of the International Congress of Mathematicians (Berlin, 1998), Vol. I), 579–604 (electronic).

[VSF] V. Voevodsky, A. Suslin, and E. M. Friedlander, *Cycles, Transfers, and Motivic Homology Theories*, Annals of Mathematics Studies 143, Princeton University Press, Princeton, NJ, 2000.

[VMW] V. Voevodsky, C. Mazza, and C. Weibel, Lectures on motivic cohomology, available online from http://math.rutgers.edu/~weibel/motiviclectures.html.

[We1] C. Weibel, Negative K-theory of varieties with isolated singularities, *J. Pure Appl. Algebra*, **34**-2–3 (1984), 331–342.

[We2] C. Weibel, Pic is a contracted functor, *Invent. Math.*, **103**-2 (1991), 351–377.

[We3] C. Weibel, Negative K-theory of surfaces, *Duke Math. J.*, **108**-1(2001), 1–35.

Algebraic Lessons from the Theory of Quantum Integrable Models

L. D. Faddeev

St. Petersburg Department of Steklov Mathematical Institute
27 Fontanka
191011 St. Petersburg
Russia
faddeev@pdmi.ras.ru

Summary. Several examples of quantum systems, obtained after discretization of space-time in the elementary models of conformal field theory, are considered. The algebraic structure of the corresponding algebra of observables and deformation of Virasoro symmetry are discussed and construction of the evolution operator is given.

Subject Classifications: 81R12, 81J40, 37K10

1 Introduction

Integrable evolution equation in $(1+1)$-dimensional space-time (integrable models) have been the subject of discussion for the last 35 years beginning with the papers [1, 2]; see the history and references in [3]. Their quantization, which makes sense due to the inherent Hamiltonian structure, was developed mostly during the 1980s on the basis of the algebraic Bethe Ansatz (see the survey in [4]) and produced beautiful algebraic structures, such as the Yang–Baxter relations and quantum groups. These objects already appeared on the kinematic level of description. In this talk I want to emphasize lesser-known features connected with dynamics, namely, the evolution operator in the Heisenberg picture. To this end, I shall use models defined on the discretized space-time. The mere existence of such integrable deformations is a result of more recent developments, and I shall give the relevant references below. It is remarkable that interesting formulas appear even in the examples corresponding in the continuum limit to the simplest models of conformal field theory. In this case the evolution operator is a mere shift along the spatial lattice site. However, even here the exact formulas will be quite instructive.

I begin by recalling the examples in the classical continuous space context, then describe space discretization and quantization, and finally produce the corresponding evolution operator. An unexpected new feature—modular doubling—will appear and I shall briefly discuss it.

2 Main examples

2.a Abelian current

The field variable $p(x)$ is defined on a circle, representing space, so that the space-time is $\mathbb{S}^1 \times \mathbb{R}$. To symbolize the circle, we shall use the periodic conditions

$$p(x+L) = p(x). \tag{1}$$

The manifold with coordinates $p(x)$ is a phase space with the Poisson bracket

$$\{p(x), p(y)\} = \gamma\delta'(x-y), \tag{2}$$

where $\delta'(x)$ is the first derivative of the delta function. The parameter γ plays the role of the coupling constant. Its relevance will become clear only after quantization. For the time being, it is a positive real number.

The function

$$P = \int_0^L p(x)dx \tag{3}$$

is a central element. The submanifold defined by the fixing the value of P is symplectic.

The Hamiltonian

$$H = \frac{1}{2\gamma}\int_0^L p^2(x)dx \tag{4}$$

gives a linear equation of motion

$$\dot{p} + p' = 0, \tag{5}$$

where dot and prime symbolize the time and space derivatives, respectively. The solution

$$p(x,t) = p(x-t), \tag{6}$$

(left mover) shows that the dynamics is just a shift along the space axis.

Let us mention in passing that the first nontrivial integrable equation of motion is given by the Hamiltonian

$$H = \frac{1}{2\gamma}\int_0^L \left(\frac{1}{3}p^3(x) + (p')^2(x)\right) dx, \tag{7}$$

and it is the famous KdV equation

$$\dot{p} + pp' + p''' = 0.$$

However, in what follows we shall discuss only the first rather trivial example of dynamics.

2.b Virasoro algebra

The combination

$$s(x) = p^2(x) + p'(x) \tag{8}$$

satisfies the Poisson bracket relation

$$\{s(x), s(y)\} = 2\gamma(s(x) + s(y))\delta'(x-y) + \gamma\delta'''(x-y), \tag{9}$$

which corresponds to the central extension of the algebra of diffeomorphisms on the circle. In the normalization used in the physics literature the corresponding central charge is given by

$$C_{\text{class}} = \frac{6\pi}{\gamma}.$$

The Hamiltonian

$$H = \frac{1}{2\gamma}\int s(x)dx,$$

which evidently coincides with (4), leads to the equation of motion of the left mover

$$\dot{s} + s' = 0. \tag{10}$$

Let us recall that the quantization of the relations (9) (see, e.g., [5, 6]) leads to the change of the central charge

$$C_{\text{quant}} = 1 + 6\left(\frac{\pi}{\gamma} + \frac{\gamma}{\pi} + 2\right) = 1 + 6\left(\tau + \frac{1}{\tau} + 2\right), \quad \tau = \frac{\gamma}{\pi}. \tag{11}$$

2.c Nonabelian current

Let $\mathcal{A}$ be a compact Lie algebra and t^a its generators with the structure constants f^{abc},

$$[t^a, t^b] = \sum_c f^{abc} t^c,$$

and normalization,

$$\text{tr}(t^a t^b) = \delta^{ab}.$$

The dynamical variable $L(x)$ (nonabelian current) has values in $\mathcal{A}$ and can be parameterized as

$$L(x) = \sum_a L^a(x) t^a$$

by periodic functions $L^a(x)$, $L^a(x+L) = L^a(x)$.

The Poisson brackets

$$\{L^a(x), L^b(y)\} = \gamma \sum_c f^{abc} L^c(x)\delta(x-y) + \gamma\delta^{ab}\delta'(x-y) \tag{12}$$

are the natural generalizations of the relation (2). The Hamiltonian

$$H = \frac{1}{2\gamma}\int_0^L \operatorname{tr} L^2 dx = \frac{1}{2\gamma}\int_0^L \sum_a (L^a)^2 dx$$

leads to the equation of motion of the left mover

$$\dot{L} + L' = 0$$

[7, 8].

Only these three examples will be used in what follows. Let us stress that they are connected with the most familiar examples of infinite-dimensional differential-geometric objects: the Virasoro algebra and Kac–Moody algebra. The variables $p(x)$, $s(x)$ and $L(x)$ have a natural geometric interpretation as connections in a vector bundle ($p(x)$ and $L(x)$) or a projective connection ($s(x)$). Thus it is only natural to consider their quantum counterparts and automorphisms.

3 Discretization and quantization

The space-time variables (x, t) will be replaced by the vertices (n, m) of the lattice. Here (n, m) are integers and the periodicity condition (1) will be replaced by the identification

$$n \equiv n + N.$$

Thus the space-time turns into $\mathbb{Z}_N \times \mathbb{Z}$. This construction can be thought of as a deformation of $\mathbb{S}^1 \times \mathbb{R}$ with the deformation parameter Δ—the length of the space-time unit of separation. The continuum contraction is obtained by the limit $\Delta \to 0$ in the expressions

$$x = n\Delta, \qquad t = m\Delta, \qquad L = N\Delta.$$

The discrete space $\mathbb{Z}_N$ will be called a chain. The dynamical variables $p(x)$, $s(x)$, $L(x)$ turn into p_n, s_n, L_n. The equation of the left mover changes into the shift

$$g_n \to g_{n-1}, \tag{13}$$

where g_n stands for p_n, s_n or L_n.

After quantization the variables g_n become generators of the algebra of observables. The shift (13) becomes an automorphism of this algebra. In the Heisenberg picture, it is given by the time evolution operator V such that

$$g_n V = V g_{n-1}.$$

In what follows we shall give the discrete deformation, quantization and construction of the operator V for all our three basic examples.

3.a Abelian current

The generators are comprised of a set of $p_n, n = 1, \dots N$, defined over discrete circle $\mathbb{Z}_n$ or on the whole $\mathbb{Z}$ with periodicity condition $p_{n+N} = p_n$. The Poisson relation (2) is replaced by the commutator one

$$\frac{i}{\hbar}[p_m, p_n] = 2\gamma(\delta_{m+1,n} - \delta_{m,n+1}). \tag{14}$$

Here $\delta_{m,n}$ is the Kronecker symbol, $\hbar$ is the Planck constant, and the RHS in (14) give the simplest deformation of the derivative of the δ-function. Relation (14) contracts to (2) in the limit $\hbar \to 0$, $\Delta \to 0$ if we put

$$p_n = 2\Delta p(x), \quad x = n\Delta.$$

It is convenient to introduce the Weyl-type variables

$$w_n = e^{ip_n}$$

instead of p_n. They have a natural interpretation as the holonomy along one lattice site for the connection $p(x)$.

The relations (14) lead to the following relations among the w_n:

$$w_{n-1}w_n = q^2 w_n w_{n-1} \tag{15}$$

and

$$w_n w_m = w_m w_n, \quad |n-m| \geq 2$$

with the factor q given by

$$q = e^{i\hbar\gamma}.$$

We see how the coupling constant γ enters together with the quantization constant $\hbar$. For real γ the factor q lives on the unit circle

$$|q| = 1, \quad \gamma \text{ real}$$

and there are two natural involutions for the generators w_n: the compact,

$$w_n^* = w_n^{-1},$$

and the noncompact,

$$w_n^* = w_n.$$

The algebra generated by w_n is a discrete quantum counterpart of the space of functions on the phase space of variables $p(x)$. It has one central element (the analogue of P from (3)),

$$Q = w_1 w_2 \dots w_N,$$

for N odd and two,

$$Q_1 = w_1 w_3 \dots w_{N-1},$$
$$Q_2 = w_2 w_4 \dots w_N,$$

for N even. The shift $w_n \to w_{n-1}$ is an automorphism for N even only if

$$Q_1 = Q_2. \tag{16}$$

Let us turn to the construction of the evolution operator V such that

$$w_n V = V w_{n-1} \tag{17}$$

following [9]. It was shown there that for N even and under the condition (16), it can be found in the form

$$V = \theta(w_1) \dots \theta(w_{N-1}), \tag{18}$$

where the function $\theta(w)$ satisfies the functional equation

$$\theta(q^2 w) = \frac{\theta(w)}{w} \tag{19}$$

up to a constant factor. Note that expression (18) is an ordered product of local factors with one factor $\theta(w_N)$ explicitly omitted.

To check the property (17) for w_n with $2 \leq n \leq N-1$ is easy. Indeed, for such w_n we have

$$w_n V = \theta(w_1) \dots w_n \theta(w_{n-1}) \theta(w_n) \dots \theta(w_{N-1})$$

and

$$V w_{n-1} = \theta(w_1) \dots \theta(w_{n-1}) \theta(w_n) w_{n-1} \dots \theta(w_{N-1}).$$

Thus the property (17) is satisfied if we have the relation

$$w_n \theta(w_{n-1}) \theta(w_n) = \theta(w_{n-1}) \theta(w_n) w_{n-1}. \tag{20}$$

Using the main exchange relation (15), we transform this into

$$\theta(q^{-2} w_{n-1}) w_n \theta(w_n) = \theta(w_{n-1}) w_{n-1} \theta(q^{-2} w_n).$$

and see that it is satisfied due to equation (19).

The check (17) for w_1 is little more elaborate. We have

$$\begin{aligned} w_1 \theta(w_1) \dots \theta(w_{N-1}) &= \theta(w_1) \theta(q^2 w_2) w_1 \theta(w_3) \dots \theta(w_{N-1}) \\ &= \theta(w_1) \theta(w_2) w_2^{-1} \theta(w_3) \dots \theta(w_{N-1}) w_1 \end{aligned}$$

and continuing taking consecutive factors w_{2m+1} or w_{2m}, $m = 1, \dots \frac{N-2}{2}$ to the right finally get

$$w_1 V = V \frac{1}{w_2 w_4 \ldots w_{N-2}} w_1 \ldots w_{N-1} = V w_N$$

if (16) is satisfied. It follows from the last equation that

$$\theta(w_1) V = V \theta(w_N),$$

and using the explicit form of V, we get the relation

$$\theta(w_1) \ldots \theta(w_{N-1}) = \theta(w_2) \ldots \theta(w_N). \tag{21}$$

From this the relation (17) for w_N follows as above.

Now we see that (21) is a particular case of the cyclic relations

$$V = b_1 b_2 \ldots b_{N-1} = b_2 b_3 \ldots b_N = \cdots = b_N b_1 \ldots b_{N-2}, \tag{22}$$

where we defined

$$b_n = \theta(w_n).$$

Furthermore, it follows from (20) that the generators b_n satisfy the Artin Braid Group relations

$$b_{n-1} b_n b_{n-1} = b_n b_{n-1} b_n \tag{23}$$

and

$$b_n b_m = b_m b_n, \quad |n - m| \geq 2. \tag{24}$$

It was shown by A. Volkov (mentioned in [10]) that the "global relations" (22) and the "local relations" (23) are equivalent if the locality condition (24) is satisfied. The variables b_n, $n = 1, \ldots N - 1$, generate the Braid Group $\mathcal{B}_{N-1}$; the variable b_N is a function of them, which can be found from one of the relations (22).

Thus we have established an intimate connection between the deformed algebra of abelian currents and the Braid Group. This is the first algebraic lesson in our exposition.

For N odd the simple formula (18) for V does not hold. One way out is to use some ordering in the algebra generated by w_n. It is easy to see that for a general monomial we have the following reaction to a shift of the origin in the product over $\mathbb{Z}_N$:

$$\begin{aligned} q^{2n_1 n_N} w_1^{n_1} w_2^{n_2} \ldots w_N^{n_N} &= q^{2n_2 n_1} w_2^{n_2} \ldots w_N^{n_N} w_1^{n_1} \\ &= \cdots = q^{2n_k n_{k-1}} w_k^{n_k} \ldots w_N^{n_N} w_1^{n_1} \ldots w_{k-1}^{n_{k-1}}. \end{aligned} \tag{25}$$

This allows us to introduce the cyclic ordering

$$(w_1^{n_1} w_1^{n_1} \ldots w_N^{n_N})_{\text{cycl}}$$

in the algebra generated by w_n for $N \geq 3$, making it equal to any term in (25). In particular, the formula for the product of local factors

$$(\theta_1(w_1)\dots\theta_N(w_N))_{\text{cycl}}$$

means the following: use the series

$$\theta_k(w) = \sum_{-\infty}^{\infty} a_{k,l} w^l$$

for each factor and in each monomial entering the product change the usual product of the powers of $w_k^{n_k}$ into the cyclic one. It is easy to see that we have the cyclic property

$$\begin{aligned}(\theta_1(w_1)\dots\theta_N(w_N))_{\text{cycl}} &= (\theta_2(w_2)\dots\theta_N(w_N)\theta_1(w_1))_{\text{cycl}} \\ &= \dots = (\theta_k(w_k)\dots\theta_N(w_N)\theta_1(w_1)\dots\theta_{k-1}(w_{k-1}))_{\text{cycl}}.\end{aligned}$$

Moreover, if we choose the "point of departure" in the product, all local factors but the first and the last one remain intact. Thus the check of property (17) for V given by the formula

$$V = (\theta(w_1)\dots\theta(w_N))_{\text{cycl}} \tag{26}$$

works for any N without complications.

A natural question appears about the relation between the definitions (26) and (18) for even N. The answer is that these formulas differ by a factor, expressed via the central elements, written formally as

$$\sum_{k=-\infty}^{\infty} (Q_1/Q_2)^k,$$

which is 0 for $Q_1 \neq Q_2$ and ∞ for $Q_1 = Q_2$. Thus the expression (26) for V is correct for odd N, and for even N one must use the expression (18).

We stop here the discussion of the abelian current and return to the explicit formula for $\theta(w)$ in the next section.

3.b Discrete Virasoro

We need to realizethe deformed version of the classical relation (9). The first version was given in [11]. Here we shall arrive at the same formula using the algebraic trick from [9]. It is easy to see that if the function $\gamma(w)$ satisfies the functional equation

$$\gamma(q^2 w) = \gamma(w)\frac{1}{1+qw}, \tag{27}$$

then

$$\theta(w) = \gamma(w)\gamma(w^{-1}) \tag{28}$$

satisfies equation (19). It was shown purely algebraically in [9] that from the equation (27) the following relation holds: let u, v be a Weyl pair

$$uv = q^2 vu. \tag{29}$$

Then we have the relations

$$\begin{aligned}\gamma(u)\gamma(v) &= \gamma(u+v),\\ \gamma(v)\gamma(u) &= \gamma(u+v+q^{-1}uv) = \gamma(u)\gamma(q^{-1}uv)\gamma(v).\end{aligned} \tag{30}$$

The last relation can be naturally called the pentagon equation and so $\gamma(u)$ can be considered as a definition of some noncommutative 3-cocycle. There will be more about this in the next section.

Now we arrive at a natural definition of the discrete Virasoro generators s_n. Take the expression (18) for the evolution operator V for even N,

$$V = \theta(w_1)\dots\theta(w_n)\theta(w_{n+1})\dots\theta(w_{N-1}),$$

and transform the generic factor $\theta(w_n)\theta(w_{n+1})$ using (28) and (30),

$$\begin{aligned}\theta(w_n)\theta(w_{n+1}) &= \gamma(w_n)\gamma(w_n^{-1})\gamma(w_{n+1})\gamma(w_{n+1}^{-1})\\ &= \gamma(w_n)\gamma(w_n^{-1} + w_{n+1} + q^{-1}w_n^{-1}w_{n+1})\gamma(w_{n+1}^{-1}).\end{aligned}$$

The expression

$$s_n = w_n^{-1} + w_{n+1} + q^{-1}w_n^{-1}w_{n+1}$$

gives the deformation of (8). It is easy to realize the corresponding contraction for

$$\hbar \to 0, \qquad \Delta \to 0.$$

We shall not write here the relations between the generators s_n generalizing the Virasoro relation (9). Instead we shall find what to substitute for the Braid Group structure of generators b_n in terms of s_n.

It is easy to see that the cyclicity (22) allows us to rewrite the evolution operator V via the variables t_n,

$$t_n = \gamma(s_n),$$

in the form

$$V = t_1\dots t_{N-1} = t_2\dots t_N = \dots = t_n\dots t_N t_1\dots t_{n-1}. \tag{31}$$

A. Volkov showed in [12] that the t_n satisfy the set of relations

$$t_{n+1}t_{n-1}t_n t_{n+1} = t_{n-1}t_{n+1}t_n, \qquad t_n t_{n-1}t_{n+1} = t_{n-1}t_n t_{n+1}t_{n-1} \tag{32}$$

and

$$t_n t_m = t_m t_n, \quad |n-m| \geq 3. \tag{33}$$

He has shown that the global cyclicity relations (31) and locality (33) lead to the relation

$$t_n t_{n-2} t_{n-1} t_n t_{n+1} = t_{n-2} t_{n-1} t_n t_{n+1} t_{n-1},$$

from which (32) follows if one adds a condition that these products are equal to $t_{n-2} t_n t_{n-1} t_{n+1}$, the origin of which is unclear until now. Needless to say, it is satisfied by our generators t_n.

From all this it is easy to see that V acts as a shift for the generators t_n,

$$t_n V = V t_{n-1}.$$

However, there is an additional bonus. It follows from (32) that the following set of relations hold:

$$t_{n+2} t_{n+1} t_n t_{n+2} = t_n t_{n+2} t_{n+1} t_n. \tag{34}$$

Here the ordering of t_n is decreasing from right to left in contrast to the increasing order we used until now. The relations (34) lead to the shift property

$$t_n W = W t_{n+2}$$

for operators W formally given by the product of t_n in the opposite order compared with that in V

$$W = t_N t_{N-1} \dots t_2 t_1$$

with some analogue of the cyclic ordering. However, the exact definition of such an ordering has not yet been found. We shall conjecture that it exists and use it in the construction of the evolution operator for the nonabelian current.

In any event, we arrive at the second lesson: the discretized and quantized Virasoro phase space is intimately connected with a new algebraic structure—the Volkov algebra.

3.c Nonabelian current

The deformation of $L(x)$ is given by matrix variables L_n with contraction defined by

$$L_n = 1 + i\Delta L(x), \quad x = n\Delta.$$

In [13, 14] the following relations were introduced to replace the Kac–Moody relations (12):

$$\begin{aligned} R_+ \overset{1}{L}_n \overset{2}{L}_n &= \overset{2}{L}_n \overset{1}{L}_n R_-; \\ \overset{1}{L}_n \overset{2}{L}_{n+1} &= \overset{2}{L}_{n+1} R_+ \overset{1}{L}_n, \\ \overset{1}{L}_n \overset{2}{L}_m &= \overset{2}{L}_m \overset{1}{L}_n, \quad |n-m| \geq 2. \end{aligned} \tag{35}$$

Here the usual notations from the RTF formulation of quantum groups [15] are used:

$$\overset{1}{L}_n = L_n \otimes I, \qquad \overset{1}{L}_n = I \otimes L_n;$$

$R_\pm$ are given in terms of the R-matrix R associated with the Lie algebra $\mathcal{A}$,

$$R_+ = R, \qquad R_- = PR_+^{-1}P,$$

where P is a permutation matrix. The Yang–Baxter relation

$$R_{12}R_{13}R_{23} = R_{23}R_{13}R_{12}$$

ensures the correctness of the relations (35) and provides the continuum classical limit (12) under the contraction $\hbar \to 0$, $\Delta \to 0$.

It is a natural place here to comment on the relation between the algebra A generated by L_n and the universal enveloping algebra of the quantum group. For this, consider the monodromy matrix

$$M = L_1 \ldots L_N.$$

It follows from (35) that M satisfies the relation

$$M^1 R_-^{-1} M^2 R_- = R_+^{-1} M^2 R_+ M^1,$$

characteristic of the quantum group U_q in the RTF formulation [15]. Furthermore, the central elements of U_q, given by the q-trace of M, are also central in A,

$$[\mathrm{tr}_q\, M, L_n] = 0.$$

This leads to the conclusion that the categories of representations of A and U_q coincide. For more on this, see [16].

Now let us turn to the discussion of the construction of the evolution operator. For this it is convenient to parametrize the elements of the matrix L_n by simpler generators—namely, to develop the analogue of the Wakimoto realization of Kac–Moody generators. We shall describe a concrete proposal of A. Volkov and myself [17] for the simplest case of $\mathcal{A} = SU(2)$, so that L_n is a 2×2 matrix. We introduce a refinement of the chain $\mathbb{Z}_N$, changing it to $\mathbb{Z}_{2N}$, so that each "physical" site n is associated with two sites $2n$, $2n-1$ on $\mathbb{Z}_{2N}$. We write the Gauss-type factorization

$$L_n = B_{2n}C_{2n-1}$$

with triangular matrices B_{2n} and C_{2n} and paramctrize them as follows:

$$B_{2n} = \begin{pmatrix} \alpha_{2n+\frac{1}{2}}^{-\frac{1}{2}} & 0 \\ 0 & \alpha_{2n+\frac{1}{2}}^{\frac{1}{2}} \end{pmatrix} \begin{pmatrix} 1 & \beta_{2n} \\ 0 & 1 \end{pmatrix},$$

$$C_{2n-1} = \begin{pmatrix} \alpha_{2n-\frac{1}{2}}^{\frac{1}{2}} & 0 \\ 0 & \alpha_{2n-\frac{1}{2}}^{-\frac{1}{2}} \end{pmatrix} \begin{pmatrix} 1 & 0 \\ \beta_{2n-1} & 1 \end{pmatrix}$$

with new generators $\beta_n, \alpha_{n+\frac{1}{2}}$ attached to the sites and edges, respectively. The new generators should satisfy the relations (we write only the nontrivial ones)

$$\alpha_{n+\frac{1}{2}}\alpha_{n-\frac{1}{2}} = q^2\alpha_{n-\frac{1}{2}}\alpha_{n+\frac{1}{2}}, \tag{36}$$

$$\beta_n\alpha_{n-\frac{1}{2}} = q^2\alpha_{n-\frac{1}{2}}\beta_n, \tag{37}$$

$$\alpha_{n+\frac{1}{2}}\beta_n = q^2\beta_n\alpha_{n+\frac{1}{2}}, \tag{38}$$

$$[\beta_{n-1}, \beta_n] = (q - q^{-1})\alpha_{n-\frac{1}{2}} \tag{39}$$

to ensure the relation (35) for L_n with the R-matrix

$$R = \begin{pmatrix} q^{\frac{1}{2}} & & & \\ & q^{-\frac{1}{2}} & q^{\frac{1}{2}} - q^{-\frac{3}{2}} & \\ & & q^{-\frac{1}{2}} & \\ & & & q^{\frac{1}{2}} \end{pmatrix}.$$

In fact, the $\alpha - \beta$ algebra is bigger than that generated by L_n. It contains a subalgebra, equivalent to an abelian current algebra, which commutes with the matrix elements of L_n. However, the reduction of the corresponding evolution operator causes no difficulties.

It follows from what was described that the evolution operator is a shift by two lattice sites of our refined chain $\mathbb{Z}_{2N}$. Exactly such a situation was discussed at the end of the previous section. To use the result mentioned there we need to find in the $\alpha - \beta$ algebra an analogue of the generators s_n. This was done in [17]. The $\alpha - \beta$ algebra is split into two commuting ones, generated by α_n and

$$t_{n-\frac{1}{2}} = q + q^2\beta_n\alpha^{-1}_{n-\frac{1}{2}}\beta_{n-1}.$$

The generators t_n have exactly the same local relations as s_n. However, in our setting they have different continuum limits.

Thus the evolution operator is factorized into two commuting factors

$$V = UW$$

and U provides the shift of α-currents, whereas W acts as

$$t_n W = W t_{n+2}.$$

From the considerations in Section 3.b, it follows that W should be given by some cyclic ordering of the local product

$$W = \gamma(t_N) \ldots \gamma(t_1).$$

However, up to now Volkov and I have not found the appropriate definition of the cyclic ordering for the generators t_n. Thus the program of the construction of the evolution operator for discretized nonabelian currents is not yet completed.

4 Functional equations and modular doubling

The functional equations (19) and (27) have the following formal solutions via the variable w:

$$\theta(w) = \sum_{k=-\infty}^{\infty} q^{-k^2} w^k, \tag{40}$$

$$\gamma(w) = \prod_{k=0}^{\infty}(1 + q^{2k+1} w) \tag{41}$$

$$= 1 + \sum_{k=1}^{\infty} q^{\frac{k(k-1)}{2}} \frac{w^n}{(q^{-1} - q) \dots (q^{-n} - q^n)} \tag{42}$$

$$= \exp \frac{1}{4} \sum_{k=1}^{\infty} \frac{(-1)^k w^k}{k(q^k - q^{-k})}. \tag{43}$$

Thus $\theta(w)$ is a θ-function, whereas γ in accordance with its three representations can be called a $q - \Gamma$-function, q-exponent or exponent of the q-dilogarithm. The last representation is relevant for the interpretation of the pentagon equation. It was shown in [18] that in the classical limit $\hbar \to 0$ the relation (30) turns into the famous five-term relation for the dilogarithm, which in turn is interpreted as a relation on the 3-cocycle for $SL(2, \mathbb{C})$. The second interpretation is closer to the relation (29), which we can trace at least to [19]. The first equality in the relation (30) is similar to a formula from [20].

The grave deficiency of solutions (40)–(43) is that the corresponding series and/or products do not converge for real values of the coupling constant γ. Even worse is the case when q is a root of unity. The way out of this first advertised in [21] is to return to the algebra generated by the noncompact generators p_n from Section 3.a. It is clear that the algebra $\mathcal{A}_q$, generated by the w_n is smaller than the algebra $\mathcal{B}$ generated by the p_n. Indeed, let us introduce variables $\tilde{w}_n$, modular dual to w_n,

$$\tilde{w}_n = w_n^{-\pi/\hbar\gamma} = e^{-ip_n/\tau},$$

where

$$\tau = \frac{\gamma\hbar}{\pi},$$

so that q obtains the traditional expression

$$q = e^{i\pi\tau}.$$

The generators $\tilde{w}_n$ have Weyl-type relations

$$\tilde{w}_{n+1}\tilde{w}_n = \tilde{q}^2 \tilde{w}_n \tilde{w}_{n+1}$$

with the modularly dual factor $\tilde{q}$,

$$\tilde{q} = e^{-i\pi/\tau},$$

and commute with w_n,

$$[w_n, \tilde{w}_m] = 0,$$

due to the Euler formula $e^{2\pi i} = 1$. Thus for the generic q, the algebra $\mathcal{B}$ is factorized into two commuting factors

$$\mathcal{B} = \mathcal{A}_q \otimes \mathcal{A}_{\tilde{q}}.$$

As is mentioned by A. Connes, (see, e.g., [22]) for real γ the algebras $\mathcal{A}_q$ and $\mathcal{A}_{\tilde{q}}$ (noncommutative tori) can be considered as factors of type II_1. We shall not discuss this in any detail. Instead let us mention an interesting case when $|\tau| = 1$. According to formula (11) it corresponds to the central charge of the Virasoro algebra in the "forbidden interval"

$$1 \leq C_{\text{quant}} \leq 25.$$

For such τ the algebra $\mathcal{B}$ has a beautiful involution

$$w_n^* = \tilde{w}_n,$$

and thus to get a selfadjoint dynamical system one should take the sum of two systems, corresponding to q and $\tilde{q}$. This is what I call "modular doubling."

What are the functions $\theta(p)$ and $\gamma(p)$ which serve for the modular doubles in the same way as $\theta(w)$ from (40) and $\gamma(w)$ from (41) work for the models in $\mathcal{A}_q$? They are quite simple:

$$\theta(p) = \exp \frac{p_n^2}{4\pi i \tau}$$

and

$$\gamma(p) = \frac{\gamma_q(w)}{\gamma_{\tilde{q}}(\tilde{w})},$$

where we attached the subindex q to $\gamma(w)$ to distinguish the factors γ corresponding to w and $\tilde{w}$. The functional equations for them,

$$\theta(p + 2\pi) = \theta(p)e^{-ip/\tau},$$
$$\theta(p + 2\pi\tau) = \theta(p)e^{-ip},$$
$$\gamma(p + 2\pi) = \frac{1}{1 + \tilde{q}e^{-ip/\tau}},$$
$$\gamma(p + 2\pi\tau) = \frac{1}{1 + qe^{ip}},$$

unify the two equations (19), (27) and their analogues for the modular double. The Braid relations for $\theta(p_n)$, the pentagon relation for $\gamma(p_n)$, and the Volkov algebra relations for $\gamma(s_n)$, where

$$s_n = \gamma(-p_n)\gamma(p_{n+1})$$

are satisfied. The discussion of this and a lot more can be found in a recent publication of Volkov [23], where a deep interrelation with noncommutative analogues of the hypergeometric identities is discussed.

Using the "noncompact" setting we do not encounter any difficulties connected with the arithmetic properties of τ. So we get one more lesson, more analytic than algebraic: discretization and quantization requires also the modular doubling of the classical system. With this we close our exposition.

Acknowledgments. The results, presented in this article, were obtained together with A. Volkov, and I thank him for successful collaboration. The work was supported by grant RFFI 02-01-00085 and the program "Nonlinear" of the Russian Academy of Sciences.

References

[1] G. S. Gardner, J. M. Greene, M. D. Kruskal, and R. M. Miura, Method for solving the Korteweg-de-Vries equation, *Phys. Rev. Lett.*, **19** (1967), 1095–1097.

[2] V. E. Zakharov and A. B. Shabat, Exact theory of two-dimensional self-focusing and one-dimensional self-modulation of waves in nonlinear media, *Soviet Phys. JETP*, **34** (1972), 62–69.

[3] L. D. Faddeev and L. A. Takhtajan, *Hamiltonian Methods in the Theory of Solitons*, Springer-Verlag, Berlin, Hiedelberg, 1987.

[4] L. D. Faddeev, How the algebraic Bethe Ansatz works for integrable models, in *Proceedings of Les Houches Summer School, Session* LXIV, NATO ASI Series, Elsevier, Amsterdam, 1998, 149–220.

[5] J.-L. Gervais and A. Neveu, The dual string spectrum in Polyakov's quantization I, *Nuclear Phys.* B, **199** (1982), 59–76.

[6] B. L. Feigin and D. B. Fuks, Verma modules over the Virasoro algebra, *Funct. Anal. Appl.*, **17** (1983), 241.

[7] E. Witten, Topological sigma model, *Comm. Math. Phys.*, **118** (1988), 411–449.

[8] S. P. Novikov, The Hamiltonian formalism and a many valued analogue of Morse theory, *Uspekhi Math. Nauk*, **37** (1982), 3–49.

[9] L. Faddeev and A. Volkov, Abelian current algebra and the Virasoro algebra on the lattice, *Phys. Lett.* B, **315** (1993), 311–318.

[10] L. Faddeev, R. Kashaev, and A. Volkov, Strongly coupled quantum discrete Liouville theory 1: Algebraic approach and duality, *Comm. Math. Phys.*, **219**-1 (2001), 199–219.

[11] L. D. Faddeev and L. A. Takhtajan, Liouville model on the lattice, in H. J. De Vega and N. Sanchez, eds., *Field Theory, Quantum Gravity, and Strings*, Lecture Notes in Physics 246, Springer-Verlag, Berlin, New York, Heidelberg, 1986, 166–179.

[12] A. Volkov, Beyond the "pentagon identity," *Lett. Math. Phys.*, **39** (1997), 393–397.

[13] A. Alekseev, L. Faddeev, M. Semenov-Tian-Shansky, and A. Volkov, The unravelling of the quantum group structure in the WZNW theory, Preprint CERN-TH-5981/91, CERN, Geneva, 1991.

[14] A. Alekseev, L. Faddeev, and M. Semenov-Tian-Shansky, Hidden quantum group inside Kac-Moody algebra, *Comm. Math. Phys.*, **149** (1992), 335–345.

[15] L. D. Faddeev, N. Yu. Reshetikhin, and L. A. Takhtajan, Quantization of Lie groups and Lie algebras, *Leningrad Math. J.*, **1** (1990), 193–225; *Algebra Anal.*, **1** (1989), 178–206.
[16] A. Alekseev, L. Faddeev, J. Fröhlich, and V. Schomerus, Representation theory of lattice current algebras, *Comm. Math. Phys.*, **191** (1998), 31–60.
[17] L. Faddeev and A. Volkov, Shift operator for nonabelian lattice current algebra, *Publ. Res. Inst. Math. Sci. Kyoto*, **40** (2004), 1113–1125.
[18] L. Faddeev and R. Kashaev, Quantum dilogarithm, *Mod. Phys. Lett.* A, **9**-5 (1994), 427–434.
[19] M.-P. Schützenberger, Une interprétation de certaines solutions de l'équation fonctionelle: $F(x+y) = F(x)F(y)$, *C. R. Acad. Sci. Paris*, **236** (1953), 352–353.
[20] I. M. Gelfand and D. B. Fairlie, The algebra of Weyl symmetrized polynomials and its quantum extension, *Comm. Math. Phys.*, **136** (1991), 487–501.
[21] L. Faddeev, Modular double of a quantum group, *Math. Phys. Stud.*, **21** (2000), 149–156.
[22] A. Connes, *Noncommutative Geometry*, Academic Press, New York, 1994.
[23] A. Yu. Volkov, Noncommutative hypergeometry, *Comm. Math. Phys.*, **258** (2005), 257–273.

Affine Structures and Non-Archimedean Analytic Spaces

Maxim Kontsevich[1] and Yan Soibelman[2]

[1] IHES
35 Route de Chartres
F-91440 Bures-sur-Yvette
France
maxim@ihes.fr

[2] Department of Mathematics
Kansas State University
Manhattan, KS 66506
USA
soibel@math.ksu.edu

Abstract. In this paper we propose a way to construct an analytic space over a non-archimedean field, starting with a real manifold with an affine structure which has integral monodromy. Our construction is motivated by the junction of the Homological Mirror conjecture and the geometric Strominger–Yau–Zaslow conjecture. In particular, we glue from "flat pieces" an analytic K3 surface. As a byproduct of our approach we obtain an action of an arithmetic subgroup of the group $SO(1, 18)$ by piecewise-linear transformations on the two-dimensional sphere S^2 equipped with naturally defined singular affine structure.

Subject Classifications: 14J32, 14G22

1 Introduction

1.1

An integral affine structure on a manifold of dimension n is given by a torsion-free flat connection with the monodromy reduced to $GL(n, \mathbf{Z})$. There are two basic situations in which integral affine structures occur naturally. One is the case of classical integrable systems described briefly in Section 3. Most interesting for us is a class of examples arising from analytic manifolds over non-archimedean fields which is discussed in Section 4. It is motivated by the approach to Mirror Symmetry suggested in [KoSo]. We recall it in Section 5.

From our point of view, manifolds with integral affine structure appear in Mirror Symmetry in two ways. One considers the Gromov–Hausdorff collapse of degenerating families of Calabi–Yau manifolds. The limiting space can be interpreted either as the contraction (see Section 5.2) of an analytic manifold over a non-archimedean field of Laurent series $\mathbf{C}((t))$, or as the base of a fibration of a Calabi–Yau manifold by Lagrangian tori (with respect to the symplectic Kähler 2-form). On a dense open subset of the limiting space one gets two integral affine structures associated with two interpretations, the non-archimedean one and the symplectic one. The mirror dual family of degenerating Calabi–Yau manifolds should have metrically the same Gromov–Hausdorff limit, with the roles of the two integral affine structures interchanged.

A very interesting question arises: how to reconstruct these families of Calabi–Yau manifolds from the corresponding manifolds with integral affine structures? This question was one of the main motivations for the present work.

1.2

Our approach to the reconstruction of analytic Calabi–Yau manifolds from real manifolds with integral affine structure can be illustrated in the following toy-model example. Let $S^1 = \mathbf{R}/\mathbf{Z}$ be a circle equipped with the affine structure induced from $\mathbf{R}$. We equip S^1 with the canonical sheaf $\mathcal{O}_{S^1}^{\mathrm{can}}$ of Noetherian $\mathbf{C}((q))$-algebras. By definition, for an open interval $U \subset S^1$, the algebra $\mathcal{O}_{S^1}^{\mathrm{can}}(U)$ consists of formal series $f = \sum_{m,n \in \mathbf{Z}} a_{m,n} q^m z^n$, $a_{m,n} \in \mathbf{C}$, such that $\inf_{a_{m,n} \neq 0}(m + nx) > -\infty$. Here $x \in \mathbf{R}$ is any point in a connected component of the preimage of U in $\mathbf{R}$, the choice of a different component $x \to x + k$, $k \in \mathbf{Z}$, corresponds to the substitution $z \mapsto q^k z$. The corresponding analytic space is the Tate elliptic curve $(E, \mathcal{O}_E)$, and there is a continuous map $\pi : E \to S^1$ such that $\pi_*(\mathcal{O}_E) = \mathcal{O}_{S^1}^{\mathrm{can}}$.

In the case of K3 surfaces one starts with S^2. The corresponding integral affine structure is well defined on the set $S^2 \setminus \{x_1, \dots, x_{24}\} \subset S^2$, where $x_1, \dots, x_{24}$ are distinct points. Similarly to the above toy-model example, one can construct the canonical sheaf $\mathcal{O}_{S^2 \setminus \{x_1,\dots,x_{24}\}}^{\mathrm{can}}$ of algebras, an open two-dimensional smooth analytic surface X' with the trivial canonical bundle (Calabi–Yau manifold), and a continuous projection $\pi' : X' \to S^2 \setminus \{x_1, \dots, x_{24}\}$ such that $\pi'_*(\mathcal{O}_{X'}) = \mathcal{O}_{S^2 \setminus \{x_1,\dots,x_{24}\}}^{\mathrm{can}}$. The problem is to find a sheaf $\mathcal{O}_{S^2}$ whose restriction to $S^2 \setminus \{x_1, \dots, x_{24}\}$ is locally isomorphic to $\mathcal{O}_{S^2 \setminus \{x_1,\dots,x_{24}\}}^{\mathrm{can}}$, an analytic compact K3 surface X, and a continuous projection $\pi : X \to S^2$ such that $\pi_*(\mathcal{O}_X) = \mathcal{O}_{S^2}$. We call this problem (in the general case) the Lifting Problem and discuss it in Section 7. Unfortunately we do not know the conditions one should impose on singularities of the affine structure in order that the Lifting Problem have a solution. We consider a special case of K3 surfaces in Sections 8–11. Here the solution is nontrivial and depends on data which are not visible in the statement of the problem. They are motivated by Mirror Symmetry and consist, roughly speaking, of an infinite collection of trees embedded into $S^2 \setminus \{x_1, \dots, x_{24}\}$ with the tail vertices belonging to the set $\{x_1, \dots, x_{24}\}$. The sheaf $\mathcal{O}_{S^2 \setminus \{x_1,\dots,x_{24}\}}^{\mathrm{can}}$ has to be modified by means of automorphisms assigned to every edge of a tree and then glued together with a certain model sheaf near each singular point x_i.

Informally speaking, we break $S^2 \setminus \{x_1, \dots, x_{24}\}$ endowed with the sheaf $\mathcal{O}_{S^2 \setminus \{x_1,\dots,x_{24}\}}^{\mathrm{can}}$ into infinitely many infinitely small pieces and then glue them back together in a slightly deformed way. The idea of such a construction was proposed several years ago independently by K. Fukaya and the first author. The realization of this idea was hindered by a poor understanding of singularities of the Gromov–Hausdorff collapse and by the lack of knowledge of certain open Gromov–Witten invariants ("instanton corrections"). The latter problem is circumvented here (and in fact solved) with the use of a certain pro-nilpotent Lie group (see Section 10).

1.3

The relationship between K3 surfaces and singular affine structures on S^2 is of very general origin. Starting with a projective analytic Calabi–Yau manifold X over a complete non-archimedean local field K, one can canonically construct a PL manifold $\mathrm{Sk}(X)$ called the skeleton of X. If X is a generic K3 surface, then $\mathrm{Sk}(X)$ is S^2. We discuss skeleta in Section 6.6. The group of birational automorphisms of X acts on $\mathrm{Sk}(X)$ by integral PL transformations. For $X = K3$ we obtain an action of an arithmetic subgroup of $\mathrm{SO}(1, 18)$ on S^2. Further examples should come from Calabi–Yau manifolds with large groups of birational automorphisms.

1.4

We have already discussed the contents of the paper. Let us summarize it. The paper is naturally divided into three parts. Part I is devoted to generalities on integral affine structures and examples, including Mirror Symmetry. Motivated by string theory we use the term A-model (respectively, B-model) for examples arising in symplectic (respectively, analytic) geometry.

In Part II we discuss the concept of singular integral affine structure, including an affine version of the Gauss–Bonnet theorem. The latter implies that if all singularities of an integral affine structure on S^2 are standard (so-called focus–focus singularities) then there are exactly 24 singular points. Part II also contains a statement of the Lifting Problem and a discussion of flat coordinates on the moduli space of complex Calabi–Yau manifolds. We expect that under mild conditions on the singular integral affine structure there exists a solution of the Lifting Problem, which is unique as long as we fix periods (see Sections 7.3 and 7.4 for more details).

The most technical Part III contains a solution of the Lifting Problem for K3 surfaces. We construct the corresponding analytic K3 surface as a ringed space. The sheaf of analytic functions is defined differently near a singular point and far from the singular set. It turns out that the "naive" candidate for the sheaf on the complement of the singular set has to be modified before we can glue it to the model sheaf near each singular point. This modification procedure involves a new set of data (we call them *lines*). We also discuss the group of automorphisms of the canonical sheaf which preserve the symplectic form. We use this group in order to modify the "naive" sheaf along each line.

The paper has two appendices. The first contains some background on analytic spaces, while the second is devoted to the Torelli theorem.

Part I

2 Z-affine structures

2.1 Definitions

Let us recall that an affine structure on manifold Y (smooth, of dimension n) is given by a torsion-free flat connection ∇ on the tangent bundle TY.

We will give below three equivalent definitions of the notion of integral affine structure.

Definition 1. *An integral affine structure on* Y *(*$\mathbf{Z}$*-affine structure for short) is an affine structure* ∇ *together with a* ∇*-covariant lattice of maximal rank* $T^{\mathbf{Z}} = (TY)^{\mathbf{Z}} \subset TY$.

It is easy to see that if Y carries a $\mathbf{Z}$-affine structure, then for any point $y \in Y$ there exist a small neighborhood U, local coordinate system $(x_1, \ldots, x_n)$ in U such that $\nabla = d$ in coordinates $(x_1, \ldots, x_n)$ and the lattice $(T_xY)^{\mathbf{Z}}$, $x \in U$, is a free abelian group generated by the tangent vectors $\partial/\partial x_i \in T_xY$, $1 \leq i \leq n$. Let us call $\mathbf{Z}$*-affine* such a coordinate system in U. (Sometimes we will call such U a $\mathbf{Z}$-affine chart.) For a covering of Y by $\mathbf{Z}$-affine charts, the transition functions belong (locally) to $\mathrm{GL}(n, \mathbf{Z}) \ltimes \mathbf{R}^n$. Explicitly, a change of coordinates is given by the formula

$$x'_i = \sum_{1 \leq j \leq n} a_{ij} x_j + b_i,$$

where $(a_{ij}) \in \mathrm{GL}(n, \mathbf{Z})$, $(b_i) \in \mathbf{R}^n$.

Hence Definition 1 is equivalent to the following.

Definition 2. *A* $\mathbf{Z}$*-affine structure on* Y *is given by a maximal atlas of charts such that the transition functions belong locally to* $\mathrm{GL}(n, \mathbf{Z}) \ltimes \mathbf{R}^n$.

In the above definition Y is just a topological manifold, a C^∞-structure on it can be reconstructed canonically from the $\mathbf{Z}$-affine structure.

We can restate the notion of $\mathbf{Z}$-affine structure in the language of sheaves of affine functions.

We say that a real-valued function f on $\mathbf{R}^n$ is $\mathbf{Z}$-affine if it has the form

$$f(x_1, \ldots, x_n) = a_1 x_1 + \cdots + a_n x_n + b,$$

where $a_1, \ldots, a_n \in \mathbf{Z}$ and $b \in \mathbf{R}$. We will denote by $\mathrm{Aff}_{\mathbf{Z},\mathbf{R}^n}$ the sheaf of functions on $\mathbf{R}^n$ which are locally $\mathbf{Z}$-affine.

Definition 3. *A* $\mathbf{Z}$*-affine structure (of dimension* n*) on a Hausdorff topological space* Y *is a subsheaf* $\mathrm{Aff}_{\mathbf{Z},Y}$ *of the sheaf of continuous functions on* Y*, such that the pair* $(Y, \mathrm{Aff}_{\mathbf{Z},Y})$ *is locally isomorphic to* $(\mathbf{R}^n, \mathrm{Aff}_{\mathbf{Z},\mathbf{R}^n})$.

The equivalence of the last two definitions follows from the observation that a homeomorphism between two open domains in $\mathbf{R}^n$ preserving the sheaf $\mathrm{Aff}_{\mathbf{Z},\mathbf{R}^n}$ is given by the same formula $x' = A(x) + b$, $A \in \mathrm{GL}(n, \mathbf{Z})$, $b \in \mathbf{R}^n$ as the change of coordinates between two $\mathbf{Z}$-affine coordinate systems.

2.2 Monodromy representation and its invariant

With a given affine structure on Y we can associate a flat affine connection ∇^{aff} (see [KN]). The corresponding parallel transport acts on tangent spaces by affine transformations. For a $\mathbf{Z}$-affine structure the monodromy of ∇^{aff} belongs to $\mathrm{GL}(n, \mathbf{Z}) \ltimes \mathbf{R}^n$, i.e., $\forall y \in Y$, we have a monodromy representation

$$\rho : \pi_1(Y, y) \to \mathrm{GL}(n, \mathbf{Z}) \ltimes \mathbf{R}^n.$$

Alternatively, we can define the monodromy representation by covering a loop in Y by $\mathbf{Z}$-affine coordinate charts and composing the corresponding transition functions.

Notice that a $\mathbf{Z}$-affine structure on Y gives rise to a class

$$[\rho] \in H^1(Y, T^{\mathbf{Z}} \otimes \mathbf{R}) = H^1(Y, T_Y^{\nabla}),$$

where $T_Y^{\nabla} \subset T_Y$ is the subsheaf of ∇-flat sections.[1] The de Rham representative of the class $[\rho]$ is given by a differential 1-form $\theta \in \Omega^1(Y, T_Y)$ such that $\theta(v) = v$ for any tangent vector v. In affine coordinates one has $\theta = \sum_i \partial/\partial x_i \otimes dx_i$. Clearly, $\nabla(\theta) = 0$.

We will later need an explicit formula for the $\mathbf{R}$-valued pairing of $[\rho]$ with a closed singular 1-chain with coefficients in the local system $(T^*)^{\mathbf{Z}} = (T^*Y)^{\mathbf{Z}}$, the dual covariant lattice in T^*Y. With any singular 1-chain c with values in $(T^*)^{\mathbf{Z}}$ we associate a real number $j(c)$ in the following way. Suppose that c is given by a continuous map $\gamma : [0,1] \to Y$ and a section $\alpha \in \Gamma([0,1], \gamma^*(T^*)^{\mathbf{Z}})$. Parallel transport via the connection ∇^{aff} gives rise to a map $\overline{\gamma} : [0,1] \to T_{\gamma(0)}Y$, $\overline{\gamma}(0) = 0$. Let $\alpha_0 = \alpha(0) \in (T^*_{\gamma(0)})^{\mathbf{Z}} \subset T^*_{\gamma(0)}Y$. We define $j(c) = \langle \alpha_0, \overline{\gamma}(1) \rangle$. We extend $j(c)$ to an arbitrary singular 1-chain c by additivity. Then the class $[\rho]$ can be calculated as $\langle [\rho], [c] \rangle = j(c)$ for any closed 1-chain $c \in C_1(Y, (T^*)^{\mathbf{Z}})$.

3 A-model construction

3.1 Integrable systems

Let (X, ω) be a smooth symplectic manifold of dimension $2n$, B_0 a smooth manifold of dimension n, $\pi : X \to B_0$ a smooth map with compact fibers, such that $\{\pi^*(f), \pi^*(g)\} = 0$ for any $f, g \in C^\infty(B_0)$. Here $\{\cdot,\cdot\}$ denotes the Poisson bracket on X. We assume that π is a submersion on an open dense subset $X' \subset X$. Such a triple (X, π, B_0) is called an *integrable system*. In applications it is typically given by a collection of smooth functions $(H_1, \dots, H_n)$ on X (these functions are called Hamiltonians) such that $\{H_i, H_j\} = 0$, $1 \le i, j \le n$. Usually the first Hamiltonian $H = H_1$ is identified with the energy of the mechanical system.

Let us consider the case when π is proper. This is a natural restriction, because in applications the energy H_1 is already a proper map $H_1 : X \to \mathbf{R}$.

[1] Here we slightly abuse notation because Y is not necessarily connected.

Let $x \in B_0$ be a point such that the restriction of π to $\pi^{-1}(x)$ is a submersion. We call such points *π-smooth.* According to Sard's theorem π-smooth points form an open dense subset of B_0. The fiber $\pi^{-1}(x)$ is a compact Lagrangian submanifold of X. The Liouville integrability theorem (see [Ar]) says that $\pi^{-1}(x)$ is a disjoint union of finitely many tori T^n_α. Moreover, for each torus T^n_α there exists a local coordinate system $(\varphi_1, \dots, \varphi_n, I_1, \dots, I_n)$ in a neighborhood W_α of T^n_α such that $\varphi_i \in \mathbf{R}/2\pi\mathbf{Z}$, $(I_1, \dots, I_n) \in \mathbf{R}^n$ and $\omega = \sum_{1\le i\le n} dI_i \wedge d\varphi_i$. These coordinates are called action-angle coordinates. The map π in action-angle coordinates is given by the projection $(\varphi_1, \dots, \varphi_n, I_1, \dots, I_n) \mapsto (I_1, \dots, I_n)$. There is an ambiguity in the choice of action-angle coordinates. In particular, the action coordinates $I = (I_1, \dots, I_n)$ are defined up to a transformation $I' = A(I) + b$, $A \in \mathrm{GL}(n, \mathbf{Z})$, $b \in \mathbf{R}^n$. Indeed, the free abelian group generated by 1-forms dI_i, $1 \le i \le n$, in each cotangent space $T^*_x B_0$ admits an invariant description. It is the free abelian group generated by the restrictions of 1-forms $\int_\gamma \omega$ to $T^*_x B_0$, where γ runs through closed singular 1-chains in $\pi^{-1}(x) \cap W_\alpha$. In this way we obtain a $\mathbf{Z}$-affine structure on $\pi(W_\alpha)$.

Let B be the set of connected components of fibers of π. Endowed with the natural topology it becomes a locally compact Hausdorff space; projection from X to B will be denoted by the same letter π. The natural continuous map $B \to B_0$ is a kind of "ramified finite covering." Let us define $B^{\mathrm{sm}} \subset B$ as the set of connected components on which π is a submersion (i.e., the set of all Liouville tori). Then B^{sm} is an open dense subset in B. Hence it carries a $\mathbf{Z}$-affine structure given by the action coordinates.

The singular part $B^{\mathrm{sing}} = B \setminus B^{\mathrm{sm}}$ consists of projections of singular fibers. Typically the codimension of B^{sing} is greater than or equal than 1. The codimension 1 stratum consists of the boundary of the image of π and of the ramification locus of the map $B \to B_0$. The structure of singularities of the integral affine structure in higher codimensions is less understood. It seems that the following property is always satisfied.

Fixed Point Property. For any $x \in B^{\mathrm{sing}}$, there is a small neighborhood U such that the monodromy representation $\pi_1((U \setminus B^{\mathrm{sing}})_\alpha) \to \mathrm{GL}(n, \mathbf{Z}) \ltimes \mathbf{R}^n$ for any connected component $(U \setminus B^{\mathrm{sing}})_\alpha$ of $U \setminus B^{\mathrm{sing}}$ has a fixed vector in $\mathbf{R}^n$ in the natural representation by affine transformations.

We will discuss this property in Section 6, devoted to compactifications.

3.1.1 Cohomological interpretation of class $[\rho]$

In Section 2.2 we introduced an invariant $[\rho] \in H^1(B^{\mathrm{sm}}, T^{\mathbf{Z}} \otimes \mathbf{R})$ of a $\mathbf{Z}$-affine structure. Here we will give an interpretation of $[\rho]$ for integrable systems.

Let us consider $X' = \pi^{-1}(B^{\mathrm{sm}})$ which is a Lagrangian torus fibration over B^{sm} (i.e., fibers are Lagrangian tori such that the fiber over $x \in B^{\mathrm{sm}}$ is isomorphic up to a shift to the torus $T^*_x B^{\mathrm{sm}}/(T^*_x B^{\mathrm{sm}})^{\mathbf{Z}}$).

Any singular closed 1-chain c on B^{sm} with values in the local system

$$(T^*_x B^{\mathrm{sm}})^{\mathbf{Z}} \simeq H_1(T^*_x B^{\mathrm{sm}}/(T^*_x B^{\mathrm{sm}})^{\mathbf{Z}}, \mathbf{Z})$$

gives a 2-chain $\overline{c}$ on X' with the boundary belonging to a finite collection of fibers $\pi^{-1}(x^{(i)})$, $1 \leq i \leq N$, of the fibration $\pi : X' \to B^{\mathrm{sm}}$. Moreover, for every point $x^{(i)}$ the part of $\partial \overline{c}$ over $x^{(i)}$ is homologous to zero in $\pi^{-1}(x^{(i)})$. Therefore, there exists a collection of 2-chains $\overline{c}_i$, $1 \leq i \leq N$, supported on $\pi^{-1}(x^{(i)}))$ such that the 2-chain $\overline{c} + \sum_{1 \leq i \leq N} \overline{c}_i$ is closed. In this way we obtain a group homomorphism $J_s : H_1(B^{\mathrm{sm}}, (T^*)^{\mathbf{Z}}) \to H_2(X', \mathbf{Z})/H_2^0(X', \mathbf{Z})$, where $H_2^0(X', \mathbf{Z}) \subset H_2(X', \mathbf{Z})$ denotes the sum of the images of $H_2(\pi^{-1}(y), \mathbf{Z})$ for $y \in B^{\mathrm{sm}}$. (It is enough to pick one base point y for any connected component of B^{sm}.) It is easy to see that $\langle [\rho], [c] \rangle = \langle [\omega], J_s([c]) \rangle$, where $[\omega]$ is the class of the symplectic form ω.

3.2 Examples of integrable systems

We describe here few examples related to the rest of the paper.

3.2.1 Flat tori

The first example is the triple (X, π, B_0), where $X = \mathbf{R}^{2n}/\Lambda$, $B_0 = \mathbf{R}^n/\Lambda'$ are tori (here $\Lambda \simeq \mathbf{Z}^{2n}$, $\Lambda' \simeq \mathbf{Z}^n$ are lattices), the projection $\pi : X \to B_0$ is an affine map of tori, and X carries a constant symplectic form. Assuming that the fibers of π are connected, we have $B_0 = B = B^{\mathrm{sm}}$. The monodromy representation is a homomorphism $\rho : \pi_1(B) \to \mathbf{R}^n \subset \mathrm{GL}(n, \mathbf{Z}) \ltimes \mathbf{R}^n$. The integral affine structure on B depends on n^2 real parameters, which are coefficients of an invertible $n \times n$ matrix expressing a basis of the lattice $\Lambda' \subset T_x B$ as a linear combination of generators of the lattice $(T_x B)^{\mathbf{Z}} \subset T_x B$, where $x \in B$ is an arbitrary point.

3.2.2 Surfaces

Let (X, ω) be a surface and $\pi : X \to B_0 = \mathbf{R}$ be an arbitrary smooth proper function with isolated critical points. Then (X, π, B_0) is an integrable system. The space B of connected components of fibers is a graph, and the $\mathbf{Z}$-affine structure on $B^{\mathrm{sm}} \subset B$ gives a length element on the edges of B.

3.2.3 Moment map

Consider a compact connected symplectic manifold (X, ω) of dimension $2n$ together with a Hamiltonian action of the torus T^n. Then one has an integrable system $\pi : X \to B_0$, where π is the moment map of the action and $B_0 = (\mathrm{Lie}(T^n))^* \simeq \mathbf{R}^n$. Furthermore, it is well known that $B = \pi(X)$ is a convex polytope and B^{sm} is the interior of B.

3.2.4 K3 surfaces

Before considering this example let us remark that one can define integrable systems in the case of complex manifolds. More precisely, assume that X is a complex manifold of complex dimension $2n$, $\omega_{\mathbf{C}}$ is a holomorphic closed nondegenerate 2-form on X, $B = B_0$ is a complex manifold of dimension n and $\pi : X \to B$ is a surjective proper holomorphic map such that generic fibers of π are connected complex Lagrangian submanifolds of X. With a complex integrable system one can associate a real one by forgetting the complex structures on X and B and taking $\omega := \mathrm{Re}(\omega_{\mathbf{C}})$ as a symplectic form on X. It is easy to see that the image of the monodromy representation belongs to $\mathrm{Sp}(2n, \mathbf{Z}) \ltimes \mathbf{R}^{2n} \subset \mathrm{GL}(2n, \mathbf{Z}) \ltimes \mathbf{R}^{2n}$.

Let (X, Ω) be a complex K3 surface equipped with a nonzero holomorphic 2-form $\omega_{\mathbf{C}} = \Omega$ and $\pi : X \to \mathbf{C}P^1$ a holomorphic fibration such that the generic fiber of π is an elliptic curve. For example, X can be represented as a surface in $\mathbf{C}P^2 \times \mathbf{C}P^1$ given by a general equation $F(x_0, x_1, x_2, y_0, y_1) = 0$ of bidegree $(3, 2)$ in homogeneous coordinates. The map π is the projection to the second factor. The holomorphic form Ω is given by

$$\Omega = i_{\mathrm{Euler}_x \wedge \mathrm{Euler}_y} \frac{dx_0 \wedge dx_1 \wedge dx_2 \wedge dy_0 \wedge dy_1}{dF},$$

where Euler_p denotes the Euler vector field along the coordinates $p = (x_i)$ or (y_i). Such an elliptic fibration gives an integrable system. Namely, we set $X := X(\mathbf{C})$, $\omega := \mathrm{Re}(\Omega)$, $B := \mathbf{C}P^1 \simeq S^2$. Generically B^{sing} is a set of $24 = \chi(X)$ points in S^2. The singularity of the affine structure near each of the 24 points is well known in the theory of integrable systems where it is called a focus–focus singularity (see, e.g., [Au, Zu]). We will discuss it in Section 6.4. Here we give a short description of this singularity. We take $\mathbf{R}^2$ with the standard integral affine structure and remove the point $(x_0, 0)$ on the horizontal axis. Then we modify the affine structure (and also the C^∞-structure!) on the ray $\{(x, 0) \mid x > x_0\}$. The new local integral affine coordinates near points of this ray will be functions y and $x + \max(y, 0)$ (see Figure 1). The monodromy of the resulting integral affine structure around removed singular point $(x_0, 0)$ is given by the transformation $(x, y) \mapsto (x + y, y)$.

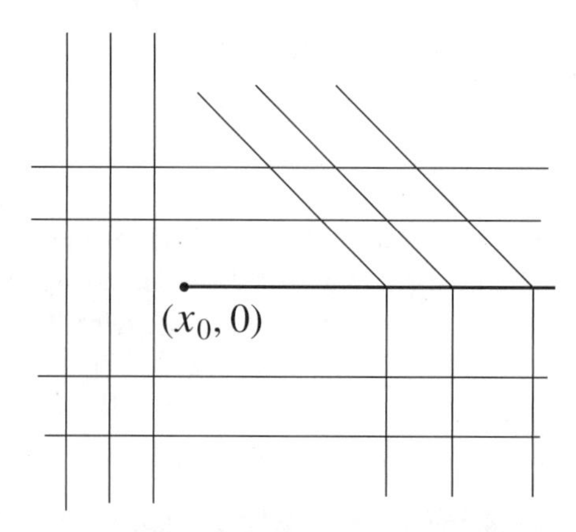

Fig. 1. Focus–focus singularity. All lines are straight in the modified **Z**-affine structure.

3.3 Families of integrable systems and PL actions

In many examples an integrable system depends on parameters. It often happens that the parameter space $\mathcal{P}$ carries a natural foliation $\mathcal{F}$ such that the fundamental group $\pi_1(\mathcal{F}_p, p)$, $p \in \mathcal{P}$, of any leaf acts on the base space B_p of the corresponding torus fibration. This action is given by piecewise-linear homeomorphisms with integral linear parts.

Let us illustrate this phenomenon in the case of the family of integrable systems associated with the K3 surface discussed above.

Here the parameter space $\mathcal{P}$ has dimension 38, which is twice the complex dimension of the space of polynomials F modulo unimodular linear transformations. On the other hand, the miniversal family of representations (up to conjugation)

$$\{\rho : \pi_1(S^2 - \{24 \text{ points}\}) \to \mathrm{SL}(2, \mathbf{Z}) \ltimes \mathbf{R}^2\},$$

such that the monodromy around each puncture is conjugate to $\left(\begin{smallmatrix} 1 & 1 \\ 0 & 1 \end{smallmatrix}\right)$, has dimension 20.

Thus we obtain a foliation $\mathcal{F}$ of $\mathcal{P}$ of rank $18 = 38 - 20$. It is defined by the following property: if we continuously vary the parameters $p \in \mathcal{P}$ along leaves of $\mathcal{F}$, then the conjugacy class of the monodromy representation ρ remains unchanged.

Notice that in the local model described above, we can move the position $(x_0, 0)$ at which we start the cut. Then we have on the sphere S^2 a set of 24 "worms" (singular points, each of them can move in its preferred direction, which is the line invariant under the local monodromy). One can show easily that any continuous deformation of $\mathbf{Z}$-affine structure, satisfying the Fixed Point Property (see Section 3.1) and preserving the conjugacy class of ρ, corresponds to a movement of worms.[2]

Moving "worms"we get a canonical identification of manifolds with integral affine structures far enough from singular points. We will see later in Section 6.4 that we also have a canonical PL identification of manifolds near singular points. Therefore, we obtain a local system along leaves of $\mathcal{F}$ with the fiber over $p \in \mathcal{P}$ being a manifold $B_p \simeq S^2$ with the above $\mathbf{Z}$-affine structure. In this way we get a homomorphism from $\pi_1(\mathcal{F}_p, p)$ to $\mathrm{Aut}_{\mathbf{Z}\mathrm{PL}}(S^2)$, where $\mathbf{Z}$PL denotes the group of integral PL transformations of S^2 equipped with the above $\mathbf{Z}$-affine structure. We will return to this action in Section 6.7, where it will be compared with another PL action on the same space.

4 B-model construction

4.1 Z-affine structure on smooth points

Here we are going to define an analog of the notion of integrable system in the framework of rigid analytic geometry. Roughly speaking, it is a triple (X, π, B),

[2] Notice that in our example $rk(\mathcal{F}) = 18$ is less than 24. This means that there are six constraints on moving worms.

where X is a variety defined over a non-archimedean field (see [Be1] and Appendix A), B is a CW complex and $\pi : X \to B$ a continuous map. More precisely, let K be a field with nontrivial valuation, X an irreducible algebraic variety over K of dimension n, $f = (f_1, \ldots, f_N)$ a collection of nonzero rational functions on X. Then we have a multivalued map

$$X(\overline{K}) \to [-\infty, +\infty]^N, \qquad x \mapsto \mathrm{val}_{\overline{K}}(f(x)) := (\mathrm{val}_{\overline{K}}(f_1(x)), \ldots, \mathrm{val}_{\overline{K}}(f_N(x))).$$

Here $\overline{K}$ is the algebraic closure of K, and $\mathrm{val}_{\overline{K}}$ denotes valuation on $\overline{K}$.

Let $\psi : [-\infty, +\infty]^N \to B$ be a continuous map such that the composition $\pi = \psi \circ \mathrm{val}(f)$ is single-valued. Our map π will always be of this form. More generally, we can take X to be a (not necessarily algebraic) compact smooth K-analytic space, and $\pi : X \to B$ to be a continuous map which factorizes as the composition of the projection $p_{\mathcal{X}} : X \to S_{\mathcal{X}}$ to the Clemens polytope $S_{\mathcal{X}}$ of some model $\mathcal{X}$ of X and a continuous map $\pi' : S_{\mathcal{X}} \to B$ (see Section 4.2.3).

Now we would like to be more precise. Let K be a complete non-archimedean field, with valuation val and the corresponding norm $|x| := \exp(-\mathrm{val}(x)) \in \mathbf{R}_{\geq 0}$. Before giving the next definition, we observe that there is a canonical continuous map $\pi_{\mathrm{can}} : (\mathbf{G}_m^{\mathrm{an}})^n \to \mathbf{R}^n$ (see Section A.2 in Appendix A). Here $\mathbf{G}_m^{\mathrm{an}}$ is a multiplicative group (considered as an analytic space over K) and the restriction of π_{can} to $(\overline{K}^{\times})^n$ is given by the formula

$$\pi_{\mathrm{can}}(z_1, \ldots, z_n) = (\log|z_1|, \ldots, \log|z_n|).$$

The sheaf $\mathcal{O}_{\mathbf{R}^n}^{\mathrm{can}} := (\pi_{\mathrm{can}})_*(\mathcal{O}_{(\mathbf{G}_m^{\mathrm{an}})^n})$ of K-algebras is called the *canonical* sheaf.

Let X be a smooth K-analytic space of dimension n and $\pi : X \to B$ a continuous map of X into a Hausdorff topological space B.

Definition 4. *We call a point $x \in B$ smooth (or π-smooth) if there is a neighborhood U of x such that the fibration $\pi^{-1}(U) \to U$ is isomorphic to a fibration $\pi_{\mathrm{can}}^{-1}(V) \to V$ for some open subset $V \subset \mathbf{R}^n$. Here the isomorphism $\pi^{-1}(U) \simeq \pi_{\mathrm{can}}^{-1}(V)$ is taken in the category of K-analytic spaces while $U \simeq V$ is a homeomorphism.*

In this case we will call π (or the triple $(\pi^{-1}(U), \pi, U)$) an analytic torus fibration.

Let B^{sm} denotes the set of smooth points of B. It is a topological subspace of B (in fact, a topological manifold of dimension n).

Theorem 1. *The space B^{sm} carries a sheaf of $\mathbf{Z}$-affine functions, which is locally isomorphic to the canonical sheaf of $\mathbf{Z}$-affine functions on $\mathbf{R}^n$.*

Proof. We start with the following lemma.

Lemma 1. *Let $V \subset \mathbf{R}^n$ be a connected open set and $\varphi \in \mathcal{O}^{\times}_{(\mathbf{G}_m^{\mathrm{an}})^n}(\pi_{\mathrm{can}}^{-1}(V))$ an invertible analytic function. Then the function $\mathrm{val}_x(\varphi(x))$ is constant along fibers of π_{can}, and it is the pullback of a $\mathbf{Z}$-affine function on $\mathbf{R}^n$.*

Proof. In order to prove the lemma we observe that any analytic function $\psi \in \mathcal{O}^{\times}_{(\mathbf{G}_m^{\mathrm{an}})^n}(\pi_{\mathrm{can}}^{-1}(V))$ can be expanded into Laurent series:

$$\psi = \sum_{I=(i_1,\ldots,i_n)\in\mathbf{Z}^n} c_I z^I, \quad c_I \in K$$

satisfying certain convergence conditions (see Section A.2).

Then for a nonzero analytic function ψ on $\pi_{\mathrm{can}}^{-1}(V)$ we introduce a real-valued function $\mathrm{Val}(\psi)(x) := \inf_{I\in\mathbf{Z}^n}(\mathrm{val}(c_I) - \langle I, x\rangle)$, $x \in V$. It is a concave, locally piecewise-linear function on V. It is easy to see that

(a) there is a dense open subset $V_1 \subset V$ such that for any $x \in V_1$ the infimum in the definition of $\mathrm{Val}(\psi)$ is achieved for a single multiindex I;
(b) $\mathrm{Val}(\psi_1\psi_2) = \mathrm{Val}(\psi_1) + \mathrm{Val}(\psi_2)$.

For an invertible function φ, we have $0 = \mathrm{Val}(1) = \mathrm{Val}(\varphi\varphi^{-1}) = \mathrm{Val}(\varphi) + \mathrm{Val}(\varphi^{-1})$. Since both $\mathrm{Val}(\varphi)$ and $\mathrm{Val}(\varphi^{-1})$ are concave, their sum can be equal to zero iff they are both affine. Moreover, they are both $\mathbf{Z}$-affine since the linear part of $\mathrm{Val}(\varphi)$ is given by the integer vector I for some single multiindex I. Finally, observe that $\mathrm{val}_x(\varphi(x)) \geq \pi_{\mathrm{can}}^*(\mathrm{Val}(\varphi))(x)$, $x \in \pi_{\mathrm{can}}^{-1}(V)$. Therefore, $\mathrm{val}_x(\varphi(x)) = \mathrm{Val}(\varphi)(\pi_{\mathrm{can}}(x))$ for invertible φ. □

Now we can finish the proof of the theorem. The above formula gives us a coordinate-free description of $\pi_{\mathrm{can}}^*(\mathrm{Val}(\varphi))$. It is easy to see that any $\mathbf{Z}$-affine function on V is of the form $\mathrm{Val}(\varphi) + c$, $c \in \mathbf{R}$, for some invertible φ (in the case of $\mathbf{R}^n$ it suffices to take monomials as φ). We can identify $\pi^{-1}(U) \to U$ with $\pi_{\mathrm{can}}^{-1}(V) \to V$ for some small open $U \subset X$ and $V \subset \mathbf{R}^n$. Then we can define $\mathrm{Val}(\varphi)$ for any invertible $\varphi \in \mathcal{O}_X(\pi^{-1}(U))$ by the above formula. Finally we define a sheaf of $\mathbf{Z}$-affine functions on B^{sm} by taking all functions of the form $\mathrm{Val}(\varphi) + c$, $c \in \mathbf{R}$. It follows from the above discussion that in this way we obtain a $\mathbf{Z}$-affine structure on B^{sm}, which is locally isomorphic to the standard one on $\mathbf{R}^n$. □

We will denote by $\mathrm{Aff}^{\mathrm{can}}_{\mathbf{Z},B^{\mathrm{sm}}}$ the sheaf of $\mathbf{Z}$-affine functions constructed in the proof.

4.2 Examples

4.2.1 Logarithmic map

This is a basic example

$$\pi = \pi_{\mathrm{can}} = \log|\cdot| : X = (\mathbf{G}_m^{\mathrm{an}})^n \to B_0 = B = \mathbf{R}^n$$

described in detail in Appendix A. For any algebraic (or analytic) subvariety $Z \subset (\mathbf{G}_m^{\mathrm{an}})^n$ of dimension $m \leq n$ its image $\pi(Z)$ is a noncompact piecewise-linear closed subset of $\mathbf{R}^n$ of real dimension m. The smooth points for $\pi_{|Z}$ are dense in $\pi(Z)$.

In particular, if Z is a curve, then $\pi(Z)$ is a graph in B with straight edges having rational directions. One can try to make a dictionary which translates the properties of the algebraic variety $Z \subset \mathbf{G}_m^n$ to the properties of the PL set $\pi(Z^{\mathrm{an}})$ which is the closure of $\pi(Z(\overline{K}))$ in $\mathbf{R}^n$. This circle of ideas is the subject of so-called "tropical geometry" (see, e.g., [Mi]).

4.2.2 Tate tori

Let $\rho : \mathbf{Z}^n \to (K^\times)^n$ be a group homomorphism such that the image of the composition $\mathrm{val} \circ \rho : \mathbf{Z}^n \to \mathbf{R}^n$ is a rank n lattice in $\mathbf{R}^n$. The group $(K^\times)^n$ acts by translations on the analytic space $(\mathbf{G}_m^{\mathrm{an}})^n$. The restriction of this action to $\mathbf{Z}^n$ (via ρ) is discrete and cocompact. The quotient is a K-analytic space X called the *Tate torus*. There is an obvious map $\pi : X \to B := \mathbf{R}^n/(\mathrm{val} \circ \rho)(\mathbf{Z}^n)$. All points of B are smooth. The space X depends on n^2 parameters taking values in $K^\times$(cf. the flat tori example in Section 3.2.1).

4.2.3 Clemens polytopes and their contractions

For any smooth projective variety X of dimension n, and and an snc model $\mathcal{X}$ of it (see Appendix A) we have a canonical projection to the corresponding Clemens polytope

$$p_{\mathcal{X}} : X^{\mathrm{an}} \to S_{\mathcal{X}}.$$

All interior points of n-dimensional simplices of $S_{\mathcal{X}}$ are $p_{\mathcal{X}}$-smooth, although there might be other smooth points too. More generally, one can compose the projection $p_{\mathcal{X}}$ with a continuous surjection $\pi' : S_{\mathcal{X}} \twoheadrightarrow B$, where B is a finite CW complex and map π' is a cell map for some cell subdivision of $S_{\mathcal{X}}$. We assume that the fibers of the composition $\pi := \pi' \circ p_{\mathcal{X}} : X^{\mathrm{an}} \to B$ are connected. This seems to be the most general case of maps from projective varieties over complete local fields to CW complexes relevant for our purposes.

4.2.4 Curves

Let X/K be a connected smooth projective curve of genus $g > 1$. After passing to a finite extension K' of K we may assume that X has a canonical model $\mathcal{X}$ with stable reduction. The graph Γ' corresponding to the special fiber $\mathcal{X}_0$ is a retraction of $(X \otimes_K K')^{\mathrm{an}}$. The quotient graph $\Gamma = \Gamma'/\mathrm{Gal}(K'/K)$ is a retraction of the analytic curve X^{an} (see [Be1]). We define $B := \Gamma$. Then B^{sm} is the complement of a finite set. As in Section 3.2.2, a $\mathbf{Z}$-affine structure on a graph is the same as a length element (i.e., a metric). Therefore, Γ is a metrized graph. Notice also that the maximal number of edges of the graph corresponding to a genus g curve is $3g-3$, which is the dimension of the moduli space of genus g curves.

Notice that if in Section 4.2.1 the subvariety Z is a curve, then its projection is a noncompact metrized graph with unbounded edges corresponding to the punctures $\overline{Z} \setminus Z$.

4.2.5 K3 surfaces

Here we will describe a particular case of the construction from Section 4.2.3 (a contraction of a Clemens polytope).

Let the field K be $\mathbf{C}((t))$ and $X \subset \mathbf{P}^3_K$ a formal family of complex K3 surfaces given by the equation

$$x_0x_1x_2x_3 + tP_4(x_0, x_1, x_2, x_3) = 0,$$

where P_4 is a generic homogeneous polynomial of degree 4, and t is a formal parameter.

The special fiber at $t = 0$ of this family is singular; it is given by the equation $x_0x_1x_2x_3 = 0$. Let us denote by $\widetilde{\mathbf{P}^3}$ the blowup of the total space of the trivial $\mathbf{P}^3$-bundle over $\mathrm{Spec}(\mathcal{O}_K)$ at 24 points p_α, $1 \le \alpha \le 24$, of the special fiber, where each p_α is a solution of the equation

$$P_4(x_0, x_1, x_2, x_3) = 0, \quad x_i = x_j = 0, \quad 0 \le i < j \le 3.$$

The closure $\mathcal{X}$ of X in $\widetilde{\mathbf{P}^3}$ is a model with simple normal crossings. The associated Clemens polytope $S_\mathcal{X}$ has 28 vertices. Four of them correspond to the coordinate hyperplanes $x_i = 0$ in $\mathbf{P}^3$, and the other 24 correspond to divisors sitting at the preimages of the points p_α. Therefore, $S_\mathcal{X}$ is the union of the boundary $\partial\Delta^3$ of the standard 3-simplex Δ^3 with 24 copies of the standard 2-simplex Δ^2. These 24 triangles Δ^2_α, $1 \le \alpha \le 24$, are decomposed into six groups of four triangles in each. All triangles from the same group have a common edge, which is identified with an edge of $\partial\Delta^3$ (tetrahedron with 24 "wings"). As we mentioned in the previous example, there is a continuous map $p : X^{\mathrm{an}} \to S_\mathcal{X}$. We are going to construct B as a retraction of $S_\mathcal{X}$.

In order to do this we observe that for an edge $e \subset \Delta^2$ and a point $a \in e$ one has the canonical retraction $p_{a,e} : \Delta^2 \to e$. Namely, let us identify the edge e with the interval $[-1, 1]$ of the real line, so that a is identified with the point $a = (a_0, 0)$, and Δ^2 is bounded by e and the segments $0 \le y \le 1 - |x|$. Then we define $p_{a,e}$ by the formulas (see Figure 2)

$$\begin{aligned} (x, y) &\mapsto (x + y, 0), & x + y &\le a_0; \\ (x, y) &\mapsto (x - y, 0), & x - y &\ge a_0; \\ (x, y) &\mapsto (a_0, 0), & &\text{otherwise.} \end{aligned}$$

Now we choose a point q_{ij}, $0 \le i < j \le 3$ in the interior of each edge e_{ij} of $\partial\Delta^3$ (here i, j are identified with the vertices of $\partial\Delta^3$). There are four "wings" Δ^2_α having e_{ij} as a common edge. Then we retract each Δ^2_α to e_{ij} by the map $p_{q_{ij},e_{ij}}$. This gives us a retraction $\pi' = \pi'_{(q_{ij})} : S_\mathcal{X} \to \partial\Delta^3$. Let $\pi := p_\mathcal{X} \circ \pi'_{(q_{ij})} : X^{\mathrm{an}} \to B$ be the composition of the projection $p_\mathcal{X} : X^{\mathrm{an}} \to S_\mathcal{X}$ with the above retraction. One can show that all points of $B := \partial\Delta^3$ are π-smooth except for the chosen six points q_{ij}, $0 \le i < j \le 3$. According to Theorem 1, we obtain a $\mathbf{Z}$-affine structure on

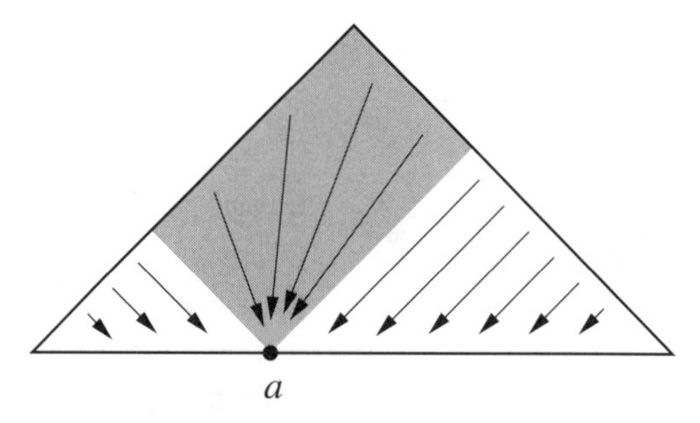

Fig. 2. Triangle contracted to one side. The dashed area maps to point a.

$S^2 \setminus \cup_{0\leq i<j\leq 3}\{q_{ij}\}$. One can show that the local monodromy around each point q_{ij} is conjugate to the matrix

$$\begin{pmatrix} 1 & 4 \\ 0 & 1 \end{pmatrix}.$$

We skip the computations here.

4.3 Stein property

A K-analytic space X is called *Stein* if the natural map

$$X \to \mathrm{Spec}^{\mathrm{an}}(\Gamma(X, \mathcal{O}_X))$$

is a homeomorphism. Here $\Gamma(X, \mathcal{O}_X)$ is considered as a topological K-algebra. This definition is equivalent to the standard one. Let us call the projection $\pi : X \to B$ *Stein* if for any $b \in B$ there exists a fundamental systems of neighborhoods U_i of x such that $\pi^{-1}(U_i) \subset X$ is a Stein domain. If π is Stein then we can reconstruct $(X, \mathcal{O}_X)$ and π from the space B endowed with the sheaf $\pi_*(\mathcal{O}_X)$ of topological K-algebras.

Proposition 1. *Let B be a contraction of the Clemens polytope $S_{\mathcal{X}}$ of some model $\mathcal{X}$ of X as in Section* 4.2.3, *and π a Stein map. Then B^{sm} is dense in B.*

Proof.[3] It suffices to prove that n-dimensional cells are dense in B; here $n = \dim X$. For any open $U \subset B$, $U \neq \emptyset$, we have $H^n_c(U, \pi_*(\Omega^n_X)) \simeq H^n_c(\pi^{-1}(U), \Omega^n_X)$.

The last group is nontrivial, because for any nonempty open $V \subset X^{\mathrm{an}}$ the integration map $\int : H^n_c(V, \Omega^n_X) \to K$ is onto. Therefore, $\dim(U) \geq n$. □

All the examples in Sections 4.2.1–4.2.5 (except Section 4.2.3) have the Stein property.

[3] We thank Ofer Gabber for suggesting this proof below.

5 Z-affine structures and mirror symmetry

5.1 Gromov–Hausdorff collapse of Calabi–Yau manifolds

We recall that a Calabi–Yau metric on a complex manifold X is a Kähler metric with vanishing Ricci curvature. If such a metric exists, then $c_1(TX) = 0 \in H^2(X, \mathbf{R})$ and hence the class of the canonical bundle $\bigwedge^{\dim X}(T^*X)$ is torsion in $\mathrm{Pic}(X)$. According to the famous Yau theorem, for any compact Kähler manifold X such that $c_1(TX) = 0 \in H^2(X, \mathbf{R})$, and any Kähler class $[\omega] \in H^2(X, \mathbf{R})$ there exists a unique Calabi–Yau metric g_{CY} with the class $[\omega]$.[4] Up to now, there is no explicitly known nonflat Calabi–Yau metric on a compact manifold.

In Mirror Symmetry one studies the limiting behavior of g_{CY} as the complex structure on X approaches a "cusp" in the moduli space of complex structures ("maximal degeneration"). A well-known conjecture of Strominger, Yau, and Zaslow (see [SYZ]) claims a torus fibration structure of Calabi–Yau manifolds near the cusp. A metric approach to the maximal degeneration (see [GW, KoSo]) explains the structure of such Calabi–Yau manifolds in terms of their Gromov–Hausdorff limits. We recall this picture below following [KoSo].

We start with the definition of a maximally degenerating family of algebraic Calabi–Yau manifolds.

Let $\mathbf{C}_t^{\mathrm{mer}} = \{f = \sum_{n \geq n_0} a_n t^n\}$ be the field of germs at $t = 0$ of meromorphic functions in one complex variable, and X_{mer} an algebraic n-dimensional Calabi–Yau manifold over $\mathbf{C}_t^{\mathrm{mer}}$ (i.e., X_{mer} is a smooth projective manifold over $\mathbf{C}_t^{\mathrm{mer}}$ with the trivial canonical class: $K_{X_{\mathrm{mer}}} = 0$). We fix an algebraic nonvanishing volume element $\Omega \in \Gamma(X_{\mathrm{mer}}, K_{X_{\mathrm{mer}}})$. The pair $(X_{\mathrm{mer}}, \Omega)$ defines a one-parameter analytic family of complex Calabi–Yau manifolds (X_t, Ω_t), $0 < |t| < \epsilon$, for some $\epsilon > 0$.

Let $[\omega] \in H^2_{DR}(X_{\mathrm{mer}})$ be a cohomology class in the ample cone. Then for every t such that $0 < |t| < \epsilon$, it defines a Kähler class ω_t on X_t. We denote by g_{X_t} the unique Calabi–Yau metric on X_t with the Kähler class $[\omega_t]$.

It follows from resolution of singularities that as $t \to 0$ one has

$$\int_{X_t} \Omega_t \wedge \overline{\Omega}_t = C(\log|t|)^m |t|^{2k}(1 + o(1))$$

for some $C \in \mathbf{C}^\times$, $k \in \mathbf{Z}$, $0 \leq m \leq n = \dim(X_{\mathrm{mer}})$.

Definition 5. *We say that X_{mer} has maximal degeneration at $t = 0$ if, in the formula above, we have $m = n$.*

Let us rescale the Calabi–Yau metric: $g_{X_t}^{\mathrm{new}} = g_{X_t} / \mathrm{diam}(X_t, g_{X_t})^{1/2}$. In this way we obtain a family of Riemannian manifolds $X_t^{\mathrm{new}} = (X_t, g_{X_t}^{\mathrm{new}})$ of diameter 1.

[4] Notice that there is a discrepancy in terminology. In the algebraic situation one usually calls Calabi–Yau a projective variety with trivial canonical class in $\mathrm{Pic}(X)$, and the polarization is not considered as part of data.

Conjecture 1. If X_{mer} has maximal degeneration at $t = 0$, then

$$\text{diam}(X_t, g_{X_t}) = -(\log |t|)^{-1} \exp(O(1)),$$

and there is a limit (B, g_B) of X_t^{new} in the Gromov–Hausdorff metric as $t \to 0$ such that we have the following:

(a) (B, g_B) is a compact metric space which contains a smooth oriented Riemannian manifold $(B^{\text{sm}}, g_{B^{\text{sm}}})$ of dimension n as a dense open metric subspace. The Hausdorff dimension of $B^{\text{sing}} = B \setminus B^{\text{sm}}$ is less than or equal to $n - 2$.
(b) B^{sm} carries a $\mathbf{Z}$-affine structure.
(c) The metric $g_{B^{\text{sm}}}$ has a potential. This means that it is locally given in affine coordinates by a symmetric matrix $(g_{ij}) = (\partial^2 F/\partial x_i \partial x_j)$, where F is a smooth function (defined modulo adding an affine function).
(d) In affine coordinates the metric volume element is constant, i.e.,

$$\det(g_{ij}) = \det(\partial^2 F/\partial x_i \partial x_j) = \text{const}$$

(real Monge–Ampère equation).

There is a more precise conjecture (see [KoSo] for the details) which says that outside of B^{sing} the space X_t^{new} is metrically close to a torus fibration with flat Lagrangian fibers (integrable system). This torus fibration can be canonically reconstructed (up to a locally constant twist) from the limiting data (a)–(d).

Conjecture 1 holds for abelian varieties (since $B = B^{\text{sm}}$ is a flat torus in this case). It is nontrivial for K3 surfaces (see [GW] for the proof). In three-dimensional case there is now substantial progress (see [LYZ]).

Definition 6. *A Monge–Ampère manifold is a triple (Y, g, ∇), where (Y, g) is a smooth Riemannian manifold with metric g, and ∇ is a flat connection on TY such that we have the following:*

(a) *∇ defines an affine structure on Y.*
(b) *Locally in affine coordinates $(x_1, \dots, x_n)$, the matrix (g_{ij}) of g is given by $(g_{ij}) = (\partial^2 F/\partial x_i \partial x_j)$ for some smooth real-valued function F.*
(c) *The Monge–Ampère equation* $\det(\partial^2 F/\partial x_i \partial x_j) = \text{const}$ *is satisfied.*

The following easy proposition is well known.

Proposition 2. *For a given Monge–Ampère manifold (Y, g_Y, ∇_Y) there is a canonically defined dual Monge–Ampère manifold $(Y^\vee, g_Y^\vee, \nabla_Y^\vee)$ such that (Y, g_Y) is identified with $(Y^\vee, g_Y^\vee)$ as Riemannian manifolds, and the local system $(TY^\vee, \nabla_Y^\vee)$ is naturally isomorphic to the local system dual to (TY, ∇_Y) (the dual local system is constructed via the metric g_Y).*

Corollary 1. *If ∇_Y defines an integral affine structure on Y with the covariantly constant lattice $(TY)^{\mathbf{Z}}$, then $\nabla_Y^\vee$ defines an integral affine structure on $Y^\vee$ such that for all $x \in Y^\vee = Y$ the lattice $(T_x Y^\vee)^{\mathbf{Z}}$ is dual to $(T_x Y)^{\mathbf{Z}}$ with respect to the Riemannian metric g_Y on Y.*

We will call *integral* a Monge–Ampère manifold with $\mathbf{Z}$-affine structure.

In Mirror Symmetry one often has a so-called dual family of Calabi–Yau manifolds associated with the given one. There is no general definition of the dual family, but there are many examples. The following conjecture (see [KoSo]) formalizes the Strominger–Yau–Zaslow picture of Mirror Symmetry.

Conjecture 2. The smooth parts of Gromov–Hausdorff limits of dual families of Calabi–Yau manifolds are dual integral Monge–Ampère manifolds.

One can say that Monge–Ampère manifolds with integral affine structures are real analogs of Calabi–Yau manifolds. Conversely, having an integral Monge–Ampère manifold $(Y, g_Y, \nabla_Y, (TY)^{\mathbf{Z}})$ one can construct a torus fibration $TY/(TY)^{\mathbf{Z}} \to Y$. It is easy to see that the total space of this fibration is in fact a Calabi–Yau manifold (typically noncompact as Y is noncompact too). Rescaling the covariant lattice we can make fibers small (of the size $O((\log|t|)^{-1})$). As we already mentioned, the extended version of Conjecture 1 says that this torus fibration is close (after a locally constant twist) to X_t^{new} outside of a "singular" subset.

5.1.1 K3 example

In the case of collapsing K3 surfaces the corresponding intergal Monge–Ampère manifold has an explicit description.

Let S be a complex surface endowed with a holomorphic nonvanishing volume form Ω_S, and $\pi : S \to C$ be a holomorphic fibration over a complex curve C such that the fibers of π are nonsingular elliptic curves.

We define a metric g_C on C as the Kähler metric associated with the $(1, 1)$-form $\pi_*(\Omega_S \wedge \overline{\Omega_S})$. Let us choose (locally on C) a basis (γ_1, γ_2) in $H_1(\pi^{-1}(x), \mathbf{Z}), x \in C$. We define two closed 1-forms on C by the formulas

$$\alpha_i = \operatorname{Re}\left(\int_{\gamma_i} \Omega_S\right), \quad i = 1, 2.$$

It follows that $\alpha_i = dx_i$ for some functions x_i, $i = 1, 2$. We define a $\mathbf{Z}$-affine structure on C, and the corresponding connection ∇, by saying that (x_1, x_2) are $\mathbf{Z}$-affine coordinates (compare with 3.2.4). One can check directly that (C, g_C, ∇) is a Monge–Ampère manifold. In a typical example of elliptic fibration of a K3 surface, one gets $C = \mathbf{C}P^1 \setminus \{x_1, \dots, x_{24}\}$, where $\{x_1, \dots, x_{24}\}$ is a set of distinct 24 points in $\mathbf{C}P^1$. M. Gross and P. Wilson (see [GW]) proved that there exists a family of K3 surfaces with Calabi–Yau metrics collapsing to $S^2 \simeq \mathbf{C}P^1$ with the intergal Monge–Ampère structure described above.

5.2 Non-archimedean picture for the space *B*

Here we would like to formulate a conjecture which relates the Gromov–Hausdorff limit with non-archimedean geometry, thus giving a purely algebraic description of

$\mathbf{Z}$-affine structure on B^{sm}. Let $\overline{\mathbf{C}_t^{\mathrm{mer}}} = \cup_{m\geq 1}\mathbf{C}_{t^{1/m}}^{\mathrm{mer}}$ be the algebraic closure of $\mathbf{C}_t^{\mathrm{mer}}$. We denote by $\pi_{\mathrm{mer}} : X(\overline{\mathbf{C}_t^{\mathrm{mer}}}) \to B$ the map which associates the limiting point (in the Gromov–Hausdorff metric) of the points $x(t^{1/m}) \in X_{t^{1/m}}(\mathbf{C})$ as $t^{1/m} \to 0$.

Let $K = \mathbf{C}((t))$ be the field of formal Laurent series. Then, by extending scalars we obtain an algebraic Calabi–Yau manifold X over K. We denote by X^{an} the corresponding smooth K-analytic space.

Conjecture 3. The map π_{mer} is well defined and extends by continuity to the map $\pi : X^{\mathrm{an}} \to B$. The set B^{sm} (defined as the maximal open subset of B on which the limiting metric is smooth) coincides with the set of π-smooth points. The two $\mathbf{Z}$-affine structures on B^{sm}, one coming from the collapse picture, the other coming from non-archimedean picture, coincide with each other.

We also make the following conjecture (or better a wish, because it is based on very thin evidence).

Conjecture 4. The map π is Stein.

Part II

6 Compactifications of Z-affine structures

6.1 Properties of compactifications

Assume that we are given a noncompact manifold B^{sm} with a $\mathbf{Z}$-affine structure. We would like to "compactify" it, i.e., to find a compact Hausdorff topological space B such that $B^{\mathrm{sm}} \subset B$ is an open dense subset. We do not require an extension of the $\mathbf{Z}$-affine structure to B. The question is: what kind of properties should one expect from such a compactification? We cannot give a complete list of such properties at the moment. Instead, we formulate two of them and illustrate the notion of compactification in the PL case. The similarity between examples in Sections 3.2 and 4.2 suggests that the class of singularities which appear in integrable systems should be more or less the same as the class of singularities appearing in non-archimedean geometry.

Let $x \in B^{\mathrm{sing}} := B \setminus B^{\mathrm{sm}}$ be a singular point of some compactification of B^{sm}. Then we require the following.

Finiteness Property. There is a fundamental system of neighborhoods $U \subset B$ of x such that the number of connected components of $U \cap B^{\mathrm{sm}}$ is finite.

Let $U \cap B^{\mathrm{sm}} = \sqcup_{1\leq i\leq N} U_i$ be the disjoint union of the connected components. Let us pick a point $x_i \in U_i$ and consider a continuous path $\gamma : [0,1] \to B$ such that $\gamma(0) = x_i, \gamma(1) = x, \gamma([0,1)) \subset U_i$. Using the affine structure we can canonically lift this path to a path $\gamma' : [0,1) \to T_{x_i}B$. We assume that the lifted path γ' extends to time $t = 1$ and is analytic at $t = 1$ (this is a technical assumption, helping to avoid some pathologies). Then we require the following.

Independence Property. The path γ with the properties as above exists, and the point $\gamma'(1) \in T_{x_i}B$ does not depend on the choice of γ.

The Independence Property implies the existence of a fixed vector for the monodromy representation restricted to $\pi_1(U \cap B^{\mathrm{sm}}, x_i)$ (this implies the Fixed Point Property from Section 3.1).

6.2 PL compactifications

Let V be a finite set, and let $S \subset 2^V$ belong to the set of $(n+1)$-element subsets of V. Then we have an n-dimensional simplicial complex $B = \cup_{Y \in S} \Delta^Y \subset \Delta^V$.

Let us choose a $\mathbf{Z}$-affine structure on the n-dimensional faces of B which is compatible with the standard affine structure, and consider all $(n-1)$-dimensional faces which enjoy the following property: they belong to exactly two n-dimensional faces. For any two such n-dimensional faces σ and τ, we choose a $\mathbf{Z}$-affine structure on $\sigma \cup \tau$ which is compatible with the already chosen $\mathbf{Z}$-affine structures on σ and τ (such a choice is equivalent to a choice of $\mathbf{Z}$-affine structure in a neighborhood of the $(n-1)$-dimensional face $\sigma \cap \tau$). In this way we obtain a $\mathbf{Z}$-affine structure on the union U of the interior points of all n-dimensional simplices and also the interior points of $(n-1)$-dimensional faces belonging to exactly two top-dimensional cells.

Proposition 3. *There exists a (unique) maximal extension of this $\mathbf{Z}$-affine structure to an open subset $U_{\max} \subset B$ containing U.*

Proof. Let us proceed inductively by codimension of faces. The induction step reduces to the obvious remark that the extension of the standard $\mathbf{Z}$-affine structure on $\mathbf{R}^n \setminus L$ to a neighborhood of point $p \in L$ in $\mathbf{R}^n$ is unique in the case when $L \subset \mathbf{R}^n$ is an affine subspace, $\dim L \leq n-2$. □

It is easy to see that $B^{\mathrm{sm}} := U_{\max}$ with $\mathbf{Z}$-affine structure on it, compactified by B, satisfies both the Finiteness and Independence Properties.

We introduce PL compactifications both as a "toy model," and also (as we hope, see Conjecture 6 in Section 6.3) as a sufficiently representative class for applications. In this case we can try to formulate additional desired properties. One of the goals is to find a good substitute for the algebro-geometric notion of a canonical singularity (which is, morally, a singularity of a noncollapsing limit of a family of Calabi–Yau manifolds with fixed Kähler class).

For a large class of maximally degenerating Calabi–Yau manifolds there is a proposal by several authors (see [GS, HZh]) for a PL compactification B conjecturally related to the Gromov–Hausdorff limit. The space B is topologically a sphere S^n; it carries two dual cell decompositions. Each of these decompositions is identified with the boundary ∂P_1 or ∂P_2 of a convex $(n+1)$-dimensional polytope. Moreover, on each n-dimensional face of each polytope we have a $\mathbf{Z}$-affine structure compatible with the natural affine structure. The assumption is that for any two open n-cells U, U' from the first and the second cell decompositions, the two induced $\mathbf{Z}$-affine structures on $U \cap U'$ coincide. This gives a $\mathbf{Z}$-affine structure on $B \setminus (\mathrm{Sk}_{n-1} \cap \mathrm{Sk}'_{n-1})$, where Sk_{n-1} and Sk'_{n-1} are the $(n-1)$-skeletons of the two CW-structures.

6.3 Some conjectures about singular sets

Our conjectures are in fact rather "wishes," i.e., they are desired properties of $B^{\mathrm{sing}} = B \setminus B^{\mathrm{sm}}$. For simplicity, we assume that B^{sing} is a stratified set (say, the CW complex) of dimension less than or equal t0 $n-1$.

Conjecture 5. We have a decomposition $B^{\mathrm{sing}} = B^{\mathrm{sing}}_{n-1} \cup B^{\mathrm{sing}}_{\leq n-2}$, where $B^{\mathrm{sing}}_{\leq n-2}$ consists of strata of dimension less than or equal to $n-2$, B^{sing}_{n-1} is the union of strata of dimension $n-1$, and locally near every point $x \in B^{\mathrm{sing}}_{n-1}$ the $\mathbf{Z}$-affine structure is modeled by the "book" $\cup_{i \in I} \mathbf{R}^{n-1} \times \mathbf{R}_{\geq 0}$. Here I is a finite set, all half-spaces have a common plane $\mathbf{R}^{n-1} \times \{0\}$ and x belongs to this plane. The $\mathbf{Z}$-affine structure on $B^{\mathrm{sm}} = \sqcup \mathbf{R}^{n-1} \times \mathbf{R}_{>0}$ is the natural one.

This conjecture gives a local model for a singular $\mathbf{Z}$-affine structure at a singular component of codimension one. Let us discuss the case of higher codimension. We start with the following definition.

Definition 7. *A $\mathbf{Z}$-affine structure with singularities on B is given by*

1. *a closed subset $B^{\mathrm{presing}} \subset B$ of a compact space B;*
2. *a $\mathbf{Z}$-affine structure on the open set $B \setminus B^{\mathrm{presing}}$.*

One can think about the closed set of "potential singularities" B^{presing} as containing the actual set of singularities B^{sing}).

Definition 8. *A continuous path $\gamma(t)$, $t \in [0,1]$, in the space of $\mathbf{Z}$-affine structures with singularities on a given compact space B is given by*

1. *a continuous path B_t^{presing} in the space of all compact subsets of B;*
2. *a $\mathbf{Z}$-affine structure on $B \setminus B_t^{\mathrm{presing}}$ for all $t \in [0,1]$.*

Notice that for each $t_0 \in (0,1)$ and $x_0 \in B \setminus B_{t_0}^{\mathrm{presing}}$ we can choose neighborhoods U_{t_0} of t_0 and U_{x_0} of x_0 such that $U_{x_0} \subset B \setminus B_t^{\mathrm{presing}}$ for all $t \in U_{t_0}$. Then we require that

3. *if U_{t_0} and U_{x_0} are sufficiently small, then the induced $\mathbf{Z}$-affine structure on U_{x_0} does not depend on $t \in U_{t_0}$.*

Notice that in the case when the homotopy type of $B \setminus B_t^{\mathrm{presing}}$ remains unchanged, the representation $\rho_t : \pi_1(B \setminus B_t^{\mathrm{presing}}) \to \mathrm{GL}(n, \mathbf{Z}) \ltimes \mathbf{R}^n$ stays the same.

We are going to give an example of a nontrivial path in the next subsection. We expect that singularities which appear in the collapse of Calabi–Yau manifolds satisfy the following.

Conjecture 6. If $B^{\mathrm{sing}} = B^{\mathrm{presing}}$ is of codimension at least two in B, then there is a continuous path $\gamma(t)$ in the space of $\mathbf{Z}$-affine structures with singularities which connects a given structure with the one coming from a PL compactification, and such that for all t we have $\mathrm{codim}(B_t^{\mathrm{presing}}) \geq 2$ and $\gamma(t)$ has the Finiteness and Independence Properties.

6.4 Standard singularities in codimension 2

Let us remove the angle $\{(x, y) \in \mathbf{R}^2 \mid 0 < x < y\}$ from $\mathbf{R}^2$. After that we identify the sides of the angle by the affine transformation $(x, y) \mapsto (x + y, y)$. In this way we introduce a new $\mathbf{Z}$-affine structure on $\mathbf{R}^2 \setminus \{(0, 0)\}$ with the monodromy around $(0, 0)$ given by the unipotent matrix (see Figure 3)

$$\begin{pmatrix} 1 & 1 \\ 0 & 1 \end{pmatrix}.$$

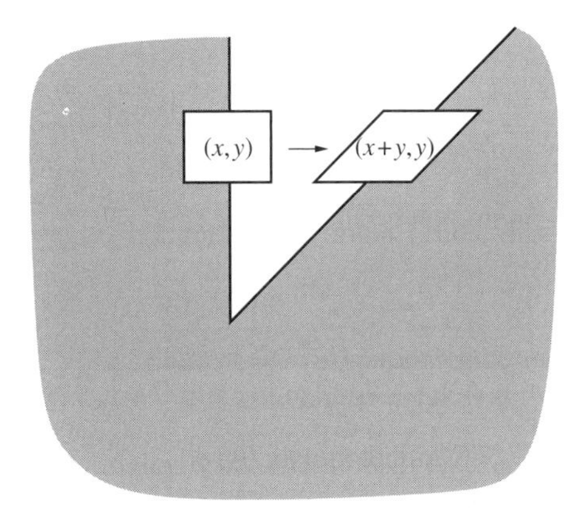

Fig. 3. Glue the two sides of the dashed area. White parallelograms are identified.

This $\mathbf{Z}$-affine structure does not admit a continuation to $\mathbf{R}^2$. Therefore, we obtain a $\mathbf{Z}$-affine structure with singularities on $\mathbf{R}^2$. We will call *standard* the singularity at $(0, 0)$.

Equivalently, we can describe this $\mathbf{Z}$-affine structure on $\mathbf{R}^2 \setminus \{(0, 0)\}$ by taking a cut along the ray $\{(x, 0) \mid x > 0\}$ in $\mathbf{R}^2$ and gluing the standard $\mathbf{Z}$-affine structure above and below the cut by means of the affine transformation $(x, y) \mapsto (x + y, y)$ (see Figure 1 in Section 3.2.4). In this description it is clear that we can start the cut at an *arbitrary* point $(x_0, 0)$ on the x-axes. The resulting singularity will also be called the standard one.

We adopt this terminology.

Remark 1. We can vary the position of $(x_0, 0)$, thus obtaining a continuous path in the space of $\mathbf{Z}$-affine structures with singularities in $\mathbf{R}^2$.

More generally, suppose that B is equipped with a $\mathbf{Z}$-affine structure which has standard singularities at the points $p_1, \ldots, p_m$. Then we can slightly move each point p_i in the direction invariant under the local monodromy around p_i. This gives a new $\mathbf{Z}$-affine structure which is $\mathbf{Z}$PL-isomorphic to the initial one.

The standard singularity is called a focus–focus singularity in the theory of integrable systems (see [Zu]). In non-archimedean geometry it appears as a singular value of some map $f : X^{\mathrm{an}} \to \mathbf{R}^2$, where X is an algebraic surface in the three-dimensional affine space $\mathbf{A}^3_K$ (see Section 8).

Let us consider the Cartesian product of $\mathbf{R}^2 \setminus \{(0,0)\}$ equipped with the above $\mathbf{Z}$-affine structure with the standard (nonsingular) $\mathbf{Z}$-affine structure on $\mathbf{R}^{n-2}$. Let us choose a continuous function $f(z_1, \dots, z_{n-2})$ and start the cuts at all points $(f(z_1, \dots, z_{n-2}), 0, z_1, \dots, z_{n-2})$. This means that we introduce the standard nonsingular $\mathbf{Z}$-affine structure in the region $y \neq 0$ as well as in the region $(y = 0, x < f(z_1, \dots, z_{n-2}))$. Near the points $(y = 0, x > f(z_1, \dots, z_{n-2}))$ we introduce a modified $\mathbf{Z}$-affine structure by declaring the functions

$$(y, x + \max(y, 0), z_1, \dots, z_{n-2})$$

to be $\mathbf{Z}$-affine coordinates. This gives an example of a "curved" singular set B^{sing} of codimension 2. Since the function f can be approximated by PL functions, the above $\mathbf{Z}$-affine structure can be deformed to a PL one.

6.5 Z-affine version of Gauss–Bonnet theorem

Let B be a connected compact oriented topological surface, and $B^{\mathrm{sing}} \subset B$ a finite set. Assume that $B^{\mathrm{sm}} = B \setminus B^{\mathrm{sing}}$ carries a $\mathbf{Z}$-affine structure such that for any $x \in B^{\mathrm{sing}}$ there exists a small neighborhood U such that $U = \cup_{i \in I} U_i$, where I is a finite set and each U_i is affine equivalent to a germ of an angle in $\mathbf{R}^2$, with x being the apex of each angle.

The aim of this section is to define a map $i_{\mathrm{loc}} : B^{\mathrm{sing}} \to \frac{1}{12}\mathbf{Z}$ (which depends only on the $\mathbf{Z}$-affine structure near B^{sing}) and to prove the following result (a kind of Gauss–Bonnet theorem).

Theorem 2. *The following equality holds:*

$$\sum_{x \in B^{\mathrm{sing}}} i_{\mathrm{loc}}(x) = \chi(B),$$

where $\chi(B)$ is the Euler characteristic of B.

We start with the construction of i_{loc}. Let us denote by $\widetilde{\mathrm{SL}(2,\mathbf{Z})}$ the preimage of $\mathrm{SL}(2,\mathbf{Z})$ in the universal covering $\widetilde{\mathrm{SL}(2,\mathbf{R})}$ of the group $\mathrm{SL}(2,\mathbf{R})$. The group $\widetilde{\mathrm{SL}(2,\mathbf{R})}$ contains $\pi_1(\mathrm{SL}(2,\mathbf{R})) \simeq \mathbf{Z}$. Let u be a generator of the latter (it also belongs to $\widetilde{\mathrm{SL}(2,\mathbf{Z})}$).

We have an exact sequence of groups

$$1 \to \mathbf{Z} \to \widetilde{\mathrm{SL}(2,\mathbf{Z})} \to \mathrm{PSL}(2,\mathbf{Z}) \to 1.$$

Notice that $\mathrm{PSL}(2,\mathbf{Z})$ is a free product $\mathbf{Z}/2 * \mathbf{Z}/3$. Moreover, in the above exact sequence $\mathbf{Z}$ is embedded into the center of $\widetilde{\mathrm{SL}(2,\mathbf{Z})}$. Notice that u is the image of $2 \in \mathbf{Z}$. One can choose representatives a_2, a_3 of the standard generators of $\mathrm{PSL}(2,\mathbf{Z})$

in such a way that $\widetilde{\mathrm{SL}(2,\mathbf{Z})}$ is generated by u, a_2, a_3 subject to the relations $a_2^2 = a_3^3, a_2^4 = a_3^6 = u$. This gives a homomorphism of groups $\phi : \widetilde{\mathrm{SL}(2,\mathbf{Z})} \to \mathbf{Z}$ such that $\phi(a_2) = 3, \phi(a_3) = 2, \phi(u) = 12$. Dividing by 12 we obtain a homomorphism $i : \widetilde{\mathrm{SL}(2,\mathbf{Z})} \to \frac{1}{12}\mathbf{Z}$ such that $i(u) = 1$.

Let us consider a topological S^1-bundle E over B, such that the fiber over $x \in B$ is the union of all affine rays emanating from x. Then the restriction of E to B^{sm} is just the spherical bundle. Let us pick $x_0 \in B^{\mathrm{sm}}$ and remove it from B together with small neighborhoods of all points B^{sing}. We denote by B_1 the topological space obtained in this way. We can trivialize the tangent bundle over B_1 (we choose a C^∞-trivialization, compatible with the $\mathrm{SL}(2,\mathbf{R})$-structure) in such a way that it extends to a continuous trivialization of the S^1-bundle E over $B\setminus\{x_0\}$. Let $\alpha \in \Omega^1(B_1)\otimes \mathrm{sl}(2,\mathbf{R})$ be a 1-form defined by means of the affine structure ∇ on B_1. Then $d\alpha + \frac{1}{2}[\alpha,\alpha] = 0$ and α defines a flat connection on the trivial $\widetilde{\mathrm{SL}(2,\mathbf{R})}$-bundle over B_1. This gives a a monodromy representation $\pi_1(B_1) \to \widetilde{\mathrm{SL}(2,\mathbf{Z})}$ defined up to conjugation. Composing it with the homomorphism i we obtain a homomorphism $\pi_1(B_1) \to \frac{1}{12}\mathbf{Z}$. Since $\frac{1}{12}\mathbf{Z}$ is an abelian group, the latter homomorphism is the composition $\pi_1(B_1) \to H_1(B_1,\mathbf{Z}) \to \frac{1}{12}\mathbf{Z}$. Let us pick small circles $[\gamma_x] \in H_1(B,\mathbf{Z})$ for each $x \in B^{\mathrm{sing}}$. Then the above homomorphism gives us a number denoted by $i_{\mathrm{loc}}(x) \in \frac{1}{12}\mathbf{Z}$.

Proof of Theorem 2. Let us pick a small circle $[\gamma_{x_0}] \in H_1(B_1,\mathbf{Z})$ around x_0. Then $\sum_{x\in B^{\mathrm{sing}}}[\gamma_x]+[\gamma_{x_0}] = 0$. The monodromy around x_0 can be easily computed via the winding number of the induced vector field (section of $E_{|\gamma_{x_0}}$) and is equal to $-\chi(B)u$. Applying the homomorphism i we obtain the result. □

Corollary 2. *Suppose that the monodromy for each point $x \in B^{\mathrm{sing}}$ is the standard one (see Section* 6.4*). Then one has two possibilities:*

(a) *$B^{\mathrm{sing}} = \emptyset$ and $B = B^{\mathrm{sm}}$ is a two-dimensional torus;*
(b) *the set B^{sing} consists of* 24 *distinct points on the sphere S^2.*

Proof. It is easy to see that for each point $x \in B^{\mathrm{sing}}$ one has $i_{\mathrm{loc}}(x) = \frac{1}{12}$. Then from the Gauss–Bonnet theorem one deduces that $\chi(B) = 2 - 2g$, where g is the genus of the Riemann surface B. Then we have

$$\frac{|B^{\mathrm{sing}}|}{12} = 2 - 2g.$$

Since the LHS is nonnegative we conclude that either $g = 1$ or $g = 0$. In the first case $|B^{\mathrm{sing}}| = 0$ and we have a $\mathbf{Z}$-affine structure on a torus. In the second case we have $|B^{\mathrm{sing}}| = 24$ and $g = 0$. □

Remark 2. This corollary was proved in [LeS] by different methods.

Similarly, for the affine structure with the monodromy at each point conjugate to

$$\begin{pmatrix} 1 & 4 \\ 0 & 1 \end{pmatrix},$$

one has $i_{\mathrm{loc}}(x) = \frac{1}{3}$, and we have six singular points on S^2 (see Section 4.2.5).

6.6 Skeleton of a non-archimedean Calabi–Yau variety

Let $X = X/K$ be a smooth proper algebraic variety over a non-archimedean field K, $\dim X = n$, and let $\Omega \in \Gamma(X, \Omega^n_X)$ be a nonzero top-degree form on X. We will associate canonically with the pair (X, Ω) a piecewise-linear compact space $\mathrm{Sk}(X, \Omega) \subset X^{\mathrm{an}}$ such that $\dim_{\mathbf{R}} \mathrm{Sk}(X, \Omega) \leq n$.

Let us assume for simplicity that $K = \mathbf{C}((t))$ and X is defined over $\mathbf{C}_t^{\mathrm{mer}} \subset K$. The analytic space X^{an} contains a dense subset X_{Div} of divisorial points corresponding to irreducible components of special fibers of all snc models $\mathcal{X}$ of X (see Appendix A):

$$X_{\mathrm{Div}} = \cup_{\mathcal{X}} i_{\mathcal{X}}(V_{S_{\mathcal{X}}}),$$

where $V_{S_{\mathcal{X}}}$ is the set of vertices of the Clemens polytope $S_{\mathcal{X}}$.

The top-degree form Ω gives rise to a map $\psi_\Omega : X_{\mathrm{Div}} \to \mathbf{Q}$. Namely, if $p : \mathcal{X} \to \mathrm{Spec}(\mathbf{C}_t^{\mathrm{mer}})$ is an snc model and $D \subset \mathcal{X}_0$ is an irreducible divisor of the special fiber, then we define

$$\psi_\Omega(D) = \frac{\mathrm{ord}_D(\Omega \wedge dt/t)}{\mathrm{ord}_D(p^*(t))}.$$

Here $\Omega \wedge dt/t$ is a meromorphic top-degree form on $\mathcal{X}/\mathbf{C}$.

It is easy to show that $\psi_\Omega(D)$ depends only on the point $i_{\mathcal{X}}(D) \in X^{\mathrm{an}}$. The function ψ_Ω is (globally) bounded from below.

Definition 9. *A divisorial point $i_{\mathcal{X}}(D)$ is called essential if*

$$\psi_\Omega(D) = \inf_{x \in X_{\mathrm{Div}}} \psi_\Omega(x).$$

Definition 10. *The skeleton $\mathrm{Sk}(X, \Omega)$ is the closure in X^{an} of the set of essential points.*[5]

Let $\mathcal{X}$ be an snc model. We will explain how to describe $\mathrm{Sk}(X, \Omega)$ in terms of $\mathcal{X}$ and Ω. In fact it is a nonempty simplicial subcomplex of $i_{\mathcal{X}}(S_{\mathcal{X}})$.

Let us call *$\mathcal{X}$-essential* a divisor $D_i \subset \mathcal{X}_0$ such that

$$\psi_\Omega(D_i) = \min_{D_j \in \mathcal{X}_0} \psi_\Omega(D_j).$$

A nonempty collection $D_{i_1}, \dots, D_{i_l}$ of divisors in $\mathcal{X}_0$ is called *$\mathcal{X}$-essential* if all D_{i_k} are $\mathcal{X}$-essential, the intersection $D_{i_1} \cap D_{i_2} \cap \dots \cap D_{i_l}$ is nonempty and does not belong to the closure of the divisor of zeros of Ω is $\mathcal{X} \setminus \mathcal{X}_0$.

Theorem 3. *The skeleton $\mathrm{Sk}(X, \Omega)$ is the image under $i_{\mathcal{X}}$ of the subcomplex $\mathrm{Sk}(\mathcal{X}, \Omega) \subset S_{\mathcal{X}}$ consisting of simplices corresponding to $\mathcal{X}$-essential collections of divisors.*

---

[5] Our notion of a skeleton should not be confused with the one introduced in [Be3]. The latter is related to the Clemens polytope $S_{\mathcal{X}}$ of an snc model $\mathcal{X}$.

Sketch of the proof. Notice that for any snc model $\mathcal{X}$ the subset $S_{\mathcal{X}}(\mathbf{Q}) \subset S_{\mathcal{X}}$ consisting of points with rational barycentric coordinates is mapped by $i_{\mathcal{X}}$ into X_{Div}. Namely, we can modify $\mathcal{X}$ by blowing up at nonempty intersections of irreducible components of the special fiber and then continue this process indefinitely. Divisorial points obtained in this way exhaust all points of $i_{\mathcal{X}}(S_{\mathcal{X}}(\mathbf{Q}))$.

We will prove that the set of essential points in X^{an} coincides with $i_{\mathcal{X}}(S_{\mathcal{X}}(\mathbf{Q}) \cap \mathrm{Sk}(\mathcal{X}, \Omega))$. First of all, a direct computation shows that ψ_{Ω} restricted to $i_{\mathcal{X}}(S_{\mathcal{X}}(\mathbf{Q}))$ achieves its absolute minimum on $i_{\mathcal{X}}(S_{\mathcal{X}}(\mathbf{Q}) \cap \mathrm{Sk}(\mathcal{X}, \Omega))$. Secondly, another straightforward computation shows that the latter set does not change under blowups of first and second type (see Section A.5 in Appendix A). This concludes the proof.□

For a Calabi–Yau manifold X we will denote $\mathrm{Sk}(X, \Omega)$ simply by $\mathrm{Sk}(X)$, as there exists only one (up to a scalar) nonzero top-degree form Ω on X and $\mathrm{Sk}(X, \lambda\Omega) = \mathrm{Sk}(X, \Omega)\ \forall \lambda \in K^{\times}$.

One can prove that the PL space $\mathrm{Sk}(X, \Omega)$ is in fact a birational invariant. Moreover, the group $\mathrm{Aut}^{\mathrm{brt}}(X)$ of birational automorphisms of X acts on the skeleton by **Z**PL transformations. In order to obtain nontrivial examples of such actions we need Calabi–Yau manifolds with large groups of birational automorphisms. An example of a **Z**PL-action is considered in the next subsection.

6.7 K3 surfaces and ZPL-actions on S^2

6.7.1 Integrable systems

Recall that in Section 3.3 we constructed a 38-dimensional space $\mathcal{P}$ parameterizing integrable systems $(X, \omega) \to B$ with $B \simeq S^2$. The space $\mathcal{P}$ carries a codimension 20 foliation $\mathcal{F}$ corresponding to small deformations of integrable systems which do not change the invariant $[\rho]$ of the local system $\rho : \pi_1(B \setminus B^{\mathrm{sing}}) \to \mathrm{SL}(2, \mathbf{Z}) \ltimes \mathbf{R}^2$. We explained that the fundamental group of a leaf of $\mathcal{F}$ acts by PL homeomorphisms of S^2. Here we are going to give a (partial) description of $\mathcal{P}$ and $\mathcal{F}$ in cohomological terms using the Torelli theorem (see Appendix B).

An algebraic polarized K3 surface $X/\mathbf{C}$ elliptically fibered over $\mathbf{C}P^1$, equipped with a holomorphic volume form Ω can be encoded by the data $(\Lambda, (\cdot, \cdot), [\omega], [\Omega], [\gamma], \mathcal{K}_X)$, where

1. $(\Lambda, (\cdot, \cdot), \mathbf{C}[\Omega], \mathcal{K}_X)$ is K3 period data;
2. $[\omega], [\gamma] \in \Lambda$, $\Omega \in \Lambda \otimes \mathbf{C}$;
3. $[\omega] \in \mathcal{K}_X$, $\gamma \in \partial\mathcal{K}_X$, $([\omega], [\Omega]) = ([\gamma], [\Omega]) = ([\gamma], [\gamma]) = 0$;
4. γ is a nonzero primitive lattice vector.

Here $[\omega]$ is the class of the polarization (projective embedding) of X, and $[\gamma]$ is dual to the class of a generic fiber of the elliptic fibration $\pi : X \to \mathbf{C}P^1$.

Perhaps one can express in cohomological terms the fact that π has exactly 24 critical values. The latter is an open condition.

Let $L \subset H_2(X, \mathbf{Z})$ be a subgroup consisting of cohomology classes of cycles which are projected into graphs in $B \setminus B^{\mathrm{sing}}$ (such cycles are circle fibrations over graphs). When we move along a leaf of $\mathcal{F}$, then the pairing of Re([Ω]) with L remains unchanged (see Section 3.1.1). Clearly, $L \subset [\gamma]^{\perp}$, and moreover, one can check that $L = [\gamma]^{\perp} \simeq \mathbf{Z}^{21}$. The pairing with Re([$\Omega$]) gives a map $\Lambda_{2,18} := [\gamma]^{\perp}/\mathbf{Z}[\gamma] \to \mathbf{R}$, where $\Lambda_{2,18}$ is the following even unimodular lattice of signature (2, 18):

$$\Lambda_{2,18} = \begin{pmatrix} 0 & 1 \\ 1 & 0 \end{pmatrix} \oplus \begin{pmatrix} 0 & 1 \\ 1 & 0 \end{pmatrix} \oplus (-E_8) \oplus (-E_8),$$

where $-E_8$ is the Cartan matrix for the Dynkin diagram E_8 taken with the minus sign.

The functional $(\mathrm{Re}[\Omega], \cdot)$ on $\Lambda_{2,18}$ can be represented as $(v_{\mathrm{Re}[\Omega]}, \cdot)$, where $v_{\mathrm{Re}[\Omega]} \in \Lambda_{2,18} \otimes \mathbf{R}$ is a vector with strictly positive squared norm. One can show that the (non-Hausdorff) space of leaves of $\mathcal{F}$ is canonically identified with the set $\{v \in \Lambda_{2,18} \otimes \mathbf{R} | (v, v) > 0\}/\operatorname{Aut}(\Lambda_{2,18})$.

The fundamental group of the leaf $\mathcal{F}_v$ corresponding to a vector $v \in \Lambda_{2,18} \otimes \mathbf{R}$ maps onto the group $\Gamma_v \subset \operatorname{Aut}(\Lambda_{2,18})$. This group is (up to conjugation) the stabilizer in $(\operatorname{Aut}(\Lambda_{2,18}), (\cdot, \cdot)_{2,18}, v)$ of the cone K_v, which is a connected component of the set

$$\{w \in \Lambda_{2,18} \otimes \mathbf{R} | (w, v) = 0, (w, w) > 0\} \setminus \bigcup_{\gamma \in \Lambda_{2,18}, (\gamma,\gamma)=-2, (\gamma, v)=0} H_\gamma$$

and $H_\gamma \in \Lambda_{2,18} \otimes \mathbf{R}$ is the hyperplane orthogonal to γ (cf. Appendix B). Let us denote by $\operatorname{Aut}_{\mathbf{Z}\mathrm{PL},v}(S^2)$ the group of piecewise-linear transformations of S^2 with integer linear part. The index v signifies the dependence of **Z**PL-structure on S^2 on v.

Conjecture 7. The homomorphism $\pi_1(\mathcal{F}_v) \to \operatorname{Aut}_{\mathbf{Z}\mathrm{PL},v}(S^2)$ arising from the monodromy of the local system along the leaf $\mathcal{F}_v$ (see Sections 3.3 and 6.4) is equal to the composition

$$\pi_1(\mathcal{F}_v) \twoheadrightarrow \Gamma_v \to \operatorname{Aut}_{\mathbf{Z}\mathrm{PL},v}(S^2),$$

where the homomorphism $\phi_v : \Gamma_v \to \operatorname{Aut}_{\mathbf{Z}\mathrm{PL},v}(S^2)$ is uniquely determined by this property.

In particular, for $v \in \Lambda_{2,18}$ the group Γ_v is a subgroup (and also a quotient group) of an arithmetic subgroup in the Lie group SO(1, 18). Also, in this case there is a Γ_v-invariant notion of a point with integer coordinates on S^2, as well of points with coordinates in $\frac{1}{N}\mathbf{Z}$ for any integer $N \geq 1$. The number $M_{v,N}$ of such points is finite, and we obtain a homomorphism from Γ_v to the mapping class group $\pi_1(\mathcal{M}^{\mathrm{unord}}_{0,M_{v,N}})$, the fundamental group of the moduli space of genus-zero complex curves with $M_{v,N}$ unordered distinct marked points. The latter group is closely related to the braid group. The conclusion is that we have constructed homomorphisms from arithmetic groups to braid groups.

One can consider the whole moduli space $\mathcal{M}_{44}$ of **Z**-affine structures on S^2 with 24 standard singularities. This space is a Hausdorff orbifold (with a natural **Z**-affine structure!) of dimension 44, and it carries a foliation of codimension 20 as before. It seems that using our main result (Theorem 5 in Part III) together with a certain natural assumption (see Conjecture 11 in Section 11.6), one can show that the action by **Z**PL transformations of S^2 of the fundamental group of leaves of the foliation on the larger space $\mathcal{M}_{44}$ is again reduced to the action of Γ_v.

6.7.2 Analytic surfaces

Let $X = (X_t)_{t\to 0}$ be a maximally degenerate K3 surface over the field $\mathbf{C}_t^{\mathrm{mer}}$ (see Section 5.1). We denote by Λ_X the quotient group $[\gamma_0]^\perp/\mathbf{Z}[\gamma_0]$, where $[\gamma_0] \in H_2(X_t, \mathbf{Z})$ is the vanishing cycle. Then $\Lambda_X \simeq \Lambda_{2,18}$. Let us assume that the monodromy acts trivially on Λ_X.

We define a natural homomorphism $\rho_X : \Lambda_X \to (\mathbf{C}_t^{\mathrm{mer}})^\times$ by the formula

$$\rho_X([\gamma]) = \exp\left(2\pi i \frac{\int_\gamma \Omega_t}{\int_{\gamma_0} \Omega_t}\right), \quad [\gamma] \in [\gamma_0]^\perp.$$

One can give a more abstract definition of ρ_X in terms of the variation of Hodge structure. It is easy to see that $(\mathrm{val}_{\mathbf{C}_t^{\mathrm{mer}}} \circ \rho_X)([\gamma]) = (v_X, [\gamma])$, where $v_X \in \Lambda_X$ is a vector such that $(v_X, v_X) > 0$, and $\mathrm{val}_{\mathbf{C}_t^{\mathrm{mer}}}$ is the standard valuation on the field $\mathbf{C}_t^{\mathrm{mer}} \subset \mathbf{C}((t))$.

Let X^{an} be the corresponding analytic K3 surface over the field $K = \mathbf{C}((t))$. We have an analytic torus fibration over $S^2 \setminus \{x_1, \dots, x_{24}\}$ which can be extended to a continuous map $X^{\mathrm{an}} \to S^2$. Let us call such an extension a *singular* analytic torus fibration with standard singularities.

Conjecture 8. For any analytic K3 surface X^{an}/K admitting an analytic torus fibration $X^{\mathrm{an}} \to S^2$ with standard singularities, one can define intrinsically the lattice $\Lambda_{X^{\mathrm{an}}}$ and the homomorphism $\rho_{X^{\mathrm{an}}} : \Lambda_{X^{\mathrm{an}}} \to K^\times$.

Notice that for K3 surfaces any birational automorphism is biregular. Hence the group of birational automorphisms $\mathrm{Aut}^{\mathrm{brt}}(X)$ acts by a $\mathbf{Z}$PL-transformation of the sphere S^2 which is equipped with a singular $\mathbf{Z}$-affine structure (see Section 6.6), i.e., we have a homomorphism

$$\mathrm{Aut}^{\mathrm{brt}}(X) = \mathrm{Aut}(X) \to \mathrm{Aut}_{\mathbf{Z}\mathrm{PL},v_X}(\mathrm{Sk}(X^{\mathrm{an}}, \Omega)) \simeq \mathrm{Aut}_{\mathbf{Z}\mathrm{PL},v_X}(S^2).$$

Conjecture 9.

(1) The image Γ_{ρ_X} of $\mathrm{Aut}(X)$ in $\mathrm{Aut}(\Lambda_X, \rho_X)$ is a subgroup of Γ_{v_X}, where $v_X := \mathrm{val}_K \circ \rho_X : \Lambda_X \to \mathbf{R}$.
(2) The homomorphism $\mathrm{Aut}(X) \to \mathrm{Aut}_{\mathbf{Z}\mathrm{PL},v_X}(S^2)$ is conjugate to the restriction to Γ_{ρ_X} of the homomorphism ϕ_{v_X} defined in the previous subsection.

6.7.3 Lattice points

Let us consider the special case when vector v is a lattice vector, i.e., $v \in \Lambda_{2,18}$. In the A-model picture, it corresponds to the integrality of the class $[\omega]$ of symplectic 2-form. In the B-model, this means that the non-archimedean field K has valuation in $\mathbf{Z} \subset \mathbf{R}$. In terms of $\mathbf{Z}$-affine structures, it means that the monodromy of the affine connection is reduced to $SL(2, \mathbf{Z}) \ltimes \mathbf{Z}^2$. The group Γ_v is a subgroup (and also a quotient group) of an arithmetic subgroup in the Lie group SO(1, 18). Also, in this case there is a Γ_v-invariant notion of a point with integer coordinates on $B \simeq S^2$, as well as points with coordinates in $\frac{1}{N}\mathbf{Z}$ for any integer $N \geq 1$. The number $M_{v,N}$ of such points is finite. It is not hard to see that $M_{v,N} = \mathrm{Area}_v + 2 = \frac{(v,v)}{2}N^2 + 2$, where Area_v is the area of B with a $\mathbf{Z}$PL structure corresponding to v. This is analogous to the Riemann–Roch formula $rk(\Gamma(X_{\mathbf{C}}, L^{\otimes N})) = \int_X \frac{c_1(L)^2}{2} + 2$ for an ample line bundle L on a complex K3 surface $X_{\mathbf{C}}$.

The action of Γ_v on S^2 gives rise to a homomorphism $\Gamma_v \to S_{M_{v,n}}$, where $S_{M_{v,n}}$ is the symmetric group. Also the action gives a homomorphism from Γ_v to the mapping class group $\pi_1(\mathcal{M}^{\mathrm{unord}}_{0,M_{v,N}})$, the fundamental group of the moduli space of genus-zero complex curves with $M_{v,N}$ unordered distinct marked points. The last group is closely related to the braid group. The conclusion is that we have constructed homomorphisms from arithmetic groups to a tower of braid groups.

One can deduce from Torelli theorem an interpretation of Γ_v as a quotient group of the fundamental group of a neighborhood U of a cusp in 19-dimensional moduli space of polarized complex algebraic K3 surfaces, where vector v corresponds to the polarization. Therefore, the homomorphism $\Gamma_v \to S_{M_{v,n}}$ gives a finite covering U' of U. One may wonder whether there exists a line bundle over U' whose direct image to U coincides with the direct image of the sheaf $L^{\otimes N}$ from the universal family of K3 surfaces. (This question is in spirit of some ideas of Andrey Tyurin; see, e.g., [Tyu].)

6.8 Further examples

There are many families of Calabi–Yau varieties with huge groups of birational automorphisms. We learned the following example from D. Panov and D. Zvonkine. For any real numbers $l_1, \ldots, l_n > 0$, we can consider the space of planar n-gons with the lengths of edges equal to $l_1, \ldots, l_n$, modulo the group of orientation-preserving motions. This space can be identified with the space of solutions of the system of equations

$$\sum l_i z_i = 0, \qquad \sum l_i z_i^{-1} = 0,$$

where $(z_1 : \cdots : z_n) \in \mathbf{C}P^{n-1}$ is a point satisfying the reality condition $|z_i| = 1$, $i = 1, \ldots, n$. Hence we obtain a singular subvariety of $\mathbf{C}P^{n-1}$ of codimension 2, depending on the parameters $l_1, \ldots, l_n$. One can check that this variety is birationally isomorphic to a nonsingular Calabi–Yau variety. For any proper subset $I \subset \{1, \ldots, n\}$, $2 \leq |I| \leq n-2$, we have a birational involution σ_I defined by the formula

$$\sigma_I^*(z_i) = \begin{cases} c/z_i & \text{if } i \in I, \\ z_i & \text{if } i \notin I, \end{cases}$$

where $c := \frac{\sum_{i \in I} l_i z_i}{\sum_{i \in I} l_i / z_i}$.

We do not know at the moment the structure of the group G_n generated by the involutions σ_I. One can easily obtain explicit formulas for the action of G_n by piecewise-linear homemorphisms of S^{n-3}. The length parameters l_i should be replaced by elements of a non-archimedean field K with "generic" norms $\lambda_i = \mathrm{val}_K(l_i) \in \mathbf{R}$.

Denote by ζ_i, $i = 1, \dots, n$, real variables which have the meaning of valuations of the variables $z_i \in K$. The sphere S^{n-3} is obtained in the following way. In $\mathbf{R}^n$ we consider the intersection of two subsets:

$$\{(\zeta_1, \dots \zeta_n) \mid \min_i(\lambda_i + \zeta_i) \text{ is achieved at least twice}\}$$

and

$$\{(\zeta_1, \dots \zeta_n) \mid \min_i(\lambda_i - \zeta_i) \text{ is achieved at least twice}\}$$

and then take the quotient by the action of $\mathbf{R}$:

$$(\zeta_1, \dots \zeta_n) \to (\zeta_1 + a, \dots, \zeta_n + a)$$

corresponding to the projectivization. For appropriately chosen $(\lambda_1, \dots, \lambda_n)$ we obtain a set which is the union of S^{n-3} with several "wings" going to infinity. The action of the involution σ_I is obtained from algebraic formulas from above, in which one replaces non-archimedean variables by real ones, addition by minimum and multiplication (division) by addition (subtraction).

7 K-affine structures

7.1 Definitions

Let B^{sm} be a manifold with $\mathbf{Z}$-affine structure. The sheaf of $\mathbf{Z}$-affine functions $\mathrm{Aff}_{\mathbf{Z}} := \mathrm{Aff}_{\mathbf{Z}, B^{\mathrm{sm}}}$ gives rise to an exact sequence of sheaves of abelian groups

$$0 \to \mathbf{R} \to \mathrm{Aff}_{\mathbf{Z}} \to (T^*)^{\mathbf{Z}} \to 0.$$

Let K be a complete non-archimedean field with a valuation map val. We give two equivalent definitions of a K-affine structure on B^{sm} compatible with a given $\mathbf{Z}$-affine structure.

Definition 11. *A K-affine structure on B^{sm} compatible with the given $\mathbf{Z}$-affine structure is a sheaf Aff_K of abelian groups on B^{sm}, an exact sequence of sheaves*

$$0 \to K^\times \to \mathrm{Aff}_K \to (T^*)^{\mathbf{Z}} \to 0,$$

together with a homomorphism Φ of this exact sequence to the exact sequence of sheaves of abelian groups

$$0 \to \mathbf{R} \to \mathrm{Aff}_{\mathbf{Z}} \to (T^*)^{\mathbf{Z}} \to 0,$$

such that $\Phi = \mathrm{id}$ on $(T^*)^{\mathbf{Z}}$ and $\Phi = \mathrm{val}$ on $K^\times$.

Since B^{sm} carries a $\mathbf{Z}$-affine structure, we have an associated $\mathrm{GL}(n, \mathbf{Z}) \ltimes \mathbf{R}^n$-torsor on B^{sm}, whose fiber over a point x consists of all $\mathbf{Z}$-affine coordinate systems at x.

Definition 12. *A K-affine structure on B^{sm} compatible with the given $\mathbf{Z}$-affine structure is a* $\mathrm{GL}(n, \mathbf{Z}) \ltimes (K^\times)^n$*-torsor on B^{sm} such that the application of* $\mathrm{val}^{\times n}$ *to* $(K^\times)^n$ *gives the initial* $\mathrm{GL}(n, \mathbf{Z}) \ltimes \mathbf{R}^n$*-torsor.*

The equivalence of the above two definitions is obvious in local $\mathbf{Z}$-affine coordinates. The reason is that the set of automorphisms of the exact sequence of groups

$$0 \to K^\times \to K^\times \times \mathbf{Z}^n \to \mathbf{Z}^n \to 0$$

which are the identity on $K^\times$ coincides with the group $\mathrm{GL}(n, \mathbf{Z}) \ltimes (K^\times)^n$.

Finally, we can formulate the Fixed Point Property for K-affine structures (see Section 3.1 for the $\mathbf{Z}$-affine case).

Fixed Point Property for K-Affine Structures. In the notation of the end of Section 3.1, for any $b \in B^{\mathrm{sing}}$ and sufficiently small neighborhood U of b the lifted monodromy representation $\pi_1(U) \to \mathrm{GL}(n, \mathbf{Z}) \ltimes (K^\times)^n$ has fixed vectors in $K^{\times n}$, and the $\mathbf{R}$-affine span of the corresponding (under the valuation map) vectors in $\mathbf{R}^n$ coincides with the set of fixed points of the monodromy representation $\pi_1(U) \to \mathrm{GL}(n, \mathbf{Z}) \ltimes \mathbf{R}^n$.

7.2 K-affine structure on smooth points

From this section until the end of the chapter (except for Section 11.7), we will assume the following.

Zero Characteristic Assumption. K is a complete non-archimedean local field such that its residue field has characteristic zero.

Let X be a K-analytic manifold of dimension n and we are given a continuous map $\pi : X \to B$, where B is a topological space. Then B^{sm} carries a $\mathbf{Z}$-affine structure (Theorem 1). Suppose that there is an open K-analytic submanifold $U \subset X$ such that $\pi^{-1}(B^{\mathrm{sm}}) \subset U$ and there is a nonwhere vanishing analytic form $\Omega \in \Gamma(U, \Omega^n_X)$. We are going to define a $\mathbf{Z}$-affine function $\mathrm{Val}(\Omega)$ similar to the definition of the function $\mathrm{Val}(\varphi)$ in Section 4.1. Namely, in local coordinates $(z_1, \dots, z_n)$ we consider the expression $\varphi := \Omega / \bigwedge_{1 \le i \le n} (dz_i/z_i)$. This is an invertible function, and we define $\mathrm{Val}(\Omega)$ as $\mathrm{Val}(\varphi)$. The independence of the choice of coordinates follows from the following lemma.

Lemma 2. *Let $(z_i)_{i=1,\dots,n}$, $(z'_i)_{i=1,\dots,n}$, be two systems of invertible coordinates on $\pi^{-1}(U)$ for some connected open $U \subset B^{\mathrm{sm}}$. Then*

$$|(\textstyle\bigwedge_{1 \le i \le n} (dz_i/z_i)) / (\bigwedge_{1 \le i \le n} (dz'_i/z'_i))|_x = 1 \quad \forall x \in \pi^{-1}(U).$$

Proof. By Lemma 1 from Section 4.1, we know that z_i', as any invertible function, can be written in the form $c_i z^{I^{(i)}}(1+o(1))$ for some nonzero $c_i \in K$ and a multiindex $I^{(i)} \in \mathbf{Z}^n$. The vectors $I^{(1)}, \dots, I^{(n)}$ form a basis of $\mathbf{Z}^n$, as follows from the condition that $z_1', \dots, z_n'$ form a coordinate system. Therefore, after applying the change of coordinates $z_i \mapsto c_i z^{I^{(i)}}$ preserving the form $\bigwedge_i dz_i/z_i$ up to sign, we may assume that $z_i' = (1+o(1))z_i$. The Jacobian matrix of the transformation $(z_i) \to (z_i')$ is the identity matrix plus terms of size $o(1)$. Therefore, its determinant has norm equal to 1. □

Now we make the following assumption.

Constant Norm Assumption. The function $\mathrm{Val}(\varphi)$ is locally constant.

Theorem 4. *If the Constant Norm Assumption is satisfied, then there is a K-affine structure on B^{sm} compatible with the $\mathbf{Z}$-affine structure* $\mathrm{Aff}^{\mathrm{can}}_{\mathbf{Z},B^{\mathrm{sm}}}$ *(see Section* 4.1).

Proof. Let us write in local coordinates $\Omega = \varphi(z_1, \dots, z_n) \bigwedge_{1\le i\le n} \frac{dz_i}{z_i}$. Define the residue $\mathrm{Res}(\Omega) \in K$ as the constant term φ_0 in the Laurent expansion $\varphi(z_1, \dots, z_n) = \sum_{I\in\mathbf{Z}^n} \varphi_I z^I$. It is easy to see that $\mathrm{Res}(\Omega)$ does not depend (up to sign) on the choice of local coordinates. For nowhere vanishing Ω satisfying the Constant Norm Assumption, we have $\exp(-\mathrm{Val}(\varphi)) = |\varphi| = |\varphi_0|$. Therefore, we have $\mathrm{Res}(\Omega) \neq 0$.

Let us return to the proof of the theorem. Let F be the sheaf of abelian groups $F \subset \pi_*(\mathcal{O}_X^\times)$ consisting of f such that $\mathrm{Val}(f) = 0$. Then we have an exact sequence of sheaves

$$0 \to K^\times/\mathcal{O}_K^\times \to \pi_*(\mathcal{O}_X^\times)/F \to (T_X^*)^{\mathbf{Z}} \to 0,$$

where $\mathcal{O}_K$ denotes the constant sheaf with fiber being the ring of integers of K. Indeed, we embed $K^\times/\mathcal{O}_K^\times$ into $\pi_*(\mathcal{O}_X^\times)/F$ as constant functions. The projection $\pi_*(\mathcal{O}_X^\times)/F \to (T_X^*)^{\mathbf{Z}}$ assigns to the function f the linear part of the corresponding $\mathbf{Z}$-affine function $\mathrm{Val}(f)$.

Notice that if $U \subset B^{\mathrm{sm}}$ is a connected domain then any $f \in \Gamma(U, F)$ can be written (noncanonically) as $f = a(1+r)$, where $a \in \mathcal{O}_K^\times$ and $r = o(1)$ in $\pi^{-1}(U)$.

We define an epimorphism of sheaves $p_\Omega : F \twoheadrightarrow \mathcal{O}_K^\times$ by the formula

$$p_\Omega(f) = p_\Omega(a(1+r)) = a\exp\left(\frac{\mathrm{Res}(\Omega\log(1+r))}{\mathrm{Res}(\Omega)}\right).$$

Here exp and log are understood as infinite convergent series. (In order to make sense of them, we use the Zero Characteristic Assumption.)

It is easy to see that p_Ω is well defined. Then the exact sequence of sheaves

$$1 \to K^\times \to \pi_*(\mathcal{O}_X^\times)/\ker(p_\Omega) \to (T_X^*)^{\mathbf{Z}} \to 1$$

defines a K-affine structure on B^{sm} compatible with $\mathrm{Aff}^{\mathrm{can}}_{\mathbf{Z},B^{\mathrm{sm}}}$. This concludes the proof of the theorem. □

Notice that the above proof gives an explicit construction of the K-affine structure. We will denote it by $\mathrm{Aff}^{\Omega}_{K,B^{\mathrm{sm}}}$. It is easy to see that this K-affine structure does not change if we make a rescaling $\Omega \mapsto c\Omega$, $c \in K^\times$.

7.3 Lifting Problem

Let K be as in Section 7.2, let $B \supset B^{\mathrm{presing}}$ be a space with singular $\mathbf{Z}$-affine structure (see Section 6.3), and assume we have an extension of the $\mathbf{Z}$-affine structure on $B \backslash B^{\mathrm{presing}}$ to a K-affine structure satisfying the Fixed Point Property (see Section 7.1). We assume that that $\mathbf{Z}$-affine structure cannot be extended to a larger open set $U \supset B \setminus B^{\mathrm{presing}}$, $U \neq B \setminus B^{\mathrm{presing}}$. Slightly abusing notation, we will denote $B \setminus B^{\mathrm{presing}}$ simply by B^{sm}. We want to have a K-analytic space X, meromorphic nonzero top-degree form Ω and a continuous proper (and maybe also Stein) map $\pi : X \to B$ such that

1. B^{presing} coincides with B^{sing}, and the $\mathbf{Z}$-affine structure on B^{sm} arising from the projection π coincides with the given one;
2. the restriction $\Omega_{|\pi^{-1}(B^{\mathrm{sm}})}$ is a nowhere vanishing analytic form which satisfies the Constant Norm Assumption;
3. the K-affine structure on B^{sm} arising from the pair (X, Ω) coincides with the initial one.

We call the problem of finding such data the *Lifting Problem.*

Remark 3. If a solution of the Lifting Problem exists, then B^{sm} is orientable. Indeed, $\mathrm{Res}(\Omega)$ is locally a constant defined up to a sign which depends on the orientation of B^{sm}. A global choice of the constant gives an orientation. For oriented B^{sm} we can rescale Ω canonically in such a way that $\mathrm{Res}(\Omega) = 1$.

Question. What restrictions on the behavior of the K-affine structure near $B^{\mathrm{presing}} = B^{\mathrm{sing}}$ should we impose in order to guarantee the existence of a solution of the Lifting Problem?

Let $B = B^{\mathrm{sm}}$ be a flat torus (see Section 3.2.1). Then the Lifting Problem has a solution (canonical up to rescaling of Ω) for any compatible K-affine structure. More precisely, the groupoid of Tate tori and isomorphisms between them is equivalent to the groupoid of K-affine structures on real flat tori.

In Sections 8–11 we are going to discuss a solution of the Lifting Problem for K3 surfaces. In that case $B^{\mathrm{sing}} \neq \emptyset$.

If we restrict ourselves only to the smooth part B^{sm} (i.e., we allow noncompact X) then there is a canonical solution of this "reduced" Lifting Problem. In other words, one can construct a smooth K-analytic space X' with an analytic top-degree form Ω' and a map $\pi' : X' \to B^{\mathrm{sm}}$ satisfying the above conditions 1–3. Let us explain this construction assuming that B^{sm} is oriented.

First, we notice that the orientation of B^{sm} gives a reduction to $\mathrm{SL}(n, \mathbf{Z}) \ltimes (K^{\times})^n$ of the structure group of the torsor defining the K-affine structure. The reduced group acts naturally by automorphisms of the fibration $\pi_{\mathrm{can}} : (\mathbf{G}_m^{\mathrm{an}})^n \to \mathbf{R}^n$ preserving the form $\bigwedge_{1 \le i \le n} \frac{dz_i}{z_i}$. The action on $(\mathbf{G}_m^{\mathrm{an}})^n$ is induced from the action on monomials. Namely, the inverse to an element $(A, \lambda_1, \dots, \lambda_n) \in \mathrm{SL}(n, \mathbf{Z}) \ltimes (K^{\times})^n$ acts on monomials as

$$z^I = z_1^{I_1} \dots z_n^{I_n} \mapsto \left(\prod_{i=1}^n \lambda_i^{I_i}\right) z^{A(I)}.$$

The action of the same element on $\mathbf{R}^n$ is given by the similar formula

$$x = (x_1, \ldots, x_n) \mapsto A(x) - (\mathrm{val}(\lambda_1), \ldots, \mathrm{val}(\lambda_n)).$$

Let $B^{\mathrm{sm}} = \cup_\alpha U_\alpha$ be an open covering by coordinate charts $U_\alpha \simeq V_\alpha \subset \mathbf{R}^n$ such that for any α, β we are given elements $g_{\alpha,\beta} \in \mathrm{SL}(n, \mathbf{Z}) \ltimes (K^\times)^n$ satisfying the 1-cocycle condition for any triple α, β, γ. Then the space X' is obtained from $\pi_{\mathrm{can}}^{-1}(V_\alpha)$ by gluing by means of the transformations $g_{\alpha,\beta}$. The form $\bigwedge_{1 \le i \le n} \frac{dz_i}{z_i}$ gives rise to a nowhere vanishing analytic top-degree form Ω' on X'. Thus we have obtained a solution of the reduced Lifting Problem. The sheaf $\pi_*(\mathcal{O}_{X'}) := \mathcal{O}_{B^{\mathrm{sm}}}^{\mathrm{can}}$ is called the *canonical sheaf*.

In the case $B^{\mathrm{presing}} \neq \emptyset$ this solution seems to be the "wrong" one, i.e., it cannot be extended to a solution $\pi : X \to B$, where X and B are compact. In the case of K3 surfaces we will show later how to modify it in order to obtain a "true" solution of the Lifting Problem.

7.4 Flat coordinates and periods

Here we are going to discuss a relation between K-affine structures and so-called flat coordinates on the moduli space of complex structures on Calabi–Yau manifolds. We assume the picture of collapse from Section 5.1.

7.4.1 Flat coordinates for degenerating complex Calabi–Yau manifolds

Let $X_{\mathrm{mer}} = (X_t)_{t \to 0}$ be a maximally degenerating algebraic Calabi–Yau manifold of dimension n over $\mathbf{C}_t^{\mathrm{mer}}$. We denote by B the Gromov–Hausdorff limit of our family (see Conjecture 1, Section 5.1). Its connected oriented open dense part B^{sm} carries a $\mathbf{Z}$-affine structure with the covariant lattice $T^{\mathbf{Z}}$.

Recall that according to the picture of collapse presented in Section 5.1 there is a canonical isotopy class of embeddings from the total space of a torus bundle $p : X_t' \to B^{\mathrm{sm}}$ to the complex manifold X_t for all sufficiently small $t \neq 0$. Let us denote by $[\gamma_0] \in H_n(X_t', \mathbf{Z})$ the fundamental class of the fiber of p. This is the homology class of a singular chain in X_t' which projects to a point by p.

Let $H_n^{\le 1}(X_t', \mathbf{Z}) \subset H_n(X_t', \mathbf{Z})$ be the subgroup generated by homology classes of chains which are projected into graphs in B^{sm}. It follows from the definition that we have an epimorphism

$$J_a : H_1(B^{\mathrm{sm}}, \textstyle\bigwedge^{n-1} T^{\mathbf{Z}}) \twoheadrightarrow H_n^{\le 1}(X_t', \mathbf{Z})/\mathbf{Z}[\gamma_0]$$

similar to the homomorphism J_s defined in the symplectic case (see Section 3.1.1). The following formula defines a homomorphism of groups:

$$P : H_n^{\le 1}(X_t', \mathbf{Z})/\mathbf{Z}[\gamma_0] \to (\mathbf{C}_t^{\mathrm{mer}})^\times, \qquad [\gamma] \mapsto \exp\left(2\pi i \frac{\int_{[\gamma]} \Omega_t}{\int_{[\gamma_0]} \Omega_t}\right).$$

We will call P the *period map*. Notice that $\mathbf{Z}[\gamma_0] := H_n^{\leq 0}(X'_t, \mathbf{Z}) \subset H_n^{\leq 1}(X'_t, \mathbf{Z})$ is the low-degree part of the limiting Hodge filtration on the homology of the Calabi–Yau manifold X_t. The nonzero complex numbers

$$\exp\left(2\pi i \frac{\int_{[\gamma_i]} \Omega_t}{\int_{[\gamma_0]} \Omega_t}\right),$$

where γ_i is a set of generators of $H_n^{\leq 1}(X'_t, \mathbf{Z})/H_n^{\leq 0}(X'_t, \mathbf{Z})$ are called *flat coordinates* in Mirror Symmetry (see, e.g., [Mor]). These are local coordinates near a point close to the "cusp" of the moduli space of complex structures (local Torelli theorem).

The orientation of B^{sm} gives rise to an isomorphism $\bigwedge^{n-1} T^{\mathbf{Z}} \simeq (T^*)^{\mathbf{Z}}$. Therefore, combining the maps J_a, P and the above isomorphism we obtain a homomorphism

$$\widetilde{P} : H^1(B^{\mathrm{sm}}, (T^*)^{\mathbf{Z}}) \to (\mathbf{C}_t^{\mathrm{mer}})^{\times}.$$

7.4.2 Non-archimedean periods

Let X^{an} be a smooth analytic Calabi–Yau manifold associated with X_{mer}. Assuming the equivalence of the Gromov–Hausdorff and non-archimedean pictures of collapse presented in Section 5, we have a continuous map $\pi : X^{\mathrm{an}} \to B$. It gives a K-affine structure on B^{sm}. The corresponding exact sequence

$$0 \to K^{\times} \to \mathrm{Aff}_K \to (T^*)^{\mathbf{Z}} \to 0$$

represents a class in $H^1(B^{\mathrm{sm}}, T^{\mathbf{Z}} \otimes K^{\times}) \simeq Ext^1((T^*)^{\mathbf{Z}}, K^{\times})$. Pairing with this class gives another homomorphism

$$P' : H_1(B^{\mathrm{sm}}, (T^*)^{\mathbf{Z}}) \to K^{\times} = H_0(B^{\mathrm{sm}}, K^{\times}).$$

Conjecture 10. The homomorphism P' is equal to the composition of $\widetilde{P}$ with the embedding $(\mathbf{C}_t^{\mathrm{mer}})^{\times} \hookrightarrow K^{\times}$.

Part III

We fix field K satisfying the Zero Characteristic Assumption.

Let B be a compact oriented surface, $B^{\mathrm{sing}} \subset B$ a finite set, and $\mathrm{Aff}_K = \mathrm{Aff}_{K,Y}$ a sheaf defining a K-affine structure on $Y := B^{\mathrm{sm}} = B \setminus B^{\mathrm{sing}}$. We assume that all singularities of the underlying $\mathbf{Z}$-affine structure are standard (see Section 6.4), and the local monodromy around each $b \in B^{\mathrm{sing}}$ acts on $(K^{\times})^2$ with a fixed point (see the Fixed Point Property at the end of Section 7.1). The main result of Part III of the article can be formulated as follows.

Theorem 5. *There exist a compact K-analytic surface X^{an}, a top-degree analytic form $\Omega = \Omega_{X^{\mathrm{an}}}$ and a continuous proper Stein map $\pi : X^{\mathrm{an}} \to B$ such that the set of π-smooth points coincides with Y and the induced K-affine structure coincides with the one given by* Aff_K.

In other words, the triple $(X^{\mathrm{an}}, \pi, \Omega)$ is a solution of the Lifting Problem.

By the Stein property it suffices to construct the sheaf $\mathcal{O}_B = \pi_*(\mathcal{O}_{X^{\mathrm{an}}})$ of K-algebras over B. We will see that outside of the finite singular set $S = \{x_1, \ldots, x_{24}\}$ the sheaf $\mathcal{O}_B$ is locally isomorphic to $\mathcal{O}_Y^{\mathrm{can}}$. In the next section we will describe the local model for the sheaf $\mathcal{O}_B$ near each singular point. It will be glued together with a modification of the canonical sheaf $\mathcal{O}_Y^{\mathrm{can}}$. This modification depends on data called *lines*. The appearance of lines is motivated by Homological Mirror Symmetry (see [Ko, KoSo]).[6] Roughly speaking, lines correspond (for mirror dual K3 surface) to collapsing holomorphic discs with boundaries belonging to fibers of the dual torus fibration (see Section 5.1 and [KoSo]). Such "bad" fibers are Lagrangian tori, but they do not correspond to objects of the Fukaya category (A-branes in the terminology of physicists). There are infinitely many such fibers and hence infinitely many lines. We will axiomatize this piece of data in Section 9. Subsequently, with each line l we will associate an automorphism of the restriction of $\mathcal{O}_Y^{\mathrm{can}}$ to l. This will give us the above-mentioned modified canonical sheaf.

8 Model near a singular point

Here we will construct an analytic torus fibration corresponding to the standard singularity (see Sections 3.2.4 and 6.4).

Let $X \subset \mathbf{A}^3$ be the algebraic surface given by the equation $(\alpha\beta - 1)\gamma = 1$ in the coordinates (α, β, γ), and X^{an} be the corresponding analytic space. We define a continuous map $f : X^{\mathrm{an}} \to \mathbf{R}^3$ by the formula $f(\alpha, \beta, \gamma) = (a, b, c)$, where $a = \max(0, \log|\alpha|_p)$, $b = \max(0, \log|\beta|_p)$, $c = \log|\gamma|_p = -\log|\alpha\beta - 1|_p$. Here $|\cdot|_p = \exp(-\mathrm{val}_p(\cdot))$ denotes the multiplicative seminorm corresponding to the point $p \in X^{\mathrm{an}}$ (see Appendix A).

Proposition 4. *The map f is proper. Moreover,*

(a) *the image of f is homeomorphic to $\mathbf{R}^2$;*
(b) *all points of the image except of* $(0, 0, 0)$ *are f-smooth.*

Proof. Here is the plan of the proof:

1. We define three open domains T_i, $i = 1, 2, 3$, in three copies of the standard two-dimensional analytic torus $(\mathbf{G}_m^{\mathrm{an}})^2$, and a continuous map $\pi_i : T_i \to \mathbf{R}^2$ such that all points of the image $U_i = \pi_i(T_i)$ are π_i-smooth (i.e., each π_i is an analytic torus fibration). The domains U_i cover $\mathbf{R}^2 \setminus \{(0, 0)\}$.
2. For each i, $1 \le i \le 3$, we construct an open embedding $g_i : T_i \hookrightarrow X^{\mathrm{an}}$.
3. We construct an embedding $j : \mathbf{R}^2 \hookrightarrow \mathbf{R}^3$ such that each open set U_i is homeomorphically identified with $f(g_i(T_i))$ and $j((0, 0)) = (0, 0, 0)$. Moreover, π_i-smooth points are mapped into f-smooth points.

[6] The main idea is that X is a component of the moduli space of certain objects (skyscraper sheaves) in the derived category $D^b(\mathrm{Coh}(X))$. These objects correspond to $U(1)$-local systems on Lagrangian tori in the Fukaya category of the mirror dual symplectic manifold.

The proposition will follow from 1–3.

Let us describe the constructions and formulas. We start with the open sets U_i, $1 \le i \le 3$. Let us fix a number $0 < \varepsilon < 1$ and define

$$U_1 = \{(x, y) \in \mathbf{R}^2 | x < \varepsilon |y|\},$$
$$U_2 = \{(x, y) \in \mathbf{R}^2 | x > 0, y < \varepsilon x\},$$
$$U_3 = \{(x, y) \in \mathbf{R}^2 | x > 0, y > 0\}$$

Clearly, $\mathbf{R}^2 \setminus \{(0,0)\} = U_1 \cup U_2 \cup U_3$. We also define a slightly modified domain U_2' as $\{(x, y) \in \mathbf{R}^2 | x > 0, y < \frac{\varepsilon}{1+\varepsilon} x\}$.

We define $T_i := \pi_{\mathrm{can}}^{-1}(U_i) \subset (\mathbf{G}_m^{\mathrm{an}})^2$, $i = 1, 3$, and $T_2 := \pi_{\mathrm{can}}^{-1}(U_2') \subset (\mathbf{G}_m^{\mathrm{an}})^2$. Then the projections $\pi_i : T_l \to U_l$ are given by the formulas

$$\pi_i(\xi_i, \eta_i) = \pi_{\mathrm{can}}(\xi_i, \eta_i) = (\log|\xi_i|, \log|\eta_i|), \quad i = 1, 3,$$
$$\pi_2(\xi_2, \eta_2) = \begin{cases} (\log|\xi_2|, \log|\eta_2|) & \text{if } |\eta_2| < 1, \\ (\log|\xi_2| - \log|\eta_2|, \log|\eta_2|) & \text{if } |\eta_2| \ge 1. \end{cases}$$

In these formulas (ξ_i, η_i) are coordinates on T_i, $1 \le i \le 3$.

We define the inclusion $g_i : T_i \hookrightarrow X$, $1 \le i \le 3$, by the following formulas:

$$g_1(\xi_1, \eta_1) = \left(\frac{1}{\xi_1}, \xi_1(1 + \eta_1), \frac{1}{\eta_1}\right),$$
$$g_2(\xi_2, \eta_2) = \left(\frac{1 + \eta_2}{\xi_2}, \xi_2, \frac{1}{\eta_2}\right),$$
$$g_3(\xi_3, \eta_3) = \left(\frac{1 + \eta_3}{\xi_3\eta_3}, \xi_3\eta_3, \frac{1}{\eta_3}\right).$$

Let us decompose $X^{\mathrm{an}} = X_- \cup X_0 \cup X_+$ according to the sign of $\log|\gamma|_p$, where $p \in X^{\mathrm{an}}$ is a point. It is easy to see that

$$f(X_-) = \{(a, b, c) \in \mathbf{R}^3 \mid c < 0, a \ge 0, b \ge 0, ab(a + b + c) = 0\},$$
$$f(X_0) = \{(a, b, c) \in \mathbf{R}^3 \mid c = 0, a \ge 0, b \ge 0, ab = 0\},$$
$$f(X_+) = \{(a, b, c) \in \mathbf{R}^3 \mid c > 0, a \ge 0, b \ge 0, ab = 0\}.$$

From this explicit description we see that f is proper and the image of f is homeomorphic to $\mathbf{R}^2$.

Let us consider the embedding $j : \mathbf{R}^2 \to \mathbf{R}^3$ given by the formula

$$j(x, y) = \begin{cases} (-x, \max(x + y, 0), -y) & \text{if } x \le 0, \\ (0, x + \max(y, 0), -y) & \text{if } x \ge 0. \end{cases}$$

One can easily check that the image of j coincides with the image of f, $j \circ \pi_i = f \circ g_i$ and $f^{-1}(j(U_i)) = g_i(T_i)$ for all $1 \le i \le 3$. This concludes the proof of the proposition. $\square$

We can derive more from the explicit formulas given in the proof.

Let us denote by $\pi : X^{an} \to \mathbf{R}^2$ the map $j^{(-1)} \circ f$. It is an analytic torus fibration outside of the point $(0,0)$. The induced $\mathbf{Z}$-affine structure on $\mathbf{R}^2 \setminus \{(0,0)\}$ is in fact the standard singular $\mathbf{Z}$-affine structure described in Sections 3.2.4 and 6.4, as follows immediately from the formulas for the projections π_i, $i = 1, 2, 3$.

Let us introduce another sheaf $\mathcal{O}^{can}$ on $\mathbf{R}^2 \setminus \{(0,0)\}$. It is defined as $(\pi_i)_*(\mathcal{O}_{T_i})$ on each domain U_i, with identifications

$$
\begin{aligned}
(\xi_1, \eta_1) &= (\xi_2, \eta_2) && \text{on } U_1 \cap U_2,\\
(\xi_1, \eta_1) &= (\xi_3, \eta_3) && \text{on } U_1 \cap U_3,\\
(\xi_2, \eta_2) &= (\xi_3\eta_3, \eta_3) && \text{on } U_2 \cap U_3.
\end{aligned}
$$

Let us consider the direct image sheaf $\pi_*(\mathcal{O}_{X^{an}})$. It is easy to see that on the sets U_1 and $U_2 \cup U_3$ this sheaf is canonically isomorphic to $\mathcal{O}^{can}$ (by identification of coordinates (ξ_1, η_1) and of the glued coordinates (ξ_2, η_2) and (ξ_3, η_3), respectively). Therefore, on the intersection $U_1 \cap (U_2 \cup U_3)$ we identify two copies of the canonical sheaf by a certain automorphism φ of $\mathcal{O}^{can}$ which preserves one coordinate (namely, the coordinate η). We will develop the theory of such transformations and their analytic continuations in Section 11. The explicit formulas for φ is

$$
\varphi(\xi, \eta) = \begin{cases} (\xi(1+\eta), \eta) & \text{on } U_1 \cap U_2,\\ (\xi(1+1/\eta), \eta) & \text{on } U_1 \cap U_3. \end{cases}
$$

We would like to now say few words about analytic volume forms. Notice that each $T_i \subset (\mathbf{G}_m^{an})^2$ carries a nowhere vanishing top-degree analytic form given by the formula $\Omega_i = \frac{d\xi_i \wedge d\eta_i}{\xi_i \eta_i}$. Then a straightforward calculation shows that Ω^{T_i} is the pullback under g_i of the nowhere vanishing on X_0^{an} analytic top-degree form

$$
\Omega = -\gamma d\alpha \wedge d\beta.
$$

The form Ω satisfies the Constant Norm Assumption; hence it gives a K-affine structure on $\mathbf{R}^2 \setminus \{(0,0)\}$. On the other hand, the sheaf $\mathcal{O}^{can}$ of algebras is also endowed with the top-degree form Ω^{can}, equal to $\frac{d\xi_i \wedge d\eta_i}{\xi_i \eta_i}$ in local coordinates.

Lemma 3. *The K-affine structure on $\mathbf{R}^2 \setminus \{(0,0)\}$ associated with Ω coincides with the one associated with Ω^{can}.*

Proof. Using definitions from Section 7.2 one sees immediately that it is enough to calculate $p_{\Omega^{can}}(1+\eta) = \exp(\mathrm{Res}(\Omega^{can} \log(1+\eta))) = 1 \in \mathcal{O}_K^{\times}$. □

In all the definitions and formulas in this section on can shift domains U_i, $i = 1, 2, 3$, by vector $(x_0, 0) \in \mathbf{R}^2$ for arbitrary $x_0 \in \mathbf{R}$, thus giving an map $X^{an} \to \mathbf{R}^2$ with singularity at the point $(x_0, 0)$.

Finally, we denote $\pi_*(\mathcal{O}_{X^{an}})$ by $\mathcal{O}_{\mathbf{R}^2}^{model}$. This will be our model for the sheaf $\mathcal{O}_B$ near each point of the singular set B^{sing}.

9 Lines on surfaces

In this section we are going to describe axiomatically the notion of collection of lines on a surface.

9.1 Data

(a) A compact oriented surface B, a finite subset $B^{\text{sing}} \subset B$.
(b) A $\mathbf{Z}$-affine structure on $Y = B^{\text{sm}} = B \setminus B^{\text{sing}}$ with the standard singularities near each $b \in B^{\text{sing}}$.
(c) A set $\mathcal{L}$ of *lines*. With each line $l \in \mathcal{L}$, there is an associated continuous map $f_l : (0, +\infty) \to Y$. We assume that $\mathcal{L}$ is decomposed into a disjoint union of two subsets $\mathcal{L} = \mathcal{L}_{\text{in}} \sqcup \mathcal{L}_{\text{com}}$. Lines belonging to $\mathcal{L}_{\text{in}}$ are called *initial*, while those in $\mathcal{L}_{\text{com}}$ are called *composite*. We assume that for any $l \in \mathcal{L}$ there exists a continuous extension $f_l : [0, +\infty) \to B$ such that $f_l(0) \in B^{\text{sing}}$ if $l \in \mathcal{L}_{\text{in}}$ and $f_l(0) \in Y = B^{\text{sm}}$ if $l \in \mathcal{L}_{\text{com}}$.
(d) A collection of covariantly constant nowhere vanishing integer-valued 1-forms $\alpha_l \in \Gamma((0, +\infty), f_l^*((T^*)^{\mathbf{Z}}), l \in \mathcal{L}$. We assume that for $l \in \mathcal{L}_{\text{in}}$ in the standard coordinates (x, y) near the singular point $f_l(0)$ we have $f_l(t) = (0, t)$ or $f_l(t) = (0, -t)$ for all sufficiently small $t > 0$, and $\alpha_l(t) = \pm f_l^*(dy)$.
(e) A map $\mathcal{L} \to \mathcal{L} \times \mathcal{L}, l \mapsto (p_{\text{left}}(l), p_{\text{right}}(l))$. (The letter p stands for "parent": one can think about these lines as "generating l in a collision.")

Notice that since the form dy is invariant with respect to the monodromy, the condition in **(d)** is coordinate-independent. The covector $\alpha_l(t)$ will be called a *direction covector* of l at time t. It gives rise to a half-plane

$$P_{l,t}^{(0)} = \{v \in T_{f_l(t)}Y | \langle \alpha_l(t), v\rangle > 0\}.$$

9.2 Axioms

To every $l 1 \mathcal{L}_{\text{in}}$ we assign a pair $(f_l(0), \text{sgn}(\alpha_l(0))) \in B^{\text{sing}} \times \{\pm 1\}$, where $\text{sgn}(\alpha_l(0))$ is a choice of sign in $\pm f_l^*(dy)$ (see data **(d)** in the previous subsection). In this way we obtain a map $r : \mathcal{L}_{\text{in}} \to B^{\text{sing}} \times \{\pm 1\}$.

Axiom 1. The map r is one-to-one.

Let $U \subset Y$ be a simply-connected domain, and assume the line l intersects U. Let $I \subset \mathbf{R}_+$ be an interval such that $f_l(I) \subset U$. Then there exists a covariantly constant closed nonzero 1-form β_U in U (with constant integer coefficients), such that $f_l^*(\beta_U) = \alpha_l$, when both sides are restricted to I.

Axiom 2. For any $t_1, t_2 \in I$, one has

$$\int_{f_l(t_1)}^{f_l(t_2)} \beta_U = t_2 - t_1.$$

Let $l_1, l_2 \in \mathcal{L}$, $t_1, t_2 > 0$ satisfy the condition $f_{l_1}(t_1) = f_{l_2}(t_2) = x \in Y$. In this case we say that the lines l_1 and l_2 have a collision at x at the times t_1 and t_2, respectively.

Axiom 3. Under the above assumptions there are only two possibilities:

3a. either $l_1 = l_2$ and $t_1 = t_2$, or
3b. the covector $\alpha_{l_1}(t_1)$ is not proportional to $\alpha_{l_2}(t_2)$. Then we may assume that $\alpha_{l_1}(t_1) \wedge \alpha_{l_2}(t_2) > 0$.

Under these conditions we require that for any coprime positive integers n_1, n_2 there exists a unique line $l \in \mathcal{L}$ such that $l_1 = p_{\text{left}}(l)$, $l_2 = p_{\text{right}}(l)$, $f_l(0) = x$ and $\alpha_l(0) = n_1\alpha_{l_1}(t_1) + n_2\alpha_{l_2}(t_2)$.

In other words, l_1 and l_2 are "parents of l," and the direction covector of l at the intersection point is a primitive integral linear combination of those for l_1 and l_2 (see Figure 4).

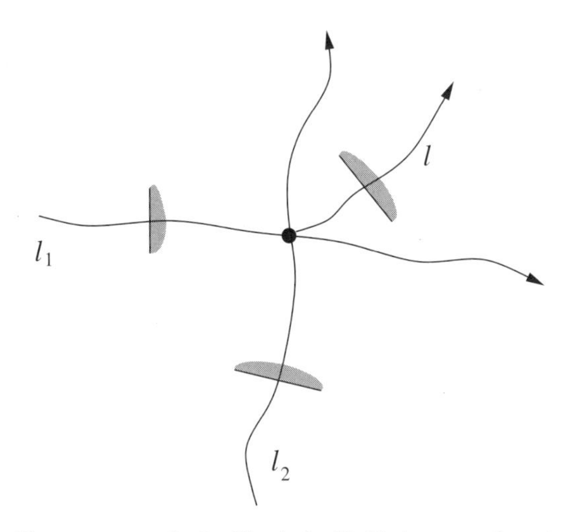

Fig. 4. A line l and its two parents l_1, l_2. The dashed half-planes are domains in tangent planes where the 1-forms α take positive values.

Axiom 4. For every line $l \in \mathcal{L}_{\text{com}}$, there exist l_1 and l_2 such that they satisfy condition **3b**.

Axiom 5. For any $x \in Y$, there are no more than two pairs $(l, t) \in \mathcal{L} \times (0, +\infty)$ such that $x = f_l(t)$. In other words, there are no more than two lines intersecting at a point in Y.

Let l_1, l_2, t_1, t_2, x mean the same as in Axiom 3, and assume that $\alpha_{l_1}(t_1) \wedge \alpha_{l_2}(t_2) > 0$. Let us consider the set $\mathcal{L}_{(x)}$ of germs of all $l \in \mathcal{L}_{\text{com}}$ starting at x (i.e., such that $f_l(0) = x$).

Axiom 6. For any finite subset $\mathcal{L}' \subset \mathcal{L}_{(x)}$ there is an orientation-preserving homeomorphism of a neighborhood of x onto a neighborhood of $(0, 0) \in \mathbf{R}^2$ such that we have the following:

6a. The germs of oriented curves which are images of l_1 and l_2 get transformed into the germs at $(0, 0)$ of the coordinate axes $(x, 0)$ and $(0, y)$, respectively.
6b. The germ of the image of $l \in \mathcal{L}'$ gets transformed into the germ of the ray $\{(n_1 t, n_2 t) \mid t > 0\}$, where $\alpha_l(0) = n_1 \alpha_{l_1}(t_1) + n_2 \alpha_{l_2}(t_2)$.

Figure 5 illustrates this axiom.

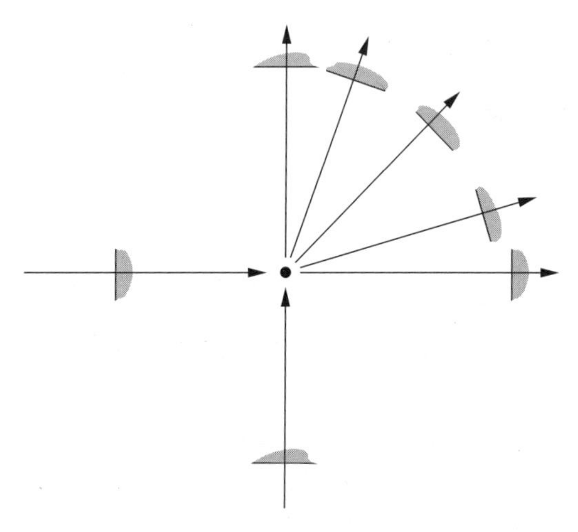

Fig. 5. Two intersecting lines and some of the new lines obtained as a result of a collision, in a canonical form.

Axiom 7. Let p_i denote either p_{left} or p_{right}. Then for any $l \in \mathcal{L}$ there exists $N \geq 1$ such that if the line $p_1(p_2(\dots p_N(l)\dots)$ is well defined, then it belongs to $\mathcal{L}_{\mathrm{in}}$.

This axiom says that any composed line $l \in \mathcal{L}_{\mathrm{com}}$ appears as a result of finitely many collisions. The tree of ancestors of a given line form a tree embedded in B; see Figure 6.

9.3 Example: Gradient lines

Here we offer a construction of the set of lines satisfying the above axioms.

Let us use the standard $\mathbf{R}^2$ as a model around each $b \in B^{\mathrm{sing}}$ in order to fix a structure of a smooth manifold on the whole surface B. Let $\widetilde{Y}$ denotes the covering of Y such that the fiber over $y \in Y$ is $(T_y^* Y)^{\mathbf{Z}} \setminus \{0\}$.

Let us fix a *generic* smooth metric on B. By pullback it gives a metric on $\widetilde{Y}$. Notice that there is a canonical closed 1-form β on $\widetilde{Y}$ such that $\beta_{|(y,\mu)} = \mu$, where

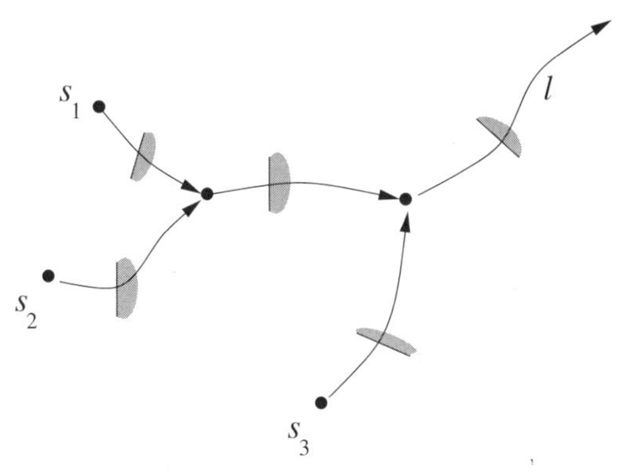

Fig. 6. The tree of ancestors of a line l starting from 3 singular points $s_1, s_2, s_3 \in B^{\mathrm{sing}}$.

$y \in Y$, $\mu \in (T_y^* Y)^{\mathbf{Z}}$. Using the metric we obtain a gradient vector field v on $\widetilde{Y}$ dual to β.

For any $s \in B^{\mathrm{sing}}$ and a choice of 1-form $\alpha(0) = \pm dy$ in local coordinates, we take the unique integral line of v starting at $(s, \alpha(0))$. The set $\mathcal{L}_{\mathrm{in}}$ will be the set of all lines obtained in this way. Each line $l \in \mathcal{L}_{\mathrm{in}}$ carries a covariantly constant closed 1-form α_l. Using Axiom 2 as a definition, we obtain a canonical parametrization of each line by the time parameter t. Since the metric is generic, a line cannot return to a point in B^{sing}.

Then we proceed inductively. If two already constructed lines $l_1, l_2 \in \mathcal{L}$ meet at $x \in Y$, we produce a new integral line l of v with the direction covector satisfying condition **3b** for any pair of coprime positive integers n_1, n_2. In this way we construct a set of lines $\mathcal{L}$ satisfying all the axioms. The only nontrivial thing to check is that for each line the values of the parameter t are in one-to-one correspondence with the interval $(0, +\infty)$. In order to see this we observe that the length of each line is infinite. Indeed, an integral curve of v cannot have a limiting point in Y (since the flow generated by v is smooth, and the lengths of tangent vectors are bounded from below because of the integrality of 1-forms).

We conclude that there exists a set $\mathcal{L}$ of lines satisfying Axioms 1–7.

10 Groups and symplectomorphisms

In this section we are going to discuss the sheaf of groups of symplectomorphisms $\mathrm{Symp} := \mathrm{Symp}(\mathcal{O}_Y^{\mathrm{can}})$ of the sheaf $\mathcal{O}_Y^{\mathrm{can}}$. Let $U \subset Y$ be an open convex subset. By definition, a *symplectomorphism* of $\mathcal{O}_Y^{\mathrm{can}}(U)$ is an automorphism of the K-algebra $\mathcal{O}_Y^{\mathrm{can}}(U)$ preserving the projection to Y and the canonical symplectic form $\Omega = \frac{d\xi \wedge d\eta}{\xi\eta}$ (the latter is understood as an element of the algebra of Kähler differential forms). To each line l we will assign a symplectomorphism of the restriction of $\mathcal{O}_Y^{\mathrm{can}}$ to l, so that the assignment will be compatible with the collision of lines. Then we are going to modify the sheaf $\mathcal{O}_Y^{\mathrm{can}}$ using symplectomorphisms associated with lines and obtain the sheaf $\mathcal{O}_Y^{\mathrm{modif}}$. This sheaf will be glued with the sheaf $\mathcal{O}_{\mathbf{R}^2}^{\mathrm{model}}$ near each point of B^{sing}.

10.1 Pro-nilpotent Lie algebra

Here it will be convenient to work in local coordinates $(x, y) = (\log|\xi|, \log|\eta|)$ on Y.

Let $(x_0, y_0) \in \mathbf{R}^2$ be a point, $\alpha_1, \alpha_2 \in (\mathbf{Z}^2)^*$ be 1-covectors such that $\alpha_1 \wedge \alpha_2 > 0$. Denote by $V = V_{(x_0,y_0),\alpha_1,\alpha_2}$ the closed angle

$$\{(x, y) \in \mathbf{R}^2 | \langle \alpha_i, (x, y) - (x_0, y_0)\rangle \geq 0,\ i = 1, 2\}.$$

Let $\mathcal{O}(V)$ be the K-algebra consisting of series $f = \sum_{n,m\in\mathbf{Z}} c_{n,m}\xi^n\eta^m$, such that $c_{n,m} \in K$ and for all $(x, y) \in V$, we have the following:

1. If $c_{n,m} \neq 0$, then $\langle (n, m), (x, y) - (x_0, y_0)\rangle \leq 0$, where we identify $(n, m) \in \mathbf{Z}^2$ with a covector in $(T_p^*Y)^{\mathbf{Z}}$.
2. $\log|c_{n,m}| + nx + my \to -\infty$ as long as $|n| + |m| \to +\infty$.

The algebra $\mathcal{O}(V)$ is a Poisson algebra with respect to the bracket $\{\xi, \eta\} = \xi\eta$.

For an integer covector $\mu = adx + bdy \in (\mathbf{Z}^2)^*$, we denote by R_μ the monomial $\xi^a\eta^b$.

Let us consider a pro-nilpotent Lie algebra $\mathbf{g} := \mathbf{g}_{\alpha_1,\alpha_2,V} \subset \mathcal{O}(V)$ consisting of series

$$f = \sum_{n_1,n_2\geq 0, n_1+n_2>0} c_{n_1,n_2} R_{\alpha_1}^{-n_1} R_{\alpha_2}^{-n_2}$$

satisfying the condition

$$\log|c_{n,m}| - n_1\langle \alpha_1, (x, y)\rangle - n_2\langle \alpha_2, (x, y)\rangle \leq 0 \quad \forall (x, y) \in V.$$

The latter condition is equivalent to $\log|c_{n,m}| - \langle n_1\alpha_1 + n_2\alpha_2, (x_0, y_0)\rangle \leq 0$.

The Lie algebra $\mathbf{g}$ admits a filtration by Lie subalgebras $\mathbf{g}^{\geq k}$, $k \in \mathbf{Z}$, $k \geq 1$, $\mathbf{g} = \mathbf{g}^{\geq 1}$, such that $\mathbf{g}^{\geq k}$ consists of the above series which satisfy the condition $n_1 + n_2 \geq k$. Clearly, $[\mathbf{g}^{\geq k_1}, \mathbf{g}^{\geq k_2}] \subset \mathbf{g}^{\geq k_1+k_2}$, and $\mathbf{g} = \varprojlim_{k\to+\infty} \mathbf{g}/\mathbf{g}^{\geq k}$.

Thus $\mathbf{g}$ is a topological complete pro-nilpotent Lie algebra over K. We denote by G the corresponding pro-nilpotent Lie group $\exp(\mathbf{g})$. It inherits the filtration by normal subgroups $G^{\geq k}$ obtained from the corresponding Lie algebras.

10.2 Lie groups G_λ

For each $\lambda \in [0, +\infty]_{\mathbf{Q}} := \mathbf{Q}_{\geq 0} \cup \infty$ define a Lie subalgebra

$$\mathbf{g}_\lambda = \left\{ \sum_{n_1,n_2} c_{m,n} R_{\alpha_1}^{-n_1} R_{\alpha_2}^{-n_2} \in \mathbf{g} \mid c_{n_1,n_2} \in K,\ \frac{n_2}{n_1} = \lambda \right\}.$$

Every $\mathbf{g}_\lambda$ is an abelian Lie algebra. It carries the induced filtration by Lie algebras $\mathbf{g}_\lambda^{\geq k} = \mathbf{g}_\lambda \cap \mathbf{g}^{\geq k}$. Denote by $G_\lambda = \exp(\mathbf{g}_\lambda)$ the corresponding pro-nilpotent group.

Lemma 4. *For any given $k \geq 1$, there exist finitely many $\lambda_1 < \lambda_2 < \cdots < \lambda_{N_k}$ such that $\mathbf{g}_\lambda/\mathbf{g}_\lambda^{\geq k} = 0$ for $\lambda \neq \lambda_i$, $1 \leq i \leq N_k$.*

Proof. Indeed, for the monomial $R_{\alpha_1}^{n_1} R_{\alpha_2}^{n_2} \in \mathbf{g}_\lambda$ which maps nontrivially to the quotient $\mathbf{g}_\lambda/\mathbf{g}_\lambda^{\geq k}$, we have $n_1 + n_2 \leq k$, $n_1/n_2 = \lambda$, where n_1, n_2 are nonnegative integers. There are finitely many such nonnegative integers n_1 and n_2. □

It follows from the lemma that we have a natural isomorphism of vector spaces $\prod_{\lambda \in [0,+\infty]_{\mathbf{Q}}} \mathbf{g}_\lambda/\mathbf{g}_\lambda^{\geq k} \to \mathbf{g}/\mathbf{g}^{\geq k}$, so the map

$$(f_\lambda)_{\lambda \in [0,+\infty]_{\mathbf{Q}}} \mapsto \sum_\lambda f_\lambda = \sum_{i=1}^{N_i} f_{\lambda_i}, \quad \text{where } f_\lambda \in \mathbf{g}_\lambda/\mathbf{g}_\lambda^{\geq k} \quad \forall \lambda \in [0, +\infty]_{\mathbf{Q}}$$

is well defined and gives rise (after taking the projective limit as $k \to +\infty$) to the isomorphism $\mathbf{g} \simeq \prod_{\lambda \in [0,+\infty]_{\mathbf{Q}}} \mathbf{g}_\lambda$.

In a similar way, we define the map $\prod_\to : \prod_{\lambda \in [0,+\infty]_{\mathbf{Q}}} G_\lambda \to G$, the product with respect to the natural order on $\mathbf{Q}$, by finite-dimensional approximations

$$\prod\nolimits_\to^{(k)} : \prod_{i=1}^{N_k} G_{\lambda_i}/G_{\lambda_i}^{\geq k} \to G/G^{\geq k}, \quad (g_1, \ldots, g_{N_k}) \mapsto g_1 \ldots g_{N_k} \quad \text{for } g_i \in G_{\lambda_i}/G_{\lambda_i}^{\geq k}.$$

Theorem 6. *The map $\prod_\to$ is a bijection of sets.*

Proof. Let $k \geq 1$ be an integer. We claim that $\prod_\to^{(k)}$ is a bijection of sets (this implies the proposition by taking the projective limit as $k \to +\infty$). We will prove the bijection by induction in k. The case $k = 1$ is obvious because all the groups under consideration are trivial.

We would like to prove that $\prod_\to^{(k+1)}$ is a bijection assuming that $\prod_\to^{(k)}$ is a bijection. Let h be an element of $G/G^{\geq k+1}$ and $\overline{h}$ its image in $G/G^{\geq k}$. By the induction assumption there exist unique $\overline{h}_i \in G_{\lambda_i}/G_{\lambda_i}^{\geq k+1}$, $1 \leq i \leq N_{k+1}$, such that $\overline{h}_1 \ldots \overline{h}_n = \overline{h}$. Let h_i, $1 \leq i \leq N_{k+1}$, be any liftings of $\overline{h}_i$ to $G_i/G_i^{\geq k}$. Then $h_1 \ldots h_{N_{k+1}} = h \pmod{G^{\geq k}}$, hence $c := h_1 \ldots h_{N_{k+1}} h^{-1}$ belongs to $G^{\geq k}/G^{\geq k+1} \subset \mathrm{Center}(G/G^{\geq k+1})$. The last inclusion holds because $[\mathbf{g}, \mathbf{g}^{\geq k}] = [\mathbf{g}^{\geq 1}, \mathbf{g}^{\geq k}] \subset \mathbf{g}^{\geq k+1}$.

Next, we observe that the isomorphism of abelian Lie algebras

$$\bigoplus_{1 \leq i \leq N_{k+1}} \mathbf{g}_{\lambda_i}^{\geq k}/\mathbf{g}_{\lambda_i}^{\geq k+1} \simeq \mathbf{g}^{\geq k}/\mathbf{g}^{\geq k+1}$$

implies an isomorphism of the corresponding abelian groups

$$\prod_{1 \leq i \leq N_{k+1}} G_i^{\geq k}/G_i^{\geq k+1} \simeq G^{\geq k}/G^{\geq k+1}.$$

Hence we can write uniquely $c = c_1 \ldots c_{N_{k+1}}$, where $c_i \in G^{\geq k}/G^{\geq k+1} \subset \mathrm{Center}(G/G^{\geq k+1})$. It follows that $\prod_\to^{(k+1)}((h_i c_i^{-1})) = h$. Also, it is now clear that this decomposition of h is unique. This concludes the proof. □

10.3 Function ord_l

For $l \in \mathcal{L}$ we will define an *order* function

$$\mathrm{ord}_l \in \Gamma((0,+\infty), f_l^*(\mathrm{Aff}_{\mathbf{Z},Y}))$$

(whose meaning will become clear later) by the following inductive procedure:

1. Let $l \in \mathcal{L}_{\mathrm{in}}$ and $t > 0$ be sufficiently small. Then in the standard affine coordinates near $s = f_l(0)$ one has $\alpha_l = \pm f_l^*(dy)$. We define $\mathrm{ord}_l = \pm f_l^*(y)$. Then $d(\mathrm{ord}_l) = \alpha_l$, and we can extend uniquely ord_l for all $t \in (0,+\infty)$.
2. Let $l \in \mathcal{L}_{\mathrm{com}}$ and l_1, l_2 be parents of l. In the notation of Axiom 3, we have $f_{l_1}(t_1) = f_{l_2}(t_2) = f_l(0)$ and $\alpha_l(0) = n_1\alpha_{l_1}(t_1) + n_2\alpha_{l_2}(t_2)$. Then we define $\mathrm{ord}_l(0) := n_1 \,\mathrm{ord}_{l_1}(t_1) + n_2 \,\mathrm{ord}_{l_2}(t_2)$. Again, using the condition $d(\mathrm{ord}_l) = \alpha_l$ and the knowledge of $\mathrm{ord}_l(0)$, we can extend ord_l for $t > 0$.

Notice that $\mathrm{ord}_l(t)$ can be thought of as an affine function on the tangent space $T_{f_l(t)}Y$ (in the induced integral affine structure). In particular, we have a half-plane $P_{l,t} \subset T_{f_l(t)}Y$ defined by the inequality $\mathrm{ord}_l(t) > 0$. The family of half-planes $P_{l,t}$ is covariantly constant with respect to ∇^{aff}.

Each half-plane $P_{l,t}$ contains $0 \in T_{f_l(t)}Y$ strictly in its interior. Recall that at the end of Section 9.1 we defined another half-plane $P_{l,t}^{(0)} \subset T_{f_l(t)}Y$. It is easy to see that $P_{l,t}^{(0)}$ is the half-plane parallel to $P_{l,t}$ such that $0 \in T_{f_l(t)}Y$ is on the boundary of $P_{l,t}^{(0)}$.

10.4 Symplectomorphisms assigned to lines

In this section we are going to assign to any line $l \in \mathcal{L}$ a symplectomorphism

$$\varphi_l \in \Gamma((0,+\infty), f_l^*(\mathrm{Symp})),$$

giving for each $t > 0$ a transformation $\varphi_l(t) : \mathcal{O}^{\mathrm{can}}_{Y,f_l(t)} \to \mathcal{O}^{\mathrm{can}}_{Y,f_l(t)}$. This symplectomorphism in local coordinates will belong to the subgroup G_λ, where λ is the slope of $\alpha_l(t)$. More precisely, we demand that $\varphi_l(t)$ be of the form

$$\varphi_l(t) = \exp\{F_{l,t}(\xi^{-a}\eta^{-b}), \cdot\},$$

where $\alpha_l(t) = a dx + b dy$, the operation $\{\cdot,\cdot\}$ is the Poisson bracket on $\mathcal{O}^{\mathrm{can}}_{Y,f_l(t)}$ and $F_{l,t}(z) \in zK[[z]]$ is an analytic function of *one* variable satisfying the following condition. Let us consider the pullback (by the exponential map) of the function $F_{l,t}(\xi^{-a}\eta^{-b})$ to a section of the sheaf $\mathcal{O}^{\mathrm{can}}$ on the $\mathbf{Z}$-affine manifold $T_{f_l(t)} \simeq \mathbf{R}^2$. Then this pullback should admit an analytic continuation from $0 \in T_{f_l(t)}$ to the half-plane $P_{l,t}$, and obey there the bound

$$|F_{l,t}(\xi^{-a}\eta^{-b})| \le \exp(-\mathrm{ord}_l(t)).$$

Let us explain the construction of $\varphi_l(t)$, leaving the justification for the following sections.

The symplectomorphisms φ_l are constructed by an inductive procedure. Let $l = l_+ \in \mathcal{L}_{\rm in}$ be (in standard affine coordinates) a line in the half-plane $y > 0$ emanating from $(0, 0)$ (there is another such line l_- in the half-plane $y < 0$). Assume that t is sufficiently small. Then we define $\varphi_l(t) \in \mathrm{Symp}_{f_l(t)}$ on topological generators ξ, η by the formula (as in Section 8)

$$\varphi_l(t)(\xi, \eta) = (\xi(1 + 1/\eta), \eta).$$

Notice that $\varphi_l(t) = \exp\{F(\eta^{-1}), \cdot\}$, where $F(z) = \sum_{n>0}(-1)^n z^n/n^2$ is convergent for $|z| < 1$.

In order to extend $\varphi_l(t)$ to the interval $(0, t_0)$, where t_0 is not small, we cover the corresponding segment of l by open charts. Notice that a change of affine coordinates transforms η into a monomial multiplied by a constant from $K^\times$. Therefore, η extends analytically in a unique way to a section in $\Gamma((0, +\infty), f_l^*((\mathcal{O}^{\rm can})^\times))$. Moreover, the norm $|\eta|$ strictly decreases as t increases, and remains strictly smaller than 1. Hence $F(\eta)$ can be canonically extended for all $t > 0$.

Each symplectomorphism $\varphi_l(t)$ is defined by a series which converges in the half-plane $P_{l,t}$. Using the exponential map associated with the affine structure as well as estimates of $\mathrm{ord}_l(t)$, we can analytically extend $\varphi_l(t)$ into a neighborhood of $f_l(t)$.

Let us now assume that l_1 and l_2 collide at $p = f_{l_1}(t_1) = f_{l_2}(t_2)$, generating the line $l \in \mathcal{L}_{\rm com}$. Then $\varphi_l(0)$ is defined with the help of the Factorization Theorem in the group G. More precisely, we set $\alpha_i := \alpha_{l_i}(t_i)$, $i = 1, 2$ and the angle V to be the intersection of the half-planes $P_{l_1,t_1} \cap P_{l_2,t_2}$. By construction, the elements $g_0 := \varphi_{l_1}(t_1)$ and $g_{+\infty} := \varphi_{l_2}(t_2)$ belong, respectively, to G_0 and $G_{+\infty}$. Then we can use the factorization Theorem 6 and write down the formula

$$g_{+\infty} g_0 = \prod\nolimits_{\rightarrow}((g_\lambda)_{\lambda \in [0,+\infty]_{\mathbb{Q}}}) = g_0 \cdots g_{1/2} \cdots g_1 \cdots g_{+\infty},$$

where $g_\lambda \in G_\lambda$ and the product on the right is in *increasing* order. There is no clash of notation because it is easy to see that the boundary factors in the decomposition from above are indeed equal to g_0 and $g_{+\infty}$. Every term g_λ with $0 < \lambda = n_1/n_2 < +\infty$ corresponds to the newborn line l with the direction covector $n_1\alpha_{l_1}(t_1) + n_2\alpha_{l_2}(t_2)$. Then we set $\varphi_l(0) := g_\lambda$. This transformation is defined by a series which is convergent in a neighborhood of p, and using the analytic continuation as above, we obtain $\varphi_l(t)$ for $t > 0$. The decomposition identity can be rewritten as

$$g_0 \cdots g_{1/2} \cdots g_1 \cdots g_{+\infty} g_0^{-1} g_{+\infty}^{-1} = \mathrm{id},$$

where each factor corresponds to half-lines at the collision point (see Figure 5), and the meaning of the identity is that the infinite composition of symplectomorphisms in the natural cyclic order on half-lines is trivial.

11 Modification of the sheaf $\mathcal{O}^{\mathrm{can}}$

11.1 Pieces of lines and convergence regions

Definition 13. *A neighborhood U of a point $x \in Y$ is* convex *if there exists an open convex $U_1 \in T_xY$, $0 \in U_1$, which is isomorphic to U by means of the exponential map* $\exp_x : T_xY \to Y$ *associated with affine structure on Y.*

For $x \in Y$ let $U \subset U'$ be convex neighborhoods of x such that U is relatively compact in U'. Let $l \in \mathcal{L}$. Then there is a natural embedding $f_l^{-1}(U) \to f_l^{-1}(U')$.

Definition 14. *A* piece *of l defined by the pair (U, U') is an element of the image of the set of connected components $\pi_0(f_l^{-1}(U))$ into $\pi_0(f_l^{-1}(U'))$ under the above embedding.*

In plain words, a piece L of l is an equivalence class of a connected interval of $l \cap U$. Two connected intervals are equivalent if they are contained in a larger connected interval of $l \cap U'$. The sole purpose of the introduction of the notion of a piece is to avoid some pathology. Namely, for any pair (U, U') as above, any $l \in \mathcal{L}$ and any $T \in \mathbf{R}_{>0}$, there is only a finite number of pieces of l in (U, U') which have points with time parameter $t \in (0, T)$.

Let L be a piece of l defined by a pair (U, U'). Then one can define an affine function $\mathrm{ord}_L \in \mathrm{Aff}_{\mathbf{Z},Y}(U')$ in the following way. Let $t > 0$ be such that $f_l(t)$ belongs to L. Since U' is convex, there is a unique continuation of $\mathrm{ord}_l(t)$ to U'. This is an affine function which does not depend on the choice of t. We will denote it by ord_L.

For any germ of a symplectomorphism $\varphi \in \mathrm{Symp}_p$ at the point $p \in Y$ we define its convergence region as the maximal convex subset $\Omega(\varphi) \subset T_pY$ such that the pullback $\exp_p^*(\varphi)$ extends to $\Omega(\varphi)$. Since the definition of φ_l (and hence its convergence region) is covariant with respect to the affine connection, we have the following result.

Proposition 5. *Let $p = f_l(t)$ belong to a line l. Then the convergence region of $\varphi_l(t)$ at p contains an open half-plane $P_{l,t}$.*

It is clear that one can define convergence regions for symplectomorphisms associated with pieces of lines, and that a similar property holds for them.

11.2 Main assumptions, and an apology

Let us suppose that our collection of lines satisfies the following assumptions.

Assumption A1. There is a smooth metric $g = g_B$ and a collection of balls $D(s, r_s)$ with centers at $s \in B^{\mathrm{sing}}$ such that each ball $D(s, r_s)$ contains exactly two lines $l_\pm \in \mathcal{L}_{\mathrm{in}}$ emanating from s.

Assumption A2. There exists $\varepsilon > 0$ such that for any $p = f_l(t) \in Y' := B \setminus \cup_{s \in B^{\text{sing}}} D(s, r_s)$ the distance in T_pY between $0 \in T_pY$ and the boundary line of $P_{l,t}$ is greater than or equal to ε.

We are going to show in Section 11 that such a collection does exist.

Assumptions A1 and A2 are very artificial; they do not hold in the physical picture which is the main motivation of the construction. It is quite possible that they can be weakened or even omitted. Their main use here is the possibility of defining a sheaf of analytic functions by simple gluing. In complex geometry it is similar to the gluing of closed Riemann surfaces with boundaries by means of real-analytic identifications of the boundaries. It is well known that one can replace real-analytic maps by smooth ones (or even by quasi-symmetric continuous maps). Maybe the rest of this section is unnecessary, and unpleasant technical arguments in Section 11.5 can be avoided.

11.3 Infinite product and its convergence

Denote by $W_{\mathcal{L}} := \cup_{l \in \mathcal{L}} f_l([0, +\infty))$ the set of all points of all lines; it has measure zero. Let p be a point of Y. We consider two convex neighborhoods $U \subset U'$ of p, such that U is relatively compact in U'.

For any two points x, y belonging $U \setminus W_{\mathcal{L}}$, and a path γ joining x and y in U, we would like to define an infinite ordered product $i^{\gamma}_{x,y}$ of transformations $\varphi_L^{\pm 1}$, where the factors correspond to the intersection points of γ with all possible pieces L relative to (U, U'). The factors in the infinite product are ordered according to the time parameter of γ; the sign corresponds to the mutual position of orientations of γ and the piece L at the intersection point.

In order to give a precise meaning to the infinite product, the neighborhood U of p should be sufficiently small. Then we will have an analytic continuation of symplectomorphisms φ_L to U, and the convergence of the infinite product. We are also going to prove that the product is independent of the choice of the path γ. In order to achieve these goals it suffices to assume the following:

C1 For any l, t such $f_l(t) \in U$, the set $(\exp_{f_l(t)})^{-1}(U)$ is contained in $P_{l,t}$.
C2 For any $C \in \mathbf{R}$, there is only a finite number of pieces L of lines in U such that $\inf_{x \in U} \operatorname{ord}_L(x) < C$.

Theorem 7. *Assume the two conditions above. Then the product defining $i^{\gamma}_{x,y}$ converges at every point of U and in fact gives an element of* Symp(U). *Moreover, the product does not depend on the choice of path γ, and for any $x, y, z \in U \setminus W_{\mathcal{L}}$ satisfies the relation $i_{x,y} i_{y,z} = i_{x,z}$.*

Proof. Condition **C1** implies that all transformations φ_L admit an analytic continuation to U. Let us introduce a decreasing filtration by positive real numbers $\mathrm{Symp}^{\geq r}(U)$, $r \in \mathbf{R}$, $r \geq 0$, on the group Symp(U) by the formula

$$\{g \in \mathrm{Symp}(U) \mid \log|\xi'/\xi - 1|, \log|\eta'/\eta - 1| < -r, \text{ where } (\xi', \eta') = g((\xi, \eta))\}.$$

This is a complete filtration, and condition **C2** implies that in any quotient $\mathrm{Symp}(U)/\mathrm{Symp}^{\geq r}(U)$ only a finite number of elements φ_L are nontrivial. Therefore, we can define the product in the quotient group.

In order to prove independence of γ, we consider the quotient group $\mathrm{Symp}(U)/\mathrm{Symp}^{\geq r}(U)$, and the finite one-dimensional CW-complex (graph) consisting of finitely many pieces L, such that $\varphi_L \neq 1$ in the quotient. For each vertex v of the graph there is a natural cyclic order on the edges incident to v. The product $\varphi_v = \prod_L \varphi_L^{\pm 1}$ taken in the cyclic order over the set of edges incident to v is equal to id (this follows from the construction of φ_l via factorizations). Since U is simply-connected, we conclude that the image of $i_{x,y}^{\gamma}$ in $\mathrm{Symp}(U)/\mathrm{Symp}^{\geq r}(U)$ does not depend on γ. Using completeness of the filtration we see that $i_{x,y} := i_{x,y}^{\gamma}$ does not depend on γ. The proof of the identity $i_{x,y} i_{y,z} = i_{x,z}$ is similar. □

Theorem 8. *Assumptions* A1 *and* A2 *imply that for any $p \in Y$ there exist neighborhoods U and U' satisfying conditions* **C1** *and* **C2**.

Proof. Assumption A1 implies that the situation near any singular point $s \in B^{\mathrm{sing}}$ is trivial, as there are only two lines near s. If we are far from B^{sing}, then obviously Assumption A2 implies **C1**.

In order to check **C2**, we prove the following lemma.

Lemma 5. *Under Assumptions* A1 *and* A2, *for any $C > 0$ the set*

$$\{(l, t) \mid \mathrm{ord}_l(t)(f_l(t)) < C\} \subset \mathcal{L} \times (0, +\infty)$$

consists of a finite number of intervals.

Proof. We proceed by induction in "complexity of the line." Let $\delta \in \mathbf{R}_{>0}$ be the infimum of $\mathrm{ord}_l(t)(f_l(t))$, where $l \in \mathcal{L}_{init}$ has a collision at time t. This is strictly positive because the number of initial lines is finite, and by Assumption A1 there are no collisions at small times. Observe that the value of $\mathrm{ord}_l(0)$ at the beginning of any composite line l is greater than or equal to the sum $\mathrm{ord}_{l_1}(t_1) + \mathrm{ord}_{l_2}(t_2)$. Therefore, the inequality in the lemma implies that the number of collisions is bounded from above. Let us observe that the length of segments of lines between intersections is also bounded since it is less than or equal to $A\,\mathrm{ord}_l$, for some absolute constant $A > 0$. Hence we have only finitely many possibilities for intersections. □

For a point $p \in Y$ which is far from B^{sing}, we choose as U a neighborhood of radius $\epsilon' \ll \epsilon$, where $\epsilon > 0$ is constant from Assumption A2. Then for any point of a line $f_l(t) \in U$, we will have the inclusion

$$U \subset \exp_{f_l(t)}\left(\frac{1}{2} P_{l,t}\right).$$

This implies that ord_L in U for the corresponding piece L is bounded below by

$$\frac{1}{2}\,\mathrm{ord}_l(t)(f_l(t)).$$

Since (by the last lemma) there exists only a finite number of pieces L intersecting such U, we obtain convergence condition **C2**. □

11.4 Construction of the modified sheaf $\mathcal{O}_B^{\mathrm{modif}}$

For any point $p \in Y$ and a neighborhood U satisfying conditions **C1** and **C2** we define a sheaf $\mathcal{O}_U^{\mathrm{modif}}$ as the identification of copies of the sheaf $(\mathcal{O}_Y^{\mathrm{can}})_{|U}$ labeled by points $x \in U \setminus W_{\mathcal{L}}$, by the isomorphisms $i_{x,y}$. It follows from formulas in Section 8 that near singular points one can canonically identify this sheaf with the restriction of the sheaf $\mathcal{O}_{\mathbf{R}^2}^{\mathrm{model}}$ to a punctured neighborhood of $(0, 0) \in \mathbf{R}^2$.

Proposition 6. *For the modified sheaf $\mathcal{O}^{\mathrm{modif}}$ one has a canonical nowhere vanishing section Ω of the associated sheaf of K-analytic 2-forms.*

The K-affine structure $\mathrm{Aff}_{K,Y}^{\Omega}$ on Y associated with Ω coincides with the initial one $\mathrm{Aff}_{K,Y}$.

Proof. The existence of Ω follows from the fact that all modifications associated with lines are symplectomorphisms. In order to finish the proof it suffices to check that the modification associated with a line does not change the K-affine structure on Y. In local coordinates we may assume that $\Omega = \frac{d\xi}{\xi} \wedge \frac{d\eta}{\eta}$ and the modification is of the form $\varphi(\xi, \eta) = (\xi f(\eta^{-1}), \eta)$, where $f(z) = 1 + \sum_{n \geq 1} c_n z^n \in K[[z]]$ is convergent in an appropriate domain. We need to check that the automorphism φ acts trivially on the quotient sheaf $\pi_*(\mathcal{O}_X^{\times})/\ker p_{\Omega}$ (see Section 7.2 for the notation). This check reduces to the calculation of

$$p_{\Omega}\left(\frac{\xi f(\eta)}{\xi}\right) = \exp\left(\frac{\mathrm{Res}(\Omega \log(\xi f(\eta)/\xi)))}{\mathrm{Res}(\Omega)}\right).$$

The latter is equal to $\exp(\mathrm{Res}(\Omega \log(f(\eta)))) = 1$ because $\log(f(\eta^{-1}))$ belongs to $\eta^{-1}K[[\eta^{-1}]]$ and therefore has no constant term. □

Thus we have a solution of the Lifting Problem under Assumption A1 and A2.

11.5 Construction of the collection of lines

We would like to show that there exists a smooth metric g and a collection of lines satisfying the Assumptions A1 and A2.

Let g_0 be an arbitrary smooth metric, flat near singular points. We define germs of lines $l \in \mathcal{L}_{\mathrm{in}}$ in such a way that for each $s \in B^{\mathrm{sing}}$ in local coordinates these lines are given by $\{(0, y)|y > 0\}$ and $\{(0, y)|y < 0\}$. The metric g will coincide with g_0 in a sufficiently small neighborhood $U = \cup_{s \in B^{\mathrm{sing}}} D(s, r_s)$ of the singular set. Hence Assumption A1 will be satisfied.

In order to construct the whole family of lines we introduce a 3-dimensional manifold $\mathcal{M}$ consisting of pairs (x, P), where $x \in B \setminus \overline{U_1}$ and P is a half-plane in $T_x B$ whose boundary contains zero. Here $U_1 := \cup_{s \in B^{\mathrm{sing}}} D(s, 2r_s)$ is a larger neighborhood of B^{sing}.

We would like to construct a smooth section $v : (x, P) \mapsto v_{(x,P)} \in T_x B$ of the pullback to $\mathcal{M}$ of the tangent bundle TB satisfying the following conditions:

1. for any $(x, P) \in \mathcal{M}$, one has $v_{(x,P)} \in \mathrm{int}(P)$;

2. for any $x \in B \setminus \overline{U}_1$, the map $(x, P) \mapsto \mathbf{R}^{\times}_{>0} \cdot v_{(x,P)}$ is an orientation-preserving diffeomorphism

$$S^1 \simeq (T_x B^* \setminus \{0\})/\mathbf{R}^{\times}_{>0} \to S^1 \simeq (T_x B \setminus \{0\})/\mathbf{R}^{\times}_{>0};$$

3. for every $l \in \mathcal{L}_{\mathrm{in}}$, there exists a smooth extension of the piece of l in U_1 to a larger piece intersecting ∂U_1 such that

$$\dot{f}_l(t) \in \mathbf{R}^{\times}_{>0} \cdot v_{(f_l(t), P_{l,t})}$$

for such $t > 0$ that $f_l(t) \in B \setminus \overline{U}_1$.

Let us associate with the section v a nowhere vanishing vector field $\hat{v}$ on $T^*(B \setminus \overline{U}_1) \setminus$ (zero section) in the following way:

- For each $(x, \alpha) \in T^*_x B$ the vector $\hat{v}(x, \alpha)$ is tangent to the horizontal distribution associated with the flat connection ∇ (the one which defines the affine structure on $B \setminus B^{\mathrm{sing}}$).
- The projection of $\hat{v}(x, \alpha)$ to B coincides with $v_{(x, P_\alpha)}$, where $P_\alpha = \{\gamma | (\alpha, \gamma) > 0\}$.

Clearly, these conditions determine $\hat{v}$ uniquely. Now we formulate the last condition:

4. There exists $r'_s > 2r_s$ such that for almost all (in the sense of Baire category) initial values $(x_0, P_0) \in \mathcal{M}$ the integral curve of $\hat{v}$ starting at (x_0, P_0) reaches the pullback of $B \setminus \cup_{s \in B^{\mathrm{sing}}} D(s, r'_s)$ in finite time.

Using the vector field $\hat{v}$ we will construct (under certain genericity assumptions) a set $\mathcal{L}$ of lines satisfying Assumption A1. Namely, the data consisting of a line l and an integer-valued 1-form α_l (see Section 9) will be an integral line of $\hat{v}$.

We are going to construct lines by induction on the number of collisions. Lines $l \in \mathcal{L}_{\mathrm{in}}$ will be constructed using condition 3. The genericity assumption mentioned after condition 4 is the assumption that no more than two lines collide and that initial values for newborn lines will be sufficiently generic. Conditions 1 and 4 plus genericity imply that one can parametrize any line $l \in \mathcal{L}$ by the new "time" $t > 0$ such that Axiom 2 is satisfied. Axiom 6 follows from condition 2. Other axioms and the Assumption A1 will be satisfied automatically.

Now we would like to discuss Assumption A2.

Proposition 7. *Suppose that the metric g and field v described above are such that for any $(x, P) \in \mathcal{M}$, there exists $C > 0$ such that*

$$(\nabla_{v_{(x,P)}} g)(n_P, n_P) \leq C g(n_P, v_{(x,P)}),$$

where n_P is the normal unit vector to P directed inward and $\nabla_{v_{(x,P)}} g$ is the covariant derivative of the metric g considered as a symmetric tensor on the cotangent bundle.

Then Assumption A2 is satisfied.

Proof. In order to satisfy Assumption A2 it suffices to find such $\varepsilon > 0$ that for any $x \in B \setminus \overline{U}_1$ and any half-plane $P_x \subset T_x B$, $0 \in \mathrm{int}(P_x)$, with the distance

$\mathrm{dist}_{g_x}(0, \partial P_x) = \varepsilon$, one has the following property: if $P_{x+\delta t v_{x,P_x}}$ is the half-plane obtained from P_x by a small covariant (using ∇^{aff}) shift δt in the direction of v_{x,P'_x} (here P'_x is parallel to P_x but $0 \in \partial P'_x$), then

$$\mathrm{dist}_{g_{x+\delta t v_{x,P'_x}}}(0, P_{x+\delta t v_{x,P'_x}}) \geq \mathrm{dist}_{g_x}(0, \partial P_x).$$

Here g_x denotes the induced flat metric on the tangent space $T_x B$. This property guarantees that the condition $\mathrm{dist}_{g_x}(0, \partial P_{l,t}) \geq \varepsilon$ will propagate along the line. For a new line obtained as a result of the collision of l_1 and l_2 at the times t_1 and t_2, respectively, one has

$$\mathrm{dist}_{g_x}(0, \partial P_{l,0}) \geq \min\{\mathrm{dist}_{g_x}(0, \partial P_{l_1,t_1}), \mathrm{dist}_{g_x}(0, \partial P_{l_2,t_2})\}$$

since $\partial P_{l,0}$ contains the intersection point $A \in \partial P_{l_1,t_1} \cap \partial P_{l_2,t_2}$; see Figure 7.

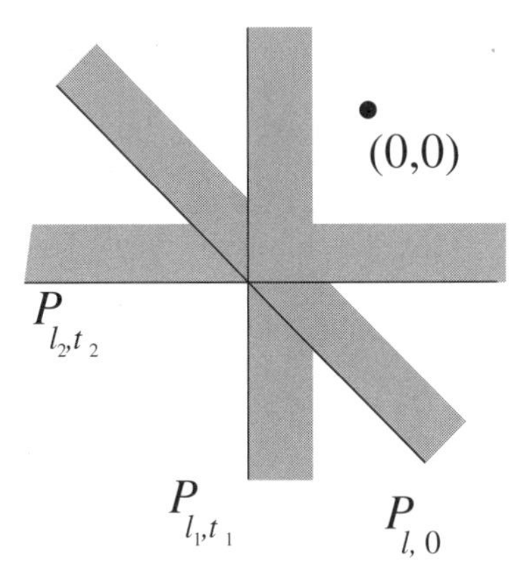

Fig. 7. Three half-planes containing zero.

One can easily see that the infinitesimal inequality above is equivalent to

$$\delta t g_x(v_{(x,P'_x)}, n_{P'_x}) + \varepsilon/2(g_{x+\delta t v_{(x,P'_x)}} - g_x)(n_{P'_x}, n_{P'_x}) \geq 0.$$

(The change in the distance consists of two summands: one corresponds to the shift along $\delta t v_{(x,P_x)}$ with the metric fixed, and the other corresponds to the change of the metric.) Taking the limit $\delta t \to 0$, we arrive at the inequality for the covariant derivative of the metric with $C = 2/\varepsilon$. □

Now our goal is to construct the field of directions v and the metric g satisfying conditions 1–4 and the inequality from the last propostion. This will conclude the construction of the set $\mathcal{L}$ of lines satisfying Assumptions A1 and A2.

Since ∂U_1 is the boundary of a convex set, we can locally model it by the graph of a function $y = f(x)$ such that $f''(x) > 0$, $f'(x_0) = 0$. We may assume that $P = P_{dy}$ is the upper half-plane. Then we take

$$v_{(x,y),P} = \partial/\partial y + \frac{(f(x) - f(x_0))/f'(x)}{f(x) - f(x_0) + f(x) - y}\partial/\partial x.$$

We extend this local model of v near ∂U_1 to $B \setminus \overline{U}_1$ in such a way that conditions 1 and 2 are satisfied. It is clear that we can satisfy conditions 3 and 4 as well by taking a small perturbation of v. In Figure 8 there is a picture of the field $(x, y) \mapsto v_{((x,y),P_{(x,y)})}$ in the case when $P_{(x_0,y_0)}$ is the upper half-plane.

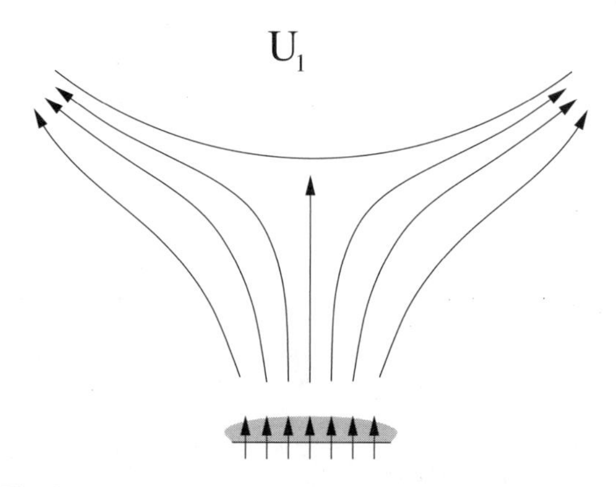

Fig. 8. Vector field near ∂U_1 for $P =$ the upper half-plane.

For an arbitrary choice of the metric g we have $g(n_P, v_{z,P}) > 0$ for all $(z, P) \in \mathcal{M}$. The problem with the inequality

$$(\nabla_{v_{(z,P)}} g)(n_P, n_P) \leq C g(n_P, v_{(z,P)})$$

arises only as the point z approaches ∂U_1. Indeed, in this case the vector $v_{(z,P)}$ can be very close to $\partial P_z \subset T_z B$.

Lemma 6. *With the above choice of v assume that the metric satisfies for any $z \in \partial U_1$ the condition*

$$(\nabla_{e_z} g)(n_z, n_z) = 0,$$

where $e_z \in T_z B$ is the unit tangent vector to ∂U_1 and n_z is the normal vector to ∂U_1 (all scalar products and lengths are taken with respect to the metric g).

Then there exists $C > 0$ such that

$$(\nabla_{v_{(z,P)}} g)(n_P, n_P) \leq C g(n_P, v_{(z,P)})$$

for all $(z, P) \in \mathcal{M}$.

Proof. We need to check that the ratio

$$\frac{(\nabla_{v_{(z,P)}} g)(n_P, n_P)}{g(n_P, v_{(z,P)})}$$

is bounded for $(z, P) \in \mathcal{M}$.

It suffices to prove the lemma assuming that U_1 is the parabolic domain $\{(x, y) \in \mathbf{R}^2 | y > x^2\}$ and P is the upper half-plane. The vector field $v_{(z,P)}$ is given for $z = (x, y)$ by the formulas

$$v_{(z,P)} = \partial/\partial y + \frac{x\partial/\partial x}{4x^2 - 2y}.$$

The denominator is equal to $g(n_P, v_{(z,P)}) = \langle dy, v_{(z,P)} \rangle \cdot \sqrt{g(\partial/\partial y, \partial/\partial y)} = \sqrt{g(\partial/\partial y, \partial/\partial y)} = \exp(O(1))$ near $(0, 0)$.

The numerator is equal to

$$\frac{x\partial/\partial x}{4x^2 - 2y} f_1(x, y) + f_2(x, y),$$

where $f_1(x, y) = (\nabla_{\partial/\partial x} g)(n_P, n_P)$ and $f_2(x, y) = (\nabla_{\partial/\partial y} g)(n_P, n_P)$ are two C^∞-functions.

By the assumption of the lemma, we have $f_1(0, 0) = 0$. Therefore, $|f_1(x, y)| \leq \text{const} \max\{|x|, |w|\}$, where $w = x^2 - y$ is a convenient local coordinate near the point $(0, 0)$. Notice also that $f_2(x, y) = O(1)$.

Now we can estimate the first summand of the numerator assuming that $|x|$ and $|w|$ are sufficiently small. As we have seen, it is bounded by

$$I := \frac{x}{x^2 + w} O(\max\{|x|, |w|\}.$$

There are three cases that we need to consider:

(a) If $0 < w < x^2$, then $I = \frac{x}{x^2} O(|x|) = O(1)$.
(b) If $x^2 \leq w < x$, then $I = \frac{x}{w} O(|x|) = O(1)$.
(c) If $x \leq w \leq 1$, then $I = \frac{x}{w} O(|w|) = O(1)$.

We see that the numerator is bounded. This concludes the proof of the lemma. □

Finally, we have the following result.

Lemma 7. *There exists a metric g satisfying the conditions of Lemma* 6.

Proof. First of all, the condition on g from Lemma 6 is the condition on a loop $g_{|T_z B}$ of scalar products on two-dimensional spaces; here $z \in \partial U_1 \simeq S^1$. We can write $g = \exp(\psi) g_0$, where $\det(g_0) = 1$ and ψ is a smooth function. Then we have

$$\nabla_{e_z}(\exp \psi g_0) = \exp(\psi) \nabla_{e_z} g_0 + \exp(\psi) \partial_{e_z}(\psi) g_0.$$

The equation of Lemma 6 gives $\partial_{e_z} \psi = -(\nabla_{e_z} g_0)(n_z, n_z)/g_0(n_z, n_z)$. The RHS of this expression is known as long as we know g_0. Hence we can say that $d\psi = \beta_{g_0}$, where β_{g_0} is a 1-form depending on the restriction $(g_0)_{|\partial U_1}$. We see that it suffices to find g_0 such that $\int_{\partial U_1} \beta_{g_0} = 0$ (then ψ and hence g exist).

Let us consider the functional $I(g_0) = \int_{S^1} \beta_{g_0}$. We can interpret a metric g_0 as a point in the Lobachevsky plane $\mathcal{H} = \mathrm{SL}(2, \mathbf{R})/\,\mathrm{SO}(2)$. More precisely, let us consider the space S of pairs (g_0, P), where g_0 is a positive quadratic form on $\mathbf{R}^2$ such that $\det(g_0) = 1$ and P is a half-plane in $\mathbf{R}^2$. (The meaning of P is the inward oriented tangent half-plane to ∂U_1 at a point $z \in \partial U_1$.) This space is naturally diffeomorphic to $S^*(\mathbf{R}^2) \times \mathcal{H}$. The latter manifold can be identified in an $\mathrm{SL}(2, \mathbf{R})$-equivariant way with the manifold consisting of pairs (x, y), where $x \in \mathcal{H}$ and y belongs to the absolute. Hence $(g_0)_{|\partial U_1}$ is (locally) a nonparametrized path in S. (It would be a global path if the bundle over S^1 given by the all metrics on S^1 with the determinant 1 was trivial.)

Next, we observe that the variation $\delta I(g_0) = \int_N \omega$, where N is a two-dimensional surface bounded by the paths defined by g_0 and $g_0+\delta g_0$, and ω is a canonical $\mathrm{SL}(2, \mathbf{R})$-invariant 2-form on S. One can show that even by a small variation of the path defined by g_0 we can make $I(g_0)$ an arbitrary real number. In particular, we can find g_0 such that $I(g_0) = 0$. This concludes the proof of Lemma 6. □

Summarizing, we have constructed a set of lines satisfying Assumptions A1 and A2. This concludes the proof of Theorem 5. Thus we have obtained a solution of the Lifting Problem, which is a K-analytic K3 surface.

11.6 Independence and uniqueness

It is natural to ask how the above construction of the K-analytic K3 surface $(X^{\mathrm{an}}, \Omega)$ depends on the choice of the set $\mathcal{L}$ of lines. We know that the "periods" of Ω (they are encoded in the initial K-affine structure) do not depend on $\mathcal{L}$ (see Sections 7.3 and 10.4). In the light of the Torelli theorem (see Appendix B) it is natural to formulate the following conjecture.

Conjecture 11. The isomorphism class of the pair $(X^{\mathrm{an}}, \Omega)$ does not depend on the choice of the set $\mathcal{L}$ of lines.

More precisely, the change in $\mathcal{L}$ corresponds to the change of the projection $\pi := \pi_{\mathcal{L}} : X^{\mathrm{an}} \to B$ (see Section 7.3).

Remark 4. For $B = S^2$ and $B^{\mathrm{sing}} = \{x_1, \dots, x_{24}\}$ with the standard singular $\mathbf{Z}$-affine structure we have constructed a K-analytic K3 surface depending on 20 parameters in $K^\times$. More precisely, we have a 20-dimensional K-analytic space of conjugacy classes of representations

$$\pi_1(S^2 \setminus B^{\mathrm{sing}}) \to \mathrm{SL}(2, \mathbf{Z}) \ltimes (K^\times)^2$$

such that the monodromy around each singular point is conjugate to the pair $(A, (1, 1))$, where $A \in \mathrm{SL}(2, \mathbf{Z})$ is equal to

$$\begin{pmatrix} 1 & 1 \\ 0 & 1 \end{pmatrix}.$$

(compare with Section 3.3).

11.7 Remark on the case of positive and mixed characteristic

Our construction of $(X^{\mathrm{an}}, \Omega)$ works even without the assumption $\mathrm{char}\, k = 0$, where k is the residue field of K. This can be explained from the point of view of the Factorization Theorem (see Section 10.4). It turns out that the symplectomorphisms which appear in the infinite product in the RHS of the Factorization Theorem are infinite series whose coefficients are integer polynomials in the coefficients of the "parent" symplectomorphisms.

For example, let $f_0(z) = 1 + \sum_{n\geq 1} c_n z^n$ and $f_\infty(z) = 1 + \sum_{n\geq 1} d_n z^n$ be two power series that are convergent for $|z| < 1$. Let us consider two symplectomorphisms: $F_0(\xi, \eta) = (\xi, \eta f_0(\xi^{-1}))$, and $F_\infty(\xi, \eta) = (\xi f_\infty(\eta^{-1}), \eta)$ and decompose $F_\infty \circ F_0$ into the infinite ordered product $\prod_\rightarrow (F_\lambda)$. Here

$$F_{p/q}(\xi, \eta) = (\xi f_{p/q}(\xi^{-p}\eta^{-q})^q, \eta f_{p/q}(\xi^{-p}\eta^{-q})^{-p}),$$

where $f_{p/q}(z) = 1 + \sum_{n\geq 1} c_n^{p/q} z^n$. Then one can check that for any coprime $p, q \in \mathbf{Z}_{>0}$ and any $n \geq 1$, one has

$$c_n^{p/q} \in \mathbf{Z}[c_1, c_2, \ldots, d_1, d_2, \ldots].$$

This implies that our construction works when one replaces K by an arbitrary commutative ring R endowed with a complete nontrivial valuation $\mathrm{val} : R \to (-\infty, +\infty]$.

11.8 Further generalizations

First of all, one can introduce a small parameter $\hbar \in K$, $|\hbar| < 1$, of noncommutativity in the picture; coordinates ξ, η will not commute but instead satisfy the relation

$$\eta\xi = \xi\eta \exp(\hbar).$$

For such a noncommutative analytic torus one can still define the sheaf $\mathcal{O}^{\hbar}_{\mathrm{can}}$ on $\mathbf{R}^2$ by the "same" formula as in the commutative case:

$$\mathcal{O}^{\hbar}_{\mathrm{can}}(U) = \left\{ \sum_{n,m\in\mathbf{Z}} c_{n,m}\xi^n\eta^m \mid \forall (x,y) \in U \sup_{n,m}(\log|c_{n,m}| + nx + my) < \infty \right\},$$

where $U \subset \mathbf{R}^2$ is connected. Also one can construct a noncommutative deformation of the model sheaf near the singular point. All arguments with the groups work as well. In this way we will obtain a kind of quantized K3 surface over a non-archimedean field.

Secondly, we believe that one can generalize our construction to higher dimensions. Instead of lines there will be codimension one walls which should be flat hypersurfaces with respect to the $\mathbf{Z}$-affine structure and carry foliations by parallel lines. Generically, on the intersection of two such foliated hypersurfaces one can "separate" variables into the product of the purely two-dimensional situation studied in the present paper, and $n-2$ dummy variables. Presumably everywhere except for a countable union of codimension 2 subsets one can use two-dimensional factorization and define gluing volume-preserving maps ϕ. One can hope that by a kind of Hartogs principle the sheaf will have a canonical extension to the whole space B.

A Analytic geometry

In this section we collect several facts and definitions about rigid analytic spaces and Clemens polytopes. Some of them are well known, the rest are borrowed from [KoT].

We always work over a complete non-archimedean local field K. The field K carries a valuation map $\mathrm{val}_K := \mathrm{val} : K \to \mathbf{R} \cup \{+\infty\}$ such that $\mathrm{val}(0) = +\infty$, $\mathrm{val}(1) = 0$, $\mathrm{val}(xy) = \mathrm{val}(x) + \mathrm{val}(y)$, $\mathrm{val}(x+y) \geq \min(\mathrm{val}(x), \mathrm{val}(y))$.

We will assume that the valuation is nontrivial. The ring

$$\mathcal{O}_K = \mathrm{val}_K^{-1}(\mathbf{R}_{\geq 0} \cup \{+\infty\})$$

is called the ring of integers of K. The residue field is defined as $k = \mathcal{O}_K / m_K$, where $m_K = \mathrm{val}_K^{-1}(\mathbf{R}_{>0} \cup \{+\infty\})$ is the maximal ideal in $\mathcal{O}_K$.

Our main example is the field $K = \mathbf{C}((t))$ of Laurent series in one variable. In this case $\mathrm{val}_K(\sum_{n \geq n_0} c_n t^n) = n_0$, as long as $c_{n_0} \neq 0$.

A.1 Berkovich spectrum

We refer the reader to [Be1] for the general definition of an analytic space and more details. In this appendix we restrict ourselves to analytic spaces associated with algebraic varieties (although we use the general definition in the paper as well).

Let $R = R/K$ be a commutative unital finitely generated K-algebra. The underlying set of the Berkovich spectrum $\mathrm{Spec}^{\mathrm{an}}(R) := \mathrm{Spec}^{\mathrm{an}}(R/K)$ can be defined in two ways. First one uses valuations (or, equivalently, multiplicative seminorms).

Definition 15 (valuations). *A point x of* $\mathrm{Spec}^{\mathrm{an}}(R/K)$ *is an additive valuation*

$$\mathrm{val}_x : R \to \mathbf{R} \cup \{+\infty\}$$

extending $\mathrm{val} := \mathrm{val}_K$*, i.e., it is a map satisfying the conditions*

- $\mathrm{val}_x(r + r') \leq \max(\mathrm{val}_x(r), \mathrm{val}_x(r'))$;
- $\mathrm{val}_x(rr') = \mathrm{val}_x(r) + \mathrm{val}_x(r')$;
- $\mathrm{val}_x(\lambda) = \mathrm{val}_K(\lambda)$

for all $r, r' \in R$ and all $\lambda \in K$.

Having a valuation and a real number $q_0 \in (0, 1)$ one can define the *multiplicative seminorm* $|a| = q_0^{\mathrm{val}_K(a)}$, $a \in R$. In particular, in the previous definition one can take seminorms $|\cdot|_x$ instead of valuations $\mathrm{val}_x(\cdot)$. The reader has noticed that in the main body of the paper, for $R = K$ we often took $|a| = e^{-\operatorname{val}(a)}$. It is easy to translate the definition of Berkovich spectrum into the language of multiplicative seminorms. We use it freely in the paper.

The second way to define X^{an} uses evaluations (characters).

Definition 16 (evaluation maps). *A point x of $\mathrm{Spec}^{\mathrm{an}}(R/K)$ is an equivalence class of homomorphisms of K-algebras*

$$\mathrm{eval}_x : R \to K_x,$$

where $K_x \supset K$ is a complete field equipped with a non-archimedean valuation which extends the valuation val_K, and such that K_x is generated by the closure of the image of eval_x.

The field K_x is determined by $x \in X^{\mathrm{an}}$ in a canonical way. We define for $r \in R$ and $x \in X^{\mathrm{an}}$ the "value" $r(x) \in K_x$ as the image $\mathrm{eval}_x(r)$.

In order to pass from the first description of $\mathrm{Spec}^{\mathrm{an}}(R/K)$ to the second, starting with a valuation val_x one defines the field K_x as the completion of the field of fractions of R/I_x, where $I_x = (\mathrm{val}_x)^{-1}(\{+\infty\})$.

Definition 17. *The topology on $\mathrm{Spec}^{\mathrm{an}}(R/K)$ is the weakest topology such that for all $r \in R$ the map*

$$\mathrm{Spec}^{\mathrm{an}}(R/K) \to \mathbf{R} \cup \{+\infty\},$$
$$x \mapsto \mathrm{val}_x(r)$$

is continuous.

An element $f \in R$ defines a function $f : \mathrm{Spec}^{\mathrm{an}}(R) \to K_x$, where K_x is the non-archimedean valuation field, which is the completion of the field of fractions of the domain $R/\ker(\mathrm{val}_x)$. Since each K_x carries a seminorm, we obtain a function $|f| : \mathrm{Spec}^{\mathrm{an}} \to \mathbf{R}_{\geq 0}$, $x \mapsto |f(x)|$.

A fundamental system of neighborhoods $U = U_x \subset \mathrm{Spec}^{\mathrm{an}}(R)$ of a point x is parametrized by the following data: a finite collection of functions

$$(f_i)_{i \in I}, (g_j)_{j \in J} \in R$$

and numbers

$$\beta_i^+, \beta_i^-, \gamma_j \in \mathbf{R}_{>0}$$

such that $\beta_i^- < |f_i(x)| < \beta_i^+$, $|g_j(x)| = 0$, The corresponding neighborhood consists of points x' such that $\beta_i^- < |f_i(x')| < \beta_i^+$, $|g_j(x')| < \gamma_j$ for all $i \in I$, $j \in J$, and $x' \in U$.

Let us assume that the elements $(f_i)_{i \in I}$, $(g_j)_{j \in J}$ generate R, i.e.,

$$R = K[(f_i)_{i \in I}, (g_j)_{j \in J}]/I,$$

where I is an ideal. Let us consider the algebra of series

$$s = \sum_{n_I \in \mathbf{Z}^I, m_J \in \mathbf{N}^J} c_{I,J} f_I^{n_I} g_J^{m_J}$$

with constants $c_{I,J} \in K$, absolutely convergent when the variables $(f_i)_{i \in I}$, $(g_j)_{j \in J}$ satisfy the above inequalities. The quotient of this algebra by the topological closure of the ideal I is the algebra $\mathcal{O}_{\mathrm{Spec}^{\mathrm{an}}(R/K)}(U)$.

As in the case of schemes we can glue $\mathrm{Spec}^{\mathrm{an}}(R/K)$ into ringed spaces called *analytic spaces* (or rigid analytic spaces). Moreover, we get a functor

$$(\mathrm{Schemes}/K) \to (K\text{-analytic spaces}),$$
$$X \mapsto (X^{\mathrm{an}}, \mathcal{O}_{X^{\mathrm{an}}}).$$

Proposition 8. *The space* X^{an}

(a) *is a locally compact Hausdorff space as long as* X *is separated;*
(b) *has the homotopy type of a finite* CW*-complex;*
(c) *is contractible if* X *has good reduction with irreducible special fiber.*

Example 1. Let $X = \mathbf{A}^1 = \mathrm{Spec}(K[x])$ be the affine line. The analytic space X^{an} contains, among others, points of the following types:

- $X(K) \hookrightarrow X(\overline{K})/\mathrm{Gal}(\overline{K}/K) \hookrightarrow X^{\mathrm{an}}$.
- For $r \in \mathbf{R}_{\geq 0}$, define

$$\left| \sum_{j=0}^{d} c_j z^j \right|_r := \max_j (|c_j| r^j).$$

  This gives an embedding $\mathbf{R}_{\geq 0} \hookrightarrow X^{\mathrm{an}}$.

We see that X^{an} contains, in a sense, both p-adic and real points.

Define the *cone* over X^{an} as

$$C_{X^{\mathrm{an}}}(\mathbf{R}) := X^{\mathrm{an}} \times \mathbf{R}_{>0}.$$

We interpret a point $\mathbf{x} = (x, \lambda)$ of $C_{X^{\mathrm{an}}}(\mathbf{R})$ as a K_x-point of X, where $K_x \supset K$ is a complete field with the $\mathbf{R}$-valued valuation

$$\mathrm{val}_{\mathbf{x}} := \lambda\, \mathrm{val}_x,$$

whose restriction to K is proportional to val_K. The set of points $\mathbf{x} \in C_{X^{\mathrm{an}}}(\mathbf{R})$ such that the valuation $\mathrm{val}_{\mathbf{x}}$ is $\mathbf{Z}$-valued is denoted by $C_{X^{\mathrm{an}}}(\mathbf{Z})$.

A.2 Algebraic torus and the logarithmic map

Here we will describe explicitly the main example for our paper. Let $X = \mathbf{G}_m^n = \mathrm{Spec}(K[z_i^{\pm 1}])$, $1 \leq i \leq n$, be an algebraic torus, and $X^{\mathrm{an}} = (\mathbf{G}_m^{\mathrm{an}})^n$ the corresponding analytic space.

First, we define an embedding $i_{\mathrm{can}} : \mathbf{R}^n \hookrightarrow X^{\mathrm{an}}$. For a real vector $(x_i)_{1 \le i \le n} \in \mathbf{R}^b$, the corresponding point $p := i_{\mathrm{can}}(x_1, \ldots, x_n) \in X^{\mathrm{an}}$ will be described in terms of valuations.

For every Laurent polynomial $f = \sum_{I \in \mathbf{Z}^n} c_I z^I$, $c_I \in K$, we set

$$\mathrm{val}_p(f) := \min_{I \in \mathbf{Z}^n} \left(\mathrm{val}(c_I) - \sum_{i=1}^{n} x_i I_i \right).$$

Next, we define a projection $\pi_{\mathrm{can}} : X^{\mathrm{an}} \to \mathbf{R}^n$ by the formula

$$\pi_{\mathrm{can}}(y) = (-\mathrm{val}_y(z_1), \ldots, -\mathrm{val}_y(z_n)) = (\log |z_1|_y, \ldots, \log |z_n|_y).$$

The fiber over a point $(x_1, \ldots, x_n) \in \mathbf{R}^n$ can be identified with the set of seminorms $|\cdot|_y$ such that $|z_i|_y = \exp(x_i)$, $1 \le i \le n$. We see that π_{can} is a kind of torus fibration. Moreover, $\pi_{\mathrm{can}} \circ i_{\mathrm{can}} = \mathrm{id}_{\mathbf{R}^n}$.

For any open connected $U \in (\mathbf{R})^n$, the K-algebra of analytic functions on $\pi_{\mathrm{can}}^{-1}(U)$ consists of series $f = \sum_{I \in \mathbf{Z}^n} c_I z^I$ with coefficients $c_I \in K$ such that for any $p = (x_1, \ldots, x_n) \in U$, we have $\log |c_I| + \sum_{i=1}^n x_i I_i \to +\infty$ when $|l| \to +\infty$. It is easy to see that $\pi_{\mathrm{can}}^{-1}(U) = \pi_{\mathrm{can}}^{-1}(\mathrm{Conv}(U))$, where $\mathrm{Conv}(U)$ is the convex hull of U.

The sheaf $(\pi_{\mathrm{can}})_*(\mathcal{O}_{X^{\mathrm{an}}}) := \mathcal{O}_{\mathbf{R}^n}^{\mathrm{can}}$ (canonical sheaf) plays an important role in the paper (see Sections 4.1, 7.3, and 8).

A.3 Clemens polytopes

Let X be a smooth proper scheme over the non-archimedean field K. We assume that K carries a discrete valuation val such that $\mathrm{val}(K^\times) = \mathbf{Z}$.

Definition 18. *A model of X is a scheme of finite type $\mathcal{X}/\mathcal{O}_K$ flat and proper over $\mathcal{O}_K$, together with an isomorphism $\mathcal{X} \times_{\mathrm{Spec}(K)} \mathrm{Spec}(\mathcal{O}_K) \simeq X$. Denote the special fiber of $\mathcal{X}$ by*

$$\mathcal{X}^0 := \mathcal{X} \times_{\mathrm{Spec}(K)} \mathrm{Spec}(k).$$

A model has no nontrivial automorphisms. Thus the stack of equivalence classes of models is in fact a set, which we denote by Mod_X. It carries a natural partial order. Namely, we say that $\mathcal{X}_1 \ge \mathcal{X}_2$ if there exists a map $\mathcal{X}_1 \to \mathcal{X}_2$ over $\mathrm{Spec}(\mathcal{O}_K)$. Such a map is automatically unique.

Definition 19. *A model $\mathcal{X}$ has* normal crossings *if the scheme $\mathcal{X}$ is regular and the reduced subscheme $\mathcal{X}^0_{\mathrm{red}}$ is a divisor with normal crossings.*

By the resolution of singularities, in the case $\mathrm{char}\, k = 0$ we know that every model is dominated by a model with normal crossings.

Definition 20. *A model $\mathcal{X}$ has simple normal crossings (snc model for short) if*

- *it has normal crossings,*
- *all irreducible components of $\mathcal{X}^0_{\mathrm{red}}$ are smooth, and*

- *all intersections of irreducible components of $\mathcal{X}^0_{\mathrm{red}}$ are either empty or irreducible.*

The set of equivalence classes of snc models will be denoted by $\mathrm{Mod}_X^{\mathrm{snc}}$. It is a filtered partially ordered set. The order is given by dominating maps of models which give the identity automorphism on the generic fiber.

It is easy to show that starting with any model with normal crossings and applying blowups centered at certain self-intersection loci of the special fiber, we can get an snc model. In what follows, we use snc models only. This choice is dictated by convenience and not by necessity. Working with snc models has the advantage that all definitions and calculations can be made very transparent. The reader can consult [Be2] for the approach in the general case, without the use of the resolution of singularities.

Let $\mathcal{X}$ be an snc model and $I = I_{\mathcal{X}}$ the set of irreducible components of $\mathcal{X}^0_{\mathrm{red}}$. Denote by $D_i \subset \mathcal{X}$ the divisor corresponding to $i \in I$. For any finite nonempty subset $J \subset I$, put

$$D_J := \bigcap_{j \in J} D_j.$$

By the snc property, the set D_J is either empty or is a smooth connected proper variety over k of dimension $\dim(D_J) = (n - |J| + 1)$. For a divisor $D_i \subset \mathcal{X}^0$ we denote by $d_i \in \mathbf{Z}_{>0}$ the order of vanishing of u at D_i, where $u \in K$ is a uniformizing element, $\mathrm{val}_K(u) = 1$. Equivalently, d_i is the multiplicity of D_i in $\mathcal{X}^0$.

Definition 21. *The Clemens polytope $S_{\mathcal{X}}$ is the finite simplicial subcomplex of the simplex Δ^I such that Δ^J is a face of $S_{\mathcal{X}}$ iff $D_J \neq \emptyset$.*

Clearly, $S_{\mathcal{X}}$ is a nonempty connected CW-complex. We will also consider the cone over $S_{\mathcal{X}}$:

$$C_{\mathcal{X}}(\mathbf{R}) := \left\{ \sum_{i \in I} a_i \langle D_i \rangle \mid a_i \in \mathbf{R}_{\geq 0}, \bigcap_{i : a_i > 0} D_i \neq \emptyset \right\} \setminus \{0\} \subset \mathbf{R}^I.$$

Analogously, we can define $C_{\mathcal{X}}(\mathbf{Z})$.

We identify $S_{\mathcal{X}}$ with the following subset of $C_{\mathcal{X}}(\mathbf{R})$:

$$\left\{ \sum_{i \in I} a_i \langle D_i \rangle \in C_{\mathcal{X}}(\mathbf{R}) \mid \sum_i a_i d_i = 1 \right\}.$$

Obviously, we can also describe $S_{\mathcal{X}}$ as a quotient of $C_{\mathcal{X}}(\mathbf{R})$:

$$S_{\mathcal{X}} = C_{\mathcal{X}}(\mathbf{R}) / \mathbf{R}_+^{\times}.$$

A.4 Simple blowups

Let $\mathcal{X}$ be an snc model, $J \subset I_{\mathcal{X}}$ a nonempty subset and $Y \subset D_J$ a smooth irreducible variety of dimension less than or equal to n. Let us assume that Y intersects transversally (in D_J) all subvarieties $D_{J'}$ of D_J (for $J' \supset J$), and that all intersections $Y \cap D_J$ are either empty or irreducible. It is obvious that the blowup $\mathcal{X}' := Bl_Y(\mathcal{X})$ of $\mathcal{X}$ with center at Y is again an snc model.

Definition 22. *For a pair of snc models $\mathcal{X}' \geq \mathcal{X}$ as above, we say that $\mathcal{X}'$ is obtained from $\mathcal{X}$ by a simple blowup. If $Y = D_J$ we say that we have a simple blowup of the first type. Otherwise (when $\dim(Y) < \dim(D_J)$), we have a simple blowup of the second type.*

Let us describe the behavior of $S_{\mathcal{X}}$ under simple blowups. To the set of vertices we add a new vertex corresponding to the divisor $\widetilde{Y}$ obtained from Y:

$$I_{\mathcal{X}'} = I_{\mathcal{X}} \sqcup \{\text{new}\}, \qquad D_{\text{new}} := \widetilde{Y}.$$

The degree of the new divisor is (for both the first and the second type)

$$d_{\text{new}} := \sum_{i \in J} d_j.$$

For blowups of the first type, we have automatically $\#J > 1$. Here is the list of faces of $S_{\mathcal{X}'}$:

(1) I' for $I' \in \text{Faces}(S_{\mathcal{X}})$, $I' \not\subset J$;
(2) $I' \sqcup \{\text{new}\}$ for $I' \in \text{Faces}(S_{\mathcal{X}})$, $I' \neq J$, $I' \cup J \in \text{Faces}(S_{\mathcal{X}})$;
(3) the vertex $\{\text{new}\}$.

For blowups of the second type, the list of faces of $S_{\mathcal{X}'}$ is

(1) I' for $I' \in \text{Faces}(S_{\mathcal{X}})$;
(2) $I' \sqcup \{\text{new}\}$ for $I' \in \text{Faces}(S_{\mathcal{X}})$, $I' \supset J$, $Y \cap D_{I'} \neq \emptyset$;
(3) the vertex $\{\text{new}\}$.

On can deduce from results [AKMW] the following.

Theorem 9 (weak factorization). *Assume that* $\operatorname{char} k = 0$. *Then for any two snc models $\mathcal{X}, \mathcal{X}'$ there exists a finite alternating sequence of simple blowups*

$$\mathcal{X} < \mathcal{X}_1 > \mathcal{X}_2 < \cdots < \mathcal{X}_{2m+1} > \mathcal{X}'.$$

Corollary 3. *The simple homotopy type of $S_{\mathcal{X}}$ does not depend on the choice of an snc model $\mathcal{X}$.*

A.5 Clemens cones and valuations

Let $\mathcal{X}$ be an snc model of X. We define a map

$$i_{\mathcal{X}} : C_{\mathcal{X}}(\mathbf{R}) \to C_{X^{\text{an}}}(\mathbf{R})$$

as follows. For $J = \{j_1, \ldots, j_k\} \subset I_{\mathcal{X}}$ such that $D_J \neq \emptyset$ let us consider a point $x \in C_{\mathcal{X}}(\mathbf{R})$,

$$x = \sum_{i=1}^{k} a_i \langle D_{j_i} \rangle, \quad a_i \in \mathbf{R}_{>0}, \quad \forall i \in \{1, \ldots, k\}$$

and an affine Zariski open subset $U \subset \mathcal{X}$ containing the generic point of D_J. One can embed $\mathcal{O}(U)$ into the algebra of formal series $K_J[[z_1, \dots, z_k]]$, where K_J is the field of rational functions on D_J and $z_i = 0$ are equations of divisors D_{j_i}, $i = 1, \dots, k$. We define a valuation v_x of $\mathcal{O}(U)$ by the formula

$$v_x\left(\sum_{n_1,\dots,n_k\geq 0} c_{n_1,\dots,n_k}\prod_{i=1}^{k} z_i^{n_i}\right) = \inf\left\{\sum a_i n_i \mid c_{n_1,\dots,n_k} \neq 0\right\}.$$

We define $i_{\mathcal{X}}(x)$ to be the image of the point $v_x \in \mathrm{Spec}^{\mathrm{an}}(\mathcal{O}(U)/K)$ in X^{an}. It is easy to check that the element $i_{\mathcal{X}}(x)$ does not depend on the choice of the open subset U.

The following proposition is obvious.

Proposition 9. *The map $i^{\mathbf{R}}_{\mathcal{X}}$ is an embedding.*

We will denote also by $i_{\mathcal{X}}$ the induced embedding $S_{\mathcal{X}} \hookrightarrow X^{\mathrm{an}}$.

A.6 Clemens cones and paths

For a model $\mathcal{X}$ we can interpret elements of $C_{X^{\mathrm{an}}}(\mathbf{Z})$ as *paths* in $\mathcal{X}$, i.e., equivalence classes of maps

$$\phi : \mathrm{Spec}(\mathcal{O}_L) \to \mathcal{X},$$

where $\mathcal{O}_L$ is the ring of integers in a field L with discrete valuation in $\mathbf{Z}$, such that the image of ϕ does not lie in $\mathcal{X}$. We define the map

$$p^{\mathbf{Z}}_{\mathcal{X}} : C_{X^{\mathrm{an}}}(\mathbf{Z}) \to C_{\mathcal{X}}(\mathbf{Z})$$

as

$$p^{\mathbf{Z}}_{\mathcal{X}}([\phi]) := \sum_i a_i \langle D_i \rangle,$$

where $a_i \in \mathbf{Z}_{\geq 0}$ is the multiplicity of the intersection of the path ϕ with the divisor D_i, $i \in I_{\mathcal{X}}$.

The following proposition can be derived from [Be1].

Proposition 10. *The map $p^{\mathbf{Z}}_{\mathcal{X}}$ extends uniquely to a continuous $\mathbf{R}^{\times}_{+}$-equivariant map $p^{\mathbf{R}}_{\mathcal{X}} : C_{X^{\mathrm{an}}}(\mathbf{R}) \to C_{\mathcal{X}}(\mathbf{R})$. The map $p^{\mathbf{R}}_{\mathcal{X}}$ is a surjection.*

We denote by $p_{\mathcal{X}} : X^{\mathrm{an}} \to S_{\mathcal{X}}$ the map induced by $p^{\mathbf{R}}_{\mathcal{X}}$.

Let $f : \mathcal{X}' \to \mathcal{X}$ be a dominating map of models. Let us denote by $m_{i,i'} \in \mathbf{Z}_{\geq 0}$ the multiplicity of a divisor $D_{i'}$, $i' \in I_{\mathcal{X}'}$, in the proper pullback of D_i, $i \in I_{\mathcal{X}}$. We define $p^{\mathbf{Z}}_{\mathcal{X}',\mathcal{X}} : C_{X^{\mathrm{an}}}(\mathbf{Z}) \to C_{\mathcal{X}}(\mathbf{Z})$ by the formulas $\sum_{i'} a_{i'}\langle D_{i'}\rangle \mapsto \sum_i m_{i,i'} a_{i'} \langle D_i \rangle$. Let $p_{\mathcal{X}',\mathcal{X}} : S_{X^{\mathrm{an}}}(\mathbf{R}) \to S_{\mathcal{X}}(\mathbf{R})$ be the corresponding map of Clemens polytopes.

Then we have the following result, which is easy to prove.

Lemma 8. *For any dominating map of models $\mathcal{X}' \to \mathcal{X}$, we have*

$$p^{\mathbf{Z}}_{\mathcal{X}} = p^{\mathbf{Z}}_{\mathcal{X}',\mathcal{X}} \circ p^{\mathbf{Z}}_{\mathcal{X}'}.$$

Corollary 4. *For dominating maps $\mathcal{X}'' \geq \mathcal{X}' \geq \mathcal{X}$, we have*

$$p_{\mathcal{X}'',\mathcal{X}} = p_{\mathcal{X}',\mathcal{X}} \circ p_{\mathcal{X}'',\mathcal{X}'}.$$

Theorem 10. *For any algebraic X the analytic space X^{an} is a projective limit over the partially ordered set of snc models $\mathcal{X}$ of Clemens polytopes $S_{\mathcal{X}}$. The connecting maps are $p_{\mathcal{X}',\mathcal{X}}$.*

With any meromorphic at $t = 0$ family of smooth complex projective varieties X_t, $0 < |t| < \epsilon$, one can associate a variety X over the field $\mathbf{C}((t))$. It is easy to see that for any snc model $\mathcal{X}$ one can canonically complete the family X_t by adding $S_{\mathcal{X}}$ as the fiber over $t = 0$. The total space is not a complex manifold but just a Hausdorff locally compact space which maps properly to the disk $\{t \in \mathbf{C} \mid |t| < \epsilon\}$. Passing to the projective limit we see that one can compactify the family X_t at $t = 0$ by X^{an}.

B Torelli theorem for K3 surfaces

Here we recall the classification theory of complex K3 sufaces (see [PSS] and its extension to the nonalgebraic case in [LP]). Let X be a complex K3 surface, i.e., smooth connected complex manifold with $\dim_{\mathbf{C}} X = 2$ which admits a nowhere vanishing holomorphic 2-form Ω, and such that $H^1(X, \mathbf{Z}) = 0$.

It is known that the group $H^2(X, \mathbf{Z})$ endowed with the Poincaré pairing $(\cdot,\cdot)$ is isomorphic to the lattice

$$\Lambda_{\mathrm{K3}} = \begin{pmatrix} 0 & 1 \\ 1 & 0 \end{pmatrix} \oplus \begin{pmatrix} 0 & 1 \\ 1 & 0 \end{pmatrix} \oplus \begin{pmatrix} 0 & 1 \\ 1 & 0 \end{pmatrix} \oplus (-E_8) \oplus (-E_8)$$

of signature $(3, 19)$.

The complex one-dimensional vector space $H^{2,0}(X) = \mathbf{C}\cdot[\Omega] \subset H^2(X,\mathbf{Z}) \otimes \mathbf{C}$ satisfies the condition $(v, v) = 0, (v, \overline{v}) > 0$ for any nonzero vector v. Finally, it is known that X admits a Kähler metric, and the Kähler cone $\mathcal{K}_X \subset H^2(X, \mathbf{R})$ of all Kähler metrics on X is an open subset of $C_X := \{[\omega] \in H^2(X, \mathbf{Z}) \otimes \mathbf{R} | ([\omega], [\Omega]) = 0, ([\omega], [\omega]) > 0\}$. In fact $\mathcal{K}_X$ is a connected component of the set $C_X \setminus \cup_{v \in H^2(X,\mathbf{Z}),(v,v)=-2,(v,[\Omega])=0} H_v$, where H_v is the hyperplane orthogonal to v.

Axiomatizing these data we arrive at the following definition.

Definition 23. *K3 period data is a quadruple $(\Lambda, (\cdot,\cdot), H^{2,0}, \mathcal{K})$ consisting of a free abelian group Λ, a symmetric pairing $(\cdot,\cdot) : \Lambda \times \Lambda \to \mathbf{Z}$, a one-dimensional complex vector subspace $H^{2,0} \subset \Lambda \otimes \mathbf{C}$ and a set $\mathcal{K} \subset \Lambda \otimes \mathbf{R}$ satisfying the following conditions:*

1. *$rk \Lambda = 22$;*

2. $(\Lambda, (\cdot, \cdot))$ *is isomorphic to* Λ_{K3}*;*
3. *for any* $v \in H^{2,0} \setminus \{0\}$ *one has* $(v, v) = 0$ *and* $(v, \overline{v}) > 0$*;*
4. *the set* $\mathcal{K}$ *is a connected component of* $C \setminus \cup_{v \in \Lambda, (v,v)=-2, (v, H^{2,0})=0} H_v$*, where* $C = \{w \in \Lambda \otimes \mathbf{R} | (w, H^{2,0}) = 0, (w, w) > 0\}$ *and* H_v *is the hyperplane orthogonal to* v.

The K3 period data form a groupoid. On the other hand, K3 surfaces also form a groupoid (morphisms are isomorphisms of K3 surfaces). Then classical global Torelli theorem can be formulated in the following way.

Theorem 11. *The groupoid of K3 surfaces is equivalent to the groupoid of K3 period data.*

In particular, the automorphism group of a K3 surface is isomorphic to the automorphism group of its period data.

More generally, one can speak about holomorphic families of K3 surfaces over complex analytic spaces. For a K3 surface over an analytic space M the period data consist of a local system of integral lattices $(\Lambda, (\cdot, \cdot))$ pointwise isomorphic to Λ_{K3}, a holomorphic line subbundle $H^{2,0}$ of $\Lambda \otimes_{\mathbf{Z}} \mathcal{O}_M$ which is isotropic with respect to the symmetric pairing $(\cdot, \cdot)$, and satisfying pointwise the condition $(v, \overline{v}) > 0$, $v \in H_x^{2,0} \setminus \{0\}$, $x \in M^{\mathrm{red}}$, and an open subset of the total space of the bundle over M^{red} with fibers $\Lambda_x \otimes \mathbf{R} \cap (H^{2,0})^{\perp}$ ($(H^{2,0})^{\perp}$ is the orthogonal complement) satisfying pointwise condition 4 from the definition of K3 period data. Then the Torelli theorem holds for families as well.

Acknowledgments. We are grateful to Ilya Zharkov and Mark Gross for useful discussions. The second author thanks the Clay Mathematics Institute for supporting him as a Fellow and IHES for excellent research and living conditions.

References

[AKMW] D. Abramovich, K. Karu, K. Matsuki, and J. Wlodarczyk, Torification and factorization of birational maps, *J. Amer. Math. Soc.*, **15** (2002), 531–572.

[Ar] V. Arnold, *Mathematical Methods of Classical Mechanics*, Springer-Verlag, New York, 1997.

[Au] M. Audin, *Spinning Tops: A Course on Integrable Systems*, Cambridge Studies in Advanced Mathematics 51, Cambridge University Press, Cambridge, UK, 1996.

[Be1] V. Berkovich, *Spectral Theory and Analytic Geometry over Non-Archimedean Fields*, AMS Mathematical Surveys and Monographs 33, American Mathematical Society, Providence, RI, 1990.

[Be2] V. Berkovich, Smooth p-adic analytic spaces are locally contractible, *Invent. Math.*, **137** (1999), 1–84.

[Be3] V. Berkovich, Smooth p-adic analytic spaces are locally contractible II, in A. Adolphson, F. Baldassarri, P. Berthelot, N. Katz, and F. Loeser, eds., *Geometric Aspects of Dwork Theory*, Walter de Gruyter, Berlin, 2004, 293–370.

[GS] M. Gross and B. Siebert, Mirror symmetry via logarithmic degeneration data I, math.AG/0309070, 2003.

[GW] M. Gross and P. Wilson, Large complex structure limits of K3 surfaces, *J. Differential Geom.*, **55** (2000), 475–546.

[HZh] C. Haase and I. Zharkov, Integral affine structures on spheres and torus fibrations of Calabi-Yau toric hypersurfaces I, math.AG/0205321, 2002.

[KN] S. Kobayashi and K. Nomizu, *Foundations of Differential Geometry*, Vol. 1, Wiley, New York, 1963.

[Ko] M. Kontsevich, *Homological algebra of mirror symmetry*, in *Proceedings of the* 1994 *International Congress of Mathematicians*, Vol. I, Birkhäuser, Zürich, 1995, 120–139.

[KoSo] M. Kontsevich and Y. Soibelman, Homological mirror symmetry and torus fibrations, in K. Fukaya, Y.-G. Oh, K. Ono, and G. Tian, eds., *Symplectic Geometry and Mirror Symmetry: Proceedings of the* 4*th KIAS Annual International Conference*, World Scientific, Singapore, 2001, 203–263.

[KoT] M. Kontsevich and Yu. Tschinkel, Non-archimedean Kähler geometry, in preparation.

[LeS] N. C. Leung and M. Symington, Almost toric symplectic four-manifolds, math.SG/0312165, 2003.

[LYZ] J. Loftin, S.-T. Yau, and E. Zaslow, Affine manifolds, SYZ geometry, and the "Y" vertex, math.DG/0405061, 2004.

[LP] E. Looijenga and C. Peters, Torelli theorems for Kähler K3 surfaces, *Compositio Math.*, **42** (1981), 145–186.

[Mi] G. Mikhalkin, Amoebas of algebraic varieties and tropical geometry, in S. Donaldson, Y. Eliashberg, and M. Gromov, eds., *Different Faces of Geometry*, International Mathematical Series, Kluwers–Plenum, New York, 2004.

[Mor] D. R. Morrison, Mathematical aspects of mirror symmetry, in J. Kollar, ed., *Complex Algebraic Geometry*, IAS/Park City Mathematics Series 3, American Mathematical Society, Providence, RI, 1997, 265–327.

[PSS] I. I. Pjateckiĭ-Šapiro and I. R. Šafarevič, A Torelli theorem for algebraic surfaces of type K3, *Math. USSR Izv.*, **5**-3 (1971), 547–588.

[SYZ] A. Strominger, S.-T. Yau, and E. Zaslow, Mirror symmetry is T-duality, *Nuclear Phys.*, **B479** (1996), 243–259.

[Tyu] A. Tyurin, On Bohr-Sommerfeld bases, *Izv. Math.*, **64**-5 (2000), 1033–1064.

[Zu] N. Zung, Symplectic topology of integrable Hamiltonian systems II: Topological classification, *Compositio Math.*, **138**-2 (2003), 125–156.

Gelfand–Zeitlin Theory from the Perspective of Classical Mechanics. II

Bertram Kostant[1] and Nolan Wallach[2]

[1] Department of Mathematics
Massachusetts Institute of Technology
Cambridge, MA 02139
USA
kostant@math.mit.edu

[2] Department of Mathematics
University of California at San Diego
San Diego, CA 92093
USA
nwallach@ucsd.edu

To Israel Gelfand:
Your friendship, personality, and incomparable mathematical insights
have made a lifelong impact on us.

Summary. In this paper, Part II, of a two-part paper we apply the results of [KW], Part I, to establish, with an explicit dual coordinate system, a commutative analogue of the Gelfand–Kirillov theorem for $M(n)$, the algebra of $n \times n$ complex matrices. The function field $F(n)$ of $M(n)$ has a natural Poisson structure and an exact analogue would be to show that $F(n)$ is isomorphic to the function field of an $n(n-1)$-dimensional phase space over a Poisson central rational function field in n variables. Instead we show that this the case for a Galois extension, $F(n, \mathfrak{e})$, of $F(n)$. The techniques use a maximal Poisson commutative algebra of functions arising from Gelfand–Zeitlin theory, the algebraic action of an $n(n-1)/2$-dimensional torus on $F(n, \mathfrak{e})$, and the structure of a Zariski open subset of $M(n)$ as an $n(n-1)/2$-dimensional torus bundle over an $n(n+1)/2$-dimensional base space of Hessenberg matrices.

Subject Classifications: 14L30, 53D17, 14M17, 14R20, 33C45

0 Part II continuation of introduction

0.6

We recall some of the notation and results in Part I, i.e., [KW]. If k is a positive integer, then $I_k = \{1, \ldots, k\}$. $M(n)$ is the algebra of all $n \times n$ complex matrices. If $m \in I_n$, then regard $M(m) \subset M(n)$ as the upper left block of all $m \times m$ matrices.

If $x \in M(n)$, then $x_m \in M(m)$ is the upper left principal $m \times m$ minor of x. Using a natural isomorphism of (the Lie algebra) $M(n)$ with its dual space, $M(n)$ becomes a Poisson manifold so that its affine ring $\mathcal{O}(M(n))$ is a Poisson algebra. For any $k \in \mathbb{Z}_+$, let $d(k) = k(k+1)/2$. The subalgebra $J(n)$ of $\mathcal{O}(M(n))$, generated by the symmetric polynomial $Gl(m)$-invariants of $M(m)$ for all $m \in I_n$, is a polynomial algebra with $d(n)$ generators and, more importantly, it is a maximal Poisson commutative subalgebra of $\mathcal{O}(M(n))$.

In Part I we showed that the Poisson vector field ξ_f on $M(n)$ corresponding to any $f \in J(n)$ is globally integrable on $M(n)$ and a choice of generators of $J(n)$ defines an abelian Lie group A of dimension $d(n-1)$ operating on $M(n)$. The orbits of A are explicitly determined in Part I, and the orbits are independent of the choice of generators. One particular choice are the functions, $p_i(x)$, $i \in I_{d(n)}$, $x \in M(n)$, where, for all $m \in I_n$, $p_{d(m-1)+k}(x)$, $k \in I_m$, are the nontrivial coefficients of the characteristic polynomial of x_m.

A suitable measure on $\mathbb{R}$ and the Gram–Schmidt process define a sequence, $\phi_k(t)$, $k \in \mathbb{Z}_+$, of orthogonal polynomials on $\mathbb{R}$. Let W_n be the span of ϕ_{m-1}, $m \in I_n$, and let $x \in M(n)$ be the matrix, with respect to this basis, of the operator of multiplication by t, followed by projection on W_n. The matrix x is Jacobi, and for $m \in I_n$ one recovers the orthogonal polynomial ϕ_m as the characteristic polynomial of x_m. In particular, the all important zeros of the orthogonal polynomials ϕ_m appear as the eigenvalues of the x_m. One motivation for our work here is to set up Poisson machinery to deal with the eigenvalues of x_m for any $x \in M(n)$. In the course of setting up this machinery we have obtained a number of new results. Some of these results have appeared in Part I [KW]. In the present paper, Part II, we will be concerned with establishing a refinement of a commutative analogue of the Gelfand–Kirillov theorem. The refinement refers to exhibiting an explicit coordinate system satisfying the Poisson commutation relations of phase space. The coordinate system emerges from the action of an algebraic group and the structure of $M(n)$, obtained in Part I, as sort of a cotangent bundle over the variety of Hessenberg matrices.

In more detail, let $M_\Omega(n)$ be the Zariski open (dense) subset of $M(n)$ defined as the set of all $x \in M(n)$ such that x_m is regular semisimple in $M(m)$ for all $m \in I_n$ and such that the spectrum of x_{m-1}, for $m > 0$, has empty intersection with the spectrum of x_m. In Part I $M_\Omega(n)$ was shown to have the following structure: It is a $(\mathbb{C}^\times)^{d(n-1)}$ bundle over a $d(n)$-dimensional base space ($d(n-1) + d(n) = n^2$). The fibers are not only the level sets for the functions in $J(n)$ but also the fibers are the orbits of A in $M_\Omega(n)$. The base space, denoted by $\mathfrak{b}_{e,\Omega(n)}$, is the intersection $\mathfrak{b}_e \cap M_\Omega(n)$, where $\mathfrak{b}_e$ is the space of Hessenberg matrices. That is, $x \in \mathfrak{b}_e$ if and only if x is of the form

$$x = \begin{pmatrix} a_{11} & a_{12} & \cdots & a_{1n-1} & a_{1n} \\ 1 & a_{22} & \cdots & a_{2n-1} & a_{2n} \\ 0 & 1 & \cdots & a_{3n-1} & a_{3n} \\ \vdots & \vdots & \ddots & \vdots & \vdots \\ 0 & 0 & \cdots & 1 & a_{nn} \end{pmatrix}.$$

Also, $\mathfrak{b}_{e,\Omega(n)}$ is Zariski dense in $\mathfrak{b}_e$. Theorem 2.5 in Part I concerning establishing a beautiful property of $\mathfrak{b}_e$ plays a major role here. In Part II a "Lagrangian" property of $\mathfrak{b}_e$ plays a key role in showing that the dual coordinates s_i, defined below, Poisson commute. Part I has three sections. A serious deficiency in $M_\Omega(n)$ in dealing with the commutative analogue of the Gelfand–Kirillov theorem is that one cannot consistently solve the characteristic polynomials of x_m for all $m \in I_n$ and all $x \in M_\Omega(n)$ to yield algebraic eigenvalue functions r_i on $M_\Omega(n)$.

0.7

We begin, in the first section of Part II, labeled Section 4, to obtain such functions on a covering space $M_\Omega(n, \mathfrak{e})$ of $M_\Omega(n)$. Initially, the covering map

$$\pi_n : M_\Omega(n, \mathfrak{e}) \to M_\Omega(n)$$

is only understood to be analytic. The covering admits, as deck transformations, a group, Σ_n, isomorphic to the direct product of the symmetric groups S_m, $m \in I_n$, and as analytic manifolds

$$M_\Omega(n, \mathfrak{e})/\Sigma_n \cong M_\Omega(n).$$

However, much more structure is needed and established in Section 4. For one thing $M_\Omega(n, \mathfrak{e})$ is a nonsingular affine variety and π_n is a finite étale morphism. For another, if $F(n)$ is the field of rational functions on $M(n)$ and $F(n, \mathfrak{e})$ is the field of rational functions on $M_\Omega(n, \mathfrak{e})$, then $F(n, \mathfrak{e})$ is a Galois extension of $F(n)$ with Σ_n as Galois group. Furthermore the affine ring $\mathcal{O}(M_\Omega(n, \mathfrak{e}))$ is the integral closure of $\mathcal{O}(M_\Omega)$ in $F(n, \mathfrak{e})$. Very significant for our purposes, there exist (eigenvalue) functions $r_i \in \mathcal{O}(M_\Omega(n, \mathfrak{e}))$, $i \in I_{d(n)}$, with the property that for any $m \in I_n$ and any $z \in M_\Omega(n, \mathfrak{e})$ the numbers $r_{d(m-1)+k}(z)$, $k \in I_m$, are the eigenvalues of x_m, where $x = \pi_n(z)$. The Poisson structure on $M_\Omega(n)$ lifts to $M_\Omega(n, \mathfrak{e})$ and one has $[r_i, r_j] = 0$ for all $i, j \in I_{d(n)}$.

0.8

Section 5 is devoted to the construction of the dual coordinates $s_j \in \mathcal{O}(M_\Omega(n, \mathfrak{e}))$, $j \in I_{d(n-1)}$. There are two key points here: (a) It is shown that the Poisson vector fields $\xi_{\mathfrak{r}_i}$ on $M_\Omega(n, \mathfrak{e})$ integrate and generate a complex algebraic torus, $A_\mathfrak{r} \cong (\mathbb{C}^\times)^{d(n-1)}$, which operates algebraically on $M_\Omega(n, \mathfrak{e}))$, and, in fact, if $M_\Omega(n, \mathfrak{e}, \mathfrak{b})$ is the π_n inverse image of $\mathfrak{b}_{\mathfrak{e},\Omega(n)}$ in $M_\Omega(n, \mathfrak{e})$, then the map

$$A_\mathfrak{r} \times M_\Omega(n, \mathfrak{e}, \mathfrak{b}) \to M_\Omega(n, \mathfrak{e}), \qquad (b, y) \mapsto \mathfrak{b} \cdot y$$

is an algebraic isomorphism. The natural coordinate system on $A_\mathfrak{r}$ then carries over to $M_\Omega(n, \mathfrak{e})$ defining functions $s_j \in \mathcal{O}(M_\Omega(n, \mathfrak{e}))$, $j \in I_{d(n-1)}$, when they are normalized so that, for all j, s_j is the constant 1 on $M_\Omega(n, \mathfrak{e}, \mathfrak{b})$. The second key point, (b), yields the Poisson commutativity $[s_i, s_j] = 0$ from the Lagrangian property of $\mathfrak{b}_e$. See Theorem 5.20 and its proof. Combining Theorems 5.14 and 5.23, one has the following.

Theorem 0.16. *The image of the map*

$$M_\Omega(n,\mathfrak{e}) \to \mathbb{C}^{n^2}, \qquad z \mapsto (r_1(z),\dots,r_{d(n)}(z),s_1(z),\dots,s_{d(n-1)}(z)) \tag{0.13}$$

is a Zariski open set Y in $\mathbb{C}^{n^2}$, and (0.13) *is an algebraic isomorphism of $M_\Omega(n,\mathfrak{e})$ with Y. Furthermore, one has the following Poisson commutation relations:*

$$\begin{aligned} &(1)\ [r_i,r_j]=0,\ i,j\in I_{d(n)},\\ &(2)\ [r_i,s_j]=\delta_{ij}s_j,\ i\in I_{d(n)},\ j\in I_{d(n-1)},\\ &(3)\ [s_i,s_j]=0,\ i,j\in I_{d(n-1)}.\end{aligned} \tag{0.14}$$

Noting that s_i vanishes nowhere on $M_\Omega(n,\mathfrak{e})$, one has $r_{(i)}\in\mathcal{O}(M_\Omega(n,\mathfrak{e}))$ for $i\in I_{d(n-1)}$, where $r_{(i)}=r_i/s_i$. Replacing r_i by $r_{(i)}$ in (2), one has the more familiar phase space commutation relation $[r_{(i)},s_j]=\delta_{ij}$. For the implication of Theorem 0.16 on the structure of the field $F(n,\mathfrak{e})$, see Theorem 5.24.

0.9

Of course, given the eigenvalue functions r_i, the dual coordinates s_j are not uniquely determined. In the present paper, they are given by the use of the algebraic group $A_\mathfrak{r}$ (defined by the r_i) and a set of Hessenberg matrices as a base space. Independently and quite differently, the papers [GKL1] and [GKL2] also deal with establishing a refined commutative analogue of the Gelfand–Kirillov theorem. A point of similarity is the use of the coordinates r_i and the necessity, thereby, to go to a covering. In [GKL1, Section 3] dual coordinates are given, denoted in that paper by Q_{nj}. It seems to be an interesting question to write down equations expressing a relation between the Q_{nj} in [GKL1] and the s_i here. The first three sections of the paper are in Part I. Part II begins with Section 4.

4 The covering $M_\Omega(n,\mathfrak{e})$ of $M_\Omega(n)$ and the eigenvalue functions r_i

4.1

We retain the general notation of Part I so that n is a positive integer and $M(n)$ is the space of all complex $n\times n$ matrices. As in (2.61), for $m\in I_n$ (see Section 1.1), let $\mathfrak{d}(m)$ be the space of all diagonal matrices in $M(m)$ and let $\mathfrak{e}(m)$ be the (connected) Zariski open subset of all regular elements in $\mathfrak{d}(m)$. That is, if $z\in\mathfrak{d}(m)$, then $z\in\mathfrak{e}(m)$ if and only if the diagonal entries of z are distinct. Consider the direct product

$$\mathfrak{e}=\mathfrak{e}(1)\times\cdots\times\mathfrak{e}(m) \tag{4.1}$$

so that if $\nu\in\mathfrak{e}$, we can write

$$\nu=(\nu(1),\dots,\nu(n)), \tag{4.2}$$

where $\nu(m) \in \mathfrak{e}(m)$. In addition, we will write

$$\nu(m) = \mathrm{diag}(\nu_{1m}, \ldots, \nu_{mm}), \tag{4.3}$$

where the numbers $\nu_{im} \in \mathbb{C}$, $i \in I_m$, are distinct. Taking notation from (2.53), let $\mathfrak{e}_{\Omega(n)}$ be the Zariski open subset of $\mathfrak{e}$ defined so that if $\nu \in \mathfrak{e}$, then $\nu \in \mathfrak{e}_{\Omega(n)}$ if and only if

$$\nu_{im} \neq \nu_{jm+1}, \quad \forall m \in I_{n-1}, \quad i \in I_m, \quad j \in I_{m+1}. \tag{4.4}$$

Of course, $\mathfrak{e}_{\Omega(n)}$ is a nonsingular variety, where

$$\dim \mathfrak{e}_{\Omega(n)} = d(n)$$

(see Section 0.1). The symmetric group S_m, as the Weyl group of $(M(m), \mathfrak{d}(m))$, operates freely on $\mathfrak{e}(m)$ and the direct product $\Sigma_n = S_1 \times \cdots \times S_n$ (a group of order $\prod_{m \in I_n} m!$) operates freely on $\mathfrak{e}_{\Omega(n)}$, where if $\sigma = (\sigma_1, \ldots, \sigma_n)$, $\sigma_m \in S_m$, is in Σ_n and $\nu \in \mathfrak{e}_{\Omega(n)}$, then, using the notation of (4.2) and (4.3),

$$\sigma \cdot \nu = (\sigma_1 \cdot \nu(1)), \ldots, \sigma_m \cdot \nu(m)) \tag{4.5}$$

and

$$\sigma_m \cdot \nu(m) = \mathrm{diag}(\nu_{\sigma_m^{-1}(1)m}, \ldots, \nu_{\sigma_m^{-1}(m)m}). \tag{4.6}$$

4.2

Recall the Zariski open set $M_\Omega(n)$ of $M(n)$ (see (2.53)). In particular, we recall that the matrices in $M_\Omega(n)$ are regular semisimple. Consider the direct product $\mathfrak{e}_{\Omega(n)} \times M_\Omega(n)$ and let

$$M_\Omega(n, \mathfrak{e}) = \{(\nu, x) \in \mathfrak{e}_{\Omega(n)} \times M_\Omega(n) \mid \nu(m) \text{ is } Gl(m)\text{-conjugate to } x_m, \ \forall m \in I_n\}. \tag{4.7}$$

It is clear that $M_\Omega(n, \mathfrak{e})$ is a Zariski closed subset of $\mathfrak{e}_{\Omega(n)} \times M_\Omega(n)$ and the maps

$$\pi_n : M_\Omega(n, \mathfrak{e}) \to M_\Omega(n), \quad \text{where } \pi_n(\nu, x) = x \tag{4.8}$$

and

$$\kappa_n : M_\Omega(n, \mathfrak{e}) \to \mathfrak{e}_{\Omega(n)}, \quad \text{where } \kappa_n(\nu, x) = \nu \tag{4.9}$$

are surjective (see Theorem 2.5) algebraic morphisms.

For $m \in I_n$ and $i \in I_m$, let ρ_{im} be the regular function on $M_\Omega(n, \mathfrak{e})$ defined so that if $z \in M_\Omega(n, \mathfrak{e})$ and $\nu = \kappa_n(z)$, then $\rho_{im}(z) = \nu_{im}$.

Remark 4.1. One notes that if $z \in M_\Omega(n, \mathfrak{e})$ and $x \in M_\Omega(n)$, then $\pi_n(z) = x$ if and only if

$$(\rho_{1m}(z), \ldots, \rho_{mm}(z)) = (\mu_{1m}(x), \ldots, \mu_{mm}(x)), \tag{4.10}$$

up to a reordering, for all $m \in I_n$, using the notation of Section 2.2.

One defines a free action of Σ_n on $M_\Omega(n, \mathfrak{e})$, operating as a group of algebraic isomorphisms, by defining

$$\sigma \cdot z = (\sigma \cdot \nu, x), \tag{4.11}$$

where $\sigma \in \Sigma_n$ and $z = (\nu, x) \in M_\Omega(n, \mathfrak{e})$.

The following well-known proposition is classical.

Proposition 4.2. *Let $m \in I_n$ and let $\gamma(m) \in \mathfrak{e}(m)$. Let $g(m) \in Gl(m)$, and put $x(m) = \operatorname{Ad} g(m)(\gamma(m))$. Then there exists an open neighborhood N of $\gamma(m)$ in $\mathfrak{e}(m)$ and a section $S \subset Gl(m)$ of the quotient map $Gl(m) \to Gl(m)/\operatorname{Diag}(m)$ (using notation in Section* 3.4*) defined on a neighborhood of $g(m)\operatorname{Diag}(m)$ such that the map*

$$S \times N \to M(m), \quad \textit{where } (g, \gamma) \mapsto \operatorname{Ad} g(\gamma) \tag{4.12}$$

is an analytic isomorphism onto an open set of $M(m)$. The elements of the image are necessarily regular semisimple elements of $M(m)$.

If U is an open subset of $M_n(\Omega)$ and $\phi : U \to \mathfrak{e}$ is an analytic map, let graph $\phi : U \to \mathfrak{e} \times M_n(\Omega)$ be the analytic map defined by putting graph $\phi(x) = (\phi(x), x)$.

Proposition 4.3. *Let $z = (\nu, x) \in M_\Omega(n, \mathfrak{e})$. Then there exist a (sufficiently small) connected open neighborhood U of x in $M_\Omega(n)$ and an analytic map $\phi : U \to \mathfrak{e}_{\Omega(n)}$ with the following properties:*

(1) graph $\phi(x) = z$, graph $\phi : U \to U_z$ *is a homeomorphism, where U_z is the image of* graph ϕ, *and $U_z \subset M_\Omega(n, \mathfrak{e})$.*
(2) $\pi_n^{-1}(U) = \sqcup_{\sigma \in \Sigma_n} \sigma \cdot U_z$.
(3) $\{\sigma \cdot U_z \mid \sigma \in \Sigma_n\}$ *are the connected components of $\pi_n^{-1}(U)$ and each component is open in $M_\Omega(n, \mathfrak{e})$.*

In particular, U is evenly covered by π_n and π_n is a covering projection (see (4.8)*).*

Proof. Statements (1) and (2) are immediate consequences of Proposition 4.2. But it is immediate from (1) and (2) that $\sigma \cdot U_z$ is connected and closed in $\pi_n^{-1}(U)$ for any $\sigma \in \Sigma_n$. Since the partition in (2) is finite, it follows that the parts are open in $\pi_n^{-1}(U)$ and hence are open in $M_\Omega(n, \mathfrak{e})$. The remaining statements are obvious. □

4.3

As in the introduction, Section 0, let $\mathfrak{b}_e = -e + \mathfrak{b}$ using the notation of Section 2.2. By Remark 2.4, Theorems 2.3 and 2.5 hold if $\mathfrak{b}_e$ replaces $e + \mathfrak{b}$. Let $\mathfrak{b}_{e,\Omega(n)} = M_\Omega(n) \cap \mathfrak{b}_e$. Then $\mathfrak{b}_{e,\Omega(n)}$ is a Zariski open subset of $\mathfrak{b}_e$ and

$$\Phi_n : \mathfrak{b}_{e,\Omega(n)} \to \Omega(n) \tag{4.13}$$

is an algebraic isomorphism by Theorem 2.5 (see Remark 2.16). In particular, $\mathfrak{b}_{e,\Omega(n)}$ is dense in $\mathfrak{b}_e$. Now let $M_\Omega(n, \mathfrak{e}, \mathfrak{b}) = \pi_n^{-1}(\mathfrak{b}_{e,\Omega(n)})$ so that, by Proposition 4.3, $M_\Omega(n, \mathfrak{e}, \mathfrak{b})$ is a covering of $\mathfrak{b}_{e,\Omega(n)}$. Now consider the restriction

$$\kappa_n : M_\Omega(n, \mathfrak{e}, \mathfrak{b}) \to \mathfrak{e}_{\Omega(n)} \tag{4.14}$$

of (4.9) to $M_\Omega(n, \mathfrak{e}, \mathfrak{b})$. Note that $\mathfrak{e}_{\Omega(n)}$ is connected since it is clearly Zariski open in $\mathfrak{e}$. One has the following.

Theorem 4.4. *The map* (4.14) *is a homeomorphism. In particular,* $M_\Omega(n, \mathfrak{e}, \mathfrak{b})$ *is connected.*

Proof. Let $\nu \in \mathfrak{e}_{\Omega(n)}$. Then by Theorem 2.5 there exists (uniquely) $x \in \mathfrak{b}_e$ such that, for any $m \in I_n$, $(\mu_{1m}(x), \ldots, \mu_{m,m}(x)) = (\nu_{1m}, \ldots, \nu_{mm})$, up to a reordering. But then $x \in \mathfrak{b}_{e,\Omega(n)}$ and, by Proposition 4.3, $(\nu, x) \in M_\Omega(n, \mathfrak{e}, \mathfrak{b})$. But then ν is in the image of (4.14). That is, (4.14) is surjective. But assume $z, z' \in M_\Omega(n, \mathfrak{e}, \mathfrak{b})$ and $\kappa_n(z) = \kappa_n(z')$. Then if $x = \pi_n(z)$ and $x' = \pi_n(z')$, one has $x, x' \in \mathfrak{b}_e$ and hence $x = x'$ by Theorem 2.5. Thus $z = z'$ so that (4.14) is injective. Hence (4.14) is bijective. But, of course, as a restriction map, (4.14) is continuous. We have only to show that its inverse is continuous.

Let β_i, $i \in I_n$, be the regular function on $\mathfrak{e}_{\Omega(n)}$ defined so that if $\nu \in \mathfrak{e}_{\Omega(n)}$, then $\beta_{d(m-1)+k}(\nu)$, $k \in I_m$, $m \in I_n$, is the elementary symmetric function of degree $m - k + 1$ in $\{\nu_{1m}, \ldots, \nu_{mm}\}$. Now let $\beta : \mathfrak{e}_{\Omega(n)} \to \mathbb{C}^{d(n)}$ be the regular algebraic map defined so that

$$\beta(\nu) = (\beta_1(\nu), \ldots, \beta_{d(n)}(\nu)). \tag{4.15}$$

One notes that if $c = \beta(\nu)$, then by (2.3), (2,4), (2.10), and (2.11),

$$(\nu_{1m}, \ldots, \nu_{mm}) = (\mu_{1m}(c), \ldots, \mu_{mm}(c)), \tag{4.16}$$

up to a reordering, for all $m \in I_n$. It follows then that

$$\beta : \mathfrak{e}_{\Omega(n)} \to \Omega(n) \tag{4.17}$$

is a surjective morphism (see (2.53)). Recalling Theorem 2.3 (where $-e$ replaces e), one has, inverting (4.13), a surjective morphism

$$\widetilde{\beta} : \mathfrak{e}_{\Omega(n)} \to \mathfrak{b}_{e,\Omega(n)}, \tag{4.18}$$

where $\beta = \Phi_n \circ \widetilde{\beta}$ noting that, for $i \in I_n$,

$$p_i(\widetilde{\beta}(\nu)) = \beta_i(\nu) \tag{4.19}$$

by (2.3), (2.4), and (2.5). But clearly $(\nu, \widetilde{\beta}(\nu)) \in M_\Omega(n, \mathfrak{e}, \mathfrak{b})$ for any $\nu \in \mathfrak{e}_{\Omega(n)}$. Hence

$$\mathfrak{e}_{\Omega(n)} \to M_\Omega(n, \mathfrak{e}, \mathfrak{b}), \qquad \nu \mapsto (\nu, \widetilde{\beta}(\nu)) \tag{4.20}$$

is an algebraic morphism. But (4.20) must be the inverse to (4.14), by the bijectivity of (4.14), since $\kappa_n((\nu, \widetilde{\beta}(\nu))) = \nu$. Hence (4.14) is a homeomorphism. □

Recalling (4.9), let $M_\nu(n, \mathfrak{e}) = \kappa_n^{-1}(\nu)$ so that one has a "fibration"

$$M_\Omega(n, \mathfrak{e}) = \sqcup_{\nu \in \mathfrak{e}_{\Omega(n)}} M_\nu(n, \mathfrak{e}) \tag{4.21}$$

of $M_\Omega(n,\mathfrak{e})$ over $\mathfrak{e}_{\Omega(n)}$ with fiber projection κ_n (see (4.9) and Proposition 4.5 below). Let $\nu \in \mathfrak{e}_{\Omega(n)}$. If $\nu \in \mathfrak{e}_{\Omega(n)}$ and $c \in \Omega(n) \subset \mathbb{C}^{d(n)}$ is defined by

$$c = \beta(\nu) \tag{4.22}$$

(see (4.17)), note that

$$M_c(n) \to M_\nu(n,\mathfrak{e}), \qquad x \mapsto (\nu, x) \tag{4.23}$$

is a homeomorphism, by the definition of $M_\Omega(n,\mathfrak{e})$. The following asserts, in particular, that the "fibers" of (4.21) are all homeomorphic.

Proposition 4.5. *One has a homeomorphism*

$$M_\nu(n,\mathfrak{e}) \cong (\mathbb{C}^\times)^{d(n-1)} \tag{4.24}$$

for any $\nu \in \mathfrak{e}_{\Omega(n)}$.

Proof. This is immediate from (4.23) and Theorem 3.23. □

Remark 4.6. Note that $M_\Omega(n,\mathfrak{e},\mathfrak{b})$ defines a cross-section of κ_n by Proposition 4.14. That is, for any $\nu \in \mathfrak{e}_{\Omega(n)}$, the intersection

$$M_\Omega(n,\mathfrak{e},\mathfrak{b}) \cap M_\nu(n,\mathfrak{e}) \tag{4.25}$$

has only one point.

Proposition 4.7. $M_\Omega(n,\mathfrak{e})$, *as a topological space (Euclidean topology), is connected.*

Proof. As a covering of the manifold of $M_\Omega(n)$, obviously $M_\Omega(n,\mathfrak{e})$ is locally connected (see Proposition 4.3) so that any connected component of $M_\Omega(n,\mathfrak{e})$ is open in $M_\Omega(n,\mathfrak{e})$. But then there exists a connected component C such that $M_\Omega(n,\mathfrak{e},\mathfrak{b}) \subset C$, by Theorem 4.4. But if $\nu \in \mathfrak{e}_{\Omega(n)}$, then $M_\nu(n,\mathfrak{e})$ is connected by (4.24). But then $M_\nu(n,\mathfrak{e}) \subset C$ by (4.25). Hence $C = M_\Omega(n,\mathfrak{e})$ by (4.21). Thus $M_\Omega(n,\mathfrak{e})$ is connected. □

4.4

The definition of variety here and throughout implies that it is Zariski irreducible. We will prove in this section that $M_\Omega(n,\mathfrak{e})$ is a nonsingular affine variety of dimension n^2. We first observe the following.

Proposition 4.8. *The nonempty Zariski open subset* $\mathfrak{e}_{\Omega(n)}$ *of* $\mathfrak{e}$ *(see Section 4.1) is a nonsingular affine variety, and the nonempty Zariski open subset* $M_\Omega(n)$ *of* $M(n)$ *is a nonsingular affine variety (see (2.53)). In particular,* $\mathfrak{e}_{\Omega(n)} \times M_\Omega(n)$ *is a nonsingular affine variety of dimension* $n^2 + d(n)$.

Proof. Recalling Section 4.1, clearly, for any $m \in I_n$, $\mathfrak{e}(m)$ is a Zariski open, nonempty subvariety of $\mathfrak{d}(m)$. It is affine since it is the complement of the zero set of the discriminant function on $\mathfrak{d}(m)$. But then $\mathfrak{e}$ is a nonsingular affine variety of dimension $d(n)$. But then $\mathfrak{e}_{\Omega(n)}$ is a nonsingular affine variety of dimension $d(n)$ since the condition (4.4) clearly defines $\mathfrak{e}_{\Omega(n)}$ as the complement of the zero set of a single regular function on $\mathfrak{e}$. But now the argument in Remark 2.16 readily characterizes $M_\Omega(n)$ as the complement in $M(n)$ of the zero set in $M(n)$ of a polynomial in $J(n)$ (see (2.30)). Thus $M_\Omega(n)$ is a nonsingular affine variety of dimension n^2. This of course proves the proposition. □

If X is an affine variety, we will denote the affine ring of X by $\mathcal{O}(X)$.

Theorem 4.9. *$M_\Omega(n, \mathfrak{e})$ is an n^2-dimensional Zariski closed affine nonsingular subvariety of the nonsingular affine variety $\mathfrak{e}_{\Omega(n)} \times M_\Omega(n)$ (see Proposition* 4.8*).*

Proof. One has

$$\mathcal{O}(\mathfrak{e}_{\Omega(n)} \times M_\Omega(n)) = \mathcal{O}(\mathfrak{e}_{\Omega(n)}) \otimes \mathcal{O}(M_\Omega(n)). \tag{4.26}$$

For $i \in I_n$, let $\beta_i \in \mathcal{O}(\mathfrak{e}_{\Omega(n)})$ be defined as in (4.15). Also let $p_i' \in \mathcal{O}(M_\Omega(n))$ be defined by putting $p_i' = p_i | M_\Omega(n)$, where $p_i \in J(n)$ is given by (2.5) (see (2.30)). For notational convenience, put $W = \mathfrak{e}_{\Omega(n)} \times M_\Omega(n)$. Now let $f_i \in \mathcal{O}(W)$ be defined by putting $f_i = \beta_i \otimes 1 - 1 \otimes p_i'$. Clearly,

$$M_\Omega(n, \mathfrak{e}) = \operatorname{Spec}[\mathcal{O}(W)/(f_1, \ldots, f_{d(n)}]. \tag{4.27}$$

But for any $z = (\nu, x) \in M_\Omega(n, \mathfrak{e})$, one has

$$(df_i)_z,\ i \in I_n, \text{ are linearly independent} \tag{4.28}$$

since $(dp_i)_x$, $i \in I_n$, are linearly independent in $T_x^*(M(n))$ by (2.55) and the definition of $M^{\mathrm{sreg}}(n)$ in Section 2.3. One also notes that $(d\beta_i)_\nu$, $i \in I_n$, are linearly independent in $T_\nu^*(\mathfrak{e})$ since $\nu(m)$ is regular in $\mathfrak{d}(m)$ for any $m \in I_n$. But then $M_\Omega(n, \mathfrak{e})$ is nonsingular at z by [M1, Chapter 3, Theorem 4, Section 4, p. 172], where X and U in the notation of that reference are equal to W here (see Proposition 4.8) and $Y = M_\Omega(n, \mathfrak{e})$. Thus $M_\Omega(n, \mathfrak{e})$ is nonsingular and has dimension n^2. But now since $M_\Omega(n, \mathfrak{e})$ is connected in the Euclidean topology, by Proposition 4.7, it is obviously connected in the Zariski topology (i.e., it is not the disjoint union of two nonempty Zariski open sets). But then, since it is nonsingular, it is irreducible as an algebraic set by [B, Corollary 17.2, p. 72]. It is then also a closed affine subvariety of W by (4.27). □

We may regard $\mathcal{O}(M_\Omega(n))$ as a module for $J(n)$ (see Section 2.4)) where p_i operates as multiplication by p_i' (using the notation in the proof of Theorem 4.9). Similarly, regard $\mathcal{O}(\mathfrak{e}_{\Omega(n)})$ as a module for $J(n)$ where p_i operates as multiplication by β_i (see (4.19)). Then Theorem 4.9 and equality (4.27) clearly imply the following.

Theorem 4.10. *As an affine ring one has the tensor product*

$$\mathcal{O}(M_\Omega(n, \mathfrak{e})) = \mathcal{O}(\mathfrak{e}_{\Omega(n)}) \otimes_{J(n)} \mathcal{O}(M_\Omega(n)) \tag{4.29}$$

4.5

Assume that Y is an affine variety and that Σ is a finite group operating as a group of algebraic isomorphisms of Y. Then Σ operates as a group of algebraic automorphisms of $\mathcal{O}(Y)$, where for any $f \in \mathcal{O}(Y)$, $\sigma \in \Sigma$ and $z \in Y$, one has

$$(\sigma \cdot f)(z) = f(\sigma^{-1} \cdot z). \tag{4.30}$$

Let Y/Σ be the set of orbits of Σ on Y. Obviously any $f \in \mathcal{O}(Y)^{\Sigma}$ defines a function of Y/Σ. It is then a classical theorem that Y/Σ has the structure of an affine variety, where

$$\mathcal{O}(Y/\Sigma) = \mathcal{O}(Y)^{\Sigma}. \tag{4.31}$$

See, e.g., [M2, Chapter 1, Section 2, Theorem 1.1, p. 27 and Amplification 1.3, p. 30]. In addition, one notes that

$$Y \to Y/\Sigma, \qquad z \mapsto \Sigma \cdot z \tag{4.32}$$

is a morphism where the corresponding cohomomorphism is the embedding

$$\mathcal{O}(Y)^{\Sigma} \to \mathcal{O}(Y). \tag{4.33}$$

Let $F(Y)$ be the quotient field of $\mathcal{O}(Y)$ and let $F(Y/\Sigma)$ be the quotient field of $\mathcal{O}(Y/\Sigma)$. Since Σ is finite, it is a simple and well-known fact that, as a consequence of (4.31),

$$F(Y/\Sigma) = F(Y)^{\Sigma} \quad \text{so that } F(Y) \text{ is a Galois extension of } F(Y/\Sigma). \tag{4.34}$$

Let $\mathrm{Clos}_{F(Y)}(\mathcal{O}(Y/\Sigma))$ be the integral closure of $\mathcal{O}(Y/\Sigma)$ in $F(Y)$.

Proposition 4.11. $\mathrm{Clos}_{F(Y)}(\mathcal{O}(Y/\Sigma))$ *is a finite module over* $\mathcal{O}(Y/\Sigma)$. *In particular* $\mathrm{Clos}_{F(Y)}(\mathcal{O}(Y/\Sigma))$ *is Noetherian. Furthermore,* $\mathcal{O}(Y)$ *is also a finite module over* $\mathcal{O}(Y/\Sigma)$ *so that* $\mathcal{O}(Y)$ *is integral over* $\mathcal{O}(Y/\Sigma)$ *and hence*

$$\mathcal{O}(Y) \subset \mathrm{Clos}_{F(Y)}(\mathcal{O}(Y/\Sigma)), \tag{4.35}$$

and one has equality in (4.35) *in case* Y *is nonsingular. Finally* (*in the sense of* [M1, *Chapter* 2, *Section* 7, *Definition* 3, *p.* 124]), *the morphism* (4.32) *is finite and the morphism*

$$Z \mapsto Y/\Sigma \tag{4.36}$$

is finite, where $Z = \mathrm{Spec}(\mathrm{Clos}_{F(Y)}(\mathcal{O}(Y/\Sigma)))$ *and* (4.36) *is defined so that the injection* $\mathcal{O}(Y/\Sigma) \to \mathrm{Clos}_{F(Y)}(\mathcal{O}(Y/\Sigma))$ *is the corresponding cohomomorphism.*

Proof. The first statement is given by [ZS, Chapter 5, Theorem 9, Section 4, p. 267]. The statement that $\mathcal{O}(Y)$ is also a finite module over $\mathcal{O}(Y/\Sigma)$ is stated as Noether's Theorem and proved as [Sm, Theorem 2.3.1, p. 26]. But now if Y is nonsingular, then $O(Y)$ is integrally closed in $F(Y)$. See, e.g., [M1, p. 197, first paragraph]. But of course $\mathrm{Clos}_{F(Y)}(\mathcal{O}(Y/\Sigma))$ is integral over $O(Y)$. Hence one has equality in (4.35). But now the finiteness of (4.32) and (4.36) follows from [M1, Chapter 2, Section 7, Proposition 5, p. 124] since Y, Y/Σ and Z are affine varieties. □

Making use of Theorem 4.9, we apply Proposition 4.11 in the case where $Y = M_\Omega(n, \mathfrak{e})$ and $\Sigma = \Sigma_n$ with, of course, the action given by (4.12). The quotient field of $\mathcal{O}(M_\Omega(n, \mathfrak{e}))$ will be denoted by $F(n, \mathfrak{e})$. The quotient field of $\mathcal{O}(M(n))$ will be denoted by $F(n)$. In the notation of Section 1.1, note that $\mathcal{O}(M(n))$ is just $P(n)$. Since $M_\Omega(n)$ is Zariski dense in $M(n)$, note that $F(n)$ is also the quotient field of $\mathcal{O}(M_\Omega(n))$.

It is clear from the definition of π_n (see (4.8)) that π_n is a morphism whose corresponding cohomomorphism maps $\mathcal{O}(M_\Omega(n))$ into $\mathcal{O}(M_\Omega(n, \mathfrak{e})^{\Sigma_n})$. Hence π_n descends to a morphism

$$M_\Omega(n, \mathfrak{e})^{\Sigma_n} \to M_\Omega(n). \tag{4.37}$$

Proposition 4.12. *The morphism* (4.37) *is an isomorphism of algebraic varieties. That is,*

$$M_\Omega(n, \mathfrak{e})^{\Sigma_n} \cong M_\Omega(n). \tag{4.38}$$

In particular, π_n is a finite morphism. Furthermore,

$$F(n, \mathfrak{e})^{\Sigma_n} \cong F(n) \tag{4.39}$$

so that, using (4.39) *to define an identification, $F(n, \mathfrak{e})$ is a Galois extension of $F(n)$ with Galois group Σ_n. In addition, using* (4.38) *to define an identification, one has*

$$\mathrm{Clos}_{F(n,\mathfrak{e})}\, \mathcal{O}(M_\Omega(n)) = \mathcal{O}(M_\Omega(n, \mathfrak{e})), \tag{4.40}$$

Proof. It is immediate from Proposition 4.3 that (4.37) is bijective. But then it is birational by [Sp, Theorem 5.1.6, p. 81] (since we are in a characteristic zero case). But $M_\Omega(n)$ is nonsingular. Thus (4.37) is an isomorphism (see, e.g., [Sp, Theorem 5.2.8, p. 85]). The rest of the statements follows from Proposition 4.11 since $M_\Omega(n, \mathfrak{e})$ is nonsingular by Theorem 4.9. □

Recall the notation of Proposition 4.3 so that $z = (\gamma, x) \in M_\Omega(n, \mathfrak{e})$. Also, U_z is a (Euclidean) open neighborhood of z in $M_\Omega(n, \mathfrak{e})$, U is a (Euclidean) open neighborhood of x in $M_\Omega(n)$, and the statements of Proposition 4.3 hold. The inverse of the homeomorphism graph $\phi : U \to U_z$ is clearly

$$\pi_n | U_z : U_z \to U. \tag{4.41}$$

Proposition 4.13. *Recalling the notation of Proposition* 4.3, *the map* (4.41) *is an analytic isomorphism. In particular* (*see* [M1, *Corollary* 2 (*p.* 182) *to Theorem* 3 *in Chapter* 3, *Section* 5]) $\pi_n : M_\Omega(n, \mathfrak{e}) \to M_\Omega(n)$ (*see* (4.8)) *is an étale morphism.*

Proof. Since π_n is a morphism, it is a holomorphic map of nonsingular analytic manifolds (see [M1, Chapter 1, Section 10, p. 58, ii]). Thus the homeomorphism (4.41) is an analytic map. It suffices to prove that

$$\mathrm{graph}\, \phi : U \to U_z \tag{4.42}$$

is analytic. To do this, first regard (4.42) as a map

$$U \to \mathfrak{e}_{\Omega(n)} \times M_\Omega(n) \tag{4.43}$$

(see (4.7)). By the definition of graph ϕ in Proposition 4.3, it is obvious that (4.43) is analytic. Hence if $g \in \mathcal{O}(\mathfrak{e}_{\Omega(n)} \times M_\Omega(n))$, then $g \circ \operatorname{graph} \phi$ is an analytic function on U. But then $f \circ \operatorname{graph} \phi$ is an analytic function on U for any $f \in \mathcal{O}(M_\Omega(n, \mathfrak{e}))$ since $\mathcal{O}(M_\Omega(n, \mathfrak{e}))$ is just the restriction of $\mathcal{O}(\mathfrak{e}_{\Omega(n)} \times M_\Omega(n))$ to $M_\Omega(n, \mathfrak{e})$. But then (4.42) is analytic since an analytic coordinate system in a Euclidean neighborhood of z in $M_\Omega(n, \mathfrak{e})$ is given by elements in $\mathcal{O}(M_\Omega(n, \mathfrak{e}))$ which are uniformizing parameters in a Zariski neighborhood of z (see [M1, Chapter 3, Section 6, p. 183]). □

Combining Propositions 4.3 and 4.13, one has the following.

Proposition 4.14. *The map π_n defines $M_\Omega(n, \mathfrak{e})$ as an analytic covering of $M_\Omega(n)$ with Σ_n as the group of deck transformations.*

4.6

Since $M_\Omega(n, \mathfrak{e})$ is locally and analytically isomorphic to $M_\Omega(n)$ (via π_n) the tensor which defines Poisson bracket of functions on $M_\Omega(n)$ lifts and defines Poisson bracket of analytic functions on $M_\Omega(n, \mathfrak{e})$. In particular $\mathcal{O}(M_\Omega(n, \mathfrak{e}))$ has the structure of a Poisson algebra. For any $f \in \mathcal{O}(M_\Omega(n))$ (noting that we regard $\mathcal{O}(M(n)) \subset \mathcal{O}(M_\Omega(n))$), let $\widehat{f} \in \mathcal{O}(M_\Omega(n, \mathfrak{e}))$ be defined by putting $\widehat{f} = f \circ \pi_n$. For $f_1, f_2 \in \mathcal{O}(M_\Omega(n))$ one then has

$$\widehat{[f_1, f_2]} = [\widehat{f_1}, \widehat{f_2}]. \tag{4.44}$$

Using the notation of Section 1.2 but now, in addition, applied to $M_\Omega(n, \mathfrak{e})$, for any $\varphi \in \mathcal{O}(M_\Omega(n, \mathfrak{e}))$, let ξ_φ be the (complex) vector field on $M_\Omega(n, \mathfrak{e})$ defined so that $\xi_\varphi \psi = [\varphi, \psi]$ for any $\psi \in \mathcal{O}(M_\Omega(n))$. It is immediate that if $f \in \mathcal{O}(M_\Omega(n)$, then $\xi_{\widehat{f}}$ is π_n-related to ξ_f so that unambiguously

$$(\pi_n)_*(\xi_{\widehat{f}}) = \xi_f. \tag{4.45}$$

Besides the (just considered) subring of $\mathcal{O}(M_\Omega(n, \mathfrak{e}))$, defined by the pullback of (the surjection) π_n, there is the subring of $\mathcal{O}(M_\Omega(n, \mathfrak{e}))$ defined by the pullback of (the surjection) κ_n (see (4.9)). Indeed, let

$$J(n, \mathfrak{e}) = \{q \circ \kappa_n \mid q \in \mathcal{O}(\mathfrak{e}_{\Omega(n)})\}. \tag{4.46}$$

From the definition of ρ_{km}, $k \in I_m$, $m \in I_m$ in Section 4.2, note that $\rho_{km} \in J(n, \mathfrak{e})$. For notational convenience let $r_i \in J(n, \mathfrak{e})$, $i \in I_{d(n)}$, be defined so that

$$r_i = \rho_{km}, \tag{4.47}$$

where

$$i = d(m-1) + k \tag{4.48}$$

so that

$$r_i \in J(n, \mathfrak{e}), \quad i \in d(n). \tag{4.49}$$

Remark 4.15. Recalling Section 4.1, note that conversely $J(n, \mathfrak{e})$ is a localization of the polynomial ring generated by the r_i, $i \in I_{d(n)}$.

Now put $\widehat{J(n)} = \{\widehat{p} \mid p \in J(n)\}$ (see Section 2.4) so that $\widehat{J(n)}$ is the polynomial ring

$$\widehat{J(n)} = \mathbb{C}[\widehat{p}_1, \ldots, \widehat{p_{d(n)}}]. \tag{4.50}$$

Let $I_{[m]} = I_{d(m)} - I_{d(m-1)}$ so that card $I_{[m]} = m$.

Proposition 4.16. *One has*

$$\widehat{J(n)} \subset J(n, \mathfrak{e}). \tag{4.51}$$

In fact, if $i \in I_{[m]}$, where $m \in I_n$, and i is written as in (4.48), then $\widehat{p_i}$ is the elementary symmetric polynomial of degree $m - k + 1$ in the functions r_j, $j \in I_{[m]}$. Indeed, if

$$P_m(\lambda) = \lambda^m + \sum_{k \in I_m} (-1)^{m-k+1} \widehat{p_{d(m-1)+k}} \lambda^{k-1},$$

then

$$P_m(\lambda) = \prod_{j \in I_{[m]}} (\lambda - r_j) \tag{4.52}$$

so that, in addition, r_j, for $j \in I_{[m]}$, satisfies the polynomial equation

$$P_m(r_j) = 0.$$

Proof. The inclusion (4.51) follows from (4.50) and (4.52). On the other hand, (4.52) follows from (2.3), (2.4), and (2.5) together with (4.10), (4.29), and (4.47). □

Since π_n is an analytic covering map (see Proposition 4.14), it follows from (2.55) and the definition of strongly regular (see Section 2.3) that the differentials $(d\widehat{p_i})_z$, $i \in I_{d(n)}$, are linearly independent at any $z \in M_\Omega(n, \mathfrak{e})$. For any $m \in I_n$, let $T_z^*(M_\Omega(n, \mathfrak{e}))^{(m)}$ be the m-dimensional subspace of the cotangent space $T_z^*(M_\Omega(n, \mathfrak{e}))$ spanned by the differentials $(d\widehat{p_i})_z$, $i \in I_{[m]}$.

Proposition 4.17. *Let $z \in M_\Omega(n, \mathfrak{e})$ and let $m \in I_n$. Then $(dr_i)_z$, $i \in I_{[m]}$, is a basis of $T_z^*(M_\Omega(n, \mathfrak{e}))^{(m)}$.*

Proof. Let V be the space of $T_z^*(M_\Omega(n, \mathfrak{e}))$ spanned by $(dr_i)_z$, $i \in I_{[m]}$. But then $\dim V \le m$. But $T_z^*(M_\Omega(n, \mathfrak{e}))^{(m)} \subset V$ by (4.51). Thus $V = T_z^*(M_\Omega(n, \mathfrak{e}))^{(m)}$ by dimension. □

In the notation above, let $T_z^*(M_\Omega(n, \mathfrak{e}))'$ (respectively, $T_z^*(M_\Omega(n, \mathfrak{e}))''$) be the $d(n-1)$-dimensional (respectively, $d(n)$-dimensional) sum of the subspaces $T_z^*(M_\Omega(n, \mathfrak{e}))^{(m)}$ over all $m \in I_{n-1}$ (respectively, $m \in I_n$). Then as an immediate consequence of Proposition 4.17, one has the following.

Proposition 4.18. *Let $z \in M_\Omega(n, \mathfrak{e})$. Then $(dr_i)_z$, $i \in I_{d(n-1)}$ (respectively, $I_{d(n)}$) is a basis of $T_z^*(M_\Omega(n, \mathfrak{e}))'$ (respectively, $T_z^*(M_\Omega(n, \mathfrak{e}))''$).*

4.7

Let $\nu \in \mathfrak{e}_{\Omega(n)}$. By definition (see (4.21)), $M_\nu(n, \mathfrak{e}) = \kappa_n^{-1}(\nu)$ so that $M_\nu(n, \mathfrak{e})$ is a Zariski closed subset of $M_\Omega(n, \mathfrak{e})$.

Remark 4.19. Note that (see (4.2), (4.3), Section 4.2, and (4.47)) $M_\nu(n, \mathfrak{e})$ may be given by the equations

$$M_\nu(n, \mathfrak{e}) = \{z \in M_\Omega(n, \mathfrak{e}) \mid r_i(z) = \nu_{km} \text{ when } i \in I_{d(n)} \text{ is put in the form (4.48)}\}. \tag{4.53}$$

Note also that $M_\nu(n, \mathfrak{e})$ is nonsingular by Proposition 4.18 (linear independence of differentials).

Proposition 4.20. *Let $\nu \in \mathfrak{e}_{\Omega(n)}$ and let $c = \beta(\nu)$ so that $c \in \Omega(n)$ (see* (4.22)*). Then the covering map π_n (see* (4.8)*) restricts to an algebraic isomorphism*

$$\pi_n : M_\nu(n, \mathfrak{e}) \to M_c(n) \tag{4.54}$$

of nonsingular affine varieties.

Proof. Recall that $M_c(n)$ is an irreducible nonsingular Zariski closed subvariety of $M(n)$ (see Theorem 3.23). The homeomorphism (4.23) can obviously be regarded as a morphism mapping $M_c(n)$ to $\mathfrak{e}_{\Omega(n)} \times M_\Omega(n)$. However the image of (4.23) is the Zariski closed subset $M_\nu(n, \mathfrak{e})$ of $\mathfrak{e}_{\Omega(n)} \times M_\Omega(n)$. Thus (4.23), as it stands, is a bijective morphism. But then $M_\nu(n, \mathfrak{e})$ is a variety (i.e., it is irreducible). Hence it is a nonsingular affine variety by Remark 4.19. Thus (4.23), as it stands, is an algebraic isomorphism. But (4.54) is just the inverse of (4.23). □

Note that (2.31) and (4.44) imply

$$[\widehat{p}, \widehat{q}] = 0 \tag{4.55}$$

for any $p, q \in J(n)$. In particular,

$$[\widehat{p_i}, \widehat{p_j}] = 0 \tag{4.56}$$

for any $i, j \in I_{d(n)}$. One consequence of Proposition 4.20 is the following.

Proposition 4.21. *Let $\nu \in \mathfrak{e}_{\Omega(n)}$ and let $z \in M_\nu(n, \mathfrak{e})$, so that $\nu = \kappa_n(z)$ (see* (4.9)*). Then $(\xi_{\widehat{p_i}})_z$, $i \in I_{d(n-1)}$, is a basis of the tangent space $T_z(M_\nu(n, \mathfrak{e}))$.*

Proof. Let $x = \pi_n(z)$ (see (4.8)) and $c = \beta(\nu)$ (see (4.22)) so that $x \in M_c(n)$. Since π_n is a local analytic isomorphism it suffices by Proposition 4.20 and (4.45) to see that $(\xi_{p_i})_x$, $i \in d(n-1)$, is a basis of $T_x(M_c(n))$. But x is strongly regular (see Section 2.3) by (2.55) since $c \in \Omega(n)$. The result then follows from Remark 2.8 and Theorems 3.4 and 3.23. □

The argument which established Theorem 3.25 may now be used to establish the following.

Theorem 4.22. *$J(n, \mathfrak{e})$ (see (4.46)) is a maximal Poisson commutative subalgebra of $\mathcal{O}(M_\Omega(n, \mathfrak{e}))$. In particular, (see (4.48)),*

$$[r_i, r_j] = 0 \tag{4.57}$$

for any $i, j \in I_{d(n)}$. Furthermore, if $f \in J(n, \mathfrak{e})$ and $\nu \in \mathfrak{e}_{\Omega(n)}$, then $f|M_\nu(n, \mathfrak{e})$ is a constant function and $\xi_f|M_\nu(n, \mathfrak{e})$ is tangent to $M_\nu(n, \mathfrak{e})$. Moreover, if we write (using Proposition 4.21)

$$\xi_f = \sum_{i \in I_{d(n-1)}} f_i \xi_{\widehat{p_i}} \tag{4.58}$$

on $M_\nu(n, \mathfrak{e})$, where $f_i \in \mathcal{O}(M_\nu(n, \mathfrak{e}))$, then all the f_i are constant on $M_\nu(n, \mathfrak{e})$. Finally,

$$\begin{aligned} &(\xi_{r_i})_z,\ i \in I_{d(n-1)}, \text{ is a basis of } T_z(M_\nu(n, \mathfrak{e})), \text{ for any } z \in M_\nu(n, \mathfrak{e}) \text{ and} \\ &(\xi_{r_i})_z = 0,\ i \in I_{[n]} = I_{d(n)} - I_{d(n-1)}, \text{ for any } z \in M_\nu(n, \mathfrak{e}). \end{aligned} \tag{4.59}$$

Proof. Let $\nu \in \mathfrak{e}_{\Omega(n)}$. The function r_i is constant on $M_\nu(n, \mathfrak{e})$ for all $i \in I_{d(n)}$ by Remark 4.19. Let $f \in J(n, \mathfrak{e})$. But then $f|M_\nu(n, \mathfrak{e})$ is a constant function by Remark 4.15. On the other hand, if $z \in M_\nu(n, \mathfrak{e})$ and W_z is the orthocomplement of $T_z(M_\nu(n, \mathfrak{e}))$ in $T^*(M_\Omega(n, \mathfrak{e}))$, then $(dr_i)_z$, $i \in I_{d(n)}$, is a basis of W_z by Proposition 4.18 and Remark 4.19. But this implies that $(df)_z \in W_z$ by Remark 4.15. In particular, $(d\widehat{p_i})_z \in W_z$ for any $i \in I_{d(n)}$ by (4.50). In fact, $(d\widehat{p_i})_z$, $i \in I_{d(n)}$, is a basis of W_z by Proposition 4.18 and the definition of $T_z^*(M_\Omega(n, \mathfrak{e}))''$ in Section 4.6.

Now for any $g \in \mathcal{O}(M_\Omega(n, \mathfrak{e}))$ the tangent vector $(\xi_g)_z$ depends only on $(dg)_z$ (see Section 1.2). But

$$(\xi_{\widehat{p_i}})_z = 0 \quad \text{for any } i \in I_{[n]} = I_{d(n)} - I_{d(n-1)} \tag{4.60}$$

by (2.7). But then $(\xi_g)_z \in T_z(M_\nu(n, \mathfrak{e}))$ if $(dg)_z \in W_z$, by Proposition 4.21. Hence $\xi_f|M_\nu(n, \mathfrak{e})$ is tangent to $M_\nu(n, \mathfrak{e})$. Furthermore, Propositions 4.17, 4.18, and 4.21 imply (4.59). Now if $g \in J(n, \mathfrak{e})$, then $g|M_\nu(n, \mathfrak{e})$ is constant. But $\xi_f|M_\nu(n, \mathfrak{e})$ is tangent to $M_\nu(n, \mathfrak{e})$. Thus $J(n, \mathfrak{e})$ is Poisson commutative. In particular, $[\xi_{\widehat{p_j}}, \xi_f] = 0$ for any $j \in I_{d(n-1)}$. But on $M_\nu(n, \mathfrak{e})$, one has

$$[\xi_{\widehat{p_j}}, \xi_f] = \sum_{i \in I_{d(n-1)}} (\xi_{\widehat{p_j}} f_i) \xi_{\widehat{p_i}}$$

by (4.58). Thus $\xi_{\widehat{p_j}} f_i = 0$, by Proposition 4.21, for all $i, j \in I_{d(n-1)}$. Hence the f_i are constants.

Now recall the definition of $M_\Omega(n, \mathfrak{e}, \mathfrak{b})$ in Section 4.3. Then $M_\Omega(n, \mathfrak{e}, \mathfrak{b})$ is a Zariski closed subset of $M_\Omega(n, \mathfrak{e})$ since, clearly, $\mathfrak{b}_{\mathfrak{e},\Omega(n)}$ is obviously closed in $M_\Omega(n)$. But $M_\Omega(n, \mathfrak{e}, \mathfrak{b})$ is irreducible since it is the image of the bijective morphism (4.20). But then (4.14) is a bijective (and hence, necessarily birational, since we are in characteristic 0) morphism of irreducible varieties. In addition $\mathfrak{e}_{\Omega(n)}$ is nonsingular (see Section 4.1). Thus (4.14) is an isomorphism of varieties. Consequently (see (4.46)), the map

$$J(n, \mathfrak{e}) \to \mathcal{O}(M_\Omega(n, \mathfrak{e}, \mathfrak{b})), \qquad g \mapsto g|M_\Omega(n, \mathfrak{e}, \mathfrak{b}) \tag{4.61}$$

is an algebra isomorphism. Consequently, given any $h \in \mathcal{O}(M_\Omega(n, \mathfrak{e}))$ there exists a unique $g \in J(n, \mathfrak{e})$ such that $g|M_\Omega(n, \mathfrak{e}, \mathfrak{b}) = h|M_\Omega(n, \mathfrak{e}, \mathfrak{b})$. But now assume that h Poisson commutes with any function in $J(n, \mathfrak{e})$. Then $h|M_\nu(n, \mathfrak{e})$ is constant, by Proposition 4.21, for any $\nu \in \mathfrak{e}_{\Omega(n)}$. But then $h = g$ on $M_\nu(n, \mathfrak{e})$ by (4.25). Thus $h = g$, by (4.21), and hence $J(n, \mathfrak{e})$ is maximally Poisson commutative in $\mathcal{O}(M_\Omega(n, \mathfrak{e}))$. □

4.8

We recall some definitions, results and notations in Part I. Generators $p_{(i)}, i \in I_{d(n)}$, of the polynomial ring $J(n)$ (see (2.30)) were defined by (3.20), recalling (2.38). In particular, two sets of generators of $J(n)$ were under consideration in Part I, namely, the $p_{(i)}$ and the p_i (see (2.5) and (2.3)), where $i \in I_{d(n)}$. From the discussion preceding (2.38), it follows that for $x \in M(n)$, and $m \in I_n$,

$$\text{span of } (dp_{(i)})_x, \text{ for } i \in I_{d(m)} = \text{span of } (dp_i)_x, \text{ for } i \in I_{d(m)} \tag{4.62}$$

and hence

$$\text{span of } (\xi_{p_{(i)}})_x, \text{ for } i \in I_{d(m)} = \text{span of } (\xi_{p_i})_x, \text{ for } i \in I_{d(m)}. \tag{4.63}$$

But then, by (4.45), for any $z \in M_\Omega(n, \mathfrak{e})$,

$$\text{span of } (\xi_{\widehat{p_{(i)}}})_z, \text{ for } i \in I_{d(m)} = \text{span of } (\xi_{\widehat{p_i}})_z, \text{ for } i \in I_{d(m)}. \tag{4.64}$$

An immediate consequence of (4.64), when $m = n - 1$, and Proposition 4.21 is the following.

Proposition 4.23. *Let $\nu \in \mathfrak{e}_{\Omega(n)}$ and let $z \in M_\nu(n, \mathfrak{e})$. Then $(\xi_{\widehat{p_{(i)}}})_z$, $i \in I_{d(n-1)}$, is a basis of $T_z(M_\nu(n, \mathfrak{e}))$.*

By definition, $\mathfrak{a}$ (see Section 3.2) is the (complex) commutative $d(n-1)$-dimensional Lie algebra of vector fields on $M(n)$ spanned by $\xi_{p_{(i)}}$, $i \in I_{d(n-1)}$. By Theorem 3.4, the Lie algebra $\mathfrak{a}$ integrates to a (complex) analytic Lie group $A \cong \mathbb{C}^{d(n-1)}$ which operates analytically on $M(n)$. Now let $\widehat{\mathfrak{a}}$ be the $d(n-1)$-dimensional complex commutative (see (4.44)) Lie algebra of vector fields on $M_\Omega(n, \mathfrak{e})$ spanned by $\xi_{\widehat{p_{(i)}}}$, $i \in I_{d(n-1)}$.

Remark 4.24. If $\nu \in \mathfrak{e}_{\Omega(n)}$, then note that by Proposition 4.23, $\widehat{\mathfrak{a}}|M_\nu(n, \mathfrak{e})$ is a commutative $d(n-1)$-dimensional Lie algebra of vector fields on $M_\nu(n, \mathfrak{e})$.

Let $\widehat{A}(\cong \mathbb{C}^{d(n-1)})$ be a simply connected Lie group with Lie algebra $\widehat{\mathfrak{a}}$. Let

$$A \to \widehat{A}, \qquad a \mapsto \widehat{a} \tag{4.65}$$

be the group isomorphism whose differential maps $\xi_{p_{(i)}}$ to $\xi_{\widehat{p_{(i)}}}$ for all $i \in I_{d(n-1)}$.

Theorem 4.25. *The Lie algebra $\widehat{\mathfrak{a}}$ integrates to an action of $\widehat{A}$ on $M_\Omega(n, \mathfrak{e})$. Furthermore, if $a \in A$ and $z \in M_\Omega(n, \mathfrak{e})$, then*

$$\pi_n(\widehat{a} \cdot z) = a \cdot x, \tag{4.66}$$

where $x = \pi_n(z)$. Moreover, $M_\nu(n, \mathfrak{e})$ is stable under the action of $\widehat{A}$ for any $\nu \in \mathfrak{e}_{\Omega(n)}$. In fact, for all ν, $\widehat{A}$ operates transitively on $M_\nu(n, \mathfrak{e})$ so that (4.21) *is the decomposition of $M_\Omega(n, \mathfrak{e})$ into $\widehat{A}$ orbits.*

Proof. Noting Remark 4.24, Theorem 4.25 is an immediate consequence of (4.45), the isomorphism (4.54), and Theorem 3.23. □

Remark 4.26. Implicit in Theorem 4.25 and its proof is the fact that if $\nu \in \mathfrak{e}_{\Omega(n)}$, then the Lie algebra $\widehat{\mathfrak{a}}|M_\nu(n, \mathfrak{e})$ of vector fields on $M_\nu(n, \mathfrak{e})$ (see Remark 4.24) integrates to the group action $\widehat{A}|M_\nu(n, \mathfrak{e})$ on $M_\nu(n, \mathfrak{e})$.

One now has an analogue of Theorem 3.5. (Actually, it is an analogue of a considerably weaker result than Theorem 3.5 in that $M_\Omega(n, \mathfrak{e})$ covers the strongly regular set $M_\Omega(n)$ and not all of $M(n)$.)

Theorem 4.27. *Let $f \in J(n, \mathfrak{e})$ (see* (4.46)*). Then the vector field ξ_f integrates to an action of $\mathbb{C}$ on $M_\Omega(n, \mathfrak{e})$. In fact, if $\nu \in \mathfrak{e}_{\Omega(n)}$, then $\xi_f|M_\nu(n, \mathfrak{e})$ is tangent to $M_\nu(n, \mathfrak{e})$. Indeed,*

$$\xi_f|M_\nu(n, \mathfrak{e}) \in \widehat{\mathfrak{a}}|M_\nu(n, \mathfrak{e}) \tag{4.67}$$

so that (see Remark 4.26*), the action of $\mathbb{C}$ stabilizes $M_\nu(n, \mathfrak{e})$.*

Proof. Clearly, $\widehat{p_{(i)}} \in J(n, \mathfrak{e})$, for $i \in I_{d(n-1)}$, by (4.50). Let $\nu \in \mathfrak{e}_{\Omega(n)}$. Then $\xi_{\widehat{p_i}}|M_\nu(n, \mathfrak{e})$, $i \in I_{d(n-1)}$, is a basis of $\widehat{\mathfrak{a}}|M_\nu(n, \mathfrak{e})$ by (the constancy of the f_i in) Theorem 4.22 and Proposition 4.23. But then one has (4.67), also by Theorem 4.22. Theorem 4.27 then follows from Remark 4.26 and Theorem 4.25. □

5 The emergence of the dual coordinates s_j, $j \in I_{d(n-1)}$

5.1

We first wish to be more explicit about the vector fields ξ_{r_j}, $j \in I_{d(n-1)}$, on $M_\Omega(n, \mathfrak{e})$. See Section 4.2 and (4.47). Fix $m \in I_n$. We have put $I_{[m]} = I_{d(m)} - I_{d(m-1)}$. For $i \in I_{[m]}$ and

$$i = d(m-1) + k \tag{5.1}$$

for $k \in I_m$, one has, on $M_\Omega(n, \mathfrak{e})$,

$$\widehat{p_{(i)}} = \frac{1}{m+1-k} \sum_{j \in I_{[m]}} r_j^{m+1-k}. \tag{5.2}$$

See (2.38), (3.20), Section 4.2, and (4.47). Thus

$$d\widehat{p_{(i)}} = \sum_{j \in I_{[m]}} r_j^{m-k} dr_j \tag{5.3}$$

and hence

$$\xi_{\widehat{p_{(i)}}} = \sum_{j \in I_{[m]}} r_j^{m-k} \xi_{r_j} \tag{5.4}$$

(see (1.14)).

Now let $z \in M_\Omega(n, \mathfrak{e})$ and let $\nu = \kappa_n(z)$ (see (4.9)) so that $z \in M_\nu(n, \mathfrak{e})$. Let $x = \pi_n(z)$ so that $x \in M_\Omega(n)$. Since the numbers (the eigenvalues of x_m) $r_j(z)$, $j \in I_{[m]}$, are distinct the Vandermonde $m \times m$ matrix

$$C_{k\ell} = r_j(z)^{m-k}, \tag{5.5}$$

where

$$j = d(m-1) + \ell \tag{5.6}$$

is invertible.

Remark 5.1. In the notation of (4.53), note that $r_j(z) = \nu_{\ell m}$.

If $m = n$, then ξ_{p_i} and $\xi_{p_{(i)}}$ vanish by (2.7) and the argument which implies (2.7). Henceforth, assume $m \in I_{n-1}$. Recalling the definition of the m-dimensional commutative Lie algebra $\mathfrak{a}(m)$ of vector fields on $M(n)$ (see Section 3.1 and (3.20)), there exists a unique basis $\eta_{j\nu}$, $j \in I_{[m]}$, of $\mathfrak{a}(m)$ such that for $i \in I_{[m]}$ and k related to i by (5.1),

$$\xi_{p_{(i)}} = \sum_{j \in I_{[m]}} r_j^{m-k}(z) \eta_{j\nu}. \tag{5.7}$$

But then (4.46), (5.4), (5.7), and the invertibilty of the Vandermonde matrix $C_{k\ell}$ imply the following.

Proposition 5.2. *Let $\nu \in \mathfrak{e}_{\Omega(n)}$ and $m \in I_{n-1}$. Then*

$$(\pi_n)_*(\xi_{r_j} | M_\nu(n, \mathfrak{e})) = \eta_{j\nu} | M_c(n), \tag{5.8}$$

where $c = \beta(\nu)$ (see (4.22)) for all $j \in I_{[m]}$.

We recall in Section 2.4 that $Z_{x,m}$ is the commutative (associative) algebra of $M(m)$ generated by x_m. Here $x = \pi_n(z)$ so that x_m is regular semisimple and hence $\dim Z_{x,m} = m$. We recall (see Section 3.1) that $G_{x,m}$ is the (algebraic) subgroup of $Gl(n)$ corresponding to $Z_{x,m}$ when $Z_{x,m}$ is regarded as a Lie subalgebra of $M(n)$. We also recall that $A(m)$ is a simply connected group corresponding to $\mathfrak{a}(m)$ and $\mathfrak{a}(m)$ integrates to an action of $A(m)$ on $M(n)$ (see Theorem 3.3). Next we recall (see (3.6)) that $\rho_{x,m}$ is the homomorphism of $A(m)$ into $G_{x,m}$ whose differential is given by

$$(\rho_{x,m})_*(\xi_{p_{(i)}}) = -(x_m)^{m-k} \tag{5.9}$$

when $i \in I_{[m]}$ has the form (5.1). But then applying $(\rho_{x,m})_*$ to (5.7), one has

$$-(x_m)^{m-k} = \sum_{j\in I_{[m]}} r_j^{m-k}(z)(\rho_{x,m})_*(\eta_{j\nu}). \tag{5.10}$$

On the other hand, put

$$h_{\nu,m} = \operatorname{diag}(\nu_{1m},\ldots,\nu_{mm}),$$

recalling (4.53), so that in the notation of Section 4.1,

$$h_{\nu,m} \in \mathfrak{e}(m). \tag{5.11}$$

Note that for the matrix units $e_{\ell\ell}$, $\ell \in I_m$, one has

$$h_{\nu,m} = \sum_{\ell\in I_m} \nu_{\ell m} e_{\ell\ell}. \tag{5.12}$$

Now let $g_z \in Gl(m)$ be such that

$$g_z h_{\nu,m} g_z^{-1} = x_m \tag{5.13}$$

(see Remark 5.1). Using the notation of Section 4.1, note that then

$$g_z \mathfrak{d}(m) g_z^{-1} = Z_{x,m}. \tag{5.14}$$

Remark 5.3. Note that g_z is unique in $Gl(m)$ modulo the maximal diagonal torus $\operatorname{Diag}(m)$ (see Section 3.4).

For $j \in I_{[m]}$, let $\varepsilon_{z,j} \in Z_{x,m}$ be the idempotent in $Z_{x,m}$ defined by putting

$$\varepsilon_{z,j} = g_z e_{\ell\ell} g_z^{-1}, \tag{5.15}$$

where ℓ is defined by (5.6). Thus (5.12) and (5.13) imply that

$$x_m = \sum_{\ell\in I_m} \nu_{\ell m} \varepsilon_{z,d(m-1)+\ell}. \tag{5.16}$$

But the $\varepsilon_{z,d(m-1)+\ell}$, $\ell \in I_m$, are orthogonal idempotents by (5.15). Hence, by Remark 5.1,

$$-(x_m)^{m-k} = -\sum_{\ell\in I_m} \nu_{\ell m}^{m-k} \varepsilon_{z,d(m-1)+\ell}. \tag{5.17}$$

Proposition 5.4. *Let $z \in M_\Omega(n,\mathfrak{e})$ and let $\nu = \kappa(z)$ (see (4.9)). Let $x = \pi_n(z)$ (see (4.8)) and let $m \in I_n$. Let $\eta_{j,\nu} \in \mathfrak{a}(m)$, $j \in I_{[m]}$, be the basis of $\mathfrak{a}(m)$ defined by (5.7) so that one has (5.8). Let $(\rho_{x,m})_* : \mathfrak{a}(m) \to Z_{x,m}$ be the Lie algebra homomorphism defined as in (3.6) and (3.7). Let $\varepsilon_{z,j}$, $j \in I_{[m]}$, be the orthogonal idempotents in $Z_{x,m}$ defined by (5.15). Then*

$$(\rho_{x,m})_*(\eta_{j,\nu}) = -\varepsilon_{z,j} \tag{5.18}$$

for any $j \in I_{[m]}$.

Proof. One has $\nu_{\ell m}^{m-k} = r_j^{m-k}(z)$ by Remark 5.1. But then (5.18) follows from the equality of the right-hand sides of (5.10) and (5.17), recalling the invertibility of the Vandermonde matrix (5.5). □

5.2

Retain the notation of Proposition 5.4. For any $\zeta \in \mathbb{C}^\times$ and $i \in I_n$, let $\delta_i(\zeta) \in \mathrm{Diag}(n)$ (see Section 3.4) be the invertible $n \times n$ diagonal matrix such that $\alpha_{jj}(\delta_i(\zeta)) = 1$ if $j \neq i$ and $\alpha_{ii}(\delta_i(\zeta)) = \zeta$, using the notation of (1.2). One notes that if $\ell \in I_m$, then $\delta_\ell(\zeta) \in \mathrm{Diag}(m)$. Also, using the relation between j and ℓ given by (5.6), one has $\gamma_{zj}(\zeta) \in G_{x,m}$, by (5.14), where we put

$$\gamma_{zj}(\zeta) = g_z \delta_\ell(\zeta) g_z^{-1}. \tag{5.19}$$

Let $q \in \mathbb{C}$. One notes that

$$\exp q e_{\ell\ell} = \delta_\ell(e^q) \tag{5.20}$$

and hence

$$\exp q \varepsilon_{z,j} = \gamma_{zj}(e^q). \tag{5.21}$$

Multiplying (5.18) by q and exponentiating (where $\exp q\eta_{j,\nu} \in A(m)$), it follows then from (3.6) that

$$\rho_{x,m}(\exp q \eta_{j,\nu}) = \gamma_{zj}(e^{-q}). \tag{5.22}$$

We can now describe the flow generated by $\xi_{\mathfrak{r}_j}$ (see (4.47) and Section 4.2) on $M_\Omega(n, \mathfrak{e})$ for any $j \in I_{d(n-1)}$ (see Theorem 4.27).

Theorem 5.5. *Let $j \in I_{d(n-1)}$ and let $m \in I_{n-1}$ be such that $j \in I_{[m]} = I_{d(m)} - I_{d(m-1)}$. Let $z \in M_\Omega(n, \mathfrak{e})$ (see (4.7)) and let $x \in M_\Omega(n)$, $\nu \in \mathfrak{e}_{\Omega(n)}$ be such that $x = \pi_n(z)$ and $\nu = \kappa_n(z)$. See (4.8) and (4.9). Let $q \in \mathbb{C}$. Then $(\nu, \mathrm{Ad}(\gamma_{zj}(e^{-q}))(x)) \in M_\nu(n, \mathfrak{e})$ (see (4.7) and (4.21)) and*

$$(\exp q \xi_{r_j}) \cdot z = (\nu, \mathrm{Ad}(\gamma_{zj}(e^{-q}))(x)), \tag{5.23}$$

where $\gamma_{zj}(e^{-q}) \in G_{x,m}$ is defined by (5.21).

Proof. Let $c = \beta(\nu)$ (see (4.22)) so that $c \in \Omega(n)$ and, by Proposition 4.20, the restriction of π_n to $M_\nu(n, \mathfrak{e})$ defines an algebraic isomorphism $M_\nu(n, \mathfrak{e}) \to M_c(n)$. But then by (5.8), one has

$$\pi_n((\exp q \xi_{r_j}) \cdot z) = (\exp q \eta_{j,\nu}) \cdot x \tag{5.24}$$

(see Theorems 3.23 and 4.27). But $(\exp q\eta_{j,\nu}) \in A(m)$ since by definition $\eta_{j,\nu} \in \mathfrak{a}(m)$. Hence, by (5.22) and Theorem 3.3, one has $(\exp q\eta_{j,\nu}) \cdot x = \mathrm{Ad}(\gamma_{zj}(e^{-q}))(x)$. But then (5.23) follows from (4.7). □

5.3

We continue with the notation of Sections 5.1 and 5.2.

Proposition 5.6. *Let $j \in I_{d(n-1)}$. Then the isomorphism $\exp q \xi_{r_j}$ of $M_\Omega(n, \mathfrak{e})$ reduces to the identity if $q \in 2\pi i \mathbb{Z}$.*

Proof. This is immediate from (5.23) since $\gamma_{zj}(e^{-q})$ is the identity matrix of $M(n)$, by (5.15) and (5.21), if $q \in 2\pi i\mathbb{Z}$. □

Let $\mathfrak{r}$ be the commutative $d(n-1)$-dimensional Lie algebra of vector fields on $M_\Omega(n,\mathfrak{e})$ with basis ξ_{r_j}, $j \in I_{d(n-1)}$ (see (4.57) and (4.59)). Let $j \in I_{d(n-1)}$ and let $A_{\mathfrak{r},j}$ be a one-dimensional complex torus with a global coordinate ζ_j defining an algebraic group isomorphism

$$\zeta_j : A_{\mathfrak{r},j} \to \mathbb{C}^\times. \tag{5.25}$$

By Proposition 5.6 (and abuse of notation), we can regard

$$\mathbb{C}\xi_{r_j} = \text{Lie } A_{\mathfrak{r},j} \tag{5.26}$$

and simultaneously have $A_{\mathfrak{r},j}$ operate on $M_\Omega(n,\mathfrak{e})$, as an integration of the vector field ξ_{r_j}, in such a fashion that if $b \in A_{\mathfrak{r},j}$, then $b = \exp q\xi_{r_j}$ if

$$\zeta_j(b) = e^q. \tag{5.27}$$

It is very easy to prove that the action of $A_{\mathfrak{r},j}$ is analytic, but what is much more important for us is to prove that this action is that of an algebraic group, operating algebraically on an affine algebraic variety.

Theorem 5.7. *Let $j \in I_{d(n-1)}$. Then the map*

$$A_{\mathfrak{r},j} \times M_\Omega(n,\mathfrak{e}) \to M_\Omega(n,\mathfrak{e}), \qquad (b,z) \mapsto b\cdot z \tag{5.28}$$

is a (algebraic) morphism.

Proof. Let $b \in A_{\mathfrak{r},j}$ and let $z \in M_\Omega(n,\mathfrak{e})$. Let $\zeta = \zeta_j(b)$ so that $\zeta \in \mathbb{C}^\times$. Recalling (4.7), write $z = (\nu, x)$, where $x \in M_\Omega(n)$ and $\nu \in \mathfrak{e}_{\Omega(n)}$. Let $m \in I_{n-1}$ be such that $j \in I_{[m]} = I_{d(m)} - I_{d(m-1)}$. Then by (5.23) and (5.27),

$$b\cdot z = (\nu, \text{Ad}(\gamma_{z,j}(\zeta^{-1}))(x)). \tag{5.29}$$

Since $\nu = \kappa_n(z)$ and $x = \pi_n(z)$ (see (4.8) and (4.9)) and both (4.8) and (4.9) are morphisms, it suffices by (5.29) to prove that the map

$$A_{\mathfrak{r},j} \times M_\Omega(n,\mathfrak{e}) \to G_{x,m}, \qquad (b,z) \mapsto \gamma_{z,j}(\zeta^{-1}) \tag{5.30}$$

is a morphism. Now $\ell \in I_m$ in (5.19) is defined so that $j = d(m-1)+\ell$. For any $g \in Gl(m)$ let $[g] \in Gl(m)/\text{Diag}(m)$ be the left coset defined by g (see Section 3.4). Of course, $Gl(m)/\text{Diag}(m)$ is an affine algebraic homogeneous space. Recalling (5.19), to prove (5.30) is a morphism, it clearly suffices to show that

$$A_{\mathfrak{r},j} \times M_\Omega(n,\mathfrak{e}) \to Gl(m)/\text{Diag}(m), \qquad (b,z) \mapsto [g_z] \tag{5.31}$$

is a morphism.

Let $E(m)$ be the set of all regular semisimple elements in $M(m)$ so that $E(m)$ has the structure of a Zariski open (and hence nonsingular) affine subvariety of $M(n)$. Then, recalling (4.1), the map

$$Gl(m) \times \mathfrak{e}(m) \to E(m), \qquad (g, \mu) \mapsto g\mu g^{-1} \tag{5.32}$$

is a surjective morphism. But now if $g \in Gl(m)$ and $\mu \in \mathfrak{e}(m)$, then $[g] \cdot \mu \in E(m)$ is well defined by putting $[g] \cdot \mu = g\mu g^{-1}$. Clearly,

$$(Gl(m)/\operatorname{Diag}(m)) \times \mathfrak{e}(m) \to E(m), \qquad ([g], \mu) \mapsto [g] \cdot \mu \tag{5.33}$$

is then also a surjective morphism. Now let

$$E(m, \mathfrak{e}) = \{(\mu, y) \in \mathfrak{e}(m) \times E(m) \mid \mu \text{ is } Gl(m)\text{-conjugate to } y\} \tag{5.34}$$

so that $E(m, \mathfrak{e})$ is a Zariski closed subset of $\mathfrak{e}(m) \times E(m)$. The argument establishing the dimension and nonsingularity in Theorem 4.9 (especially using the independence of the differentials dp_i, $i \in I_{[m]}$, at all points in $E(m)$) can obviously be modified to apply here and prove that

$$E(m, \mathfrak{e}) \text{ is a nonsingular } m^2\text{-dimensional Zariski closed subset of } \mathfrak{e}(m) \times E(m). \tag{5.35}$$

But now (5.33) may be augmented to define the map

$$(Gl(m)/\operatorname{Diag}(m)) \times \mathfrak{e}(m) \to E(m, \mathfrak{e}), \qquad ([g], \mu) \mapsto (\mu, [g] \cdot \mu). \tag{5.36}$$

But (5.33) readily implies that (5.36) is a surjectve morphism so that, for one thing, $E(m, \mathfrak{e})$ is irreducible. Hence $E(m, \mathfrak{e})$ is a nonsingular variety. But (5.36) is obviously bijective and hence birational. But then (5.36) is an algebraic isomorphism. Let

$$E(m, \mathfrak{e}) \to (Gl(m)/\operatorname{Diag}(m)) \times \mathfrak{e}(m) \tag{5.37}$$

be the inverse isomorphism. Projecting on the first factor defines a morphism $\sigma : E(m, \mathfrak{e}) \to Gl(m)/\operatorname{Diag}(m)$, where, for $g \in Gl(m)$ and $\mu \in \mathfrak{e}(m)$,

$$\sigma((\mu, [g] \cdot \mu)) = [g]. \tag{5.38}$$

But now, since (4.8) and (4.9) are morphisms, it follows that $\tau : A_{\mathfrak{r},j} \times M_{\Omega}(n, \mathfrak{e}) \to E(m, \mathfrak{e})$ is a morphism, where, using the noataion of (4.3), $\tau((b, z)) = (\nu(m), x_m)$. But

$$\sigma \circ \tau((b, z)) = [g_z] \tag{5.39}$$

by (5.13), since clearly $h_{\nu,m} = \nu(m)$. See (4.3) and (5.12). This proves that (5.31) is a morphism. $\square$

5.4

Let $m \in I_{n-1}$ and let $\mathfrak{r}(m)$ be the span of the vector fields ξ_{r_i}, $i \in I_{[m]}$, and so that, as defined in Section 5.3,

$$\mathfrak{r} = \mathfrak{r}(1) \oplus \cdots \oplus \mathfrak{r}(m-1). \tag{5.40}$$

By (4.57) and (4.59), $\mathfrak{r}(m)$ is a commutative Lie algebra of dimension m, and as we have already noted, $\mathfrak{r}$ is a commutative Lie algebra of dimension $d(n-1)$. Let (see (5.26))

$$\begin{aligned} A_{\mathfrak{r}}(m) &= A_{\mathfrak{r},d(m-1)+1} \times \cdots \times A_{\mathfrak{r},d(m-1)+m}, \\ A_{\mathfrak{r}} &= A_{\mathfrak{r}}(1) \times \cdots \times A_{\mathfrak{r}}(n-1) \end{aligned} \tag{5.41}$$

so that as algebraic groups

$$\begin{aligned} A_{\mathfrak{r}}(m) &\cong (\mathbb{C}^{\times})^{m}, \\ A_{\mathfrak{r}} &\cong (\mathbb{C}^{\times})^{d(n-1)}. \end{aligned} \tag{5.42}$$

In addition, by (5.26),

$$\begin{aligned} \mathfrak{r}(m) &= \operatorname{Lie} A_{\mathfrak{r}}(m), \\ \mathfrak{r} &= \operatorname{Lie} A_{\mathfrak{r}}. \end{aligned} \tag{5.43}$$

As an immediate consequence of Theorem 5.7 and commutativity, one has the following.

Theorem 5.8. *Let $m \in I_{n-1}$. Then the Lie algebras $\mathfrak{r}(m)$ and $\mathfrak{r}$, respectively, integrate to an algebraic action of $A_{\mathfrak{r}}(m)$ and $A_{\mathfrak{r}}$ on $M_{\Omega}(n,\mathfrak{e})$.*

The following result is a refinement of Theorem 4.25. We are now dealing with the "eigenvalue" vector fields ξ_{r_i} themselves on $M_{\Omega}(n,\mathfrak{e})$ rather than the more crude "eigenvalue symmetric function" vector fields $\xi_{\widehat{p(i)}}$.

Theorem 5.9. *Let $\nu \in \mathfrak{e}_{\Omega(n)}$. Then $M_{\nu}(n,\mathfrak{e})$ is stable under the algebraic group $A_{\mathfrak{r}}$. Furthermore $A_{\mathfrak{r}}$ operates simply and transitively on $M_{\nu}(n,\mathfrak{e})$. In particular, the disjoint union* (4.21) *is the $A_{\mathfrak{r}}$-orbit decomposition of $M_{\Omega}(n,\mathfrak{e})$.*

Proof. By Theorems 4.22 and 4.27, one has

$$\mathfrak{r}|M_{\nu}(n,\mathfrak{e}) = \widehat{\mathfrak{a}}|M_{\nu}(n,\mathfrak{e}). \tag{5.44}$$

Hence

$$A_{\mathfrak{r}}|M_{\nu}(n,\mathfrak{e}) = \widehat{A}|M_{\nu}(n,\mathfrak{e}). \tag{5.45}$$

But then $M_{\nu}(n,\mathfrak{e})$ is stable under $A_{\mathfrak{r}}$ and $A_{\mathfrak{r}}$ operates transitively on $M_{\nu}(n,\mathfrak{e})$ by Theorem 4.25. The only question concerns the simplicity of this action.

Let $b \in A_{\mathfrak{r}}$. By definition, there exist $q_j \in \mathbb{C}$, $j \in I_{d(n-1)}$, such that

$$b = \exp q_1 \xi_{r_1} \cdots \exp q_{d(n-1)} \xi_{r_{d(n-1)}}. \tag{5.46}$$

But then if $c = \beta(\nu)$ (see (4.22)), it follows from Proposition 5.2 that there exist $\eta_{j\nu} \in \mathfrak{a}$, $j \in I_{d(n-1)}$, such that if $a \in A$ is defined by putting

$$a = \exp q_1 \eta_{1,\nu} \cdots \exp q_{d(n-1)} \eta_{d(n-1),\nu}, \tag{5.47}$$

then

$$b|M_\nu(n, \mathfrak{e}) = \widehat{a}|M_\nu(n, \mathfrak{e}), \tag{5.48}$$

recalling Theorem 4.25. Also, Proposition 5.2 implies that $a = a(1) \cdots a(n-1)$, where for $m \in I_{n-1}$, $a(m) \in A(m)$ is given by

$$a(m) = \prod_{j \in I_{[m]}} \exp q_j \eta_{j,\nu}. \tag{5.49}$$

Now assume that $b|M_\nu(n, \mathfrak{e})$ has a fixed point. But then by the commutativity of $A_{\mathfrak{r}}$ and the transitivity of $A_{\mathfrak{r}}$ on $M_\nu(n, \mathfrak{e})$, it follows that $b|M_\nu(n, \mathfrak{e})$ reduces to the identity. We must prove

$$q_j \in 2\pi i \mathbb{Z}, \quad \forall j \in I_{d(n-1)} \tag{5.50}$$

by (5.27) and (5.41). But now $a|M_c(n)$ reduces to the identity by (4.66) and (5.48). See Proposition 4.20. But then $a \in D_c$ (see Section 3.64). Hence $a(m) \in D_c(m)$ for any $m \in I_{n-1}$ by Theorem 3.28. But Theorem 3.28 also asserts that if $z \in M_\nu(n, \mathfrak{e})$ and $x = \pi_n(z)$, then also $a(m) \in \operatorname{Ker} \rho_{x,m}$. But by (5.15) and (5.18) (see also (5.19) and (5.22)), this implies (5.50) since $m \in I_{n-1}$ is arbitrary. □

5.5

In the introduction, Section 0, we defined $\mathfrak{b}_e \subset M(n)$. In common parlance (for some people), $\mathfrak{b}_e$ is the space of all $n \times n$ Hessenberg matrices. In Section 4.3 we defined $\mathfrak{b}_{e,\Omega(n)}$ to be the intersection $M_\Omega(n) \cap \mathfrak{b}_e$ so that $\mathfrak{b}_{e,\Omega(n)}$ is a Zariski open subvariety of $\mathfrak{b}_e$ (and hence a nonsingular variety) and a closed subvariety of $M_\Omega(n)$. We also defined $M_\Omega(n, \mathfrak{e}, \mathfrak{b}) = \pi_n^{-1}(\mathfrak{b}_{e,\Omega(n)})$ (see (4.8) and Section 4.3) so that $M_\Omega(n, \mathfrak{e}, \mathfrak{b})$ is a Zariski closed subset of $M_\Omega(n, \mathfrak{e})$. Sharpening Theorem 4.4, we shall need the following.

Theorem 5.10. *The restriction (see* (4.9))

$$\kappa_n : M_\Omega(n, \mathfrak{e}, \mathfrak{b}) \to \mathfrak{e}_{\Omega(n)} \tag{5.51}$$

is an algebraic isomorphism so that $M_\Omega(n, \mathfrak{e}, \mathfrak{b})$ is a closed nonsingular subvariety of $M_\Omega(n, \mathfrak{e})$.

Proof. Clearly, (5.51) is a morphism since it the restriction of (4.9) to a Zariski closed subset of $M_\Omega(n, \mathfrak{e})$. On the other hand, it is bijective by Theorem 4.4. However, the inverse of (5.51) is a morphism. See (4.20). □

An easy consequence of Theorem 5.10 is the following.

Theorem 5.11. *The image of the map*

$$M_\Omega(n, \mathfrak{e}, \mathfrak{b}) \to \mathbb{C}^{d(n)}, \qquad y \mapsto (r_1(y), \ldots, r_{d(n)}(y)) \tag{5.52}$$

is a Zariski open set in $\mathbb{C}^{d(n)}$ *of the form* $(\mathbb{C}^{d(n)})_q$, *where* q *is a nonzero polynomial on* $\mathbb{C}^{d(n)}$ *and (5.52) is an algebraic isomorphism of* $M_\Omega(n, \mathfrak{e}, \mathfrak{b})$ *with* $(\mathbb{C}^{d(n)})_q$.

Proof. Recalling the definition of $\mathfrak{e}_{\Omega(n)}$ in Section 4.1, Theorem 5.11 follows immediately from Theorem 5.10 and the definition of r_i in (4.47) and ρ_{km} in Section 4.2. □

On the other hand, we establish the following product structure for $M_\Omega(n, \mathfrak{e})$.

Theorem 5.12. *The map*

$$A_\mathfrak{r} \times M_\Omega(n, \mathfrak{e}, \mathfrak{b}) \to M_\Omega(n, \mathfrak{e}), \qquad (b, y) \mapsto b \cdot y \tag{5.53}$$

is an algebraic isomorphism.

Proof. The map (5.53) is bijective by (4.21), (4.25), and Theorem 5.9. But then (5.53) is a bijective morphism of nonsingular algebraic varieties by Theorems 5.8 and 5.10. Hence (5.53) is an algebraic isomorphism. □

By definition (see (5.41)) any element $b \in A_\mathfrak{r}$ can be uniquely written

$$b = (b_1, \ldots, b_{d(n-1)}), \tag{5.54}$$

where $b_j \in A_{\mathfrak{r},j}$. We now extend the domain of the function ζ_j on $A_{\mathfrak{r},j}$ to all of $A_\mathfrak{r}$ so that, in the notation of (5.54),

$$\zeta_j(b) = \zeta_j(b_j). \tag{5.55}$$

Thus $\zeta_j \in \mathcal{O}(A_r)$ and the map

$$A_\mathfrak{r} \to (\mathbb{C}^\times)^{d(n-1)}, \qquad b \mapsto (\zeta_1(b), \ldots, \zeta_{d(n-1)}(b)) \tag{5.56}$$

is an isomorphism of algebraic groups. For $i \in d(n-1)$, let λ_{r_i} be the left invariant vector field on $A_\mathfrak{r}$ whose value at the identity of $A_\mathfrak{r}$ corresponds to (abuse of notation) ξ_{r_i}. Thus, by (5.27), in the coordinates ζ_j of $A_\mathfrak{r}$, one has

$$\lambda_{r_i} = \zeta_i \frac{\partial}{\partial \zeta_i}$$

and hence

$$\lambda_{r_i}\zeta_j = \delta_{ij}\zeta_i. \tag{5.57}$$

But now, by Theorem 5.12, every $z \in M_\Omega(n,\mathfrak{e})$ can be uniquely written $z = b \cdot y$ where $b \in A_\mathfrak{r}$ and $y \in M_\Omega(n,\mathfrak{e},\mathfrak{b})$. Hence, by Theorem 5.12, one has a well-defined function $s_j \in \mathcal{O}(M_\Omega(n,\mathfrak{e}))$, $j \in I_{d(n-1)}$, where

$$s_j(z) = \zeta_j(b^{-1}). \tag{5.58}$$

Furthermore, Theorems 5.8, 5.9, and 5.12 and (5.55) also clearly imply the following.

Theorem 5.13. *Let $\nu \in \mathfrak{e}_{\Omega(n)}$. Then the map*

$$M_\nu(n,\mathfrak{e}) \to (\mathbb{C}^\times)^{d(n-1)}, \qquad z \mapsto (s_1(z),\ldots,s_{d(n-1)}(z))$$

is an algebraic isomorphism.

The action of $A_\mathfrak{r}$ on $M_\Omega(n,\mathfrak{e})$ of course introduces, contragrediently, an action of $A_\mathfrak{r}$ on $\mathcal{O}(M_\Omega(n,\mathfrak{e}))$ so that if $b \in A_\mathfrak{r}$, $f \in \mathcal{O}(M_\Omega(n,\mathfrak{e}))$, and $z \in M_\Omega(n,\mathfrak{e})$, then $(b\cdot f)(z) = f(b^{-1}\cdot z)$. It is immediate from (5.58) that

$$b \cdot s_j = \zeta_j(b)s_j, \quad j \in I_{d(n-1)}. \tag{5.59}$$

But then with regard to Poisson bracket on $M_\Omega(n,\mathfrak{e})$, at this stage, we can say, for $i \in I_{d(n)}$ and $j \in I_{d(n-1)}$,

$$[r_i, s_j] = \delta_{ij}s_j \tag{5.60}$$

since, by differentiating (5.59) and applying (5.57), clearly,

$$\xi_{r_i}s_j = \delta_{ij}s_j, \tag{5.61}$$

recalling (4.59).

Combining Theorems 5.11, 5.12, and 5.13 and (5.61), one sees that the s_i, r_j, $i \in I_{d(n-1)}$, $j \in I_{d(n)}$, form a system of uniformizing parameters on $M_\Omega(n,\mathfrak{e})$.

Theorem 5.14.

For any $z \in M_\Omega(n,\mathfrak{e})$, the n^2 differentials $(dr_i)_z$, $(ds_j)_z$, $i \in I_{d(n)}$, $j \in I_{d(n-1)}$, are a basis of the cotangent space $T^*_z(M_\Omega(n,\mathfrak{e}))$. (5.62)

Furthermore, the image of the map

$$M_\Omega(n,\mathfrak{e}) \to \mathbb{C}^{n^2}, \qquad z \mapsto (r_1(z),\ldots,r_{d(n)}(z), s_1(z),\ldots,s_{d(n-1)}(z)) \tag{5.63}$$

is a Zariski open set Y in $\mathbb{C}^{n^2}$ and (5.63) is an algebraic isomorphism of $M_\Omega(n,\mathfrak{e})$ with Y.

Proof. Let $z \in M_\Omega(n,\mathfrak{e})$. The $(dr_i)_z$, $i \in I_{d(n)}$, are linearly independent by Proposition 4.18 (double prime statement). On the other hand, $(ds_j)_z$, $i \in I_{d(n)}$, $j \in I_{d(n-1)}$, are clearly linearly independent by (5.61). But (5.61) together with (4.57) implies that $(dr_i)_z$ are independent of the $(ds_j)_z$. By dimension this proves (5.62). The image of (5.63) is the just the product of the image of (5.52) and (5.59) by Theorem 5.12. Here we are using the constancy of the r_i on $M_\nu(n, \mathfrak{e})$ (see (4.53)). But then the image Y of (5.63) is Zariski open in $\mathbb{C}^{n^2}$ by Theorems 5.11 and 5.13. But the morphism (5.63) is bijective by Theorems 5.11, 5.12, and 5.13. Since both Y and $M_\Omega(n, \mathfrak{e})$ are nonsingular varieties it follows (a version of Zariski's Main Theorem) that (5.63) is an algebraic isomorphism. □

5.6

It is our main objective now to prove the s_i, $i \in I_{d(n-1)}$, Poisson commute among themselves. Localizing this problem, letting $z \in M_\Omega(n, \mathfrak{e})$ and $i, j \in I_{d(n-1)}$ it is enough to show that

$$[s_i, s_j](z) = 0. \tag{5.64}$$

Let $\nu_o = \kappa_n(z)$ so that $z \in M_{\nu_o}(n, \mathfrak{e})$. By (4.25) there exists a unique element z_o in $M_\Omega(n, \mathfrak{e}, \mathfrak{b}) \cap M_{\nu_o}(n, \mathfrak{e})$. Let $x_o = \pi_n(z_o)$ (see (4.8)). Recalling the definition of $M_\Omega(n, \mathfrak{e}, \mathfrak{b})$ and $\mathfrak{b}_{e,\Omega(n)}$ in Section 4.3, it follows that $x_o \in \mathfrak{b}_{e,\Omega(n)}$ and, by Proposition 4.14, π_n defines $M_\Omega(n, \mathfrak{e}, \mathfrak{b})$ as an analytic covering space of $\mathfrak{b}_{e,\Omega(n)}$ (see also Theorem 5.10). Thus there exists an open connected (in the Euclidean sense) neighborhood V_{x_o} of x_o in $\mathfrak{b}_{e,\Omega(n)}$ such that V_{x_o} is evenly covered by π_n. Let V_{z_o} be the connected component of $\pi_n^{-1}(V_{x_o})$ which contains z_o. Thus V_{z_o} is an open connected neighborhood of z_o in $M_\Omega(n, \mathfrak{e}, \mathfrak{b})$ and

$$\pi_n : V_{z_o} \to V_{x_o} \tag{5.65}$$

is an analytic isomorphism.

Now, recalling the analytic isomorphism (4.13), let $c_o = \Phi_n(x_o)$ (see (2.8)) and let V_{c_o} be the open connected neighborhood of c_o in $\Omega(n)$ defined by putting $V_{c_o} = \Phi_n(V_{x_o})$. Finally, let $V_{\nu_o} = \kappa_n(V_{z_o})$ so that (see (5.51)) V_{ν_o} is open connected neighborhood of ν_o in $\mathfrak{e}_{\Omega(n)}$.

Lemma 5.15. *The map*

$$\beta \circ \kappa_n : V_{z_o} \to V_{c_o} \tag{5.66}$$

is an analytic isomorphism (*see* (4.15) *for the definition of* β).

Proof. One readily notes that $\beta \circ \kappa_n$ restricted to V_{z_o} is the same as $\Phi_n \circ \pi_n$ restricted to V_{z_o} (a commutative diagram). But (5.65) is an analytic isomorphism and (4.13) is an analytic isomorphism. □

Let $x = \pi_n(z)$. Let $W_z = A_\mathfrak{r} \cdot V_{z_o}$ and let $W_x = \pi_n(W_z)$.

Proposition 5.16. *W_z is an open connected neighborhood of z in $M_\Omega(n, \mathfrak{e})$. Furthermore,*

$$W_z = \sqcup_{\nu \in V_{\nu_o}} M_\nu(n, \mathfrak{e}). \tag{5.67}$$

In addition, W_x is an open connected neighborhood of x in $M_\Omega(n)$ and

$$W_x = \sqcup_{c \in V_{c_o}} M_c(n) \tag{5.68}$$

(*see Section* 2.2). *Finally,*

$$\pi_n : W_z \to W_x \tag{5.69}$$

is an analytic isomorphism.

Proof. Since V_{z_o} is open and connected in $M_\Omega(n, \mathfrak{e}, \mathfrak{b})$ it follows from Theorem 5.12 that W_z is open and connected in $M_\Omega(n, \mathfrak{e})$. Furthermore, (5.67) follows from Theorems 5.9 and 5.12. Also $z \in W_z$ since $\nu_o \in V_{\nu_o}$. Since a covering map is an open map it follows that W_x is an open connected neighborhood of x in $M_\Omega(n)$. Also (5.68) follows from (5.67) and Proposition 4.20. Since (5.69) is both open and continuous, to prove that it is an analytic isomorphism it suffices to prove that it is bijective. For this, of course, one must see that it is injective. But this clearly follows from Proposition 4.20, Lemma 5.15, and the bijectivity of (5.51). □

Proposition 5.16 enables us to transport holomorphic functions on W_z to W_x. Using the notation of Section 1.2, for $i \in I_{d(n)}$, $j \in I_{d(n-1)}$, let $r_i', s_j' \in \mathcal{H}(W_x)$ be defined so that on W_z, $r_i' \circ \pi_n = r_i$ and $s_j' \circ \pi_n = s_j$. Recalling the definition of the Poisson structure on $M_\Omega(n, \mathfrak{e})$ (see Section 4.6), it follows that (5.69) is an isomorphism of Poisson manifolds. Hence, to prove (5.64), it suffices to prove

$$[s_i', s_j'](x) = 0. \tag{5.70}$$

Obviously, for $i \in I_{d(n-1)}$,

$$(\pi_n)_*(\xi_{r_i} | W_z) = \xi_{r_i'}. \tag{5.71}$$

If $\mathfrak{r}'$ is the commutative Lie algebra span of $\xi_{r_i'}$, $i \in I_{d(n-1)}$, then, recalling Proposition 4.20, Theorem 5.9, (5.67), and (5.68), $\mathfrak{r}'$ integrates to a group $A_{\mathfrak{r}'}$, which operates on W_x, and which admits an isomorphism

$$A_{\mathfrak{r}} \to A_{\mathfrak{r}'}, \qquad b \mapsto b' \tag{5.72}$$

such that for any $w \in W_z$ and $b \in A_{\mathfrak{r}}$,

$$\pi_n(b \cdot w) = b' \cdot \pi_n(w). \tag{5.73}$$

In addition, Theorems 5.9 and 5.12 imply the following.

Proposition 5.17. (5.68) *is the orbit decomposition of* $A_{\mathfrak{r}'}$ *on* W_x *and* $A_{\mathfrak{r}'}$ *operates simply* (*as well as transitively*) *on* $M_c(n)$ *for every* $c \in V_{c_o}$. *Furthermore* (*see* (5.65)),

$$W_x = \sqcup_{b \in A_r} b' \cdot V_{x_o}. \tag{5.74}$$

5.7

We retain the notation of the previous section and recall some notation from Section 1.2. If $O \subset M(n)$ is an adjoint orbit of $Gl(n)$ and $y \in O$, then $O_y = O$. If O is an orbit of regular semisimple elements, then

$$\dim O = 2d(n-1), \tag{5.75}$$

and if q_k, $k = 1, \ldots, n$, is the constant value that the $Gl(n)$-invariant (see Section 2.1) $p_{d(n-1)+k}$ takes on O, then O is clearly determined by those values. That is,

$$O = \{y \in M(n) \mid p_{d(n-1)+k}(y) = q_k\}. \tag{5.76}$$

It follows then from (2.9) that if (see Section 2.2)

$$\mathbb{C}^{d(n)}(O) = \{c \in \mathbb{C}^{d(n)} \mid c_{d(n-1)+k} = q_k,\ k \in I_n\},$$

one has

$$O = \sqcup_{c \in \mathbb{C}^{d(n)}(O)} M_c(n). \tag{5.77}$$

Let $\mathcal{S}$ be the set of all $Gl(n)$-adjoint orbits of regular semisimple elements. Since any $y \in M_\Omega(n)$ (see (2.53)) is regular semisimple, one has $O_y \in \mathcal{S}$ for any $y \in W_x$. Let

$$\mathcal{S}(W_x) = \{O \in \mathcal{S} \mid W_x \cap O \neq \emptyset\}$$

and for any $O \in \mathcal{S}(W_x)$ let $V_{c_o}(O) = \mathbb{C}^{d(n)}(O) \cap V_{c_o}$ so that

$$O \cap W_x = \sqcup_{c \in V_{c_o}(O)} M_c(n) \tag{5.78}$$

by (5.68) and (5.77).

Now, recall from Section 1.2 that any adjoint orbit has the structure of a symplectic manifold. If $O \in \mathcal{S}$, then as one knows O is closed in $M(n)$. If $O \in \mathcal{S}(W_x)$, then $O \cap W_x$ is open in O and hence $O \cap W_x$ is a symplectic manifold.

Proposition 5.18. *Let $O \in \mathcal{S}(W_x)$. Then the group $A_{\mathfrak{r}'}$ (see Proposition 5.17) stabilizes $O \cap W_x$ and operates as a group of symplectomorphisms on $O \cap W_x$.*

Proof. For any $i \in I_{d(n-1)}$, the vector field $\xi_{r_i'}|O \cap W_x$ is tangent to $O \cap W_x$ and is a Hamiltonian vector field on $O \cap W_x$ by Proposition 1.3 and especially (1.19). Thus $\mathfrak{r}'|O \cap W_x$ is a Lie algebra of Hamiltonian vector fields on $O \cap W_x$. But by (5.78) and Proposition 5.17, $O \cap W_x$ is stabilized by the integrated group $A_{\mathfrak{r}'}$. Hence $A_{\mathfrak{r}'}$ operates as a group of symplectomorphisms of $O \cap W_x$. □

Assume X is a submanifold of W_x. Let $\mathcal{S}(X) = \{O \in \mathcal{S} \mid O \cap X \neq \emptyset\}$ so that $\mathcal{S}(X) \subset \mathcal{S}(W_x)$. We will say X is Lagrangian in W_x if $\dim X = d(n)$ and $O \cap X$ is a Lagrangian submanifold of the symplectic manfold O, for any $O \in \mathcal{O}(X)$.

Proposition 5.19. *Assume X is a Lagrangian submanifold of W_x. Let $b \in A_{\mathfrak{r}}$. Then $\mathcal{S}(X) = \mathcal{S}(b' \cdot X)$ and $b' \cdot X$ is again a Lagrangian submanifold of W_x.*

Proof. Obviously, $\dim b' \cdot X = d(n)$. Let $O \in \mathcal{S}(X)$. Then, of course, $O \cap X = (O \cap W_x) \cap X$ and, by definition of being Lagrangian, obviously $(O \cap W_x) \cap X$ is Lagrangian in $O \cap W_x$. But b' operates as a symplectomorphism of $O \cap W_x$ by Proposition 5.18. Thus $b' \cdot ((O \cap W_x) \cap X) = (O \cap W_x) \cap b' \cdot X$ is again Lagrangian in $O \cap W_x$. That is, $O \in \mathcal{S}(b' \cdot X)$ and $b' \cdot (O \cap X) = O \cap b' \cdot X$ is Lagrangian in O. By using b^{-1} one readily reverses the argument to show that if $O \in \mathcal{S}(b' \cdot X)$, then $O \in \mathcal{S}(X)$ and $O \cap b' \cdot X$ is Lagrangian in O. □

The following result will be seen to be the key point in proving (5.70) and consequently (5.64).

Theorem 5.20. *Retain the notation of* (5.65) (*or* (5.74)). *Then V_{x_o} is a Lagrangian submanifold of W_x.*

Proof. We will use results in [K2]. These are stated for complex semisimple Lie groups but their extension to the reductive group $Gl(n)$ is immediate and we will apply the results for that case. Let $N \subset Gl(n)$ be the maximal unipotent subgroup where Lie N is the Lie subalgebra of all strictly upper triangular matrices. Retaining notation in [KW], we have put $\mathfrak{u} = \text{Lie}\, N$. See (3.51)). Let $\mathfrak{s}$ be defined by [K2, (1.1.5)] so that if $\mathfrak{s}_e = -e + \mathfrak{s}$, using the notation of Section 2.2, then $\mathfrak{s}_e \subset \mathfrak{b}_e$ (see Section 4.3) and, as asserted by [K2, Theorem 1.1], (a) $\mathfrak{s}_e$ is a cross-section for the adjoint action of $Gl(n)$ on the set of all regular elements in $M(n)$. On the other hand, (b) [K2, Theorem 1.2] asserts that $\mathfrak{b}_e$ is stable under Ad N and the map

$$N \times \mathfrak{s}_e \to \mathfrak{b}_e, \qquad (u, w) \mapsto \text{Ad}\, u(w) \tag{5.79}$$

is an isomorphism of affine varieties.

To prove that V_{x_o} is Lagrangian in W_x, we first observe that $\dim V_{x_o} = d(n)$. This is clear since V_{x_o} is open in $\mathfrak{b}_e$ (see Section 5.6). Now let $O \in \mathcal{S}(V_{x_o})$. It remains to show that $O \cap V_{x_o}$ is Lagrangian in O. But now, by (a) above, $O \cap \mathfrak{s}_e$ consists of a single point y_o, and hence by (b), one must have

$$O \cap \mathfrak{b}_e = \text{Ad}\, N(y_o). \tag{5.80}$$

Since O is closed in $M(n)$, this implies that Ad $N(y_o)$ is closed in $\mathfrak{b}_e$ and one has

$$O \cap V_{x_o} = \text{Ad}\, N(y_o) \cap V_{x_o} \tag{5.81}$$

since $V_{x_o} \subset \mathfrak{b}_e$. But V_{x_o} is open in $\mathfrak{b}_e$ and hence (5.81) implies that $O \cap V_{x_o}$ is open in Ad $N(y_o)$. Consequently, to prove that $O \cap V_{x_o}$ is Lagrangian in O, it suffices to prove that Ad $N(y_o)$ is Lagrangian in O. But $\dim \text{Ad}\, N(y_o) = \dim N$ by (5.79), and $\dim N = d(n-1)$, which is half the dimension of O. On the other hand, if $y \in \text{Ad}\, N(y_o)$ and $u, v \in \mathfrak{u}$ (see (3.51)), we must show that $\omega_y(\eta^u, \eta^v) = 0$ (see Section 1.2 and, more specifically, (1.10)). But $\omega_y(\eta^u, \eta^v) = B(y, [u, v])$ by (1.10). But $B(y, [u, v]) = 0$ since $y \in \mathfrak{b}_e$, and one notes that $\mathfrak{b}_e$ is B-orthogonal to $[\mathfrak{u}, \mathfrak{u}]$. □

By Theorem 5.12, one has the following disjoint union:

$$M_\Omega(n, \mathfrak{e}) = \sqcup_{b\in A_\mathfrak{r}} b \cdot M_\Omega(n, \mathfrak{e}, \mathfrak{b}). \tag{5.82}$$

On the other hand, we note that the components in (5.82) are level sets of the functions s_j, $j \in I_{d(n-1)}$. Indeed, for any $\tau \in (\mathbb{C}^\times)^{d(n-1)}$, let $\tau_i \in \mathbb{C}^\times$, $i \in I_{d(n-1)}$, be defined so that

$$\tau = (\tau_1, \ldots, \tau_{d(n-1)}) \tag{5.83}$$

and for $\tau \in (\mathbb{C}^\times)^{d(n-1)}$, let

$$M_\Omega(n, \mathfrak{e}, \tau) = \{y \in M_\Omega(n, \mathfrak{e}) \mid s_i(y) = \tau_i, \ \forall i \in I_{d(n-1)}\} \tag{5.84}$$

so that

$$M_\Omega(n, \mathfrak{e}) = \sqcup_{\tau\in(\mathbb{C}^\times)^{d(n-1)}} M_\Omega(n, \mathfrak{e}, \tau). \tag{5.85}$$

The following proposition is an immediate consequence of (5.58).

Proposition 5.21. *The partitions* (5.74) *and* (5.82) *of* $M_\Omega(n, \mathfrak{e})$ *are the same. That is, for any* $b \in A_\mathfrak{r}$,

$$b \cdot M_\Omega(n, \mathfrak{e}, \mathfrak{b}) = M_\Omega(n, \mathfrak{e}, \tau), \tag{5.86}$$

where for $i \in I_{d(n-1)}$, $\tau_i = \zeta_i(b^{-1})$.

Remark 5.22. Since V_{z_o} is open in $M_\Omega(n, \mathfrak{e}, \tau)$ (see Section 5.6), note from the definition W_z in Section 5.6 one has the disjoint union

$$W_z = \sqcup_{b\in A_\mathfrak{r}} b \cdot V_{z_o} \tag{5.87}$$

and hence by (5.86),

$$b \cdot V_{z_o} = \{w \in W_z \mid s_i(w) = \zeta_i(b^{-1}), \ \forall i \in I_{d(n-1)}\}. \tag{5.88}$$

We can now prove one of the main theorems of the paper.

Theorem 5.23. *The uniformizing parameters* r_i, s_j, $i \in I_{d(n)}$, $j \in I_{d(n-1)}$, *of* $M_\Omega(n, \mathfrak{e})$ *(see Theorem* 5.14*) satisfy the following Poisson commutation relations:*

$$\begin{aligned} &(1)\ [r_i, r_j] = 0,\ i, j \in I_{d(n)},\\ &(2)\ [r_i, s_j] = \delta_{ij} s_j,\ i \in I_{d(n)},\ j \in I_{d(n-1)},\\ &(3)\ [s_i, s_j] = 0,\ i, j \in I_{d(n-1)}. \end{aligned} \tag{5.89}$$

Proof. (1) and (2) have already been proved. See (4.57) and (5.60). We therefore have only to prove (3). We use the notation of Section 5.6, where we have to prove (5.64). But as we have observed, this comes down to proving (5.70). Recalling the isomorphism (5.69) and the definitions of s_i' and b' for $b \in A_\mathfrak{r}$ in Section 5.6, one has the disjoint union

$$W_x = \sqcup_{b\in A_\mathfrak{r}} b' \cdot V_{x_o} \tag{5.90}$$

and

$$b' \cdot V_{x_o} = \{y \in V_x \mid s_i(y) = \zeta_i(b^{-1}),\ \forall i \in I_{d(n-1)}\} \tag{5.91}$$

by (5.87) and (5.88).

But now, by (5.90), there exists $b \in A_{\mathfrak{r}}$ such that $x \in b' \cdot V_{x_o}$. For notational simplicity put $X = b' \cdot V_{x_o}$. But then X is a Lagrangian submanifold of V_x by Proposition 5.19 and Theorem 5.20. Obviously, $O_x \in \mathcal{S}(X)$ (using the notation of Proposition 5.19). Thus $O_x \cap X$ is a Lagrangian submanifold of the symplectic manifold O_x and $x \in O_x \cap X$. Let $v \in T_x(O_x \cap X)$. But since the s_i' are constant on X by (5.91) one has $vs_i' = 0$ for all $i \in I_{d(n-1)}$. On the other hand, $(\xi_{s_i'})_x \in T_x(O_x)$ by Proposition 1.3. Thus $\omega_x((\xi_{s_i'})_x, v) = 0$, by (1.17) and (1.19), for all v in the Lagrangian subspace $T_x(O_x \cap X)$ of $T_x(O_x)$. But by the isotropic maximality of $T_x(O_x \cap X)$, with respect to ω_x, one must have $(\xi_{s_i'})_x \in T_x(O_x \cap X)$. But then $(\xi_{s_i'})_x s_j' = 0$ for all $i, j \in I_{d(n-1)}$ since s_j' is constant on $O_x \cap X$. This proves (5.70). □

5.8

The Poisson bracket in $P(n)$ (see Section 1.1) or, as denoted in Section 4.4, $\mathcal{O}(M(n))$, extends in the obvious way to the quotient field, $F(n)$ (see Section 4.5). Similarly, the Poisson structure in $\mathcal{O}(M_\Omega(n, \mathfrak{e})$ extends to the function field $F(n, \mathfrak{e})$ (see Section 4.5), which, we recall, is a Galois extension of $F(n)$. See Proposition 4.12.

The Gelfand–Kirillov theorem is the statement that the quotient division ring of the universal enveloping algebra of $M(n)$ is isomorphic to the quotient division ring of a Weyl algebra over a (central) rational function field. A natural commutative analogue is the statement that $F(n)$, as a Poisson field, is isomorphic to the rational function field of a classical phase space over (a Poisson central) function field. Using the eigenvalue functions, r_i, and the commutative algebraic group $A_{\mathfrak{r}}$ we now find that the statement is explicitly true for the Galois extension $F(n, \mathfrak{e})$ of $F(n)$.

Theorem 5.24. *For $i \in I_{d(n-1)}$ one has $s_i^{-1} \in \mathcal{O}(M_\Omega(n, \mathfrak{e}))$ (see (5.58)) so that (see Section 4.2 and (4.47)) $r_{(i)} \in \mathcal{O}(M_\Omega(n, \mathfrak{e}))$, where*

$$r_{(i)} = r_i / s_i. \tag{5.92}$$

Then $F(n, \mathfrak{e})$ is the rational function field in n^2 variables,

$$F(n, \mathfrak{e}) = \mathbb{C}(r_{(1)}, \ldots, r_{(d(n-1))}, s_1, \ldots, s_{d(n-1)}, r_{d(n-1)+1}, \ldots, r_{d(n-1)+n}). \tag{5.93}$$

Furthermore, one has the Poisson commutation relations: $r_{d(n-1)+k}$ Poisson commutes with all every element in $F(n, \mathfrak{e})$ for $k \in I_n$ and, for $i, j \in I_{d(n-1)}$,

$$\begin{aligned} &(1)\ [r_{(i)}, r_{(j)}] = 0,\\ &(2)\ [r_{(i)}, s_j] = \delta_{ij},\\ &(3)\ [s_i, s_j] = 0. \end{aligned} \tag{5.94}$$

Proof. As a function field, one has

$$F(n, \mathfrak{e}) = \mathbb{C}(r_1, \ldots, r_{d(n-1)}, s_1 \ldots, s_{d(n-1)}, r_{d(n-1)+1}, \ldots, r_{d(n-1)+n}) \tag{5.95}$$

by Theorem 5.14. For $i \in I_{d(n-1)}$ one has $s_i^{-1} \in \mathcal{O}(M_\Omega(n, \mathfrak{e}))$ since s_i vanishes nowhere on $M_\Omega(n, \mathfrak{e})$ by (5.58). As a function field (5.93) follows immediately from (5.95). The commutation relations (5.94) follow easily from (5.89). □

Involved in the paper are two groups that operate on $M_\Omega(n, \mathfrak{e})$ and then contragrediently on $\mathcal{O}(M_\Omega(n, \mathfrak{e}))$. In both cases, the latter action extends to an action on the field $F(n, \mathfrak{e})$. The first group is the Galois group Σ_n (see Section 4.5 and Proposition 4.12). It is immediate from Proposition 4.14 that Σ_n preserves Poisson bracket. The second group is the $d(n-1)$-dimensional complex torus, $A_\mathfrak{r}$. Since the action of $A_\mathfrak{r}$ is algebraic (see Theorem 5.8) the group $A_\mathfrak{r}$ stabilizes $\mathcal{O}(M_\Omega(n, \mathfrak{e}))$. In fact, since the action is algebraic, as one knows, for any $f \in \mathcal{O}(M_\Omega(n, \mathfrak{e}))$,

$$A_\mathfrak{r} \cdot f \text{ spans a finite-dimensional subspace of } \mathcal{O}(M_\Omega(n, \mathfrak{e})). \tag{5.96}$$

In addition, the action of $A_\mathfrak{r}$ obviously extends to an action on the field $F(n, \mathfrak{e})$. Furthermore, this action also preserves the Poisson bracket (an easy consequence of (5.46) and (5.96)). The following theorem explicitly determines the action of the two groups. If $\alpha \in \mathbb{Z}^{d(n-1)}$ and $j \in I_{d(n-1)}$, let $\alpha_j \in \mathbb{Z}$ be such that $\alpha = (\alpha_1, \ldots, \alpha_{d(n-1)})$. For $\alpha \in \mathbb{Z}$, let ζ^α be the character on the torus $A_\mathfrak{r}$ defined by putting (see (5.55) and (5.56))

$$\zeta^\alpha = \zeta_1^{\alpha_1} \cdots \zeta_{d(n-1)}^{\alpha_{d(n-1)}}.$$

Also, let $s^\alpha \in F(n, \mathfrak{e})$ be defined by putting

$$s^\alpha = s_1^{\alpha_1} \cdots s_{d(n-1)}^{\alpha_{d(n-1)}}.$$

Theorem 5.25. *Let $m \in I_n$ and let $j \in I_{[m]} = I_{d(m)} - I_{d(m-1)}$. Then for any $\sigma \in \Sigma_n$, there exists $k \in I_{[m]}$ such that*

$$\sigma \cdot r_j = r_k. \tag{5.97}$$

Furthermore, if $m \in I_{n-1}$, then $\sigma \cdot s_j = s_k$ and $\sigma \cdot r_{(j)} = r_{(k)}$. In addition Σ_n normalizes the torus $A_\mathfrak{r}$. In fact, in the preceding notation,

$$\sigma A_{\mathfrak{r},j} \sigma^{-1} = A_{\mathfrak{r},k} \tag{5.98}$$

(see Section 5.3 *and* (5.41)*).*

Next, for any rational function $f \in \mathbb{C}(r_1, \ldots, r_{d(n)})$ and any $\alpha \in \mathbb{Z}^{d(n-1)}$, one has

$$b \cdot (f s^\alpha) = \zeta^\alpha(b) f s^\alpha \tag{5.99}$$

for any $b \in A_\mathfrak{r}$.

Proof. (5.97) follows immediately from the definition of the action of Σ_n. See (4.5), (4.6), (4.10), (4.11), the definition of ρ_{im} in Section 4.2, and (4.47). Since σ preserves the Poisson bracket structure in $M_\Omega(n, \mathfrak{e})$, it carries the vector field ξ_{r_j} to ξ_{r_k} and hence transforms the corresponding flows (where, by (4.59), we may assume $m \in I_{n-1}$) so as to yield (5.98). But σ also stabilizes $M_\Omega(n, \mathfrak{e}, \mathfrak{b})$ since, by definition (see Section 5.5), $M_\Omega(n, \mathfrak{e}, \mathfrak{b}) = \pi_n^{-1}(\mathfrak{b}_{e,\Omega(n)})$, But then $\sigma \cdot s_j = s_k$ by (5.58) and (5.98). It is, of course, then immediate that $\sigma \cdot r_{(j)} = r_{(k)}$.

Equation (5.99) follows obviously from the definition of $A_\mathfrak{r}$ (see (5.41)) and (5.59). □

Acknowledgments. The research of the first author was supported in part by NSF grant DMS-0209473 and in part by the KG&G Foundation. The research of the second author was supported in part by NSF grant MTH 0200305.

References

[B] A. Borel, *Linear Algebraic Groups*, W. A. Benjamin, New York, 1969.

[GKL1] A. Gerasimov, S. Kharchev, and D. Lebedev, Representation Theory and quantum inverse scattering method: The open Toda chain and the hyperbolic Sutherland model, *Internat. Math. Res. Notes*, **17** (2004), 823–854.

[GKL2] A. Gerasimov, S. Kharchev, and D. Lebedev, On a class of integrable systems connected with $GL(N, \mathbb{R})$, *Internat. J. Mod. Phys.* A, **19**-supplement (2004), 823–854.

[K2] B. Kostant, On Whittaker vectors and representation theory, *Invent. Math.*, **48** (1978), 101–184.

[KW] B. Kostant and N. Wallach, Gelfand-Zeitlin theory from the perspective of classical mechanics I, in J. Bernstein, V. Hinich, and A. Melnikov, eds., *Studies in Lie Theory: A. Joseph Festschrift*, Progress in Mathematics, Birkhäuser Boston, Cambridge, MA, 2005; also available online from http://www.arXiv.org/pdf/math.SG/0408342.

[M1] D. Mumford, *The Red Book of Varieties and Schemes*, Lecture Notes in Mathematics 1358, Springer-Verlag, New York, 1995.

[M2] D. Mumford and J. Fogarty, *Geometric Invariant Theory*, Ergebnisse der Mathematik und ihrer Grenzgebiete 2 Folge 34, Springer-Verlag, Berlin, 1965.

[Sm] L. Smith, *Polynomial Invariants of Finite Groups*, Research Notes in Mathematics 6, A. K. Peters, Wellesley, MA, 1995.

[Sp] T. Springer, *Linear Algebraic Groups*, 2nd ed., Progress in Mathematics 9, Birkhäuser Boston, Cambridge, MA, 1998.

[ZS] O. Zariski and P. Samuel, *Commutative Algebra*, Vol. 1, Van Nostrand, New York, 1958.

Mirror Symmetry and Localizations

Chien-Hao Liu[1], Kefeng Liu[2,3], and Shing-Tung Yau[1]

[1] Department of Mathematics
Harvard University
Cambridge, MA 02138
USA
chienliu@math.harvard.edu, yau@math.harvard.edu
[2] Department of Mathematics
University of California at Los Angeles
Los Angeles, CA 90095-1555
USA
liu@math.ucla.edu
[3] Center of Mathematical Sciences
Zhejiang University
Hangzhou
People's Republic of China
liu@cms.zju.edu.cn

Summary. We describe the applications of localization methods, in particular the functorial localization formula, in the proofs of several conjectures from string theory. The functorial localization formula pushes the computations on complicated moduli spaces to simple moduli spaces. It is a key technique in the proof of the general mirror formula, the proof of the Hori–Vafa formula for explicit expressions of basic hypergeometric series of homogeneous manifolds, and the proof of the Mariño–Vafa formula for Hodge integrals. The proposal of Strominger–Yau–Zaslow of mirror symmetry will also be discussed.

Subject Classifications: 14J32, 14D21, 14N35

1 Introduction

The main purpose of this article is to explain the applications of a variation of the localization formula of Atiyah–Bott in solving various conjectures from string theory. We call this variation the *Functorial Localization Formula*. We will also discuss the role of the SYZ proposal in mirror symmetry.

We start with a review of the Atiyah–Bott localization formula [A-B]. Recall the definition of equivariant cohomology group for a manifold X with a torus T action:

$$H_T^*(X) = H^*(X \times_T ET),$$

where ET is the universal bundle of T.

Example. We know $ES^1 = S^\infty$. If S^1 acts on $\mathbf{P}^n$ by

$$\lambda \cdot [Z_0, \ldots, Z_n] = [\lambda^{w_0} Z_0, \ldots, \lambda^{w_n} Z_n],$$

then

$$H_{S^1}(\mathbf{P}^n; \mathbb{Q}) \cong \mathbb{Q}[H, \alpha]/\langle (H - w_0\alpha) \cdots (H - w_n\alpha) \rangle,$$

where α is the generator of $H^*(BS^1, \mathbb{Q})$.

Atiyah–Bott Localization Formula. *For $\omega \in H_T^*(X)$ an equivariant cohomology class, we have*

$$\omega = \sum_E i_{E*}\left(\frac{i_E^*\omega}{e_T(E/X)}\right),$$

where E runs over all connected components of the T fixed point set.

This formula is very effective in the computations of integrals on manifolds with torus T symmetry. The idea of localization is fundamental in many subjects of geometry. In fact, Atiyah and Witten proposed to formally apply this localization formula to loop spaces and the natural S^1-action, from which one gets the Atiyah–Singer index formula. In fact, the Chern characters can be interpreted as equivariant forms on loop space, and the $\hat{A}$-class is the inverse of the equivariant Euler class of the normal bundle of X in its loop space LX:

$$e_T(X/LX)^{-1} \sim \hat{A}(X),$$

which follows from the normalized infinite product formula

$$\left(\prod_{n \neq 0}(x + n)\right)^{-1} \sim \frac{x}{\sin x}.$$

K. Liu observed in [Liu] that the normalized product

$$\prod_{m,n}(x + m + n\tau) = 2q^{\frac{1}{8}}\sin(\pi x) \cdot \prod_{j=1}^{\infty}(1 - q^j)(1 - e^{2\pi i x}q^j)(1 - e^{-2\pi i x}q^j),$$

where $q = e^{2\pi i \tau}$, also has deep geometric meaning. This formula is the Eisenstein formula. It can be viewed as a double loop space analogue of the Atiyah–Witten observation. This formula gives the basic Jacobi θ-function. As observed by K. Liu, formally this gives the $\hat{A}$-class of the loop space, and the Witten genus which is defined to be the index of the Dirac operator on the loop space:

$$e_T(X/LLX) \sim \hat{W}(X),$$

where LLX is the double loop space, the space of maps from $S^1 \times S^1$ into X. $\hat{W}(X)$ is the Witten class. See [Liu] for more details.

The variation of the localization formula to be used in various situations is the following.

Functorial Localization Formula. *Let X and Y be two manifolds with torus action. Let $f : X \to Y$ be an equivariant map. Given $F \subset Y$ a fixed component, let $E \subset f^{-1}(F)$ be those fixed components inside $f^{-1}(F)$. Let $f_0 = f|_E$; then for $\omega \in H_T^*(X)$ an equivariant cohomology class, we have the following identity on F:*

$$f_{0*}\left[\frac{i_E^*\omega}{e_T(E/X)}\right] = \frac{i_F^*(f_*\omega)}{e_T(F/Y)}.$$

This formula will be applied to various settings to prove conjectures from physics. It first appeared in [L-L-Y1, I]. It is used to push computations on complicated moduli spaces to simpler moduli spaces. A K-theory version of the functorial localization formula also holds [L-L-Y1, II]; interesting applications are expected.

Remark. Consider the diagram

$$\begin{array}{ccc} H_T^*(X) & \xrightarrow{f_*} & H_T^*(Y) \\ \downarrow i_E{}^* & & \downarrow i_F{}^* \\ H_T^*(E) & \xrightarrow{f_{0*}} & H_T^*(F). \end{array}$$

The functorial localization formula is like Riemann–Roch with the inverted equivariant Euler classes of the normal bundle as "weights," in a way similar to the Todd class for the Riemann–Roch formula. In fact, if we formally apply this formula to the map between the loop spaces of X and Y, equivariant with respect to the rotation of the circle, we do formally get the differentiable Riemann–Roch formula. We believe this can be done rigorously by following Bismut's proof of the index formula which made rigorous the above argument of Atiyah–Witten.

This formula will be used in the following setups:

(1) We call the mirror principle the proof of the mirror formulas and its generalizations. The mirror principle implies all of the conjectural formulas for toric manifolds and their Calabi–Yau submanifolds from string theory. In this case we apply the functorial localization formula to the map from the nonlinear moduli space to the linearized moduli space. This transfers the computations of integrals on complicated moduli space of stable maps to computations on rather simple spaces like projective spaces. From this, the proof of the mirror formula and its generalizations become conceptually clean and simple.

In fact, the functorial localization formula was first found and used in Lian–Liu–Yau's proof of the mirror conjecture.

(2) The proof of the Hori–Vafa conjecture and its generalizations for Grassmannians and flag manifolds. This conjecture predicts an explicit formula for the basic hypergeometric series of a homogeneous manifold in terms of the basic series of a simpler manifold such as the product of projective spaces. In this case we use the functorial localization formula twice to transfer the computations on the complicated moduli spaces of stable curves to the computations on Quot-schemes. The first is a map from moduli space of stable maps to a product of projective spaces, and the

other one is a map from the Quot-scheme into the same product of projective spaces. A key observation we had is that these two maps have the same image.

This approach was first sketched in [L-L-Y1, III]; the details for Grassmannians were carried out in [L-L-L-Y] and [B-CF-K]. The most general case of flag manifolds was carried out in [ChL-L-Y2].

(3) The proof of a remarkable conjecture of Mariño–Vafa on Hodge integrals by C.-C. M. Liu, K. Liu, and Z. Zhou [L-L-Z1]. This conjecture gives a closed formula for the generating series of a class of triple Hodge integrals for all genera and any number of marked points in terms of the Chern–Simons knot invariant of the unknot. This formula was conjectured by M. Mariño and C. Vafa in [M-V] based on the duality between large N Chern–Simons theory and string theory. Many Hodge integral identities, including the ELSV formula for Hurwitz numbers [ELSV] and the λ_g conjecture [Ge-P, Fa-P2], can be obtained by taking various limits of the Mariño–Vafa formula [L-L-Z2]. The Mariño–Vafa formula was proved by applying the functorial localization formula to the branch morphism from the moduli space of relative stable maps to a projective space.

2 Mirror principle

There have been many discussions of the mirror principle in the literature. Here we only give a brief account of the main ideas of the setup and proof of the mirror principle. We will use two very interesting examples to illustrate the algorithm.

The goal of the mirror principle is to compute the characteristic numbers on moduli spaces of stable maps in terms of certain hypergeometric type series. This was motivated by mirror symmetry in string theory. The most interesting case is the counting of the numbers of curves which corresponds to the computations of Euler numbers. More generally, we would like to compute the characteristic numbers and classes induced from the general Hirzebruch multiplicative classes such as the total Chern classes. The computations of integrals on moduli spaces of those classes pulled back through evaluation maps at the marked points and the general Gromov–Witten invariants can also be considered as part of the mirror principle. Our hope is to develop a "black-box" method which makes easy the computations of the characteristic numbers and the Gromov–Witten invariants.

The general setup of the mirror principle is as follows. Let X be a projective manifold, $\mathcal{M}_{g,k}(d, X)$ the moduli space of stable maps of genus g and degree d with k marked points into X, modulo the obvious equivalence. The points in $\mathcal{M}_{g,k}(d, X)$ are triples $(f; C; x_1, \ldots, x_k)$, where $f : C \to X$ is a degree d holomorphic map and $x_1, \ldots, x_k$ are k distinct smooth points on the genus g curve C. The homology class $f_*([C]) = d \in H_2(X, \mathbb{Z})$ is identified as an integral n-tuple $d = (d_1, \ldots, d_n)$ by choosing a basis of $H_2(X, \mathbb{Z})$, dual to the Kähler classes.

In general, the moduli space may be very singular, and may even have different, dimension for different components. To define integrals on such singular spaces, we need the virtual fundamental cycle of Li-Tian [L-T], and also Behrend–Fantechi [B-F], which we denote by $[\mathcal{M}_{g,k}(d, X)]^v$. This is a homology class of the expected dimension

$$2(c_1(TX)[d] + (\dim_{\mathbb{C}} X - 3)(1 - g) + k)$$

on $\mathcal{M}_{g,k}(d, X)$.

Let us consider the case $k = 0$ first. Note that the expected dimension of the virtual fundamental cycle is 0 if X is a Calabi–Yau 3-fold. This is the most interesting case for string theory.

The starting data of the mirror principle are as follows. Let V be a concavex bundle on X, which we defined as the direct sum of a positive and a negative bundle on X. Then V induces a sequence of vector bundles V_d^g on $\mathcal{M}_{g,0}(d, X)$ whose fiber at $(f; C; x_1, \ldots, x_k)$ is given by $H^0(C, f^*V) \oplus H^1(C, f^*V)$. Let b be a multiplicative characteristic class. So far, for all applications in string theory, b is the Euler class.

The problem of the mirror principle is to compute

$$K_d^g = \int_{[\mathcal{M}_{g,0}(d,X)]^v} b(V_d^g).$$

More precisely, we want to compute the generating series

$$F(T, \lambda) = \sum_{d,g} K_d^g \lambda^g \, e^{d \cdot T}$$

in terms of certain hypergeometric type series. Here λ, $T = (T_1, \ldots, T_n)$ are formal variables.

The most famous formula in the subject is the Candelas formula as conjectured by P. Candelas, X. de la Ossa, P. Green, and L. Parkes [CdGP]. This formula changed the history of the subject. More precisely, the Candelas formula considers the genus 0 curves, that is, we want to compute the so-called *A-model potential* of a Calabi–Yau 3-fold M given by

$$\mathcal{F}_0(T) = \sum_{d \in H_2(M;\mathbb{Z})} K_d^0 e^{d \cdot T},$$

where $T = (T_1, \ldots, T_n)$ are considered as the coordinates of the Kähler moduli of M, and K_d^0 is the genus zero, degree d invariant of M which gives the numbers of rational curves of all degree through the multiple cover formula [L-L-Y1]. The famous mirror conjecture asserts that there exists a mirror Calabi–Yau 3-fold M' with *B-model potential* $\mathcal{G}(t)$, which can be computed by period integrals, such that

$$\mathcal{F}(T) = \mathcal{G}(t),$$

where t denotes the coordinates of complex moduli of M'. The map $t \mapsto T$ is called the *mirror map*. In the toric case, the period integrals are explicit solutions to the GKZ-system, that is, the Gelfand–Kapranov–Zelevinsky hypergeometric series. While the A-series are usually very difficult to compute, the B-series are very easy to get. This is the magic of the mirror formula. We will discuss the proof of the mirror principle which includes the proof of the mirror formula.

The key ingredients for the proof of the mirror principle consist of

1. linear and nonlinear moduli spaces;
2. Euler data and hypergeometric (HG) Euler data.

More precisely, the nonlinear moduli is the moduli space $M_d^g(X)$ which is the moduli space of stable maps of degree $(1, d)$ and genus g into $\mathbf{P}^1 \times X$. A point in $M_d^g(X)$ consists of a pair (f, C), where $f : C \to \mathbf{P}^1 \times X$ with C a genus g (nodal) curve, modulo obvious equivalence. The linearized moduli W_d for toric X were first introduced by Witten and used by Aspinwall–Morrison to do approximate computations.

Example. Consider the projective space $\mathbf{P}^n$ with homogeneous coordinates $[z_0, \ldots, z_n]$. Then the linearized moduli W_d is defined as projective space with homogeneous coordinates $[f_0(w_0, w_1), \ldots, f_n(w_0, w_1)]$, where the $f_j(w_0, w_1)$ are homogeneous polynomials of degree d.

This is the simplest compactification of the moduli space of degree d maps from $\mathbf{P}^1$ into $\mathbf{P}^n$. The following lemma is important. See [L-L-Y1, IV] for its proof. The $g = 0$ case was given in [Gi] and in [L-L-Y1, I].

Lemma 1. *There exists an explicit equivariant collapsing map*

$$\varphi : M_d^g(\mathbf{P}^n) \longrightarrow W_d.$$

For a general projective manifold X, the nonlinear moduli $M_d^g(X)$ can be embedded into $M_d^g(\mathbf{P}^n)$. The nonlinear moduli $M_d^g(X)$ is very "singular" and complicated, but the linear moduli W_d is smooth and simple. The embedding induces a map of $M_d^g(X)$ to W_d. The functorial localization formula pushes the computations onto W_d. Usually, mathematical computations are done on the moduli of stable maps, while physicists have tried to use the linearized moduli to approximate the computations. So the functorial localization formula connects the computations of mathematicians and physicists. In some sense the mirror symmetry formula is more or less the comparison of computations on nonlinear and linearized moduli.

A mirror principle has been proved to hold for balloon manifolds. A projective manifold X is called a balloon manifold if it admits a torus action with isolated fixed points, and if the following conditions hold. Let

$$H = (H_1, \ldots, H_k)$$

be a basis of equivariant Kähler classes such that

1. the restrictions $H(p) \neq H(q)$ for any two fixed points $p \neq q$;
2. the tangent bundle T_pX has linearly independent weights for any fixed point p.

This notion was introduced by Goresky–Kottwitz–MacPherson.

Theorem 2. *The mirror principle holds for balloon manifolds and for any concavex bundles.*

Remarks.

1. All toric manifolds are balloon manifolds. For $g = 0$ we can identify the hypergeometric series explicitly. Higher genus cases need more work to identify such series.

2. For toric manifolds and $g = 0$, the mirror principle implies all of the mirror conjectural formulas from string theory.
3. For Grassmannian manifolds, the explicit mirror formula is given by the Hori–Vafa formula to be discussed in Section 3.
4. The case of direct sum of positive line bundles on $\mathbf{P}^n$, including the Candelas formula, has two independent approaches, by Givental and by Lian–Liu–Yau.

Now we briefly discuss the proof of the mirror principle. The main idea is to apply the functorial localization formula to φ, the collapsing map and the pullback class $\omega = \pi^* b(V_d^g)$, where $\pi : \mathcal{M}_d^g(X) \to M_{g,0}(d, X)$ is the natural projection.

Such classes satisfy a certain induction property. To be precise we introduce the notion of *Euler Data*, which naturally appears on the right-hand side of the functorial localization formula, $Q_d = \varphi_!(\pi^* b(V_d^g))$, which is a sequence of polynomials in equivariant cohomology rings of the linearized moduli spaces with simple quadratic relations. We also consider their restrictions to X.

From the functorial localization formula, we prove that by knowing the Euler data Q_d we can determine the K_d^g. On the other hand, there is another much simpler Euler data, the *HG Euler data* P_d, which coincides with Q_d on the "generic" part of the nonlinear moduli. We prove that the quadratic relations and the coincidence on the generic part determine the Euler data uniquely up to a certain degree. We also know that Q_d always has the right degree for $g = 0$. We then use the mirror transformation to reduce the degrees of the HG Euler data P_d. From these we deduce the mirror principle.

Remarks.

1. Both the denominator and the numerator in the HG series, the generating series of the HG Euler data, are equivariant Euler classes. In particular, the denominator is exactly from the localization formula. This is easily seen from the functorial localization formula.
2. The quadratic relation of Euler data, which naturally comes from gluing and functorial localization on the A-model side, is closely related to special geometry, and is similar to the Bershadsky–Cecotti–Ooguri–Vafa *holomorphic anomaly* equation on the B-model side. Such a relation can determine the polynomial Euler data up to a certain degree.
   It is an interesting task to use special geometry to understand the mirror principle computations, especially the mirror transformation as a coordinate change.
3. The Mariño–Vafa formula to be discussed in Section 4 is needed to determine the hypergeometric Euler data for higher genus computations in the mirror principle. The Mariño–Vafa formula comes from the duality between Chern–Simons theory and Gromov–Witten theory. This duality and the matrix model for Chern–Simons theory indicate that the mirror principle may have a matrix model description.

Let us use two examples to illustrate the algorithm of the mirror principle.

Example. Consider the Calabi–Yau quintic in $\mathbf{P}^4$. In this case,

$$P_d = \prod_{m=0}^{5d} (5\kappa - m\alpha)$$

where α can be considered as the weight of the S^1 action on $\mathbf{P}^1$, and κ denotes the generator of the equivariant cohomology ring of W_d.

The starting data of the mirror principle in this case is $V = \mathcal{O}(5)$ on $X = \mathbf{P}^4$. The hypergeometric series, after taking $\alpha = -1$, is given by

$$HG[B](t) = e^{Ht} \sum_{d=0}^{\infty} \frac{\prod_{m=0}^{5d}(5H+m)}{\prod_{m=1}^{d}(H+m)^5} e^{dt},$$

where H is the hyperplane class on $\mathbf{P}^4$ and t is a formal parameter.

We introduce the series

$$\mathcal{F}(T) = \frac{5}{6}T^3 + \sum_{d>0} K_d^0 e^{dT}.$$

The algorithm is as follows. Take the expansion in H:

$$HG[B](t) = H\{f_0(t) + f_1(t)H + f_2(t)H^2 + f_3(t)H^3\},$$

from which we have the famous *Candelas Formula*: With $T = f_1/f_0$,

$$\mathcal{F}(T) = \frac{5}{2}\left(\frac{f_1}{f_0}\frac{f_2}{f_0} - \frac{f_3}{f_0}\right).$$

Example. Let X be a toric manifold and $g = 0$. Let $D_1, \dots, D_N$ be the T-invariant divisors in X. The starting data consist of $V = \oplus_i L_i$ with $c_1(L_i) \geq 0$ and $c_1(X) = c_1(V)$. Let us take $b(V) = e(V)$ the Euler class. We want to compute the A-series

$$A(T) = \sum K_d^0 e^{d\cdot T}.$$

The HG Euler series which is the generating series of the HG Euler data can be easily written down as

$$B(t) = e^{-H\cdot t} \sum_d \prod_i \prod_{k=0}^{\langle c_1(L_i), d\rangle} (c_1(L_i) - k) \frac{\prod_{\langle D_a, d\rangle < 0} \prod_{k=0}^{-\langle D_a, d\rangle - 1} (D_a + k)}{\prod_{\langle D_a, d\rangle \geq 0} \prod_{k=1}^{\langle D_a, d\rangle} (D_a - k)} e^{d\cdot t}.$$

Then the mirror principle implies that there are explicitly computable functions $f(t)$, $g(t)$, which define the mirror map, such that

$$\int_X (e^f B(t) - e^{-H\cdot T} e(V)) = 2A(T) - \sum T_i \frac{\partial A(T)}{\partial T_i},$$

where $T = t + g(t)$. From this equation we can easily solve for $A(T)$.

In general, we want to compute

$$K^g_{d,k} = \int_{[\mathcal{M}_{g,k}(d,X)]^v} \prod_{j=1}^{k} ev_j^* \omega_j \cdot b(V_d^g),$$

where $\omega_j \in H^*(X)$ and ev_j denotes the evaluation map at the jth marked point. We form a generating series with t, λ and ν formal variables,

$$F(t, \lambda, \nu) = \sum_{d,g,k} K^g_{d,k} e^{dt} \lambda^{2g} \nu^k.$$

The ultimate mirror principle we want to prove is to compute this series in terms of certain explicit HG series. It is easy to show that those classes in the integrand can still be combined to induce Euler data. Actually the Euler data really encode the geometric structure of the stable map moduli.

We only use one example to illustrate the higher genus mirror principle.

Example. Consider an open toric Calabi–Yau manifold $\mathcal{O}(-3) \to \mathbf{P}^2$. Here $V = \mathcal{O}(-3)$. Let

$$Q_d = \sum_{g \geq 0} \varphi_!(\pi^* e_T(V_d^g))\lambda^{2g}.$$

Then it can be shown that the corresponding HG Euler data is given explicitly by

$$P_d J(\kappa, \alpha, \lambda) J(\kappa - d\alpha, -\alpha, \lambda),$$

where P_d is exactly the genus 0 HG Euler data and J is the generating series of Hodge integrals with summation over all genera. J may be considered as the degree 0 Euler data. In fact, we may say that the computations of Euler data include computations of all Gromov–Witten invariants, and even more. Some closed formulas can be obtained. We have proved that the mirror principle holds in such a general setting. The remaining task is to determine the explicit HG Euler data.

Finally, we mention some recent works. First, we have constructed the refined linearized moduli space for higher genus, the *A-twisted moduli stack* $\mathcal{AM}_g(X)$ of genus g curves associated to a smooth toric variety X, induced from the gauged linear sigma model studied by Witten.

This new moduli space is constructed as follows. A morphism from a curve of genus g into X corresponds to an equivalence class of triples $(L_\rho, u_\rho, c_m)_{\rho,m}$, where each L_ρ is a line bundle pulled back from X, u_ρ is a section of L_ρ satisfying a nondegeneracy condition, and the collection $\{c_m\}_m$ gives conditions to compare the sections u_ρ in different line bundles L_ρ, (cf.[Cox]). $\mathcal{AM}_g(X)$ is the moduli space of such data. It is an Artin stack, fibered over the moduli space of quasi-stable curves [ChL-L-Y1]. We hope to use this refined moduli to do computations for the higher genus mirror principle.

On the other hand, motivated by recent progress in open string theory, we are also trying to develop an open mirror principle. Open string theory predicts formulas for the counting of holomorphic discs with boundary inside a Lagrangian submanifold, more generally of the counting of the numbers of open Riemann surfaces with boundary in a Lagrangian submanifold. The linearized moduli space for such data is being constructed which gives a new compactification of such moduli spaces.

3 Hori–Vafa formula

In [H-V], Hori and Vafa generalize the world-sheet aspects of mirror symmetry to being the equivalence of $d = 2$, $N = (2, 2)$ supersymmetric field theories (i.e., without imposing the conformal invariance on the theory). This leads them to a much broader encompassing picture of mirror symmetry. See [HKKPTVVZ] for full explanations. Putting this in the framework of abelian gauged linear sigma models (GLSM) [Wi1] enables them to link many $d = 2$ field theories together. The generalization of this setting to nonabelian GLSM [Wi1, Section 5.3] leads them to the following conjecture, when the physical path integrals are interpreted appropriately mathematically.

Conjecture 3 (*Hori–Vafa* [H-V, *Appendix* A]). The hypergeometric series for a given homogeneous space (e.g., a Grassmannian manifold) can be reproduced from the hypergeometric series of simpler homogeneous spaces (e.g., a product of projective spaces). Similarly for the twisted hypergeometric series that are related to the submanifolds in homogeneous spaces.

In other words, different homogeneous spaces (or some simple quotients of them) can give rise to generalized mirror pairs.

Some progress towards this conjecture has been made for general flag manifolds by using hyper-Quot schemes in [ChL-L-Y2]. The derivation of the formula for flag manifolds is rather complicated, involving many technical new ingredients like restrictive flag manifolds. A main object to be understood in the above conjecture is the fundamental hypergeometric series $HG[\mathbf{1}]^X(t)$ associated to the flag manifold X. Recall that in the computations of the mirror principle, the existence of linearized moduli made the computations for toric manifolds easy.

An outline of how this series may be computed was given in [L-L-Y1, III] via an extended mirror principle diagram. To make clear the main ideas we will only focus on the case of Grassmannian manifolds in this article. The main problem for the computation is that there is no known good linearized moduli for Grassmannians or general flag manifolds. To overcome this difficulty we use the Grothendieck quot scheme to play the role of the linearized moduli. The method gives a complete proof of the Hori–Vafa formula in the Grassmannian case.

Let $ev : \mathcal{M}_{0,1}(d, X) \to X$ be the evaluation map on the moduli space of stable maps with one marked point, and c the first Chern class of the tangent line at the marked point. The fundamental hypergeometric series for the mirror formula is given by the push-forward

$$ev_* \left[\frac{1}{\alpha(\alpha - c)} \right] \in H^*(X)$$

or, more precisely, the generating series

$$HG[1]^X(t) = e^{-tH/\alpha} \sum_{d=0}^{\infty} ev_* \left[\frac{1}{\alpha(\alpha - c)} \right] e^{dt}.$$

Assume the linearized moduli exists. Then the functorial localization formula applied to the collapsing map: $\varphi : M_d \to N_d$, immediately gives the expression as the denominator of the hypergeometric series.

Example. $X = \mathbf{P}^n$, then we have $\varphi_*(1) = 1$, and functorial localization immediately gives us

$$ev_*\left[\frac{1}{\alpha(\alpha - c)}\right] = \frac{1}{\prod_{m=1}^{d}(x - m\alpha)^{n+1}},$$

where the denominators of both sides are equivariant Euler classes of normal bundles of the fixed points. Here x denotes the hyperplane class.

For $X = \mathrm{Gr}(k, n)$ or general flag manifolds, no explicit linearized moduli is known. Hori–Vafa conjectured a formula for $HG[1]^X(t)$ by which we can compute this series in terms of those of projective spaces.

Hori–Vafa Formula for Grassmannians. *We have*

$$\begin{aligned}&HG[1]^{\mathrm{Gr}(k,n)}(t)\\&= \frac{e^{(k-1)\pi\sqrt{-1}\sigma/\alpha}}{\prod_{i<j}(x_i - x_j)} \cdot \prod_{i<j}\left(\alpha\frac{\partial}{\partial x_i} - \alpha\frac{\partial}{\partial x_j}\right)\Bigg|_{t_i = t + (k-1)\pi\sqrt{-1}} HG[1]^{\mathbf{P}}(t_1, \ldots, t_k),\end{aligned}$$

where $\mathbf{P} = \mathbf{P}^{n-1} \times \cdots \times \mathbf{P}^{n-1}$ *is the product of k copies of projective space, σ is the generator of the divisor classes on* $\mathrm{Gr}(k, n)$ *and x_i the hyperplane class of the ith copy of* $\mathbf{P}^{n-1}$:

$$HG[1]^{\mathbf{P}}(t_1, \ldots, t_k) = \prod_{i=1}^{k} HG[1]^{\mathbf{P}^{n-1}}(t_i).$$

Now we describe the ideas of the proof of the above formula. As mentioned above we use another smooth moduli space, the Grothendieck Quot-scheme Q_d to play the role of the linearized moduli, and apply the functorial localization formula. Here is the general setup:

To start, note that the Plücker embedding $\tau : \mathrm{Gr}(k, n) \to \mathbf{P}^N$ induces an embedding of the nonlinear moduli M_d of $\mathrm{Gr}(k, n)$ into that of $\mathbf{P}^N$. The composition of this map with the collapsing map gives us a map $\varphi : M_d \to W_d$ into the linearized moduli space W_d of $\mathbf{P}^N$. On the other hand, the Plücker embedding also induces a map $\psi : Q_d \to W_d$. We have the following three crucial lemmas proved in [L-L-L-Y].

Lemma 4. *The above two maps have the same image in W_d:* $\mathrm{Im}\,\psi = \mathrm{Im}\,\varphi$. *And all the maps are equivariant with respect to the induced circle action from* $\mathbf{P}^1$.

Just as in the mirror principle computations, our next step is to analyze the fixed points of the circle action induced from $\mathbf{P}^1$. In particular we need the distinguished fixed point set to get the equivariant Euler class of its normal bundle. The distinguished fixed point set in M_d is $\mathcal{M}_{0,1}(d, \mathrm{Gr}(k, n))$ with the equivariant Euler class of its normal bundle given by $\alpha(\alpha - c)$, and we know that φ restricts to ev.

Lemma 5. *The distinguished fixed point set in Q_d is a union: $\cup_s E_{0s}$, where each E_{0s} is a fiber bundle over* $\mathrm{Gr}(k, n)$ *with fiber given by a flag manifold.*

It is a complicated endeavor to determine the fixed point sets E_{0s} and the weights of the circle action on their normal bundles. The situation for the flag manifold case is much more involved. See [L-L-L-Y] and [ChL-L-Y2] for details.

Now let p denote the projection from E_{0s} onto $\mathrm{Gr}(k, n)$. The functorial localization formula, applied to φ and ψ, gives us the following.

Lemma 6. *We have the equality on* $\mathrm{Gr}(k, N)$*:*

$$ev_* \left[\frac{1}{\alpha(\alpha - c)}\right] = \sum_s p_* \left[\frac{1}{e_T(E_{0s}/Q_d)}\right],$$

where $e_T(E_{0s}/Q_d)$ is the equivariant Euler class of the normal bundle of E_{0s} in Q_d.

Finally, we compute $p_*[\frac{1}{e_T(E_{0s}/Q_d)}]$. There are two different approaches; the first one is by direct computations in [L-L-L-Y], and the other one is by using the well-known Euler sequences for universal sheaves [B-CF-K]. The second method has the advantage of being more explicit. Note that

$$e_T(TQ|_{E_{0s}} - TE_{0s}) = e_T(TQ|_{E_{0s}})/e_T(TE_{0s}).$$

Both $e_T(TQ|_{E_{0s}})$ and $e_T(TE_{0s})$ can be written down explicitly in terms of the universal bundles on the flag bundle $E_{0s} = \mathrm{Fl}(m_1, \dots, m_k, S)$ over $\mathrm{Gr}(r, n)$. Here S is the universal bundle on the Grassmannian.

The push-forward by p from $\mathrm{Fl}(m_1, \dots, m_k, S)$ to $\mathrm{Gr}(r, n)$ is done by an analogue of the family localization formula of Atiyah–Bott, which is given by a sum over the Weyl groups along the fiber which labels the fixed point sets.

In any case, the final formula of degree d is given by

$$p_* \left[\frac{1}{e_T(E_{0s}/Q_d)}\right] = (-1)^{(r-1)d} \sum_{\substack{(d_1,\dots,d_r) \\ d_1+\cdots+d_r=d}} \frac{\prod_{1\le i<j\le r}(x_i - x_j + (d_i - d_j)\alpha)}{\prod_{1\le i<j\le r}(x_i - x_j) \prod_{i=1}^{r} \prod_{l=1}^{d_i} (x_i + l\alpha)^n}.$$

Here $x_1, \dots, x_r$ are the Chern roots of S^*. As a corollary of our approach, we have the following.

Corollary 7. *The Hori–Vafa conjecture holds for Grassmann manifolds.*

This corollary was derived in [B-CF-K] by using the idea and method and also the key results in [L-L-L-Y]. The explicit form of the Hori–Vafa conjecture for general flag manifolds and its justifications require further study in the future.

4 Mariño–Vafa formula

To compute the mirror formula for higher genus, we need to compute Hodge integrals, which are defined as follows. Let $\mathcal{M}_{g,h}$ be the moduli space of stable curves of genus g with h marked points. The Hodge bundle $\mathbb{E}$ is the rank g vector bundle over $\mathcal{M}_{g,h}$ whose fiber over $[(C, x_1, \ldots, x_h)] \in \mathcal{M}_{g,h}$ is $H^0(C, \omega_C)$. The λ classes are defined by

$$\lambda_j = c_j(\mathbb{E}) \in H^{2i}(\mathcal{M}_{g,h}; \mathbb{Q}).$$

The cotangent line $T^*_{x_i}C$ of C at the ith marked point x_i gives a line bundle $\mathbb{L}_i$ over $\mathcal{M}_{g,h}$. The ψ classes are defined by

$$\psi_i = c_1(\mathbb{L}_i) \in H^2(\mathcal{M}_{g,h}; \mathbb{Q}).$$

Hodge integrals are intersection numbers of λ classes and ψ classes.

We next introduce a particular form of Hodge integrals. Given a partition

$$\mu = (\mu_1 \geq \cdots \geq \mu_h > 0),$$

define $\ell(\mu) = h$, and $|\mu| = \mu_1 + \cdots + \mu_h$. Given a triple (g, μ, τ), where g is a nonnegative integer, μ is a partition, and $\tau \in \mathbb{Z}$, we define the *one-partition Hodge integral* as follows:

$$G_{g,\mu}(\tau) = \frac{-\sqrt{-1}^{|\mu|+\ell(\mu)}}{|\operatorname{Aut}(\mu)|}(\tau(\tau+1))^{\ell(\mu)-1} \prod_{i=1}^{\ell(\mu)} \frac{\prod_{a=1}^{\mu_i-1}(\mu_i\tau + a)}{(\mu_i - 1)!}$$
$$\cdot \int_{\mathcal{M}_{g,\ell(\mu)}} \frac{\Lambda_g^\vee(1)\Lambda_g^\vee(-\tau-1)\Lambda_g^\vee(\tau)}{\prod_{i=1}^{\ell(\mu)}(1-\mu_i\psi_i)},$$

where

$$\Lambda_g^\vee(u) = u^g - \lambda_1 u^{g-1} + \cdots + (-1)^g \lambda_g.$$

The one-partition Hodge integral can be simplified in special cases:

- $g = 0$: $\Lambda_0^\vee(u) = 1$.

$$\begin{aligned}
\int_{\mathcal{M}_{0,h}} &\frac{1}{\prod_{i=1}^h (1-\mu_i\psi_i)} \\
&= \sum_{k_1+\cdots+k_h=h-3} \mu_1^{k_1}\cdots\mu_h^{k_h} \int_{\mathcal{M}_{0,h}} \psi_1^{k_1}\cdots\psi_h^{k_h} \\
&= \sum_{k_1+\cdots+k_h=h-3} \mu_1^{k_1}\cdots\mu_h^{k_h} \frac{h!}{k_1!\cdots k_h!} \\
&= |\mu|^{h-3}.
\end{aligned}$$

- $\tau = 0$: $G_{g,\mu}(0) = 0$ if $\ell(\mu) > 1$, and

$$G_{g,(d)}(0) = \sqrt{-1}^{d+1} \int_{\mathcal{M}_{g,1}} \frac{\lambda_g}{1 - d\psi} = \sqrt{-1}^{d+1} d^{2g-2} b_g,$$

where

$$b_g = \begin{cases} 1, & g = 0, \\ \int_{\mathcal{M}_{g,1}} \lambda_g \psi^{2g-2}, & g > 0. \end{cases}$$

To state Mariño–Vafa's conjecture on one-partition Hodge integrals, we introduce some generating functions.

We first define generating functions of one-partition Hodge integrals. Introduce variables λ and $p = (p_1, p_2, \ldots)$. Given a partition μ, let

$$p_\mu = p_{\mu_1} \cdots p_{\mu_{\ell(\mu)}}.$$

Define generating functions to be

$$\begin{aligned} G_\mu(\lambda; \tau) &= \sum_{g=0}^{\infty} \lambda^{2g-2+\ell(\mu)} G_{g,\mu}(\tau), \\ G(\lambda; \tau; p) &= \sum_\mu G_\mu(\lambda; \tau) p_\mu, \\ G^\bullet(\lambda; \tau; p) &= \exp(G(\lambda; \tau; p)) = \sum_\mu G^\bullet_\mu(\lambda; \tau) p_\mu. \end{aligned}$$

We next define generating functions of symmetric group representations. Let χ_μ denote the character of the irreducible representation of the symmetric group $S_{|\mu|}$ indexed by μ with $|\mu| = \sum_j \mu_j$, and let C_μ denote the conjugacy class of $S_{|\mu|}$ indexed by μ. Introduce

$$\begin{aligned} V_\mu(\lambda) = &\prod_{1 \le a < b \le \ell(\mu)} \frac{\sin[(\mu_a - \mu_b + b - a)\lambda/2]}{\sin[(b-a)\lambda/2]} \\ &\cdot \frac{1}{\prod_{i=1}^{\ell(v)} \prod_{v=1}^{\mu_i} 2 \sin[(v - i + \ell(\mu))\lambda/2]}, \end{aligned}$$

which has an interpretation in terms of *quantum dimension* in Chern–Simons knot theory. Define

$$\begin{aligned} R^\bullet_\mu(\lambda; \tau) &= \sum_{|\nu| = |\mu|} \frac{\chi_\nu(C_\mu)}{z_\mu} e^{\sqrt{-1}(\tau + \frac{1}{2})\kappa_\nu \lambda/2} V_\nu(\lambda), \qquad (1) \\ R^\bullet(\lambda; \tau; p) &= \sum_\mu R^\bullet_\mu(\lambda; \tau) p_\mu, \end{aligned}$$

where

$$z_\mu = |\operatorname{Aut}(\mu)|\mu_1 \cdots \mu_{\ell(\mu)}, \qquad \kappa_\mu = |\mu| + \sum_i (\mu_i^2 - 2i\mu_i).$$

Define

$$R(\lambda; \tau; p) = \log(R^\bullet(\lambda; \tau; p)).$$

Conjecture 8 (*Mariño–Vafa* [M-V]).

$$G(\lambda; \tau; p) = R(\lambda; \tau; p). \tag{2}$$

The Mariño–Vafa formula (2) provides a highly nontrivial link between geometry (Hodge integrals) and combinatorics (representations of symmetric groups). Note that for each fixed partition μ, the Mariño–Vafa formula gives a closed and finite formula for $G_\mu(\lambda; \tau)$, a generating function for all genera.

We now outline the proof of the Mariño–Vafa formula due to C.-C. M Liu, K. Liu, and J. Zhou [L-L-Z1]. There is another approach due to A. Okounkov and R. Pandharipande [O-P].

At $\tau = 0$, both sides of the Mariño–Vafa formula can be greatly simplified:

$$G(\lambda; 0; p) = -\sum_{d=1}^{\infty} \frac{\sqrt{-1}^{d+1} p_d}{\lambda d^2} \sum_{g=0}^{\infty} b_g (\lambda d)^{2g}, \tag{3}$$

$$R(\lambda; 0; p) = -\sum_{d=1}^{\infty} \frac{-\sqrt{-1}^{d+1} p_d}{2d \sin(\lambda d/2)} \tag{4}$$

They are equal by a previous result [Fa-P1]:

$$\sum_{g=0}^{\infty} b_g t^{2g} = \frac{t/2}{\sin(t/2)}. \tag{5}$$

Note that both sides of the Mariño–Vafa formula (2) are valid for $\tau \in \mathbb{C}$. It follows from the expression (1) that

$$R^\bullet_\mu(\lambda; \tau) = \sum_{|\nu|=|\mu|} R^\bullet_\nu(\lambda; 0) z_\nu \Phi^\bullet_{\nu\mu}(\sqrt{-1}\lambda\tau), \tag{6}$$

where

$$\Phi^\bullet_{\nu\mu}(\lambda) = \sum_\chi \lambda^{-\chi+\ell(\nu)+\ell(\mu)} \frac{H^\bullet_{\chi,\nu,\mu}}{(-\chi+\ell(\nu)+\ell(\mu))!} = \sum_\eta \frac{\chi_\eta(C_\nu)}{z_\nu} \frac{\chi_\eta(C_\mu)}{z_\mu} e^{\kappa_\eta \lambda/2}$$

is the generating function of disconnected double Hurwitz numbers $H^\bullet_{\nu,\mu,\chi}$. The convolution equation (6) is equivalent to the following *cut-and-join equation*:

$$\frac{\partial R}{\partial \tau} = \frac{\sqrt{-1}\lambda}{2} \sum_{i,j=1}^{\infty} \left((i+j) p_i p_j \frac{\partial R}{\partial p_{i+j}} + ij p_{i+j} \left(\frac{\partial R}{\partial p_i} \frac{\partial R}{\partial p_j} + \frac{\partial^2 R}{\partial p_i \partial p_j} \right) \right). \tag{7}$$

In the symmetric group S_d, a transposition can *cut* an $(i+j)$-cycle into an i-cycle and an j-cycle:

$$(s,t)(s,s_2,\dots,s_i,t,t_2,\dots,t_j)=(s,s_2,\dots,s_i)(t,t_2,\dots t_j).$$

This corresponds to the *cut* operator

$$(i+j)p_ip_j\frac{\partial}{\partial p_{i+j}}.$$

A transposition can also *join* an i-cycle and a j-cycle to form an $(i+j)$-cycle:

$$(s,t)(s,s_2,\dots,s_i)(t,t_2,\dots t_j)=(s,s_2,\dots,s_i,t,t_2,\dots t_j).$$

This corresponds to the *join* operator

$$ijp_{i+j}\frac{\partial}{\partial p_i}\frac{\partial}{\partial p_j}.$$

The Mariño–Vafa formula will follow from the initial values (3), (4), (5), the cut-and-join equation (7) of $R(\lambda;\tau;p)$, and the following cut-and-join equation of $G(\lambda;\tau;p)$.

Theorem 9 (Liu–Liu–Zhou [L-L-Z1]).

$$\frac{\partial G}{\partial\tau}=\frac{\sqrt{-1}\lambda}{2}\sum_{i,j=1}^{\infty}\left((i+j)p_ip_j\frac{\partial G}{\partial p_{i+j}}+ijp_{i+j}\left(\frac{\partial G}{\partial p_i}\frac{\partial G}{\partial p_j}+\frac{\partial^2 G}{\partial p_i\partial p_j}\right)\right). \tag{8}$$

The cut-and-join equation (8) is equivalent to the following convolution equation:

$$G^\bullet_\mu(\lambda;\tau)=\sum_{|\nu|=|\mu|}G^\bullet_\nu(\lambda;0)z_\nu\Phi^\bullet_{\nu,\mu}(\sqrt{-1}\lambda\tau). \tag{9}$$

Theorem 9 is proved by applying functorial localization to the branch morphism

$$\mathrm{Br}:\mathcal{M}_g(\mathbf{P}^1,\mu)\to\mathrm{Sym}^r\,\mathbf{P}^1\cong\mathbf{P}^r,$$

where $\mathcal{M}_g(\mathbf{P}^1,\mu)$ is the moduli space of relative stable maps from a genus g curve to $\mathbf{P}^1$ with fixed ramification type $\mu=(\mu_1,\dots,\mu_h)$ at ∞, and

$$r=2g-2+|\mu|+\ell(\mu)$$

is the virtual dimension of $\mathcal{M}_g(\mathbf{P}^1,\mu)$. Note that the $\mathbb{C}^*$-action on $\mathbf{P}^1$ induces $\mathbb{C}^*$-actions on the domain and the target of Br, and Br is $\mathbb{C}^*$-equivariant. This is similar to the setup of the mirror principle, with a different linearized moduli.

We end this section with some applications of the Mariño–Vafa formula, following [L-L-Z2]. We have

$$G_{g,\mu}(\tau) = \sum_{k=\ell(\mu)-1}^{2g-2+|\mu|+\ell(\mu)} G_{g,\mu}^k \tau^k,$$

where

$$G_{g,\mu}^{2g-2+|\mu|+\ell(\mu)} = \frac{-\sqrt{-1}^{|\mu|+\ell(\mu)}}{|\operatorname{Aut}(\mu)|} \frac{\mu_i^{\mu_i}}{\mu_i!} \int_{\mathcal{M}_{g,\ell(\mu)}} \frac{\Lambda_g^\vee(1)}{\prod_{i=1}^{\ell(\mu)}(1-\mu_i\psi_i)},$$

$$G_{g,\mu}^{\ell(\mu)-1} = \frac{-\sqrt{-1}^{|\mu|+\ell(\mu)}}{|\operatorname{Aut}(\mu)|} \frac{\lambda_g}{\prod_{i=1}^{\ell(\mu)}(1-\mu_i\psi_i)}.$$

The part corresponding to $G_{g,\mu}^{2g-2+|\mu|+\ell(\mu)}$ in $R(\lambda;\tau;p)$ reduces to the Burnside formula of Hurwitz numbers $H_{g,\mu}$. We obtain the ELSV formula [ELSV]:

$$\frac{1}{|\operatorname{Aut}(\mu)|} \frac{\mu_i^{\mu_i}}{\mu_i!} \int_{\mathcal{M}_{g,\ell(\mu)}} \frac{\Lambda_g^\vee(1)}{\prod_{i=1}^{\ell(\mu)}(1-\mu_i\psi_i)} = \frac{H_{g,\mu}}{(2g-2+|\mu|+\ell(\mu))!}. \tag{10}$$

Extracting the part corresponding to $G_{g,\mu}^{\ell(\mu)-1}(\tau)$ from $R(\lambda;\tau;p)$, we obtain

$$\sum_{g=0}^\infty \lambda^{2g} \int_{\mathcal{M}_{g,n}} \frac{\lambda_g}{\prod_{i=1}^n (1-\mu_i\psi_i)} = |\mu|^{n-3} \frac{|\mu|\lambda/2}{\sin(|\mu|\lambda/2)}. \tag{11}$$

The identity (11) is true for any partition of length n, so it may be viewed as an identity of polynomials in $\lambda, \mu_1, \ldots, \mu_n$. This gives us the values of all λ_g-integrals:

$$\int_{\mathcal{M}_{g,n}} \psi_1^{k_1} \cdots \psi_n^{k_n} \lambda_g = \binom{2g+n-3}{k_1, \ldots, k_n} \frac{2^{2g-1}-1}{2^{2g-1}} \frac{|B_{2g}|}{(2g)!}. \tag{12}$$

The identity (12) was first proved in [Fa-P2].

The following identities proved in [Fa-P1] are also consequences the Mariño–Vafa formula:

$$\int_{\mathcal{M}_g} \lambda_{g-2}\lambda_{g-1}\lambda_g = \frac{1}{2(2g-2)!} \frac{|B_{2g-2}|}{2g-2} \frac{|B_{2g}|}{2g},$$

$$\int_{\mathcal{M}_{g,1}} \frac{\lambda_{g-1}}{1-\psi_1} = b_g \sum_{i=1}^{2g-1} \frac{1}{i} - \frac{1}{2} \sum_{\substack{g_1+g_2=g \\ g_1,g_2>0}} \frac{(2g_1-1)!(2g_2-1)!}{(2g-1)!} b_{g_1} b_{g_2}.$$

5 Mirror symmetry

In the previous sections we discussed the localization method to understand the counting function of Gromov–Witten invariants. These formulas are rather difficult to predict. They were motivated by important concepts of duality. A very important duality is called mirror symmetry. The counting function of Gromov–Witten invariants appears as instanton contribution to the II_A theory of one Calabi–Yau manifold M. The ability to compute it came from the symmetry that the II_A theory of M is isomorphic to the II_B theory of another Calabi–Yau manifold $\widehat{M}$ which is "mirror" to M. The II_B theory can be computed by deformation of complex structures, which can in turn be computed by studying the periods of holomorphic differentials.

However, the construction of $\widehat{M}$ has not been explained in a fundamental way, except for some special cases. About eight years ago, Strominger, Zaslow, and Yau, based on the newly developed brane theory, proposed a geometric construction of $\widehat{M}$. The program is still being pursued vigorously and it is closely related to the (more algebraic) homological mirror conjecture of Kontsevich and Fukaya.

We now explain the construction of SYZ and some of the important questions to be answered.

Motivated by understanding supersymmetric cycles in Calabi–Yau manifolds, Becker–Becker–Strominger [B-B-S] considered the concept of Lagrangian subvarieties V of a CY manifold M so that the holomorphic three-form, when restricted to the subvariety, is a (complex) constant (with norm one) multiple of the volume form of the subvariety. They consider a pair (V, L), where L is a $U(1)$ flat line bundle over V. These branes play an important role in understanding questions of duality, as supersymmetric cycles are protected when coupling constants of the theory change.

Soon it was found that such subvarieties V had been studied by Harvey–Lawson earlier based on their interest on understanding examples of area-minimizing subvarieties in Euclidean space. They were called special Lagrangian cycles by them. Later McLean [Mc] proved that the local moduli of a special Lagrangian submanifold V in a Calabi–Yau manifold are parametrized by harmonic one-form on V. The space of the harmonic one-form also parametrizes flat $U(1)$ bundles over V. Hence one can put an almost complex structure on the moduli space of the pair (V, L). This was observed by Strominger–Yau–Zaslow [S-Y-Z] and was proposed there to study this moduli space as an interesting complex manifold. In particular, when the first Betti number of V is equal to three, this complex manifold is three dimensional.

Based on the theory of branes, it was proposed by SYZ that if V is a three-dimensional torus, we can replace V by its dual V^*, the moduli space of flat $U(1)$ line bundles over V, and obtain a new complex manifold $\widehat{M}$. In general, we shall need to make instanton corrections to the complex structure on $\widehat{M}$. We proposed $\widehat{M}$ to be the mirror manifold of M.

The foliation defined by the special Lagrangian torus have singular leaves. We expect that in the large radius limit, there is a map $f : M \longrightarrow S^3$ so that outside a trivalent graph $G \subset S^3$ the fibers are nonsingular special Lagrangian tori. If these tori are linear we call the picture semiclassical. Leung–Yau–Zaslow [Le-Y-Z] has studied this mirror construction quite extensively. Many interesting predictions for mirror symmetry hold for this semiclassical setting.

The first explicit construction of Ricci flat metric for the semiclassical setting was due to Greene–Shapere–Vafa–Yau [G-S-V-Y]. It is an interesting question to find instanton corrections to their metric to obtain the Ricci flat metric on the K3 surface. M. Gross and P. Wilson [G-W] studied this problem based on perturbation of the semiflat Ricci flat metric. Unfortunately, we still have little information about the instantons which are holomorphic disks whose boundaries give nontrivial homology classes on the Lagrangian torus.

If we consider the domain which parametrizes the special Lagrangian tori in M, assuming we are in the semiflat situation, there is a Weil–Petersson metric on $S^3 \setminus G$. The form of such metrics was worked out by N. Hitchin [Hi], namely, that it is a Hessian metric defined on an affine flat manifold. (This kind of metric was introduced by Cheng–Yau [C-Y] in 1980 as an analogue of a Kähler metric for flat affine structures. Under some assumptions, Cheng–Yau also proved existence and uniqueness theorems for such a metric.) Hence there is a flat affine structure on $S^3 \setminus G$ and in order for the torus to be defined, the monodromy group must be a subgroup of $SL(3, \mathbb{Z})$. It is believed that there is a well-defined volume form on $S^3 \setminus G$ so that in suitable flat coordinates, the metric has the form $\sum \frac{\partial^2 u}{\partial x_i \partial x_j} dx_i dx_j$ and $\det(\frac{\partial^2 u}{\partial x_i \partial x_j}) = 1$. The existence of such a metric on a tube domain is related to the existence of Ricci flat Kähler flat metrics if we look at the complexified coordinates $x_j + \sqrt{-1} y_j$. (This ansatz was first proposed by E. Calabi.) However, its existence and the behavior near the triple singular point of G is nontrivial. This was worked out recently by Loftin–Yau–Zaslow [Lo-Y-Z].

Potentially, the construction of SYZ geometry can be reduced to the following data:

1. Construction of the flat affine structure in $S^3 \setminus G$ whose holonomy group is a subgroup of $SL(3, \mathbb{Z})$.
2. Construction of Cheng–Yau-type Hessian metric with a given flat volume form.
3. Construction of a map from $S^3 \setminus G$ to the moduli space of flat tori that is compatible with the holonomy group mentioned above.

Once this construction is carried out, one can construct the mirror manifold $\widehat{M}$ in the large radius limit. A very important verification of the SYZ construction of the mirror conjecture is to understand the deformation of the complex structure of $\widehat{M}$ and relate it to periods of the holomorphic three-form. It should reflect the counting of holomorphic curves of M.

Under reasonable topological assumptions, M. Gross studied the SYZ construction for the quintic in $\mathbf{P}^4$ and computed the Hodge diagram of the mirror manifold. He concludes that the picture is consistent. W.-D. Ruan [Ru] studied the Lagrangian fibration for Calabi–Yau manifolds that are constructed from toric manifolds.

The mirror correspondence is supposed to map the even cohomology of M to the odd cohomology of $\widehat{M}$. We propose to construct this map in terms of the SYZ construction in the following manner.

For the map $f : M \longrightarrow S^3$ and its mirror $\widehat{f} : \widehat{M} \longrightarrow S^3$, we can form a nine-dimensional variety by forming their fiber product $M \underset{S^3}{\times} \widehat{M} \longrightarrow S^3$. The general fiber of this map is given by $T^3 \times (T^3)^*$, which admits the standard Poincaré (complex) line bundle L so that L restricted to $T^3 \times \{l\}$ is given by l. We assume that L can be extended to be a line bundle (or a sheaf) over $M \underset{S^3}{\times} \widehat{M}$.

Let $\Pi_1 : M \underset{S^3}{\times} \widehat{M} \longrightarrow M$ and $\Pi_2 : M \underset{S^3}{\times} \widehat{M} \longrightarrow M$. Then we can define the mirror map on the cohomology level by taking any even degree cohomology class ω in M, and mapping it to $(\Pi_2)_*[(\Pi_1^*\omega)\exp(c_1(L))]$, which gives odd cohomology in $\widehat{M}$.

This assertion should be easier to verify in the semiclassical picture when we have flat affine constructions. When one counts instanton corrections, one should be able to map quantum cohomology of M to $H^1(T_{\widehat{M}})$, the deformation space of complex structures of $\widehat{M}$.

In [Le-Y-Z], we discuss how to map special Lagrangian cycles in $\widehat{M}$ to stable holomorphic sheaves over M. It would be important to prove this picture rigorously (which in turn depends on a rigorous construction of $\widehat{M}$). The potential construction of special Lagrangian cycles in $\widehat{M}$ can give a way to construct holomorphic cycles in M. Therefore it becomes an important question to understand which odd-dimensional cohomology classes in $\widehat{M}$ admit special Lagrangian cycles. The Hodge conjecture on M may suggest that an integral multiple of each odd-dimensional cohomology class in $\widehat{M}$ should be representable by special Lagrangian cycles.

There are several directions in which we may want to generalize the above pictures.

1. When M is Calabi–Yau, we can look for fiber spaces, which are holomorphic. Hence $f : M \longrightarrow N$ is holomorphic, the general fiber T is polarized Calabi–Yau and the space N is a Fano variety or a variety with negative Kodaira dimension. We can replace each fiber T by its mirror manifold $\widehat{T}$. Hopefully, one can complete the process to form a new compact Kähler manifold $\widehat{M}$ which is still Calabi–Yau. Obviously there are conditions one needs to impose in order for such an assertion to hold.
   The new manifold $\widehat{M}$ should reflect a great deal about the geometry of M. One can still define the transfer map from M to $\widehat{M}$. Since everything is now complex, it maps even cohomology of M to even cohomology of $\widehat{M}$.
   At least when T is a complex torus, it may give an isomorphism of derived category of M to a derived category of $\widehat{M}$, the Chow rings of M to Chow rings of $\widehat{M}$.
2. When M is a more general Kähler–Einstein manifold, the notion of special Lagrangian does not make sense. However, we can replace it by Lagrangian cycles which are area minimizing among all Lagrangian cycles. Hence we are looking for Lagrangian cycles whose mean curvature one-form is harmonic.

Many interesting questions in geometry can be motivated by such pictures.

References

[A-B] M. F. Atiyah and R. Bott, The moment map and equivariant cohomology, *Topology*, **23**-1 (1984), 1–28.

[B-B-S] K. Becker, M. Becker, and A. Strominger, Fivebranes, membranes and non-perturbative string theory, *Nuclear Phys.*, **B456** 1–2 (1995), 130–152.

[B-CF-K] A. Bertram, I. Ciocan-Fontanine, and B. Kim, Two proofs of a conjecture of Hori and Vafa, *Duke Math. J.*, **126**-1 (2005), 101–136.

[B-F] K. Behrend and B. Fantechi, The intrinsic normal cone, *Invent. Math.*, **128** (1997), 45–88.

[Ca] E. Calabi, A construction of nonhomogeneous Einstein metrics, in *Differential Geometry*, Proceedings of of Symposia in Pure Mathematics 27, American Mathematical Society, Providence, RI, 1975, 17–24.

[CdGP] P. Candelas, X. C. de la Ossa, P. S. Green, and L. Parkes, An exactly soluble superconformal theory from a mirror pair of Calabi-Yau manifolds, *Phys. Lett.* B, **258**-1–2 (1991), 118–126.

[Cox] D. A. Cox, The functor of a smooth toric variety, *Tôhoku Math. J.*, **47**-2 (1995), 251–262.

[C-V] S. Cecotti and C. Vafa, On the classification of $N = 2$ supersymmetric theories, *Comm. Math.*, **158** (1993), 569–644.

[C-Y] S. Y. Cheng and S.-T. Yau, The real Monge–Ampère equation and affine flat structures, in *Proceedings of the* 1980 *Beijing Symposium on Differential Geometry and Differential Equations*, Vol. 1, Science Press, Beijing, 1982, 339–370.

[ELSV] T. Ekedahl, S. Lando, M. Shapiro, and A. Vainshtein, Hurwitz numbers and intersections on moduli spaces of curves, *Invent. Math.*, **146**-2 (2001), 297–327.

[Fa-P1] C. Faber and R. Pandharipande, Hodge integrals and Gromov-Witten theory, *Invent. Math.*, **139**-1 (2000), 173–199.

[Fa-P2] C. Faber and R. Pandharipande, Hodge integrals, partition matrices, and the λ_g conjecture, *Ann. Math.* (2), **157**-1 (2003), 97–124.

[F] K. Fukaya, Multivalued Morse theory, asymptotic analysis, and mirror symmetry, in M. Lyubich and L. Takhtajan, eds., *Graphs and Patterns in Mathematics and Theoretical Physics*, Proceeding of Symposia in Pure Mathematics 73, American Mathematical Society, Providence, RI, 2005.

[Gi] A. Givental, Equivariant Gromov-Witten invariants, *Internat. Math. Res. Notices*, **13** (1996), 613–663.

[G-S-V-Y] B. Greene, A. Shapere, C. Vafa and S.-T. Yau, Stringy cosmic strings and non-compact CY manifold, *Nuclear Phys.*, **B337** (1990), 1–3.

[Gr] M. Gross, Topological mirror symmetry, *Invent. Math.*, **144** (2001), 75–137.

[Ge-P] E. Getzler and R. Pandharipande, Virasoro constraints and the Chern classes of the Hodge bundle, *Nuclear Phys.*, **B530**-3 (1998), 701–714.

[G-W] M. Gross and P. M. H. Wilson, Large complex structure limits of K3-surfaces, *J. Differential Geom.*, **55**-3 (2000), 475–546.

[H-L] R. Harvey and H. B. Lawson, Jr., Calibrated geometries, *Acta Math.*, **148** (1982), 47–157.

[Hi] N. J. Hitchin, The moduli space of special Lagrangian submanifolds, *Ann. Scuola Norm. Sup. Pisa Cl. Sci.*, **25** (1997), 503–515.

[H-I-V] K. Hori, A. Iqbal, and C. Vafa, D-branes and mirror symmetry, hep-th/0005247, 2000.

[HKKPTVVZ] K. Hori, S. Katz, A. Klemm, R. Pandharipande, R. Thomas, C. Vafa, R. Vakil, and E. Zaslow, *Mirror Symmetry*, Clay Mathematics Institute Monographs 1, American Mathematical Society, Providence, RI, 2003.

[H-V] H. Hori and C. Vafa, Mirror symmetry, hep-th/0002222, 2000.

[K-S] M. Kontsevich and Y. Soibelman, Homological mirror symmetry and torus fibration, in K. Fukaya, Y.-G. Oh, K. Ono, and G. Tian, eds., *Symplectic Geometry and Mirror Symmetry*, World Scientific, Singapore, 2001, 203–263.

[Le-Y-Z] N. C. Leung, S.-T. Yau, and E. Zaslow, From special Lagrangian to Hermitian–Yang–Mills via Fourier–Mukai transform, *Adv. Theor. Math. Phys.*, **4**-6 (2002), 1319–1341.

[Liu] K. Liu, Modular invariance and characteristic numbers, *Comm. Math. Phys.*, **174** (1995), 29–42.

[L-L-L-Y] B. Lian, C.-H. Liu, K. Liu, and S.-T. Yau, The S^1 fixed points in Quot-schemes and mirror principle computations, in S. D. Cutkosky, D. Edidin, Z. Qin, and Q. Zhang eds., *Vector Bundles and Representations Theory*, Contemporary Mathematics 322, American Mathematical Society, Providence, RI, 2003, 165–194.

[L-L-Y1] B. Lian, K. Liu, and S.-T. Yau, Mirror principle I, II, III, IV, *Asian J. Math.*, **1** (1997), 729–763; *Asian J. Math.*, **3** (1999), 109–146; *Asian J. Math.*, **3** (1999), 771–800; math.AG/0007104, 2000.

[L-L-Y2] B. Lian, K. Liu, and S.-T. Yau, A survey of mirror principle, in E. D'Hoker, D. H. Phong, and S.-T. Yau, eds., *Mirror Symmetry* IV: *Proceedings of the Conference on Strings, Duality, and Geometry*, AMS/IP Studies in Advanced Mathematics 33, American Mathematical Society, Providence, RI, 2002, 3–10.

[ChL-L-Y1] C.-H. Liu, K. Liu, and S.-T. Yau, On A-twisted moduli stack for curves from Witten's gauged linear sigma models, *Comm. Anal. Geom.*, **12**-1–2 (2004), 233–280.

[ChL-L-Y2] C.-H. Liu, K. Liu, and S.-T. Yau, S^1-fixed-points in hyper-quot-schemes and an exact mirror formula for flag manifolds from the extended mirror principle diagram, math.AG/0401367, 2004.

[L-L-Z1] C.-C. M. Liu, K. Liu, and J. Zhou, A proof of a conjecture of Mariño-Vafa on Hodge integrals, *J. Differential Geom.*, **65**-2 (2003), 289–340.

[L-L-Z2] C.-C. M. Liu, K. Liu, and J. Zhou, Mariño-Vafa formula and Hodge integral identities, math.AG/0308015, 2003.

[L-T] J. Li and G. Tian, Virtual moduli cycles and Gromov-Witten invariants of algebraic varieties, *J. Amer. Math. Soc.*, **11** (1998), 119–174.

[Lo-Y-Z] J. Loftin, S.-T. Yau, and E. Zaslow, Affine manifolds, SYZ geometry and the "Y" vertex, math.DG/0405061, 2004.

[Mc] R. C. Mclean, Deformation of calibrated submanifolds, *Comm. Anal. Geom.*, **6** (1998), 705–774.

[M-V] M. Mariño and C. Vafa, Framed knots at large N, in *Orbifolds in Mathematics and Physics: Proceedings of a Conference on Orbifolds in Mathematics and Physics, May* 4–8, 2001, *Madison, Wisconsin*, Contemporary Mathematics 310, American Mathematical Society, Providence, RI, 2002, 185–204.

[O-P] A. Okounkov and R. Pandharipande, Hodge integrals and invariants of the unknots, *Geom. Topol.*, **8** (2004), 675–699.

[Ru] W.-D. Ruan, Lagrangian torus fibrations of Calabi-Yau hypersurfaces in toric varieties and SYZ mirror symmetry conjecture, in *Mirror Symmetry* IV, AMS/IP Studies in Advanced Mathematics 33, American Mathematical Society, Providence, RI, 2002, 33–55.

[S-Y-Z] A. Strominger, S.-T. Yau, and E. Zaslow, Mirror symmtry in T-duality, *Nuclear Phys.*, **B479** (1996), 243–259.

[Wi1] E. Witten, Phases of $N = 2$ theories in two dimensions, *Nuclear Phys.*, **B403** (1993), 159–222.

[Wi2] E. Witten, The Verlinde algebra and the cohomology of the Grassmannian, in S.-T. Yau ed., *Geometry, Topology, and Physics for Raoul Bott*, International Press, Somerville, MA, 1994, 357–422.

Character Sheaves and Generalizations

G. Lusztig

Department of Mathematics
Massachusetts Institute of Technology
Cambridge, MA 02139
USA
gyuri@math.mit.edu

Dedicated to I. M. Gelfand on the occasion of his 90th birthday.

Subject Classification: 20G45

1

Let $\mathbf{k}$ be an algebraic closure of a finite field $\mathbf{F}_q$. Let $G = GL_n(\mathbf{k})$. The group $G(\mathbf{F}_q) = GL_n(\mathbf{F}_q)$ can be regarded as the fixed point set of the Frobenius map $F : G \to G$, $(g_{ij}) \mapsto (g_{ij}^q)$. Let $\bar{\mathbf{Q}}_l$ be an algebraic closure of the field of l-adic numbers, where l is a prime number invertible in $\mathbf{k}$. The characters of irreducible representations of $G(\mathbf{F}_q)$ over an algebraically closed field of characteristic 0, which we take to be $\bar{\mathbf{Q}}_l$, have been determined explicitly by J. A. Green [G]. The theory of character sheaves [L2] tries to produce some geometric objects over G from which the irreducible characters of $G(\mathbf{F}_q)$ can be deduced for any q. This allows us to unify the representation theories of $G(\mathbf{F}_q)$ for various q. The geometric objects needed in the theory are provided by intersection cohomology.

Let X be an algebraic variety over $\mathbf{k}$, let X_0 be a locally closed irreducible, smooth subvariety of X and let $\mathcal{E}$ be a local system over X_0 (we say "local system" instead of "$\bar{\mathbf{Q}}_l$-local system"). Deligne, Goresky, and MacPherson attach to this datum a canonical object $IC(\bar{X}_0, \mathcal{E})$ (intersection cohomology complex) in the derived category $\mathcal{D}(X)$ of $\bar{\mathbf{Q}}_l$-sheaves on X; this is a complex of sheaves which extends $\mathcal{E}$ to X (by 0 outside the closure $\bar{X}_0$ of X_0) in the most economical possible way so that local Poincaré duality is satisfied. We say that $IC(\bar{X}_0, \mathcal{E})$ is irreducible if $\mathcal{E}$ is irreducible.

Now take $X = G$ and take $X_0 = G_{rs}$ to be the set of regular semisimple elements in G. Let T be the group of diagonal matrices in G. For any integer $m \geq 1$ invertible in $\mathbf{k}$ we have an unramified $n!m^n$-fold covering

$$\pi_m : \{(g, t, xT) \in G_{rs} \times T \times G/T; x^{-1}gx = t^m\} \to G_{rs}, \quad (g, t, xT) \mapsto g.$$

An irreducible local system $\mathcal{E}$ on G_{rs} is said to be admissible if it is a direct summand of the local system $\pi_{m!}\bar{\mathbf{Q}}_l$ for some m as above. The character sheaves on G are the complexes $IC(G, \mathcal{E})$ for various admissible local systems $\mathcal{E}$ on G_{rs}.

We show how the irreducible characters of $G(\mathbf{F}_q)$ can be recovered from character sheaves on G. If A is a character sheaf on G, then its inverse image F^*A under F is again a character sheaf. There are only finitely many A (up to isomorphism) such that F^*A is isomorphic to A. For any such A we choose an isomorphism $\phi : F^*A \xrightarrow{\sim} A$ and we form the characteristic function $\chi_{A,\phi} : G(\mathbf{F}_q) \to \bar{\mathbf{Q}}_l$ whose value at g is the alternating sum of traces of ϕ on the stalks at g of the cohomology sheaves of A. Now ϕ is unique up to a nonzero scalar; hence $\chi_{A,\phi}$ is unique up to a nonzero scalar. It turns out that

(a) $\chi_{A,\phi}$ *is (up to a nonzero scalar) the character of an irreducible representation of* $G(\mathbf{F}_q)$ *and* $A \mapsto \chi_{A,\phi}$ *gives a bijection between the set of (isomorphism classes of) character sheaves on* G *that are isomorphic to their inverse image under* F *and the irreducible characters of* $G(\mathbf{F}_q)$.

(This result is essentially contained in [L1, L3].) The main content of this result is that the (rather complicated) values of an irreducible character of $G(\mathbf{F}_q)$ are governed by a geometric principle, namely by the procedure which gives the intersection cohomology extension of a local system.

2

More generally, assume that G is a connected reductive algebraic group over $\mathbf{k}$. The definition of the $IC(G, \mathcal{E})$ given above for GL_n makes sense also in the general case. The complexes on G obtained in this way form the class of *uniform* character sheaves on G. Consider now a fixed $\mathbf{F}_q$-rational structure on G with Frobenius map $F : G \to G$. The analogue of property 1(a) does not hold in general for (G, F). It is still true that the characteristic functions of the uniform character sheaves that are isomorphic to their inverse image under F are linearly independent class functions $G(\mathbf{F}_q) \to \bar{\mathbf{Q}}_l$. However, they do not form a basis of the space of class functions. Moreover, they are in general not irreducible characters of $G(\mathbf{F}_q)$ (up to a scalar); rather, each of them is a linear combination with known coefficients of a "small" number of irreducible characters of $G(\mathbf{F}_q)$ (where "small" means "bounded independently of q"); this result is essentially contained in [L1, L3].

It turns out that the class of uniform character sheaves can be naturally enlarged to a larger class of complexes on G.

For any parabolic P of G, U_P denotes the unipotent radical of P. For a Borel B in G, the images under $c^B : G \to G/U_B$ of the double cosets BwB form a partition $G/U_B = \cup_w (BwB/U_B)$.

An irreducible intersection cohomology complex $A \in \mathcal{D}(G)$ is said to be a character sheaf on G if it is G-equivariant and if for some/any Borel B in G, $c^B_! A$ has the following property:

(*) *Any cohomology sheaf of this complex restricted to any BwB/U_B is a local system with finite monodromy of order invertible in* $\mathbf{k}$.

Then any uniform character sheaf on G is a character sheaf on G. For $G = GL_n$ the converse is also true, but for general G this is not so.

Consider again a fixed $\mathbf{F}_q$-rational structure on G with Frobenius map $F : G \to G$. The following partial analogue of property 1(a) holds (under a mild restriction on the characteristic of $\mathbf{k}$):

(a) *The characteristic functions of the various character sheaves A on G (up to isomorphism) such that $F^*A \xrightarrow{\sim} A$ form a basis of the vector space of class functions $G(\mathbf{F}_q) \to \bar{\mathbf{Q}}_l$.*

3

We now fix a parabolic P of G. For any Borel B of P, let $\tilde{c}^B : G/U_P \to G/U_B$ be the obvious map. Now P acts on G/U_P by conjugation.

An irreducible intersection cohomology complex $A \in \mathcal{D}(G/U_P)$ is said to be a parabolic character sheaf if it is P-equivariant and if for some/any Borel B in P, $\tilde{c}^B_! A$ has property 2(*). When $P = G$, we recover the definition of character sheaves on G.

Consider now a fixed $\mathbf{F}_q$-rational structure on G with Frobenius map $F : G \to G$ such that P is defined over $\mathbf{F}_q$. Then G/U_P has a natural $\mathbf{F}_q$-rational structure with Frobenius map F. The following generalization of 2(a) holds (under a mild restriction on the characteristic of $\mathbf{k}$):

(a) *The characteristic functions of the various parabolic character sheaves A on G/U_P (up to isomorphism) such that $F^*A \xrightarrow{\sim} A$ form a basis of the vector space $\mathcal{V}$ of $P(\mathbf{F}_q)$-invariant functions $G(\mathbf{F}_q)/U_P(\mathbf{F}_q) \to \bar{\mathbf{Q}}_l$.*

The proof is given in [L5]. It relies on a generalization of property 2(a) to not necessarily connected reductive groups which will be contained in the series [L6].

If $h : G(\mathbf{F}_q) \to \bar{\mathbf{Q}}_l$ is the characteristic function of a character sheaf as in 2(a), then by summing h over the fibers of $G(\mathbf{F}_q) \to G(\mathbf{F}_q)/U_P(\mathbf{F}_q)$ we obtain a function $\bar{h} \in \mathcal{V}$. It turns out that each function $\bar{h}$ is a linear combination of a "small" number of elements in the basis of $\mathcal{V}$ described above. (The fact such a basis of $\mathcal{V}$ exists is not a priori obvious.)

The parabolic character sheaves on G/U_P are expected to be a necessary ingredient in establishing the conjectural geometric interpretation of Hecke algebras with unequal parameters given in [L4].

4

In this section G denotes an abelian group with a given family $\mathfrak{F}$ of automorphisms such that

(i) if $F \in \mathfrak{F}$ and $n \in \mathbf{Z}_{>0}$, then $F^n \in \mathfrak{F}$;
(ii) if $F \in \mathfrak{F}$, $F' \in \mathfrak{F}$, then there exist $n, n' \in \mathbf{Z}_{>0}$ such that $F^n = F'^{n'}$;
(iii) for any $F \in \mathfrak{F}$, the map $G \to G, x \mapsto F(x)x^{-1}$ is surjective with finite kernel.

For $F \in \mathfrak{F}$ and $n \in \mathbf{Z}_{>0}$, the homomorphism

$$N_{F^n/F} : G \to G, \qquad x \mapsto xF(x)\dots F^{n-1}(x),$$

restricts to a surjective homomorphism $G^{F^n} \to G^F$. (If $y \in G^F$, we can find $z \in G$ with $y = F^n(z)z^{-1}$, by (i), (iii).) We set $x = F(z)z^{-1}$. Then $x \in G^{F^n}$ and $N_{F^n/F}(x) = y$.) Let X be the set of pairs (F, ψ), where $F \in \mathfrak{F}$ and $\psi \in \mathrm{Hom}(G^F, \bar{\mathbf{Q}}_l^*)$. Consider the equivalence relation on X generated by $(F, \psi) \sim (F^n, \psi \circ N_{F^n/F})$. Let G^* be the set of equivalence classes. We define a group structure on G^*. We consider two elements of G^*; we represent them in the form (F, ψ), (F', ψ'), where $F = F'$ (using (ii)) and we define their product as the equivalence class of $(F, \psi\psi')$; one checks that this product is independent of the choices. This makes G^* into an abelian group. The unit element is the equivalence class of $(F, 1)$ for any $F \in \mathfrak{F}$. For $F \in \mathfrak{F}$ we define an automorphism $F^* : G^* \to G^*$ by sending an element of G^* represented by (F^n, ψ) with $n \in \mathbf{Z}_{>0}$, $\psi \in \mathrm{Hom}(G^{F^n}, \bar{\mathbf{Q}}_l^*)$ to $(F^n, \psi \circ F)$ (here $\psi \circ F$ is the composition $G^{F^n} \xrightarrow{F} G^{F^n} \xrightarrow{\psi} \bar{\mathbf{Q}}_l^*$); one checks that this is well defined. For any $F \in \mathfrak{F}$ the map $\mathrm{Hom}(G^F, \bar{\mathbf{Q}}_l^*) \to G^*$, $\psi \mapsto (F, \psi)$ is

(a) *a group isomorphism of* $\mathrm{Hom}(G^F, \bar{\mathbf{Q}}_l^*)$ *onto the subgroup* $(G^*)^{F^*}$ *of* G^*.

(This follows from the surjectivity of $N_{F^n/F} : G^{F^n} \to G^F$.)

5

Assume now that G is an abelian, connected (affine) algebraic group over $\mathbf{k}$. We define the notion of character sheaf on G.

Let $\mathfrak{F}$ be the set of Frobenius maps $F : G \to G$ for various rational structures on G over a finite subfield of $\mathbf{k}$. (These maps are automorphisms of G as an abstract group.) Then properties 4(i)–(iii) are satisfied for $(G, \mathfrak{F})$; hence the abelian group G^* is defined as in Section 4. We will give an interpretation of G^* in terms of local systems on G. Let $F \in \mathfrak{F}$. Let $L : G \to G$ be the Lang map $x \mapsto F(x)x^{-1}$. Consider the local system $E = L_!\bar{\mathbf{Q}}_l$ on G. Its stalk at $y \in G$ is the vector space E_y consisting of all functions $f : L^{-1}(y) \to \bar{\mathbf{Q}}_l$. We have $E_y = \oplus_{\psi \in \mathrm{Hom}(G^F, \bar{\mathbf{Q}}_l^*)} E_y^\psi$, where

$$E_y^\psi = \{f \in E_y;\ f(zx) = \psi(z)f(x)\ \forall z \in G^F, x \in L^{-1}(y)\}.$$

We have a canonical direct sum decomposition $E = \oplus_\psi E^\psi$, where E^ψ is a local system of rank 1 on G whose stalk at $y \in G$ is E_y^ψ (ψ as above). There is a unique isomorphism of local systems $\phi : F^*E^\psi \xrightarrow{\sim} E^\psi$ which induces identity on the stalk at 1. This induces for any $y \in G$ the isomorphism $E_{F(y)}^\psi \to E_y^\psi$ given by $f \mapsto f'$, where $f'(x) = f(F(x))$. If $y \in G^F$, this isomorphism is multiplication by $\psi(y)$. Thus the characteristic function $\chi_{E^\psi,\phi} : G^F \to \bar{\mathbf{Q}}_l$ is the character ψ.

Let $n \in \mathbf{Z}_{>0}$. Let $L' : G \to G$ be the map $x \mapsto F^n(x)x^{-1}$. Conider the local system $E' = L'_!\bar{\mathbf{Q}}_l$ on G. Its stalk at $y \in G$ is the vector space E'_y consisting of all functions $f' : L'^{-1}(y) \to \bar{\mathbf{Q}}_l$. We define $E_y \to E'_y$ by $f \mapsto f'$, where $f'(x) = f(N_{F^n,F}x)$ (note that $N_{F^n/F}(L'^{-1}(y)) \subset L^{-1}(y)$). This is induced by a morphism of local systems $E \to E'$ which restricts to an isomorphism $E^\psi \xrightarrow{\sim} E'^{\psi'}$, where $\psi' = \psi \circ N_{F^n/F} \in \mathrm{Hom}(G^{F^n}, \bar{\mathbf{Q}}_l^*)$.

From the definitions we see that, if $\psi, \psi' \in \mathrm{Hom}(G^F, \bar{\mathbf{Q}}_l^*)$, then for any $y \in G$ we have an isomorphism $E_y^\psi \otimes E_y^{\psi'} \xrightarrow{\sim} E_y^{\psi\psi'}$ given by multiplication of functions on $L^{-1}(y)$. This comes from an isomorphism of local systems $E^\psi \otimes E^{\psi'} \xrightarrow{\sim} E^{\psi\psi'}$.

A *character sheaf* on G is by definition a local system of rank 1 on G of the form E^ψ for some (F, ψ) as above. Let $\mathcal{S}(G)$ be the set of isomorphism classes of character sheaves on G. Then $\mathcal{S}(G)$ is an abelian group under tensor product. The arguments above show that $(F, \psi) \mapsto E^\psi$ defines a (surjective) group homomorphism $G^* \to \mathcal{S}(G)$. This is in fact an isomorphism. (It is enough to show that, if (F, ψ) is as above and $\psi' \in \mathrm{Hom}(G^F, \bar{\mathbf{Q}}_l^*)$ is such that the local systems $E^\psi, E^{\psi'}$ are isomorphic, then $\psi = \psi'$.) As we have seen earlier, each $E^\psi, E^{\psi'}$ has a unique isomorphism ϕ, ϕ' with its inverse image under $F : G \to G$ which induces the identity at the stalk at 1. Then we must have $\chi_{E^\psi,\phi} = \chi_{E^{\psi'},\phi'}$ hence $\psi = \psi'$. Note that for $F \in \mathfrak{F}$, the map $F^* : G^* \to G^*$ corresponds under the isomorphism $G^* \xrightarrow{\sim} \mathcal{S}(G)$ to the map $\mathcal{S}(G) \to \mathcal{S}(G)$ given by the inverse image under F. Using this and 4(a), we see that, for $F \in \mathfrak{F}$, the map $\mathrm{Hom}(G^F, \bar{\mathbf{Q}}_l^*) \to \mathcal{S}(G)$, $\psi \mapsto E^\psi$ is a group isomorphism of $\mathrm{Hom}(G^F, \bar{\mathbf{Q}}_l^*)$ onto the subgroup of $\mathcal{S}(G)$ consisting of all character sheaves on G that are isomorphic to their inverse image under F. We see that in this case the analogue of 1(a) holds.

From the definitions, we see that

(a) *if* $\mathcal{L}_1 \in \mathcal{S}(G)$ *and* $m : G \times G \to G$ *is the multiplication map, then* $m^*\mathcal{L}_1 = \mathcal{L}_1 \otimes \mathcal{L}_1$.

In the case where $G = \mathbf{k}$, our definition of character sheaves on G reduces to that of the Artin–Schreier local systems on $\mathbf{k}$.

6

In this section we assume that G is a unipotent algebraic group over $\mathbf{k}$ of "exponential type," that is, such that the exponential map from Lie G to G is well defined (and an isomorphism of varieties.) In this case we can define character sheaves on G using Kirillov theory. Namely, for each G-orbit in the dual of Lie G we consider the local system $\bar{\mathbf{Q}}_l$ on that orbit extended by 0 on the complement of the orbit. Taking the Fourier–Deligne transform we obtain (up to shift) an irreducible intersection cohomology complex on Lie G (since the orbit is smooth and closed, by Kostant–Rosenlicht). We can view it as an intersection cohomology complex on G via the exponential map. The complexes on G thus obtained are by definition the character sheaves of G. Using Kirillov theory (see [K]), we see that in this case the analogue of 1(a) holds.

Assume, for example, that G is the group of all matrices

$$[a, b, c] = \begin{pmatrix} 1 & a & b \\ 0 & 1 & c \\ 0 & 0 & 1 \end{pmatrix}$$

with entries in $\mathbf{k}$ and that $2^{-1} \in \mathbf{k}$. Consider the following intersection cohomology complexes on G:

(i) the complex which on the center $\{(0, b, 0); b \in \mathbf{k}\}$ is the local system $\mathcal{E} \in \mathcal{S}(\mathbf{k}), \mathcal{E} \neq \bar{\mathbf{Q}}_l$ extended by 0 to the whole of G;
(ii) the local system $f^*\mathcal{E}$, where $f[a, b, c] = (a, c)$ and $\mathcal{E} \in \mathcal{S}(\mathbf{k}^2)$.

The complexes (i)–(ii) are the character sheaves of G.

7

In this section we assume that G is a connected unipotent algebraic group over $\mathbf{k}$ (not necessarily of exponential type). We expect that in this case there is again a notion of character sheaf on G such that over a finite field, the characteristic functions of character sheaves form a basis of the space of class functions and each characteristic function of a character sheaf is a linear combination of a "small" number of irreducible characters. Thus here the situation should be similar to that for a general connected reductive group rather than that for GL_n. We illustrate this in one example. Assume that $\mathbf{k}$ has characteristic 2. Let G be the group consisting of all matrices of the form

$$\begin{pmatrix} 1 & a & b & c \\ 0 & 1 & d & b+ad \\ 0 & 0 & 1 & a \\ 0 & 0 & 0 & 1 \end{pmatrix}$$

with entries in $\mathbf{k}$; we also write $[a, b, c, d]$ instead of the matrix above. (This group can be regarded as the unipotent radical of a Borel in $Sp_4(\mathbf{k})$.)

Let $\mathcal{E}_0 \in \mathcal{S}(\mathbf{k})$ be the local system on $\mathbf{k}$ associated in Section 5 to $\mathbf{F}_q$ and to the homomorphism $\psi_0 : \mathbf{F}_q \to \bar{\mathbf{Q}}_l^*$ (composition of the trace $\mathbf{F}_q \to \mathbf{F}_2$ and the unique injective homomorphism $\mathbf{F}_2 \to \bar{\mathbf{Q}}_l^*$).

Consider the following intersection cohomology complexes on G:

(i) the complex which on the center $\{[0, b, c, 0]; (b, c) \in \mathbf{k}^2\}$ is the local system $\mathcal{E} \in \mathcal{S}(\mathbf{k}^2), \mathcal{E} \neq \bar{\mathbf{Q}}_l$ (see Section 5) extended by 0 to the whole of G;

(ii) the complex which on $\{[a_0, b, c, 0]; (b, c) \in \mathbf{k}^2\}$ (with $a_0 \in \mathbf{k}^*$ fixed) is the local system $pr_c^*\mathcal{E}$, where $\mathcal{E} \in \mathcal{S}(\mathbf{k})$, $\mathcal{E} \neq \bar{\mathbf{Q}}_l$ (see Section 5) extended by 0 to the whole of G;

(iii) the complex which on $\{[0, b, c, d_0]; (b, c) \in \mathbf{k}^2\}$ (with $d_0 \in \mathbf{k}^*$ fixed) is the local system $f^*\mathcal{E}_0$, where $f[0, b, c, d_0] = \alpha b + \alpha^2 d_0 c$ (with $\alpha \in \mathbf{k}^*$ fixed) extended by 0 to the whole of G;

(iv) the complex which on $\{[a_0, b, c, d_0]; (b, c) \in \mathbf{k}^2\}$ (with $a_0, d_0 \in \mathbf{k}^*$ fixed) is the local system $f^*\mathcal{E}_0$, where $f[a_0, b, c, d_0] = a_0^{-2}d_0^{-1}c$ extended by 0 to the whole of G;

(v) the local system $f^*\mathcal{E}$ on G, where $f[a, b, c, d] = (a, d) \in \mathbf{k}^2$ and $\mathcal{E} \in \mathcal{S}(\mathbf{k}^2)$.

By definition, the character sheaves on G are the complexes in (i)–(v) above. Note that there are infinitely many subvarieties of G which appear as supports of character sheaves (this in contrast with the case of reductive groups). There is a symmetry that exchanges the character sheaves of type (ii) with those of type (iii). Namely, define $\xi : G \to G$ by

$$[a, b, c, d] \mapsto [d, c + ab + a^2 d, b^2 + dc + abd, a^2].$$

Then ξ is a homomorphism whose square is $[a, b, c, d] \mapsto [a^2, b^2, c^2, d^2]$; moreover, ξ^* interchanges the sets (ii) and (iii) and it leaves stable each of the sets (i), (iv), and (v).

Now G has an obvious $\mathbf{F}_q$-structure with Frobenius map $F : G \to G$. We describe the irreducible characters of $G(\mathbf{F}_q)$:

(I) We have q^2 one-dimensional characters $U \to \bar{\mathbf{Q}}_l^*$ of the form $[a, b, c, d] \mapsto \psi_0(xa + yd)$ (one for each $x, y \in \mathbf{F}_q$).

(II) We have $q - 1$ irreducible characters of degree q of the form $[0, b, c, 0] \mapsto q\psi_0(xb)$ (all other elements are mapped to 0), one for each $x \in \mathbf{F}_q - \{0\}$.

(III) We have $q - 1$ irreducible characters of degree q of the form $[0, b, c, 0] \mapsto q\psi_0(xc)$ (all other elements are mapped to 0), one for each $x \in \mathbf{F}_q - \{0\}$.

(IV) We have $4(q - 1)^2$ irreducible characters of degree $q/2$, one for each quadruple $(a_0, d_0, \epsilon_1, \epsilon_2)$, where

$$a_0 \in \mathbf{F}_q^*, \quad d_0 \in \mathbf{F}_q^*, \quad \epsilon_1 \in \mathrm{Hom}(\{0, a_0\}, \pm 1), \quad \epsilon_2 \in \mathrm{Hom}(\{0, d_0\}, \pm 1),$$

namely,

$$[a, b, c, d] \mapsto (q/2)\epsilon_1(a)\epsilon_2(d)\psi_0(a_0^{-2}d_0^{-1}(ba + ba_0 + c)),$$

if $a \in \{0, a_0\}$, $d \in \{0, d_0\}$; all other elements are sent to 0.

A character of type (II) is obtained by inducing from the subgroup $\{[a, b, c, d] \in G(\mathbf{F}_q); d = 0\}$ the one-dimensional character $[a, b, c, 0] \mapsto \psi_0(xb)$, where $x \in \mathbf{F}_q - \{0\}$. A character of type (III) is obtained by inducing from the commutative subgroup $\{[a, b, c, d] \in G(\mathbf{F}_q); a = 0\}$ the one-dimensional character $[0, b, c, d] \mapsto \psi_0(xc)$, where $x \in \mathbf{F}_q - \{0\}$. A character of type (IV) is obtained by inducing from the subgroup $\{(a, b, c, d) \in G(\mathbf{F}_q); a \in \{0, a_0\}\}$ (where $a_0 \in \mathbf{F}_q - \{0\}$ is fixed) the one-dimensional character $[a, b, c, d] \mapsto \epsilon_1(a)\psi_0(fd + a_0^{-2}d_0^{-1}(ba + ba_0 + c))$, where $f \in \mathbf{F}_q$ is chosen so that $\psi_0(f d_0) = \epsilon_2(d_0)$. (The induced character does not depend on the choice of f.)

Consider the matrix expressing the characteristic functions of character sheaves A such that $F^*A \cong A$ (suitably normalized) in terms of irreducible characters of $G(\mathbf{F}_q)$. This matrix is square and a direct sum of diagonal blocks of size 1×1 (with entry 1) or 4×4 with entries $\pm 1/2$, representing the Fourier transform over a two-dimensional symplectic $\mathbf{F}_2$-vector space. There are $(q-1)^2$ blocks of size 4×4 involving the irreducible characters of type IV.

We see that in our case the character sheaves have the desired properties. We also note that in our case $G(\mathbf{F}_q)$ has some irreducible character whose degree is not a power of q (but $q/2$) in contrast with what happens in the situation in Section 6.

8

Let ϵ be an indeterminate. For $r \geq 2$ let $\mathcal{A}_r = \mathbf{k}[\epsilon]/(\epsilon^r)$. Let $G = GL_n(\mathcal{A}_r)$. Let B (respectively, T) be the group of upper triangular (respectively, diagonal) matrices in G. Then G is in a natural way a connected affine algebraic group over $\mathbf{k}$ of dimension $n^2 r$ and B, T are closed subgroups of G. On G we have a natural $\mathbf{F}_q$-structure with Frobenius map $F : G \to G$, $(g_{ij}) \mapsto (g_{ij}^{(q)})$, where for $a_0, a_1, \dots, a_{r-1}$ in $\mathbf{k}$ we set $(a_0 + a_1\epsilon + \cdots + a_{r-1}\epsilon^{r-1})^{(q)} = a_0^q + a_1^q\epsilon + \cdots + a_{r-1}^q\epsilon^{r-1}$. The fixed point set of $F : G \to G$ is $GL_n(\mathbf{F}_q[\epsilon]/(\epsilon^r))$. For $i \neq j$ in $[1, n]$, we consider the homomorphism $f_{ij} : \mathbf{k} \to T$ which takes $x \in \mathbf{k}$ to the diagonal matrix with ii-entry equal to $1 + \epsilon^{r-1}x$, jj-entry equal to $1 - \epsilon^{r-1}x$ and other diagonal entries equal to 1. Since T is connected and commutative, the group $\mathcal{S}(T)$ is defined (see Section 5). Let $\mathcal{L} \in \mathcal{S}(T)$. We will assume that $\mathcal{L}$ is *regular* in the following sense: for any $i \neq j$ in $[1, n]$, $f_{ij}^*\mathcal{L}$ is not isomorphic to $\bar{\mathbf{Q}}_l$.

Let $\pi : B \to T$ be the obvious homomorphism. Consider the diagram

$$G \overset{a}{\leftarrow} Y \overset{b}{\rightarrow} T,$$

where

$$Y = \{(g, xB) \in G \times G/B; x^{-1}gx \in B\}, \quad a(g, xB) = g, \quad b(g, xB) = \pi(x^{-1}gx).$$

Then $b^*\mathcal{L}$ is a local system on Y and we may consider the complex $a_! b^*\mathcal{L}$ on G.

As in Section 5, we can find an integer $m_0 > 0$ such that, for any $m \in \mathcal{M} = \{m_0, 2m_0, 3m_0, \dots\}$, $\mathcal{L}$ is associated to $(\mathbf{F}_{q^m}, \psi_m)$, where $\psi_m \in \mathrm{Hom}(T^{F^m}, \bar{\mathbf{Q}}_l^*)$. We can regard ψ_m as a character $B(\mathbf{F}_{q^m}) \to \bar{\mathbf{Q}}_l^*$ via $\pi : B \to T$; inducing this from $B(\mathbf{F}_{q^m})$ to $G(\mathbf{F}_{q^m})$ we obtain a representation of $G(\mathbf{F}_{q^m})$ whose character is denoted by c_m. It is easy to see (using the regularity of $\mathcal{L}$) that this character is irreducible.

For $m \in \mathcal{M}$, there is a unique isomorphism $(F^m)^*\mathcal{L} \xrightarrow{\sim} \mathcal{L}$ of local systems on T which induces the identity on the stalk of $\mathcal{L}$ at 1. This induces an isomorphism $(F^m)^*(b^*\mathcal{L}) \xrightarrow{\sim} b^*\mathcal{L}$ (where $F : Y \to Y$ is $(g, xB) \mapsto (F(g), F(x)B)$) and an isomorphism $(F^m)^*(a_! b^*\mathcal{L}) \xrightarrow{\sim} a_! b^*\mathcal{L}$ in $\mathcal{D}(G)$. Let $\chi_m : G^{F^m} \to \bar{\mathbf{Q}}_l$ be the characteristic function of $a_! b^* \mathcal{L}$ with respect to this isomorphism. From the definitions we see that $\chi_m = c_m$. This shows that $a_! b^* \mathcal{L}$ behaves like a character sheaf except for the fact that it is not clear that it is an intersection cohomology complex.

We conjecture that

(a) *if $\mathcal{L}$ is regular, then $a_! b^* \mathcal{L}$ is an intersection cohomology complex on G.*

(The conjecture also makes sense and is expected to be true when GL_n is replaced by any reductive group, and G by the corresponding group over $\mathcal{A}_r$.) Thus one can expect that there is a theory of character sheaves for G, as far as generic principal series representations and their twisted forms are concerned. But one cannot expect a complete theory of character sheaves in this case (see Section 13).

In Sections 9–12, we prove the conjecture in the special case where $G = GL_2(\mathbf{k})$ and $r = 2$.

9

Let $\mathcal{A} = \mathcal{A}_2 = \mathbf{k}[\epsilon]/(\epsilon^2)$. Let V be a free $\mathcal{A}$-module of rank 2. Let G be the group of automorphisms of the $\mathcal{A}$-module V. This is the group of all automorphisms of the four-dimensional $\mathbf{k}$-vector space V that commute with the map $\epsilon : V \to V$ given by the $\mathcal{A}$-module structure. Hence G is an algebraic group of dimension 8 over $\mathbf{k}$. Let ${}^0\tilde{G}$ be the set of all pairs (g, V_2), where $g \in G$ and V_2 is a free $\mathcal{A}$-submodule of V of rank 1 such that $gV_2 = V_2$. For $k = 1, 2$, let X_k be the set of all $\mathcal{A}$-submodules of V that have dimension k as a $\mathbf{k}$-vector space. Let $\tilde{G}$ be the set of all triples (g, V_1, V_2), where $g \in G$, $V_1 \in X_1$, $V_2 \in X_2$, $V_1 \subset V_2$, $gV_1 = V_1$, $gV_2 = V_2$ and the scalars by which g acts on V_1 and V_2/V_1 coincide. We can regard ${}^0\tilde{G}$ as a subset of $\tilde{G}$ by $(g, V_2) \mapsto (g, \epsilon V_2, V_2)$. Note that $\tilde{G}$ is naturally an algebraic variety over $\mathbf{k}$ and ${}^0\tilde{G}$ is an open subset of $\tilde{G}$.

The group of units $\mathcal{A}'$ of $\mathcal{A}$ is an algebraic group isomorphic to $\mathbf{k}^* \times \mathbf{k}$. Hence $\mathcal{S}(\mathcal{A}')$ is defined. Let $\mathcal{L}_1 \in \mathcal{S}(\mathcal{S}')$, $\mathcal{L}_2 \in \mathcal{S}(\mathcal{S}')$. Let $\mathcal{L} = \mathcal{L}_1 \boxtimes \mathcal{L}_2 \in \mathcal{S}(\mathcal{A}' \times \mathcal{A}')$, $\mathcal{E} = \mathcal{L}_2 \otimes \mathcal{L}_1^* \in \mathcal{S}(\mathcal{A}')$. Define $f : {}^0\tilde{G} \to \mathcal{A}' \times \mathcal{A}'$ by $f(g, V_2) = (\alpha_1, \alpha_2)$, where $\alpha_1 \in \mathcal{A}'$ is given by $gv = \alpha_1 v$ for $v \in V_2$ and $\alpha_2 \in \mathcal{A}'$ is given by $gv' = \alpha_2 v'$ for $v' \in V/V_2$. Let $\tilde{\mathcal{L}} = f^*(\mathcal{L}_1 \boxtimes \mathcal{L}_2)$, a local system on ${}^0\tilde{G}$. Define $f_i : {}^0\tilde{G} \to \mathcal{A}'$ $(i = 1, 2)$ by $f_1(g, V_2) = \alpha_1\alpha_2$, $f_2(g, V_2) = \alpha_1$, where α_1, α_2 are as above. Then $\tilde{\mathcal{L}} = f_1^* \mathcal{L}_1 \otimes f_2^* \mathcal{L}$. (We use 5(a).)

We shall assume that $\mathcal{L}$ is *regular* in the following sense: the restriction of $\mathcal{E}$ to the subgroup $\mathcal{T} = \{1 + \epsilon c; c \in \mathbf{k}\}$ of $\mathcal{A}'$ is not isomorphic to $\bar{\mathbf{Q}}_l$.

Lemma 10.

(a) *$\tilde{G}$ is an irreducible, smooth variety and $\tilde{G} - {}^0\tilde{G}$ is a smooth irreducible hypersurface in $\tilde{G}$.*
(b) *We have $IC(\tilde{G}, \tilde{\mathcal{L}})|_{\tilde{G} - {}^0\tilde{G}} = 0$.*

Note that $f_1 : {}^0\tilde{G} \to \mathcal{A}'$ extends to the whole of $\tilde{G}$ by $f_1(g, V_1, V_2) = \det_{\mathcal{A}}(g : V \to V)$. Hence $f_1^* \mathcal{L}_1$ extends to a local system on $\tilde{G}$ and we have $IC(\tilde{G}, \tilde{\mathcal{L}}) =$

$f_1^*\mathcal{L}_1 \otimes IC(\tilde{G}, f_2^*\mathcal{E})$. Hence to prove (b) it is enough to show that $IC(\tilde{G}, f_2^*\mathcal{E})$ is zero on $\tilde{G} - {}^0\tilde{G}$.

Let Z (respectively, H) be the fiber of the second projection $\tilde{G} \to X_1$ (respectively, $\tilde{G} - {}^0\tilde{G} \to X_1$) at $V_1 \in X_1$. Since G acts transitively on X_1 it is enough to show that Z is smooth, irreducible, H is a smooth, irreducible hypersurface in Z and $IC(Z, f_2^*\mathcal{E})$ is zero on H. (The restriction of f_2 to Z is again denoted by f_2.)

Let e_1, e_2 be a basis of V such that $V_1 = \mathbf{k}\epsilon e_1$. The subspaces $V_2 \in X_2$ such that $V_1 \subset V_2$ are exactly the subspaces $V_2^{z',z''} = \mathbf{k}\epsilon e_1 + \mathbf{k}(z'e_1 + z''\epsilon e_2)$, where $(z', z'') \in \mathbf{k}^2 - \{0\}$. An element $g \in G$ is of the form

$$ge_1 = a_0e_1 + b_0e_2 + a_1\epsilon e_1 + b_1\epsilon e_2,$$
$$ge_2 = c_0e_1 + d_0e_2 + c_1\epsilon e_1 + d_1\epsilon e_2,$$

where $a_i, b_i, c_i, d_i \in \mathbf{k}$ satisfy $a_0d_0 - b_0c_0 \neq 0$.

The condition that $g\epsilon e_1 \in \mathbf{k}\epsilon e_1$ is $b_0 = 0$. The condition that $gV_2^{z',z''} = V_2^{z',z''}$ is that $z'b_1 + z''d_0 = a_0z''$ if $z' \neq 0$ (no condition if $z' = 0$). The condition that the scalars by which g acts on V_1 and $V_2^{z',z''}/V_1$ coincide is $a_0 = d_0$ if $z' = 0$ (no condition if $z' \neq 0$).

We see that we may identify Z with

$$\{(a_0, c_0, d_0, a_1, b_1, c_1, d_1; z', z'') \in \mathbf{k}^7 \times (\mathbf{k}^2 - \{0\})/\mathbf{k}^*;$$
$$a_0 \neq 0, d_0 \neq 0, z'b_1 = z''(a_0 - d_0)\}$$

and H with the subset defined by $z' = 0$. In this description it is clear that Z is irreducible, smooth and H is a smooth, irreducible hypersurface in Z. The function f_2 takes a point with $z' \neq 0$ to $a_0 + \epsilon(a_1 + z''z'^{-1}c_0)$. To prove the statement on intersection cohomology, we may replace Z by the open subset $z'' \neq 0$ containing H. Thus we may replace Z by

$$Z_1 = \{(a_0, c_0, d_0, a_1, b_1, c_1, d_1; z) \in \mathbf{k}^7 \times \mathbf{k}; a_0 \neq 0, d_0 \neq 0, zb_1 = a_0 - d_0\}$$

and H by the subset defined by $z = 0$. The function f_2 is defined on $Z_1 - H$ by

$$a_0 + \epsilon(a_1 + z^{-1}c_0) = (a_0 + \epsilon a_1)(1 + \epsilon z^{-1}c_0a_0^{-1}).$$

Thus $f_2 = f_3f_4$, where f_3 (respectively, f_4) is defined on $Z_1 - H$ by $a_0 + \epsilon a_1$ (respectively, $1 + \epsilon z^{-1}c_0a_0^{-1}$). Hence $f_2^*\mathcal{E} = f_3^*\mathcal{E} \otimes f_4^*\mathcal{E}$. Now f_3 extends to Z_1 hence $f_3^*\mathcal{E}$ extends to a local system on Z_1. We have $IC(Z_1, f_3^*\mathcal{E} \otimes f_4^*\mathcal{E}) = f_3^*\mathcal{E} \otimes IC(Z_1, f_4^*\mathcal{E})$. It is enough to show that $IC(Z_1, f_4^*\mathcal{E})$ is zero on H. We make the change of variable $c = c_0a_0^{-1}$. Then Z_1 becomes

$$Z_1 = \{(a_0, c, a_1, b_1, c_1, d_1; z) \in \mathbf{k}^7 \times \mathbf{k}; a_0 \neq 0, a_0 - zb_1 \neq 0\},$$

H is the subset defined by $z = 0$ and $f_4 : Z_1 - H \to \mathcal{A}'$ is given by $1 + \epsilon z^{-1}c$. Let $\tilde{Z}_1 = \{(a_0, c, a_1, b_1, c_1, d_1; z) \in \mathbf{k}^7 \times \mathbf{k}\}$ and let H_1 be the subset of $\tilde{Z}_1$ defined by $z = 0$. Then Z_1 is open in $\tilde{Z}_1$ and f_4 is well defined on $\tilde{Z}_1 - H_1$ by $1 + \epsilon z^{-1}c$.

Hence $f_4^*\mathcal{E}$ is well defined on $\tilde{Z}_1 - H_1$. It is enough to show that $IC(\tilde{Z}_1, f_4^*\mathcal{E})$ is zero on H_1. Let $H' = \{(c,z) \in \mathbf{k}^2; z = 0\}$ and define $f' : \mathbf{k}^2 - H' \to \mathcal{A}'$ by $f'(c,z) = 1 + \epsilon z^{-1}c$. It is enough to show that $IC(\mathbf{k}^2, f'^*\mathcal{E})$ is zero on H'. Let P be the projective line associate to $\mathbf{k}^2$. Then H' defines a point $x_0 \in P$. Since f' is constant on lines, it defines a map $h : P - \{x_0\} \to \mathcal{A}'$. Since P is one-dimensional we have $IC(P, h^*\mathcal{E}) = \mathcal{F}$, where $\mathcal{F}$ is a constructible sheaf on P whose restriction to $P - \{x_0\}$ is $h^*\mathcal{E}$. It is enough to show that

(c) the stalk of $\mathcal{F}$ at x_0 is 0;
(d) $H^i(P, \mathcal{F}) = 0$ for $i = 0, 1$.

(Indeed, (c) implies that $IC(\mathbf{k}^2, f'^*\mathcal{E})$ is zero at $(c, 0)$ with $c \neq 0$ and (d) implies that $IC(\mathbf{k}^2, f'^*\mathcal{E})$ is zero at $(0, 0)$.)

Consider the standard $\mathbf{F}_q$-rational structures an $\mathbf{k}^2$, X, $\mathcal{A}'$ and let F be the corresponding Frobenius map. We may assume that $\mathcal{E}$ is associated as in Section 5 to $(\mathbf{F}_q, \psi)$, where $\psi \in \mathrm{Hom}(\mathcal{A}'^F, \bar{\mathbf{Q}}_l^*)$. For any $m \in \mathbf{Z}_{>0}$ there is a unique isomorphism $\phi_m : (F^m)^*\mathcal{E} \xrightarrow{\sim} \mathcal{E}$ which induces the identity on the stalk of $\mathcal{E}$ at 1. The characteristic function of $\mathcal{E}$ with respect to this isomorphism is $a' \mapsto \psi(N_{F^m/F}(a'))$, $a' \in \mathcal{A}'^{F^m}$. Since, by assumption, $\mathcal{E}|_{\mathcal{T}}$ is not isomorphic to $\bar{\mathbf{Q}}_l$, $\psi|_{\mathcal{T}^F}$ is not a trivial character. Hence $\psi \circ N_{F^m/F} : \mathcal{A}'^{F^m} \to \bar{\mathbf{Q}}_l^*$ is nontrivial on $\mathcal{T}^{F^m}$. Now ϕ_m induces an isomorphism $\phi'_m : (F^m)^*h^*\mathcal{E} \xrightarrow{\sim} h^*\mathcal{E}$. We show that

(e) $\sum_{x \in P^{F^m} - \{x_0\}} \mathrm{tr}(\phi'_m, (h^*\mathcal{E})_x) = 0$.

An equivalent statement is

$$\sum_{(c,z) \in \mathbf{F}_{q^m} \times \mathbf{F}_{q^m}^*} (\psi \circ N_{F^m/F})(1 + \epsilon z^{-1}c) = 0,$$

which follows from the fact that $\psi \circ N_{F^m/F} : \mathcal{A}'^{F^m} \to \bar{\mathbf{Q}}_l^*$ is nontrivial on $\mathcal{T}^{F^m}$. Introducing (e) in the trace formula for Frobenius, we see that

(f) $\sum_{i=0}^{2} (-1)^i \mathrm{tr}(\phi'_m, H^i(P, \mathcal{F})) = \mathrm{tr}(\phi'_m, \mathcal{F}_{x_0})$,

where $\mathcal{F}_{x_0}$ is the talk of $\mathcal{F}$ at x_0 and ϕ'_m is in fact equal to $\phi_1'^m$ (for $m = 1, 2, 3, \dots$). By Deligne's purity theorem, $H^i(P, \mathcal{F})$ together with ϕ'_1 is pure of weight i; by Gabber's theorem [BBD], $\mathcal{F}_{x_0}$ together with ϕ'_1 is mixed of weight ≤ 0. Hence from (f) we deduce that $H^1(P, \mathcal{F}) = 0$, $H^2(P, \mathcal{F}) = 0$ and $\dim H^0(P, \mathcal{F}) = \dim \mathcal{F}_{x_0}$. By the hard Lefschetz theorem [BBD] we have $\dim H^0(P, \mathcal{F}) = \dim H^2(P, \mathcal{F})$. It follows that $H^0(P, \mathcal{F}) = 0$, hence $\mathcal{F}_{x_0} = 0$. This proves (c), (d). The lemma is proved.

Lemma 11. *Define* $\rho : {}^0\tilde{G} \to G$ *by* $(g, V_2) \mapsto g$. *Let* $K = \rho_!\tilde{\mathcal{L}}$. *Let* G_0 *be the open dense subset of* G *consisting of all* $g \in G$ *such that* $g : \epsilon V \to \epsilon V$ *is regular, semisimple. Let* $\rho_0 : \rho^{-1}(G_0) \to G_0$ *be the restriction of* ρ. *Then* $\rho_{0!}\tilde{\mathcal{L}}$ *is a local system on* G_0. *We have* $\dim \operatorname{supp} \mathcal{H}^i K < \dim G - i$ *for any* $i > 0$.

The first assertion of the lemma follows from the fact that ρ_0 is a double covering. To prove the second assertion it is enough to show that, for $i > 0$, the set G_i consisting of the points $g \in G$ such that $\dim \rho^{-1}(g) = i$ and $\oplus_j H_c^j(\rho^{-1}(g), \tilde{\mathcal{L}}) \neq 0$ has codimension $> 2i$ in G.

Consider the fiber $\rho^{-1}(g)$ for $g \in G$. We may assume that, with respect to a suitable $\mathcal{A}$-basis of V, g can be represented as an upper triangular matrix $\left(\begin{smallmatrix} a & b \\ 0 & c \end{smallmatrix}\right)$ with a, c in $\mathcal{A}'$ and $b \in \mathcal{A}$. (Otherwise, $\rho^{-1}(g)$ is empty.) There are five cases:

Case 1. $a - d \in \mathcal{A}'$. Then $\rho^{-1}(g)$ consists of two points.
Case 2. $a - d \in \epsilon\mathcal{A}$, $b \in \mathcal{A}'$. Then $\rho^{-1}(g)$ is an affine line.
Case 3. $a - d \in \epsilon\mathcal{A} - \{0\}$, $b \in \epsilon\mathcal{A}$. Then $\rho^{-1}(g)$ is a disjoint union of two affine lines.
Case 4. $a = d$, $b \in \epsilon\mathcal{A} - \{0\}$. Then $\rho^{-1}(g)$ is an affine line.
Case 5. $a = d$, $b = 0$. Then $\rho^{-1}(g)$ is an affine line bundle over a projective line.

In Case 2, we may identify $\rho^{-1}(g), \tilde{\mathcal{L}}|_{\rho^{-1}(g)}$ with $P - \{x_0\}, \mathcal{F}|_{P-\{x_0\}}$ in the proof of Lemma 10. Then the argument in that proof shows that $H_c^j(\rho^{-1}(g), \tilde{\mathcal{L}}) = 0$ for all j. We see that G_1 consists of all g as in Cases 3 and 4, hence G_1 has codimension 3 in G. We see that G_2 consists of all g as in Case 5, hence G_2 has codimension 6 in G. The lemma is proved. Note that without the assumption that $\mathcal{L}$ is regular, the last assertion of the lemma would not hold. (There would be a violation coming from g in Case 2.)

12

We show that

(a) $\rho_!\tilde{\mathcal{L}} = IC(G, \rho_{0!}\tilde{\mathcal{L}})$.

Define $\tilde{\rho} : \tilde{G} \to G$ by $\tilde{\rho}(g, V_1, V_2) = g$. Clearly, $\tilde{\rho}$ is proper. Let $j : {}^0\tilde{G} \to G$ be the inclusion. We have $\rho = \tilde{\rho} \circ j$, hence $\rho_!\tilde{\mathcal{L}} = \tilde{\rho}_!(j_!\tilde{\mathcal{L}})$. By Lemma 10, we have $j_!\tilde{\mathcal{L}} = IC(\tilde{G}, \tilde{\mathcal{L}})$ hence $\rho_!\tilde{\mathcal{L}} = \tilde{\rho}_! IC(\tilde{G}, \tilde{\mathcal{L}})$. Since $\tilde{\rho}$ is proper, $\tilde{\rho}_!$ commutes with the Verdier duality $\mathfrak{D}$. Hence $\mathfrak{D}(\rho_!\tilde{\mathcal{L}}) = \tilde{\rho}_!\mathfrak{D}\, IC(\tilde{G}, \tilde{\mathcal{L}})$. Hence $\mathfrak{D}(\rho_!\tilde{\mathcal{L}})$ equals $\tilde{\rho}_! IC(\tilde{G}, \tilde{\mathcal{L}}^*)$ up to a shift. Now the same argument that shows $j_!\tilde{\mathcal{L}} = IC(\tilde{G}, \tilde{\mathcal{L}})$ shows also $j_!\tilde{\mathcal{L}}^* = IC(\tilde{G}, \tilde{\mathcal{L}}^*)$. Hence, up to shift, $\mathfrak{D}(\rho_!\tilde{\mathcal{L}})$ equals $\tilde{\rho}_! j_!\tilde{\mathcal{L}}^* = \rho_!\tilde{\mathcal{L}}^*$. Now the argument in Lemma 11 can also be applied to $\tilde{\mathcal{L}}^*$ instead of $\tilde{\mathcal{L}}$ and yields $\dim \operatorname{supp} \mathcal{H}^i \rho_!\tilde{\mathcal{L}}^* < \dim G - i$ for any $i > 0$. Thus, $\rho_!\tilde{\mathcal{L}}$ satisfies the defining properties of $IC(G, \rho_{0!}\tilde{\mathcal{L}})$, hence it is equal to it. This proves (a).

We see that conjecture 8(a) holds for $n = 2, r = 2$.

13

If G is a connected affine algebraic group over $\mathbf{k}$ which is neither reductive nor nilpotent, one cannot expect to have a complete theory character sheaves for G. Assume for example that G is the group of all matrices

$$[a, b] = \begin{pmatrix} a & b \\ 0 & 1 \end{pmatrix}$$

with entries in $\mathbf{k}$. The group $G(\mathbf{F}_q)$ (for the obvious $\mathbf{F}_q$-rational structure) has $(q-1)$ one-dimensional representations and one $(q-1)$-dimensional irreducible representation. The character of a one-dimensional representation can be realized in terms of an intersection cohomology complex (a local system on G), but that of the $(q-1)$-dimensional irreducible representation appears as a difference of two intersection cohomology complexes, one given by the local system $\bar{\mathbf{Q}}_l$ on the unipotent radical of G and one supported by the unit element of G. A similar phenomenon occurs for G as in Section 9 and for a (q^2-1)-dimensional irreducible representation of $G(\mathbf{F}_q)$.

Acknowledgment. This work was supported in part by the National Science Foundation.

References

[BBD] A. Beilinson, J. Bernstein, and P. Deligne, Faisceaux pervers, in *Analyse et topologie sur les espaces singuliers*, Vol. 1, Astérisque 100, Société Mathématique de France, Paris, 1982.

[G] J. A .Green, The characters of the finite general linear groups, *Trans. Amer. Math. Soc.*, **80** (1955), 402–447.

[K] D. Kazhdan, Proof of Springer's hypothesis, *Israel J. Math.*, **28** (1977), 272–286.

[L1] G. Lusztig, *Characters of Reductive Groups over a Finite Field*, Annals of Mathematics Studies 107, Princeton University Press, Princeton, NJ, 1984.

[L2] G. Lusztig, Character sheaves I–V, *Adv. Math.*, **56** (1985), 193–237, **57** (1985), 226–265, **57** (1985), 266–315, **59** (1986), 1–63, **61**-2 (1986), 103–155.

[L3] G. Lusztig, Green functions and character sheaves, *Ann. Math.*, **131** (1990), 355–408.

[L4] G. Lusztig, *Hecke Algebras with Unequal Parameters*, CRM Monographs Series 18, American Mathematical Society, Providence, RI, 2003.

[L5] G. Lusztig, Parabolic character sheaves I, *Moscow Math. J.*, **4** (2004), 153–179.

[L6] G. Lusztig, Character sheaves on disconnected groups I–III, *Representation Theory*, **7** (2003), 374–403; **8** (2004), 72–124; **8** (2004), 125–144.

Symplectomorphism Groups and Quantum Cohomology

Dusa McDuff

Department of Mathematics
Stony Brook University
Stony Brook, NY 11794-3651
USA
dusa@math.sunysb.edu
http://www.math.sunysb.edu/~dusa

To my teacher, Israel Moiseevich Gelfand, on the occasion of his 90th birthday.

Summary. We discuss the question of what quantum methods (J-holomorphic curves and quantum homology) can tell us about the symplectomorphism group and its compact subgroups. After describing the rather complete information we now have about the case of the product of two 2-spheres, we describe some recent results of McDuff–Tolman concerning the symplectomorphism group of toric manifolds. This leads to an interpretation of the relations in the quantum cohomology ring of a symplectic toric manifold in terms of the Seidel elements of the generating circles of the torus action.

Key words: symplectomorphism group, Seidel representation, quantum cohomology, toric manifold, toric automorphism group

Subject Classifications: 53D35, 57R17

1 Introduction

1.1 Overview

The group $\mathrm{Symp}(M, \omega)$ of symplectomorphisms of a symplectic manifold (M, ω) is an interesting but largely unknown group. The manifold for which we have the most information is $S^2 \times S^2$ with its family of symplectic forms $\omega^\lambda := \lambda\pi_1^*(\sigma) + \pi_2^*(\sigma)$, where $\lambda \geq 1$. We begin by discussing recent results due to Abreu, McDuff, and Anjos–Granja on the homotopy type of the corresponding family of groups

$$\mathcal{G}^\lambda := \mathrm{Symp}(S^2 \times S^2, \omega^\lambda).$$

In all cases that have so far been calculated, the homotopy groups of $\mathcal{G}^\lambda$ are generated by its compact subgroups. These appear as the automorphism groups of the different toric structures on $S^2 \times S^2$.

As a first step towards generalizing these results, one can look at the relation of the toric automorphism group $\mathrm{Aut}(M, T)$ of a toric manifold (M, T) to $\mathrm{Symp}(M, \omega)$. Recent work by McDuff–Tolman gives examples where the inclusion of $\mathrm{Aut}(M, T)$ does not induce an injection on π_1. Nevertheless, our work suggests that the map $\pi_1(\mathrm{Aut}(M, T)) \to \pi_1 \mathrm{Symp}(M, \omega)$ is noninjective only in cases where the manifold M has very special structure. After discussing such questions, we explain a simple way of understanding the corrections needed to Batyrev's formula for the quantum cohomology ring of a non-Fano toric manifold. These come from the Seidel elements of the generating circles of the torus action. In the nef case they give a new perspective on Givental's change of variable formula that relates the I- and J-functions in his proof of the mirror conjecture.

This paper is rather narrowly focussed; more general information on symplectomorphism groups may be found in the survey articles [16, 17]. For background material on symplectic topology see McDuff–Salamon [18, 19].

1.2 Preliminaries

Consider a closed manifold M of dimension $2n$ and its symplectomorphism group $\mathrm{Symp}(M, \omega)$, consisting of all diffeomorphisms that preserve the symplectic form. The identity component $\mathrm{Symp}_0(M, \omega)$ contains a normal subgroup, the Hamiltonian group $\mathrm{Ham}(M, \omega)$, made up of the time-1 maps of the flows $\phi_t^H, t \geq 0$, generated by time-dependent Hamiltonian functions $H : M \times [0, 1] \to \mathbb{R}$. If $H^1(M; \mathbb{R}) = 0$, then $\mathrm{Ham}(M, \omega) = \mathrm{Symp}_0(M, \omega)$; in general it is a subgroup of codimension equal to $\dim H^1(M; \mathbb{R})$. In all cases, the group $\mathcal{G} := \mathrm{Ham}(M, \omega)$ may be considered as a Fréchet Lie group whose Lie algebra consists of all normalized Hamiltonians:

$$\mathrm{Lie}\,\mathcal{G} = \left\{ F : M \to \mathbb{R} \,\middle|\, \int_M F\omega^n = 0 \right\}.$$

As Reznikov pointed out in [22] the formula

$$\langle F, G \rangle := \int_M FG\omega^n$$

defines a nondegenerate form on $\mathrm{Lie}\,\mathcal{G}$ that is invariant under the adjoint action of $\mathcal{G}$, and so is analogous to a Killing form. Thus, although $\mathcal{G} := \mathrm{Ham}(M, \omega)$ is an infinite-dimensional group, its Lie algebra behaves like the Lie algebra of a compact Lie group, and one might hope that this is reflected in the topological properties of $\mathcal{G}$. Investigating this is one of the ideas behind this paper.

We first observe that the above invariant form may be used to define an analog of Chern–Weil theory for Hamiltonian bundles (i.e., bundles with fiber M and structural group $\mathcal{G} = \mathrm{Ham}$). Guillemin–Lerman–Sternberg pointed out in [8] that given any such bundle $M \to P \to B$ the fiberwise symplectic form ω has a closed extension Ω. One can normalize the class $[\Omega]$ by requiring that

$$\pi_!([\Omega]^{n+1}) = \int_M [\Omega]^{n+1} = 0 \in H^2(B),$$

where $\int_M$ denotes the integral over the fiber of $\pi : P \to B$. Such a form Ω defines an Ehresmann connection on the bundle $P \to B$ whose horizontal spaces are the Ω-orthogonal complements to the fiber. It turns out that the holonomy of this connection is Hamiltonian. Moreover, given vector fields $v, w \in T_b B\mathcal{G}$ with horizontal lifts $v^\sharp, w^\sharp$, the function $\Omega(v^\sharp, w^\sharp)(x)$ restricts on each fiber $M_b := \pi^{-1}(b)$ to an element $F(v, w) \in \mathrm{Lie}(\mathcal{G})$ that represents the curvature $\widetilde{\Omega}(v, w)$ of this connection at (v, w). By making finite-dimensional approximations, one can make sense of this construction on the universal Hamiltonian bundle

$$(M, \omega) \to (M_{\mathcal{G}}, \Omega) \to B\mathcal{G}.$$

Any Ad-invariant polynomial $\mathcal{I}^k : \mathrm{Lie}(\mathcal{G})^{\otimes k} \to \mathbb{R}$ therefore gives rise to a characteristic class $c_k^{\mathcal{I}}$ in $H^*(B\mathcal{G})$, namely the class represented by the closed real-valued $2k$-form $\mathcal{I}^k \circ \widetilde{\Omega}^k$. Just as in the case of $U(n)$ we may define $\mathcal{I}^k$ by using the Killing form, namely

$$\mathcal{I}^k(F_1 \otimes \cdots \otimes F_k) := \int_M F_1 \cdots F_k \omega^n.$$

We claim that up to a constant $c_k^{\mathcal{I}}$ equals the class defined by the fiberwise integral

$$\mu_k := \int_M [\Omega]^{n+k} \in H^{2k}(B\mathcal{G}). \tag{1.1}$$

The classes $c_k^{\mathcal{I}}$ are variants of the ones defined by Reznikov [22], while the μ_k were considered by Januszkiewicz–Kędra in [9]. The following proof is taken from Kędra–McDuff [10].

Lemma 1.1. *This class $c_k^{\mathcal{I}}$ is a nonzero multiple of μ_k.*

Proof. Let $v_1, \dots, v_{2k}$ be vector fields on $B\mathcal{G}$ with horizontal lifts $v_1^\sharp, \dots, v_{2k}^\sharp$. Then if the w_j are tangent to the fiber at $x \in M_{\mathcal{G}}$, we find

$$\Omega^{n+k}(w_1, \dots, w_{2n}, v_1^\sharp, \dots, v_{2k}^\sharp)(x) = \sum_\sigma \varepsilon(\sigma) \binom{n+k}{n} \times F_{1,\sigma}(x) \cdots F_{k,\sigma}(x) \omega^n(w_1, \dots, w_{2n}),$$

where, for each permutation σ of $\{1, \dots, 2k\}$, $\varepsilon(\sigma)$ denotes its signature and

$$F_{j,\sigma}(x) := \Omega(v_{\sigma(2j-1)}^\sharp, v_{\sigma(2j)}^\sharp)(x) = \widetilde{\Omega}(v_{\sigma(2j-1)}, v_{\sigma(2j)})(x).$$

Therefore, $(\pi_! \Omega^{n+k})(v_1, \dots, v_{2k}) = c\mathcal{I}^k \circ \widetilde{\Omega}^k(v_1, \dots, v_{2k})$, as claimed. □

2 The group of symplectomorphisms of $(S^2 \times S^2, \omega^\lambda)$

Consider the group $\mathcal{G}^\lambda := \mathrm{Symp}(S^2 \times S^2, \omega^\lambda)$ defined above. We need only consider the range $\lambda \geq 1$ since $\mathcal{G}^{1/\lambda}$ is isomorphic to $\mathcal{G}^\lambda$ (because $\omega^{1/\lambda}$ is a scalar multiple of ω^λ). We shall think of $S^2 \times S^2$ as a trivial S^2-bundle over S^2 where the base is identified with the first (i.e., the larger) factor.

Gromov proved in [7] that

$$\mathcal{G}^1 \simeq SO(3) \times SO(3), \qquad \mathcal{G}^\lambda \not\simeq \mathcal{G}^1 \quad \text{if } \lambda > 1,$$

where $\simeq$ denotes homotopy equivalence. Abreu [1] calculated $H^*(\mathcal{G}^\lambda; \mathbb{Q})$ for $1 < \lambda \leq 2$; his calculation was completed by Abreu–McDuff [2] to all λ. The following theorem combines this with some results from McDuff [15] and very recent work by Abreu, Granja, and Kitchloo.

Theorem 2.1.

(i) *The homotopy type of* $\mathcal{G}^\lambda$ *is constant on the intervals* $k < \lambda \leq k+1$, $k \geq 1$.
(ii) $H^*(\mathcal{G}^\lambda; \mathbb{Q}) \cong \Lambda(t, x, y) \otimes \mathbb{Q}[w_k]$ *when* $k < \lambda \leq k+1$. *Here* $\deg t = 1$, $\deg x = \deg y = 3$, *and* $\deg w_k = 4k$.
(iii) $H^*(B\mathcal{G}^\lambda; \mathbb{Q}) = \mathbb{Q}[T, X, Y]/T(X - Y + T^2) \dots (k^4 X - k^2 Y + T^2)$, *where* T, X, Y *are appropriate deloopings of* t, x, y, *respectively.*

We now explain the relation between statements (ii) and (iii) and their connection to the different toric structures on $S^2 \times S^2$. The generator t in (ii) is dual to the element $\tau \in \pi_1(\mathcal{G}^\lambda)$ represented by the circle action α on $S^2 \times S^2 \cong \mathbb{P}(\mathcal{O}(2) \oplus \mathbb{C})$ given by rotating the fiber of the line bundle $\mathcal{O}(2) \to \mathbb{CP}^1$. (Here we are identifying $S^2 \times S^2$ with the second Hirzebruch surface $\mathbb{P}(\mathcal{O}(2) \oplus \mathbb{C})$, where $\mathcal{O}(2)$ and $\mathbb{C}$ are bundles over $S^2 = \mathbb{CP}^1$. We denote by J_1 the corresponding complex structure on $S^2 \times S^2$.)

The generators x, y in (ii) are dual to the 3-spheres ξ, η in $\mathcal{G}^\lambda$ given by the inclusion of each factor of $SO(3) \times SO(3)$ in $\mathcal{G}^\lambda$. Thus $x + y$ is dual to the diagonal copy $\xi + \eta$ of $SO(3)$ in $SO(3) \times SO(3)$. We claim that that this commutes with the circle action α. Indeed, one can identify $S^2 \times S^2$ with the toric manifold $\mathbb{P}(\mathcal{O}(2) \oplus \mathbb{C})$ in such a way that its toric (or Kähler) automorphism group $K_1 := \mathrm{Aut}(J_1)$ coincides with the product $SO(3) \times S^1$ of α with the diagonal copy of $SO(3)$: see [2]. In this realization α has the formula

$$\tau_t(z, w) \mapsto (z, R_{t,z} w),$$

where $R_{t,z} : S^2 \to S^2$ is the rotation through angle $2\pi t$ with axis through the point z and its antipode. Thus α fixes the points on the diagonal and antidiagonal in $S^2 \times S^2$ and commutes with the diagonal $SO(3)$ action.

If $1 < \lambda \leq 2$, then one can show that α does not commute with the individual $SO(3)$-factors ξ, η even up to homotopy. More precisely, one can show that the remaining generator $w_1 \in H^4(\mathcal{G}^\lambda)$ is dual to the element $[\xi, \tau] \in \pi_4(\mathcal{G}^\lambda)$ given by the Samelson product:

$$\begin{array}{ccc} S^3 \times S^1 & \xrightarrow{(\xi,\tau)\mapsto \xi\tau\xi^{-1}\tau^{-1}} & \mathcal{G}^\lambda \\ \downarrow & & =\downarrow \\ S^4 := S^3 \times S^1/S^3 \vee S^1 & \xrightarrow{[\xi,\tau]} & \mathcal{G}^\lambda. \end{array}$$

Since Samelson products deloop to Whitehead products, the element w_1 is not transgressive but rather gives rise to the relation $T(X - Y + T^2) = 0$ in $H^*(B\mathcal{G}^\lambda)$.

When $\lambda > 2$ there is another Hirzebruch structure on $S^2 \times S^2$ that supports a Kähler structure in the class $[\omega^\lambda]$, namely the complex structure J_2 coming from the identification of $S^2 \times S^2$ with $\mathbb{P}(\mathcal{O}(4) \oplus \mathbb{C})$. The autmorphism group $\mathrm{Aut}(J_2)$ of this structure is again isomorphic to $SO(3) \times S^1$ but its image K_2 in $\mathcal{G}^\lambda$ contains the rational homotopy classes $\xi + 4\eta$, τ. Therefore, the class τ can be represented in $\mathcal{G}^\lambda$ by a circle action in K_1 that commutes with $\xi + \eta$ and by a circle action in K_2 that commutes with $\xi + 4\eta$. Hence by the linearity of the Samelson product $[\xi, \tau]$ now vanishes in $\pi_*(\mathcal{G}^\lambda)$. Therefore, one can define the higher product $[\xi, \xi, \tau]$ and it turns out that this does not vanish when $2 < \lambda \le 3$. Similarly, when $k < \lambda \le k + 1$ there are $k + 1$ toric structures on $S^2 \times S^2$ and w_k is dual to a kth order Samelson product of the form $[\xi, \dots, \xi, \tau]$.

2.1 The integral homotopy type of $\mathcal{G}^\lambda$

We now explain some recent work by Anjos [3] and Anjos–Granja [4] that gives a beautiful description of the full homotopy type of $\mathcal{G}^\lambda$ in the first interesting case, namely $1 < \lambda \le 2$. This description arises naturally from the geometry that is the basis of the proof of Theorem 2.1. The arguments go back, of course, to Gromov's original paper on J-holomorphic curves. Recall that an almost complex structure J on a symplectic manifold (M, ω) is said to be *tamed* by ω if $\omega(v, Jv) > 0$ for all nonzero $v \in TM$. Further, a map $f : S^2 \to M$ is said to be *J-holomorphic* if $df \circ j = J \circ df$, where j is the standard complex structure on S^2. For short, one sometimes calls such f a J-sphere.

Gromov looked at the contractible space

$$\mathcal{J}^\lambda$$

of all almost complex structures J on $M := S^2 \times S^2$ that are tamed by ω^λ. He proved that when $\lambda = 1$ every $J \in \mathcal{J}^1$ has the same pattern of holomophic curves as does the product structure $J_0 := j \times j$. In particular, there are two foliations by J-spheres, one consisting of spheres in the class $A := [S^2 \times pt]$ and the other of spheres in the class $B := [pt \times S^2]$. For each such J one can think of these spheres as providing "coordinates" on $M = S^2 \times S^2$, i.e., there is a diffeomorphism $\psi_J : M \to M$ that takes the standard foliations of M by the spheres $S^2 \times pt$, $pt \times S^2$ to the two foliations by J-spheres. This diffeomorphism ψ_J is not quite symplectic, but has a canonical homotopy to an element $\psi_J' \in \mathcal{G}^1$. Moreover, ψ_J' is independent of choices modulo the subgroup $SO(3) \times SO(3) = \mathrm{Aut}(J_0)$. It follows that the sequence of maps

$$\mathrm{Aut}(J_0) \longrightarrow \mathcal{G}^1 \longrightarrow \mathcal{J}$$

is a fibration up to homotopy, that is split by the map $J \mapsto \psi'_J$. Since $\mathcal{J}$ is contractible, we arrive at Gromov's result that $\mathrm{Aut}(J_0) \simeq \mathcal{G}^1$.

When $1 < \lambda \leq 2$ it is no longer true that all elements in $\mathcal{J}^\lambda$ have the same pattern of J-spheres. There is now a second model, namely the pattern formed by the J_1-curves. In this case there is only one foliation, by spheres in the class $B = [pt \times S^2]$ of the smaller sphere and there is an isolated curve in the class $A - B$ of the antidiagonal (the rigid curve $\mathbb{P}(\mathcal{O}(2) \oplus 0)$ in the Hirzebruch surface). In this case there can be no A-curves, by positivity of intersections: if A_1, A_2 are represented by distinct connected J-curves, then one has $A_1 \cdot A_2 \geq 0$. Thus

$$\mathcal{J}^\lambda = U_0 \cup U_1,$$

where, for $i = 0, 1$, the set U_i consists of all J whose J-spheres are like those of J_i. This decomposition has the following properties:

- U_1 is a codimension 2-submanifold of $\mathcal{J}^\lambda$. In other words, there is a neighborhood NU_1 of U_1 in $\mathcal{J}^\lambda$ such that

$$NU_1 \setminus U_1 \to U_1$$

  is an S^1-bundle.
- $\mathrm{Aut}(J_1) = SO(3) \times S^1$.
- There are homotopy equivalences

$$\mathcal{G}^\lambda / \mathrm{Aut}(J_0) \simeq U_0, \qquad \mathcal{G}^\lambda / \mathrm{Aut}(J_1) \simeq U_1.$$

The following theorem shows that $\mathcal{G}^\lambda$ is homotopic to an amalgamated free product of the two Lie groups $\mathrm{Aut}(J_0)$ and $\mathrm{Aut}(J_1)$.

Theorem 2.2. *$\mathcal{G}^\lambda$ is homotopy equivalent to the pushout of the diagram*

$$\begin{array}{ccc} SO(3) & \longrightarrow & \mathrm{Aut}(J_0) \\ \downarrow & & \\ \mathrm{Aut}(J_1) & & \end{array}$$

in the category of topological groups.

To prove this it suffices to show that the pushout of the quotients

$$\begin{array}{ccc} \mathcal{G}^\lambda / SO(3) & \longrightarrow & \mathcal{G}^\lambda / \mathrm{Aut}(J_0) \\ \downarrow & & \\ \mathcal{G}^\lambda / \mathrm{Aut}(J_1) & & \end{array}$$

in the homotopy category is contractible. But the above remarks show that this is equivalent to the pushout of the diagram

$$\begin{array}{ccc} NU_1 \setminus U_1 & \longrightarrow & U_0 \\ \downarrow & & \\ U_1. & & \end{array}$$

Since all the maps here are cofibrations (inclusions), the pushout is simply the contractible set $U_0 \cup U_1 = \mathcal{J}^\lambda$.

It is very likely that this work can be extended to all $\lambda > 1$, although the resulting pushout diagrams will be more complicated. As already mentioned when $k < \lambda \leq k+1$, there are $k+1$ different integrable complex structures $J_0, \dots, J_k$ that are tamed by ω^λ, and the corresponding Lie groups $\mathrm{Aut}(J_i)$ define the structure of the generators and relations for the rational homotopy type of $\mathcal{G}^\lambda$ and $B\mathcal{G}^\lambda$. One can show (using a gluing argument) that there is a corresponding stratification of $\mathcal{J}^\lambda$. Therefore, all the ingredients are in place except that one has to find a pushout diagram (or other categorical construction) that corresponds to a higher order Whitehead product.

Similar results are true for the nontrivial S^2-bundle over S^2, i.e., the one-point blowup of $\mathbb{CP}^2$, and also, by recent work of Lalonde–Pinsonnault [13], for the one-point blowup of $S^2 \times S^2$ in the range $1 < \lambda \leq 2$. Here are two open problems:

- What happens with the many-point blowup of $\mathbb{CP}^2$?
- What is the homotopy type of $\mathrm{Symp}(T^2 \times S^2, \lambda\sigma_0 + \sigma_1)$ for $\lambda > 0$?

Lalonde and Pinsonnault are working on the first of these problems. Some information on the second may be found in McDuff [15] and Buse [6]. In particular, the homotopy type of the group is constant for λ in the range $0 < \lambda \leq 1$. However, it is not yet known what this group is, even rationally.

3 Higher-dimensional toric manifolds

A $2n$-dimensional symplectic manifold (M, ω) is said to be toric if it admits a Hamiltonian action of an n-torus $T := T^n$. Such a manifold (M, T) can always be realised as the symplectic reduction $M := \mathbb{C}^N /\!\!/ T'$ at some element $\nu \in (\mathfrak{t}')^* \cong \mathbb{R}^{N-n}$ of a high-dimensional Euclidean space $\mathbb{C}^N$ by the action of a $(N-n)$-dimensional subtorus T' of the standard N-torus $T^N \subset U(N)$. Further, T can be identified with the quotient T^N/T' and the toric autmorphism group $\mathrm{Aut}(M, T)$ is the quotient by T' of the centralizer of T' in $U(N)$:

$$\mathrm{Aut}(M, T) := \mathrm{Cent}(T')/T'.$$

Here is an example. Let M be the symplectic quotient $\mathbb{C}^5 /\!\!/ T'$, where T' is a 2-torus whose generators ξ_1, ξ_2 act with the weights $(1, 1, 1, 0, 0)$ and $(5, 1, 0, 1, 1)$, respectively. Quotienting out by ξ_2 gives the vector bundle $\mathcal{O}_5 \oplus \mathcal{O}_1 \oplus \mathbb{C} \to \mathbb{CP}^1$, and quotienting that by ξ_1 gives the projectivization. Thus M is a bundle over $\mathbb{CP}_1$ with fiber $\mathbb{CP}^2$. M can also be written as the (ordinary) quotient $S^3 \times_{S^1} \mathbb{CP}^2$, where S^1 acts on $\mathbb{CP}^2$ via the circle $\mathrm{diag}(\lambda^3, \lambda^{-1}, \lambda^{-2})$, $\lambda \in S^1$, in $PSU(3)$. Since this circle contracts in $PSU(3)$, M is diffeomorphic to the product, and it is shown in McDuff–Tolman [21] that if we choose any of its toric Kähler forms, it is symplectomorphic to a product $(\mathbb{CP}^1 \times \mathbb{CP}^2, \omega^\mu)$, where

$$\omega^\mu := \mu\sigma_{\mathbb{CP}^1} + \sigma_{\mathbb{CP}^2}, \quad \mu > 3.$$

(Here we assume that the forms $\sigma_{\mathbb{CP}^i}$ are normalized to have integral 1 over the projective line.)

Observe that the centralizer of T' is isomorphic to $S^1 \times S^1 \times S^1 \times U(2)$, so that $\mathrm{Aut}(M, T) = T^3 \times U(2)/T'$. Thus $\pi_1(\mathrm{Aut}(M, T))$ has rank 2.

The product $(\mathbb{CP}^1 \times \mathbb{CP}^2, \omega^\mu)$ has many toric structures, denoted $(M_{a,b}, T_{a,b})$, that are obtained similarly but with ξ_2 allowed to have any weights $(a, b, 0, 1, 1)$ such that

$$a \geq b \geq 0, \qquad a + b = 3k, \qquad 3\mu > 2a - b.$$

Thus for $1 < \mu \leq 2$ the possibilities for (a, b) are $(0, 0)$, $(2, 1)$ and $(3, 3)$, while for $2 < \mu \leq 3$ one must add to these the pairs $(3, 0)$, $(4, 2)$, $(5, 4)$, and $(6, 6)$.

It would be very interesting to understand the relation of the different Lie subgroups $\mathrm{Aut}(M_{a,b}, T_{a,b})$ to the homotopy groups of $\mathcal{G}^\mu := \mathrm{Symp}(M, \omega^\mu)$. One cannot hope for such complete information as in the four-dimensional case since the analysis there was based on our exhaustive knowledge of the J-curves. Nevertheless, the more elementary aspects of the four-dimensional case do generalize. For example, the group $\pi_1(\mathcal{G}^\mu)$ for $\mu > 1$ still contains an element of infinite order that does not appear in $\mathcal{G}^1$. This was discovered by Seidel [24]. This element is realised as the circle action corresponding to a suitable facet of the moment polytope of $(M_{2,1}, T_{2,1})$ and hence appears in $\mathrm{Aut}(M_{2,1}, T_{2,1})$.

In general, given a symplectic toric manifold (M, T, ω^μ), one can try to understand the induced map

$$\pi_* \mathrm{Aut}(M, T) \to \pi_*(\mathcal{G}^\mu)$$

where $\mathcal{G}^\mu := \mathrm{Symp}_0(M, \omega^\mu)$? For example, is it always injective, at least rationally? How does it vary with the cohomology class of $[\omega^\mu]$?

The following result from [21] applies to generic low-dimensional toric manifolds.

Proposition 3.1. *If* $\dim M \leq 6$ *and* $\mathrm{Aut}(M, T) = T$, *then* $\pi_1(T)$ *injects into* $\pi_1(\mathcal{G}^\mu)$ *for all forms* ω^μ.

The proof is elementary, using only the geometry of the moment polytope. One might hope that this result would generalize to all dimensions, but that is not so; it already fails in dimension 8. One important reason for this lack of injectivity is demonstrated in the example $(M_{a,b}, T_{a,b})$ discussed above. We saw earlier that $\pi_1(\mathrm{Aut}(M, T))$ has rank 2 when $a > b > 0$. On the other hand this automorphism group is contained in the group $\mathcal{G}_{\mathrm{fib}}$ of diffeomorphisms of M that commute with the projection to $\mathbb{CP}^1$ and restrict to symplectomorphisms on each fiber $pt \times \mathbb{CP}^2$. Since $\mathrm{Symp}_0(\mathbb{CP}^2) \simeq PSU(3)$ (by Gromov), $\mathcal{G}_{\mathrm{fib}}$ deformation retracts to the product of the group of orientation preserving diffeomorphisms of the base with the group of fiberwise diffeomorphisms that fix each fiber:

$$\mathcal{G}_{\mathrm{fib}} \simeq SO(3) \times \mathrm{Map}(S^2, PSU(3)).$$

Hence $\pi_1(\mathcal{G}_{\mathrm{fib}})$ has rank 1. Although the elements of $\mathcal{G}_{\mathrm{fib}}$ are not symplectomorphisms, a standard Moser-type argument shows that any compact subset of $\mathcal{G}_{\mathrm{fib}}$ can be homotoped into $\mathcal{G}^\mu$ for sufficiently large μ. It follows that

$$\pi_1(\mathrm{Aut}(M_{2,1}, T_{2,1})) \to \pi_1(\mathcal{G}^\mu)$$

is not injective for sufficiently large μ. By directly constructing a homotopy, one can show that in fact it is not injective for all $\mu > 1$: see [21].

Note that $\mathrm{Aut}(M, T)$ is always a maximal connected compact subgroup of $\mathrm{Symp}_0(M)$ because its subgroup T is a maximal connected abelian subgroup of $\mathrm{Symp}_0(M)$. Therefore, by analogy with what happens in the finite-dimensional case (where a simple G deformation retracts to its maximal compact subgroup), one would expect that at the very least the homotopy carried by $\mathrm{Aut}(M, T)$ would not completely disappear in $\mathrm{Ham}(M, \omega)$. The next result from Kędra–McDuff [10] gives some supporting evidence.

Theorem 3.2. *Let (M, ω) be a symplectic manifold of dimension $2n$ and set $\mathcal{G} := \mathrm{Ham}(M, \omega)$. Suppose given a nonconstant homomorphism $\alpha : S^1 \to \mathcal{G}$ that represents the zero element in $\pi_1(\mathcal{G})$ and so extends to a map $\widetilde{\alpha} : D^2 \to \mathcal{G}$. Define $\rho \in \pi_3(\mathcal{G}) \otimes \mathbb{Q}$ by*

$$\begin{aligned} S^3 := (D^2 \times S^1)/((D^2 \times \{1\}) \vee (\partial D^2 \times S^1)) &\to \mathcal{G}, \qquad (3.1)\\ (z, t) &\mapsto [\widetilde{\alpha}(z), \alpha(t)], \end{aligned}$$

where the bracket $[\phi, \psi]$ denotes the commutator $\phi\psi\phi^{-1}\psi^{-1}$. Then $\rho \neq 0$ and is independent of the choice of extension $\widetilde{\alpha}$. Moreover, ρ transgresses to an element $\overline{\rho} \in \pi_4(B\mathcal{G}) \otimes \mathbb{Q}$ with nonzero image in $H_4(B\mathcal{G})$.

This is proved by showing that the characteristic class μ_2 defined in (1.1) is nontrivial on $\overline{\rho}$. As an example, if α is a nonzero element in the kernel of the map $\pi_1(\mathrm{Aut}(M_{2,1}, T_{2,1})) \to \pi_1(\mathcal{G}^\mu)$, then ρ is represented in the $PSU(3)$-factor of

$$\mathrm{Aut}(M_{0,0}, T_{0,0}) \cong PSU(3) \times SO(2).$$

We conclude with several remarks.

- Because the element ρ above is detected by a characteristic class that exists on all simply connected symplectic manifolds, it is very robust and persists under small variations of the class $[\omega]$ of the form. This should be contrasted with the elements w_k of Theorem 2.1 that disappear under appropriate perturbations of $[\omega]$. Thus these elements are quite different in nature even though they are constructed in ostensibly similar ways, i.e., via commutators.
- In the case of the toric manifolds $(M_{a,b}, T_{a,b}, \omega^\mu)$ it would be interesting to work out the relation between the groups $\mathcal{G}^\mu$ and the fiberwise diffeomorphism group $\mathcal{G}_{\mathrm{fib}}$. There is an analogous question in the case $(S^2 \times S^2, \omega^\lambda)$. But here one can use the existence of J-spheres in the fiber class B to define natural maps $\mathcal{G}^\lambda \to \mathcal{G}^{\lambda'}$ whenever $\lambda < \lambda'$ and can show that

$$\lim_{\lambda \to \infty} \mathcal{G}^\lambda \simeq \mathcal{G}_{\mathrm{fib}}.$$

In the six-dimensional case such maps do not seem to exist. Nevertheless, one still should be able to make sense of the limit as $\mu \to \infty$ and to investigate its relation to the group $\mathcal{G}_{\text{fib}}$. In this case, of course, there is no symmetry between μ and $1/\mu$ and so one could similarly consider the limit as $\mu \to 0$. This should be related to the group of fiberwise diffeomorphisms of the fibration $\mathbb{CP}^1 \times \mathbb{CP}^2 \to \mathbb{CP}^2$.

- An interesting question in the six-dimensional case is the extent to which J-spheres can give useful information. One can no longer use their geometry; for example, spheres in the class $[\mathbb{CP}^1 \times pt]$ need not form the leaves of a foliation. Nevertheless, one can get information from quantum cohomology as in Seidel [24]. Such methods allow one to show that the elements in $\pi_*(\mathcal{G}_{\text{fib}}) \otimes \mathbb{Q}$ coming from $\pi_*(\Omega^2(PSU(3))) \otimes \mathbb{Q}$ do not appear in $\mathcal{G}^\mu$ for $\mu \leq 1$ though they are there when $\mu > 1$. Buse [6] has a different approach to these problems that uses equivariant Gromov–Witten invariants.

4 Quantum cohomology of toric manifolds

Consider a compact symplectic manifold (M, ω) and its Hamiltonian group $\mathcal{G} := \mathrm{Ham}(M, \omega)$. One very useful tool in understanding $\pi_1(\mathcal{G})$ is Seidel's representation

$$\mathcal{S} : \pi_1(\mathcal{G}) \to \mathrm{Units}(\mathrm{QH}^*(M))$$

of $\pi_1(\mathcal{G})$ in the group of multiplicative units in the quantum cohomology of M. Here we use the coefficients $\Lambda := \Lambda^{\mathrm{univ}}[q, q^{-1}]$ for quantum cohomology $\mathrm{QH}^*(M) := H^*(M) \otimes \Lambda$, where q is a variable of degree 2 and Λ^{univ} is a generalized Laurent series ring in a variable t of degree 0:

$$\Lambda^{\mathrm{univ}} := \left\{ \sum_{\kappa \in \mathbb{R}} r_\kappa t^\kappa \;\middle|\; r_\kappa \in \mathbb{Q},\ \#\{\kappa < c \mid r_\kappa \neq 0\} < \infty\ \forall c \in \mathbb{R} \right\}.$$

Thus typical elements in $\mathrm{QH}^*(M)$ have the form $\sum_{d,\kappa} a_{d,\kappa} q^d t^\kappa$ where $a_{d,\kappa} \in H^*(M)$ satisfy the same finiteness condition as do the r_κ.

Given an element $\phi \in \pi_1(\mathcal{G})$, the element $\mathcal{S}(\phi)$ is constructed as follows. First consider the Hamiltonian fibration

$$(M, \omega) \longrightarrow (P_\phi, \Omega) \xrightarrow{\ \pi\ } S^2$$

whose clutching function is a representing loop $\{\phi_t\}$ for ϕ. Thus P_ϕ is the union of two copies of $D^2 \times M$ whose boundaries are identified via $\{\phi_t\}$. It carries two canonical cohomology classes, $u \in H^2(P_\phi)$ which is the class of the coupling form and $c_1^{\mathrm{Vert}} \in H^2(P_\phi)$, the first Chern class of the vertical tangent bundle.

As mentioned in §1.2, the fiberwise form ω has a closed extension Ω that we may assume to be symplectic by adding to it the pullback of a suitable area form on the base. Choose an Ω-tame almost complex structure $\widetilde{J}$ on P_ϕ that preserves the tangent bundle to the fiber and projects under π to the standard complex structure on S^2. Then, because the fibers of π are $\widetilde{J}$-holomorphic, every $\widetilde{J}$-sphere $f : S^2 \to P_\phi$ that represents a section class $\widetilde{A}$ in P_ϕ (i.e., a class such that $\widetilde{A} \cap [M] = 1$) may be parametrized as a section. Denote by $\mathcal{M}(\widetilde{A}, \widetilde{J})$ the space of all such sections. In good cases this is a manifold with boundary of codimension ≥ 2, so that its intersection with a fixed fiber $[M]$ represents a homology class in M which we denote $\alpha_{\widetilde{A}} := [\mathcal{M}(\widetilde{A}, \widetilde{J})] \cap [M]$. In general, this homology class is defined using Gromov–Witten invariants in P:

$$\alpha_{\widetilde{A}} \cdot_M \beta = \mathrm{GW}^P_{\widetilde{A},3}([M], [M], i_*(\beta)) \quad \forall \beta \in H_*(M).$$

Here $[M] \in H_*(P)$ denotes the homology class of a fiber and $i : M \to P$ is the inclusion of a fiber. Then we define

$$\mathcal{S}(\phi) := \sum_{\widetilde{A}} \mathrm{PD}(a_{\widetilde{A}}) \otimes q^{c_1^{\mathrm{Vert}}(\widetilde{A})} t^{u(\widetilde{A})}.$$

One can show that this sum satisfies the requisite finiteness condition and so represents an element in $\mathrm{QH}^*(M) = H^*(M) \otimes \Lambda$ of degree 0. Further, the image $\mathcal{S}(0)$ of the constant loop is the multiplicative unit $1 \in H^*(M) \subset \mathrm{QH}^*(M)$, and a gluing argument shows that

$$\mathcal{S}(\phi + \psi) = \mathcal{S}(\phi) * \mathcal{S}(\psi),$$

where $*$ denotes quantum multiplication. (Proofs in various contexts may be found in Seidel [23], Lalonde–McDuff–Polterovich [12], McDuff [14] and McDuff–Salamon [19]. We denote the group operation in $\pi_1(\mathcal{G})$ by $+$ since this is an abelian group.)

In general it is not easy to calculate $\mathcal{S}(\phi)$. The following result is proved in McDuff–Tolman [20]. It applies when ϕ is represented by a circle action $t \mapsto \phi_t$. Denote by $K : M \to \mathbb{R}$ the normalized moment map of this action (i.e., $\int_M K\omega^n = 0$), and by

$$X_{\max} := K^{-1}(\max K)$$

the maximal fixed point set. Further, if J is an S^1-invariant and ω-tame almost complex structure then we say that (M, J) is Fano (respectively, nef) if $c_1(TM, J)$ is positive (respectively, nonnegative) on every J-sphere.

Theorem 4.1. *Suppose that ϕ_K is represented by a circle action with normalized moment map K. Suppose further that the weights of the linearized action at the maximal fixed point component $X_{\max}$ are 0 or -1. Then*

$$\mathcal{S}(\phi_K) = \mathrm{PD}[X_{\max}] \otimes q^{-1} t^{-\max K} + \sum_{\kappa > -\max K, d} a_{d,\kappa} \otimes q^d t^\kappa.$$

If (M, J) is Fano all the higher order terms $a_{d,\kappa} \in H^*(M)$ vanish, while if (M, J) is nef they vanish unless $\deg(a_{d,\kappa}) \leq 2$.

Again following [20], we now explain what this theorem tells us about the quantum cohomology of a symplectic toric manifold (M, T). We begin by reviewing the structure of the usual cohomology ring. Denote the Lie algebra of T by $\mathfrak{t}$ and its dual by $\mathfrak{t}^*$. Let $\Phi : M \to \mathfrak{t}^*$ be the normalized moment map for the T-action, i.e., each of its components is mean normalized. The image of Φ is a convex polytope $\Delta \subset \mathfrak{t}^*$, and we denote its facets (the codimension one faces) by $D_1, \dots, D_N$ and the outward primitive integral normal vectors by $\eta_1, \dots \eta_N \in \mathfrak{t}$. Let Σ be the set of subsets $I = \{i_1, \dots, i_k\} \subseteq \{1, \dots, N\}$ for which $D_{i_1} \cap \dots \cap D_{i_k} \neq \emptyset$. Define two ideals in $\mathbb{Q}[x_1, \dots, x_N]$:

$$P(\Delta) = \left\langle \sum (\xi, \eta_i) x_i \;\middle|\; \xi \in \mathfrak{t}^* \right\rangle \quad \text{and} \quad SR(\Delta) = \langle x_{i_1} \cdots x_{i_k} \mid \{i_1, \dots, i_k\} \notin \Sigma \rangle.$$

A subset $I \subseteq \{1, \dots, N\}$ is called *primitive* if I is not in Σ but every proper subset is. Clearly,

$$SR(\Delta) = \langle x_{i_1} \cdots x_{i_k} \mid \{i_1, \dots, i_k\} \subseteq \{1, \dots, N\} \text{ is primitive} \rangle.$$

The following result is well known.

Proposition 4.2. *The map that sends x_i to the Poincaré dual of $\Phi^{-1}(D_i)$ (which we shall also denote by $x_i \in H^2(M)$) induces an isomorphism*

$$\mathbb{Q}[x_1, \dots, x_N]/(P(\Delta) + SR(\Delta)) \cong H^*(M, \mathbb{Q}). \tag{4.1}$$

Moreover, there is a natural isomorphism between $H_2(M; \mathbb{Z})$ and the set of tuples $(a_1, \dots, a_N) \in \mathbb{Z}^N$ such that $\sum a_i \eta_i = 0$, under which the pairing between such an element of $H_2(M, \mathbb{Z})$ and x_i is a_i.

The linear functional η_i is constant on D_i; let $\eta_i(D_i)$ denote its value. Under the isomorphism of (4.1) (extended to real coefficients),

$$[\omega] = \sum_i \eta_i(D_i) x_i \quad \text{and} \quad c_1(M) = \sum_i x_i. \tag{4.2}$$

Note also that each element η_i lies in the integer lattice of $\mathfrak{t}$ and so corresponds to a circle action λ_i on M. By Theorem 4.1,

$$\mathcal{S}(\lambda_i) = y_i \otimes q^{-1} t^{-\eta_i(D_i)} \in \text{Units}(\text{QH}^*(M)),$$

where the element y_i has the form $x_i + \sum_{\kappa > 0} a_{d,\kappa} q^d t^\kappa$.

We are now ready to examine the quantum cohomology of a toric variety. Given any face of Δ, let $D_{j_1}, \dots, D_{j_\ell}$ be the facets that intersect to form this face. The *dual cone* is the set of elements in $\mathfrak{t}$ which can be written as a positive linear combination of $\eta_{j_1}, \dots, \eta_{j_\ell}$. Every vector in $\mathfrak{t}$ lies in the dual cone of a unique face of Δ. Therefore, given any subset $I = \{i_1, \dots, i_k\} \subseteq \{1, \dots, N\}$ there is a unique face of Δ so that $\eta_{i_1} + \dots + \eta_{i_k}$ lies in its dual cone. Let $D_{j_1}, \dots, D_{j_\ell}$ be the facets that intersect to form this unique face. Then there exist unique positive integers $m_1, \dots, m_\ell$ so that

$$\eta_{i_1} + \dots + \eta_{i_k} - m_1 \eta_{j_1} - \dots - m_\ell \eta_{j_\ell} = 0.$$

Batyrev showed that if I is primitive the sets I and $J = \{j_1, \dots, j_\ell\}$ are disjoint. Let $\beta_I \in H_2(M, \mathbb{Z})$ be the class corresponding to the above relation. By (4.2), we see that

$$c_1(\beta_I) = k - m_1 - \cdots - m_\ell,$$

and

$$\omega(\beta_I) = \eta_{i_1}(D_{i_1}) + \cdots + \eta_{i_k}(D_{i_k}) - m_1\eta_{j_1}(D_{j_1}) - \cdots - m_\ell\eta_{j_\ell}(D_{j_\ell}).$$

Since $\eta_{i_1} + \cdots + \eta_{i_k} = m_1\eta_{j_1} + \cdots + m_\ell\eta_{j_\ell}$, the corresponding circle actions are also equal. Using the fact that the Seidel representation is a homomorphism, we find

$$\begin{aligned} y_{i_1} * \cdots * y_{i_k} &\otimes q^{-k} t^{-\eta_{i_1}(D_{i_1}) - \cdots - \eta_{i_k}(D_{i_k})} \\ &= {y_{j_1}}^{m_1} * \cdots * {y_{j_\ell}}^{m_\ell} \otimes q^{-m_1 - \cdots - m_\ell} t^{-m_1\eta_{j_1}(D_{j_1}) - \cdots - m_\ell\eta_{j_\ell}(D_{j_\ell})}. \end{aligned}$$

Therefore,

$$y_{i_1} * \cdots * y_{i_k} - {y_{j_1}}^{m_1} * \cdots * {y_{j_\ell}}^{m_\ell} \otimes q^{c_1(\beta_I)} t^{\omega(\beta_I)} = 0.$$

Since $x_1, \ldots, x_N$ generate $H^*(M)$, the natural homomorphism

$$\Theta : \mathbb{Q}[x_1, \ldots, x_N] \otimes \Lambda \to \mathrm{QH}^*(M)$$

which takes x_i to the Poincaré dual of $\Phi^{-1}(D_i)$ is surjective. To compute $\mathrm{QH}^*(M)$, we need to find the kernel of Θ. It is not hard to check that there is

$$Y_i = x_i + \text{ higher order terms } \in \mathbb{Q}[x_1, \ldots, x_N] \otimes \Lambda \tag{4.3}$$

such that $\Theta(Y_i) = y_i$. Define an ideal $SR_Y(\Delta) \subset \mathbb{Q}[x_1, \ldots, x_N] \otimes \Lambda$ by

$$\begin{aligned} &SR_Y(\Delta) \\ &= \langle Y_{i_1} \cdots Y_{i_k} - {Y_{j_1}}^{m_1} \cdots {Y_{j_\ell}}^{m_\ell} \otimes q^{c_1(\beta_I)} t^{\omega(\beta_I)} \mid I = \{i_1, \ldots, i_k\} \text{ is primitive}\}\rangle, \end{aligned} \tag{4.4}$$

where the Y_i are as in (4.3). Note that $SR_Y(\Delta)$ depends on the Y_i. Additionally, even if y_i is known, it is not in general possible to describe Y_i without prior knowledge of the ring structure on $\mathrm{QH}^*(M)$. On the other hand, SR_Y is clearly contained in the kernel of Θ. Moreover, Batyrev shows that $\omega(\beta_I) > 0$ for all primitive I. Hence we conclude the following.

Proposition 4.3. *Let* $\mathrm{QH}^*(M)$ *denote the small quantum cohomology of the toric manifold* (M, ω) *with coefficients* $\Lambda = \Lambda^{\mathrm{univ}}[q, q^{-1}]$. *The map which sends* x_i *to the Poincaré dual of* $\Phi^{-1}(D_i)$ *induces an isomorphism*

$$\mathbb{Q}[x_1, \ldots, x_N] \otimes \Lambda / \langle P(\Delta) + SR_Y(\Delta)\rangle \cong \mathrm{QH}^*(M).$$

In the Fano case, Theorem 4.1 states that the higher order terms in Y_i vanish. Therefore, we recover the formula for the small quantum cohomology of a Fano toric variety given by Batyrev and proved by Givental.

In the nef case there may be higher order terms in the Seidel elements y_i. However, these terms only involve cohomology classes $a_{d,\kappa}$ of degree ≤ 2. Therefore, $a_{d,\kappa}$

either lifts to the unit 1 in $\mathbb{Q}[x_1, \dots, x_N] \otimes \Lambda$ or to some linear combination of the x_i that is unique modulo the additive relations $P(\Delta)$. Hence we do not need to know the quantum multiplication in M in order to define the Y_i. The rest of the information needed to define the relations $P(\Delta)$ and $SR_Y(\Delta)$ is contained explicitly in Δ.

Thus in the nef case, once one knows the Seidel elements $\Psi(\Lambda_i)$, $i = 1, \dots, N$, there is an easy formula for the quantum cohomology ring based on the combinatorics of the moment polytope Δ. This substitution of the Y_i for the x_i in the Stanley–Reisner ring SR_Y should be related to Givental's change of variable formulae as discussed in [5, 11.2.5.2].

It should also be possible to calculate the Y_i by an explicit formula from the polytope. In [20] we show that its terms are generated by certain chains of edges in the polytope, and that the coefficient of each such term is determined locally, i.e., by a neighborhood of the chain. However we do not attempt to calculate these coefficients. Even in nef examples in four dimensions, the chains can be quite complicated.

Observe finally that here we are working with the stripped down coefficient ring $\Lambda^{\text{univ}}[q, q^{-1}]$. However, as described in McDuff–Salamon [19, Chapter 11.4], it is possible to obtain a similar description of the quantum cohomology for (M, T) with coefficients in the usual Novikov ring (the completed group ring of $H_2(M; \mathbb{Z})/\text{tor}$) by varying the cohomology class of $[\omega]$.

Acknowledgment. The author is partly supported by the NSF grant DMS 0305939.

References

[1] M. Abreu, Topology of symplectomorphism groups of $S^2 \times S^2$, *Invent. Math.*, **131** (1998), 1–23.
[2] M. Abreu and D. McDuff, Topology of symplectomorphism groups of rational ruled surfaces, *J. Amer. Math. Soc.*, **13** (2000), 971–1009.
[3] S. Anjos, The homotopy type of symplectomorphism groups of $S^2 \times S^2$, *Geom. Topol.*, **6** (2002), 195–218.
[4] S. Anjos and G. Granja, Homotopy decomposition of a group of symplectomorphisms of $S^2 \times S^2$, *Topology*, **43**-3 (2004), 599–618.
[5] D. Cox and S. Katz, *Mirror Symmetry and Algębraic Geometry*, Mathematical Surveys 68, American Mathematical Society, Providence, RI, 1999.
[6] O. Buse, in preparation.
[7] M. Gromov, Pseudo holomorphic curves in symplectic manifolds, *Invent. Math.*, **82** (1985), 307–347.
[8] V. Guillemin, E. Lerman, and S. Sternberg, *Symplectic Fibrations and Multiplicity Diagrams*, Cambridge University Press, Cambrigde, UK, 1994.
[9] T. Januszkiewicz and J. Kędra, Characteristic classes of smooth fibrations, SG/0209288, 2002.
[10] J. Kędra and D. McDuff, Homotopy properties of Hamiltonian group actions, *Geom. Topol.*, **9** (2005), 121–162.
[11] F. Lalonde and D. McDuff, Symplectic structures on fiber bundles, *Topology*, **42** (2003), 309–347.

[12] L. Lalonde, D. McDuff, and L. Polterovich, Topological regidity of Hamiltoian loops and quantum homology, *Invent. Math.*, **135** (1999), 69–85.

[13] F. Lalonde and M. Pinsonnault, The topology of the space of symplectic balls in rational 4-manifolds, *Duke Math J.*, **122**-2 (2004), 347–397.

[14] D. McDuff, Quantum homology of fibrations over S^2, *Internat. J. Math.*, **11** (2000), 665–721.

[15] D. McDuff, Symplectomorphism groups and almost complex structures, *Enseign. Math.*, **38** (2001), 1–30.

[16] D. McDuff, Lectures on groups of symplectomorphisms, *Rend. Circ. Mat. Palermo Ser.* II, **72** (2004), 43–78.

[17] D. McDuff, A survey of topological properties of groups of symplectomorphisms, in U. L. Tillmann, ed. *Topology, Geometry and Quantum Field Theory: Proceedings of* 2002 *Symposium in Honor of G. B. Segal*, Cambridge University Press, Cambridge, UK, 2004.

[18] D. McDuff and D. Salamon, *Introduction to Symplectic Topology*, 2nd ed., Oxford University Press, Oxford, UK, 1998.

[19] D. McDuff and D. Salamon, *J-Holomorphic Curves and Symplectic Topology*, Colloquium Publications 52, American Mathematical Society, Providence, RI, 2004.

[20] D. McDuff and S. Tolman, Topological properties of Hamiltonian circle actions, SG/0404338, 2004.

[21] D. McDuff and S. Tolman, The symplectomorphism group of toric manifolds, in preparation.

[22] A. G. Reznikov, Characteristic classes in symplectic topology, *Selecta Math.*, **3** (1997), 601–642.

[23] P. Seidel, π_1 of symplectic automorphism groups and invertibles in quantum cohomology rings, *Geom. Funct. Anal.*, **7** (1997), 1046–1095.

[24] P. Seidel, On the group of symplectic automorphisms of $\mathbb{C}P^m \times \mathbb{C}P^n$, *Amer. Math. Soc. Transl.* (2), **196** (1999), 237–250.

Algebraic Structure of Yang–Mills Theory

M. Movshev[1] and A. Schwarz[2]

[1] Institut Mittag-Leffler
Auravägen 17
S-182 60 Djursholm
Sweden
[2] Department of Mathematics
University of California at Davis
Davis, CA 95616
USA
asschwarz@ucdavis.edu

To I. M. Gelfand with admiration.

Subject Classifications: 81T70, 81T13, 16E40

In the present paper we analyze algebraic structures arising in Yang–Mills theory. The paper should be considered as a part of a project started with [15] and devoted to maximally supersymmetric Yang–Mills theories. In this paper we collect those of our results which hold without the assumption of supersymmetry and use them to give rigorous proofs of some results of [15]. We consider two different algebraic interpretations of Yang–Mills theory—in terms of A_∞-algebras and in terms of representations of Lie algebras (or associative algebras). We analyze the relations between these two approaches and calculate some Hochschild (co)homology of the algebras in question.

1 Introduction

Suppose $\mathfrak{g}$ is a Lie algebra equipped with a nondegenerate inner product $\langle .,. \rangle$. We consider a Yang–Mills field A as a $\mathfrak{g}$-valued one-form on a complex D-dimensional vector space $\mathbf{V}$ equipped with a symmetric bilinear inner product $(.,.)$. (All vector spaces in this paper are defined over the complex numbers.) We write this form as $A = \sum_{i=1}^{D} A_i dx^i$, where $x^1, \ldots, x^D$ is an orthogonal coordinate system on $\mathbf{V}$ which is fixed for the rest of the paper.

We assume that the field A interacts with bosonic and fermionic matter fields ϕ, ψ which are functions on the vector space $\mathbf{V}$ with values in $\Phi \otimes \mathfrak{g}$, $\Pi\mathbf{S} \otimes \mathfrak{g}$, respectively. The symbol Π stands for the change of parity. In other words matter fields transform according to the adjoint representation of $\mathfrak{g}$. The linear space Φ is equipped with a symmetric inner product $(.,.)$, the linear space $\mathbf{S}$ is equipped with a symmetric bilinear map $\Gamma : \mathrm{Sym}^2(\mathbf{S}) \to \mathbf{V}$. An important example is 10D SUSY Yang–Mills theory, where $D = 10$, $\boldsymbol{\Phi} = 0$, $\mathbf{V}$ and $\mathbf{S}$ are the spaces of the vector and spinor representations of SO(10), and Γ stands for the SO(10)-intertwiner $\mathrm{Sym}^2\,\mathbf{S} \to \mathbf{V}$.

We will always consider the action functional S as a holomorphic functional on the space of fields; to quantize one integrates $\exp(-S)$ over a real slice in this space. (For example, if $\mathfrak{g} = \mathfrak{gl}(n)$, one takes $\mathfrak{u}(n)$-valued gauged fields as a real slice in the space of gauge fields.) All considerations are local; in other words our fields are polynomials or power series on $\mathbf{V}$. This means that the action functionals are formal expressions (integration over $\mathbf{V}$ is ill-defined). However, we work with the equations of motion, which are well-defined. It is easy to get rid of this nuisance and make the definitions completely rigorous.

Choosing once and for all an orthonormal basis in Φ and some basis in $\mathbf{S}$ we can identify ϕ, ψ with $\mathfrak{g}$-valued fields $(\phi_1, \dots, \phi_{d'})$ and (ψ^α). The Lagrangian in these bases takes the form

$$L = \frac{1}{4}\sum_{i,j=1}^{D}\langle F_{ij}, F_{ij}\rangle + \sum_{i=1}^{D}\sum_{j=1}^{d'}\langle \nabla_i\phi_j, \nabla_i\phi_j\rangle + \sum_{i=1}^{D}\sum_{\alpha\beta}\langle \Gamma^i_{\alpha\beta}\nabla_i\psi^\alpha, \psi^\beta\rangle - U(\phi, \psi), \tag{1}$$

where ∇_i stands for the covariant derivative built out of A_i, $F_{ij} = \partial_i A_j - \partial_j A_i + [A_i, A_j]$ denotes the gauge field strength, and $\Gamma^i_{\alpha\beta}$ is the matrix of the linear map Γ in the chosen bases; U is a $\mathfrak{g}$-invariant potential. The corresponding action functional S_{cl} is gauge invariant and can be extended to a solution of the BV master equation in the standard way:

$$S = S_{cl} + \int_{\mathbf{V}}\left(\sum_{i=1}^{D}\langle \nabla_i c, A^{*i}\rangle + \sum_{\alpha}\langle [c, \psi^\alpha], \psi^*_\alpha\rangle + \sum_{j=1}^{d'}\langle [c, \phi_j], \phi^{*j}\rangle + \frac{1}{2}\langle [c, c], c^*\rangle\right) dx^1 \dots dx^D. \tag{2}$$

Here c stands for a Grassmann odd ghost field, and A^{*i}, ψ^*_α, ϕ^{*j}, c^* are antifields for A_i, ψ^α, ϕ_j, c. (The parity of antifields is opposite to the parity of fields.)

The BV action functional S determines a vector field Q on the space of fields, where Q obeys $Q^2 = 0$. The space of solutions of the equations of motion in the BV formalism coincides with the zero locus of Q. Using Q we introduce a structure of L_∞-algebra on the space of fields (see [1] or [15, Appendix C]). (Recall that the Taylor coefficients of a vector field Q obeying $Q^2 = 0$ at a point belonging to the zero locus of Q specify an L_∞-algebra. A point in the space of fields where all fields vanish belongs to the zero locus of Q; we construct the L_∞-algebra using the Taylor expansion of Q at this point.) The equations of motion can be identified with the Maurer–Cartan equations for the L_∞-algebra.

The L_∞-algebra $\mathcal{L}$ we constructed depends on the choice of Lie algebra $\mathfrak{g}$ and other data: potential, inner products on the spaces $\mathbf{V}$, Φ, and the bilinear map Γ on

S. When we need to emphasize such dependence we will do it by an appropriate subscript: e.g., $\mathcal{L}_{\mathfrak{gl}_n}$ shows that the Lie algebra $\mathfrak{g}$ is $\mathfrak{gl}_n$. We will assume that the potential $U(\phi, \psi)$ in our algebraic approach is a polynomial (or, more generally, a formal power series) in ϕ and ψ. If $\mathfrak{g} = \mathfrak{gl}_n$ it has the form

$$U(\phi, \psi) = \mathrm{tr}(P(\phi, \psi)),$$

where $P(\phi, \psi)$ is a noncommutative polynomial in the matrix fields $\phi_1, \ldots, \phi_{d'}, \psi^\alpha$. In this case we construct the A_∞-algebra $\mathcal{A}$ in a such a way that the L_∞-algebra $\mathcal{L}_{\mathfrak{gl}_n}$ is built in a standard way from the A_∞-algebra $\mathcal{A} \otimes \mathrm{Mat}_n$. We can say that working with the A_∞-algebra $\mathcal{A}$ we are working with all algebras $\mathcal{L}_{\mathfrak{gl}_n}$ at the same time. (Moreover, we can say that we are working with the gauge theories of all classical gauge groups at the same time.)

We mentioned already that for a Q-manifold X (a supermanifold equipped with an odd vector field Q obeying $Q^2 = 0$) one can construct an L_∞-algebra on the vector space $\Pi T^*_{x_0} X$ for every point $x_0 \in X$ in the zero locus of Q. In the finite-dimensional case, we can identify the L_∞-algebra with a formal Q-manifold. On the other hand the algebra of functions on a formal Q-manifold X (= the algebra of formal power series) is a differential commutative algebra. This algebra by definition is dual to the L_∞-algebra $\mathcal{L}$.

Similar definitions can be given for A_∞-algebras. The algebra of functions on a formal noncommutative manifold X is defined as the topological algebra of formal noncommutative power series. More precisely, if W is a $\mathbb{Z}_2$-graded topological vector space we can consider the tensor algebra $T(W) = \bigoplus_{n \geq 1} W^{\otimes n}$. This algebra has an additional $\mathbb{Z}$-grading with nth graded component $W^{\otimes n}$ and a descending filtration $K^n = \bigoplus_{i \geq n} W^{\otimes i}$. The algebra of formal power series $\widehat{T(W)}$ is defined as the completion of $T(W)$ with respect to this filtration. By definition the completion $\widehat{T(W)}$ consists of infinite series in the generators which become finite upon projection to $T(W)/K^n$ for every n. The elements of $\widehat{T(W)}$ are infinite sums of monomials formed by elements of a basis of W. The algebra $\widehat{T(W)}$ is $\mathbb{Z}_2$-graded; the filtration on $T(W)$ generates a filtration on the completion. $\widehat{T(W)}$ can be considered as the inverse limit of the spaces $T(W)/K^n$; we equip $\widehat{T(W)}$ with the topology of inverse limit. (The topology on $T(W)/K^n$ is defined as the strongest topology compatible with the linear structure.) A formal noncommutative Q-manifold is by definition a topological algebra $\widehat{T(W)}$ equipped with a continuous odd differential Q obeying $Q^2 = 0$. We say that the formal Q-manifold $(\widehat{T(W)}, Q)$ specifies a structure of A_∞-coalgebra $\mathcal{H}$ on the space ΠW. We are saying that the differential topological algebra $(\widehat{T(W)}, Q)$ is (bar)-dual to the A_∞-coalgebra $\mathcal{H}$. One says also that the differential algebra $(\widehat{T(W)}, Q)$ is obtained from the A_∞-coalgebra $\mathcal{H}$ by means of the bar-construction; we denote it by $\mathrm{Bar}\,\mathcal{H}$. The homology of $(\widehat{T(W)}, Q)$ is called the Hochschild homology of $\mathcal{H}$. Notice that in this definition Hochschild homology is $\mathbb{Z}_2$-graded. We also obtain a structure of A_∞-algebra $\mathcal{A} = \mathcal{H}^*$ on ΠW^*. In the finite-dimensional case the notion of A_∞-algebra on the vector space V is equivalent to the notion of A_∞-coalgebra on the vector space V^*. However, in the infinite-dimensional case it is much simpler to use A_∞-coalgebras. We will consider the case when the space W is equipped with a descending filtration F^n; then we can extend F^n to a filtration of $T(W)$ and $\widehat{T(W)}$ is defined by means of this filtration. (See Section 3.2 in the appendix for definitions.)

Another way of algebraization of Yang–Mills theory is based on consideration of the equations of motion (the equations are treated as defining relations in an associative algebra). We analyze the relations between two ways of algebraization and study some properties of the algebras at hand. In particular we calculate some Hochschild homology.

The paper is organized as follows. In Section 1.1 we formulate our main results. In Section 2.1 we give proofs in the case of Yang–Mills theory reduced to a point, and in Sections 2.2 and 2.3 we consider the more general case of Yang–Mills theory reduced to any dimension. In Sections 2.4, 2.5, and 2.6 we make some homological calculations that allow us to apply general results to the case of maximally supersymmetric theories.

All proofs in the paper are rigorous. However, our exposition in Section 2.1 is sometimes sketchy; the exposition of more general results in Sections 2.2 and 2.3 is more formal.

Notation

Denote by $\langle a_1, \ldots, a_n\rangle$ the span of the vectors $a_1, \ldots, a_n$ in some linear space.

Denote by $\mathbb{C}\langle a_1, \ldots, a_n\rangle$ the free algebra without a unit on the generators $a_1, \ldots, a_n$. If $\langle a_1, \ldots, a_n\rangle = W$, then an alternative notation for $\mathbb{C}\langle a_1, \ldots, a_n\rangle$ is

$$T(W) = \bigoplus_{n\geq 1} W^{\otimes n}. \tag{3}$$

All algebras in this paper are nonunital algebras, unless the opposite is explicitly stated.

Any algebra has a canonical filtration $F^n = \{\sum a_1 \ldots a_n\}$.

Suppose A is an algebra with a unit and augmentation (i.e., a homomorphism $\varepsilon : A \to \mathbb{C}$). Denote $I(A) = \operatorname{Ker} \varepsilon$.

Suppose A is an algebra (unit is irrelevant). Denote by $\underline{A} = A + \mathbb{C}$ the algebra with the following multiplicative structure: $(a, \alpha)(b, \beta) = (ab + \alpha b + \beta a, \alpha\beta))$. In this construction we formally adjoin a unit to A equal to $(0, 1)$. The algebra $\underline{A}$ has an augmentation $\varepsilon(a, \alpha) = \alpha$ and $I(\underline{A}) = A$.

Suppose the $\mathbb{Z}_2$-grading of the Hochschild homology comes from a $\mathbb{Z}$-grading. This happens in the case when the algebra A has no differential, or is $\mathbb{Z}$-graded. In the $\mathbb{Z}$-graded case by our definition the zeroth Hochschild homology of the algebra without unit is equal to zero (sometimes it is called the reduced homology). Sometimes it will be convenient for us to define $H_0(A) = \mathbb{C}$, so $H_\bullet(A)$ would become an A_∞-coalgebra with counit and coaugmentation. The completion will be denoted by $\overline{H_\bullet(A)}$. It is easy to see that it is equal to the standard unreduced Hochschild homology of the unital algebra $\underline{A}$, which is denoted by $H_\bullet(\underline{A}, \mathbb{C})$ and defined in [12]. We will use this notation also in the $\mathbb{Z}_2$-graded case.

1.1 Main results

Let us reduce the theory to a point, i.e., consider fields that do not depend on x^i, $i = 1, \dots, D$ (the case of x-dependent fields will be considered later). The BV action functional becomes a (super)function and takes the form

$$
\begin{aligned}
S = &\frac{1}{4}\sum_{i,j=1}^{D}\langle [A_i, A_j], [A_i, A_j]\rangle + \frac{1}{2}\sum_{i=1}^{D}\sum_{j=1}^{d'}\langle [A_i, \phi_j], [A_i, \phi_j]\rangle \\
&+ \frac{1}{2}\sum_{i,j=1}^{D}\sum_{\alpha\beta}\Gamma^i_{\alpha\beta}\langle [A_i, \psi^\alpha], \psi^\beta\rangle - U(\phi, \psi) \qquad (4)\\
&+ \sum_{i=1}^{D}\langle [A_i, c], A^{i*}\rangle + \sum_{j=1}^{d'}\langle [c, \phi_j], \phi^{*j}\rangle + \sum_{\alpha}\langle [c, \psi^\alpha], \psi^*_\alpha\rangle + \frac{1}{2}\langle [c, c], c^*\rangle .
\end{aligned}
$$

Here A_i, ϕ_i, ψ^α, c^* are elements of $\mathfrak{g}$, and A^{*i}, ϕ^{*i}, ψ^*_α and c are elements of $\Pi\mathfrak{g}$. The vector field Q corresponding to S is given by the formulas

$$
\begin{aligned}
Q(A_i) &= [c, A_i],\\
Q(\phi_j) &= [c, \phi_j],\\
Q(\psi^\alpha) &= [c, \psi^\alpha],\\
Q(c) &= \frac{1}{2}[c, c],\\
Q(c^*) &= \sum_{i=1}^{D}[A_i, A^{*i}] + \sum_{j=1}^{d'}[\phi_j, \phi^{*j}] + \sum_{\alpha}\{\psi^\alpha, \psi^*_\alpha\} + [c, c^*],\\
Q(A^{*m}) &= -\sum_{i=1}^{D}[A_i, [A_i, A_m]] - \sum_{k=1}^{d'}[\phi_k[\phi_k, A_m]] \qquad (5)\\
&\quad + \frac{1}{2}\sum_{\alpha\beta}\Gamma^m_{\alpha\beta}\{\psi^\alpha, \psi^\beta\} - [c, A^{*m}],\\
Q(\phi^{*j}) &= -\sum_{i=1}^{D}[A_i[A_i, \phi_j]] - \frac{\partial U}{\partial \phi_j} - [c, \phi^{*j}],\\
Q(\psi^{*s}_\alpha) &= -\sum_{i=1}^{D}\sum_{\beta}\Gamma^i_{\alpha\beta}[A_i, \psi^\beta] - \frac{\partial U}{\partial \psi^\alpha} - [c, \psi^*_\alpha].
\end{aligned}
$$

Let us consider the case $\mathfrak{g} = \mathfrak{gl}(n)$. In this case all fields are matrix-valued functions. In order to pass from the L_∞ to the A_∞ construction we need to assume that the functions $u_i(x)$ are equal to matrix polynomials.

We can construct an A_∞-algebra $\mathcal{A}_0$ such that the L_∞-algebra $\mathcal{L}_{\mathfrak{gl}(n)}$ can be obtained as the L_∞-algebra corresponding to the A_∞-algebra algebra $\mathcal{A}_0 \otimes \mathrm{Mat}_n$. The construction is obvious—we consider $A_i, \phi_j, c, A^{*i}, \phi^{*j}, c^*$ in (5) not as matrices but as formal generators. Then Q determines the derivation $\hat{Q}$ in the algebra $\widehat{T(W)}$ of formal power series with respect to the free generators. (We consider $A_i, \phi_j, \psi^*_\alpha, c^*$ as even elements and $A^{*i}, \phi^{*j}, \psi^\alpha, c$ as odd ones.) The space W can be considered as the direct sum of the spaces V, $\mathbf{\Phi}$, $\Pi\mathbf{S}$, $\Pi\mathbf{C}$, $\Pi\mathbf{V}$, $\Pi\mathbf{\Phi}$, $\mathbf{S}^*$, $\mathbf{C}$.

The derivation $\hat{Q}$ obeys $\hat{Q}^2 = 0$; hence it specifies a structure of A_∞-coalgebra on ΠW. The potential $U(\phi, \psi)$ is a linear combination of cyclic words in the alphabet ϕ_j, ψ^α. In other words $U(\varphi, \psi)$ is an element of $\operatorname{Cyc} W = T(W)/[T(W), T(W)]$ or, if we allow infinite sums, an element of the completion of this space. The linear space $[T(W), T(W)]$ is spanned by $\mathbb{Z}_2$-graded commutators. Notice that for every $w \in W^*$ one can define the derivative $\partial/\partial w : \operatorname{Cyc} W \to T(W)$; this map can be extended to completions. The derivative $\partial U/\partial \phi_j$ in the definition of the operator Q should be understood in this way. The derivation $\hat{Q}$ specifies not only a structure of A_∞-coalgebra on ΠW but also a structure of the A_∞-algebra on ΠW^*. Let us consider for simplicity the case of the bosonic theory; writing the potential in the form

$$U(\phi) = \sum_k c^{j_1,\dots,j_k} \phi_1 \dots \phi_k,$$

we can represent the operations in the A_∞-algebra in the following way:

$$\begin{aligned}
m_k(\mathbf{p}^{j_1}, \dots, \mathbf{p}^{j_k}) &= -c^{j_1,\dots,j_{k+1}} \mathbf{p}^*_{j_{k+1}} \quad (k \geq 2),\\
m_2(\mathbf{a}^{i_1}, \mathbf{a}^*_{i_2}) &= m_2(\mathbf{a}^*_{i_2}, \mathbf{a}^{i_1}) = -\bar{\mathbf{c}}^* \delta^{i_1}_{i_2},\\
m_2(\mathbf{p}^{j_1}, \mathbf{p}^*_{j_2}) &= m_2(\mathbf{p}^*_{j_2}, \mathbf{p}^{j_1}) = -\bar{\mathbf{c}}^* \delta^{j_1}_{j_2},\\
m_3(\mathbf{a}^{i_1}, \mathbf{a}^{i_2}, \mathbf{a}^{i_3}) &= -(\delta^{i_1 i_2} \mathbf{a}^*_{i_3} - 2\delta_{i_1 i_3} \mathbf{a}^*_{i_2} + \delta^{i_2 i_3} \mathbf{a}^*_{i_1}),\\
m_3(\mathbf{a}^{i_1}, \mathbf{a}^{i_2}, \mathbf{p}^j) &= -\delta_{i_1 i_2} \mathbf{p}^*_j,\\
m_3(\mathbf{a}^{i_1}, \mathbf{p}^j, \mathbf{a}^{i_2}) &= 2\delta_{i_1 i_2} \mathbf{p}^*_j,\\
m_3(\mathbf{p}^j, \mathbf{a}^{i_1}, \mathbf{a}^{i_2}) &= -\delta_{i_1 i_2} \mathbf{p}^*_j,\\
m_3(\mathbf{p}^{j_1}, \mathbf{p}^{j_2}, \mathbf{a}^i) &= -\delta_{j_1 j_2} \mathbf{a}^*_i,\\
m_3(\mathbf{p}^{j_1}, \mathbf{a}^i, \mathbf{p}^{j_2}) &= 2\delta_{j_1 j_2} \mathbf{a}^*_i,\\
m_3(\mathbf{a}^i, \mathbf{p}^{j_1}, \mathbf{p}^{j_2}) &= -\delta_{j_1 j_2} \mathbf{a}^*_i.
\end{aligned} \tag{6}$$

(We are using a basis of ΠW^* that is dual to the basis of W.) There is an additional set of equations relating $\mathbf{c}$ with the rest of the algebra. They simply assert that $\mathbf{c}$ is a unit.

The ideal $I(c) \subset T(W)$ generated by the element c is closed under the differential $\hat{Q}$. Denote by BV_0 the quotient differential algebra $T(W)/I(c)$. To define a filtration on BV_0 we introduce first of all a grading on W assuming that

$$\begin{aligned}
&\deg(c) = 0, \qquad \deg(A_i) = \deg(\phi_i) = 2, \qquad \deg(\psi^\alpha) = 3,\\
&\deg(\psi^*_\alpha) = 5, \qquad \deg(A^{*i}) = \deg(\phi^{*i}) = 6, \qquad \deg(c^*) = 8.
\end{aligned}$$

The corresponding multiplicative grading on $T(W)$ in general does not descend to BW_0, but the decreasing filtration on $T(W)$ generated by the grading descends to BV_0 and is compatible with the differential Q in the case when $\deg(U) \geq 8$. (The grading on W induces a grading on the space $\mathrm{Cyc}(\Phi + \Pi S)$ of cyclic words and the corresponding filtration F^k; the notation $\deg(U) \geq 8$ means that $U \in F^8$.)

We always impose the condition $\deg(U) \geq 8$ in considering BV_0; under this condition we can consider the completion of BV_0 as a filtered differential algebra $\widehat{BV_0}$ that can be identified with the quotient algebra $\widehat{T(W)}/\widehat{I(c)}$. Notice that Q is a polynomial vector field; hence, instead of the algebra $\widehat{T(W)}$ of formal power series, we can work with the tensor algebra $T(W) = \bigoplus_{k \geq 0} W^{\otimes k}$. However, without additional assumptions on the potential U, the results of our paper are valid only for the completed algebra $\widehat{T(W)}$. The differential algebra $(\widehat{T(W)}, \hat{Q})$ is dual to the A_∞-algebra $\mathcal{A}$. It will be more convenient for us to work with A_∞-coalgebras. The motivation is that we would like to avoid dualization in the category of infinite-dimensional vector spaces as much as possible. However, there is an involutive duality functor on the category of finite-dimensional or graded vector spaces. It implies that the category of finite-dimensional A_∞-algebras is dual to the category of A_∞-coalgebras. The same statement is true for the category of A_∞-(co)algebras equipped with an additional grading. There is a topological version of such a duality which is not an autoequivalence of the appropriate category; rather it is an equivalence between two different categories.

Another approach to algebraization of Yang–Mills theory is based on consideration of the equations of motion. We will illustrate it in the case of Yang–Mills theory reduced to a point. We consider its equations of motion,

$$\sum_{i=1}^{D}[A_i,[A_i,A_m]] + \sum_{k=1}^{d'}[\phi_k[\phi_k,A_m]] - \frac{1}{2}\sum_{\alpha\beta}\Gamma^m_{\alpha\beta}\{\psi^\alpha,\psi^\beta\} = 0, \quad m = 1,\ldots,D, \tag{7}$$

$$\sum_{i=1}^{D}[A_i[A_i.\phi_j]] + \frac{\partial U}{\partial \phi_j} = 0, \quad j = 1,\ldots,d', \tag{8}$$

$$\sum_{\beta}\sum_{i=1}^{D}\Gamma^i_{\alpha\beta}[A_i,\psi^\beta] + \frac{\partial U}{\partial \psi^\alpha} = 0, \tag{9}$$

as defining the relations of an associative algebra with generators A_i, ϕ_j, ψ^α. This algebra will be denoted by YM_0. (The algebras $\mathcal{A}$ and YM_0 depend on the choice of potential U; hence more accurate notations would be $\mathcal{A}_0^U$ and YM_0^U). One can say that YM_0 is a quotient of the tensor algebra $T(W_1)$ with generators A_i, ϕ_j, ψ^α with respect to some ideal. The grading on $W_1 = \mathbf{V} + \Phi + \Pi\mathbf{S}$ generates a grading on YM_0 if $\deg(U) = 8$. It generates a decreasing filtration compatible with the algebra structure on YM_0 if $\deg(U) \geq 8$; in this case we can introduce an algebra structure on the completion $\widehat{YM}_0$. The graded algebra associated with the filtered algebra YM_0

will be denoted by YM'_0; it can be described as the algebra YM_0 corresponding to the component of U having degree 8.

Theorem 1. *If the potential U has degree ≥ 8, then the differential algebra $(\widehat{BV_0}, \hat{Q})$ is quasi-isomorphic to the algebra $\widehat{YM_0}$. If the potential U has degree 8, we can say also that the differential algebra (BV_0, Q) is quasi-isomorphic to the algebra YM_0.*

Proof. See Section 2.1 for the proof. □

The above constructions can be included into the following general scheme.

Let us consider the $\mathbb{Z}_2$-graded vector spaces $W_1 = V$ with basis $e_1, \dots, e_n$, $W_2 = \Pi V^*$ with dual basis $e^{*1}, \dots, e^{*n}$, and the one-dimensional spaces W_0 with odd generator c, $W_3 = \Pi W_0^*$ with even generator c^*. Take $L \in \mathrm{Cyc}\, W_1$.

Define a differential Q on the free algebra $T(W)$ by the rule

$$\begin{aligned} Q(e_i) &= [c, e_i], \\ Q(\bar{e}^i) &= \frac{\partial L}{\partial e^i} + [c, e^{*i}], \\ Q(c) &= -\frac{1}{2}[c, c], \\ Q(c^*) &= \sum_i [e_i, \bar{e}^i] + [c, c^*]. \end{aligned} \tag{10}$$

Define $W^{\mathrm{red}} = W_1 + W_2 + W_3$. Denote the algebra $T(W)$ with the differential Q defined by formula (10) as $T(W)^L$. Denote by $I(c)$ the ideal generated by c. It is easy to see that it is a differential ideal. Write $BV^L = T(W)^L/I(c)$.

It is easy to see that $BV_0 = BV^L$ for L defined in (1), where one should disregard $\langle , \rangle$ signs. (We consider $A_i, \phi_j, \dots$ as free variables and L as a $\mathbb{Z}_2$-graded cyclic word.) The algebra YM^L is defined as the quotient of $T(W_1)$ with respect to the ideal generated by $\partial L/\partial e_i$. There exists a natural homomorphism of BV^L onto YM^L that sends $e_i \to e_i, e^{*i} \to 0, c^* \to 0$. If this homomorphism is a quasi-isomorphism, we say that L is regular. Theorem 1 gives a sufficient condition for regularity.

Theorem 1 can be generalized to unreduced Yang–Mills theory or to the theory reduced to d dimensions, $0 < d \leq D$. Our considerations will be local; this means that for the theory with gauge group $\mathfrak{g}$ reduced to d dimensions, we consider $\mathfrak{g}$-valued fields $A_i, A^{*i}, \phi_j, \phi^{*j}, c, c^*$ that are formal functions of the first d variables. They span the space $W_d \otimes \mathfrak{g}$, where $W_d = W \otimes \mathbb{C}[[x^1, \dots, x^d]]$. The space W_d is equipped with a filtration F^s, which induces a filtration on $W_d \otimes \mathfrak{g}$ in a trivial way. The group F^s consists of all power series with coefficients in W with Taylor coefficients vanishing up to degree s. The filtration F^s defines a topology in the standard way.

The solutions of the equations of motion in the BV formalism correspond to the zeros of a vector field Q defining a structure of A_∞-coalgebra on the space ΠW_d; we denote this coalgebra by bv_d. We can also work with the A_∞-algebra defined on the space $\Pi W^* \otimes \mathbb{C}[x_1, \dots, x_d]$.

The role of the algebra YM_0 in the case at hand is played by the truncated Yang–Mills algebra $T_d YM$. We consider a set of differentials $\partial_k : \underline{YM}_0 \to \underline{YM}_0$, $k = 1, \dots, d$. The differentials are defined by the formula

$$\begin{aligned} \partial_k(A_i) &= \delta_{ki}, && 1 \le i \le d, \\ \partial_k(A_i) &= 0, && i > d, \\ \partial_k(\phi_j) &= 0 && \text{for all } j. \end{aligned} \tag{11}$$

Definition 2. *We define* $\underline{T_d YM}$ *as* $\bigcap_{k=1}^{d} \operatorname{Ker} \partial_k$. *We assume that* ∂_k, $k = 1, \dots, d$, *is generic and that the matrix* Γ *is nondegenerate. We define* $T_d YM$ *in the standard way as* $I(\underline{T_d YM})$.

The precise meaning of the assumptions in this definition can be explained in the following way.

Definition 3. *We say that the set of differentials* ∂_k, $k = 1, \dots, d$, *is generic if the restriction of the bilinear form from* $\mathbf{V}$ *to* $T_d YM \cap \mathbf{V}$ *is nondegenerate. If the set of generators includes fermions, then we require that there be a subspace* $V' \subset \mathbf{V}$ *of codimension one such that* $V' \supset T_d YM \cap \mathbf{V}$ *and the bilinear form* $(s_1, s_2)_v \stackrel{\text{def}}{=} [\Gamma(s_1, s_2) \to \mathbf{V}/V']$ *is nondegenerate.*

Definition 4. *We say that the matrix* Γ *is nondegenerate if there is at least one generic differential from Definition* 3.

If we do not assume genericity we use an alternative notation $T_\mu YM$ for $T_d YM$ which will be adopted throughout the main part of the paper. The algebra $T_d YM$ is filtered and we can define its completion $\widehat{T(W_d)}$. The ideal $I(c) \subset \widehat{T(W_d)}$ generated by the element c is closed under the differential $\hat{Q}$. Denote by $\widehat{BV}_d$ the quotient algebra $\widehat{T(W_d)}/\widehat{I(c)}$. We use the notation $\widehat{BV}_\mu$ for the quotient $\widehat{T_\mu(W)}/I(c)$ without the assumption of genericity of the family ∂_k, $k = 1, \dots, d$.

Theorem 5. *The differential algebra* $(\widehat{BV}_d, \hat{Q})$ *is quasi-isomorphic to* $\widehat{T_d YM}$.

Proof. See Proposition 46. □

Corollary 6. *The algebra* $\widehat{T_d YM}$ *is quasi-isomorphic to the dual of the coalgebra* bv_d.

Proof. See Proposition 47, where the A_∞-coalgebra bv_d is denoted as bv_μ. □

Let us analyze the structure of the algebra $T_d YM$ for $d \ge 1$.

Definition 7. *Define the algebra* $K(q_1, \dots, q_n | p^1, \dots, p^n; \psi^1, \dots, \psi^{n'})$ *as the quotient algebra* $\mathbb{C}\langle q_1, \dots, q_n, p^1, \dots, p^n, \psi^1, \dots, \psi^{n'}\rangle / I(\omega)$, *where the ideal* $I(\omega)$ *is generated by the element*

$$\omega = \sum_{i=1}^{n} [q_i, p^i] - \frac{1}{2} \sum_{j=1}^{n'} \{\psi^j, \psi^j\}. \tag{12}$$

Theorem 8. *The algebra* $\widehat{T_1 YM}$ *is isomorphic to the algebra* $\widehat{K}$.

Theorem 9. *The algebra $\widehat{T_d YM}$ for $d \geq 2$ is isomorphic to the completion of the free algebra $T(\mathcal{H} + \mathcal{S} + \mathcal{G})$, where $\mathcal{H}$ stands for the space of all polynomial harmonic two-forms on $\mathbb{C}^d$ and $\mathcal{S}$ stands for the space of harmonic spinors and $\mathcal{G}$ stands for the space of harmonic polynomials on $\mathbb{R}^d$ with values in $\Phi + \mathbb{C}^{D-d}$.*

Proof. The proofs of statements that are more general than Theorems 8 and 9 are given in Examples 1, 2, and 4 and Propositions 50 and 52. □

Notice that the above theorems have a physical interpretation.

Theorems 1 and 5 can be interpreted as the statement that the BV formalism is equivalent to the more traditional approach to the theory of gauge fields. (One can relate this theorem to the calculation of BV homology in [2].) Theorem 8 is related to the Hamiltonian formalism in gauge theory when we neglect the dependence of the fields on all spatial variables. A solution to the equation of motion in such a theory is characterized by a point of a phase space; the degeneracy of the Lagrangian leads to the constraint (12) on the phase space variables.

Theorems 8 and 9 mean that there exists a one-to-one correspondence between solutions of the full Yang–Mills equation of motion and solutions of the linearized version of this equation.

The phase space dynamics specifies an action of the one-dimensional Lie algebra $\mathfrak{a}$ on the algebra K. More precisely, we define an action of exterior derivative H, corresponding to the generator of $\mathfrak{a}$, by the rule

$$H(q_i) = p^i, \tag{13}$$

$$H(Q_i) = P^i, \tag{14}$$

$$H(p^i) = -\sum_{i=1}^{D-1}[q_i,[q_i,q_m]] - \sum_{j=1}^{d'}[Q_j,[Q_j,q_m]] + \frac{1}{2}\sum_{\alpha\beta}\Gamma^m_{\alpha\beta}\{\psi^\alpha,\psi^\beta\}, \tag{15}$$

$$H(P^j) = -\sum_{i=1}^{D-1}[q_i,[q_i,Q_j]] - \frac{\partial U}{\partial Q_j}, \tag{16}$$

$$H(\psi^\alpha) = -\sum_{i=1}^{D-1}\sum_{\beta}\Gamma^i_{\alpha\beta}[q_i,\psi^\beta] - \frac{\partial U}{\partial \psi^\alpha}. \tag{17}$$

Definition 10. *Suppose A is an algebra with a unit and $\mathfrak{g}$ is a Lie algebra which acts upon A via derivations. Let*

$$\rho : \mathfrak{g} \to \operatorname{Der} A \tag{18}$$

be the homomorphism to the Lie algebra of derivations that corresponds to the action. Let $U(\mathfrak{g})$ be the universal enveloping algebra of $\mathfrak{g}$. Denote by $U(\mathfrak{g}) \ltimes A$ the algebra defined on the space $U(\mathfrak{g}) \otimes A$ in the following way. It contains $U(\mathfrak{g}) \otimes 1$ and $1 \otimes A$ as subalgebras. For $g \in U(\mathfrak{g})$ and $a \in A$, $ga = g \otimes a$. If g is a linear generator of $\mathfrak{g}$, then $ga - ag = \rho(g)a$. We call $U(\mathfrak{g}) \ltimes A$ the semidirect product.

Theorem 11. *Suppose the Γ-matrices used in the definition of YM_0 are nondegenerate in the sense of Definition 4. Then the algebra $\underline{YM}_0$ is isomorphic to the semidirect product*

$$\underline{YM}_0 = U(\mathfrak{a}) \ltimes \underline{K}(q_1, \dots, q_{D-1}, Q_1, \dots, Q_{d'} | p^1, \dots, p^{D-1}, P^1, \dots, P^{d'}; \psi^\alpha). \tag{19}$$

In the above formula $U(\mathfrak{a})$ is the universal enveloping algebra of the abelian one-dimensional Lie algebra $\mathfrak{a}$ spanned by the element H, which acts on $\widehat{\underline{K}}$ as an outer derivation.

Formula (19) remains valid for the completed algebras $\widehat{\underline{YM}_0}$ and $\widehat{\underline{K}}$ if we replace the semidirect product with its completion with respect to the multiplicative decreasing filtration F^n which coincides with the intrinsic filtration on $\widehat{\underline{K}}$ and is determined by the condition $H \in F^2$.

Proof. See Section 2.1 for the proof. □

Using general results about Hochschild homology (the main reference is [12]; see also the appendix) for algebras with one relation and on the homology of a cross-product, we get the following theorem.

Theorem 12. *The Hochschild homology $\underline{H}_*(K(q_1, \dots, q_n | p^1, \dots, p^n; \psi^1, \dots, \psi^{n'}))$ is isomorphic to*

$$\underline{H}_0(K) = \mathbb{C}, \tag{20}$$

$$\underline{H}_1(K) = \langle [q_1], \dots, [q_n], [p_1], \dots, [p_n], [\psi^1], \dots, [\psi^{n'}] \rangle, \tag{21}$$

$$\underline{H}_2(K) = \langle r \rangle, \tag{22}$$

$$\underline{H}_i(K) = 0 \quad \textit{for } i \geq 3. \tag{23}$$

The symbols $[q_1], \dots, [q_n], [p_1], \dots, [p_n], [\psi^1], \dots, [\psi^{n'}]$ are in one-to-one correspondence with the generators $q_1, \dots, q_n, p_1, \dots, p_n, \psi^1, \dots, \psi^{n'}$. There is a nondegenerate even skew-symmetric pairing on $\underline{H}_1(K)$ which depends on the choice of a generator r in $\underline{H}_2(K)$. The statement of the proposition holds if one replaces the algebra K by its completion $\widehat{K}$.

Proof. See Section 2.1 for the proof. □

Theorem 13. *The Hochschild homology $\underline{H}_*(\widehat{YM_0})$ is isomorphic to*

$$\begin{aligned} \underline{H}_0(\widehat{YM_0}) &= W_0 = \langle c \rangle, \\ \underline{H}_1(\widehat{YM_0}) &= W_1 = \Phi + \Pi \mathbf{S}, \\ \underline{H}_2(\widehat{YM_0}) &= W_2 = \Pi W_1^*, \\ \underline{H}_3(\widehat{YM_0}) &= W_3 = \langle c^* \rangle. \end{aligned} \tag{24}$$

There is a graded commutative duality pairing

$$\underline{H}_i(\widehat{YM_0}) \otimes \underline{H}_{3-i}(\widehat{YM_0}) \to \mathbb{C}$$

which depends on the choice of a generator c^* in W_3.

The algebra YM_0 has the same Hochschild homology.

Proof. See Section 2.1 for the proof. □

The duality of A_∞-algebras is closely related to the Koszul duality of quadratic algebras (see, e.g., [11]).

Let **S** be a spinor representation of Spin(10). Let

$$\mathcal{S} = \mathrm{Sym}(\mathbf{S}) / \sum_{\alpha\beta} \Gamma^i_{\alpha\beta} u^\alpha u^\beta,$$

where $\Gamma^i_{\alpha\beta}$ are spinor Γ-matrices and $u^1, \ldots, u^{16}$ is a basis of **S**. The algebra $\mathcal{S}$ can be considered as the algebra of polynomial functions on the space of pure spinors (spinors in $\mathbf{S}^*$ satisfying $\sum_{\alpha\beta} \Gamma^i_{\alpha\beta} u^\alpha u^\beta = 0$). Denote

$$B_0 = \mathcal{S} \otimes \Lambda(\mathbf{S}) \tag{25}$$

with linear generators of $\Lambda(\mathbf{S})$ denoted by $\theta^1, \ldots, \theta^{16}$. Define a differential on the algebra B_0 by the rule

$$Q(\theta^\alpha) = u^\alpha. \tag{26}$$

We call the differential algebra (B_0, Q) the Berkovits algebra. From here until the end of the section, we assume that YM_0 is built from the following data: $D = 10$, $d' = 0$, **S** is an irreducible spinor representation $\mathfrak{s}_l$ of Spin(10), where $\Gamma^i_{\alpha\beta}$ are the Γ-matrices associated with the spinor representation **S**. (This means that YM_0 is obtained from 10D SUSY YM theory reduced to a point.)

We checked in [15] that YM_0 maps to the Koszul dual of the Berkovits algebra (B_0, Q) (see [15] and references therein about Koszul duality). In this paper we prove a statement which was formulated in [15] without proof.

Theorem 14. *The Koszul dual to the algebra (B_0, Q) is quasi-isomorphic to YM_0.*

Proof. To prove this fact we should know the homology of the Berkovits algebra. A heuristic calculation of this homology was given in [6]. We present a rigorous calculation in Section 2.6. □

Theorem 15. *The Berkovits algebra (B_0, Q) is quasi-isomorphic to the A_∞-algebra $\mathcal{A}$ obtained from the $D = 10$ SUSY YM action functional reduced to a point.*

This statement was formulated in [15]. It follows from Theorem 14 and from the relation between Koszul duality and bar-duality.

Let us now consider the d-dimensional Berkovits algebra (B_d, Q).

Definition 16. *The Berkovits algebra* (B_d, Q) *is defined as the algebra of polynomial functions of the pure spinor u, the odd spinor* $\theta = (\theta^1, \dots, \theta^{16})$*, and the commuting coordinates* $x^1, \dots, x^d$ $(d \leq 10)$*, equipped with the differential*

$$Q = \sum_{\alpha=1}^{16} u^\alpha \frac{\partial}{\partial \theta^\alpha} + \sum_{\alpha\beta=1}^{16} \sum_{i=1}^{d} \Gamma^i_{\alpha\beta} u^\alpha \theta^\beta \frac{\partial}{\partial x^i}.$$

The algebra B_d is a quadratic algebra.

Theorem 17. *The Koszul dual to the differential algebra* (B_d, Q) *is quasi-isomorphic to the truncated Yang–Mills algebra* $T_d YM$.

Proof. See Propositions 73 and 74. □

2 Proofs

2.1 Algebras $\widehat{K}$ and $\widehat{YM_0}$

Proof of Theorem 11. We need to rewrite relations (7), (8), (9) in a slightly different form. Introduce the notation $q_i = A_{i+1}$ $(i \geq 1)$, $Q_i = \phi_i (i \geq 1)$, $p^i = [A_1, A_{i+1}] (i \geq 1)$, $P^j = [A_1, \phi_j]$ $(j \geq 1)$. Commutation with A_1 preserves the algebra generated by $q_i, Q_i, p^i, P^j, \psi^\alpha$. Denote the operation of commutation with A_1 by H: $[A_1, x] = H(x)$. Then by definition we have (13), (14).

In the new notation, when $m = 0$ relation (7) becomes (12). To prove this, we use the nondegeneracy of the Γ-matrices. It allows us to set $\Gamma^1_{\alpha\beta} = \delta_{\alpha\beta}$ by appropriate choice of bases in **S** and in **V**. Relations (7) when $m > 0$ become (15), relations (8) become (16), and relations (8) become (17). We can see that in this representation the algebra $\underline{YM}_0$ is the semidirect product of two algebras: the universal enveloping algebra $U(\mathfrak{a})$ of an abelian algebra with one generator H and of the algebra $\underline{K}$.

The action of the universal enveloping algebra is given as outer derivation H (the letter H stands for the Hamiltonian) by the formulas (13), (14), (12), (15), (16), (17).

If H acts on K via formulas (13), (14), (12), (15), (16), (17), then the map $p : \underline{YM}_0 \to U(\mathfrak{a}) \ltimes \underline{K}$, defined by the formulas

$$\begin{aligned} &p(A_1) = H, \qquad p(A_{i+1}) = q_i, \quad 1 \leq i \leq D, \\ &p(\phi_j) = Q_j, \quad 1 \leq j \leq d', \qquad p(\psi^\alpha) = \psi^\alpha, \end{aligned}$$

is well defined and agrees on filtrations. This is because formulas (13), (14), (12), (15), (16), (17) imply (7), (8), (9)). As a result, p is a continuous isomorphism that can be extended to completions. □

We need to formulate basic theorems on how to compute the Hochschild homology of some algebras.

Proposition 18.

(a) *Suppose* $\mathfrak{g}$ *is a Lie algebra. Then* $H_1(\mathfrak{g}, \mathbb{C}) \cong \mathfrak{g}/[\mathfrak{g}, \mathfrak{g}]$, *where* $[\mathfrak{g}, \mathfrak{g}]$ *is the ideal of* $\mathfrak{g}$ *consisting of elements of the form* $[a, b]$, *where* $a, b \in \mathfrak{g}$.

(b) *Suppose* $\mathfrak{g} = F/R$, *where* F *is a free Lie algebra and* R *is an ideal of relations. Then* $H_2(\mathfrak{g}, \mathbb{C}) \cong R \cap [F, F]/[F, R]$.

Proof. See [12] for the proof. □

Corollary 19. *Suppose* $\mathfrak{g}$ *is a positively graded Lie algebra.*

(a) *Let* $V \subset \mathfrak{g}$ *be a minimal generating subspace. Then the canonical map* $V \to \mathfrak{g}/[\mathfrak{g}, \mathfrak{g}] \cong H_1(\mathfrak{g}, \mathbb{C})$ *is an isomorphism.*
(b) *Suppose* $\mathfrak{g} = F/R$, *where* F *is a free algebra and* R *is an ideal. Assume also that the minimal linear subspace of relations* L *which generates the ideal* R *is a subspace of* $[F, F]$. *Then the canonical map* $L \to R/[F, R] = R \cap [F, F]/[F, R] \cong H_2(\mathfrak{g}, \mathbb{C})$ *is an isomorphism.*

Proposition 20. *Let* A *be an algebra complete with respect to the decreasing multiplicative filtration* F^s, $s \geq 1$, *such that* $\bigcap F^s = 0$. *There is an isomorphism between* $H_1(A)$ *and a minimal linear space* $X \subset A$ *such that the subalgebra generated by* X *is dense (in the sense of the topology generated by the filtration) in* A.

Definition 21. *Suppose* A *is an algebra with a unit and* M *is a bimodule. If* $i \geq 0$ *is the minimal number such that the Hochschild homology* $H_{i+k}(A, M) = 0$ *for all* M, $k > 0$, *then one says that the homological dimension of* A *is equal to* i.

Proposition 22 ([8]). *Suppose the positively graded algebra* A *is the quotient of a free algebra* T *by the ideal generated by one element* r. *If* $r \neq aba$, *where* $a, b \in T$ *and* $\deg(a) > 0$, *then the homological dimension of* $\underline{A}$ *is equal to* 2 *and* $H_1(A)$ *is isomorphic to a minimal set of generators and* $H_2(A) = \langle r \rangle$.

Proposition 23 ([12]). *There is an isomorphism* $H_1(T(W)) = W$. *All other homologies of the free algebra* $T(W)$ *are trivial.*

Proposition 24. *Let* A *be a positively graded algebra and* $\widehat{A}$ *its completion with respect to the filtration associated with the grading. Then* $H_*(\widehat{A}) = \widehat{H_*(A)}$, *where* $\widehat{H_*(A)}$ *stands for the completion of* $H_*(A)$ *by means of the filtration associated with the grading.*

Proof. Obvious. □

Proof of Theorem 12. We can use Proposition 22 for the computation of homology groups. The algebra has only one relation in degree 2; therefore, $H_2(K) = \langle r \rangle$ is one dimensional. There is a comultiplication map

$$\Delta : H_2 \to H_2 \otimes H_0 + H_1 \otimes H_1 + H_0 \otimes H_2. \tag{27}$$

The image of r in the middle component gives the matrix of the pairing. It is nondegenerate, and after the inversion one gets

$$[q_i] * [p^j] = -[p^j] * [q_i] = \delta_i^j, \\ [\psi^i] * [\psi^j] = [\psi^j] * [\psi^i] = \delta^{ij} \tag{28}$$

and all other products are equal to zero. The vanishing of $H_i(K) = 0$ for $i \geq 3$ is a corollary of Proposition 22. The completion of K is associated with the grading; therefore, the statement of the proposition for the completed algebra follows from Proposition 24. □

Proof of Theorem 13. By definition we have isomorphisms $\underline{H}_*(\widehat{YM}_0) = H_*(\underline{YM}_0, \mathbb{C})$ and $\underline{H}_*(\widehat{K}) = H_*(\underline{K}, \mathbb{C})$. The algebra $\widehat{\underline{YM}}_0$ contains a dense semidirect product $U(\mathfrak{a}) \ltimes \widehat{\underline{K}}$ and the algebra $\underline{YM}_0$ is equal to a semidirect product.

Let A be one of these algebras. Introduce an increasing filtration $G^n A$ defined as follows. The algebra $U(\mathfrak{a}) = \mathbb{C}[H]$ has a grading such that $\deg(H) = 1$; denote by $G^n U(\mathfrak{a})$ the associated increasing filtration. Write $G^n(A) = G^n U(\mathfrak{a}) \otimes B$, where B is either $\underline{K}$ or $\widehat{\underline{K}}$. This filtration induces a filtration of the bar-complex. It leads to a spectral sequence which is usually attributed to Serre and Hochschild.

The next computations will be carried out for the case of the semidirect product $U(\mathfrak{a}) \ltimes \underline{K} = \underline{YM}_0$.

The E^2-term of the spectral sequence is $E^2_{ij} = H_i(U(\mathfrak{a}), \underline{H}_j(K)) = H_i(\mathfrak{a}, \underline{H}_j(K))$. (Here $\underline{H}_i(\mathfrak{a}, \dots)$ stands for the homology of the Lie algebra $\mathfrak{a}$ with coefficients in some module.) We have convergence, $E^2_{ij} \Rightarrow H_{i+j}(U(\mathfrak{a}) \ltimes \underline{K}, \mathbb{C})$. The homology with trivial coefficients of the algebra $\underline{K}$ is computed in Theorem 12. We have

$$\underline{H}_1(\widehat{K}) = \langle [q_1], \dots, [q_d], [p^1], \dots, [p^d], [Q_1], \dots, [Q_{d'}], [P^1], \dots, [P^{d'}], [\psi^\alpha]] \rangle.$$

The action of the algebra $\mathfrak{a}$ or (what is the same) the action of its generator H on the homology of K is easy to describe. It is trivial for $\underline{H}_0(K)$ for obvious reasons. It is trivial on $\underline{H}_2(K)$ because if one applies the differential of a free algebra $\mathbb{C}\langle q_1, \dots, q_{D-1}, p^1, \dots, p^{D-1}, Q_1, \dots, Q_{d'}, P^1, \dots, P^{d'}, \psi^\alpha \rangle$ defined by the formulas (13)–(17) to the LHS of equation (12), one gets zero. The action of H on $[q_i], [Q_j], [\psi^\alpha] \in \underline{H}_1(K)$ is zero and $H[p^i] = -[q_i]$, $H[P^i] = -[Q_i]$. (This follows from the formulas (13)–(17) and the restriction on the degree of the potential.)

The differential $d_A : C_1(\mathfrak{a}, \underline{H}_1(K)) = \underline{H}_1(K) \xrightarrow{H} \underline{H}_1(K) = C_0(\mathfrak{a}, \underline{H}_1(K))$(Here C_i denotes the group of i-chains of the abelian Lie algebra $\mathfrak{a}$). This is the only nontrivial differential in the E_1-term. The spectral sequence degenerates in the E^2-term because all linear spaces which higher differentials can hit are equal to $\{0\}$. This implies that the classes $[A_1], \dots, [A_D], [\phi_1], \dots, [\phi_{d'}], [\psi^\alpha]$ form a basis of $H_1(U(\mathfrak{a}) \ltimes \underline{K}, \mathbb{C})$. We have a nondegenerate pairing on E^2 between $E^2_{i,j}$ and $E^2_{1-i,2-j}$. It comes from the diagonal $E^2_{1,2} \to E^2_{i,j} \otimes E^2_{1-i,2-j}$. This diagonal is the tensor product of the diagonal from Theorem 12 and the diagonal in the homology of an abelian Lie algebra. It indicates that $H_2(U(\mathfrak{a}) \ltimes \underline{K}, \mathbb{C})$ is dual to $H_1(U(\mathfrak{a}) \ltimes \underline{K}, \mathbb{C})$ and the relations (7), (8) are a minimal set of relations, $H_3(U(\mathfrak{a}) \ltimes \underline{K}, \mathbb{C}) = \mathbb{C}$.

All homology groups $H_i(U(\mathfrak{a}) \ltimes \underline{K})$, $i \geq 4$, are equal to zero because all contributors to these groups in the spectral sequence vanish.

This proves the statement for the algebra $U(\mathfrak{a}) \ltimes \underline{K} = \underline{YM}_0$. We cannot apply the proof directly to $\widehat{\underline{YM}}_0$ because it is not a semidirect product. We have the following argument in this case:

The algebra $\widehat{\underline{YM}}_0$ is filtered with Gr $\widehat{\underline{YM}}_0 = \widehat{\underline{YM}}'_0$. To define $\widehat{\underline{YM}}'_0$, we need to take the potential U of $\widehat{\underline{YM}}_0$ and extract its degree 8 homogeneous part. The algebra $\underline{YM}'_0$ is graded and its completion with respect to the decreasing completion associated with the grading is equal to $\widehat{\underline{YM}}'_0$. By Proposition 20 the Hochschild cohomology of $\widehat{\underline{YM}}'_0$ is equal to the completion of the homology of $\underline{YM}'_0$. The latter are finite dimensional. They coincide with the homology of $\underline{YM}'_0$. The spectral sequence associated with the filtration $F^n\widehat{\underline{YM}}_0$ degenerates at the E^1-term. The proof follows from this fact. □

Definition 25. *Introduce a multiplicative filtration on $T(W)$ by extending the filtration from the generating space W. After completing the algebra $T(W)$ we get $\widehat{T(W)}$. Define a continuous differential on the algebra $\widehat{T(W)}$ by the formula* (5). *Under the assumptions of Theorem* 1, *the differential* (5) *leaves the subalgebra $T(W) \subset \widehat{T(W)}$ invariant. Denote the resulting differential algebra by $(T(W), Q)$.*

Define a map

$$p : \widehat{BV}_0 \to \widehat{YM}_0 \tag{29}$$

by its values on topological generators:

$$\begin{aligned} p(A_i) &= A_i, \\ p(\phi_j) &= \phi_j, \\ p(\psi^\alpha) &= \psi^\alpha, \\ p(A^{*i}) &= p(\phi^{*j}) = p(\psi^*_\alpha) = p(c^*) = 0, \end{aligned} \tag{30}$$

then extend it to the entire algebra using the properties of homomorphism and continuity. The maps $p : BV_0 \to YM_0$ and $p : BV'_0 \to YM'_0$ are defined by the same formulas.

Proof of Theorem 1. We will use Theorems 13 and their corollaries. The Hochschild homologies of $\widehat{YM}_0$ and of YM_0 were calculated in Theorem 13. To calculate the homology of the differential algebra $\widehat{BV}_0$, we start with $\widehat{BV}_0$ considered as an algebra without differential.

It is easy to check that this is a completed free algebra with free topological generators:

$$A_i,\ \phi_i,\ \psi^\alpha,\ A^{*i},\ \phi^{*i},\ \psi^*_\alpha,\ c^*.$$

It follows from Propositions 23 and 24 that $H_1(\widehat{BV}_0) = W_1 \oplus W_1^* \oplus \mathbb{C}$ is the space spanned by free topological generators, $H_i(\widehat{BV}_0) = 0$ for $i \neq 1$.

The Hochschild homology of the differential algebra $(\widehat{BV}_0, Q)$ is not $\mathbb{Z}$-graded, but it is $\mathbb{Z}_2$-graded. The calculation of the completed Hochschild homology of $(\widehat{BV}_0, Q)$ is based on the general Lemma 83.

In our case H is the coalgebra that corresponds to the differential Q on $\widehat{BV}_0$. This coalgebra is filtered (in the sense described in the appendix) with $F^1(H) = H$,

$F^2(H) = 0$. The A_∞-coalgebra H has differential equal to zero. We obtain that as a $\mathbb{Z}_2$-graded vector space the Hochschild homology of $(\widehat{BV}_0, Q)$ is isomorphic to the Hochschild homology of $\widehat{BV}_0$.

More precisely, this homology is spanned by the following cocycles, where the map ι is the identification of the generators of BV_0 and the homology classes in $\mathrm{Bar}(BV_0, Q)$:

$$\iota(A^{*m}) = -A^{*m} + \sum_{i=1}^{D}(A_i|[A_i, A_m] - [A_i, A_m]|A_i) + \sum_{k=1}^{d'}(\phi_k|[\phi_k, A_m] - [\phi_k, A_m]|\phi_k) + \frac{1}{2}\sum_{\alpha\beta}\Gamma^m_{\alpha\beta}\psi^\alpha|\psi^\beta, \tag{31}$$

$$\iota(\phi^{*j}) = -\phi^{*j} + \sum_{i=1}^{D}(A_i|[A_i.\phi_j] - [A_i.\phi_j]|A_i) + \frac{\widetilde{\partial U}}{\partial\phi_j}, \tag{32}$$

$$\iota(\psi^*_\alpha) = -\psi^{*s}_\alpha + \sum_{i=1}^{D}\sum_{\beta}(\Gamma^i_{\alpha\beta}A_i|\psi^\beta - \psi^\beta|A_i) + \frac{\widetilde{\partial U}}{\partial\psi^\alpha}, \tag{33}$$

$$\iota(c^*) = \sum_i(A_i|(A^{*i} - \iota(A^{*i})) - (A^{*i} - \iota(A^{*i}))|A_i) + \sum_{j=1}^{d'}(\phi_j|(\phi^{*j} - \iota(\phi^{*j})) - (\phi^{*j} - \iota(\phi^{*j}))|\phi_j) + \sum_\alpha(\psi^\alpha|(\psi^*_\alpha - \iota(\psi^*_\alpha)) + (\psi^*_\alpha - \iota(\psi^*_\alpha))|\psi^\alpha). \tag{34}$$

The map ι is the identity map on A_i, ϕ_j, ψ^α. We need to explain what $\widetilde{\ }$ means in the formulas (33) and (32). The tensor algebra $T(V)$ generated by the linear space V can be considered as the free product of the algebra V with zero multiplication and the algebra spanned by the $\otimes$ symbol (the multiplication in such an algebra is trivial). Consider the algebra spanned by the bar symbol $|$ with zero multiplication. Then $\mathrm{Bar}\,T(V)$ is a subspace of the free product $V \circ \langle\otimes\rangle \circ \langle|\rangle$, where $\circ$ denotes free product. Define a derivation on such an algebra by the rule $d(v) = 0$, $v \in W$, $d_|(\otimes) = |$, $d_|(|) = 0$. The partial derivatives of U in formulas (32), (33) are elements of the tensor algebra $T(\phi_1, \ldots, \phi_{d'}, \psi^\alpha) \subset \mathrm{Bar}\,T(\phi_1, \ldots, \phi_{d'}, \psi^\alpha)$. Then $\frac{\widetilde{\partial U}}{\partial\phi_j} = d_|\frac{\partial U}{\partial\phi_j}$ and $\frac{\widetilde{\partial U}}{\partial\psi^\alpha} = d_|\frac{\partial U}{\partial\psi^\alpha}$, where $d_|$ is extended by continuity to $\widehat{\mathrm{Bar}}\,\widehat{T(V)}$.

Simple direct calculation using formulas (31)–(34) shows that the homomorphism $(\widehat{BV}_0, Q) \to \widehat{YM}_0$ induces an isomorphism on Hochschild homology. □

Definition 26. *Denote the minimal model (see* [10] *for definition) for the coalgebra* $\widehat{\mathrm{Bar}}(\widehat{BV}_0, d)$ *by* $\widehat{bv}_0$. *The linear space of* $\widehat{bv}_0$ *coincides with* $W_1 + W_2 + W_3$ *and the structure maps coincide with* (5) *(the variable* c *is set to zero). The coalgebras* bv_0 *and* bv'_0 *are noncomplete and graded versions of the coalgebra* $\widehat{bv}_0$.

Proposition 27. *Theorem* 1 *together with Lemma* 83 *can be rephrased as the statement that the* A_∞*-coalgebra* bv_0 *is bar-dual to the algebra* $\widehat{YM}_0$.

2.2 Truncated Yang–Mills algebra

We need to describe the notation adopted in this section.

Define a bigraded vector space $W = \bigoplus_{i=0}^{3} \bigoplus_{j=0}^{8} W_i^j$,

$$\begin{aligned} W_0^0 &= \langle c \rangle, & W_1^2 &= \mathbf{V} + \Phi, & W_1^3 &= \mathbf{S}, \\ W_3^8 &= \langle c^* \rangle, & W_2^6 &= \mathbf{V}^* + \Phi^*, & W_2^5 &= \mathbf{S}^*. \end{aligned} \tag{35}$$

We define

$$W_i = \bigoplus_{j=0}^{8} W_i^j, \qquad W^j = \bigoplus_{i=0}^{3} W_i^j. \tag{36}$$

We refer to

$$W = \bigoplus_{i=0}^{3} W_i \text{ as homological} \quad \text{and} \quad W = \bigoplus_{j=0}^{8} W^j \text{ as additional} \tag{37}$$

gradings. We also have a filtration,

$$F^n(W)_i = \bigoplus_{j \geq k}^{8} W_i^j. \tag{38}$$

The algebra $\widehat{YM}_0$ is filtered by the multiplicative filtration $F^n(\widehat{YM}_0)$. On the generating space W_1 it is given by formula (38), which determines it uniquely. This filtration was alluded to at the end of the proof of Theorem 13.

A continuous differential ∂_i of $\underline{\widehat{YM}_0}$ is defined by the formulas

$$\partial_i(A_j) = \delta_{ij}, \tag{39}$$

$$\partial_i(\phi_k) = \partial_i(\psi^\alpha) = 0. \tag{40}$$

We use δ_{ij} for the Kronecker δ symbol. All such differentials span the D-dimensional vector space $\mathbf{V}^*$. They can be arranged into one differential $\partial : \underline{\widehat{YM}_0} \to \mathbf{V}^* \otimes \underline{\widehat{YM}_0}$, $\partial(a) = \partial_1(a), \ldots, \partial_d(a)$. A choice of projection

$$\mu : \mathbf{V}^* \to V \to 0 \tag{41}$$

on the space V specifies a differential ∂_μ of YM_0 with values in the bimodule $V \otimes \underline{\widehat{YM}_0}$.

Denote

$$\underline{\widehat{TYM}_\mu} = \operatorname{Ker} \partial_\mu \quad \text{and} \quad \widehat{TYM}_\mu = I(\underline{\widehat{TYM}_\mu}).$$

The algebra $\widehat{TYM}_\mu$ is filtered by $\widehat{TYM}_\mu \cap F^n$, where $F^n = F^n(\widehat{YM}_0)$ is the filtration on $\widehat{YM}_0$. Similar constructions hold for the algebras YM_0 and YM'_0. As a result, we can define $T_\mu YM$ and $T_\mu YM'$.

Define the algebra $\underline{\widehat{E_\mu YM}} = \underline{\widehat{YM}_0} \otimes \Lambda^\bullet(V)$. The differential is defined by the rule

$$\begin{aligned} d(a) &= \partial_\mu(a) \in YM_0 \otimes V \subset YM_0 \otimes \Lambda^\bullet(V), \quad & a \in YM_0, \\ d(v) &= 0, & v \in \Lambda^\bullet(V). \end{aligned} \tag{42}$$

The differential can be extended uniquely to the algebra using the Leibniz rule. Denote

$$\widehat{E_\mu YM} = I(\underline{\widehat{E_\mu YM}}).$$

A similar construction works for the algebras YM_0 and YM'_0. We can define the algebras $E_\mu YM$ and $E_\mu YM'$. Define a multiplicative decreasing filtration $F^s(\underline{\widehat{E_\mu YM}}) = F^s$ on $\underline{\widehat{E_\mu YM}}$ which extends the filtration on $F^s(\underline{\widehat{YM}_0})$. It is uniquely determined by the condition $V \subset F^1$. It defines a filtration on $\widehat{E_\mu YM}$ for which we keep the same notation. A similar filtration exists on $E_\mu YM$. The algebra $E_\mu YM'$ is graded, the grading is a multiplicative extension of the grading on YM'_0, the grading of the space $V \subset E_\mu YM'$ is equal to one. In the case of the algebras $\widehat{E_\mu YM}$, $E_\mu YM$, the differential preserves the filtration. In the case of $E_\mu YM'$ the differential preserves the grading.

Lemma 28. *Suppose B is an algebra with a unit generated by elements $B_1, \dots, B_n$. Assume that we are given k commuting differentials ∂_s, $s = 1, \dots, k$, $k \le n$, of the algebra B such that $\partial_s B_j = \delta_{sj}$. Then there exists an increasing filtration G^i, $i = 0, 1, \dots,$ such that*

(a) $\bigcup_i G^i = B$.
(b) $G^i G^j \subset G^{i+j}$.
(c) $B_i \in G^1$.

Proof. Define the filtration G^i inductively. By definition $G^0 = \bigcap_{s=1}^k \operatorname{Ker} \partial_{si}$, then

$$G^{i+1} = \{x | \partial_s(x) \in G^i \text{ for all } s,\ 1 \le s \le k\}. \tag{43}$$

Property (b) follows from the Leibniz rule. By definition $B_i \in G^1$; hence (c) follows. Since B_i are generators, (b) and (c) imply (a). □

Consider the algebra $\mathrm{Gr}_G(B) = \bigoplus_{i=0}^\infty G^{i+1}B/G^iB$. Denote the image of the elements $B_1, \dots, B_k$ in G^1/G^0 by $\hat{B}_1, \dots, \hat{B}_k$.

Lemma 29. *The algebra $\mathrm{Gr}_G(B)$ has the following properties:*

(a) *The elements $\hat{B}_1, \dots, \hat{B}_k$ commute in $\mathrm{Gr}_G(B)$.*
(b) *The elements $\hat{B}_1, \dots, \hat{B}_k$ commute with G^0B.*
(c) *The subalgebra of $\mathrm{Gr}_G(B)$ generated by G^0B and $\hat{B}_1, \dots, \hat{B}_k$ is isomorphic to $\mathbb{C}[\hat{B}_1, \dots, \hat{B}_k] \otimes G^0B$.*
(d) *The elements $\hat{B}_1, \dots, \hat{B}_k$ and G^0B generate $\mathrm{Gr}_G(B)$.*
(f) *We have an isomorphism $\mathrm{Gr}_G(B) = \mathbb{C}[\hat{B}_1, \dots, \hat{B}_k] \otimes G^0B$.*

Proof.

(a) $\partial_s[B_i, B_j] = [\delta_{si}, B_j] + [B_i, \delta_{sj}] = 0$; therefore, $[B_i, B_j] \in G^0$, hence $[\hat{B}_i, \hat{B}_j] = 0$.
(b) Similarly, $\partial_s[B_i, m] = [\delta_{si}, m] = 0$ for $m \in G^0 \subset \operatorname{Ker} \partial_s$ for $s = 1, \dots, k$, Therefore, $[B_i, m] \in G^0$ and $[\hat{B}_i, \hat{m}] = 0$.
(c) Denote the subalgebra generated by G^0B and $\hat{B}_1, \dots, \hat{B}_k$ by C. There is a surjective map

$$\mathbb{C}[\hat{B}_1, \dots, \hat{B}_k] \otimes G^0B \to C \tag{44}$$

and an inclusion $G^0 \subset C$. Denote the kernel of the map (44) by I. Then

$$I \cap G^0B = 0. \tag{45}$$

Suppose $0 \neq a = \sum a_{i_1,\dots,i_k} \hat{B}_1^{i_1} \dots \hat{B}_k^{i_k} \in I$. Since a is a polynomial in $\hat{B}_1, \dots, \hat{B}_k$, there are $i_1, \dots, i_k$ such that there is no $a_{i'_1,\dots,i'_k} \neq 0$ with $i_1 < i'_1, \dots, i_k < i'_k$. This means that an element $\partial_1^{i_1} \dots \partial_k^{i_k} a \neq 0$ belongs to I and is independent of $\hat{B}_1, \dots, \hat{B}_k$, which contradicts (45).
(d) We are going to prove the statement by induction on the index i in $\mathrm{Gr}^i_G(B)$. If $i = 0$, then there is nothing to prove. Suppose we have an element $\hat{a} \in \mathrm{Gr}^{i+1}_G(B)$. Let $a \in G^{i+1}$ be its representative in B. Then $\partial_s a = b_s \in G^i$ and by the inductive assumption $\hat{b}_s = \hat{b}_s(\hat{B}_1, \dots, \hat{B}_k) \in C$. The elements b_s satisfy $\partial_i \hat{b}_s = \partial_s \hat{b}_i$. This implies that there is $b \in B$ such that $\partial_s \hat{b} = \hat{b}_s$. Consider the difference $a - b = c$, $\partial_s \hat{c} = 0$, hence $\partial_s c \in G^{i-1}$ for every s; therefore, $c \in G^i$ and $\hat{a} = \hat{b}$. □

Lemma 30. *Suppose an algebra B satisfies the conditions of Lemma 28. Denote by V the linear space with the basis $[\partial_1], \dots, [\partial_k]$. Define the structure of a complex with differential $\sum_{s=1}^k [\partial_s]\partial_s$ on the linear space $\Lambda^i(V) \otimes B$. Then the cohomology of this complex is concentrated in degree 0 and is isomorphic to G^0.*

Proof. Define a filtration on the complex $H_i = B \otimes \Lambda^i(V)$ by $G^j H_i = G^{i+j} \otimes \Lambda^i(V)$. The adjoint quotients of this filtration are isomorphic to

$$G^j H_i / G^{j-1} H_i = \Lambda^i(V) \otimes \mathrm{Sym}^{(i+j)}(V) \otimes G^0. \tag{46}$$

The differential coincides with the de Rham differential. Its cohomology is isomorphic to G^0 in degree zero and to 0 in higher degrees. The spectral sequence corresponding to the filtration G^i collapses at the E_1-term and converges to the cohomology we are looking for. □

Lemma 31.

(a) *The embeddings $(\underline{TYM_\mu}, 0) \to (\underline{E_\mu YM}, d)$ and $(\underline{TYM'_\mu}, 0) \to (\underline{E_\mu YM'}, d)$ are quasi-isomorphisms.*
(b) *The map*

$$(\widehat{\underline{TYM_\mu}}, 0) \to (\widehat{\underline{E_\mu YM}}, d) \tag{47}$$

is a filtered quasi-isomorphism of algebras (see the appendix for the definition).

Proof.

(a) The algebras YM_0 and YM'_0 satisfy Lemma 30; hence the proof follows.
(b) The morphism (47) is compatible with the filtrations which exist on its range and domain. It induces a map of spectral sequences associated with the filtrations. Let us analyze the E^2-term.
The algebra $A = \mathrm{Gr}_F(\widehat{YM}_0)$ is isomorphic to YM'_0, where in the relations (8), (9) we drop the potential U. The algebra YM'_0 is finitely generated and graded, and it carries no topology. The algebra A satisfies the conditions of Lemma 30. This implies that there is a quasi-isomorphism $(\mathrm{Gr}\,\widehat{TYM}_\mu, 0) \to (\Lambda(V) \otimes \underline{\widehat{YM}_0}, d)$.
We see that the map (47) induces an isomorphism of the E^1-terms of the corresponding spectral sequences. Since the range and domain are complete with respect to the filtrations, we conclude that the map (47) is a quasi-isomorphism. The property that it induces is the quasi-isomorphism of adjoint quotients, and it is the same as the filtered property. □

Corollary 32. *The map* $(\widehat{TYM}_\mu, 0) \to (\widehat{E_\mu YM}_0, d)$ *is a filtered quasi-isomorphism.* $(TYM_\mu, 0) \to (E_\mu YM, d)$ *and* $(TYM'_\mu, 0) \to (E_\mu YM', d)$ *are quasi-isomorphisms.*

Remark 33. The algebra $\widehat{E_\mu YM}$ is topologically finitely generated. This implies that the canonical filtration is comparable with the filtration $F^s(\widehat{E_\mu YM})$. This implies that the algebra $\widehat{E_\mu YM}_0$ is complete with respect to the canonical filtration $I^n_{\widehat{E_\mu YM}_0}$ and $\widehat{TYM}_\mu$ with respect to $\widehat{TYM}_\mu \cap I^n_{\widehat{E_\mu YM}_0}$.

2.3 Construction of BV_μ

Let V be the vector space generated by the symbols ∂_i, $i = 1, \ldots, d$, $\mathrm{Sym}(V) = \bigoplus_{i=0}^\infty \mathrm{Sym}^i(V)$. Denote the decreasing filtration of $\mathrm{Sym}(V)$ associated with the grading by $F^n(\mathrm{Sym}(V))$. The vector spaces W_i, $i = 0, \ldots, 3$, were defined by the formulas (36),

$$\begin{aligned} \overline{W}_i &= W_i \otimes \mathrm{Sym}(V), \\ \overline{W} &= \overline{W}_0 + \overline{W}_1 + \overline{W}_2 + \overline{W}_3. \end{aligned} \tag{48}$$

The last direct sum decomposition is called the homological grading on $\overline{W}$.

Define two filtrations on $\overline{W}$. The first one is

$$F^n(\overline{W}) = \sum_{i+j \geq n} F^i(W) \otimes F^j(\mathrm{Sym}(V)), \tag{49}$$

where $F^i(W)$ was defined in (38). The second is

$$\tilde{F}^n(\overline{W}) = W \otimes F^n(\mathrm{Sym}(V)). \tag{50}$$

We have

$$\tilde{F}^n(\overline{W}) \subset F^n(\overline{W}) \subset \tilde{F}^{n-8}(\overline{W}) \tag{51}$$

with finite-dimensional quotients. Denote the completion of $\overline{W}$ by $\widehat{W} = \widehat{W}_0 + \widehat{W}_1 + \widehat{W}_2 + \widehat{W}_3$.

The filtrations $\tilde{F}^n(\overline{W})$ and $F^n(\overline{W})$ define two multiplicative filtrations on $T(\overline{W}[1])$, which we denote by $\tilde{F}^n(T(\overline{W}[1]))$ and $F^n(T(\overline{W}[1]))$. The algebra $T(\overline{W}[1])$ acquires a homological grading by multiplicative extension of the homological grading from $\overline{W}$.

Define an additional grading on $\overline{W}$:

$$\overline{W}^k = \bigoplus_{i+j=k} W^i \otimes \mathrm{Sym}^j(V). \tag{52}$$

The algebra $T(\overline{W}[1])$ acquires an additional grading by multiplicative extension of the additional grading from $\overline{W}$.

Proposition 34. *The completions of $T(\overline{W}[1])$ with respect to the filtrations $\tilde{F}^n(\overline{W})$ and $F^n(\overline{W})$ coincide and we denote it by $\widehat{T(\overline{W}[1])}$. Similarly, the two completions of the space of generators of $\overline{W}$ coincide.*

Proof. The filtration satisfies the inclusions $\tilde{F}^n(T(\overline{W}[1])) \subset F^n(T(\overline{W}[1])) \subset \tilde{F}^{n-k}(T(\overline{W}[1]))$ for some finite k with finite-dimensional quotients. This is a simple corollary of equation (51). We see that the filtrations are commensurable. It is a simple exercise to show that the completions are equal. □

The operators ∂_i act by multiplication on the set of generators of the algebra $T(\overline{W}[1])$ (recall that it is a free $\mathrm{Sym}(V)$-module). We extend the action of ∂_i on $\widehat{T(\overline{W}[1])}$ as a continuous derivation, which we denote by the same symbol ∂_i.

There is a linear map $\mu : W_1 \to V$. It is an extension by zero from $\mathbf{V}$ to $W_1 = \mathbf{V} + \Pi\mathbf{S} + \Phi$ of the map μ defined in (41). We used the identification of $\langle A_1, \dots, A_D \rangle$ and $\langle A_1, \dots, A_D \rangle^*$ provided by the canonical bilinear form. The algebra $\widehat{T(\overline{W}[1])}$ admits a continuous action of the outer derivation ∇_i defined by the formula

$$\nabla_i x = \mu(A_i)x + [A_i, x]. \tag{53}$$

The commutator is defined as

$$[\nabla_i, \nabla_i] = \mu(A_i)A_j - \mu(A_j)A_i + [A_i, A_j]. \tag{54}$$

Definition 35. *Define a continuous differential in the algebra $\widehat{T(\overline{W}[1])}$ by the formulas* (5), *where* $\deg(U) \geq 8$. *Define a differential on the algebra $T(\overline{W}[1])$ by the formula* (5), *where* $\deg(U) \geq 8$, *and we impose some finiteness conditions on the potential. We denote this algebra by $(T(\overline{W}[1]), Q^\mu)$. The filtration on the algebras $\widehat{T(\overline{W}[1])}$ and $T(\overline{W}[1])$ is preserved by the differential.*

If one assumes that the potential satisfies $\deg(U) = 8$, *we denote the algebra $(T(\overline{W}[1]), Q^\mu)$ by $(T(\overline{W}[1])', Q^\mu)$. In this algebra the differential has degree zero with respect to the additional grading.*

One can extend Q^μ uniquely to the entire set of generators, using commutation properties with ∂_i.

Proposition 36. *The differential Q^μ satisfies $(Q^\mu)^2 = 0$. The differential has degree -1 with respect to the homological grading.*

Observe that though the elements ∇_i do not belong to the algebra $\widehat{T(\overline{W}[1])}$, all RHS expressions in the formulas (5) do.

Define an odd differential ε of the algebra $T(\overline{W}[1])$ by the formula $\varepsilon(c) = 1$, the value of ε on all other generators being equal to zero. It can be extended by continuity to $\widehat{T(\overline{W}[1])}$.

Proposition 37. *The commutator $\{Q^\mu, \varepsilon\}$ is a differential P of the algebra $\widehat{T(\overline{W}[1])}$ which on generators is equal to the identity transformation. The same holds for $T(\overline{W}[1])$.*

Proof. Direct computation. □

This implies that P is an invertible linear transformation on $\widehat{T(\overline{W}[1])}$ and on $T(\overline{W}[1])$, and $H = \varepsilon/P$ is a contracting homotopy. We are interested in a modification of the algebras $(\widehat{T(\overline{W}[1])}, Q^\mu)$, $(T(\overline{W}[1]), Q^\mu)$, $(T(\overline{W}[1])', Q^\mu)$. Denote by $\widehat{I(c)}$ the closure of the ideal generated by c. A simple observation is that $\widehat{I(c)}$ is a differential ideal. The ideal $\widehat{I(c)}$ is not closed under the action of ∂_i, however.

Definition 38. *Denote by $\widehat{BV_\mu}$ the quotient algebra $\widehat{T(\overline{W}[1])}/\widehat{I(c)}$. Similarly, define BV_μ and BV'_μ. The former algebra is filtered and the latter is graded.*

Remark 39. In contrast with the algebras $\widehat{BV_0}$, BV_0, BV'_0 the algebras $\widehat{BV_\mu}$, BV_μ, BV'_μ have nontrivial components with negative homological grading.

Proposition 40. *There is a morphism of differential graded algebras $(\widehat{BV_\mu}, Q^\mu) \to (\widehat{E_\mu YM}, d)$ defined by the formulas*

$$\begin{aligned} p(\partial_i c) &= [\partial_i], \\ p(A_i) &= A_i, \\ p(\phi_i) &= \phi_i, \\ p(\psi^\alpha) &= \psi^\alpha, \end{aligned} \tag{55}$$

which preserves the filtrations (so it is continuous). The map is zero on the rest of the generators. The formulas (55) *define a map $(BV_\mu, Q^\mu) \to (E_\mu YM, d)$ which preserves the filtrations and a map $(BV'_\mu, Q^\mu) \to (E_\mu YM', d)$ of degree zero.*

The filtrations $F^n \widehat{T(\overline{W}[1])}$ and $\tilde{F}^n \widehat{T(\overline{W}[1])}$ of $\widehat{T(\overline{W}[1])}$ induce similarly named filtrations of $\widehat{BV_\mu}$, denoted $F^n \widehat{BV_\mu}$ and $\tilde{F}^n \widehat{BV_\mu}$. The filtrations F^n and $\tilde{F}^n$ are also defined on BV_μ and BV'_μ. In the latter case the filtration F^n coincides with the decreasing filtration associated with the grading.

Define $\mathrm{Gr}\,\widehat{BV}_\mu = \prod_{n\geq 0} \mathrm{Gr}^n\,\widehat{BV}_\mu$ as

$$\mathrm{Gr}^n\,BV_\mu = \tilde{F}^n\widehat{BV}_\mu/\tilde{F}^{n+1}\widehat{BV}_\mu. \tag{56}$$

Proposition 41. *The algebra* $\mathrm{Gr}\,\widehat{BV_\mu}$ *is the completion of the free algebra with the same space of generators as* $\widehat{BV_\mu}$. *The differential* Q *is defined by the formulas* (5), *except that in the formula* (53) *one needs to replace*

$$\mu(A_i)x + [A_i, x] \Rightarrow [A_i, x]. \tag{57}$$

Similarly, $\mathrm{Gr}\,BV_\mu$ *and* $\mathrm{Gr}\,BV'_\mu$ *coincide with* BV_μ *as algebras. In the definition of* Q^μ *one has to alter* ∇_i *according to the rule* (57).

Proof. Obvious. □

The nonreduced bar-complex of an A_∞-coalgebra H with a counit ε is by definition the bar-complex of H as if it had no counit. A simple theorem asserts that in the presence of a counit it is always contractible.

Definition 42. *The algebras* $(\widehat{T(\overline{W}[1])}, Q^\mu)$ *and* $(\mathrm{Gr}\,T(\overline{W}[1]), \mathrm{Gr}\,Q)$ *can be thought of as nonreduced bar-complexes of* A_∞*-coalgebras with a counit and coaugmentation, which we denote by* $(\widehat{\overline{W}}, Q^\mu)$ *and* $(\widehat{\overline{W}}, \mathrm{Gr}(Q^\mu))$; *the corresponding coideals are denoted as* $\widehat{bv_\mu}$ *and* $\widehat{\mathrm{Gr}\,bv_\mu}$. *Similarly, we have the* A_∞*-coalgebras* $(\overline{W}, Q^\mu)$, $(\overline{W}, \mathrm{Gr}\,Q^\mu)$; $(\overline{W}', Q^\mu)$, $(\overline{W}', \mathrm{Gr}\,Q^\mu)$ *with coaugmentation coideals* (bv_μ, Q^μ), $(\mathrm{Gr}\,bv_\mu, \mathrm{Gr}\,Q^\mu)$ *and* (bv'_μ, Q^μ), $(\mathrm{Gr}\,bv'_\mu, \mathrm{Gr}\,Q^\mu)$, *respectively.*

There is a general construction of the tensor product of A_∞-coalgebras. Its description when one of the tensor factors is an ordinary coalgebra is very simple. Suppose H, G are two A_∞-algebras and G has only one nontrivial operation $\Delta_2 = \Delta : G \to G^{\otimes 2}$. Define the mapping $\nu_n \to G^{\otimes n}$ by the formula

$$\nu_n^G = (\Delta \otimes \mathrm{id} \otimes \ldots \mathrm{id}) \circ \cdots \circ \Delta. \tag{58}$$

The tensor product of H and G has its underlying vector space equal to $H \otimes G$. The operations $\Delta_i^{H\otimes G}$ are defined by the formula

$$\Delta_n^{H\otimes G}(a \otimes b) = T\Delta_n^H(a) \otimes \nu_n^G(b) \tag{59}$$

where the operator T is a graded permutation which defines an isomorphism $H^{\otimes n} \otimes G^{\otimes n} \cong (H \otimes G)^{\otimes n}$. We need to draw attention to the fact that although Δ_n^H is the nth operation in the coalgebra H, the map ν_n^G is not such for the coalgebra G, but rather the nth iteration of the binary operation. As you can see, this construction is not symmetric.

Observe that on the category of A_∞-algebras a similar operation corresponds to the extension of the ring of scalars.

It turns out that the algebras $\widehat{\mathrm{Gr}\, bv_\mu}$, $\mathrm{Gr}\, bv_\mu$, $\mathrm{Gr}\, bv'_\mu$ have an alternative description in terms of the finite-dimensional A_∞-coalgebras $\widehat{bv}_0$, bv_0, bv'_0 introduced in Definition 26.

The Künneth formula asserts the following.

Proposition 43. *There is a quasi-isomorphism of algebras* $\widehat{\mathrm{Bar}}(H \otimes G) \to \widehat{\mathrm{Bar}}(H) \hat{\otimes} \widehat{\mathrm{Bar}}(G)$.

Proof. See [12] for the proof in the case of algebras. The coalgebra case is similar.□

Proposition 44. *There is an isomorphism of coalgebras* $\underline{\widehat{\mathrm{Gr}\, bv_\mu}} = W \otimes \underline{\widehat{\mathrm{Sym}(V)}}$, $\underline{\mathrm{Gr}\, bv_\mu} = W \otimes \underline{\mathrm{Sym}(V)}$, $\underline{\mathrm{Gr}\, bv'_\mu} = W' \otimes \underline{\mathrm{Sym}(V)}$.

Proof. All nontrivial interactions between the bv_0 and $\mathrm{Sym}(V)$ parts inside bv_μ stem from the $\mu(A_i)x$ part in the formula (53) which we kill in passing from $\widehat{bv_\mu}$ to $\mathrm{Gr}\, \widehat{bv_\mu}$.

In the case at hand, $H = \widehat{bv_0}$, $G = \mathrm{Sym}(V)$. The symmetric algebra $\mathrm{Sym}(V)$ has diagonal Δ which on $v \in V$ is equal to $\Delta(v) = v \otimes 1 + 1 \otimes v$. The arguments remain valid in the noncomplete and graded cases. □

Proposition 45. *The map* $p : (\widehat{BV_\mu}, Q) \to (\widehat{E_\mu YM}, d)$ *defined in equations* (55) *is a quasi-isomorphism. The map* $p : (BV'_\mu, Q) \to (E_\mu YM', d)$ *is also a quasi-isomorphism.*

Proof. Define a filtration $\tilde{F}^n E_\mu YM_0$ by the formula

$$\tilde{F}^n E_\mu YM = \bigoplus_{k \geq n} \Lambda^k(V) \otimes \widehat{YM}_0. \tag{60}$$

The filtrations $\tilde{F}^n \widehat{E_\mu YM}$ are compatible with the map $p : \widehat{BV}_\mu \to \widehat{E_\mu YM}$. The map p induces a map of spectral sequences associated with the filtrations.

The E_0 term of the spectral sequence associated with $\widehat{BV}_\mu$ coincides with $\mathrm{Gr}\, \widehat{BV}_\mu$.

Proposition 43 implies a series of quasi-isomorphisms:

$$\widehat{\mathrm{Gr}\, BV_\mu} \overset{\mathrm{def}}{=} \widehat{\mathrm{Bar}}(\mathrm{Gr}\, \widehat{bv_\mu}) = \widehat{\mathrm{Bar}}(\widehat{bv_0} \otimes \widehat{\mathrm{Sym}}(V)) \xrightarrow{k} \widehat{\mathrm{Bar}}(\widehat{bv_0}) \otimes \widehat{\mathrm{Bar}}(\widehat{\mathrm{Sym}}(V)), \tag{61}$$

The map $\mathrm{Gr}(p)$ factors through the map k: $\mathrm{Gr}(p) = p' \circ k$, where p' is

$$\widehat{\mathrm{Bar}}(bv_0) \otimes \widehat{\mathrm{Bar}}(\mathrm{Sym}(V)) \xrightarrow{p'} \widehat{YM}_0 \otimes \Lambda(V). \tag{62}$$

The map p' in the formula (62) is the tensor product of two quasi-isomorphisms. The first one is from Theorem 1; the second one is the classical quasi-isomorphism $\widehat{\mathrm{Bar}}(\widehat{\mathrm{Sym}(V)}) \to \Lambda(V)$.

These considerations imply that there is an isomorphism $p : H^\bullet \mathrm{Gr}\, BV_\mu \to H^\bullet \mathrm{Gr}\, E_\mu YM$. This means that we have an isomorphism of spectral sequences associated with the filtration $\tilde{F}^n$ starting with the E_1 term.

It implies that the map p induces a quasi-isomorphism of completed complexes: $p : \widehat{BV}_\mu \to \widehat{E_\mu YM}$. The proof goes through in the graded case. (All the tools which are needed for the proof are collected in the appendix in the segment devoted to homogeneous A_∞-(co)algebras.) The obstacle to the proof in the noncomplete case is the absence of a quasi-isomorphism $\mathrm{Bar}\, bv_0 = BV_0 \to YM_0$. □

Proposition 46. *$(\widehat{BV_\mu}, Q)$, $\widehat{T_\mu YM}$ and (BV'_μ, Q), $T_\mu YM'$ are pairs of quasi-isomorphic algebras.*

Proof. By Proposition 45, the algebra $(\widehat{BV}_\mu, Q)$ is quasi-isomorphic to $(\widehat{E_\mu TYM}, Q)$. By Lemma 31 the algebra $(\widehat{E_\mu TYM}, Q)$ is quasi-isomorphic to $\widehat{T_\mu YM}$. The proof for the second pair is similar. □

Proposition 47. *The Hochschild homology $H_i(\widehat{T_\mu YM}, \mathbb{C})$ as an A_∞-coalgebra is isomorphic to the A_∞-coalgebra $\widehat{bv_\mu}$. The same isomorphism holds in the graded case.*

Proof. There is a series of quasi-isomorphisms

$$\widehat{T_\mu YM} \overset{a}{\cong} \widehat{E_\mu YM} \overset{b}{\cong} \widehat{\mathrm{Bar}}(\widehat{bv_\mu}). \tag{63}$$

Theorem 80 asserts that if all quasi-isomorphisms in equation (63) are filtered, then we have a quasi-isomorphism

$$\widehat{\mathrm{Bar}}\, \widehat{T_\mu YM} \cong \widehat{\mathrm{Bar}}\, \widehat{\mathrm{Bar}}(\widehat{bv_\mu}). \tag{64}$$

Lemma 31 asserts that the quasi-isomorphism a is filtered. The proof of Proposition 45 shows that the quasi-isomorphism b is filtered. Lemma 83 claims that for any algebra complete with respect to the canonical filtration we have a quasi-isomorphism

$$\widehat{bv_\mu} \cong \widehat{\mathrm{Bar}}\, \widehat{\mathrm{Bar}}(\widehat{bv_\mu}). \tag{65}$$

By definition, the homology of the algebra $\widehat{T_\mu YM}$ is the homology of the bar-complex $\widehat{\mathrm{Bar}}\, T_\mu YM$. By the result of [13], there is a quasi-isomorphism of A_∞-coalgebras $H(\widehat{T_\mu YM})$ and $\widehat{\mathrm{Bar}}\, \widehat{T_\mu YM}$. Quasi-isomorphisms (64) and (65) finish the proof. The proof in the graded case is similar. □

Proposition 48. *The differential Q_1^μ in the A_∞-coalgebra $\widehat{bv_\mu}$ is defined on $\underline{\widehat{\mathrm{Sym}}(V)}$-generators by the formulas*

$$\begin{aligned}
Q_1^\mu(\phi_k) &= 0,\\
Q_1^\mu(A_i) &= -\mu(A_i)c,\\
Q_1^\mu(\psi^\alpha) &= 0,\\
Q_1^\mu(c) &= 0,
\end{aligned}$$

$$
\begin{aligned}
Q_1^{\mu}(c^*) &= \sum_{i=1}^{D} \mu(A_i)A^{*i}, \\
Q_1^{\mu}(A^{*m}) &= \sum_{i=1}^{D} -\mu(A_i)\mu(A_i)A_m + \mu(A_m)\mu(A_i)A_i, \\
Q_1^{\mu}(\phi^{*j}) &= \sum_{i=1}^{D} -\mu(A_i)\mu(A_i)\phi_j, \\
Q_1^{\mu}(\psi_{\alpha}^{*}) &= \sum_{i=1}^{D}\sum_{\beta} -\Gamma_{\alpha\beta}^{i}\mu(A_i)\psi^{\beta}.
\end{aligned}
\tag{66}
$$

The same formulas hold in the graded case. The homological grading on the complex $\widehat{W}$ is defined as follows: the component of homological degree i is equal to $\widehat{W}_i$.

Proof. Direct inspection. □

2.4 Examples of computations

Example 1. The first trivial example is when $V = 0$ and $\mu = 0$. In this case the differential Q_1 in (66) is equal to zero and we get that $\underline{H_\bullet(\widehat{YM}_0)} = W$, where the graded space W is defined by formula (36). This is a tautological result.

The second example is when $\dim(V) = 1$. We have two options: restriction of the bilinear form $(.,.)$ to the kernel of the map (41) is (a) *invertible*; (b) *degenerate*.

Example 2. Let us analyze case (a). Below is an explicit description of the complex (66):

$$
\begin{array}{cccc}
L\otimes c^* \xrightarrow{t} & L\otimes A^{*1} & & \\
 & L\otimes A^{*2} \xrightarrow{t^2} & L\otimes A_2 & \\
 & \cdots & & \\
 & L\otimes A^{*D} \xrightarrow{t^2} & L\otimes A_D & \\
 & L\phi^{*1} \xrightarrow{t^2} & L\otimes \phi_1 & \\
 & \cdots & & \\
 & L\otimes \phi^{*d'} \xrightarrow{t^2} & L\otimes \phi_{d'} & \\
 & L\otimes \psi_{\alpha}^{*} \xrightarrow{t} & L\otimes \psi^{\alpha} & \\
 & \cdots & & \\
 & & L\otimes A_1 \xrightarrow{t} & L\otimes c \\
\hline
3 & 2 & 1 & 0
\end{array}
\tag{67}
$$

Here $L = \widehat{\mathbb{C}[t]}$. The cohomology classes of this complex are

(a) In dimension 0, it is the space of constants.

(b) In dimension 1 the space is spanned by $A_2, \dots, A_D, tA_2, \dots, tA_D, \phi_1, \dots, \phi_{d'}, t\phi_1, \dots, t\phi_{d'}, \psi^\alpha$.
(c) In dimension 2, the space is spanned by A^{*1}.
(d) In dimension 3, the space of cocycles is zero.

This computation enables us to identify the algebra $T_\mu YM$ with the subalgebra $K \subset YM_0$ defined in Theorem 11. The connection is $[q_i] = A_{i+1}$, $[p^i] = tA_{i+1}$, $[P^j] = \phi_j$, $[Q^j] = t\phi_j$, $[\psi^\alpha] = \psi^\alpha$. The symbol $[a]$ denotes the homology class of a generator a. The cocycle A^{*1} corresponds to the single relation $\sum_{i=1}^{D-1}[q_i, p^i] + \sum_{i=1}^{d'}[Q_i, P^i] - \frac{1}{2}\sum_\alpha\{\psi^\alpha, \psi^\alpha\}$.

The algebra K has homological dimension 2. There is up to a constant only one homology class, which we denote $\int \in H_2(K)$. The algebra $\underline{K}$ is the universal enveloping algebra of a Lie algebra $\mathfrak{k}$ with the same set of generators and relations. We have an isomorphism $H^2(K) = H^2(\mathfrak{k}, \mathbb{C})$. It can be used to define a symplectic structure on the moduli space of representations of $\mathfrak{k}$ in a semisimple Lie algebra $\mathfrak{g}$, equipped with an invariant dot-product $(.,.)_\mathfrak{g}$. It is well known what the tangent space to a point ρ of the moduli of representations of a Lie algebra $\mathfrak{m}$ is. It is equal to $H^1(\mathfrak{m}, \mathfrak{g})$. In our case it is equal to $H^1(\mathfrak{k}, \mathfrak{g})$. If we have two elements $a, b \in H^1(\mathfrak{k}, \mathfrak{g})$, the cohomological product $a \cup b \in H^2(\mathfrak{k}, \mathfrak{g} \otimes \mathfrak{g})$. The composition with $(.,.)_\mathfrak{g}$ gives an element of $H^2(\mathfrak{k}, \mathbb{C})$, whose value on the homology class $\int$ is equal to the value of the symplectic dot-product $\omega(a, b)$. In more condensed notation, we can write

$$\omega(a, b) = \int (a, b)_\mathfrak{g}. \tag{68}$$

Proposition 49. *The symplectic form $\omega(a, b)$ defined on the moduli space* $\mathrm{Mod}_\mathfrak{k}(\mathfrak{g})$ *is nondegenerate and closed.*

Proof. There is a different description of the space $\mathrm{Mod}_\mathfrak{k}(\mathfrak{g})$. Consider the linear space $(\mathfrak{g} + \mathfrak{g})^{\times(D-1)} + (\mathfrak{g} + \mathfrak{g}) \otimes \Phi^* + \Pi\mathfrak{g} \otimes \mathbf{S}^*$. We can identify the vector space $\mathfrak{g}+\mathfrak{g}$ with $\mathfrak{g}+\mathfrak{g}^*$, by means of the invariant bilinear form $(.,.)_\mathfrak{g}$. The space $\mathfrak{g}+\mathfrak{g}^*$ is a symplectic manifold. The space $\Pi\mathfrak{g}$ is an odd-dimensional symplectic manifold with symplectic form equal to $(.,.)_\mathfrak{g}$. The Lie algebra $\mathfrak{g}$ acts on this space by symplectic vector fields. Define the set of functions $f_i = (e_i, \sum_{k=1}^{D-1}[q_k, p^k] + \sum_{k=1}^{d'}[Q_k, P^k] - \frac{1}{2}\sum_\alpha\{\psi^\alpha, \psi^\alpha\})$. It is easy to see that this set of functions defines a set of Hamiltonians for the generators e_i of the Lie algebra $\mathfrak{g}$. The symplectic reduction with respect to the action of $\mathfrak{g}$ gives rise precisely to the manifold we are studying. The statement of the proposition follows from the general properties of Hamiltonian reduction. □

Example 3. Now we want to discuss case (b), where the restriction of the bilinear form to the kernel of the map μ is degenerate. It is easy to see that the null space of the form is one dimensional. Without loss of generality we may assume that $\mu(A_0) = t$, $\mu(A_1) = it$ (i is the imaginary unit) and the map μ on the rest of the generators is equal to zero. It is convenient to make a change of coordinates

$$\begin{aligned} v &= \frac{1}{\sqrt{2}}(A_1 + iA_2), \\ u &= \frac{1}{\sqrt{2}}(A_1 - iA_2), \\ u^* &= \frac{1}{\sqrt{2}}(A^1 + iA^{*2}), \\ v^* &= \frac{1}{\sqrt{2}}(A^1 - iA^{*2}). \end{aligned} \tag{69}$$

In this notation, the differential Q_1 looks like

$$\begin{array}{cccc} \mathbb{C}[t]\otimes c^* \overset{\sqrt{2}t}{\to} & \mathbb{C}[t]\otimes u^* & & \\ & \mathbb{C}[t]\otimes v^* \overset{-2t^2}{\to} & \mathbb{C}[t]\otimes v & \\ & \mathbb{C}[t]\otimes A^{*3} \overset{0}{\to} & \mathbb{C}[t]\otimes A_3 & \\ & \dots & & \\ & \mathbb{C}[t]\otimes A^{*D} \overset{0}{\to} & \mathbb{C}[t]\otimes A_D & \\ & \mathbb{C}[t]\phi^{*1} \overset{0}{\to} & \mathbb{C}[t]\otimes \phi_1 & \\ & \dots & & \\ & \mathbb{C}[t]\otimes \phi^{*d'} \overset{0}{\to} & \mathbb{C}[t]\otimes \phi_{d'} & \\ & \mathbb{C}[t]\otimes \psi^*_\alpha \overset{Gt}{\to} & \mathbb{C}[t]\otimes \psi^\alpha & \\ & \dots & & \\ & & \mathbb{C}[t]\otimes u \overset{\sqrt{2}t}{\to} & \mathbb{C}[t]\otimes c \\ \hline 3 & 2 & 1 & 0 \end{array} \tag{70}$$

where G is a linear map $\mathbf{S}^* \to \mathbf{S}$. In the degenerate case, not much can be said about G. If G is built from spinorial Γ-matrices, G has a kernel with dimension equal to $1/2\dim(\mathbf{S})$. An important observation is that the complex (70) has infinite homology groups in dimensions $1, 2$. The homology in dimension 3 is trivial and the zeroth homology is one dimensional. To simplify formulas for truncated Yang–Mills algebra in this case we get rid of fermions. After the change of variables (69), the relations (7), (8), (9) become

$$-[v[v,u]] + \sum_{i=3}^{D}[A_i,[A_i,v]] + \sum_{k=1}^{d'}[\phi_k[\phi_k,v]] = 0,$$

$$-[u[u,v]] + \sum_{i=3}^{D}[A_i,[A_i,u]] + \sum_{k=1}^{d'}[\phi_k[\phi_k,u]] = 0,$$

$$[u[v,A_m]] + [v[u,A_m]] + \sum_{i=3}^{D}[A_i,[A_i,A_m]] + \sum_{k=1}^{d'}[\phi_k[\phi_k,A_m]] = 0,$$

$$m = 3,\dots,D,$$

$$[u[v,\phi_j]]+[v[u,\phi_j]]+\sum_{k=3}^{D}[A_k[A_k.\phi_j]]+\frac{\partial U}{\partial \phi_j}=0, \qquad j=1,\dots,d'. \tag{71}$$

The generators of the algebra $T_\mu YM$ with $\mathrm{rk}(\mu)=1$ and $\mathrm{ind}(\mu)=0$ are

$$\begin{aligned} &v,\\ p&=[u,v],\\ A_m^n&=\mathrm{Ad}^n(u)A_m, \quad n\geq 0, \quad m=3,\dots,D,\\ \phi_j^n&=\mathrm{Ad}^n(u)\phi_j, \quad n\geq 0, \quad j=1,\dots,d'. \end{aligned} \tag{72}$$

As in the case of the first example, the algebra YM_0 is the semidirect product of an abelian one-dimensional Lie algebra and the algebra $T_\mu YM$. The relations in $T_\mu YM$ and the action of the generator of the abelian Lie algebra (Hamiltonian) can be read off from equations (71). The action of the Hamiltonian is given by the formulas

$$\begin{aligned} H(p)&=\sum_{i=3}^{D}[A_i^0,A_i^1]+\sum_{k=1}^{d'}[\phi_k^0,\phi_k^1],\\ H^m(A_i^0)&=A_i^m,\\ H^m(\phi_i^0)&=\phi_i^m \end{aligned}$$

The relations are

$$\begin{aligned} &[v,p]+\sum_{i=3}^{D}[A_i^0,[A_i^0,v]]+\sum_{k=1}^{d'}[\phi_k^0[\phi_k^0,v]]=0,\\ &A_0^{*m}=[p,A_m^0]+2[v,A_m^1]+\sum_{i=3}^{D}[A_i^0,[A_i^0,A_m^0]]+\sum_{k=1}^{d'}[\phi_k^0[\phi_k^0,A_m^0]]=0,\\ &\quad m=3,\dots,D,\\ &\phi_0^{*j}=[p,\phi_j^0]+2[v,\phi_j^1]+\sum_{k=3}^{D}[A_k^0[A_k^0.\phi_j^0]]+\frac{\partial U}{\partial \phi_j^0}=0, \quad j=1,\dots,d',\\ &A_n^{*m}=H^n(A_0^{*m}), \quad m=3,\dots,D, \quad n\geq 1,\\ &\phi_n^{*j}=H^n(\phi_0^{*j}), \qquad j=1,\dots,d', \quad n\geq 1. \end{aligned} \tag{73}$$

Proposition 50. *There is an isomorphism of $T_\mu YM$ and the quotient algebra $\mathbb{C}\langle v,p,A_k^n,\phi_j^n,\psi_\alpha\rangle/(I)$, where the ideal is generated by relations* (73). *There is an isomorphism $\mathbb{C}[H]\ltimes T_\mu YM\cong YM_0$.*

Example 4. Suppose $V=\mathbf{V}$ and $\mu=\mathrm{id}$. We need to compute the cohomology of complexes

$$
\begin{array}{ccccccc}
 & & \mathbf{S}^* \otimes \underline{\widehat{\mathrm{Sym}}(\mathbf{V})} & \overset{\not{D}}{\to} & \mathbf{S} \otimes \underline{\widehat{\mathrm{Sym}}(\mathbf{V})} & & \\
\langle c^* \rangle \otimes \underline{\widehat{\mathrm{Sym}}(\mathbf{V})} & \overset{d}{\to} & \Pi\mathbf{V}^* \otimes \underline{\widehat{\mathrm{Sym}}(\mathbf{V})} & \overset{d*d}{\to} & \mathbf{V} \otimes \underline{\widehat{\mathrm{Sym}}(\mathbf{V})} & \overset{d}{\to} & \langle c \rangle \otimes \widehat{\mathrm{Sym}}(\mathbf{V}) \\
 & & \Phi^* \otimes \underline{\widehat{\mathrm{Sym}}(\mathbf{V})} & \overset{\times||v||^2}{\to} & \Phi \otimes \underline{\widehat{\mathrm{Sym}}(\mathbf{V})} & & \\
\hline
3 & & 2 & & 1 & & 0
\end{array}
\tag{74}
$$

It is easy to see that the homology of the complex (74) coincides with the completion of the homology of a similar complex with $\underline{\widehat{\mathrm{Sym}}}$ stripped of the completion sign. Therefore, we will examine only the noncompleted version. It is particularly easy to compute the $\Phi^* - \Phi$ part of the cohomology. It is equal to zero in all dimensions but one, where it is $\Phi \otimes \mathrm{Sym}(\mathbf{V})/(||v||^2)$. By $(||v||^2)$ we denote the homogeneous ideal of functions equal to zero on the quadric q given by the equation $||v||^2 = 0$.

Denote

$$
\begin{aligned}
T &= \{(a, b) \in \mathbf{V} \times \mathbf{V} \mid ||a||^2 = 0,\ (a, b) = 0\}, \\
X &= \{(a, \tilde{b}) \in T | b \text{ is defined up to addition of a multiple of } a\}.
\end{aligned}
\tag{75}
$$

Then X is the quotient bundle of the tangent bundle T to the quadric by the one-dimensional subbundle L. L consists of all vector fields that are tangent to the projection $q\{0\} \to \tilde{q} \subset \mathbb{P}^{D-1}$. The space of global sections of X is precisely the first cohomology of the complex (74) in the $\mathbf{V} \otimes \mathrm{Sym}(\mathbf{V})$-term. The zeroth cohomology in the $\langle c \rangle \otimes \mathrm{Sym}(\mathbf{V})$-term is one dimensional for obvious reasons. The vanishing of the third and the second cohomology will be proved in Proposition 52 under more general assumptions.

There is a standard "adjoint" Dirac operator $\not{D}^* : \mathbf{S} \otimes \mathrm{Sym}(\mathbf{V}) \to \mathbf{S}^* \otimes \mathrm{Sym}(\mathbf{V})$. Together $\not{D}$ and $\not{D}^*$ satisfy

$$
\begin{aligned}
\not{D}\not{D}^* &= ||.||^2, \\
\not{D}^*\not{D} &= ||.||^2,
\end{aligned}
\tag{76}
$$

where $||.||^2$ is the operator of multiplication on the quadric. Equations (76) imply that there is no second cohomology in the $\mathbf{S}^* \otimes \mathrm{Sym}(\mathbf{V})$-term. There is a similar geometric interpretation of cohomology in the $\mathbf{S} \otimes \mathrm{Sym}(\mathbf{V})$-term. Suppose $\mathbf{S}$ is a spinor representation of $\mathrm{Spin}(n)$; upon restriction of $\mathbf{S}$ to $\mathrm{Spin}(n-2)$, $\mathbf{S}$ splits into two nonisomorphic spinor representations $\mathbf{S}_1, \mathbf{S}_2$; choose the one of the two which contains the highest weight vector of $\mathbf{S}$ as a representation of $\mathrm{Spin}(n)$. The Levi subgroup of the stabilizer of a point l of the quadric q is equal to $\mathrm{SO}(n-2)$. One can induce a vector bundle C on q from the representation $\mathbf{S}_1$ of $\mathrm{Spin}(n-2)$. It is not hard to see that the first cohomology of the complex (74) in the term $\mathbf{S} \otimes \mathrm{Sym}(\mathbf{V})$ is isomorphic to the direct sum of the space of global sections of C.

It is useful to use the Borel–Weil theorem to compute the spaces of global sections of these bundles.

As an illustration, let us carry out such a computation in the case $D = 10, d' = 0, N = 1$.

The Dynkin graph of the group $\mathrm{Spin}(10) = \mathrm{Spin}(\mathbf{V})$ is

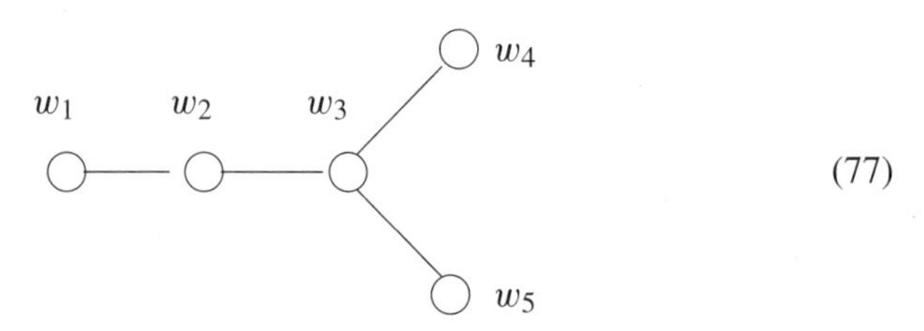

(77)

We will encode a representation which is labeled by the Dynkin diagram above by an array $[w_1, w_2, w_3, w_4, w_5]$. Our convention is that the spinor representation $\mathbf{S}$ is equal to the irreducible representation with highest weight $[0, 0, 0, 1, 0]$, the tautological representation in $\mathbb{C}^{10} = \mathbf{V}$ is equal to $[1, 0, 0, 0, 0]$, the exterior square of the latter representation is equal to $[0, 1, 0, 0, 0]$. In the case of the $N = 1$, $D = 10$ super-Yang–Mills theory the cohomology is equal to

$$\begin{aligned} &\bigoplus_{i\geq 0}[i, 1, 0, 0, 0] \quad \text{harmonic two-forms,} \\ &\bigoplus_{i\geq 0}[i, 0, 0, 1, 0] \quad \text{harmonic spinors.} \end{aligned} \tag{78}$$

An interesting feature of the algebra $T_{\mathrm{d}}YM$ is that its second homology vanishes. As a result we conclude that the algebra $\widehat{T_{\mathrm{d}}YM}$ is the completed free algebra. It is useful to exhibit the set of free generators of such an algebra. Before doing this introduce some notation. The space $\mathbf{V}$ is a representation of $SO(10)$ and a basis vector A_D can be taken to be the highest weight vector. The element $(A_D)^{\otimes i}$ is a highest weight vector in the ith symmetric power $\mathrm{Sym}^i(\mathbf{V})$. Let W be an irreducible representation of $\mathrm{Spin}(D)$ with highest weight vector w. Then the vector $(A_D)^i \otimes w \in \mathrm{Sym}^i(\mathbf{V}) \otimes W$ will be a highest weight vector and generates an irreducible subrepresentation of $\mathrm{Spin}(D)$ in $\mathrm{Sym}^i(\mathbf{V}) \otimes W$; denote the projection on such a representation by p. For example, the representation $[i, 1, 0, 0, 0]$ is isomorphic to the image of $p : \mathrm{Sym}^i(\mathbb{C}^{10}) \otimes \Lambda^2(\mathbb{C}^{10}) \to \mathrm{Sym}^i(\mathbb{C}^{10}) \otimes \Lambda^2(\mathbb{C}^{10})$.

Denote $\mathrm{Ad}(x)(y) = [x, y]$, $F_{ij} = [A_i, A_j]$. Introduce elements $\mathrm{Ad}(A_{(i_1})\dots\mathrm{Ad}(A_{i_{k-1}})F_{i_k)j}$, where ( ) denotes symmetrization. This element belongs to $\mathrm{Sym}^{k-1}(\mathbf{V}) \otimes \Lambda^2(\mathbf{V})$. Similarly, the elements $\mathrm{Ad}(A_{(i_1})\dots\mathrm{Ad}(A_{i_k})\phi_j$ belong to $\mathrm{Sym}^k(\mathbf{V}) \otimes \Phi$ and $\mathrm{Ad}(A_{(i_1})\dots\mathrm{Ad}(A_{i_{k-1}})\psi^\alpha$ belong to $\mathrm{Sym}^k(\mathbf{V}) \otimes \mathbf{S}$. The elements

$$\begin{aligned} &p(\mathrm{Ad}(A_{(i_1})\dots\mathrm{Ad}(A_{i_{k-1}})F_{i_k)j}), \\ &p(\mathrm{Ad}(A_{(i_1})\dots\mathrm{Ad}(A_{i_k)})\psi^\alpha), \\ &p(\mathrm{Ad}(A_{(i_1})\dots\mathrm{Ad}(A_{i_k)})\phi_j) \end{aligned} \tag{79}$$

form a topological free set of generators of the algebra $T_{\mathrm{d}}YM$. One can check this by looking at the image of these elements in the first homology.

We would like to elucidate some general features of the complex (66). Its structure depends on the map μ.

Definition 51. *Let $\mu(b)$ be the image of the tensor $q \in \mathrm{Sym}^2(\mathbf{V})$ inverse to the scalar product under the map μ (see* (41)*). Write* $\mathrm{ind}(\mu) = \mathrm{rk}\, \mu(b)$ *and* $\mathrm{rk}(\mu) = \dim V$.

There are three classes of maps:

(a) $\mu = 0$;
(b) $\mathrm{ind}(\mu) = 0, \mu \neq 0$;
(c) $\mathrm{ind}(\mu) = 1, \mathrm{rk}(\mu) = 1$;
(d) all other cases.

The importance of such a division is justified by the following proposition.

Proposition 52.

(a) *If condition* (a) *is satisfied, the algebra $T_\mu YM$ has homological dimension* 3 *and coincides with YM_0.*
(b) *If condition* (b) *is satisfied, the algebra $T_\mu YM$ has homological dimension* 2 *and has an infinite number of generators and relations.*
(c) *If condition* (c) *is satisfied, then the algebra $T_\mu YM$ has homological dimension* 2*, and has a finite number of generators and one relation.*
(d) *If the condition* (d) *is satisfied, the algebra $T_\mu YM$ is a completed free algebra with an infinite number of generators.*

Proof.

(a) The proof is a tautology.
If $\mu \neq 0$ the third homology is equal to zero. Indeed, the space of four-chains is equal to zero. On the space of three-chains (equal to $\mathrm{Sym}(V) \otimes c^*$) the differential is injective. This implies that if $\mu \neq 0$ the homological dimension of all algebras in question is less than or equal to 2.
(b) The case when the dimension d (the number of generators of A_k) is less than or equal to three and $\mathrm{ind}(\mu) = 0$ was covered by Example 2. We may assume that $d \geq 4$. Then the restriction of the map $Q_1 : \Pi\mathbf{V}^* \otimes \widehat{\mathrm{Sym}}(V) \to \mathbf{V} \otimes \widehat{\mathrm{Sym}}(V)$ contains a free module with at least $[d/2]$ generators ($[d/2]$ is the dimension of the maximal isotropic subspace in the space $\Pi\mathbf{V}^*$ equipped with the standard bilinear form). This implies that the image $\mathrm{Im}[Q_1 : \overline{W}_3 \to \overline{W}_2]$ cannot cover $\mathrm{Ker}\, Q_1 \cap \overline{W}_2$ and the second cohomology is infinite dimensional.
(c) This was worked out as the first nontrivial example of computation of cohomology.
(d) If $\mathrm{ind}(\mu) \geq 1$ one can choose a subspace $V' \subset V$ of codimension 1 such that the image of $\mu(b)$ in $\mathrm{Sym}^2(V/V')$ is nonzero. Choose some complement V'' to V' such that $V = V' + V''$. Define the linear space F^1 as the ideal in $\widehat{\mathrm{Sym}}(V)$ generated by V''. Introduce the multiplicative descending filtration F^i of $\widehat{\mathrm{Sym}}(V)$ generated by F^1. Define the filtration of $\widehat{\overline{W}}$ as $F^s(\widehat{\overline{W}}) = W \otimes F^s$. The E_2-term of the spectral sequence associated with such a filtration is equal to the cohomology of (67), where $L = \widehat{\mathbb{C}[t]} \hat{\otimes} \widehat{\mathrm{Sym}}(V')$. The element t is a generator of V''. The space of the second homology of the E_2-term is the free $\widehat{\mathrm{Sym}}(V')$-module generated by A^{*1} (we can assume this without loss

of generality). Similarly, the first homology of the E_2-term is a free $\widehat{\mathrm{Sym}}(V')$ module of rank $D+d'-1+l$. The differential in E_2 is a $\widehat{\mathrm{Sym}}(V')$ homomorphism and it is injective on the second homology, if we can prove that it is nonzero on A^{*1}. A simple analysis shows that $Q_1 A^{*1} \neq 0$ in E_2 if the map μ satisfies condition (d). □

2.5 Fano manifolds

Let M be a smooth manifold of dimension n. Denote by Ω^p the sheaf of holomorphic p-forms on M. Let us fix a line bundle $\mathcal{L}$. In the most interesting situations M is a projective manifold and $\mathcal{L}$ is obtained from the tautological line bundle $\mathcal{O}(1)$ on projective space by means of restriction to M. We will use the notation $\mathcal{O}(1)$ for $\mathcal{L}$ and $\mathcal{O}(-1)$ for the dual bundle $\mathcal{L}^*$ also in the general case. We identify line bundles with invertible sheaves. For any sheaf $\mathcal{F}$ denote by $\mathcal{F}(i)$ the sheaf $\mathcal{F} \otimes \mathcal{O}(1)^{\otimes i}$, where the tensor product is taken over the structure sheaf $\mathcal{O}$.

The Serre algebra $\mathcal{S}$ is defined by the formula

$$\mathcal{S} = \bigoplus_{i=0}^{\infty} \mathcal{S}_i = \bigoplus_{i=0}^{\infty} H^0(M, \mathcal{O}(i)). \tag{80}$$

It can be embedded in the differential algebra $\mathcal{B}$ (Koszul–Serre algebra) in the following way. As an algebra

$$\mathcal{B} = \mathcal{S} \otimes \Lambda(\mathcal{S}_1), \tag{81}$$

where $\Lambda(\mathcal{S}_1)$ is the exterior algebra of $\mathcal{S}_1$. The algebra $\mathcal{B}$ is $\mathbb{Z}$-graded: an element $a \in \mathcal{S}_i \otimes \Lambda^j(\mathcal{S}_1)$ has degree $\deg(a) = j$. Let v_α, $\alpha = 1, \ldots, s$, be a basis of $\mathcal{S}_1 \subset \mathcal{S}$ and let θ_α be the corresponding basis of $\mathcal{S}_1 \subset \Lambda(\mathcal{S}_1)$. The algebra $\mathcal{B}$ carries a differential d of degree -1. If $a \in \mathcal{S}$, then $d(a) = 0$, $d(\theta_\alpha) = v_\alpha$. It can be extended to $\mathcal{B}$ by the Leibniz rule.

There is an additional grading on $\mathcal{B}$, which we denote by Deg. An element $a \in \mathcal{S}_i \otimes \Lambda^j(\mathcal{S}_1)$ has degree $\mathrm{Deg}(a) = i + j$. The differential has degree zero with respect to the additional grading. According to the definition in the appendix such an algebra is called homogeneous. We can split $\mathcal{B}$ into a sum

$$\mathcal{B} = \bigoplus_{i,j} \mathcal{B}_{j,i} \tag{82}$$

such that $a \in \mathcal{B}_{j,i}$ has the degrees $\deg(a) = j$, $\mathrm{Deg}(a) = i$. A line bundle $\mathcal{L}$ is called ample if for some positive n the tensor power $\mathcal{L}^{\otimes n}$ defines an embedding of the manifold M into $\mathbf{P}(H^0(M, \mathcal{L}^{\otimes n})^*)$. We assume that

$$\begin{gathered} M \text{ is an algebraic smooth manifold of dimension } n, \\ \text{the canonical bundle } \Omega^n \text{ is isomorphic to } \mathcal{O}(-k), \quad k > 0, \\ \mathcal{O}(1) = \mathcal{L} \text{ is ample.} \end{gathered} \tag{83}$$

The constant k is called the index. Manifolds satisfying the above assumptions are Fano manifolds (i.e., the anticanonical line bundle is ample). Conversely, every Fano manifold can be equipped with the above structure. (We can always take $k = 1$.)

The goal of this section is to illuminate some properties of the cohomology of the following complexes:

$$Q_i^\bullet = (0 \to \Lambda^i(\mathcal{S}_1) \to \Lambda^{i-1}(\mathcal{S}_1) \otimes \mathcal{S}_1 \to \cdots \to \mathcal{S}_i \to 0) = \bigoplus_{j=0}^{i} \mathcal{B}_{j,i}. \tag{84}$$

Some preliminaries on $H^\bullet(M, \mathcal{O}(i))$

Proposition 53 (Kodaira [9]). *Suppose $\mathcal{L}$ is an ample bundle over a complex manifold N. Then*

$$H^i(N, \Omega^j \otimes \mathcal{L}) = 0 \quad \textit{if } i + j > n.$$

Corollary 54. *Under the assumptions* (83),

$$\begin{aligned}
H^i(M, \mathcal{O}(l)) &= 0 && \textit{for } 0 < i < n \textit{ and any } l,\\
H^0(M, \mathcal{O}(-l)) &= 0 && \textit{for } l > 0,\\
H^0(M, \mathcal{O}(l)) &= 0 && \textit{for } l > -k,\\
H^0(M, \mathcal{O}) &= \mathbb{C}, &&\\
H^n(M, \mathcal{O}(-l)) &= H^0(M, \mathcal{O}(l-k))^*, && l \geq k.
\end{aligned} \tag{85}$$

Proof. The proof is straightforward: use Theorem 53 and Serre duality. □

Theorem 55. *Suppose M is an n-dimensional Fano manifold of index k. Let $\mathcal{B}$ be the differential algebra* (81). *There exists a nondegenerate pairing*

$$H^{j,i}(\mathcal{B}) \otimes H^{s-n-1-j,s-k-i}(\mathcal{B}) \to H^{s-n-1,s-k}(\mathcal{B}),$$

where $s = \dim \mathcal{S}_1$.

Proof. There is a short exact sequence of vector bundles over M (Euler sequence):

$$0 \to \mathcal{O}(-1) \to \mathcal{S}_1^* \xrightarrow{m^*} T(-1) \to 0, \tag{86}$$

where $\mathcal{S}_1^*$ is the trivial bundle with fiber $\mathcal{S}_1^*$. The first map is the tautological embedding. The vector bundle $T(-1)$ is the quotient $\mathcal{S}_1^*/\mathcal{O}(-1)$

The ith exterior power of the dual of this sequence gives rise to the complex of vector bundles

$$N_i = (0 \to E^i \to \Lambda^i(\mathcal{S}_1) \to \Lambda^{i-1}(\mathcal{S}_1)(1) \to \cdots \to \Lambda^1(\mathcal{S}_1)(i-1) \to \mathcal{O}(i) \to 0), \tag{87}$$

which is acyclic everywhere except in degree zero. The zeroth cohomology is equal to $\Lambda^i(T(-1)) = E^i$. Since $E^s = 0$, the complex N_s has a particularly simple form.

After tensoring the resolution $N_s^\bullet(i)$ by the Dolbeault complex $\Omega^{0,\bullet}$ one can compute the diagonal cohomology of the corresponding bicomplex (hypercohomology) which we denote by $\mathbb{H}^\bullet(N_s^\bullet(i))$. By the acyclicity of $N_s^\bullet(i)$, we have the equality

$$\mathbb{H}^\bullet(N_s^\bullet(i)) = 0. \tag{88}$$

There is a spectral sequence of the bicomplex $N_s^\bullet(i) \otimes \Omega^{0,\bullet}$ whose $E_1^{p,q}$-term coincides with

$$E_1^{p,q} = \Lambda^p(\mathcal{S}_1) \otimes H^q(\mathcal{O}(i-p)).$$

According to equation (85) and Corollary 54, all nonzero entries of the E_1-term are concentrated on horizontal segments:

$$E_1^{p,0} = \Lambda^p(\mathcal{S}_1) \otimes \mathcal{S}_{i-p}, \qquad 0 \le p \le i, \tag{89}$$

$$E_1^{p,n} = \Lambda^p(\mathcal{S}_1) \otimes \mathcal{S}^*_{p-i-k}, \quad i+k \le p \le s. \tag{90}$$

All entries $E_1^{p,q}$ not mentioned in the above table are equal to zero.

Definition 56. *Let $K^\bullet = \cdots \to K^i \to K^{i+1} \to \ldots.$ The complex $K^\bullet[l]$ is a complex with shifted grading:*

$$K^i[l] = K^{i+l}. \qquad \textit{The differential in the new complex is equal to } (-1)^l d.$$

We have an obvious equality of complexes:

$$(E_1^{\bullet,0}, d) = Q_i^\bullet, \tag{91}$$

$$(E_1^{\bullet,n}, d) = Q_{s-k-i}^{\bullet *}[-s]. \tag{92}$$

The second isomorphism depends on a choice of a linear functional on the space $\Lambda^s(\mathcal{S}_1) \cong \mathbb{C}$. The symbol $*$ means dualization. The spectral sequence converges to zero in the $(n+1)$st term. The differential d_{n+1} on $E_2^{p,q} = E_{n+1}^{p,q}$ induces an isomorphism

$$d_{n+1} : E_2^{p,n} \to E_2^{p-n-1,0}, \tag{93}$$

which is a map of $E_2^{\bullet,0}$ modules, because the spectral sequence is multiplicative.

The isomorphism (93) can be interpreted as a nondegenerate pairing:

$$(.,.) : H^l(Q_i) \otimes H^{s-n-1-l}(Q_{s-k-i}) \to \mathbb{C} \tag{94}$$

It satisfies $(ab, c) = (a, bc)$, because d_{n+1} is a map of $E_2^{\bullet,0}$ modules. The pairing can be recovered from the functional $\lambda(a) = (a, 1)$ by the rule $(a, b) = \lambda(ab)$. The functional is not equal to zero only on $H^{s-n-1}(Q_{s-k})$. The proof follows from this.

A direct inspection of the complex $Q_i^\bullet$ shows that $H^i(Q_i^\bullet) = 0$ for $i \ne 0$. This implies by the duality (94) that

$$H^{i,i}(\mathcal{B}) = H^{s-n-1-i,s-k-i}(\mathcal{B}) = 0 \quad \text{for } i \ne 0. \qquad \square \tag{95}$$

The duality (94) also implies that the cohomology of $Q_i^\bullet$ is not equal to zero only in the range $0 \leq \deg \leq s - n - 1$; therefore, we have proved the following proposition.

Proposition 57. *Under the assumptions of Theorem* 55,

$$H^{j,i}(\mathcal{B}) \neq 0 \quad \textit{only for}$$
$$0 \leq j \leq s - n - 1 \quad \textit{and} \quad j \leq i$$
$$\textit{and by duality}, \quad i \leq j + n + 1 - k.$$

Notice that an analogue of Theorem 55 can be proved for Calabi–Yau manifolds.

Proposition 58. *Suppose an n-dimensional smooth algebraic manifold has $\Omega^n = \mathcal{O}$, $H^i(M, \mathcal{O}) = 0$ for $0 < i < n$ and $\mathcal{L} = \mathcal{O}(1)$ is ample. Then there is a nondegenerate pairing*

$$H^{j,i}(\mathcal{B}) \otimes H^{s-n-1-j,s-i}(\mathcal{B}) \to H^{s-n-1,s}(\mathcal{B})$$

for the differential graded algebra $\mathcal{B}$ defined in (81).

Proof. The proof goes along the same lines as Theorem 55. □

2.6 Berkovits algebra

In this section we will be dealing with some structures built from the 16-dimensional spinor representation $\mathbf{S} = \mathfrak{s}_l$ of the group Spin(10) defined over the complex numbers. This group is a double cover of the group of all linear transformations of the linear space $\mathbf{V}$, $\dim(\mathbf{V}) = 10$ that preserve a nondegenerate form $(.,.)$ and have determinant equal to one. The Dynkin diagram D_5 that corresponds to the Lie algebra SO($\mathbf{V}$) can be found in the picture (77). Our convention is that the representation $\mathbf{S}$ is equal to the irreducible representation with highest weight $[0, 0, 0, 1, 0]$.

Let $\mathcal{S}_i$ be $[0, 0, 0, i, 0]$. There is a structure of algebra on

$$\bigoplus_{i \geq 0} \mathcal{S}_i = \mathcal{S} \tag{96}$$

induced by the tensor product of representations and projection on the leading component.

According to Cartan [5] there is a compact Kähler 10-dimensional homogeneous space $O\mathrm{Gr}(5, 10)$ of the group $O(\mathbf{V})$—the Grassmannian of maximal (five-dimensional) isotropic subspaces of $\mathbf{V}$. This space is called the isotropic Grassmannian. It has two connected components. They are isomorphic as Kähler manifolds but not as homogeneous spaces. An element $e \in O(\mathbf{V})$ with $\det(e) = -1$ swaps the spaces. Let us describe one of the connected components, which we denote $\mathcal{Q}$ and will call the space of pure spinors. Fix $W_0 \in O\mathrm{Gr}(5, 10)$; then the other isotropic subspace W_1 belongs to the same component $\mathcal{Q}$ if $\dim(W_0 \cap W_1)$ is odd.

The complex group Spin($\mathbf{V}$) (in fact SO($\mathbf{V}$)) acts transitively on $\mathcal{Q}$; the corresponding stable subgroup P is a parabolic subgroup. To describe the Lie algebra $\mathfrak{p}$ of

P, we notice that the Lie algebra $\mathfrak{so}(\mathbf{V})$ of SO($\mathbf{V}$) can be identified with $\Lambda^2(\mathbf{V})$ (the space of antisymmetric tensors ρ_{ab}, where $a, b = 0, \ldots, 9$). The vector representation $\mathbf{V}$ of SO($\mathbf{V}$) restricted to the group $GL(5, \mathbb{C}) = GL(W) \subset SO(\mathbf{V})$ is equivalent to the direct sum $W \oplus W^*$ of the vector and covector representations of $GL(W)$, where $\dim(W) = 5$. The direct sum $W + W^*$ carries a canonical symmetric bilinear form. The Lie algebra of SO($\mathbf{V}$) as a vector space can be decomposed as $\Lambda^2(W) + \mathfrak{p}$, where $\mathfrak{p} = (W \otimes W^*) \oplus \Lambda^2(W^*)$ is the Lie algebra of P. Using the language of generators we can say that the Lie algebra $\mathfrak{so}(10, \mathbb{C})$ is generated by skew-symmetric tensors m_{ab}, n^{ab} and by k_a^b, where $a, b = 1, \ldots, 5$. The subalgebra $\mathfrak{p}$ is generated by k_a^b and n^{ab}. The corresponding commutation relations are

$$[m, m'] = [n, n'] = 0, \tag{97}$$

$$[m, n]_a^b = m_{ac} n^{cb}, \tag{98}$$

$$[m, k]_{ab} = m_{ac} k_b^c + m_{cb} k_a^c, \tag{99}$$

$$[n, k]_{ab} = n^{ac} k_c^b + n^{cb} k_c^a. \tag{100}$$

Proposition 59 (Borel–Weil–Bott theorem [4]). *Suppose $\mathcal{L}$ is an invertible homogeneous line bundle over the Kähler compact homogeneous space M of a semisimple group G. Then $H^i(M, \mathcal{L})$ can be nonzero only for one value of i. For this value $H^i(M, \mathcal{L})$ is an irreducible representation.*

Corollary 60. *$H^i(\mathcal{Q}, \mathcal{O}) = 0$ for $i > 0$.*

Since $H^1(\mathcal{Q}, \mathcal{O}) = 0$ all holomorphic topologically trivial line bundles are holomorphically trivial. The corollary and the Hodge decomposition imply that $H^2(\mathcal{Q}, \mathcal{O}) = H^0(\mathcal{Q}, \Omega^2) = 0$ and $H^2(\mathcal{Q}, \mathbb{Z}) = \mathrm{Pic}(\mathcal{Q})$.

Proposition 61.

(a) *The group* $\mathrm{Pic}(\mathcal{Q}) = \mathbb{Z}$.
(b) *The group* $\mathrm{Pic}(\mathcal{Q})$ *has a very ample generator, which we denote by* $\mathcal{O}(1) = \mathcal{L}$, *such that* $H^0(\mathcal{Q}, \mathcal{O}(1)) = \mathbf{S}$.
(c) *The canonical class of* $\mathcal{Q}$ *is isomorphic to* $\mathcal{O}(-8)$.

Proof. We saw that that the Levi subgroup of the parabolic group P contains a center isomorphic to $\mathbb{C}^\times$, so the singular cohomology $H^1(P, \mathbb{Z}) = \mathbb{Z}$. Transgression arguments imply that $H^2(\mathcal{Q}, \mathbb{Z}) = \mathbb{Z}$. This proves that the Picard group of Q is equal to $\mathbb{Z}$.

Denote by $\widetilde{GL(W)} \subset \mathrm{Spin}(\mathbf{V})$ the double cover of $GL(W)$. The restriction of $\mathbf{S}^*$ to $\widetilde{GL(W)}$ is isomorphic to

$$[\mathbb{C} + \Lambda^2(W) + \Lambda^4(W)] \otimes \det{}^{-1/2}(W). \tag{101}$$

By the Borel–Weil theorem this implies that the ample generator of $\mathrm{Pic}(\mathcal{Q})$, which we denote by $\mathcal{O}(1)$, has its space of global sections isomorphic to $\mathbf{S}$.

Consider the representation of $\widetilde{GL(W)}$ in $\Lambda^2(W)$. It is easy to see that

$$\det(\Lambda^2(W)) = \det{}^4(W) = (\det{}^{1/2}(W))^8. \tag{102}$$

We can interpret $\Lambda^2(W)$ as an isotropy representation of the parabolic subgroup P in the tangent space of $\mathcal{Q}$ at a point which is fixed by P. This implies that the canonical class K is isomorphic to $\mathcal{O}(-8)$. □

By the Borel–Weil–Bott theorem the algebra $\bigoplus_{n\geq 0} H^0(Q, \mathcal{O}(n))$ is equal to $\mathcal{S}$.

It is possible to write a formula for $\mathcal{S}$ in terms of generators and relations. To do this observe that

$$\mathrm{Sym}^2(\mathbf{S}) = \mathcal{S}_2 \oplus \mathbf{V}.$$

Denote by

$$\Gamma : \mathbf{V} \to \mathrm{Sym}^2(\mathbf{S})$$

the inclusion of representations. We use the same letter for the projection

$$\Gamma : \mathrm{Sym}^2(\mathbf{S}) \to \mathbf{V}.$$

To distinguish these two maps we will always specify the arguments. The first map $\Gamma(v)$ has a vector argument v. The second map $\Gamma(s_1, s_2)$ has two spinor arguments s_1, s_2.

Proposition 62 (Cartan [5, 3]).

(a) *Denote by* $A_1, \ldots, A_{10}$ *a basis of* $\mathbf{V}$. *Then the algebra* $\mathcal{S}$ *defined in* (96) *can be described through generators and relations:*

$$\mathcal{S} = \mathrm{Sym}(\mathbf{S})/(\Gamma(A_1), \ldots, \Gamma(A_{10})).$$

(b) *The space* Q *can be identified with all points* $\lambda \in \mathbb{P}(\mathbf{S}^*)$ *such that* $\Gamma(\lambda, \lambda) = 0$.

Consider the complex

$$\mathrm{Kos}^\bullet(\mathbf{S})(i) = (0 \to \Lambda^i(\mathbf{S}) \to \Lambda^{i-1}(\mathbf{S}) \otimes \mathrm{Sym}^1(\mathbf{S}) \to \cdots \to \mathrm{Sym}^i(\mathbf{S}) \to 0); \tag{103}$$

it is a classical Koszul complex.

The cohomological grading is the degree in the exterior algebra. The sum $\mathrm{Kos}^\bullet(\mathbf{S}) = \bigoplus_i \mathrm{Kos}^\bullet(\mathbf{S})(i)$ is an algebra. It contains $\mathrm{Sym}(\mathbf{S}) = \bigoplus_{i\geq 0} \mathrm{Sym}^i(\mathbf{S})$ as a subalgebra. The algebra $\mathcal{S}$ is a module over $\mathrm{Sym}(\mathbf{S})$. Then

$$B_0 = \mathrm{Kos}^\bullet(\mathbf{S}) \underset{\mathrm{Sym}(\mathbf{S})}{\otimes} \mathcal{S} = \bigoplus_i Q^\bullet(i) \tag{104}$$

is called the (reduced) Berkovits algebra. The complex $\mathrm{Kos}^\bullet(\mathbf{S})$ is a free resolution of $\mathbb{C}$ over $\mathrm{Sym}(\mathbf{S})$. This implies that we have an identity

$$H^n(Q^\bullet(i)) = \mathrm{Tor}^{n,i}_{\mathrm{Sym}(\mathbf{S})}(\mathbb{C}, \mathcal{S}). \tag{105}$$

(The left upper index corresponds to the cohomology index; the right index corresponds to the homogeneity index.) There is a symmetry

$$\mathrm{Tor}^{n,i}_{\mathrm{Sym}(\mathbf{S})}(\mathbb{C}, \mathcal{S}) = \mathrm{Tor}^{n,i}_{\mathrm{Sym}(\mathbf{S})}(\mathcal{S}, \mathbb{C}). \tag{106}$$

One way to compute the groups $H^n(Q^\bullet(i))$ is to construct a minimal resolution of $\mathcal{S}$ as a module over $\mathrm{Sym}(\mathbf{S})$ (instead of $\mathbb{C}$ as $\mathrm{Sym}(\mathbf{S})$ module). Then the generators of the modules in the resolution will coincide with the cohomology classes of the complexes $Q^\bullet(i)$. Such a resolution was constructed in [7] (though $\mathrm{Spin}(\mathbf{V})$ equivariance in their approach is not apparent). Another option is to compute the cohomology of $Q^\bullet(i)$, $i = 0, \dots, 4$ by brute force using a computer. This has been (partly) done in [6]. In all these approaches, due to the duality proved in Theorem 55 the only nontrivial task is the computation of $H^\bullet(Q^\bullet(4))$.

We are going to construct a (partial) free resolution of $\mathrm{Sym}(\mathbf{S})$ module $\mathcal{S}$ whose graded components are schematically depicted in the picture (107):

$$\begin{array}{ccccccccccc}
\dots & \dots & \dots & \dots & \dots & \dots & \dots & \dots & \dots & \dots & \dots \\
\{0\} & \to & M_3^3 & \stackrel{\delta_3}{\to} & M_3^2 & \stackrel{\delta_2}{\to} & M_3^1 & \stackrel{\delta_1}{\to} & M_3^0 & \stackrel{\delta_0}{\to} & \mathcal{S}_3 \\
 & & \{0\} & \to & M_2^2 & \stackrel{\delta_2}{\to} & M_2^1 & \stackrel{\delta_1}{\to} & M_2^0 & \stackrel{\delta_0}{\to} & \mathcal{S}_2 \\
 & & & & \{0\} & \to & M_1^1 & \stackrel{\delta_1}{\to} & M_1^0 & \stackrel{\delta_0}{\to} & \mathcal{S}_1 \\
 & & & & & & \{0\} & \to & M_0^0 & \stackrel{\delta_0}{\to} & \mathcal{S}_0.
\end{array} \tag{107}$$

By definition $M^i = \bigoplus_{j \geq 0} M^i_j$ and $M^0 = \mathrm{Sym}(\mathbf{S})$.

Since the algebra $\mathcal{S}$ is quadratic with ideal of relations $I = \bigoplus_{j \geq 1} I_j$ generated by $\mathbf{V} = I_2$, we have $M^1_2 = \mathbf{V}$, $M^1_1 = 0$ and $M^1 = \mathbf{V} \otimes \mathrm{Sym}(\mathbf{S})$. Denote by A_i, $i = 1, \dots, 10$, the basis of $\mathbf{V}$ and by u^α, $\alpha = 1, \dots, 16$, the basis of $\mathbf{S}$. The map δ_1 is defined by the formula $\delta_1(A_i) = \sum_{\alpha\beta=1}^{16} \Gamma_{i\alpha\beta} u^\alpha u^\beta$.

The linear space $M^2_3 = B$ is the kernel of the surjection $\mathbf{V} \otimes \mathbf{S} \to I_3$. In the case of interest, the representation content of I_3 is $[1, 0, 0, 1, 0]$, and the representation content of $\mathbf{V} \otimes \mathbf{S}_1$ is $[0, 0, 0, 0, 1] \oplus [1, 0, 0, 1, 0]$. This implies that $B = [0, 0, 0, 0, 1] = \mathbf{S}^*$. Denote a basis of the vector space $\mathbf{S}^*$ by $\psi_\alpha, \alpha = 1, \dots, 16$. The map δ_2 is given on generators by

$$\begin{aligned}
&\delta_2 : \mathbf{S}^* \to \mathbf{S} \otimes \mathbf{V}, \\
&\delta_2(\psi_\beta) = \sum_{\alpha=1}^{16} \sum_{i=1}^{10} \Gamma_{\alpha\beta i} A_i u^\alpha.
\end{aligned} \tag{108}$$

We conclude that the module M^2 is equal to $\mathbf{S}^* \otimes \mathrm{Sym}(\mathbf{S}) + \tilde{M}^2$, where $\tilde{M}^2$ is a free module with generators of degree greater than 3. We will see later that $\tilde{M}^2 = 0$. Now we need to prove a weaker statement: $\tilde{M}^2_4 = 0$. Indeed, it is equal to the cohomology of the complex

$$\mathbf{S}^* \otimes \mathbf{S} \stackrel{\delta_2}{\to} \mathbf{V} \otimes \mathrm{Sym}^2(\mathbf{S}) \stackrel{\delta_1}{\to} I_4 \to 0$$

in the term $\mathbf{V} \otimes \mathrm{Sym}^2(\mathbf{S})$. We have the following representation content:

$$
\begin{aligned}
I_4 &= [1,0,0,2,0] \oplus [2,0,0,0,0], \\
\mathbf{V} \otimes \mathrm{Sym}^2(\mathbf{S}) &= [0,0,0,0,0] \oplus [0,0,0,1,1] \oplus [0,1,0,0,0] \\
&\quad \oplus [1,0,0,2,0] \oplus [2,0,0,0,0], \\
\mathbf{S}^* \otimes \mathbf{S} &= [0,0,0,0,0] \oplus [0,0,0,1,1] \oplus [0,1,0,0,0].
\end{aligned}
\tag{109}
$$

Since the map δ_1 is surjective we need to check that $\mathbf{S}^* \otimes \mathbf{S} \xrightarrow{\delta_2} \mathbf{V} \otimes \mathrm{Sym}^2(\mathbf{S})$ is the inclusion. This can be readily checked by applying the map δ_2 to the highest vectors of each representation in the decomposition of $\mathbf{S}^* \otimes \mathbf{S}$. Let us extend the partial resolution of $\mathbf{S}$ to an arbitrary full resolution. Using this resolution, we can compute $\mathrm{Tor}^{i,j}_{\mathrm{Sym}(\mathbf{S})}(\mathbb{C}, \mathcal{S})$. Simple computations give the following answer:

$$
\begin{aligned}
&\mathrm{Tor}^{0,i}_{\mathrm{Sym}(\mathbf{S})}(\mathbb{C}, \mathcal{S}) = \mathbb{C} \quad \text{if } i = 0 \quad \text{and } \{0\} \quad \text{if } i \neq 0, \\
&\mathrm{Tor}^{1,i}_{\mathrm{Sym}(\mathbf{S})}(\mathbb{C}, \mathcal{S}) = \mathbf{V} \quad \text{if } i = 2 \quad \text{and } \{0\} \quad \text{if } i \neq 1, \\
&\mathrm{Tor}^{2,i}_{\mathrm{Sym}(\mathbf{S})}(\mathbb{C}, \mathcal{S}) = \mathbf{S}^* \quad \text{if } i = 3 \quad \text{and } \{0\} \quad \text{if } i < 3 \quad \text{or} \quad i = 4, \\
&\mathrm{Tor}^{3,i}_{\mathrm{Sym}(\mathbf{S})}(\mathbb{C}, \mathcal{S}) = \{0\} \quad \text{if } i < 5.
\end{aligned}
\tag{110}
$$

Using equation (105), the general duality theorem (Theorem 55), and Proposition 57, we prove the following.

Proposition 63. *The cohomology of the algebra B_0 is*

$$
\begin{aligned}
H^{0,0} &= \mathbb{C}, \\
H^{1,2} &= \mathbf{V}, \\
H^{2,3} &= \mathbf{S}^*, \\
H^{3,5} &= \mathbf{S}, \\
H^{4,6} &= \mathbf{V}, \\
H^{5,8} &= \mathbb{C}, \\
H^{p,q} &= 0 \quad \textit{for all } p, q \textit{ not listed above.}
\end{aligned}
$$

As we know from [15] the Koszul dual to the algebra $\mathcal{S} = F(\hat{Q})$ is the algebra $\mathcal{S}^!$) with generators $\lambda_\alpha, \alpha = 1, \ldots, 16$, which span a linear space $\mathbf{S}^*$ and the relations

$$
\sum_{\alpha\beta=1}^{16} \Gamma^{\alpha\beta}_{m_1 \ldots m_5} \{\lambda_\alpha, \lambda_\beta\} = 0. \tag{111}
$$

This algebra is the universal enveloping algebra $U(\mathbb{L})$ of a Lie algebra $\mathbb{L}$ with the same set of generators and relations.

The Cartan–Eilenberg complex of a positively graded Lie algebra $\mathfrak{g}$ is the exterior algebra $\Lambda(\mathfrak{g}^\dagger) = \Lambda^\bullet(\mathfrak{g}^\dagger)$; the dual complex is denoted by $\Lambda(\mathfrak{g}) = \Lambda^\bullet(\mathfrak{g})$. The sign $\dagger$ denotes dualization in the category of graded vector spaces (see the appendix). The differential $\Lambda^\bullet(\mathfrak{g}^\dagger)$ is given by the formula

$$(d\nu)(x) = \nu([x_1, x_2]) \tag{112}$$

where $\nu \in \Lambda^1(\mathfrak{g}^\dagger)$ is a linear generator. Denote by $H^\bullet(\mathfrak{g}, \mathbb{C})$ the homology of this complex; the homology of the dual complex $\Lambda_\bullet(\mathfrak{g})$ will be denoted as $H_\bullet(\mathfrak{g}, \mathbb{C})$ (See [12] for details about the (co)homology of Lie algebras). For any positively graded Lie algebra $\mathfrak{g}$, there is a canonical quasi-isomorphism $\Lambda_\bullet(\mathfrak{g}) \to \underline{\mathrm{Bar}}(U(\mathfrak{g}))$ and the dual quasi-isomorphism $\underline{\mathrm{Bar}}(U(\mathfrak{g}))^\dagger \to \Lambda(\mathfrak{g}^\dagger)$ (see [12] for details).

By one of the properties of the Koszul duality transformation (see [11]) there is an inclusion $U(\mathfrak{g})^! \subset H^\bullet(\mathfrak{g}, \mathbb{C})$, for any quadratic Lie algebra.

We need the following proposition.

Proposition 64 ([3]). *For any compact homogeneous Kähler manifold G/P of a reductive group G and an ample line bundle α on it the Serre algebra $\bigoplus_{n\geq 0} H^0(G/P, \alpha^{\otimes n})$ is Koszul.*

Since the Koszul relation is reflexive for the case at hand, we have an isomorphism:

$$U(\mathbb{L})^! = H^\bullet(\mathbb{L}, \mathbb{C}). \tag{113}$$

Proposition 65. *There is a quasi-isomorphism $\rho : \Lambda^\bullet(\mathbb{L}^\dagger) \to \mathcal{S}$, which maps the linear functional $\lambda^{*\alpha}$ on $\mathbb{L}$ into the generator u^α of $\mathcal{S}$. This map is zero on the subspace $\bigoplus_{i\geq 2} \mathbb{L}_i^* \subset \Lambda^1(\mathbb{L}^*)$.*

Proof. The only statement which needs to be checked is that ρ commutes with differentials. This is obvious, however. □

Our next goal is to relate the Berkovits algebra (B_d, Q) with the classical BV approach to YM theory.

Let $\mu : \mathbf{V} \to V$ be a surjective linear map. We assume that the linear space $\mathbf{V}$ has an orthonormal basis $A_1, \dots, A_{10}$ and that the linear space V has a basis generated by the symbols $\frac{\partial}{\partial x^1}, \dots, \frac{\partial}{\partial x^d}$. Let $\mathrm{Sym}(V^*)$ be the symmetric algebra on the space dual to V. The space V^* has a basis $x^1, \dots, x^d$. Then $\mu(A_i)$ defines a linear functional on V^*, which can be extended to a derivation of $\mathrm{Sym}(V^*)$. Introduce the algebra

$$B_\mu = B_0 \otimes \mathrm{Sym}(V^*) \tag{114}$$

and a differential on it by the rule

$$Q = \sum_{\alpha=1}^{16} u^\alpha \frac{\partial}{\partial \theta^\alpha} + \sum_{\alpha\beta=1}^{16} \sum_{i=1}^{10} \Gamma^i_{\alpha\beta} u^\alpha \theta^\beta \mu(A_i). \tag{115}$$

Recall that the differential algebra (B_d, Q) was defined in the introduction in Definition 16; (B_μ, Q) is a minor generalization of it.

Consider a graded extension $\mathbb{M}_\mu$ of the Lie algebra $\mathbb{L}$. The linear space $\mathbb{L}_1 = \mathbf{S}^*$. Then $\mathbb{M}_{\mu 0} = \mathbf{S}^*$ and this space has a basis $\tau_\alpha, \alpha = 1, \dots, 16$. The linear space V has a basis $\xi_1, \dots, \xi_d$ and $\mathbb{M}_{\mu 1} = \mathbf{S}^* + V$. The linear space V is by definition dual to the linear space generated by $x^1, \ldots, x^d$ of linear coordinates on d-dimensional linear space. For $i \geq 3$ we have $\mathbb{M}_{\mu i} = \mathbb{L}_i$.

The parity of elements of $\mathbb{M}$ is reduction of grading modulo 2. The Lie algebra $\mathbb{M}$ is equipped with a differential defined by the formulas

$$\begin{aligned} d &: \mathbb{M}_{\mu 1} \to \mathbb{M}_{\mu 0}; \qquad \mathbf{S}^* + V \overset{\mathrm{id},0}{\to} \mathbf{S}^*, \\ d &: \mathbb{M}_{\mu 2} \to \mathbb{M}_{\mu 1}; \qquad \mathbf{V} \overset{0,\mu}{\to} \mathbf{S}^* + V, \\ d &: \mathbb{M}_{\mu i} \to \mathbb{M}_{\mu i-1}; \qquad i \geq 3, \quad d = 0. \end{aligned} \tag{116}$$

The commutation relations in the algebra $\mathbb{M}$ are those of the semidirect product $\mathbb{L} \ltimes (\mathbf{S}^* + V)$, where $\mathbf{S}^* \subset \mathbb{M}_{\mu 0}$, $V \subset \mathbb{M}_{\mu 1}$. The linear space $\mathbf{S}^* + V$ is an abelian ideal. The action of $\mathbb{L}$ on $\mathbf{S}^* + V$ is given by the rule

$$\begin{aligned} [\theta_\alpha, \tau_\beta] &= 2\mu \sum_{i=1}^{10} \Gamma^i_{\alpha\beta} A_i, \\ [\theta_\alpha, \xi_i] &= 0. \end{aligned} \tag{117}$$

One can consider a version of the Cartan–Eilenberg complex for the differential graded Lie algebra $(\mathbb{M}_\mu, d)$; in the complex $\Lambda^\bullet(\mathbb{M}^\dagger_\mu)$, formula (112) becomes

$$(D\nu)(x) = \nu([x_1, x_2]) + \nu(d(x_3)). \tag{118}$$

There is a map

$$\chi : (\Lambda(\mathbb{M}^\dagger_\mu), D) \to (B_\mu, Q). \tag{119}$$

On the generators, the map is

$$\begin{aligned} \chi(\xi^{*i}) &= x^i, \\ \chi(\lambda^{*\alpha}) &= u^\alpha, \\ \chi(\tau^\alpha) &= \theta^\alpha, \\ \chi &\text{ is zero on the rest of the generators.} \end{aligned} \tag{120}$$

To make this map a map of complexes, one has to modify slightly the grading on B_μ:

$$\begin{aligned} \tilde{x^i} &= 2, \\ \tilde{u^\alpha} &= 2, \\ \tilde{\theta^\alpha} &= 1. \end{aligned} \tag{121}$$

The grading on $\Lambda(\mathbb{M}^\dagger_\mu)$ is the standard cohomological grading.

Proposition 66. *The map χ is well defined and is a quasi-isomorphism of the algebras $(\Lambda(\mathbb{M}^\dagger_\mu), D)$ and (B_μ, Q).*

Proof. We leave the proof of the first statement as an exercise for the reader. The algebra $\mathbb{M}_\mu$ carries an action of $\mathbb{C}^\times$ which commutes with the differential D. It manifests itself in a grading. In this grading $\mathrm{DEG}(\lambda_\alpha) = 1$. This condition allows us to uniquely extend the grading to the entire algebra. The induced grading on $\Lambda(\mathbb{M}_\mu)$ has the following feature: all graded components become finite-dimensional complexes bounded from both sides. Such a grading can be pushed onto B_μ. The simple observation is that the map χ is surjective. A filtration of $\Lambda(\mathbb{M}_\mu)$ which leads to the Serre–Hochschild spectral sequence, based on the extension

$$0 \to \mathbf{S}^* + V \to \mathbb{M}_\mu \to \mathbb{L} \to 0 \tag{122}$$

can be pushed onto the algebra B_μ. The E_2-terms of the corresponding spectral sequences are isomorphic to the algebra B_μ. Therefore, the limiting terms of the spectral sequence, which converge strongly, must coincide. □

Proposition 67. *The universal enveloping algebra $U(\mathbb{M}_\mu)$ is Koszul dual to B_μ.*

Proof. The proof is a straightforward application of the definitions. □

In order to avoid confusion, when we talk about differential Lie algebras, by (co)homology we always mean cohomology of the Cartan–Eilenberg complex. However, the linear space of the algebra itself carries a differential which we call the intrinsic differential. The cohomology of such a differential will be called the intrinsic cohomology.

The algebra $\mathbb{M}_\mu$ carries an intrinsic differential. It allows us to reduce the space of the algebra without affecting its cohomology. Introduce two subalgebras $E_\mu\mathbb{M} = \bigoplus_{i\geq 1} E_\mu\mathbb{M}_i$ and $T_\mu\mathbb{M} = \bigoplus_{i\geq 2} T_\mu\mathbb{M}$ of $\mathbb{M}_\mu$:

$$\begin{aligned} E_\mu\mathbb{M}_1 &= V,\\ E_\mu\mathbb{M}_i &= \mathbb{L}_i, \quad i \geq 2, \end{aligned} \tag{123}$$

the differential is a restriction of the differential to $E_\mu\mathbb{M}$,

$$\begin{aligned} T_\mu\mathbb{M}_2 &= \mathrm{Ker}[\mu : \mathbb{L}_2 \to V],\\ T_\mu\mathbb{M}_i &= \mathbb{L}_i, \quad i \geq 3,\\ d &= 0. \end{aligned} \tag{124}$$

Proposition 68. *The algebras $E_\mu\mathbb{M}$, $T_\mu\mathbb{M}$ quasi-isomorphically embed into the algebra $\mathbb{M}_\mu$.*

Proof. Obvious. □

Corollary 69. $H^\bullet(T_\mu\mathbb{M}, \mathbb{C}) = H^\bullet(E_\mu\mathbb{M}, \mathbb{C}) = H^\bullet(B_\mu)$, *where the first two groups are the cohomology of Lie algebras. The last group is intrinsic to the differential of the algebra B_μ.*

We need to identify the algebras $T_\mu\mathbb{M}$ and $E_\mu\mathbb{M}$. Assume that $\mu = 0$. In [15] we indicated that the algebra $\mathbb{L}$ contains a homomorphic image of the Lie algebra $\mathbb{S}YM$.

The universal enveloping algebra $U(\mathbb{S}YM)$ is isomorphic to YM_0 with $D = 10$, $d' = 0$, $N = 1$ and the potential U equal to zero. The Lie algebra $\mathbb{S}YM$ is generated by

$$\begin{aligned} A_i &= \sum_{\alpha\beta} \Gamma_i^{\alpha\beta}\{\lambda_\alpha, \lambda_\beta\}, \\ \psi^\alpha &= \sum_\beta \sum_{m=1}^{10} \Gamma^{\alpha\beta m}[\lambda_\beta, A_m] \end{aligned} \tag{125}$$

with relations (7), (8), (9).

Proposition 70. *The Lie algebra* $\mathbb{S}YM$ *is isomorphic to the algebra* $\bigoplus_{i\geq 2}\mathbb{L}_i$.

Proof. By construction $\mathbb{S}YM$ maps into $\bigoplus_{i\geq 2}\mathbb{L}_i$. We need to check that the set of generators of $\bigoplus_{i\geq 2}\mathbb{L}_i$ coincides with $A_1, \ldots, A_{10}, \psi^1, \ldots, \psi^{16}$ and there are no relations other than (7), (8), (9) with $D = 10$, $d' = 0$, $N = 1$, and $U = 0$. To do so, we take advantage of Proposition 18 and Corollary 19. The space $H^\bullet(E_0\mathbb{M}, \mathbb{C})$ is quasi-isomorphic to $H^\bullet(B_0)$, which was computed in Proposition 63. According to this proposition, $H^1(E_0\mathbb{M}, \mathbb{C}) = \mathbf{V} + \mathbf{S}^*$ and $H^2(E_0\mathbb{M}, \mathbb{C}) = \mathbf{V} + \mathbf{S}$. It implies that elements $A_1, \ldots, A_{10}$ which span $\mathbf{V}$ and $\psi^1, \ldots, \psi^{16}$ which span $\mathbf{S}$ are indeed the generators of the algebra $E_0\mathbb{M}$. The relations (7) (if we think of them as elements of a free algebra) transform as the representation $\Pi\mathbf{V}^*$, (7) transform as the representation $\mathbf{S}^*$ are indeed the generators of the ideal of relations of the algebra $E_0\mathbb{M}$. □

If $\mu = 0$ then $\mathbb{M}_\mu$ is quasi-isomorphic to $T_0\mathbb{M} = \bigoplus_{i\geq 2}\mathbb{L}_i$.

Proposition 71. *The Koszul dual to* B_0 *is quasi-isomorphic to* $U(\mathbb{S}YM)$.

Proof. According to Proposition 70, $U(\mathbb{S}YM)$ is quasi-isomorphic to $U(\bigoplus_{i\geq 2}\mathbb{L}_i)$. By the remark from the previous paragraph, $\bigoplus_{i\geq 2}\mathbb{L}_i$ is quasi-isomorphic to $T_0\mathbb{M}$, which by Proposition 68 is quasi-isomorphic to $\mathbb{M}_0$. By Proposition 67 $\mathbb{M}_0$ is Koszul dual to B_0. □

Corollary 72. *There is a quasi-isomorphism* $bv_0^* \cong B_0$.

Proof. We already established a quasi-isomorphism between $\Lambda(E_0\mathbb{M})$ and B_0. We have a canonical quasi-isomorphism $\underline{\mathrm{Bar}}(E_0\mathbb{M})^\dagger \to \Lambda(E_0\mathbb{M}^\dagger)$. On the other hand, according to Theorem 1, we have a quasi-isomorphism $\mathrm{Bar}\, YM \cong bv_0$. Dualization of the last quasi-isomorphism gives the necessary quasi-isomorphism. □

Proposition 73. *There is a quasi-isomorphism*

$$B_\mu \cong bv_\mu^*. \tag{126}$$

Proof. There is an obvious identification $U(T_\mu\mathbb{M}) \cong T_\mu YM$ (U stands for the universal enveloping algebra) which comes from the identification $T_0\mathbb{M} \cong \mathbb{S}YM$. Then by Lemma 31 and Propositions 45 and 66, we have a quasi-isomorphism (126). □

Proposition 74. *The Koszul dual to* B_μ *is quasi-isomorphic to* $U(T_\mu\mathbb{M})$.

Proof. According to Proposition 67, the Koszul dual to B_μ is equal to $U(\mathbb{M}_\mu)$. The result follows from the previous observation and Proposition 68. □

3 Appendix: Dualization in the category of linear spaces: Bar duality

3.1 Dual spaces

Suppose we have an inverse system of finite-dimensional vector spaces $\cdots \to N^{i+1} \xrightarrow{i} N^i \to \cdots \to N^0$ $(i \geq 0)$, where all maps i are surjective. Let $N = \lim_i N^i$. There is a canonical map $N \to N^i$ which in our case is surjective. Denote the kernel of this map by J^i. It is clear that $J^{i-1} \subset J^i$ and the set of linear spaces J^i completely determines the inverse system, and for every linear space W with a decreasing filtration J^i such that

$$\bigcap_{\infty}^{i=0} J^i = \{0\}, \tag{127}$$

$$\dim(J^i/J^{i-1}) < \infty, \tag{128}$$

we have $W \subset \lim_i W/L_i = \widehat{W}$. Define a direct system of finite-dimensional vector spaces $M^0 \to M^{-1} \to \ldots M^n \to M^{n-1} \to \ldots$ as $M^n = N^{*-n}$. We call such a direct system the dual to N^n and denote it by N^{*n}. It should be clear how to define N^{**n} and that it is equal to N^n. Observe that $\operatorname{colim}_n N^{*n}$ has an increasing filtration by spaces $F^i = N^{i*}$. Denote $M = \operatorname{colim}_n N^{*n}$. The filtration satisfies

$$M = \bigcup_i F^i, \tag{129}$$

$$\dim(F^{i+1}/F^i) < \infty. \tag{130}$$

Skipping all mentionings of limits we can say that there is a dualization invertible functor from the category of complete linear spaces W with decreasing filtration, J_i $(i \leq 0)$, where J_i satisfies (127), (128) and linear spaces M equipped with increasing filtration, and F^i such that F^i satisfies (129) and (128). We will refer to such a duality as *topological*.

Definition 75. *Denote by $U = \bigoplus_{i \in \mathbb{Z}} U_i$ a graded vector space with* $\dim(U_i) < \infty$. *One can define a dualization functor on the category of such vector spaces. Indeed, by definition $U^\dagger = \bigoplus_{i \in \mathbb{Z}} U^*_{-i}$, and the grading of U^*_i is equal to $-i$. Observe that the functor $U^\dagger$ is autoduality in the category of graded linear spaces. A vector space dual in this sense to U will be called the* algebraic *dual.*

Given a graded vector space $U = \bigoplus_{i \geq 0} U_i$ one can define two linear spaces with filtrations:

(a) $U = \bigoplus_{i \geq 0} U_i$ and a filtration is defined as $F^n = \bigoplus_{0 \leq i \leq n} U_i$.

(b) $\widehat{U} = \prod_{i \geq 0} U_i$ and a filtration is defined as $J_n = \prod_{n \leq i} U_i$. In the future the sign $\widehat{\ }$ will always means completion of the space W with respect to a decreasing filtration.

In the future, if we write U^* for $U = \bigoplus_{i \geq 0} U_i$, we will always mean the topological dual.

3.2 A_∞-algebras

Let us consider a $\mathbb{Z}_2$-graded vector space W and the corresponding tensor algebra $T(W) = \bigoplus_{n \geq 1} W^{\otimes n}$. The tensor algebra $T(W)$ is $\mathbb{Z}$-graded, but it also has a $\mathbb{Z}_2$-grading coming from the $\mathbb{Z}_2$-grading of W. We say that the differential (= an odd derivation having zero square) Q on $T(W)$ specifies a structure of A_∞-coalgebra on $V = \Pi W$.

One can describe the structure of A_∞-coalgebra on $V = \Pi W$ as a sequence of linear maps $\Delta_1 : V \to V$, $\Delta_2 : V \to V \otimes V$, $\Delta_n : V \to V^{\otimes n}$. Using the Leibniz rule we can extend $\Delta_1, \Delta_2, \dots$ to a derivation Q of $(T(W)$; the condition $Q^2 = 0$ implies some conditions on $\Delta_1, \Delta_2, \dots$. The map Δ_1 is a differential ($\Delta_1^2 = 0$); the map Δ_2 can be interpreted as comultiplication. If $\Delta_n = 0$ for $n \geq 3$, we obtain a structure of associative coalgebra on V.

One says that the differential algebra $(T(W), Q)$ is (bar-)dual to the A_∞-coalgebra (V, m) or that $(T(W), Q)$ is obtained from (V, Δ) by means of the bar-construction. We will use the notation $\mathrm{Bar}(V, \Delta)$ for this differential algebra. The Hochschild homology of (V, Δ) is defined as the homology of $(T(W), Q)$.

An A_∞-map of coalgebras is defined as a homomorphism of corresponding differential tensor algebras (i.e., as an even homomorphism $\mathrm{Bar}(V, \Delta) \to \mathrm{Bar}(V', \Delta')$ commuting with the differentials).

One can describe an A_∞-map $\phi(V, \Delta) \to (V', \Delta')$ by means of a sequence of maps $\varphi_n : V \to V'^{\otimes n}$, where $V = \Pi W$, $V' = \Pi W'$.

The map φ_1 commutes with the differentials ($\Delta_1' \varphi_1 = \varphi_1 \Delta_1$); hence it induces a homomorphism of the homology of (V, Δ_1) into the homology of (V', Δ_1'). If the induced homomorphism is an isomorphism one says that the A_∞-map is a quasi-isomorphism.

One can define an A_∞-algebra structure on the $\mathbb{Z}_2$-graded vector space V as an odd coderivation Q of the tensor coalgebra $T(W) = \bigoplus_{n \geq 1} W^{\otimes n}$ obeying $Q^2 = 0$. (Here $W = \Pi V$).

Equivalently, an A_∞ algebra (V, m) can be defined as a $\mathbb{Z}_2$-graded vector space V equipped with a series of operations

$$m_1 : V \to V, m_2 : V^{\otimes 2} \to V, \dots, m_n : V^{\otimes n} \to V,$$

obeying certain conditions.

One says that the differential coalgebra $(T(W), Q)$ is bar-dual to the the A_∞ algebra (V, m) or that $(T(W), Q)$ is obtained from (V, m) by means of the bar-construction. We will use the notation $\mathrm{Bar}(V, m)$ for the differential coalgebra

$(T(W), Q)$. The Hochschild homology of (V, m) is defined as the homology of $(T(W), Q)$.

Usually one considers bar-duality for $\mathbb{Z}$-graded A_∞-algebras and A_∞-coalgebras.

The algebra $\mathrm{Bar}(V, \Delta) = (T(W), Q)$ that is dual to a $\mathbb{Z}$-graded A_∞-coalgebra (V, Δ) will be considered as a $\mathbb{Z}$-graded differential algebra. The space W is equal to $V[-1]$; in other words W coincides with V with grading shifted by -1. Similarly, the $\mathbb{Z}$-graded coalgebra $\mathrm{Bar}(V, m)$ dual to the graded A_∞-algebra (V, m) will be considered as a graded differential coalgebra.

Let us consider the case of A_∞-(co)algebras (V, m) and (V', m) having an additional positive grading. This is an auxiliary grading which has no correlation with the internal (homological) $\mathbb{Z}_2(\mathbb{Z})$-grading. We assume that all structure maps m_k have degree zero with respect to this additional grading. The same applies to A_∞-morphisms. Such A_∞-(co)algebras will be called homogeneous. We will use the abbreviations h.morphism, h.quasi-isomorphism, etc., for homogeneous morphism, homogeneous quasi-isomorphism, etc.

Theorem 76. *Two homogeneous* $\mathrm{Bar}(V, \Delta)$ *A_∞-(co)algebras are h.quasi-isomorphic iff their dual algebras (coalgebras) are h.quasi-isomorphic.*

Theorem 77. *If the A_∞-morphism $f : (V, m) \to (V', m')$ of homogeneous (co)algebras induces an isomorphism of Hochschild homology, then f is an h.quasi-isomorphism.*

For quadratic algebras bar-duality is closely related to Koszul duality. Let A be a quadratic algebra $A = \bigoplus_{n>0} A^n$, $\dim A^n < \infty$, and $B = \bigoplus_{n>0} A^{*n}$ the dual graded coalgebra.

Proposition 78. *The differential graded algebra bar-dual to the coalgebra B is quasi-isomorphic to the Koszul dual $A^!$ if A is a Koszul algebra.*

We will consider duality in the more general situation when the A_∞-coalgebra (V, Δ) whose descending filtration F^k satisfies $F^1 = V, \cap_{k\geq 1} F^k = 0$ and V is complete with respect to the filtration. Then we can introduce the corresponding filtration F^k on $T(W)$. We define $F^p(T(W))$ by the formula

$$\sum_{\sum_{r=1}^k n_r \geq p} F^{n_1} \otimes \cdots \otimes F^{n_k}. \tag{131}$$

We assume that the structure of A_∞-coalgebra is compatible with the filtration. This means that

$$\Delta_k(F^s) \subset \sum_{n_1+\cdots+n_k \geq s} F^{n_1} \otimes \cdots \otimes F^{n_k}, \quad n_k \geq 1. \tag{132}$$

In the language of tensor algebras we require that $Q(F^k(T(W))) \subset F^k(T(W))$. In particular, for the filtered A_∞-coalgebra we have $\Delta_1(F^s) \subset F^s$; hence we can consider the homology of (F^s, Δ_1).

The bar dual to the filtered A_∞-coalgebra (V, Δ) is defined as the topological differential algebra $(\widehat{T(W)}, Q)$ obtained from $(T(W), Q)$ by means of completion with respect to the filtration F^k.

The A_∞-maps of filtered coalgebras should agree with the filtrations; they can be considered as continuous homomorphisms of the dual topological differential algebras.

Representing the A_∞-map $\phi : (V, \Delta_k) \to (V', \Delta'_k)$ as a series of maps $\varphi_k : V \to V'^{\otimes k}$ and using that $\Delta'_1\varphi_1 = \varphi_1\Delta_1$, ϕ_1 induces a homomorphism of homology of $(F^k/F^{k+1}, \Delta_1)$ into the homology of $(F'^k/F'^{k+1}, \Delta'_1)$; if all of these homomorphisms are isomorphisms, we say that ϕ is a filtered quasi-isomorphism. There are nonfiltered quasi-isomorphisms between filtered objects.

We also introduce a notion of filtered A_∞-algebra (V, m) fixing a decreasing filtration F^p on V, $p \geq 1$, that satisfies the conditions

$$\mu_k : F^{s_1} \otimes \vdots \otimes F^{s_k} \to F^{s_1+\cdots+s_k}, \qquad k \geq 1, \tag{133}$$

$$\bigcap_s F^s = 0, \qquad F^1 = V \tag{134}$$

and V is complete with respect to such a filtration. (Notice that the notion of filtered A_∞-algebra is not dual to the notion of filtered A_∞-coalgebra (a filtration that is dual to a decreasing filtration is an increasing filtration). The differential coalgebra $\mathrm{Bar}(V, m)$ corresponding to the filtered A_∞-algebra can be considered as a filtered coalgebra (see formula (132) for the filtration). Its completion $\widehat{\mathrm{Bar}}(V, m)$ can also be regarded as a filtered differential topological coalgebra.

Let f be an A_∞-morphism of filtered A_∞-algebras $(V, m) \to (V', m')$ that is compatible with the filtrations. It induces a map $f_* : \mathrm{Bar}(V, m) \to \mathrm{Bar}(V', m')$ of the corresponding dual coalgebras that can be extended to a map $\hat{f}_* : \widehat{\mathrm{Bar}(V, m)} \to \widehat{\mathrm{Bar}(V', m')}$.

We need the following statements proved in [14].

Theorem 79. *If the filtered A_∞-coalgebras are quasi-isomorphic, then the dual topological algebras are quasi-isomorphic.*

Theorem 80. *Let (V, Δ) and (V', Δ') be two filtered A_∞-algebras. Then quasi-isomorphism of the corresponding topological differential coalgebras $(\widehat{T(W)}, Q)$ and $(\widehat{T(W')}, Q')$ implies quasi-isomorphism of the A_∞-algebras (V, Δ) and (V', Δ').*

Let (V, Δ) and (V', Δ') be two filtered A_∞-algebras. Then filtered quasi-isomorphism of (V, Δ) and (V', Δ') implies quasi-isomorphism of the corresponding topological differential coalgebras $(\widehat{T(W)}, Q)$ and $(\widehat{T(W')}, Q')$.

Theorem 81. *Let (V, Δ) and (V', Δ') be two filtered A_∞-coalgebras. Then filtered quasi-isomorphism of the corresponding topological differential algebras $(\widehat{T(W)}, Q)$ and $(\widehat{T(W')}, Q')$ implies quasi-isomorphism of the A_∞-coalgebras (V, Δ) and (V', Δ').*

Theorem 82. *If the map $f_* : \widehat{\mathrm{Bar}}(V, m) \to \widehat{\mathrm{Bar}}(V', m)$ is a quasi-isomorphism, then the original map f is also a quasi-isomorphism.*

Lemma 83. *For any A_∞ filtered coalgebra H, there is an A_∞ morphism*

$$\widehat{\mathrm{Bar}}\,\widehat{\mathrm{Bar}}(H) \stackrel{\psi}{\to} H \tag{135}$$

of A_∞-coalgebras. The morphism ψ is a quasi-isomorphism.

Similarly, for any A_∞ filtered algebra A, there is an $A_\infty i$-morphism

$$A \stackrel{\phi}{\to} \widehat{\mathrm{Bar}}\,\widehat{\mathrm{Bar}}(A) \tag{136}$$

of A_∞-coalgebras. The morphism ϕ is a quasi-isomorphism.

Acknowledgments. We would like to thank A. Bondal, A. Gorodentsev, M. Kontsevich, D. Piontkovsky, A. Rudakov, and E. Witten for stimulating discussions. Part of this work was done when one or both of the authors were staying at the Mittag-Leffler Institute; we appreciate the hospitality of this institution.

The work of both authors was partially supported by NSF grant DMS-0204927.

References

[1] M. Alexandrov, A. Schwarz, O. Zaboronsky, and M. Kontsevich, The geometry of the master equation and topological quantum field theory, *Internat. J. Modern Phys.* A, **12**-7 (1997), 1405–1429.

[2] G. Barnich, F. Brandt, and M. Henneaux, Local BRST cohomology in gauge theories, *Phys. Rep.*, **338** (2000), 439–569.

[3] R. Bezrukavnikov, Koszul property and Frobenius splitting of Schubert varieties, alg-geom/9502021, 1995.

[4] R. Bott, Homogeneous vector bundles, *Ann. Math.* (2), **66** (1957), 203–248.

[5] É. Cartan, *The Theory of Spinors*, foreword by R. Streater, reprint of the 1966 English translation, Dover Books on Advanced Mathematics, Dover, New York, 1981.

[6] M. Cederwall, B. Nilsson, and D. Tsimpis, Spinorial cohomology and maximally supersymmetric theories, *J. High Energy Phys.*, **2**-9 (2002), 009.

[7] A. Corti and M. Reid, Weighted Grassmannians, in M. C. Beltrametti, ed., *Algebraic Geometry: A Volume in Memory of Paolo Francia*, Walter de Gruyter, Berlin, New York, 2002, 141–164.

[8] W. Dicks, A survey of recent work on the cohomology of one-relator associative algebras, in J. L. Bueso, P. Jara, and B. Torrecillas, eds., *Ring Theory*, Lecture Notes in Mathematics 1328, Springer-Verlag, Berlin, 1988, 75–81.

[9] Ph. Griffiths and J. Harris, *Principles of Algebraic Geometry*, Pure and Applied Mathematics, Wiley–Interscience, New York, 1978.

[10] M. Kontsevich and Y. Soibelman, Homological mirror symmetry and torus fibrations, in *Symplectic Geometry and Mirror Symmetry* (*Seoul*, 2000), World Scientific, River Edge, NJ, 2001, 203–263.

[11] C. Löfwall, On the subalgebra generated by the one-dimensional elements in the Yoneda extalgebra, in L. L. Avramov and K. B. Tchakerian, *Algebra: Some Current Trends*, Lecture Notes in Mathematics 1352, Springer-Verlag, Berlin, 1988, 291–338.
[12] S. MacLane, *Homology*, Die Grundlehren der mathematischen Wissenschaften 114, Academic Press, New York, Springer-Verlag, Berlin, Göttingen, Heidelberg, 1963.
[13] S. Merkulov, Strong homotopy algebras of a Kähler manifold, *Internat. Math. Res. Notices*, **1999**-3 (1999), 153–164.
[14] M. Movshev, Bar duality, in preparation.
[15] M. Movshev and A. Schwarz, On maximally supersymmetric Yang-Mills theories, *Nuclear Phys.* B, **681**-3 (2004), 324–350.

Seiberg–Witten Theory and Random Partitions

Nikita A. Nekrasov[1, 2] and Andrei Okounkov[3]

[1] Institut des Hautes Études Scientifiques
35 Route de Chartres
F-91440 Bures-sur-Yvette
France
nikita@ihes.fr
[2] ITEP
Moscow 117259
Russia
(on leave of absence)
[3] Department of Mathematics
Princeton University
Princeton, NJ 08544
USA
okounkov@princeton.edu

Summary. We study $\mathcal{N} = 2$ supersymmetric four-dimensional gauge theories, in a certain $\mathcal{N} = 2$ supergravity background, called the Ω-background. The partition function of the theory in the Ω-background can be calculated explicitly. We investigate various representations for this partition function: a statistical sum over random partitions, a partition function of the ensemble of random curves, and a free fermion correlator.

These representations allow us to derive rigorously the Seiberg–Witten geometry, the curves, the differentials, and the prepotential.

We study pure $\mathcal{N} = 2$ theory, as well as the theory with matter hypermultiplets in the fundamental or adjoint representations, and the five-dimensional theory compactified on a circle.

Subject Classifications: 81T60, 81T13, 81T45

1 Introduction

Supersymmetric gauge theories are interesting theoretical laboratories. They are rich enough to exhibit most of the quantum field theory phenomena, yet they are rigid enough to contain a lot of exactly calculable information [1, 2, 3, 4]. They embed easily into string theory, and provide an exciting arena in the search for string/gauge dualities [5, 6, 7, 8].

In the past few years a lot of progress has been made in understanding some of this rich structure using direct field theoretic techniques (see [9, 10] and references therein), in the case of the theories with extended supersymmetry, and in [8] in the case of $\mathcal{N}=2$ SUSY broken down to $\mathcal{N}=1$ by the superpotential.

In particular, a connection between four-dimensional gauge theories and two-dimensional conformal field theories seemed to appear. Some earlier indications for such a connection were observed in the study of $\mathcal{N}=4$ super-Yang–Mills theory [11, 12].

In this paper we shall make this connection more transparent in the case of $\mathcal{N}=2$ theories.

We shall also derive, by purely field theoretic means, via direct instanton calculus, the solution for the low-energy effective theories, proposed by Seiberg and Witten in 1994 [13] and further generalized in [4, 14].

1.1 Notation

Throughout the paper, we denote color indices by lowercase Latin letters $l, m, n = 1, \dots, N$, the vevs of the Higgs field by

$$\mathbf{a} = \operatorname{diag}(a_1, \dots, a_N), \quad a = \frac{1}{N}\sum_l a_l, \quad \tilde{a}_l = a_l - a,$$

and the dual vevs by

$$\xi = \operatorname{diag}(\xi_1, \dots, \xi_N).$$

The vector

$$\rho_l = \frac{1}{N}\left(l - \frac{N+1}{2}\right).$$

will occur often.

For the gauge group $G = U(N)$, the group of gauge transformations on $\mathbf{R}^4$ which extend smoothly to $\mathbf{S}^4$ will be denoted by $\mathcal{G}$. Its normal subgroup $\mathcal{G}_\infty$, which consists of the gauge transformations trivial at ∞, will play a special role.

1.2 Organization of the paper

Section 2 is addressed to physicists who want to jump quickly onto the subject. It contains previously unpublished details on the noncommutative regularizations of the theories we consider, as well as a systematic introduction to Ω-backgrounds. In this paper we shall not touch upon the recently revived [15] C-backgrounds of [16, 17], postponing the explanations of the relations between Ω and C to some future work.

Section 3 starts by setting the stage for the mathematical problem and reviews the ingredients needed for its solution. We present the formula for the partition function of the pure $\mathcal{N}=2$ gauge theory in the Ω-background, as a sum over random partitions. We also begin to formulate the same sum in terms of random paths, which arise as the boundaries of the Young diagrams representing the random partitions. The mathematically oriented reader can skip Section 2 and proceed directly to Section 3. (We also recommend [18, 19] for orientation.)

Section 4 attacks the problem of the calculation of the prepotential. Physically it has to do with the limit on which the Ω-background approaches flat space. In the random path representation, this limit is the quasiclassical limit, which can be evaluated using the saddle point method. This is exactly the idea of our derivation. We extensively discuss the equations on the minimizing path, and their solution. We find that the solution is most simply described in terms of the Seiberg–Witten curve.

Section 5 explains the fermionic representation of the partition function in the special Ω-background, preserving twice as many supersymmetries as compared to the generic Ω-background. Even though the formalism of free fermions is well known to mathematicians under the name of the infinite wedge representations of $\mathbf{gl}(\infty)$ algebra, we supply the necessary details. We find that a certain transform of the partition function can be written as a matrix element (current conformal block) of the exponentials in the $\widehat{U(N)}$ currents on the sphere $\mathbf{S}^2$.

Section 6 begins our quest for generalizations. We discuss the softly broken $\mathcal{N} = 4$ theory, i.e., $\mathcal{N} = 2$ gauge theory with the matter hypermultiplet in the adjoint representation. As most of the steps are similar to the pure gauge theory case, we move faster and write down the expression for the partition function in terms of partitions, paths, and chiral fermions, which this time live on an elliptic curve, determined by the microscopic gauge coupling. Again we perform the saddle point evaluation, and find that the prepotential is encoded, as conjectured by Donagi and Witten [20], in the spectral curves of the elliptic Calogero–Moser integrable system.

Section 7 considers gauge theories with matter in the fundamental representation, and gauge theories with the tower of Kaluza–Klein states, coming from the compactification of the five-dimensional theory on a circle.

Section 8 presents our conclusions and the discussion of unsolved problems.

Note added in proof

H. Nakajima informed us that he and K. Yoshioka found another proof [21] of our main theorem of Section 4, which relates the partition function, prepotential and the Seiberg–Witten curves. They use blowup techniques.

2 $\mathcal{N} = 2$ gauge theory, deformations, and backgrounds

In this section we recall the construction of [9, 10] and also provide quite a few new details. We consider $\mathcal{N} = 2$ supersymmetric gauge theory with gauge group $U(N)$. We study the Euclidean path integral with fixed vev of the adjoint Higgs field on $\mathbf{R}^4$ with certain nonminimal couplings. Most of the interesting dynamics in such theories is associated with instanton [22] effects [23].

2.1 Lagrangian, fields, couplings

The simplest way to write down the action of four-dimensional super-Yang–Mills theory with extended supersymmetry is to use dimensional reduction of the higher-dimensional minimal supersymmetric theory.

In particular, starting with the six-dimensional $\mathcal{N} = 1$ SYM, we arrive at

$$
\begin{aligned}
L_{\text{flat}} = \frac{1}{4g_0^2} \int \sqrt{G} d^4x \, \text{Tr} \Big\{ & -F_{IJ}F^{IJ} - 2D_I\phi D^I\bar{\phi} - [\phi, \bar{\phi}]^2 \\
& - i\bar{\lambda}_{\mathbf{i}}^{\dot{\alpha}} \sigma^I_{\alpha\dot{\alpha}} D_I \lambda^{\alpha\mathbf{i}} + \frac{i}{2}(\phi\epsilon^{\mathbf{ij}}[\bar{\lambda}_{\dot{\alpha}\mathbf{i}}, \bar{\lambda}_{\mathbf{j}}^{\dot{\alpha}}] - \bar{\phi}\epsilon_{\mathbf{ij}}[\lambda^{\alpha\mathbf{i}}, \lambda^{\mathbf{j}}_{\alpha}]) \Big\} \\
& + \frac{\vartheta_0}{2\pi} \int \text{Tr}\, F \wedge F.
\end{aligned}
\tag{2.1}
$$

Here $A_I, \phi, \bar{\phi}$ are the components of the six-dimensional gauge field, which decompose as the four-dimensional gauge field and an adjoint complex Higgs field (or two real Higgs fields); $\lambda_{\alpha\mathbf{i}}$ are $\mathcal{N} = 2$ gluions—a pair of four-dimensional Weyl spinors, transforming in the adjoint representation of the gauge group.

The indices $\alpha, \beta = 1, 2$ correspond to the doublet of $SU(2)_L$, $\dot{\alpha}, \dot{\beta} = 1, 2$ is that for $SU(2)_R$, while $\mathbf{i}, \mathbf{j} = 1, 2$ are the internal indices, which reflect the $SU(2)_I$ R-symmetry of the theory. They are raised and lowered using the SU(2) invariant tensor $\varepsilon_{12} = -\varepsilon_{21} = 1 = \varepsilon^{21} = -\varepsilon^{12}$. Space-time Lorentz indices will be denoted throughout the paper by the uppercase Latin letters $I, J, \ldots = 1, 2, 3, 4$. The Pauli tensors, relating the spinor and vector indices, are $\sigma_{I\alpha\dot{\alpha}}$.

In (2.1) we have used the bare coupling constant g_0 and the bare theta angle ϑ_0. The bare coupling corresponds to some high energy cutoff scale μ.

The action (2.1) is the limit of the six-dimensional action of the theory on a six manifold $\mathbf{T}^2 \times \mathbf{R}^4$ with the standard flat product metric.

2.2 Ω-background

However, in going from six to four dimensions one may have started with a nontrivial six-dimensional metric. In particular, by reducing on the two-torus one may have considered $\mathbf{R}^4$ bundles with nontrivial flat SO(4) connections, such as the space $\mathcal{N}_6$ with the metric

$$
ds^2 = Adzd\bar{z} + g_{IJ}(dx^I + V^I dz + \bar{V}^I d\bar{z})(dx^J + V^J dz + \bar{V}^J d\bar{z}) \tag{2.2}
$$

with $V^I = \Omega^I_J x^J$, $\bar{V}^I = \bar{\Omega}^I_J x^J$, and the area A of the torus to be sent to zero. For $[\Omega, \bar{\Omega}] = 0$ the metric (2.2) is flat.

However, for $\Omega \neq 0$, the background (2.2) will break all supersymmetries. Indeed, the spinors $\epsilon^{\mathbf{i}}_{\alpha}, \bar{\epsilon}_{\dot{\alpha}\mathbf{i}}$ generating supersymmetries of (2.1) are the components of a six-dimensional Weyl spinor. In order to generate the symmetry of the theory on a curved background the spinor must be covariantly constant. In our case this means that the spinor, viewed as four-dimensional Weyl spinor, should be invariant under the holonomies around the two cycles of the two-torus, which is possible only for discrete choices of $\Omega, \bar{\Omega}$.

2.2.1 Ω-background in the physical formalism

Fortunately there exists a continuous deformation of the $\mathcal{N} = 2$ theory preserving some fermionic symmetry (which also deforms along the way). The trick is to use the R-symmetry, which is manifest in the four-dimensional theory—the SU(2) group, which acts on the internal index $\mathbf{i} = 1, 2$ of the gluinos λ.

Namely, in addition to the nontrivial metric (2.2), we turn on a Wilson loop in the R-symmetry group, which should compensate some part of the metric induced holonomy on the spinors.

As a result, in the limit $A \to 0$ with $\Omega, \bar{\Omega}$ fixed the action (2.1) gets extra terms:

$$\Delta L = \Omega^I_J L^{(1)J}_I + \bar{\Omega}^I_J \bar{L}^{(1)J}_I + \Omega^I_J \bar{\Omega}^K_L L^{(2)JL}_{IK}, \tag{2.3}$$

where

$$L^{(1)J}_I = \int d^4x \sqrt{G} (x^J (\mathrm{Tr}\, F_{IK} D^K \bar{\Phi} + \epsilon^{\mathbf{ij}} \mathrm{Tr}\, \bar{\lambda}_{\dot{\alpha}\mathbf{i}} D_I \bar{\lambda}^{\dot{\alpha}}_{\mathbf{j}}) + G^{JK} \bar{\sigma}^{\mathbf{ij}}_{IK} \mathrm{Tr}[\bar{\lambda}_{\dot{\alpha}\mathbf{i}}, \bar{\lambda}^{\dot{\alpha}}_{\mathbf{j}}]), \tag{2.4}$$

where the tensor $\bar{\sigma}^{\mathbf{ij}}_{IK} = -\bar{\sigma}^{\mathbf{ij}}_{KI} = \bar{\sigma}^{\mathbf{ji}}_{IK} = \frac{1}{2}\varepsilon_{IKJL}\bar{\sigma}^{\mathbf{ij}JL}$ is the 't Hooft projector: $\mathbf{so}(4) = \mathbf{su}_L(2) \oplus \mathbf{su}_R(2) \to \mathbf{su}_R(2) = ((\frac{1}{2}) \otimes (\frac{1}{2}))_{\mathrm{sym}}$:

$$\sigma^{\dot{\beta}}_{IJ\dot{\alpha}} = \frac{1}{4}(\sigma^{\beta\dot{\beta}}_I \sigma_{J\beta\dot{\alpha}} - \sigma^{\beta\dot{\beta}}_J \sigma_{I\beta\dot{\alpha}})$$

and, for future use,

$$\sigma^{\beta}_{IJ\alpha} = \frac{1}{4}(\sigma_{I\alpha\dot{\alpha}} \sigma^{\beta\dot{\alpha}}_J - \sigma_{J\alpha\dot{\alpha}} \sigma^{\beta\dot{\alpha}}_I).$$

Finally,

$$L^{(2)JL}_{IK} = \int d^4x \sqrt{G} x^J x^L G^{MN} \mathrm{Tr}\, F_{IM} F_{KN}. \tag{2.5}$$

2.2.2 Ω-background in the twisted formalism

Supersymmetric gauge theories with extended supersymmetry can be formulated in a way which guarantees the existence of a nilpotent symmetry in any curved background. This formulation, sometimes called twisted, or topological, or cohomological, makes use of the R-symmetry of the theory. The coupling to the curved metric is accompanied by the coupling to the R-symmetry gauge field, which is taken to be equal to the corresponding projection of the spin connection [24, 25]. In this way, the fermions of the pure gauge theory become a one-form ψ, a self-dual two-form χ, and a scalar η. The bosons A_I and ϕ are not sensitive to the twisting.

The advantage of such a formulation of the theory is the clear geometric meaning of all the terms in the action. It is well known that $\mathcal{N} = 2$ super-Yang–Mills in the twisted formulation provides an integral representation for the Donaldson invariants of four-manifolds (for $G = \mathrm{SO}(3)$). In our study the gauge theory lives on $\mathbf{R}^4$, which is boring topologically. However, we study something different from the Donaldson theory, due to the Ω-background. It corresponds to the K-equivariant version of Donaldson invariants, $K = \mathrm{Spin}(4)$ being the group of rotations.

In general, if the metric $G = G_{IJ}dx^I dx^J$ of the Euclidean space-time manifold has some isometries, one can deform the standard Donaldson–Witten [24, 26] action by coupling it to the isometry vector fields V and $\bar{V}$, which should commute $[V, \bar{V}] = 0$. Explicitly, the action of the theory is given by

$$\begin{aligned} L = \frac{1}{2g_0^2}\Big(&-\frac{1}{2}\operatorname{Tr} F \star F + \operatorname{Tr}(D_A\phi - \iota_V F) \star (D_A\bar{\phi} - \iota_{\bar{V}} F) \\ &+ \frac{1}{2}\operatorname{Tr}([\phi, \bar{\phi}] + L_V\bar{\phi} - L_{\bar{V}}\phi)^2 \operatorname{vol}_g\Big) \\ &+ \operatorname{Tr}(\chi(D_A\psi)^+ + \eta D_A^*\psi + \chi \star L_V\chi + \eta \wedge \star L_V\eta + \psi \star L_{\bar{V}}\psi) \\ &+ \operatorname{Tr}(\chi \star [\phi, \chi] + \eta \star [\phi, \eta] + \psi \star [\bar{\phi}, \psi]) \\ &+ \frac{\vartheta_0}{2\pi}\operatorname{Tr} F \wedge F. \end{aligned} \tag{2.6}$$

On $\mathbf{R}^4$, we take, as above,

$$V^I = \Omega^I_J x^J, \qquad \bar{V}^I = \bar{\Omega}^I_J x^J \tag{2.7}$$

with

$$\Omega^{IJ} = \begin{pmatrix} 0 & \epsilon_1 & 0 & 0 \\ -\epsilon_1 & 0 & 0 & 0 \\ 0 & 0 & 0 & \epsilon_2 \\ 0 & 0 & -\epsilon_2 & 0 \end{pmatrix}, \qquad \bar{\Omega}^{IJ} = \begin{pmatrix} 0 & \bar{\epsilon}_1 & 0 & 0 \\ -\bar{\epsilon}_1 & 0 & 0 & 0 \\ 0 & 0 & 0 & \bar{\epsilon}_2 \\ 0 & 0 & -\bar{\epsilon}_2 & 0 \end{pmatrix}, \tag{2.8}$$

where $\Omega^{IJ} = G^{JK}\Omega^I_K$, etc.

2.3 On supersymmetry

For completeness, we list here the formulae for the supersymmetries of the actions (2.1), (2.3), (2.6).

2.3.1 Supersymmetry of the physical theory

The supercharges transform in the representation $(\mathbf{2}, \mathbf{1}, \mathbf{2}) \oplus (\mathbf{1}, \mathbf{2}, \mathbf{2})$ of the Lorentz $\times$ R-symmetry group $SU(2)_L \times SU(2)_R \times SU(2)_I$. Introducing the corresponding infinitesimal parameters $\zeta^{\mathbf{i}}_\alpha$ and $\bar{\zeta}^{\mathbf{i}}_{\dot{\alpha}}$, the supersymmetry transformations can be written

$$\begin{aligned} \delta A_I &= -i\bar{\lambda}^{\dot{\alpha}}_{\mathbf{i}}\sigma_{I\alpha\dot{\alpha}}\zeta^{\alpha\mathbf{i}} - i\lambda^{\alpha\mathbf{i}}\sigma_{I\alpha\dot{\alpha}}\bar{\zeta}^{\dot{\alpha}}_{\mathbf{i}}, \\ \delta\lambda_{\alpha\mathbf{i}} &= \sigma^{IJ}_{\alpha\beta}\zeta^{\beta\mathbf{i}}F_{IJ} + i\zeta^{\mathbf{i}}_\alpha D + i\sqrt{2}\sigma^I_{\alpha\dot{\alpha}}D_I B\epsilon^{\mathbf{ij}}\bar{\zeta}^{\dot{\alpha}}_{\mathbf{j}}, \\ \delta\bar{\lambda}^{\mathbf{i}}_{\dot{\alpha}} &= \sigma^{IJ}_{\dot{\alpha}\dot{\beta}}\bar{\zeta}^{\dot{\beta}}_{\mathbf{i}}F_{IJ} - i\bar{\zeta}_{\dot{\alpha}\mathbf{i}}D + i\sqrt{2}\sigma^I_{\alpha\dot{\alpha}}D_I\bar{B}\epsilon_{\mathbf{ij}}\zeta^{\alpha\mathbf{j}}, \\ \delta B &= \sqrt{2}\zeta^{\alpha\mathbf{i}}\lambda_{\alpha\mathbf{i}}, \\ \delta\bar{B} &= \sqrt{2}\bar{\zeta}^{\dot{\alpha}}_{\mathbf{i}}\bar{\lambda}^{\mathbf{i}}_{\dot{\alpha}} \end{aligned} \tag{2.9}$$

with the auxiliary field $D = [B, \bar{B}]$.

2.3.2 Twisted superalgebra

The twisted formulation of the theory is achieved by replacing the Lorentz group $K = SU(2)_L \times SU(2)_R \in K \times SU(2)_I$ by another subgroup of $K \times SU(2)_I$, namely, $SU(2)_L \times SU(2)_d$, with $SU(2)_d$ being diagonally embedded into $SU(2)_R \times SU(2)_I$. In other words, the internal index $\mathbf{i}$ is identified with another $SU(2)_R$ index, $\dot{\alpha}$. The fields are redefined according to

$$\psi_I = \lambda_{\alpha\dot{\beta}}\sigma_I^{\alpha\dot{\beta}}, \qquad \chi_{IJ} = \sigma_{IJ\dot{\alpha}\dot{\beta}}\bar{\lambda}^{\dot{\alpha}\dot{\beta}}, \qquad \eta = \epsilon_{\dot{\alpha}\dot{\beta}}\bar{\lambda}^{\dot{\alpha}\dot{\beta}}. \tag{2.10}$$

Similarly, the supersymmetry parameters $\zeta_\alpha^{\mathbf{i}}$ become ζ^I, $\bar{\zeta}_{\dot{\alpha}\mathbf{i}}$ become ζ^{IJ} and ζ. Of course, ζ^{IJ} is self-dual.

The supersymmetry algebra becomes

$$\begin{aligned}
\delta A_I &= -i\zeta\psi_I - i\zeta_{IJ}\psi^J - i\zeta_I\eta,\\
\delta\psi_I &= +\zeta^J F_{IJ}^- + i\zeta_I D + i\zeta D_I\phi,\\
\delta\chi_{IJ} &= \zeta F_{IJ}^+ - i\zeta_{IJ} D + i(\zeta_{[I} D_{J]}\bar{\phi})^+,\\
\delta\eta &= \zeta^{IJ} F_{IJ} - i\zeta D + \zeta^I D_I\bar{\phi},\\
\delta\phi &= \zeta^I\psi_I,\\
\delta\bar{\phi} &= \zeta\eta + \zeta^{IJ}\chi_{IJ}.
\end{aligned} \tag{2.11}$$

The geometric meaning of the transformations δ is the following. The space of fields A_I, ϕ, ψ_I represents the $\mathcal{G}$-equivariant de Rham complex of the space $\mathcal{A}$ of gauge fields on $\mathbf{R}^4$, together with the ingredients needed to construct a Mathai-Quillen representative of the Euler class of a certain infinite-dimensional bundle over $\mathcal{A}$, and the projection form [27] associated with the projection $\mathcal{A} \to \mathcal{A}/\mathcal{G}$.

The space $\mathbf{R}^4$ is hyper-Kähler, and possesses an action of the group $\mathbf{R}^4$ of translations.

The transformation generated by ζ is the $\mathcal{G}$-equivariant de Rham differential. The transformations generated by ζ^{IJ} correspond to the $\mathcal{G}$-equivariant $\bar{\partial}_{\mathcal{I},\mathcal{J},\mathcal{K}}$ differentials, corresponding to the three complex structures on $\mathcal{A}$ induced from the complex structures on $\mathbf{R}^4$. Finally, ζ^I correspond to the operators ι_{∂_I} of contraction with the vector fields on $\mathcal{A}$, induced by the vector fields

$$\frac{\partial}{\partial x^I}$$

generating translations of $\mathbf{R}^4$.

2.3.3 Supersymmetry of the Ω-background

The transformations which generate the symmetry of the Ω-background utilize the rotational symmetries of $\mathbf{R}^4$. To the vector field

$$V = \Omega^I_J x^J \frac{\partial}{\partial x^I}$$

there is an associated vector field on $\mathcal{A}$ and the associated operation of contraction on the $\mathcal{G}$-equivariant de Rham complex. In terms of the transformations δ of the twisted theory these are simply the transformation (2.11) with the space-dependent transformation parameter $\zeta^I = V^I(x) = \Omega^I_J x^J$.

The theory (2.3) in invariant under the transformation generated by

$$(\zeta, \zeta^I, \zeta^{IJ}) = (\zeta, V^I(x)\zeta, 0).$$

The supercharge generating this transformation will be denoted by $\tilde{Q}$.

2.4 Noncommutative deformation

The theory (2.1) allows yet another deformation, which we shall implicitly use to simplify our calculations. Let $\Theta^{IJ} = -\Theta^{JI}$ be a constant Poisson tensor on $\mathbf{R}^4$. The theory on $\mathbf{R}^4$ can be deformed to that on the noncommutative space, $\mathbf{R}^4_\Theta$. A naive way to define this deformation is to replace all the products of functions (or components of various tensor fields) in (2.1) by the so-called Moyal product:

$$f \star g(x) = \exp \frac{i}{2}\Theta^{IJ} \frac{\partial}{\partial \xi^I}\frac{\partial}{\partial \eta^J}\Big|_{\eta=\xi=0} f(x+\xi)g(x+\eta). \tag{2.12}$$

A more conceptual definition goes as follows (cf. [28, 29]). Consider the theory (2.1) dimensionally reduced to zero space-time dimensions. We get some sort of supersymmetric matrix model. Replace the matrices by operators in the Hilbert space $\mathcal{H}$. Explicitly, we get the theory of six bosonic operators,

$$X^I, \ \Phi, \ \bar{\Phi}; \quad I = 1, \ldots, 4, \tag{2.13}$$

and eight fermionic ones (we use the twisted formulation),

$$\Psi^I, \ \eta, \ \chi^{IJ} = \frac{1}{2}\varepsilon^{IJKL}\chi_{KL}; \quad I, J = 1, \ldots, 4, \tag{2.14}$$

where we lower indices using some Euclidean metric g_{IJ}:

$$\chi_{KL} = g_{KI} g_{LJ} \chi^{IJ},$$

The action reads as follows:

$$\begin{aligned} L = -\frac{1}{4\mathbf{g}^2} \mathrm{Tr}_{\mathcal{H}} & (g_{IK} g_{JL}([X^I, X^J][X^K, X^L] + \chi^{IJ}[X^K, \Psi^L] + \chi^{IJ}[\Phi, \chi^{KL}]) \\ & + 2 g_{IJ}(\eta[X^I, \Psi^J] + [\bar{\Phi}, X^I][\Phi, X^J] + \Psi^I[\bar{\Phi}, \Psi^J]) \\ & + ([\Phi, \bar{\Phi}]^2 + \eta[\Phi, \eta])) \\ + \mathrm{Tr}_{\mathcal{H}} & \left(i B_{IJ}[X^I, X^J] + \frac{\vartheta}{8\pi}\varepsilon_{IJKL}[X^I, X^J][X^K, X^L] + L_0 \right). \end{aligned} \tag{2.15}$$

The action (2.15) leads to the equations of motion which do not depend on the parameters entering the last line in (2.15), i.e., B_{IJ}, ϑ_0, L_0.

A special class of extrema of the action (2.15) with all the fermionic fields set to zero is achieved on the operators X^I, Φ, $\bar{\Phi}$ obeying

$$\begin{aligned} &[X^I, X^J] = i\Theta^{IJ}\mathbf{I}, \quad \Theta^{IJ} = 2g_0^2 g^{IK} g^{JL} B_{KL}, \\ &[X^I, \Phi] = 0 = [X^I, \bar{\Phi}] = [\Phi, \bar{\Phi}]. \end{aligned} \tag{2.16}$$

Let us fix a standard set of operators $\mathbf{x}^I$ in $\mathcal{H}$ obeying

$$[\mathbf{x}^I, \mathbf{x}^J] = i\Theta^{IJ}\mathbf{1}. \tag{2.17}$$

The algebra (2.17) may be represented reducibly in $\mathcal{H}$. In general, for nondegenerate Θ^{IJ},

$$\mathcal{H} = \mathcal{H}_0 \otimes W, \tag{2.18}$$

where $\mathcal{H}_0$ is an irreducible representation of (2.17), and W is a multiplicity space, which we shall assume to be a finite-dimensional Hermitian vector space, of complex dimension N. From now on, $\mathbf{x}^I$ will denote the operators on $\mathcal{H}_0$. The corresponding operators on $\mathcal{H}$ will be denoted as $\mathbf{x}^I \otimes \mathbf{1}_N$. Expanding,

$$X^I = \mathbf{x}^I \otimes \mathbf{1}_N + i\Theta^{IJ} A_J(\mathbf{x}), \qquad \Phi = \phi(\mathbf{x}) = \mathbf{1} \otimes \mathbf{a} + \phi_\infty(\mathbf{x}), \qquad \Psi^I = \Theta^{IJ}\psi_J(\mathbf{x}), \tag{2.19}$$

where $\phi_\infty(x) \to 0$, $x \to \infty$, we arrive at the naive formulation, with the metric on $\mathbf{R}^4$, given by

$$\begin{aligned} G^{IJ} &= g_{KL}\Theta^{IK}\Theta^{JL}, \\ \vartheta_0 &= \vartheta\, \mathrm{Pf}(\Theta), \\ \frac{1}{g_0^2} &= \frac{1}{\mathbf{g}^2}\,\mathrm{Pf}(\Theta)\sqrt{\det g}. \end{aligned} \tag{2.20}$$

In (2.19), both $A_J = A_{J|ln} \otimes E^{ln}$ and ϕ are valued in the $N \times N$ matrices, anti-Hermitian operators in W. To arrive at the ordinary gauge theory, we should take the limit $\Theta \to 0$, while keeping G_{IJ}, g_0^2, ϑ_0 finite. The curvature F of the gauge field A is given by

$$[X^I, X^J] = i\Theta^{IJ} + \Theta^{IK}\Theta^{JL}F_{KL}. \tag{2.21}$$

2.4.1 Combining Ω and Θ

The universal gauge theory (2.15) can also be subject to the nontrivial Ω-background:

$$
\begin{aligned}
L = -\frac{1}{4\mathbf{g}^2}\,\mathrm{Tr}_{\mathcal{H}}(g_{IK}g_{JL}([X^I, X^J][X^K, X^L] \\
+ \chi^{IJ}[X^K, \Psi^L] + \chi^{IJ}([\Phi, \chi^{KL}] + 2\Omega^K_M \chi^{ML})) \\
+ 2g_{IJ}(([\bar{\Phi}, X^I] + \bar{\Omega}^I_L X^L)([\Phi, X^J] + \Omega^J_K X^K) \\
+ \eta[X^I, \Psi^J] + \Psi^I([\bar{\Phi}, \Psi^J] + \bar{\Omega}^J_L \Psi^L)) \\
+ ([\Phi, \bar{\Phi}]^2 + \eta[\Phi, \eta])) \\
+ \mathrm{Tr}_{\mathcal{H}}(iB_{IJ}[X^I, X^J] + \frac{\vartheta}{8\pi}\varepsilon_{IJKL}[X^I, X^J][X^K, X^L] + L_0).
\end{aligned}
\tag{2.22}
$$

The vacua of the theory (2.22) are given by the operators which solve slightly different equations than (2.16). Namely, the condition on Φ is now

$$
[X^I, \Phi] = \Omega^I_J X^J, \qquad [X^I, \bar{\Phi}] = \bar{\Omega}^I_J X^J, \tag{2.23}
$$

which is consistent with $[X^I, X^J] = i\Theta^{IJ}$ only under certain conditions on $\Omega, \Theta, \bar{\Omega}$. Namely, assuming nondegeneracy of Θ, with $\omega = \Theta^{-1}$, the matrices

$$
\mathcal{E}_{IJ} = \Omega^K_J \omega_{KI}, \qquad \bar{\mathcal{E}}_{IJ} = \bar{\Omega}^K_J \omega_{KI} \tag{2.24}
$$

must be symmetric. Then the solution to (2.16) is given by

$$
X^I = \mathbf{x}^I, \qquad \Phi = \frac{1}{2}\mathcal{E}_{IJ}\mathbf{x}^I\mathbf{x}^J, \qquad \bar{\Phi} = \frac{1}{2}\bar{\mathcal{E}}_{IJ}\mathbf{x}^I\mathbf{x}^J. \tag{2.25}
$$

Now, the gauge field A_I and the Higgs field ϕ are introduced via

$$
X^I = \mathbf{x}^I \otimes \mathbf{1}_N + i\Theta^{IJ}A_J(\mathbf{x}), \qquad \Phi = \frac{1}{2}\mathcal{E}_{IJ}X^I X^J \otimes \mathbf{1}_N + \phi(\mathbf{x}). \tag{2.26}
$$

2.4.2 Supersymmetry of the noncommutative theory in Ω-background

We shall only write down the supercharge of interest to us:

$$
\begin{aligned}
\tilde{Q}X^I &= \Psi^I, & \tilde{Q}\Psi^I &= [\Phi, X^I] + \Omega^I_J X^J, \\
\tilde{Q}\Phi &= 0, & \tilde{Q}\chi^{IJ} &= H^{IJ}, \\
\tilde{Q}H^{IJ} &= [\Phi, \chi^{IJ}] - (\Omega^I_K \chi^{JK} - \Omega^J_K \chi^{IK})^+, \\
\tilde{Q}\bar{\Phi} &= \eta, & \tilde{Q}\eta &= [\Phi, \bar{\Phi}].
\end{aligned}
\tag{2.27}
$$

Here H^{IJ} is an auxiliary field, which is equal to $[X^I, X^J]^+$ on-shell.

2.4.3 Observables in the supersymmetric gauge theory: $\Omega = 0$

In the ordinary $\mathcal{N} = 2$ supersymmetric gauge theory a special class of observables play an important role. They are distinguished by the property that a certain supercharge annihilates them. Of particular interest for application to Donaldson theory are the observables which are constructed out of the invariant polynomials in the Higgs field ϕ by means of the descent procedure. In terms of the twisted fields, these observables appear as follows. Let $P(\phi)$ be any G-invariant polynomial on the Lie algebra $\mathbf{g} = \mathrm{Lie}(G)$ of the gauge group G. Then

$$\begin{aligned}
\mathcal{O}_P^{(0)}(x) &= P(\phi(x)),\\
\mathcal{O}_P^{(1)}(C) &= \oint_C \frac{\partial P}{\partial \phi^a}\psi^a,\\
\mathcal{O}_P^{(2)}(\Sigma) &= \int_\Sigma \frac{\partial P}{\partial \phi^a}F^a + \frac{1}{2}\frac{\partial^2 P}{\partial\phi^a\partial\phi^b}\psi^a \wedge \psi^b,\\
&\ \vdots\\
\mathcal{O}_P^{(4)}(X) &= \int_X \frac{1}{2}\frac{\partial^2 P}{\partial\phi^a\partial\phi^b}F^a \wedge F^b\\
&\quad + \cdots + \frac{1}{24}\frac{\partial^4 P}{\partial\phi^a\partial\phi^b\partial\phi^c\partial\phi^d}\psi^a \wedge \psi^b \wedge \psi^c \wedge \psi^d.
\end{aligned} \tag{2.28}$$

Here $x, C, \Sigma, \ldots, X$ represent a $0, 1, 2, \ldots, 4$-cycles in the space-time manifold, respectively. The main idea behind (2.28) is to use the fact that the supercharge Q acts on $P(\Phi(x)+\cdots)$ as the de Rham differential. To compare the observables of the ordinary gauge theory to those of the noncommutative gauge theory, we shall utilize the generating function (form):

$$\mathcal{O}_P = P\left(\phi(x) + \psi_I(x)dx^I + \frac{1}{2}F_{IJ}(x)dx^I \wedge dx^J\right) \in \Omega^*(\text{space-time}). \tag{2.29}$$

Its main property is

$$d_x\langle \mathcal{O}_P(x)\ldots\rangle = 0.$$

The noncommutative gauge symmetry does not allow, naively, for local observables as in the first line of (2.28). However, it has as many gauge invariant observables as does the ordinary theory. For the gauge group $U(N)$ the invariant polynomials can be expressed as polynomials in the single trace operators

$$P_n(\phi) = \mathrm{Tr}\,\phi^n. \tag{2.30}$$

It is convenient to introduce the character

$$P_\beta(\phi) = \mathrm{Tr}\,e^{i\beta\phi} = \sum_{n=0}^{\infty}\frac{(i\beta)^n}{n!}P_n. \tag{2.31}$$

The analogue of (2.28)–(2.29) is the following closed form on $\mathbf{R}^4$:

$$\mathcal{O}_{P_\beta}(x, dx) = \beta^2 e^{-\beta\omega}\int d^4\vartheta d^4\kappa e^{-i\beta(\kappa_I x^I + \vartheta_I dx^I)}\,\mathrm{Tr}_{\mathcal{H}}\,e^{i\beta\Phi(\vartheta,\kappa)}, \tag{2.32}$$

where

$$\Phi(\vartheta,\kappa) = \Phi+\kappa_I X^I+\vartheta_I \Psi^I+\frac{1}{2}\vartheta_I\vartheta_J[X^I, X^J], \qquad \omega = \frac{1}{2}\omega_{IJ}dx^I\wedge dx^J. \quad (2.33)$$

The closedness of $\mathcal{O}_{P_\beta}$ is proved with the help of the symmetry acting on ϑ, κ:

$$\delta\vartheta = \kappa, \quad \delta\kappa = 0. \quad (2.34)$$

2.4.4 Observables: $\Omega \neq 0$

The chiral observables get deformed when $\Omega \neq 0$. First of all, in the ordinary gauge theory, the generating form $\mathcal{O}_P$ becomes equivariantly closed under the correlator:

$$(d + \iota_V)\mathcal{O}_P = 0. \quad (2.35)$$

In the noncommutative gauge theory we get the same statement, but now we need to modify the symmetry δ to its equivariant analogue,

$$\delta\vartheta_I = \kappa_I, \qquad \delta\kappa_I = -\Omega_I^J\vartheta_J, \quad (2.36)$$

and change (2.32) to

$$\begin{aligned} \mathcal{O}_{P_\beta;\Omega} &= \prod_{\alpha=1,2}\frac{e^{\beta\epsilon_\alpha}-1}{\epsilon_\alpha}e^{\beta\omega_\Omega}\int d^4\vartheta d^4\kappa e^{-i\beta(\kappa_I x^I+\vartheta_I dx^I)}\,\mathrm{Tr}_{\mathcal{H}}\,e^{i\beta\Phi(\vartheta,\kappa)}, \\ \omega_\Omega &= \frac{1}{2}(\omega_{IJ}dx^I\wedge dx^J - \mathcal{E}_{IJ}x^I x^J). \end{aligned} \quad (2.37)$$

In what follows we consider either the 0-observable or the integrated 4-observable. In the first case, we set $x = 0$ (to be at the fixed point), and get the integral over ϑ, κ of $\mathrm{Tr}_{\mathcal{H}}\,e^{i\beta\Phi(\vartheta,\kappa)}$. In the second case, we get the integral over ϑ, κ of $\mathrm{Tr}_{\mathcal{H}}\,e^{i\beta\Phi(\vartheta,\kappa)}$ with an extra δ-invariant Gaussian factor. Actually, (2.36) implies that the integral over ϑ, κ is localized at $\kappa = \vartheta = 0$, so that both the 0- and 4-observables are equal, up to an ϵ_α-dependent factor

$$\begin{aligned} \mathcal{O}^{(0)}_{P_\beta;\Omega} &= \prod_{\alpha=1,2}(e^{\beta\epsilon_\alpha}-1)\,\mathrm{Tr}_{\mathcal{H}}\,e^{i\beta\Phi}, \\ \int_{\mathbf{R}^4}\mathcal{O}^{(4)}_{P_\beta;\Omega} &= \prod_{\alpha=1,2}\frac{e^{\beta\epsilon_\alpha}-1}{\epsilon_\alpha}\,\mathrm{Tr}_{\mathcal{H}}\,e^{i\beta\Phi}. \end{aligned} \quad (2.38)$$

The observables (2.38) are most natural in the five-dimensional gauge theory compactified on the circle of the circumference β. The periodicity of $\mathcal{O}_{P_\beta}$ in Φ has a simple origin there—large gauge transformations [14]. We shall return to this theory in Section 7.

3 Gauge theory partition function

In the ordinary gauge theory the natural object of study is the Euclidean path integral over the field configurations, such that

$$\phi(x) \to \mathbf{a}, \quad x \to \infty. \tag{3.1}$$

We are aiming to calculate the path integral in the Ω-background, with the boundary conditions (3.1) (and, in fact, using noncommutative deformation):

$$Z(\mathbf{a}, \epsilon_1, \epsilon_2; \Lambda) = \int DAD\psi D\eta D\chi D\Phi D\bar{\Phi} e^{-\int_{\mathbf{R}^4} L\sqrt{g} d^4x}. \tag{3.2}$$

In the noncommutative gauge theory the condition (3.1) is phrased differently. Think of Φ as an element of the Lie algebra of the group $\mathcal{G}_\Theta \approx U(\mathcal{H})$ of gauge transformations. (Note that normally one takes $\mathcal{G}_\Theta = PU(\mathcal{H})$.) The identification $\mathcal{H} = \mathbf{C}^N \otimes \mathcal{H}_0$ allows one to view Φ as an $N \times N$ matrix of operators in $\mathcal{H}_0$. The condition (3.1) is now stated as

$$\Phi = \mathbf{a} \otimes \mathbf{1}_{\mathcal{H}_0} + \frac{1}{2} \mathbf{1}_N \otimes \mathcal{E}_{IJ} \mathbf{x}^I \mathbf{x}^J + \varphi(\mathbf{x}). \tag{3.3}$$

where $\varphi(x) \to 0$ as $x \to \infty$ faster than any power of x.

The integral (3.2) needs a UV cutoff μ. The bare couplings g_0 and ϑ_0 are renormalized, thus generating an effective scale Λ,

$$\Lambda^{2N} \sim \mu^{2N} e^{-\frac{8\pi^2}{g_0^2} + 2\pi i \vartheta_0}. \tag{3.4}$$

Remark. On the $U(1)$ factor. The gauge theory with gauge group $U(N)$ has an interesting pattern of the renormalization group flow. The ordinary gauge theory has a decoupled $U(1)$ factor, which does not exhibit any renormalization of its coupling, and an interacting SU(N) part, with the famous phenomenon of asymptotic freedom, reflected in (3.4). When the theory is deformed by Θ, the perturbative loop calculations are altered in a Θ-dependent way, as it introduces, among other things, an energy scale. In particular, the $U(1)$ factor coupling constant starts running, in the energy range

$$\mu \gg \Lambda \gg \frac{1}{\mu\Theta},$$

where the formula (3.4) with $N = 1$ holds (this result is a simple generalization of [30], which can also be justified using [1, 31]). The $U(1)^{N-1} \subset \mathrm{SU}(N)$ gauge couplings experience different renormalization group flow, as they are affected by the loops of the charged W-bosons. In principle, we should introduce two distinct low-energy scales, one for the SU(N) part of the gauge group, another for $U(1)$: Λ, Λ_0. As we shall eventually remove noncommutativity, it makes sense to keep $\Lambda_0 \approx \mu$. We shall encounter the ambiguity related to the $U(1)$ factor in the analysis of the Seiberg–Witten curves. In the paper we concentrate on the SU(N) dynamics, and mostly set $a = 0$.

Remark. The theory (2.1) has an anomalous $U(1)_{gh}$ symmetry (gh for ghost). Under this symmetry, the adjoint Higgs field Φ has charge $+2$, $\bar{\Phi}$ has charge -2, the fermions ψ have charge $+1$, and χ, η have charge -1. In the background of the instanton of charge k the path integral measure transforms under the ghost $U(1)$ with the charge $-4kN$. The Ω deformation breaks $U(1)_{gh}$. It can be restored by assigning to $\Omega, \bar{\Omega}$ the charges $+2, -2$, respectively.

3.1 Partition function as a sum over partitions

The partition function $Z(\mathbf{a}, \epsilon_1, \epsilon_2; \Lambda)$ can be explicitly evaluated as a sum over instantons. Moreover, the Ω-background lifts instanton moduli, leaving only a finite number of isolated points on the appropriately compactified instanton moduli space as a full set of supersymmetric minima of the action. The evaluation of (3.2) is then reduced to the calculation of the ratios of the bosonic and fermionic determinants near each critical point. These points are labeled by colored partitions.

Consequently, (3.2) is given by the sum over N-tuples of partitions [9, 10]. (See Appendix B for the notations and definitions related to partitions.) Explicitly,

$$Z(\mathbf{a}; \epsilon_1, \epsilon_2, \Lambda) = Z^{\text{pert}}(\mathbf{a}; \Lambda, \epsilon_1, \epsilon_2) \sum_{\mathbf{k}} \Lambda^{2N|\mathbf{k}|} Z_{\mathbf{k}}(\mathbf{a}; \epsilon_1, \epsilon_2), \tag{3.5}$$

where

$$\begin{aligned} Z_{\mathbf{k}}(\mathbf{a}; \epsilon_1, \epsilon_2) &= \prod_{l,n;i,j} \frac{a_l - a_n + \epsilon_1(i-1) + \epsilon_2(-j)}{a_l - a_n + \epsilon_1(i - \tilde{k}_{nj} - 1) + \epsilon_2(k_{li} - j)} \\ &= \prod_{l,n;i,j} \frac{a_l - a_n + \epsilon_1(-i) + \epsilon_2(j-1)}{a_l - a_n + \epsilon_1(\tilde{k}_{lj} - i) + \epsilon_2(j - k_{ni} - 1)} \\ &= \frac{1}{\epsilon_2^{2N|\mathbf{k}|}} \prod_{(l,i)\neq(n,j)} \frac{\Gamma(k_{li} - k_{nj} + \nu(j-i+1) + b_{ln})\Gamma(\nu(j-i) + b_{ln})}{\Gamma(k_{li} - k_{nj} + \nu(j-i) + b_{ln})\Gamma(\nu(j-i+1) + b_{ln})}, \end{aligned} \tag{3.6}$$

$$b_{ln} = \frac{a_l - a_n}{\epsilon_2}, \qquad \nu = -\frac{\epsilon_1}{\epsilon_2}, \tag{3.7}$$

and

$$Z^{\text{pert}}(\mathbf{a}; \epsilon_1, \epsilon_2, \Lambda) = \exp\left(\sum_{l,n} \gamma_{\epsilon_1,\epsilon_2}(a_l - a_n; \Lambda)\right), \tag{3.8}$$

where the function $\gamma_{\epsilon_1,\epsilon_2}$ is defined in Appendix A.

Remark. The product over i, j in (3.6) in $Z_{\mathbf{k}}$ is infinite and needs a precise definition. Here it is: Fix a pair (l, n). Consider all the factors in $Z_{\mathbf{k}}$ corresponding to l, n. Split the set of indices (i, j) into four groups:

$$\begin{aligned} \mathbf{Z}_+^2 &= S_{++} \cup S_{+-} \cup S_{-+} \cup S_{--}, \\ S_{++} &= \{(i,j) \mid 1 \le i \le \ell(\mathbf{k}_l),\ 1 \le j \le k_{n1}\}, \\ S_{+-} &= \{(i,j) \mid 1 \le i \le \ell(\mathbf{k}_l),\ k_{n1} < j\}, \\ S_{-+} &= \{(i,j) \mid \ell(\mathbf{k}_l) < i,\ 1 \le j \le k_{n1}\}, \\ S_{--} &= \{(i,j) \mid \ell(\mathbf{k}_l) < i,\ k_{n1} < j\}. \end{aligned} \tag{3.9}$$

The set S_{++} is finite, the set S_{--} contributes 1. The set S_{+-} contributes

$$\begin{aligned} Z^{ln}_{+-;\mathbf{k}} &= \prod_{i=1}^{\ell(\mathbf{k}_l)} \prod_{j=k_{n1}+1}^{\infty} \frac{a_{ln} + \epsilon_1(i-1) + \epsilon_2(-j)}{a_{ln} + \epsilon_1(i-1) + \epsilon_2(k_{li} - j)} \\ &:= \prod_{i=1}^{\ell(\mathbf{k}_l)} \prod_{j=1}^{k_{li}} \frac{1}{a_{ln} + \epsilon_1(i-1) + \epsilon_2(k_{li} - k_{n1} - j)}, \end{aligned} \tag{3.10}$$

and similarly for S_{-+}. See also [32].

The special case $\epsilon_2 = -\epsilon_1 = \hbar$, i.e., $\nu = 1$ deserves special attention. In this case, the expression for the partition function simplifies to

$$\begin{aligned} Z(\mathbf{a}; \hbar, \Lambda) &= \sum_{\mathbf{k}} \Lambda^{2N|\mathbf{k}|} Z_{\mathbf{k}}(\mathbf{a}; \hbar), \\ Z_{\mathbf{k}}(\mathbf{a}; \hbar) &= Z^{\mathrm{pert}}(\mathbf{a}; \hbar)\mu^2_{\mathbf{k}}(\mathbf{a}, \hbar), \\ \mu^2_{\mathbf{k}}(\mathbf{a}, \hbar) &= \prod_{(l,i)\neq(n,j)} \left(\frac{a_l - a_n + \hbar(k_{l,i} - k_{n,j} + j - i)}{a_l - a_n + \hbar(j-i)}\right), \\ Z^{\mathrm{pert}}(\mathbf{a}; \hbar) &= \exp\left(\sum_{l,n} \gamma_\hbar(a_l - a_n; \Lambda)\right), \end{aligned} \tag{3.11}$$

where the function $\gamma_\hbar(x; \Lambda)$ is defined in Appendix A.

3.2 Plancherel measure on partitions

In the case when $N = 1$ and $\nu = 1$, the sum in (3.5) is over a single partition $\mathbf{k}$ and the weight $\mu_{\mathbf{k}}(\mathbf{a}, \hbar)$ reduces to

$$\mu(\mathbf{k}) = \prod_{i<j} \left(\frac{k_i - k_j + j - i}{j - i}\right) = \prod_{\square\in\mathbf{k}} \frac{1}{h(\square)}, \tag{3.12}$$

where the second product is over all squares $\square$ in the diagram of the partition $\mathbf{k}$ and $h(\square)$ denotes the corresponding hook-length.

The weight $\mu(\mathbf{k})^2$ is known as the Plancherel measure on partitions because of the relation

$$\mu(\mathbf{k}) = \frac{\dim R_{\mathbf{k}}}{|\mathbf{k}|!}, \tag{3.13}$$

where $R_{\mathbf{k}}$ is the irreducible representation of the symmetric group corresponding to the partition $\mathbf{k}$. As in the classical Plancherel theorem, the Fourier transform on the symmetric group an isometry of the L^2-spaces with respect to the Haar (i.e., counting) measure on the group and the Plancherel measure on the set of its irreducible representations. Observe that from (3.13) it follows that

$$\sum_{|\mathbf{k}|=n} \mu^2(\mathbf{k}) = \frac{1}{n!}. \tag{3.14}$$

The Plancherel measure is the most fundamental and natural measure on the set of partitions. In many aspects, the set of partitions equipped with the Plancherel measure is the proper discretization of the Gaussian Unitary Ensemble (GUE) of random matrices [33]. In particular, the integrable structures of random matrices are preserved and, in fact, become more natural and transparent for the Plancherel measure; see [34, 35]. Similarly, Plancherel random partitions play a central role in the Gromov–Witten theory of target curves, extending the role played by random matrices in the case of the point target; see [36].

A pedestrian explanation of the relation between Plancherel measure and GUE can be obtained by rewriting the weight (3.12) as the product of the Vandermonde determinant in the variables $k_i - i$ and a multinomial coefficient, which, of course, is the discrete analogue of the Gaussian weight.

As we will see below in Section 5, for $N > 1$ and $\nu = 1$ the partition function is again related to the Plancherel measure, but now periodically weighted with period N. Finally, the case $\nu \neq 1$ leads to the Jack polynomial analogue of the Plancherel measure; see [37].

4 Prepotential

In this section we study the limit $\epsilon_1, \epsilon_2 \to 0$ of the partition function (3.2). In this limit, according to field theory arguments [9] the partition function must behave as

$$Z(\mathbf{a}; \epsilon_1, \epsilon_2, \Lambda) = \exp\left(-\frac{1}{\epsilon_1\epsilon_2}\mathcal{F}(\mathbf{a}; \epsilon_1, \epsilon_2, \Lambda)\right) \tag{4.1}$$

where $\mathcal{F}$ is analytic in ϵ_1, ϵ_2 for $\epsilon_{1,2} \to 0$. We prove the conjecture of [9], which identifies $\mathcal{F}_0(\mathbf{a}, \Lambda) \equiv \mathcal{F}(\mathbf{a}; 0, 0, \Lambda)$ with the Seiberg–Witten prepotential of the low-energy effective theory.

4.1 Quasiclassical limit and SW curve

By setting $\nu = 1$ we deduce from the field theory prediction (4.1) that for $\hbar \to 0$ the partition function (3.11) has an asymptotic expansion,

$$Z(\mathbf{a};\hbar,\Lambda) = \exp\left(\sum_{g=0}^{\infty} \hbar^{2g-2}\mathcal{F}_g(\mathbf{a},\Lambda)\right), \tag{4.2}$$

where

$$\mathcal{F}_0(\mathbf{a},\Lambda) = -\frac{1}{2}\sum_{l,n}(a_l - a_n)^2\left(\log\left(\frac{a_l - a_n}{\Lambda}\right) - \frac{3}{2}\right) + \sum_{k=1}^{\infty}\Lambda^{2kN} f_k(\mathbf{a}) \tag{4.3}$$

is the so-called prepotential of the low-energy effective theory. The coefficients f_k for $k = 1, 2$ were computed directly from instanton calculus in [38, 39] for $k \leq 5$ in [9].

The prepotential was identified in [3, 4, 40] with the prepotential of the periodic Toda lattice. The arguments for this identification were indirect. In [41] the coefficients f_k for $k \leq 5$ were computed for the Toda prepotential, in agreement with the later results of [9].

The purpose of this paper is to give a proof of this relation to all orders in the instanton expansion.

More precisely, we shall show the following:

The set $(a_1, \ldots, a_N; \frac{1}{2\pi i}\frac{\partial\mathcal{F}_0}{\partial a_1}, \ldots, \frac{1}{2\pi i}\frac{\partial\mathcal{F}_0}{\partial a_N})$ coincides with the set of periods of the differential

$$dS = \frac{1}{2\pi i} z \frac{dw}{w} \tag{4.4}$$

on the curve $\mathcal{C}_u$ defined by

$$\Lambda^N\left(w + \frac{1}{w}\right) = P_N(z), \quad P_N(z) = z^N + u_1 z^{N-1} + u_2 z^{N-2} + \cdots + u_N. \tag{4.5}$$

Remark. Normally the factor $2\pi i$ is included in the definition of the prepotential [18]. We chose not to do this to get real expressions for $\mathcal{F}_0$ for real a_ls and Λ.

For generic $u = (u_1, u_2, \ldots, u_N)$, the curve (4.5) is a smooth hyperelliptic curve of genus $N - 1$. The differential dS has poles at two points $\infty_\pm$ over $x = \infty$. The homology group

$$H_1(\mathcal{C}_u - \{\infty_+, \infty_-\}; \mathbf{Z})$$

is isomorphic to $\mathbf{Z}\oplus\mathbf{Z}^{2(N-1)}$, the first summand being generated by the small loop σ_∞ around ∞_+ (the loop around ∞_- is in the same homology class). The intersection pairing identifies the dual space with

$$H_1(\mathcal{C}_u, \{\infty_+, \infty_-\}; \mathbf{Z}),$$

the relative cycle dual to σ_∞ being the path ℓ^∞ connecting ∞_-, ∞_+.

Note that in the N-parameter family of curves (4.5), there is a set of singular curves. The identification between the periods and $(a, \frac{\partial\mathcal{F}_0}{\partial a})$ is up to the action of the monodromy group. We need to fix this identification in some open domain of the parameter space.

Consider the region $\mathcal{U}_\infty$ of the parameter space where $u_s \gg \Lambda^s$. In this domain the curve (4.5) can be approximately described as follows. Let $\alpha_1, \ldots, \alpha_N$ be the zeroes of the polynomial $P_N(x)$:

$$P_N(x) = \prod_{l=1}^{N}(x - \alpha_l). \tag{4.6}$$

For small Λ, for each l we can unambiguously find $\alpha_l^{\pm} \approx \alpha_l$ such that

$$P_N(\alpha_l^{\pm}) = \pm 2\Lambda^N. \tag{4.7}$$

These are the branch points of the two-fold covering $\rho : \mathcal{C}_u \to \mathbf{P}^1, \mathbf{P}^1 \ni x$.

Let $\mathbf{a}_l$ be the 1-cycles on $\mathcal{C}_u$ which are the lifts of the cycles on the x-plane which surround the cuts going from α_l^- to α_l^+. Let $\mathbf{b}_l$ be the relative 1-cycle represented by the path on $\mathcal{C}_u$ which starts at ∞_-, goes to α_l^+, and then goes on the second sheet to ∞_+. The cycles $\mathbf{a}_l$, and $\mathbf{b}_l$ have the canonical intersection pairing:

$$\mathbf{a}_l \cap \mathbf{b}_m = \delta_{lm}. \tag{4.8}$$

The position of the $\mathbf{a}$-cycles and $\mathbf{b}$-cycles on the curve $\mathcal{C}_u$ is illustrated in Figure 1.

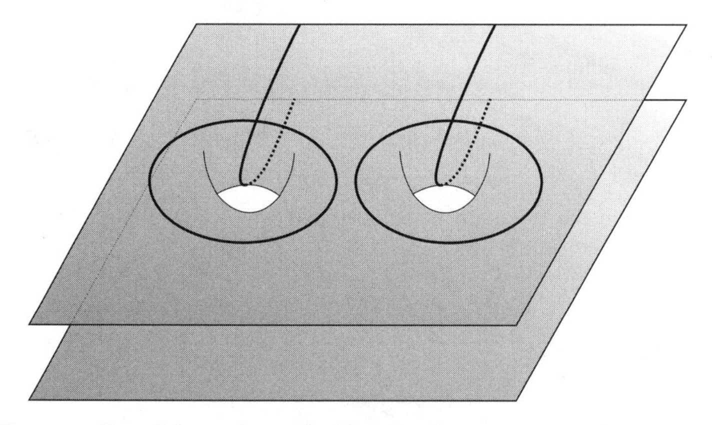

Fig. 1. The curve $\mathcal{C}_u$ and the cycles on it. The closed ones are $\mathbf{a}_l$s. The noncompact ones are $\mathbf{b}_l$s.

The cycles σ_∞ and ℓ^∞ are related to $\mathbf{a}_l, \mathbf{b}_m$ via

$$\sigma_\infty = \sum_l \mathbf{a}_l, \qquad \ell^\infty = \sum_l \mathbf{b}_l. \tag{4.9}$$

The periods a_l and $\frac{\partial \mathcal{F}_0}{\partial a_l}$ are defined via

$$a_l = \oint_{\mathbf{a}_l} dS, \qquad \frac{\partial \mathcal{F}_0}{\partial a_l} = 2\pi i \int_{\mathbf{b}_l(\mu)} dS, \tag{4.10}$$

where $\mathbf{b}_l(\mu)$ is the regularized contour of integration, which connects $w|_{z=\mu}$ and $w^{-1}|_\mu$ instead of ∞_+ and ∞_-.

The divergent (with μ) part of the periods is easy to calculate:

$$\frac{1}{2\pi i}\frac{\partial \mathcal{F}_0}{\partial a_l} = 2N\mu - u_1 \log \mu + \text{finite part}. \tag{4.11}$$

At the same time,

$$\oint_{\sigma_\infty} dS = a = \sum_l a_l = -u_1. \tag{4.12}$$

As we said above, in what follows we set $a = 0 = u_1$. Also, to avoid worrying about the cutoff, we work with the absolute cycles $\mathbf{b}_l - \mathbf{b}_{l+1}$.

4.2 Partitions and their profiles

The standard geometric object associated to a partition is its diagram. In this paper, we will draw partition diagrams in what is sometimes referred to as the Russian form (as opposed to the traditionally competing French and English traditions of drawing partitions). For example, Figure 2 shows the diagram of the partition (8, 6, 5, 3, 2, 2, 1, 1). The upper boundary of the diagram of $\mathbf{k}$ is the graph of a piecewise-linear function $f_{\mathbf{k}}(x)$, which we will call the *profile* of the partition $\mathbf{k}$. Explicitly,

$$f_{\mathbf{k}}(x) = |x| + \sum_{i=1}^{\infty}[|x - k_i + i - 1| - |x - k_i + i| + |x + i| - |x + i - 1|]. \tag{4.13}$$

The profile is plotted in bold in Figure 2.

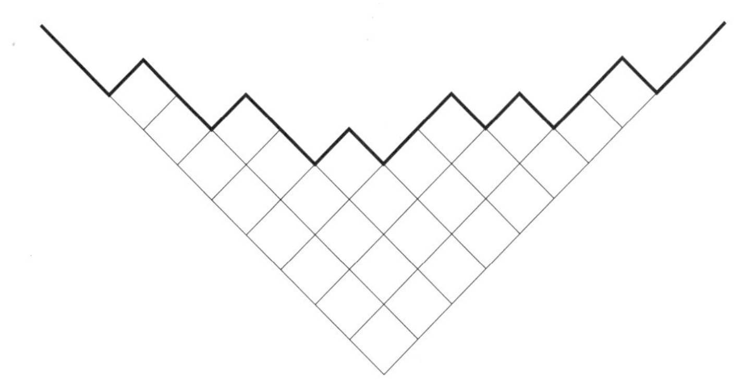

Fig. 2. Диаграмма Юнга (Young diagram).

For general $\epsilon_2 > 0 > \epsilon_1$, it is convenient to extend the definition of $f_{\mathbf{k}}(x)$ by scaling the two axes by $-\epsilon_1$ and ϵ_2, respectively. For example, for $(\epsilon_1, \epsilon_2) = (-1, \frac{1}{2})$,

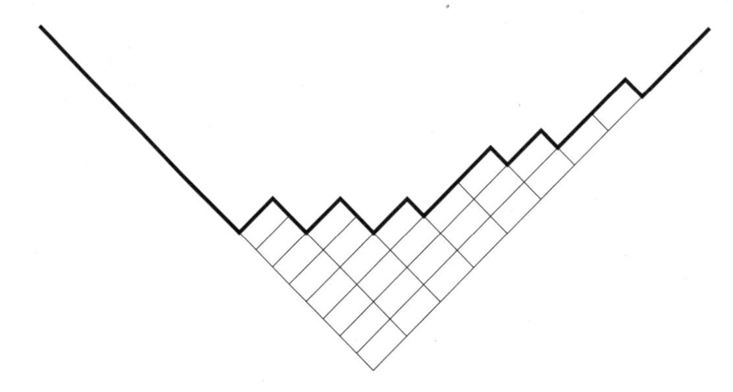

Fig. 3. Squeezed Young diagram.

the scaled diagram of the same partition (8, 6, 5, 3, 2, 2, 1, 1) and the corresponding profile $f_{\mathbf{k}}(x|\epsilon_1, \epsilon_2)$ will look as Figure 3.

We have

$$\begin{aligned} f_{\mathbf{k}}(x|\epsilon_1,\epsilon_2) &= |x| + \sum_{i=1}^{\infty}[|x+\epsilon_1-\epsilon_2 k_i-\epsilon_1 i| - |x-\epsilon_2 k_i-\epsilon_1 i| \\ &\qquad - |x+\epsilon_1-\epsilon_1 i| + |x-\epsilon_1 i|] \\ &= |x| + \sum_{j=1}^{\infty}[|x+\epsilon_2-\epsilon_1\tilde{k}_j-\epsilon_2 j| - |x-\epsilon_1\tilde{k}_j-\epsilon_2 j| \\ &\qquad - |x+\epsilon_2-\epsilon_2 j| + |x-\epsilon_2 j|]. \end{aligned} \tag{4.14}$$

By construction, the profile of a partition satisfies

$$\begin{aligned} &f'_{\mathbf{k}}(x|\epsilon_1,\epsilon_2) = \pm 1, \\ &f_{\mathbf{k}}(x|\epsilon_1,\epsilon_2) \geq |x|, \\ &f_{\mathbf{k}}(x|\epsilon_1,\epsilon_2) = |x| \quad \text{for } |x| \gg 0. \end{aligned} \tag{4.15}$$

We also define the profile of a charged partition

$$f_{a;\mathbf{k}}(x|\epsilon_1,\epsilon_2) = f_{\mathbf{k}}(x-a|\epsilon_1,\epsilon_2).$$

The charge and the size are easily recovered from $f_{a;\mathbf{k}}(x)$:

$$\begin{aligned} a &= \frac{1}{2}\int_{\mathbf{R}} dx x f''_{a;\mathbf{k}}(x|\epsilon_1,\epsilon_2) = -\frac{1}{2}\int_{\mathbf{R}} dx f'_{a;\mathbf{k}}(x|\epsilon_1,\epsilon_2), \\ |\mathbf{k}| &= \frac{a^2}{2\epsilon_1\epsilon_2} - \frac{1}{4\epsilon_1\epsilon_2}\int dx x^2 f''_{a;\mathbf{k}}(x|\epsilon_1,\epsilon_2) \\ &= \frac{1}{2\epsilon_1\epsilon_2}\left(a^2 - \int dx (f_{a;\mathbf{k}}(x|\epsilon_1,\epsilon_2) - |x|)\right). \end{aligned} \tag{4.16}$$

Here and in what follows, we denote by

$$\fint_D g(x)dx = \lim_{L\to\infty,\delta\to 0} \int_{D\cap[-L,L]\setminus \mathrm{sing}_\delta(g)} g(x)dx$$

the principal value integral over a domain $D \subset \mathbf{R}$, where $\mathrm{sing}_\delta(g)$ denotes the δ-neighborhood of the singularities of $g(x)$.

For a colored partition $\mathbf{k}$ and a vector $\mathbf{a}$, we define

$$f_{\mathbf{a};\mathbf{k}}(x|\epsilon_1, \epsilon_2) = \sum_{l=1}^{N} f_{a_l;\mathbf{k}_l}(x|\epsilon_1, \epsilon_2). \tag{4.17}$$

For example, for $\epsilon_2 = -\epsilon_1 = \hbar$, $a_1 = -a_2 = 11\hbar$ and the partition

$$\{(7, 4, 3, 3, 2, 1), (8, 7, 4, 4, 3, 1)\},$$

the corresponding profile looks as in Figure 4.

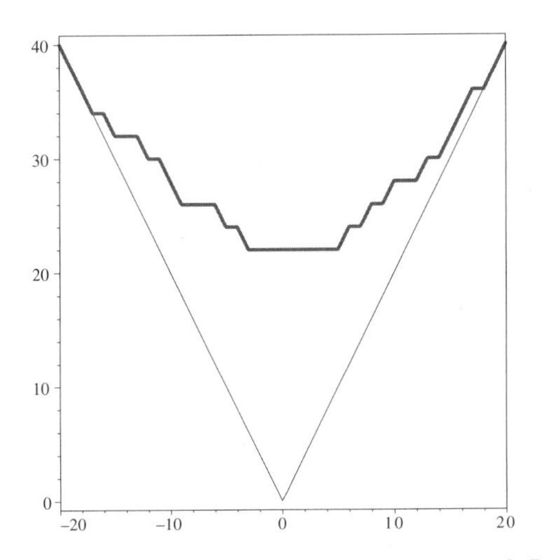

Fig. 4. The profile of the colored partition $\{(7, 4, 3, 3, 2, 1), (8, 7, 4, 4, 3, 1)\}$.

4.3 Vacuum expectation values and resolvents

The profile function is natural from the gauge theory point of view: the vacuum expectation values of the single trace operators $\mathrm{Tr}\,\phi^n$ have a simple expression in terms of the profiles

$$\langle \mathrm{Tr}\,\phi^n \rangle_{\mathbf{a}} = \frac{1}{Z(\mathbf{a}; \epsilon_1, \epsilon_2, \Lambda)} \sum_{\mathbf{k}} \Lambda^{2N|\mathbf{k}|} Z_{\mathbf{k}}(\mathbf{a}; \epsilon_1, \epsilon_2, \Lambda)\mathcal{O}_n[\mathbf{k}], \tag{4.18}$$

where

$$\mathcal{O}_n[\mathbf{k}] = \frac{1}{2} \int_{\mathbf{R}} dx x^n f''_{\mathbf{a};\mathbf{k}}(x|\epsilon_1, \epsilon_2). \tag{4.19}$$

The second derivative f'' of a partition profile is a compactly supported distribution on $\mathbf{R}$ which for random partitions plays a role similar to the role of the spectral measure for random matrices. In particular, it is convenient to introduce the resolvent $R(z|\epsilon_1, \epsilon_2)$ of a colored partition $\mathbf{k}$ by

$$R(z|\epsilon_1, \epsilon_2) = \frac{1}{2} \int_{\mathbf{R}} dx \frac{f''_{\mathbf{a};\mathbf{k}}(x|\epsilon_1, \epsilon_2)}{z - x}. \tag{4.20}$$

Note that

$$R(z|\epsilon_1, \epsilon_2) = \frac{N}{z} + \frac{a}{z^2} + O\left(\frac{1}{z^3}\right), \quad z \longrightarrow \infty. \tag{4.21}$$

In our context, the limit shapes f of partitions will be convex piecewise analytic functions for which the second derivative $f''(x)$ will be positive and compactly supported.

4.4 Real vs. complex

In the gauge theory the vacuum expectation values a_l of the Higgs field are, in general, complex. In this case the profile function $f_{\mathbf{a},\mathbf{k}}$ does not make sense. However, the resolvent (4.20) of a colored partition is well defined and can be effectively used in the analysis below. The formula (4.19) is replaced by

$$\mathcal{O}_n[\mathbf{k}] = \frac{1}{2\pi i} \oint z^n R(z|\epsilon_1, \epsilon_2) dz, \tag{4.22}$$

where the contour goes around $z = \infty$.

Complex values of a_l can be reached from the real values by analytic continuation. By the same token, it is enough to analyze the problem for values of a_l in any open set of $\mathbf{R}^N$. In particular, it is enough to analyze the problem in the asymptotic domain $\mathcal{U}_\infty$, which is what we will do next.

4.5 The thermodynamic limit

Our strategy in extracting the $\epsilon_1, \epsilon_2 \to 0$ limit of the sum (3.5) is the following. The typical size of the partition $\mathbf{k}$ contributing to the sum is of order $|\mathbf{k}| \sim \frac{1}{\epsilon_1 \epsilon_2}$, and is so large that the sum over the partitions can be approximated by an integral over the space of continuous Young diagrams $f(x)$. A continuous Young diagram, by definition, is a function satisfying the following weakening of condition (4.15):

$$
\begin{aligned}
f(x) &= |x|, \quad |x| \gg 0, \\
|f(x) - f(y)| &\leq |x - y|, \\
\int_{\mathbf{R}} dx f'(x) &= 0, \\
\int_{\mathbf{R}} dx(f(x) - |x|) &< \infty.
\end{aligned} \tag{4.23}
$$

Note that the condition $f'(x) = \pm 1$ is replaced by the weaker Lipschitz condition.

We will show that this path integral is dominated by a unique saddle point (in fact, a strict maximum). In more mathematical language, this means the following. By associating to every partition its profile, we get from (3.5) a measure on the space of Lipschitz functions. As $\epsilon_1, \epsilon_2 \to 0$, these measures concentrate around a single point, that is, converge to the delta measure at a single function. This function is the limit shape of our random partition. We will construct this limit shape explicitly and show that it has a very simple and direct relation to the Seiberg–Witten geometry.

We begin by observing that the Plancherel measure $\mu(\mathbf{k})$ can be written in terms of the profile $f_{\mathbf{k}}(x|\hbar)$ as follows:

$$
\mu(\mathbf{k}) = \exp\left(-\frac{1}{8}\int_{x \neq y} f''_{\mathbf{k}}(x|\hbar) f''_{\mathbf{k}}(y|\hbar)\gamma_{\hbar}(x - y; \Lambda) dx dy\right). \tag{4.24}
$$

More generally, we have

$$
\begin{aligned}
&Z_{\mathbf{k}}(\mathbf{a}; \epsilon_1, \epsilon_2, \Lambda) \\
&= \exp\left(-\frac{1}{4}\int dx dy f''_{\mathbf{a},\mathbf{k}}(x|\epsilon_1, \epsilon_2) f''_{\mathbf{a},\mathbf{k}}(y|\epsilon_1, \epsilon_2)\gamma_{\varepsilon_1,\varepsilon_2}(x - y, \Lambda)\right),
\end{aligned} \tag{4.25}
$$

as can be easily checked with the help of the main difference equation. Denoting the right-hand side of (4.25) by $Z_f(\epsilon_1, \epsilon_2, \Lambda)$, we have

$$
Z(\mathbf{a}; \epsilon_1, \epsilon_2, \Lambda) = \sum_{f \in \Gamma_{\mathbf{a}}^{\text{discrete}}} Z_f(\epsilon_1, \epsilon_2, \Lambda), \tag{4.26}
$$

where the summation is over the set $\Gamma_{\mathbf{a}}^{\text{discrete}}$ of paths of the form $f = f_{\mathbf{a},\mathbf{k}}$.

When $\epsilon_1, \epsilon_2 \to 0$ with ν fixed, the size of a typical partition $\mathbf{k}_l$ in (4.26) grows like $|\mathbf{k}_l| \sim \frac{1}{\epsilon_1\epsilon_2}$. In this limit, the sum (4.26) looks like an integral over the space $\Gamma_{\mathbf{a}}$ of paths f of the form

$$
f(x) = \sum_{l=1}^{N} f_l(x - a_l) \tag{4.27}
$$

with each f_l satisfying (4.23). For this integral, $\epsilon_1\epsilon_2$ plays the role of the Planck constant. Our strategy, therefore, is to find a saddle point (in fact, the minimum) of the action

$$
\mathcal{E}_{\Lambda}(f) = \frac{1}{4}\int_{y<x} dx dy f''(x) f''(y)(x - y)^2 \left(\log\left(\frac{x - y}{\Lambda}\right) - \frac{3}{2}\right) \tag{4.28}
$$

on the space $\Gamma_{\mathbf{a}}$ of paths (4.27). The action (4.28) is the leading term as $\epsilon_1, \epsilon_2 \to 0$ of the action in (4.25):

$$Z_f(\mathbf{a}; \epsilon_1, \epsilon_2, \Lambda) \sim \exp\left(\frac{1}{\epsilon_1\epsilon_2}\mathcal{E}_\Lambda(f)\right). \tag{4.29}$$

Integrating by parts, we rewrite (4.28) as

$$\mathcal{E}_\Lambda(f) = -\frac{1}{2}\fint_{x<y}(N + f'(x))(N - f'(y))\log\left(\frac{y-x}{\Lambda}\right)dxdy, \tag{4.30}$$

which, for $N = 1$, reproduces the result of Logan–Schepp–Kerov–Vershik [42, 43, 44, 45]. Thus we have

$$\mathcal{F}_0(\mathbf{a}, \Lambda) = -\operatorname{Crit}_{f\in\Gamma_{\mathbf{a}}}\mathcal{E}_\Lambda(f). \tag{4.31}$$

4.5.1 Profiles vs. eigenvalue densities

In [9] an expression for $Z_k(\mathbf{a}, \epsilon_1, \epsilon_2) = \sum_{\mathbf{k}, |\mathbf{k}|=k} Z_{\mathbf{k}}$ was given in terms of a certain contour integral over k eigenvalues ϕ_I, $I = 1, \ldots, k$. This expression follows straightforwardly from the ADHM construction of the moduli space of instantons, and, therefore, easily generalizes to the case of SO, Sp gauge groups. It is, therefore, quite remarkable that the formula (4.29) can be obtained directly from the contour integral expression, avoiding the actual evaluation of the integral (which is, of course, needed to get correctly the $\mathcal{F}_g$s with $g > 0$).

$$Z_k(\mathbf{a}, \epsilon_1, \epsilon_2) = \oint\prod_{I=1}^{k}\left[\frac{\epsilon_1+\epsilon_2}{2\pi\epsilon_1\epsilon_2}\frac{d\phi_I}{P(\phi_I)P(\phi_I+\epsilon_1+\epsilon_2)}\right]\prod_{I\neq J}\frac{\phi_{IJ}(\phi_{IJ}+\epsilon_1+\epsilon_2)}{(\phi_{IJ}+\epsilon_1)(\phi_{IJ}+\epsilon_2)}, \tag{4.32}$$

where

$$\begin{aligned} P(x) &= \prod_{l=1}^{N}(x - a_l), \\ \phi_{IJ} &= \phi_I - \phi_J. \end{aligned} \tag{4.33}$$

We now multiply Z_k by Λ^{2kN} and sum over $k = 0, 1, \ldots,$ to get Z^{inst}. Take the limit $\epsilon_1, \epsilon_2 \to 0$. The typical k which will contribute most to the sum will be of order $k \sim \frac{1}{\epsilon_1\epsilon_2}$. So we introduce the density of eigenvalues,

$$\rho(x) = \epsilon_1\epsilon_2\sum_{I=1}^{k}\delta(x - \phi_I), \tag{4.34}$$

which is normalized in a k-independent way that nevertheless guarantees its finiteness in the limit we are taking. Now, the difference from the ordinary 't Hooft-like limit of ordinary matrix integrals is the presence of an equal number of the ϕ_{IJ} terms in the numerator and denominator of the measure (4.32). This will change qualitatively the density dependence of the effective potential on the eigenvalues, and the resulting equilibrium distribution of the eigenvalues. In particular, in our limit the former superymmetric matrix integral (4.32) scales as $\exp kF$ as opposed to 't Hooft's $\exp k^2 F$.

Nevertheless, we have a sharp peak in the measure, which justifies the application of the saddle point method. Indeed, by expanding in ϵ_1, ϵ_2 we map the measure in (4.32) onto

$$\Lambda^{2kN} Z_k(\mathbf{a}, \epsilon_1, \epsilon_2) \sim \exp\left(\frac{1}{\epsilon_1 \epsilon_2} \mathbf{E}_\Lambda[\rho]\right), \tag{4.35}$$

where

$$\mathbf{E}_\Lambda[\rho] = -⨍_{x \neq y} dx dy \frac{\rho(x)\rho(y)}{(x-y)^2} - 2\int dx \rho(x) \log\left(\frac{P(x)}{\Lambda^N}\right). \tag{4.36}$$

Up to the perturbative piece

$$\frac{1}{2}\sum_{l,n}(a_l - a_n)^2 \log\left(\frac{a_l - a_n}{\Lambda}\right),$$

the energy (4.36) coincides with the action (4.28) if we identify

$$\boxed{f(x) - \sum_{l=1}^{N} |x - a_l| = \rho(x).} \tag{4.37}$$

The same method applies to other theories considered in [9, 46].

Thus we have learned yet another interpretation of the limiting profile of the colored partition. In the rest of the paper we shall work with limit profiles, as the equations on the extremum are identical to those following from (4.36).

4.5.2 Surface tension

Given a function $f(x)$ as in (4.27), how can we extract a_ls? This is easy to do in the region $\mathcal{U}_\infty$ where

$$a_l \ll a_{l+1},$$

in which case the supports of the functions $f_{a_l,\mathbf{k}_l}$ do not intersect. Introduce parameters $\xi_1, \ldots, \xi_N$ which will play the role of dual variables to the charges a_l. From (4.16), we have the following.

Proposition 1. *For $a_l \ll a_{l+1}$, we have*

$$\sum_l \xi_l a_l = -\frac{1}{2}⨍_{\mathbf{R}} \sigma(f'(x))dx,$$

where $\sigma(y)$ *is a concave, piecewise-linear function on* $[-N, N]$ *such that*

$$\sigma'(y) = \xi_l, \quad y \in [-N + 2(l-1), -N + 2l]$$

and

$$\sigma(-N) = -\sigma(N) = -\sum_l \xi_l.$$

An example of the graph of σ *is plotted in Figure* 5.

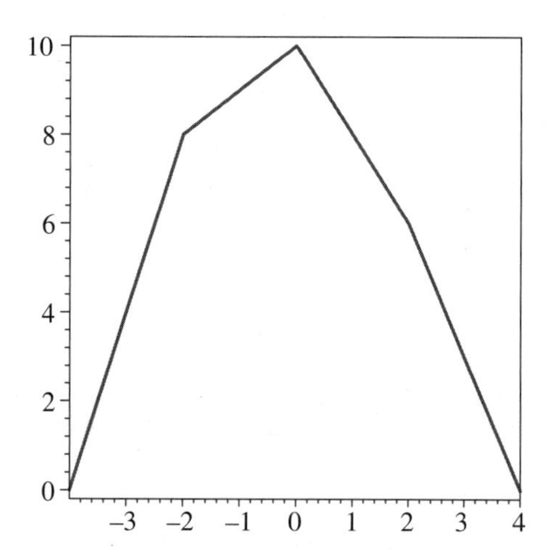

Fig. 5. Surface tension for $\xi = (4, 1, -2, -3)$.

In Section 5, we define a dual partition function $Z^D(\xi, \hbar, \Lambda)$. We will see that the dual partition function Z^D can be interpreted in terms of a periodically weighted (with period N) Plancherel measure on partitions. The periodic weights in this formalism are precisely e^{ξ_l} and the function σ becomes the corresponding surface tension function.

This story has many parallels and direct connections to the theory of periodically weighted planar dimers developed in [47].

It is a general principle that singularities of surface tension (usually referred to as "cusps") correspond to flat regions ("crystal facets") of the corresponding action minimizing shapes. In our case, the singularities of the function σ correspond to gaps between the supports of the functions $f_{\star,l}(x - a_l)$ for the minimizer $f_\star(x)$.

4.6 Equations for the limiting shape

In order to minimize $\mathcal{E}(f)$ with fixed a_ls, we introduce Lagrange multipliers ξ_l, which we order as

$$\xi_1 > \cdots > \xi_N, \tag{4.38}$$

and look for the maximizer of the total action

$$\mathcal{S}_\Lambda(f) = -\mathcal{E}_\Lambda(f) + \frac{1}{2}\int \sigma(f'(x))x \tag{4.39}$$

for fixed values of the ξ_ls. To get back the minimum of $\mathcal{E}(f)$ with fixed a_l we shall later perform the Legendre transform with respect to the ξ_ls. Note that the surface tension term $\frac{1}{2}\int \sigma(f')$ in (4.39) under the conditions $a = 0$ and (4.38) can be made arbitrarily large by making the individual profiles f_l sufficiently separated from each other. The price one pays for this is the decrease in the "energy" term $-\mathcal{E}_\Lambda(f)$, and thus there is a competition between the two terms in (4.39) which leads to the global maximum.

The action (4.39) is a concave functional. In fact, the first term in (4.39) is strictly concave (with proper boundary conditions), which can be seen by rewriting it as a certain Sobolev norm; see [44]. Therefore, any critical point of the action $\mathcal{S}(f)$ is automatically a global minimizer. It is, therefore, enough to look at the first variation of (4.39). Because of the singularities of σ, this first variation will involve one-sided derivatives.

Taking the first variation and integrating once by parts, we find the following equation:

$$\fint_{y\neq x} dy(y-x)\left(\log\left|\frac{y-x}{\Lambda}\right| - 1\right) f''(y) = \sigma'(f'(x)). \tag{4.40}$$

This equation should be satisfied for any point x for which $f'(x)$ is a point of continuity of σ'. When

$$f'(x) \in \{-N + 2l \mid l = 1, \dots, N-1\},$$

then considering the left and right derivatives separately, we obtain the inequalities

$$\mathbf{X}f(x) \in (\sigma'(f'(x) - 0), \sigma'(f'(x) + 0)), \tag{4.41}$$

where, by definition,

$$[\mathbf{X}f](x) = \fint_{y\neq x} dy(y-x)\left(\log\left|\frac{y-x}{\Lambda}\right| - 1\right) f''(y). \tag{4.42}$$

The transform $\mathbf{X}f$ is closely related to the standard Hilbert transform

$$[\mathbf{H}g](x) = \frac{1}{\pi}\fint_{y\neq x} dy \frac{g(y)}{y-x}. \tag{4.43}$$

Indeed,

$$[\mathbf{X}f]'' = \pi\mathbf{H}(f''). \tag{4.44}$$

With $\xi_0 = +\infty, \xi_{N+1} = -\infty$, conditions (4.40) and (4.41) can be recast in the following form.

Proposition 2. *A function $f_\star(x)$ is a critical point of $\mathcal{S}(f)$ iff*

$$\begin{aligned} \mathbf{X} f_\star(x) &= \xi_l && \text{whenever } -N+2l-2 < f_\star{}'(x) < -N+2l, \\ \xi_l > \mathbf{X} f_\star(x) &> \xi_{l+1} && \text{whenever } f_\star{}'(x) = -N+2l, \quad l = 0, \dots, N. \end{aligned} \tag{4.45}$$

In other words, the function

$$\varphi(x) = f_\star'(x) + \frac{1}{\pi i}[\mathbf{X} f_\star]'(x) \tag{4.46}$$

defines a map from $\mathbf{R}$ to the boundary of the domain Δ depicted in Figure 6. We now describe the solution to (4.45) in great detail.

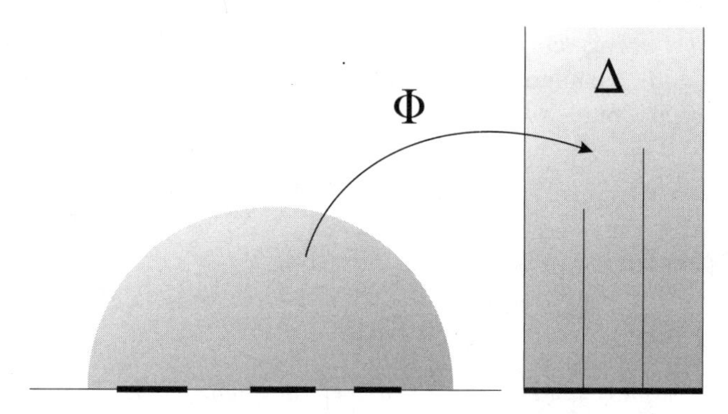

Fig. 6. Conformal map for $N = 3$.

4.7 Construction of the maximizer

Let $\Phi(z)$ be the conformal map from the upper half-plane to the domain Δ which is the half-strip

$$\Delta = \{\varpi \mid |\Re(\varpi)| < N, \Im(\varpi) > 0\} \tag{4.47}$$

with vertical slits along

$$\{\Re(\varpi) = -N + 2l, \Im(\varpi) \in [0, \eta_l]\}, \quad l = 1, \dots, N-1. \tag{4.48}$$

The positive reals η_l, $l = 1 \dots N-1$, are the parameters of the function Φ.

We normalize Φ by the condition that it maps infinity to infinity and

$$\Phi(z) = N + \frac{2N}{\pi i} \log \frac{\Lambda}{z} + O\left(\frac{1}{z}\right), \quad z \to \infty. \tag{4.49}$$

This fixes Φ up to an overall shift $x \mapsto x + \text{const}$. This ambiguity is related to the overall shift of the limit shape (and hence, to the overall charge of our colored partition) and is immaterial.

4.7.1 Construction of the conformal map

The map Φ can, of course, be found using the Schwarz–Christoffel formula, but it is easier to construct it as the following sequence of elementary conformal maps.

We take $P_N(z)$ to be a monic real polynomial

$$P_N(z) = z^N + \cdots$$

such that all roots $\alpha_l^{\pm}$ of the equation

$$P_N(z)^2 - 4\Lambda^{2N} = \prod_{l=1}^{N}(z - \alpha_l^+)(z - \alpha_l^-) \tag{4.50}$$

are real. Let w be the smaller root of the equation

$$\Lambda^N\left(w + \frac{1}{w}\right) = P_N(z). \tag{4.51}$$

The function

$$w \mapsto \Lambda^N\left(w + \frac{1}{w}\right)$$

is known as the Zhukowski function and it maps the open disk $|w| < 1$ to the exterior of the segment $[-2\Lambda^N, 2\Lambda^N]$. It also maps reals to reals, and therefore the smaller root of (4.51) maps the upper half-plane to the disk $|w| < 1$ with slits along the real axis. It remains to take the logarithm to obtain the map Φ. Concretely,

$$\Phi(z) = \frac{2}{\pi i}\log(w) + N, \tag{4.52}$$

where w is the smaller root of the equation (4.51) and we take the branch of the logarithm satisfying

$$\Im \log(w) \to 0, \quad z \to +\infty.$$

Since

$$w \sim \frac{\Lambda^N}{z^N}, \quad z \to \infty,$$

we get the normalization condition (4.49).

4.7.2 Example: $N = 2$

As an example, consider the case $N = 2$. A real quadratic polynomial

$$P_2(z) = z^2 + \cdots$$

maps the upper half-plane to the entire complex plane with a cut along the ray

$$\left[\min_{x\in\mathbf{R}} P_2(x), +\infty\right).$$

In particular, if $\min_{\mathbf{R}} P_2(x) < -2\Lambda^2$, that is, if both roots of

$$P_2(x) = -2\Lambda^2$$

are real, then the segment $[-2\Lambda^2, 2\Lambda^2]$ is contained in this cut.

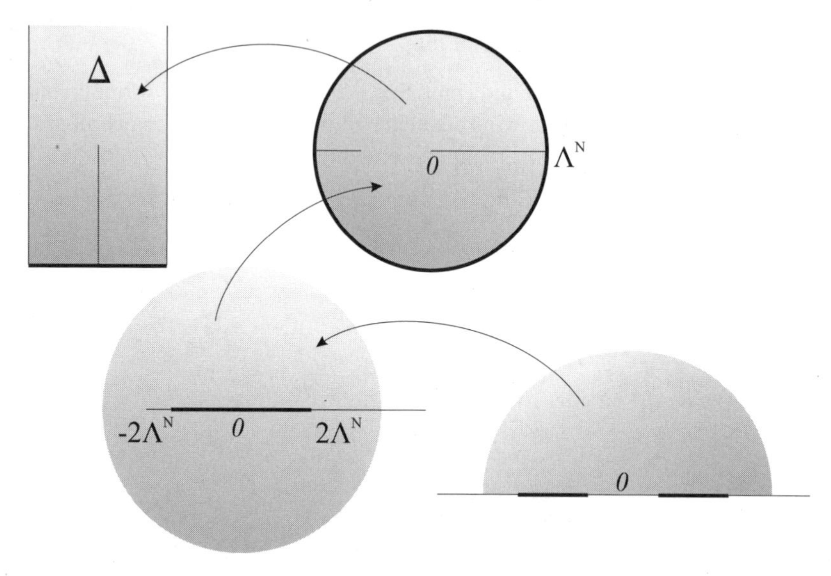

Fig. 7. A sequence of maps for $N = 2$.

4.7.3 Gaps and bands

For general N, the only difference from the $N = 2$ case is that the map P_N winds the upper half-plane $N/2$ times around the complex plane. Correspondingly, the root of (4.51) lives on a certain cover of the disk $|w| < 1$, which then gets unfolded by the logarithm.

Observe that the conformal map $\Phi(z)$ has a well-defined extension to the boundary $\mathbf{R}$ of the upper half-plane

$$\varphi(x) = \Phi(x + i0),$$

where the notation is consistent with (4.6).

We label the roots of (4.50) so that the union of the intervals $[\alpha_l^-, \alpha_l^+]$ is the preimage of the base of the half-strip Δ under the map φ, that is,

$$\varphi^{-1}([-N, N]) = \bigcup_{l=1}^{N} [\alpha_l^-, \alpha_l^+]. \tag{4.53}$$

These intervals are plotted in bold in the above figures. We will call these intervals the *bands* of Φ and the complementary intervals the *gaps* of Φ.

Proposition 3. *We claim that for a choice of ξ_ls in (4.38), we can find the corresponding values of η_l, $l = 1, \ldots, N$, such that the function $f_\star$ satisfying*

$$\boxed{f_\star'(x) = \Re\varphi(x),} \tag{4.54}$$

is the maximizer of the action $S(f)$. The bands and gaps of Φ will correspond to the curves and flat parts of the limit shape $f_\star$, respectively. They will also correspond to *the bands and gaps in the spectrum of the Lax operator for the periodic Toda chain, as will be explained in the next section.*

Let us apply the Schwarz reflection principle to any vertical part of the boundary of Δ and the corresponding piece of the real axis in the domain of Φ. Taking the resulting function modulo 2, we obtain an N-fold covering map

$$\Phi \bmod 2 : \mathbf{C} \setminus \bigcup_{l=1}^{N} [\alpha_l^-, \alpha_l^+] \to \{\Im \varpi > 0\} / \bmod 2 \tag{4.55}$$

from the complex plane minus the bands to the half-infinite cylinder, shown schematically in Figure 8. Note that this map has square-root branching precisely over the points $i\eta_l$, $l = 1, \dots, N-1$.

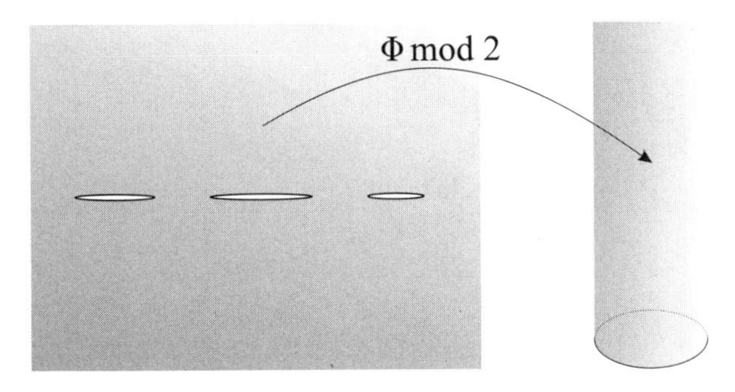

Fig. 8. Covering half-cylinder.

It follows that $\Phi'(z)$ extends to an analytic function in the complement of the bands. By the reflection principle, the value of $\Phi'(z)$ on the other side of the cut is precisely $-\Phi'(z)$. Also note that $\Phi'(z)$ is real (respectively, purely imaginary) on bands (respectively, gaps). We conclude that

$$\Phi'(z) = \frac{1}{\pi i} \int_{\mathbf{R}} \frac{\Re \varphi'(x)}{x - z} dx = -\frac{2}{\pi i} R_{f\star}(z), \tag{4.56}$$

where $R_{f_\star}(z)$ is ($\epsilon_1, \epsilon_2 \to 0$ limit of) the resolvent of the limit profile $f_\star$, defined by (4.54). We can, therefore, identify

$$l\text{th band} \equiv [\alpha_l^-, \alpha_l^+] = \operatorname{supp} f_l''(x - a_l), \quad l = 1, \dots, N. \tag{4.57}$$

4.7.4 Periods

It follows from (4.56) that

$$a_l = \frac{1}{2\pi i} \oint_{\mathbf{a}_l} z R_{f_\star}(z) dz = \oint_{\mathbf{a}_l} dS, \tag{4.58}$$

where we used (4.52), and, in agreement with (4.4), we have defined the Seiberg–Witten differential

$$\boxed{dS = \frac{1}{2\pi i} z \frac{dw}{w}.} \tag{4.59}$$

Taking the average of $\Phi'(z)$ on the two sides of the cut in (4.56), we get

$$\mathbf{H}\Re\varphi'(x) = -\Im\varphi'(x). \tag{4.60}$$

Using (4.44), we conclude from (4.60) that

$$[\mathbf{X} f_\star]'(x) = -\pi\Im\varphi(x) + \text{const.}$$

We claim that this constant is, in fact, zero. Indeed, we have

$$[\mathbf{X} f_\star]'(x) = -\int \log\left(\frac{|y-x|}{\Lambda}\right) f''(y)dy = 2N \log\left(\frac{\Lambda}{|x|}\right) + O\left(\frac{1}{x}\right), \quad x \to \infty. \tag{4.61}$$

Comparing this to (4.49), we conclude that

$$[\mathbf{X} f_\star]'(x) = -\pi\Im\varphi(x). \tag{4.62}$$

Since, clearly, $\Im\varphi(x) \geq 0$ and $\Im\varphi(x)$ vanishes on the bands, it follows from (4.62) that $\mathbf{X} f_\star$ is monotone decreasing and constant on the bands. It is also clear that the function (4.54) is monotone increasing and constant on the gaps, where it takes the values

$$-N + 2l, \quad l = 0, \ldots, N.$$

This verifies the conditions (4.45) with the ξ_l being the values of $\mathbf{X} f_\star$ on the bands.

To recover the ξ_l, we just integrate $[\mathbf{X} f_\star]'$ along the gap, where

$$i\Im\varphi(x) = \varphi(x) + \text{const.}$$

We find

$$\xi_{l+1} - \xi_l = -\pi \int_{\alpha_l^+}^{\alpha_{l+1}^-} \Im\varphi(x)dx = -\pi i \int_{\alpha_l^+}^{\alpha_{l+1}^+} x d\varphi(x),$$

integrating by parts and using the vanishing of $\Im\varphi(x)$ at the endpoints of a gap. In terms of the Seiberg–Witten differential and the cycles $\mathbf{b}_l$, the last relation reads

$$\xi_l - \xi_{l+1} = 2\pi i \oint_{\mathbf{b}_l - \mathbf{b}_{l+1}} dS. \tag{4.63}$$

This, together with the overall constraint $\sum \xi_l = 0$, fixes the values of the ξ_l.

4.7.5 Completeness

We now show that by a suitable choice of the slit-lengths η_l, we can achieve any value of the parameters (4.38). The period map

$$\mathbf{R}_{>0}^{N-1} \ni (\eta_1, \ldots, \eta_{N-1}) \mapsto (\xi_1 > \cdots > \xi_N), \tag{4.64}$$

is a continuous (in fact, analytic) map of open sets of $\mathbf{R}^{N-1}$. Because of the uniqueness of the maximizer $f_\star$, this map is one-to-one. Hence, if we can additionally show that it maps boundary to boundary, it will follow that this map is onto.

It is clear that if $\eta_l \to 0$ for some l, then $\xi_l - \xi_{l+1} \to 0$. Suppose that for some $n = 1, \ldots, N$,

$$\eta_n = \max_l \eta_l \to +\infty.$$

It is clear from the electrostatic interpretation of the conformal map Φ that the tip of the nth slit will not be screened by other slits, that is, for some constants $\delta_1, \delta_2 > 0$, we have

$$\int_{x \in [\alpha_l^+, \alpha_l^-], \Im\Phi(x) > (\eta_n - \delta_1)} dx > \delta_2,$$

whence

$$\xi_l - \xi_{l+1} = \int_{[\alpha_l^+, \alpha_l^-]} \Im\Phi(x)dx > \delta_2(\eta_n - \delta_1) \to \infty.$$

4.7.6 Periods and the prepotential

We now perform a check. We consider the partials

$$\frac{\partial \mathcal{S}(f_\star)}{\partial \xi_l} \tag{4.65}$$

and relate them to the A-periods a_l (4.58) of the differential dS.

We have

$$\frac{\partial}{\partial \xi_l}\mathcal{S}(f_\star) = \left[\frac{\partial}{\partial \xi_l}\mathcal{S}\right](f_\star) \tag{4.66}$$

because any infinitesimal change in $f_\star$ can only decrease the value of $\mathcal{S}(f_\star)$, which forces the variation of $\mathcal{S}$ due to the change in $f_\star$ to vanish (this is essentially the usual argument about the variation of the critical value of the action with respect to the parameters of the action). The rest is trivial:

$$\begin{aligned} \delta\mathcal{S}(f_\star) &= \frac{1}{2}\int [\delta\sigma](f_\star{}')dx = -\frac{1}{2}\int [\delta\sigma]'(f_\star{}')x f_\star{}''(x)dx \\ &= -\frac{1}{2}\sum_l \delta\xi_l \int_{\alpha_l^-}^{\alpha_l^+} x f_\star{}''(x)dx \\ &= -\frac{1}{4}\sum_l \delta\xi_l \oint_{\mathbf{a}_l} z d\Phi(z), \end{aligned} \tag{4.67}$$

which is exactly what we wanted, given (4.52), (4.59). Thus

$$\boxed{\begin{aligned} \mathcal{F}_0(\mathbf{a}, \Lambda) &= -\mathcal{E}_\Lambda(f_\star), \\ R_{f_\star}(z)dz &= d\log(w), \\ a_l &= \oint_{\mathbf{a}_l} dS, \\ \xi_l &= 2\pi i \oint_{\mathbf{b}_l} dS. \end{aligned}} \tag{4.68}$$

The integral in the last line is to be understood with the cutoff. Again, in the SU(N) theory, where only the differences $a_l - a_m$, and $\xi_l - \xi_m$ make sense, the cutoff never show up. It is the Cheshire cat smile of the noncommutative regularization.

4.8 Lax operator

The Seiberg–Witten curves arising from conformal maps Φ can be parameterized as the spectral curves in the periodic Toda chain corresponding to *real* initial conditions.

Consider the infinite Toda chain with particle coordinates q_i and momenta p_i. Make it periodic by imposing the constraints

$$q_{i+N} = q_i - N\log\Lambda \tag{4.69}$$

for all i. The Lax operator of this periodic Toda chain is a discrete Schrödinger operator of the form

$$L(w) = \begin{pmatrix} p_1 & e^{q_1-q_2} & & & w\Lambda^N e^{q_N-q_1} \\ e^{q_1-q_2} & p_2 & e^{q_2-q_3} & & \\ & e^{q_2-q_3} & \ddots & & \\ & & & \ddots & e^{q_{N-1}-q_N} \\ w^{-1}\Lambda^N e^{q_N-q_1} & & & e^{q_{N-1}-q_N} & p_N \end{pmatrix}, \tag{4.70}$$

where w is the Bloch–Floquet multiplier.

The integrals of motions are summarized by the spectral curve, which is the curve defined by the characteristic polynomial

$$\det(z - L(w)) = P(z) - \Lambda^N\left(w + \frac{1}{w}\right).$$

Here $P(z)$ is a monic polynomial of degree N. Observe that all roots of the polynomials

$$P(z) \pm 2\Lambda^N = \det(z - L(\mp 1)) \tag{4.71}$$

are real because $L(\pm 1)$ is a real symmetric matrix. This fact plays an important role in the dynamics of the periodic Toda lattice.

The bands and gaps of the map Φ are precisely the bands and gaps in the spectrum of the associated periodic discrete Schrödinger operator L on $\mathbf{Z}$:

$$[Lf](i) = e^{q_{i-1}-q_i} f(i-1) + p_i f(i) + e^{q_i - q_{i+1}} f(i+1), \quad i \in \mathbf{Z}.$$

Indeed, by (4.69), L commutes with the translation operator T,

$$[Tf](i) = f(i+N),$$

and (4.70) is the restriction of L to the N-dimensional w-eigenspace of T. In a band, the Bloch–Floquet multiplier w is a complex number of absolute value 1 and hence L has a bounded eigenfunction.

It can be shown that all curves of the form (4.51) with N real ovals arise in this way. This is similar to the result of [47] that the spectral curves of periodically weighted planar dimers parameterize Harnack (also known as maximal) plane curves. Incidentally, they are also M-curves, not just because they are used in the M-theory construction of the gauge theory [7] but also, and mostly,[1] because they have exactly real N ovals.

For example, taking the lattice at rest leads to Chebyshev polynomials. In this case all gaps shrink to points.

4.9 An SU(3) example

Here is an example of a limit shape $f_\star$. Take

$$P(z) = z^3 - 4z.$$

To visualize the curve $w + \frac{1}{w} = P(z)$, let's look at the plots of $\Re(w)$ and $\Im(w)$ for $z \in [-3, 3]$ plotted in Figure 9 in bold and normal, respectively. The three parts of this curve correspond to the three bands of the corresponding limit shape $f_\star$ plotted in Figure 10.

4.10 Higher Casimirs

As in [10] we could deform the theory by adding arbitrary higher Casimirs to the microscopic prepotential:

$$\mathcal{F}_{UV} = 2\pi i \left[\frac{\tau_0}{2} \operatorname{Tr} \Phi^2 + \sum_{\mathbf{n}} \tau_{\mathbf{n}} \prod_{J=1}^{\infty} \left(\frac{1}{J} \operatorname{Tr} \Phi^J \right)^{n_J} \right]. \tag{4.72}$$

The deformations by the single trace operators are especially simple, as they would lead to the modification of the action $\mathcal{S}(f)$ by the purely surface term

$$\mathcal{S}(f; \tau_{\mathbf{n}}) = -\mathcal{E}_\Lambda(f) + \frac{1}{2} \int dx \sigma(f') + \frac{1}{2} \int dx f''(x) \sum_{k=1}^{\infty} \tau_k \frac{x^{k+1}}{k+1}. \tag{4.73}$$

Note that τ_1 shifts $\log(\Lambda)$. Presumably the critical point of (4.73) would be a solution of some generalization of the Whitham equations [48, 40, 18, 49, 50, 51].

[1] We should apologize to M-theorists for the fact that M-manifolds were introduced a long time before 11d SUGRA was invented.

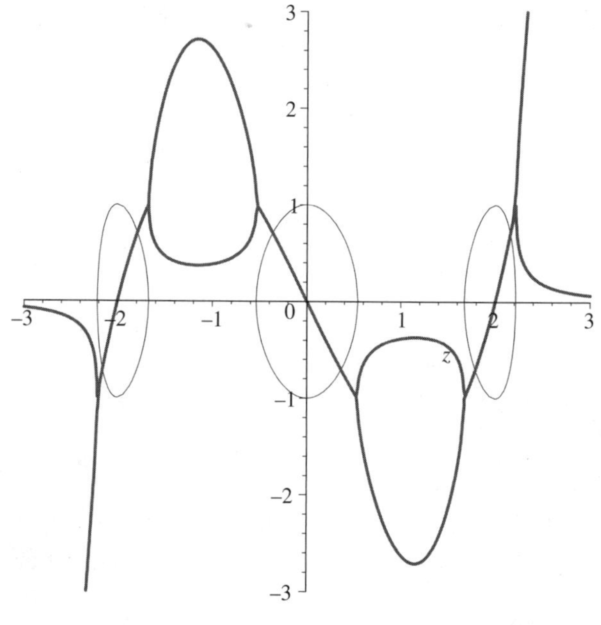

Fig. 9.

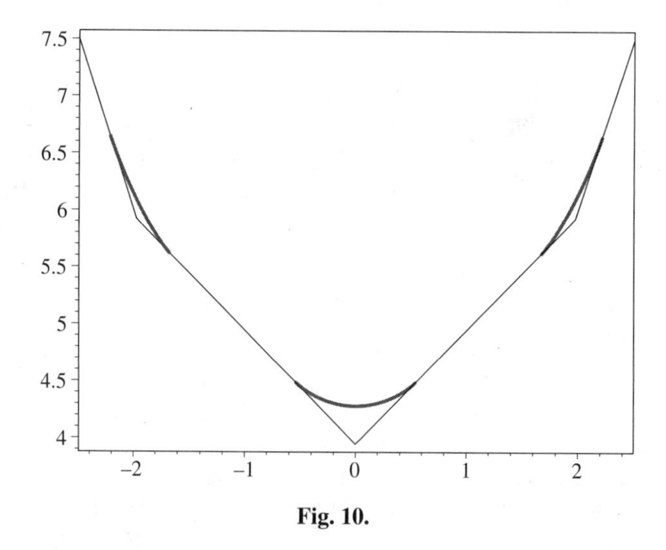

Fig. 10.

5 Dual partition function and chiral fermions

In this section we show that a certain transform of the partition function (3.2) has a natural fermionic representation. The physical origin of these fermions is still not completely clear. One way to understand them is to invoke the Chern–Simons/closed string duality of Gopakumar–Vafa [52], and then the fermionic representation of large

N topological gauge theories [53]. Another possible origin is through the M-theory five-brane realization of the gauge theory. Mathematically this story is very much related to the Heisenberg algebra representation in the cohomology of the moduli space of torsion-free sheaves on $\mathbf{CP}^2$, studied in [54].

5.1 Dual partition function

Let $\xi_1, \ldots, \xi_N$ be complex parameters,

$$\sum_l \xi_l = 0.$$

Consider

$$Z^D(\xi; p; \hbar, \Lambda) = \sum_{\substack{p_1,\ldots,p_N \in \mathbf{Z},\\ \sum_l p_l = p}} Z(\hbar(p_l + \rho_l); \hbar, \Lambda) \exp\left(\frac{i}{\hbar}\sum_l p_l \xi_l\right). \tag{5.1}$$

Clearly,

$$Z^D(\xi; p + N; \hbar, \Lambda) = Z^D(\xi; p; \hbar, \Lambda) \tag{5.2}$$

(shift all p_ls by 1). Thus there are essentially N partition functions one could consider. They are labeled by the level 1 integrable highest weights of $\widehat{\mathbf{sl}}_N$. Moreover,

$$Z^D(\xi; p + 1; \hbar, \Lambda) = Z^D(\xi^+; p; \hbar, \Lambda), \tag{5.3}$$

where

$$\xi^+ = (\xi_2, \xi_3, \ldots, \xi_N, \xi_1). \tag{5.4}$$

To extract from Z^D the partition function of interest we perform a contour integral. In the search for the prepotential we are actually interested in the extremely high frequency Fourier modes of Z^D, as we want $\mathcal{F}_0(\mathbf{a}; \Lambda)$ as a function of finite $a_l = \hbar(p_l + \rho_l)$, with $\hbar \to 0$. This means that the inverse Fourier transform can be evaluated using the saddle point, which we already analyzed.

Clearly,

$$Z^D(\xi; p; \hbar, \Lambda) = \exp \sum_{g=0}^{\infty} \hbar^{2g-2} \mathcal{F}_g^D(\xi; p, \Lambda), \tag{5.5}$$

where $\mathcal{F}_0^D$ is in fact p-independent, and is given by the Legendre transform of $\mathcal{F}_0$:

$$\mathcal{F}_0^D(\xi; \Lambda) = i \sum_l \xi_l a_l + \mathcal{F}_0(\mathbf{a}; \Lambda), \quad \xi_l = i \frac{\partial \mathcal{F}_0}{\partial a_l}; \tag{5.6}$$

i.e., ξ_l must be given by the $\mathbf{b}_l$ periods of the differential dS.

Remark. Note that in this section ξ_l differs by the factor i from the ξ_l of the previous section.

5.2 Free fermions

Introduce N free chiral fermions $\psi^{(l)}$:

$$\begin{aligned}\psi^{(l)}(z) &= \sum_{r\in\mathbf{Z}+\frac{1}{2}} \psi_r^{(l)} z^{-r}\left(\frac{dz}{z}\right)^{\frac{1}{2}}\\ \tilde{\psi}^{(l)}(z) &= \sum_{r\in\mathbf{Z}+\frac{1}{2}} \tilde{\psi}_r^{(l)} z^{-r}\left(\frac{dz}{z}\right)^{\frac{1}{2}}\\ \{\psi_r^{(l)}, \tilde{\psi}_s^{(m)}\} &= \delta_{lm}\delta_{r+s},\end{aligned} \tag{5.7}$$

which can also be packaged into a single chiral fermion Ψ

$$\begin{gathered}\Psi_r, \tilde{\Psi}_r, \quad r\in\mathbf{Z}+\frac{1}{2},\\ \{\Psi_r, \tilde{\Psi}_s\} = \delta_{r+s},\\ \Psi(z) = \sum_{r\in\mathbf{Z}+\frac{1}{2}} \Psi_r z^{-r}\left(\frac{dz}{z}\right)^{\frac{1}{2}}, \qquad \tilde{\Psi}(z) = \sum_{r\in\mathbf{Z}+\frac{1}{2}} \tilde{\Psi}_r z^{-r}\left(\frac{dz}{z}\right)^{\frac{1}{2}}\end{gathered} \tag{5.8}$$

in the standard fashion [55]:

$$\Psi_{N(r+\rho_l)} = \psi_r^{(l)}, \qquad \tilde{\Psi}_{N(r-\rho_l)} = \tilde{\psi}_r^{(l)}. \tag{5.9}$$

The operators Ψ_r, $\tilde{\Psi}_s$ act in the standard fermionic Fock space $\mathcal{H}$ (sometimes called an infinite wedge representation). It splits as a sum of Fock subspaces with fixed $U(1)$ charge, defined below:

$$\mathcal{H} = \oplus_{p\in\mathbf{Z}}\mathcal{H}[p].$$

Introduce affine $\widehat{U(N)}_1$ currents, which act within $\mathcal{H}[p]$ for any p:

$$J^{ln}(z) =: \psi^{(l)}\tilde{\psi}^{(n)} := \sum_{k\in\mathbf{Z}} \frac{dz}{z^{k+1}} \sum_{r\in\mathbf{Z}+\frac{1}{2}} : \psi_r^{(l)}\tilde{\psi}_{k-r}^{(n)} : . \tag{5.10}$$

Here we normal order with respect to the vacuum $|0\rangle$, which is annihilated by

$$\begin{aligned}\Psi_r|0\rangle &= 0, \quad r>0,\\ \tilde{\Psi}_s|0\rangle &= 0, \quad s>0,\end{aligned} \tag{5.11}$$

which is equivalent to

$$\begin{aligned}\psi_r^{(l)}|0\rangle &= 0, \quad r>0,\\ \tilde{\psi}_s^{(l)}|0\rangle &= 0, \quad s>0.\end{aligned} \tag{5.12}$$

The normal ordered product is simply

$$
\begin{aligned}
:\Psi_r \tilde{\Psi}_s : &= \begin{cases} \Psi_r \tilde{\Psi}_s, & s > 0, \\ -\tilde{\Psi}_s \Psi_r, & r > 0, \end{cases} \\
:\psi_r^{(l)} \tilde{\psi}_s^{(m)} : &= \begin{cases} \psi_r^{(l)} \tilde{\psi}_s^{(m)}, & s > 0, \\ -\tilde{\psi}_s^{(m)} \psi_r^{(l)}, & r > 0. \end{cases}
\end{aligned}
\tag{5.13}
$$

One can also introduce vacua with different overall $U(1)$ charges:

$$
\begin{aligned}
\Psi_r |p\rangle &= 0, \qquad r > p, \\
\tilde{\Psi}_s |p\rangle &= 0, \qquad s > -p, \\
|p\rangle &\in \mathcal{H}[p].
\end{aligned}
\tag{5.14}
$$

It is also useful to work with Ψ and the corresponding $\widehat{U(1)}_1$ currents,

$$
\mathcal{J}(z) =: \Psi \tilde{\Psi} : (z) = -\frac{1}{N} \sum_{l,m} J^{lm}(z^{-N}) z^{l-m}, \tag{5.15}
$$

and Virasoro generators,

$$
L_0 = \sum_{r \in \mathbf{Z} + \frac{1}{2}} r : \Psi_r \tilde{\Psi}_{-r}. \tag{5.16}
$$

Note the following:

$$
L_0 |p\rangle = \frac{p^2}{2} |p\rangle. \tag{5.17}
$$

5.2.1 Bosonization

It is sometimes convenient to work with the chiral boson

$$
\begin{aligned}
\phi(z) &= q - i \mathcal{J}_0 \log(z) + i \sum_{n \neq 0} \frac{1}{n} \mathcal{J}_n z^{-n}, \\
\partial \phi &\equiv z \partial_z \phi = -i \mathcal{J}, \\
[q, J_0] &= i, \\
\Psi(z) &=: e^{i\phi(z)} :, \qquad \tilde{\Psi}(z) =: e^{-i\phi(z)} : .
\end{aligned}
\tag{5.18}
$$

We will also use a truncated boson, with the zero mode q removed:

$$
\varphi(z) = -i \mathcal{J}_0 \log(z) + i \sum_{n \neq 0} \frac{1}{n} \mathcal{J}_n z^{-n}. \tag{5.19}
$$

The space $\mathcal{H}[p]$ is actually an irreducible representation of the Heisenberg algebra generated by $\mathcal{J}$:

$$
\mathcal{H}[p] = \mathrm{Span}_{n_1, \ldots, n_k > 0} \, \mathcal{J}_{-n_1} \cdots \mathcal{J}_{-n_k} |p\rangle, \quad \mathcal{J}_0 |_{\mathcal{H}[p]} = p. \tag{5.20}
$$

We also recall that

$$
\Psi : \mathcal{H}[p] \to \mathcal{H}[p+1], \qquad \tilde{\Psi} : \mathcal{H}[p] \to \mathcal{H}[p-1]. \tag{5.21}
$$

5.3 Dual partition function as a current correlator

We claim that

$$Z^D(\xi; p; \hbar, \Lambda) = \langle p| e^{\frac{1}{\hbar}\oint \mathrm{Tr}\, E_+(z)J(z)} e^{\oint \frac{1}{\hbar}\mathrm{Tr}\, H(z)J(z)} \Lambda^{2L_0} e^{-\frac{1}{\hbar}\oint \mathrm{Tr}\, E_-(z)J(z)} |p\rangle, \tag{5.22}$$

where the matrices $E_\pm$, H are given by

$$\begin{aligned} E_+(z) &= zE^{N,1} + \sum_{l=2}^{N} E^{l-1,l}, \\ H(z) &= \sum_l \xi_l E^{l,l}, \\ E_-(z) &= z^{-1}E^{1,N} + \sum_{l=2}^{N} E^{l,l-1}. \end{aligned} \tag{5.23}$$

5.4 Affine algebras and arbitrary gauge groups

Note that (5.22) can be written using the Chevalley generators e_i, f_i, h_i of the affine Lie algebra $\widehat{\mathbf{sl}}_N$:

$$\begin{aligned} Z^D(\xi; p; \hbar, \Lambda) &= (u_\hbar, \Lambda^{2h_0} e^{\mathbf{H}_\xi} u_\hbar)_{V_{\omega_p}}, \\ u_\hbar &= \exp\left(\frac{N}{\hbar}\sum_{l=1}^{N} f_{l-1}\right) v_0, \\ \mathbf{H}_\xi &= \frac{1}{\hbar}\sum_{l=1}^{N-1}(\xi_l - \xi_{l+1})h_l, \end{aligned} \tag{5.24}$$

where V_{ω_p} is the integrable highest weight module with highest weight vector v_0 (annihilated by the simple roots e_l) and highest weight ω_p, $p = 1, \ldots, N$.

The formula (5.24) has an obvious generalization to any simple Lie algebra $\widehat{\mathbf{g}}$. Conjecturally, it gives the partition function in the Ω-background in the $\mathcal{N} = 2$ gauge theory with gauge group S-dual to G (Langlands dual).

5.5 Dual partition function and $\mathbf{gl}(\infty)$

We now return to the A_{N-1} case.

In the language of the single fermion Ψ, the formula (5.22) reads

$$Z^D(\xi; p; \hbar, \Lambda) = \langle p| e^{\frac{\mathcal{J}_1}{\hbar}} e^{\mathbf{H}_\xi} \Lambda^{2L_0} e^{\frac{\mathcal{J}_{-1}}{\hbar}} |p\rangle, \tag{5.25}$$

where $\mathbf{H}_\xi$ is a diagonal matrix

$$\mathbf{H}_\xi = \frac{1}{\hbar}\sum_r \xi_{(r+\frac{1}{2}) \bmod N} : \Psi_r \tilde{\Psi}_{-r} : .$$

Clearly, $[\mathbf{H}_\xi, L_0] = 0$. The formula (5.25) expresses the dual partition function as an average with the Plancherel measure of the N-periodic weight $e^{\mathbf{H}_\xi}$.

Remark (on the Toda equation). The function Z^D obeys the Toda equation (cf. [56]):

$$4\partial^2_{\log(\Lambda)} \log(Z^D(p)) = \frac{Z^D(p+1)Z^D(p-1)}{(Z^D(p))^2}. \tag{5.26}$$

In fact, the formula (5.25) identifies Z^D as the tau-function of the Toda lattice hierarchy, thanks to the results of [57, 36, 58], with the specific parameterization of the times. We hope to return to this property of the partition function in a future publication.

5.6 The $U(1)$ case

In order to understand (5.25), (5.22), we first consider the case $N = 1$. As the Plancherel measure is a-independent in this case, we can study both the partition function and its dual with ease. Let us represent the states in the fermionic Fock space as semi-infinite functions of Ψ_r. The charge p vacuum $|p\rangle$ in this representation corresponds to the product

$$|p\rangle = \overrightarrow{\prod_{r>-p}} \Psi_r \equiv \Psi_{-p+\frac{1}{2}} \Psi_{-p+\frac{3}{2}} \cdots . \tag{5.27}$$

To every partition $\mathbf{k}$, there corresponds the so-called *charge p partition state*:

$$\begin{aligned} |p; \mathbf{k}\rangle &= \overrightarrow{\prod_{i=1,2,\ldots}} \Psi_{-p-\frac{1}{2}+i-k_i} = \overrightarrow{\prod_{1\le i\le n}} \Psi_{-p-\frac{1}{2}+i-k_i} \overleftarrow{\prod_{1\le i\le n}} \tilde{\Psi}_{p+\frac{1}{2}-i} |p\rangle \\ &= \overrightarrow{\prod_{1\le i\le n}} \Psi_{-p-\frac{1}{2}+i-k_i} |p-n\rangle. \end{aligned} \tag{5.28}$$

These states form a complete basis in the space $\mathcal{H}[p]$. The important fact, responsible for (5.25), is the variant of the boson–fermion correspondence:

$$\exp \frac{1}{\hbar} \mathcal{J}_{-1} |p\rangle = \sum_{\mathbf{k}} \frac{\mu(\mathbf{k})}{\hbar^{|\mathbf{k}|}} |p; \mathbf{k}\rangle, \tag{5.29}$$

where the factor $\mu(\mathbf{k})$ is the Plancherel measure (3.12). It follows that

$$Z(a, \hbar, \Lambda) = \Lambda^{\frac{a^2}{\hbar^2}} \langle 0| e^{-\frac{1}{\hbar}\mathcal{J}_1} \Lambda^{2L_0} e^{\frac{1}{\hbar}\mathcal{J}_{-1}} |0\rangle, \tag{5.30}$$

which for $a = \hbar p$ can also be written as

$$Z(a, \hbar, \Lambda) = \langle p| e^{-\frac{1}{\hbar}\mathcal{J}_1} \Lambda^{2L_0} e^{\frac{1}{\hbar}\mathcal{J}_{-1}} |p\rangle. \tag{5.31}$$

To make use of (5.29), we *blend* N partitions $\mathbf{k}_l, l = 1, \ldots, N$, into a single one, $\mathbf{K}$. Let $p_1, \ldots, p_N$ be some integers. Consider the following countable set of distinct integers:

$$\{N(p_l + k_{li} - i) + l - 1 | l = 1, \ldots, N,\ i \in \mathbf{N}\} = \{p + K_I - I | I \in \mathbf{N}\}, \tag{5.32}$$

where the sequence $K_1 \geq K_2 \geq \cdots$ is defined by (5.32) and the condition that it stabilizes to zero. Likewise, p is determined from (5.32) and depends only on the p_l:

$$p = \sum_l p_l. \tag{5.33}$$

Let

$$a_l = \hbar(p_l + \rho_l).$$

The size of the blended partition $\mathbf{K}$ is expressed in terms of the a_l (see Appendix A):

$$|\mathbf{K}| = \frac{1 - N^2}{24} - \frac{p^2}{2} + N \sum_l \frac{\tilde{a}_l^2}{2\hbar^2} + N \sum_{l,i} k_{li} \tag{5.34}$$

It is now straightforward to evaluate $\mu_{\mathbf{K}} := \mu(\mathbf{K})$,

$$\mu_{\mathbf{K}}^2 = Z_{\mathbf{k}}(a; \hbar), \tag{5.35}$$

and arrive at (5.25).

6 $\mathcal{N} = 2$ theory with adjoint hypermultiplet

In a sense the most interesting $\mathcal{N} = 2$ theory is the theory with a massive hypermultiplet in the adjoint representation. This theory is ultraviolet finite, and is thus characterized by the microscopic coupling $\tau_0 = \frac{\vartheta_0}{2\pi} + \frac{4\pi i}{g_0^2}$, and by the mass $\mathbf{m}$ of the hypermultiplet. It is convenient to use the nodal parameter $q = e^{2\pi i \tau_0}$ to count instantons.

In this case the prepotential of the low-energy effective theory is expected to have the following expansion:

$$\begin{aligned}\mathcal{F}_0(\mathbf{a}, \mathbf{m}, q) = \pi i \tau_0 \sum_l a_l^2 & \\ - \frac{1}{2} \sum_{l \neq n} [& (a_l - a_n)^2 \log(a_l - a_n) \\ & - (a_l - a_n + \mathbf{m})^2 \log(a_l - a_n + \mathbf{m})] \\ + \sum_{k=1}^{\infty} q^k f_k(\mathbf{a}, \mathbf{m}) &\end{aligned} \tag{6.1}$$

The previous calculations of the coefficients f_k for low values of k were done in [59]. One of the remarkable features of the prepotential of the low-energy effective theory of the theory with an adjoint hypermultiplet is its relation to the elliptic Calogero–Moser system [60]. Indeed, in [20] an ansatz for the family of curves encoding the prepotential was proposed, using a version of the Hitchin system [61]. This very construction of the elliptic Calogero–Moser system was found earlier [62]. One of the advantages of this realization is its simple extension to Lie groups other than $SU(N)$ (see, e.g., [63]). The coefficients f_k of the prepotential of the Calogero–Moser system were calculated for $k = 1, 2$ in, e.g., [41].

6.1 Partition function

The partition function of the theory with an adjoint hypermultiplet in the Ω-background is explicitly calculable (in [9] the countour integral representation was given; the poles of the integral are exactly the same as those of the contour integral for the pure theory) and the answer is ($\mu = \mathbf{m}/\hbar$):

$$\begin{aligned} Z(\mathbf{a}, \mathbf{m}; \hbar, q) &= \exp \sum_{g=0}^{\infty} \hbar^{2g-2} \mathcal{F}_g(\mathbf{a}, \mathbf{m}; q) \\ &= q^{\frac{1}{2\hbar^2} \sum_l a_l^2 - \frac{1}{24} N(\mu^2 - 1)} \sum_{\mathbf{k}} q^{|\mathbf{k}|} Z_{\mathbf{k}}(\mathbf{a}, \mathbf{m}; \hbar, q), \end{aligned} \tag{6.2}$$

where

$$\begin{aligned} Z_{\mathbf{k}}(\mathbf{a}, \mathbf{m}; \hbar, q) &= Z^{\mathrm{pert}}(\mathbf{a}, \mathbf{m}; \hbar, q) \mu_{\mathbf{k}}^2(\mathbf{a}, \hbar) \\ &\quad \times \prod_{(l,i) \neq (n,j)} \frac{(a_l - a_n + \mathbf{m} + \hbar(j-i))}{(a_l - a_n + \mathbf{m} + \hbar(k_{l,i} - k_{n,j} + j - i))}, \end{aligned} \tag{6.3}$$

$$Z^{\mathrm{pert}}(\mathbf{a}, \mathbf{m}; \hbar, q) = \exp \sum_{l,n} \gamma_\hbar(a_l - a_n; q) - \gamma_\hbar(a_l - a_n + \mathbf{m}; q).$$

6.2 Abelian theory

Let us consider the $N = 1$ case first. Let $\mu = \frac{\mathbf{m}}{\hbar}$. In the $N = 1$ case, we are to sum over all partitions $\mathbf{k}$:

$$Z(a, \mathbf{m}; \hbar, q) = q^{\frac{a^2}{2\hbar^2} + \frac{1}{24}(\mu^2 - 1)} e^{-\gamma_\hbar(\mathbf{m}; q)} \sum_{\mathbf{k}} q^{|\mathbf{k}|} \prod_{\square \in \mathbf{k}} \left(\frac{h(\square)^2 - \mu^2}{h(\square)^2} \right). \tag{6.4}$$

The fact that (6.2) reduces to (6.4) for $N = 1$ is a consequence of a simple identity between the Chern characters:

$$\sum_{i<j} e^{\hbar(k_i - k_j + j - i)} - e^{\hbar(j-i)} - e^{\mathbf{m} + \hbar(k_i - k_j + j - i)} + e^{\mathbf{m} + \hbar(j-i)} = \sum_{\square \in \mathbf{k}} e^{\mathbf{m} + \hbar h(\square)} - e^{\hbar h(\square)}. \tag{6.5}$$

Another expression for (6.4) will be useful immediately:

$$Z(a,\mathbf{m};\hbar,q) = q^{\frac{a^2}{2\hbar^2}} e^{-\gamma_\hbar(\mathbf{m};q)} \mathcal{Z}(\mu), \qquad \mathcal{Z}(\mu) = q^{\frac{1}{24}(\mu^2-1)} \sum_{\mathbf{k}} q^{|\mathbf{k}|} \mathcal{Z}_{\mathbf{k}}(\mu),$$

$$\mathcal{Z}_{\mathbf{k}}(\mu) = \prod_{i=1}^{\ell(\mathbf{k})} \frac{(\ell(\mathbf{k})+k_i-i+\mu)!(\ell(\mathbf{k})+k_i-i-\mu)!(\ell(\mathbf{k})-i)!^2}{(\ell(\mathbf{k})-i+\mu)!(\ell(\mathbf{k})-i-\mu)!(\ell(\mathbf{k})+k_i-i)!^2} \times \frac{\det_{1\le i,j\le \ell(\mathbf{k})} \| \frac{1}{k_i-k_j+j-i+\mu} \|}{\det_{1\le i,j\le \ell(\mathbf{k})} \| \frac{1}{j-i+\mu} \|}. \tag{6.6}$$

Proposition 4. *The analogue of the formula* (5.31) *for the theory with adjoint matter is*

$$\mathcal{Z}(\mu) = q^{\frac{\mu^2}{24}} \operatorname{Tr}_{\mathcal{H}_0} q^{L_0 - \frac{1}{24}} \mathcal{V}_\mu(1), \tag{6.7}$$

where

$$\mathcal{V}_\mu(z) =: e^{i\mu\varphi(z)} := \exp\left(-\mu \sum_{n>0} \mathcal{J}_{-n} \frac{z^n}{n}\right) z^{\mu \mathcal{J}_0} \exp\left(\mu \sum_{n>0} \mathcal{J}_n \frac{z^{-n}}{n}\right), \tag{6.8}$$

$\varphi(z)$ *is the bosonic field* (5.19), *and* $\mathcal{H}_\lambda$ *is the charge* λ *subspace of the fermionic Fock space.*

Proof. We represent the trace (6.7) as a sum over partition states: $|p;\mathbf{k}\rangle$, where $p = \frac{a}{\hbar}$ is fixed. We start by evaluating

$$\langle \mathbf{k} | \mathcal{V}_\mu(1) | \mathbf{k} \rangle = \frac{\det_{1\le i,j\le \ell(\mathbf{k})} \| G_{k_i-i+\ell(\mathbf{k}), k_j-j+\ell(\mathbf{k})} \|}{\det_{1\le i,j\le \ell(\mathbf{k})} \| G_{-i+\ell(\mathbf{k}), -j+\ell(\mathbf{k})} \|}, \tag{6.9}$$

where we employ Wick's theorem and where for $i, j \in \mathbf{Z}_+$,

$$\begin{aligned} G_{ij} &= (-)^{i-j} \frac{(i+\mu)!(j-\mu)!}{i!j!(i-j+\mu)} \frac{\sin \pi\mu}{\pi\mu} \\ &= \operatorname{Coeff}_{x^i y^j} \left(\frac{1-y}{1-x}\right)^\mu \frac{1}{1-xy} \\ &= \langle 0 | \tilde{\Psi}_{j+\frac{1}{2}} \mathcal{V}_\mu(1) \Psi_{-\frac{1}{2}-i} | 0 \rangle. \end{aligned} \tag{6.10}$$

The second line of (6.10) follows from

$$[\mu - y\partial_y + x\partial_x] \left\{ \left(\frac{1-y}{1-x}\right)^\mu \frac{1}{1-xy} \right\} = \mu(1-y)^{\mu-1}(1-x)^{-\mu-1},$$

while the last line is most easily derived from the current action on the fundamental fermions Ψ, $\tilde{\Psi}$,

$$\begin{aligned}
&\langle 0|\tilde{\Psi}(y^{-1})\mathcal{V}_{\mu}(1)\Psi(x)|0\rangle \\
&\quad = \exp\left(\mu\sum_{n>0}\frac{1}{n}(x^n - y^n)\right) \times \langle 0|\tilde{\Psi}(y^{-1})\Psi(x)|0\rangle \\
&\quad = \left(\frac{1-y}{1-x}\right)^{\mu}\frac{(dx)^{\frac{1}{2}}(dy)^{\frac{1}{2}}}{1-xy},
\end{aligned} \tag{6.11}$$

from which (6.10) follows by employing the expansion (5.8). □

Remark. Note that an immediate consequence of the equalities (6.4) and (6.7) is the identity

$$\begin{aligned}
\prod_{n>0}(1-q^n)^{\mu^2-1} &= \sum_{\mathbf{k}} q^{|\mathbf{k}|}\prod_{\square\in\mathbf{k}}\frac{h(\square)^2-\mu^2}{h(\square)^2}, \\
\mathcal{Z}(\mu) &= \eta(q)^{\mu^2-1}
\end{aligned} \tag{6.12}$$

(see Appendix C for notation), which interpolates between the Dyson–Macdonald formulas for the root systems A_n. Indeed, when $\mu = n$, the exponent on the left in (6.12) equals

$$\dim \mathrm{SU}(\mu) = \mu^2 - 1,$$

while on the right we have a sum over all partitions with no hooks of length μ, which one can easily identify with the sum over the weight lattice of $\mathrm{SU}(n)$ entering the Dyson–Macdonald formula.

6.2.1 Higher Casimirs and the Gromov–Witten theory of elliptic curves

Just as in [10] we can consider deforming the theory by the higher Casimirs. And as in [10] these are represented in the free fermion formalism by the $\mathcal{W}$-generators, and the trace (6.7) becomes the $W_{1+\infty}$ character in the presence of the "vertex" operator $\mathcal{V}_{\frac{\mathbf{m}}{\hbar}}$.

For $\mathbf{m} = 0$ this trace becomes *exactly* the $W_{1+\infty}$ character, as studied in [64, 65]. As shown in [36], this trace also coincides with the all-genus partition function of the Gromov–Witten theory of an elliptic curve (see also [66]). It would be nice to find a Gromov–Witten dual of the "vertex" operator $\mathcal{V}_{\mu}$, and establish the precise relation of (6.12) to [36, 65, 64].

Note that the "flow to the pure $\mathcal{N} = 2$ theory," which is represented by the limit $\mathbf{m} \to \infty$, $q \to 0$, such that $\Lambda^2 = \mathbf{m}^2 q$ stays finite, leads to the dual Gromov–Witten theory of $\mathbf{CP}^1$ with $\log(\Lambda^2)$ playing the role of the Kähler class.

6.3 Nonabelian case

Using (6.4) it is simple to express the partition function (6.2) in terms of free fermions:

$$Z(\mathbf{a};\hbar,\mathbf{m},q) = \mathrm{Tr}_{\mathcal{H}^{(N)}_{\frac{\mathbf{a}}{\hbar}}}(q^{L_0}\mathcal{V}_{\mu}(1)), \tag{6.13}$$

where

$$\mathcal{H}_{\mathbf{p}}^{(N)} = \otimes_{l=1}^{N} \mathcal{H}[\mathbf{p}_l]$$

stands for the Fock space of N free fermions, with specified charges $(\mathbf{p}_1, \ldots, \mathbf{p}_N)(U(1)^N$ (weight, in the language of affine Lie algebras).

We also have an expression for the dual partition function:

$$Z^D(\xi, p; \hbar, \mathbf{m}, q) = \mathrm{Tr}_{\mathcal{H}_p}(q^{L_0} \mathcal{V}_\mu(1) e^{\mathbf{H}_\xi}). \tag{6.14}$$

Of course, the modular properties of the partition functions (6.6), (6.12), (6.14) reflect the S-duality of the $\mathcal{N} = 4$ theory [11].

6.4 Path representation

As in the pure gauge theory case it is extremely useful to represent the partition function as a sum over the profiles—the paths $f(x)$. The result is

$$\begin{aligned} Z(\mathbf{a}, \mathbf{m}; \hbar, q) &= \sum_{f \in \Gamma_{\mathbf{a}}^{\text{discrete}}} Z_f(\mathbf{a}, \mathbf{m}; \hbar, q), \\ Z_f(\mathbf{a}, \mathbf{m}; \hbar, q) &= \exp\Bigg(-\frac{1}{4} \int dx dy f''(x) f''(y) \gamma_\hbar(x - y; q) \\ &\quad + \frac{1}{4} \int dx dy f''(x) f''(y) \gamma_\hbar(x - y + \mathbf{m}; q) \\ &\quad + \frac{1}{4\hbar^2} \log(q) \int dx x^2 f''(x) \Bigg). \end{aligned} \tag{6.15}$$

Interchanging x and y, one sees that this is symmetric with respect to $\mathbf{m} \mapsto -\mathbf{m}$.

6.5 Variational problem for the prepotential

Again, as in the case of the pure $\mathcal{N} = 2$ gauge theory we are looking for the extremum of a certain action functional on the space $\Gamma_{\mathbf{a}}$ of profiles $f(x)$ with fixed partial charges. By adding the surface tension term to the action we get a minimization problem on the space of all profiles, i.e., functions that satisfy

$$f(x) - N|x| = 0, \quad |x| \gg 0,$$

and the Lipschitz condition $|f'(x)| \leq N$. The Lipschitz condition in our case will be satisfied automatically, since all extrema will be convex. We also impose the total zero charge condition, $a = 0$, i.e.,

$$\int_{\mathbf{R}} x f''(x) dx = 0. \tag{6.16}$$

In this section we set $\mathbf{m} = im$.

6.5.1 Path energy and surface tension

Let

$$L(x) = \frac{1}{2} x \log \left(\frac{x^2}{x^2 + m} \right) - m \arctan \frac{x}{m}. \tag{6.17}$$

We have

$$\begin{aligned} L'(x) &= \frac{1}{2} \log \left(\frac{x^2}{x^2 + m^2} \right), \\ L''(x) &= \frac{1}{x} - \frac{1}{2} \left(\frac{1}{x + im} + \frac{1}{x - im} \right). \end{aligned} \tag{6.18}$$

We introduce a generalization of the energy functional (4.28):

$$\begin{aligned} \mathcal{E}_{q,m}(f) = &-\frac{1}{2} \fint_{x<y} (N + f'(x))(N - f'(y)) L'(y - x) dx dy \\ &- \frac{i\pi\tau}{2} \int_{\mathbf{R}} x^2 f''(x) dx, \end{aligned} \tag{6.19}$$

where the first term is the leading $\hbar \to 0$ asymptotics of (6.15) and the second term comes from the weight $q^{|\mathbf{k}|}$. In the limit $q \to 0$, $m \to \infty$, so that $\Lambda = mq^{\frac{1}{2N}}$ stays finite, $\mathcal{E}_{q,m} \to \mathcal{E}_\Lambda$. (This reflects the flow of the theory with adjoint hypermultiplet to the pure $\mathcal{N} = 2$ super-Yang–Mills theory.)

The functional we want to maximize is

$$\mathcal{S}_{q,m}(f) = -\mathcal{E}_{q,m}(f) + \frac{1}{2} \int \sigma(f'(x)) dx, \tag{6.20}$$

where the last term is the usual surface tension term. Again, the functional (6.20) is convex; therefore, every local maximizer is also a global maximizer.

6.5.2 Variational equations

Varying $f'(x)$ in (6.20), and integrating by parts, we obtain the equation

$$\fint_{y \neq x} f''(y) L(y - x) dy - \sigma'(f'(x)) + 2\pi i \tau x = \text{const}, \tag{6.21}$$

which should be satisfied at any point of continuity of $\sigma'(f'(x))$. The constant in (6.21) is the Lagrange multiplier corresponding to the constraint (6.16). At the points of discontinuity of $\sigma'(f'(x))$, the equality (6.21) should be replaced by the corresponding two-sided inequalities.

The function σ' is piecewise constant, therefore differentiating (6.21) with respect to x we get the equation

$$\int f''(y) L'(y - x) dy = 2\pi i \tau. \tag{6.22}$$

6.6 The spectral curve

The facets of the limit shape correspond to the singularities of the surface tension σ, therefore the maximizer f will have $N-1$ of them. This means that the support of f'' will be N intervals (bands),

$$\operatorname{supp} f'' = \bigcup_{l=1,\dots,N} \operatorname{supp} f_l'', \tag{6.23}$$

and, as in (4.57), we denote

$$\operatorname{supp} f_l'' = [\alpha_l^-, \alpha_l^+]. \tag{6.24}$$

Consider the resolvent $R(z) \equiv R_f(z; 0, 0)$ introduced in (4.20). It has jump discontinuities along the bands with the jump equal to $\pi i f''(x)$. Set, by definition (cf. [67, 68]),

$$G(z) = R\left(z - \frac{im}{2}\right) - R\left(z + \frac{im}{2}\right).$$

This function is analytic in the domain U which is the Riemann sphere with $2N$ symmetric cuts

$$U = \bar{\mathbf{C}} \setminus \bigcup_{l=1}^{N} \left[\alpha_l^- \pm \frac{im}{2}, \alpha_l^+ \pm \frac{im}{2}\right].$$

We have

$$G(z) = O(z^{-2}), \quad z \to \infty,$$

and hence we can consider the multivalued function

$$\int_\infty^z G(y)dy.$$

The period of $\int G(y)dy$ around any cut is equal to the integral of the jump of $G(z)$ along the cut. By construction, the jump of $G(z)$ is $\pm \pi i f''(x)$ and as $\int_{\alpha_l^-}^{\alpha_l^+} f''(y)dy = 2$, we have the well-defined function

$$\begin{aligned} F(z) &= \frac{1}{2\pi i} \int_\infty^z G(y)dy \bmod \mathbf{Z} \\ &= \frac{1}{4\pi i} \int_{\mathbf{R}} dy f''(y) \log\left(\frac{z - \frac{im}{2} - y}{z + \frac{im}{2} - y}\right) \bmod \mathbf{Z}. \end{aligned} \tag{6.25}$$

We will show that F is an N-fold branched cover of the standard cylinder

$$C_\tau = \{\varpi \mid |\Im(\varpi)| < -i\tau/2\} \bmod \mathbf{Z}$$

by the domain U. Clearly, F maps the infinity of U to the origin $\varpi = 0$ of C_τ.

The following proposition describes the behavior of the function $F(z)$ on both sides of the two cuts corresponding to any band $[\alpha_l^-, \alpha_l^+]$.

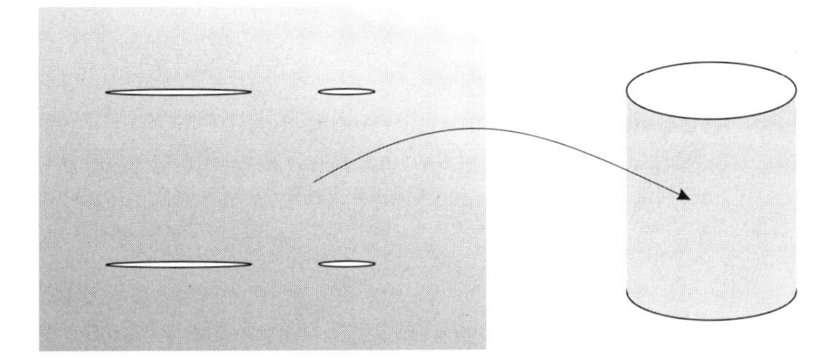

Fig. 11. Covering finite cylinder.

Proposition 5. *For any* $x \in [\alpha_l^-, \alpha_l^+]$, *we have*

$$F\left(x + \frac{im}{2} \pm i0\right) - F\left(x - \frac{im}{2} \mp i0\right) = \tau. \tag{6.26}$$

Proof. This is, in fact, a reformulation of (6.22). Indeed, let $x_\pm$ denote the two arguments of F in (6.26). We need to show that

$$\tau = \int_{x_-}^{x_+} G(y)dy = \frac{1}{4\pi i} \int_{x_-}^{x_+} dz \int_{\mathbf{R}} \left(\frac{1}{y - z + \frac{im}{2}} - \frac{1}{y - z - \frac{im}{2}} \right) f''(y)dy. \tag{6.27}$$

Since the integrand is nonsingular on the domain of integration, we can change the order of integration, which gives

$$\frac{1}{2\pi i} \int L'(y - x) f''(y)dy = \tau, \tag{6.28}$$

which is precisely (6.22). □

Since $f''(x)$ is real, we have

$$R(\bar{z}) = \overline{R(z)},$$

from which it follows that

$$G(\bar{z}) = -\overline{G(z)}, \qquad F(\bar{z}) = \overline{F(z)}. \tag{6.29}$$

Proposition (6.26) now implies that

$$\Im F\left(x \pm \frac{im}{2}\right) = \mp i\tau/2, \quad x \in [\alpha_l^-, \alpha_l^+].$$

which means that F maps the boundary ∂U to the boundary ∂C_τ. It follows that F maps U onto C_τ and, moreover, since F has degree N on the boundary, the map

$$F : U \to C_\tau$$

is an N-fold branched cover.

Proposition (6.26) also gives the analytic continuation of the function $F(x)$ across the cuts of U. The Riemann surface $\mathcal{C}$ of the function F is obtained by taking countably many copies of the domain U and identifying the point

$$x + \frac{im}{2} \pm i0, \quad x \in [\alpha_l^-, \alpha_l^+],$$

on one sheet with the point

$$x - \frac{im}{2} \mp i0$$

on the next sheet.

By construction, F extends to the map from $\mathcal{C}$ to the infinite cylinder

$$F : \mathcal{C} \to \mathbf{C} \bmod \mathbf{Z},$$

satisfying the property

$$F^{-1}(\varpi + \tau) = F^{-1}(\varpi) + im. \tag{6.30}$$

More geometrically, this means that the curve $\mathcal{C}$ is imbedded into the total space $\mathcal{A}$ of the affine bundle over the elliptic curve $E_\tau = \mathbf{C}/\mathbf{Z} \oplus \tau\mathbf{Z}$ which is obtained from C_τ by identifying its boundaries:

$$\mathcal{C} \subset \mathcal{A} = \{(z, \varpi) | z \in \mathbf{C}, \varpi \in \mathbf{C}\}/(z, \varpi) \sim (z, \varpi + 1) \sim (z + \mathbf{m}, \varpi + \tau). \tag{6.31}$$

The map F is the restriction to $\mathcal{C}$ of the projection $\mathcal{A} \to E_\tau$. The coordinate z is the coordinate on the fiber of $\mathcal{A}$.

This implies, in turn, that $\mathcal{C}$ is the spectral curve of the elliptic Calogero–Moser system [60]. Indeed, the latter is the spectral curve of the Lax operator (see Appendix C):

$$\begin{aligned} \mathcal{C}: \quad & \mathrm{Det}_{l,n}(L(\varpi) - z) = 0, \\ & L_{ln}(\varpi) = \delta_{ln}\left(p_n + \mathbf{m}\frac{1}{2\pi i}\log(\theta_{11}(\varpi))'\right) \\ & \qquad + \frac{\mathbf{m}}{2\pi i}(1 - \delta_{ln})\frac{\theta_{11}(\varpi + q_l - q_n)\theta_{11}'(0)}{\theta_{11}(\varpi)\theta_{11}(q_l - q_n)}, \end{aligned} \tag{6.32}$$

and is naturally viewed as the holomorphic curve in $\mathcal{A}$ representing the homology class of $N[E_\tau]$ under the identification $\mathcal{A} \approx \mathbf{C} \times E_\tau$: $(z, \varpi) = (p + \mathbf{m}\log(\theta_{11}(\varpi))', \varpi)$. This condition, together with the behavior near $\varpi = 0$, fixes uniquely the N-parameter family of curves $\mathcal{C}$. The coordinates $u = (u_1, \ldots, u_N)$ on the base of the family of curves can be defined using the characteristic polynomial of the Lax operator (6.32). An alternative set of (local) coordinates is given by the so-called action variables, which can be defined as a set of periods of the complexified Liouville one-differential $z d\varpi$ on $\mathcal{C}$. In our construction these periods are linear combinations of a_ls, ξ_ls, and m; see below.

Remark. For $N = 1$, the function $F^{-1}(\varpi)$ is, up to normalization, the classical Weierstraß function $\zeta(\varpi)$.

The limit shape f is reconstructed from the fact that $f'(x) \bmod 2\mathbf{Z}$ is equal to the jump of the function $F(x)$ across the cut of U.

6.6.1 Periods

We shall now study the periods of the differential

$$dS = z d\varpi|_{\mathcal{C}} = z dF(z) = \frac{1}{2\pi i} z \left(R\left(z - \frac{im}{2}\right) - R\left(z + \frac{im}{2}\right)\right) dz \tag{6.33}$$

on the Riemann surface $\mathcal{C}$.

For $l = 1, \ldots, N$ let $\mathbf{a}_l^{\pm}$ denote the 1-cycle, which circles around the cut $[\alpha_l^- \pm \frac{im}{2}, \alpha_l^+ \pm \frac{im}{2}]$. Clearly,

$$\oint_{\mathbf{a}_l^{\pm}} dS = a_l \pm \frac{im}{2}. \tag{6.34}$$

For $l = 1, \ldots, N-1$, let β_l be the cycle on $\mathcal{C}$ joining some points $x + \frac{im}{2}$ and $y + \frac{im}{2}$, where

$$x \in (\alpha_l^-, \alpha_l^+), \qquad y \in (\alpha_{l+1}^-, \alpha_{l+1}^+),$$

above both cuts on some sheet of $\mathcal{C}$ and below both cuts on the next sheet of $\mathcal{C}$.

By construction of $\mathcal{C}$, we have

$$\begin{aligned}\oint_{\beta_l} z dF(z) &= \int_{x+\frac{im}{2}+i0}^{y+\frac{im}{2}+i0} z dF(z) + \int_{y-\frac{im}{2}-i0}^{x-\frac{im}{2}-i0} z dF(z) \\ &= \int_{y-\frac{im}{2}-i0}^{y+\frac{im}{2}+i0} z dF(z) - \int_{x-\frac{im}{2}-i0}^{x+\frac{im}{2}+i0} z dF(z).\end{aligned}$$

By the same principle as in the proof of Proposition (6.26), we obtain

$$\begin{aligned}\int_{x-\frac{im}{2}-i0}^{x+\frac{im}{2}+i0} z dF(z) &= \frac{1}{4\pi i} \int_{\mathbf{R}} f''(t) dt \\ &\quad \times \left[\left(t + \frac{im}{2}\right) \log \frac{x - t + i0}{x - t - im} + \left(t - \frac{im}{2}\right) \log \left(\frac{x - t - i0}{x - t + im}\right)\right] \\ &= \frac{1}{2\pi i} \int_{\mathbf{R}} (x L'(t - x) + L(t - x)) f''(t) dt \\ &= \frac{1}{2\pi i} \xi_{l+1} + \text{const}\end{aligned} \tag{6.35}$$

thanks to (6.21), (6.28), and the fact that $\sigma'(f'(x)) = \xi_l$ because x lies on the lth band. We summarize:

$$\boxed{\begin{aligned} \xi_l - \xi_{l+1} &= 2\pi i \oint_{\beta_l} dS, \\ a_l \pm \frac{im}{2} &= \oint_{\mathbf{a}_l^{\pm}} dS, \\ \mathcal{F}(\mathbf{a}, \mathbf{m}, q) &= -\mathcal{E}_{q,\mathbf{m}}(f_\star). \end{aligned}} \tag{6.36}$$

7 On another matter

The results of the previous chapters can be generalized in various ways. One can incorporate hypermultiplets in arbitrary representations. One can add a tower of Kaluza–Klein states. One can study quiver gauge theories. We shall only sketch the results for fundamental matter, or for five-dimensional theories. The other cases will be treated in future publications.

7.1 Hypermultiplets in the fundamental representation

In this section $\mathbf{m}$ denotes the vector of masses:

$$\mathbf{m} = \mathrm{diag}(m_1, \ldots, m_k)$$

We will denote by $\mathbf{f} = 1, \ldots, k$ the flavor index.

The partition function of the theory with k matter hypermultiplets in the fundamental representation of $U(N)$ is given by the following sum over colored partitions [9]:

$$\begin{aligned} Z(\mathbf{a}, \mathbf{m}; \hbar, \Lambda) &= \sum_{\mathbf{k}} \Lambda^{(2N-k)|\mathbf{k}|} Z_{\mathbf{k}}(\mathbf{a}, \mathbf{m}; \hbar, \Lambda), \\ Z_{\mathbf{k}}(\mathbf{a}, \mathbf{m}; \hbar, \Lambda) &= Z^{\mathrm{pert}}(\mathbf{a}, \mathbf{m}; \hbar, \Lambda) \prod_{l,i,\mathbf{f}} \frac{\Gamma\left(\frac{m_{\mathbf{f}}+a_l}{\hbar} + k_{li} + 1 - i\right)}{\Gamma\left(\frac{m_{\mathbf{f}}+a_l}{\hbar} + 1 - i\right)} \mu_{\mathbf{k}}^2(\mathbf{a}, \hbar), \\ Z^{\mathrm{pert}}(\mathbf{a}, \mathbf{m}; \hbar, \Lambda) &= \exp\left(\sum_{l,n} \gamma_\hbar(a_l - a_n; \Lambda) + \sum_{l,\mathbf{f}} \gamma_\hbar(a_l + m_{\mathbf{f}}; \Lambda) \right). \end{aligned} \tag{7.1}$$

As before, the partition function (7.1) can be written as a sum over profiles $f = f_{\mathbf{a},\mathbf{k}}$ as follows:

$$\begin{aligned} Z(\mathbf{a}, \mathbf{m}; \hbar, \Lambda) &= \sum_{f} Z_f(\mathbf{a}, \mathbf{m}; \hbar, \Lambda), \\ Z_f(\mathbf{a}, \mathbf{m}; \hbar, \Lambda) &= \exp\left(-\frac{1}{4} \int dx dy f''(x) f''(y) \gamma_\hbar(x - y; \Lambda) \right. \\ &\quad \left. + \frac{1}{2} \sum_{\mathbf{f}} \int dx f''(x) \gamma_\hbar(x + m_{\mathbf{f}}; \Lambda) \right). \end{aligned} \tag{7.2}$$

Observe that the action in (7.2) is the old action (4.25) evaluated at a modified function $\tilde{f}$ such that

$$\tilde{f}'' = f'' - \sum_{\mathbf{f}} \delta(x + m_{\mathbf{f}}) \tag{7.3}$$

Therefore, the variational problem for the limit shape f is essentially the variational problem solved in Section 4, the only modification being that the maximizer $\tilde{f}_\star$ is required to have corners at the points $x = -m_{\mathbf{f}}$.

We will solve this problem under the assumption that the $m_{\mathbf{f}}$ are real and $|m_{\mathbf{f}}| \gg 0$. Without loss of generality, we can assume that the number k of flavors is exactly $2N-1$ since letting some $m_{\mathbf{f}}$ go to infinity reduces the number of labels.

Let $\tilde{\Delta}$ be the half-strip

$$\{|\Re(\varpi)| < N, \Im(\varpi) > 0\},$$

but now with two kinds of vertical slits. As before, we have $N-1$ slits going from the points

$$-N + 2l + i\eta_l, \quad l = 1, \dots, N-1,$$

down to the real axis. Additionally, we introduce $2N-1$ slits from the points

$$-N + l + i\tilde{\eta}_l, \quad l = 1, \dots, 2N-1,$$

going up to infinity. Let $\tilde{\Phi}$ be the conformal map from the upper half-plane to the domain $\tilde{\Delta}$ as in Figure 12.

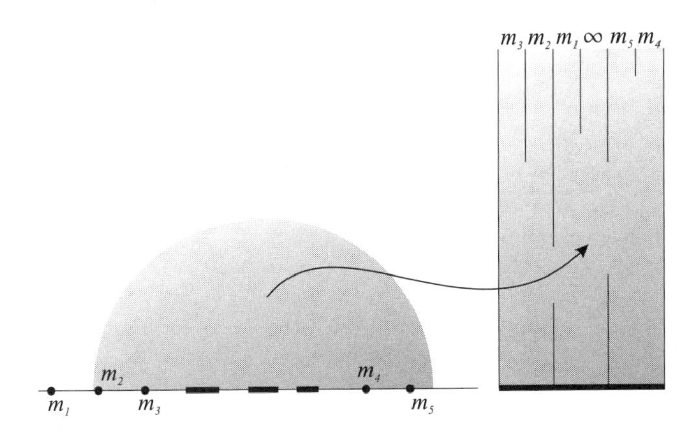

Fig. 12. The map $\tilde{\Phi}$ for $N = 3$.

The numbers η_l are, as before, the parameters of this map, as well as the number Λ. The numbers $\tilde{\eta}_l$ are uniquely fixed by the requirement that

$$\Phi^{-1}(\infty) = \{m_1, \dots, m_{2N-1}, \infty\}.$$

Let $P(z)$ be a monic degree N polynomial with all real roots. Suppose that Λ is sufficiently small and the $m_{\mathbf{f}}$ are large enough. If

$$m_1 < \cdots < m_r \ll 0 \ll m_{r+1} < \cdots < m_{2N-1},$$

then, as before, one checks that the map

$$\tilde{\Phi}(z) = \frac{2}{\pi i} \ln \frac{w}{\sqrt{\Lambda}} - N + r + 1, \tag{7.4}$$

where w is the smaller root of the equation

$$w + \frac{\Lambda}{w} = \frac{P(z)}{\sqrt{Q(z)}}, \quad Q(z) = \prod_{\mathbf{f}=1}^{2N-1} (z + m_{\mathbf{f}}), \tag{7.5}$$

is the required conformal map. Its construction is illustrated in Figure 13, which is a modification of Figure 7. The branch of the logarithm in (7.4) is chosen so that

$$\Im \log(w) \to 0, \quad z \to \infty.$$

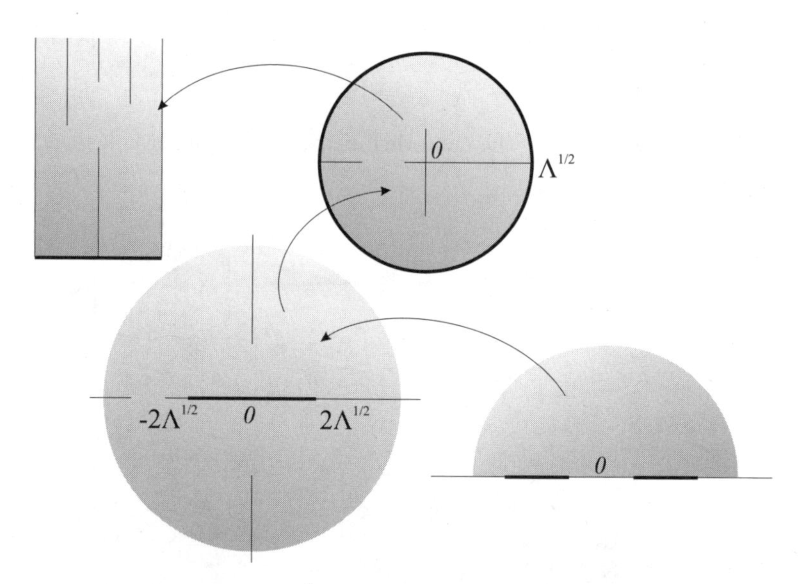

Fig. 13. The map $\tilde{\Phi}$ as a composition of maps for $N = 2$.

We now claim that the maximizer $\tilde{f}_\star$ is given by the same formula as before,

$$\tilde{f}_\star(x)' = \Re\tilde{\Phi}(x), \tag{7.6}$$

which simply means that the nontrivial part of the maximizer $f_\star$ is determined by the formula

$$f_\star(x)' = \Re\tilde{\Phi}(x), \quad |x| < \min\{|m_{\mathbf{f}}|\}, \tag{7.7}$$

whereas $f_\star(x) = N|x|$ outside of this interval.

Indeed, $\tilde{f}_\star(x)'$ clearly has the required jumps at the points $x = -m_{\mathfrak{f}}$. The equation

$$(\mathbf{X}\tilde{f}_\star)' = -\pi \Im \Phi$$

is checked in the exact same way as before. In particular, the asymptotics

$$\tilde{\Phi}(z) \sim \frac{1}{\pi i} \ln \frac{\Lambda}{z} + \text{real constant}, \quad z \to \infty, \tag{7.8}$$

agrees with the fact that from (4.61) we have

$$(\mathbf{X}\tilde{f}_\star)'(z) = \ln \Lambda - \ln |z| + O\left(\frac{1}{z}\right), \quad z \to \infty. \tag{7.9}$$

This verifies the predictions of [3, 4]. If $2N = k$, then the curve will be covering an elliptic curve and again can be explicitly constructed.

7.2 Five-dimensional theory on a circle

Consider five-dimensional pure supersymmetric gauge theory with eight supercharges, compactified on the circle $\mathbf{S}^1$ of circumference β. In addition, put the Ω-twist in the noncompact four dimensions. The resulting theory is a deformation of (2.1) by β-dependent terms.

The theory was analyzed in [14, 9] and the result of the calculation of the partition function is the following:

$$\begin{aligned} Z(\mathbf{a}; \hbar, \beta, \Lambda) &= \sum_{\mathbf{k}} (\beta\Lambda)^{2N|\mathbf{k}|} Z_{\mathbf{k}}(\mathbf{a}; \hbar, \beta) \\ Z_{\mathbf{k}}(\mathbf{a}; \hbar, \beta) &= Z^{\text{pert}}(\mathbf{a}; \hbar, \beta)\mu_{\mathbf{k}}^2(\mathbf{a}, \beta, \hbar), \\ \mu_{\mathbf{k}}^2(\mathbf{a}, \beta, \hbar) &= \prod_{(l,i)\neq(n,j)} \frac{\sinh \frac{\beta}{2}(a_l - a_n + \hbar(k_{l,i} - k_{n,j} + j - i))}{\sinh \frac{\beta}{2}(a_l - a_n + \hbar(j - i))}, \\ Z^{\text{pert}}(\mathbf{a}; \hbar, \beta) &= \Lambda^{\frac{1-N^2}{12}} \exp \sum_{l,n} \gamma_\hbar(a_l - a_n|\beta; \Lambda). \end{aligned} \tag{7.10}$$

Note that we included a power of β in Λ in (7.10) in order to have a simple four-dimensional limit $\beta \to 0$, Λ finite. This "renormalization" group relation was actually derived in [6, 14].

7.2.1 Path representation

The partition function (7.10) is easily written as a path sum:

$$Z(\mathbf{a}; \hbar, \beta, \Lambda) = \sum_{f \in \Gamma_{\mathbf{a}}} \exp\left(-\frac{1}{4} \fint_{x \neq y} f''(x) f''(y) \gamma_\hbar(x - y|\beta; \Lambda)\right). \tag{7.11}$$

7.2.2 Prepotential

From (7.11) we derive, as before, the equations on the critical path, which, in turn, determine the prepotential. Assuming $\beta > 0$, we have

$$\mathcal{F}_0(\mathbf{a}, \beta, \Lambda) = \frac{1}{4}\mathcal{E}_\beta(f_\star), \tag{7.12}$$

where $f_\star$ is the maximizer and the functional $\mathcal{E}_\beta$ is defined by

$$\mathcal{E}_\beta(f) = \fint_{x<y} dxdy(N + f'(x))(N - f'(y)) \log \frac{2}{\beta\Lambda} \sinh \frac{\beta|x-y|}{2}. \tag{7.13}$$

Again, we modify the functional (7.12) by a surface energy term to obtain the following total action functional:

$$\mathcal{S}(f) = \frac{1}{4}\mathcal{E}_\beta(f) + \frac{1}{2}\int_{\mathbf{R}} \sigma(f')dx. \tag{7.14}$$

We now introduce a five-dimensional analogue of the transform $\mathbf{X}f$ (see Appendix A for the definition of the function $\gamma_0(x;\beta)$):

$$[\mathbf{X}_\beta f](x) = \frac{1}{2}\fint_{y\neq x} dy \operatorname{sgn}(y-x)\gamma_0'(|y-x|;\beta)f''(y). \tag{7.15}$$

Note that the relation

$$\gamma_0(x;\beta)'' = \log \frac{2}{\beta\Lambda} \sinh \frac{\beta x}{2} \tag{7.16}$$

implies that

$$[\mathbf{X}_\beta f]'(x) = -\frac{1}{2}\fint_{y\neq x} dy \log\left(\frac{2}{\beta\Lambda} \sinh \frac{\beta|x-y|}{2}\right) f''(y). \tag{7.17}$$

The equations on the minimizer for (7.14) are formulated exactly as in Proposition 2, (4.45), with the replacement of $\mathbf{X}f$ by $\mathbf{X}_\beta f$.

The solutions to these equations can be constructed as follows. Let $\Phi(z)$ be the conformal map from the horizontal strip

$$0 < \Im z < \frac{\pi}{\beta} \tag{7.18}$$

to the slit half-strip $\tilde{\Delta}$ with $N-1$ slits of the form (4.48) and an additional vertical semi-infinite slit

$$\Re\varpi = 0, \quad \Im\varpi > \tilde{\eta},$$

as in Section 7.1 above. We normalize Φ by requiring that it map the two ends of the horizontal strip to the two ends of $\tilde{\Delta}$. This fixes it up to precomposing with an overall shift by a real constant.

The exponential map

$$z \mapsto X = e^{\beta z}$$

maps the strip (7.18) to the upper half-plane, mapping $\mathbf{R}$ to $\mathbf{R}_{>0}$. Therefore, we can use the results of the previous section to conclude that Φ has the form

$$\Phi(z) = N + \frac{2}{\pi i} \log w,$$

where w is the smaller root of the equation

$$X^{-\frac{N}{2}} P_N(X) = (\beta\Lambda)^N \left(w + \frac{1}{w}\right), \quad X = e^{\beta z}. \tag{7.19}$$

Here P_N is a monic polynomial of degree N and $\log w$ is normalized, as usual, by the requirement that $\Im \log w \to 0$ as $z \to +\infty$ along the real axis.

In the form (7.19), the parameter $\tilde{\eta}$ of the map Φ becomes a function of Λ and other parameters. The equation (7.19) defines the spectral curve of the relativistic Toda chain, as predicted in [14].

Applying the Schwarz reflection principle to any vertical part of the boundary of $\tilde{\Delta}$ and taking the resulting function modulo 1, we obtain a $2N$-fold map from the horizontal cylinder of circumference $2\pi/\beta$ with N cuts along the real axis to the unit half-cylinder, as shown in Figure 14. Here each cut covers 2-to-1 the finite boundary of the half-cylinder, while both ends of the cylinder cover N-to-1 each the infinite end of the half-cylinder.

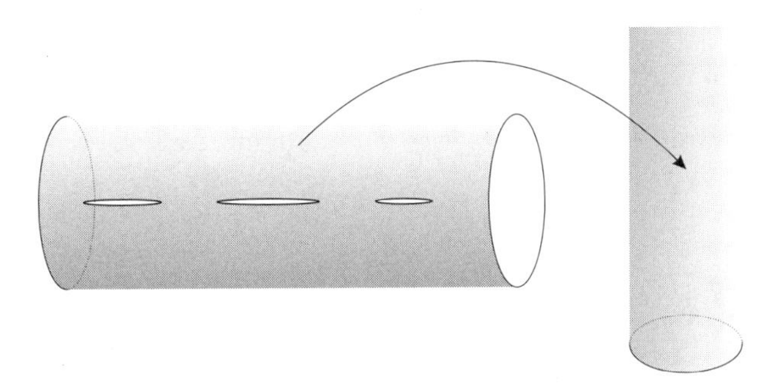

Fig. 14. The map Φ mod 1.

As before, it follows that $\Phi'(z)$ extends to a $2\pi i/\beta$-periodic function in the entire complex plane with N cuts along the real axis, periodically repeated. The values of $\Phi'(z)$ on the two sides of the cut are real and opposite in sign. We have

$$\Phi(x + i0) = \pm \left(i \frac{N\beta x}{\pi} + N\right) + \frac{2N}{\pi i} \log \beta\Lambda, \quad x \to \pm\infty, \tag{7.20}$$

whence

$$\Phi'(x + i0) = \mp \frac{N\beta}{\pi i}, \quad x \to \pm\infty. \tag{7.21}$$

It follows that the function $\Phi'(z)$ satisfies the integral equation

$$\Phi'(z) = \frac{\beta}{2\pi i} \int_{\mathbf{R}} \Re\Phi'(x) \coth\left(\beta \frac{(x-z)}{2}\right) dx. \tag{7.22}$$

Moreover, by comparing the asymptotics, we can integrate (7.22) to

$$\Phi(z) = N + \frac{i}{\pi} \int_{\mathbf{R}} \Re\Phi'(x) \log\left(\frac{2}{\beta\Lambda} \sinh \frac{\beta(z-x)}{2}\right) dx. \tag{7.23}$$

We claim that the maximizer $f_\star$ is given in terms of the map Φ by the usual formula

$$f_\star'(x) = \Re\Phi(x + i0).$$

Indeed, from (7.23) and (7.17), we conclude that

$$(\mathbf{X}_\beta f_\star)'(x) = -\frac{\pi}{2} \Im\Phi(x). \tag{7.24}$$

The difference from the previously considered cases lies in the fact that to reconstruct the values of ξ_ls (and a_ls) we have to integrate the multivalued differential

$$dS_5 = \frac{1}{\beta} \log(X) \frac{dw}{w}$$

on the curve (7.19), as indeed proposed in [14]. Also note that the domain where the map Φ is defined is different. This leads to the new form of the Seiberg–Witten differential.

7.2.3 Free-field representation

Again we start with $N = 1$ case. Let $Q = e^{\beta\hbar}$. In this case we are to sum over all partitions:

$$\begin{aligned} Z(a, \hbar, \beta, \Lambda) &= \sum_{\mathbf{k}} (-1)^{|\mathbf{k}|} (\beta\Lambda)^{2|\mathbf{k}|} \prod_{\square \in \mathbf{k}} \left(\frac{\beta}{2 \sinh\left(\frac{\beta\hbar h(\square)}{2}\right)} \right)^2 \\ &= \exp \sum_{n=1}^{\infty} \frac{(\beta\Lambda)^{2n}}{4n \sinh^2\left(\frac{\beta\hbar n}{2}\right)}. \end{aligned} \tag{7.25}$$

The free-field representation of the sum (7.25) is most simply done using the vertex operators

$$\Gamma_\pm(\beta, \hbar) = \exp \sum_{\pm n > 0} \frac{1}{n(1 - Q^n)} \mathcal{J}_n. \tag{7.26}$$

Then

$$Z(a, \hbar, \beta, \Lambda) = \langle 0 | \Gamma_+(\beta, \hbar) (\beta\Lambda)^{2L_0} \Gamma_-(\beta, \hbar) | 0 \rangle, \tag{7.27}$$

and for $N > 1$, we would get

$$Z^D(\xi; p; \beta, \hbar, \Lambda) = \langle p | \Gamma_+(\beta, \hbar) (\beta\Lambda)^{2L_0} e^{\mathbf{H}_\xi} \Gamma_-(\beta, \hbar) | p \rangle. \tag{7.28}$$

8 Conclusions and future directions

8.1 Future directions and related work

In this paper we extensively analyzed the partition functions of various $\mathcal{N} = 2$ supersymmetric gauge theories, subject to the Ω-background of $\mathcal{N} = 2$ supergravity. In all cases the partition function was identified with the statistical sum of an ensemble of random partitions with various Boltzmann weights. The grand canonical ensemble corresponds to a fixed theta angle in the gauge theory. In the thermodynamic limit the statistical sums are dominated by a saddle point, a master partition, from which one can easily extract various important characteristics of the low-energy effective gauge theory, such as the prepotential of the effective action, which is identified with the free energy per unit volume in the dual statistical model, provided we relate the volume to the parameters ϵ_1, ϵ_2 of the Ω-background, via

$$V = \frac{1}{\epsilon_1 \epsilon_2}.$$

The particle number is clearly the instanton charge. So the gauge theory partition function corresponds to the grand canonical ensemble, with $\log(\Lambda)$ being the chemical potential. The average particle number per unit volume, the density, is easy to determine:

$$\begin{aligned}\rho = \left\langle \frac{k}{V} \right\rangle &= \frac{\epsilon_1 \epsilon_2}{2N Z^{\text{inst}}(\mathbf{a}, \epsilon_1, \epsilon_2, \Lambda)} \frac{\partial}{\partial \log(\Lambda)} Z^{\text{inst}}(\mathbf{a}, \epsilon_1, \epsilon_2, \Lambda) \\ &\underset{V\to\infty}{\longrightarrow} \frac{1}{2N} \frac{\partial}{\partial \log(\Lambda)} \mathcal{F}_0^{\text{inst}}(\mathbf{a}, \Lambda) = u_2 - \sum_l \frac{a_l^2}{2}\end{aligned} \tag{8.1}$$

(recall (4.5)). In the region $\mathcal{U}_\infty$, the right-hand side of (8.1) is clearly very small; it is of order

$$\rho \sim \frac{\Lambda^{2N}}{a^{2(N-1)}} \sim M_W^2 e^{-\frac{8\pi^2}{g_{\text{eff}}^2(M_W)}},$$

where M_W is the typical W-boson mass, and $g_{\text{eff}}(M_W)$ is the effective coupling at this scale. The dependence of the instanton density on the coupling is the typical one in the dilute gas approximation (cf. [23, 69]). In the semiclassical region, there are very few instantons per unit volume.

Of course, a cautious reader may wonder about the dimensionality of the instanton density (which normally should be $(\text{mass})^4$). The point is that the noncommutative regularization which we used in the calculations introduces a scale $\sim \sqrt{\Theta}$, which is responsible for the dimensional transmutation we see in (8.1). Indeed, the physical volume, occupied by k instantons sitting on top of each other, is roughly $k\Theta^2$. Such instanton "foam" [70], or perhaps, liquid, may well be related to more complicated phenomenological pictures, say, instanton liquid [71]. Moreover, as we go deeper into the moduli space of vacua, the diluteness of the instanton gas also must cease to hold. However, the analytic properties of the partition function (3.2) are powerful enough to uniquely fix it by the instanton gas expansion in $\mathcal{U}_\infty$.

Note that one of the interpretations of our result is the existence of the master field in the $\mathcal{N} = 2$ gauge theories, which, however, works only for special (chiral) observables, unlike the master field of the large N gauge theories [72]. The latter, however, is much more elusive.

In our story, the master field is constructed as follows. First of all, the labeled partitions $\mathbf{k}$ we sum over are in one-to-one correspondence with the (noncommutative) gauge fields, describing certain arrangements of the nearly point-like instantons, sitting nearly on top of each other. The "master" partition with the profile $f_\star$ corresponds to the statistically most favorable configuration.

We extensively exploited in this paper the fact that the Ω-background acts as a box, similarly, in some respects, to the AdS space for the supergravity theory. One may wonder: Why couldn't we use the more traditional ways of regularizing infrared divergences of the gauge theory? In fact, the number of options is rather limited. One may study gauge theories on compact four-manifolds. To preserve supersymmetry one needs to turn on certain nonminimal couplings, which effectively twist the gauge theory. However, in these approaches one cannot learn directly the properties of the gauge theory in infinite volume, as all vacua are averaged over, and either most of them do not contribute to the correlation functions of the chiral operators (which is the case for simple-type manifolds), or (for manifolds with $b_2^+ \leq 1$) the contribution of the various vacua is related to the prepotential in a complicated fashion [73, 74]. It is out of such attempts to regulate the $\mathcal{N} = 2$ theory that one naturally arrives at the concept of the Ω-background.

The topic which we completely neglected in our paper is the application of our formalism to theories with $\mathcal{N} = 1$ supersymmetry. Recently, there has been a lot of excitement related to the exact calculations of the effective *super*potentials of the theories, obtained from $\mathcal{N} = 2$ theories by superpotential deformations [8, 75, 76]. It should be straightforward to apply our techniques in these setups as well. One encouraging feature of our formalism in the case of pure $\mathcal{N} = 2$ and softly broken $\mathcal{N} = 4$ theories is the striking similarity of the expression we have for the partition functions and the matrix models calculating the effective superpotentials of the $\mathcal{N} = 1$ gauge theories with an additional adjoint chiral multiplet, or with three adjoint chiral multiplets, with specific superpotential.

The similarity is the measure—in the first case, it is the regularized Vandermonde determinant of the infinite matrix $\hat{\Phi}$ with the eigenvalues $a_l + \hbar(k_{li} - i)$, and in the second case, it is the ratio of the determinants:

$$\frac{\mathrm{Det}'(\mathrm{ad}\,\hat{\Phi})}{\mathrm{Det}(\mathrm{ad}\,\hat{\Phi} + \mathbf{m})}.$$

The fact that the eigenvalues concentrate (for small $\hbar$) around a_ls also has a matrix model counterpart. Namely, to get the theory with $U(1)^N$ gauge symmetry, one adds to the $U(N)$ theory a tree-level superpotential with extrema near $a_1, \ldots, a_N$. In the limit of vanishing superpotential, one is left essentially with the Haar measure on matrices, together with the prescription to distribute eigenvalues near the extrema of the phantom of the superpotential [8].

The differences are also easy to see: on our case the matrix is infinite, but the eigenvalues are discrete and are summed over. In the Dijkgraaf–Vafa case, the matrix is finite, but the eigenvalues are continuous and integrated over. We have $\hbar \to 0$, they have $\hat{N} \to \infty$. The issue is tantalizing and is under investigation [77].

Oh, East is East, and West is West, and never the twain shall meet....

Another issue which we didn't touch much upon is the string theory dual of our calculation. There are several points one may want to stress.

First of all, the summation over partitions that we encountered is very similar to the summation over partitions one encounters in the two-dimensional $U(\hat{N})$ Yang–Mills theory, in the large $\hat{N}$ limit. In particular, when working on a two-sphere, one finds a master partition [78], and moreover, for the values of the 't Hooft coupling constant the profile of the master partition has facets, just like ours.

Of course, technically speaking, the two-dimensional YM theory corresponds to the $U(1)$ theory in our case, and the 't Hooft coupling there corresponds to the higher Casimir coupling τ_3 in our game. But the phenomena have similar origin. Now, the two-dimensional Yang–Mills theory has a dual closed string representation [79], and so does our four-dimensional partition function. And again, just like $1/\hat{N}$ played the role of the string coupling constant, $\hbar$ plays this role in the four-dimensional story.

In the $U(1)$ case all this is more or less well known [36, 10] by now. However, the full string dual of the $U(N)$ theory is yet to be discovered (see [10] for discussion; see also [80, 81]).

We should add that from the point of view of Gromov–Witten theorists the dual type A topological string theory, whatever it is, must be rather simple. Most of them should have one-dimensional target spaces. The rough correspondence between the gauge theory and the dual GW theory states that the $U(N)$ theory with g adjoint hypermultiplets (which is not asymptotically free/conformal for $g > 1$) is dual to the GW theory of N copies of the Riemann surface of genus g. The words "N copies" still do not quite have a formal meaning.

In the large $\hat{N}$ description of the two-dimensional Yang–Mills theory a prominent role was played by the formalism of free fermions [53]. Of course, the chiral fermions of [53] are ours $\Psi, \tilde{\Psi}$. (Somehow four-dimensional gauge theories do not see the antichiral sector.)

One of the exciting problems is to better understand the nature of these chiral fields in the string realizations of $\mathcal{N} = 2$ gauge theories. The conjecture of [9] was that they arise as the modes of the chiral two-form propagating on the worldvolume of the NS5/M5 brane, trapped by the Ω-background. In the bosonized form, these can also be mapped to the truncated version of the Kodaira–Spencer field, propagating along the Seiberg–Witten curve. We are planning to investigate these issues in the future.

Yet another exciting area of research revolving around our partition functions is their two-dimensional (anyonic) interpretation in the $\nu \neq 1$ case. The connections to the Jack polynomials, which are the eigenfunctions of the Sutherland many-body Hamiltonians with $\nu(\nu - 1)$ coupling constant, suggest strongly some relations to the physics of the quantum Hall effect, and also to the theory of analytic maps [82]. The latter also seem to be responsible for the higher Casimir deformation (4.73).

The Seiberg–Witten curve itself seems to arise as some quasiclassical object in the theory of some many-body ($\hat{N} \to \infty$) system. Indeed, the space z, w where it is embedded can be viewed as a one-particle phase space. The particles are basically free for $\nu = 1$, and have some phantom interaction for $\nu \neq 0, 1$, which only reveals itself in the generalized Pauli exclusion principle (Haldane exclusion principle). They fill some sort of Fermi sea, bounded by the Seiberg–Witten curve. Perhaps, the relation to random dimers [47] will also prove useful in understanding these issues. Another possibly useful feature of our energy functional is its seimple relation to the two-dimensional local theory of a free boson. Indeed, the bilocal part of (4.36) is nothing but the induced boundary action, with $\rho(x)$ being the boundary condition in the theory of the free boson, living on the upper half-plane. Of course, when the subleading in $\hbar$ corrections are taken into account the theory will cease to be free. It is also interesting to point out that our extremizing configuration $f_{\star}(x)$ corresponds to some sort of multiple D-brane boundary state, with N D-branes located at $x \sim a_l$, $l = 1, \ldots, N$.

It is of course very tempting to develop these pictures further, connect instanton gas/liquid/crystal to the (Luttinger?) liquid of the dual anyons, and learn more about the properties of five-branes from all this. Perhaps the results of [83, 84] will prove useful along this route.

Finally, the theory with adjoint matter should be closely related to conformal field theory on an elliptic curve. We have uncovered some of the relations in the partition function being interpreted as a trace in the representation of a current algebra. We should also note that the elliptic Calogero–Moser system, whose spectral curve, as we showed above, encodes the quasiclassical/thermodynamic limit of the partition function, allows a nonstationary generalization, related to the KZB equations, which, conjecturally, governs our full partition function [85, 86, 87, 88, 89, 90].

8.2 Summary of the results

In this paper we have advanced in the study of the vacuum structure of gauge theories. We considered $\mathcal{N} = 2$ supersymmetric gauge theories in four dimensions, and have subjected them to the so-called Ω-background. In this background one can calculate exactly the partition function in any instanton sector. The result has the form of a statistical sum of the ensemble of random partitions. The limit where the Ω-background approaches flat space is the most interesting for the applications, as there one is supposed to learn about the properties of the supersymmetric gauge theory in flat space-time. In terms of random partitions this is a thermodynamic, or quasiclassical limit. The flat space limit of the free energy coincides with the prepotential of the low-energy effective theory.

We have evaluated the sum over instantons by applying the saddle point method, and thus have succeeded (to our knowledge, for the first time in the literature) in producing the all-instanton direct calculation of the prepotential.

In all cases considered: pure gauge theory, theory with matter hypermultiplets, five-dimensional theory compactified on a circle—the saddle point corresponds to some *master partition*, which is the analogue of the *eigenvalue distribution* in the theory of random matrices. And in all cases one can encode the solution in some family of algebraic curves, endowed with a meromorphic (sometimes multivalued) differential, whose periods contain the information about the prepotential.

We have also found interesting representation of the full partition function (which, in addition to the prepotential, also contains certain higher gravitational couplings $\mathcal{F}_g$ of the gauge theory) as a partition function of the theory of chiral fermions/bosons on a sphere (for the pure gauge theory), on a torus (for the theory with adjoint matter), or some q-analogue thereof (for the five-dimensional theory).

We have also uncovered numerous puzzling relations between various seemingly unrelated topics which leave a lot of work for the future.

Disclaimer

Opinions presented in this paper do not necessarily reflect the authors' point of view. There are about $\sim 2\pi\imath$ misprints in this paper. We are, however, confident that most of them cancel each other.

A The function $\gamma_\hbar(x;\Lambda)$

A.0.1 Free case: $\epsilon_2 = -\epsilon_1 = \hbar$

The function $\gamma_\hbar(x;\Lambda)$ is characterized by the following properties:

1. Asymptotic expansion for $\hbar \to 0$:

$$\gamma_\hbar(x;\Lambda) = \sum_{g=0}^{\infty} \hbar^{2g-2}\gamma_g(x). \tag{A.1}$$

2. Finite-difference equation:

$$\gamma_\hbar(x+\hbar;\Lambda) + \gamma_\hbar(x-\hbar;\Lambda) - 2\gamma_\hbar(x;\Lambda) = \log\left(\frac{x}{\Lambda}\right). \tag{A.2}$$

The conditions (A.1)–(A.2) specify $\gamma_\hbar(x;\Lambda)$ uniquely up to a linear function in x. All the terms $\gamma_g(x)$, $g > 0$ are uniquely determined:

$$\begin{aligned}
\gamma_0(x) &= \frac{1}{2}x^2\log\left(\frac{x}{\Lambda}\right) - \frac{3}{4}x^2,\\
\gamma_1(x) &= -\frac{1}{12}\log\left(\frac{x}{\Lambda}\right),\\
\gamma_2(x) &= -\frac{1}{240}\frac{1}{x^2},\\
&\vdots\\
\gamma_g(x) &= \frac{B_{2g}}{2g(2g-2)}\frac{1}{x^{2g-2}}, \quad g > 1,
\end{aligned} \tag{A.3}$$

where the B_n are the usual Bernoulli numbers:

$$\frac{t}{e^t - 1} = \sum_{n=0}^{\infty} \frac{B_n}{n!} t^n.$$

The function $\gamma_\hbar(x; \Lambda)$ is closely related to the gamma function:

$$\gamma_\hbar\left(x + \frac{\hbar}{2}; \Lambda\right) - \gamma_\hbar\left(x - \frac{\hbar}{2}; \Lambda\right) = \log\left(\frac{1}{\sqrt{2\pi}} \hbar^{\frac{x}{\hbar}} \Gamma\left(\frac{1}{2} + \frac{x}{\hbar}\right)\right). \tag{A.4}$$

Another definition of the function $\gamma_\hbar(x; \Lambda)$ is through zeta-regularization:

$$\gamma_\hbar(x; \Lambda) = \frac{d}{ds}\Big|_{s=0} \frac{\Lambda^s}{\Gamma(s)} \int_0^\infty \frac{dt}{t} t^s \frac{e^{-tx}}{(e^{\hbar t} - 1)(e^{-\hbar t} - 1)}. \tag{A.5}$$

Remark. Note that

$$\gamma_\hbar(0; \Lambda) = -\frac{1}{12}.$$

The function $\gamma_\hbar(x; \Lambda)$ for $\Lambda = 1$ arises as the free energy of the $c = 1$ string with $\hbar$ being the string coupling and x the cosmological constant. It is also related to the free energy of the topological type A string on the conifold (see also below). The adepts of the applications of matrix models in SUSY gauge theories [8] praise yet another property of $\gamma_\hbar(x; \Lambda)$:

$$\log(\mathrm{Vol}\, U(N)) = \gamma_1(N; 1). \tag{A.6}$$

A.0.2 Anyon case: General ϵ_1, ϵ_2

$$\gamma_{\epsilon_1,\epsilon_2}(x; \Lambda) = \frac{d}{ds}\Big|_{s=0} \frac{\Lambda^s}{\Gamma(s)} \int_0^\infty \frac{dt}{t} t^s \frac{e^{-tx}}{(e^{\epsilon_1 t} - 1)(e^{\epsilon_2 t} - 1)}. \tag{A.7}$$

In the case $\epsilon_2 = -\epsilon_1 = \hbar$, this reduces to $\gamma_\hbar(x; \Lambda)$:

$$\gamma_\hbar(x; \Lambda) = \gamma_{-\hbar,\hbar}(x; \Lambda).$$

The *main difference equation*:

$$\gamma_{\epsilon_1,\epsilon_2}(x; \Lambda) + \gamma_{\epsilon_1,\epsilon_2}(x - \epsilon_1 - \epsilon_2; \Lambda) - \gamma_{\epsilon_1,\epsilon_2}(x - \epsilon_1; \Lambda) - \gamma_{\epsilon_1,\epsilon_2}(x - \epsilon_2; \Lambda) = \log\left(\frac{\Lambda}{x}\right). \tag{A.8}$$

Reflection:

$$\gamma_{-\epsilon_1,\epsilon_2}(x; \Lambda) = -\gamma_{\epsilon_1,\epsilon_2}(x - \epsilon_1; \Lambda) - \frac{2x + \epsilon_1}{2\epsilon_2} \log(\Lambda). \tag{A.9}$$

For $\nu \in \mathbf{Q}$, the function $\gamma_{\epsilon_1,\epsilon_2}(x; \Lambda)$ can be related to $\gamma_\hbar(x; \Lambda)$. Suppose

$$\epsilon_1 = -\frac{\hbar}{p}, \qquad \epsilon_2 = \frac{\hbar}{q}, \qquad p, q \in \mathbf{N}. \tag{A.10}$$

Then

$$\gamma_{\epsilon_1,\epsilon_2}(x; \Lambda) = \sum_{i=0}^{p-1}\sum_{j=0}^{q-1} \gamma_\hbar\left(x + \frac{\hbar}{pq}(pj - qi); \Lambda\right). \tag{A.11}$$

A.0.3 Trigonometric analogue of $\gamma_{\epsilon_1,\epsilon_2}(x; \Lambda)$

The natural generalization of (A.7) is the function

$$\begin{aligned}\gamma_{\epsilon_1,\epsilon_2}(x|\beta; \Lambda) = {}& \frac{1}{2\epsilon_1\epsilon_2}\left(-\frac{\beta}{6}\left(x + \frac{1}{2}(\epsilon_1 + \epsilon_2)\right)^3 + x^2\log(\beta\Lambda)\right) \\ & + \sum_{n=1}^{\infty} \frac{1}{n}\frac{e^{-\beta n x}}{(e^{\beta n\epsilon_1} - 1)(e^{\beta n \epsilon_2} - 1)},\end{aligned} \tag{A.12}$$

which obeys the *main Q-difference equation*:

$$\begin{aligned}&\gamma_{\epsilon_1,\epsilon_2}(x|\beta; \Lambda) + \gamma_{\epsilon_1,\epsilon_2}(x - \epsilon_1 - \epsilon_2|\beta; \Lambda) \\ &\qquad - \gamma_{\epsilon_1,\epsilon_2}(x - \epsilon_1|\beta; \Lambda) - \gamma_{\epsilon_1,\epsilon_2}(x - \epsilon_2|\beta; \Lambda) \\ &\quad = -\log\left(\frac{2}{\beta\Lambda}\sinh\frac{\beta x}{2}\right).\end{aligned} \tag{A.13}$$

We shall only use in this paper the special case $\epsilon_2 = -\epsilon_1 = \hbar$, where we get the function

$$\begin{aligned}\gamma_\hbar(x; \beta; \Lambda) &= \frac{\beta x^3}{12\hbar^2} - \frac{x^2}{2\hbar^2}\log(\beta\Lambda - \frac{\beta x}{24} + \sum_{n=1}^{\infty}\frac{1}{n}\frac{e^{-\beta n x}}{(e^{-\beta n\hbar} - 1)(e^{\beta n\hbar} - 1)} \\ &= \sum_{g=0}^{\infty} \gamma_g(x; \beta)\hbar^{2g-2}, \\ \gamma_0(x; \beta) &= -\frac{x^2}{2}\log\beta\Lambda + \beta\frac{x^3}{12} - \frac{1}{\beta^2}\mathrm{Li}_3(e^{-\beta x}), \\ \gamma_1(x; \beta) &= -\frac{1}{12}\log\left(2\sinh\frac{\beta x}{2}\right), \\ \gamma_g(x; \beta) &= \frac{B_{2g}\beta^{2g-2}}{2g(2g-2)}\mathrm{Li}_{3-2g}(e^{-\beta x}).\end{aligned} \tag{A.14}$$

Note that up to terms of instanton degree zero the function $\gamma_\hbar(x|\beta; \Lambda)$ coincides with the all-genus free energy of the type A topological string on the resolved conifold, with βx being the Kähler class of the $\mathbf{P}^1$, and $\beta\hbar$ the string coupling [91]. Another Gromov–Witten interpretation is via the local $\mathbf{F}_1$, with the Kähler class of the base $\mathbf{P}^1$ being $\log(\beta\Lambda)$ (considered to be big), and the fiber $\mathbf{P}^1$ with the Kähler class βx [6].

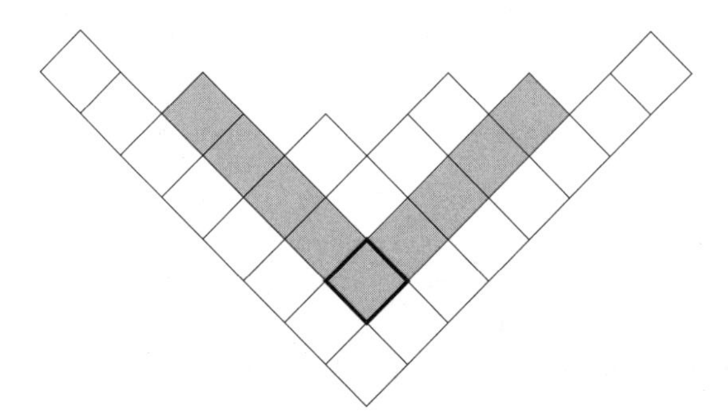

Fig. 15. Box, hook, ...

B Partitions, charges, colors

Partition: A nonincreasing sequence of nonnegative integers, stabilizing at some point at zero:

$$\mathbf{k} : k_1 \geq k_2 \geq \cdots \geq k_n > 0 = k_{n+1} = k_{n+2} = \cdots .$$

$n \equiv \ell(\mathbf{k})$ is called the *length* of $\mathbf{k}$, $|\mathbf{k}| = \sum_i k_i$, is called the *size* of the partition, and the k_i are called the *parts* of the partition. The parts of a partition are labeled by $i, j = 1, 2, \ldots$. For the partition $\mathbf{k}$,

$$\square = (i, j) \in \mathbf{k} \Longleftrightarrow 1 \leq i, \quad 1 \leq j \leq k_i .$$

Dual partition: $\tilde{\mathbf{k}}$,

$$(i, j) \in \tilde{\mathbf{k}} \Longleftrightarrow (j, i) \in \mathbf{k},$$

i.e., $\tilde{k}_i = \#\{j | i \leq k_j\}$.

Hook-length $h_{i,j} = h(\square)$ of the (i, j) box in the Young diagram of the partition $\mathbf{k}$:

$$h_{i,j} = \tilde{k}_j + k_i - i - j + 1.$$

Colored partition: $\mathbf{k}$, the N-tuple of partitions:

$$\mathbf{k} = (\mathbf{k}_1, \ldots, \mathbf{k}_N);$$

individual partitions are denoted as

$$\mathbf{k}_l = (k_{l,1} \geq k_{l,2} \geq \cdots \geq k_{l,n_l} > k_{l,n_l+1} = 0 = \cdots),$$
$$|\mathbf{k}| = \sum_{l,i} k_{li} .$$

Charged partition: $(p; \mathbf{k})$, the set of nonincreasing integers $\kappa_i = k_i + p$, where

$$\mathbf{k} = (k_1 \geq k_2 \geq \cdots)$$

is a partition and $p \in \mathbf{Z}$. The limit $\kappa_\infty \equiv p$ is called the *charge*.

Blending of colored partitions: Given a vector $\mathbf{p} = (p_1, \ldots, p_N)$, with

$$\sum_l p_l = 0$$

and an N-tuple of partitions $\mathbf{k}$, we define the *blended* partition K as follows:

$$\{K_i - i \mid i \in \mathbf{N}\} = \{N(k_{li} - i + p_l) + l - 1 \mid l = 1, \ldots, N,\ i \in \mathbf{N}\}. \tag{B.1}$$

B.0.4 Power-sums

To the charged partition $(p; \mathbf{k})$ it is useful to associate the following shifted-symmetric generating function, analytic in t, with a single pole at $t = 0$, defined for $\Re t > 0$ by the series

$$\mathbf{p}_{p;\mathbf{k}}(t) = \sum_{i=1}^{\infty} e^{t(p+k_i-i+\frac{1}{2})}. \tag{B.2}$$

The expansion of $\mathbf{p}_{p;\mathbf{k}}(t)$ near $t = 0$ contains information about the charge, the size of $\mathbf{k}$, etc.:

$$\mathbf{p}_{p;\mathbf{k}}(t) = \frac{1}{t} + p + t\left(\frac{p^2}{2} + |\mathbf{k}| - \frac{1}{24}\right) + \cdots. \tag{B.3}$$

Given N partitions $\mathbf{k}_l$, $l = 1, \ldots, N$, and the charges $p_l \in \mathbf{Z}$, we associate to them the generating function

$$\mathbf{p}_{\mathbf{p};\mathbf{k}}(t) = \sum_{l,i} e^{t\left(N(p_l+k_{l,i}-i)+l-\frac{1}{2}\right)} = \sum_l e^{Nt\rho_l} \mathbf{p}_{p_l,\mathbf{k}_l}(Nt), \tag{B.4}$$

which corresponds to the blended partition $\mathbf{K}$ of charge

$$p = \sum_l p_l = \sum_l \tilde{p}_l \tag{B.5}$$

and size

$$|\mathbf{K}| = \sum_I K_I = N \sum_l \left(\frac{1}{2}\tilde{p}_l^2 + |\mathbf{k}_l|\right) - \frac{1}{2}p^2 - \frac{N^2-1}{24}, \tag{B.6}$$

where

$$\tilde{p}_l = p_l + \rho_l.$$

The function $f_{p;\mathbf{k}}(x)$ is related to $\mathbf{p}_{p;\mathbf{k}}(t)$ by the integral transformation

$$f''_{p;\mathbf{k}}(x) = -\frac{1}{\pi i} \int_{\mathbf{R}} \mathrm{d}t e^{-itx} \sin \frac{t\hbar}{2} \mathbf{p}_{p;\mathbf{k}}(it\hbar). \tag{B.7}$$

The resolvent (4.20) is related to $\mathbf{p}_{p;\mathbf{k}}(t)$ by another integral transformation:

$$R(z|\epsilon_1, \epsilon_2) = -\frac{1}{\pi i} \int_0^{\infty} \mathrm{d}t e^{-tz} \sinh \frac{t\hbar}{2} \mathbf{p}_{p;\mathbf{k}}(t\hbar). \tag{B.8}$$

C Theta-function

For completeness, we list here the relevant formulae for the odd theta function, which we use in (6.32):

$$\begin{aligned}\theta_{11}(\varpi;\tau) &= \sum_{n\in\mathbf{Z}} e^{\pi i\tau(n+\frac{1}{2})^2+2\pi i(\varpi+\frac{1}{2})(n+\frac{1}{2})},\\ \theta_{11}(\varpi+1;\tau) &= -\theta_{11}(\varpi;\tau),\\ \theta_{11}(\varpi+\tau;\tau) &= -e^{-\pi i(2\varpi+\tau)}\theta_{11}(\varpi;\tau).\end{aligned} \tag{C.1}$$

From these formulae one easily concludes that the Lax operator $L(\varpi)$ (6.32) is the meromorphic Higgs operator in the rank N vector bundle over the elliptic curve E_τ, twisted by the one-dimensional affine bundle, which makes its spectrum to live in the affine bundle as well.

In (6.12), we use the Dedekind eta-function

$$\eta(q) = q^{\frac{1}{24}}\prod_{n=1}^{\infty}(1-q^n), \quad q = e^{2\pi i\tau}. \tag{C.2}$$

Acknowledgments. We thank A. Braverman, R. Kenyon, A. Losev, G. Moore, A. Morozov and S. Smirnov for useful discussions. This research was partly supported by NSF grant DMS-0100593 (AO), by the Packard Foundation (AO), by РФФИ grant 01-01-00549 (NN), by grant 00-15-96557 (NN) for scientific schools, and by the Clay Mathematical Institute (NN).

AO thanks Institut Henri Poincaré for hospitality. NN thanks Institute for Advanced Study, New High Energy Theory Center at Rutgers University, Kavli Institute for Theoretical Physics at University of California Santa Barbara, Forschungsinstitut für Mathematik, ETH Zurich, University of São Paulo, and ICTP, Trieste, for hospitality during the preparation of the manuscript.

References

[1] V. Novikov, M. Shifman, A. Vainshtein, and V. Zakharov, *Nuclear Phys.* B, **229** (1983), 381; *Nuclear Phys.* B, **260** (1985), 157–181; *Phys. Lett.* B, **217** (1989), 103–106.
[2] N. Seiberg, *Phys. Lett.* B, **206** (1988), 75–80.
[3] N. Seiberg and E. Witten, hep-th/9407087; hep-th/9408099.
[4] A. Klemm, W. Lerche, S. Theisen, and S. Yankielowicz, hep-th/9411048.
P. Argyres and A. Faraggi, hep-th/9411057.
A. Hanany and Y. Oz, hep-th/9505074.
[5] A. Klemm, W. Lerche, P. Mayr, C. Vafa, and N. Warner, hep-th/9604034.
[6] S. Katz, A. Klemm, and C. Vafa, hep-th/9609239.
[7] E. Witten, hep-th/9703166.
[8] R. Dijkgraaf and C. Vafa, hep-th/0206255; hep-th/0207106; hep-th/0208048.
R. Dijkgraaf, S. Gukov, V. Kazakov, and C. Vafa, hep-th/0210238.
R. Dijkgraaf, M. Grisaru, C. Lam, C. Vafa, and D. Zanon, hep-th/0211017.
M. Aganagic, M. Marino, A. Klemm, and C. Vafa, hep-th/0211098.
R. Dijkgraaf, A. Neitzke, and C. Vafa, hep-th/0211194.

[9] N. Nekrasov, hep-th/0206161.
[10] A. Losev, A. Marshakov, and N. Nekrasov, hep-th/0302191.
[11] C. Vafa and E. Witten, hep-th/9408074.
[12] J. A. Minahan, D. Nemeschansky, C. Vafa, N. P. Warner, hep-th/9802168.
T. Eguchi and K. Sakai, hep-th/0203025; hep-th/0211213.
[13] N. Seiberg and E. Witten, hep-th/9908142; *J. High Energy Phys.*, **9909** (1999), 032.
[14] N. Nekrasov, hep-th/9609219.
A. Lawrence and N. Nekrasov, hep-th/9706025.
[15] H. Ooguri and C. Vafa, hep-th/0302109; hep-th/0303063.
N. Seiberg, hep-th/0305248.
[16] M. Bershadsky, S. Cecotti, H. Ooguri, and C. Vafa, hep-th/9309140.
[17] R. Gopakumar and C.Vafa, hep-th/9809187; hep-th/9812127.
[18] A. Marshakov, *Seiberg-Witten Theory and Integrable Systems*, World Scientific, Singapore, 1999.
H. Braden and I. Krichever, eds., *Integrability: The Seiberg-Witten and Whitham Equations*, Gordon and Breach, New York, 2000.
[19] R. Donagi, alg-geom/9705010.
[20] R. Donagi and E. Witten, hep-th/9510101.
[21] H. Nakajima and K. Yoshioka, math.AG/0306198.
[22] A. Belavin, A. Polyakov, A. Schwarz, and Yu. Tyupkin, *Phys. Lett.* B, **59** (1975), 85–87.
[23] S. Coleman, The uses of instantons, in *Aspects of Symmetry: Selected Erice Lectures*, Cambridge University Press, Cambridge, UK, 1985, 265–350.
[24] E. Witten, *Comm. Math. Phys.*, **117** (1988), 353.
[25] E. Witten, hep-th/9403195.
[26] M. Atiyah and L. Jeffrey, *J. Geom. Phys.*, **7** (1990), 119–136.
[27] M. Atiyah and R. Bott, *Philos. Trans. Roy. Soc. London Ser. A*, **308** (1982), 524–615.
E. Witten, hep-th/9204083.
S. Cordes, G. Moore, and S. Rangoolam, hep-th/9411210.
[28] N. Ishibashi, H. Kawai, Y. Kitazawa, and A. Tsuchiya, *Nuclear Phys.* B, **498** (1997), 467; hep-th/9612115.
[29] N. Seiberg, hep-th/0008013.
[30] S. Minwala, M. van Raamsdonk, and N. Seiberg, hep-th/9912072.
[31] N. Nekrasov and A. S. Schwarz, hep-th/9802068; *Comm. Math. Phys.*, **198** (1998), 689.
[32] A. Okounkov, hep-th/9702001.
[33] K. Johansson, The longest increasing subsequence in a random permutation and a unitary random matrix model, *Math. Res. Lett.*, **5**-1–2 (1998), 63–82.
J. Baik, P. Deift, and K. Johansson, On the distribution of the length of the longest increasing subsequence of random permutations, *J. Amer. Math. Soc.*, **12**-4 (1999), 1119–1178.
A. Okounkov, Random matrices and random permutations, *Internat. Math. Res. Notices*, **20** (2000), 1043–1095.
A. Borodin, A. Okounkov, and G. Olshanski, Asymptotics of Plancherel measures for symmetric groups, *J. Amer. Math. Soc.*, **13**-3 (2000), 481–515.
K. Johansson, Discrete orthogonal polynomial ensembles and the Plancherel measure, *Ann. Math.* (2), **153**-1 (2001), 259–296.
[34] A. Okounkov, Infinite wedge and random partitions, *Selecta Math.* (*N.S.*), **7**-1 (2001), 57–81.
[35] P. van Moerbeke, Integrable lattices: random matrices and random permutations, in P. M. Bleher and A. R. Its, eds., *Random Matrix Models and Their Applications*, Mathematical Sciences Research Institute Publications 40, Cambridge University, Press, Cambridge, UK, 2001, 321–406.

[36] A. Okounkov and R. Pandharipande, math.AG/0207233; math.AG/0204305.

[37] S. V. Kerov, Interlacing measures, in *Kirillov's Seminar on Representation Theory*, American Mathematical Society Translations Series 2, American Mathematical Society, Providence, RI, 1998, 35–83.
S. V. Kerov, Anisotropic Young diagrams and symmetric Jack functions, *Функционал. Анал. и Приложен.*, **34**-1 (2000), 51–64, 96 (in Russian); *Functional Anal. Appl.*, **34**-1 (2000), 41–51.
A. M. Vershik, Hook formulae and related identities, *Записки сем. ЛОМИ*, **172** (1989), 3–20 (in Russian).
S. V. Kerov, Random Young tableaux, *Теор. вероят. и ее применения*, **3** (1986), 627–628 (in Russian).

[38] N. Dorey, T. J. Hollowood, V. Khoze, and M. Mattis, hep-th/0206063 and references therein.

[39] N. Dorey, V. V. Khoze, and M. P. Mattis, hep-th/9607066.

[40] A. Gorsky, I. Krichever, A. Marshakov, A. Mironov, and A. Morozov, *Phys. Lett.* B, **355** (1995), 466–474; hep-th/9505035.

[41] G. Chan and E. D'Hoker, hep-th/9906193.
E. D'Hoker, I. Krichever, and D. Phong, hep-th/9609041.

[42] B. F. Logan and L. A. Shepp, A variational problem for random Young tableaux, *Adv. Math.*, **26**-2 (1977), 206–222.

[43] S. V. Kerov and A. M. Vershik, Asymptotics of the Plancherel measure of the symmetric group and the limit shape of the Young diagrams, *ДАН СССР*, **233**-6 (1977), 1024–1027 (in Russian).

[44] S. V. Kerov and A. M. Vershik, Asymptotic behavior of the maximum and generic dimensions of irreducible representations of the symmetric group, *Функционал. Анал. и Приложен.*, **19**-1 (1985), 25–36 (in Russian).

[45] S. V. Kerov, Gaussian limit for the Plancherel measure of the symmetric group, *C. R. Acad. Sci. Paris Sér.* I *Math.*, **316**-4 (1993), 303–308.

[46] G. Moore, N. Nekrasov, and S. Shatashvili, hep-th/9712241; hep-th/9803265.

[47] R. Kenyon, A. Okounkov, and S. Sheffield, Dimers and amoebae, to appear.
R. Kenyon and A. Okounkov, in preparation.

[48] I. Krichever, hep-th/9205110; *Comm. Math. Phys.*, **143** (1992), 415.

[49] A. Gorsky, A. Marshakov, A. Mironov, and A. Morozov, *Nuclear Phys.* B, **B527** (1998), 690–716; hep-th/9802007.

[50] M. Mariño and G. Moore, hep-th/9802185.
J. Edelstein, M. Mariño, and J. Mas, hep-th/9805172.

[51] K. Takasaki, hep-th/9901120.

[52] R. Gopakumar and C. Vafa, hep-th/9802016; hep-th/9811131.

[53] M. Douglas, hep-th/9311130; hep-th/9303159.

[54] H. Nakajima, *Lectures on Hilbert Schemes of Points on Surfaces*, University Lecture Series, American Mathematical Society, Providence, RI, 1999.

[55] M. Jimbo and T. Miwa, Solitons and infinite dimensional Lie algebras, *Publ. RIMS Kyoto Univ.*, **19** (1983), 943–1001.

[56] K. Ueno andK. Takasaki, *Adv. Stud. Pure Math.*, **4** (1984), 1.

[57] A. Orlov, nlin.SI/0207030; nlin.SI/0305001.

[58] T. Takebe, Representation theoretical meaning of the initial value problem for the Toda lattice hierarchy I, *Lett. Math. Phys.*, **21**-1 (1991), 77–84.

[59] T. Hollowood, hep-th/0201075; hep-th/0202197.

[60] F. Calogero, *J. Math Phys.*, **12** (1971), 419.
J. Moser, *Adv. Math.*, **16** (1975), 197–220.
M. Olshanetsky and A. Perelomov, *Invent. Math.*, **31** (1976), 93; *Phys. Rev.*, **71** (1981), 313.
I. Krichever, *Funk. An. Appl.*, **14** (1980), 45.
[61] N. Hitchin, *Duke Math. J.*, **54**-1 (1987).
[62] N. Nekrasov and A. Gorsky, hep-th/9401021.
N. Nekrasov, hep-th/9503157.
[63] J. Hurtubise and E. Markman, math.AG/9912161.
[64] H. Awata, M. Fukuma, S. Odake, and Y.-H. Quano, hep-th/9312208.
H. Awata, M. Fukuma, Y. Matsuo, and S. Odake, hep-th/9408158.
R. Dijkgraaf, hep-th/9609022.
[65] S. Bloch and A. Okounkov, alg-geom/9712009.
[66] R. Dijkgraaf, Mirror symmetry and elliptic curves, in R. H. Dijkgraaf, C. Faber, and G. B. M. van der Geer, eds., *The Moduli Space of Curves*, Progress in Mathematics, Birkhäuser Boston, Cambridge, MA, 1994, 149–163.
[67] J. Hoppe, Quantum theory of a massless relativistic surface and a two-dimensional bound state problem, *Elementary Particle Res. J. (Kyoto)*, **80** (1989).
[68] V. Kazakov, I. Kostov, and N. Nekrasov, D-particles, matrix integrals and KP hierarchy, *Nuclear Phys.* B, **557** (1999), 413–442; hep-th/9810035.
[69] A. M. Polyakov, *Gauge Fields and Strings*, Contemporary Concepts in Physics 3, Harwood Academic Publishers, Chur, Switzerland, 1987.
[70] H. Braden and N. Nekrasov, hep-th/9912019.
[71] E. Shuryak, *Z. Phys.* C, **38** (1988), 141–145, 165–172; *Nuclear Phys.* B, **319** (1989), 521–541; *Nuclear Phys.* B, **328** (1989), 85–102.
[72] D. Gross and R. Gopakumar, hep-th/9411021.
R. Gopakumar, hep-th/0211100.
[73] G. Moore and E. Witten, hep-th/9709193.
[74] A. Losev, N. Nekrasov, and S. Shatashvili, hep-th/9711108; hep-th/9801061.
[75] H. Kawai, T. Kuroki, and T. Morita, hep-th/0303210.
[76] F. Cachazo, M. Douglas, N. Seiberg, and E. Witten, hep-th/0211170.
[77] N. Nekrasov and S. Shatashvili, in preparation.
[78] M. Douglas and V. Kazakov, hep-th/9305047.
[79] D. Gross, hep-th/9212149.
D. Gross and W. Taylor, hep-th/9301068; hep-th/9303046.
[80] A. Iqbal, hep-th/0212279.
[81] M. Aganagic, M. Mariño, and C. Vafa, hep-th/0206164.
[82] P. Wiegmann and A. Zabrodin, hep-th/9909147.
I. K. Kostov, I. Krichever, M. Mineev-Weinstein, P. Wiegmann, and A. Zabrodin, hep-th/0005259.
[83] V. Pasquier, hep-th/9405104.
R. Caracciollo, A. Lerda, and G. R. Zemba, hep-th/9503229.
J. Minahan, A. P. Polychronakos, hep-th/9404192; hep-th/9303153.
A. P. Polychronakos, hep-th/9902157.
E. Langmann, math-ph/0007036; math-ph/0102005.
[84] S. Girvin, cond-mat/9907002.
[85] V. Knizhnik and A. Zamolodchikov, *Nuclear Phys.* B, **247** (1984), 83–103.
[86] D. Bernard, *Nuclear Phys.* B, **303** (1988), 77–93; *Nuclear Phys.* B, **309** (1988), 145–174.
[87] A. S. Losev, *Coset Construction and Bernard Equation*, Preprint TH-6215-91, CERN, Geneva, 1991.

[88] G. Felder, hep-th/9609153.
[89] D. Ivanov, hep-th/9610207.
[90] M. A. Olshanetsky, Painlevé type equations and Hitchin systems, in *Seiberg-Witten Theory and Integrable Systems*, World Scientific, Singapore, 1999.
[91] C. Vafa, hep-th/0008142.

Quantum Calabi–Yau and Classical Crystals

Andrei Okounkov[1], Nikolai Reshetikhin[2], and Cumrun Vafa[3]

[1] Department of Mathematics
Princeton University
Princeton, NJ 08544
USA
okounkov@math.princeton.edu
[2] Department of Mathematics
University of California at Berkeley
Evans Hall #3840
Berkeley, CA 94720-3840
USA
reshetik@math.berkeley.edu
[3] Jefferson Physical Laboratory
Harvard University
Cambridge, MA 02138
USA
vafa@string.harvard.edu

Summary. We propose a new duality involving topological strings in the limit of the large string coupling constant. The dual is described in terms of a classical statistical mechanical model of crystal melting, where the temperature is the inverse of the string coupling constant. The crystal is a discretization of the toric base of the Calabi–Yau with lattice length g_s. As a strong piece of evidence for this duality we recover the topological vertex in terms of the statistical mechanical probability distribution for crystal melting. We also propose a more general duality involving the dimer problem on periodic lattices and topological A-model string on arbitrary local toric threefolds. The (p, q) 5-brane web, dual to Calabi–Yau, gets identified with the transition regions of rigid dimer configurations.

Subject Classifications: 81T45, 81T30, 14J32, 82B23

1 Introduction

Topological strings on Calabi–Yau threefolds have been a fascinating class of string theories, which have led to insights into the dynamics of superstrings and supersymmetric gauge theories. They have also been shown to be equivalent in some cases to noncritical bosonic strings. In this paper we ask how the topological A-model which "counts" holomorphic curves inside the Calabi–Yau behaves in the limit of large values of the string coupling constant $g_s \gg 1$. We propose a dual description which is given in terms of a discrete statistical mechanical model of a three-dimensional real crystal with boundaries, where the crystal is located in the toric base of the Calabi–Yau threefold, with the "atoms" separated by a distance of g_s. Moreover, the temperature T in the statistical mechanical model corresponds to $1/g_s$. Heating up the crystal leads to melting of it. In the limit of large temperature, or small g_s, the Calabi–Yau geometry emerges from the geometry of the molten crystal!

In the first part of this paper we focus on the simplest Calabi–Yau, namely, $\mathbf{C}^3$. Even here there are a lot of nontrivial questions to answer. In particular, the computation of topological string amplitudes when we put D-branes in this background is nontrivial and leads to the notion of a topological vertex [1, 2] (see also the recent paper [3]). Moreover, using the topological vertex one can compute an all order amplitude for topological strings on arbitrary local Calabi–Yau manifolds. This is an interesting class to study, as it leads to nontrivial predictions for instanton corrections to gauge and gravitational couplings of a large class of $N = 2$ supersymmetric gauge theories in four dimensions via geometric engineering [4] (for recent progress in this direction see [5, 6]). It is also the same class which is equivalent (in some limits) to noncritical bosonic string theories. In this paper we will connect the topological vertex to the partition function of a melting corner with fixed asymptotic boundary conditions. Furthermore we find an intriguing link between dimer statistical mechanical models and noncompact toric Calabi–Yau threefolds. In particular, the dimer problems in two dimensions naturally get related to the study of configurations of the (p, q) 5-brane web, which is dual to noncompact toric Calabi–Yau threefolds.

The organization of this paper is as follows: In Section 2 we will motivate and state the conjecture. In Section 3 we check aspects of this conjecture and derive the topological vertex from the statistical mechanical model. In Section 4 we discuss dimer problems and its relation to topological strings on Calabi–Yau.

Our proposal immediately raises many questions, which are being presently investigated. Many of them will be pointed out in the paper. One of the most interesting physical questions involves the superstring intepretation of the discretization of space. On the mathematical side, we expect that our statistical mechanical model should have a deep meaning in terms of the geometry of the target space based on the interpretation of its configurations as torus fixed points in the Hilbert scheme of curves of the target threefold. Also, our 3d model naturally extends the random 2d partition models that arise in $N = 2$ supersymmetric gauge theory [5] and Gromov-Witten theory of target curves [7]. For some mathematical aspects of the topological vertex, see [8, 9, 3].

2 The conjecture

2.1 Hodge integrals and 3d partitions

Consider topological A-model strings on a Calabi–Yau threefold. For simplicity, let us consider the limit when the Kähler class of the Calabi–Yau is rescaled by a factor that goes to infinity. As explained in [10], in this limit the genus g amplitude is given by

$$\frac{\chi}{2}\int_{\overline{\mathcal{M}}_g} c_{g-1}^3(\mathcal{H})$$

where χ is the Euler characteristic of the Calabi–Yau threefold, $\overline{\mathcal{M}}_g$ is the moduli space of Riemann surfaces, $\mathcal{H}$ is the Hodge bundle over $\overline{\mathcal{M}}_g$, and c_{g-1} denotes its $(g-1)$st Chern class. For genus 0 and 1, there is also some Kähler dependence (involving the volume and the second Chern class of the tangent bundle), which we subtract out to get a finite answer. Consider

$$Z = \exp\left[\frac{\chi}{2}\sum_g g_s^{2g-2}\int_{\overline{\mathcal{M}}_g} c_{g-1}^3(\mathcal{H})\right].$$

It has been argued physically [11] and derived mathematically [12] that

$$Z = f^{\chi/2},$$

where

$$f = \prod_n \frac{1}{(1-q^n)^n}$$

and

$$q = e^{-g_s}.$$

By the classical result of MacMahon, the function f is the generating function for 3d partitions, that is,

$$f = \sum_{\text{3d partitions}} q^{\#\text{ boxes}},$$

where, by definition, a 3d partition is a 3d generalization of 2d Young diagrams and is an object of the kind seen on the left in Figure 3. This fact was pointed out to one of us as a curiosity by R. Dijkgraaf shortly after [11] appeared.

2.2 Melting of a crystal and Calabi–Yau threefold

It is natural to ask whether there is a deeper reason for this correspondence. What could three-dimensional partitions have to do with the A-model topological string on a Calabi–Yau threefold? The hint comes from the fact that we are considering the limit of large Kähler class and in this limit the Calabi–Yau looks locally like $\mathbf{C}^3$s glued together. It is then natural to view this torically, as we often do in the topological string, in the context of mirror symmetry, and write the Kähler form as

$$\sum_{i=1,2,3} dz_i \wedge d\bar{z}_i \sim \sum_{i=1,2,3} d|z_i|^2 \wedge d\theta_i,$$

where the $|z_i|^2$ span the base of a toric fibration of $\mathbf{C}^3$. Note that $|z_i|^2 = x_i$ parameterize the positive octant $O^+ \subset \mathbf{R}^3$. If we assign Euler characteristic "2" to each $\mathbf{C}^3$ patch, the topological string amplitudes on it get related to the MacMahon function.

Then it is natural to think that the octant is related to three-dimensional partitions, in which the boxes are located at $\mathbf{Z}^3$ lattice points inside O^+. Somehow the points of the Calabi–Yau, in this case $\mathbf{C}^3$, become related to integral lattice points on the toric base.

The picture we propose is the following. We identify the highly quantum Calabi–Yau with the frozen crystal, that is, the crystal in which all atoms (indexed by lattice points in O^+) are in place. We view the excitations as removing lattice points as in Figure 1. The rule is that we can remove lattice points only if there are no pairs of atoms on opposite sides. This gives the same rule as 3d partitions. Note that if one holds the page upside-down, one sees a 3d partition in Figure 1, namely, the partition from Figure 3.

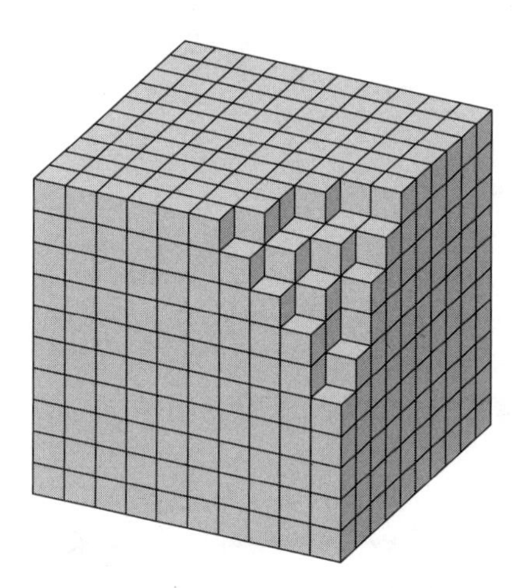

Fig. 1. A melting crystal corner.

Removing each atom contributes the factor $q = e^{-\mu/T}$ to the Boltzmann weight of the configuration, where μ is the chemical potential (the energy of the removal of an atom) and T is the temperature. We choose units in which $\mu = 1$. To connect this model to the topological string on $\mathbf{C}^3$ we identify $g_s = 1/T$. In particular, in the $g_s \to 0$, that is, the $q \to 1$ limit the crystal begins to melt away. Rescaled in all direction by a factor of $1/T = g_s$, the crystal approaches a smooth limit shape which has been studied from various viewpoints [15, 16]. This is plotted in Figure 2.

The analytic form of this limit shape is encoded in terms of a complex Riemann surface, which in this case is given by

$$F(u, v) = e^{-u} + e^{-v} + 1 = 0 \tag{2.1}$$

defined as a hypersurface in $\mathbf{C}^2$ with a natural 2-form $du \wedge dv$, where u, v are periodic variables with period $2\pi i$. In coordinates e^{-u} and e^{-v} this is simply a straight line in $\mathbf{C}^2$. Consider the following function of the variables $U = \mathrm{Re}(u)$ and $V = \mathrm{Re}(v)$

$$R(U, V) = \frac{1}{4\pi^2} \iint_0^{2\pi} \log |F(U + i\theta, V + i\phi)| d\theta d\phi. \tag{2.2}$$

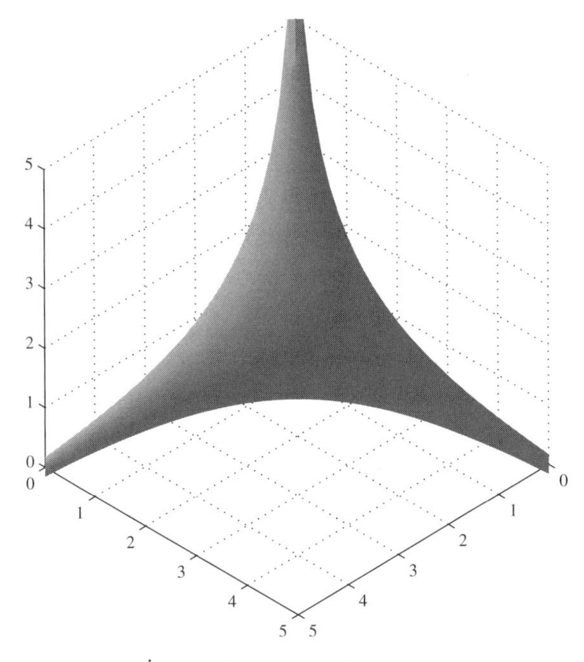

Fig. 2. The limit shape of a 3d partition.

This function is known as the *Ronkin function* of F. In terms of the Ronkin function, the limit shape can be parameterized as follows:

$$(x_1, x_2, x_3) = (U + R, V + R, R), \qquad R = R(U, V).$$

Note that

$$U = x_1 - x_3, \qquad V = x_2 - x_3.$$

The projection of the curved part of the limit shape onto the (U, V)-plane is the region bounded by the curves

$$\pm e^{-U} \pm e^{-V} + 1 = 0,$$

excluding the case when both signs are positive. This region is the *amoeba* of the curve (2.1), which, by definition, is its image under the map $(u, v) \mapsto (U, V)$. In different coordinates, this is the planar region actually seen in Figure 2.

2.3 Mirror symmetry and the limit shape

Consider topological strings on $\mathbf{C}^3$. One can apply mirror symmetry in this context by dualizing the three phases of the complex parameters according to T-duality. This has been done in [13]. One introduces dual variables Y_i which are periodic, with period $2\pi i$ and are related to the z_i by

$$|z_i|^2 = \mathrm{Re}(Y_i).$$

(This is quantum mechanically modified by addition of an i-independent large positive constant to the right-hand side.) The imaginary part of the Y_i is "invisible" to the original geometry, just as is the phase of the z_i to the Y_i variables. They are T-dual circles. But we can compare the data between $\mathbf{C}^3$ and the mirror on the base of the toric variety which is visible to both. In mirror symmetry the Kähler form of $\mathbf{C}^3$ gets mapped to the holomorphic three-form which in this case is given by

$$\Omega = \prod_{i=1}^{3} dY_i \exp[W],$$

where

$$W = \sum e^{-Y_i}.$$

Note that shifting $Y_i \to Y_i + r$ shifts $W \to e^{-r} W$. The analog of rescaling in the mirror is changing the scale of W. Let us fix the scale by requiring $W = 1$; this will turn out to be the mirror statement to rescaling by g_s to get a limit shape. In fact, we will see below that the limit shape corresponds to the toric projection of the complex surface in the mirror given simply by $W(Y_i) = 1$. Let us define

$$u = Y_1 - Y_3, \qquad v = Y_2 - Y_3.$$

Then

$$W = e^{-Y_3} F(u, v),$$

where

$$F(u, v) = e^{-u} + e^{-v} + 1;$$

we will identify $F(u, v)$ with the Riemann surface of the crystal melting problem. In this context also according to the mirror map the points in the u, v space get mapped to the points on the toric base (i.e., O^+) satisfying

$$x_1 - x_3 = \mathrm{Re}(u) = U,$$
$$x_2 - x_3 = \mathrm{Re}(v) = V.$$

We now wish to understand the interpretation of the limit shape from the viewpoint of the topological string. We propose that the boundary of the molten crystal which is a 2-cycle on the octant should be viewed as a special Lagrangian cycle of the A-model with one hidden circle in the fiber. Similarly, in the B-model mirror it should be viewed as the B-model holomorphic surface, which in the case at hand gets identified with $W = 1$. (Recall that in the LG models B-branes can be identified with $W = \text{const.}$ [14].) On this surface we have

$$e^{-Y_3} F(u, v) = 1.$$

If we take the absolute value of this equation, to find the projection onto the base, we find

$$e^{-\mathrm{Re}(Y_3)} |F(u, v)| = 1.$$

If we take the logarithm of this relation, we have

$$-\operatorname{Re}(Y_3) + \log|e^{-u} + e^{-v} + 1| = 0 \rightarrow x_3 = \log|e^{-u} + e^{-v} + 1|.$$

However, we have a fuzziness in mapping this to the toric base: $u = \operatorname{Re}(u) + i\theta$ and $v = \operatorname{Re}(v) + i\phi$ and so a given value for $\operatorname{Re}(u)$ and $\operatorname{Re}(v)$ does not give a fixed value of x_3. That depends in addition on the angles θ, ϕ of the mirror torus which are invisible to the A-model toric base. It is natural to take the average values as defining the projection to the base, i.e.,

$$x_3 = \frac{1}{4\pi^2} \int d\theta d\phi \log|F(U + i\theta, V + i\phi)|,$$

which is exactly the expression for the limit shape. We thus find some further evidence that the statistical mechanical problem of crystal melting is rather deeply related to the topological string and mirror symmetry on Calabi–Yau. Moreover, we can identify the points of the crystal with the discretization of points of the base of the toric Calabi–Yau.

To test this conjecture further we will have to first broaden the dictionary between the two sides. In particular, we ask what is the interpretation of the topological vertex for the statistical mechanical problem of crystal melting? For the topological vertex we fix a 2d partition on each of the three legs of the toric base. There is only one natural interpretation of what this could mean in the crystal melting problem: This could be the partition function of the melting crystal with three fixed asymptotic boundary shapes for the molten crystal, dictated by the corresponding partition. We will show that this is indeed the case in the next section.

3 Melting corner and the topological vertex

3.1 Transfer matrix approach

The grand canonical ensemble of 3d partitions weighted with $q^{\#\,\text{boxes}}$ is the simplest model of a melting crystal near its corner. We review the transfer matrix approach to this model following [16]. This approach can be easily generalized to allow for certain inhomogeneity and periodicity, which is useful in the context of the more general models discussed in Section 4.

We start by cutting the 3d partition into diagonal slices by planes $x_2 - x_1 = t$; see Figure 3. This operation makes a 3d partition a sequence $\{\mu(t)\}$ of ordinary partitions indexed by an integer variable t. Conversely, given a sequence $\{\mu(t)\}$, it can be assembled into a 3d partition provided it satisfies the following *interlacing* condition. We say that two partitions μ and ν interlace, and write $\mu \succ \nu$ if

$$\mu_1 \geq \nu_1 \geq \mu_2 \geq \nu_2 \geq \cdots.$$

It is easy to see that a sequence of slices $\{\mu(t)\}$ of a 3d partition satisfies

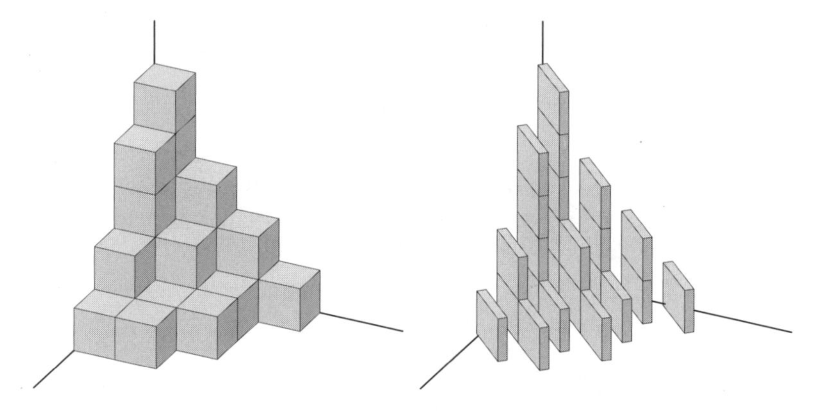

Fig. 3. A 3d partition and its diagonal slices.

$$\mu(t) \prec \mu(t+1), \quad t < 0,$$

and the reverse relation for $t \geq 0$.

There is a well-known map from partitions to states in the NS sector of the complex fermionic oscillator. Let a_i and b_i be the areas of pieces one gets by slicing a 2d partition first diagonally and then horizontally (resp. vertically) above and the below the diagonal, respectively. Formally,

$$a_i = \mu_i - i + \frac{1}{2}, \qquad b_i = \mu_i^t - i + \frac{1}{2},$$

where i ranges from 1 to the number of squares on the diagonal of μ. In mathematical literature, these coordinates on partitions are known as (modified) Frobenius coordinates. The fermionic state associated to μ is

$$|\mu\rangle = \prod_{j=1}^{d} \psi_{a_i}^* \psi_{b_i} |0\rangle.$$

Note that

$$q^{L_0}|\mu\rangle = q^{\#\,\text{boxes}}|\mu\rangle = q^{|\mu|}|\mu\rangle,$$

where we denote the total number of boxes of μ by $|\mu|$. We can write a bosonic representation of this state by the standard bosonization procedure.

Consider the operators

$$\Gamma_{\pm}(z) = \exp\left(\sum_{\pm n > 0} \frac{z^n J_n}{n}\right),$$

where the J_n denote the modes of the fermionic current $\psi^*\psi$. The operators $\Gamma_{\pm}(z)$ can be identified with annihilation and creation parts of the bosonic vertex operator $e^{\phi(z)}$. The relevance of these operators for our problem lies in the following formulas:

$$\Gamma_-(1)|\mu\rangle = \sum_{\nu \succ \mu} |\nu\rangle,$$
$$\Gamma_+(1)|\mu\rangle = \sum_{\nu \prec \mu} |\nu\rangle. \tag{3.1}$$

To illustrate their power we will now derive, as in [16], the MacMahon generating function for 3d partitions

$$Z = \sum_{\text{3d partitions } \pi} q^{\#\text{ of boxes}}.$$

By the identification (3.1) of the transfer matrix, we have

$$Z = \left\langle \left(\prod_{t=0}^{\infty} q^{L_0} \Gamma_+(1) \right) q^{L_0} \left(\prod_{t=-\infty}^{-1} \Gamma_-(1) q^{L_0} \right) \right\rangle. \tag{3.2}$$

Now we commute the operators q^{L_0} to the outside, splitting the middle one in half. This yields

$$Z = \left\langle \prod_{n>0} \Gamma_+(q^{n-\frac{1}{2}}) \prod_{n>0} \Gamma_-(q^{-n-\frac{1}{2}}) \right\rangle. \tag{3.3}$$

Now we commute the creation operators through annihilation operators, using the commutation relation

$$\Gamma_+(z)\Gamma_-(z') = (1 - z/z')^{-1} \Gamma_-(z')\Gamma_+(z).$$

The product of resulting factors gives directly the MacMahon function

$$Z = \prod_{n>0} (1 - q^n)^{-n} = f.$$

Same ideas can be used to get more refined results, such as, for example, the correlation functions, see [16]. We will generalize below the above computation to the case when 3d partitions have certain asymptotic configuration is in the direction of the three axes. Before doing this, we will review some relevant theory of symmetric functions.

3.2 Skew Schur functions

Skew Schur functions $s_{\lambda/\mu}(x_1, x_2, \dots)$ are certain symmetric polynomials in the variables x_i indexed by a pair of partitions μ and λ such that $\mu \subset \lambda$, see [17]. Their relevance for us lies in the well-known fact (see, e.g., [18]) that

$$\prod_i \Gamma_-(x_i)|\mu\rangle = \sum_{\lambda \supset \mu} s_{\lambda/\mu}(x)|\lambda\rangle. \tag{3.4}$$

When $\mu = \emptyset$, this specializes to the usual Schur functions. The Jacobi–Trudi determinantal formula continues to hold for skew Schur functions:

$$s_{\lambda/\mu} = \det(h_{\lambda_i - \mu_j + j - i}). \tag{3.5}$$

Here h_k is the complete homogeneous function of degree k—the sum of all monomials of degree k. They can be defined by the generating series

$$\sum_{n \geq 0} h_n t^n = \prod_i (1 - t x_i)^{-1} = \exp\left(\sum_{n>0} \frac{t^n}{n} \sum_i x_i^n\right). \tag{3.6}$$

The formula (3.5) is very efficient for computing the values of skew Schur functions, including their values at the points of the form

$$q^{\nu+\rho} = \left(q^{\nu_1 - 1/2}, q^{\nu_2 - 3/2}, q^{\nu_3 - 5/2}, \dots\right), \tag{3.7}$$

where ν is a partition. In this case the sum over i in (3.6) becomes a geometric series and can be summed explicitly.

There is a standard involution in the algebra of symmetric functions which acts by

$$s_\lambda \mapsto s_{\lambda^t},$$

where λ^t denotes the transposed diagram. It continues to act on skew Schur functions in the same manner,

$$s_{\lambda/\mu} \mapsto s_{\lambda^t/\mu^t}.$$

It is straightforward to check that

$$s_{\lambda/\mu}(q^{\nu+\rho}) = (-1)^{|\lambda|-|\mu|} s_{\lambda^t/\mu^t}(q^{-\nu-\rho}). \tag{3.8}$$

Finally, the following property of skew Schur functions will be crucial in making the connection to the formula for the topological vertex from [1]. The coefficients of the expansion

$$s_{\lambda/\mu} = \sum_\nu c^\lambda_{\mu\nu} s_\nu \tag{3.9}$$

of skew Schur functions in terms of ordinary Schur functions are precisely the tensor product multiplicities, also known as the Littlewood-Richardson coefficients.

3.3 Topological vertex and 3d partitions

Topological vertex in terms of Schur functions

Our goal now is to recast the topological vertex [1] in terms of Schur functions. The basic ingredient of the topological vertex involves the expectation values of the $U(\infty)$ Chern–Simons Hopf link invariant $W_{\mu\lambda} = W_{\lambda\mu}$ in representations μ and λ. The expression of the Hopf link invariant $W_{\mu\lambda}$ in terms of the Schur functions is the following:

$$W_{\mu\lambda} = W_\mu s_\lambda(q^{\mu+\rho}), \tag{3.10}$$

where

$$W_\mu = q^{\kappa(\mu)/2} s_{\mu^t}(q^\rho). \tag{3.11}$$

Here

$$\kappa(\lambda) = \sum_i \left[\left(\lambda_i - i + \tfrac{1}{2}\right)^2 - \left(-i + \tfrac{1}{2}\right)^2\right] = 2 \sum_{\square=(i,j)\in\lambda} (j-i) \tag{3.12}$$

is the unique up to scalar quadratic Casimir such that

$$\kappa(\emptyset) = \kappa(\square) = 0.$$

Using (3.9) it is straightforward to check that the topological vertex $C(\lambda, \mu, \nu)$ in the standard framing has the following expression in terms of the skew Schur functions:

$$C(\lambda, \mu, \nu) = q^{\kappa(\lambda)/2+\kappa(\nu)/2} s_{\nu^t}(q^\rho) \sum_\eta s_{\lambda^t/\eta}(q^{\nu+\rho}) s_{\mu/\eta}(q^{\nu^t+\rho}). \tag{3.13}$$

The lattice length

To relate the crystal melting problem to the topological vertex, we first have to note that the topological vertex refers to computations in the A-model corresponding to placing Lagrangian branes on each leg of $\mathbf{C}^3$, assembled into representations of $U(\infty)$ and identified with partitions. If we place the brane at a fixed position and put it in a representation μ, the effect of moving the brane from a position l to the position $l+k$ (in string units) affects the amplitude by multiplication by

$$\exp(-k|\mu|).$$

Now consider the lattice model, where we fix the asymptotics at a distance $L \gg 1$ to be fixed to be a fixed partition μ. Then, if we change $L \to L+K$, the amplitude gets weighted by $q^{K|\mu|}$. Since we have already identified $q = e^{-g_s}$, comparing these two expressions, we immediately deduce that

$$k = K g_s.$$

In other words, the distance in the lattice computation times g_s is the distance as measured in string units. This is satisfactory as it suggests that as $g_s \to 0$ the lattice spacing in string units goes to zero and the space becomes continuous.

In defining the topological vertex one gets rid of the propagator factors above (which will show up in the gluing rules). Similarly, in the lattice model when we fix the asymptotic boundary condition to be given by fixed 2d partitions we should multiply the amplitudes by $q^{-L|\mu|}$ for each fixed asymptote at lattice position L. Actually this is not precisely right: we should rather counterweight it with $q^{-(L+\frac{1}{2})|\mu|}$. To see this note that if we glue two topological vertices with lattice points L_1 and L_2 along the joining edge, the number of points along the glued edge is $L_1 + L_2 + 1$. Putting the $\frac{1}{2}$ in the above formula gives a symmetric treatment of this issue in the context of gluing.

Framing

The topological vertex also comes equipped with a framing [1] for each edge, which we now recall. Toric Calabi–Yau's come with a canonical direction in the toric base. In the case of $\mathbf{C}^3$, it is the diagonal line $x_1 = x_2 = x_3$ in O^+. One typically projects vectors on this toric base along this direction, to a two-dimensional plane (the U, V plane in the context we have discussed). At each vertex there are integral projected 2d vectors along the axes which sum up to zero. The topological vertex framing is equivalent to picking a 2d projected vector on each axis whose cross product with the integral vector along the axis is $+1$ (with a suitable sense of orientation). If v_i is a framing vector for the ith axis, and e_i denotes the integral vector along the ith axis, then the most general framing is obtained by

$$v_i \to v_i + n_i e_i,$$

where n_i is an integer. We now interpret this choice in our statistical mechanical model: In describing the asymptotes of the 3d partition we have to choose a slicing along each axis. We use the framing vector, together with the diagonal direction $x_1 = x_2 = x_3$, to define a slicing 2-plane for that edge. The standard framing corresponds to choosing the framing vector in cyclic order: On the x_1-axis, we choose x_3, on the x_3-axis we choose x_2 and on the x_2-axis we choose x_1. This together with the diagonal line determines a slicing plane on each axis. Note that all different slicings will have the diagonal line on them. This line passes through the diagonal of the corresponding 2d partition.

Before doing any detailed comparison with the statistical mechanical model with fixed asymptotes we can check whether framing dependence of the topological vertex can be understood. This is indeed the case. Suppose we compute the partition function with fixed 2d asymptotes and with a given framing (i.e., slicing). Suppose we shift the framing by n_i. This will still cut the asymptotic diagram along the same 2d partition. Now, however, the total number of boxes of the 3d partition has changed. The diagonal points of the partition have not moved as they are on the slicing plane for each framing. The farther a point is from the diagonal the more it has moved. Indeed the net number of points added to the 3d partition is given by

$$n_i \sum_{k,l \in \mu} (k - l) = n_i \kappa(\mu).$$

Thus the statistical mechanical model will have the extra Boltzmann weight $q^{n_i \kappa(\mu)}$. This is precisely the framing dependence of the vertex. Encouraged by this observation we now turn to computing the topological vertex in the standard framing from the crystal point of view.

3.4 The perpendicular partition function

Definition

Now our goal is to find an exact match between the formula (3.13) and the partition function $P(\lambda, \mu, \nu)$ for 3d partitions whose asymptotics in the direction of the three coordinates axes is given by three given partitions λ, μ, and ν. This generating function, which we call the *perpendicular partition function*, will be defined and computed presently.

Consider three-dimensional partitions π inside the box

$$[0, N_1] \times [0, N_2] \times [0, N_3].$$

Let the boundary conditions in the planes $x_i = N_i$ be given by three partitions λ, μ, and ν. This means that, for example, the facet of π in the plane $x_1 = N_1$ is the diagram of the partition λ oriented so that λ_1 is its length in the x_2 direction. The other two boundary partitions are defined in the cyclically symmetric way. See Figure 4, in which $\lambda = (3, 2)$, $\mu = (3, 1)$ and $\nu = (3, 1, 1)$.

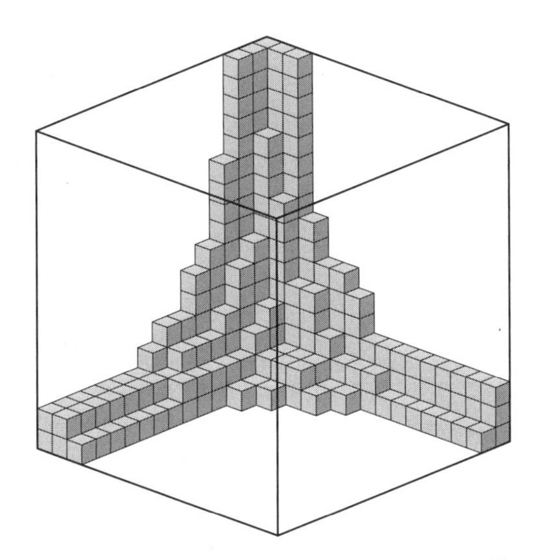

Fig. 4. A 3d partition ending on three given 2d partitions.

Let $P_{N_1,N_2,N_3}(\lambda, \mu, \nu)$ be the partition function in which every π is weighted by $q^{\mathrm{vol}(\pi)}$. It is obvious that the limit

$$P(\lambda, \mu, \nu) = \lim_{N_1,N_2,N_3 \to \infty} q^{-N_1|\lambda|-N_2|\mu|-N_3|\nu|} P_{N_1,N_2,N_3}(\lambda, \mu, \nu) \tag{3.14}$$

exists as a formal power series in q. What should the relation of this to the topological vertex be? From what we have said before this should be the topological vertex itself up to framing factors. Let us also fix the framing factor, to compare it to the canonical framing. First of all, we need to multiply P by

$$q^{\frac{-1}{2}(|\lambda|+|\mu|+|\nu|)}.$$

This is related to our discussion of the gluing algorithm (of shifting $N_i \to N_i + \frac{1}{2}$) and splitting the point of gluing between the two vertices. Second, this perpendicular slicing is not the same as canonical framing. We need to rotate the perpendicular slicing to become the canonical framing. This involves the rotation of the partition along its first column by one unit and the sense of the rotation is to increase the number of points. For each representation, this gives

$$\frac{1}{2}\sum_i \lambda_i(\lambda_i - 1) = \frac{1}{2}(\|\lambda\|^2 - |\lambda|)$$

extra boxes which we have to subtract off and gives the additional weight. Here

$$\|\lambda\|^2 = \sum \lambda_i^2.$$

Combining with the previous factor, we get a net factor of

$$q^{\frac{1}{2}(\|\lambda\|^2 + \|\mu\|^2 + \|\nu\|^2)}.$$

Moreover, we should normalize as usual by dividing by the partition function with trivial asymptotic partition, which is the MacMahon function. We thus expect

$$\prod_n (1 - q^n)^n q^{\frac{1}{2}(\|\lambda\|^2 + \|\mu\|^2 + \|\nu\|^2)} P(\lambda, \mu, \nu) = C(\lambda, \mu, \nu).$$

We will see below that this is true up to $g_s \to -g_s$ and an overall factor that does not affect the gluing properties of the vertex (as they come in pairs) and it can be viewed as a gauge choice for the topological vertex.

Transfer matrix formula

Recall that in the transfer matrix setup, one slices the partition diagonally. Compared with the perpendicular cutting, the diagonal cutting adds extra boxes to the partition and, as a result, it increases its volume by $\binom{\lambda}{2} + \binom{\mu^t}{2}$, where, by definition,

$$\binom{\lambda}{2} = \sum_i \binom{\lambda_i}{2}. \tag{3.15}$$

Also, for the transfer matrix method it is convenient to let $N_3 = \infty$ from the very beginning, that is, to consider

$$P_{N_1,N_2}(\lambda, \mu, \nu) = \lim_{N_3 \to \infty} q^{-N_3|\nu|} P_{N_1,N_2,N_3}(\lambda, \mu, \nu). \tag{3.16}$$

The partition function $P_{N_1,N_2}(\lambda, \mu, \nu)$ counts 3d partitions π with given boundary conditions on the planes $x_{1,2} = N_{1,2}$ inside the container which is a semi-infinite cylinder with base

$$[0, N_1] \times [0, N_2] \setminus \nu;$$

see Figure 5. In other words, $P_{N_1,N_2}(\lambda, \mu, \nu)$ counts *skew 3d partitions* in the sense of [16].

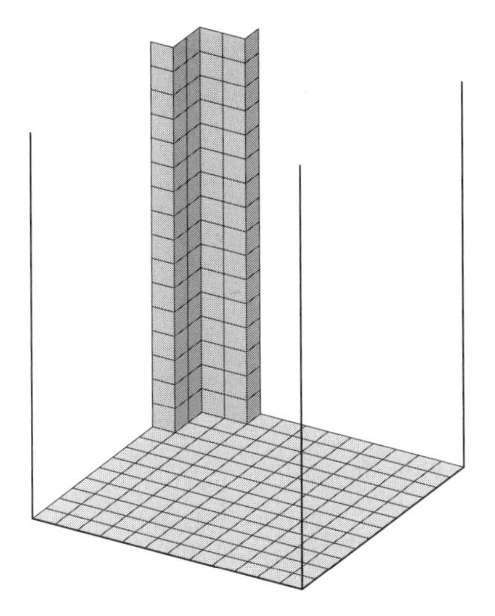

Fig. 5. The container for skew 3d partitions.

The main observation about skew 3d partitions is that their diagonal slices interlace in the pattern dictated by the shape ν. This gives the following transfer matrix formula for (3.16), which is a direct generalization of (3.2):

$$P_{N_1,N_2}(\lambda,\mu,\nu) = q^{-\binom{\lambda}{2}-\binom{\mu^t}{2}} \times \left\langle \lambda^t \left| \left(\prod_{\substack{N_1-1\\ \text{terms}}} q^{L_0}\Gamma_\pm(1) \right) q^{L_0} \left(\prod_{\substack{N_2-1\\ \text{terms}}} \Gamma_\pm(1) q^{L_0} \right) \right| \mu \right\rangle, \tag{3.17}$$

where the pattern of pluses and minuses in $\Gamma_\pm$ is dictated by the shape of ν.

Transformation of the operator formula

Now we apply to the formula (3.17) the following three transformations:

- Commute the operators q^{L_0} to the outside, splitting the middle one in half. The operators $\Gamma_\pm(1)$ will be conjugated to the operators $\Gamma_\pm(q^{\cdots})$ in the process.
- Commute the raising operators Γ_- to the left and the lowering operators Γ_+ to the right. There will be some overall, ν-dependent multiplicative factor $Z(\nu)$ from this operation.
- Write the resulting expression as a sum over intermediate states $|\eta\rangle\langle\eta|$.

The result

We obtain

$$P(\lambda,\mu,\nu) = Z(\nu) q^{-\binom{\lambda}{2}-\binom{\mu^t}{2}-|\lambda|/2-|\mu|/2} \sum_\eta s_{\lambda^t/\eta}(q^{-\nu-\rho}) s_{\mu/\eta}(q^{-\nu^t-\rho}). \tag{3.18}$$

In order to determine the multiplicative factor $Z(\nu)$, we compute

$$P(\emptyset, \emptyset, \nu) = P(\nu, \emptyset, \emptyset)$$

using formula (3.18). We get

$$Z(\nu) = \frac{q^{-\binom{\nu}{2}-|\nu|/2} s_{\nu^t}(q^{-\rho})}{\prod_{n>0}(1-q^n)^n}. \tag{3.19}$$

Since

$$\frac{\kappa(\mu)}{2} = \binom{\mu}{2} - \binom{\mu^t}{2}, \tag{3.20}$$

comparing (3.18) with (3.13), we obtain

$$\boxed{C(\lambda, \mu, \nu; 1/q) = q^{\frac{\|\lambda^t\|+\|\mu^t\|+\|\nu^t\|}{2}} \prod_{n>0}(1-q^n)^n P(\lambda, \mu, \nu),} \tag{3.21}$$

where

$$\frac{\|\lambda\|^2}{2} = \sum_i \frac{\lambda_i^2}{2} = \binom{\lambda}{2} + \frac{|\lambda|}{2}. \tag{3.22}$$

This is exactly what we aimed for; see Section 3.4, up to the following immaterial details. The difference between $\|\lambda^t\|$ and $\|\lambda\|$ is irrelevant since the gluing formula pairs λ with λ^t. Also, since the string amplitudes are even functions of g_s, they are unaffected by the substitution $q \mapsto 1/q$. These differences can be viewed as a gauge choice for the topological vertex.

3.5 Other generalizations

One can use the topological vertex to glue various local $\mathbf{C}^3$ patches and obtain the topological A-model amplitudes. Thus it is natural to expect that there is a natural lattice model. There is a natural lattice [26] for the crystal in this case, obtained by viewing the Kähler form divided by g_s as defining the first Chern class of a line bundle and identifying the lattice model with holomorphic sections of this bundle. This is nothing but geometric quantization of Calabi–Yau with the Kähler form playing the role of symplectic structure and g_s playing the role of $\hbar$. The precise definition of lattice melting problem for this class is investigated in [26]. The fact that we have already seen the emergence of topological vertex in the lattice computation corresponding to $\mathbf{C}^3$ would lead one to expect that asymptotic gluings suitably defined should give the gluing rules of the statistical mechanical model. It is natural to expect that this idea also works the same way in the compact case, by viewing the Calabi–Yau as a noncommutative manifold with noncommutativity parameter being $g_s k$, where k is the Kähler form.

It is also natural to embed this in superstring [27], where g_s will be replaced by the graviphoton field strength. The large g_s in this context translates to strong graviphoton field strength, for which it is natural to expect discretization of spacetime. This is exciting as it will potentially give a novel realization of the superstring target space as a discrete lattice.

4 Periodic dimers and toric local CY

4.1 Dimers on a periodic planar bipartite graph

A natural generalization of the ideas discussed here is the planar dimer model, see [19] for an introduction. Let Γ be a planar graph ("lattice") which is periodic and bipartite. The first condition means that it is lifted from a finite graph in the torus $\mathbf{T}^2$ via the standard covering map $\mathbf{R}^2 \to \mathbf{T}^2$. The second condition means that the vertices of Γ can be partitioned into two disjoint subsets ("black" and "white" vertices) such that edges connect only white vertices to black vertices. Examples of such graphs are the standard square or honeycomb lattices.

By definition, a dimer configuration on Γ is a collection of edges $D = \{e\}$ such that every vertex is incident to exactly one edge in D. Subject to suitable boundary conditions, the partition function of the dimer model is defined by

$$Z = \sum_{D} \prod_{e \in D} w(e),$$

where $w(e)$ is a certain (Boltzmann) weight assigned to a given edge. For example, one can take both weights and boundary conditions to be periodic, in which case Z is a finite sum. One can also impose boundary conditions at infinity by saying that the dimer configuration should coincide with a given configuration outside some ball of large radius. The relative weight of such a configuration is a well-defined finite product, but the sum Z itself is infinite. In order to make it convergent, one introduces a factor of $q^{\mathrm{volume}(D)}$, defined in terms of the height function; see below. Other boundary conditions can be given by cutting a large but finite piece out of the graph Γ and considering dimers on it.

Simple dimer models, such as equal weight square or honeycomb grid dimer models with simple boundary conditions were first considered in the physics literature many years ago [20, 21]. A complete theory of the dimer model on a periodic weighted planar bipartite graph was developed in [24, 25]. It has some distinctive new features due to spectral curve being a general high genus algebraic curve. We will now quote some results of [24, 25] and indicate their relevance in our setting.

4.2 Periodic configurations and spectral curve

Consider a dimer configuration D on a torus or, equivalently, a dimer configuration in the plane that repeat, periodically, like a wallpaper pattern. There are finitely many such configurations and they will play a special role for us, namely, they will describe the possible facets of our CY crystal. We will now introduce a certain refined counting of these configurations.

Given two configurations D_1 and D_2 on a torus $\mathbf{T}^2$, their union is a collection of closed loops on $\mathbf{T}^2$. These loops come with a natural orientation by, for example, going from white to black vertices along the edges of the first dimer and from black to white vertices along the edges of the second dimer. Hence, they define an element (by summing over all classes of the loops)

$$h = (h_1, h_2) \in H_1(\mathbf{T}^2, \mathbf{Z})$$

of the first homology group of the torus. Fixing any configuration D_0 as our reference point, we can associate $h = h(D)$ to any other dimer configuration and define

$$F(z, w) = \sum_D (-1)^{Q(h)} z^{h_1} w^{h_2} \prod_{e \in D} w(e), \tag{4.1}$$

where $Q(h)$ is any of the four theta-characteristics, for example, $Q((h_1, h_2)) = h_1 h_2$. The ambiguity in the definition of (4.1) comes from the choice of the reference dimer D_0, which means overall multiplication by a monomial in z and w, the choice of the basis for $H_1(\mathbf{T}^2, \mathbf{Z})$, which means $SL(2, \mathbf{Z})$ action, and the choice of the theta-characteristic, which means flipping the signs of z and w.

The locus $F(z, w) = 0$ defines a curve in the toric surface corresponding to the Newton polygon of F. It is called the *spectral curve* of the dimer problem for the given set of weights. It is the spectral curve of the Kasteleyn operator on Γ, the variables z and w being the Bloch–Floquet multipliers in the two directions. In our situation, the curve $F(z, w)$ will be related to the mirror Calabi–Yau threefold.

4.3 Height function and "empty" configurations

Now consider dimer configurations in the plane. The union of two dimer configurations D_1 and D_2 again defines a collection of closed oriented loops. We can view it as the boundary of the level sets of a function h, defined on the cells (also known as faces) of the graph Γ. This function h is well-defined up to a constant and is known as the *height function*.

It is instructive to see how for dimers on the hexagonal lattice this reproduces the combinatorics of the 3d partitions, the height function giving the previously missing 3rd spatial coordinate; see Figure 6. The "full corner" or "empty room" configuration, which was our starting configuration describing the fully quantum $\mathbf{C}^3$ in the language of the dimers becomes the unique, up-to translation, configuration in which the periodic dimer patterns can come together. Each rhombus corresponds to one dimer (the edge of the honeycomb lattice inside it).

For general dimers, there are many periodic dimer patterns and there are (integer) moduli in how they can come together. For example, for the square lattice, which is the case corresponding to the $\mathcal{O}(-1) \oplus \mathcal{O}(-1) \to \mathbf{P}^1$ geometry, the periodic patterns are the brickwall patterns and there is one integer degree of freedom in how they can be patched together. One possible such configuration is shown in Figure 7. The arrows in Figure 7 point from white vertices to black ones to help visualize the difference between the four periodic patterns.

All of these "empty" configurations can serve as the initial configuration, describing the fully quantum toric threefold, for the dimer problem. When lifted into 3d via the height function, each empty configuration follows a piecewise linear function, which is the boundary of the polyhedron defining the toric variety. In particular, the number of "empty room" moduli matches the Kähler moduli of the toric threefold and changes in its combinatorics correspond to the flops.

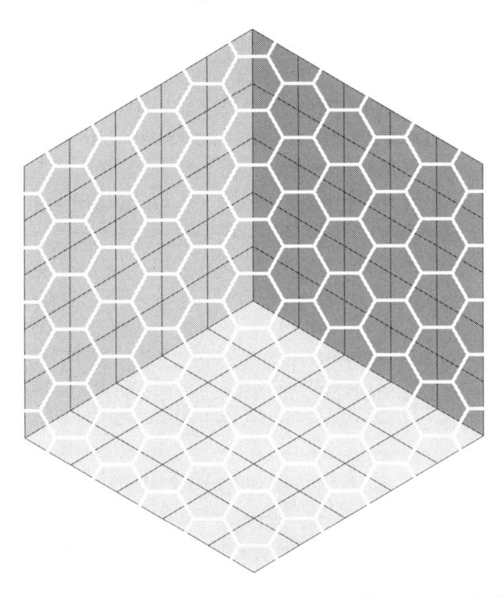

Fig. 6. The "empty room" configuration of honeycomb dimers.

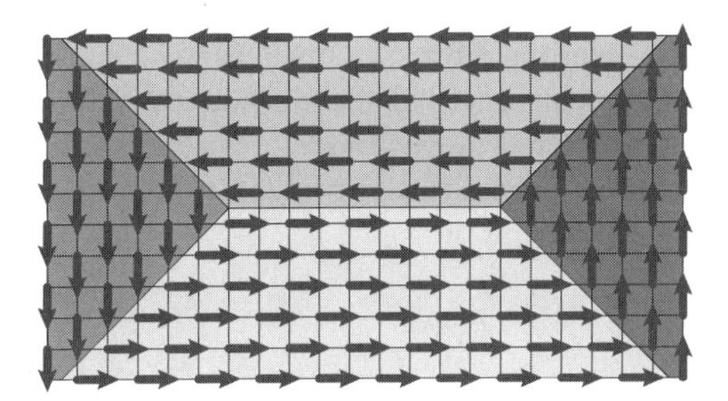

Fig. 7. An "empty room" configuration of square dimers.

Readers familiar with (p, q) 5-brane webs [22] and their relation to toric Calabi–Yau [23] immediately see a dictionary: A 5-brane configuration is identified as the transition line from one rigid dimer configuration to other, which in toric language is related to which $\mathbf{T}^2$ cycle of Calabi–Yau shrinks over it. For example, the $\mathbf{C}^3$ geometry is mapped to three 5-branes (of types $(1, 0)$, $(0, 1)$, $(-1, -1)$) on the 2-plane meeting at a point where each region gets identified with a particular dimer configuration. A similar description holds for arbitrary 5-brane webs. In this context, $F(z, w) = F(e^{-u}, e^{-v})$ is identified with the mirror geometry [13].[1]

[1] More precisely as in [13, 14] this corresponds to a LG theory with $W = e^{-Y_3}(F(u, v))$ or to a noncompact CY given as a hypersurface: $\alpha\beta - F(u, v) = 0$.

4.4 Excitations and limit shape

Now we can start "removing atoms" from the "full crystal" configuration, adding the cost of q to each increase in the height function. The limit shape that develops is controlled by the *surface tension* of the dimer problem. This is a function of the slope measuring how much the dimer likes to have height function with this slope. Formally, it is defined as the $n \to \infty$ limit of the free energy per fundamental domain for dimer configurations on the $n \times n$ torus $\mathbf{T}^2$ restricted to lie in a given homology class or, equivalently, restricted to have a certain slope when lifted to a periodic configuration in the plane.

One of the main results of [24] is the identification of this function with the Legendre dual of the *Ronkin function* of the polynomial (4.1) defined by

$$R(U, V) = \frac{1}{4\pi^2} \iint_0^{2\pi} \log |F(U + i\theta, V + i\phi)| d\theta d\phi. \tag{4.2}$$

The Wulff construction implies that the Ronkin function itself is one of the possible limit shapes, the one corresponding to its own boundary conditions. Note that this is exactly what one would anticipate from our general conjecture if we view $F(z, w)$ as the describing the mirror geometry. In fact, following the same type of argument as in the $\mathbf{C}^3$ case discussed before would lead us to the above Ronkin function.

As an example consider the dimers on the hexagonal lattice with 1×1 fundamental domain. In this case the edge weights can be gauged away and (4.2) becomes the Ronkin function considered in Section 2.2. For the square lattice with 1×1 fundamental domain, there is one gauge invariant combination of the four weights and the spectral curve takes the form

$$F(z, w) = 1 + z + w - e^{-t} zw,$$

where t is a parameter related to the size of $\mathbf{P}^1$ in the $\mathcal{O}(-1) \oplus \mathcal{O}(-1) \to \mathbf{P}^1$ geometry. The spectral curve is a hyperbola in $\mathbf{C}^2$ and (the negative of) its Ronkin function is plotted in Figure 8. Note how one can actually see the projection of the mirror curve!

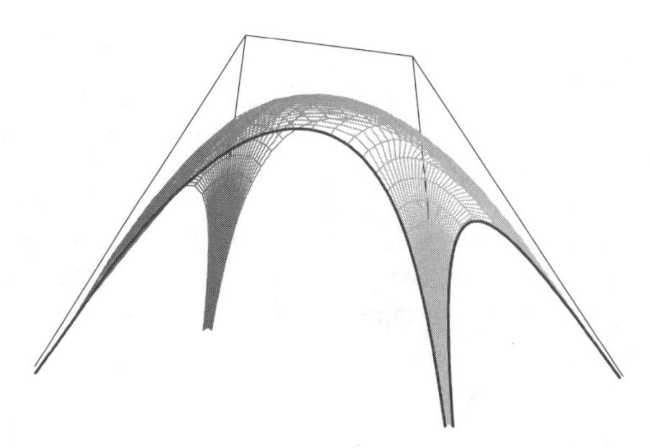

Fig. 8. The Ronkin function of a hyperbola.

All possible limit shapes for given set of dimer weights are maximizers of the surface tension functional and in this sense are very similar to minimal surfaces. An analog of the Weierstraß parameterization for them in terms of analytic data was found in [25]. It reduces the solution of the Euler–Lagrange PDEs to solving equations for finitely many parameters, essentially finding a plane curve of given degree and genus satisfying certain tangency and periods conditions.

In the limit of extreme weights and large initial configurations, the amoebas and Ronkin functions degenerate to the piecewise-linear toric geometry. In this limit, it is possible to adjust parameters to reproduce the topological vertex formula for the GW invariants of the toric target obtained in [1]. We expect that the general case will reproduce the features of the background dependence (i.e., holomorphic anomaly) in the A-model [10]. This issue is presently under investigation [28].

Acknowledgments. We would like to thank the Simons workshop in Mathematics and Physics at Stony Brook for a very lively atmosphere which directly led to this work. We would also like to thank M. Aganagic, R. Dijkgraaf, S. Gukov, A. Iqbal, S. Katz, A. Klemm, M. Mariño, N. Nekrasov, H. Ooguri, M. Rocek, G. Sterman, and A. Strominger for valuable discussions. We thank J. Zhou for pointing out an inaccuracy in the formula (3.13) in the first version of this paper.

A. O. was partially supported by NSF grant DMS-0096246 and fellowships from the Packard foundation. N. R. was partially supported by NSF grant DMS-0307599. The research of C. V. is supported in part by NSF grants PHY-9802709 and DMS-0074329.

References

[1] M. Aganagic, A. Klemm, M. Mariño, and C. Vafa, The topological vertex, *Comm. Math. Phys.*, **254** (2005), 425–478.
M. Aganagic, R. Dijkgraaf, A. Klemm, M. Mariño, and C. Vafa, Topological strings and integrable hierarchies, hep-th/0312085, 2003.

[2] A. Iqbal, All genus topological string amplitudes and 5-brane webs as Feynman diagrams, hep-th/0207114, 2002.

[3] D. E. Diaconescu and B. Florea, Localization and gluing of topological amplitudes, hep-th/0309143, 2003.

[4] S. Katz, A. Klemm, and C. Vafa, Geometric engineering of quantum field theories, *Nuclear Phys.* B, **497** (1997), 173–195.
S. Katz, P. Mayr, and C. Vafa, Mirror symmetry and exact solution of 4D $N = 2$ gauge theories I, *Adv. Theor. Math. Phys.*, **1** (1998), 53–114.

[5] N. Nekrasov and A. Okounkov, Seiberg–Witten theory and random partitions, in P. Etinghof, V. Retakh, and I. Singer, eds., *The Unity of Mathematics*, Birkhäuser Boston, Cambridge, MA, 2005 (this volume), 525–596.

[6] A. Iqbal and A. K. Kashani-Poor, $SU(N)$ geometries and topological string amplitudes, hep-th/0306032, 2003.

[7] A. Okounkov and R. Pandharipande, Gromov-Witten theory, Hurwitz theory, and completed cycles, math.AG/0204305, 2002.
A. Okounkov and R. Pandharipande, The equivariant Gromov-Witten theory of P^1, math.AG/0207233, 2002.

A. Okounkov and R. Pandharipande, Virasoro constraints for target curves, math.AG/0308097, 2003.

[8] C.-C. M. Liu, K. Liu, and J. Zhou, On a proof of a conjecture of Marino-Vafa on Hodge integrals, *Math. Res. Lett.*, **11**-2 (2004), 259–272.

C.-C. M. Liu, K. Liu, and J. Zhou, A proof of a conjecture of Marino-Vafa on Hodge integrals, *J. Differential Geom.*, **65** (2004), 289–340.

C.-C. M. Liu, K. Liu, and J. Zhou, Mariño-Vafa formula and Hodge integral identities, math.AG/0308015, 2003.

[9] A. Okounkov and R. Pandharipande, Hodge integrals and invariants of the unknot, *Geom. Topol.*, **8** (2004), 675–699.

[10] M. Bershadsky, S. Cecotti, H. Ooguri, and C. Vafa, Kodaira-Spencer theory of gravity and exact results for quantum string amplitudes, *Comm. Math. Phys.*, **165** (1994), 311–428.

[11] R. Gopakumar and C. Vafa, M-theory and topological strings I, hep-th/9809187, 1998.

[12] C. Faber and R. Pandharipande, Logarithmic series and Hodge integrals in the tautological ring (with an appendix by D. Zagier), math.AG/0002112, 2000.

[13] K. Hori and C. Vafa, Mirror symmetry, hep-th/0002222, 2000.

[14] K. Hori, A. Iqbal, and C. Vafa, D-branes and mirror symmetry, hep-th/0005247, 2000.

[15] R. Cerf and R. Kenyon, The low-temperature expansion of the Wulff crystal in the 3D Ising model, *Comm. Math. Phys.*, **222**-1 (2001), 147–179.

[16] A. Okounkov and N. Reshetikhin, Correlation function of Schur process with application to local geometry of a random 3-dimensional Young diagram, *J. Amer. Math. Soc.*, **16**-3 (2003), 581–603.

A. Okounkov and N. Reshetikhin, Random skew plane partitions and the Pearcey process, math.CO/0503508, 2005.

[17] I. G. Macdonald, *Symmetric Functions and Hall Polynomials*, Clarendon Press, Oxford, UK, 1995.

[18] V. Kac, *Infinite-Dimensional Lie Algebras*, Cambridge University Press, Cambridge, UK, 1990.

[19] R. Kenyon, An introduction to the dimer model, math.CO/0310326, 2003; also available online from http://topo.math.u-psud.fr/~kenyon/papers/papers.html.

[20] P. W. Kasteleyn, The statistics of dimers on a lattice, *Physica*, **27** (1961), 1209–1225.

[21] M. Fisher and H. Temperley, Dimer problem in statistical mechanics: An exact result, *Philos. Mag.*, **6** (1961), 1061–1063.

[22] O. Aharony, A. Hanany, and B. Kol, Webs of (p, q) 5-branes, five dimensional field theories and grid diagrams, *J. High Energy Phys.*, **9801** (1998), 002.

[23] N. C. Leung and C. Vafa, Branes and toric geometry, *Adv. Theoret. Math. Phys.*, **2** (1998), 91–118.

[24] R. Kenyon, A. Okounkov, and S. Sheffield, Dimers and amoebae, to appear; math-ph/0311005, 2003.

[25] R. Kenyon and A. Okounkov, Planar dimers and Harnack curves, math.AG/0311062, 2003; Limit shapes and the complex Burgers equation, math-ph/0507007, 2005.

[26] A. Iqbal, N. Nekrasov, A. Okounkov, and C. Vafa, Quantum foam and topological strings, hep-th/0312022, 2003.

[27] H. Ooguri, A. Strominger, and C. Vafa, Black hole attractors and the topological string, *Phys. Rev.* D, **70** (2004), 106007.

[28] R. Kenyon, A. Okounkov, and C. Vafa, work in progress.

Gelfand–Tsetlin Algebras, Expectations, Inverse Limits, Fourier Analysis

A. M. Vershik

St. Petersburg Department of Steklov Institute of Mathematics
Russian Academy of Sciences
27 Fontanka
St. Petersburg 191011
Russia
vershik@pdmi.ras.ru

Summary. This text mainly follows my talk at the conference "Unity of Mathematics" (Harvard, September 2003), devoted to the 90th birthday of I. M. Gelfand. I introduce some new notions that are related to several old ideas of I. M. and try to give a draft of the future development of this area, which includes the representation theory of inductive families of groups and algebras and Fourier analysis on such groups. I also include a few reminiscences about I. M. as my guide.

Subject Classifications: 22D20, 43A30

0 Historical excursus: I. M. Gelfand as my correspondence advisor

The first substantial series of mathematical works that I studied as a student was the series of papers by Gelfand, Raikov, and Shilov (GIMDARGESH, as I called it to myself) on commutative normed rings and subsequent papers on generalized Fourier analysis. This theory became a mathematical inspiration for me; I was struck by its beauty and naturalness, universality and depth.

Before this I hesitated whether I should join the Department of Algebra—I attended the course of Z. I. Borevich on group theory and the course of D. K. Faddeev on Galois theory—or the Department of Mathematical Analysis, where my first advisor G. P. Akilov worked; in the latter case I could choose complex analysis (V. I. Smirnov, N. A. Lebedev) or real and functional analysis (G. M. Fikhtengolts, L. V. Kantorovich). But now the choice was clear: the functional analysis. At the same time I was more interested in the works of the Moscow (Gelfand) school of functional analysis focused on noncommutative problems than in the works of the Leningrad school, which was oriented towards the theory of functions and operator theory.

Since then the works of I. M. Gelfand and his school in various fields have become a kind of mathematical guidebook for me. My master's thesis was devoted to the theory of generalized functions; this topic equally interested the Leningrad mathematicians (L. V. Kantorovich, G. P. Akilov). Later, following G. P. Akilov's advice, I began to study representation theory, which was at the time absolutely unrepresented in Leningrad. While I was a postgraduate student, I. M. Gelfand popularized problems concerning measure theory in infinite-dimensional spaces, inspired by the theory of distributions, the notion of generalized random processes, and quantum physics. These problems were communicated to us by D. A. Raikov, who, following I. M. Gelfand's advice, worked in the new theory of locally convex and nuclear spaces, which we also studied in G. P. Akilov's seminar.

In Leningrad, measure theory in linear topological spaces was studied in the late 1950s–early 1960s by V. N. Sudakov and myself. At the time everybody believed that the theory of generalized functions and measure theory in infinite-dimensional spaces would require one to overstep the limits of conventional Banach functional analysis, which would be replaced by the theory of nuclear spaces (Minlos–Sazonov and Gelfand–Kostyuchenko theorems, quasi-invariant measures, etc.). However, it soon became clear that measure theory in linear spaces is a natural part of general measure theory, and Banach analysis continued to be the traditional language of functional analysis. After a while, the interest in all these problems gradually died away.

V. A. Rokhlin'a arrival at Leningrad thoroughly changed the mathematical landscape in the Department of Mathematics. In particular, he organized seminars on ergodic theory and topology. V. A. became my principal advisor during my postgraduate studies and several subsequent years. I seriously studied the theory of dynamical systems and general measure theory, and both my dissertations were devoted to these problems. But representation theory continued to fascinate me equally. Even earlier, in his talks on problems of functional analysis at the All-Union Conference on Functional Analysis and the 3rd Mathematical Congress (1956), I. M. spoke about von Neumann factors and Wiener measure as subjects that were possibly related and underestimated at the time. Later, in the 1960s, I began to study factors and relations of the theory of C^*-algebras, introduced by Gelfand and Naimark, with the theory of dynamical systems; this became the subject of my research for several years.

Except for several short discussions with I. M. in the mid- and late 1960s and the acquaintance by correspondence via V. A. Rokhlin (and possibly via Yu. V. Linnik), our close acquaintance took place in the spring of 1972. After a session of his seminar, I began to talk to him about my work (joint with A. A. Shmidt) on the limit statistics of cycles of random permutations; and the next day, at his home, about my plans to study the representations of the symmetric groups. Though at first he said that with them everything is clear and started to talk enthusiastically about the theory of symmetric functions, later he agreed that not everything is that clear and advised me to look at the paper by E. Thoma on the characters of the infinite symmetric group, which was of great interest for me. This paper played an important role in our subsequent studies of this group with my pupil S. V. Kerov. One of our principal contributions was an explanation and a new proof of Thoma's result in terms of representation theory (asymptotics of Young diagrams). And in that conversation I. M. approved wholeheartedly of my ideas, which I later called asymptotic representation theory; and even when he retold them to D. Kazhdan, who appeared a little later, he referred to the theorems on the asymptotic behavior of Young diagrams, characters, etc., which were only conjectured at the time (many of them were proved later in joint papers with S. V. Kerov), as if they were results already obtained. Those results I talked about that were already proved related rather to probability theory (Poisson–Dirichlet measures) and the theory of dynamical systems than to representation theory. Other groups besides the symmetric groups and their representations were not discussed in those conversations. I took leave of him and was about to depart for Leningrad.

Suddenly, on the day of my departure, I. M., having found out, in a rather complicated way, the phone number of my friends with whom I stayed at Moscow, called me and asked me to come to him immediately. He also invited M. I. Graev, and during our long walk told me about the problem of constructing the noncommutative integral of representations for semisimple groups, and especially for $SL(2, \mathbb{R})$. He had earlier offered this problem to other pupils of his, but he said that he had no doubt that it "fitted" me. I was slightly surprised, because I supposed that he could not know to what extent I was acquainted with the representation theory of Lie groups, and in particular that of $SL(2)$; as I have mentioned above, we did not discuss these matters at all.

But I. M. was right—this problem was offered to me at a very appropriate moment. Several years before this conversation, at the youth seminar organized by L. D. Faddeev and myself, we studied Gelfand's volumes on generalized functions and other useful things, which were not widespread in Leningrad. And in the early 1970s, apart from my studies of ergodic theory, I gave a course and seminars focused on the representation theory of groups and algebras, tensor products, and factors. Apparently, I. M. had heard about it, but I did not ask him. Thus his problem appeared at an appropriate moment. We coped with it within several months (the end 1972—the beginning 1973). The first paper in *Uspekhi* (*Russian Math. Surveys*) appeared in a volume dedicated to Kolmogorov in 1973, and this was the beginning of our collaboration with I. M. and M. I. Graev, which lasted with gaps of about ten years and which I am going to describe one day in more detail. That first (the best, in I. M.'s and my opinion) paper of this series was devoted to the "integral" of representations of $SL(2, \mathbb{R})$ and touched upon many topics that are still current; in that paper we rediscovered several constructions that had recently appeared (Araki's Gaussian construction, cohomology in groups without Kazhdan's property, etc.), gave the first explicit formulas for the nonzero cohomology of semisimple groups of rank 1, and constructed irreducible nonlocal representations of current groups with values in finite-dimensional Lie groups. I. M. repeatedly (and the last time—at this conference (Harvard 2003)) expressed his wish to continue our joint work in this direction. We had no doubt that this series of papers would have various applications, which has already been repeatedly confirmed, and that work would be continued.

This paper is devoted to a subject from another line, which also goes back to I. M.'s works. Having worked for many years with inductive families of semisimple algebras, S. V. Kerov and the author at once appreciated the importance of the notion that we called the Gelfand–Tsetlin algebras; this notion is a generalization of the well-known and still popular construction of the Gelfand–Tsetlin bases for the unitary and orthogonal groups. These algebras serve as a basis for harmonic analysis and Fourier analysis on noncommutative groups. They play an especially important role in the representation theory of locally finite groups, symmetric groups, and, more generally, inductive limits of groups and algebras. Our joint works with A. Okounkov (see [1] and [2]) show how applying these algebras, and especially a natural basis in them (the Young–Jucys–Murphy basis), allows one to reconstruct the representation theory of the symmetric groups in a completely different basis. In my talk and in this paper I draw attention to yet another idea, closely related to the previous one; namely, to the idea of inverse limits of algebras with respect to conditional expectations. For the symmetric group, this question will be considered in detail in a joint work with N. V. Tsilevich (in preparation). On the other hand, inverse limits of finite-dimensional algebras generalize von Neumann's theory of complete and noncomplete tensor products [3], and I remember one of my first visits to Gelfand's seminar in the late 1950s, when this von Neumann paper was being discussed and commented on by the head of the seminar. In this paper I do not touch upon one subject that I mentioned in the talk, namely, the results on representations of the group of infinite matrices over a finite field, which we intensively studied with S. V. Kerov during the last several years. It will be considered in other publications under preparation.

1 Definition of a generalized expectation on a subalgebra

Let A be a C^*-algebra over $\mathbf{C}$ with involution $*$, and let B be its involution C^*-subalgebra. All algebras in the paper are assumed to be algebras with identity, and all subalgebras are assumed to contain this identity. Here we mainly consider finite-dimensional algebras, but the definitions below are valid for the general case.

Definition 1. *A linear operator $P : A \longrightarrow B$ is called a* conditional mathematical expectation, *or* expectation[1] *for short, of the algebra A onto the subalgebra B if*

1. $P(b) = b$ *and* $P(b_1 a b_2) = b_1 P(a) b_2$ *for all* $a \in A$ *and* $b, b_1, b_2 \in B$;
2. $P(a^*) = a^*,\ P1 = 1$;

and

3. $P \geq 0$, *which means that for all* $a \in A$, $P(aa^*)$ *is positive, i.e., belongs to the real cone in B generated by elements of the form* bb^*.

We will say that P is a generalized expectation *if only the first and second conditions hold, and P is a* true expectation, *or* expectation, *if condition* 3 *also holds.*

The notion of ("conditional"!) expectation is well known and has been used in many situations; for commutative algebras, it coincides with the ordinary notion of

[1] The word "conditional" is the traditional one, but I prefer to omit it below, as well as the word "mathematical," violating the old tradition. The reason is that the "unconditional expectation" is simply the "conditional expectation" onto the algebra of scalars $\mathbf{C}$ (the conditions are trivial); thus if we fix a subalgebra B, we do not need to use the word "conditional," because it is clear what the "conditions" are.

(mathematical) conditional expectation on a sigma-subfield or subalgebra. A fruitful example of *generalized*, i.e., *nonpositive* expectation appeared, I believe, only recently, in the very concrete situation of the group algebra of the symmetric group (see below), and this is the reason for considering this notion in full generality. Sometimes people require that an expectation P should be not only positive but even *totally positive*, but we will not put emphasis on this.

Note also that it is clear from the definition that the set of all expectations in an algebra A to a subalgebra B is a convex set.

In the main part of the paper, our attention will be focused not on a single generalized expectation for some pair $B \subset A$, but on *sequences of generalized expectations in an inductive family of algebras*.

It is not difficult to describe all expectations for finite-dimensional semisimple C^*-algebras over $\mathbf{C}$, which are the sums of several copies of full matrix algebras $M_n(\mathbf{C})$, as well as to describe generalized conditional expectations for these algebras. Recall that for a general pair (A, B), where $A = \sum_{j=1}^{m} A_j$ is a finite-dimensional C^*-algebra, $B = \sum_{i=1}^{k} B_i$ is a C^*-subalgebra, and $A_j = M_{k_j}(\mathbf{C})$, $j = 1, \dots, m$, $B_i = M_{n_i}(\mathbf{C})$, $i = 1, \dots, k$, are their decompositions into simple algebras, one can define a *bipartite multigraph* in which the first (upper) part of the vertices is indexed by the subalgebras B_i, $i = 1, \dots, k$, and the second (lower) part of the vertices is indexed by the subalgebras A_j, $j = 1, \dots, m$, and the multiplicity of an edge (i, j) is equal to the number of copies of the subalgebra B_i as a subalgebra of A_j. We will use this construction in the theorem below (claim 2). For the sake of clarity, we consider the multiplicity-free case when each B_i belongs to at most one A_j; a pair (i, j) is called admissible if it is an edge, or $B_i \subset A_j$. In order to determine the pair (A, B) uniquely up to isomorphism, we must fix this bipartite multigraph and positive integers in each upper vertex (the dimensions of the B_i).

Theorem 1.

1. *First, assume that* $A = M_n(\mathbf{C})$ *and that its subalgebra* B *is also a full matrix algebra* $B = M_m(\mathbf{C})$ *(that is, the multigraph reduces to two vertices and one edge). Then there exists a unique expectation* $P(a) = pap$*, where* $a \in A$ *and* p *is the natural orthogonal projection determined by the identity of the algebra* B.
2. *Suppose that* A *is a finite-dimensional semisimple algebra and* B *is a semisimple subalgebra as above. Then every conditional expectation* $P : A \longrightarrow B$ *is the sum*

$$P = \sum_{i,j} P_{i,j}$$

over all admissible pairs (i, j) *of generalized expectations from claim* 1: $P_{i,j} : A_j \longrightarrow B_i$, $P_{i,j}(a) = \lambda_{i,j} p_{i,j} a p_{i,j}$, *where* $\lambda_{i,j}$ *are real numbers (for a true expectation, nonnegative real numbers) such that* $\sum_j \lambda_{i,j} = 1$ *for every* i.

The proof of claim 1 is obvious; in order to prove claim 2, it suffices to separate the restrictions of P to each A_j by the linearity of P and then apply claim 1 and condition 2 from the definition of expectation ($P1 = 1$).

Thus a real matrix $\{\lambda_{i,j}\}$ satisfying the condition $\sum_j \lambda_{i,j} = 1$ for every i is a parameter on the set of generalized conditional expectations for a fixed semisimple finite-dimensional algebra and a subalgebra; for true expectations, we have an additional condition $\lambda_{i,j} \geq 0$, and $\{\lambda_{i,j}\}$ is a *Markovian matrix on the bipartite graph.* For this reason, in the general case we will say that the matrix $\lambda_{i,j}$ is a generalized Markovian matrix. It is clear that the set of (generalized) expectations for a finite-dimensional pair $B \subset A$ is always nonempty.

The conjugate operator to a generalized expectation P is an operator P^* from the space A^* conjugate to A to B^*. If P is a true (positive) expectation, then P^* maps each state (= positive normalized functional) on B to some state on A. But since P is not a homomorphism of algebras, it does not map traces (characters) to traces. We may consider more refined properties of expectations in regard to this fact, e.g., call an expectation *central* if the image of each trace is a trace, etc. We will not discuss this topic here.

The following natural question arises. Suppose that P_B is an expectation for a pair of finite-dimensional algebras A, B. Let us regard A as a vector space. The problem is to describe the $*$-algebra $E = \langle A, P_B \rangle$ generated by the left action of A and P_B in $\mathrm{END}(A)$. We give the answer to this question in terms of the decomposition of E into simple algebras.

Theorem 2. *Let $\Gamma(L_B, L_A)$ be the bipartite graph corresponding to the pair (A, B), where L_A (L_B) are the vertices of Γ corresponding to the decomposition of A (B), respectively. Then the diagram of the triple of algebras $(B \subset A \subset E)$ is the graph $\Gamma(L_B, L_A, L_E)$, where the bipartite part $\Gamma(L_A, L_E)$ is the* reflection *of $\gamma(L_B, L_A)$, which means that $L_E \equiv L_B$ and the edges between the vertices of (L_A, L_E) are the same as the corresponding edges in $\Gamma(L_B, L_A)$. This means, in particular, that the algebra $E = \langle A, P_B \rangle$ does not depend on the choice of the expectation P_B, but only on the subalgebra B itself, so that we can denote it by $E(A, B)$.*

The proof of this theorem uses Theorem 1 (the structure of expectations); we will not give examples and details here. Theorem 1 was firstly proved by V. Jones [12]; see also [13].

2 Two classes of examples for group algebras

For the group algebras (over **C**) of finite groups, we present two types of expectations related to the group structure. Since a linear map in the group algebra is determined by its values on the group, we can state the question in terms of the group.

1. The first type of examples relates to the case when the value of the expectation at an element of the group (regarded as a subset of the group algebra) again belongs to the group.

In this case we can formulate a purely group-theoretical question concerning a group analogue of expectation.

Assume that G is a finite group and H is a subgroup. When does there exist a map p from G onto H such that

$$p(h) = h, \qquad p(h_1 g h_2) = h_1 p(g) h_2, \qquad p(e_G) = e_H$$

for all $h, h_1, h_2 \in H$ and $g \in G$, where e_H and e_G are the identity elements in G and H, respectively?

If such a map p exists, we say that it is a *virtual projection* of the group G to the subgroup H.

Theorem 3. *The following two conditions are equivalent:*

1. *There exists a virtual projection $p : G \to H$.*
2. *There exists a set $K \subset G$ such that*
   (a) *K is invariant under the inner automorphisms generated by the elements of H, that is, for every $h \in H$, for every $k \in K$, $hkh^{-1} \in K$;*
   (b) *the intersection of the set K with any left (equivalently, right) coset of H in G has only one element; in other words, for all $k, k' \in K$, $k \neq k'$, we have $k^{-1}k' \notin H$.*

Proof. The proof is straightforward, and we only supplement it with some comments. Condition (b) means that the group G can be partitioned into left cosets of the subgroup H, each of them containing exactly one element of the set K; thus $G \cong H \times K$, and for every $g \in G$ there is a unique left decomposition $g = hk$ with $h \in H$, $k \in K$; condition (a) gives the right decomposition with the same h but another $k' \in K : g = k'h$. We assert that there is a bijection between the set of all virtual projections $p : G \to H$ and the set of all subsets K that satisfy these conditions. Namely, if K enjoys properties (a) and (b) above, then the corresponding virtual projection p is given by the formula

$$p(g) = h$$

for the element $g = hk = k'h$; and vice versa: if p is a virtual projection, then the set $K = p^{-1}(e_H) \subset G$ enjoys properties (a) and (b). □

Remark 1. It is clear from the construction that the set K is the union of orbits of the group of inner automorphisms of H. If O is one of these orbits in K, then its characteristic function commutes with H. In the case of the symmetric group, K is a single orbit.

Remark 2. The set K above can be described in the following terms (E. Vinberg's observation): $\bar{K} = \{k \in G : k$ belongs to the center of the group $H \cap k^{-1}Hk\}$. Then our K is a subset of $\bar{K}$, which is H-invariant and intersects each left (and, automatically, right) coset of the subgroup H at one point.

For different groups, it may happen that such a set K either is nonunique, or does not exist at all.

It is an interesting question for what pairs $H \subset G$ a virtual projection (in terms of Theorem 3, a set K with properties (a) and (b)) does exist. In the trivial example G is the direct product of two groups: $G = H \times K$.

As a nontrivial example, consider the symmetric groups $G = S_n$ and $H = S_{n-1}$ with the ordinary embedding; then K is the set of transpositions (i, n), where i runs over $1, 2, \ldots, n$. The map $p : G \to H$ determined by this decomposition is a virtual projection; it simply deletes the element n from a permutation. This projection was defined in [4] (see also [5]) and called the *virtual projection*. It is easy to check that for $n > 4$, the virtual projection and the corresponding set K are unique; for $n = 3, 4$, there are several possibilities to choose such a set K.

Let us extend a virtual projection by linearity to an operator P in the group algebra:

$$P : C(G) \longrightarrow C(H) \subset C(G).$$

Lemma 1. *The linear operator P defined above is a generalized expectation of the algebra $C(G)$ to $C(H)$ in the sense of Definition* 1.

An important remark: in general, the generalized expectation P does not satisfy the positivity condition 3 from Definition 1; for example, in the case of the symmetric group (see above), this operator is not positive, because, e.g., the signature of a permutation can change under this projection. Thus P is not an expectation, but a generalized expectation.

Thus we have defined a particular class of generalized expectations on group algebras, which arise from virtual projections on groups. A very interesting problem is to describe pairs (G, H) for which a virtual projection, and hence the corresponding generalized expectation on the group algebra, does exist. For an abelian group, it is easy to describe all virtual projections (they exist for all pairs (G, H) and determine true expectations), but even for metabelian groups I do not know the answer.

For some classes of groups, such as free groups, "local groups" (see [6]), Coxeter groups, presumably the following recipe works: suppose that $G_n = \langle \sigma_1, \sigma_2, \ldots, \sigma_n \rangle$ and $G_n \supset G_{n-1} = \langle \sigma_1, \sigma_2, \ldots, \sigma_{n-1} \rangle$. There exists a normal form of each element of G_n as a word in the alphabet $\sigma_1, \ldots, \sigma_n$ such that the deletion of the letter σ_n in this normal form is a virtual projection of G_n onto G_{n-1}. This is true for free, locally free, and symmetric groups (such a normal form does exist).

2. The second type of example is closer to the classical definitions, because it leads to true (positive) expectations. Again let G and H be a finite group and its subgroup, respectively; now we allow the values of expectations at the elements of the group not only to belong to the group, but also to be equal to zero. Define a projection

$$P : C(G) \longrightarrow C(H) \subset C(G)$$

as follows: P is the linear extension to the whole group algebra of the following map on the group: $P(h) = h$ for all $h \in H$, and $P(g) = 0$ if $g \in G$, $g \notin H$. This definition makes sense for an arbitrary group and a subgroup. Obviously, P is a (positive) expectation. For some reason, we call it the *Plancherel expectation*. This definition leads, in particular, to Fourier analysis on the symmetric groups, which will be the subject of the joint paper with N. Tsilevich which is now in preparation.

It is easy to formulate the analogue of Lemma 2 for algebras: the set of all generalized expectations $P : A \longrightarrow B$ is in a one-to-one correspondence with the set of subspaces T of A satisfying the following properties:

1. T is a closed complement to the subspace B of the vector space A;
2. $BTB \subset T$.

The correspondence is as follows: $T = \ker P$.

Because of the convexity of the set of expectations, we can consider convex combinations of these two types of examples. For the symmetric group, such deformations are related to the contents of the papers [4, 5].

3 Gelfand–Tsetlin (GZ-) algebras

Now we introduce the central notion of the theory of inductive families of algebras (not only finite-dimensional). This notion follows the idea of the classical papers by Gelfand and Tsetlin [7, 8], in which a particular basis was defined for the orthogonal SO(n) and unitary SU(n) groups. This basis appears only if we consider not just one group, say SO(n) or SU(n), but the whole inductive family $\mathrm{SO}(2) \subset \mathrm{SO}(3) \subset \cdots \subset \mathrm{SO}(n)$ or $\mathrm{SU}(1) \subset \mathrm{SU}(2) \subset \cdots \subset \mathrm{SU}(n)$ simultaneously. Since the restrictions of irreducible representations of the group SO(n) to the subgroup SO($n-1$) (and similarly with SU) are multiplicity-free, this inductive family determines a basis (Gelfand–Tsetlin basis), which is unique up to scalar multiples (see below). But even more important is the notion of Gelfand–Tsetlin algebras, which was introduced for a general inductive family of algebras in our papers with S. Kerov (a detailed exposition is given in [9]) and independently, but not in the same spirit, in [10]). I do not know any papers about Gelfand–Tsetlin algebras even in the classical case (that of the universal enveloping algebras of semisimple Lie algebras) apart from the paper [11], which concerns a completely different problem. The most important problem is to define reasonable multiplicative generators of the Gelfand–Tsetlin algebras in terms of the initial algebras; having such generators, one can create the representation theory of an inductive family of algebras in a very natural way. The realization of this plan allows one to define an analogue of the Fourier transform for algebras with inductive family of subalgebras inside it. For the symmetric group, these generators were defined (independently of GZ-algebras) by A. Young and in more recent times by Jucys and Murphy (YJM-generators). The consistent development of the representation theory of the symmetric groups was given in [1, 2]. For other groups (even for the orthogonal and unitary groups), this is still not done. Below we consider only complex $*$-representations of algebras over $\mathbf{C}$.

Definition 2 (Gelfand–Tsetlin algebra). *Suppose we are given a finite or infinite family A_k, $k = 0, \ldots, n$ (here n can be finite or infinite) of semisimple algebras over $\mathbf{C}$, $A_0 = \mathbf{C}$, $A_k \subset A_{k+1}$. Assume for the sake of clarity that the multiplicity of the restriction of an irreducible representation of A_k to A_{k-1} for $k = 1, \ldots, n-1$ is equal to one or zero (the so-called simple spectrum). By definition, the Gelfand–Tsetlin algebra GZ_n is the algebra generated by the* centers, *which we denote by $\zeta(A_k) \subset A_k$, $k = 0, \ldots, n$:*

$$\mathrm{GZ}_n = \langle \zeta(A_1), \ldots, \zeta(A_n) \rangle.$$

(The notation $\langle \dots \rangle$ *stands for the subalgebra of* A_n *generated by the contents of the brackets.)*

It is clear from this definition that all GZ_k are abelian algebras and the family of algebras $\{\mathrm{GZ}_k\}_1^n$ is an inductive family of subalgebras in A_n (the centers do not form an inductive family); the definition and the assumption on the simplicity of the spectrum also imply that GZ_n is a maximal abelian subalgebra of A_n. Moreover, from the definition we can conclude that there is a particular basis (defined up to scalars) in the algebra GZ_n, which we call the GZ-basis; and, consequently, there is a particular basis in each irreducible representation of A_n—this is what people usually called the Gelfand–Tsetlin basis. In the case of the groups $\mathrm{SO}(n)$ and $\mathrm{SU}(n)$, this is just the classical Gelfand–Tsetlin basis [7, 8]. It leads to the well-known notion of Gelfand–Tsetlin patterns.

The elements of the GZ-basis of the algebra GZ_n in the general case are defined as those elements such that each of them has a nonzero image in only one irreducible representation. All such elements are defined uniquely (up to scalars). We may say that there is a bijection between this basis and paths in the graph of the Bratteli diagram of the algebra A_n (see below). As we have mentioned above, a nontrivial problem is to describe the GZ-algebra, as well as the GZ-basis, using some multiplicative generators of $\mathrm{GZ}(A_n)$, not in terms of representations, but in intrinsic terms of the initial definition of the algebras A_n (or groups in the case when A_n is a group algebra). This problem leads to what we called the *Fourier analysis* of inductive families of algebras (groups).

We want to emphasize that the notion of GZ_n-subalgebra of an algebra A_n does depend on the structure of the inductive family $A_i, i = 1, \dots, n$, and not only on the algebra A_n itself; so if we choose another inductive family inside A_n, then GZ_n can also change. The development of these ideas for the symmetric groups can be found in [1, 2]. The assumption on the simplicity of the spectrum is assumed to be satisfied in all further considerations.

The analysis of examples of Gelfand–Tsetlin algebras in the case of groups, and especially of the GZ_n subalgebras of $C(S_N)$, allows us to formulate the following theorem.

Theorem 4. *Suppose that* $G_1 \subset G_2 \subset \cdots \subset G_n$ *is a finite sequence of finite groups. Suppose that the restriction of irreducible representations of* G_k *to* G_{k-1}, $k = 1, \dots, n$, *is multiplicity-free and there exists a virtual projection of* G_k *to* G_{k-1}, $k = 1, \dots, n$. *Then the family of sets* $\{X_k = \ker P_k,\ k = 1, \dots, n\}$ *generates (as multiplicative generators) the subalgebra* GZ_n*; here* P_k *is the generalized expectation* $C(G_k) \longrightarrow C(G_{k-1})$ *corresponding to the virtual projection* $p_k : G_k \to G_{k-1}$ *(see the previous section).*

Proof. Using Remark 1 after Theorem 2, we can prove that the center of $C(G_k)$ belongs to the algebra generated by GZ_{k-1} and the set X_k. □

In the case of the symmetric group, the set X_k is determined by the YJM-elements.

4 The inverse limit of an inductive family of algebras and GZ-algebras, and martingales

Suppose now we have a countable sequence $A_n, n = 0, 1, \ldots, A_0 = \mathbf{C}, A_n \subset A_{n+1}$, of C^*-algebras that form an inductive family of algebras and define the inductive limit

$$A_\infty = \underset{\rightarrow}{\lim\operatorname{ind}}\, A_i$$

with respect to the embedding of algebras.

In the same spirit we can define the inductive limit of the Gelfand–Tsetlin algebras

$$\mathrm{GZ}_\infty = \underset{\rightarrow}{\lim\operatorname{ind}}\, \mathrm{GZ}_n;$$

under our assumptions, it is again a maximal abelian subalgebra of A_∞.

Using Theorem 4 from the previous section, we can define multiplicative generators of $\mathrm{GZ}_\infty = \lim\operatorname{ind} \mathrm{GZ}_n$ for the case of group algebras. In particular, this gives a description of a multiplicative basis for the GZ-algebra of the infinite symmetric group.

An inductive family $\{A_n\}$ of *finite-dimensional algebras* determines a $\mathbf{Z}_+$-graded graph Y (the Bratteli diagram). The vertices of level $n \geq 0$ correspond to the simple subalgebras of the algebra A_n (at the zero level we have one vertex $\mathbf{0}$), and two adjacent levels Y_n and Y_{n-1} form precisely the bipartite graph that was mentioned in Section 2. The set of all maximal paths (finite if the number of algebras is finite, or infinite) from the vertex $\mathbf{0}$ to the end is called the set of *tableaux* and is denoted by $T(Y)$ (recall that a path is a sequence of edges, and in the multiplicity-free case a path is also a sequence of vertices). Now let us choose a sequence of generalized expectations of these algebras at each level:

$$P_n : A_n \longrightarrow A_{n-1}, \quad n = 1, 2, \ldots.$$

Lemma 2. *The restriction of the generalized expectation P_n to the Gelfand–Tsetlin algebra GZ_n sends it to GZ_{n-1}; thus this restriction is an expectation of GZ_n to GZ_{n-1}.*

Proof. Each expectation sends the center of the algebra onto the center of the subalgebra: $P_n(\zeta(A_n)) = \zeta(A_{n-1})$. Indeed, let $z \in \zeta(A_n)$ and $b \in A_{n-1}$; then $P_n(zb) = P_n(z)b = P_n(bz) = bP_n z$. At the same time $P_n(\zeta(A_{n-1})) = \zeta(A_{n-1})$. Consequently, $P_n(\mathrm{GZ}_n) = \mathrm{GZ}_{n-1}$ by definition. □

Now let us define the projective limit

$$A^\infty = \underset{\leftarrow}{\lim\operatorname{proj}}\{A_n, P_n\}$$

with respect to the sequence of generalized expectations. It is obvious from the definition that the following lemma holds.

Lemma 3. *A^∞ is a left and right A_∞-bimodule (but not an algebra in general).*

Indeed, all algebras A_n act from the left and from the right on all A_m, $m > n$; thus these actions extend to the projective limit. This definition makes sense for a general inductive family with an arbitrary system of expectations.

By Lemma 3, we can also correctly define the inverse (projective) limit of the algebras $\{\mathrm{GZ}_n\}$:

$$\mathrm{GZ}^\infty = \lim_{\leftarrow}\mathrm{proj}\{\mathrm{GZ}_n, P_n\}.$$

This is not an algebra either, but a module over GZ_∞. The interpretation of this limit will be given below.

Suppose now that all algebras A_n, $n = 1, 2, \dots$, are finite-dimensional semisimple algebras. Since (generalized) expectations are determined by systems of (generalized) Markovian matrices, the projective module is determined by the system of matrices Λ_n, $n = 1, 2, \dots$, where Λ_n determines the expectation of A_n to A_{n-1}.

Let us fix such a system of generalized (or true) Markovian matrices Λ_n, $n = 1, 2, \dots$. The size of Λ_n is $m_n \times m_{n-1}$, where m_k is the number of simple subalgebras in the algebra A_k. We denote this system of matrices by $\mathbf{L} = \{\Lambda_n,\ n = 1, 2, \dots\}$, and in order to emphasize the dependence of the projective limit on the expectations, we will sometimes write

$$A_{\mathbf{L}}^\infty = \lim_{\leftarrow}\mathrm{proj}\{A_n, P_n\}$$

and

$$\mathrm{GZ}_{\mathbf{L}}^\infty = \lim_{\leftarrow}\mathrm{proj}\{A_n, P_n\}.$$

In the case of abelian algebras, as well as in the case of GZ-algebras, such an inverse limit is well known by another name, at least when all matrices Λ_n are true Markovian matrices. We will shortly explain this link.

First of all, as usual, the system of Markovian matrices $\mathbf{L}$ determines a Markov measure $\mu_{\mathbf{L}}$ on the space of tableaux $T(Y)$ (see above). Thus we have a measure space (more precisely, a Lebesgue space) $(T(Y), \mathbf{A}\mu_{\mathbf{L}})$, where $\mathbf{A}$ is the sigma-field generated by elementary cylindrical sets (an elementary cylindrical set of order n is the set of all paths with common fragment of length n). Second, in $\mathbf{A}$ we have an increasing sequence of finite sigma-subfields of cylindrical sets of order n. Following the general definition of *martingales*, we can now define the vector space $\mathbf{M}_{\mathbf{L}}$ of martingales over this increasing sequence of sigma-subfields, each of them being a sequence $\{f_n\}_n$ of measurable functions such that f_n is $\mathbf{A}_n$-measurable and the expectation of f_n on the sigma-field $\mathbf{A}_{n-1}$ is equal to f_{n-1}.

It is clear from the definition that *this space of martingales is exactly the inverse limit* $\mathrm{GZ}_{\mathbf{L}}^\infty$ defined above.

This is the reason for calling the elements of the inverse limit $A_{\mathbf{L}}^\infty$ of algebras noncommutative martingales. This opens a wide range of generalizations of the martingale theory to this noncommutative case.

If we have a generalized expectation, then we need to consider martingales with respect to nonpositive measures, which, as far as I know, have never been considered.

In the group case there is a distinguished Markov measure—the so-called Plancherel measure on the space of tableaux $T(Y)$; namely, if $G = \mathrm{limind}\, G_n$ is a locally finite group with simple spectrum (like $S_\infty = \text{limind}\, S_n$), then, using one of the expectations defined in the previous section, we obtain the Plancherel measure on $T(Y)$, which is the inverse limit of the Plancherel measures on the spaces of finite tableaux. Martingales with respect to the Plancherel measure play an important role as a special kind of modules over the group algebras of the group G.

Our last remark concerns the link with von Neumann's theory of infinite tensor products: if our algebra A_∞ is the *infinite tensor product of algebras of matrices* (e.g., of order 2), the so-called Glimm algebras, then each incomplete tensor product of Hilbert spaces in the sense of [3] is generated by the inverse limit of algebras with respect to some sequence of expectations. In this spirit, the scheme of this section allows us to generalize von Neumann's theory to an arbitrary inductive limit of finite-dimensional algebras instead of Glimm algebras.

References

[1] A. Okounkov and A. Vershik, A new approach to representation theory of symmetric group, *Selecta Math.*, **2**-4 (1996), 581–605.

[2] A. Vershik and A. Okounkov, A new approach to representation theory of symmetric group-II, *Zap. Nauchn. Semin. POMI*, **307** (2004) (in Russian).

[3] J. von Neumann, On infinite direct products, *Compositio Math.*, **6** (1938), 1–77.

[4] S. Kerov, G. Olshanski, and A. Vershik, Harmonic analysis on the infinite symmetric group: A deformation of the regular representation, *C. R. Acad. Sci. Paris Ser.* I *Math.*, **316**-8 (1993), 773–778.

[5] S. Kerov, G. Olshanski, and A. Vershik, Harmonic analysis on the infinite symmetric group, to appear; math.RT/0312270, 2003.

[6] A. Vershik, Dynamic theory of growth in groups: Entropy, boundaries, examples, *Russian Math. Surveys*, **55**-4 (2000), 667–733.

[7] I. M. Gelfand and M. L. Tsetlin, Finite-dimensional representations of the group of unimodular matrices, *Dokl. Akad. Nauk SSSR (N.S.)*, **71** (1950), 825–828 (in Russian); in I. M. Gelfand, *Collected Papers*, Vol. II, Springer-Verlag, Berlin, 1987, 653–656 (in English).

[8] I. M. Gelfand and M. L. Tsetlin, Finite-dimensional representations of groups of orthogonal matrices, *Dokl. Akad. Nauk SSSR (N.S.)*, **71** (1950), 1017–1020 (in Russian); in I. M. Gelfand, *Collected Papers*, Vol. II, Springer-Verlag, Berlin, 1987, 657–661 (in English).

[9] A. M. Vershik and S. V. Kerov, Locally semisimple algebras: Combinatorial theory and the K-functor, in *Itogi Nauki i Tekhniki*, **26**, VINITI, Moscow, 1985, 3–56; *J. Soviet Math.*, **38** (1987), 1701–1733 (in English).

[10] S. Stratila and D. Voiculescu, Representations of AF-Algebras and of the Group $U(\infty)$, Lecture Notes in Mathematics 486, Springer-Verlag, Berlin, 1975.

[11] E. Vinberg, Some commutative subalgebras of a universal enveloping algebra, *Izv. Akad. Nauk SSSR Ser. Mat.*, **54**-1 (1990), 3–25 (in Russiam); *Math. USSR-Izv.*, **36**-1 (1991), 1–22 (in English).

[12] V. F. R. Jones, Index for subfactors, *Invent. Math.*, **72** (1983), 1–25.

[13] F. M. Goodman, P. de la Harpe, and V. F. R. Jones, *Coxeter Graphs and Towers of Algebras*, Springer-Verlag, Berlin, New York, Heidelberg, 1989.